ELECTRICAL TECHNOLOGY

ELECTRICAL TECHNOLOGY

BY

H. COTTON
M.B.E., D.Sc., M.I.E.E.
Emeritus Professor of Electrical Engineering, University of Nottingham

SEVENTH EDITION
(M.K.S. UNITS)

CBS Publishers & Distributors Pvt. Ltd.
New Delhi • Bengaluru • Chennai • Kochi • Kolkata • Mumbai
Hyderabad • Nagpur • Patna • Pune • Vijayawada

ISBN: 81-239-0928-4

First Indian Edition: 1984
Reprint: 1999, 2000, 2002, 2003, 2004, 2005

This edition has been published in India by arrangement with
M/s Pitman, London

Published by **Satish Kumar Jain** and produced by **Varun Jain** for
CBS Publishers & Distributors Pvt. Ltd.,
4819/XI Prahlad Street, 24 Ansari Road, Daryaganj, New Delhi - 110002
delhi@cbspd.com, cbspubs@airtelmail.in • www.cbspd.com
Ph.: 23289259, 23266861, 23266867 • Fax: 011-23243014

Corporate Office: 204 FIE, Industrial Area, Patparganj, Delhi - 110 092
Ph: 49344934 • Fax: 011-49344935
E-mail: publishing@cbspd.com • publicity@cbspd.com

Branches:

- ***Bengaluru:*** 2975, 17th Cross, K.R. Road, Bansankari 2nd Stage, Bengaluru - 70 • Ph: +91-80-26771678/79 • Fax: +91-80-26771680 E-mail: cbsbng@gmail.com, bangalore@cbspd.com
- ***Chennai:*** No. 7, Subbaraya Street, Shenoy Nagar, Chennai - 600030 Ph: +91-44-26681266, 26680620 • Fax: +91-44-42032115 E-mail: chennai@cbspd.com
- ***Kochi:*** Ashana House, 39/1904, A.M. Thomas Road, Valanjambalam, Ernakulum, Kochi • Ph: +91-484-4059061-65 Fax: +91-484-4059065 • E-mail: cochin@cbspd.com
- ***Kolkata:*** 6-B, Ground Floor, Rameshwar Shaw Road, Kolkata - 700014 Ph: +91-33-22891126/7/8 • E-mail: kolkata@cbspd.com
- ***Mumbai:*** 83-C, Dr. E. Moses Road, Worli, Mumbai - 400018 Ph: +91-9833017933, 022-24902340/41 • E-mail: mumbai@cbspd.co

Representatives:

- Hyderabad: 0-9885175004
- Patna: 0-9334159340
- Jharkhand: 0-9811541605
- Nagpur: 0-9021734563
- Pune: 0-9623451994
- Uttarakhand: 0-9716462459

Printed at:
J.S. Offset Printers, Delhi (India)

PREFACE
TO THE SEVENTH EDITION

THIS book was first published thirty-two years ago as a volume in Pitman's "Engineering Degree Series." It was designed to cover the requirements of the syllabus of the subject "Electrical Technology" in the Regulations of the University of London for the External Degree in Engineering. This indicates that its content was adequate at that time. Now, the first edition goes little beyond the requirements of the subject "Applied Electricity" in the Regulations for Part One of most University Degrees, or of the Ordinary National Certificate in Electrical Engineering. This is an indication of the magnitude of the progressively increasing demands which have been made on students in the last thirty years.

As far as this book is concerned these demands were met from time to time by additions to the text, but, after a time this process tended towards discontinuity and a major revision was indicated. The development of the subject was, in the past editions, based exclusively on the c.g.s. system of units, and the recent recognition and adoption by important bodies, such as our own Institution of Electrical Engineers, of the rationalized m.k.s. system, converted a major revision from a desirability to a necessity.

Considering first of all the increases in the content of the book, these have been made necessary by the vast developments in all branches of Electrical Practice since the end of the first World War, and by the demand by industry for personnel with a greatly increased knowledge of fundamental principles. For this reason all the chapters, with one exception, have been enlarged, some very considerably, particularly those dealing with Electrostatics, Electromagnetics, and Circuit Theory, both d.c. and a.c. The chapters dealing with Electrical Plant and Apparatus have also been enlarged, with the exception of that on Converting Machinery. The characteristics of the Alternator and the Synchronous Motor are discussed at much greater length, including the phenomena of rotor oscillation, the treatment of the Induction Motor is also greatly enlarged, and that of the various forms of Commutator Motor is expanded to such an extent that it almost justifies the publication of a separate book. There are also appreciable additions to the chapters on Mercury-arc Rectifiers, on Phase Advancing, and on Illumination. Special d.c. machines of the cross-field type are also considered. All these additions have resulted in a book of materially increased size.

Perhaps the greatest difficulty associated with this new edition was the method of presentation of the rationalized m.k.s. system. There appeared to be two alternatives, (*a*) completely to ignore the c.g.s. system, (*b*) to assume a knowledge of the c.g.s. system, to build up the m.k.s. treatment on this as a framework, and then to drop the framework as quickly as possible. The first method is, at first sight, the better, and so it is, in the case of an elementary text written for students who are studying the subject *de novo*, and for whom, in consequence, one particular name for a unit will present no more difficulty than any other name. Thus it will be no more difficult to define a magnetic flux in webers than in maxwells, while it will be much easier to think of a capacitance in coulombs per volt than in centimetres. Unfortunately the m.k.s. system is, as yet, little used in the Schools and it is therefore still the rule that boys entering the Universities and Technical Colleges know only the c.g.s. system and have acquired such facility in its use that their mental processes, as far as the study of Electricity is concerned, are naturally conditioned by this system.

Again, the present book is not intended to be an elementary text, it is believed that it will be of value, not only to examination candidates, but also to engineers in practice. At the present stage of the process of transition from the old to the new system the great majority of these work in terms of the c.g.s. system, or the so-called English system, or even a mixture of both, and they may experience undue difficulty in the reading of a book which ignores the system with which they are so familiar and in which, because of rationalization, many important everyday formulae are completely altered in form.

For these reasons the second alternative has been adopted.

In view of the considerable elapsed time since the publication of the first edition, and the large amount of new material which has been incorporated, the preparation of this new edition has entailed a great deal of labour. I am therefore glad to record my thanks to my former colleagues in the Department of Electrical Engineering at the University of Nottingham, particularly to my friend Mr. S. A. Swann. Acknowledgments are made to the firms who have very readily loaned figures or supplied information. Tables showing the relationship between c.g.s. and m.k.s. systems of units are given in the Appendix. These are reproduced from *The M.K.S. System of Units* (Pitman, 1953) by kind permission of the author, Mr. T. McGreevy.

At the end of most of the chapters there is a list of books for further reading. As the literature of Electrical Engineering is now so vast, the selection has been confined mainly to books which I have formerly recommended to my students for the purpose of ancillary reading.

H. COTTON

ST. IVES
CORNWALL

PREFACE
TO THE FIRST EDITION

THIS text-book has been written with two main objects in view; first, to cover the ground of the more important examination syllabuses in Electrical Technology; and second, to be sold at a price which is within the means of all students. These two aims are diametrically opposite, but it is hoped that both have been fulfilled through the elimination of the more elementary portions of the subject, which should be thoroughly understood by students reaching this stage, and by the non-inclusion of half-tone blocks, which, although undoubtedly attractive, are of no real educational value.

Although the book is written primarily for examination purposes, the practical side has been kept in view, and it is hoped that it will appeal to practical engineers as well as to students.

The fundamental principles of Electrical Technology, both direct and alternate working, are covered, as are also certain principles of design, but no detailed designs are worked out, as these are beyond the scope of the book. The mathematical knowledge required of the reader is, on the whole, elementary, except in the chapter on "Electrical Oscillations," in which differential equations are used of necessity. There is a large number of worked examples in the text, and most of the chapters have questions to be answered; these are mainly drawn from examination papers set by the London University and by the City and Guilds of London Institute.

The great majority of the diagrams have been specially drawn, but a few have been kindly lent by Messrs. Pitman, and others by various manufacturing firms. These are acknowledged in the text.

H. COTTON

UNIVERSITY COLLEGE
NOTTINGHAM

CONTENTS

PAGE

Preface to Seventh Edition v
Preface to First Edition vii

CHAPTER 1

THE M.K.S. SYSTEM OF UNITS 1

Mechanical units—The values of μ_0 and ε_0—Magnetic flux—Rationalization—Further reading—Exercises

CHAPTER 2

ELECTROSTATICS 6

Electrification—Coulomb's law—The electric field—Field due to straight uniformly charged wire—Field at a point on the axis of a uniformly charged ring—Gauss's theorem—Applications of Gauss's theorem—Field inside a charged hollow spherical conductor—Coulomb's theorem—Equipotential surfaces—Potential due to a point charge—Charged sphere—Capacitance—Systems of capacitors—Capacitors in parallel—Potential coefficients—Energy stored in an electric field—Energy of a charged system—Mechanical force on a dielectric—Loss of energy—Charge and discharge of a capacitor—Oscillatory charge and discharge of a capacitor—The equations of Poisson and Laplace—Dielectrics—Dielectric phenomena—Dielectric breakdown—Breakdown of air—Further reading—Exercises

CHAPTER 3

THE ELECTRIC CURRENT—SYSTEMS OF CONDUCTORS . . 65

The Electric Current

Resistance—Variation of resistance with temperature—Power—Resistance calculation by the Laplace equation

Systems of Conductors

Series and parallel connections—Kirchhoff's laws—Delta-star transformation—Superposition theorem—Thévenin's theorem—Nonlinear resistors—Exercises

CHAPTER 4

ELECTROMAGNETISM 85

The ampere—Magnetic flux density—Magnetic flux—Field intensity due to a long straight wire—Flux density inside a straight wire—Field due to a circular current—Intensity due to a straight wire of finite length—Intensity due to a solenoid—Intensity due to a short solenoid—Force between two parallel current-carrying conductors—The magnetic force—The unit magnetic pole—Flux inside a toroid—The magnetic circuit—Magnetic leakage—Magnetic circuits in parallel—Energy density—Lifting power of an electromagnet—Ferromagnetism—Magnetization curves—Practical calculations—The Curie temperature—Magnetic reversals—Coercive force and retentivity—Permanent magnets—Theory of magnetism—Determination of magnetization curves—Magnetic potentiometer method—The magnetic potentiometer—Further reading—Exercises

CHAPTER 5 PAGE

ELECTROMOTIVE FORCE 135

Dynamically-induced e.m.f.—Statically-induced e.m.f.—Self-induced e.m.f.—Coefficients of self- and mutual induction—Rise and decay of current in an inductive circuit—Relative permeability not constant—Mutual inductance—Inductance of a pair of parallel conductors—Exercises

CHAPTER 6

CONSTRUCTION OF D.C. MACHINES 151

Essential features—The field magnet—The armature—The commutator and brush gear—Armature windings—Example of a simple lap winding—Example of a simple wave winding—Action of the commutator—Multi-turn windings—Numerical example—Equalizer rings—Selective commutation—E.M.F. equation—Fluctuation in the induced e.m.f.—Further reading—Exercises

CHAPTER 7

ARMATURE REACTION AND COMMUTATION IN D.C. GENERATORS 185

Armature reaction—Position of brushes—Armature ampere-turns—Commutation — Reactance voltage — Interpoles — High-resistance brushes—Compensating winding—Special armature windings—Armature m.m.f. diagram—Exercises

CHAPTER 8

CHARACTERISTICS OF D.C. GENERATORS 197

Types of generators—Series-wound generator—Separately-excited generator—Shunt-wound generator—Conditions for self-excitation—The compound-wound generator—Efficiency of d.c. machines—Exercises

CHAPTER 9

CROSS-FIELD DYNAMOS 211

The Rosenberg constant-current dynamo—The amplidyne—The magnicon—Further reading

CHAPTER 10

THE D.C. MOTOR 220

Back e.m.f.—Speed of a d.c. motor—Conditions for maximum power—Torque—Armature reaction and commutation—Characteristics of d.c. motors: the series motor—The shunt-wound motor—Compound-wound motor—Motor starters—Starters for series motors—Timer of acceleration—Contactor starters—Speed control of d.c. motors—Calculation of mass of flywheel—Modified Ward-Leonard control—Exercises

CHAPTER 11

OPERATION OF D.C. GENERATORS IN SERIES AND IN PARALLEL . 251

Series generators in series—Series generators in parallel—Shunt generators in series—Shunt generators in parallel—Compound generators in parallel—Exercises

CHAPTER 12 PAGE

TESTING OF D.C. MACHINES 259

Type of tests—Indirect methods—Separation of losses—The "retardation" or "running-down" test—Regenerative methods—Series motor tests—Exercises

CHAPTER 13

DISTRIBUTION 275

Effect of supply voltage on the size of distributors and feeders—Effect of pressure on the efficiency of transmission—Three-wire system—Balancers—Boosters—Negative booster—Battery boosters—Calculation of feeders—Calculation of the drop of volts in distributors—Further reading—Exercises

CHAPTER 14

ELECTROLYSIS—SECONDARY CELLS—PRIMARY CELLS . 295

Phenomena at the electrodes—Quantitative laws of electrolysis—Polarization—Accumulators—Testing—The alkaline accumulator—Primary cells—Further reading—Exercises

CHAPTER 15

A.C. CIRCUITS 307

The simple alternator—Effective value—Average and effective values of a sinusoidal wave—Graphical representation of alternating quantities—Phase difference—The addition of alternating e.m.f.'s, or alternating currents—Pure resistance circuit—Purely inductive circuit—Resistance and inductance in series—Circuit containing capacitance only—General series circuit—Electrical resonance—Properties of a resonant circuit—The pass-band of a resonant circuit—Circuits in parallel—Graphical solution of a branched circuit—Resonance in a branched circuit—Circuit with mutual inductance—The symbolic method—Application to electric circuit calculations—Solution of the differential equation of the *R, L, C* circuit—Starting conditions—Further reading—Exercises

CHAPTER 16

POLYPHASE CURRENTS 359

Relation of output to number of phases—Two-phase working—Three-phase working—Volt-amperes and reactive volt-amperes—Calculation of balanced three-phase networks—Measurement of power in three-phase circuits—Corrections to be applied to wattmeter readings—Polyphase vector diagrams—Calculation of unbalanced three-phase circuits—Reduction of star connection to equivalent mesh connection—The drop of volts in a three-phase line—Symmetrical components of unbalanced voltages and currents—Further reading—Exercises

CHAPTER 17

THE TRANSFORMER 393

Transformation ratio—Primary no-load current—Transformer on load—Effect of resistance and reactance in the windings—Equivalent resistance and reactance—Kapp regulation diagram—Equivalent circuit—Percentage resistance, reactance, and impedance—"All-day" efficiency—Relation between copper and iron loss—Division of load between two transformers in parallel—Effect of an inequality in the turns ratios—The transformer as a mutually-inductive circuit—Testing of transformers—Methods of cooling transformers—Heating-time curve—The auto-transformer—Use of transformers with measuring instruments—Transformation from three to two phase—Three-phase transformation—Tests on three-phase transformers—Further reading—Exercises

CHAPTER 18 PAGE

ROTATING MAGNETIC FIELDS 443

Ferrari's principle—Two-phase supply—Three-phase supply—Rotating field of a three-phase machine—The three-phase power-factor meter

CHAPTER 19

THE ALTERNATOR 451

The stator—Frequency—Stator windings–single phase—Polyphase windings—Fractional-pitch windings—The e.m.f. equation—Notes on the choice of a stator winding—Further reading—Exercises

CHAPTER 20

THE ALTERNATOR ON LOAD 472

Rotating field produced by a polyphase winding—Non-stepped m.m.f. distribution—Fourier analysis of the m.m.f. wave—M.M.F. of a fractional-pitch winding—Effect of the rotor construction—Calculation of leakage reactance—Predetermination of voltage regulation—Separation of armature reaction and leakage reactance effects—Transient reactance of synchronous machines—Further reading—Exercises

CHAPTER 21

THE SYNCHRONOUS MOTOR 502

Torque—Single-phase machine—Three-phase machine—General properties of the motor—Vector diagram of motor—Operation with fixed excitation—Operation at constant power—The Blondel diagram—Conditions for stability—V characteristics—Torque-angle characteristic—Phase swinging or hunting—Further reading—Exercises

CHAPTER 22

THE OPERATION OF ALTERNATORS 528

Automatic voltage regulators—The Tirrill regulator—Parallel operation of alternators—Conditions necessary for successful parallel operation—Effect of a change in excitation—Effect of a change in steam supply—Effect of load on the synchronizing power—Solution by symbolic method—Division of load between two machines—Period of free oscillation of a synchronous machine—Cyclic irregularity—Damped oscillations—Further reading—Exercises

CHAPTER 23

THE INDUCTION MOTOR 550

Slip—Phase-wound and Cage-type resistors—Relation between slip and rotor I^2R loss—Torque—Starting torque—Methods of starting squirrel-cage motors—Operation over the normal range—Modes of operation of the motor—More accurate analysis of the torque-slip relationship—Another equivalent circuit—The resistance of the Cage rotor—Crawling of induction motors—Magnetic locking—The circle diagrams of the induction motor—The effect of magnetic leakage on the operation of the induction motor—Testing of induction motors—Squirrel-cage motors with high starting torque—Control of induction motors—The induction motor run as a synchronous motor—The induction motor run as a generator—The single-phase induction motor—Rotating coil in an alternating field—Further reading—Exercises

CHAPTER 24 PAGE

CONVERTING MACHINERY 624

Methods of conversion—The rotary convertor—Voltage and current ratios—Waveform of the armature current—Armature reaction in polyphase convertors—Methods of obtaining six-phase current—Methods of starting rotary convertors—Voltage regulation of rotary convertors—Parallel operation of rotary convertors—Further reading—Exercises

CHAPTER 25

SINGLE-PHASE COMMUTATOR MOTORS 644

The e.m.f. of a single-phase commutator motor—The single-phase series motor—The torque of a single-phase series motor—Commutation—Constructional features—Conditions necessary for a good power factor—The circle diagram—The effect of losses on the diagram—Another vector diagram—The repulsion motor—Mechanics of torque production—The compensated repulsion motor—The doubly-fed or series repulsion motor—Vector diagram of the repulsion motor—Further reading—Exercises

CHAPTER 26

THE POLYPHASE COMMUTATOR MOTOR 665

Action of the commutator—Induced e.m.f. equation—The three-phase series motor—The three-phase shunt commutator motor—The expedor—The susceptor—The stator-fed motor with induction regulator control—Commutation in the stator-fed machine—The Schrage motor—Theory of the Schrage motor—Cascade control with commutator machine as the rear machine—Further reading—Exercises

CHAPTER 27

THE MERCURY-ARC RECTIFIER OR CURRENT CONVERTOR . . 707

The mechanism of conduction—Influence of the vapour pressure—Arc volt drop—Constructions—Single-phase and three-phase rectification—Voltage and current relationships—Nature of the anode currents—Effect of transformer reactance—Voltage regulation—Star-connection of the transformer primary—Phase-equalizing circuits—Grid control of mercury-arc rectifiers—Inverted operation—The single-anode mercury-arc rectifier—Hot-cathode rectifiers—Metal rectifiers—Further reading

CHAPTER 28

COMPLEX WAVEFORMS 757

Effective value of a complex wave—Influence of the nature of the circuit on the shape of the current wave—Selective resonance—Power conveyed by complex waves—Effect of iron on the shape of the current wave—Waveform of an alternator—Determination of the harmonics in field form—Elimination of harmonics by special arrangements of the stator windings—Suppression of tooth ripples—Effect of the interconnection of the phases—Effect of waveform on iron losses—Effect of waveform on the operation of a synchronous motor or on two alternators in parallel—Effect of waveform on the drop in a transformer—Effect of harmonics on the form of the rotating field produced by a three-phase winding—Experimental determination of waveform—The cathode-ray oscillograph—Further reading—Exercises

CHAPTER 29

PHASE ADVANCING 787

Methods of obtaining an improved power factor—Economical limit of power-factor correction—Further reading—Exercises

CHAPTER 30 PAGE

ILLUMINATION 799

Radiant energy—Colour—Relative luminous efficiency of radiation—Radiant efficiency and luminous efficiency—Definitions—Notes on some of the units—Laws of illumination—Measurement of candle power—The illumination photometer—Photoelectric methods—Determination of m.h.-s.c.p. and m.s.c.p.—Incandescent lamps—Construction—Life of incandescent lamps—Effect of variations in voltage on operation—Starting characteristics—Discharge lamps—The electron volt—Wave-length of radiated emission—The hydrogen discharge—The sodium discharge—The mercury discharge—Ionization—The sodium-vapour lamp—The mercury-vapour lamp—The high-pressure mercury lamp—Calculation of an illumination scheme—Further reading—Exercises

Appendix 851

Index 857

PLATES

PLATE | | *facing page*

I. Bitter Patterns 124

II. Rotor Slot and Teeth of a Turbo-alternator with Hollow Conductors 464

III. End of Stator Winding of a Turbo-alternator with Hollow Conductors 465

IV. Torque-speed Curves for Different Induction Motor Rotor Constructions 480

V. Stator of a Scherbius Machine 481

CHAPTER 1

THE M.K.S. SYSTEM OF UNITS

It has been said that no science has inherited such a bewildering array of units as the science of electromagnetism. For example there are in everyday use the c.g.s. absolute electrostatic system of units, the c.g.s. absolute electromagnetic system, the system of practical units, and the system of international units. The student and practising engineer have to make calculations in terms of all of these, and frequently it is necessary, in a calculation, to convert from one set to another.

It would therefore appear illogical to make use of still another system, but the justification is based on the belief that it will bring order to what is at present chaos. The c.g.s. systems are based on so-called absolute units of length (the centimetre), mass (the gramme), and time (the second). The practical units, in which there is so much vested interest that their abandonment is quite out of the question, are the volt, the ampere and the units derived from them. Briefly, it can be said that the main function of the m.k.s. system is to combine the c.g.s. and the practical systems so that the practical units, the volt, ampere, watt, joule, coulomb, ohm, henry, and farad become both practical and absolute units. Only in the case of magnetism is it necessary to introduce new units.

Mechanical Units. We have for the new fundamental units of length, mass and time—

Length, $L = 1$ metre $= 10^2$ cm
Mass, $M = 1$ kilogramme $= 10^3$ g
Time, $T = 1$ second, as in the c.g.s. system

Velocity

$$v = L/T = LT^{-1}$$

Denote the m.k.s. unit by v and the c.g.s. unit by v', then

$$v/v' = LT^{-1}/[L'(T')^{-1}] = 10^2 \times 1/(1 \times 1) = 10^2$$

Angular Velocity

$$\omega = 1 \text{ radian/sec in both m.k.s. and c.g.s. systems}$$

Acceleration

$$a = LT^{-2}$$

$$\therefore a/a' = LT^{-2}/[L'(T')^{-2}] = 10^2 \times 1/(1 \times 1) = 10^2$$

Force

$$F = \text{mass} \times \text{acceleration} = MLT^{-2}$$

$$\therefore F/F' = MLT^{-2}/[M'L'(T')^{-2}] = 10^3 \times 10^2 \times 1/(1 \times 1 \times 1) = 10^5$$

The m.k.s. unit is the newton,

$\therefore$ 1 newton $= 10^5$ dynes.

Work and Energy

$$W = \text{force} \times \text{distance} = ML^2T^{-2}$$

$$\therefore W/W' = ML^2T^{-2}/[M'L'(T')^{-2}] = 10^3 \times 10^4 \times 1/(1 \times 1 \times 1) = 10^7$$

The m.k.s. unit of work is the joule, and the c.g.s. unit the erg, so that this is a restatement of the relationship

$$1 \text{ joule} = 10^7 \text{ ergs}$$

Power. Power is the rate of doing work,

$$\therefore P = \text{work/time} = ML^2T^{-3}$$

$$\therefore P/P' = 10^3 \times 10^4 \times 1/(1 \times 1 \times 1) = 10^7$$

The m.k.s. unit of power is the watt, so that this is a re-statement of the relationship

$$1 \text{ watt} = 10^7 \text{ erg/sec}$$

Torque. Torque is given by the quotient of power and angular velocity. Using the symbol M we have $M/M' = P/P' = 10^7$. The m.k.s. unit is the newton-metre,

$$\therefore 1 \text{ newton-metre} = 10^7 \text{ dyne-cm}$$

$$= [10^7/(981 \times 454)]\,[1/(2{\cdot}54 \times 12)]$$

$$= 0{\cdot}7373 \text{ lb-ft}$$

The Values of μ_0 and ε_0. Although it is possible to define a mechanical system in terms of the three fundamental quantities, length, mass, and time, it is necessary to introduce a fourth quantity in order that an electrical system may be defined. One possible fourth fundamental quantity is the absolute permeability of empty space. This is taken to be unity in the c.g.s. system, being based on the fact that, in terms of lines of force, B and H are numerically equal. In the m.k.s. system the permeability of free space, or preferably, the magnetic space constant, denoted by μ_0, is not unity. Similarly, the permittivity of free space, or preferably, the electric space constant, ε_0, is not unity. These two quantities are related as follows,

$$1/\varepsilon_0\mu_0 = c^2$$

where c is the velocity with which electromagnetic phenomena are propagated in free space. The value of c is, very approximately

$$c = 3 \times 10^{10} \text{ cm/sec} = 3 \times 10^{8} \text{ m/sec}$$

$$\therefore\ \varepsilon_0\mu_0 = 1/c^2 = 10^{-16}/9$$

It appears to be impossible to state separately the magnitudes of ε_0 and μ_0. Since we can give any particular unit any magnitude we like, provided that this value is reflected in the magnitude of all the other units, we can give a value to one of μ_0 and ε_0. The values chosen for M, L and T, and the retention of the practical units, impose the value 10^{-7} for μ_0, and therefore

$$\varepsilon_0 = (10^{-16} \times 10^{7})/9 = 10^{-9}/9 = (10/9) \times 10^{-10}$$

Thus $\quad \varepsilon_0 = 1{\cdot}113 \times 10^{-10}$

(giving c the more accurate value of $2{\cdot}998 \times 10^{8}$ m/sec).

The reason for this particular value of μ_0 is as follows. The dimensional equations for e.m.f. and current in the c.g.s. electromagnetic system are

$$[E] = \mu^{\frac{1}{2}}M^{\frac{1}{2}}L^{\frac{3}{2}}T^{-2}$$

$$[I] = \mu^{-\frac{1}{2}}M^{\frac{1}{2}}L^{\frac{1}{2}}T^{-1}$$

The unit of time remains the same, namely the second. Let the unit length, mass, and permeability be changed in the ratio 10^{l}, 10^{m} and 10^{p} respectively. Then the units of e.m.f. and current are changed in the ratios $10^{(3l+m+p)/2}$ and $10^{(l+m-p)/2}$ respectively.

Now $\quad$ one volt $= 10^{8}$ c.g.s. electromagnetic units (abvolts)

and $\quad$ one ampere $= 10^{-1}$ c.g.s. electromagnetic units (abamps)

Hence, if the volt and the ampere are to be retained we must have

$$\left.\begin{aligned} \tfrac{1}{2}(3l + m + p) &= 8 \\ \text{and} \quad \tfrac{1}{2}(l + m - p) &= -1 \end{aligned}\right\}$$

Thus there are only two equations for three unknowns, and one of the three must be chosen arbitrarily, the only conditions being that, since the metre system is to be retained, all three values must be integers. Two solutions which fulfil these requirements are

$$l = 2,\ m = 3,\ p = 7$$

$$l = 9,\ m = -11,\ p = 0$$

The first of these provides the basis of the unrationalized m.k.s. system, the units of length and mass thus being the metre and kilogramme respectively. The new unit for μ is 10^{7}, which means that a medium whose permeability is 10^{7} times that of free space is of unit permeability in the m.k.s. system. Consequently the permeability of free space is $\mu_0 = 10^{-7}$.

The objection to the use of μ_0 as the connecting link between the electrical and mechanical units is that it is not possible to make a standard of μ_0 for use in standardizing laboratories. Consequently, after much discussion it was decided to use the ampere, defined from its magnetic effect, as the fourth fundamental quantity.

In the c.g.s. system, when we make use of the relationship

$$B = \mu H$$

we tend to ignore the dimensions of μ because we say that $B = H$ when $\mu = 1$. In the new system we are not able to ignore the space constant μ_0 and consequently we have to use the relationship

$$B = \mu\mu_0 H$$

where μ is the relative permeability, as before, and $\mu_0 = 10^{-7}$.

Similarly for the relationship between dielectric flux density D, and electric field strength E, we have

$$D = \varepsilon\varepsilon_0 E$$

It is very essential to realize that B and H can never be equal to one another and that D and E can never be equal to one another. Hence the force on a current-carrying conductor in a magnetic field is BIl, *not* HIl. The capacitance of a capacitor is $DA/(4\pi \times \text{voltage})$, *not* $EA/(4\pi \times \text{voltage})$.

Magnetic Flux. The magnetic line of flux has no real existence, but the conception is so widespread that it is doubtful if it would be practicable to express a flux in terms of anything else. The m.k.s. unit of magnetic flux is the *weber* and it is equal to 10^8 maxwells. This new unit tidies up the expressions for induced e.m.f. because the familiar term 10^{-8} disappears. Thus, for the instantaneous e.m.f., e, induced in a coil of T turns by a rate of flux change $d\Phi/dt$ we have

$$e = -\, Td\Phi/dt \text{ volts}$$

This gives rise to new values of B and H. Thus the m.k.s. unit of flux density is

$$\begin{aligned} B' &= \text{webers/metre}^2 \\ &= 10^8/10^4 = 10^4 B \\ H &= B/\mu\mu_0 = 10^{-4}B'/(\mu \times 10^{-7}) \\ &= 10^3 \times B'/\mu \\ &= 10^3 H' \end{aligned}$$

Thus one m.k.s. unit of m.m.f. gradient H is equal to 10^{-3} oersted.

Rationalization. Before discussing the effects of the new magnitudes of the fundamental units on the expressions, and for the purpose of magnetic and electric calculations, it is necessary to consider the process known as rationalization.

When trying to form a mental picture of electrostatic and electromagnetic phenomena we imagine lines of force in some medium, and in the c.g.s. and the unrationalized m.k.s. systems we picture 4π lines radiating from a unit charge or from a unit pole. This involves fractional numbers of lines, a matter of little moment, but it also introduces π into expressions other than those expressions related to the circle or the sphere, where its appearance would be quite natural. For example, the capacitance of a parallel-plate capacitor with homogeneous dielectric is

$$C = \varepsilon\varepsilon_0 A/4\pi d$$

while that of a sphere of radius r is

$$C = \varepsilon\varepsilon_0 r$$

Similarly with magnetic calculations.

Rationalization is merely the removal of the quantity π from situations where it does not naturally belong, e.g. in the expression for the capacitance of a parallel-plate capacitor, and its transference to situations where it does belong, e.g. to the expression for the capacitance of a sphere. This is accomplished by replacing

ε_0 by $4\pi\varepsilon_{r0}$ where $\varepsilon_{r0} = 1{\cdot}113 \times 10^{-10}/4\pi = 8{\cdot}854 \times 10^{-12}$

and μ_0 by $\mu_{r0}/4\pi$ where $\mu_{r0} = 10^{-7} \times 4\pi = 1{\cdot}257 \times 10^{-6}$*

It is important to note that rationalization is not something peculiar to the m.k.s. system: it could, if desired, be applied to any of the other systems, e.g. the c.g.s. system.

For further information on this subject of fundamental importance, reference should be made to the following books and papers.

Further Reading

"Symposium of papers on the m.k.s. system of units." *J. Instn Elect. Engrs.* **97**, Pt. 1 (1950).

McGreevy, T. *The M.K.S. System of Units.* Pitman, London, 1953.

Bradshaw, E. *Electrical Units.* Chapman & Hall London, 1952.

Carr, L. H. A. "The rationalized m.k.s. system of units." *M. V. Gaz.* (March, 1952).

* N.B. In subsequent discussions the double suffix is dropped, ε_0 and μ_0 being the rationalized values unless othe wise stated.

CHAPTER 2

ELECTROSTATICS

IT is now known that the atom of any given element, instead of being indivisible, is made up of components which, by virtue of their properties can be regarded as particles. These particles are all entirely different in their nature from the atom of which they form a part. The particles are electrons, protons and neutrons.

An electron is a particle of negative electricity, which when removed from an atom, exhibits none of the properties of ordinary matter. All electrons are exactly alike, even if the atoms from which they are derived are different. The mass of an electron is $9{\cdot}11 \times 10^{-28}$ g.

The proton is the opposite of the electron, being a particle of positive electricity. The positive charge on the proton is exactly equal to the negative charge on the electron, but the mass of the proton is so very much greater than that of the electron that to all intents and purposes the mass of an atom can be said to be the sum of the masses of all the protons in it. The mass of a proton is $1{\cdot}63 \times 10^{-24}$ g.

Neutrons are particles of the same mass as the protons, but they have no charge. Protons and neutrons together constitute the nucleus of the atom. The electrons are external to the nucleus.

The structure of an atom is according to the Bohr theory, that of a solar system in miniature,* the central sun the nucleus, and the planets a series of electrons rotating about the nucleus in a series of orbits. The simplest atom is that of hydrogen, which has a single proton for its nucleus and a single electron rotating round the nucleus. The radius of the orbit is of the order of 10,000 times the diameter of the electron. In more complex atoms the nucleus may have protons and neutrons together. Thus, in the case of helium, the nucleus has 2 protons and 2 neutrons, while, as planets, there are 2 electrons. A more complicated case is that of the copper atom in which the nucleus has 29 protons and 35 neutrons, while, as planets, there are 29 electrons.

In the case of all elements the atom is electrically neutral, since there are exactly as many electrons as protons. The atomic weight is (neglecting the masses of the electrons) given by the number of protons and neutrons in the atom. The atomic number of an element is equal to the number of protons in the nucleus, or, what amounts to the same thing, the number of planetary electrons. It is this number which determines the chemical properties of the atom. So far, the greatest known atomic number is 92, that for uranium, and therefore, if we assume that there must be an element to each atomic number, there must be at least 92 elements.

* This is not in accordance with the most recent views, but it is sufficiently accurate for the purpose of a mental picture.

Electrification. Since the aggregate of the positive electricity of the protons of any atom is exactly equal to the aggregate of negative electricity of the electrons, it follows that any normal atom is not electrified from the point of view of having any electrical influence external to itself. If some of the outer electrons are removed then the atom remains an atom, but its chemical nature may be altered, and, what is immediately obvious, the balance of positive to negative electricity will be upset. There will, in fact, be a surplus of positive electricity, and, if this removal of some of the electrons has been common to all the atoms in a body, this body will be charged positively, or will have acquired a positive charge.

If, on the other hand, an electron is added instead of removed there will be a preponderance of negative electricity, and if this addition has been made to all the atoms of a body this body will be charged negatively.

The process by which an originally neutral atom becomes positively charged through the removal of electrons is called ionization, and the atom itself is called an ion.

Summarizing, we can say that positive electrification is the result of a deficiency of electrons, while negative electrification is the result of an excess of electrons. The total deficiency, or excess, of electrons is called the charge.

Coulomb's Law. It is well known that bodies with like charges repel one another, and that bodies with unlike charges attract. In the case of bodies which are so small that their charges can be regarded as point charges, the magnitude of this force in air, or more strictly *in vacuo*, is, in the c.g.s. system

$$f = q_1 q_2 / d^2 \text{ dynes}$$

where q_1 and q_2 are measured in electrostatic units of charge, and d is in cm. In the unrationalized m.k.s. system we have

$$f = q_1 q_2 / \varepsilon_0 d^2 \text{ newtons}$$

where q_1 and q_2 are in coulombs, d is in metres, and ε_0 is the permittivity of free space. We have seen that

$$\varepsilon_0 \mu_0 = 1/c^2 = 1{\cdot}113 \times 10^{-17}$$

and $$\mu_0 = 10^{-7}$$

$$\therefore \ \varepsilon_0 = 1{\cdot}113 \times 10^{-10}$$

To effect rationalization we modified the values of the space constants as follows

$$\mu_0 = 4\pi \times 10^{-7}$$

Hence, to retain the relationship $\varepsilon_0 \mu_0 = 1/c^2$, we have

$$\varepsilon_0 = 8{\cdot}854 \times 10^{-12}$$

The Coulomb's law relationship thus becomes

$$f = q_1 q_2 / 4\pi\varepsilon_0 d^2 \text{ newtons}$$

Example. Calculate the electrostatic force of repulsion between two α-particles separated a distance of 10^{-10} cm, given that the charge on the α-particle is $3{\cdot}2 \times 10^{-19}$ coulomb.

$$f = q^2 / 4\pi\varepsilon_0 d^2$$

$$q = 3{\cdot}2 \times 10^{-19} \text{ coulomb}$$

$$d = 10^{-10} \text{ cm} = 10^{-12} \text{ metre}$$

$$\therefore f = 3{\cdot}2^2 \times 10^{-38}/(4\pi \times 8{\cdot}854 \times 10^{-12} \times 10^{-24})$$

$$= 9{\cdot}18 \times 10^{-4} \text{ newton}$$

Since one newton $= 10^5$ dynes and 1 g $= 981$ dynes,

$$f = 91{\cdot}8 \text{ dynes} = 0{\cdot}094 \text{ g}$$

In a medium other than air, then, provided that the medium is an insulator, i.e. not a conductor of electricity, we have

$$f = q_1 q_2 / 4\pi\varepsilon\varepsilon_0 d^2 \text{ newtons}$$

where ε is the relative permittivity of the medium.

The Electric Field. Imagine two neighbouring charges, q_1 and q_2: there will be a force of mutual attraction or repulsion (according to the signs of the charges, i.e. whether unlike or like) which will be directed along the straight line joining the charges. The magnitude of this force will be given by Coulomb's Law. At the point occupied by the charge q_2 we can imagine that the electrical state of the space is caused by the other charge q_1. This electrical state, or stress, is the result of the electric field caused by q_1. The property of an electric field that causes it to exert a force on a charge introduced into it is called the field intensity, E. It is given by the relationship

$$E = f/q \text{ newton/coulomb}$$

Imagine then, a point charge q and a unit charge distant d from it, then the force on the unit charge is

$$f = q/4\pi\varepsilon\varepsilon_0 d^2 \text{ newtons}$$

But the force on a unit charge is numerically equal to E, so that the field intensity at a point distant d metres from a point charge of q coulombs is

$$E = q/4\pi\varepsilon\varepsilon_0 d^2 \text{ newtons/coulomb.}$$

We shall now see that the newton per coulomb is the same as the volt per metre.

The field intensity is a vector quantity, and therefore we can imagine a line, the familiar line of force, drawn in such a way that, at any point on it, the tangent is directed along the direction of the intensity. Imagine a point charge of q coulombs placed on such a line of force, then the force acting on it will be qE newtons directed along the tangent at that point. The work done in moving the charge a distance dl along the line of force is thus,

$$dW = qEdl$$

For any finite distance the work done is

$$W = \int qEdl \text{ joules}$$

Now let the difference of potential between the points at the beginning and end of the path taken by q be $V_1 \sim V_2$ volts. Then, if V_1 is the higher potential and the charge is moved against the potential gradient, q coulombs will have been moved through a potential difference of $(V_1 - V_2)$ volts, the work done therefore being

$$W = q(V_1 - V_2) \text{ joules}$$

Hence

$$q\int Edl = q(V_1 - V_2)$$

$$\therefore \; Edl = -dV$$

the significance of the negative sign being that the work done is positive when the charge is moved "up" the potential gradient, that is from a point of lower to one of higher potential

$$\therefore \; E = -dV/dl \text{ volts/metre}$$

Example 1. An electron of charge $1{\cdot}6 \times 10^{-19}$ coulomb is placed in a field of 10^5 newtons/coulomb. If the force on the electron is directed vertically downwards and the mass of the electron is $9{\cdot}1 \times 10^{-28}$ g, calculate the total downward force acting on the electron.

Electric force

$$f_e = Ee$$

$$= 10^5 \times 1{\cdot}6 \times 10^{-19} = 1{\cdot}6 \times 10^{-14} \text{ newtons}$$

where e is the charge of the electron.

Gravitational force

$$f_g = mg$$

$$m = 9{\cdot}1 \times 10^{-28} \text{ g} = 9{\cdot}1 \times 10^{-31} \text{ kg}$$

$$g = 981 \text{ cm/sec}^2 = 9{\cdot}81 \text{ m/sec}^2$$

$$\therefore \; f_g = 9{\cdot}1 \times 10^{-31} \times 9{\cdot}81 = 89{\cdot}3 \times 10^{-31} \text{ newtons}$$

$$\therefore \; f_e + f_g \simeq 1{\cdot}6 \times 10^{-14} \text{ newtons}$$

the gravitational force being negligibly small.

Example 2. If the electron can move freely in the field of Example 1 for a distance of 2 cm, and if it starts with an initial velocity of zero, what velocity will it attain, what will be the time taken, and what will be its kinetic energy.

Neglecting the gravitational force, we have for the acceleration a

$$a = f_e/m = 1{\cdot}6 \times 10^{-14}/(9{\cdot}1 \times 10^{-31})$$

$$= 1{\cdot}76 \times 10^{16} \text{ m/sec}^2$$

Distance $d = 2 \times 10^{-2}$ metre

$$\therefore v = \sqrt{(2ad)} = (2 \times 1{\cdot}76 \times 10^{16} \times 2 \times 10^{-2})^{\frac{1}{2}}$$
$$= 2{\cdot}66 \times 10^{7} \text{ m/sec}$$

$$\text{Time } t = v/a = 2{\cdot}66 \times 10^{7}/(1{\cdot}76 \times 10^{16}) = 1{\cdot}51 \times 10^{-9} \text{ sec}$$

$$\text{Kinetic energy} = \tfrac{1}{2}mv^2 = \tfrac{1}{2} \times 9{\cdot}1 \times 10^{-31} \times (2{\cdot}66 \times 10^{7})^2$$
$$= 3{\cdot}21 \times 10^{-16} \text{ joules}$$

Field due to a Straight Uniformly Charged Wire. The intensity, in *vacuo*, at a distance d from a point charge of q coulombs is $q/4\pi\varepsilon_0 d^2$

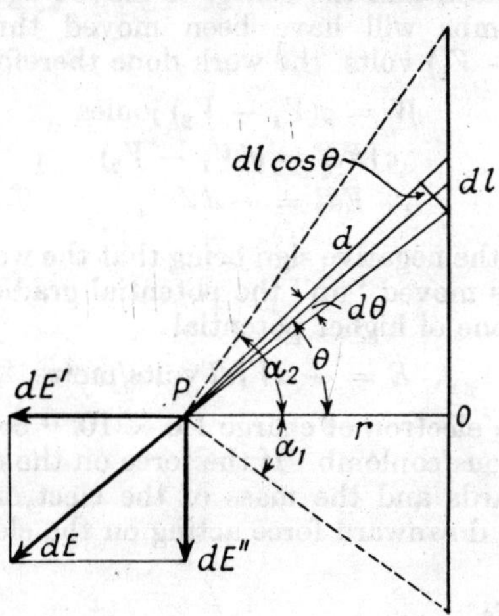

Fig. 2.1. Intensity due to a Straight Uniformly Charged Wire

newton per coulomb. If the wire has a charge of λ coulombs per metre (Fig. 2.1), then the intensity dE at point P due to an element dl is

$dE = \lambda dl/4\pi\varepsilon_0 d^2$ *in vacuo*

or $dE = \lambda dl/4\pi\varepsilon\varepsilon_0 d^2$ in a medium of relative permittivity ε

This will be directed *away* from the element, as shown in Fig. 2.1.

Changing the variable we have

$$dl \cos\theta = d \,.\, d\theta$$
$$\therefore dl = d \sec\theta d\theta$$

Again $d = r\sec\theta$

$$\therefore dE = d\lambda \sec\theta d\theta/(4\pi\varepsilon\varepsilon_0 d \,.\, r\sec\theta) = (\lambda/4\pi\varepsilon\varepsilon_0 r)d\theta$$

The radial component of dE is

$$dE' = dE\cos\theta$$
$$= (\lambda/4\pi\varepsilon\varepsilon_0 r)\cos\theta d\theta$$

If the ends of the wire are at angular distances α_1 and α_2 from the point P then, adopting the convention that α_2 is positive and α_1 negative in the figure, the total radial component becomes

$$E' = \frac{\lambda}{4\pi\varepsilon\varepsilon_0 r} \times \int_{\alpha_1}^{\alpha_2} \cos\theta d\theta$$

$$E' = [\lambda/(4\pi\varepsilon\varepsilon_0 r)] (\sin\alpha_2 - \sin\alpha_1)$$

Denoting the longitudinal component by E'' we have

$$E'' = \frac{\lambda}{4\pi\varepsilon\varepsilon_0 r}\int_{\alpha_1}^{\alpha_2} \sin\theta d\theta$$

If the wire is so long in comparison with the distance r that α_1 and α_2 both approach $\pi/2$ we have

$$E' = \lambda/2\pi\varepsilon\varepsilon_0 r, \text{ and } E'' = 0$$

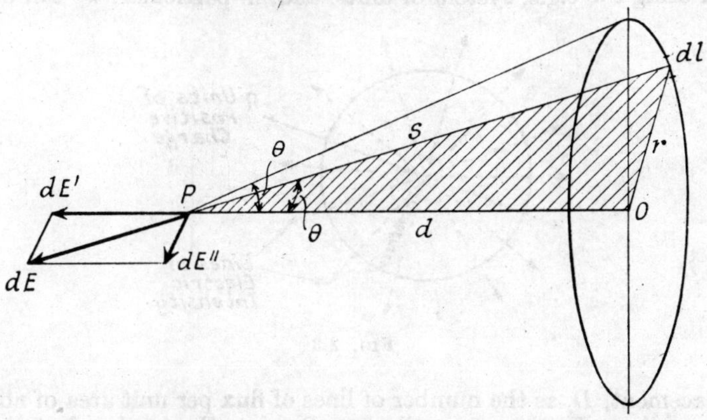

FIG. 2.2. INTENSITY ON THE AXIS OF A UNIFORMLY CHARGED RING

Field at a Point on the Axis of a Uniformly Charged Ring. Again let the charge be λ coulombs per metre, then intensity at point P (Fig. 2.2) due to an element dl is

$$dE = \lambda dl/4\pi\varepsilon\varepsilon_0 s^2 \text{ in the general case}$$

Axial component

$$dE' = dE\cos\theta = \lambda\cos\theta\, dl/4\pi\varepsilon\varepsilon_0 s^2$$

$\therefore$ Total axial component

$$E' = \frac{\lambda\cos\theta}{4\pi\varepsilon\varepsilon_0 s^2}\int dl$$

$$= (\lambda/4\pi\varepsilon\varepsilon_0)(d/s^3)2\pi r$$

Now $2\pi r\lambda = q$, the total charge on the ring

$$\therefore\ E' = qd/[4\pi\varepsilon\varepsilon_0(r^2 + d^2)^{\frac{3}{2}}]$$

If d is very great in comparison with r then

$$E' = q/4\pi\varepsilon\varepsilon_0 d^2$$

which means that, at such a distance, the ring acts like a point charge.

As in the analogous magnetic case we see that, by considering diametrical elements in pairs, the total quadrature component E'' is zero.

Gauss's Theorem. Imagine a point charge of q coulombs of positive electricity situated at the centre of a sphere of radius r metres (Fig. 2.3), then the intensity at any point on the surface of the sphere is

$$E = q/4\pi\varepsilon\varepsilon_0 r^2 \text{ newton/coulomb}$$

We can represent the properties of electric fields by lines of force as when using the c.g.s. system of units, and, in particular, we can define

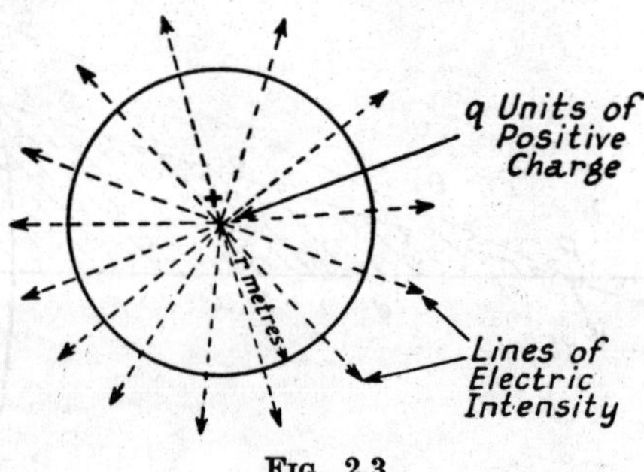

FIG. 2.3

displacement, D, as the number of lines of flux per unit area of normal cross section. In the magnetic case $B = \mu\mu_0 H$; in the electric case $D = \varepsilon\varepsilon_0 E$. Obviously, in the case of Fig. 2.3, all the lines of force are normal to the spherical surface, and therefore the total flux is

$$\Psi = D \times 4\pi r^2 = (\varepsilon\varepsilon_0 q/4\pi\varepsilon\varepsilon_0 r^2) \times 4\pi r^2 = q$$

Now if we imagine a closed surface of any shape whatsoever enclosing the charge q then, provided this surface does not pass through material of different ε, the flux across this surface must be the same as that across the surface of the sphere, namely q. Again, if the charge is distributed it can be regarded as made up of a number of elementary charges small enough to be regarded as point charges. Hence, in a homogeneous medium we see that if a charge q of any distribution whatsoever is enclosed within a surface, the total flux crossing this surface is q. This is known as Gauss's Law.

Applications of Gauss's Theorem. (a) **Field due to a Long Straight Charged Conductor.** For an element of length dl the lines of force will be radially outwards, so that a suitable gaussian surface is a concentric cylinder of any radius r, and the same axial length dl (Fig. 2.4). With a charge of λ coulombs per unit length the total charge within the surface is

$$q = \lambda dl$$

$\therefore$ Flux per unit of length

$$\Psi = \lambda dl$$

Surface area crossed by this flux

$$a = 2\pi r dl$$

$\therefore$ Intensity at radial distance r,

$$E = D/\varepsilon\varepsilon_0 = \Psi/\varepsilon\varepsilon_0 a$$
$$= \lambda/2\pi\varepsilon\varepsilon_0 r, \text{ as before}$$

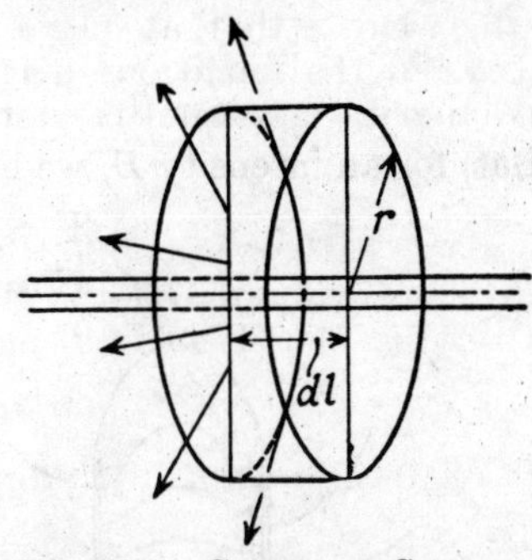

FIG. 2.4. GAUSSIAN SURFACE ROUND AN ELEMENT OF A STRAIGHT WIRE

(b) **Field due to a Charged Sphere.** Let the surface density of charge be σ coulombs per square metre, then total charge on a sphere of radius r is

$$q = 4\pi r^2\sigma \text{ coulombs}$$

A convenient gaussian surface for a point distant R metres from the centre is obviously a concentric sphere of radius R, and of surface $4\pi R^2$. The total flux will be radial, because of radial symmetry, and of magnitude

$$\Psi = q = 4\pi r^2\sigma$$

Hence intensity at a radial distance R

$$E = (4\pi r^2\sigma/\varepsilon\varepsilon_0) \div 4\pi R^2$$
$$= (r/R)^2(\sigma/\varepsilon\varepsilon_0)$$

If we express E in terms of the total charge q, we have

$$E = q/4\pi\varepsilon\varepsilon_0 R^2$$

which, being independent of r, the radius of the sphere, shows that, to points external to it, the sphere behaves as though its charge were concentrated at the centre.

(c) **Field between Two Charged Parallel Plates.** The distribution of flux between two oppositely charged parallel plates is indicated in Fig. 2.5 (a). At the middle the lines of force are normal to the surfaces and the intensity is uniform, but towards the edges the lines bulge outwards as shown. In order to obtain a uniform field in practice, it is therefore usual to surround one surface with a guard plate, separated from the plate itself by a small annular gap, the two being in electrical connection (Fig. 2.5 (b)). Since it is thus possible to realize a uniform field in practice consider the uniform portion of Fig. 2.5 (a). The gaussian surface is a pillbox of lid area dA (say).

Let σ be the surface density at the uniform portion, then the charge enclosed by the surface is

$$q = \sigma dA$$
$$\therefore \Psi = \sigma dA$$

But, for an intensity E, we have

$$\Psi = DdA = \varepsilon\varepsilon_0 EdA$$
$$\therefore E = \sigma/\varepsilon\varepsilon_0$$

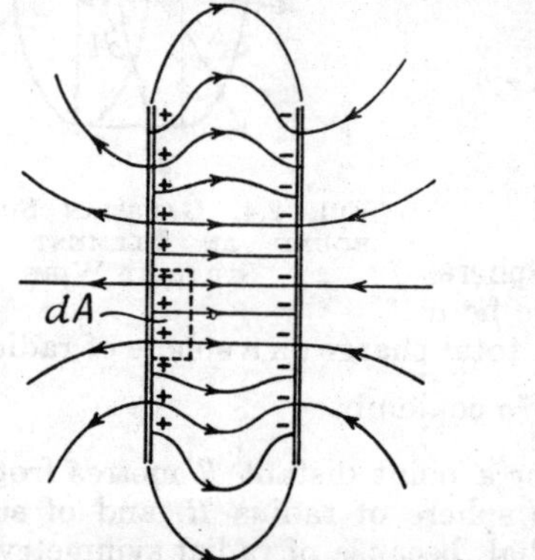

(a) Field Distribution (b) Plate with Guard Plate

FIG. 2.5. OPPOSITELY CHARGED PARALLEL PLATES

Field Inside a Charged Hollow Spherical Conductor. Fig. 2.6 shows a hollow sphere charged to a surface density σ. Consider a cone with apex at any point P and cutting off small areas ω_1 and ω_2. The charges on the two areas are thus $\sigma\omega_1$ and $\sigma\omega_2$ and the total intensity due to these at point P is

$$E = \sigma\omega_1/4\pi\varepsilon\varepsilon_0 r_1^2 - \sigma\omega_2/4\pi\varepsilon\varepsilon_0 r_2^2$$

The projection of area ω_1 perpendicular to the axis of the cone is $\omega_1 \cos\phi_1$, and therefore if the solid angle of the cone is δs, we have

$$\omega_1 \cos\phi_1/\delta s = r_1^2/1$$
$$\therefore \delta s = \omega_1 \cos\phi_1/r_1^2$$

Similarly $$\delta s = \omega_2 \cos\phi_2/r_2^2$$

Also $$\phi_1 = \phi_2$$

$$\therefore \omega_1/r_1^2 = \omega_2/r_2^2$$
$$\therefore E = 0$$

Similarly we can subdivide the whole surface into pairs of small areas such as ω_1 and ω_2 and, since each pair will produce zero intensity at P, it follows that the intensity is zero for the whole sphere. Now P

is any point whatsoever, and therefore the intensity at all points inside the sphere is zero. We have seen that intensity can be represented in terms of lines of force, and therefore it follows that there can be no lines of force inside a closed charged sphere. This applies, in general, to a closed surface of any shape.

Coulomb's Theorem. Consider a small area ω of a charged surface, Fig. 2.7, and let the surface density be σ. The total charge on the area is $\sigma\omega$. Hence, by Gauss's theorem the total flux radiating from this surface is $\sigma\omega$.

If the surface under consideration is a conducting surface, then we

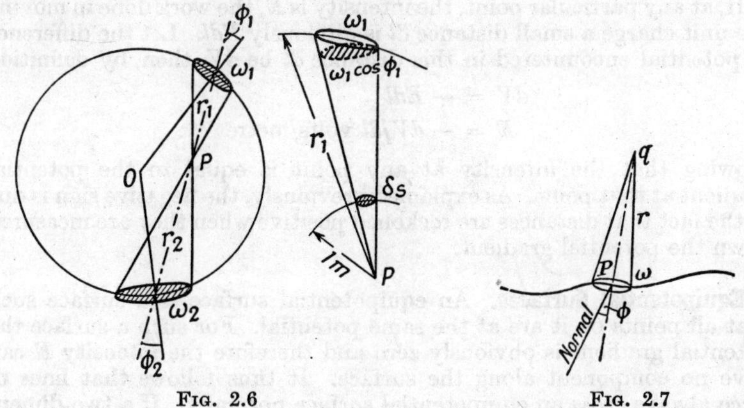

Fig. 2.6 Fig. 2.7

have seen that no flux can exist inside, and consequently the whole of the flux passes outward normally. The induction density at the surface is therefore

$$\sigma\omega \div \omega = \sigma$$

and this, expressed in newtons per coulomb, is the intensity at a point close to the surface. It is in a direction normal to the surface.

Potential. If a mass is moved from one point to another in the gravitational field, work will be done, either against, or by, the force of gravity, according as the vertical distance of the mass above the earth is increased or decreased. Also, provided the starting point and finishing point are the same, the amount of work done will be the same, no matter what path may be taken. Before the operation the mass will have, in general, a finite potential energy, and the work done during the positional change will be added to, or deducted from, this initial energy. Hence, the total potential energy of the mass is a function of the position of the mass with respect to the earth. The case of the electric field round a charged body is analogous to that of the gravitational field round the earth. If a small charge is introduced into the electric field it will be repelled or attracted by the body, and any change of position will, in general, necessitate that work shall be done against, or by, the force acting on the charge. We can thus regard potential in the electric field in exactly the same way as we regard potential in the gravitational field.

In general, the potential at any point in an electric field is defined as the work done, in joules, in moving a positive charge of one coulomb from a point at infinity to that point. The potential difference between two points in the field is the work done, in joules, in moving a positive charge of one coulomb from the point of lower potential to the point of higher potential. The potential difference is independent of the route taken by the positive charge.

If, at any particular point, the intensity is E, the work done in moving the unit charge a small distance dl is obviously Edl. Let the difference of potential encountered in this distance dl be dV, then, by definition

$$dV = -Edl$$

$$E = -dV/dl \text{ volts/metre}$$

showing that the intensity at any point is equal to the potential gradient at that point. As explained previously, the negative sign is due to the fact that distances are reckoned positive when they are measured down the potential gradient.

Equipotential Surfaces. An equipotential surface is a surface such that all points on it are at the same potential. For such a surface the potential gradient is obviously zero and therefore the intensity E can have no component along the surface. It thus follows that lines of force always cross an equipotential surface normally. If a two-dimensional drawing is made of the lines of force in any particular electric field, then the equipotential lines (which in such a drawing are the intersections of the plane of the paper with the equipotential surface) must cut the lines of force at right angles.

Potential due to a Point Charge. Let the intensity at a point P (Fig. 2.8 (a)) be E and let a point charge of one coulomb be moved a

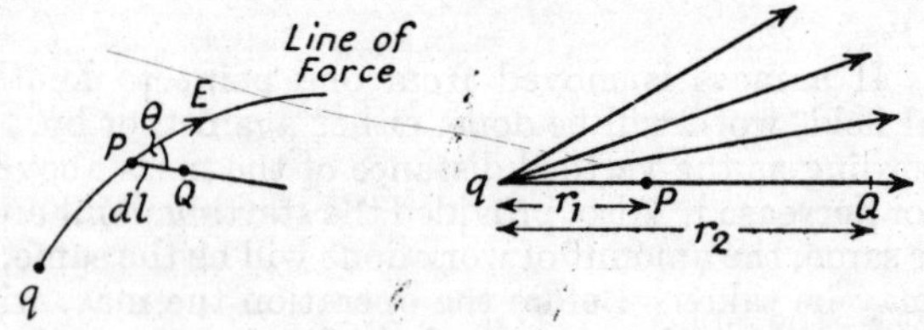

Fig. 2.8

distance dl in any direction θ to the direction of the intensity at P, to a new point Q. Then the potential difference between P and Q is

$$dV = -E\cos\theta \,.\, dl$$

If the field is due to a point charge q and the unit charge is moved along a radial line of force, Fig. 2.8 (*b*), $\theta = 0$ and

$$dV = -\,Edl$$

Let the limits of movement of the unit charge along this line of force be radial distances r_1 and r_2. At any radial distance r

$$E = q/4\pi\varepsilon\varepsilon_0 r^2$$

Hence the potential difference $V_{r_1r_2}$ between the points P and Q is

$$V_{r_1r_2} = -\int Edl, \text{ and } dl = dr$$

$$= -\int_{r_2}^{r_1} \frac{q}{4\pi\varepsilon\varepsilon_0} . dr/r^2$$

$$= (q/4\pi\varepsilon\varepsilon_0)(1/r_1 - 1/r_2)$$

The absolute potential at a point at radial distance r_1 is obtained by putting $r_2 = \infty$, since the absolute potential is defined as the work done in bringing a unit charge from infinity to that point. Hence, for the absolute potential at any radial distance r, we have

$$V_r = q/4\pi\varepsilon\varepsilon_0 r$$

The potential gradient is therefore

$$G = -\,dV_r/dr = q/4\pi\varepsilon\varepsilon_0 r^2 = E \text{ volts/metre}$$

Thus the potential gradient at any point is equal to the intensity at that point.

Charged Sphere. Let the sphere be of radius a. At all points inside the sphere the potential is the same as at the surface, and therefore since we can regard the charge as being concentrated at the centre, we have for this potential

$$V_a = q/4\pi\varepsilon\varepsilon_0 a$$

For any point distant r ($> a$) from the centre, we have

$$V_r = q/4\pi\varepsilon\varepsilon_0 r$$

and if we consider the case of a sphere in air

$$V_r = q/4\pi\varepsilon_0 r$$

For radii > 0 but $\not> a$ the potential is constant, since a is a constant, and therefore the gradient is zero. For points outside the sphere we have

$$E = q/4\pi\varepsilon_0 r^2$$

showing that the sphere is a surface of maximum potential. The variations of potential and of potential gradient with respect to radial distance are shown in Fig. 2.9.

This is a case of very great practical importance because there is a limit to the potential gradient which can be imposed on air at atmospheric pressure. This limit is 3×10^6 volts per metre, and above this the air becomes conducting owing to the phenomenon of ionization. Now at the surface

$$V_a = q/4\pi\varepsilon_0 a \text{ and } E_a = q/4\pi\varepsilon_0 a^2$$
$$\therefore\ V_a = aE_a$$

Substituting the limiting value 3×10^6 for G_a ($= E_a$) we have the relationship between the maximum possible potential to which a sphere can be raised and the radius of the sphere.

$$V_{max} = 3 \times 10^6\, a \text{ volts, where } a \text{ is in metres}$$

Thus if we make a equal to one metre, the maximum potential to which it can be raised is three million volts.

If we make a very small then the value of V_{max} decreases accordingly, and this accounts for the well-known action of points which, at

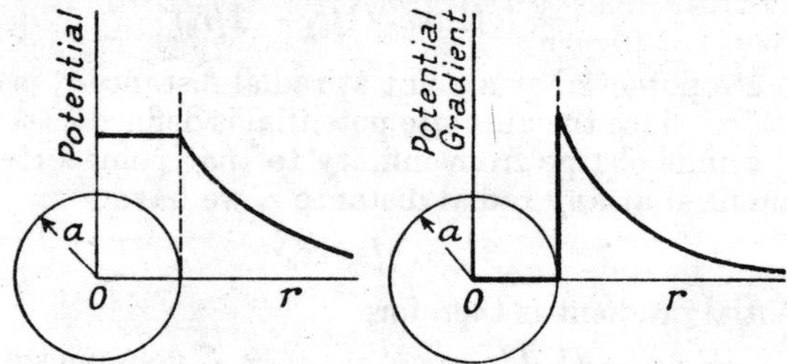

FIG. 2.9. POTENTIAL AND POTENTIAL GRADIENT DUE TO A CHARGED SPHERE

quite small potentials, set up such intense local potential gradient that ionization results.

For a sphere of radius a the surface area is $4\pi a^2$ and therefore $E_a = \sigma/\varepsilon_0$. If the medium is of permittivity ε then $E_a = \sigma/\varepsilon\varepsilon_0$.

Capacitance. Consider a system composed of two insulated conductors disposed in such a way that, if a charge of electricity is imparted to one of them, all the lines of force emanating from this charge will end on the other conductor. Then, if the charge is $+ q$ on one conductor, there will be an induced charge of $- q$ on the other. At any point in the field the intensity is given by

$$E = - dV/dl$$

and therefore the potential difference between the two conductors will be given by

$$V = - \int E dl$$

But, from Coulombs' law we know that V is directly proportional to q, since V is proportional to E. For any particular system the ratio of q to V is a constant, depending only on the size and shape of the electric field and the medium in which the lines of force exist. Denoting the proportionality constant by C we have

$$C = q/V$$

This proportionality constant is called the capacitance, and we see that the capacitance of a system is the ratio of its charge to its potential.

This also applies to an isolated conductor, but in this case we have to take infinity as the reference level of potential. With q expressed in coulombs and V in volts, C is given in farads. The farad is a very large unit and in practice it is more convenient to express the capacitance of an actual system in microfarads (μF), one microfarad being equal to 10^{-6} farad.

(*a*) Isolated Conductor. In the case of a charged isolated sphere of radius a we have seen that the potential is

$$V = q/4\pi\varepsilon_0 a \textit{ in vacuo}$$

or, in general

$$V = q/4\pi\varepsilon\varepsilon_0 a$$

$$\therefore\ C = q/V = 4\pi\varepsilon\varepsilon_0 a \text{ farads}$$

We note that the capacitance is proportional to the first power of the linear dimension a. In general, for systems which are geometrically similar, the capacitance varies in direct proportion to the linear dimensions.

When dealing with capacitance calculations it is of interest to compare the simplicity of the m.k.s. system with the complications of the c.g.s. systems in which there are two distinct systems, with difficult transformations from one to the other to be memorized. The coulomb is equal to 10^{-1} c.g.s. electromagnetic absolute unit of quantity while the volt is equal to 10^8 absolute units

$$\therefore\ \text{the farad} = 10^{-1}/10^8 = 10^{-9} \text{ absolute e.m. units}$$

When capacitance is calculated from the dimensions of a conductor the result is given in electrostatic units, namely in centimetres. The absolute unit is an electromagnetic unit and therefore, to convert from centimetres to farads, it is necessary to make use of the relationships between electrostatic and electromagnetic units.

$$\begin{aligned} \text{E.m. unit of capacitance} &= \text{e.s. unit} \times c^2 \\ &= \text{e.s. unit} \times 9 \times 10^{20} \\ \text{Farads} &= \text{e.m. units} \times 10^{-9} \\ \therefore\ \mu\text{F} &= \text{e.m. units} \times 10^{-15} \\ &= \text{e.s. units} \times 10^{-15} \times 9 \times 10^{20} \\ &= \text{e.s. units} \times 9 \times 10^{5} \end{aligned}$$

Example. Calculate the capacitance of an isolated sphere of 10 cm radius in free space

$$\begin{aligned} C &= 4\pi\varepsilon\varepsilon_0 a \times 10^6\ \mu\text{F} \ (a \text{ in metres}) \\ &= 4\pi \times 1 \times 8{\cdot}854 \times 10^{-12} \times 10^{-1} \times 10^6 \\ &= 1{\cdot}113 \times 10^2 \times 10^{-12} \times 10^{-1} \times 10^6 = 1{\cdot}113 \times 10^{-5}\ \mu\text{F} \end{aligned}$$

In terms of the c.g.s. system we have

$$\begin{aligned} C &= a/(9 \times 10^5) \ (a \text{ in centimetres}) \\ &= 10/(9 \times 10^5) = 1{\cdot}11 \times 10^{-5}\ \mu\text{F} \end{aligned}$$

(*b*) CONCENTRIC SPHERES. The p.d. between two points of radial distances r_1 and r_2 from a point charge q has been shown to be

$$V_{r_1r_2} = (q/4\pi\varepsilon\varepsilon_0)(1/r_1 - 1/r_2)$$

Let there be conducting spheres at these radial distances, then since the spherical surfaces coincide with equipotential surfaces no difference will be made to the distribution of electric intensity external to the inner sphere. Hence, denoting the radii of the inner and outer spheres by a and b respectively we have for the p.d. between them

$$V = (q/4\pi\varepsilon\varepsilon_0)(1/a - 1/b)$$
$$\therefore C = q/V = 4\pi\varepsilon\varepsilon_0/(1/a - 1/b)$$

where ε is the relative permittivity of the medium in the space between the two spheres.

(*c*) PARALLEL PLATES. If we assume that the electric field is uniform over the plate area to be considered, the intensity between the plates is $\sigma/\varepsilon\varepsilon_0$. Now σ is the surface density of the charge, so that for a plate area A we have

$$\sigma = q/A \quad \therefore E = q/\varepsilon\varepsilon_0 A$$

Let d be the distance between the plates, then the potential difference between them is

$$V = Ed = qd/\varepsilon\varepsilon_0 A$$
$$\therefore C = q/V = \varepsilon\varepsilon_0 A/d \text{ farads}$$

If the medium is air, for which $\varepsilon = 1$

$$C = \varepsilon_0 A/d \text{ farads}$$

For a given capacitance C and separation d we thus have

$$A = Cd/\varepsilon_0$$

It is of interest to calculate the plate area necessary for a capacitance of 1 farad. Suppose that the separation is as small as 1 mm $= 10^{-3}$ m

$$A = 1 \times 10^{-3}/(8{\cdot}854 \times 10^{-12}) = 1{\cdot}13 \times 10^8 \text{ m}^2$$

Suppose the plates are square, then the side will be $1{\cdot}065 \times 10^4$ m, or 7·7 miles. This example shows why it is necessary to express practical capacitances in microfarads or, in the case of very small capacitances, in micromicrofarads ($\mu\mu$F).

EXAMPLE. A parallel-plate capacitor has plates 3 m² and separation 2 cm immersed in a liquid for which $\varepsilon = 2{\cdot}5$. A p.d. of 5000 V is maintained between the plates. Calculate capacitance, charge per plate, surface density, and intensity.

$$\begin{aligned} C &= \varepsilon\varepsilon_0 A/d \\ &= 2{\cdot}5 \times 8{\cdot}854 \times 10^{-12} \times 3/(2 \times 10^{-2}) \text{ farad} \\ &= 3{\cdot}32 \times 10^{-9} \text{ farad} \\ &= 3{\cdot}32 \times 10^{-3}\ \mu\text{F} \end{aligned}$$

$$q = CV = 3{\cdot}32 \times 10^{-9} \times 5 \times 10^3$$
$$= 1{\cdot}66 \times 10^{-5} \text{ coulomb}$$
$$\sigma = q/A = 1{\cdot}66 \times 10^{-5} \div 3$$
$$= 5{\cdot}53 \times 10^{-6} \text{ coulomb/m}^2$$
$$E = V/d = 5000/(2 \times 10^{-2}) = 2{\cdot}5 \times 10^5 \text{ V/m}$$

Alternatively

$$E = \sigma/\varepsilon\varepsilon_0 = 5{\cdot}53 \times 10^{-6}/(2{\cdot}5 \times 8{\cdot}854 \times 10^{-12})$$
$$= 2{\cdot}5 \times 10^5 \text{ V/m}$$

(*d*) Coaxial Cylinders. We have seen that the intensity at a radial distance r from an isolated charged conductor having λ coulombs per unit length is

$$E = \lambda/2\pi\varepsilon\varepsilon_0 r$$

Let the radius of the conductor be a, and let there be a concentric cylinder of radius b round it. Then as the surface of this cylinder will coincide with the original equipotential at radius b, the intensity distribution over the distance $(b - a)$ will remain as before.

Hence p.d. between the cylinders

$$V = \int E dr$$
$$= \int_a^b \frac{\lambda}{2\pi\varepsilon\varepsilon_0} \cdot \frac{dr}{r}$$
$$= (\lambda/2\pi\varepsilon\varepsilon_0) \log_e(b/a)$$

quantity per unit length,

$$q = \lambda \text{ coulombs}$$
$$\therefore C = q/V = 2\pi\varepsilon\varepsilon_0/\log_e(b/a)$$
$$= 2\pi\varepsilon\varepsilon_0/2{\cdot}303 \log_{10}(b/a) \text{ farads/metre}$$

Now we have seen that the intensity is equal to the voltage gradient

$$g = E = \lambda/2\pi\varepsilon\varepsilon_0 r \text{ at any radius } r$$

But
$$\lambda = 2\pi\varepsilon\varepsilon_0 V/\log_e(b/a)$$
$$\therefore g = V/r \log_e(b/a) = V/2{\cdot}303 r \log_{10}(b/a)$$

At the surface of the inner conductor r has the minimum value of a and therefore the gradient has a maximum value of

$$g_m = V/a \log_e(b/a) = V/2{\cdot}303\, a \log_{10}(b/a)$$

Example. A concentric cable has a core diameter of 0·927 in. It is insulated with impregnated paper of thickness 0·65 in., and permittivity 3·5, and covered with a lead sheath. If the working pressure is

66,000 V (alternating) calculate the capacitance of the cable in μF per mile, and the maximum potential gradient.

$$a = 0{\cdot}464 \text{ in.}$$
$$b = 0{\cdot}464 + 0{\cdot}65 = 1{\cdot}114 \text{ in.}$$
$$\therefore\ b/a = 1{\cdot}114/0{\cdot}464 = 2{\cdot}41$$
$$\therefore\ \log_{10} b/a = 0{\cdot}382$$
$$\therefore\ C = [2\pi \times 3{\cdot}5 \times 8{\cdot}854 \times 10^{-12}/(2{\cdot}303 \times 0{\cdot}382)] \times 10^6\ \mu\text{F/m}$$
$$= 2{\cdot}204 \times 10^{-4}\ \mu\text{F/m}$$
$$= 2{\cdot}204 \times 10^{-4} \times 1609\ \mu\text{F/mile}$$
$$= 3{\cdot}55 \times 10^{-1}\ \mu\text{F/mile}$$
$$a = 0{\cdot}464 \text{ in.} = 1{\cdot}18 \times 10^{-2} \text{ m}$$
$$\therefore\ g_m = \sqrt{2} \times 6{\cdot}6 \times 10^4/(2{\cdot}303 \times 1{\cdot}18 \times 10^{-2} \times 0{\cdot}382)$$
$$= 9 \times 10^6 \text{ V/m}$$

The factor $\sqrt{2}$ is introduced into the numerator because, with an alternating voltage from a public supply, the maximum voltage is $\sqrt{2}$ times the stated voltage.

(*e*) PARALLEL CONDUCTORS. Fig. 2.10 shows two parallel conductors, each of radius r, separated by a distance d, and it is assumed that d is

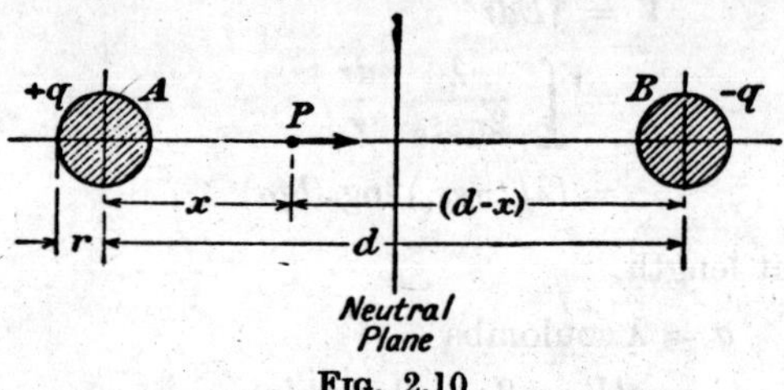

FIG. 2.10

large compared with r. If conductor A has a charge of $+\,q$ along its length, this will induce a charge $-\,q$ on conductor B. Hence, for any point P we have, for a charge λ coulombs per unit length, intensity due to conductor $A = \lambda/2\pi\varepsilon\varepsilon_0 x$ and intensity due to conductor $B = \lambda/[2\pi\varepsilon\varepsilon_0(d - x)]$. Hence, total intensity at P,

$$E = (\lambda/2\pi\varepsilon\varepsilon_0)[1/x + 1/(d - x)]$$

For the p.d. between the wires we have

$$V = \int E dx = \frac{\lambda}{2\pi\varepsilon\varepsilon_0}\int_r^{d-r}\left(\frac{1}{x} + \frac{1}{d - x}\right) dx$$
$$= (\lambda/2\pi\varepsilon\varepsilon_0)\Big[\log_e x - \log_e(d - x)\Big]_r^{d-r}$$
$$= (\lambda/\pi\varepsilon\varepsilon_0)\log_e[(d - r)/r]$$

Hence, capacitance per unit length

$$C = \lambda/V = \pi\varepsilon\varepsilon_0/\log_e[(d - r)/r]$$
$$\simeq \pi\varepsilon\varepsilon_0/\log_e(d/r) \text{ farads/metre}$$
$$\simeq \pi\varepsilon\varepsilon_0/2{\cdot}303 \log_{10}(d/r) \text{ farads/metre}$$

EXAMPLE. The 132-kV British Grid lines have the following data: $d = 12$ ft, $2r = 0{\cdot}77$ in. Calculate the capacitance per mile of a line having two conductors only.

$$d/r = 12 \times 12/0{\cdot}385 = 374$$
$$\therefore \log_{10}(d/r) = 2{\cdot}5729$$
$$\therefore C = \pi \times 1 \times 8{\cdot}854 \times 10^{-12}/(2{\cdot}303 \times 2{\cdot}5729)$$
$$= 4{\cdot}695 \times 10^{-12} \text{ farad/m}$$
$$= 4{\cdot}695 \times 10^{-6}\ \mu\text{F/m}$$
$$= 4{\cdot}695 \times 10^{-6} \times 1609\ \mu\text{F/mile}$$
$$= 7{\cdot}55 \times 10^{-3}\ \mu\text{F/mile}$$

To determine the voltage gradient it is convenient to regard the centre plane as a neutral plane so that, if we denote the potential of one conductor with respect to this plane as V', we have

$$V' = V/2 = (\lambda/2\pi\varepsilon\varepsilon_0) \log_e[(d - r)/r]$$

At any point P the potential with respect to the neutral plane is

$$V_x' = \int_x^{d/2} \frac{\lambda}{2\pi\varepsilon\varepsilon_0}\left(\frac{1}{x} + \frac{1}{d - x}\right)dx$$
$$= (\lambda/2\pi\varepsilon\varepsilon_0) \log_e[(d - x)x]$$

Hence, for the voltage gradient at any point P we have

$$g = dV_x'/dx$$
$$= -(\lambda/2\pi\varepsilon\varepsilon_0)[d/\{x(d - x)\}]$$

But $\lambda/2\pi\varepsilon\varepsilon_0 = V'/\log_e[(d - r)/r]$

$$\therefore g = V'[d/\{x(d - x)\}][1/\log_e\{(d - r)/r\}]$$

The gradient is a maximum at the conductor surface, where $x = r$, and since $(d - r) \simeq d$, we have

$$g = V'[d/\{x(d - x)\}]/\log_e(d/r) = V'[d/\{x(d - x)\}]/2{\cdot}3 \log_{10}(d/r)$$

and $g_m = V'[2{\cdot}3r \log_{10}(d/r)]$

EXAMPLE: $V = 200{,}000$ V (steady), $d = 10$ ft, $2r = 1$ in.

$$V' = 100{,}000 \text{ V}$$
$$d = 120 \text{ in. and } r = 0{\cdot}5 \text{ in.}$$
$$\therefore \log_{10}[d/r]$$
$$= \log_{10} 240 = 2{\cdot}3802$$
$$\therefore g = 100{,}000 \times [120/x(120 - x)] \times (1/2{\cdot}3 \times 2{\cdot}38)$$
$$= (18{,}300/x) \times [120/(120 - x)]$$

The units depend on the unit of length adopted. As the dimensions and spacings are given in inches, and it is probable that these units will be used for some considerable time for practical cases, we have, with x in inches,

$$g = 18{,}300/x \text{ volts/inch}$$

At the surface of the conductor $x = r = 0{\cdot}5$,

$$\therefore\ g_m = 18{,}300/0{\cdot}5 = 36{,}600 \text{ V/in.}$$

The variations in G across the space between the conductors are shown in Fig. 2.11.

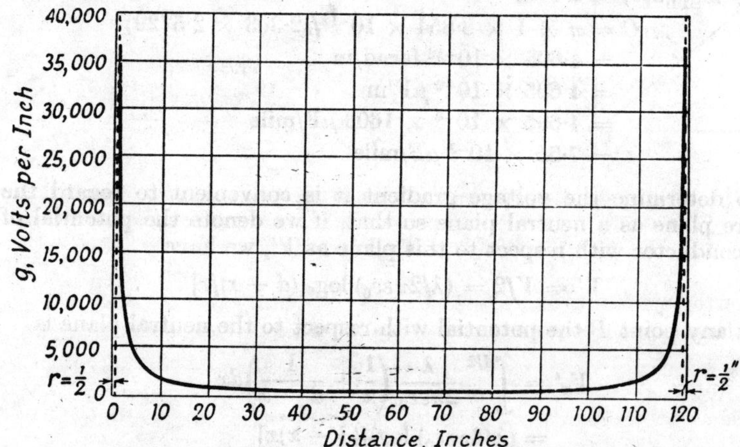

FIG. 2.11. POTENTIAL GRADIENT BETWEEN PARALLEL CONDUCTORS

Systems of Capacitors. If a number of capacitors of individual capacitances C_1, C_2, C_3 are connected in series, then the total capacitance is given by

$$1/C = 1/C_1 + 1/C_2 + 1/C_3 + \ldots\ldots$$

The charge on each capacitor is the same, so that if V_1, V_2, V_3, etc., are the p.d.'s on the individual capacitors, and V the total applied p.d., we have

$$q = CV = C_1V_1 = C_2V_2 = C_3V_3, \text{ etc.}$$

$$\therefore\ V_1 = q/C_1$$

and $$V = q/C$$

$$\therefore\ V_1/V = C/C_1$$

Similarly $$V_2/V = C/C_2;\ \ V_3/V = C/C_3;\ \text{etc.}$$

This is of great importance in connection with the compound dielectrics used in the manufacture of electrical machinery and other apparatus. Take, for example, the slot of a high-voltage alternator.

The armature conductors are at a high potential with respect to the core, and conductors and core are separated by dielectric, with the result that the system constitutes a capacitor. If the dielectric is homogeneous, then the potential gradient will be uniform, but if the dielectric consists of layers of different materials with different permittivities, then the gradient will no longer be uniform, but will be different for each material. Fig. 2.12 shows three such cases, the total thickness of

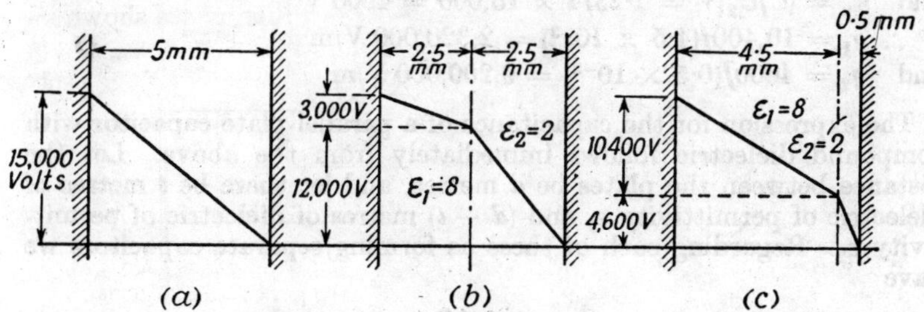

FIG. 2.12. POTENTIAL GRADIENT IN HOMOGENEOUS AND COMPOSITE DIELECTRICS

insulation being 5 mm in each case, and the total p.d. 15,000 V. In case (*a*) the material is homogeneous so that g is uniform and equal to

$$g = 15{,}000/5 \times 10^{-3} = 3 \times 10^{6} \text{ V/m}$$

In case (*b*) there are two materials of permittivities 8 and 2, but of equal thicknesses. We can regard this system as the equivalent of two capacitors in series, the capacitances being

$$C_1 = \varepsilon_1\varepsilon_0 A/d = 8\varepsilon_0 A/(0{\cdot}25 \times 10^{-2}) = 32\varepsilon_0 A \times 10^{2}$$

$$C_2 = \varepsilon_2\varepsilon_0 A/d = 2\varepsilon_0 A/(0{\cdot}25 \times 10^{-2}) = 8\varepsilon_0 A \times 10^{2}$$

Hence total capacitance

$$C = C_1C_2/(C_1 + C_2) = [(32 \times 8)/(32 + 8)]\varepsilon_0 A \times 10^{2} = (32/5)\varepsilon_0 A \times 10^{2}$$

$$\therefore\ V_1 = (C/C_1)V = [32/(5 \times 32)] \times 15{,}000 = 3000 \text{ V}$$

$$\therefore\ g_1 = 3000/(2{\cdot}5 \times 10^{-3}) = 1{,}200{,}000 \text{ V/m}$$

$$V_2 = (C/C_2)V = [32/(5 \times 8)] \times 15{,}000 = 12{,}000 \text{ V}$$

$$\therefore\ g_2 = 12{,}000/(2{\cdot}5 \times 10^{-3}) = 4{,}800{,}000 \text{ V/m}$$

This shows that the gradient across the material of low permittivity is greater than that across the material with the high permittivity.

Now take the more likely case of Fig. 2.12 (*c*). Then

$$C_1 = \varepsilon_1\varepsilon_0 A/d_1 = 8\varepsilon_0 A/(4{\cdot}5 \times 10^{-3}) = 1{\cdot}778 \times 10^3\varepsilon_0 A$$
$$C_2 = \varepsilon_2\varepsilon_0 A/d_2 = 2\varepsilon_0 A/(0{\cdot}5 \times 10^{-3}) = 4 \times 10^3\varepsilon_0 A$$
$$\therefore C = C_1C_2/(C_1 + C_2) = [(1{\cdot}78 \times 4 \times 10^6)/(5{\cdot}78 \times 10^3)]\varepsilon_0 A$$
$$= 1{\cdot}23 \times 10^3\varepsilon_0 A$$
$$\therefore V_1 = (C/C_1)V = 1{\cdot}23/1{\cdot}78 \times 15{,}000 = 10{,}400 \text{ V}$$
$$\text{and } V_2 = (C/C_2)V = 1{\cdot}23/4 \times 15{,}000 = 4600 \text{ V}$$
$$\therefore g_1 = 10{,}400/(4{\cdot}5 \times 10^{-3}) = 2{,}320{,}000 \text{ V/m}$$
$$\text{and } g_2 = 4600/(0{\cdot}5 \times 10^{-3}) = 9{,}200{,}000 \text{ V/m}$$

The expression for the capacitance of a parallel-plate capacitor with compound dielectric follows immediately from the above. Let the distance between the plates be d metres, and let there be t metres of dielectric of permittivity ε_1 and $(d - t)$ metres of dielectric of permittivity ε_2. Regarding each of these as forming separate capacitors we have

$$C_1 = \varepsilon_1\varepsilon_0 A/t$$
$$C_2 = \varepsilon_2\varepsilon_0 A/(d - t)$$

Hence, since C_1 and C_2 are in series, we have

$$C = C_1C_2/(C_1 + C_2)$$
$$= \varepsilon_0 A\left[\frac{(\varepsilon_1/t) \times \varepsilon_2/(d - t)}{\varepsilon_1/t + \varepsilon_2/(d - t)}\right]$$
$$= \varepsilon_1\varepsilon_2\varepsilon_0 A/[\varepsilon_1(d - t) + \varepsilon_2 t]$$
$$= \frac{\varepsilon_0 A}{[(d - t)/\varepsilon_2 + t/\varepsilon_1]}$$

Another practical example involving two capacitors in series is an insulated cable with compound dielectric.

Let r_1 = radius of core

r_2 = radius of inner dielectric of relative permittivity ε_1

r_3 = radius of outer dielectric of relative permittivity ε_2

Then regarding each dielectric as a separate capacitor we have, Fig. 2.13,

$$C_1 = 2\pi\varepsilon_1\varepsilon_0/\log_e(r_2/r_1) \text{ farads/metre}$$
$$C_2 = 2\pi\varepsilon_2\varepsilon_0/\log_e (r_3/r_2) \text{ farads/metre}$$
$$\therefore V_2/V_1 = C_1/C_2 = \varepsilon_1 \log_e (r_3/r_2)/\varepsilon_2 \log_e (r_2/r_1)$$
$$= \varepsilon_1 \log_{10} (r_3/r_2)/\varepsilon_2 \log_{10} (r_2/r_1)$$

As a numerical example, let $r_1 = 3$ mm, $r_2 = 6$ mm, $r_3 = 10$ mm. $\varepsilon_1 = 7$, $\varepsilon_2 = 5$.

$$\therefore \varepsilon_1 \log_{10} (r_3/r_2) = 7 \log_{10} (10/6) = 1{\cdot}553$$
$$\varepsilon_2 \log_{10} (r_2/r_1) = 5 \log_{10} (6/3) = 1{\cdot}505$$

$$\therefore\ V_2/V_1 = 1{\cdot}553/1{\cdot}505 = 1{\cdot}03$$

$$\therefore\ V/V_1 = 2{\cdot}03/1$$

$$V_1 = V/2{\cdot}03$$

Thus suppose V is 10,000 V, then

$$V_1 = 10{,}000/2{\cdot}03 = 4870\ \text{V}$$

and $$V_2 = 10{,}000 - 4870 = 5130\ \text{V}$$

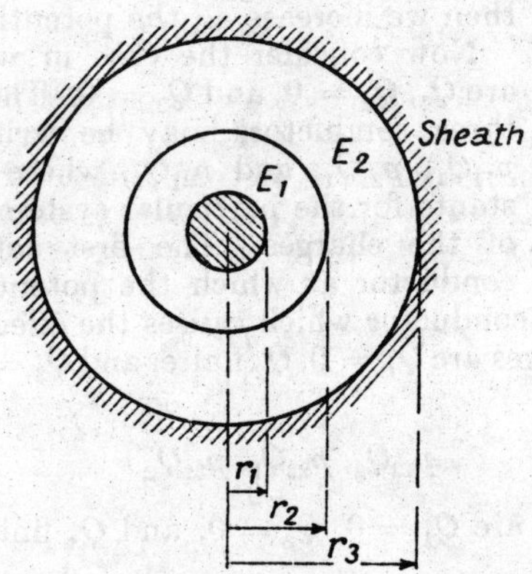

FIG. 2.13. CABLE WITH COMPOSITE DIELECTRIC

Capacitors in Parallel. If a number of capacitors are connected in parallel, then for the total capacitance we have

$$C = C_1 + C_2 + C_3 + \ldots$$

Each capacitor is subjected to the whole of the applied p.d.

Potential Coefficients. When two conductors are disposed in simple geometrical arrangements we have seen that the capacitance may be easily calculated in certain cases. The methods adopted in the previous sections are not suitable in general, since there may be more than two conductors, and therefore what is required is a method suited to the general case. This is provided by the method of potential coefficients originally introduced by Maxwell.* Thse coefficients are particularly valuable in the solution of transmission-line problems.

If an isolated conductor is given a charge Q, then its potential is raised to some value V, and we have seen that the ratio Q/V is called the capacitance of the conductor. When there are several neighbouring conductors the potential of each will no longer be determined solely by its own charge and isolated capacitance, because its charge

* Clerk Maxwell, *Electricity and Magnetism*, Vol. I, Chap. III. J. Jeans, *Mathematical Theory of Electricity and Magnetism*, Cambridge, 1925. L. Page and N. I. Adams, *Principles of Electricity*, Van Nostrand.

distribution will be modified by the presence of the neighbouring conductors.

Consider the group of three conductors in Fig. 2.14. Let their charges be Q_1, Q_2 and Q_3, and their potentials V_1, V_2 and V_3. If their charges are Q_1', Q_2' and Q_3', their potentials will have new values V_1', V_2' and V_3'. Now let the charges be $Q_1 + Q_1'$, $Q_2 + Q_2'$, and $Q_3 + Q_3'$, then, by the principle of superposition, the potentials will be $V_1 + V_1'$, $V_2 + V_2'$, and $V_3 + V_3'$. In general, if we increase all the charges n times, then we increase all the potentials n times.

FIG. 2.14

Now consider the case in which the charges are Q_1, $Q_2 = 0$, and $Q_3 = 0$. The potentials of the three conductors may be written in the form $p_{11}Q_1$, $p_{21}Q_1$, and $p_{31}Q_1$, where the p's are constants for the particular system and independent of the charges. The first suffix indicates the conductor at which the potential is considered, and the second, the conductor which causes the effect.

Hence, if the charges are $Q_1 = 0$, Q_2 finite, and $Q_3 = 0$, the potentials can be written,

$$p_{12}Q_2,\ p_{22}Q_2,\ p_{32}Q_2$$

while if the charges are $Q_1 = 0$, $Q_2 = 0$, and Q_3 finite, the potentials become

$$p_{13}Q_3,\ p_{23}Q_3,\ p_{33}Q_3$$

If all the conductors have finite charges, then

$$V_1 = p_{11}Q_1 + p_{12}Q_2 + p_{13}Q_3$$
$$V_2 = p_{21}Q_1 + p_{22}Q_2 + p_{23}Q_3$$
$$V_3 = p_{31}Q_1 + p_{32}Q_2 + p_{33}Q_3$$

The magnitudes of the p's depend on the geometry of the system and on the relative permittivity of the medium.

If we require the charges in terms of the potentials then we can write a similar set of equations—

$$Q_1 = q_{11}V_1 + q_{12}V_2 + q_{13}V_3$$
$$Q_2 = q_{21}V_1 + q_{22}V_2 + q_{23}V_3$$
$$Q_3 = q_{31}V_1 + q_{32}V_2 + q_{33}V_3$$

where the q's are new constants which can be calculated in terms of the p's. When the suffixes of the q's are the same, as in q_{11}, q_{22}, etc., these constants are called coefficients of capacitance. If the suffixes are not the same, as in q_{12}, q_{23}, etc., they are called coefficients of induction because, for example, $q_{12}V_2$ means the charge on conductor 1, due to the potential V_2 of conductor 2, i.e. the charge induced on conductor 1, by the potential V_2 of conductor 2.

The results are general for any number of conductors n, the equations for the potentials being

$$\begin{aligned} V_1 &= p_{11}Q_1 + p_{12}Q_2 + p_{13}Q_3 + \cdots + p_{1n}Q_n \\ V_2 &= p_{21}Q_1 + p_{22}Q_2 + p_{23}Q_3 + \cdots + p_{2n}Q_n \\ &\cdot \\ &\cdot \\ &\cdot \\ V_n &= p_{n1}Q_1 + p_{n2}Q_2 + p_{n3}Q_3 + \cdots + p_{nn}Q_n \end{aligned}$$

By means of Green's reciprocity theorem it can be shown that $p_{12} = p_{21}$, or in general $p_{nm} = p_{mn}$. In words, this means that the potential of any conductor n due to unit charge on m is the same as the potential of m due to unit charge on n.

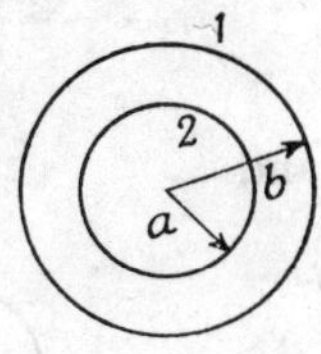

FIG. 2.15

EXAMPLE 1. CAPACITANCE OF CONCENTRIC SPHERES. Denote the spheres as 1 and 2, Fig. 2.15, and let the radii be a and b. The potentials of the spheres in terms of the charges are

$$V_1 = p_{11}Q_1 + p_{12}Q_2 \quad \cdot \quad \cdot \quad \cdot \quad \cdot \quad (1)$$

$$V_2 = p_{21}Q_1 + p_{22}Q_2 \quad \cdot \quad \cdot \quad \cdot \quad \cdot \quad (2)$$

Put $\qquad Q_1 = 1 \text{ and } Q_2 = 0$

then $\qquad V_1' = p_{11} = 1/4\pi\varepsilon\varepsilon_0 b$

Now put $\qquad Q_1 = 0 \text{ and } Q_2 = 1$

then $\qquad V_2' = p_{22} = 1/4\pi\varepsilon\varepsilon_0 a$

Let $Q_1 = Q$, then we know that $Q_2 = -Q$, and therefore, subtracting equation (2) from (1) we have

$$\begin{aligned} V = V_1 - V_2 &= (p_{11} - p_{12} - p_{21} + p_{22})Q \\ &= \left(\frac{1}{b} - \frac{1}{b} - \frac{1}{b} + \frac{1}{a}\right)Q/4\pi\varepsilon\varepsilon_0 \\ &= \frac{b-a}{ab} \cdot \frac{Q}{4\pi\varepsilon\varepsilon_0} \end{aligned}$$

$\therefore$ from $C = Q/V$

$$C = \frac{4\pi\varepsilon\varepsilon_0 ab}{b-a}$$

This is, of course, the capacitance between the inner and outer spheres. The outer sphere has an additional capacitance of $4\pi\varepsilon\varepsilon_0 b$.

EXAMPLE 2. ELECTRIC SHIELDING. Let a conductor 1, Fig. 2.16, be entirely surrounded by a conductor 2 which is maintained at zero potential. We require the effect of conductor 3 on the potential of conductor 1. We have

$$Q_1 = q_{11}V_1 + q_{12}V_2 + q_{13}V_3 \quad . \quad . \quad . \quad (1)$$
$$Q_2 = q_{21}V_1 + q_{22}V_2 + q_{23}V_3 \quad . \quad . \quad . \quad (2)$$
$$Q_3 = q_{31}V_1 + q_{32}V_2 + q_{33}V_3 \quad . \quad . \quad . \quad (3)$$

But $V_2 = 0$, and if $Q_1 = 0$ then $V_1 = 0$ also, because the potential inside a hollow conductor is that of the conductor itself unless there is a charged body in it.

$$\therefore q_{13}V_3 = 0 \text{ from equation (1), i.e. } q_{13} = 0$$
$$\therefore Q_1 = q_{11}V_1$$

and $$Q_3 = q_{33}V_3 \text{ (since } q_{13} = q_{31})$$

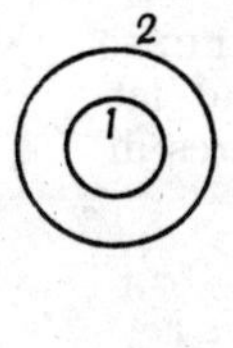

FIG. 2.16

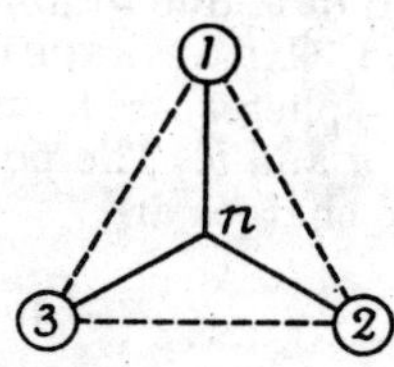

FIG. 2.17

The electrical condition of conductor 1 is thus entirely independent of the electric field due to conductor 3, and similarly the field of conductor 1 has no effect on conductor 3.

EXAMPLE 3. CAPACITANCE TO NEUTRAL OF A THREE-PHASE OVERHEAD TRANSMISSION LINE WITH SYMMETRICALLY SPACED CONDUCTORS. The conductors are shown as 1, 2, and 3 in Fig. 2.17. For simplicity the conductors are regarded as being so far from the earth that their capacitances to earth may be neglected. With balanced conditions the neutral n will be at the centre of the system, as shown, and we have

$$Q_1 = q_{11}V_1 + q_{12}V_2 + q_{13}V_3 + q_{1n}V_n$$
$$Q_2 = q_{21}V_1 + q_{22}V_2 + q_{23}V_3 + q_{2n}V_n$$
$$Q_3 = q_{31}V_1 + q_{32}V_2 + q_{33}V_3 + q_{3n}V_n$$
$$Q_n = q_{n1}V_1 + q_{n2}V_2 + q_{n3}V_3 + q_{nn}V_n$$

Now the potential at the neutral is zero in a balanced system. Q_n is zero, and $V_1 + V_2 + V_3 = 0$,

$$\therefore Q_1 = q_{11}V_1 + q_{12}(V_2 + V_3)$$
$$= (q_{11} - q_{12})V_1$$

$\therefore$ Capacitance to neutral

$$= Q_1/(V_1 - V_n) = Q_1/V_1 = q_{11} - q_{12}$$

Now consider conductors 1 and 2 alone.

Let $$Q_1 = Q \text{ and } Q_2 = -Q$$

Then
$$2Q = V_1(q_{11} - q_{21}) + V_2(q_{12} - q_{22})$$
$$= (q_{11} - q_{12})(V_1 - V_2)$$
because $q_{22} = q_{11}$, by symmetry, and $q_{12} = q_{21}$

$\therefore$ Capacitance between conductors 1 and 2
$$= Q/(V_1 - V_2) = (q_{11} - q_{12})/2$$

This expression is one-half that for the capacitance to neutral, showing that the capacitance to neutral of one conductor of a three-phase symmetrically spaced transmission line is twice the capacitance between any pair of the conductors.*

Energy Stored in an Electric Field. Since the lines of force of an electric field start on one conductor and end on another, it will be realized that any electrostatic field is associated with a capacitor. Now the shapes of the lines of force depend on the shape, size, and relative positions of the conductors and therefore on the capacitance of the capacitor formed by them. Consider for simplicity a parallel-plate capacitor. The two plates attract one another with a certain force f, and if their separation is increased by a small amount dx, work equal to fdx will be expended. As there is no possibility of heat production, the whole of this work will be stored in the extended electric field in the form of potential energy. Thus the electric field is the seat of potential energy, just as a magnetic field is a seat of potential energy.

Let the p.d. at any instant between the plates be v, then v is the work done in taking a unit positive charge along one of the lines of force against the potential gradient. Hence, if the charge on the capacitor is increased by a small amount dq, the work done must be vdq. Now the p.d. v is associated with a charge q according to the law

$$q = Cv$$
$$\therefore dq = Cdv$$

$\therefore$ Work done by increasing the charge by dq

$$vdq = Cv\,.\,dv$$

Hence work done in increasing the p.d. from zero to V

$$= \int_0^V Cvdv$$
$$= \tfrac{1}{2}CV^2$$

If C and V are expressed in electrostatic units, then the work done will be in ergs. If C and V are in practical units—viz. C in farads and V in volts—then the work done will be in joules. Thus the stored energy in the electric field is $\frac{1}{2}CV^2$ joules. Compare this with the expression $\frac{1}{2}LI^2$ for the stored energy in a magnetic field.

* The method of potential coefficients is particularly powerful in the case of transmission-line problems. (*See* L. V. Bewley, *Travelling Waves on Transmission Systems*, Ch. 9 & 10. Chapman & Hall, London, 1951.)

EXAMPLE. A capacitor of capacitance 3 μF is charged to a p.d. of 10,000 V. Calculate the stored energy in ergs, ft.lb, joules, and watt-hours.

$$V = 10{,}000 \text{ V} = 10{,}000/300$$
$$= 33{\cdot}3 \text{ e.s. c.g.s. units}$$
$$C = 3 \times 10^{-6} \text{ F}$$
$$= 3 \times 10^{-6} \times 9 \times 10^{11} \text{ e.s. c.g.s. units}$$
$$= 27 \times 10^{5} \text{ e.s. c.g.s. units}$$

(i) $\therefore$ Ergs $= \frac{1}{2} \times (27 \times 10^5) \times (33{\cdot}3)^2$
$= 1{\cdot}5 \times 10^9$

(ii) Joules $= \frac{1}{2} \times (3 \times 10^{-6}) \times (10^4)^2$
$= 150$

Alternatively, since one joule $= 10^7$ ergs, we have

Joules $= 1{\cdot}5 \times 10^9 \times 10^{-7}$
$= 150$ as before

(iii) One joule = one watt-second
$\therefore$ One watt-hour $= 60^2$ joules
$\therefore$ Watt-hours $= 150/3600 = 0{\cdot}042$

(iv) One erg = one dyne-cm
$= 1/(981 \times 454) \times 1/(2{\cdot}54 \times 12)$ ft-lb
$\therefore$ Ft-lb $= 1{\cdot}5 \times 10^9/(981 \times 454 \times 2{\cdot}54 \times 12)$
$= 110$

Energy of a Charged System. If there are several neighbouring charged bodies we have for the total energy

$$W = \tfrac{1}{2}\Sigma(QV)$$

Consider the simple case of two charged bodies, and let the energy, expressed in terms of their charges, be W_Q, then

$$W_Q = \tfrac{1}{2}Q_1V_1 + \tfrac{1}{2}Q_2V_2$$
$$= \tfrac{1}{2}Q_1(p_{11}Q_1 + p_{12}Q_2) + \tfrac{1}{2}Q_2(p_{21}Q_1 + p_{22}Q_2)$$
$$= \tfrac{1}{2}(p_{11}Q_1^2 + 2p_{12}Q_1Q_2 + p_{22}Q_2^2)$$

Now let W_V be the energy expressed in terms of their potentials (W_Q and W_V are, of course, numerically equal), then

$$W_V = \tfrac{1}{2}(q_{11}V_1^2 + 2q_{12}V_1V_2 + q_{22}V_2^2)$$

Mechanical Force on a Dielectric. If a system changes its configuration under its own forces, as for example if the dielectric, or part of the dielectric, is free to move, then work is done at the expense of the stored energy. In other words the system tends to adjust itself in such a way as to reduce the stored energy to a minimum. Consider a capacitor of such characteristics that when one conductor moves a distance dx under the force f acting on it, the charges on both conductors remain constant, then work done, in terms of the charges,

$$dW_Q = -fdx$$

i.e.

$$f = -dW_Q/dx$$

The coefficients p in the expression for W_Q will be functions of x, while the factors Q will be constant. It often happens that the potentials are kept constant instead of the charges, and the force f must then be calculated differently. Imagine a movement dx with constant potentials effected by a movement with constant charges, followed by a modification of the charges sufficient to restore the potentials to their original values. Let the energy added in this latter adjustment be dW then,

$$dW_V = dW_Q + dW$$

During the first step, the potentials change by

$$dV_1 = [\partial(p_{11}Q_1 + p_{12}Q_2)/\partial x]dx$$
$$dV_2 = [\partial(p_{21}Q_1 + p_{22}Q_2/\partial x]dx$$

Hence, in the second, or adjusting, step we must add charges dQ_1 and dQ_2, obeying the expressions

$$dV_1 = p_{11}dQ_1 + p_{12}dQ_2$$
$$= -[(\partial p_{11}/\partial x)Q_1 + (\partial p_{12}/\partial x)Q_2]dx$$

and

$$dV_2 = p_{21}dQ_1 + p_{22}dQ_2$$
$$= -[(\partial p_{21}/\partial x)Q_1 + (\partial p_{22}/\partial x)Q_2]dx$$

The increments of charge will cause the system energy to increase by

$$\begin{aligned} dW &= V_1dQ_1 + V_2dQ_2 \\ &= (p_{11}Q_1 + p_{12}Q_2)dQ_1 + (p_{21}Q_1 + p_{22}Q_2)dQ_2 \\ &= (p_{11}dQ_1 + p_{21}dQ_2)Q_1 + (p_{12}dQ_1 + p_{22}dQ_2)Q_2 \\ &= -[(\partial p_{11}/\partial x)Q_1^2 + 2(\partial p_{12}/\partial x)Q_1Q_2 + (\partial p_{22}/\partial x)Q_2^2[dx \\ &= -(2dW_Q/dx)dx \\ &= -2dW_Q \end{aligned}$$

$$\begin{aligned} \therefore dW_V &= dW_Q - 2dW_Q \\ &= -dW_Q \\ &= fdx \end{aligned}$$

$$\therefore f = dW_V/dx$$

Hence during motion with constant potentials, the potential energy increases by an amount equal to the mechanical work done by the system. The sources of constant potential therefore supply an amount of energy equal to twice the gain of potential energy. A gravitational system is analogous to an electrostatic system with constant charges, mass corresponding to charge and here, work done by the systen equals the loss of potential energy.

Loss of Energy. This consideration can be extended as follows Consider again the parallel-plate capacitor. We have

$$C = \varepsilon\varepsilon_0 A/d \text{ farads}$$

The plates attract one another and therefore we can imagine one o: them displaced a small distance δd under the influence of this attraction The work done can be regarded as negative, since it is with, and no against the force. Hence, denoting the force by f we have for th work done,

$$\delta W = -f\delta d$$

This is obviously equal to the change in the potential energy stored i the dielectric. The capacitance is a function of the separation d, so w can write

$$\delta C = -(\varepsilon\varepsilon_0 A/d^2)\delta d$$

But

$$W = \tfrac{1}{2}CV^2$$

$$\therefore \delta W = \tfrac{1}{2}\delta C V^2$$

$$= -\tfrac{1}{2}(\varepsilon\varepsilon_0 A\delta d/d^2)V^2$$

$$\therefore f = \tfrac{1}{2}\varepsilon\varepsilon_0 A V^2/d^2 \text{ newtons}$$

If the force of attraction is measured, as in the Kelvin attracted-disc electrometer, then we can calculate the p.d. between the plates, since

$$V = d(2f/\varepsilon\varepsilon_0 A)^{\frac{1}{2}}$$

Again energy density

$$\delta W/\delta(\text{Vol.}) = \tfrac{1}{2}[\varepsilon\varepsilon_0 A\delta d/\{(A\delta d)d^2\}]\,.\,V^2$$

$$= \tfrac{1}{2}\varepsilon\varepsilon_0 V^2/d^2$$

$$= \tfrac{1}{2}\varepsilon\varepsilon_0 E^2$$

In the magnetic case we define the induction B as

$$B = \mu\mu_0 H$$

Similarly in the electric field we have

$$D = \varepsilon\varepsilon_0 E$$

where D is the electrostatic flux density, or displacement density

$$\therefore \text{Energy density} = \left.\begin{aligned}&\tfrac{1}{2}\varepsilon\varepsilon_0 E^2\\ =\ &\tfrac{1}{2}D^2/\varepsilon\varepsilon_0\\ =\ &\tfrac{1}{2}DE\end{aligned}\right\} \text{joules/metre}^3$$

EXAMPLE. 1. The dielectric of a parallel-plate capacitor is withdrawn in a direction parallel to one side of the plates, while the p.d., V, is kept constant. Calculate the force acting on the dielectric (Fig. 2.18).

As there is one variable, namely x, we have

$$f = dW_V/dx$$
$$W_V = \tfrac{1}{2}\varepsilon\varepsilon_0 E^2 bdx + \tfrac{1}{2}\varepsilon_0 E^2 bd(l - x)$$
$$\therefore\ dW_V/dx = \tfrac{1}{2}\varepsilon_0 E^2 bd(\varepsilon - 1)$$
$$= \tfrac{1}{2}\varepsilon_0 (V^2/d^2) bd(\varepsilon - 1)$$
$$= (\varepsilon_0/2d) V^2 b(\varepsilon - 1) \text{ newtons}$$

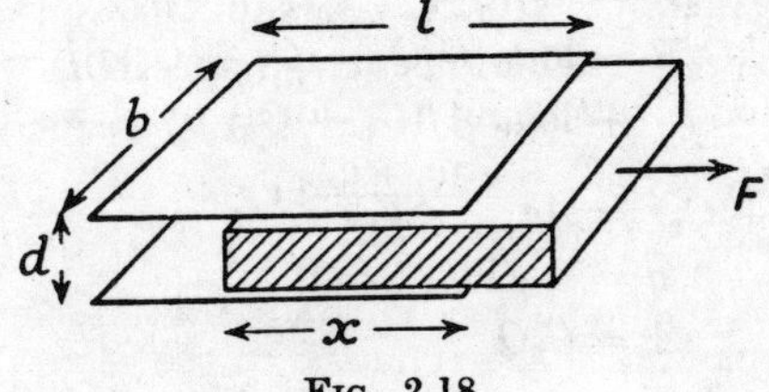

Fig. 2.18

Example. 2. The maximum intensity that can be imposed in air at atmospheric pressure without breakdown is 30 kV/cm. Calculate the corresponding energy density.

$$E = 30{,}000 \text{ V/cm} = 3 \times 10^6 \text{ V/m}$$
$$\varepsilon = 1$$

$\therefore$ Energy density

$$= \tfrac{1}{2} \times 8{\cdot}854 \times 10^{-12} \times 9 \times 10^{12} = 40 \text{ joules/m}^3$$

Division of Stored Energy. If a capacitor of capacitance C_1 is given a charge Q the stored energy is $Q^2/2C_1$. If now an uncharged capacitor

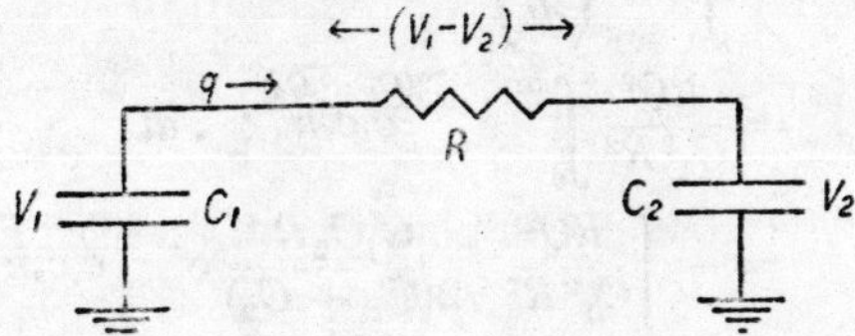

Fig. 2.19

of capacitance C_2 is connected in parallel with C_1 the charge Q is distributed over a capacitance of $(C_1 + C_2)$, the new energy being

$$Q^2/2(C_1 + C_2)$$

Hence there is a loss of energy of

$$\tfrac{1}{2}Q^2[1/C_1 - 1/(C_1 + C_2)]$$

Consider the case of a charged capacitor of capacitance C_1 joined by a resistance R to an uncharged capacitor of capacitance C_2, as in Fig. 2.19.

Let the original charge on C_1 be Q. After a quantity q has been transferred from C_1 to C_2 the p.d.s. will be

$$V_1 = (Q - q)/C_1$$
$$V_2 = q/C_2$$
$$V_1 - V_2 = Ri = R(dq/dt)$$
$$\therefore\ (Q - q)/C_1 - q/C_2 = R(dq/dt)$$

or
$$dt/C_1C_2R = dq/[C_2Q - (C_1 + C_2)q]$$

Integrating,
$$t/C_1C_2R = \log_e [C_2Q - (C_1 + C_2)q]/ - (C_1 + C_2) + \log_e A/(C_1 + C_2)$$

$$\therefore\ C_2Q - q(C_1 + C_2) = Ae^{-\frac{(C_1 + C_2)}{C_1C_2R}.t}$$

Now
$$q = 0 \text{ when } t = 0$$
$$\therefore\ A = C_2Q$$
$$\therefore\ q = [C_2Q/(C_1 + C_2)](1 - e^{-\frac{C_1 + C_2}{C_1C_2R}.t})$$

and, as
$$i = dq/dt$$
$$i = [C_2Q/(C_1 + C_2)] . [(C_1 + C_2)/C_1C_2R]e^{-\frac{C_1 + C_2}{C_1C_2R}}$$
$$= (Q/C_1R)e^{-\frac{C_1 + C_2}{C_1C_2R}.t}$$

Rate of change of energy
$$= i^2R$$
$$= (dq/dt)^2R$$

Hence for the total energy lost in the resistance R we have

$$\int_o^\infty i^2R\,dt = \int_o^\infty R\left(\frac{dq}{dt}\right)^2 . dt$$
$$= \frac{RQ^2}{C_1{}^2R^2}\int_o^\infty e^{-\frac{2(C_1 + C_2)}{C_1C_2R}} . dt$$
$$= -\left[\frac{RQ^2}{C_1{}^2R^2} \cdot \frac{C_1C_2R}{2(C_1 + C_2)} e^{-\frac{2(C_1 + C_2)}{C_1C_2R}.t}\right]_0^\infty$$
$$= [RQ^2/C_1{}^2R^2] . [C_1C_2R/2(C_1 + C_2)]$$
$$= Q^2C_2/[2C_1(C_1 + C_2)]$$

This is independent of the value of R, thereby proving the important principle that when an uncharged capacitor is connected to a charged capacitor the loss of energy resulting from the re-distribution of charge is independent of the resistance of the connector. The energy lost is, of course, converted into heat energy, as indicated by the fact that it was computed from the expression

$$\int_o^\infty i^2R\,dt$$

Charge and Discharge of a Capacitor. Let a circuit be made up of a battery of e.m.f. E, a resistance R ohms, a capacitor of capacitance C farads, and a switch S, as shown in Fig. 2.20. At the instant S is closed the p.d. across the capacitor will be zero, but this p.d. will gradually increase until eventually it is equal to E. While the p.d. is increasing, the charge q on the capacitor plates will be increasing, and the whole of this quantity will be derived from the battery. This means that while the p.d. is changing, the battery will be delivering current. This current, however, will not flow across the dielectric, but the quantity of electricity delivered by it will be stored by the capacitor. Such a current is called a displacement current, to distinguish it from a current in the ordinary sense which flows right round a circuit.

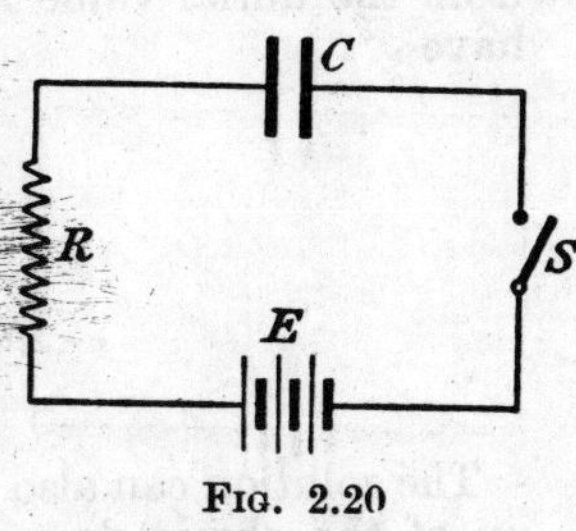

Fig. 2.20

Let v = p.d. across capacitor terminals at any instant

The displacement current at any instant is given by

$$i = (dq/dt = (d/dt)(CV) = C(dv/dt)$$

But we also have

$$i = (\text{drop in volts along } R)/R = (E - v)/R$$

$$\therefore (E - v)/R = C.(dv/dt)$$

or

$$v = E - CR(dv/dt)$$

The solution of this is

$$v = E(1 - e^{-t/CR})$$

$$\therefore q = EC(1 - e^{-t/CR})$$

the capacitor thus acquiring a p.d. equal to E in infinite time.

Now let the terminals of the charged capacitor be joined by a wire of resistance R, then the initial conditions are now $v = E$ and $i = 0$. The current i will now decrease instead of increasing, so that we have

$$i = -dq/dt = -C(dv/dt)$$

Also the drop in volts along the resistance R is now equal to the capacitor p.d., so that

$$v = iR \text{ or } i = v/R$$

$$\therefore v/R = -C(dv/dt)$$

or

$$v = -CR(dv/dt)$$

The solution of this is

$$v = Ee^{-t/CR}$$

since the value of v when $t = 0$ is E

$$\therefore q = CEe^{-t/CR}$$

Suppose we wish to find the time taken for the capacitor p.d. to drc from the initial value E to some lower value V, then at this p.d. w have

$$q' = CV = CEe^{-t'/CR}$$

$$\therefore V/E = e^{-t'/CR}$$

$$\therefore E/V = e^{t'/CR}$$

$$\therefore \log_e (E/V) = t'/CR$$

$$\therefore t' = CR \log_e (E/V)$$

The solution can also be made as follows: the applied e.m.f. is made up of the ohmic drop in the resistor and the p.d. at the capacitor terminals

$$\therefore E = iR + q/C$$

$$= iR + \int i dt/C$$

Differentiating with respect to t, we have

$$di/dt + i/RC = 0$$

$$\therefore i = (E/R)e^{-t/CR}$$

Now immediately after a circuit is switched on to a supply, or, for that matter, immediately after any abrupt change in circuit conditions, the current flowing will consist of two components, namely (1) a steady-state component representing the state which the circuit ultimately attains, and (2) a transient component which enables the circuit gradually to reach the steady-state condition. The transient results from the ability of the circuit to store energy and its nature depends on the manner in which this energy is stored. In the case of a circuit possessing resistance only there is no stored energy and therefore no transient, with the result that the steady-state condition of $I = E/R$ is attained immediately after applying the voltage E.

In the above equation for the charging of the RC circuit the two components are

$$\text{Steady-state current} = 0$$

$$\text{Transient current} = (E/R)e^{-t/CR}$$

If we put $t = 0$, then $i = E/R$, showing that at the moment of applying the voltage the capacitor (if uncharged) acts like a short-circuit.

Now let the capacitor be discharged through the resistor.

The e.m.f. equation is

$$0 = iR + q/C$$

$$0 = iR + \int i dt/C$$

$$\therefore di/dt + i/RC = 0$$

Since, initially $q = CE$, the solution is

$$i = (E/R)e^{-t/CR}$$

This is wholly a transient, the steady-state current again being zero.

$$\therefore e = iR$$
$$= Ee^{-t/CR}$$

Rate of discharge of the capacitor

$$-de/dt = E \times (1/CR)e^{-t/CR}$$

When $t = 0$, $\quad -de/dt = E/CR$

In CR seconds the voltage falls to

$$e = Ee^{-CR/CR}$$
$$= Ee^{-1} = 0{\cdot}368E$$

The term CR is called the *time constant*.

Now find the time t_2 for the voltage to fall to $E/2$

$$\therefore E/2 = E \,.\, e^{-t_2/CR}$$
$$1/2 = e^{-t_2/CR}$$
$$t_2 = CR \log_e 2 = 0{\cdot}692CR$$

RC Circuit with Shunted Capacitor. The circuit is shown in Fig. 2.21.

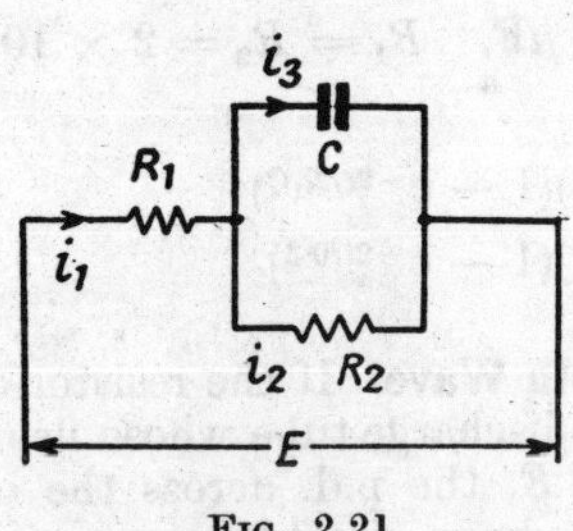

Fig. 2.21

$$i_3 = Cdv/dt; \quad v = \text{the capacitor p.d.}$$

$\therefore$ at any time t $\quad i_2 = v/R_2$

Now $\quad i_1 = i_c + i_2$

$$= C\,dv/dt + v/R_2$$

$\therefore$ p.d. across R_1

$$R_1 i_1 = R_1(v/R_2 + C\,dv/dt)$$
$$= (R_1/R_2)v + R_1C\,dv/dt$$

Equating voltages

$$\begin{aligned} E &= \text{p.d. across } R_1 + \text{p.d. across } R_2 \\ &= (R_1/R_2)v + R_1C\, dv/dt + v \\ &= (1 + R_1/R_2)v + R_1C\, dv/dt \\ &= av + R_1C\, dv/dt, \text{ say} \end{aligned}$$

$$E - av = R_1C\, dv/dt$$

$$dv/(E - av) = dt/R_1C$$

$$\therefore\ 1/-a \log_e (E - av) = t/R_1C + K$$

$$\log_e (E - av) = -at/R_1C + \log_e A \text{ (another constant)}$$

$$E - av = Ae^{-at/R_1C}$$

When $t = 0, v = 0 \ \therefore\ A = E$

$$\therefore\ v = (E/a)(1 - e^{at/R_1C})$$

We can determine the initial and final conditions as follows—

(1) When $t = 0$ the capacitor can be regarded as a short-circuit

$$\therefore\ (i_1)_{t=0} = E/R_1, \text{ and } v = 0$$

(2) When $t = \infty$, the circuit having reached the steady-state condition, the capacitor can be regarded as an open-circuit

$$\therefore\ (i_1)_{t=\infty} = (i_2)_{t=\infty} = E/(R_1 + R_2)$$

and
$$v = R_2 i_1 = E \times [R_2/(R_1 + R_2)]$$

EXAMPLE. $C = 0{\cdot}15\ \mu\text{F}, \quad R_1 = R_2 = 2 \times 10^6\ \Omega$

$$\therefore\ a = 2$$

$$\begin{aligned} \therefore\ v &= (E/2)(1 - e^{-2t/R_1C}) \\ &= (E/2)(1 - e^{-2t/0{\cdot}3}) \end{aligned}$$

Generation of a Sawtooth Wave. If the resistor of the RC circuit of Fig. 2.20 is shunted by a discharge tube whose firing voltage E_d is less than the supply voltage E, the p.d. across the capacitor will be of approximately triangular shape, reaching a maximum in an approximately linear manner and then suddenly falling to zero, this process being repeated at time intervals decided by the circuit constants. Denoting the voltages at R and C by v_R and v_C respectively we have

$$v_R = Ri = Ee^{-t/CR}$$

$$v_C = E - v_R = E(1 - e^{-t/CR})$$

Thus v_C increases from zero and provided it is not allowed to become too high the increase will be approximately linear. When v_C reaches the firing voltage E_d the tube will discharge and act like a short-circuit, thus suddenly reducing the voltage v_C, not to zero, but to a value v'

which is unable to maintain the discharge. For the time taken to reach the value E_d after the first application of the e.m.f. E, we have

$$E_d = E(1 - e^{-t_d/CR})$$

$$\therefore\ t_d = CR \log_e [E/(E - E_d)]$$

At $t = t_d$ the capacitor p.d. suddenly falls, not to zero, but to the above value v', and therefore time for the next rise in this p.d. to the

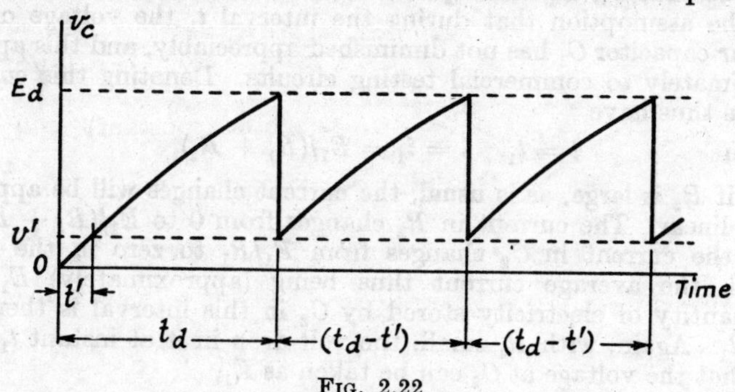

FIG. 2.22

actual value E_d will be less than before, by the time taken for the p.d. to rise from zero to v'. This is clearly

$$t' = CR \log_e [E/(E - v')]$$

Hence time of second and all successive teeth

$$t_d - t' = CR \log_e [(E - v')/(E - E_d)]$$

The variations of voltage with time are thus as shown in Fig. 2.22.

***RC* Circuit with Shunted Capacitor Supplied from a Charged Reservoir Capacitor.** This is an important circuit as it corresponds to that sometimes used for the application of an impulse voltage to a test object when this object is essentially a capacitor. The circuit is that of Fig. 2.23 in which the capacitor C_1 charged initially to a voltage of E_1 is suddenly connected to the rest of the circuit, usually by the breakdown of a spark-gap. The following solution is mainly concerned with the general conditions and, although approximate, is often of a sufficient accuracy. It avoids a very lengthy exact solution.

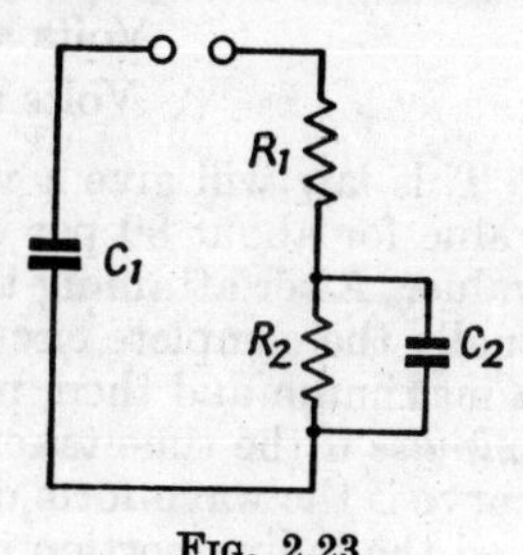

FIG. 2.23

Initially the capacitor C_2 is uncharged and, in this state, it acts as though short-circuited. Hence at the moment the spark-gap breaks down

$$i_0 = E_1/R_1$$

This current flows in resistor R_1 and in capacitor C_2, the current in R_2 at this instant being zero.

It is obvious that the current in C_1 and R_1 will rise to a maximum and eventually die away exponentially. The voltage at capacitor C_2 will also rise to a maximum and then die away. Let it reach a maximum when $t = t_1$, then since, at a maximum $dv_2/dt = 0$ the current in $C_2 = 0$ so that C_1, R_1 and R_2 now carry the same current. We now make the assumption that during the interval t_1 the voltage of the reservoir capacitor C_1 has not diminished appreciably, and this applies approximately to commercial testing circuits. Denoting this current by i_1 we thus have

When $$t = t_1, \quad i = i_1 = E_1/(R_1 + R_2)$$

Again, if R_2 is large, as is usual, the current changes will be approximately linear. The current in R_2 changes from 0 to $E_1/(R_1 + R_2)$ in t_1 sec., the current in C_2 changes from E_1/R_1 to zero in the same interval, the average current thus being (approximately) $E_1/2R_1$. The quantity of electricity stored by C_2 in this interval is therefore $E_1t_1/2R_1$. Again, with R_1 small, the volt-drop in it at instant t_1 is so small that the voltage at C_2 can be taken as E_1,

$$C_2E_1 = E_1t_1/2R_1$$

$$\therefore t_1 = 2C_2R_1$$

This is (approximately) the time taken for the voltage at C_2 to attain its maximum value, and the interesting fact to notice is that, in spite of the very large value of R_2, it is the value of R_1 which decides, almost entirely, the length of the time interval t_1.

We have assumed that E_1 does not sensibly diminish, and we have, with most test objects, the condition that R_2 is very large, so that, as a further approximation we can regard the circuit as consisting of C_1, C_2 and R_1 in series. The current therefore rises according to the law

$$i = (E_1/R_1)e^{-t/C_2R_1} *$$

$$\text{Volts across } R_1 = E_1e^{-t/C_2R_1}$$

$$\therefore \text{Volts across } C_2 = E_1(1 - e^{-t/C_2R_1})$$

This law will give a value for v_{c_2} which is very close to the correct value for about 80 per cent of the time taken to attain its maximum value. After attaining this maximum the voltage dies away and eventually the complete circuit is discharged. Thus the voltage v_{c_2} rises to a maximum and then, more gradually, falls to zero. It is therefore an *impulse* if the time taken is very short. The shape of the voltage-time curve is the wave-form of the impulse, the rising portion the wavefront, and the falling portion the wave tail.

For the wave tail we can again make an approximation by neglecting R_1 and C_2, thereby simplifying the circuit to C_1 in series with R_2, the initial condition being that $v \simeq E_m$. Thus for the tail we have

$$v_{c_2} \text{ or } v_{R_2} \simeq E_m(1 - e^{-t/C_1R_2})$$

* The argument applies only to the important practical case in which $C_1 \gg C_2$.

Example. $E_1 = 10^6$ V, $R_1 = 500\ \Omega$, $R_2 = 44{,}500\ \Omega$, $C_1 = 0{\cdot}005\ \mu$f, $C_2 = 100\ \mu\mu$f. The stages in the development of the impulse applied to the test object are illustrated in Fig. 2.24

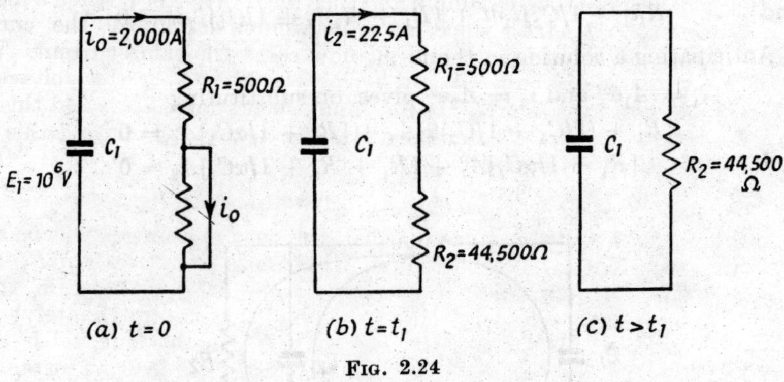

Fig. 2.24

(*a*) Initially, $t = t_0$ (Fig. 2.24(*a*)). The capacitor C_2 acting as a short circuit, the reservoir capacitor discharges into the resistance R_1, giving

$$i_0 = 10^6/500 = 2{,}000 \text{ A}$$

(*b*) Wave front:

$$t_1 = 2C_2R_1 = 2 \times 100 \times 10^{-12} \times 500$$
$$= 10^{-7} \text{ sec}$$

The *initial* rate of rise of the wave front is given by

$$v_{c_2} = E_1(1 - \mathrm{e}^{-t/C_2R_1})$$
$$= 10^6(1 - \mathrm{e}^{-t/5\times 10^{-8}})$$

where t is small, and this will hold, very nearly, up to $t \ngtr 0{\cdot}8t_1$, i.e. up to 8×10^{-8} sec.

(*c*) At the wave crest the current in C_2 is zero, the conditions being given (Fig. 2.24 (*b*)). The capacitor C_1, in view of its large capacitance, will have experienced little fall in voltage, so that the approximate condition is

$$i_2 = 10^6/(500 + 44{,}500) = 22{\cdot}2 \text{ A}$$
$$\text{crest voltage } E_m = i_2R_2 = 22{\cdot}2 \times 44{,}500$$
$$= 0{\cdot}99 \times 10^6 \text{ V}$$

(*d*) The wave tail

$$= E_m(1 - \mathrm{e}^{-t/C_1R_2})$$
$$= 0{\cdot}99 \times 10^6(1 - \mathrm{e}^{-t/(2{\cdot}225\times 10^{-4})})$$

The application of Kirchhoff's Laws affords a neat solution of the above problem.* From the loops 1 and 2 in Fig. 2.25 we have, subsequent to the moment $t = 0$.

$$R_1 i_1 + (1/C_1 + 1/C_2)\int i_1 dt + R_1 i_2 + 1/C_1 \int i_2 dt = 0$$

and $$R_1 i_1 + 1/C_1 \int i_1 dt + (R_1 + R_2) i_2 + 1/C_1 \int i_2 dt = 0$$

Anticipating a solution in the form

$i_1 = A_1 e^{\alpha t}$ and $i_2 = A_2 e^{\alpha t}$ gives, on substituting

$$[R_1 + (1/C_1 + 1/C_2)\alpha]A_1 + [R_1 + 1/\alpha C_1]A_2 = 0$$

and $$[R_1 + 1/\alpha C_1]A_1 + [R_1 + R_2 + 1/\alpha C_1]A_2 = 0$$

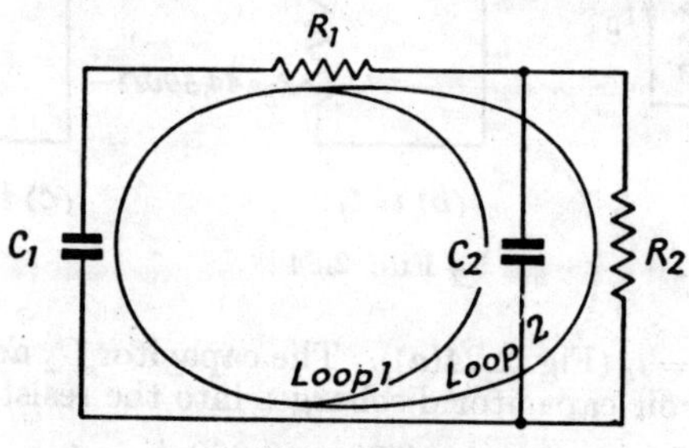

Fig. 2.25

From these two equations we derive expressions for the ratios A_2/A_1, namely

$$A_2/A_1 = -\frac{R_1 + (1/C_1 + 1/C_2)\alpha}{R_1 + 1/\alpha C_1}$$

and $$A_2/A_1 = -\frac{R_1 + 1/\alpha C_1}{R_1 + R_2 + 1/\alpha C_1}$$

Equating these two expressions for the same ratio gives

$$[R_1 + (1/C_1 + 1/C_2)\alpha]/(R_1 + 1/\alpha C_1) = (R_1 + 1/\alpha C_1)/(R_1 + R_2 + 1/\alpha C_1)$$

Hence α is given by the equation

$$C_1 C_2 R_1 R_2 \alpha^2 + (C_1 R_1 + C_1 R_2 + C_2 R_2)\alpha + 1 = 0$$

say $$a\alpha^2 + b\alpha + 1 = 0$$

giving $$\alpha = (-b \pm \sqrt{b^2 - 4a})/2a$$

The two values of α give the *modes* of the circuit, the meaning of the term "mode" being very well expressed by the following extract from the paper quoted. "If all the sources of e.m.f. in a network are removed by a switching operation and the network contains elements capable of

* J. Willis. "The solution of transient problems in terms of the modes of the circuit." *Bull. Elec. Eng. Educ.*, No. 2.

storing electrical energy, then the subsequent currents in the network are of a transitory nature. These currents exhibit certain modes of behaviour in the form of natural frequencies and rates of decay. A charged capacitor switched across a resistor is discharged exponentially with a time constant defined by the product of the resistance and the capacitance of the circuit; a capacitor discharging through a coil, if the resistance of the coil is small enough, does so in an oscillatory manner, the frequency of the oscillation and its rate of decay being determined by the values of the circuit resistance, inductance and capacitance. More complicated networks have their currents characterized by several normal modes. The possession of these modes is a fundamental property of the network and the transient response of the network is wholly expressed in them. These modes are, in fact, the exponential terms which make up the solutions of the differential equations which express the behaviour of the network currents after the switching operation."

The circuit considered has two modes, given by the solution of the equations for α. Denoting them by α_1 and α_2 we have two sets of solutions, namely

$$i_1 = A_1' e^{\alpha_1 t}; \quad i_2 = A_2' e^{\alpha_1 t}$$

and

$$i_1 = A_1'' e^{\alpha_2 t}; \quad i_2 = A_2'' e^{\alpha_2 t}$$

Since the differential equations of the circuit are linear the above solutions can be added together, giving

$$i_1 = A_1' e^{\alpha_1 t} + A_1'' e^{\alpha_2 t}$$
$$i_2 = A_2' e^{\alpha_1 t} + A_2'' e^{\alpha_2 t}$$

We shall again see that the currents in the above circuit have two modes, which, in the case of the wave tail of Fig. 2.29 are denoted by the symbols m and n.

Oscillatory Charge and Discharge of a Capacitor. If a circuit contains resistance and capacitance in series, then rise of current during charging of the capacitor, and the decay of current during discharging, are both exponential in character. There are no current oscillations of any kind. Now consider a circuit having resistance, inductance, and capacitance in series. At any instant the quantity of electricity stored in the capacitor is given by

$$q = \int i \, . \, dt$$

The p.d. across the capacitor terminals is equal to

$$\frac{q}{C} = \frac{1}{C}\int i \, . \, dt$$

Hence, if at any instant the e.m.f. impressed on the circuit is e, we have

$$Ri + L\frac{di}{dt} + \frac{1}{C}\int i \, . \, dt = e$$

Put $e = E_{max} \sin \omega t$, and differentiate with respect to t. Then

$$R(di/dt) + L(d^2i/dt^2) + i/C = \omega E_{max} \cos \omega t$$

The solution of this is

$$i = A \sin (\omega t - \alpha) + Be^{-\frac{R}{2L}\cdot t} \sin (\omega' t - \beta)$$

where $$A = \frac{E_{max}}{\sqrt{[R^2 + (L\omega - 1/C\omega)^2]}}$$

$$\omega' = \sqrt{(1/LC - R^2/4L^2)}$$

$$\alpha = \tan^{-1} L\omega/R$$

B, α and β, are constants, and e is the base of natural logarithms.

The first term of the expression for i represents a forced oscillation which obeys the laws that will be considered in detail in Chapter 22. The second term represents a free oscillation, sometimes called a "transient," of frequency

$$f' = \omega'/2\pi = (1/2\pi)\sqrt{(1/LC - R^2/4L^2)}$$

This oscillation has a damping term $e^{-\frac{R}{2L}\cdot t}$ showing that its amplitude dies away exponentially as shown in Fig. 2.26. The term $R/2L$ is called the "attenuation constant." It will be seen that the resistance R has some influence on the frequency of the free oscillation, but if $R^2/4L^2$ is small compared with $1/LC$, as for example, in the oscillatory circuits of wireless apparatus, then the frequency of the free oscillation becomes

$$f' \simeq (1/2\pi)\sqrt{(1/LC)}$$

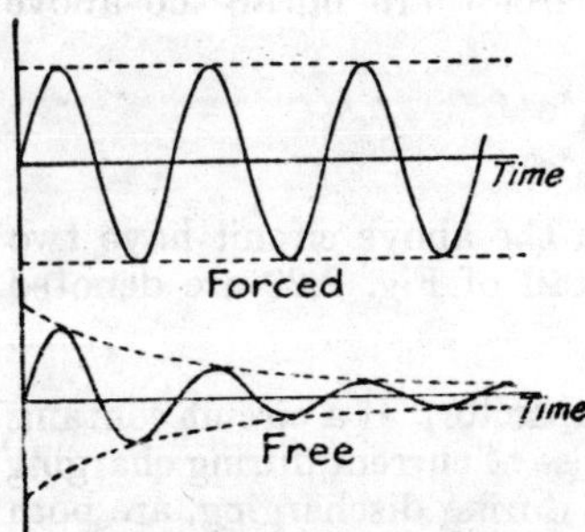

Fig. 2.26. Illustrating Free and Forced Oscillations

Now suppose that the circuit, instead of having an external e.m.f. impressed on it, is allowed to discharge itself after the capacitor has previously been charged up. The e.m.f. equation then becomes

$$R(di/dt) + L(d^2i/dt^2) + i/C = 0$$

or $$d^2i/dt^2 + (R/L)(di/dt) + i/LC = 0$$

To solve this, put $i = e^{\lambda t} y$

$$\therefore\ d^2i/dt^2 = [(d^2y/dt^2) + 2\lambda(dy/dt) + \lambda^2 y]e^{\lambda t}$$

$$(R/L)(di/dt) = (R/L)[(dy/dt) + \lambda y]e^{\lambda t}$$

$$i/LC = (1/LC)ye^{\lambda t}$$

$$\therefore \; d^2y/dt^2 + (2\lambda + R/L)(dy/dt) + [\lambda^2 + (R/L) \cdot \lambda + 1/LC]y = 0$$

Now put $\lambda = -R/2L$ so that $i = ye^{-\frac{R}{2L} \cdot t}$

$$\therefore \; d^2y/dt^2 + (1/LC - R^2/4L^2)y = 0$$

There are three cases—

(a) Let $1/LC > R^2/4L^2$. Put $(1/LC - R^2/4L^2) = \omega^2$

$$\therefore \; d^2y/dt^2 + \omega^2 y = 0$$

$$\therefore \; y = A\cos\omega t + B\sin\omega t$$

$$\therefore \; i = (A\cos\omega t + B\sin\omega t)e^{-\frac{R}{2L} \cdot t}$$

$$= \alpha e^{-\frac{R}{2L} \cdot t}\cos(\omega t + \phi)$$

where α and ϕ are constants. This represents a simple harmonic oscillation whose amplitude $\alpha e^{-\frac{R}{2L} \cdot t}$ continually diminishes as t increases, as shown in Fig. 2.27. The period

$$\frac{2\pi}{\omega} = \frac{2\pi}{\sqrt{(1/LC)}(1 - R^2C/4L)^{\frac{1}{2}}}$$

If the resistance were zero, the period would be $2\pi\sqrt{(LC)}$, showing that the presence of resistance lengthens the period in the ratio

$$\frac{1}{(1 - R^2C/4L)^{\frac{1}{2}}}$$

If R is small, then $R^2C/4L$ can be neglected as a small quantity of the second order, and the change in period is then negligible.

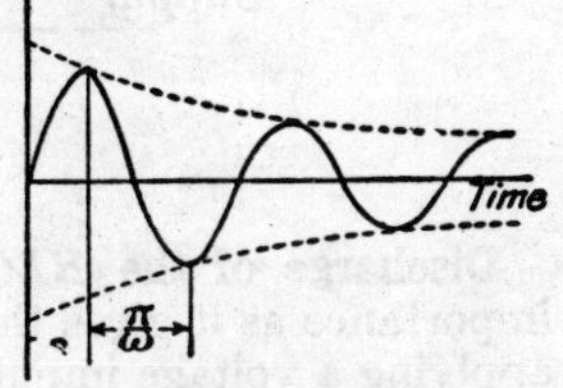

Fig. 2.27. Discharge Current of a Capacitor

To find the points of maximum amplitude, we have

$$di/dt = 0$$

This equation has the form

$$C\cos\omega t + D\sin\omega t = 0$$

$$\therefore \; \tan\omega t = -C/D$$

The solution of this is an arithmetic progression whose common difference is π. The corresponding increments of time are therefore π/ω, and the ratio of consecutive amplitudes in positive and negative directions is

$$e^{\frac{R}{2L} \cdot \frac{\pi}{\omega}}$$

The log of this, namely

$$(\pi R/2L\omega)\log_{10} e$$

is called the *logarithmic decrement.*

(b) Let $1/LC < R^2/4L^2$

Put
$$R^2/4L^2 - 1/LC = m^2$$
$$\therefore\ d^2y/dt^2 - m^2y = 0$$
$$\therefore\ y = Ae^{mt} + Be^{-mt}$$
and
$$i = Ae^{-\left(\frac{R}{2L} - m\right)t} + Be^{-\left(\frac{R}{2L} + m\right)t}$$

Since $R/2L > m$, both terms diminish asymptotically to zero, and the current is aperiodic instead of oscillatory.

(c) Let
$$1/LC = R^2/4L^2$$
Then
$$d^2y/dt^2 = 0;\ \therefore\ y = At + B$$
$$\therefore\ i = (At + B)e^{-\frac{R}{2L}.t}$$

The current in this case approaches zero asymptotically, and there is, at the most, one passage through the zero position.

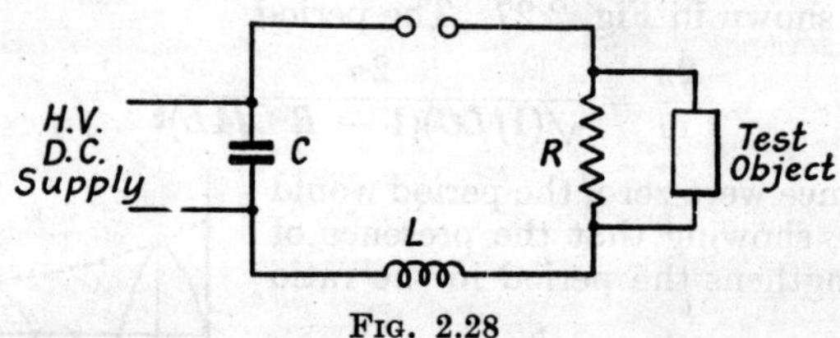

Fig. 2.28

Discharge of the *RLC* Circuit. This is a case of great practical importance as it gives the circuit characteristics during one method of applying a voltage impulse to a test object. For this purpose the test object is connected across the resistor, as shown in Fig. 2.28, and a gradually increasing unidirectional voltage applied to the capacitor. The capacitor voltage at the moment of its discharge into the circuit is pre-determined by giving the adjustable gap the required setting. Solving for i, but in a somewhat different manner from the above we have
$$d^2i/dt^2 + (R/L)di/dt + i/LC = 0$$

Let $i = A'e^{\alpha t}$ be a solution
$$\therefore\ di/dt = \alpha A'e^{\alpha t};\quad d^2i/dt^2 = \alpha^2 A'e^{\alpha t}$$
$$\therefore\ \alpha^2 + (R/L)\alpha + 1/LC = 0$$
$$\therefore\ \alpha = -R/2L \pm \sqrt{R^2/4L^2 - 1/LC}$$

Denote the two values of α by α_1 and α_2; also put $R/2L = a$ and $\sqrt{R^2/4L^2 - 1/LC} = b$
$$\therefore\ \alpha_1 = -a + b$$
and
$$\alpha_2 = -a - b$$
$$\therefore\ i = Ae^{-(a-b)t} + Be^{-(a+b)t}$$

When $t = 0$, $i = 0$ $\quad\therefore\ P = -A$.

PLATE V. STATOR OF A SCHERBIUS MACHINE
(*Brown-Boveri*)

Again when $t = 0$ the condition of zero current means that there is no voltage drop along R. Hence initial capacitor voltage

$$E = L(di/dt)_{t=0}$$

$$\therefore A = E/2Lb$$

Hence $$i = E/2Lb[e^{-(a-b)t} - e^{-(a+b)t}]$$

Hence voltage across test object

$$v_R = Ri$$

$$= RE/2Lb[e^{-(a-b)t} - e^{-(a+b)t}]$$

Put $$RE/2Lb = E'$$

$$a - b = m$$

$$a + b = n$$

$$\therefore v_R = E'(e^{-mt} - e^{-nt})$$

Since E, R, L and b are known from the current constants, E' is also known. Assuming that the circuit parameters give conditions (*b*) in the preceding section, we see that v_R is the difference between two exponentially decaying voltages and consequently the nature of v_R is that it is initially zero, it rises to a maximum, and then falls away exponentially. The shape of the voltage impulse applied to the test object is therefore as indicated in Fig. 2.29. Let the time taken to

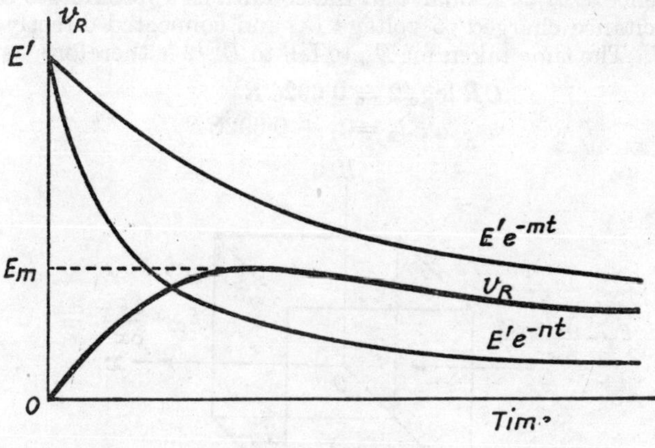

Fig. 2.29

reach the maximum value be t_1 and the time taken for the *tail* to fall to one-half of the maximum be t_2, then the wave is called a t_1/t_2 wave. The British Standard impulse wave is a 1/50 wave, which means that the peak value is reached in one microsecond, and the tail falls to one-half of the peak value in 50 microseconds. We calculate t_1 as follows:

hen $t = t_1$ $dv_R/dt = 0$

$$\therefore 0 = E'(-me^{-mt_1} + ne^{-nt_1})$$
$$me^{-mt_1} = ne^{-nt_1}$$
$$m/n = e^{-(n-m)t_1}$$
$$\therefore t_1 = [\log_e (m/n)]/(n - m)$$

It is difficult to calculate t_2 accurately but, for practical purposes, an approximate solution is sufficient. In all impulse-test equipment the capacitance C and resistance R are such that

$$CR^2 \gg 4L$$

with the result that m and n simplify to

$$m = 1/CR$$
$$n = R/L - 1/CR$$

If we use these values in the expression for t_1 we have

$$n/m = CR^2/L - 1$$
$$n - m = R/L - 2/CR$$
$$t_1 = [\log_e (CR^2/2 - 1)]/(R/L - 2/CR)$$

The peak voltage E_m occurs when $t = t_1$, hence

$$E_m = E'(e^{-mt_1} - e^{-nt_1})$$

After this value has been reached dv_R/dt is small and therefore di/dt is small. Hence $L\,di/dt$ is small and the conditions approximate to those of a capacitance charged to voltage E_m and connected directly to the resistor R. The time taken for E_m to fall to $E_m/2$ is therefore, as before

$$CR \log_e 2 = 0{\cdot}692CR$$
$$\therefore t_2 = t_1 + 0{\cdot}692CR$$

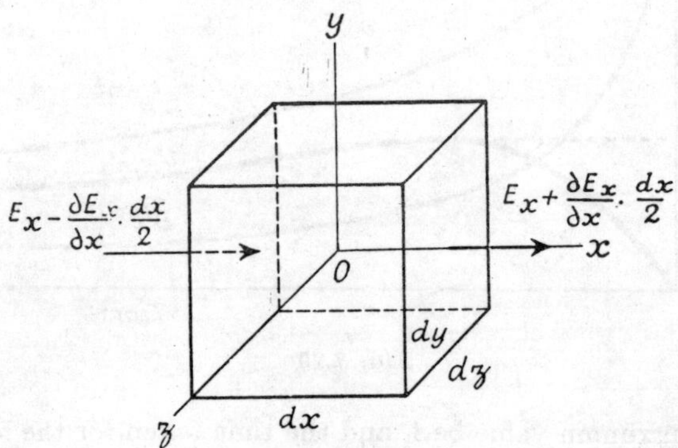

FIG. 2.30

The Equations of Poisson and Laplace. Consider an elementary cube of dielectric $dxdydz$ and let the origin be at the centre, Fig. 2.30. Let the components of the electric force be E_x, E_y, and E_z and the volume density of charge ρ. The surface integral of outward normal components of E consists of

$$[E_x + (\partial E_x/\partial x)(dx/2)]dy \,.\, dz - [E_x - (\partial E_x/\partial x)(dx/2)]dy\, dz$$
$$= (\partial E_x/\partial x)\, dx \,.\, dy \,.\, dz \text{ in the } x \text{ direction}$$

Similarly $\quad (\partial E_y/\partial y)\, dx \,.\, dy \,.\, dz$ in the y direction

and $\quad (\partial E_z/\partial z)\, dx \,.\, dy \,.\, dz$ in the z direction

The enclosed charge is $\quad \rho\, dx\, dy\, dz$

Hence, by Gauss's theorem

$$(\partial E_x/\partial x + \partial E_y/\partial y + \partial E_z/\partial z)\, dx\, dy\, dz = \rho\, dx\, dy\, dz/\varepsilon\varepsilon_0$$
$$\therefore\ \partial E_x/\partial x + \partial E_y/\partial y + \partial Ez/\partial z = \rho/\varepsilon\varepsilon_0$$

If $\rho = 0$, we have

$$\partial E_x/\partial x + \partial E_y/\partial y + \partial E_z/\partial z = 0$$

and in this form the equation is known as the equation of continuity.

Now $\quad E_x = -\,\partial V_x/\partial x;\ E_y = -\,\partial V_y/\partial y;\ E_z = -\,\partial V_z/\partial z$

and therefore for the cases of ρ finite and ρ zero we have the two equations—

$$\partial^2 V/\partial x^2 + \partial^2 V/\partial y^2 + \partial^2 V/\partial z^2 = -\,\rho/\varepsilon\varepsilon_0 \quad \text{Poisson's equation}$$

and $$\partial^2 V/\partial x^2 + \partial^2 V/\partial y^2 + \partial^2 V/\partial z^2 = 0 \quad \text{Laplace's equation}$$

EXAMPLE. THE CYLINDRICAL CAPACITOR. If the cylinders are

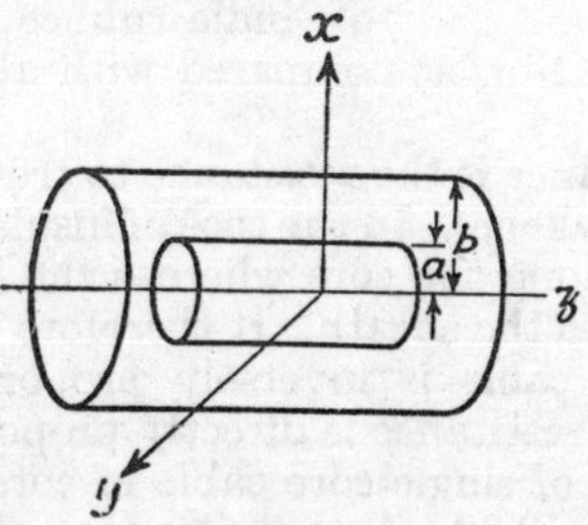

FIG. 2.31

infinitely long there is no variation of V in the z direction (Fig. 2.31), and therefore Laplace's equation becomes

$$\partial^2 V/\partial x^2 + \partial^2 V/\partial y^2 = 0$$

which can be transformed to

$$\partial^2 V/\partial r^2 + (1/r)\partial V/\partial r = 0$$

there being no θ term because of radial symmetry.

The solution of this is

$$V_r = k_1 + k_2 \log_e r$$

Denoting the potentials of the inner and outer cylinders by V and 0 respectively, we have

when $r = a \quad V_r = V \quad \therefore \quad V = k_1 + k_2 \log_e a$

when $r = b \quad V_r = 0 \quad \therefore \quad 0 = k_1 + k_2 \log_e b$

This gives

$$k_2 = -V/\log_e (b/a)$$
$$k_1 = +V \log_e b/\log_e (b/a)$$
$$\therefore V_r = (V/\log_e b/a)(\log_e b - \log_e r)$$
$$= \frac{V \log_e (b/r)}{\log_e (b/a)}$$

Dielectrics. An insulator is a material which has a very high specific resistance in comparison with a good conductor of electricity such as copper or aluminium. For this reason conductors which have to be maintained at a high potential with respect to earth are either entirely sheathed in insulating material or are supported by insulators. Even so, there is always a certain amount of leakage to earth, although the leakage current which passes through the insulating material is exceedingly small in comparison with the useful current carried by the conductor. Thus the specific resistance of india-rubber is of the order of 2×10^{12} to 10×10^{12} Ω-m, as compared with about $1{\cdot}7 \times 10^{-8}$ Ω-m for copper.

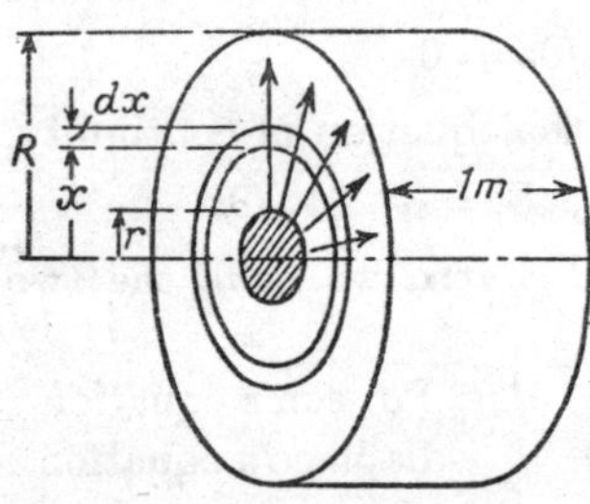

Fig. 2.32

The insulation resistance is the resistance to true leakage current (i.e. not including surface leakage). In the case of insulated cables, the useful current flows axially along the core whereas the leakage current flows radially from the core to the sheath. It therefore follows that the insulation resistance of a cable is inversely proportional to its length, whereas the conductor resistance is directly proportional to the length.

Consider unit length of single-core cable of core radius r and insulation radius R, as in Fig. 2.32. An annular ring of radius x will have a surface area of $2\pi x \times 1 = 2\pi x$ square metres, and since the flow of leakage current is radial as shown, such a ring of thickness dx will offer to the leakage current a path of length dx and area $2\pi x$.

Hence, for the resistance of the ring, we have

$$dR = \rho dx/2\pi x$$

Hence insulation resistance/metre length of cable

$$R = \frac{\rho}{2\pi}\int_r^R \frac{dx}{x}$$

$$(\rho/2\pi)\log_e(R/r)$$
$$= (2{\cdot}3\rho/2\pi)\log_{10}(R/r)$$

Hence, for a length l metres of cable, we have for the insulation resistance

$$R = (2{\cdot}3\rho/2\pi l)\log_{10}(R/r)$$

EXAMPLE. $r = 1{\cdot}3 \times 10^{-2}$ metre, $R = 4 \times 10^{-2}$ metre, $\rho = 8 \times 10^{12}$ Ω-m = 8×10^6 MΩ-m. Find the insulation resistance/mile of cable.

$$\log_{10}(R/r) = \log(4/1{\cdot}3) = 0{\cdot}488$$
$$l = 5280 \times 12 \times 2{\cdot}54 \times 10^{-2}$$
$$= 16{\cdot}2 \times 10^2 \text{ metres}$$
$$\therefore R = 2{\cdot}3 \times 8 \times 10^6 \times 0{\cdot}488/(2\pi \times 16{\cdot}2 \times 10^2)$$
$$= 883 \text{ M}\Omega\text{/mile, or, more correctly,}$$
$$883 \text{ M}\Omega\text{-miles}$$

The following are the important electrical properties of insulating materials—

(i) Specific resistance; this has already been considered above.
(ii) Dielectric strength or breakdown strength.
(iii) Dielectric loss when subjected to alternating electric stresses.
(iv) Relative permittivity, or dielectric constant.

Certain mechanical properties are also essential, depending on the use to which the material is put. Thus for the insulating materials used in the manufacture of electrical machinery, the following are essential—

(v) Sufficient mechanical strength to withstand vibration, and the bending and abrasion experienced during the manufacture of the machine.
(vi) Good heat conductivity.
(vii) Ability to withstand the maximum temperature attained, without change in physical properties or in chemical composition. Table 2.1 gives some of the important characteristics of a selection of common dielectric materials.*

The breakdown strength naturally depends upon the thickness of the material but it is not, as might be expected, proportional to the thickness. Thus if the breakdown strength of a certain sample is given as x kilovolts per mm, this value x refers to a definite thickness of sample. The breakdown strength is also dependent on the shape of the electrodes to which the voltage is applied, the frequency and wave form of the voltage (when alternating), the time of application, and on the time taken to bring up the voltage from zero to the breakdown value, also on temperature and moisture content of test piece. It will therefore

* Insulating materials are discussed very fully by R. H. Robinson, *The Inside of Electrical Machines*, Benn, London; D. F. Milner, *The Insulation of Electrical Apparatus*, McGraw Hill, New York; H. Warren, *Electrical Insulating Materials*, Benn, London, and H. Schwarzer, *Theory of Dielectrics*, Wiley, New York.

Table 2.1. Properties of Dielectric Materials

Material	Dielectric strength in kV/mm at 50 c/s	Specific Resistance when dry at 25°C, MΩ/cm cube	Permittivity	Affected by Moisture	Safe Temperature in °C	Uses
Asbestos	3–4·5	$1 \cdot 6 \times 10^5$	—	Absorbent	500 or more	Covering of wires in very highly rated machines.
Bakelite	20–25	—	5–6	No	200	Bakelized paper made up in form of boards. Many uses.
Bitumen (vulcanized)	14	—	4·5	No	About 60	Low-voltage mining cables. Cable-box filling compound.
Cotton	3–4	1000 upwards according to dryness	—	Absorbent	90	Covering for wires.
Ebonite	10–40	$2 - 100 \times 10^8$	2–3	Slightly	40	Covers for resistance boxes, etc.
Empire cloth	10–20	As cotton	2	Absorbent if varnish layer is cracked	90	Wrapping for groups of wires, e.g. armature coils.
Fibre	5	As cotton	4–6	Ditto	90	In sheet form, slot linings.
Glass	5–12	5×10^6 to infinity	3–8	No, except on surface	Room temperatures only or may crack	Not used in electrical practice except to a small extent for transmission-line insulators.
Guttapercha	10–20	5 to 25×10^8	3–5	No	40	Covering for submarine cables.
India rubber	10–25	2 to 10×10^8	2–3	Slightly	40	Cable insulation.
Marble	6	400	8	Somewhat absorbent	Room temperatures	Formerly used for switchboards.
Mica	40–150	5 to 100×10^6	3–8	No	500 or more	Not generally used for machines in the pure form.
Micanite	30	10 to 6000×10^6	6–8	No	130 when under pressure	Commutator segments. Slot linings for high-voltage machines. Bushes.
Paper	4–10	As cotton	2	Absorbent	90	Cable insulation when oil impregnated. Covering for transformer conductors.
Paraffin wax	8	3×10^{10}	2	No	Under 50	
Porcelain	9–20	1 to 1000×10^6	4–7	Not when vitreous and glazed	Room temperatures	Insulators for overhead lines.
Shellac	5–20	9×10^9	2·5–3·5	No	Under 60	Insulating varnishes. Cement for manufacture of micanite.
Slate	3	40	7	Rather absorbent	Room temperatures	For face-plates of starters, etc., not now used for switchboards.

be seen that figures of breakdown value mean very little unless the test conditions are very completely specified.*

Provided that the test conditions are identical and only the thickness of the material is varied, the relationship between breakdown voltage V and thickness t is given approximately by Baur's Law, viz.

$$V = at^{\frac{2}{3}}$$

where a is a constant depending on the material and also on the thickness t.

The reason for the decrease in the electric strength with increase in thickness is probably due to the fact that a thick material cannot get rid of internally produced heat so well as a thin material, a smaller applied voltage per unit thickness therefore being required to produce a given maximum temperature. This is in agreement with the thermal theory of breakdown given by Professor Miles Walker.† (*See* Dielectric Breakdown, p. 59.)

Dielectric loss when the material is subjected to alternating voltage is of little importance so long as the voltage is low, but it becomes of very considerable importance in high-tension work. For example, insulated cables are now made for working voltages of 100,000 and more, and the dielectric loss is a limiting factor in their design. Imagine a capacitor charged by the application of a steady voltage, then there is an initial rush of current which follows the exponential law previously given, but, after this, instead of falling to zero the current persists at a low value for an appreciable time. The current has been described as "soaking in" the dielectric. Similarly, when the capacitor is discharged, there is an initial rush of current, and the p.d. very quickly falls to zero. If, however, the discharging connection is removed and the capacitor left for some time, it will be found that a charge has been acquired, and this charge in turn can be removed by the process of discharging. With some dielectrics this can be repeated several times. The first of these phenomena is called *absorption*, and all dielectrics are absorptive to a certain degree. The charge recovered or retained after an initial discharge is called the *residual charge*.

Dielectric Phenomena. The high electrical conductivity of metals is explained by the fact that the metallic atoms have only one, or two, valency electrons, and these are so far removed from the nucleus that they are effectively screened from it by the inner, completed, electron shells. In consequence, they can easily detach themselves, say as the result of thermal agitation, and they are able to wander at random about the interspaces between the more stable portions of the atoms, which thereby, in virtue of losing their valency electrons, have become positive ions. If a p.d. is applied to the ends of the conductor

* The standard test conditions can be found in the *Electrical Engineers' Data Book*, Vol. 2.

† *J. Instn Elect. Engrs.*, **49**, p. 71.

an axial drift of electrons towards the end at higher potential is superposed on the random motions, and it is this axial component of electron motion which constitutes the electric current.

All liquid or solid insulators are compounds and therefore the ultimate particles of which they are composed are the molecules, and in many cases the molecules are very heavy and complex. From the electrical point of view we can regard the molecules as collections of positive and negative charges. Outside the molecule the positive charges will act as though there were only one positive resultant charge situated at some

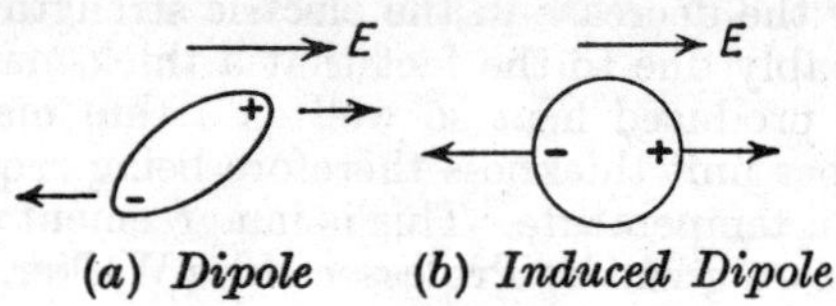

FIG. 2.33

particular point, and similarly with all the negative charges. There are two cases—

(1) The positive and negative resultant charges occupy different positions, the molecule therefore being a dipole, Fig. 2.33 (*a*). In an unstressed dielectric of this nature the dipoles have random orientations and the total strain can be regarded as zero. If a stress, in the form of an electric intensity E is applied then, in general, there will be a turning moment acting on each dipole, and all the dipoles will try to orient themselves along the direction of the electric force. Work has to be done against the frictional forces between neighbouring dipoles.

(2) The positive and negative resultant charges act at the same point. In such a case the molecule in its normal state is neutral, but on the application of an electric force, the charges are attracted in opposite directions. As a result they suffer displacement, work being done against the internal "binding forces" which act somewhat after the manner of a spring joining the charges together. In this stressed condition the molecule becomes an induced dipole, Fig. 2.33 (*b*).

In either case it will be seen that the result of the application of an electric field is a displacement of all the positive charges in the direction of the field and a displacement of the negative charges in the opposite direction. The magnitude of the displacement will obviously increase as the field intensity is increased (*a*) by causing a greater angular movement in the case of permanent dipoles and (*b*) by increasing the separation in the case of induced dipoles. Thus the arrangement of the molecules on two kinds of dielectric under stressed conditions will be rather as in Fig. 2.34. The dotted planes in the two diagrams indicate that in both cases there is a boundary of negative charge adjacent to the positive electrode, and one of positive charge adjacent to the negative electrode. If we imagine an isolated positive charge placed inside the dielectric the total force acting on it due to the whole of the molecular charges within these boundaries is zero, but the effects of the

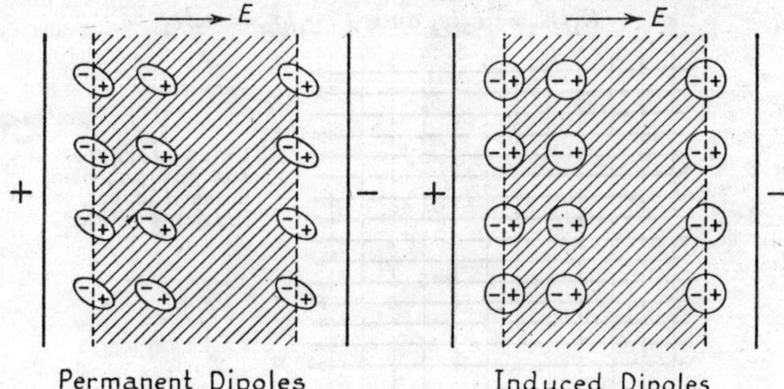

Fig. 2.34. Dielectric under Condition of Stress

two charged boundaries are to produce an attraction in the opposite direction to the applied field. Thus there is a phenomenon somewhat analogous to the demagnetizing effect of a magnet.

Now consider a parallel-plate air capacitor, and let a sheet of dielectric be inserted. We can imagine that all the lines of force penetrate the dielectric initially and thereby produce the above kind of electric displacement resulting in the induced boundary charges. These will set up an electric field in the opposite direction, in virtue of which an additional field, in opposition to the original field will be set up. The result will be a weakening of the field in the dielectric, but not in the air, as shown in Fig. 2.35. If charges in the inserted material were perfectly free to move, as in the case of a metal insertion, such motion would continue until the surface charges on the metal were equal and opposite to those on the capacitor plates. The field inside the conductor would be zero, its net charge would be zero, but its surfaces would have induced charges of opposite signs. There is no such freedom in the case of a dielectric insertion and consequently the rotation of the permanent dipoles or the electrical elongation of the induced dipoles is not sufficient to set up an induced field equal and opposite to the inducing field.

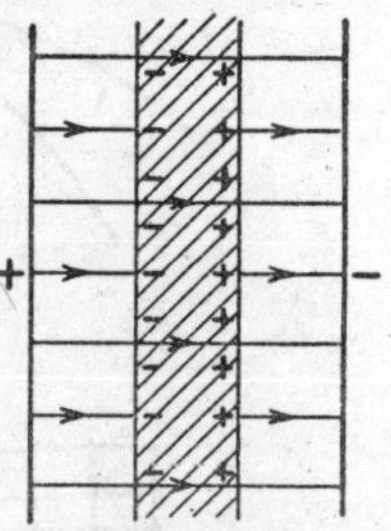

Fig. 2.35. Field in a Composite Dielectric

Now in the magnetic circuit, if there are two contiguous portions in series and of the same section, the inductions B in the two are equal. The analogous electrical quantity is the displacement D, and therefore in the above case of a capacitor with dielectric insertion the displacements in the air and in the dielectric are equal. This applies in the general case in which, instead of air there may be another dielectric. For two dielectrics we therefore have

$$D_1 = \varepsilon_1\varepsilon_0 E_1 \text{ and } D_2 = \varepsilon_2\varepsilon_0 E_2$$

also

$$D_1 = D_2$$

$$\therefore\ E_1/E_2 = \varepsilon_2/\varepsilon_1 \text{ and } \therefore\ g_1/g_2 = \varepsilon_2/\varepsilon_1$$

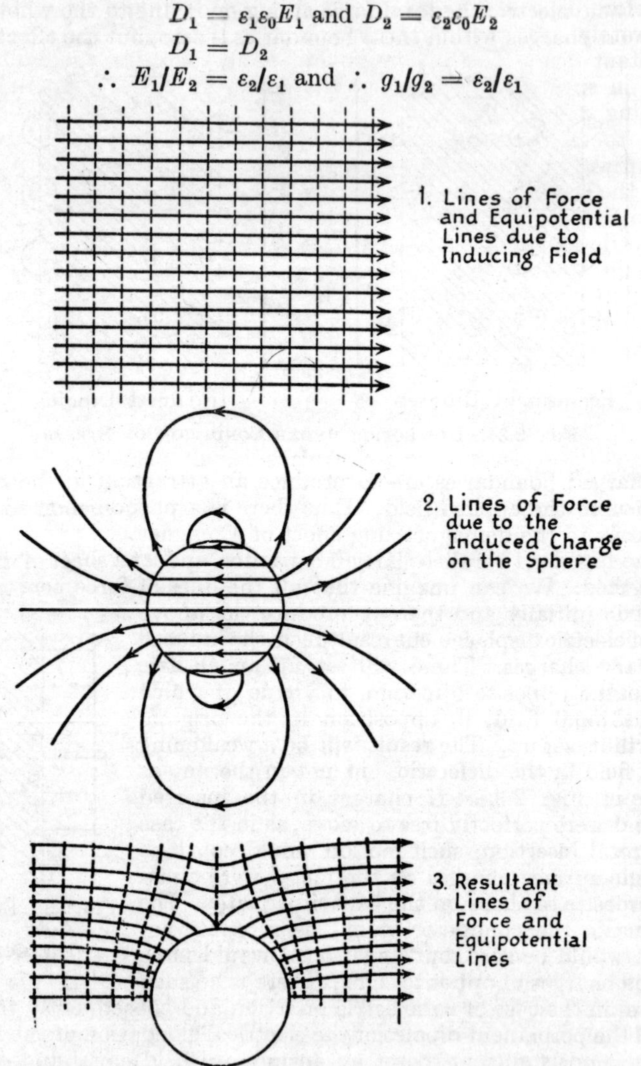

FIG. 2.36. LINES OF FORCE AND EQUIPOTENTIAL LINES

This gives a more direct way in which to solve the problem of the capacitor in Fig. 2.12.

Now imagine a spherical conductor placed between the capacitor plates. Inside the conductor there will be an induced field which completely neutralizes the inducing field, and this inducing field will be the result of induced charges at the surface of the sphere. We can therefore draw the original field and the induced field of the sphere separately, and then superpose them to determine the nature of the resultant field. This is done in Fig. 2.36. Drawing the equipotential lines in such a way as to cut the lines of force at right angles, and starting well away from the sphere with equally spaced equipotential lines to represent equal increments of potential, we see that the potential gradient is no longer uniform but is considerably increased near the surface of the sphere, particularly at the ends of the diameter drawn in the directions of the field. The effect is similar in the case of a dielectric sphere of permittivity ε greater than that of the surrounding medium, but for the reasons given in connection with the system of Fig. 2.31, the field inside the sphere is not zero and consequently the deformation of the lines of force and of the equipotential lines is not so great. The increase in potential gradient is therefore not so great.

Dielectric Breakdown. This is an extremely complex subject and can only be considered briefly in a text of this kind. It might appear at first sight that if the intensity E is sufficiently high in the case of the systems of Fig. 2.34 the forces acting on the molecular charges, in the cases of both permanent and induced dipoles, can be made so great that the molecule itself will be disrupted. That this is not the case is proved by the fact that breakdown takes place with field strengths much lower than those necessary to overcome the internal binding forces of the molecule. Consequently the breakdown of the molecule is not a purely electrical phenomenon.

In the case of air and other gaseous dielectrics breakdown is due to the ionization resulting from collisions of fast-moving electrons with neutral particles, and this is discussed in Chapter 27. This phenomenon may take place in a solid or liquid dielectric after the molecules have been split up by other causes, but the conduction current in dielectrics of importance in the electrical engineering industry is so very small that there can be, relatively, very few electrons. Furthermore their mean free paths are so small that the velocities attained may not be high enough to cause ionization.

According to Professor Miles Walker's theory of thermal breakdown, the action of a solid dielectric is governed by the fact that all such dielectrics have a very marked temperature coefficient of electrical conductivity, or, what amounts to the same thing, a decrease in resistance with increase in temperature. When a dielectric is stressed by the application of an alternating field, losses analogous to the hysteresis losses in ferromagnetic materials are produced in virtue of the frictional resistance which has to be overcome in causing the molecular deformation or molecular re-orientation. These losses result in the production of heat. Consequently the resistance decreases, causing an increase in current and therefore in heat production, and so the process becomes cumulative. A stable state will be reached if there are facilities for heat dissipation, but if there are no such facilities the temperature will eventually attain a sufficiently high value to cause chemical breakdown.

There is no doubt as to the importance of the thermal effect, but it is to be noted that breakdown never occurs simultaneously through the whole mass of the dielectric, but only in one or more very localized regions, and, naturally, the reasons for this localization have a distinct bearing on the phenomena.

There is no solid or liquid dielectric which is absolutely homogeneous, since there are invariably impurities or inclusions in the material. These are of two general kinds, voids and minute particles. The voids are tiny volumes of air which can easily be trapped where a dielectric is in the form of a wrapping, such as the paper of a paper-insulated cable. As an example, imagine such a void in the insulation of a cable. Let the potential gradient be 2×10^7 V/m adjacent to the void, and let the relative permittivity be 3·0. Then for the potential gradient across the void we have

$$2 \times 10^7 \times 3 = 6 \times 10^7 \text{ V/m}$$
$$= 6 \times 10^4 \text{ V/mm}$$

This is much greater than air can withstand and, in consequence, an electric discharge takes place with consequent disintegration of the walls of the void. This results in a concentration of energy, with consequent local heating.

The inclusions in the form of minute particles can be of two forms, insulating and conducting. If we regard these as spherical we have seen that there is a local concentration of potential gradient with conducting particles such as metal, carbon, or water, and a smaller concentration with insulating particles of greater permittivity than the surrounding dielectric. Particles of both kinds are therefore undesirable in liquid dielectrics such as the oil of transformers or high-voltage switches. The concentration of potential gradient causes local concentration of energy loss which may result in local disintegration of the oil, with consequent formation of solid residue which will adhere to the original inclusion. The inclusion will therefore tend to grow along the direction of the line of force, as indicated in a general manner in Fig. 2.37. If this phenomenon is taking place throughout the whole of the mass of dielectric then, if we trace any particular line of force from one electrode to the other, it is clear that the total distance through the dielectric in its original form becomes progressively less and the potential gradient therefore becomes progressively greater. This ultimately results in breakdown.

In the case of drops of water the effect must be extremely complex for the following reasons—

(1) The water is a relatively good conductor and thus phenomena such as the above take place.

(2) Water has a relative permittivity of about 80, and this also leads to concentration of potential gradient.

(3) The inclusion of water lowers the electrical resistance very

considerably and therefore increases the conduction current. This increases the number of free electrons and so increases the tendency to ionization.

(4) If a dielectric occupies an unsymmetrical position with respect to the electrodes we have seen that it is attracted in such a way as to increase the capacitance: this can be stated in general terms as follows: if the dielectric or the plates of a capacitor are free to move their movements will be such as to increase the capacitance. The relevance of this to the drop of water is, that since ε for water is so high, the drops will move into the regions of maximum electric gradient because, by so doing, they will be obeying the above law. The drops will then elongate themselves in the direction of the electric force, thus behaving in a similar manner to the gradually growing solid inclusions of Fig. 2.30. In obedience to the same law, bubbles of air, for which $\varepsilon = 1$, move away from the regions of high potential gradient.

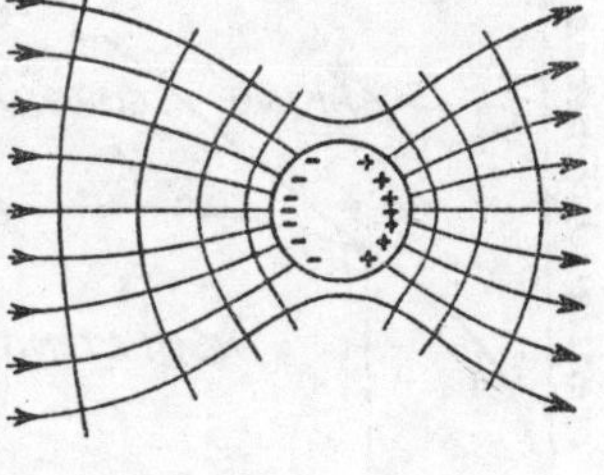

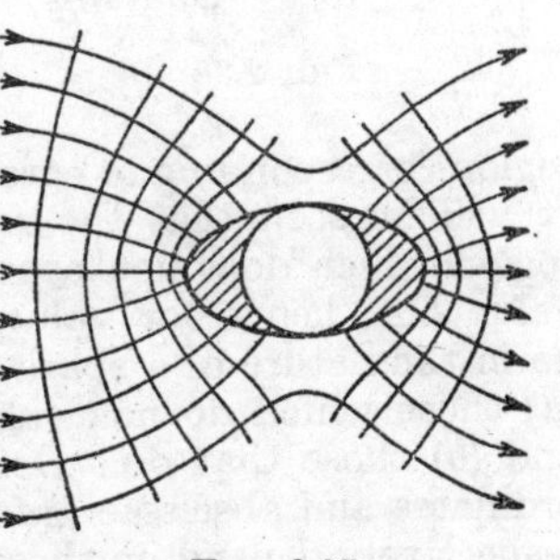

Fig. 2.37

Breakdown of Air. Breakdown of a dielectric has been defined by Wheatcroft* as "the change from a smaller to a greater conductance. Usually it is an abrupt change from a very small value to a value where the current, I, is limited only by the resistance of the external circuit. When the current at the critical point is small and the resistance between the generator and the electrodes not excessive, the generator voltage E differs little from V, the potential difference at the electrodes. If this is so the breakdown may be defined for practical purposes as $dV/dI = 0$; in other words, by a voltage maximum in the volt-ampere characteristic." While the gas is in an insulating state its properties appear to be dependent on the existence of some external ionizing agent, such as ultra-violet radiations or X-rays. Such agencies will initiate conduction but this conduction will cease with the removal of the agency. If the voltage is raised to such a value that, on the removal of this agency, the conduction continues, then this represents the conditions of breakdown.

The practical problem is to decide what shape and spacing of a pair of conductors, between which a given potential difference is maintained, will ensure that breakdown does not take place. There are two kinds of discharge (*a*) the glow, or corona discharge, and (*b*) the self-maintained discharge. The glow discharge, although a state of self-maintained conduction, is also a state of stable conduction,

* E. L. E. Wheatcroft, *Gaseous Electrical Conductors*.

with the result that a system can experience corona discharge for an indefinite period. Glow discharge is, as the name indicates, characterized by a luminosity which is distributed over the surface of the electrodes—the conductors of a transmission line for example. At high gas densities, atmospheric pressure or above, and with fairly uniform electric field intensities, between two metal spheres for example, the self-maintained state is inherently unstable, and conduction results in a spark which, unless the circuit is automatically interrupted, may be followed by an arc maintained by the full voltage of the system.

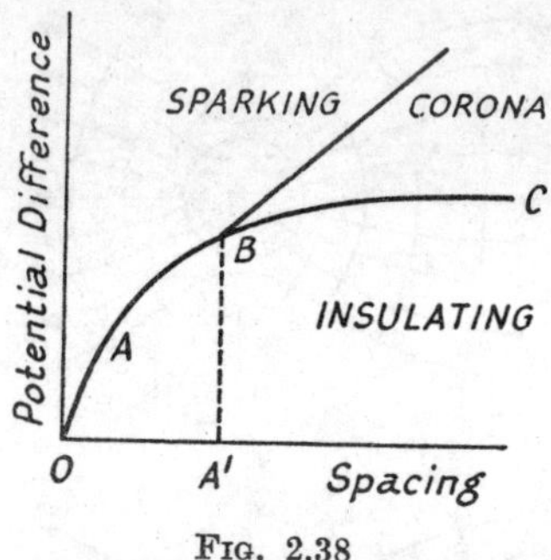

Fig. 2.38

For a given pair of electrodes the relationship between the potential difference and the spacing is given by characteristics like those of Fig. 2.38. Curve *ABC* is the boundary between the regions of insulation and of self-maintained conduction. This second region is also divided into two, the right-hand region corresponding to corona discharge and the left-hand to sparking. It will be seen that the region for corona discharge is an enclosed region which does not commence until the spacing reaches a critical value *OA*, hence for values less than *OA* the breakdown can only be in the nature of a spark. It follows that air-gaps are of two kinds (*a*) those which do not exhibit corona before spark-over takes place and (*b*) those that do. Definite characteristics, that is those having ordinates and abscissae graduated in actual volts and actual spacings respectively depend on these factors—

(*a*) shape of the electrodes and, to some extent on the condition of the surface, i.e. whether clean or dirty, smooth or pitted;

(*b*) on the spacing;

(*c*) on the gas pressure, temperature, and humidity;

(*d*) on the nature of the applied voltage, e.g. whether direct, alternating, or in the nature of a suddenly applied impulse. To produce breakdown a voltage applied for a short time must be appreciably greater than a sustained voltage, and this applies particularly if the duration is so short that the voltage becomes an impulse. The reason is that there is then no time for corona formation which, with its ion production, paves the way for a subsequent spark discharge.

Corona loss, being associated with a self-maintained current, results in a loss of energy,* and therefore it is usual with very high voltage apparatus and transmission lines to design for operation below the conditions for corona, at any rate in normal atmospheric conditions, although it is not unusual for corona to take place regularly where very-high-voltage lines are taken to high altitudes.

* F. W. Peck, *Dielectric Phenomena*, p .122.

Further Reading

STARLING, S. G. *Electricity and Magnetism for Degree Students.* Longmans, London, 1953.

PAGE, L. and ADAMS, N. I. *Principles of Electricity.* Van Nostrand, New York, 1949.

SMYTHE, W. R. *Static and Dynamic Electricity.* McGraw Hill, New York, 1950.

JEANS, Sir James. *Mathematical Theory of Electricity and Magnetism.* Cambridge University Press, Cambridge, 1925.

CARTER, G. W. *The Simple Calculation of Electrical Transients.* Cambridge University Press, Cambridge, 1944.

GUILLEMIN, E. A. *Communications Networks,* Vol. 1. J. Wiley, New York, 1931.

WARE, L. A., and TOWN, G. R. *Electrical Transients.* Macmillan, New York, 1954.

Examples on Chapter 2

1. What velocity is acquired by an electron which, starting from rest, travels through a p.d. of 600 V.

Ans. $1{\cdot}47 \times 10^7$ m/sec.

2. What energy is acquired by an electron which travels through a p.d. of 1000 V.

Ans. $15{\cdot}9 \times 10^{-7}$ joules.

3. A potential gradient of 3×10^6 V/m is maintained between two parallel plates 1 cm apart. An electron starts from rest at the negative plate and travels under the influence of the potential gradient to the positive plate. Given that the mass of the electron is $9{\cdot}1 \times 10^{-28}$ g, and the charge $1{\cdot}603 \times 10^{-19}$ coulomb, calculate (*a*) the force acting on the electron, (*b*) the ratio of the electric force to the gravitational force, (*c*) the acceleration, (*d*) the time taken to reach the positive plate.

Ans. (*a*) $4{\cdot}803 \times 10^{-13}$ newtons, (*b*) $5{\cdot}38 \times 10^{16}$, (*c*) $1{\cdot}03 \times 10^8$ m/sec.^{2} (*d*) $1{\cdot}95 \times 10^{-10}$ sec.

4. A particle has a charge of $4{\cdot}8 \times 10^{-10}$ e.s.u. It is in equilibrium when placed between two horizontal metal plates 2 cm apart, between which there is a p.d. of 3000 V. Calculate the mass of the particle.

Ans. $2{\cdot}45 \times 10^{-15}$ g.

5. Charges of $+ 20 \times 10^{-9}$ and $- 20 \times 10^{-9}$ coulomb are placed 10 cm apart. Calculate the potentials at (*a*) a point in the line of the charges 4 cm from the negative charge, (*b*) at a point in the line of the charges 4 cm outside the positive charge, (*c*) at a point at the apex of an equilateral triangle with the 10 cm spacing as base.

Ans. (*a*) − 1500 V, (*b*) + 3220 V, (*c*) zero.

6. Two metal spheres, *A* and *B*, each of 5 cm radius, are arranged with their centres 1 m apart. If *A* is given a charge of $+ 15 \times 10^{-9}$ coulomb, and *B* a charge of $- 30 \times 10^{-9}$ coulomb, calculate the p.d. between them, assuming that the proximity of the two spheres does not result in any appreciable re-distribution of the charges.

Ans. 7290 V.

7. The radius of the orbit of the single electron of the hydrogen atom is, according to the Rutherford-Bohr conception of the structure of the atom, $5{\cdot}28 \times 10^{-9}$ cm. What is the potential gradient of the field in which the electron rotates?

Ans. $5{\cdot}18 \times 10^7$ V/m.

8. An air capacitor of capacitance $0{\cdot}005\ \mu$F is connected to a direct voltage of 500 V, disconnected, and then immersed in oil of relative permittivity 2·5 Find the energy stored in the capacitor before and after immersion, and account for the difference. (L.U.)

Ans. Before, $6{\cdot}25 \times 10^{-8}$ J; after, $2{\cdot}5 \times 10^{-8}$ J.

9. In a parallel-plate capacitor with solid dielectric the plates are a tenth of a millimetre apart, and when charged the surface density is 5×10^{-10} coulombs/cm^2. If the relative permittivity is 2·2, what is the voltage between the plates? (L.U.)

Ans. 25·7 V.

10. Calculate the capacitance of a spherical capacitor if the diameter of the inner sphere is 20 cm and that of the outer sphere 30 cm, the space between them being filled with a liquid of relative permittivity 2. (L.U.)

Ans. $0{\cdot}667 \times 10^{-4}$ μF.

11. A capacitor is made up of two parallel metal discs separated by three layers of dielectric of equal thickness but having relative permittivities of 2, 3, and 4 respectively. If the metal discs are 6 in. diameter and the distance between them is 0·3 in., calculate the potential gradient in each dielectric and the total energy stored in each when a p.d. of 1000 V is applied between the discs. (L.U.)

Ans. 1815, 1210, 907·5 V/cm. $13{\cdot}51 \times 10^{-6}$, $8{\cdot}98 \times 10^{-6}$, $6{\cdot}76 \times 10^{-6}$ J.

12. A concentric cylindrical capacitor has radii of 1 and 2 cm respectively for the inner and outer (inner surface) conductors. If the space between them is filled with oil of relative permittivity 2·5 and breakdown strength 60,000 V/cm what is the maximum p.d. which can be applied to the conductors?

Ans. 41,600 V.

13. An alternating p.d. of r.m.s. value 33,000 V is applied to a capacitor of 5×10^{-6} F. Given that the applied p.d. rises to a peak value of $\sqrt{2}$ times the r.m.s. value, calculate the stored energy in (*a*) ergs, (*b*) joules, (*c*) ft-lb, (*d*) watt-hours, at the instant the p.d. is a maximum.

Ans. (*a*) $6{\cdot}7 \times 10^{10}$, (*b*) $6{\cdot}7 \times 10^{3}$, (*c*) 4940, (*d*) 1·87.

14. A piece of metal foil of area 1 cm^2 and of weight 10^{-4} g is placed on an insulated metal plate of area 10^4 cm^2. What charge must be imparted to the plate in order that the foil may be just repelled against the force of gravity?

Ans. $4{\cdot}16 \times 10^{-7}$ coulomb.

15. Three capacitors of 2, 4 and 8 μF respectively are connected in series. If a p.d. of 500 V is applied, calculate the p.d. across each.

Ans. 286, 143 and 71 V.

16. A 10 μF capacitor is in series with a 10 MΩ resistor and a p.d. of 100 V is applied to the whole. If zero time is taken at the moment of application calculate the instantaneous charging currents and charges in the capacitor at 0·05, 10 and 100 sec.

Ans. 100, 60, 14·0 μA.
400, 630, 1000 μ coulomb.

CHAPTER 3

THE ELECTRIC CURRENT—SYSTEMS OF CONDUCTORS

The Electric Current

The electric current in a conductor is the axial drift of the free electrons along the length of the conductor. Let there be n free electrons per unit volume and let the mean velocity of the axial drift be v. Then in time dt the mean distance of translation is vdt. If the cross-section is a then an elementary volume of length vdt, Fig. 3.1, will contain $navdt$ electrons,

∴ Number of electrons crossing normal section in dt seconds

$$= navdt$$

∴ Total quantity of electricity crossing normal section in dt seconds

$$= naevdt$$

$$\therefore dq = naevdt$$

$$\therefore \text{Current } i = dq/dt = naev$$

$$\therefore \text{Current density } J = i/a = nev$$

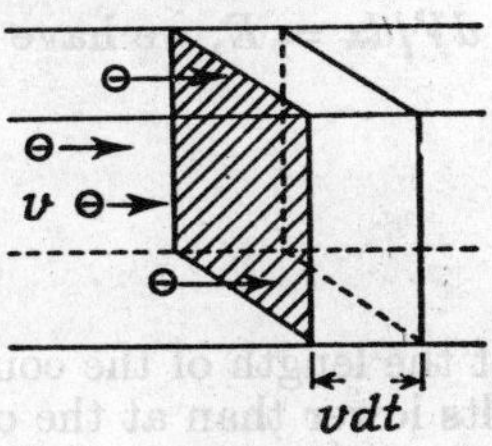

Fig. 3.1

Example. **Given that there are $8{\cdot}5 \times 10^{28}$ free electrons in one cubic metre of copper, calculate the velocity of the axial drift when a copper conductor is worked at 1000 A/in.^{2}**

$$v = J/ne$$

$$J = 1000 \text{ A/in.}^2$$

$$= (1000/6{\cdot}45) \times 10^4 = 1{\cdot}55 \times 10^6 \text{ A/m}^2$$

$$e = 1{\cdot}6 \times 10^{-19} \text{ coulomb}$$

$$\therefore v = 1{\cdot}55 \times 10^6/(8{\cdot}5 \times 10^{28} \times 1{\cdot}6 \times 10^{-19})$$

$$= 1{\cdot}14 \times 10^{-4} \text{ m/sec}$$

This example shows that, contrary to a common, but mistaken view, the velocity of electron drift is very small, even when conductors are worked at a high current density.

Resistance. **In order that the axial translation may take place there must be between the ends of the wire a p.d. which sets up an electric intensity inside the wire. If m is the mass of the electron we then have**

$$m(dv/dt) = Ee$$

for a medium in which there is no resistance to motion. In practice, at normal temperatures, there is some resistance because the ions of the metal are not free to move in the same way as the free electrons. If the resistance is proportional to the velocity v

$$m(dv/dt) = Ee - \lambda v$$

where λ is a constant.

Under steady-state conditions, i.e. with zero acceleration,

$$\lambda v = Ee$$
$$v = E(e/\lambda) \propto E$$

But the current density is proportional to v

$$\therefore\ J \propto E = gE, \text{ say,}$$

where g is a constant for a given material operated at a given temperature. It is called the *specific conductance*. Putting $i/a = J$ and $-dV/dx = E$, we have

$$i/a = -g(dV/dx)$$
$$i = -ga(dV/dx)$$
$$idx = -gadV$$

Let the length of the conductor be l and the potential at one end be V volts lower than at the other end, then

$$\int_0^l idx = -ga\int_0^V dV$$
$$li = gaV$$
$$i = (ga/l)V$$

The quantity l/ga is termed the electrical resistance, R, of the conductor. It is usual to replace g by

$$\rho = 1/g = \text{the } \textit{specific resistance} \text{ or}$$
$$\therefore\ R = \rho l/a \text{ and } I = V/R,$$

substituting I for i when there is no variation of current with time. If we substitute $l = 1$ and $a = 1$ we see that ρ is the resistance of a unit cube of the material to electrons flowing parallel to one edge. It can therefore be expressed as ohms per centimetre cube, per millimetre cube, or per metre cube, according to the unit in which length is measured. Now

$$\rho = Ra/l$$
$$= [\text{Resistance}] \times [L^2]/[L] \text{ dimensionally}$$
$$= [\text{Resistance}] \times [L]$$

Thus it is more logical to express ρ as so many ohms–centimetre, ohm–inch, or ohm–metre. In wire tables ρ is commonly given in centimetres or inch values. We have

$$\rho_{\text{metre}} = \rho_{\text{cm}} \times 10^{-2}$$
$$\rho_{\text{inch}} = \rho_{\text{cm}} \times (2{\cdot}54)^{-1}$$

For example, for copper at normal room temperature

$$\begin{aligned}\rho &= 1{\cdot}72 \times 10^{-6}\ \Omega\text{-cm}\\ &= 1{\cdot}72 \times 10^{-8}\ \Omega\text{-m}\\ &= 0{\cdot}677 \times 10^{-6}\ \Omega\text{-in.}\end{aligned}$$

Variation of Resistance with Temperature. Most conductors experience an increase in resistance with increase in temperature. In a general way this can be regarded as due to the thermal agitation of the ions. These do not occupy fixed positions, but oscillate about a mean position, the amplitude increasing with the temperature. Thus, as the temperature is increased, the chances of any particular free electron avoiding collision are diminished, and the resistance to axial motion thereby increased. In general we can write

$$\rho = \rho_0 + \alpha t + \beta t^2 + \ldots$$

where ρ_0 is the resistivity at 0°C. For many conductors, over their operating range of temperature, it is possible to neglect terms in powers of t greater than 1, and then

$$\rho = \rho_0 + \alpha t$$

The constant α is called the *temperature coefficient of resistivity* for the particular material. Let the suffixes 1 and 2 refer to t_1°C and t_2°C respectively and let a conductor be heated from one initial temperature t_1 to a final temperature t_2, then

$$\rho_1 = \rho_0(1 + \alpha_0 t_1)$$
$$\rho_2 = \rho_0(1 + \alpha_0 t_2)$$
$$\therefore\ \rho_2 = \rho_1[(1 + \alpha_0 t_2)/(1 + \alpha_0 t_1)]$$

Now let α_1 be the temperature coefficient corresponding to the initial temperature t_1, then

$$\rho_2 = \rho_1[1 + \alpha_1(t_2 - t_1)]$$
$$\therefore\ 1 + \alpha_1(t_2 - t_1) = (1 + \alpha_0 t_2)/(1 + \alpha_0 t_1)$$

giving finally
$$\alpha_1 = \alpha_0/(1 + \alpha_0 t_1)$$

In tables it is usual to give values of α for some particular initial temperature, often 0°C. The above expression enables the value to be calculated for any other initial temperature.

EXAMPLE. A copper coil of resistance 100 Ω at 40°C is heated to

100°C. Given that $\alpha_0 = 0{\cdot}00427$ per °C, calculate the resistance when hot.

$$\alpha_{40} = 0{\cdot}00427/(1 + 40 \times 0{\cdot}00427) = 0{\cdot}00364$$

$$\therefore\ R_{100} = R_{40}[1 + \alpha_{40}(100 - 40)]$$
$$= 100(1 + 0{\cdot}00364 \times 60)$$
$$= 121{\cdot}8\,\Omega$$

Material	Specific Resistance at 60°F or 15·6°C		Temperature Coefficient	
	Microhm-metre	Microhm-inch	Per °C at 15·6°C	Per °F at 60°F
Copper (annealed)	$1{\cdot}696 \times 10^{-2}$	0·668	$0{\cdot}401 \times 10^{-2}$	$0{\cdot}223 \times 10^{-2}$
Copper (hard drawn)	$1{\cdot}731 \times 10^{-2}$	0·681	0·401 ,,	0·223 ,,
Silver (annealed)	$1{\cdot}560 \times 10^{-2}$	0·614	0·376 ,,	0·209 ,,
Aluminium	$2{\cdot}91 \times 10^{-2}$	1·146	0·38 ,,	0·21 ,,
German Silver (50 % Cu, 30 % Ni, 20 % Zn)	0·30	11·8	0·04 ,,	0·02 ,,
Manganin (84 % Cu, 12 % Mn, 4 % Ni)	0·44	17·3	0·00 ,,	0·00 ,,
Eureka (60 % Cu, 40 % Ni)	0·49	19·3	0·00 ,,	0·00 ,,
Graphite	30	1200	0·05 ,,	0·03 ,,

Power. Under steady-state conditions the number of electrons crossing normal cross-section at the beginning of a conductor is equal to the number crossing at the end of the conductor. Let the current be i, then in any time interval dt the quantity of electricity crossing the cross-sections is

$$dq = idt$$

Let V be the p.d. in volts between the two ends, then the work done in moving a single electron from one end to the other is

$$\int Ee\,dx = e\int \frac{dV}{dx}\,.\,dx = Ve$$

Hence work done in moving *navdt* electrons is

$$dW = Venavdt$$
$$= Vdq$$
$$= Vidt$$
$$\therefore\ dW/dt = Vi$$

Now dW/dt is the rate of doing work, that is the power. Hence, if we put $i = I$, for steady-state conditions

$$\left.\begin{aligned} P &= VI \text{ joules} \\ &= I^2R \text{ joules} \\ &= V^2/R \text{ joules} \end{aligned}\right\}$$

Resistance Calculation by the Laplace Equation. As an example of the method find the resistance of the medium between the rectangular equipotential surfaces represented in plan by AB and CD, Fig. 3.2. From the nature of the problem it is clear that plane polar coordinates are more suitable than rectangular coordinates, and so we write the equation in the form

$$r^2\nabla^2 V = r^2(\partial^2 V/\partial r^2) + r(\partial V/\partial r) + \partial^2 V/\partial\theta^2 = 0$$

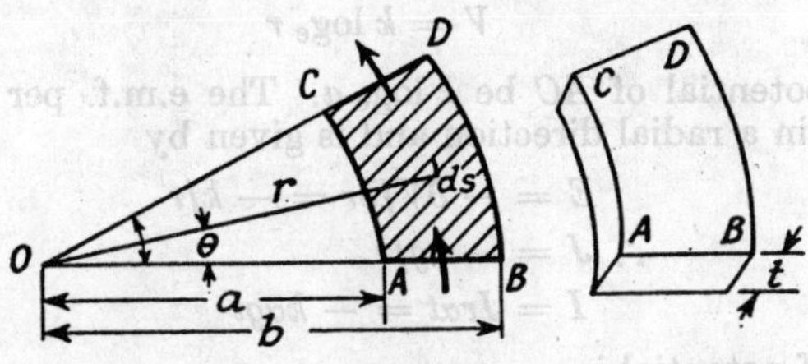

FIG. 3.2

This must hold true everywhere in the field of current flow. Along the boundaries AB and CD, we have $\partial V/\partial r = 0$ for any value of α, and therefore the requirements are satisfied by

$$\partial^2 V/\partial\theta^2 = 0$$

This gives

$$V = k\theta$$

taking the potential of AB as zero. The equipotential surfaces are therefore given by

$$\theta = \text{a constant}$$

The direction of the e.m.f. at any point is normal to the equipotential surface through that point, and is therefore along the circular arc of radius r passing through the point. Calling the arc length s, the e.m.f. per unit length in the direction of current flow at any point (r, θ) is

$$E = -\,\partial V/\partial s = -\,\partial(ks/r)/\partial s = -\,k/r$$

The current density J is given by

$$J = gE = -\,kg/r$$

Hence, total current leaving AB is

$$I = \int_a^b J t dr = -\,kgt\int_a^b \frac{dr}{r} = -\,kgt\log_e\frac{b}{a}$$

where t is the thickness of the conductor; also total potential fall from AB to CD is

$$V(0) - V(\alpha) = -k\alpha$$

Hence the resistance between equipotential surfaces at AB and CD is

$$\alpha/gt\log_e(b/a)$$

As a further example suppose that we require the resistance between AC and BD to a current flow which is radially outwards from O. In this case $\partial V/\partial\theta = 0$ and the equation becomes

$$r^2(\partial^2 V/\partial r^2) + r(\partial V/\partial r) = 0$$

or

$$\partial[r(\partial V/\partial r)]/\partial r = 0$$

The solution is

$$V = k \log_e r$$

Let the potential of AC be $k \log_e a$. The e.m.f. per unit length is everywhere in a radial direction and is given by

$$E = -\partial V/\partial r = -k/r$$

$$\therefore J = -kg/r$$

and

$$I = Jr\alpha t = -k\alpha gt$$

The fall of potential is

$$V(a) - V(b) = k \log_e(a/b) = -k \log_e(b/a)$$

The resistance between the curved equipotentials AC and BD is thus

$$\frac{[\log_e(b/a)]}{\alpha gt}$$

Note that these two results could be obtained more quickly by direct elementary methods, and the problem was presented as another illustration of the manner in which the Laplace equation can be applied. The power of the method lies in its application to more complex boundary conditions than could be dealt with by elementary methods.

Systems of Conductors

Series and Parallel Connections. If a number of resistances R_1, R_2, R_3, etc., are connected in series, then the total resistance is given by

$$R = R_1 + R_2 + R_3 + \ldots \text{ ohms}$$

If an e.m.f. E is applied to the whole circuit, the current is given by

$$I = E/R$$

and this current flows through each of the resistances. The voltage divides itself between the various resistances in such a way that the "drop" along any one section is proportional to the resistance of that section. Thus denoting the drops by V_1, V_2, V_3, etc., we have

$$V_1 = (R_1/R)E;\ V_2 = (R_2/R)E;\ V_3 = (R_3/R)E;\ \text{etc.}$$

If the various resistances are all connected in parallel, then for the total resistance we now have

$$1/R = 1/R_1 + 1/R_2 + 1/R_3 + \ldots$$

or, denoting the reciprocal of resistance by the conductance G, the practical unit being the *mho*, we have

$$G = G_1 + G_2 + G_3 + \ldots \text{ mhos}$$

The total current produced by an applied voltage E is now

$$I = E/R = EG$$

while the portions of this current carried by the various branches are given by

$$I_1 = (G_1/G)I;\; I_2 = (G_2/G)I;\; I_3 = (G_3/G)I;\; \text{etc.}$$

Kirchhoff's Laws. Circuits which do not come within the category of simple series or parallel circuits, such as the above, can be solved by means of Kirchhoff's Laws. These are as follows—

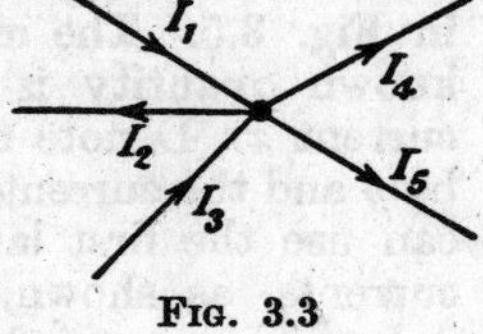

Fig. 3.3

First Law. The algebraic sum of the currents at any junction of conductors is zero. In other words, the sum of the currents flowing towards a junction is equal to the sum of the currents flowing away from the junction. Thus in Fig. 3.3 the currents I_1 and I_3 are flowing towards the junction, while I_2, I_4, and I_5 are flowing away from the junction. From this first law we therefore have for this particular junction

$$I_1 + I_3 = I_2 + I_4 + I_5$$

Second Law. In any closed circuit the algebraic sum of the potential drops in the various parts of the circuit is equal to the electromotive force acting round the circuit. To illustrate the law three possible cases are illustrated in Fig. 3.4.

(*a*) In Fig. 3.4 (*a*) the closed circuit does not contain a battery or other source of e.m.f., and therefore the algebraic sum of the various

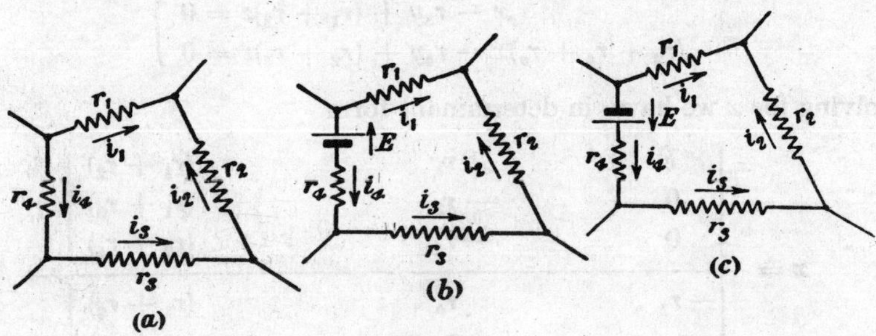

Fig. 3.4

drops in volts is equal to zero. Taking a clockwise direction round the closed mesh as the positive direction, we therefore have

$$i_1r_1 - i_2r_2 - i_3r_3 - i_4r_4 = 0$$

(*b*) In Fig. 3.4 (*b*) one arm contains a cell of e.m.f. E, and this e.m.f. tends to circulate current in the clockwise or positive direction

$$\therefore\; i_1r_1 - i_2r_2 - i_3r_3 - i_4r_4 = E$$

(c) In Fig. 3.4 (c) this cell is reversed, so that it tends to circulate current in the negative direction round the mesh. Its e.m.f. must therefore be reckoned negative, and we have

$$i_1r_1 - i_2r_2 - i_3r_3 - i_4r_4 = -E$$

It is to be noted that in cases (b) and (c) the resistance r_4 must include the internal resistance of the cell.

As an example of the application of these two laws, take the case of the well-known Wheatstone network, as shown in Fig. 3.5. The most important unknown quantity is the galvanometer current x. Denote the battery current by y and the current in r_1 by z, then we can use the first law to fix the other currents, as shown, in terms of three unknowns, x, y, and z. To obtain three equations we now apply the second law to three closed circuits, and we will choose $EBAC$, BAD, and DAC. We then have

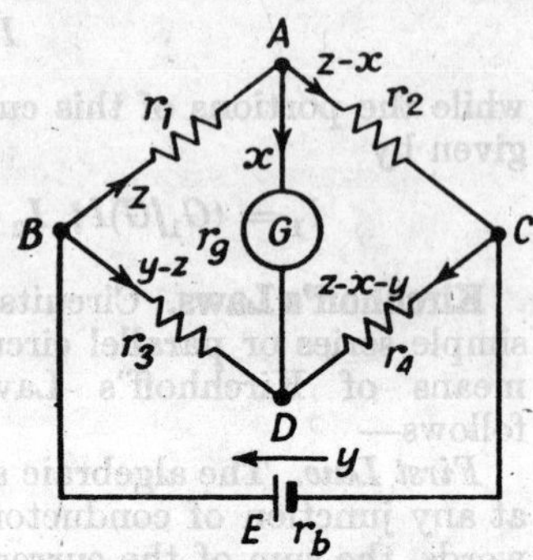

Fig. 3.5. Wheatstone Bridge Network

$$\left.\begin{aligned} r_1z + r_2(z-x) + r_by &= E \\ r_1z + r_gx - r_3(y-z) &= 0 \\ r_2(z-x) + r_4(z-x-y) - r_gx &= 0 \end{aligned}\right\}$$

These reduce to

$$\left.\begin{aligned} -r_2x + r_by + (r_1+r_2)z &= E \\ r_gx - r_3y + (r_1+r_3)z &= 0 \\ -(r_2+r_4+r_g)x - r_4y + (r_2+r_4)z &= 0 \end{aligned}\right\}$$

Solving for x we have, in determinant form

$$x = \frac{\begin{vmatrix} E & r_b & (r_1+r_2) \\ 0 & -r_3 & (r_1+r_3) \\ 0 & -r_4 & (r_2+r_4) \end{vmatrix}}{\begin{vmatrix} -r_2 & r_b & (r_1+r_2) \\ r_g & -r_3 & (r_1+r_3) \\ -(r_2+r_4+r_g) & -r_4 & (r_2+r_4) \end{vmatrix}}$$

Evaluating the numerator we have

$$E[-r_3(r_2+r_4) + r_4(r_1+r_3)]$$

which reduces to

$$E(r_1r_4 - r_2r_3)$$

Hence, denoting the denominator by Δ, we have

$$x = E(r_1r_4 - r_2r_3)/\Delta$$

$$\therefore\ x = 0 \text{ when } r_1r_4 = r_2r_3$$

or

$$r_1/r_2 = r_3/r_4$$

the well-known condition for a balance.

As a numerical example, let $E = 2$ V, $r_1 = 10\ \Omega$, $r_2 = 20\ \Omega$, $r_3 = 30\ \Omega$, $r_4 = 40\ \Omega$, $r_g = 50\ \Omega$, and $r_b = 1\ \Omega$.
Then we have

$$x = \frac{\begin{vmatrix} 2 & 1 & 30 \\ 0 & -30 & 40 \\ 0 & -40 & 60 \end{vmatrix}}{\begin{vmatrix} -20 & 1 & 30 \\ 50 & -30 & 40 \\ -110 & -40 & 60 \end{vmatrix}} = \frac{\gamma}{\Delta} \text{ say}$$

Then

$$\gamma = [(2 \times -30 \times 60) + (1 \times 40 \times 0) + (30 \times 0 \times -40)] - [(0 \times -30 \times 30) + (-40 \times 40 \times 2) + (60 \times 0 \times 1)]$$

$$\Delta = [(-20 \times -30 \times 60) + (1 \times 40 \times -110) + (30 \times 50 \times -40)] - [(-110 \times -30 \times 30) + (-40 \times 40 \times -20) + (60 \times 50 \times 1)]$$

$$\therefore\ x = \frac{\gamma}{\Delta} = \frac{-400}{-162{,}400}$$

$$= 0{\cdot}00246 \text{ A}$$

Delta-star Transformation. In some cases a network can be solved readily by means of what is called a delta-star transformation, and it

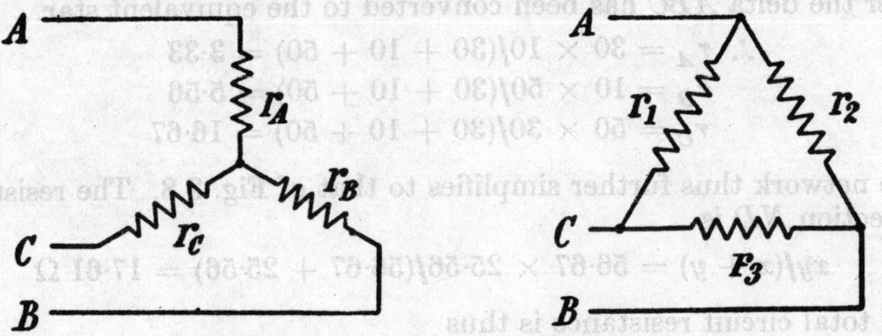

Fig. 3.6. Star-Delta Conversion

is to be noted that this method is applicable to a.c. as well as to d.c. circuits. Consider the two networks of Fig. 3.6. If these are identical then the resistance between any pair of lines will be the same when the third line is opened. Hence

when A is open $r_B + r_C = r_3(r_1 + r_2)/(r_1 + r_2 + r_3)$
when B is open $r_C + r_A = r_1(r_2 + r_3)/(r_1 + r_2 + r_3)$
when C is open $r_A + r_B = r_2(r_3 + r_1)/(r_1 + r_2 + r_3)$

The solution of these equations is

$$r_A = r_1 r_2/(r_1 + r_2 + r_3)$$
$$r_B = r_2 r_3/(r_1 + r_2 + r_3)$$
$$r_C = r_3 r_1/(r_1 + r_2 + r_3)$$

As an example, the previous Wheatstone's Bridge network is given in

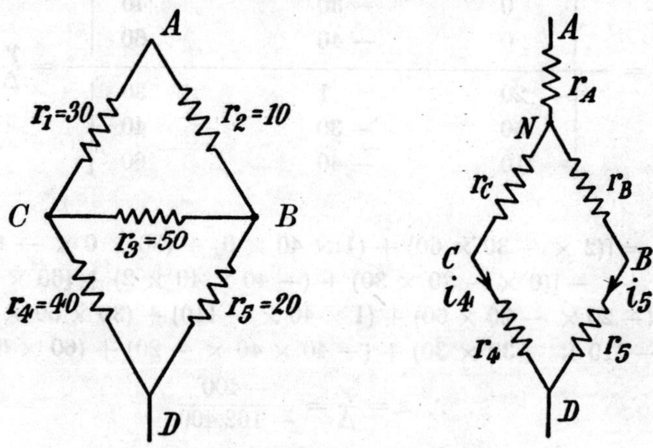

FIG. 3.7

Fig. 3.7, the right-hand figure of which gives the equivalent network after the delta ABC has been converted to the equivalent star.

$$\therefore r_A = 30 \times 10/(30 + 10 + 50) = 3{\cdot}33$$
$$r_B = 10 \times 50/(30 + 10 + 50) = 5{\cdot}56$$
$$r_C = 50 \times 30/(30 + 10 + 50) = 16{\cdot}67$$

The network thus further simplifies to that of Fig. 3.8. The resistance of section ND is

$$xy/(x + y) = 56{\cdot}67 \times 25{\cdot}56/(56{\cdot}67 + 25{\cdot}56) = 17{\cdot}61\ \Omega$$

The total circuit resistance is thus

$$1 + 3{\cdot}33 + 17{\cdot}61 = 21{\cdot}94\ \Omega$$

and the current delivered by the cell is

$$2/21{\cdot}94 = 0{\cdot}0912 \text{ A}$$

Current $i_4 = (25{\cdot}56/82{\cdot}23) \times 0{\cdot}0912 = 0{\cdot}0283$ A

and $i_5 = (56{\cdot}67/82{\cdot}23) \times 0{\cdot}0912 = 0{\cdot}0629$ A

The p.d. of point B with respect to $D = 0{\cdot}0629 \times 20 = 1{\cdot}258$ V.

The p.d. of point C with respect to $D = 0{\cdot}0283 \times 40 = 1{\cdot}132$ V.

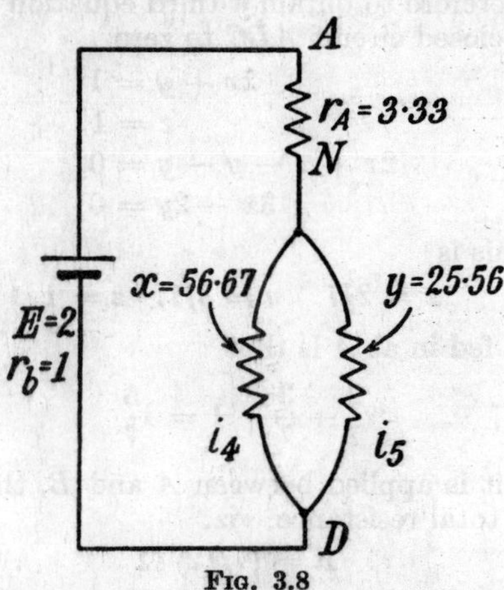

Fig. 3.8

Thus point B is at a higher potential than C, the difference being

$$1{\cdot}258 - 1{\cdot}132 = 0{\cdot}126 \text{ V}$$

The galvanometer current is therefore

$$0{\cdot}126/50 = 0{\cdot}00252 \text{ A}$$

The difference between the two values represents slide-rule errors, and shows that when an accurate result is desired it is preferable to use log tables.

As a second example, take the case of a triangular pyramid, $ABCD$, built up of six wires whose resistances are as follows: $AB = 1\Omega$,

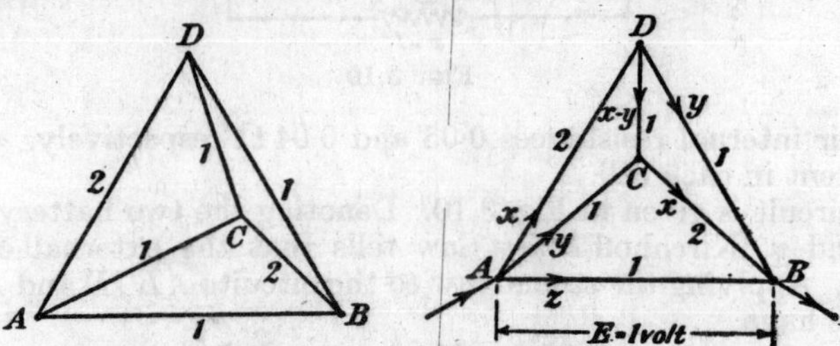

Fig. 3.9. Network illustrating Kirchhoff's Laws

$AC = 1\ \Omega$, $CB = 2\ \Omega$, $AD = 2\ \Omega$, $DC = 1\ \Omega$, and $DB = 1\ \Omega$. Calculate its resistance between the points A and B.

The pyramid is shown in Fig. 3.9, along with the equivalent "flat" circuit. Imagine that an e.m.f. of 1 V is applied between the points A and B, then, inserting symbols for the currents in the various resistances, we obtain the distribution shown, since it is clear that the currents in AD and CB will be equal, and also the currents in AC and DB. The drop of volts is equal to E (i.e. 1 V) along ($AD + DB$), along ($AC + CB$), and along AB. The first two of these really give the same condition, and therefore to obtain a third equation we will equate the total drop in the closed circuit ADC to zero.

$$\therefore\ 2x + y = 1$$
$$z = 1$$
$$2x + x - y - y = 0$$

or

$$3x - 2y = 0$$

The solution of this is

$$x = 2/7, \quad y = 3/7, \quad z = 1$$

The total current fed in at A is thus

$$\frac{2}{7} + \frac{3}{7} + 1 = 1\frac{5}{7}$$

Hence, as one volt is applied between A and B, the reciprocal of the current gives the total resistance, viz.

$$R = (7/12)\ \Omega$$

As a third example, consider the following. Two cells are connected in parallel and supply a circuit of 1 Ω. Their e.m.f.'s are 2·05 and 2·15 V,

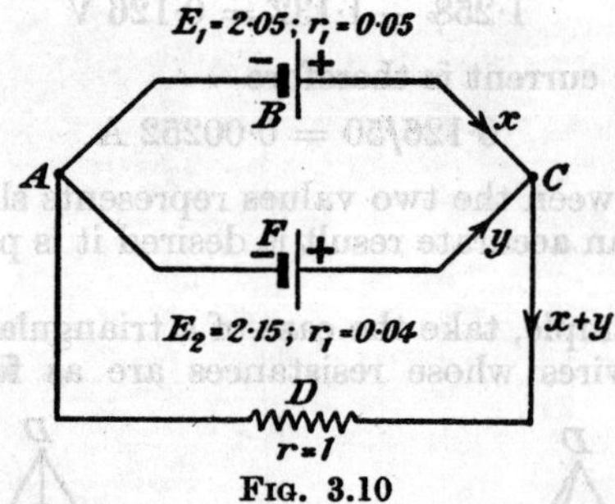

Fig. 3.10

and their internal resistances 0·05 and 0·04 Ω respectively. Calculate the current in each cell.

The circuit is given in Fig. 3.10. Denoting the two battery currents by x and y, Kirchhoff's first law tells that the external current is $(x + y)$. Applying the second law to the circuits $ABCD$ and $AFCD$ in turn, we have

$$0{\cdot}05x + 1(x + y) = 2{\cdot}05$$
$$0{\cdot}04y + 1(x + y) = 2{\cdot}15$$

Hence

$$1{\cdot}05x + y = 2{\cdot}05$$

and

$$x + 1{\cdot}04y = 2{\cdot}15$$

The solution is

$$x = -\ 0{\cdot}19\ \text{A}, \quad y = 2{\cdot}25\ \text{A}$$

Superposition Theorem. Suppose that a single e.m.f., E_1, acts in a network, then, in accordance with Ohm's Law and the two Kirchhoff Laws each element of the network will carry a current which can be calculated, and that current distribution will be unique. Suppose now that the points at which E_1 was applied are short-circuited, and that

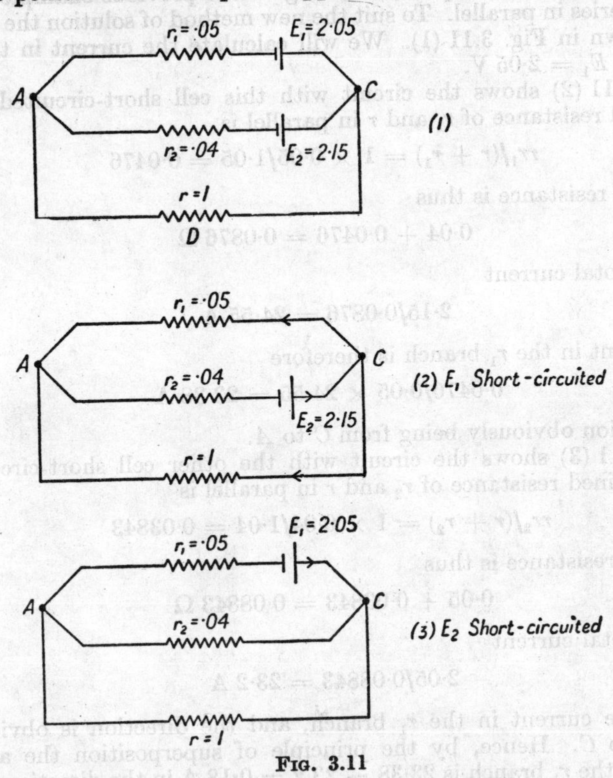

FIG. 3.11

two other points are opened and a new e.m.f. E_2 introduced, then there will be a second distribution of current which will also be unique. If, now both E_1 and E_2 exist simultaneously, then (assuming network elements which are linear, that is, obey the simple Ohm's Law) since each current distribution due to one e.m.f. alone is unique, the resulting distribution due to E_1 and E_2 acting simultaneously is given by superposing the two distributions due to E_1 and E_2 acting alone. Clearly the argument can be applied to a network of linear conductors in which any number of e.m.f.'s act simultaneously. A general statement of the law of superposition is therefore as follows: If there are several e.m.f.'s acting simultaneously in any network of linear conductors, the current in any particular conductor is given by the algebraic superposition of the current for that conductor due to each e.m.f.

acting in turn, when all the other pairs of points at which e.m.f.'s act are short-circuited. If any source of e.m.f. is of finite resistance then its resistance must be added to the resistance of the appropriate branch of the network when this source is imagined to be short-circuited.

As a numerical example consider again the previous example of the two batteries in parallel. To suit the new method of solution the circuit is re-drawn in Fig. 3.11 (1). We will calculate the current in the cell of e.m.f., $E_1 = 2{\cdot}05$ V.

Fig. 3.11 (2) shows the circuit with this cell short-circuited. The combined resistance of r_1 and r in parallel is

$$rr_1/(r + r_1) = 1 \times 0{\cdot}05/1{\cdot}05 = 0{\cdot}0476$$

The total resistance is thus

$$0{\cdot}04 + 0{\cdot}0476 = 0{\cdot}0876\ \Omega$$

and the total current

$$2{\cdot}15/0{\cdot}0876 = 24{\cdot}55\ \text{A}$$

The current in the r_1 branch is therefore

$$0{\cdot}0476/0{\cdot}05 \times 24{\cdot}55 = 23{\cdot}38\ \text{A}$$

the direction obviously being from C to A.

Fig. 3.11 (3) shows the circuit with the other cell short-circuited. The combined resistance of r_2 and r in parallel is

$$rr_2/(r + r_2) = 1 \times 0{\cdot}04/1{\cdot}04 = 0{\cdot}03843$$

The total resistance is thus

$$0{\cdot}05 + 0{\cdot}03843 = 0{\cdot}08843\ \Omega$$

and the total current

$$2{\cdot}05/0{\cdot}08843 = 23{\cdot}2\ \text{A}$$

This is the current in the r_1 branch, and the direction is obviously from A to C. Hence, by the principle of superposition the actual current in the r_1 branch is $23{\cdot}38 - 23{\cdot}2 = 0{\cdot}18$ A in the direction from C to A.

Thévenin's Theorem. Fig. 3.12 shows part of a network of conductors brought out to two terminals A and B. In general, there will be a

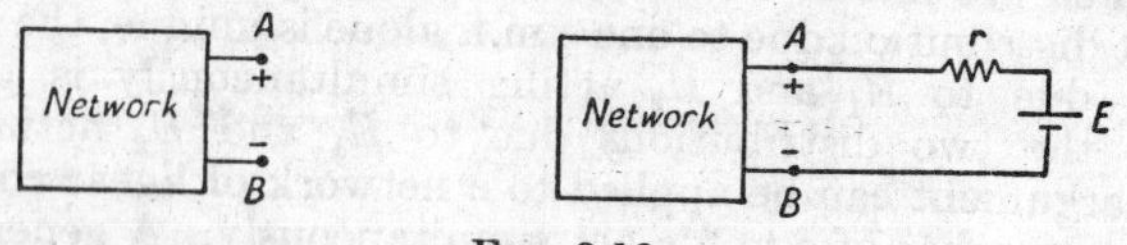

FIG. 3.12

certain p.d., V, acting between A and B. Now suppose that an additional resistance r in which an e.m.f. E, is acting, is connected to

A and B as in the second figure, and furthermore let the e.m.f., E, be equal to V the open-circuit voltage across A and B, and in opposition to it. Then, clearly no current will flow in this additional resistance, r. According to the principle of superposition this current of zero amperes in r can be assumed to be the algebraic sum of (*a*) the current which will flow when the e.m.f., E, acts alone, all the other e.m.f.'s being short-circuited, and (*b*) the current which will flow when all the other e.m.f.'s act and E is short-circuited.

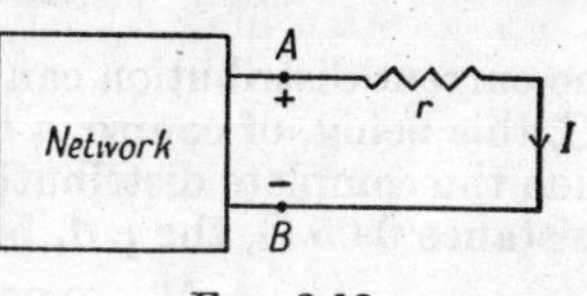

Fig. 3.13

Now let r_s be the short-circuit resistance of the original network with respect to the terminals A and B, namely, the resistance of this network between A and B when the e.m.f.'s are short-circuited. Then Thévenin's Theorem states that the current in any resistance r connected between the points AB (Fig. 3.13) is given by

$$I = E/(r + r_s)$$

where r_s is as stated above.

Example 1. Take the case of the network of Fig. 3.14. We require the current in all five conductors. First of all imagine that the conductor AB is removed, leaving the simple network of Fig. 3.14 (*b*). To

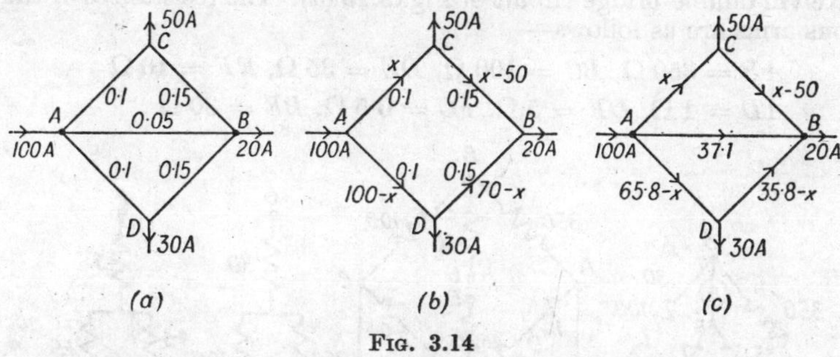

Fig. 3.14

solve this denote by x the current along AC. Then the currents in the other resistances are as shown in the figure. In determining this current distribution we have made use of Kirchhoff's first law, and therefore to determine x we equate the total e.m.f. round the closed loop to zero, since there is no e.m.f. acting anywhere in the loop.

$$\therefore\ 0{\cdot}1x + 0{\cdot}15(x - 50) - 0{\cdot}15(70 - x) - 0{\cdot}1(100 - x) = 0$$

$$\therefore\ x = 56\ \text{A}$$

The p.d. between A and B is therefore

$$0{\cdot}1 \times 56 + 0{\cdot}15 \times 6 = 6{\cdot}5\ \text{V}$$

The short-circuit resistance of the network of Fig. 3.14 (*b*) between the points A and B is

$$r_s = 0{\cdot}25 \times 0{\cdot}25/(0{\cdot}25 + 0{\cdot}25) = 0{\cdot}125\ \Omega$$

Hence, by Thévenin's Theorem, the current in the resistance AB, when network is complete, is

$$I = E/(r + r_s)$$
$$= 6{\cdot}5/(0{\cdot}05 + 0{\cdot}125)$$
$$= 37{\cdot}1\ \text{A}$$

The current distribution can now be solved. Let x be the current along AC, this being, of course, a different current from that of Fig. 3.14 (b). Then the complete distribution is given in Fig. 3.14 (c). Since AB is of resistance 0·05 Ω, the p.d. between A and B is

$$0{\cdot}05 \times 37{\cdot}1 = 1{\cdot}85\ \text{V}$$

Hence for the path ACB we have

$$0{\cdot}1x + 0{\cdot}15(x - 50) = 1{\cdot}85$$
$$0{\cdot}25x = 9{\cdot}35$$
$$x = 37{\cdot}4\ \text{A}$$

This shows that the current in CB and DB will be in the reverse direction to the arrows.

EXAMPLE 2. Calculate the current flowing in the galvanometer of the Kelvin double-bridge circuit of Fig. 3.15 (a). The resistances of the various arms are as follows—

$$AB = 350\ \Omega,\ BC = 100\ \Omega,\ DE = 35\ \Omega,\ EF = 10\ \Omega$$
$$AD = 1\ \Omega,\ DF = 5\ \Omega,\ FC = 0{\cdot}5\ \Omega,\ BE = 30\ \Omega$$

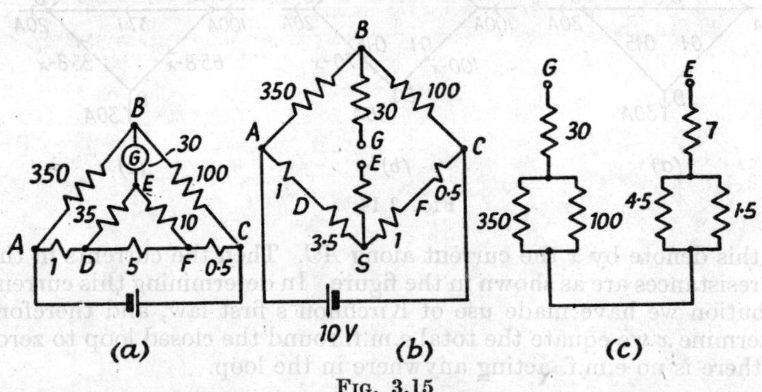

FIG. 3.15

The battery has an e.m.f. of 10 V and negligible internal resistance.

The delta mesh DEF is first transformed into the star network; the resistances of the various branches are

$$R_{ES} = R_{ED}R_{EF}/(R_{ED} + R_{EF} + R_{DF})$$
$$= 35 \times 10/(35 + 10 + 5) = 7\ \Omega$$
$$R_{FS} = 10 \times 5/50 = 1\ \Omega$$
$$R_{SD} = 35 \times 5/50 = 3{\cdot}5\ \Omega$$

The Kelvin bridge circuit can now be replaced by the Wheatstone-bridge circuit of Fig. 3.15 (*b*). To determine the current in the galvanometer circuit we have first to break the circuit at *GE* and determine the voltage across the break.

Open-circuit voltage across *GE*

$$= V_{AB} - V_{AS}$$
$$= 10 \times 350/450 - 10 \times 4{\cdot}5/6$$
$$= 7{\cdot}77 - 7{\cdot}50 = 0{\cdot}27 \text{ V}$$

Next the resistance "looking in" at *GE* is calculated, the battery being short-circuited: this is most easily done by drawing the bridge in the form of Fig. 3.15 (*c*).

Resistance "looking in"

$$= 30 + [(350 \times 100)/450] + [(4{\cdot}5 \times 1{\cdot}5)/6] + 7$$
$$= 116\ \Omega$$

∴ Galvanometer current

$$= 0{\cdot}27/116 = 2{\cdot}33 \times 10^{-3} \text{ A}$$

Nonlinear Resistors. A nonlinear resistor is one which does not obey the Ohm's law relationship (constancy of the ratio potential difference/

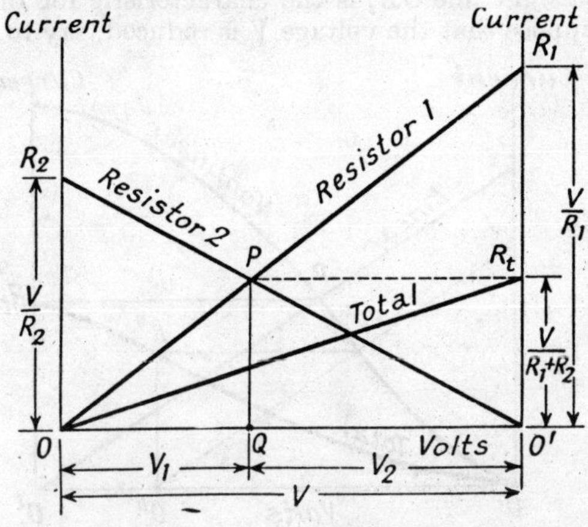

FIG. 3.16. GRAPHICAL SOLUTION OF TWO LINEAR RESISTORS IN SERIES

current). Strictly speaking, most resistors are nonlinear, a copper wire, for example, since the above ratio increases with increasing current unless steps are taken to maintain the temperature constant. However, the term is reserved for those cases in which resistance changes very appreciably with charges of current. The commonest example is the metal-filament lamp. When cold this lamp has a resistance much less than that at normal working temperature, with the result that, at the moment of switching on there is a current rush of several times the normal current.

If the current-voltage characteristic of a nonlinear resistance is known then, if the resistor has nothing in series with it, the current corresponding to any given applied p.d. can be determined immediately by reference to the characteristic. The important case is therefore that of a nonlinear resistor which is a component of a series circuit, for example in series with a linear resistor. Before considering this, take the case of two linear resistors R_1 and R_2 in series. Let the total applied p.d. be V, then this will divide between the two in direct proportion to their resistances. The current-voltage graphs will be straight lines through the origin, and it is convenient to have an origin O for R_1, and O' for R_2, where $OO' = V$. Then the intercept on the vertical arm through O is V/R_2, and that on the vertical arm through O' is V/R_1, as shown in Fig. 3.16. The intercept P obviously gives the only solution to the circuit, and therefore OQ is the p.d. V_1 and QO' the p.d. V_2. The current is PQ and is given by

$$I = V_1/R_1 = V_2/R_2 = V/(R_1 + R_2)$$

Draw the horizontal PR_t. Then

$$PQ/R_2O = QO'/OO' = (V - V_1)/V = (R_1 + R_2 - R_1)/(R_1 + R_2)$$

$$\therefore PQ = (V/R_2)[R_2/(R_1 + R_2)] = V/(R_1 + R_2) = I$$

Hence the straight line OR_t is the characteristic for the two resistors in series. Suppose that the voltage V is reduced, say to one-half of the

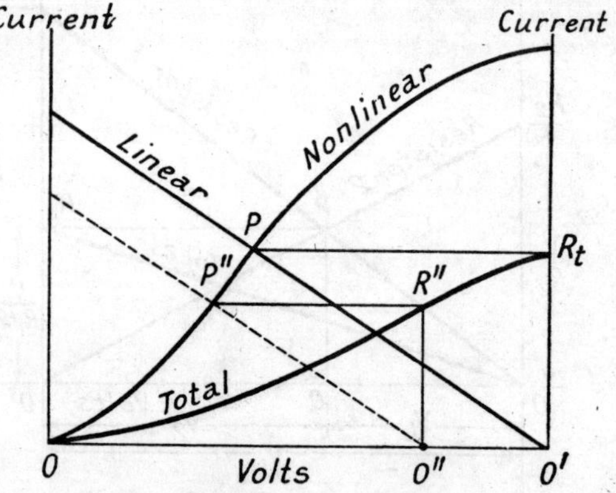

FIG. 3.17. GRAPHICAL SOLUTION FOR LINEAR AND NON-LINEAR RESISTORS IN SERIES

previous value, then the ordinates of all the characteristics will be reduced in proportion. For example, the characteristic for resistor R_2 will become a dotted characteristic as in Fig. 3.16.

This is the basis of the method of determination of the total characteristic for a nonlinear resistor in series with a linear resistor, or for two nonlinear resistors in series. Consider the former case. The characteristic for the nonlinear resistor is drawn from the origin O, and that for the linear resistor from the origin O', where $OO' = V$, as before (*see* Fig. 3.17). A horizontal PR_t is drawn, and the intercept $O'R_t$ gives the current when the full p.d. V is applied. To find other points on the total characteristic take a point O'' where

$$OO''/OO' = (m - 1)/m$$

where m can be the number of any convenient number of subdivisions of OO', say 10. Erect $O''R''$ and draw the reduced voltage characteristic for the linear resistor as shown dotted. Project horizontally from P'' to intersect the perpendicular through O'' at R''. Then R'' is the point on the total characteristic corresponding, to an applied p.d. of $(m - 1)/m$ of V.

Now locate O''' where $OO'''/OO' = (m - 2)/m$ and proceed as before, when a new point R''' will be obtained; and so on.

Examples on Chapter 3

1. A copper wire of cross-section $2{\cdot}1 \times 10^{-6}$ m^2 carries a current of 20 A. Calculate the mean velocity of the electron drift.

Ans. 6×10^{-4} m/sec.

2. Four conductors AB, BC, CD, and DA of resistances 0·5, 1·0, 1·5 and 2·0 Ω respectively are arranged in a square and an additional resistance of 2·5 Ω connected across BD. If a current of 1 A flows in at A and out at C, calculate the current in the additional resistance.

Ans. 0·0667 A.

3. A circle of radius r is made from a piece of wire of uniform cross-section, and a straight piece of identical wire is joined to it across a diameter. If current enters at one of the junctions and leaves at a point on the circumference midway between the two junctions, show that the resistance of the arrangement is equal to $\pi r R(\pi + 6)/4(\pi + 4)$, where R is the resistance per unit length.

4. Two batteries, one of 6 V e.m.f. and internal resistance 0·5 Ω, the other of 10 V e.m.f. and internal resistance 1 Ω are joined in parallel. They supply an external resistance of 12 Ω. Calculate the current in each battery.

Ans. – 2·270, 2·865 A.

5. Twelve equal wires, each of 0·1 Ω resistance are joined so as to form a hollow cube. A p.d. of 2 V is applied to the ends of one of the wires. Calculate the total current.

Ans. 34·29 A.

6. The four arms of a Wheatstone bridge have the following resistances: $AB = 100$, $BC = 10$, $CD = 4$, $DA = 50$ Ω. A galvanometer of 20 Ω resistance is connected across BD. Calculate the current through the galvanometer when a p.d. of 10 V is maintained across AC. (L.U.)

Ans. 0·00513 A.

7. The four arms of a Wheatstone bridge have the following resistances: $AB = 6$, $BC = 9{\cdot}8$, $CD = 6$, $DA = 24$ Ω. A galvanometer of 12 Ω resistance is connected across BD. A cell of 2 V and internal resistance 8 Ω is connected across AC. Calculate the galvanometer current.

Ans. 0·027 A.

8. Determine to the nearest milliampere, the currents in each of the five sections of the following network when 1 V is applied between the points a and d: $ab = 2\ \Omega$, $bd = 8\ \Omega$, $dc = 6\ \Omega$, $ca = 4\ \Omega$, and $cb = 10\ \Omega$. (L.U.)

***Ans.* 111, 97, – 105, – 91, – 14 μA respectively.**

9. A hollow triangular pyramid is made up of six wires as follows: base $AB = 1$, $BD = 2$, $DA = 3\ \Omega$; edges, joining at the apex C, $CA = 1{\cdot}5$, $CB = 2$, $CD = 2{\cdot}5\ \Omega$. Calculate the network resistance between points A and B.

***Ans.* 0·671 Ω.**

10. A regular hexagon is formed from six wires each of resistance 1 Ω. The corners are joined to the centre by six more wires each of resistance 1 Ω. Calculate the resistance of the network (*a*) when current enters at one corner and leaves at the opposite corner, (*b*) when it leaves at the adjacent corner.

***Ans.* (*a*) 0·8 Ω; (*b*) 0·55 Ω.**

11. A Wheatstone's network $ABCD$ is made up of five resistors as follows: $AB = 4$, $BC = 6$, $CD = 7$, $DA = 3$, $BD = 5\ \Omega$. In series with this combination and connected at point A is a resistor of 2 Ω. Using the delta-star transformation calculate the resistance of the whole between the outer end of the resistor 2 Ω and the point C.

***Ans.* 7·0 Ω.**

12. Two batteries, one of 12 V e.m.f. and resistance 2 Ω, the other of 6 V e.m.f. and resistance 1 Ω, are in parallel. They are joined by a resistor of such value that it takes 3 A in the direction of normal output of the batteries. What is its resistance, and what are the battery currents?

***Ans.* 2 Ω; 3 A and 0 A.**

13. Three resistors AB, BC, CD are in series and point A is connected to one terminal of a battery of 10 V and negligible resistance. The magnitudes are $AB = 6$, $BC = 10$, $CD = 2\ \Omega$. A 4-Ω resistor is connected between B and D, an 8-Ω resistor between C and D, and a 4-Ω resistor between C and D. Calculate the currents in AB, BC, and CD.

***Ans.* 1·1, 0·25, 0·138 A.**

CHAPTER 4

ELECTROMAGNETISM

The Ampere. As stated previously, the ampere is now regarded in the M.K.S. system as the fourth fundamental unit. Its function, as a unit, is to link the electrical and mechanical systems, and consequently it must be defined with this in mind. The definition is as follows—"The ampere is that current which, when flowing in each of two very long parallel conductors of negligible cross-section, and separated by a distance of one metre from one another in vacuo produces between these conductors a force of 2×10^{-7} newtons per metre of their length."

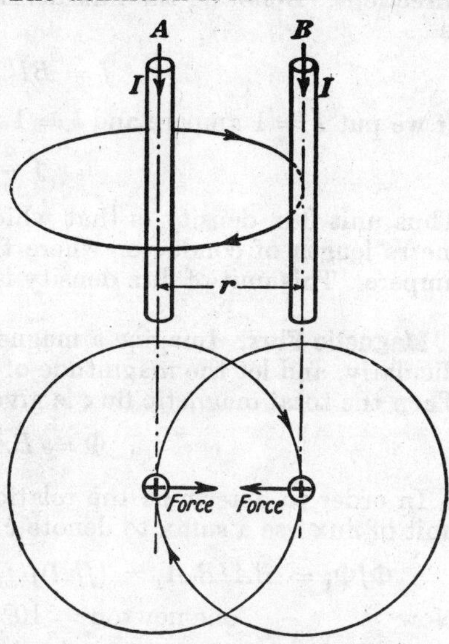

FIG. 4.1. MECHANICAL FORCE BETWEEN PARALLEL CONDUCTORS

Consider two parallel conductors separated a distance r as in Fig. 4.1. It is an experimental fact that the force per unit length between them in a medium such as air, is given by

$$f \propto I^2/r$$

or, if the currents in the two wires are different, say I_1 and I_2, then

$$f \propto I_1 I_2/r$$

This is a statement of Ampère's Law. Ampères' experiments were performed in air, but it has since been shown that the force is dependent on the nature of the medium, being proportional to the relative permeability, μ, of the medium. Hence we have, more generally

$$f \propto \mu I_1 I_2/r$$

By introducing a suitable constant we can convert the above into an equation. The constant we use is $2\ \mu_0$, where μ_0 is the magnetic space constant in the unrationalized m.k.s. system, so that

$$f = 2\mu\mu_0 I_1 I_2/r \text{ newtons/metre length}$$

It is from this expression, remembering that $\mu_0 = 10^{-7}$, that we derive the definition of the ampere given above. In the rationalized system, we give μ_0 the value

$$\mu_0 = 4\pi \times 10^{-7}$$

$$\therefore\ f = 2\ \mu\mu_0 I_1 I_2/4\pi r \text{ newtons/metre length}$$

When the medium is air we take $\mu = 1$, so that

$$f = 2\,\mu_0 I_1 I_2/4\pi r \text{ newtons/metre length}$$

Magnetic Flux Density. Considering again the two parallel wires of Fig. 4.1 we can regard one of them as setting up a magnetic field which, reacting on the other conductor, sets up a side thrust whose direction is towards the first conductor when the currents are in the same direction in space, and away from it when the currents are in opposite directions. Denoting the flux density by B the equation for the force is

$$f = BIl \text{ newtons}$$

If we put $I = 1$ ampere and $l = 1$ metre then we have, numerically

$$f = B$$

Thus unit flux density is that which gives a force of one newton per metre length of conductor where the current in the conductor is one ampere. This unit of flux density is the weber per square metre.

Magnetic Flux. Imagine a magnetic flux entering an area A perpendicularly, and let the magnitude of B be uniform over the whole of A. Then the total magnetic flux is given by

$$\Phi = BA \text{ webers}$$

In order to determine the relationship of this unit to the old c.g.s. unit of flux use a suffix to denote c.g.s. quantities. Then

$$\Phi/\Phi_1 = BA/B_1A_1 = (f/Il)/(f_1/I_1l_1) \times A/A_1 = fAI_1l_1/f_1A_1Il$$

Now one newton $= 10^5$ dynes, $\therefore f/f_1 = 10^{-5}$

One square metre $= 10^4$ cm^2, $\therefore A/A_1 = 10^{-4}$

$I_1 = 10^{-1}I$, since the c.g.s. unit of current is equal to 10 amperes.

One metre $= 10^2$ cm, $\therefore l/l_1 = 10^{-2}$

$$\therefore \quad \Phi/\Phi_1 = \frac{fAI_1l_1}{f_1A_1Il} = 10^{-5} \times 10^{-4} \times 10^{-1} \times 10^2 = 10^{-8}$$

$\therefore$ Webers/c.g.s. units (maxwells) $= 10^{-8}$

Webers $=$ maxwells $\times 10^{-8}$

Field Intensity due to a Long Straight Wire. Fig. 4.2 shows a portion of a straight current-carrying conductor: under the old conception the conductor is surrounded by circular magnetic lines of force having their centres on the axis of the wire. Consider again the equation of force

$$f = BIl \text{ dynes, in the c.g.s. system}$$

or

$$f = BI \text{ dyne/cm}$$

If the medium is air, for which we take $\mu = 1$, then $B = H$ numerically, so that

$$f = HI \text{ dyne/cm, or } H = f/I$$

If we now make $H = 1$, we have

$$f = I, \text{ numerically}$$

Thus we could define the c.g.s. unit of field intensity as the current per centimetre. Similarly in the rationalized m.k.s. system we define

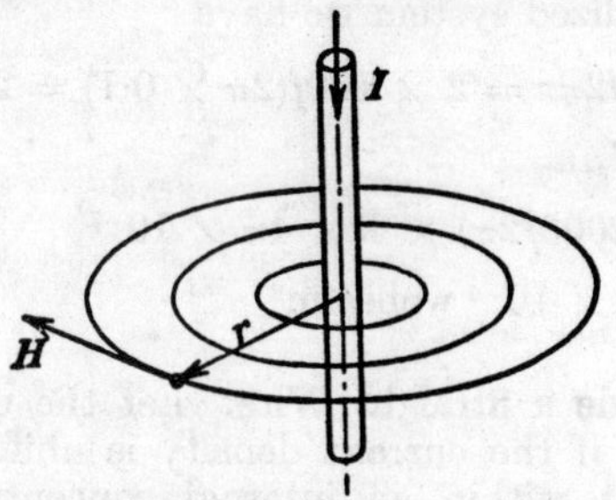

FIG. 4.2. MAGNETIC FIELD OF A STRAIGHT CONDUCTOR

the field intensity as the ampere per metre. Now the circumference of a circle of radius r is $2\pi r$, and therefore at a radius r the number of amperes per metre is $I/2\pi r$

$$\therefore H_r = I/2\pi r$$

This gives the magnitude of H in the m.k.s. *rationalized* system.

Now regard the problem in a different way. The lateral force exerted on one another by two parallel conductors carrying a current I is

$$f = 2\mu\mu_0 I^2 l/r \text{ where } \mu_0 = 10^{-7}$$

also
$$f = BIl$$
$$\therefore B = 2\,\mu\mu_0 I/r$$

But
$$H_u = B/\mu\mu_0$$
$$\therefore H_u = 2I/r$$

which gives the magnitude of H in the m.k.s. *unrationalized* system.

Arguing in the same way, but using the rationalized system, we have

$$f = 2\,\mu\mu_0 I^2 l/4\pi r$$
$$f = BIl$$
$$\therefore B = 2\,\mu\mu_0 I/4\pi r$$

Again in the rationalized system

$$H_r = B/\mu\mu_0$$
$$\therefore H_r = I/2\pi r$$

as before. The unit is now the ampere per metre.

It is therefore necessary for the student to realize that rationalization can alter the numerical values of the intensity. This is analogous to the use in the c.g.s. system of H in oersteds and H in ampere-turns per centimetre length. The magnitudes of flux and flux density are not affected.

From now on we will drop the suffixes u and r and use the symbol H to refer to the field intensity in the rationalized system.

EXAMPLE. Calculate the field intensity and flux density at a distance of 0·1 m from a long straight wire carrying a current of 200 A.

Using the rationalized system we have

$$H = I/2\pi r = 2 \times 100/(2\pi \times 0{\cdot}1) = 2000/2\pi$$

$$B = H\mu\mu_0$$

$$= (2000/2\pi) \times 1 \times 4\pi \times 10^{-7}$$

$$= 4 \times 10^{-4} \text{ weber/m}^2$$

Flux Density inside a Straight Wire. Let the conductor radius be a centimetre, then, if the current density is uniform over the cross-section, the current within an internal concentric circle of radius $r_1 (< a)$ will be

$$Ir_1 = I(r_1{}^2/a^2)$$

and the field intensity at a distance r_1 from the axis will be

$$H = I(r_1{}^2/a^2)(1/2\pi r_1)$$

$$= Ir_1/2\pi a^2$$

$$\therefore B = \mu\mu_0 H$$

$$= (Ir_1/2\pi a^2) \times 1 \times 4\pi \times 10^{-7}$$

$$= (2Ir_1/a^2) \times 10^{-7} \text{ weber/square metre}$$

The flux density inside the conductor is therefore a linear function of the distance from the axis. Just at the surface, where $r_1 = a$, we have

$$B_a = (2I/a) \times 10^{-7} \text{ weber/square metre}$$

EXAMPLE. A conductor of circular cross-section, carrying 1000 A, has a diameter of 2 cm. Plot the variation of flux density with distance from the centre

$$a = 1 \text{ cm} = 10^{-2} \text{ metre}$$

$$\therefore B_a = 2 \times 1000/0{\cdot}01 \times 10^{-7}$$

$$= 2 \times 10^{-2} \text{ weber/m}^2$$

The distribution is shown in Fig. 4.3.

Field due to a Circular Current. Fig. 4.4 shows a circular conductor carrying current. Consider an element ds of conductor, the field

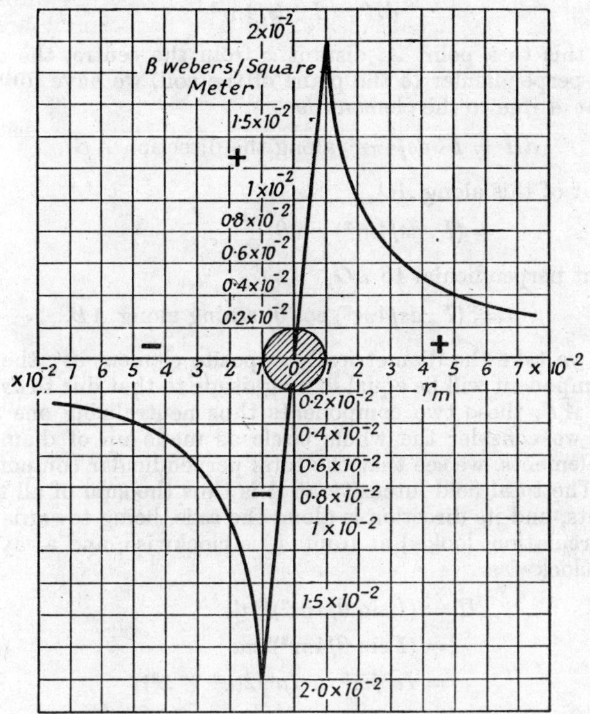

FIG. 4.3. MAGNETIC FIELD EXTERNAL TO A STRAIGHT CONDUCTOR

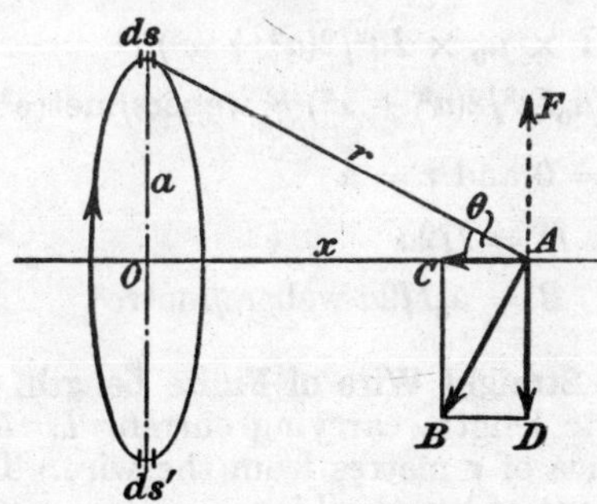

FIG. 4.4. CIRCULAR CONDUCTOR CARRYING CURRENT

intensity at a distance r at right angles to the tangent at ds is given by Laplace's law.

$$dH = I \cdot ds/r^2$$

the equation being of the same form for the c.g.s. and the unrationalized m.k.s. systems. For the rationalized m.k.s. system we have

$$dH = I \, . \, ds/4\pi r^2$$

Applying this to a point A, distant x from the centre, the direction AO being perpendicular to the plane of the coil, we have for the field strength at A due to the element ds

$$dH = I \, . \, ds/4\pi r^2, \text{ along the direction } AB$$

Component of this along AO

$$= (I \, . \, ds/4\pi r^2) \sin \theta$$

Component perpendicular to AO

$$= (I \, . \, ds/4\pi r^2) \cos \theta, \text{ acting along } AD$$

Now if we take the diametrically opposite element ds', the perpendicular component will be equal in magnitude to that due to ds but will act along AF, these two components thus neutralizing one another. Hence, if we consider the whole circle as made up of diametrically opposite elements, we see that the total perpendicular component will be zero. The total field intensity at A is thus the sum of all the axial components, and its direction is along the axis, being towards O if the current circulation, looked at from A, is clockwise, and away from O if counterclockwise.

$$\therefore \; H = (I \sin \theta/4\pi r^2)\Sigma \, ds$$
$$= (I \sin \theta/4\pi r^2)2\pi a$$
$$= Ia^2/2r^3 \text{ or } Ia^2/2(a^2 + x^2)^{\frac{3}{2}}$$

The flux density at point A is thus for air

$$B = \mu\mu_0 H$$
$$= 1 \times \mu_0 \times Ia^2/2(a^2 + x^2)^{3/2}$$
$$= \mu_0 Ia^2/2(a^2 + x^2)^{3/2} \text{ webers/metre}^2$$

At O, the centre, $x = 0$ and $r = a$

$$\therefore \; H = I/2a$$

and

$$B = \mu_0 I/2a \text{ webers/metre}^2$$

Intensity due to a Straight Wire of Finite Length. Fig. 4.5 shows a straight wire of finite length, carrying current I. A point P is at a perpendicular distance of r metres from the wire. The field intensity at P due to an element of length dl is

$$dH = (Idl/4\pi d^2) \sin \beta$$

using rationalized units, and is directed into the paper and perpendicular to it. This condition of field direction applies no matter where the element dl may be located, and consequently for the magnitude of the total intensity at P we have

$$H = \frac{1}{4\pi}\int_{l_1}^{l_2} \frac{I \sin\beta}{d^2} . dl$$

Now

$$dl \sin\beta = AB, \text{ the projection of } dl$$
$$= d \, . \, d\theta$$

again $\quad d = r \sec\theta$

Hence, if we replace the limits l_1 and l_2 by α_1 and α_2, we have

$$H = \frac{1}{4\pi}\int_{\alpha_1}^{\alpha_2} \frac{Id}{dr \sec\theta} d\theta$$
$$= I/(4\pi r) \, [\sin\theta]_{\alpha_1}^{\alpha_2}$$
$$= (I/4\pi r) \, (\sin\alpha_2 - \sin\alpha_1)$$

Hence, in air, we have

$$B = (\mu_0 I/4\pi r) \, (\sin\alpha_2 - \sin\alpha_1) \text{ webers/metre}^2$$

If the wire is very long $\sin\alpha_2$ tends towards the value 1, and $\sin\alpha_1$ towards the value -1, showing that for a very long wire

$$B = \mu_0 I/2\pi r \text{ webers/metre}^2$$

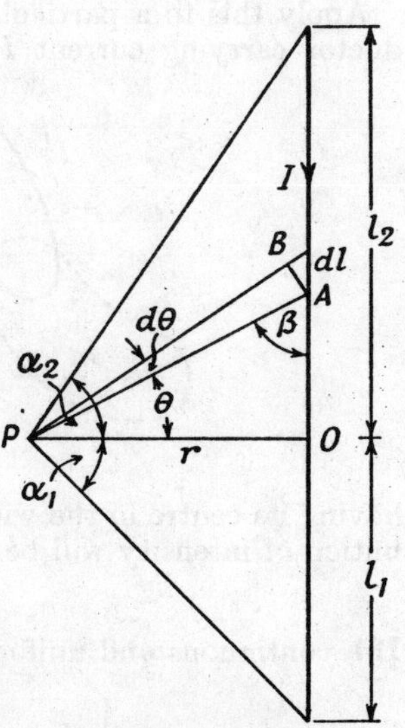

FIG. 4.5. INTENSITY DUE TO A STRAIGHT WIRE OF FINITE LENGTH

EXAMPLE. A straight conductor is carrying 1000 A and a point P is sited in such a way that $\alpha_2 = 60°$ and $\alpha_1 = -30°$, the negative sign being taken when α_1 is in the opposite sense to α_2, as in Fig. 4.5. Calculate the intensity at P if its perpendicular distance from the conductor is 0·1 m.

$$H = [1000/(4\pi \times 0{\cdot}1)] \, [\sin 60° - \sin(-30°)]$$
$$= [1000/(4\pi \times 0{\cdot}1)] \, [0{\cdot}866 + 0{\cdot}5]$$
$$= 13{,}660/4\pi \text{ rationalized m.k.s. units}$$
$$= 1088 \text{ A/m}$$

Intensity due to a Solenoid. Imagine a coil of one turn, carrying a current I' ampere (Fig. 4.6). Along a certain closed path, as indicated by the dotted line, this coil will set up a definite distribution of magnetic intensity. Now suppose that the coil of one turn is replaced by one of two turns very close together, the current now being $I'/2$. It is quite obvious that, except for points very close to the turns, the magnetic effect of the two-turn coil will be the same as that of the one-turn coil, and consequently it will set up the same distribution of magnetic intensity along the same path. It follows that we can replace an actual coil by an equivalent coil, the criterion for equivalence being that the distribution of magnetic intensity throughout the region in which one is interested, is unchanged.

Apply this to a particular case. Fig. 4.7 shows a long straight conductor carrying current I'. Along a circular path of any radius r,

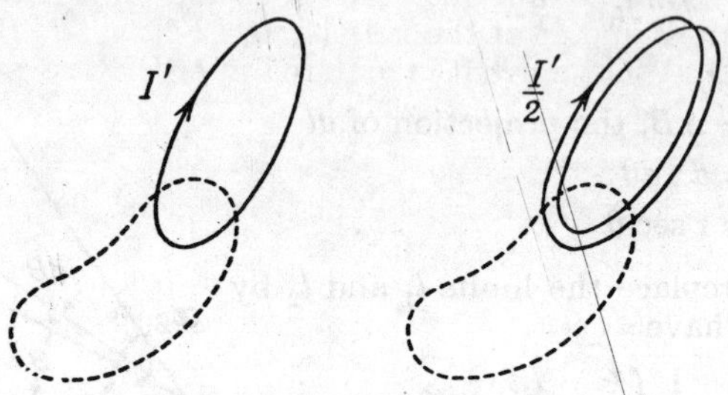

FIG. 4.6

having its centre in the wire, as shown by the dotted circle, the distribution of intensity will be uniform, and of value

$$H = I'/2\pi r.$$

If a continuous and uniform solenoid is wound with the circular path

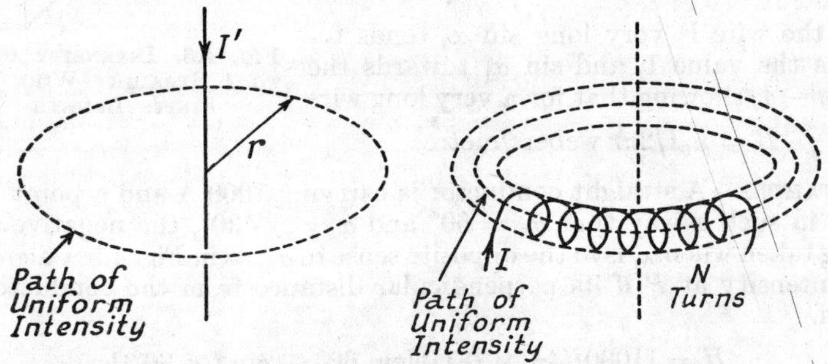

FIG. 4.7. COMPARISON OF STRAIGHT CONDUCTOR AND CLOSED, UNIFORMLY WOUND SOLENOID

as its axis, this solenoid, when carrying current I, will set up a uniform distribution of intensity along this path. For the solenoid to be equivalent to the straight wire we have

$$NI = I', \text{ where } N = \text{number of turns}$$

But $I' = H \,.\, 2\pi r$

$= Hl$, where l is the length of the path

$$\therefore NI = Hl$$

or $H = NI/l$ ampere/metre

If the solenoid is filled with a medium of relative permeability μ, then

$$B = \mu\mu_0 H$$

$$B = \mu\mu_0 NI/l \text{ webers/square metre}$$

The above formulae apply in the case of a solenoid which is long in relation to the diameter of a turn since, in such a case, the intensity in the region outside the solenoid is so small that, for practical calculations, it can be neglected.

EXAMPLE. A solenoid of 360 turns is wound on a continuous ring of iron, the mean diameter of the ring being 12 cm. The relative permeability is 1250. What current is required in order that the flux density in the iron shall be 15,000 maxwells/cm²?

(*a*) Conversion

$l =$ mean circumference of ring $= 12\pi$ cm $= 0{\cdot}12\pi$ metre

$B = 15{,}000$ maxwells/cm²

$= 15{,}000 \times 10^{-8} \times 10^{4} = 1{\cdot}5$ weber/m²

(*b*) Substitution

$$B = \mu\mu_0 NI/l$$

$$\therefore I = Bl/\mu\mu_0 N$$

$$= 1{\cdot}5 \times 0{\cdot}12\pi/(1{\cdot}25 \times 10^{3} \times 4\pi \times 10^{-7} \times 360)$$

$$= 1 \text{ A}$$

Intensity due to a Short Solenoid. Let the solenoid be of length l and let the radius of each turn be a, and let it be uniformly wound with N turns. Then there will be N/l turns per unit length and in an element of length, dx, Fig. 4.8, there will be $N\,dx/l$ turns. Now a very short element of length can be regarded as a coil of very short axial length, carrying a current $NI\,dx/l$ ampere, and consequently the field intensity at any point on the axis due to this element will be given by the type of expression derived on p. 89. Thus at the centre, O, of the solenoid we have

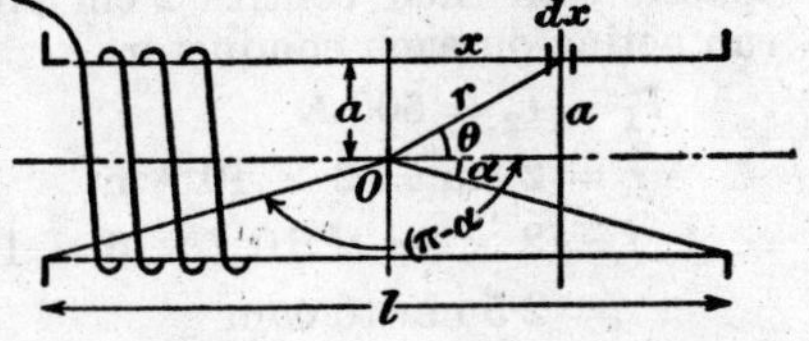

FIG. 4.8. SELF-INDUCTION OF A SOLENOID

$$dH = NI(dx/l)[(\sin\theta)/4\pi r^2]\,2\pi a$$

$$= (NI/2)[a\,(\sin\theta)/lr^2]\,dx$$

$$= (NI/2)[(\sin^3\theta)/al]\,dx$$

Now
$$x = a\cot\theta$$

$$\therefore dx = -\,(a/\sin^2\theta)\,d\theta$$

$$\therefore dH = -\,(NI/2)(\sin\theta d\theta/l)$$

$$\therefore H = -\frac{NI}{2l}\int_{\pi-\alpha}^{\alpha}\sin\theta\,.\,d\theta$$

$$= NI/2l\,[\cos\theta]_{\pi-\alpha}^{\alpha}$$

$$= (NI/l)\cos\alpha$$

If the solenoid is very long, the angle α becomes very small and $\cos \alpha$ tends towards unity. The expression then becomes

$$H = NI/l, \text{ as before}$$

Now suppose that the solenoid is cut into two equal halves at the centre line. Then for each half solenoid N is halved and l is halved, and therefore, as I remains the same as before, the field intensity remains the same. Thus the intensity inside a half solenoid is the same as that at the centre of the parent solenoid.

At the point O we can imagine that the intensity is due to the effects of the two halves separately, and therefore the contribution of each half, *at this particular point*, is $NI/2l$. If then we remove one-half, so that point O becomes the end, and not the middle point, the actual intensity is $NI/2l$. We therefore see that just at the end of a solenoid the intensity has one-half of the value inside the solenoid.

Force between Two Parallel Current-carrying Conductors. We have already seen that, for currents I_1 and I_2, the magnitude of the lateral force exerted by one conductor on the other is, for a medium such as air

$$f = 2\mu_0 I_1 I_2 / 4\pi r \text{ newtons/metre}$$

The force is one of attraction if the currents are in the same direction, and one of repulsion if they are in opposite directions.

EXAMPLE. Two long parallel conductors, each carrying 500 A, are spaced with their centres 2 cm apart. Calculate the force per metre run acting on each conductor.

$$I_1 = I_2 = 500 \text{ A}$$

$$r = 2 \text{ cm} = 2 \times 10^{-2} \text{ m}$$

$$\therefore f = 2 \times 4\pi \times 10^{-7} \times 5 \times 10^2 \times 5 \times 10^2 / (4\pi \times 2 \times 10^{-2})$$

$$= 2{\cdot}5 \text{ newtons/m}$$

In cases where parallel conductors carrying heavy conductors are installed very close to one another, the force set up may be very considerable. One very important example is the system of end connections (the overhang) of the stator winding of a very large alternator. The spacing is, of necessity, very small, and under short-circuit conditions the current may rise suddenly to several times the normal full-load value, thereby setting up sudden and violent mechanical stresses on the windings. To prevent damage it is necessary to adopt a very complete system of mechanical supports.

The Magnetic Force. A magnetic intensity of magnitude H at any particular point sets up at that point a magnetic flux density of

$$B = \mu\mu_0 H$$

If there is a uniform distribution of flux density B, over an area A, at right-angles to the direction of magnetization, then the product BA is called the magnetic flux, Φ

Thus

$$\Phi = BA$$

The fundamental concept in which the whole of the preceding discussion has been based is the side thrust exerted by a magnetic field on a current-carrying conductor. The defining equation for flux density is thus

$$f = BIl$$

from which we see that unit flux density, the weber per square metre, is that flux density which results in a side thrust of one newton per metre on a long straight conductor carrying one ampere, and arranged at right angles to the direction of magnetization.

The flux, being the product of B and A when B is uniformly distributed, or, in general, $\int BdA$, is therefore expressed in webers.

Any effect presupposes a cause. For example, we say that the establishment of an electric current round a circuit is the result of the application of an e.m.f. Under steady-state conditions there is the Ohm's Law relationship

$$\text{current} = \text{e.m.f./resistance}$$

Similarly the establishment of a magnetic flux presupposes the presence of a cause which, by analogy, is called the magnetomotive force, m.m.f. The direct result of the m.m.f. is the establishment, round closed paths, of distributions of magnetic intensity H, and it is this field intensity which produces the quantity known as flux density B. Except in certain simple cases the intensity H is not constant, but varies from point to point along the closed path. The m.m.f. F is equal to $\oint Hdl$, the integral $\oint$ meaning that the integration must be taken all the way round a closed circuit. We have seen that the definition of field intensity is the ampere per metre, and consequently it follows that $\oint Hdl$ has the dimensions of a current.

Consider two simple cases,

(*a*) Straight conductor, at radius r

$$H = I/2\pi r$$

and is uniform round the circular path of radius r. Hence m.m.f.

$$F = \oint Hdl = I/2\pi r \times 2\pi r = I$$

(*b*) Long solenoid

$$H = NI/l$$

We assume that the intensity has this value along the whole length l, and that the value at all points along the path outside the solenoid is negligibly small

$$\therefore\ F = \oint Hdl = NI/l \times l = NI$$

Thus the magnetomotive force of a current-carrying coil is the product of the turns and the amperes, that is, the ampere-turns.

The Unit Magnetic Pole. When using the c.g.s. system of units the science of electromagnetics is based on the conception of the unit magnetic pole. A unit pole is defined as one which repels a like and equal pole situated 1 cm from it with a force of one dyne. It follows from this that when magnetic flux is defined in terms of "lines of force," there must be 4π such lines (maxwells) radiating from a unit pole. Suppose we wish to retain the unit pole and the conception of lines of force, how many maxwell lines radiate from the rationalized m.k.s. unit pole?

When calculating the m.m.f. of a simple system, such as a straight conductor, or a solenoid, we make use of the relationship

m.m.f. = work done in taking a unit pole once round a closed magnetic path embracing the current

and work done = current × number of lines of force cut by the current during such a tour

∴ m.m.f. = current × number of lines cut

Let there be x lines emanating from the rationalized m.k.s. unit pole, then in taking such a pole right down the inside of a solenoid and then back to the starting point by any outside path, each of the x lines cuts each of the N turns, so that the number of cuttings is Nx. Hence

$$F = INx$$

But

$$F = IN$$

$$\therefore x = 1$$

Thus the total flux radiating from a rationalized m.k.s. unit pole is *one weber.*

Actually the conception of the unit magnetic pole is now of little value because an isolated magnetic pole is physically unrealizable, and also because, in the m.k.s. system the science of electromagnetism is based on the force exerted by one current-carrying conductor on a parallel current-carrying conductor.

Flux inside a Toroid. Consider a toroid wound on a former of rectangular cross-section, so that each individual turn is of rectangular shape. Let the axial length be l and the inner and outer radii r_1 and r_2, Fig. 4.9. The intensity at a radius x is

$$H_x = F/2\pi x = NI/2\pi x$$

$$\therefore B_x = \mu\mu_0 H_x$$

$$= \mu\mu_0(NI/2\pi x)$$

∴ Total flux inside the toroid

$$\Phi = \int B_x da = \int_{r_1}^{r_2} B_x l dx$$

$$= \frac{\mu\mu_0 NIl}{2\pi}\int_{r_1}^{r_2}\frac{dx}{x}$$

$$= (\mu\mu_0 NIl/2\pi)\log_e(r_2/r_1)$$

EXAMPLE. A toroid of 1000 turns is wound on a circular non-magnetic former of rectangular cross-section, of inner and outer radii 0·1 and 0·13 m, respectively, and of axial length 0·2 m. Calculate the flux produced by a current of 1 A.

$$\Phi = (\mu\mu_0 NIl/2\pi)\log_e(r_2/r_1)$$
$$= (1 \times 4\pi \times 10^{-7} \times 10^3 \times 1 \times 0{\cdot}2 \times \log_e 1{\cdot}3) \div 2\pi$$
$$= 1{\cdot}045 \times 10^{-5} \text{ webers}$$

The Magnetic Circuit. The path of the magnetic flux is spoken of as the magnetic circuit. Just as the flow of electric current in an electric

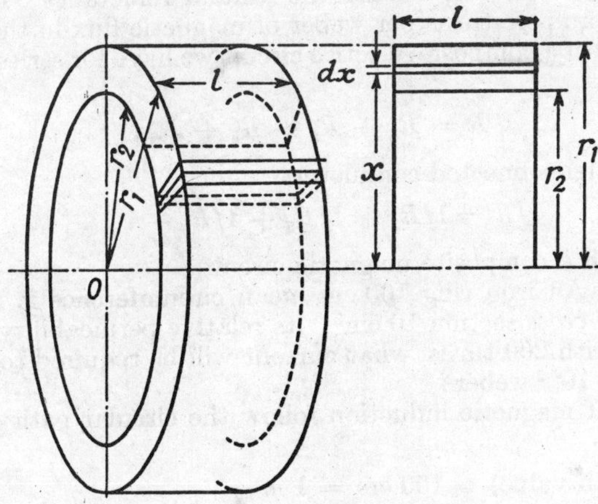

FIG. 4.9. TOROID

circuit necessitates the presence of an e.m.f., so the establishment of a magnetic flux requires the presence of a m.m.f. Every line of magnetic induction is a closed line, and although there is nothing in the nature of a flow of magnetism comparable with the axial flow of electrons which constitutes the electric current in a metallic conductor, there is, nevertheless, a useful analogy between the magnetic and electric cases.

Under steady-state conditions the current in an electric circuit is given by the Ohm's law relationship

$$I = E/R$$

also $$R = \rho l/a$$

or $$R = l/ga$$

where $$g = 1/\rho$$

the reciprocal of the specific resistance ρ.

Now consider the case of a long solenoid. We have seen that

$$B = \mu\mu_0 NI/l$$
$$\therefore \Phi = Ba$$
$$= \mu\mu_0 NIa/l$$
$$= NI \div (l/\mu\mu_0 a)$$

Now $NI = F$, the magnetomotive force, and therefore we see that, in terms of the analogy between the magnetic and electric circuits, the quantity $l/\mu\mu_0 a$ corresponds to the electric resistance l/ga. It is called the magnetic reluctance S. Hence

$$\text{Flux} = (\text{m.m.f.})/(\text{reluctance}),\ \Phi = F/S$$

No name is in common use for the unit of reluctance; it is clearly expressed in ampere-turns per weber of magnetic flux in the circuit.

In the case of a composite electric circuit we have for series-connected conductors

$$R = R_1 + R_2 + R_3 + \ldots$$

and for parallel-connected conductors

$$1/R = 1/R_1 + 1/R_2 + 1/R_3 + \ldots$$

Similarly with a composite magnetic circuit.

EXAMPLE. An iron ring 100 cm mean circumference is made from round iron of cross-section 10 cm²; its relative permeability is 500. If it is wound with 200 turns, what current will be required to produce a flux of $0{\cdot}1 \times 10^{-2}$ weber?

The lines of magnetic induction follow the circular path of the iron so that

$$l \text{ (mean value)} = 100 \text{ cm} = 1 \text{ m}$$
$$a = 10 \text{ cm}^2 = 10^{-3} \text{ m}^2$$
$$\therefore S = l/\mu\mu_0 a = 1/(500 \times 4\pi \times 10^{-7} \times 10^{-3})$$
$$= 10^7/6{\cdot}28 = 1{\cdot}592 \times 10^6$$
$$\Phi = 10^{-3} \text{ weber}$$
$$\therefore \quad F = \Phi S = 10^{-3} \times 10^7/6{\cdot}28 = 1{\cdot}592 \times 10^3$$

But
$$F = NI$$
$$\therefore I = F/N = 10^4/(6{\cdot}28 \times 200) \simeq 7{\cdot}96 \text{ A}$$

Magnetic Leakage. In the above example we assumed that all the lines of magnetic induction followed the closed path of the iron ring. In practice such an ideal state is never achieved, since a portion of the total flux will follow paths different from that of the intended circuit. The shapes of these paths, and the amount of flux in them, depend on the geometry of the magnetic circuit and also on the value of the relative permeability μ. Fig. 4.10 shows four cases. The flux taking the intended path, the main flux, is shown by continuous lines, that taking other paths, the leakage flux, is shown by dotted lines. The relative permeability of the main path is important because, if its value is low, there will be a tendency for flux to take short paths through the air, for which the relative permeability is unity.

An exact calculation of a magnetic circuit with distributed leakage

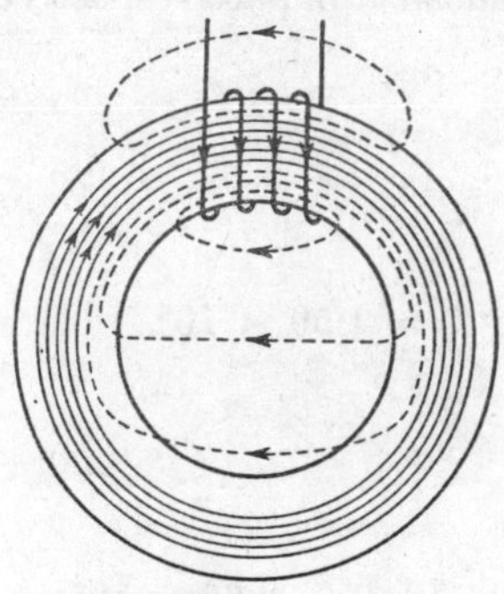

1. Ferromagnetic Ring

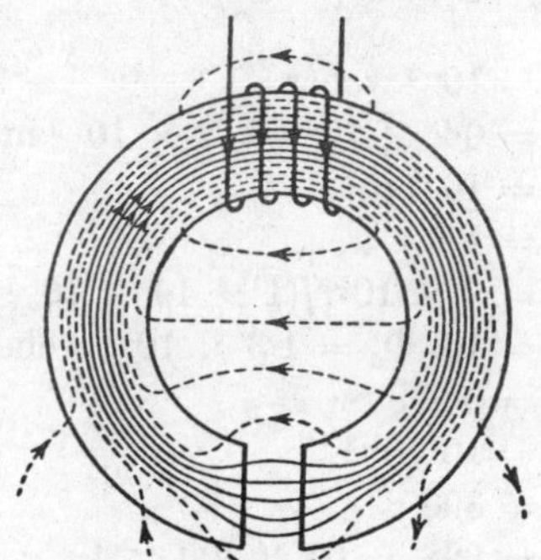

2. Ferromagnetic Ring with Air-gap.

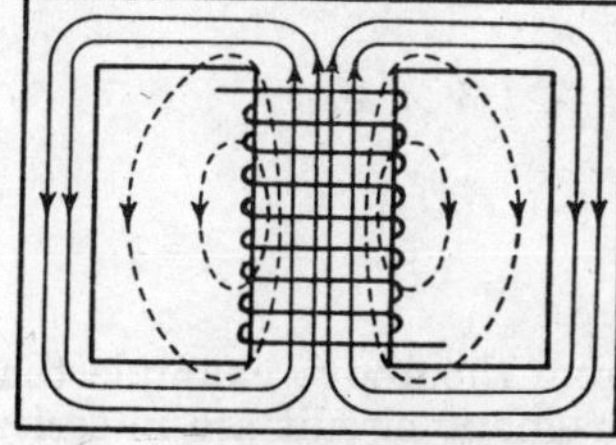

3. Ironclad Electromagnet

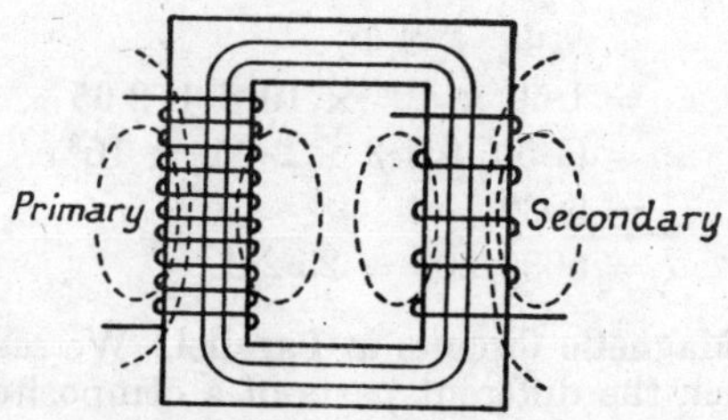

4. Simple A.C. Transformer

FIG. 4.10. EXAMPLES OF MAGNETIC LEAKAGE

The leakage flux is indicated by the dotted lines. Note that the numbers of lines representing the useful and leakage fluxes do not represent *in any way* the relative magnitudes.

would be exceedingly difficult, and for practical purposes it is assumed that the total flux is carried by that part of the circuit on which the magnetizing coil is wound, and that the main flux, or useful flux is carried by that part of the circuit in which the flux is actually utilized,

Let Φ_t = total flux

Φ = useful flux

Φ_f = leakage flux

Then $\Phi_t = \Phi + \Phi_f$

The ratio Φ_t/Φ is called the leakage factor.

EXAMPLE. The iron ring in the previous example has a saw cut 2 mm wide made in it. Find the new current required to produce the flux of $0{\cdot}1 \times 10^{-2}$ weber in the air-gap, given that the leakage factor is 1·3, and that the relative permeability of the iron under the new conditions is 300.

Let suffixes a and i refer to the air-gap and the iron path respectively, then

$$\Phi_a = 10^{-3} \text{ weber}$$
$$l_a = 0{\cdot}2 \times 10^{-2} = 2 \times 10^{-3} \text{ m}$$
$$a_a = 10^{-3} \text{ m}^2$$
$$\mu_a = 1$$
$$\therefore\ S_a = 2 \times 10^{-3}/(1 \times 4\pi \times 10^{-7} \times 10^{-3}) = 1{\cdot}59 \times 10^6$$
$$\Phi_i = 1{\cdot}3\ \Phi_a = 1{\cdot}3 \times 10^{-3} \text{ weber}$$
$$l_i = 0{\cdot}998 \text{ m}$$
$$a_i = 10^{-3} \text{ m}^2$$
$$\mu_i = 300$$
$$\therefore\ S_i = 998 \times 10^{-3}/(300 \times 4\pi \times 10^{-7} \times 10^{-3}) = 2{\cdot}65 \times 10^6$$

Since the iron and air paths carry different fluxes we are not able to add together the values of S_a and S_i to give the total reluctance. We therefore proceed as follows—

$$F = F_a + F_i$$
$$= S_a\Phi_a + S_i\Phi_i$$
$$= 1{\cdot}59 \times 10^6 \times 10^{-3} + 2{\cdot}65 \times 10^6 \times 1{\cdot}3 \times 10^{-3}$$
$$= (1{\cdot}59 + 1{\cdot}3 \times 2{\cdot}65) \times 10^3$$
$$= 5030$$
$$\therefore\ I = 5030/200 = 25{\cdot}2 \text{ A}$$

Magnetic Circuits in Parallel. We see from the above example that when the different parts of a composite magnetic circuit are in series,

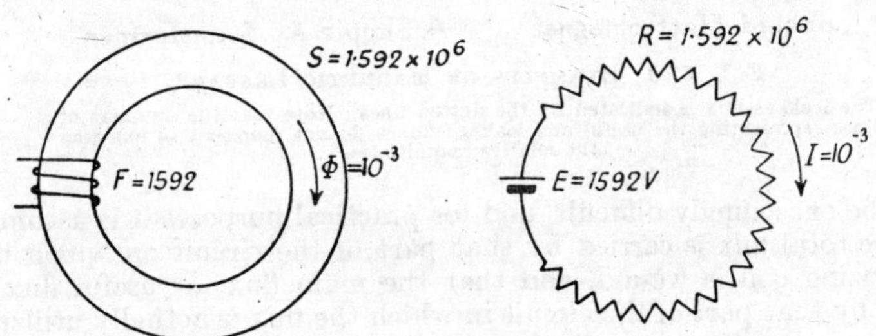

FIG. 4.11. ANALOGOUS MAGNETIC AND ELECTRIC CIRCUITS

that is, when the closed path of the main flux passes through all the parts, the total m.m.f. required by the complete circuit is equal to the sum of the m.m.f. required by the separate parts. In this respect the magnetic circuit is somewhat analogous to the electric circuit.

Consider first the simple iron ring of the example on p. 98, $F = 10^4/6{\cdot}28 = 1592$, $S = 10^7/6{\cdot}28 = 1{\cdot}592 \times 10^6$, $\Phi = 10^{-3}$. The analogous electric circuit is a simple closed circuit of $R = 1{\cdot}592 \times 10^6\,\Omega$, acted on by an e.m.f. $E = 1592$ V, and therefore carrying a current of $I = 10^{-3}$ A, as illustrated in Fig. 4.11. In the case of the iron ring with air-gap (p. 100) we regard the leakage flux Φ_l as parallel to the air-gap flux Φ_a, and for the total iron flux Φ_i we have

$$\Phi_i = \Phi_a + \Phi_l$$

The m.m.f. required to produce Φ_l is thus the same as that required for Φ_a, so that we need calculate for one only. In the example this was for Φ_a since we had the data necessary for the direct calculation of S_a, and therefore of $F_a = \Phi_a S_a$. Since F_l also equals F_a, we have

$$\Phi_l S_l = \Phi_a S_a = 10^{-3} \times 1{\cdot}592 \times 10^6 = 1592$$

$$\therefore\; S_l = \Phi_a S_a/\Phi_l$$

$$= 1592\,(1{\cdot}3 \times 10^{-3} - 10^{-3}) = 5{\cdot}3 \times 10^6$$

The magnetic circuit and the analogous current are illustrated in Fig. 4.12. In order to simplify calculation the assumption is made that the whole of the magnetic leakage flux is concentrated in the

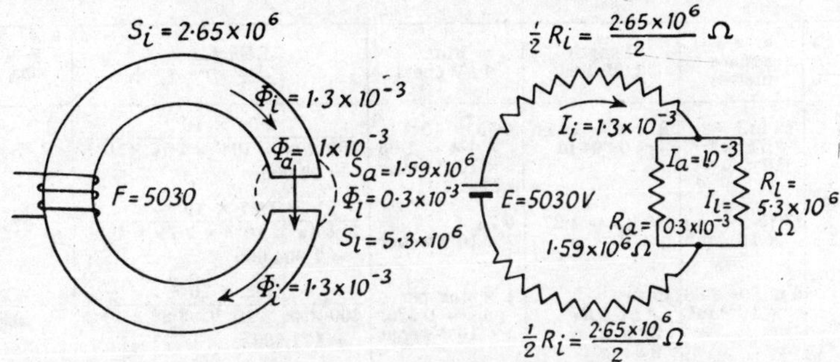

FIG. 4.12. ANALOGOUS MAGNETIC AND ELECTRIC CIRCUITS WHEN THERE IS LEAKAGE IN THE MAGNETIC CIRCUIT

immediate neighbourhood of the air-gap, each leakage path therefore including the whole of the iron path. It is this which leads to the simple branched circuit with S_a and S_l in parallel, as in the electrical analogue. Actually the leakage flux is distributed, as shown by the dotted lines of Fig. 4.13, but to try to take this into account in a calculation would be quite uneconomical in the very great majority of practical problems, as, for example, the design of the magnetic circuit of a generator or motor.

It must be realized that if there are two or more magnetic paths in parallel then each path requires the same m.m.f. as the others, so that it is sufficient to consider one path only. In many practical cases there is the further simplification that there are two identical paths only.

As an example, consider the ironclad electromagnet of Fig. 4.14. A total of 750,000 maxwells is required across the air gap, the leakage factor is 1·4, and the relative permeability of the iron under these

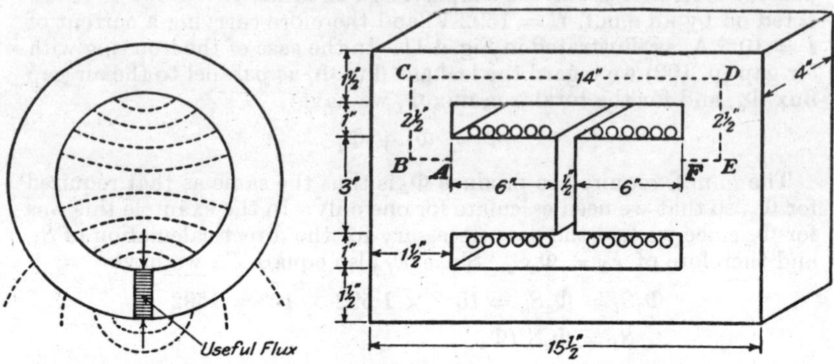

FIG. 4.13. MAGNETIC LEAKAGE

FIG. 4.14. IRONCLAD ELECTROMAGNET

conditions 600. There are two magnetic paths in parallel, one of them being indicated by the dotted path. The calculation is conveniently made in tabular form, as below. N.B. 1 weber = 10^8 maxwells.

Part of Circuit	area a, square metres	Length l, Metres	Flux Φ, Webers	Reluctance $S = l/\mu\mu_0 a$	$F = \Phi S$
Poles .	12 in.^{2} = 7·75 × 10^{-3} m^2	2 × 6 = 12 in. = 0·304 m	0·75 × 10^{-2} × 1·4 = 1·05 × 10^{-2} weber	$\dfrac{3{\cdot}04 \times 10^{-1}}{600 \times 4\pi \times 10^{-7} \times 7{\cdot}75 \times 10^{-3}}$ = 52,000	546
Air gap .	7·75 × 10^{-3} m^2	0·5 in. = 1·27 × 10^{-2} m	0·75 × 10^{-2} weber	$\dfrac{1{\cdot}27 \times 10^{-2}}{1 \times 4\pi \times 10^{-7} \times 7{\cdot}75 \times 10^{-3}}$ = 1,304,000	9780
Yoke .	6 in.^{2} = 3·88 × 10^{-3} m^2	Path *ABCDEF* 20·5 in. = 0·52 m	½ × flux per pole = 0·525 × 10^{-2} weber	$\dfrac{5{\cdot}2 \times 10^{-1}}{600 \times 4\pi \times 10^{-7} \times 3{\cdot}88 \times 10^{-3}}$ = 177,900	934
				Total F	= 11,260 ampere turns

In the previous calculations, the reluctances of the various parts of the magnetic path have been calculated, and these have depended on the value of the relative permeability, μ. In practice, μ is not known unless the induction density B is known, in fact, with all ferromagnetic materials μ is function of B (*see* p. 105). In this respect the magnetic circuit is different from the electric circuit, in which the quantity analogous to the relative permeability, namely the specific conductance g (= $1/\rho$) is independent of the current, provided that a steady thermal state has been reached, the temperature therefore remaining steady. It will also have been noticed that the magnitude of the reluctance in m.k.s. measure depends on whether the units are unrationalized or rationalized. For these reasons the method usually adopted in practical magnetic calculations is the "ampere-turn per unit length" method: this is explained on p. 106.*

Energy Density. When an electric circuit is such that a magnetic flux is set up by a flow of current, the circuit possesses self-inductance. It is shown in Chap. 5 that the self-inductance is given by the expression $L =$ (number of turns) $\times$ (flux per ampere) henrys.

Thus if the geometry of the circuit is such that the quotient flux per ampere can be calculated, then the self-inductance of the circuit can be calculated. It is also shown that if a current I ampere flows in a circuit of self-inductance L, then, in order to overcome the back e.m.f. of self-inductance, work equal to

$$W = \tfrac{1}{2}LI^2 \text{ joules}$$

has to be performed. This work is entirely separate from the work which has to be done to maintain the I^2R loss in the winding, and it is customary to associate it with energy stored in the magnetic field. Consider a simple case, such as a continuous ring with a uniform magnetizing winding, and of cross-section sufficiently small to warrant the assumption that B is uniform. We have seen that the flux in such a ring is given by

$$\Phi = NI\mu\mu_0 a/l$$

$$\therefore \Phi/I = N\mu\mu_0 a/l$$

$$\therefore L = N^2\mu\mu_0 a/l$$

$$\therefore W = \tfrac{1}{2}\,.\,(NI)^2\mu\mu_0 a/l$$

$\therefore$ Energy density

$$W/al = \tfrac{1}{2}(NI/l)^2\mu\mu_0$$

But $NI/l = H$ and $NI\mu\mu_0/l = B$

$\therefore$ Energy density

$$= \tfrac{1}{2}BH = \tfrac{1}{2}B^2/\mu\mu_0$$

$$\text{or } \tfrac{1}{2}\mu\mu_0 H^2 \text{ joules/metre}^3$$

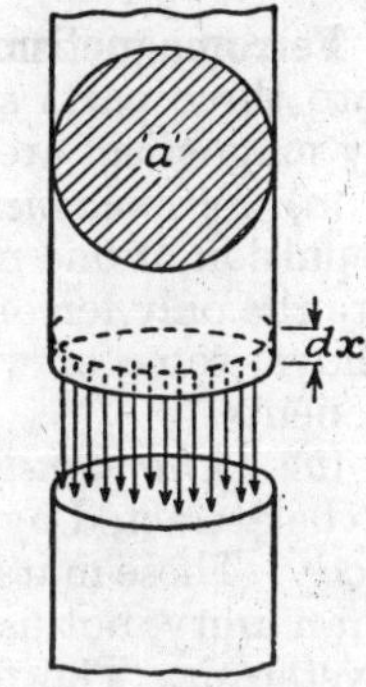

Fig. 4.15

Lifting Power of an Electromagnet. Consider two poles arranged directly opposite one another, as in Fig. 4.15. Let each have an area a square metres, and let f newtons be the force of attraction between them. Let one pole be moved a very short distance δx against the force f, then the work done is obviously $f\delta x$ newton-metres or joules.†

* L. R. Blake, "The design of electromagnets." *Electronic Eng.* **25** (Sept. and Oct., 1953).

† The current in the exciting coil is assumed to be maintained constant during the movement of δx.

But the volume of the magnetic field between the poles is increased by $a\delta x$ cubic metres, and therefore the energy stored is increased by

$$\tfrac{1}{2}(B^2/\mu\mu_0) \times a\delta x \text{ joules}$$

$$\therefore f\delta x = \tfrac{1}{2}(B^2/\mu\mu_0) \times a\delta x$$

$$\therefore f = B^2a/2\mu\mu_0 \text{ newtons}$$

EXAMPLE. In a telephone receiver the size of each of the poles is 1·2 cm × 0·2 cm, and the flux between each pole and the diaphragm is 300 lines. With what force is the diaphragm attracted to the poles? (*Lond. Univ.*).

$$\Phi = 300 \times 10^{-8} = 3 \times 10^{-6} \text{ webers}$$

$$a = 1{\cdot}2 \times 0{\cdot}2 = 0{\cdot}24 \text{ cm}^2 = 0{\cdot}24 \times 10^{-4} \text{ m}^2$$

$$\therefore B = 3 \times 10^{-6}/(0{\cdot}24 \times 10^{-4}) = 12{\cdot}5 \times 10^{-2} \text{ webers/m}^2$$

$$\mu = 1 \text{ for air}$$

$$\mu_0 = 4\pi \times 10^{-7}$$

$$\therefore f = (12{\cdot}5)^2 \times 10^{-4} \times 0{\cdot}24 \times 10^{-4}/(2 \times 1 \times 4\pi \times 10^{-7})$$

$$= 0{\cdot}1494 \text{ newton per pole}$$

$$= 0{\cdot}299 \text{ newton altogether}$$

$$= 0{\cdot}299 \times 10^5/981 = 30{\cdot}5 \text{ grammes weight}$$

Ferromagnetism. The science of magnetism is conveniently divided into three parts since the reactions of materials which are influenced by magnetism are of three kinds—

(*a*) *Ferromagnetism*, characterized by the strong attraction or repulsion of one magnetized body on another. Iron, nickel and cobalt are the only ferromagnetic elements, but there are many ferromagnetic alloys, some of which do not even possess iron as one of their components.

(*b*) *Paramagnetism*, a quality of some substances which causes them to be attracted by the poles of an electromagnet or strongly magnetized body. These materials have a relative permeability only slightly greater than unity, not usually greater than 1·001, and they do not exhibit any hysteresis. The ferromagnetic materials are all solids, but the paramagnetic may be solid, e.g. the elements sodium and potassium, liquid, e.g. oxygen, or gaseous, e.g. oxygen gas.

(*c*) *Diamagnetism*, the property of being repelled by a magnet. In these substances the magnetization is in the opposite direction to the magnetizing field, and so the permeability is less than unity. Many of the metals are diamagnetic, e.g. copper, gold and bismuth, and most non-metals, e.g. carbon, both graphite and diamond, carbon-dioxide and nitrogen in gaseous form.

Magnetization Curves. If an iron specimen, e.g. a closed ring, is subjected to a progressively increasing magnetizing force H, and the resulting induction density B plotted against H, the familiar "magnetization curve" will be obtained, (Fig. 4.16), since

$$B = \mu\mu_0 H$$
$$\mu = B/\mu_0 H$$

and from this relationship the curve of relative permeability may be plotted. The process of magnetization up to a high value of B can be

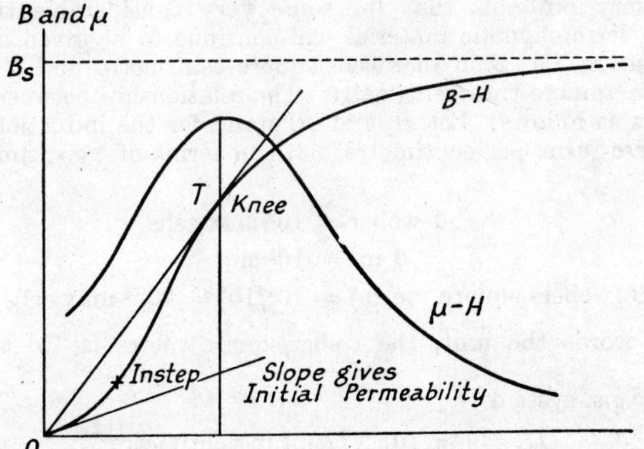

Fig. 4.16. Characteristics of Ferromagnetic Material

conveniently divided into three stages at what are called the "instep" and the "knee" of the curve.

(*a*) *The Initial Stage.* In the neighbourhood of the origin the graph is a straight line through the origin, and the slope gives the initial permeability. The value of H at which the graph begins to bend over varies over very wide limits, from one material to another.

(*b*) *The Middle Portion,* between the instep and the knee. The slope is the greatest and consequently the point of maximum permeability is within this region. If curves for decreasing, as well as increasing values of H are obtained it will be found that these differ markedly, and it is within this region that these irreversible changes are the greatest. These changes and the associated losses in energy are discussed in the next paragraph.

(*c*) *The Upper Portion beyond the Knee.* At the point I, at which the tangent to the magnetization curve passes through the origin, μ has its maximum value. For increases in H beyond this particular value there is a progressive (but not linear) decrease in μ, which ultimately tends towards unity.

In high fields B-H approaches a limit called the saturation, B_s. This upper portion of the curve is represented with fair accuracy by Fröhlich's equation

$$B - H = H/(a + bH)$$

where a is a "hardness" constant, and is a measure of the value of H

necessary to attain a given fraction of saturation, while $b = 1/B_s$. When H can be neglected in comparison with B, the Kennelly equation can be used,

$$1/\mu\mu_0 = a + bH$$

This gives a linear relationship between $1/\mu$ and H. The ratio $1/\mu\mu_0$ is called the *reluctivity*.

It is very probable that for some very considerable time data regarding ferromagnetic material will continue to be given in the old c.g.s. notation, i.e. B in maxwells/square centimetre and H in c.g.s. units or in ampere-turn/centimetre. The relationship between the two systems is as follows: Let B_c and at_c stand for the induction density and ampere-turns per centimetre, both in terms of c.g.s. units; then since

$$1 \text{ weber} = 10^8 \text{ maxwells}$$

and

$$1 \text{ m}^2 = 10^4 \text{ cm}^2$$

$$B \text{ (webers/square metre)} = 10^4/10^8 = 10^{-4} \text{ maxwells}$$

In other words the unit, the weber/square metre is 10^4 times the maxwell.

In the c.g.s. system

$$H_c = (4\pi/10)(NI/l),\ l \text{ in centimetre}$$

and

$$\therefore\ at_c = NI/l \text{ ampere-turns/centimetre}$$

$$= (10/4\pi)H_c, \text{ or } H_c = (4\pi/10)at_c$$

Now the rationalized m.k.s. H is

$$H = at = NI/l \text{ ampere-turns/metre}$$

$$\therefore\ H = 10^2\, at_c$$

For example, one particular point on the magnetization curve of a specimen of annealed iron is

$$H = 80 \text{ AT/m}$$

$$B = 0{\cdot}45 \text{ webers/m}^2$$

$$\therefore\ B_c = 0{\cdot}45 \times 10^4 = 4500 \text{ maxwell/cm}^2 \text{ or gauss}$$

$$at_c = H/100 = 0{\cdot}8$$

$$\therefore\ H_c = (4\pi/10) \times 0{\cdot}8 = 1{\cdot}0 \text{ (very approx.)}$$

$$\therefore\ \mu = 4500/1 = 4500 \text{ (relative permeability)}$$

Typical magnetization curves for a variety of commonly used magnetic materials are given in Fig. 4.17.

Practical Calculations. In practical magnetic calculations it is not usual to make use of the reluctance of the circuit. All that is necessary is to calculate for each part of the circuit the value of B, and then to refer to the magnetization curves for the particular materials of which

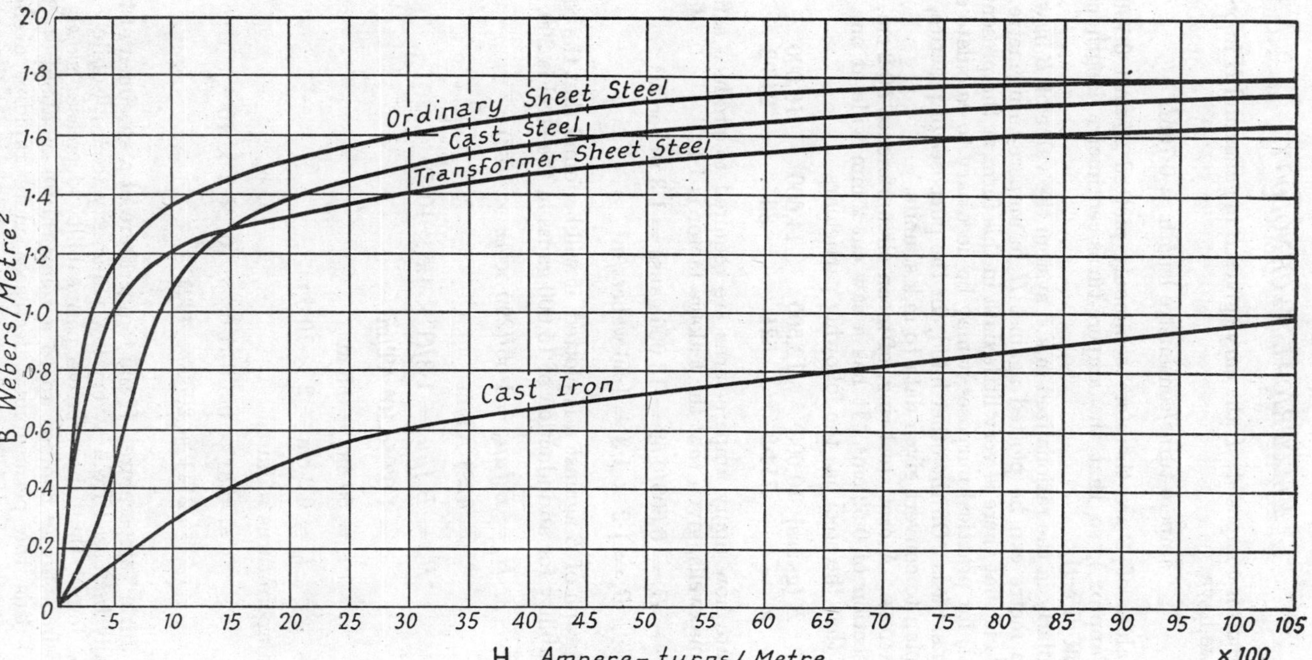

FIG. 4.17. MAGNETIZATION CURVES

the circuit is composed. If the curve gives B in gauss and H in oersteds then we have

$$\text{ampere-turns/centimetre length} = 0{\cdot}796H$$

The values of B are therefore commonly plotted against $0{\cdot}796H$ (or $0{\cdot}8H$ approx.), so that the ampere-turns/centimetre length can be read off directly.

Similarly in the rationalized m.k.s. system the value of B in webers/square metre can be plotted against H in ampere-turns/metre. The trouble is that there is very little data in this form at the present time so that, for practical purposes it may be necessary to calculate on the old c.g.s. basis. On the other hand, for the purposes of practice, it is a good plan to convert given data to m.k.s. units.

Example. A cast steel ring, of mean circumference 50·2 cm, has a cross-section of 0·52 cm². It has a saw cut 2 mm wide at one place. Given that the steel has the following particulars

B (gauss)	10,000	12,500	14,600	16,000
μ	714·3	521	365	246·2

Calculate how many ampere-turns are required to produce a flux of 6,500 maxwells, given that the leakage factor is 1·2.

$$B_a = 6{,}500/0{\cdot}52 = 13{,}000 \text{ gauss} = 1{\cdot}3 \text{ weber/m}^2$$

$$\therefore\ B_s = 1{\cdot}2 \times 1{\cdot}3 = 1{\cdot}56 \text{ webers/m}^2$$

If the graph of B against μ is plotted, it will be found that the relative permeability for an induction of 15,600 gauss in the steel is 290.

$$\therefore\ H_s = B_s/\mu\mu_0 = 1{\cdot}56/(290 \times 4\pi \times 10^{-7})$$

$$= 4281 \text{ AT/m}$$

$$H_a = B_a/\mu\mu_0 = 1{\cdot}3/(1 \times 4\pi \times 10^{-7})$$

$$= 1{\cdot}03 \times 10^6 \text{ AT/m}$$

$$l_s = 50 \text{ cm} = 0{\cdot}5 \text{ m}$$

$$l_a = 0{\cdot}2 \text{ cm} = 2 \times 10^{-3} \text{ m}$$

$\therefore$ ampere-turns required

$$= 4281 \times 0{\cdot}5 + 1{\cdot}03 \times 10^6 \times 2 \times 10^{-3}$$

$$= 2141 + 2060 = 4201$$

The Curie Temperature. If the temperature of a specimen of iron is gradually increased from a very low value, say from absolute zero, it will be found that the relative permeability decreases slowly, and almost in a linear manner up to a temperature of about + 300°C. Beyond this the permeability decreases more rapidly, eventually becoming unity. If H is kept constant there will be a definite temperature, called the Curie temperature, at which this happens. For one particular sample of annealed iron the Curie temperature with $H = 8 \times 10^5$ AT/m was 760°C. At the Curie temperature iron ceases to be ferromagnetic, and becomes paramagnetic.

Magnetic Reversals. If the magnetizing force applied to a specimen of iron is increased from zero to some maximum value, and the force is gradually reduced again to zero, it will be found that the new *B-H* curve for decreasing values of *H* lies somewhat above the original curve, and that when *H* is zero again *B* is finite. This effect is called *hysteresis*, since the values of *B* apparently lag behind those of *H*. The finite value of *B* when *H* is zero, *OR* in Fig. 4.18, is a measure of the

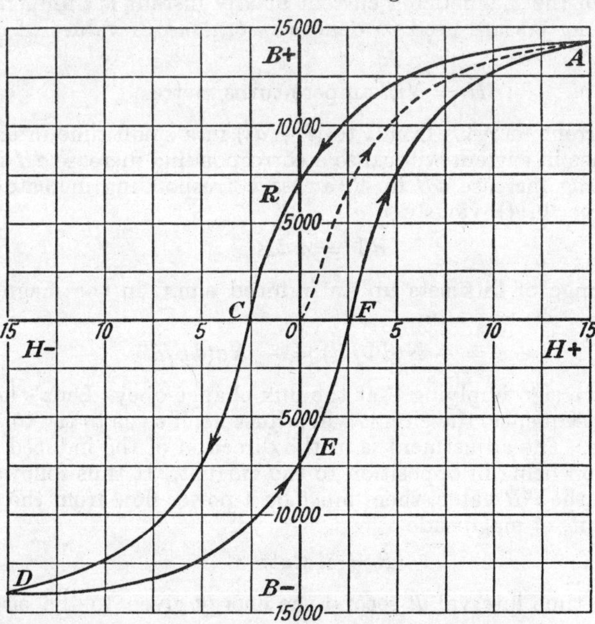

FIG. 4.18. HYSTERESIS LOOP FOR WROUGHT IRON
To convert to rationalized m.k.s. units multiply ordinates by 10^{-4}, abscissae by $10^3/4\pi$.

residual magnetism or *retentivity*. In order that the iron may be demagnetized it is necessary to apply a negative magnetizing force represented by *OC*. This negative force is called the *coercive force*. If now *H* is increased in this negative direction to its previous maximum value, the curve will reach a point *D*, at which the induction is equal to the previous maximum value, B_{max}. If, finally, *H* is gradually reduced to zero, reversed, and increased to its maximum value in the original direction, the curve *DEFA* will be traced out, the complete curve forming a closed loop, usually called a hysteresis loop. It will be seen that to take the iron through the various states represented by this loop, an alternating magnetizing force has been applied. If these alternations are maintained, the value of B_{max} being the same during each cycle, then the iron will continue to go through the same series of changes.*

The unit of retentivity has the dimensions metre per square metre, and of coercive force the ampere-turn per metre.

The hysteresis loop can be regarded as a magnetic indicator diagram, and from this analogy we should expect that during each cycle an amount of energy represented by the area enclosed by the loop is expended. That this is true can be proved as follows.

Consider an iron ring of mean circumference l metres and cross section a square metres. Let the magnetizing coil have N turns. If the value of the magnetizing current at any instant is i ampere (note that small letters are used to denote *instantaneous* values of current and voltage), then

$$H = Ni/l \text{ ampere-turns/metre}$$

Let the current increase from i to $(i + di)$ in a small time interval dt. This increase in current will cause a corresponding increase dH in H, a corresponding increase dB in B, and a corresponding increase $d\Phi$ in the total flux Φ. Obviously

$$d\Phi = a\,dB$$

Now a change of flux sets up an induced e.m.f. in the magnetizing winding of

$$e = -N(d\Phi/dt) = -Na(dB/dt)$$

The negative sign implying that the flux change obeys Lenz's law and that, in consequence, the circuit will adjust itself so as to try to oppose this change. The adjustment is in the direction of the induced e.m.f., this direction being in opposition to the current. It thus follows that, apart from the i^2R value, there must be a power flow from the source to the circuit, of magnitude

$$P = ei \text{ watts}$$

During the time interval dt second the energy given to the circuit is thus

$$\begin{aligned} dW_h &= Pdt \\ &= Nia(dB/dt)\,dt \\ &= Nia\,dB \end{aligned}$$

But

$$\begin{aligned} H &= Ni/l \\ \therefore Nia &= Hla \\ &= Hv \text{ where } v = \text{volume in cubic metres} \\ \therefore dW_h &= HdB\,v \\ \therefore W_h &= v\oint HdB \text{ joules} \end{aligned}$$

* In general, a sample of iron will not follow a closed hysteresis cycle if the cycle is only carried out once. The iron must be brought to what is called the cyclic state, by being subjected to a series of cycles, during which the maximum value of H, both positive and negative, is kept constant.

Now $\oint HdB$ is the area enclosed by the hysteresis loop, showing that

hysteresis loss in joules/cubic metre = area of hysteresis loop

In all classes of electrical machinery some part of the magnetic circuit is taken through rapid reversals of magnetism. Hysteresis loss therefore takes place, and in order that the efficiency and temperature rise of a machine may be calculable, it is necessary to know this loss. If the shape of the loop is known then the loss can be calculated by determining the area of the loop, but this method, although suitable for testing specimens of iron, is not so suitable for calculations on machines. Again, it does not follow that the magnetic properties of a material as incorporated into an actual machine will be the same as those of a sample tested in the laboratory, because these properties are dependent on the magnetic history of the material—this including how it has been handled during manufacture, for example, whether it has been filed or machined, whether holes or slots have been punched out of sheet material, and so on.

In practice it is therefore common to use empirical methods of pre-determining certain losses. The loss rises rapidly as the maximum induction density is increased. For example, a certain iron having a maximum relative permeability of 4500 has the relationship

$$W_h \propto B^3_{\text{max}} \text{ when } B_{\text{max}} < 0{\cdot}1 \text{ webers/m}^2$$

For the range 0·1 to 1·5 weber/m², the range of special importance in the design of electrical machinery,

$$W_h \propto B_{\text{max}}{}^{1\cdot 6}$$

or

$$W_h = \eta B_{\text{max}}{}^{1\cdot 6} \text{ ergs/cm}^3$$

which is known as the law of Steinmetz. The coefficient η is the Steinmetz coefficient and is approximately constant for any one material over the range for which the law holds. This coefficient is usually tabulated in c.g.s. units: the values can be converted to m.k.s. units as follows. Let the suffix c denote c.g.s. values, as before, then

$$\begin{aligned} W_h &= \eta_c(B_{\text{max }c})^{1\cdot 6} \text{ ergs/cm}^3 \\ &= \eta_c(B_{\text{max}} \times 10^4)^{1\cdot 6} \text{ ergs/cm}^3 \\ &= \eta_c(B_{\text{max}} \times 10^4)^{1\cdot 6} \times 10^{-7} \text{ joules/cm}^3 \\ &= \eta_c \,.\, 10^6 \,.\, 10^{-7} \,.\, (10^4)^{1\cdot 6} B_{\text{max}}{}^{1\cdot 6} \text{ joules/metre}^3 \end{aligned}$$

But $W_h = \eta B_{\text{max}}{}^{1\cdot 6}$

$$\therefore \eta = 2{\cdot}512 \times 10^5 \, \eta_c$$

In certain pieces of apparatus, iron having a steady unidirectional magnetization has superposed on this an alternating magnetization. In such cases the magnetic cycles are no longer symmetrical and the Steinmetz law does not apply, in fact, there is no empirical relationship which satisfactorily fits this case.

The following data* is of value—

* Based on material given by Miles Walker in *Specification and Design of Dynamo-electric Machinery*, p. 48. Longmans, London.

Material	η_c ergs/cm³	η joules/metre³ $= 2{\cdot}512 \times 10^5\, \eta_c$
Hard cast steel . .	0·028	7034
Cast steel . . .	0·003 to 0·012	753·6 to 3014
Cast iron . . .	0·011 to 0·016	2763 to 4019
Very soft iron . . .	0·002	502
Good dynamo sheet steel .	0·002	502
Silicon iron (0·2%Si) .	0·0021	527·6
Silicon iron (4.8%Si) .	0·00076	191

Example. A cylinder of iron, of volume 10,000 cm³, revolves for 30 min at a speed of 3000 r.p.m. in a two-pole field of such strength that the induction density in the iron is 8000 gauss. If the hysteresis coefficient of the iron is 0·003 erg/cm³, the specific heat of iron 0·11, the density of iron 7·8 g/cm³, the loss due to eddy currents is equal to that due to hysteresis, and 30 per cent of the heat produced is lost by radiation, find the temperature rise of the iron.

$$\text{Mass of iron} = 78{,}000 \text{ g}$$

If θ°C is the temperature rise, heat retained by the iron

$$= 78{,}000 \times 0{\cdot}11 \times \theta \text{ calories}$$

$\therefore$ Total heat produced

$$= (78{,}000 \times 0{\cdot}11 \times 100/70)\theta$$
$$= 12{,}250\ \theta \text{ calories}$$
$$W_h = \eta B_{\max}{}^{1\cdot6}\, vft \text{ joules}$$

where $\eta = 0{\cdot}003 \times 2{\cdot}512 \times 10^5 = 753{\cdot}6 \text{ joules/m}^3$

$$B_{\max} = 8000 \times 10^{-4} = 0{\cdot}8 \text{ weber/m}^2$$
$$v = 10{,}000 \times 10^{-6} = 10^{-2} \text{ m}^3$$

and $f = 50$ c/sec,

since in a two-pole field there is only one magnetic reversal in each revolution

$$t = 30 \times 60 = 1800 \text{ sec}$$
$$\therefore\ W_h = 753{\cdot}6 \times (0{\cdot}8)^{1\cdot6} \times 10^{-2} \times 50 \times 1800$$
$$= 47{\cdot}5 \times 10^4 \text{ joules}$$
$$\therefore \text{ Total loss} = 95 \times 10^4 \text{ joules}$$
$$\therefore\ 4{\cdot}2 \times 12{,}250\ \theta = 95 \times 10^4$$
$$\theta = 95 \times 10^4/(1{\cdot}225 \times 10^4 \times 4{\cdot}2) = 18^\circ\text{C}$$

Coercive Force and Retentivity. Referring to Fig. 4.18, it will be seen that on the descending portion of the curve there is a particular value of B, namely OR, which represents the induction density when the magnetizing force has been reduced to zero. This value is called the retentivity or remanence. In order to proceed down the descending portion it is necessary that the magnetizing force shall be reversed, and the particular value *OC* which just reduces the induction density to zero is called the coercive force or coercivity.

When considering the phenomenon of hysteresis we imagined a sample initially unmagnetized, so that the first portion of the magnetization process was the following of the magnetization curve *OA* which commences at the origin. In view of the property of coercivity it might be wondered how a specimen, once magnetized, could be brought back to the unmagnetized state. This is accomplished by taking the sample through a number of cycles in which the maximum value of the magnetizing force is progressively decreased. The iron is then taken round a kind of spiral curve of progressively decreasing B_{max}, as indicated in Fig. 4.19.

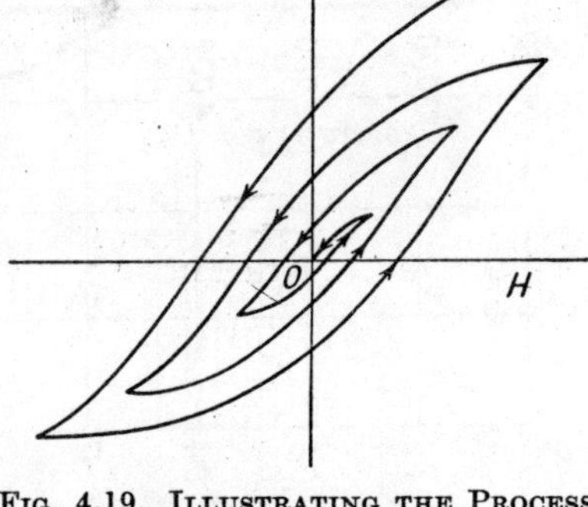

Fig. 4.19. Illustrating the Process of Demagnetization

The suitability of a ferromagnetic material for any particular application is largely determined by the shape of its hysteresis loop, and materials can be classified broadly into three groups—

(1) Those with a very steep rise to the knee, thus giving a high permeability. Largely as a result of the work of B. W. Eliven it was found that alloys having a large percentage of nickel, 35 to 90 per cent, have permeabilities much larger than those of iron. These are called "Permalloys," and the hysteresis loop of 78 permalloy, compared with the loop of soft iron and taken to the same value of B_{max}, is given in Fig. 4.20. The magnetic properties of these materials depend, not only on the percentage of nickel, but also on the heat treatment. Tests show that the maximum permeability (corresponding to the knee) is a function of the percentage of nickel for three kinds of heat treatment. Magnetic annealing consists in cooling from a higher temperature down to about 500°C, maintaining this temperature for a short time and at the same time acting on the specimen with a magnetizing force of several oersteds. The sample is then cooled to room temperature. The effect is very marked with a nickel content of 65 to 70 per cent, as indicated by the graph, but the most surprising result is the change in shape of the hysteresis loop. This becomes rectangular with vertical sides, as shown in Fig. 4.21, while the knee becomes almost a right-angled bend.

Besides nickel, molybdenum, chromium and copper have been used in the manufacture of ferromagnetic alloys, one of these, "Mumetal," having 5 per cent copper, 2 per cent chromium, 75 per cent nickel, and 18 per cent iron. These alloys have high initial permeabilities and, at

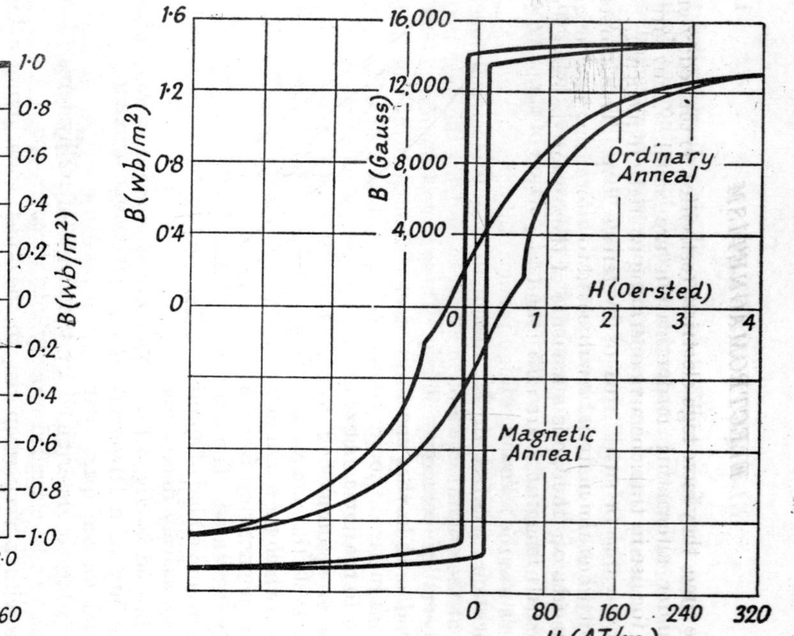

Fig. 4.20. Comparison of the Hysteresis Loops of Permalloy and Soft Iron

Fig. 4.21. Rectangular Hysteresis Loop of Permalloy 65 obtained by Annealing in presence of Magnetic Field

(*Bozorth and Dillinger*)

the same time, they have high electrical restivities, so that eddy-current losses under alternating magnetization are small. They are thus suitable for use in transformers operating at high frequencies.

(2) Those with a small area of hysteresis loop. This property is important when the material is subjected to alternating magnetization, since the area of the loop is a measure of the hysteresis loss. Many of the materials in group 1 are also found in group 2. Table 4.1* gives the properties of a number of high permeability materials.

Of particular interest in the design of heavy electrical machines, with magnetic reversals up to 50 or 60 per second, are the sheet materials having a small percentage of silicon. These have a Steinmetz coefficient much smaller than that for sheet steel without silicon and their cost is not prohibitive. The very-low-loss alloys described in Group 1 are much too expensive to be used in commercial machines and power-frequency transformers. For example, if we call the cost per pound of annealed cast iron unity, the costs of some of the others are of the order: Permalloy 28, Mumetal 27, Permendur 69, 1 per cent silicon iron 7, 4·5 per cent silicon iron 8.

(3) Those with a very large coercive force. These materials are used for the production of permanent magnets, and we shall see that it is not only the coercive force which is important, but also the area under that portion of the hysteresis loop which lies in the second quadrant, namely between the residual induction and the coercive force: the portion *RC* in Fig. 4.18. This portion is called the demagnetization curve.

Up to about 1930 the carbon steels were used almost exclusively for the making of permanent magnets. These do not have a large area under the demagnetization curve, and, in addition, the very high percentage of iron leads to ageing, i.e. a gradual diminution in the strength of the magnet. With permanent magnets used in electrical instrument manufacture it is obvious that ageing should be eliminated, and it has been found that the best way to achieve this is to reduce the iron content as far as possible. The evolution of magnet steels is briefly as follows, (*a*) alloying with tungsten; (*b*) alloying with chromium; (*c*) addition of cobalt to the chrome alloys, in one case up to as much as 36 per cent cobalt; (*d*) dispersion-hardened materials containing no carbon and in which the hardness is caused by the presence of very small particles of one kind dispersed in a matrix having a different composition from that of the other particles.

The properties of some permanent magnetic materials are given in Table 4.2.†

The demagnetization curves for a number of these materials are given in Fig. 4.22. It is interesting to compare their retentivity and coercivity with those of a low-loss material. Thus for Alnico *V* the values are 12,500 gauss or 1·25 webers/m^2, and 550 oersteds or 43,780 AT/m;

* From Bell Telephone System, Monograph *B*-1445. *See* also F. Brailsford, *Magnetic Materials*, Methuen Monographs, 1951, and L. F. Bates, *Modern Magnetism*. Cambridge University Press, London, 1951.

† From Bell Telephone System Monograph B-1445.

Table 4.1. Some Properties of High-permeability Materials

Name	Composition %	Heat Treatment*	Initial Relative Perme-ability μ	Maximum Relative Perme-ability μ_m	Coercive Force H_c (amp-turns/metre)	Saturation Induction B_s (webers/metre²)	Electrical Resistivity ρ ($\mu\Omega$ – m)	Curie Point θ (°C)	Density δ (g/cm³)	Uses
							× 10⁻²			
Iron	99·9Fe	900°C	200	5,000	80	2·15	10	770	7·87	Relays
Silicon-iron	4Si, 96Fe	800°C	400†	7,000	48	1·97	59	690	7·60	Power Transformers
Silicon-iron	3·3Si, 97Fe	Cold-rolled 800°C	600†	10,000	16	2·00	50	700	7·65	Power Transformers
Hypersil‡	3·3Si, 97Fe	Cold-rolled 1200°C, H_2	1,500†	40,000	8	2·00	48	700	7·65	Power Transformers
Sendust	9Si, 5Al, 85Fe	Cast	30,000	120,000	4	1·00	80	500	7·1	
45 Permalloy	45Ni, 54Fe	1050°C	2,500	25,000	24	1·60	50	440	8·17	Transformers, relay
Hypernick	50Ni, 50Fe	1200°C, H_2	4,000	80,000	4	1·60	35	500	8·25	Transformers
Radiometal	47Ni, 3Cu, 49Fe	1050°C	2,500	25,000	24	1·56	55	—	8·3	Transformers
78 Permalloy	78Ni, 21Fe	1050°C 600°C	8,000	100,000	4	1·00	16	580	8·60	Sensitive relays
4-79 Permalloy	79Ni, 4Mo, 16Fe	1100°C	20,000	80,000	4	0·87	57	420	8·72	Transformers
Mumetal	75Ni, 2Cr, 5Cu 18Fe	1100°C, H_2	20,000	110,000	2·4	0·72	60	430	8·58	Transformers
1040 Alloy	72Ni, 14Cu, 3Mo	1100°C, H_2	40,000	100,000	1·6	0·60	56	290	8·76	
Supermalloy	79Ni, 5Mo, 15Fe	1300°C, H_2	100,000	800,000	0·32	0·80	60	400	8·87	Transformers
Permendur	50Co 50Fe	800°C	800	5,000	160	2·45	7	980	8·3	Electromagnets
V-Permendur	49Co, 2V, 49Fe	800°C	800	4,500	160	2·40	26	980	8·2	Telephone dia-phragms
Powdered Per-malloy	81Ni, 2Mo, 17Fe	Pressed 650°C	125	130	—	—	10^6	480	7·8	Loading coils

* "H" indicates heat treatment in pure hydrogen. † Measured at $B = 20$ rather than $B = 0$. ‡ Also called Corosil.

TABLE 4.2. SOME PROPERTIES OF PERMANENT-MAGNET MATERIALS

Name	When First Used	Typical Composition* (%)	Coercive Force H_c (amp.-turns/metre)	Retentivity B_r (webers/m²)	Energy Product (joules/m³)	Fabrication	Heat Treatment (°C)
Tungsten steel	1855	6W, 0·7C, 0·3Mn	5,200	1·05	2,400	Forged	Q850°C
Low-chrome steel	1916	0·9Cr, 0·6C, 0·4Mn	4,000	1·00	1,600	Forged	Q800
High-chrome steel	1916	3·5Cr, 1C, 0·4Mn	5,200	0·95	2,400	Forged	Q825
KS magnet steel†	1917	36Co, 7W, 3·5Cr, 0·9C	16,400	1·00	7,200	Forged	Q950
Cobalt chrome steel‡	1921	16Co, 9Cr, 1C, 0·3Mn	16,400	0·80	4,800	Forged	Q975
Remalloy§	1931	12Co, 17Mo (or W)	20,000	1·05	8,800	Forged	Q1200, B675
Mishima alloys‖	1931	25Ni, 12Al	40,000	0·70	11,200	Cast	Q1200, B650
Oxide magnet	1933	15Co, 28O	72,000	0·20	4,800	Sintered	A1000‡‡
Alnico 2¶	1934	12Co, 17Ni, 10Al, 6Cu	44,800	0·73	13,600	Cast	A1200, B650
Magniflex**	1935	20Ni, 60Cu	36,000	0·53	8,800	Cold-rolled	B600
New KS‖‖	1935	35Co, 18Ni, 6Al, 8Ti	72,000	0·57	14,400	Cast	B650
Platinum cobalt alloy	1936	77Pt, 23Co	212,000††	0·45	30,400	Forged	Q1200, B650
Vicalloy	1938	52Co, 11V	16,000	1·15	12,000	Cold-rolled§§	B600
Alnico 5 (Alcomax)	1940	24Co, 14Ni, 8Al, 3Cu	44,000	1·25	36,000	Cast	A1300, B600‡‡

* Balance is iron.
† Called Koerzit in Germany.
‡ Also alloys containing 3, 6, 9Co with variable Cr.
§ Also called Comol.
‖ Also called Alnico 3; Alnico 1 and 4 contain Co.
¶ Called Alnico in England and Derstit in Germany.
** Also called Cunife 1. Cunife 2 and Cunico contain cobalt.
†† H for $B = 0$.
‡‡ Must be cooled in strong magnetic field.
§§ Cold-rolled in grooved rolls. Higher values in drawn wire, lower values in rolled sheet.
‖‖ Composition given is Alnico 12. Nipermag and Alnico 7 contain smaller amounts of titanium.

Q Quenched from centigrade temperature indicated, in oil or water.
A Cooled in air from temperature indicated.
B Aged or baked, cooling rate not important.

Forged alloys may also be hot-rolled or cast to size.
Cast alloys (unless forgeable) must be ground to size.
Cold-rolled alloys are usually hot-rolled before cold rolling.

the corresponding values for electrical sheet material with 1 per cent silicon are 7200 gauss or 7·2 webers/m² and 0·7 oersted or 55·7 AT/m respectively.

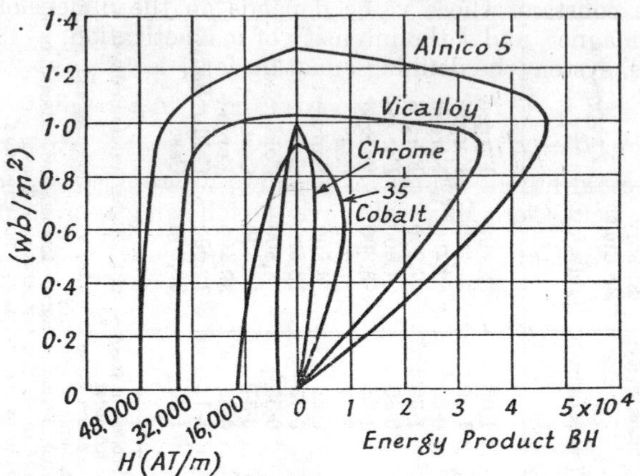

FIG. 4.22. DEMAGNETIZATION AND ENERGY PRODUCT CURVES FOR SEVERAL PERMANENT-MAGNET MATERIALS

Permanent Magnets. Imagine a magnetized bar *NS*, Fig. 4.23, with a very small cavity in the direction of the lines of induction. Then, since no lines will emerge from the iron into the cavity, the walls of this cavity will not exhibit any form of polarity. The flux through the cavity will therefore be that corresponding to the magnetizing force *H* set up by the current, and not to the total flux density *B*, and, consequently, if an isolated north pole *P** could be placed within the cavity,

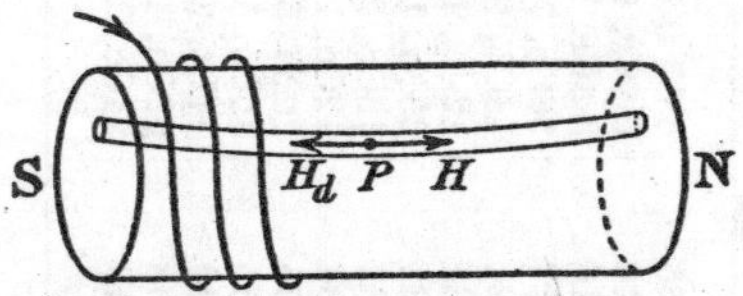

FIG. 4.23. DEMAGNETIZING FORCE

it would be acted on by a force of *H* dynes. But it is clear that the isolated pole would be repelled by the *N* pole of the magnetized bar and attracted by the *S* pole, these two together setting up a force H_d which is in direct opposition to *H*. In other words, the fact of the bar having polarized ends is responsible for the setting up of a self-demagnetizing effect.

* The concept of the magnetic pole, which forms the basis of the classical theory of magnetism, is no longer used to any extent. but it is nevertheless convenient in certain cases.

If H_{int} is the value of the resultant magnetizing force in the interior of the bar, and H is the value when the bar is absent then

$$H_{int} = H - H_d = H - KJ$$

where K is a constant whose value depends on the dimensions and shape of the magnet, and J the intensity of magnetization.

In the c.g.s. system the defining equation for J is

$$B = H + 4\pi J, \text{ with } B \text{ in gauss and } H \text{ in oersteds}$$

$$\therefore J = (B - H)/4\pi$$

For an ellipsoid having semi-axes a and c, where c is the long axis parallel to the field, Clerk Maxwell* gave the following expression for K

$$K = 4\pi(1/e^2 - 1)\{(1/2e) \log [(1 + e)/(1 - e)] - 1\}$$

where $e = [1 - (a/c)^2]^{1/2}$

In the case of a rod of length l and diameter d, we replace it by a slender ellipsoid having

$$e = [1 - (d/l)^2]^{1/2}$$

The relationship between l/d and K is then as follows—

l/d	K
10	0·251
50	0·01817
200	0·00157

This shows that with short magnets a very large proportion of the applied magnetizing force is required to overcome the demagnetizing effect.

Now let the suffix m denote magnitudes expressed in rationalized m.k.s. units. Then

$$\begin{aligned} J &= B_m \times 10^4/4\pi - (\mu_0 \times 10^7/4\pi) \\ &\quad \times (H_m \times 4\pi/1000) \times 1/4\pi \\ &= 10^4/4\pi(B_m - \mu_0 H_m) \\ &= (10^4/4\pi)J_m \end{aligned}$$

Again, from

$$H_{int} = H - KJ$$

$$(4\pi/1000)H_{m \cdot int} = (4\pi/1000)H_m - K \times (10^4/4\pi)J_m$$

$$\therefore H_{m \cdot int} = H_m - 10^7/(4\pi)^2 KJ_m$$

But $$H_{m \cdot int} = H_m - K_m J_m$$

$$\therefore K_m = [10^7/(4\pi)^2]K = 63{,}300K$$

We now have for the above ratios of l/d

l/d	K (m.k.s. unit)
10	15,900
50	1,150
200	100

* James Clerk Maxwell. *A Treatise on Electricity and Magnetism*, Vol. II, p. 65, Oxford University Press, Oxford, 1873.

The significance of this in connection with permanent magnets is as follows: the demagnetizing effect in a closed ring is zero, but, in order that the magnetic flux may be available, all permanent magnets must have a gap, thus requiring two polarized ends, and so giving rise to self-demagnetization. Consider the portion of the hysteresis loop shown in Fig. 4.24, and let the magnetization be taken to the point P. On removal of the magnetizing force the flux density will decrease to OQ in the case of a closed ring, but, in a practical form of permanent magnet, the self-demagnetizing effect will cause a further reduction to, say, R. If the magnet is jarred, there will be a further reduction in flux density, and, without any special ageing process, most of the magnetization may be ultimately lost.

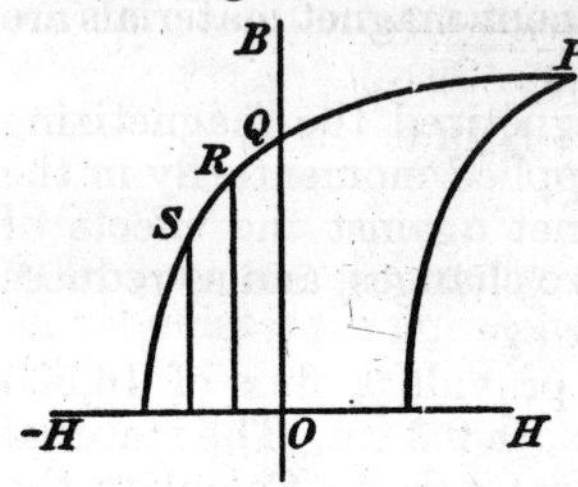

FIG. 4.24. DEMAGNETIZING FORCE

A better form of permanent magnet is thus a nearly closed ring, with the air-gap only large enough to enable the flux in the gap to be utilized.

Since modern permanent-magnet materials are very expensive, it is necessary that the volume used for any particular magnet should be the minimum possible. The criterion, which is known as Evershed's criterion,* is that the product BH shall be a maximum.

Consider a ring magnet having a gap in the form of a radial saw cut. Let the suffix a refer to the air-gap and the suffix s to the steel. Then since the magnetic paths are all closed paths, and there is no external magnetizing solenoid, the total m.m.f. must be zero,

$$\therefore \int H_a dl_a + \int H_s dl_s = 0$$

$$\therefore H_a l_a = -H_s l_s$$

$$\therefore l_s = (H_a/H_s)l_a \text{ numerically}$$

The flux through the air-gap is

$$\Phi_a = B_a a_a = \mu_0 H_a a_a$$

The flux through the steel is

$$\Phi_s = B_s a_s$$

Again $\Phi_s = \lambda \Phi_a$, where λ is the leakage factor

$$\therefore B_s a_s = \lambda \mu_0 H_a a_a$$

$$\therefore a_s = \lambda \mu_0 H_a a_a / B_s$$

$\therefore$ Volume of steel

$$= a_s l_s = (\lambda \mu_0 H_a a_a / B_s) \times (H_a/H_s) l_a$$

$$= \lambda \mu_0 a_a l_a H_a^2 / B_s H_s$$

* *J. Inst. Elect. Engrs.*, **58**, (1920) 780; **63** (1925) 725.

Now for a stable state the numerator is a constant, and therefore the volume of steel is a minimum when the denominator B_sH_s is a maximum. We have already seen, p. 103, that the product $BH/2$ gives the energy density in joules/cubic metre, and this product is therefore called the *energy product*. The demagnetization curves with the corresponding energy product curves for a number of permanent-magnet materials are given in Fig. 4.22.

When a permanent magnet has been magnetized the magnetizing field is removed and a relatively weak field applied momentarily in the opposite direction. This stabilizes the magnet against the effects of stray fields, mechanical shocks, or temperature changes, and so reduces the tendency for the gap field to reduce with age.

EXAMPLE. A permanent magnet has to provide a flux of 16,500 maxwells in an air-gap of section 3 cm² and length 0·3 cm. The material to be used is cobalt steel and the leakage factor is 4. Calculate the dimensions of the magnet.

$$\Phi_a = 1{\cdot}65 \times 10^4 \times 10^{-8} = 1{\cdot}65 \times 10^{-4} \text{ webers}$$

$$a_a = 3 \times 10^{-4} \text{ m}^2$$

$$\therefore B_a = 0{\cdot}55 \text{ weber/m}^2$$

$$l_a = 0{\cdot}3 \text{ cm} = 3 \times 10^{-3} \text{ m}$$

$\therefore$ Gap volume

$$v = 3 \times 10^{-4} \times 3 \times 10^{-3} = 9 \times 10^{-7} \text{ m}^3$$

$\therefore$ Gap energy

$$= v\, B_a^2/2\,\mu_0 = [0{\cdot}55^2/(2 \times 4\pi \times 10^{-7})] \times 9 \times 10^{-7}$$

$$= 0{\cdot}55^2 \times 9/8\pi = 0{\cdot}109 \text{ joules}$$

Since

$$\int H_a dl_a + \int H_s dl_s = 0$$

$$H_a l_a = H_s l_s \text{ with uniform cross-section}$$

$$\therefore \frac{\text{Energy in steel}}{\text{Energy in gap}} = \frac{\frac{1}{2}B_sH_sl_sa_s}{\frac{1}{2}B_aH_al_aa_a}$$

$$= B_sa_s/B_aa_a$$

$$= \Phi_s/\Phi_a$$

$$= 4$$

From Fig. 4.22, we see that for cobalt steel the energy product has a maximum value of $9{\cdot}4 \times 10^3$ m.k.s. units and therefore

$$\tfrac{1}{2} \times 9{\cdot}4 \times 10^3 \times \text{core volume} = 0436$$

$$\therefore \text{core volume} = (2 \times 0{\cdot}436/9{\cdot}4) \times 10^{-3}$$

$$= 0{\cdot}928 \times 10^{-4} \text{ m}^3$$

$$\text{Core flux} = 4 \times 1{\cdot}65 \times 10^{-4}$$

$$= 6{\cdot}6 \times 10^{-4} \text{ webers}$$

The induction corresponding to maximum BH product is 6100 gauss, or 0·61 weber/m^2,

$\therefore$ Core section $= 6{\cdot}6 \times 10^{-4} \div 0{\cdot}61 = 10{\cdot}8 \times 10^{-4}$ m^2, or 10·8 cm^2

$$\therefore \text{ core length} = (\text{core volume})/(\text{section})$$
$$= 0{\cdot}928 \times 10^{-4}/(10{\cdot}8 \times 10^{-4})$$
$$= 8{\cdot}6 \times 10^{-2} \text{ m, or } 8{\cdot}6 \text{ cm}$$

Theory of Magnetism. Reasoning from the magnetic effects of a current-carrying wire discovered by Oersted, Ampère suggested that the magnetic property of iron and similar materials might be due to circulations of current round each molecule. Hence, macroscopically, a mass of iron would be an assemblage of molecular magnets. About fifty years later Weber anticipated the modern theory of atomic structure by suggesting that the atoms of magnetic particles had associated with them a negative charge, and that, round this, a light particle carrying a positive charge rotated. Thus each molecule of iron was regarded as a permanent magnet. Furthermore, each molecular magnet was subject to a frictional force which tended to prevent any displacement from an established orientation. This explained the phenomenon of magnetic saturation and also, in a way, that of residual magnetism. Maxwell modified Weber's theory by discarding the idea of molecular friction, and introducing the conception of more than one fixed position of equilibrium. The modern theory of magnetism gives a clearer picture of the source of atomic magnetism, and accounts for the properties of magnetic materials by a grouping of molecules into so-called *domains*.

According to the modern theory the atom of an element consists of a nucleus associated with a number of negative charges (electrons). The nucleus contains positive charges (protons), and the total positive content exactly balances the negative content of the electrons in the case of a normal atom. The nucleus has other constituents in addition to the protons, but as these do not take part in any of the physical changes which normally interest the electrical engineer, they need not be considered. The properties which are the concern of the electrical engineer can be explained by means of the Rutherford-Bohr atom model. It is assumed that the nucleus is a collection of particles and that it has a positive charge, while the electrons are particles of negative electricity which revolve in planetary manner round the nucleus. This is about as far as the astronomical analogy goes, since the planetary electrons themselves are grouped in a definite manner. These groups are called *shells* and are designated by the letters K, L, M, N, etc. There are limits to the number of electrons per shell, as follows

K shell, 2 electrons
L shell, 8 electrons
M shell, 18 electrons
N shell, 32 electrons

for the first four shell's. A shell can contain the maximum number or any smaller number of electrons. Remembering that, for each added electron, there must also be a proton and also other additions to the nucleus, the atomic arrays of the first few elements are as follows.

	K shell	L shell
H	1 electron	—
He	2 electrons	—
Li	2 electrons	1 electron
Be	2 electrons	2 electrons
B	2 electrons	3 electrons

and so on. When the L shell has its full compliment of 8 electrons we have

Ne K shell, 2 electrons, L shell, 8 electrons

To add a new electron it is necessary to start a new shell. We then have the sodium atom,

Na K shell, 2 electrons, L shell, 8, M shell, 1

If we proceed through the periodic table in this manner we shall find that in the case of all *metallic* atoms the highest shell has only one or two electrons. Thus with the alkaline metals we have

Li	one electron in the L shell
Na	one electron in the M shell
K	one electron in the N shell

With some other metals we have

Al	one electron in the M shell
Fe	two electrons in the N shell
Cu	one electron in the N shell

These lone electrons, the valency electrons, play a very important part in deciding the physical and chemical properties of the elements.

The shells themselves can be divided into sub-shells and, from the electron array for the iron atom (Fig. 4.25), we see that there are

		electrons
K	shell . . .	2
L	sub-shell 1 . . .	2
	sub-shell 2 . . .	6
M	sub-shell 1 . . .	2
	sub-shell 2 . . .	6
	sub-shell 3 . . .	6
N	shell . . .	2

Thus there are 26 electrons altogether, and therefore there are 26 protons in the nucleus.

Now, when an electron travels round a closed path of any shape it will constitute an electric current of that shape, since the electric current is merely a translation of electric charge. Now a current round a closed path, e.g. a circle, sets up a magnetic force along the axis, and consequently all the electrons will set up their individual magnetic forces. When atoms combine to form molecules and the molecules aggregate to form a finite mass of the material the effects of these orbital motions may neutralize one another, and may thus make no contribution to the properties of the material.

In addition to the orbital translations of the electrons as a whole there are, however, rotations, or spins of the electrons about their own axes. An electron, regarded as a particle, is of finite, although extremely

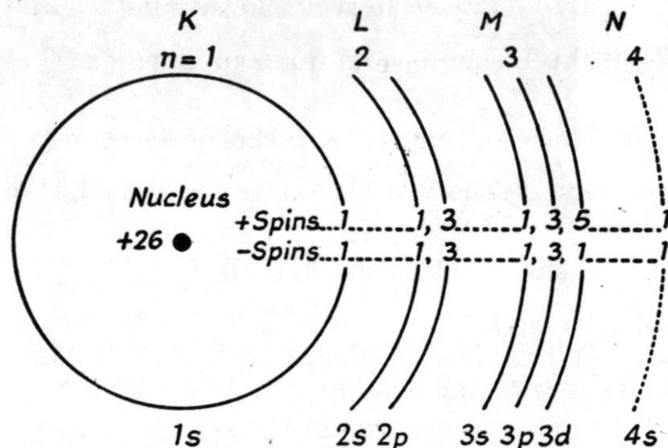

FIG. 4.25. ELECTRON ARRAY FOR IRON ATOM
(*Bozorth*)

small, size and consequently any element of the electronic charge at a finite distance from the axis will be taken round a circular path and will therefore set up a magnetic force directed along the axis of rotation. Thus each spinning electron will set up a force directed along its axis, and the direction of the force will depend on the direction of spin.* In any atom some of the electrons spin in one direction and some in the other, and the effects of all the spins may thus balance out. If they do not balance, then there will be a resultant magnetic force and the atom will correspond to one of Ampère's elementary magnets. This is the case with iron, as can be seen from Fig. 4.25. In the K, L, and N shells the positive and negative spins balance out; in the M shell there are nine electrons with positive spin but only five with negative spin.

The properties of a mass of iron, as distinct from those of a single atom, are due to the assemblage of molecules into groups called *domains*. Within the domain all the elementary magnets have a parallel alignment, and when the state of magnetization of the material is changed all the elementary magnets move together. The difference between the unmagnetized and the magnetically saturated states is thus represented in a very elementary manner, in Fig. 4.26. Actually magnetization results not only from changes in the orientation of the

* There is an experimental basis for the conclusion that ferromagnetism is the result of unbalanced electron spins, and not of orbital motions. *See* S. J. Barnett, *Reviews of Modern Physics*, **7**, (1935) 129.

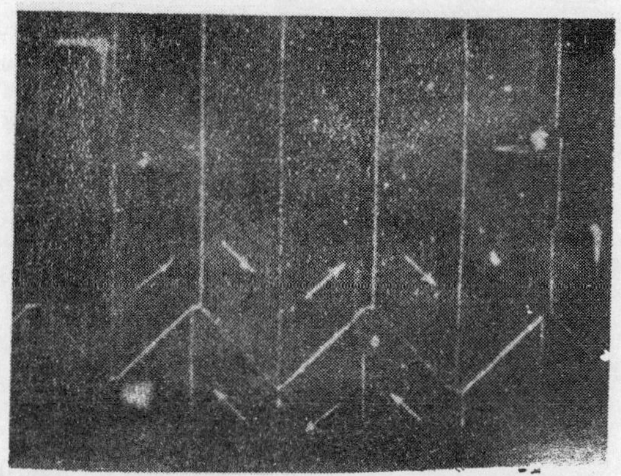

PLATE I. BITTER PATTERNS

(*Above*) Bitter pattern of a portion of a single crystal of pure iron in the demagnetized state.

(*Bates*)

(*Below*) Simple domain pattern in a single crystal of silicon iron, the direction of magnetization being as indicated.

(*Bozorth*)

domains, but also from increases in size of some of the domains, at the expense of their neighbours. This is called *boundary displacement.* We have seen that the process of magnetization takes place in three stages, namely, up to the instep, from instep to knee, and finally beyond the knee. In each portion one type of process predominates

(1) up to the instep, reversible boundary displacement;
(2) instep to knee, irreversible boundary displacement;
(3) beyond the knee, reversible domain rotation.

The evidence for the existence of domains is based largely, but not

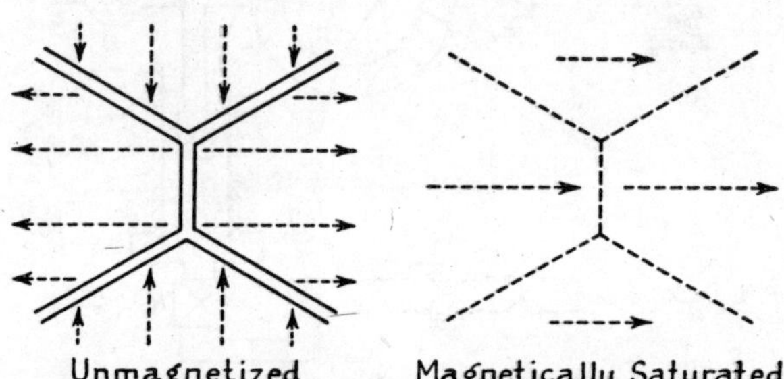

FIG. 4.26. TO ILLUSTRATE THE DOMAIN THEORY

entirely, on the Bitter powder patterns.* A suspension of Fe_2O_3, having particles of the order of 10^{-4} cm diameter, is applied to a polished, magnetized surface which is then examined under the microscope. It is, in a sense, a refinement of the well-known iron filing technique, the particles being so exceedingly small that they can indicate by their change in position very minute local inhomogenieties in the state of magnetization. By using a colloidal suspension of particles small enough to show Brownian movement, Elmore showed that the movement of the domain boundaries during magnetization could be made visible to the eye. A typical Bitter pattern is shown in Plate I†: it is a photograph of the colloidal magnetite on the central portion of a surface of a single crystal of pure iron in the demagnetized state. The figure shows the usual fir-tree patterns found on such a surface in the demagnetized state, together with some closure domains in the neighbourhood of imperfections or inclusions on the surface. The sharp boundary on the left of the photograph is the boundary between two main domains.

Determination of Magnetization Curves. Probably the most important method of determining a *B-H* curve is the Ballistic method, so called because a ballistic galvanometer is used. The classical form of specimen is that of a ring and the radial thickness is preferably kept

* Bitter, F., *Phys. Rev.* **38** (1931) 1903; **41** (1932) 507, and *Mech. Eng.* **55** (1933) 287.

† Supplied by Professor L. F. Bates, University of Nottingham.

small compared with the diameter, so that H will not vary appreciably over the cross section.* The ring has two windings, one, the magnetizing winding, distributed over the whole circumference as evenly as possible; the other a secondary winding having 30 or 40 turns. The magnetizing coil is connected to a direct-current supply through an adjustable resistance R (Fig. 4.27). An ammeter, A_1, measures the current, and a reversing switch, K_1, is included, so that the magnetizing current can

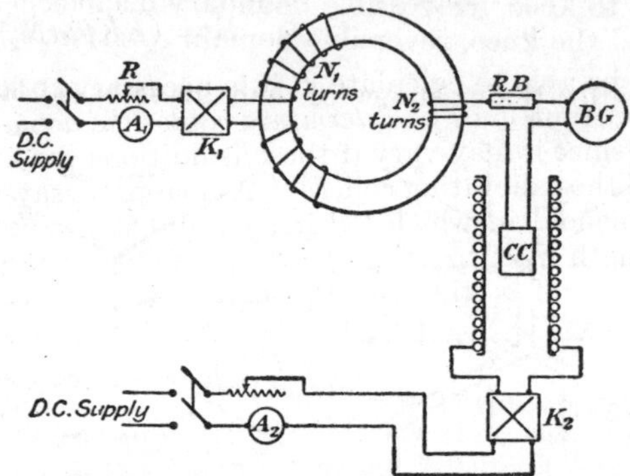

FIG. 4.27. BALLISTIC METHOD OF DETERMINING MAGNETIZATION CURVES

be reversed. The secondary is connected to the ballistic galvanometer, and a resistance box, RB, and a calibrating coil, CC, are included in this circuit. The coil CC is placed in the middle of a large solenoid whose equipment is similar to that of the magnetizing coil for the specimen.

Let N_1 = number of turns on magnetizing coil
l = mean circumference of specimen in metres
a = cross section in square metres

Then for any magnetizing current I, the magnetizing force

$$H = N_1 I/l$$

If, for any value of I, the switch K_1 is reversed, the flux density B inside the ring is reversed, and it therefore changes from $+B$ to $-B$, a change of $2B$. Hence, the change of flux in the ring is $2Ba$, and if this change takes place in a small interval of time δt, then

$$\text{Rate of change of flux} = 2Ba/\delta t$$

Hence, if the secondary coil has N_2 turns,

$$\text{e.m.f. induced in the secondary, } e = N_2(2Ba/\delta t)$$

If the total resistance of the galvanometer circuit is R, the average current through the galvanometer will therefore be

$$2BaN_2/R\delta t$$

Hence, quantity of electricity = Average current $\times\ \delta t$
$= 2BaN_2/R$

* *See* the theory of the toroid, p. 97.

Now the "throw" of a Ballistic galvanometer, when corrected for logarithmic decrement, is proportional to the quantity of electricity which passes through it, provided that this quantity has passed through in a small interval of time, a condition which holds in this case. Hence, if θ is the corrected throw and K the galvanometer constant,

$$\text{Quantity of electricity} = K \times \theta$$

$$\therefore 2BaN_2/R = K\theta$$

$$B = (RK/2aN_2)\theta$$

In order to be able to calculate B it is necessary to know the constant K. This constant *must be determined under the actual conditions of the experiment*, since it may vary if these conditions vary. It is determined by means of the calibrating coil, CC. A current, I_2 say, is passed through the large solenoid in which CC hangs, and it produces through CC a field of strength N_3I_2/l_2, where N_3 and l_2 refer to the solenoid. If CC has a cross section of a_2 the flux through it will be $\mu_0N_3I_2a_2/l_2$, and on reversing the switch K_2 there will be a change of flux through CC of

$$2\mu_0N_3I_2a_2/l_2$$

Hence, if CC has N_4 turns, the quantity of electricity which will pass through the galvanometer on reversing K_2 will, by the same reasoning as before, be

$$2\mu_0N_3N_4I_2a_2/l_2R$$

If the corrected galvanometer throw when K_2 is reversed is ϕ, then

$$2\mu_0N_3N_4I_2a_2/l_2R = K\phi$$

$$K = 2\mu_0N_3N_4I_2a_2/l_2R\phi$$

Substituting this in the expression for B, we have

$$B = (\mu_0N_3N_4I_2a_2/N_2l_2a\phi)\theta$$

All the terms in the expression in the brackets are known, and therefore B can be calculated from observed values of θ. If it is not required to determine a hysteresis loop, the magnetizing current I is given a series of gradually increasing values, and the throw observed on reversing K_1. It will, of course, be necesary to make a preliminary test in order to be sure that the throw is of suitable magnitude when K_1 is reversed with I at its highest value. Before taking observations it will then be necessary to demagnetize the iron, this being most easily accomplished by applying a gradually decreasing alternating potential to the magnetizing coil.

If it is required to take the iron through a complete loop, then the procedure is somewhat different from the above. The current is suddenly increased in small steps and the galvanometer throw noted at each increase. Obviously for any given current the throw from which the corresponding flux density is calculated is the *sum* of all the previous throws, since the throw is only proportional to the change in flux and not to the total flux. When the maximum current is reached, the current is reduced in small steps and at last brought to zero. Then K_1 is reversed and the current increased in steps again until the same maximum value has been attained. This procedure is followed until the complete loop has been traced out. Reversed galvanometer throws

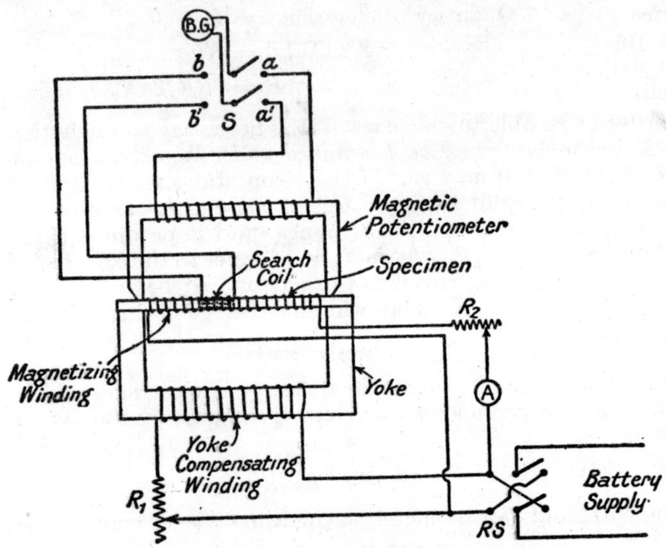

FIG. 4.28. THE ILLIOVICI PERMEAMETER

will obviously be experienced when B is being reduced. These are reckoned negative, and the total throw from which B is calculated at any step is the algebraic sum of all the throws up to that point.

Magnetic Potentiometer Method. The great disadvantage of a ring specimen from the practical point of view is that the specimen itself is expensive, whether solid or whether made up of laminations, and, in addition, a separate magnetizing winding, put on by hand, is required for each one.

Practical methods involving samples, as distinct from materials in bulk, therefore require a more convenient shape, as, for example, a short rectangular bar.* In the Illiovici Permeameter the specimen is used to close the magnetic circuit of a massive yoke, as shown in Fig. 4.28. Both specimen and yoke carry separate windings, the function of the yoke winding being to provide the ampere-turns necessary for the yoke and the two gaps. The specimen is, of course, clamped down to the

* Illiovici, A. "A new universal permeameter," *Bull. Soc. Int. Elec.* **3.** (1913), 581. *See* also: Spooner, T. *Properties and Testing of Magnetic Materials.* McGraw Hill, New York. Wall, T. F. *Applied Magnetism.* Longmans, London. Golding, E. W. *Electrical Measurements and Measuring Instruments.* Pitman, London.

yoke, but even then the gap ampere-turns may not be negligible, and furthermore will be different for every test. Compensation being thus provided for yoke and gaps, it is known that ampere-turns of the winding on the specimen will be required by the specimen only. The interest of the method lies in the method of ensuring that the yoke winding carries just sufficient current to compensate for the yoke and gaps. The appliance used is the magnetic potentiometer, which employs a thin strip or rod of flexible insulating material wound uniformly with a helix of thin wire. This coil is connected to a ballistic galvanometer. An e.m.f. will be induced in this coil either by a change in the magnetizing current producing the magnetic field in which the coil is situated, or by a rapid movement of one end of the helix from one point in the field to another. It can be shown that the galvanometer throw is proportional to the change in magnetic potential, this potential being given by $\int H dl$.*

In the above test the magnetic potentiometer is applied to the ends of the specimen, as shown in the figure, and is kept in that position. In making the test the current in the ammeter A is adjusted to give the desired value of H and the throw-over switch S placed on contacts aa'. The current in the yoke compensating coil is now adjusted by means of R_1, until, on reversing both currents by means of RS, no galvanometer deflection is noted. When this adjustment is made, the magnetic potential between the two ends of the search coil will be zero, and therefore the m.m.f. of the coil on the specimen will be required to overcome the reluctance of the specimen only. The value of H in the specimen is now known to be

$$H = NI/l \text{ ampere-turns/metre}$$

The switch S is now changed over to contacts bb', thus connecting the search coil to the galvanometer, and RS reversed. The galvanometer throw is noted, and the value of B calculated as for the previously described ballistic test.

The Magnetic Potentiometer. A method of testing employing the magnetic potentiometer is described on p. 128. This device consists simply of a uniform helix of thin wire wound in a special way on a flexible strip or rod of insulating material. The method of winding is as follows† "The winding, No. 40 S.W.G., d.s.c. copper wire was wound from the centre towards one end. It was lightly coated with shellac and then wound back towards the other end to an equal distance from the centre. It was again lightly coated with shellac and wound back to the centre again, so that the two ends of the windings emerged from the former. . . . The feet of the former were filed flat until the material was almost down to the wire; the final removal of the material was done by careful rubbing on sand-paper until bare

* *See* p. 130.

† Bates, L. F., "The Magnetic Potentiometer Study of Permanent Magnets." *Phil. Mag.*, Ser. 7. **34** (1945) 297.

wire was just visible." This description makes it quite clear that it is not merely a simple coil of fine wire wound on a flexible former, as is sometimes implied.

The manipulation of the potentiometer is very simple: the ends of the winding are connected to a ballistic galvanometer and the former is bent so that the two ends can be applied to the points between which the magnetic potential difference ($\int H dl$) is required. The potentiometer is then suddenly removed, and the reading noted. The potentiometer is calibrated during a set of measurements either in a standard solenoid or in the uniform field of an electromagnet with plain pole-pieces. The former method is the more reliable.*

The theory of the magnetic potentiometer is as follows—

We have seen that the quantity of electricity induced by the reversal of a flux Ba in a search coil is

$$Q = 2BaN/R$$

Hence, if the flux is reduced to zero, *not reversed*, as in the case of the magnetic potentiometer,

$$Q = BaN/R$$

This, of course, applies to a search coil in which the flux Ba is uniform throughout the whole length. Now regard the magnetic potentiometer as a special kind of search coil having N' turns per unit length. Then for any element of length dl, we have

$$dq = BaN'dl/R$$

$$\therefore\ Q = \int dq = \frac{N'a}{R}\int B dl = \frac{\mu_0 N'a}{R}\int H\, dl$$

since the search coil is on a core for which $\mu = 1$

This is proportional to the magnetic potential between the points of application.

Magnetic Moment. In the c.g.s. system a pole of strength m placed in a field of strength H will be acted on by a force mH. Consequently a bar magnet placed transversely across the lines of force, as in Fig. 4.29 (a) will be acted on by a turning moment of

$$T = mlH \text{ dyne-cm}$$

The product ml is called the magnetic moment.

In the rationalized m.k.s. system a flux of Φ webers leaving the end of a magnet corresponds to a pole strength Φ.† A flux density, in air,

* Bates, L. F., *Modern Magnetism*, p. 63. Cambridge U.P., London.

† In the rationalized m.k.s. system the satisfactory treatment of the magnetic pole presents considerable difficulty, particularly when the classical conception is entirely abandoned, and the science of electromagnetics developed *de novo*. This difficulty is frankly stated by McGreevey in *The M.K.S. System of Units*, Chap. VI, Pitman, London, 1954. In problems such as this, the retention of the magnetic pole conception gives a very simple solution.

of B weber/metre^2 corresponds to a field strength of $H = B/\mu_0$ ampere turns/metre, Fig. 4.29 (*b*).

$$\therefore\ T = \Phi l B/\mu_0 \text{ newton-metres}$$

The magnetic moment is now Φl and it is expressed in weber-metres.

EXAMPLE. A bar magnet of magnetic length 10 cm. is made from round bar 1 cm. dia. The intensity of magnetization at the ends is

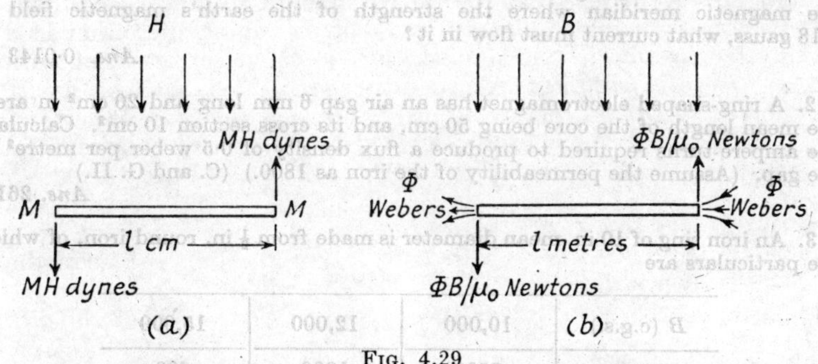

FIG. 4.29

1·5 wb/m^2. If the magnet is suspended at right-angles to the lines of force of a magnetic field of density 10^{-3} wb/m^2, calculate (1) pole strength, (2) magnetic moment (3) torque.

$$a = (\pi/4)d^2 = (\pi/4) \times (10^{-2})^2 = 7{\cdot}854 \times 10^{-5} \text{ m}^2$$

$$\therefore\ \Phi = Ja = 1{\cdot}5 \times 7{\cdot}854 \times 10^{-5} = 11{\cdot}78 \times 10^{-5} \text{ wb}$$

$$\text{magnetic moment} = \Phi l = 11{\cdot}78 \times 10^{-5} \times 10^{-1} = 11{\cdot}78 \times 10^{-6} \text{ wb-m}$$

$$T = \Phi l B/\mu_0 = \Phi l \times B/\mu_0$$
$$= (11{\cdot}78 \times 10^{-6} \times 10^{-3}) \div (4\pi \times 10^{-7})$$
$$= 9{\cdot}36 \times 10^{-3} \text{ newton-metres}$$

Suppose that the bar magnet is a linear electromagnet, energized by a magnetizing coil of N turns carrying a current of I amperes. Then

$$\Phi = \text{MMF}/S$$
$$= NI/(l/a\mu\mu_0)$$
$$= NIa\mu\mu_0/l \text{ webers}$$
$$\therefore\ T = \Phi l H$$
$$= (NIa\mu\mu_0/l) \times l \times (B/\mu_0)$$
$$= BNIa\mu \text{ newton-metres}$$

If the solenoid is wound on a core of relative permeability unity this reduces to

$$T = BNIa \text{ newton-metres}$$

FURTHER READING

In addition to those mentioned at the end of Chapter 2, the following books may, with advantage, be consulted.

M.I.T. Electrical Engineering Staff. *Magnetic Circuits and Transformers.* Wiley, New York, 1944.
Spooner, T. *Properties and Testing of Magnetic Materials.* McGraw Hill, New York, 1927.
Bates, L. F. *Modern Magnetism.* Cambridge University Press, London, 1951.
Bozorth, R. M. *Ferromagnetism.* D. van Nostrand, 1951.

Examples on Chapter 4

1. A solenoid 1 m long and of small diameter has 1,000 turns. If it is placed in the magnetic meridian where the strength of the earth's magnetic field is 0·18 gauss, what current must flow in it?

Ans. 0·0143 A.

2. A ring-shaped electromagnet has an air gap 6 mm long and 20 cm² in area, the mean length of the core being 50 cm, and its cross section 10 cm². Calculate the ampere-turns required to produce a flux density of 0·5 weber per metre² in the gap. (Assume the permeability of the iron as 1800.) (C. and G. II.)

Ans. 2610.

3. An iron ring of 10 in. mean diameter is made from ½ in. round iron, of which the particulars are

B (c.g.s.)	10,000	12,000	15,000
μ	2500	1800	600

If it has a saw-cut $\frac{1}{16}$ in. wide and its leakage factor is 1·25, calculate how many ampere-turns are required to produce a flux density in the gap of 1·1 weber per metre²?

Ans. 2195.

4. A circular lifting magnet has a diameter of 50 in., the inner circular pole is 16 in. in diameter, and the outer annular pole has the same area as the inner one. The mean length of the magnetic path through the magnet is 60 in., and the relative permeability of the steel can be taken as 1000. If it is lifting steel plates of negligible reluctance, and the plates are separated ¼ in. from the magnet poles, find the ampere-turns necessary to produce a flux density in the poles of 1 weber/m².

Ans. 11,350.

5. Calculate the force of attraction between the magnet and plates in the above example.

Ans. $1{\cdot}03 \times 10^5$ newtons.

6. Find the force of attraction in pounds weight between the square-faced ends of two rods of iron placed in a solenoid. The area of the iron rods is 6 cm², the relative permeability of the iron is 1000, and the magnetic force produced in the iron by the solenoid is 14 c.g.s. units. (L.U.)

Ans. 105.

7. A smooth core armature, working in a four-pole field magnet, has a gap (from iron to iron) of 0·5 in. The area of surface of each pole is 1 ft². The flux from each pole is 7×10^{-2} webers. Find (*a*) the mechanical force with which the pole attracts the armature; (*b*) the amount of energy expressed in joules that is stored in the four gaps.

Ans. (*a*) $2{\cdot}08 \times 10^4$ newtons; (*b*) 1065 joules.

8. The armature core of a dynamo is a cylinder of length 12 in. and internal and external diameters 7 and 12 in. respectively. It rotates at 1000 r.p.m. in a

four-pole field, and from each pole a flux of $1 \cdot 55 \times 10^{-2}$ webers emerges. If the c.g.s. value of the Steinmetz coefficient for the sheet iron used in the core is 0·0025, and 80 per cent of the core is iron, find the hysteresis loss in watts.

Ans. 81.

9. The ascending and descending values of B and H for a half-cycle are as follows—

Ascending—

H	151	159	239	318	358	AT/m
B	0	0·2	0·58	0·7	0·73	Wb/m²

Descending—

H	199	79·6	0	− 79·6	− 151	AT/m
B	0·7	0·61	0·53	0·38	0	Wb/m²

Plot the curve and calculate the energy dissipated in hysteresis in joules per cycle. Calculate also the hysteresis coefficient.

Ans. (*a*) $4 \cdot 5 \times 10^{-4}$; (*b*) 729 joules/m³.

10. Define the terms "Magneto-motive Force," "Magnetic Flux," and "Magnetic Reluctance," and prove the relation which holds between these quantities for a magnetic circuit.

Estimate the number of ampere-turns necessary to produce a flux of 1×10^{-3} webers round an iron ring of 6 cm² cross section and 20 cm mean diameter, having an air gap 2 mm wide across it. The relative permeability of the iron may be taken to be 1200. Neglect the leakage flux outside the 2 mm air gap. (L.U.)

Ans. 3370.

11. Define hysteresis. Prove that when any material is subjected to cyclic changes of magnetism, a loss of energy is involved proportional to the area of the hysteresis loop. Calculate the loss of energy caused by hysteresis in 1 hr in 50 kg of iron when subjected to cyclic magnetic changes. The frequency is 25, the area of the hysteresis loop represents 2400 ergs/cm³, the density of the iron is 7·8 g/cm³. (L.U.)

Ans. 138,000 joules.

12. Calculate the strength of the magnetic field at the centre of a single-turn square core of 1 ft side when carrying a current of 20 A. Prove the formula employed. (L.U.)

Ans. $7 \cdot 41 \times 10^{-5}$ wb/m².

13. A long straight conductor is of rectangular section in which the width $2b$ is large compared with the thickness. Show that the m.m.f. gradient at a point distant d from the centre and situated in the plane of the width is given by

$$H = (1/4\pi b) \cdot \log_e[(d + b)/(d - b)] \text{ ampere-turns/metre}$$

14. Two parallel conductors are spaced 0·3 m apart, and carry currents of 100 A in opposite directions. Calculate the magnetic flux density at a point 0·15 m from one and 0·25 m from the other.

Ans. $1 \cdot 6 \times 10^{-4}$ wb/m².

15. A single turn of insulated wire, 30 cm mean dia. rests on a similar turn, the distance between the centres being 5 mm. A current of 30 A flows through one, and a current in the opposite direction of 50 A flows through the other. Find the direction and magnitude of the resultant force. (L.U.)

Ans. 0·0565 newton.

16. A solenoid is 2 ft long and 1 in. diameter and is uniformly wound with 600 turns of insulated wire. Calculate the strength of the magnetic field at the centre when the current is 2 A. If a secondary coil of 50 turns is wound round the central part of the solenoid and is connected to a ballistic galvanometer through a resistance which makes the total resistance of the circuit 10,000 Ω, calculate the quantity of electricity discharged through the galvanometer on reversing the current of 2 A in the primary winding. Estimate the error due to assuming the solenoid infinitely long. (L.U.)

Ans. $2{\cdot}47 \times 10^{-3}$ wb/m², $1{\cdot}25 \times 10^{-8}$ coulomb, 0·09%.

17. An iron ring of 3·5 cm² cross-sectional area and mean length 100 cm is wound with a magnetizing winding of 100 turns. A secondary coil with 200 turns of wire is connected to a ballistic galvanometer having a constant of 1 micro-coulomb per scale division, the total resistance of the secondary circuit being 2000 Ω. On reversing a current of 10 A in the magnetizing coil the galvanometer gave a throw of 100 scale divisions. Calculate the flux density in the specimen and the value of the (relative) permeability at this flux density. (C. and G.)

Ans. 1·43 wb/m², 1136.

18. The demagnetization curve for a sample of permanent-magnet steel after hardening and ageing is as follows

Permanent flux density in wb/metre²	0·65	0·59	0·52	0·43	0·31	0·14
Demagnetizing ampere-turns per metre	4000	12000	20000	280000	36000	44000

The air-gap flux density in the moving-coil instrument where this steel is to be used is to be 0·09 wb/m², the length of the single gap is to be 0·12 cm, and the area of the gap 10 cm². To ensure the necessary permeance, the ratio of the area of the gap divided by its length, to the area of cross-section of the magnet divided by its length is to be 300. Assuming the leakage flux to be equal to the useful flux, and regarding all leakage as being concentrated at the pole shoes, calculate the necessary length and cross-sectional area of the magnet. (L.U.)

Ans. 0·115 m, $3{\cdot}2 \times 10^{-4}$ m².

CHAPTER 5

ELECTROMOTIVE FORCE

Production of an Electromotive Force. An electromotive force can be produced in the following ways: (*a*) by chemical action as in a voltaic cell; (*b*) by the heating of a thermo-junction; (*c*) by electro-magnetic action. The third method is by far the most important, and it has two subdivisions. First, an e.m.f. can be produced by the motion of a conductor in a magnetic field; such an e.m.f. is called a "dynamically-induced" e.m.f. Second, an e.m.f. is produced when the flux which threads, or links with a coil, changes. In this case there is no motion of the coil relative to the magnetic field, and therefore the e.m.f. so produced is "statically-induced."

Dynamically-induced E.M.F. In Fig. 5.1 three conductors, *A*, *B*, and *C*, are shown in cross section in a magnetic field, and the arrows attached

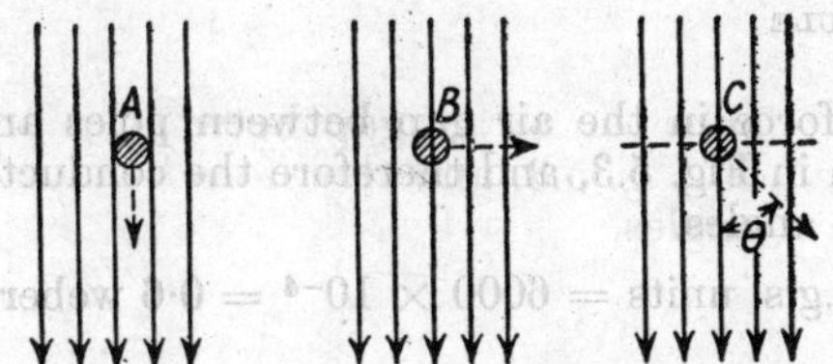

FIG. 5.1. DYNAMICALLY-INDUCED E.M.F.

to them indicate their directions of motion. Conductor *A* is moving in the direction of the lines of force and therefore, since it does not *cut* any of them, no e.m.f. is induced in it. Conductor *B* is moving in a direction perpendicular to its own length and perpendicular to the lines of force, and consequently, since it cuts the lines of force, it has an e.m.f. induced in it, of magnitude

$$E = Blv \text{ volts}$$

where B is the field strength in webers/square metre, l the length of conductor in metres, and v its velocity in metres/second. The direction of this induced e.m.f. is given by Fleming's Right-hand Rule (Fig. 5.2). Hold the thumb and first finger of the right hand at right angles and bend the second finger so as to point at right angles to the plane of these two. Then if the first finger is pointed in the direction of the field, and the thumb in the direction of motion, the second finger will point in the direction of the induced e.m.f.

Conductor C in Fig. 5.1 is moving at an angle θ to the direction of the field. In this case the magnitude of the induced e.m.f. is proportional to the component of the velocity perpendicular to the direction of the field. The modified form of the e.m.f. equation is, therefore,

$$E = Blv \sin \theta \text{ volts}$$

EXAMPLE. A conductor 12 in. long on the periphery of an armature of diameter 18 in. rotates at 1000 r.p.m. If the field strength under the poles is 6000 maxwell/cm², find the e.m.f. induced in the conductor.

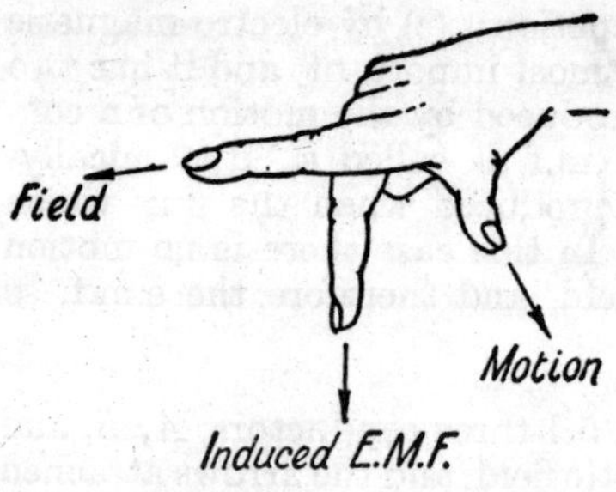

FIG. 5.2. FLEMING'S RIGHT-HAND RULE

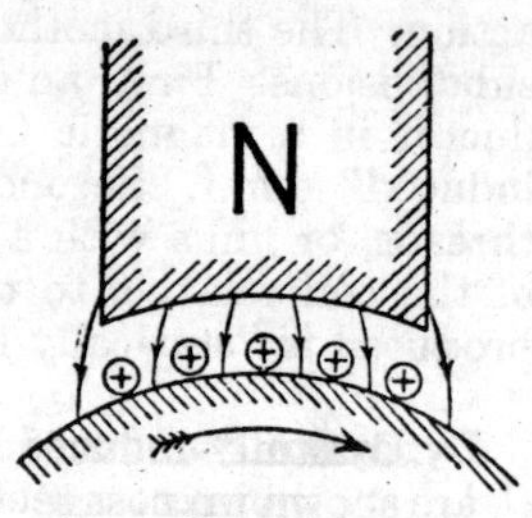

FIG. 5.3

The lines of force in the air gap between poles and armature are radial, as shown in Fig. 5.3, and therefore the conductor cuts the lines of force at right angles.

$$B = 6000 \text{ c.g.s. units} = 6000 \times 10^{-4} = 0{\cdot}6 \text{ weber/m}^2$$

$$l = 12 \text{ in.}$$

$$= 30{\cdot}5 \text{ cm} = 0{\cdot}305 \text{ m}$$

$$v = \pi \times \text{diameter} \times \text{revolutions/sec}$$

$$= \pi \times 18 \times 2{\cdot}54 \times (1000/60) = 2390 \text{ cm/sec} = 23{\cdot}9 \text{ m/sec}$$

$$\therefore E = 0{\cdot}6 \times 0{\cdot}305 \times 23{\cdot}9 = 4{\cdot}4 \text{ V}$$

Statically-induced E.M.F. A statically-induced e.m.f. may be (*a*) "mutually," (*b*) "self" induced. Consider first of all its production by the process of mutual induction. In Fig. 5.4 there are two coils, *A* and *B*, placed close together. *A* has a battery and switch connected to it, and *B* is connected to a galvanometer. If *A* carries either zero current or a finite steady current, there will be no deflection of the galvanometer, thus showing that there is no e.m.f. in the coil *B*. If, when the current in *A* is zero, the switch is suddenly closed, there will be a momentary deflection of the galvanometer, but not a permanent deflection so long as the switch is kept closed and the current in *A* is not varied. If now the switch is suddenly opened there will be another momentary deflection, this time in the reverse direction, and immediately afterwards the galvanometer needle will return to the zero position. These experiments show that an e.m.f. is induced in *B* whenever the current in *A* is *changing*, but not when it is steady. Also the direction of the e.m.f. induced by an increase in current is opposite to that induced by a decrease in current. These phenomena are explained by the following laws—

1. Whenever the number of lines of force linking with a circuit

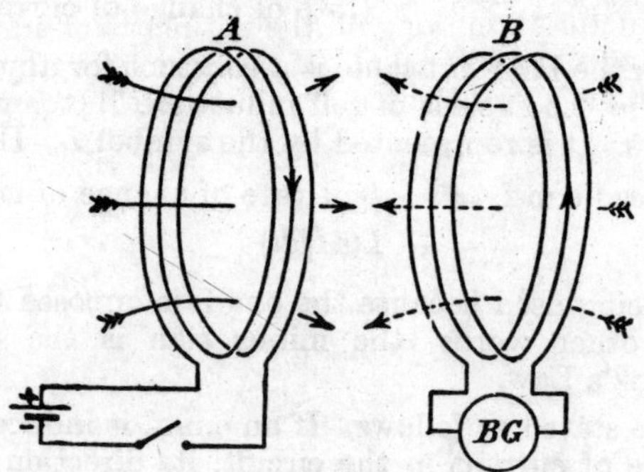

Fig. 5.4. Mutually-induced E.M.F.

changes, an e.m.f. is induced in the circuit proportional to the rate of change of flux.

This is Faraday's law of electromagnetic induction.

2. The direction of the induced e.m.f. is such that the current set up by it tends to stop the motion or change producing it.

This is known as Lenz's Law, and it follows from the fact that in order to set up an induced current some energy must be expended.

Self-induced E.M.F. In the case of a mutually-induced e.m.f., the e.m.f. is set up by a change in flux through a coil when the flux is produced by a neighbouring current. Obviously an e.m.f. will be produced whenever the flux changes, no matter how the flux may be produced. Thus if a single coil carries a current it will produce a magnetic flux, and if the current changes the flux will change. The change in flux will induce an e.m.f. in the coil, which in this case is called a self-induced e.m.f., because it is set up by a change in its own current instead of by a change in a neighbouring current. If the current increases, the self-induced e.m.f. will oppose the current; whereas if the current decreases, it will act in the same direction as the current.

Coefficients of Self- and Mutual-Induction. Consider a solenoid of N turns, length l metres, and cross-section a square metres. If it is carrying current, it will produce a flux; and if the current changes, the flux will change. We then have

$$\text{Self-induced e.m.f.} = -\,(\text{Rate of change of flux}) \times N \text{ volts}$$

Now $$\text{Flux} = (\text{Flux set up by one ampere}) \times I$$

the quantity in the brackets being a constant (if the permeability of the core is constant), which can be determined from the data of the circuit. Therefore we can write

(Rate of change of flux) = (Flux set up by one ampere)
× (rate of change of current)

∴ (Self-induced e.m.f.) = − [(Flux set up by one ampere) × N]
× (rate of change of current)

The expression in the large brackets is a constant for any given circuit and it is called the "coefficient of self-induction." It is given in practical units, henrys. It is represented by the symbol L. Hence we have

$$\text{Self-induced e.m.f.} = -L \times \text{rate of change of current}$$

or

$$= -L(dI/dt)$$

the minus sign being used because the e.m.f. is opposed to the change of current; in other words, the minus sign is the mathematical equivalent of Lenz's Law.

The law can be stated as follows. If an e.m.f. is induced in a circuit through a change of current in the circuit, its direction is such as to oppose the change of current. Thus, if the change is an increase in current, the e.m.f. will act in the opposite direction to the current; if the change is a decrease in current, the induced e.m.f. will act in the same direction as the current.

It is very easy to determine L for a simple circuit such as a solenoid. We have, for any current I,

$$\text{flux} = (\text{m.m.f.})/(\text{reluctance}) = NI\mu\mu_0 a/l$$

if the core is magnetic and of relative permeability μ. Hence, flux per ampere

$$= N\mu\mu_0 a/l$$

$$\therefore L = (\text{flux per ampere}) \times N$$

$$= N^2\mu\mu_0 a/l \text{ henrys}$$

The "coefficient of mutual induction" of a coil B relative to a coil A is obtained as follows—

Let A and B have N_1 and N_2 turns respectively,

(Mutually induced e.m.f. in B)
= [(Flux through B due to one ampere in A) × N_2]
× (rate of change of current in A)

The expression in the large brackets is the coefficient of mutual induction M, of B with respect to A. The flux produced by A, and therefore that portion of it which links with B, is proportional to N_1. We thus see that the coefficient of mutual induction of two coils is proportional to the product of their numbers of turns, whereas the coefficient of self-induction of a single coil is proportional to the square of its turns.

Rise and Decay of Current in an Inductive Circuit. If a coil possessing no self-induction, that is, a coil which does not produce a magnetic field when current is passed through it, has a p.d. of V volts applied to its terminals, the current produced reaches the Ohm's Law value of V/R instantaneously. If the coil possesses self-induction, the current theoretically takes infinite time to reach this value. With a non-inductive coil the applied p.d. has to overcome the ohmic drop IR only, but in the case of an inductive coil it has, in addition, to overcome the back e.m.f. set up by self-induction, this back e.m.f. only becoming zero when the current is steady. Now the self-induced e.m.f. is $-L\,di/dt$, where i is the instantaneous value of the current, and the applied p.d. therefore has to possess a component equal and opposite to this, as well as a component equal to the ohmic drop iR. Hence, when the current is increasing we have

$$V = Ri + L\,di/dt$$

If we multiply both sides by idt we have an energy equation

$$Vidt = Ri^2dt + Lidt(di/dt)$$

$Vidt$ is the total energy supplied to the coil in time dt, Ri^2dt is the energy converted into heat due to the ohmic resistance of the coil, and $Lidt(di/dt)$ is the energy required to establish the magnetic field whose presence is the cause of the coil's self-induction. The solution of the original equation is

$$i = \frac{V}{R}\left(1 - e^{-\frac{R}{L}}\right)$$

$$= I(1 - e^{-\frac{R}{L}t})$$

where I is the final value of the current.

EXAMPLE 1. A closed iron ring of mean diameter 12 cm is made from round iron bar of diameter 2 cm. It has a uniform winding of 1000 turns. Calculate the current required to produce an induction of 1·5 webers/m², given that the relative permeability is then 1250. Hence calculate the self inductance.

$$l = \pi \times 12 \times 10^{-2}\ \text{m}$$

$$a = \pi/4 \times 2^2 \times 10^{-4}\ \text{m}^2$$

$$\therefore S = \pi \times 12 \times 10^{-2}/(1250 \times 4\pi \times 10^{-7} \times \pi/4 \times 4 \times 10^{-4})$$

$$= 0{\cdot}96 \times 10^7/4\pi$$

$$\Phi = 1{\cdot}5 \times (\pi/4) \times 2^2 \times 10^{-4} = 1{\cdot}5\pi \times 10^{-4}\ \text{weber}$$

$$\therefore NI = \Phi S$$

$$= 1{\cdot}5\pi \times 10^{-4} \times 0{\cdot}96 \times 10^7/4\pi = 360$$

$$\therefore I = 360/1000 = 0{\cdot}36\ \text{A}$$

$$\therefore \text{(Flux per ampere)}$$

$$= 1{\cdot}5\pi \times 10^{-4}/0{\cdot}36 = 1{\cdot}31 \times 10^{-3}\ \text{weber}$$

$$\therefore L = 1{\cdot}31 \times 10^{-3} \times 10^3 = 1{\cdot}31\ \text{H}$$

EXAMPLE 2. A coil of self-inductance 0·75 H and resistance 4 Ω carries a current which is gradually increasing. At a certain instant the

current is 20 A and is increasing at the rate of 10 A/sec. Calculate the p.d. at the coil terminals, the rate at which energy is being transferred to the magnetic field, and the amount of energy stored at that particular instant.

$$V = Ri + L(di/dt)$$
$$= 4 \times 20 + 0{\cdot}75 \times 10$$
$$= 87{\cdot}5 \text{ V}$$

Rate of supply of energy to the field,

$$Li(di/dt) = 0{\cdot}75 \times 20 \times 10$$
$$= 150 \text{ W}$$

Energy stored $\quad \frac{1}{2}Li^2 = \frac{1}{2} \times 0{\cdot}75 \times 20^2$
$$= 150 \text{ joules}$$

EXAMPLE 3. A constant p.d. of 1 V is applied to a coil of resistance 1 Ω and inductance 1 H; plot the curve of current against time.

The final value of the current is $I = E/R = 1$ A, hence, for the current at any instant we have

$$i = I(1 - e^{-\frac{R}{L}t})$$
$$= 1 - e^{-t}$$

when $t = 0{\cdot}1$ sec, $i = 1 - e^{-0.1} = 1 - 2{\cdot}7183^{-0.1} = 0{\cdot}095$

when $t = 0{\cdot}2$ sec, $i = 1 - e^{-0.2} = 1 - 2{\cdot}7183^{-0.2} = 0{\cdot}181$, and so on.

(The curve is as shown in Fig. 5.5.)

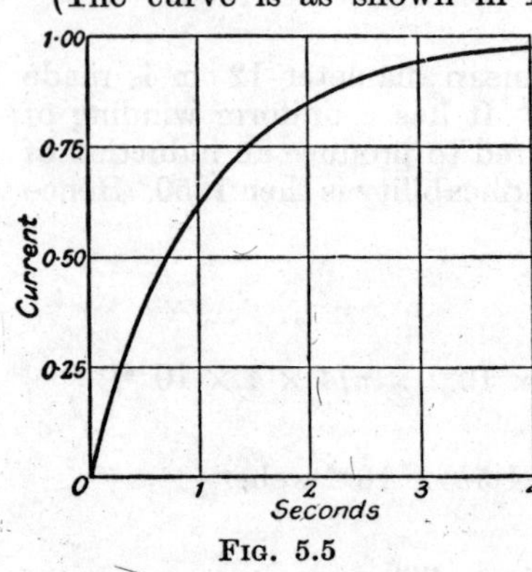

FIG. 5.5

After an interval of time of L/R seconds the current reaches the value $i = I(1 - e^{-1}) = 0{\cdot}6321I$.

This is a definite fraction of I, and the ratio L/R is therefore called the "time-constant" of the coil. The time-constant can thus be defined as the time required for the current to reach 0·6321 of its final value.

From the energy equation we saw that the energy imparted to the magnetic field in time dt was

$$Lidt(di/dt)$$

Hence, when the current has attained its final value I, the energy of the field will be

$$\int_0^I Lidi = \tfrac{1}{2}LI^2$$

Now L is the flux per ampere $\times N$ and therefore the energy of the field is

$$\tfrac{1}{2} \times \text{current} \times \text{total flux} \times N$$

We have now to consider the decay of the current when the applied p.d. is removed. Suppose, for instance, that the coil is suddenly short-circuited and the source of p.d. at the same time disconnected. Then, since the current now decreases, its rate of change di/dt is negative, and the e.m.f. equation becomes

$$O = Ri + L\,di/dt$$

the solution of which is

$$i = \left(\frac{V}{R}\right)e^{-\frac{R}{L}t} = I\,.\,e^{-\frac{R}{L}t}$$

This shows that the curve dies away exponentially, as shown in Fig. 5.6. The curves of rise and decay are complementary, for if they are

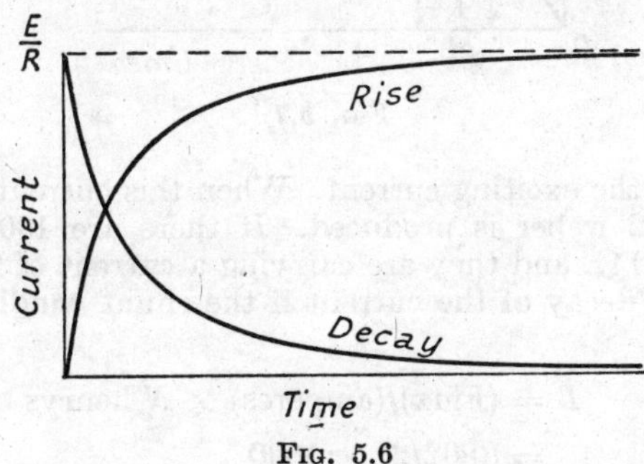

FIG. 5.6

drawn with the same origin as in Fig. 5.6 the sum of the ordinates at any instant is equal to the Ohm's Law value of the current I.

If both sides of the e.m.f. equation are multiplied by idt, another energy equation is obtained, namely,

$$O = Ri^2dt + Li(di/dt)\,.\,dt$$

This shows that, when the current is decaying, the whole of the energy for the production of heat in the winding is drawn from the stored energy of the magnetic field, and eventually, when the current is zero, the field also is of zero strength.

Consider again the curve giving the rate of rise of current. At any instant the rate of rise of current is

$$\frac{di}{dt} = \frac{(V/R - i)}{L/R}$$

if time is reckoned from the instant of zero current. The denominator L/R is called the time constant. If the curve of rise of current is known, L/R can be determined graphically as follows.

At any point P, Fig. 5.7, draw the tangent PT to cut the horizontal AB at T, where OA = final current = V/R. Then

$$\tan \theta = di/dt$$

$$QP = V/R - i$$

$$\therefore\ QT = L/R \text{ in seconds}$$

Example. A shunt dynamo gives an air-gap flux which is practically

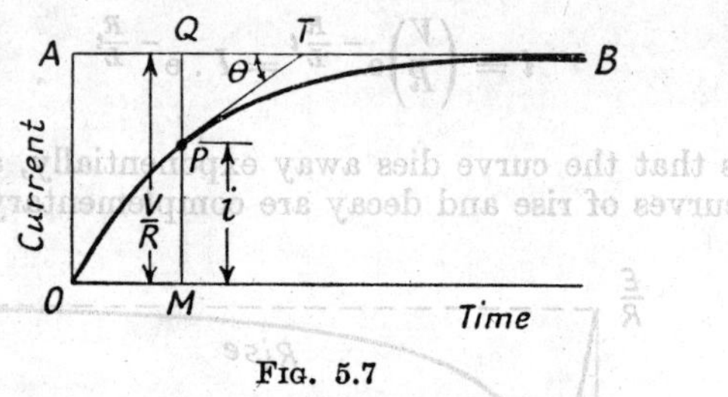

Fig. 5.7

proportional to the exciting current. When this current is 2 A, an air-gap flux of 0·02 weber is produced. If there are 1000 shunt turns of resistance 100 Ω, and they are carrying a current of 2 A, determine the equation of decay of the current if the shunt winding is suddenly short-circuited.

$$L = (\text{Flux})/(\text{amperes}) \times N \text{ henrys}$$

$$= (0{\cdot}02/2) \times 1000$$

$$= 10 \text{ H}$$

$$R = 100$$

Initial current $I = 2$ A

$$\therefore\ i = I \times e^{-\frac{R}{L}}$$

$$= 2e^{-10t}$$

Relative Permeability Not Constant. In the above discussion of the effect of self-induction we have assumed that the magnetization characteristic of the circuit, namely, the graph of flux against current, is a straight line, or, what amounts to the same thing, the permeability of the magnetic circuit is constant. If the path of the lines of flux is through iron, this is no longer the case, particularly if the iron is taken up to saturation point, the result being that the coefficient of self-induction is no longer a constant, but is a function of the current. To determine the self-induction it is, therefore, necessary to calculate the flux per ampere for a series of values of the current, the flux per ampere being given by the gradient to the curve of flux against current. A numerical example will make this clear.

A ring of iron whose magnetic characteristics are given by the following figures—

B	2500	5000	7500	$10{,}000 \times 10^{-4}$
μ	1250	1200	1000	800

has a mean circumference of 3 m and cross-section of 10^{-2} m². It has a magnetizing coil of 250 turns. Plot a curve of self-induction against magnetizing current.

The four values of H $(= B/\mu\mu_0)$ are equal to 160, 334, 600, and 1000 respectively.

The total AT $(= at \times l)$ are, therefore, 480, 1002, 1800, and 3000 respectively. Dividing these values of AT by the number of turns we obtain the currents, whose values are, therefore, 1·92, 4·01, 7·2, and 12·0 respectively.

The total fluxes $(\Phi = BA)$ are $2{\cdot}5 \times 10^{-3}$, 5×10^{-3}, $7{\cdot}5 \times 10^{-3}$, and 10^{-2} webers. The curve of Φ against I can now be plotted, and is given in Fig. 5.8. Tangents to the curve are drawn at the points corresponding to the above values of the current, and their gradients, in webers/ampere, are $1{\cdot}218 \times 10^{-3}$, $9{\cdot}14 \times 10^{-4}$, $6{\cdot}35 \times 10^{-4}$, and $4{\cdot}26 \times 10^{-4}$ respectively.

We now apply the general expression for the self-induction, and the corresponding values are—

$$\begin{aligned}
&\text{when } I = 1{\cdot}92;\ L = 1{\cdot}218 \times 10^{-3} \times 2{\cdot}5 \times 10^{2} = 0{\cdot}305 \text{ H}\\
&\text{when } I = 4{\cdot}01;\ L = 9{\cdot}14 \times 10^{-4} \times 2{\cdot}5 \times 10^{2} = 0{\cdot}229 \text{ H}\\
&\text{when } I = 7{\cdot}2\ ;\ L = 6{\cdot}35 \times 10^{-4} \times 2{\cdot}5 \times 10^{2} = 0{\cdot}159 \text{ H}\\
&\text{when } I = 12{\cdot}0\ ;\ L = 4{\cdot}26 \times 10^{-4} \times 2{\cdot}5 \times 10^{2} = 0{\cdot}107 \text{ H}
\end{aligned}$$

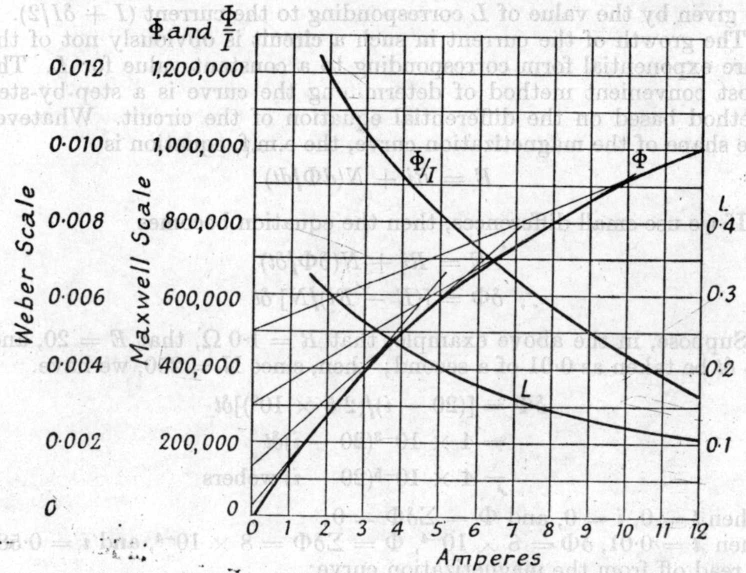

FIG. 5.8

If the current in such a circuit varies by a small amount, from, say, I to $(I + \delta I)$, then the self-induced e.m.f. will be calculated from the value of L corresponding to the mean current, namely, $(I + \frac{1}{2} . \delta I)$. On the other hand, if the current I is suddenly changed by a small

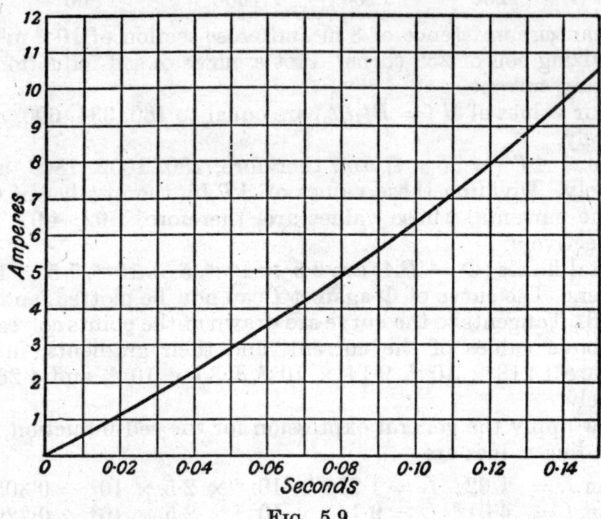

Fig. 5.9

amount to $I + \delta I$, then the *mean* value of the self-induced e.m.f. will be given by the value of L corresponding to the current $(I + \delta I/2)$.

The growth of the current in such a circuit is obviously not of the pure exponential form corresponding to a constant value for L. The most convenient method of determining the curve is a step-by-step method based on the differential equation of the circuit. Whatever the shape of the magnetization curve, the e.m.f. equation is

$$E = Ri + N(d\Phi/dt)$$

If we use small differences, then the equation becomes

$$E = Ri + N(\delta\Phi/\delta t)$$

$$\therefore \; \delta\Phi = [(E - Ri)/N]\, \delta t$$

Suppose, in the above example, that $R = 1{\cdot}0\ \Omega$, that $E = 20$, and let δt be taken as 0·01 of a second; then, since $N = 250$, we have

$$\begin{aligned} \delta\Phi &= [(20 - i)/(2{\cdot}5 \times 10^2)]\delta t \\ &= 4 \times 10^{-3}(20 - i)\delta t \\ &= 4 \times 10^{-5}(20 - i) \text{ webers} \end{aligned}$$

When $t = 0$, $i = 0$, and $\Phi = \Sigma\delta\Phi = 0$;
when $t = 0{\cdot}01$, $\delta\Phi = 8 \times 10^{-4}$, $\Phi = \Sigma\delta\Phi = 8 \times 10^{-4}$, and $i = 0{\cdot}58$, as read off from the magnetization curve;

when $t = 0{\cdot}02$, $\delta\Phi = 7{\cdot}77 \times 10^{-4}$, $\Phi = (8 + 7{\cdot}77) \times 10^{-4} = 15{\cdot}77 \times 10^{-4}$ and $i = 1{\cdot}1$;

when $t = 0{\cdot}03$, $\delta\Phi = 7{\cdot}56 \times 10^{-4}$, $\Phi = (15{\cdot}77 + 7{\cdot}56) \times 10^{-4} = 23{\cdot}33 \times 10^{-4}$

and $i = 1{\cdot}7$, and so on.

In this way the curve of i against t can be determined as far as desired, and obviously the accuracy of the determination depends solely on the smallness of the intervals δt. The curve of Fig. 5.9 gives the curve of growth of current as so determined up to a time of 0·15 sec after switching on. It will be seen that this curve is, up to $t = 0{\cdot}13$ sec, concave upwards instead of concave downwards, as in the case of a circuit for which L is constant. The reason for this is that the value of L is continually falling, as is indicated by the curve of L against I, and the growth-of-current curve will only assume the normal form when L has become sensibly constant.

Mutual Inductance. Suppose there are two coils A and B, whose coefficients of self-induction are L_A and L_B respectively, and whose coefficient of mutual induction is M. If they are traversed by currents I_A and I_B respectively, then

The energy due to self-induction is clearly

$$\tfrac{1}{2}L_A I_A^2 + \tfrac{1}{2}L_B I_B^2$$

The energy due to mutual induction is

$$\int i_A M(di_B/dt)dt + \int i_B M(di_A/dt)dt$$

using the symbol i for the value of the current at any instant

$$= M(\int i_A di_B + \int i_B di_A)$$

$$= M\int d(i_A i_B) = MI_A I_B$$

The total energy is, therefore,

$$\tfrac{1}{2}L_A I_A^2 + MI_A I_B + \tfrac{1}{2}L_B I_B^2$$

We can write the voltage equations for the two coils as follows, remembering that (*a*) the applied p.d. V_A to coil A has to supply not only the resistance drop in this coil, but also the self- and mutually-induced e.m.f.s in it, and (*b*) the applied p.d. to coil B is zero.

$$V_A = L_A(di_A/dt) + M(di_B/dt) + R_A i_A$$

$$0 = L_B(di_B/dt) + M(di_A/dt) + R_B i_B$$

Suppose that we simplify the problem by assuming that R_A and R_B are both very small, as is often the case with transformers, for example, then

$$V_A = L_A(di_A/dt) + M(di_B/dt)$$

$$0 = L_B(di_B/dt) + M(di_A/dt)$$

Solving for i_A we have

$$(di_B/dt) = -(M/L_B)(di_A/dt)$$
$$\therefore\ V_A = L_A(di_A/dt) - (M^2/L_B)(di_A/dt)$$
$$\therefore\ (di_A/dt) = V_A/(L_A - M^2/L_B)$$

Integrating this, we have for the relationship between i_A and time t

$$i_A = [V_A/(L_A - M^2/L_B)]t$$

This shows that the primary current rises in a linear manner with respect to time when a steady p.d. is applied to it.

For the secondary induced current i_B, we have

$$(di_B/dt) = -(M/L_B)(di_A/dt)$$
$$\therefore\ i_B = -(M/L_B)i_A$$
$$= -(M/L_B)[V_A/(L_A - M^2/L_B)]t$$

This shows that the current i_B is also linear with respect to time, but when i_A is reckoned positive then i_B is negative.

We can deduce the expression for the energy in an alternative way, as follows. The whole of the energy is derived from the source to which the primary is connected, and therefore we have

$$\text{Total energy} = \int_0^t V_A i_A dt$$
$$= \int_0^t V_A \times \left(\frac{V_A}{L_A - M^2/L_B}\right) t \,.\, dt$$
$$= \tfrac{1}{2}[V_A^2/(L_A - M^2/L_B)]t^2$$
$$= \tfrac{1}{2}[V_A/(L_A - M^2/L_B)]^2 t^2 (L_A - M^2/L_B)$$
$$= \tfrac{1}{2}L_A i_A^2 - \tfrac{1}{2}L_B i_B^2$$
$$= \tfrac{1}{2}L_A i_A^2 + \tfrac{1}{2}L_B i_B^2 - L_B i_B^2$$

But since $\quad i_B = -(M/L_B)i_A$

$$L_B i_B^2 = +L_B \times (M^2/L_B^2)i_A^2$$
$$\therefore\ \text{Total energy} = \tfrac{1}{2}L_A i_A^2 + M i_A i_B + \tfrac{1}{2}L_B i_B^2$$

In the treatment the symbol i is used instead of I since the currents both vary with respect to time.

Inductance of a Pair of Parallel Conductors. The lines of force due to either conductor are concentric circles round the conductor and also concentric circles inside the conductor (Fig. 5.10). Consider first of all the flux inside the conductor. The current inside a line of force of radius x is $I \times x^2/r^2$, and therefore the field strength at a distance x from the centre

$$H_x = Ix/2\pi r^2$$
$$\therefore\ B_x = \mu\mu_0 H_x$$
$$= \mu_0 Ix/2\pi r^2,\ \mu \text{ being unity}$$

Hence, the flux through a cylindrical shell of radial thickness dx and axial length 1 m is

$$B_x dx = (\mu_0 I/2\pi r^2)x\,dx$$

But the flux links with only x^2/r^2 of the conductor, so that the linkage of the shell is

$$(\mu_0 I/2\pi r^2)x\,dx(x^2/r^2) = (\mu_0 I/2\pi r^4)x^3\,dx$$

$\therefore$ Total linkage inside the conductor

$$\frac{\mu_0 I}{2\pi r^4}\int_0^r x^3\,dx = \frac{\mu_0 I}{8\pi}$$

Now consider the flux between the two conductors. At any distance

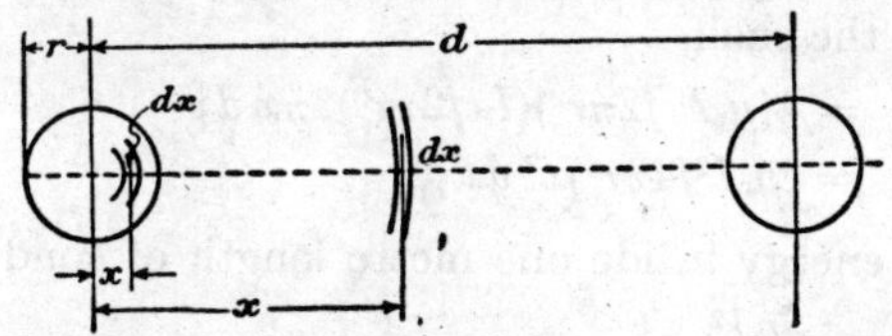

Fig. 5.10

x the field strength is again $I/2\pi x$, and therefore the flux through a cylindrical shell of thickness dx and axial length 1 m is $B_x dx = (\mu_0 I/2\pi x)dx$. Hence, linkage

$$\frac{\mu_0 I}{2\pi}\int_r^d \frac{dx}{x} = \frac{\mu_0 I}{2\pi}\log_e\frac{d}{r}$$

Hence, the total linkage per conductor, reckoning the fluxes inside and outside the conductor, is

$$\mu_0 I/8\pi + [\mu_0 I/2\pi]\log_e(d/r)$$

The total linkage for both conductors is, therefore

$$\mu_0 I/4\pi + [\mu_0 I/\pi]\log_e(d/r)$$
$$= (\mu_0 I/\pi)(\log_e(d/r) + \tfrac{1}{4})$$

$\therefore$ linkage per ampere

$$= (\mu_0/\pi)[\log_e(d/r) + \tfrac{1}{4}]$$

and therefore, as the two conductors correspond to only one turn we have

$$L = (\mu_0/\pi)[\log_e(d/r) + \tfrac{1}{4}] \text{ henrys/metre}$$

If the log is reduced to base 10, and the inductance in henrys/mile, we have, finally

$$L = (1609/\pi) \times 4\pi \times 10^{-7}[2{\cdot}303\log_{10}(d/r) + \tfrac{1}{4}]$$
$$= 6{\cdot}436 \times 10^{-4}[2{\cdot}303\log_{10}(d/r) + \tfrac{1}{4}]$$

When the spacing is large it is possible to neglect the second term in the brackets and we then have, approximately

$$L = 14{\cdot}8 \times 10^{-4} \log_{10} (d/r) \text{ henrys/mile}$$

It is of interest to consider the effect of the flux inside the conductor in another way. At any radial distance x we have seen that

$$B_x = \mu_0 Ix/2\pi r^2$$

The corresponding value of H is

$$H_x = Ix/2\pi r^2$$

Now we have seen that energy density is equal to $\frac{1}{2}BH$. Hence, since the volume of an elementary cylindrical shell of radius x and axial length 1 m is $2\pi x\, dx$,

stored energy in the shell

$$= \tfrac{1}{2}(\mu_0 Ix/2\pi r^2)(Ix/2\pi r^2)2\pi x\, dx$$
$$= (\mu_0 I^2/4\pi r^4)x^3\, dx$$

$\therefore$ Total stored energy inside one metre length of conductor is

$$\frac{\mu_0 I^2}{4\pi r^4}\int_0^r x^3\, dx = \frac{\mu_0 I^2}{16\pi}$$

But energy $= \frac{1}{2}LI^2$

$$\therefore\ L = \mu_0/8\pi \text{ per conductor}$$
$$= \mu_0/4\pi \text{ for both conductors, as before.}$$

Examples on Chapter 5

1. A conductor 12 in. long rotates about one end at 1000 r.p.m. in a plane perpendicular to a magnetic field of strength 0·5 weber/m². Find the e.m.f. induced in it.

Ans. 2·42 V.

2. Define the coefficient of inductance of a magnetic circuit and show that it is proportional to the square of the magnetizing turns and inversely to the reluctance. Show how to find, by observation of the curve of rise of current when a steady voltage is applied to the inductance, the change of reluctance due to saturation of the iron. (L.U.)

3. A coil of resistance 10 Ω and inductance 1 H has a current which increases uniformly at the rate of 10,000 A/sec. Find the value of the necessary applied p.d. (*a*) when the current is 10 A, (*b*) when it is 50 A.

Ans. (*a*) 10,100; (*b*) 10,500 V.

4. Plot a curve giving the value at each instant of an electric current which varies in the following way—At time 0 it is 4 A. It increases at the rate of 10 A/sec for 2 sec; it then remains constant for 2 sec; it then decreases at the rate of 4 A/sec for 6 sec; it then follows the law $12 \sin 2\pi/20 \,.\, t$. Plot curves showing the voltage at the terminals of a resistance of 3 Ω and at the terminals of an inductance of 4 H respectively when this current is passed through them. (L.U.)

5. A certain choke coil has 1500 turns. When 4 A are passed through it, the total magnetic flux threading the coil is 6,000,000 maxwells. The resistance of the coil

is 20 Ω. Find an expression for the current in the coil immediately after it is switched on to a d.c. supply yielding 100 V. Assume that the self-induction of the coil is constant during the rise of current. (L.U.)

Ans. $i = 5(1 - e^{-0{\cdot}89t})$.

6. A certain electromagnet has a winding of resistance 5 Ω and inductance 4H. After how many seconds, after the application of a steady p.d., will the current rise to one half of its final value.

Ans. 0·575 sec.

7. Define the unit of inductance. Obtain an expression by which the inductance of an anchor ring of D cm mean diameter and A cm^2 cross section wound with n turns of wire may be calculated approximately, if the permeability (μ) of the iron is assumed constant. A field magnet coil wound with 1500 turns of wire produces a flux of 2·8 megalines when carrying a current of 2 A; estimate the inductance of the coil in henrys. (L.U.)

Ans. 21 H

8. What methods may be used for limiting the voltage rise of an inductive coil when it is disconnected from a d.c. supply? Estimate the discharge resistance necessary to prevent the voltage rise exceeding 300 V when a coil of large inductance having a resistance of 100 Ω is disconnected from a supply at 200 V. (L.U.)

Ans.—50 Ω.

9. Prove, for any simple case, that the mechanical work done by, or on, a coil traversed by a steady current I due to a movement from a position in which the total magnetic flux through the coil is N_1, to another position, in which the total flux is N_2, is $I(N_1 - N_2)$. State precisely the units in which the different quantities are measured, and what determines whether the work done is positive or negative.

The coefficient of mutual induction M, in henrys, between two coils, one of which is fixed, and the other movable about an axis, is given by $M = a + b\theta$, where a and b are known constants, and θ is the deflection in degrees measured from some zero position of the moving coil. Calculate the torque between the two coils for any deflection θ, when the coils are traversed by two currents I_1 and I_2 amperes. State the unit of torque in which the result is expressed. (L.U).

10. A transmission line consists of a pair of $\frac{1}{4}$ in. conductors spaced 8 ft apart. Calculate the inductance of the loop formed by joining the two conductors at one end.

Ans. 4·25 mH/mile.

11. A copper wire of circular section carries a current of I amperes. Show that the magnetic energy within the wire is equal to $\mu_0 I^2/16\pi$ joules per metre length.

12. A mutual inductor consists of two coils for which $L_1 = 1{\cdot}0$ H, $L_2 = 2{\cdot}0$ H, and $M = 1{\cdot}2$ H. A p.d. of 10 V is suddenly applied to coil 1. Calculate the magnetic energy after one second.

Ans. 179 J.

13. An iron ring 1 ft mean diameter is made of round iron 1 in. diameter, and is uniformly wound with 500 turns of copper wire 0·05 in. diameter. A second winding of 1000 turns of wire 0·025 in. diameter is uniformly wound over the first. Assuming the iron to have a (relative) permeability of 800, calculate the self-inductance of each winding and the mutual inductance between them. (L.U.)

Ans. 0·133, 0·532, 0·266 H.

14. A toroidal coil is uniformly wound with 250 turns. The core is of marble with a rectangular cross-section $2\frac{1}{2}$ in. deep by 2 in. wide; the outer diameter of the toroid is 6 in., and the inner diameter is 2 in. Find the inductance of the toroid, proving any formula used. (L.U.)

Ans. 0·344 mH.

15. There are two neighbouring coils. The primary has a resistance of 10 Ω and self inductance 0·2 H, the secondary a resistance of 5 Ω and self inductance of 0·8 H. The mutual inductance is 0·1 H. Calculate the voltages of self induction and the voltages of mutual induction in the two coils at the instant a p.d. of 200 V is applied to the primary.

***Ans.* Voltages of self induction, 534 and 266 V;**
Voltages of mutual induction, 334 and 266 V.

16. A circuit consisting of a non-inductive resistor in series with an air-cored inductor has 277·2 Ω resistance and 4 H inductance. A voltage of 1000 V d.c. is suddenly applied. After 0·03 sec. the applied e.m.f. is reversed and simultaneously one-half of the resistance is removed, the remainder of the circuit being closed on itself.

Construct the curve of current against time for a total time of 0·1 sec. and from the curve deduce the current at 0·09 sec. (A.M.E.M.E.)

***Ans.* 0·38 A.**

17. A circuit consisting of a resistor of 6 Ω in series with an inductor of 3 H has an e.m.f. of 60 V suddenly applied to it. Calculate *(a)* the initial rate of rise of current, *(b)* the rate of rise of current at the instant the current has reached 5 A, *(c)* the instantaneous current 0·2 sec. after the application of the e.m.f.

***Ans.* 20 A/sec.; 10 A/sec.; 3·25 A.**

18. A resistor of R_2 Ω is in series with an inductor of R_1 Ω and L_1 H. An e.m.f. E is suddenly applied. Calculate the p.d. across the inductor terminals after an elapsed time of t sec.

Ans. $v = (E/R_1 + R_2)(R_1 + R_2 e^{-(R_1+R_2)t/2})$

CHAPTER 6

CONSTRUCTION OF D.C. MACHINES

Essential Features. In order to set up a dynamically-induced e.m.f. three things are required: a magnetic field, a conductor, and motion of the conductor relative to the field. In d.c. dynamo-electric machinery the magnetic field is produced by the "field-magnet," the conductors in which the e.m.f. is induced are placed on the "armature," and the necessary motion is obtained by the rotation of the armature within the field magnet. We shall see that the internally induced e.m.f. is an alternating e.m.f., that is, alternately positive and negative. The e.m.f. acting round the external circuit to which the armature winding is connected is a unidirectional e.m.f., this resulting from the fact that the field system is stationary in space. Connection between the external circuit and the armature winding is made by means of fixed "brushes" which press on a rotating member, the "commutator" which is, in effect, an extension of the armature winding which is less expensive to wear away than the winding itself. It is often stated that the commutator acts as a rectifier, but this is not the case.

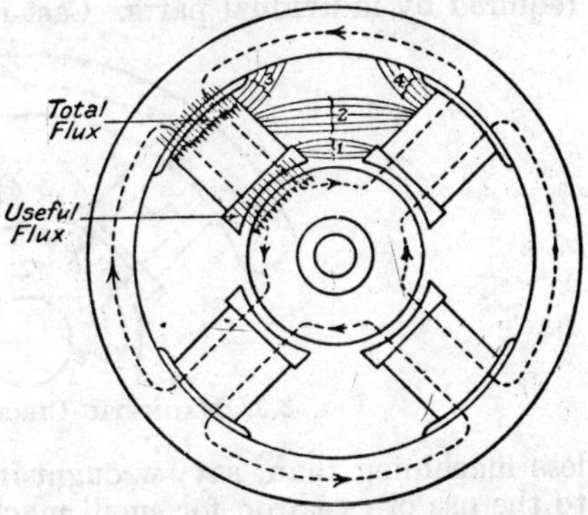

The Field Magnet. Practically all modern direct-current machines have a field magnet consisting of a circular yoke with inwardly projecting poles, the polarity of which is alternately N. and S. A typical four-pole frame is shown in Fig. 6.1. The dotted lines indicate the paths of the main magnetic fluxes and the full lines the leakage paths. It will be seen that the flux per pole divides at the yoke, so that the cross section of the yoke carries only half the flux per pole. Similarly the cross-section of the armature core carries half the flux which enters the armature from the air gap.

Fig. 6.1. Fluxes in a Typical Field Magnet

The yoke of a large machine is now almost invariably fabricated steel, because this material has roughly twice the permeability of cast iron. Only half the weight of fabricated steel need therefore be used, an important consideration in large machines. At the present time the main poles are nearly always (*a*) laminated throughout, i.e. built up of sheet material, or (*b*) solid steel pole cores with laminated pole shoes.

When laminated, the pole is secured to the yoke by a bolt or bolts screwed into a bar passing through the laminations, as shown in Fig. 6.2.

In the case of small machines, cast iron is often used for both yoke and poles, because here the main consideration is cheapness, both from

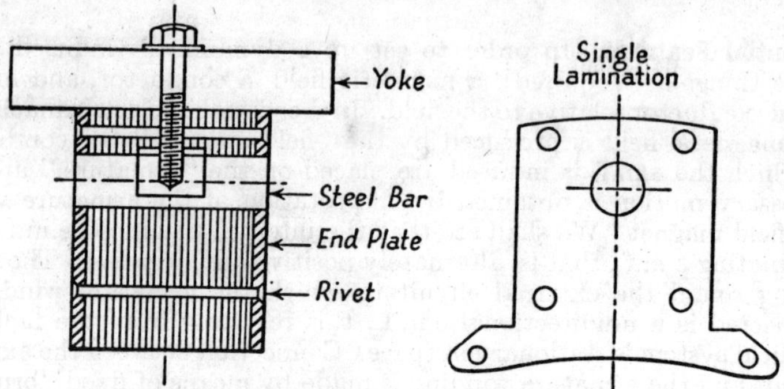

FIG. 6.2. CONSTRUCTION OF A LAMINATED POLE

the point of view of the materials used and also the amount of machining required by individual parts. Cast-iron poles obviously require much

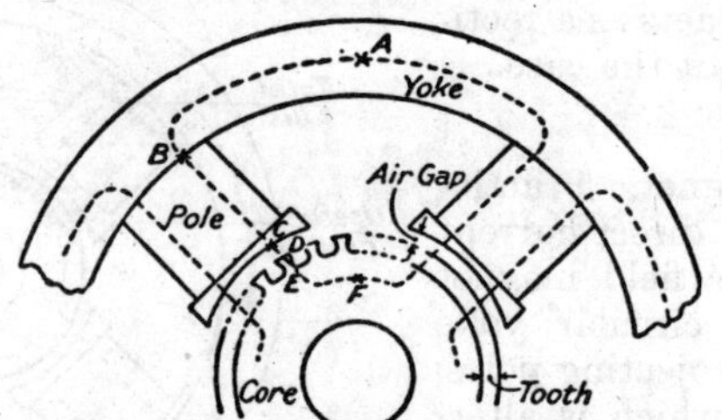

FIG. 6.3. MAGNETIC CIRCUIT OF D.C. MACHINE

less machining than, say, wrought-iron poles. The extra weight due to the use of cast iron for small machines is not so important as with large machines.

The calculation of the ampere-turns per pole for a given magnetic circuit is made as follows. The mean path of the flux (*see* Fig. 6.3) is divided up between the various parts of the circuit. The ampere-turns per pole have to overcome the reluctance of half the mean circuit, i.e. the path A to F, Fig. 6.3.

Let l_y = mean magnetic length of yoke = AB

l_p = length of pole = BC

l_g = length of air gap = CD

l_t = length of a tooth = DE

l_c = mean length in core = EF

Let corresponding suffixes refer to the areas a_y, a_p, etc.

Denote the flux entering the armature by Φ and the leakage factor by λ, then the total flux through the pole core is $\lambda\Phi$. The fluxes carried by the various parts of the circuit are thus

$$\begin{aligned} \text{Yoke } & \tfrac{1}{2}\lambda\Phi \\ \text{Pole } & \lambda\Phi \\ \text{Air gap } & \Phi \\ \text{Teeth under one pole } & \Phi \\ \text{Armature core } & \Phi/2 \end{aligned}$$

The calculation of the necessary number of field ampere-turns per pole for the yoke, poles, and armature core is straightforward and needs no comment. The calculations are more complex for the gap and teeth, and as most of the ampere-turns are needed for these two portions the accuracy must be sufficiently high. It must be realized that magnetic calculations cannot be made with a very high order of accuracy, and a designer will always provide more ampere-turns than the calculated value, in order to allow for contingencies.

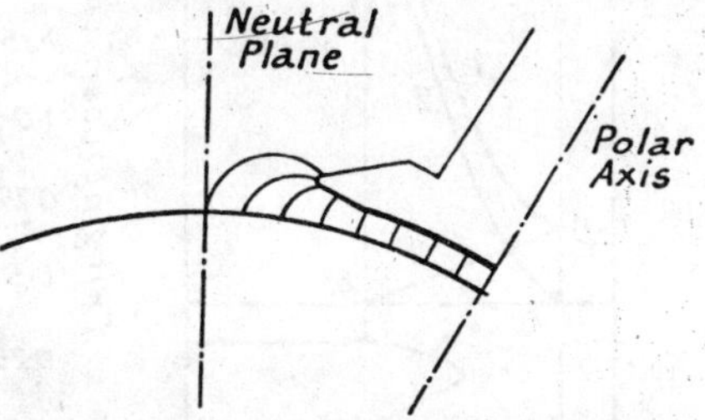

FIG. 6.4. FLUX FRINGE AT END OF POLE SHOE

(1) THE AIR-GAP. The first step is to make an allowance for the flux fringes at the pole shoes (Fig. 6.4). There are two general ways of doing this—

(*a*) A careful plot of the flux is made, the armature arc, namely, one-half of the pole pitch, less one-half the width of the interpole shoe, being divided into a number of equal paths, say 20.* The lines of force are drawn in as carefully as possible, assistance with this being afforded by drawing the equipotential lines at the same time and arranging so that lines of force and equipotential intersect at right-angles. As the metal surfaces are equipotentials except in the case of high magnetic saturation, the lines of force must leave these surfaces normally. It is a good plan to start at the point of maximum induction density and plot outwards. When completed, the map will consist of a number of "curvilinear squares," in each of which all four corner angles are right angles. As the potential drop is the same for all the lines of force the induction densities at the various points chosen are inversely proportional to the lengths of the lines of force, which are therefore measured carefully. Calling the induction at the centre point B_m the graph of induction against distance round the armature is drawn, as in Fig. 6.5, curve 1. Now some of the flux fringe passes to the tip of the adjoining pole, a fact which cannot be taken into account in the flux plot without leading to some complication. A simple correction is to

* The technique of field mapping is explained fully in S. S. Attwood's *Electric and Magnetic Fields.*

join point B to the point A, where the lengthening of the air-gap begins. The ordinates of the line BA are then deducted from curve 1, giving the corrected curve 2. The area under this curve is obtained by planimeter and the ratio B_{av}/B_m thus obtained.

A second method, due to Carter*, is to regard the fringing as increasing the effective width of the pole shoe. Let w_{ps} be the pole shoe width,

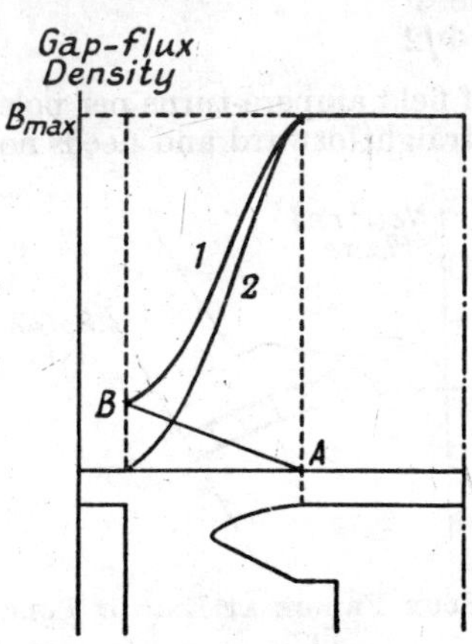

Fig. 6.5. Graphical Method of Allowing for Flux Fringe

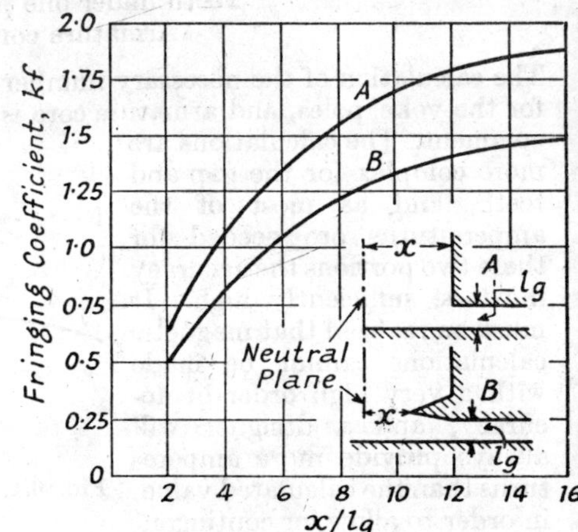

Fig. 6.6. Graph giving Fringing Coefficient, k_f

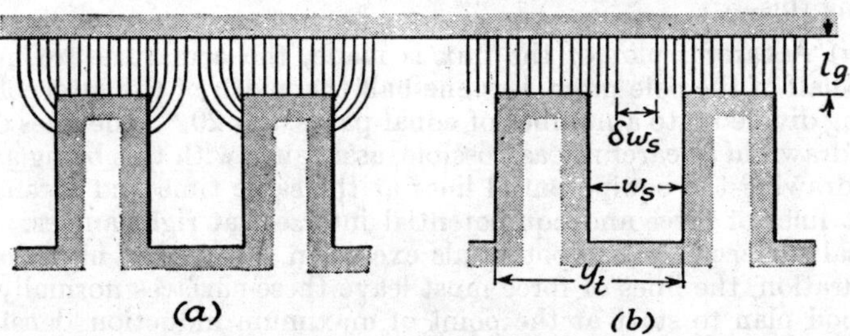

Fig. 6.7. Effect of Armature Slots

and l_g the gap length at the tip of the pole shoe, then the effective width is

$$w_{ps} + 2\,k_f l_g$$

where k_f is a fringing coefficient whose value can be determined from the graph of Fig. 6.6.

The next step is to calculate the effect of the armature slots, since these cause the flux to "tuft" into the tooth tops and flanks. For the purpose of calculation it is assumed that the actual distribution, Fig.

* Carter, F. W., *J. Inst. Elect. Engrs.*, **29** (1900) 925.

6.7 (*a*), is replaced by an equivalent radial distribution, Fig. 6.7 (*b*), in which a portion of the slot width, δw_s carries no flux. This is tantamount to reducing the slot width from w_s to δw_s. Over a whole slot pitch y_s, the width of a group of lines of force is thus reduced from y_s to $(y_s - \delta w_s)$.

With a uniform gap density each group of lines of force would have occupied the width y_s, and thus the effect of armature slots has reduced the air-gap area in the ratio

$$(y_s - \delta w_s)/y_s$$

Denoting the axial gap length by l, the uncorrected gap area is

$$l(w_{ps} + 2k_f l_g)$$

and the corrected value is therefore

$$l(w_{ps} + 2k_f l_g)[(y_s - \delta w_s)/y_s]$$

The ratio $(y_s - \delta w_s)/y_s$ is called the *contraction* coefficient. Denoting it by k_c we have for the corrected gap area,

$$A_g = k_c l(w_{ps} + 2k_f l_g)$$

If the armature has ventilating ducts then a packet of laminations between two ducts must be regarded as a wide tooth, and the ducts as

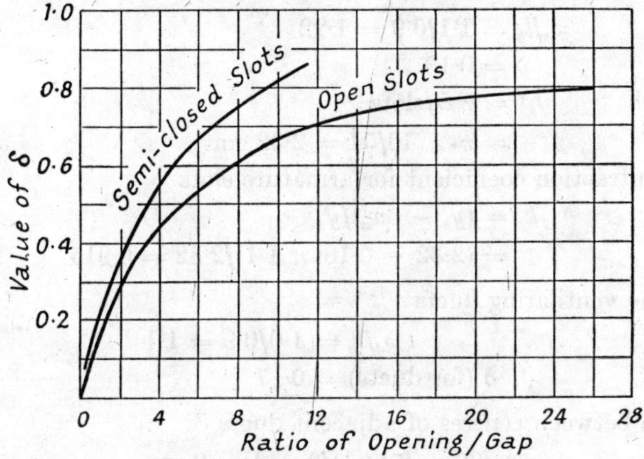

FIG. 6.8. CONTRACTION COEFFICIENT δ FOR OPEN SLOTS

a slot. There must be another contraction coefficient k'_c calculated for this, giving

$$A_g = k_c k_c' l(w_{ps} + 2k_f l_g)$$

If the pole face is slotted, as for example in a machine with a pole-face compensating winding, then another contraction coefficient must be determined in the same way, For open slots, as is usual with d.c. machines, the contraction coefficient δ can be determined from the graph of Fig. 6.8.

EXAMPLE. A 4-pole machine having a flux per pole of 0·042 weber has an armature of 40 cm, diameter with 54 slots each of width 1·1 cm. The axial length is 26 cm and there are two ventilating ducts, each 1 cm wide. The ratio of pole-shoe width to pole pitch is 0·7, and the radial gap length is 0·9 cm. Calculate the ampere-turns per pole required for the air-gap.

$$\text{Diameter at centre of gap} = D + l_g = 40{\cdot}9 \text{ cm}$$

$$\therefore \text{ Pole pitch } y = 40{\cdot}9\pi/4 = 32{\cdot}2 \text{ cm}$$

$$\text{Width of pole shoe } w_{ps} = 0{\cdot}7 \times 32{\cdot}2 = 22{\cdot}5 \text{ cm}$$

$$\therefore x \text{ (Fig. 6.6)} = \tfrac{1}{2}(32{\cdot}2 - 22{\cdot}5) = 4{\cdot}85 \text{ cm}$$

$$\therefore x/l_g = 4{\cdot}85/0{\cdot}9 = 5{\cdot}4$$

$$\therefore k_f = 1{\cdot}05 \text{ (from graph)}$$

$\therefore$ Corrected pole width

$$w_{ps} + 2k_f l_g = 22{\cdot}5 + 2 \times 1{\cdot}05 \times 0{\cdot}9 = 24{\cdot}5 \text{ cm}$$

Net iron length of core, allowing a factor 0·92 for core disc insulation

$$l_i = (26 - 2 \times 1) \times 0{\cdot}92$$
$$= 22 \text{ cm}$$

For the armature slots

$$w_s/l_g = 1{\cdot}1/0{\cdot}9 = 1{\cdot}22$$
$$\therefore \delta = 0{\cdot}18$$

Slot pitch $\quad y_s{}^* = \pi D/\text{slots}$

$$= \pi \times 40/54 = 2{\cdot}32 \text{ cm}$$

$\therefore$ Contraction coefficient for armature slots

$$k_c = (y_s - \delta w_s)/y_s$$
$$= (2{\cdot}32 - 0{\cdot}18 \times 1{\cdot}1)/2{\cdot}32 = 0{\cdot}915$$

For the ventilating ducts

$$w_d/l_g = 1{\cdot}0/0{\cdot}9 = 1{\cdot}1$$
$$\therefore \delta \text{ (for ducts)} = 0{\cdot}17$$

Distance between centres of adjacent ducts

$$(26 - 2 \times 1)/3 + 1 = 9 \text{ cm}$$

$\therefore$ Contraction coefficient for ducts—

$$k_c' = (9 - 0{\cdot}17 \times 1)/9 = 0{\cdot}98$$

$$\therefore A = 0{\cdot}915 \times 0{\cdot}98 \times 26 \times 24{\cdot}5$$
$$= 572 \text{ cm}^2 = 0{\cdot}0572 \text{ m}^2$$

$$B_g = \Phi_g/A_g = 0{\cdot}42/0{\cdot}572$$
$$= 0{\cdot}735 \text{ weber/m}^2$$

* $y_s = y_t$, the tooth pitch.

$$\therefore H_g = 0{\cdot}735/(4\pi \times 10^{-7})$$
$$= 5{\cdot}85 \times 10^5$$
$$l_g = 9 \times 10^{-1} \text{ cm} = 9 \times 10^{-3} \text{ m}$$

$\therefore$ ampere-turns per pole required for the air-gap

$$= 5{\cdot}85 \times 10^5 \times 9 \times 10^{-3}$$
$$= 5265$$

(2) THE ARMATURE TEETH. The roots of armature teeth, owing to their small cross-section, are sometimes highly saturated, to values of 2·5 webers/m², or even more. In such cases some of the flux from the gap enters the flanks of the teeth, the fluxes at the various tooth sections thus being different. Let w_t be the tooth width at any section, Φ_i, the tooth flux per tooth pitch, and Φ_a the slot flux, Fig. 6.9. The

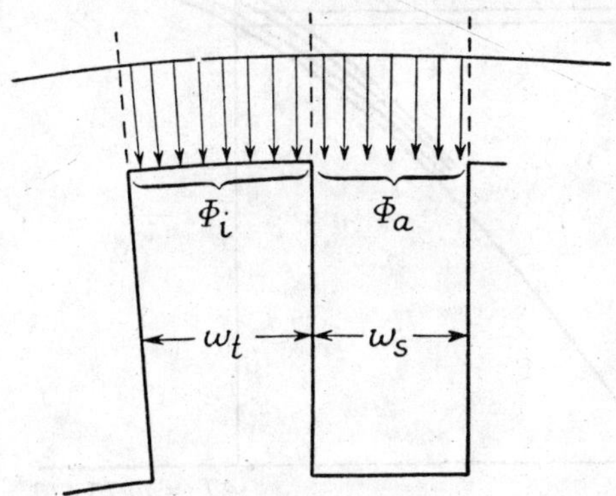

FIG. 6.9. SATURATION OF ARMATURE TEETH

uncorrected tooth density, that is, on the assumption that the tooth carries the whole flux $\Phi_i + \Phi_a$ throughout, is

$$B_t' = (\Phi_i + \Phi_a)/l_i w_t = \Phi_i/l_i w_t + \Phi_a/l_i w_t$$
$$= \Phi_i/l_i w_t + (\Phi_i/l_i w_s)(w_s/w_t) = B_t + B_g k_t$$

where k_t is the ratio air-area/iron-area for the particular sections of tooth under consideration. The gap density can be written

$$B_g = \mu_0 H$$
$$\therefore B_t' = B_t + \mu_0 k_t H$$

From the above expression a set of curves of B_t against B_t' can be constructed by taking a series of values of B_t, reading off from the magnetization curve the corresponding values of H, and then adding the product $\mu_0 k_t H$. In this way, by taking a series of values of k_t, a family of B_t/B_t' curves can be drawn. Thus, suppose that for a particular sheet material we have $H = 10{\cdot}75 \times 10^4$ AT/m, when

$B = 2{\cdot}3$ weber/m², then, by taking for k_t the values 0 (for smooth core), 0·5, 1·0, 1·5, etc., we have for the corresponding values of B_t'

$k_t = 0 \quad \therefore B_t' = 2{\cdot}3 + 0 = 2{\cdot}3$ webers/m²

$k_t = 0{\cdot}5 \quad \therefore B_t' = 2{\cdot}3 + 0{\cdot}5 \times 4\pi \times 10^{-7} \times 10{\cdot}75 \times 10^4 = 2{\cdot}367$

$k_t = 1{\cdot}0 \quad \therefore B_t' = 2{\cdot}3 + 4\pi \times 10^{-7} \times 10{\cdot}75 \times 10^4 = 2{\cdot}435$

$k_t = 1{\cdot}5 \quad \therefore B_t' = 2{\cdot}3 + 1{\cdot}5 \times 4\pi \times 10^{-7} \times 10{\cdot}75 \times 10^4 = 2{\cdot}502$

and so on.

We next take other values of B_t (from the original curve) and proceed in the same way. The process is illustrated by Fig. 6.10, which shows

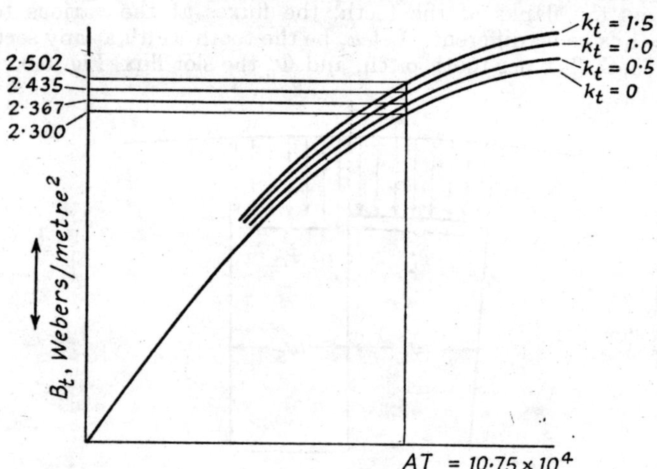

FIG. 6.10. POINTS CORRESPONDING TO $B_t = 2{\cdot}3$

the points corresponding to $B_t = 2{\cdot}3$. A set of such curves for a grade of sheet steel commonly used for d.c. armatures is given in Fig. 6.11.

EXAMPLE. An armature of diameter 130 cm and length 35 cm has four ducts, each 1 cm wide. It has 114 slots, each of dimensions 1·6 × 4·2 cm. The gap density is 1·155 webers/m². If the stampings have the characteristics of Fig. 6.11, calculate the ampere-turns per pole for the teeth.

Net iron length of core $l_i = (35 - 4 \times 1) \times 0{\cdot}92$
$= 28{\cdot}5$ cm

Armature diameter at top of tooth $= D = 130$ cm
Armature diameter at middle of tooth $= D - l_t = 125{\cdot}8$ cm
Armature diameter at root of tooth $= D - 2l_t = 121{\cdot}6$ cm
Periphery at top of tooth $= \pi \times 130 = 408$ cm
Periphery at middle of tooth $= \pi \times 125{\cdot}8 = 395$ cm
Periphery at root of tooth $= \pi \times 121{\cdot}6 = 382$ cm

For the tooth pitches at these positions we have

$$y_{t1} = 408/114 = 3{\cdot}58 \text{ cm} \quad \therefore k_{t1} = 1{\cdot}6/(3{\cdot}58 - 1{\cdot}6) = 0{\cdot}81$$
$$y_{t2} = 395/114 = 3{\cdot}46 \text{ cm} \quad \therefore k_{t2} = 1{\cdot}6/(3{\cdot}46 - 1{\cdot}6) = 0{\cdot}86$$
$$y_{t3} = 382/114 = 3{\cdot}34 \text{ cm} \quad \therefore k_{t3} = 1{\cdot}6/(3{\cdot}34 - 1{\cdot}6) = 0{\cdot}92$$

For the apparent inductions at the three sections,

$$B'_{t1} = B_g(y_{t1}/w_{t1}) = 1{\cdot}155 \times (3{\cdot}58/1{\cdot}98) = 2{\cdot}088 \text{ weber/m}^2$$
$$B'_{t2} = B_g(y_{t1}/w_{t2}) = 1{\cdot}155 \times (3{\cdot}58/1{\cdot}86) = 2{\cdot}223 \text{ weber/m}^2$$
$$B'_{t3} = B_g(y_{t1}/w_{t3}) = 1{\cdot}155 \times (3{\cdot}58/1{\cdot}74) = 2{\cdot}376 \text{ weber/m}^2$$

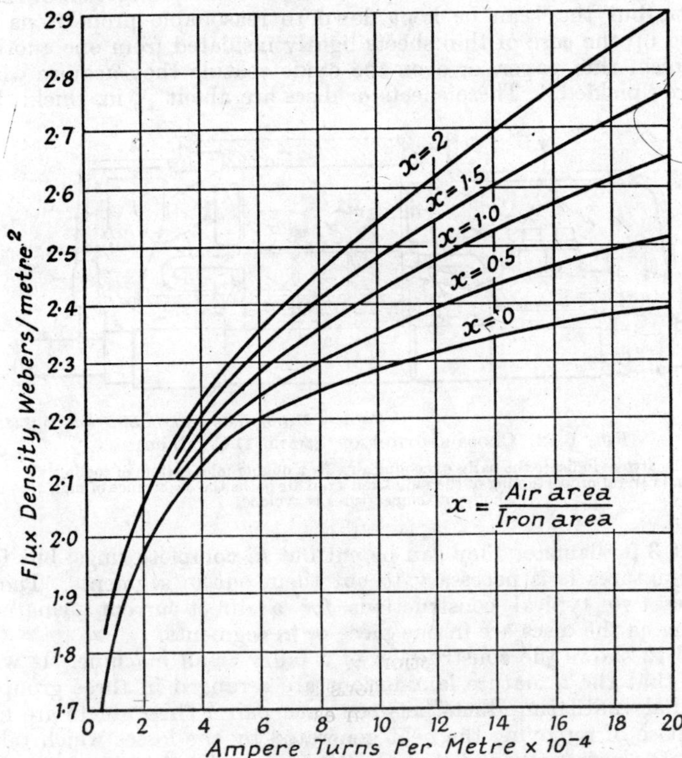

FIG. 6.11. B_t CURVES CORRESPONDING TO STEEL OF D.C. ARMATURE GRADE

From the curves of Fig. 6.11 the corresponding ampere-turns per metre length are, for the particular values of k_t,

$$(AT)_{t1} = 2{\cdot}7, \quad (AT)_{t2} = 4{\cdot}9, \quad (AT)_{t3} = 8{\cdot}9$$

Using Simpson's rule, we have for the ampere-turns for the teeth

$$1/6[(2{\cdot}7 + 4 \times 4{\cdot}9 + 8{\cdot}9) \times 10^4] \times 4{\cdot}2 \times 10^{-2} = 2180$$

The Armature. The armature consists of the core and the winding. The core obviously has to be magnetic in order to provide a path of low reluctance to the lines of force, and since iron is a good conductor of electricity, the rotation of a solid core in a strong magnetic field would result in very heavy eddy currents (or Foucault currents) being set up. These eddy currents would necessitate such a heavy expenditure of energy, and would cause such a large amount of heat to be produced, that it is imperative to eliminate them as far as possible. Since iron is a conductor it is impossible entirely to eliminate the production of these currents, but they can be kept down to reasonable proportions by building up the core of thin sheets lightly insulated from one another by varnish, thin paper, or even the oxide coating they acquire when they are "pickled." These sheets or discs are about $\frac{1}{40}$ in. thick. Up to about 3 ft diameter they can be cut out in complete rings, but for large armatures it is necessary to cut them out in segments. There are thus two typical constructions for a direct-current armature according as the discs are in one piece or in segments.

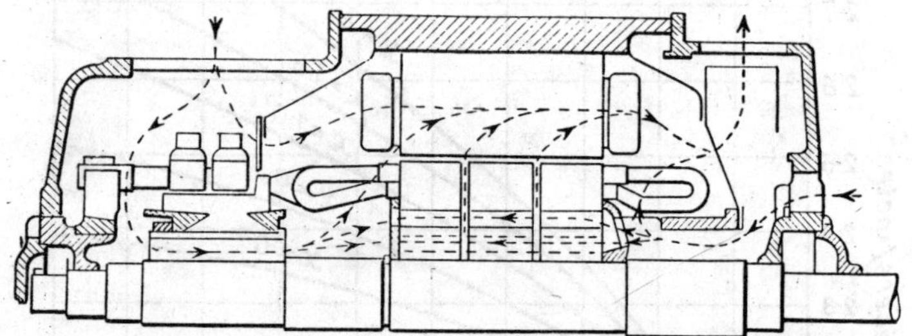

(Courtesy Metropolitan Vickers Electrical Co. Ltd.)

FIG. 6.12. CROSS-SECTION OF SMALL D.C. MACHINE

The arrows indicate the paths of cooling air. By a double-inlet system of cooling, and careful proportioning of the radial and axial air paths the occurrence of high hot spot temperatures is avoided.

Fig. 6.12 shows the construction of a fairly small machine. It will be seen that the armature laminations are arranged in three groups, with radial ventilating ducts between each pair. These ducts are for the purpose of removing the heat generated by the losses which take place in the armature, and it is therefore essential that air from the ends of the machine shall be able to reach them. This is effected by a series of longitudinal ducts which intersect the radial ducts. In the machine shown, the ventilating air is pulled through by means of the fan shown at the right-hand end, the paths of the air currents being as indicated.

The construction of a generator of large capacity is shown in Fig. 6.13. In this case the ratio of diameter to length is much greater than

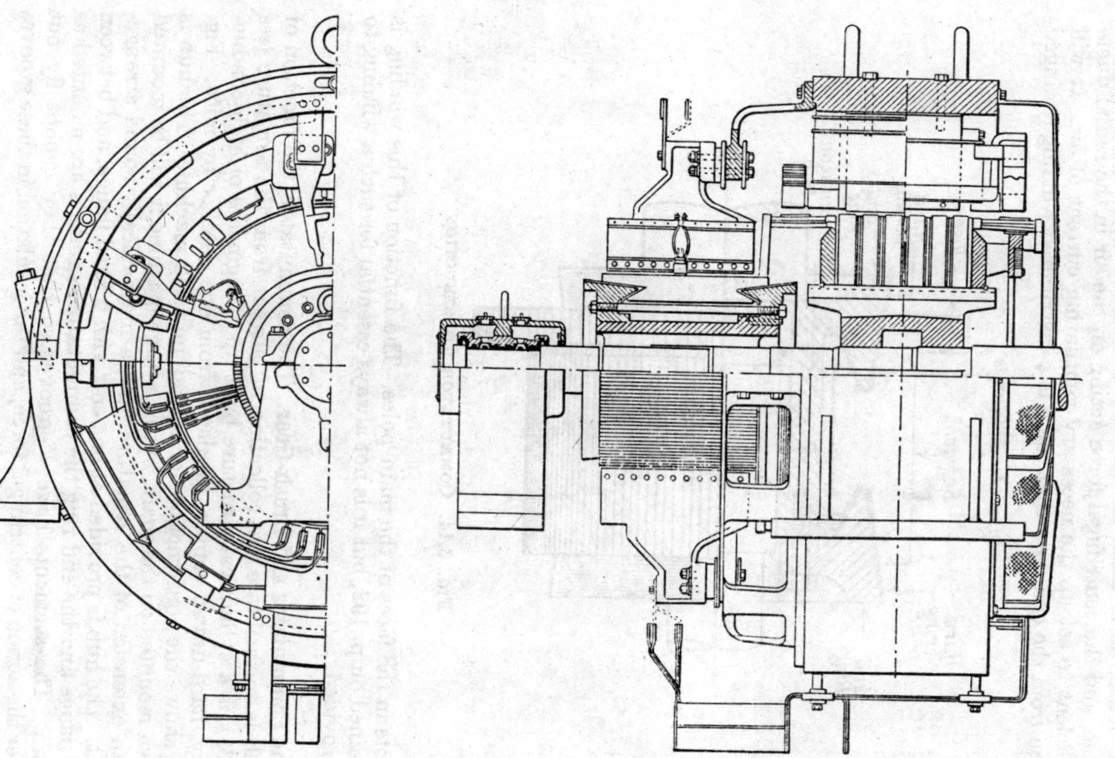

(Courtesy *Metropolitan Vickers Electrical Co. Ltd.*)

FIG. 6.13. CROSS-SECTIONAL VIEWS OF LARGE DIRECT-CURRENT GENERATOR

in the case of the machine of Fig. 6.12. If the machine is housed in a power house, as is usual for this type, then an open construction is possible, and the centrifugal force acting on the air in the radial ducts is sufficient to set up the necessary ventilating current of air. It will be seen from the end view that there is a distributed winding arranged

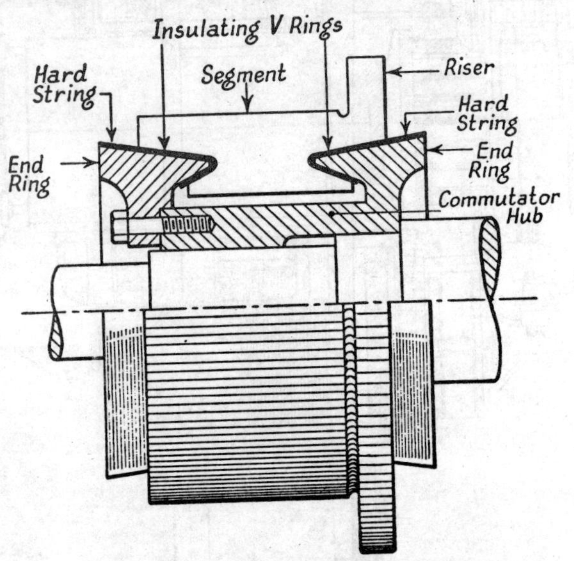

Fig. 6.14. Commutator Construction

in slots in the faces of the main poles. The function of this winding is explained on p. 192, but it is not always essential for such a winding to be provided.

The Commutator and Brush-Gear. The commutator, the function of which is to facilitate the collection of current from the armature (*see* p. 168), is a cylindrical structure built up of segments of high-conductivity, hard-drawn copper, insulated from one another by mica. Fig. 6.14 shows the construction for a medium-sized machine. The hub is shown mounted on the shaft, but in a large machine it will be mounted on an extension of the armature hub, a mechanically sound arrangement. The hub is provided at one end with a fixed flange, and between this flange and the end ring the commutator segments are mounted as shown. The segments have *V*-grooves so that they cannot fly out under the action of centrifugal force, and the insulation in these grooves is in the form of conical micanite rings.

The brushes, which are almost invariably of some form of carbon, are housed in brush holders. There are several forms of holder, the box type probably being the best. Fig. 6.15 shows a typical holder of this type. The holder is mounted on a spindle which passes through the hole shown, and the brush slides in a rectangular slot. It is pressed on the commutator by a spring whose tension can be adjusted by placing the small lever in one of the notches shown. At the top of the brush a flexible copper "pig-tail" is mounted, and this conveys the current

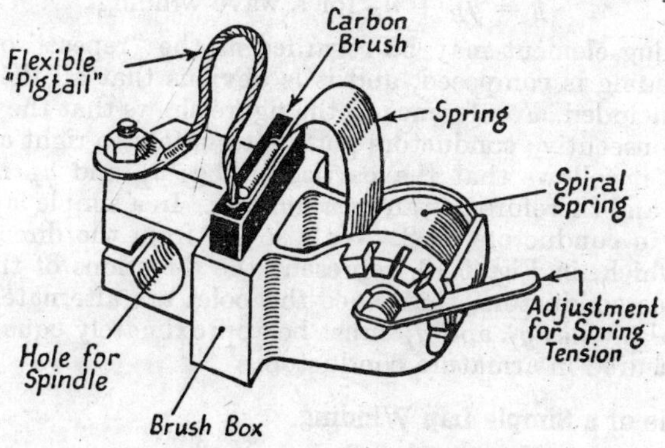

FIG. 6.15. BOX-TYPE BRUSHHOLDER

collected by the brush to the holder. There may be several brushes per spindle, according to the magnitude of the current to be collected.

Armature Windings. All modern windings are what are called lap or wave windings, or modifications of these. The differences in the two types are illustrated in Fig. 6.16, and it will be seen that these

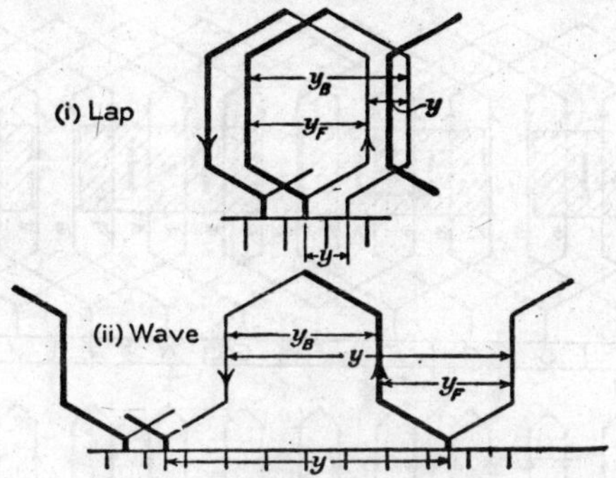

FIG. 6.16. WINDING PITCHES

differences consist merely in the arrangement of the end connections at the front of the armature. The various "pitches" are also shown in this figure. The length of the end connection at the back, measured in terms of armature conductors, is called the back pitch y_B; the length of the front connection is the front pitch y_F. The two conductors which join two commutator bars which are consecutive in the scheme of the winding, constitute a winding element, and the distance between the first and last conductors in an element is called the resultant pitch y. Hence,

$$y = y_B - y_F \text{ for a lap winding,}$$
$$y = y_B + y_F \text{ for a wave winding.}$$

The winding element may be regarded as the "repeat" of which the whole winding is composed, and it is obvious that all the conductors must be included. A reference to the figure shows that the end connections of consecutive conductors point alternately to right and left, and from this it follows that the partial pitches y_B and y_F must be *odd* numbers, and therefore y is an *even* number. In a simple lap winding y, reckoned in conductors, is 2. We also see from the directions of the arrows, which, in Fig. 6.16, represent the directions of the e.m.f.s in the various conductors, that since the poles are alternately N and S, the partial pitches y_B and y_F must be approximately equal to the pole pitch measured in armature conductors.

Example of a Simple Lap Winding.

Let No. of poles, $2p = 4$

No. of conductors, $Z = 24$

$\therefore$ pole pitch $= 24/4 = 6$

Now the partial pitches must both be about the same as the pole pitch, but they must be odd numbers. Again, their difference must be equal to 2. Suitable values are, therefore,

$$y_B = 7;\ y_F = 5$$

The complete winding shown in the form of a developed view is given

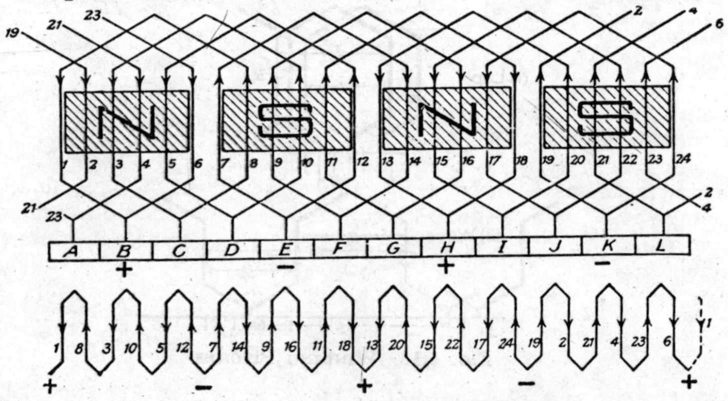

FIG. 6.17. SIMPLE LAP WINDING

in Fig. 6.17. The connections to the commutator bars are also shown, and it will be seen that there are half as many bars as conductors.

It is now necessary to find the brush positions, and this requires that the directions of the e.m.f.s should be inserted. As diagrams of armature windings are liable to be somewhat confusing, it is a good plan to draw the equivalent ring or spiral winding. The ring winding, now obsolete, consists simply of a solenoid wound on an annular ring, with equidistant tappings taken to the commutator bars. As there are no cross-overs of the end connections it is very easy to understand. The equivalent ring is shown in the bottom half of the figure, and the directions of the e.m.f.s are obtained by reference to the actual diagram. Thus, if the e.m.f.s in the conductors under the N poles are assumed to be, say, downwards, then those in the conductors under the S poles will be upwards. There will, of course, be some conductors in the inter-polar gaps, but if there is any doubt as to the direction of the e.m.f. in any particular conductor it need not be inserted. Tracing the general trend of the e.m.f.s through the equivalent ring, we find that there are two meeting points of two e.m.f.s and two separating points of two e.m.f.s. These are obviously the positions at which the + and − brushes respectively should be placed. It may happen that, according to the position of the poles relative to the winding, some of the brush positions may apparently come at the back of the armature. The brush, in such a case, must obviously be placed opposite to this position, but at the front.

The brushes of the same polarity are connected together, and therefore the armature winding is divided into four paths in parallel. In general, the number of parallel paths through a lap winding is equal to the number of poles; the terminal e.m.f. is equal to the e.m.f. induced in one parallel path; the current delivered to the external circuit is equal to the current in each armature conductor multiplied by the number of parallel paths.

Example of a Simple Wave Winding. Sometimes called a series winding. In the simple lap winding all the conductors can be included in the winding if the number of conductors N is an even number. Any even number of conductors will not necessarily do for a wave winding, for when a number of winding elements equal to the number of pairs of poles has been passed through, the winding returns practically to the starting point, namely, two conductors in front of, or behind, the starting point. Hence, the product of the resultant pitch and the number of pairs of poles must be two greater or less than the number of conductors.

$$y \times p = Z \pm 2$$

$$y = (Z \pm 2)/p$$

$$\therefore \text{ Mean pitch, i.e. } (y_B + y_F)/2 = y/2 = (Z \pm 2)/2p$$

As an example, take again $2p = 4$ and let Z be in the neighbourhood of 24 as before. We see that $Z = 24$ will not do, because the quotient of (24 ± 2) divided by 4 is not a whole number. $Z = 26$ will do, and we then have $6\frac{1}{2}$ for the pole pitch. Suitable values for the partial pitches are therefore

$$y/2 = (26 \pm 2)/4 = 6 \text{ or } 7$$

Hence, $y_B = y_F = 7$, if we take $y = 14$

If we take $y = 12$, then one of the partial pitches must be equal to 7 and the other to 5, and in such a case it is usual to make y_B the smaller, because the value of y_B fixes the width of the armature coils, and therefore the amount of copper used in the coil end when multi-turn coils are used. The values $y_B = 5$ and $y_F = 7$ are used in the winding shown in Fig. 6.18. It is assumed that the winding is in front of the poles and that it is moving from left to right, the induced e.m.f.s being therefore downwards in front of the N poles and upwards in front of the S poles. The equivalent ring winding is now drawn, and it will be noticed that the conductors in this winding are numbered, not

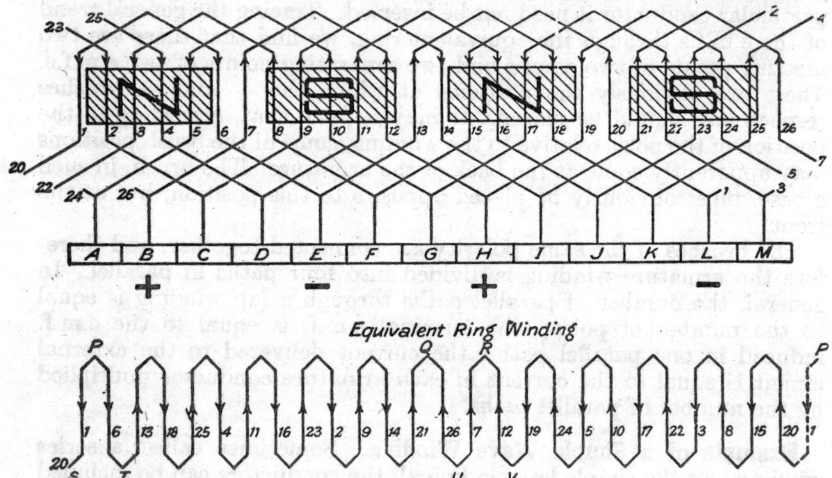

FIG. 6.18. SIMPLE WAVE WINDING

consecutively, but in the order in which they are connected together in the scheme of the winding. On inserting the directions of the e.m.f.s as obtained from the actual winding, we see that the winding divides itself electrically into two halves, namely, the portion of the winding lying between points *P* and *Q* on the equivalent diagram, and that lying between points *R* and *P*. In the first part, the e.m.f.s induced in the various conductors act in series with a general trend from left to right, while in the second part the general trend is from right to left. Hence, there can only be two parallel paths through the winding, and therefore only two brush sets are required, one positive, the other negative. The brush positions are found as follows. The point *P* is the separating point of the e.m.f.s in the two halves of the winding, and it therefore corresponds to the position of the negative brush. But it is at the back of the winding and not at the commutator end. We thus have two alternative positions for the negative brush, namely, points *S* and *T* on the equivalent winding, which corresponds to the points on the actual winding from which tappings are taken to the commutator segments *L* and *E* respectively. For the placing of the positive brush, we see from the diagram that there are two meeting points of e.m.f.s, namely, Q and R, and both of these are at the back of the winding. They are separated by one loop only, that composed of conductors 26 and 7, and therefore the middle point U of the loop is the required position. This corresponds to the tapping brought out to the commutator segment B. It appears, therefore, that the e.m.f.s in these two conductors oppose the e.m.f.s acting in the parallel paths in which they lie, but on referring to the actual winding it will be seen that they are each situated almost in the middle of an interpolar gap, the e.m.f.s induced in them being therefore very small.

For one positive brush position, we thus see that there are two alternative positions for the negative brush. Now consider the two conductors 26 and 7 situated between the points Q and R. The winding is moving across the pole faces so that its position relative to the poles as represented by the figure is only instantaneous. We see that conductor 26 is just about to move from the influence of a S to that of a N pole, and the e.m.f. in it is therefore on the point of reversing. But conductor 7 is already past the point where its e.m.f. is reversed, so that its e.m.f. will not change in direction but will gradually increase in magnitude. This means that in a very short interval of time the point R will become the meeting point of two e.m.f.s, and since it is at the back of the winding, there will be two alternative positions for the positive brush, namely, point U, which has already been considered, and the point V, corresponding to commutator segment H. Similarly, when the armature has moved a little farther forward the segment I adjacent to H will become the second alternative position for the positive brush. Hence, if one positive brush is making contact with segment B and not either of the adjacent segments, then the second positive brush, if used, should be in contact with both of the segments H and I.

If brushes are placed in both alternative positions for both positive and negative, the effect is merely to short-circuit one loop of the winding between two brushes of the same polarity. Thus, whether only two or four brushes are used, the number of parallel paths still remains at two. Also these loops during the very short period of short-circuit occupy a position almost symmetrical with respect to one of the poles so that the actual e.m.f. acting round the loop and tending to produce circulating currents through the loops and connections to brushes of like polarity is very small.

The e.m.f. developed by a simple wave winding is therefore equal to that induced in one-half of the total number of armature conductors all connected in series, whereas the current delivered to the external circuit is twice the current in an individual armature conductor.

Series-parallel Windings. These are windings in which the number of parallel paths, $2a$, through the winding is greater than 2 but less than $2p$, the number of poles. Thus a 6-pole machine will have 6 parallel paths with a lap winding, and 2 with a wave, but it can have 4 with a series-parallel winding. The winding equation is

$$\text{Mean pitch } y/2 = (Z \pm 2a)/2p$$

where a is any even number greater than 1 and less than $2p$. We have already seen that there are half as many commutator bars as coil sides Z, so that the amount $y/2$ is equal to the pitch y_K in commutator bars.

$$\therefore y_K = (Z \pm 2a)/2p$$

The number of coils is equal to $Z/2$ and there are two cases:

(*a*) If y_K and $Z/2$ have no common factor the winding will close on itself once only, like the simple wave winding.

(*b*) If y_K and $Z/2$ have a common factor, the whole winding will consist of a number of independent closed windings, the number being equal to this common factor.

EXAMPLE 1. Let $Z = 58$, $2p = 6$ poles, $a = 2$

$$\therefore y_K = \frac{58 \pm 4}{6} = 9, \text{ taking the minus sign}$$

Hence we can make

$$y_B = y_F = 9$$

also since y_K and $Z/2 (= 29)$ have no common factor the winding will consist of a single coil, that is, it will be singly re-entrant.

EXAMPLE 2. Let $Z = 56$, $2p = 6$, $a = 2$

$$\therefore y_K = \frac{56 \pm 4}{6} = 10$$

since we must take the positive sign.

The partial pitches must now be unequal and we can make $y_B = 9$ and $y_F = 11$. y_K and $Z/2$ $(= 28)$ have a common factor of 2, and therefore the winding will consist of two entirely separate windings. It will therefore be doubly re-entrant.

Action of the Commutator. Fig. 6.19 shows a small portion of a lap winding moving from left to right. The conductors under the influence of a N pole have induced e.m.f.s which act downwards, while those under a S pole have e.m.f.s which act upwards. In the figure the conductors marked 1, 2, and 3 are at the top of a slot, and those marked 1′, 2′, and 3′ at the bottom. Consider first the coil formed by joining conductors 1 and 1′ by the front connector. No. 1 is clearly nearer to the S pole than to the N and so its e.m.f. acts upwards. Conductor 1′ is under the N pole and so its e.m.f. acts downwards. Thus as far as the loop formed by these two conductors is concerned the two e.m.f.s act in the same sense and so the nose of the coil at A is neither a meeting point nor a separating point of two e.m.f.s.

Examining in the same way the coil formed by conductors 2 and 2′ we see that although conductor 2′ has a downward e.m.f., conductor 2 is so near to the middle of the interpolar gap that its induced e.m.f. will be zero or negligible. Hence point B also is neither a meeting point nor a separating point of two e.m.f.s.

The e.m.f.s in conductor 3 and 3′ forming the third coil in the figure both act downwards and therefore point C is a meeting point of two e.m.f.s. In other words point C is a point of maximum potential in the winding. Now when the armature winding has moved a distance equal to BC, the point B will be where C is at present and the coil 22′ will occupy the position at present occupied by coil 33′. Hence the point B will now be the point of maximum e.m.f., but only in virtue of its new position with respect to the poles. We thus see that the points of maximum potential, and similarly the points of minimum potential are fixed in space, and therefore the potential distribution as a whole

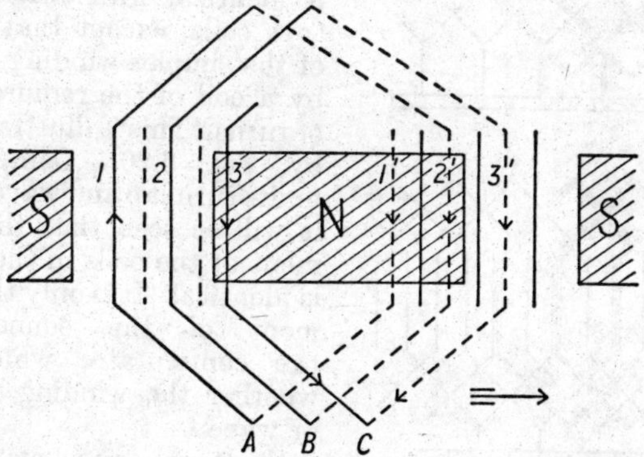

FIG. 6.19. TO ILLUSTRATE THE FUNCTION OF THE COMMUTATOR

is fixed in space. But we have made no reference whatever to the commutator, and therefore this member has nothing whatever to do with the fact that the potential distribution round the winding is fixed in space.

As shown before, the brushes are placed at meeting points and separating points of two e.m.f.s, but strictly speaking, these points are at the noses of the coils. If a d.c. armature were constructed without a commutator, but the insulations were stripped at the noses so that brushes could press on the exposed copper the machine could function as a dynamo or motor, but it would not be long before the coils were worn through and a completely new armature winding required. Thus, in fact, the commutator is merely an extension of the armature windings, and its function is to withstand the wear and tear of current collection. From time to time it can be turned or ground to a true cylindrical surface, and when this is no longer possible a new commutator can be fitted.

Multi-turn Windings. The single-turn windings, described above, are used whenever possible. Such windings necessitate half as many commutator bars as armature conductors. If the voltage to be generated is high, particularly in a small machine, it will obviously be impossible to house only two conductors in a slot, one at the top and the other at the bottom, since this would necessitate too many slots.

In such a case, six or eight conductors may be placed in one slot, those at the top being taped together, and also those at the bottom. The coils so formed are still single-turn coils, since they each consist of three or four single turns, insulated from one another, and mounted together for convenience.

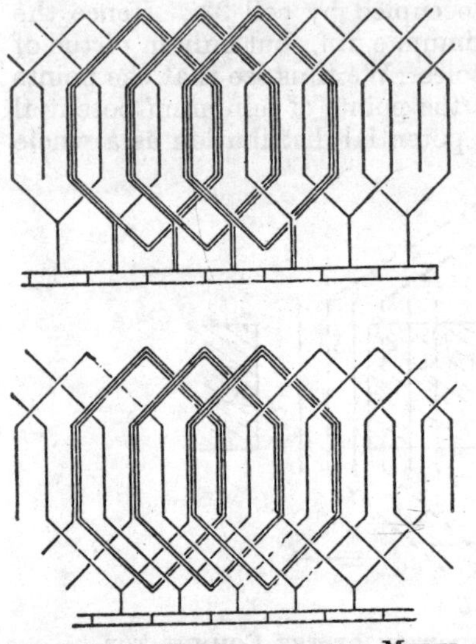

Fig. 6.20. Windings with Multi-turn Coils

In the case of small machines it is necessary to adopt multi-turn coils. The winding in such a case is identical with that for single-turn coils, except that each loop of the simpler winding is replaced by a coil of the required number of turns. This is illustrated clearly by Fig. 6.20. Comparing the multi-turn lap and wave windings, it will be seen that the arrangement of the coils in the two cases is identical. It is only the arrangement of the connections to the commutator which decides whether the winding will be lap or wave.

Since the construction of an actual winding is by no means obvious from an examination of the developed diagrams, such as those given in Figs. 6.17 and 6.18, Fig. 6.21 has been drawn. This figure shows the actual arrangement of a four-pole lap-wound armature, having as many elements as the winding in Fig. 6.17, namely, 24. Each coil consists of several turns, and there are two coil sides per slot, each coil side consisting of only one element. There are, therefore, half as many

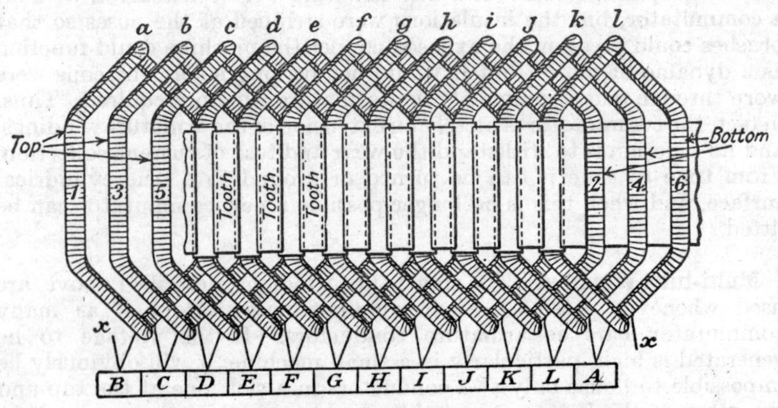

Fig. 6.21. Four-pole Lap-wound Armature

slots as elements, that is, 12 slots in the example. It would, of course, be possible to tape up coils a and b into a composite coil with two elements per coil side, similarly with coils c and d, and so on. In such a case the number of slots would be only one-half the number required when there are only two elements per slot. Similarly, if coils a, b, and c were taped up into a composite coil, likewise d, e, and f, and so on, the winding would still be essentially the same, although it would now only require one-third as many slots as when there are only two elements per slot. For a clear understanding of the completed winding it is therefore necessary to realize that (a) the winding is arranged in two layers, and (b) that what appears to be a single coil on the armature may in reality consist of several quite independent coils all taped up together. This is indicated by a comparison of the number of commutator bars and of slots.

Numerical Example. The actual disposition of an armature winding in the slots is decided, in practice, by

(a) the pitch of the commutator segments;
(b) the total ampere-conductors per slot;
(c) the voltage between adjacent commutator segments.

The segment pitch should not be less than 0·5 cm, even on a small machine, nor greater than 1·0 cm on a large machine. This shows that the range for this particular dimension is very small. There is some variation in the total ampere-conductors per slot, the upper limit for even a very large machine being about 1000. For a machine of 100 kW it will be of the order of 550. The mean voltage between adjacent commutator segments should not exceed 15 V.

We will now consider a numerical example and show how the above requirements are fulfilled. Suppose that a shunt dynamo has to be designed to give an output of 80 kW at 230 V, the speed being 500 r.p.m. Then output current = 80,000/230 = 348 A. Allowing for the drop of volts in the armature and interpoles, and for the exciting current, we will assume round values of

240 V for the induced e.m.f.

350 A for the total armature current.

The designer will know immediately how many poles the machine will have, and what will be the diameter of the armature. His preliminary calculations will then give the core length, from which the next step will be the flux per pole. We will therefore assume that the designer has decided on the following—

$$\text{No. of poles} = 4$$

$$\text{Flux per pole} = 6{\cdot}4 \times 10^{-2} \text{ weber}$$

$$\text{Armature diameter} = 53 \text{ cm}$$

$$\text{No. of slots} = 75$$

$$\text{Commutator diameter} = 35 \text{ cm}$$

It is proved, p. 178, that the induced e.m.f. of a d.c. machine is given by

$$E = \Phi ZN(P/A) \text{ volts}$$

where Φ = flux per pole, in webers
Z = No. of armature conductors
N = speed in revolutions per second
P = No. of poles
A = No. of parallel paths through the armature winding

$$\therefore Z = (E/\Phi N)(A/P)$$

(*a*) Lap winding: $A = P$

$$\therefore Z = (240/6{\cdot}4 \times 10^{-2})(60/500)$$
$$= 450$$

As this is divisible by 75, the number of slots, giving a quotient of 6, we see that 6 conductors per slot are possible. We will now check the various requirements.

First the pitch of the commutator segments. If we use single-turn coils

$$\text{No. of bars} = Z/2 = 225$$
$$\text{Comm. periphery} = \pi \times 35 = 110 \text{ cm}$$
$$\therefore \text{Segment pitch} = 110/225 = 0{\cdot}5 \text{ cm, almost exactly.}$$

Second, the total ampere-conductors per slot.

$$\text{Conductor current} = (\text{total current})/(\text{parallel paths})$$
$$= 350/4 = 87{\cdot}5 \text{ A}$$
$$\therefore \text{Slot loading} = 6 \times 87{\cdot}5$$
$$= 525 \text{ ampere-conductors}$$

Voltage between adjacent segments.

Since the brushes have an angular distance equal to the pole pitch and there are 4 poles, the full terminal voltage of 230 V will exist over one-quarter of the commutator periphery, that is over 225/4, say 56 bars

$\therefore$ average voltage between segments

$$= 230/56 = 4{\cdot}1 \text{ V}$$

The above calculations show that a satisfactory winding will be a lap winding having single-turn coils and having six coil sides per slot, three at the top, and three at the bottom.

(*b*) Wave winding: $A = 2$; $P = 4$

$$\therefore Z = (240/6{\cdot}4 \times 10^{-2})(60/500)(2/4)$$
$$= 225$$

This, when divided by the number of slots gives the odd number of 3, which is not possible. If we adopt 2 the number of slots will be too large, and if we adopt 4 the number will be rather low. However, for the purpose of comparison we will adopt 4 conductors per slot, the number of slots thus being 57 (it is preferably odd for a wave winding), and the actual number of conductors thus

$$Z = 4 \times 57 = 228$$

With single-turn coils the number of segments will be

$$228/2 = 114$$

$\therefore$ Segment pitch = 110/114 = 1·0 very nearly, which is much too large for a small machine.

The current per conductor for a wave winding

$$= 350/2 = 175 \text{ A}$$

$\therefore$ Slot loading = 4 × 175 = 700 ampere-conductors, and this is too large.

The full voltage is again built up over one-quarter of the commutator periphery, i.e. over 114/4, say 28 segments

$\therefore$ average voltage between segments

$$= 230/28 = 8{\cdot}2 \text{ V}$$

Thus in every respect the lap winding is the better for the particular specification of 80 kW, at 230 V, and 500 r.p.m.

Now suppose that the specification were to be 80 kW at 460 V, and 500 r.p.m. Then clearly the same field system and the same size armature core would do. Considering first of all the lap winding we could double the voltage by doubling the numbers of conductors, thereby making Z equal to 900, and having 12 per slot. This would be possible because, with the current being halved, the conductor section is halved and the slot will hold all 12 conductors and still have the previous slot loading of 515 ampere-conductors.

If single-turn coils are attempted then the number of segments will be doubled, i.e. 450, and so the segment pitch will be only 0·25 cm. This is much too low, and the difficulty can now be met by having coils of two turns each, thereby retaining the previous number of segments, namely 225, and the segment pitch of 0·5 cm.

As the voltage is doubled, but the number of segments kept as before, the mean voltage between segments now becomes

$$2 \times 4{\cdot}1 = 8{\cdot}2 \text{ V}$$

Such a machine would be perfectly satisfactory. The disadvantages neither of them serious, are: firstly, a two-turn coil gives a higher reactance voltage than a single-turn coil (*see* p. 188), secondly, a conductor current of nearly 44 amperes requires a rectangular copper strip and this makes a rather awkward nose at the end remote from the commutator because the two bends do not fit very snugly together.

Now consider a wave winding for the new specification. We now have an induced e.m.f. of about 480 V so that

$$Z = (480/6{\cdot}4 \times 10^{-2})(60/500)(2/4)$$
$$= 450$$

This is exactly divisible by 75, the quotient being 6, showing that we can adopt 75 slots and 6 conductors per slot. With single turn coils the number of segments will be 225 and the segment pitch 0·5 cm. The reader can prove for himself that the slot loading and voltage between bars are also satisfactory. Hence for the higher voltage of 460 V the wave winding is more satisfactory than the lap.

With very small machines of only a few kW output, then multi-turn coils are essential, but as the conductor current is very small, round wire can be used, and this presents no difficulties in winding.

Equalizer Rings. In a lap winding any path from brush to brush consists of conductors under the influence of two adjacent poles only and not all the poles. If the fluxes from all the poles are equal, the e.m.f.s induced in all the paths will be equal, and when there is no current in the external circuit there will be zero current in the armature. In practice it is impossible to obtain absolutely equal fluxes from each pole, the result being that the e.m.f.s in the various parallel paths are not equal. The resulting e.m.f. acting round the armature may give rise to large currents even with no external load. When the armature is delivering current the circulating current is superposed on the load current and produces unequal heating in the armature.

Referring, again, to the lap winding of Fig. 6.17, if we take the conductors lying between the first + brush and the first − brush positions, we see straight away from the equivalent ring winding that these are the conductors numbered 1, 8, 3, 10, 5, and 12. The poles influencing these conductors are the first north and south poles. Similarly with the other three parallel paths, the conductors in them are influenced by two poles only. It therefore follows that if the magnetic fluxes entering the armature from each of the poles are not exactly equal, the e.m.f.s induced in the various parallel paths will not be equal. Since the armature winding forms a closed circuit independent of any external connections to the brushes, and since, with respect to the armature itself, these paths are *in series*, we see that the e.m.f.s set up in the armature will not exactly balance out. Hence, a parasitic current will be set up in the armature winding, this current being independent of, and superposed on, the current due to the external load. In order to illustrate this point, the equivalent ring winding of the lap winding previously considered is reproduced in a modified form in Fig. 6.22. The e.m.f.s in the four paths are indicated as E_1, E_2, E_3, and E_4, their directions being indicated by the arrows. The resultant e.m.f. acting round the winding in a given direction, say, clockwise, is

$$(E_2 + E_4) - (E_1 + E_3)$$

If these four e.m.f.s are exactly equal, their resultant will, of course, be zero, and no parasitic currents will be set up round the winding. If they are not exactly equal, but have a resultant equal to, say, e, then a circulating current will be set up, its magnitude being equal to

$$i = e/R$$

where R is the resistance of all the armature conductors and their end-connections reckoned in series, and not the armature resistance as measured from brush to brush. This current will be quite independent of the external load on the machine, in fact, it will flow even when the brushes are lifted from the commutator.

When the brushes are in place and the machine is delivering current to an external load, the conditions are not quite so simple. In the simplified scheme of Fig. 6.22 it appears that since the four e.mf.s, E_1,

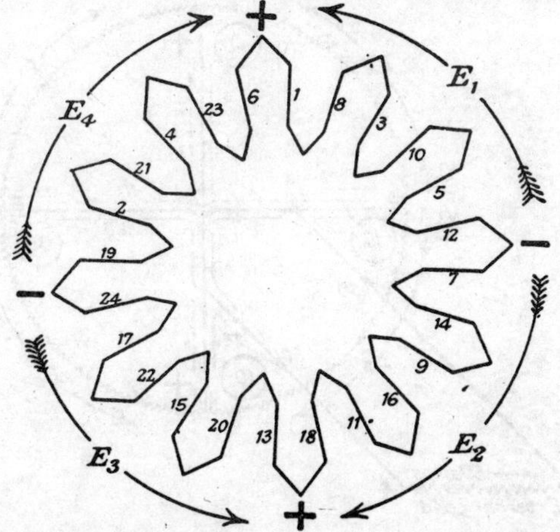

FIG. 6.22. BALANCE OF E.M.F.S IN A LAP WINDING

E_2, E_3, and E_4, are not equal, the two positive brushes can be at slightly different potentials, and also the two negative brushes. Actually such differences of potential between brushes of like polarity cannot exist, because such brushes are joined together externally by copper straps of negligible resistance. These connections are shown in Fig. 6.23, in which the winding is again represented as a ring winding for simplicity. It is obvious that currents will flow along these connections, and since their magnitudes are only limited by the resistance of the armature and the brush-contact resistances, they may assume very large proportions. In the figure they are indicated as 10 and 15 A, while the load current is represented as 50 A. Now the load current divides into two halves of 25 A each at the brush connections, and these halves divide again into quarters of 12½ A each, through each armature path, these currents all being indicated in the figure. Hence, when we superpose the circulating current through the brush connections on to the true load currents, we find that the currents collected by the four brush sets are 40, 35, 10, and 15 A, as indicated by the numbers in small circles. Actually each brush should collect 25 A. Obviously the overloaded brushes will give trouble; their temperature rise will become much too high, and they will probably spark badly, no matter how their position may be adjusted. It is to be noted, however, that in the case of lap (not wave) windings, any very excessive unbalancing of the current in the different armature paths is to some extent automatically checked by the effect of armature reaction, to be considered later.

It will be seen that the conditions represented by Figs. 6.22 and 6.23 are very different. Without the connections between brushes of the same polarity the brush potentials are incorrect, but when these connections are made, as in Fig. 6.23, the brush potentials are corrected,

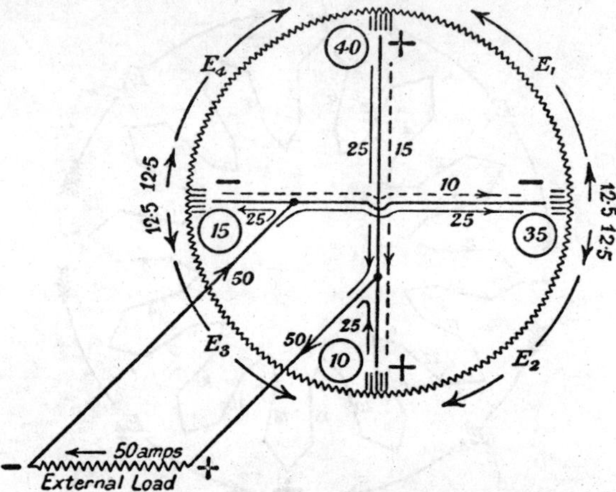

Fig. 6.23. Effect of Brush Connections on Distribution of Current in a Lap Winding

but at the expense of incorrect proportioning of the current between the various brush sets. Now, these connections put the potentials right, and therefore if similar connections could be made without involving the brushes at all, the potentials and current distribution *at the brushes* would be corrected. The clue is given by the fact that the brushes joined together are all the same potential, and therefore all that is necessary is to join together by a permanent connection (i.e. not involving the brushes) points on the winding which should be at the same potential. Such points are very easy to find from the wiring diagram. Thus, in Fig. 6.17 conductors 1 and 13 occupy exactly the same positions with respect to the two north poles, so that their potentials are the same. Also, conductors 8 and 20 are at the same potential, and since conductors 1 and 8 form one loop, while 13 and 20 form another loop, the middle points of their end connections, namely, the apexes of the end connections joining 1 to 8 and 13 to 20, could be connected together. Similarly other pairs which could be joined together can be easily found. Actually the number of points to be joined by any one connection is equal to half the number of poles. Such connections are called equalizing connections, and they are usually carried out by rings joined to tappings taken off from the back of the armature. Fig. 6.24 shows the elementary scheme of equalizer connections, the number of rings in this case being three. A very large armature may have as many as 8 or 10 equalizer rings. A more complete

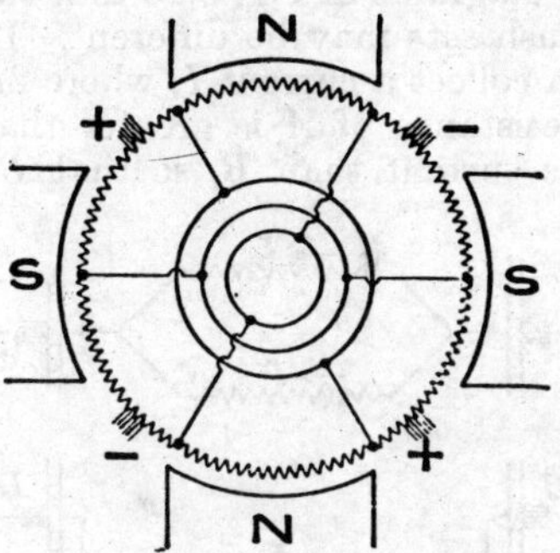

FIG. 6.24. EQUALIZER CONNECTIONS

diagram of equalizer connections is given in Fig. 6.25. Notice that there are three tappings to each equalizer, three being half the number of poles.

Selective Commutation. Since each path in a wave-wound armature consists of conductors distributed under all the poles, inequalities in

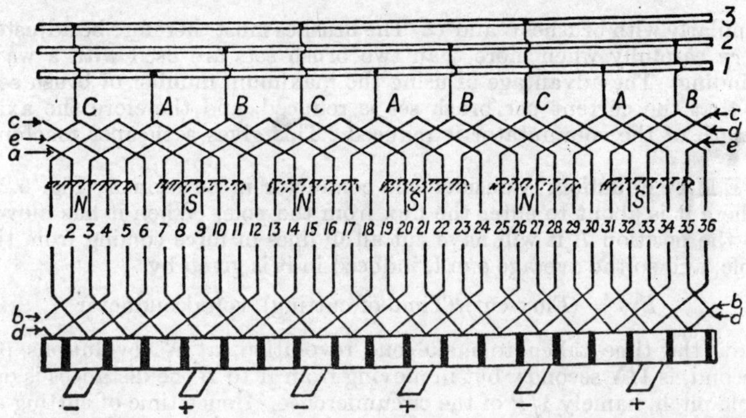

FIG. 6.25. SIX-POLE LAP WINDING SHOWING CONNECTIONS TO EQUALIZER RINGS, 36 CONDUCTORS

the fluxes from the poles do not produce inequalities in the total e.m.f.s induced in each of the two paths. As a result, equalizer connections are not required with a simple wave winding. If more than two brush sets are used, then trouble with the brushes may be produced by what is called "selective commutation." The use of more than two brushes does not divide the armature into more than two parallel paths, but the current collected from the armature is divided between the brushes of like polarity. If there are differences in contact resistance, then it is easy to see, from the diagrams in Fig. 6.26 that the currents picked off by the individual brush sets may be different. Thus the two brushes A and B should each collect a current I, where $2I$ is the load current, but if the contact resistance of A is greater than that of B, A will obviously collect less current than B, so that B will be overloaded.

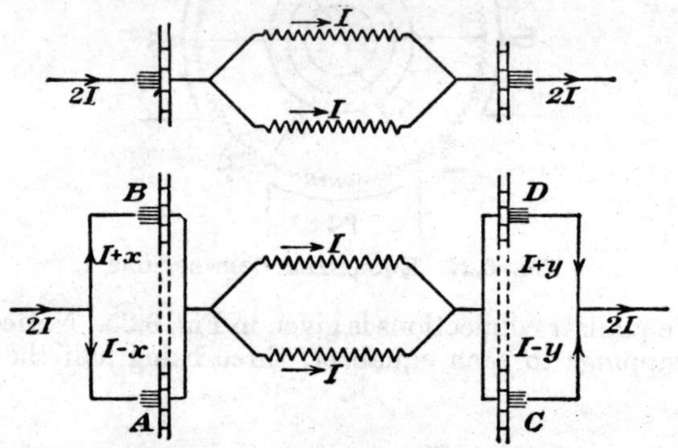

FIG. 6.26. SELECTIVE COMMUTATION

Similarly with brushes C and D. The brushes must therefore be adjusted very carefully when more than two brush sets are used with a wave winding. The advantage of using the maximum number of brush sets is that the current per brush set is reduced, and therefore the axial length of the commutator is reduced. This gives a cheaper machine.

E.M.F. Equation. Consider one conductor in position A, Fig. 6.27, where it is about to enter the flux from the pole. When it has moved to the position B it will have cut all Φ lines of force coming from the pole, and so the average e.m.f. induced in it is given by

$$E_{av} = (\text{Flux cut})/(\text{Time of cutting}) \text{ volts/conductor}$$

Now the time taken to make one revolution, at N revolutions per second, is $1/N$ second: but in moving from A to B the distances is one pole pitch, namely $1/P$ of the circumference. Hence time of cutting all Φ lines is $1/NP$ second,

$$\therefore\ E_{av} = \Phi \div 1/NP = \Phi NP \text{ volts/conductor}$$

But there are Z conductors arranged in A parallel paths, the number in series per path thus being Z/A. Again all these conductors occupy during a movement of one pole pitch all the possible positions in the field, so that their summation gives, not the average e.m.f. for the whole winding, but the actual value which appears at the brushes

$$\therefore\ E = \Phi NP(Z/A) \text{ or } E = \Phi ZN(P/A) \text{ volts}$$

Another derivation of this equation is as follows—

Fig. 6.28 shows a portion of a multipolar machine. The flux through

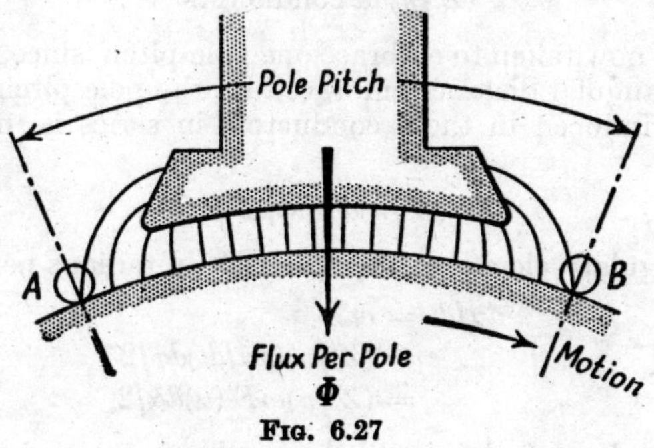

FIG. 6.27

the armature core is Φ/2 and any particular conductor is associated with this flux. The opposite conductor with which it forms a coil is

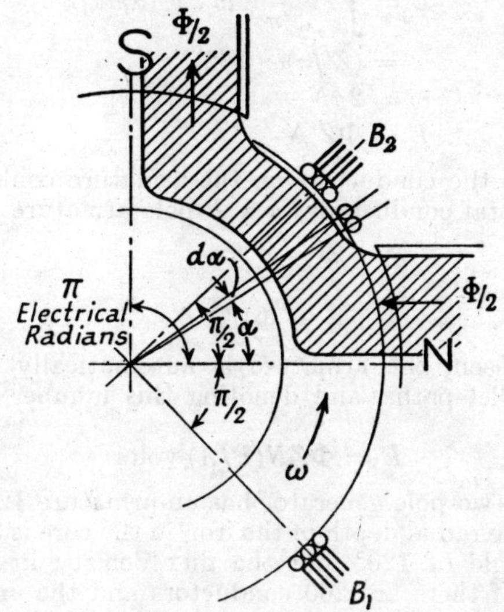

FIG. 6.28. PERTAINING TO THE E.M.F. EQUATION

associated with the flux $\Phi/2$ in the core opposite, so that we can regard the flux associated with one coil as the flux per pole Φ.

$$\therefore \text{ e.m.f. per coil} = d\Phi/dt$$

$$\therefore \text{ e.m.f. per conductor} = (d/dt)(\Phi/2)$$

Now the flux through any section of the core varies with the angular position α, and can be written

$$\phi = F(\alpha)$$

Let there be Z' conductors per pole, then within the distance $d\alpha$ there are

$$(Z'/\pi)d\alpha \text{ conductors}$$

The angle π is now taken to embrace one pole-pitch, since, in a two-pole machine, the angular distance embraced by the pole pitch is π radians.

The e.m.f. induced in these conductors in series is, neglecting the negative sign,

$$e = (Z'/\pi)d\alpha(d\phi/dt)/2$$

If ω is the angular velocity of the armature in radians per second

$$d\alpha/dt = \omega$$

$$\therefore\ e = (Z'/\pi)(d\alpha/dt)d\phi/2$$

$$= (Z'/\pi)\omega F'(\alpha)d\alpha/2$$

Hence, the total e.m.f. due to all the conductors in series between the brush positions B_1 and B_2 is

$$E = \int_{-\pi/2}^{\pi/2} \frac{Z'}{\pi} \cdot \omega \cdot F'(\alpha)d\alpha/2$$

$$= (Z'/\pi)\omega(\Phi/2)$$

Now $$\omega = 2\pi N$$

$$\therefore\ E = \Phi Z'N$$

Suppose that all the conductors on the armature could be connected in series, then total conductors for a P-pole armature

$$Z = Z'P$$

and we should have

$$E = \Phi Z'NP$$

As we have seen, the armature is automatically divided into a number of parallel paths, and denoting this number by A we have finally

$$E = \Phi ZN(P/A) \text{ volts}$$

Example. A two-pole generator has an armature 12 in. long, 12 in. diameter, and the radial depth of the iron in the core is 3 in. Each pole subtends an angle of 120° and the flux density in the air gap is 0·5 weber/m². If there are 200 conductors and the armature rotates at 1000 r.p.m., find (*a*) the e.m.f. induced; (*b*) the induction in the core.

Each pole subtends one-third of the periphery,

$$\therefore \text{ Area of air gap} = 12(\pi \times 12/3) \times 2{\cdot}54^2 \times 10^{-4} \text{ m}^2$$

$$= 9{\cdot}75 \times 10^{-2} \text{ m}^2$$

$$\therefore \text{ Flux per pole, } \Phi = 9{\cdot}75 \times 10^{-2} \times 0{\cdot}5 = 4{\cdot}875 \times 10^{-2} \text{ webers}$$

Since $$P = 2, A \text{ also must be } 2$$

$$\therefore\ E = \Phi ZN = 4{\cdot}875 \times 10^{-2} \times 200 \times (1000/60)$$

$$= 162 \text{ V}$$

The flux through the cross section of the core is one-half the useful flux, i.e. $2 \cdot 44 \times 10^{-2}$ webers.

$$\text{Cross section of core} = \text{axial length} \times \text{radial depth}$$
$$= 36 \text{ in.}^2 \text{ or } 2 \cdot 33 \times 10^{-2} \text{ m}^2 \text{ gross}$$

A portion of this will be taken up by insulation between core discs and, if 10 per cent is allowed for this, the net section will be

$$0 \cdot 9 \times 2 \cdot 33 \times 10^{-2} = 2 \cdot 1 \times 10^{-2} \text{ m}^2$$

$\therefore$ Flux density in core

$$= 2 \cdot 44 \times 10^{-2}/(2 \cdot 1 \times 10^{-2})$$
$$= 1 \cdot 16 \text{ webers/m}^2$$

Fluctuations in the Induced E.M.F. An elementary armature, consisting of a single coil with a two-part commutator, gives an e.m.f. which falls periodically to zero. As the number of coils is increased the magnitude of the voltage fluctuation, in relation to the minimum voltage decreases, becoming very small when the number of coils is large. Consider the case in which the distribution of flux density round the air-gap is sinusoidal. Reckoning angular distances from the neutral plane we have, at any angular position θ, (Fig. 6.29)

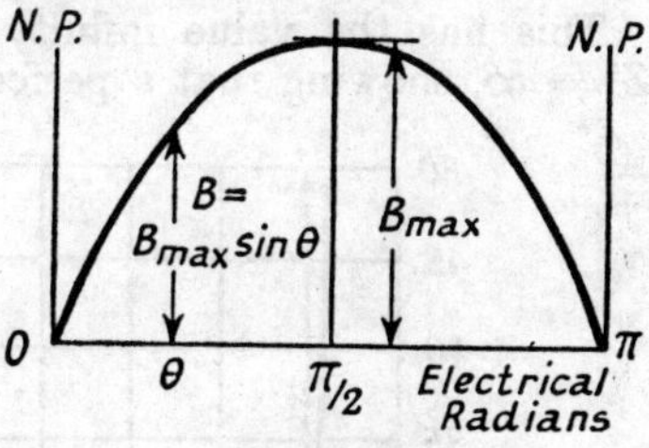

FIG. 6.29

$$B = B_m \sin \theta$$

The instantaneous e.m.f. induced in a conductor at this position is therefore

$$e = Blv = B_m lv \sin \theta$$

Let there be Z' conductors (or conductor groups in the case of a machine with several turns per coil) per pole pair, then, electrically, two pole pitches correspond to 2π radians, so that the spacing of consecutive conductors is $2\pi/Z'$. The instantaneous e.m.f.s in successive conductors or groups are therefore

$$e_1 = B_m lv \sin \theta$$
$$e_2 = B_m lv \sin (\theta + 2\pi/Z')$$
$$e_3 = B_m lv \sin (\theta + 2 \,.\, 2\pi/Z'), \text{ etc.}$$

A complete half wave of air-gap flux density has an angular spread of π and therefore embraces $Z'/2$ conductors. The final term in the series is thus

$$e_{Z'/2} = B_m lv \sin [\theta + (Z'/2 - 1)(2\pi/Z')]$$
$$= B_m lv \sin [\theta + (\pi - 2\pi/Z')]$$

The total e.m.f. contribution of one pole is thus

$$\Sigma(e) = B_m lv [\sin \theta + \sin (\theta + 2\pi/Z') + \ldots$$
$$+ \sin \{\theta + (\pi - 2\pi/Z')\}]$$

This expression has a minimum value when $\theta = 0$ and a maximum value when $\theta = \pi/Z'$, the total e.m.f. therefore fluctuating between these values. We have

$$E_{\min} = B_m lv \{\sin(2\pi/Z') + \sin(4\pi/Z') + \ldots + \sin(\pi - 2\pi/Z')\}$$
$$= B_m lv \cot(\pi/Z')$$
$$E_{\max} = B_m lv \{\sin(\pi/Z') + \sin(3\pi/Z') + \ldots + \sin(\pi - \pi/Z')\}$$
$$= B_m lv \operatorname{cosec}(\pi/Z')$$

The percentage e.m.f. fluctuation is given by

$$100 \times \frac{(E_{\max} - E_{\min})}{E_{\min}}$$
$$= 100 \times \frac{\operatorname{cosec}(\pi/Z') - \cot(\pi/Z')}{\cot(\pi/Z')}$$

This has the value infinity when $Z' = 2$ and the value 0 when $Z' = \infty$, showing that a perfectly uniform voltage is not practicable.

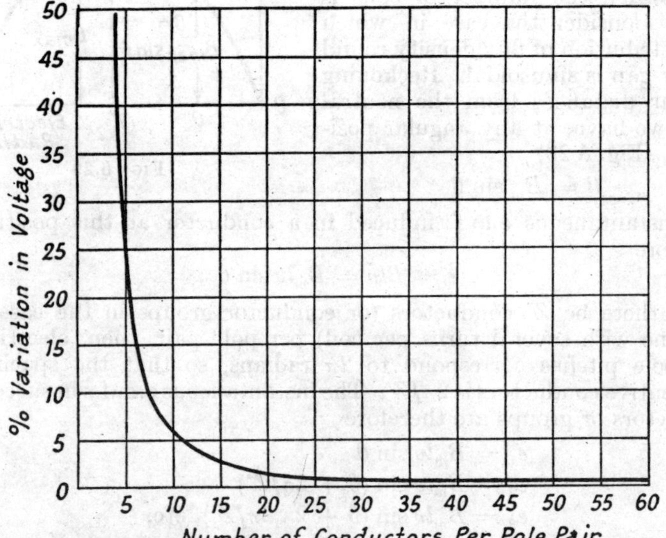

FIG. 6.30. DEPENDENCE OF VOLTAGE FLUCTUATION ON THE NUMBER OF CONDUCTORS PER POLE PAIR

The value of Z' to give any particular value of the percentage fluctuation can be calculated, the simplest plan being to give Z' a series of values and to plot the corresponding percentage fluctuation. This is done in Fig. 6.30. Thus suppose that we wish to limit the percentage variation to 0·5, then Z' must be not less than 34.

Further Reading

CLAYTON, A. E. *Performance and Design of D.C. Machines.* Pitman, London, 1938.
HAWKINS, C. C. *The Dynamo.* Pitman, London, 1922.
WALKER, M. *Specification and Design of Dynamo Electric Machinery.* Longmans, London, 1917.
SMITH, S. P. and SAY, M. G. Electrical Engineering Design Class Manual. Oxford U.P., Oxford, 1950.
An excellent statement of the theory of d.c. armature windings is given in
SMITH, S. P. *Notes on Theory and Design of Continuous Current Machines.* Benn, London, 1920.
An exhaustive account is also given in the following paper—
SMITH, S. P. "Theory of armature windings." *J. Inst. Elect. Engrs.* **59.** No. 261 (1916), 18.

Examples on Chapter 6

1. The two-circuit armature of a four-pole generator has 51 slots, each slot containing 20 conductors. What will be the voltage generated in the machine when driven at 1500 r.p.m., assuming the useful flux per pole to be $0{\cdot}7 \times 10^{-2}$ wb? (L.U.)

Ans. 357 V.

2. State Faraday's Law of magneto-electric induction, and find *from first principles* the voltage of a four-pole lap-wound continuous current generator having a pole flux of 2×10^6 maxwells, and 560 conductors running at a speed of 980 r.p.m. How is the terminal voltage affected by load? (L.U.)

Ans. 183 V.

3. Calculate the resistance of an armature from the following data—

Number of slots = 150.
Conductors per slot = 8.
Mean length of one turn = 250 cm.
Cross section of each conductor = $1{\cdot}0 \times 0{\cdot}25$ cm.
Number of parallel paths through armature = 6.
Specific resistance of copper at working temperature = 2×10^{-6} Ω-cm.

(If the resistance of one parallel path is calculated, the actual resistance will be one-sixth of this.)

Ans. 0·033 Ω.

4. A four-pole armature is built up of 400 core discs each $\frac{1}{40}$ in. thick; the discs are 12 in. external and 6 in. internal diameter. There are 600 conductors and the armature is wave wound. If the flux density in the armature is one wb/m², and the speed is 1000 r.p.m., what will be the e.m.f. generated?

Ans. 778 V.

5. The exciting winding of a dynamo has 2000 turns per pole and the exciting current is 2 A. The armature, which is lap wound, has 500 conductors, rotates at 1200 r.p.m., and gives 100 V. Calculate the reluctance of the magnetic circuit.

Ans. The reluctance of an individual magnetic path is $1{\cdot}6 \times 10^{-6}$.

6. A 40 h.p., 500 V traction motor has four poles. There are only two brush arms. Assuming that there are 41 slots and 123 commutator bars, make a diagram sufficient to explain how the armature winding may be arranged and connected to the commutator. It is sufficient to show the connections of three coils. (L.U.)

7. What is the object of equalizing connections on an a.c. generator, and to what class of winding are they applied? Give a diagram to show the way in which these equalizers are connected up to the armature winding. (L.U.)

8. A 4-pole generator with 1,200 conductors generates 250 V on open-circuit when driven at 500 r.p.m. The pole shoes have a bore of 14 in. and the ratio pole arc/pole pitch is 0·7, while the length of the shoes is 8 in. Find the mean flux density in the air gap. (I.E.E.)

***Ans.* 0·632 Wb/m^2.**

9. A 300 kW, 300 r.p.m. d.c. generator has 8 poles, and the armature has a series-parallel winding with 4 parallel paths between terminals. The no-load speed is 3 per cent greater than the full-load speed. If there are 684 armature conductors calculate the flux per pole on no load.

***Ans.* 7·3 × 10^{-2} webers.**

10. The armature of the machine in question 9 has single-turn coils, the total length of each being 1·60 m. The conductor cross-section is 0·37 × 10^{-4} m^2 and the specific resistance at the working temperature 2 × 10^{-2} $\mu\Omega$-m. Calculate the resistance of the winding.

***Ans.* 0·0185 Ω.**

11. Find possible windings for a 4-pole, 2-circuit armature. The number of conductors in series must not be greater than 400 nor less than 360; the number of slots is to be between 40 and 45, and the number of commutator segments within 126 ± 4. (C. and G.)

***Ans.* 738, 750, or 774 conductors.**

CHAPTER 7

ARMATURE REACTION AND COMMUTATION IN D.C. GENERATORS

Armature Reaction. Armature reaction is the effect of the magnetomotive force set up by the armature currents, on the distribution of flux under the main poles. Consider an armature rotating in a clockwise direction in a bi-polar field, as shown in Fig. 7.1. Let the brushes make contact with those conductors which lie in the geometric neutral plane (g.n.p.). Then all the conductors under the N pole carry currents whose directions are inwards, while the currents in the conductors under the S pole are outwards. This distribution of armature currents is magnetically equivalent to a solenoid carrying current, and we therefore see that the armature sets up a m.m.f. directed along the brush axis, in this case from right to left. The main m.m.f. is downwards, and if we represent these two m.m.f.s by the vectors OA and OB respectively we have OC for the resultant m.m.f. The magnetic neutral plane (m.n.p.) is perpendicular to OC, but on no load, when there is no armature reaction, m.n.p. is coincident with g.n.p. The effect of armature reaction is thus to shift the m.n.p. round in the direction of rotation. The magnitude of the shift obviously depends upon the length of OA, and therefore on the magnitude of the armature current. The lines of force produced by the armature reaction assuming that this reaction exists independently of the main field, take the path across the pole faces, as shown by the dotted lines. Each line of force crosses the air gap twice, and, comparing the directions of the armature and main fluxes, we see that the field strength in the gaps is weakened under the leading pole tips and strengthened under the trailing pole tips. The distributions of gap flux density on load and on no load are shown in Fig. 7.2, curves I and II. If the armature teeth are not saturated, the weakening of the field under the leading pole tips will be equal to the strengthening under the trailing pole tips and the total flux will be unaltered. The areas under the two curves in Fig. 7.2 will thus be equal. If the teeth are strongly saturated on no-load, the ampere-turns producing the armature cross-magnetizing field, as it is called, will not be able to increase the

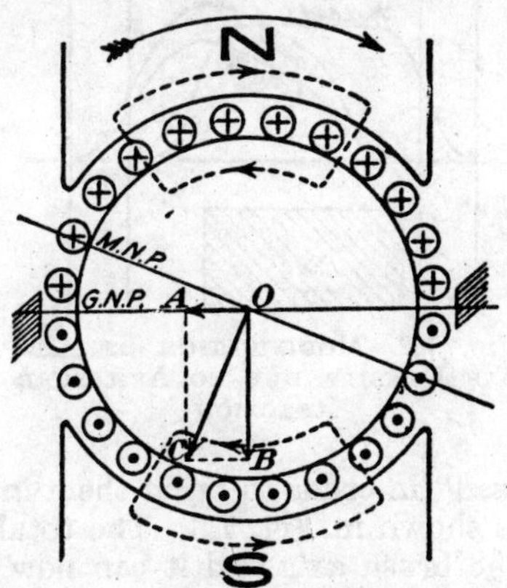

FIG. 7.1. ARMATURE REACTION IN A GENERATOR

flux through the teeth under the trailing half of the pole to the extent that they are able to demagnetize the teeth under the leading half. The result will be that, on the whole, there will be a certain amount of demagnetizing action as well as distortion of the main field.

Position of Brushes. Since the meeting and separating points of two e.m.f.s occur at the *magnetic* neutral planes, and the brushes are placed at these points, it is obvious that the brushes must be given a "forward

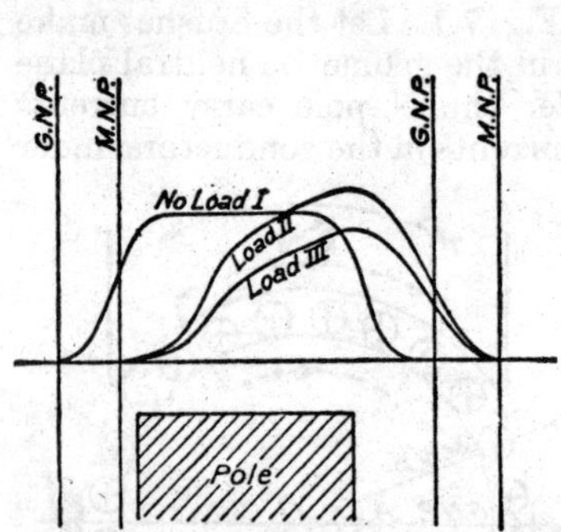

FIG. 7.2. MODIFICATION OF AIR-GAP DENSITY DUE TO ARMATURE REACTION

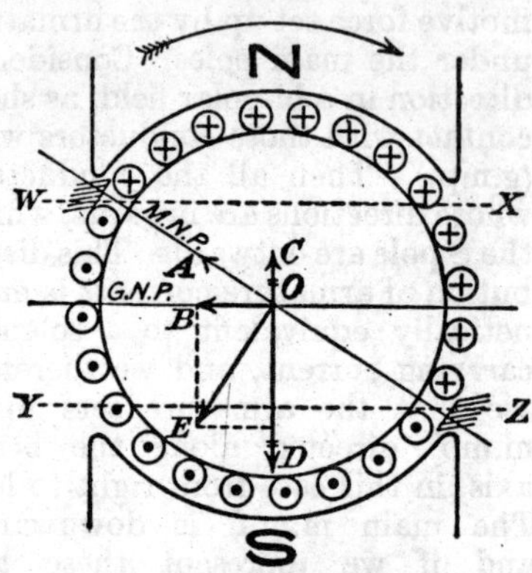

FIG. 7.3. ARMATURE REACTION IN A GENERATOR

lead" in order to bring them into the resultant m.n.p. This condition is shown in Fig. 7.3. The total armature m.m.f., OA, still acts along the brush axis and it can now be resolved into two components, the cross-magnetizing component, OB, acting along the g.n.p., and a demagnetizing component, OC, acting in opposition to the main m.m.f., OD. The resultant, OE, of all these m.m.f.s is of course perpendicular to the m.n.p. We see that when the brushes are given a forward lead the armature exerts a definite demagnetizing as well as distorting effect. The armature distorting, or cross-magnetizing ampere-turns are those lying above the horizontal WX and below the horizontal YZ; the demagnetizing ampere-turns are those lying between WX and YZ, and they are comprised within an angular distance of twice the brush lead. The distribution of air-gap density is now represented by curve III in Fig. 7.2.

Armature Ampere-turns. It is obvious that the armature demagnatizing ampere-turns must be neutralized by adding extra ampere-turns to the main field winding. It is therefore necessary to be able to calculate the armature ampere-turns per pole. Fig. 7.4 shows a portion of a multipolar armature the brushes of which have been given a forward lead of $\theta°$ mechanical. There are thus $(2\theta/360)Z$ demagnetizing conductors in each interpolar gap, that is $(\theta/360)Z$ on either side of each pole tending to demagnetize it. These will act as though they were joined together as shown, and they will produce $(\theta/360)Z$ demagnetizing *turns* per pole. Hence, if I is the current flowing in each conductor, demagnetizing ampere-turns per pole,

$$(AT)_d = (\theta/360)ZI$$

The total ampere-conductors per pole, both cross and demagnetizing,

$$= ZI/P$$

∴ Total armature ampere-turns per pole

$$= ZI/2P$$

∴ Cross-magnetizing ampere-turns per pole,

$$(AT)_c = ZI(1/2P - \theta/360)$$

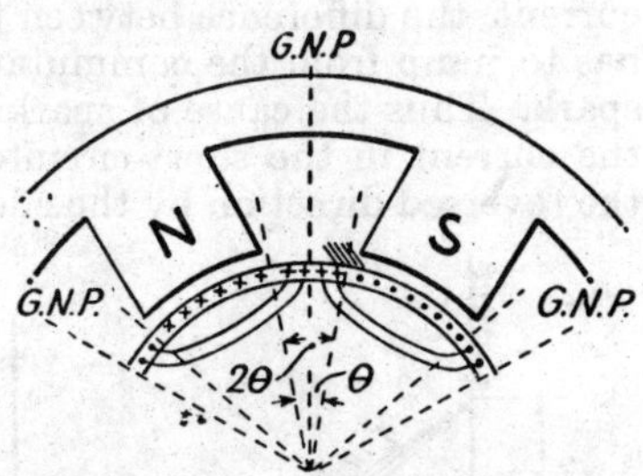

Fig. 7.4. Calculation of Armature Reaction

Example. A four-pole generator has a wave-wound armature with 722 conductors, which delivers 50 A on full load. If the brush lead is 8°, calculate the armature demagnetizing and cross-magnetizing ampere-turns per pole.

Current per conductor

$$I = I_a/2 = 50/2 = 25 \text{ A}$$

$$(AT)_d = (8/360) \times 722 \times 25$$
$$= 400$$

$$(AT)_c = 722 \times 25(1/8 - 8/360)$$
$$= 1850$$

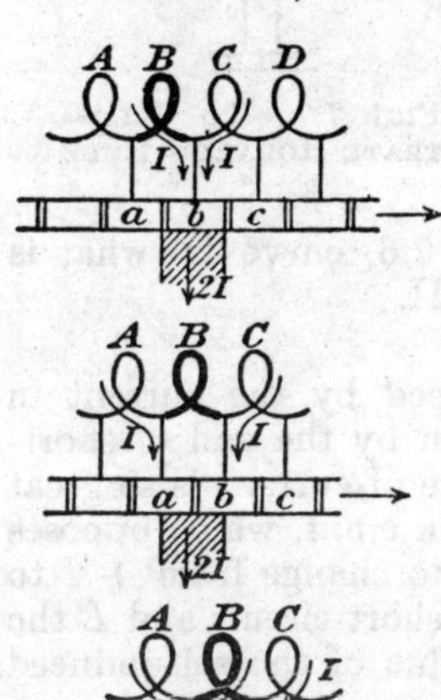

Fig. 7.5. Commutation

Commutation. Whenever a brush spans two commutator segments the winding element connected to those segments is short-circuited. By commutation we mean the changes that take place in a winding element during the period of short-circuit by a brush. These changes are illustrated in Fig. 7.5, the winding elements being drawn as portions of a ring winding for simplicity. In (*a*) the element B is on the point of being short-circuited and it is carrying, in a direction from left to right, half the current delivered by the armature to the brush; (*b*) shows the element in the middle of its short-circuit period, and it will be seen that it is possible for the currents to reach the brush without passing through this element; in (*c*) the same element B is shown immediately after short-circuit and in this position it is, or *should be*, carrying the full current in a direction from right to left. We thus see

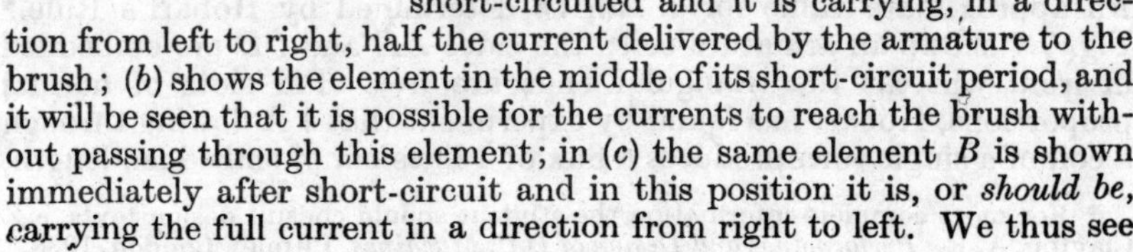

that during the period of short-circuit by a brush the current in a short-circuited element should be reversed *and brought up to its full value in the reversed direction.* If the current in B has not attained its full value in the position shown in (c), then, since element C is carrying the full current, the difference between the currents through elements C and B has to jump from the commutator bar b to the brush in the form of a spark. Thus the cause of sparking at the commutator is the failure of the current in the short-circuited elements to reach the full value in the reversed direction by the end of short-circuit. The curve of current

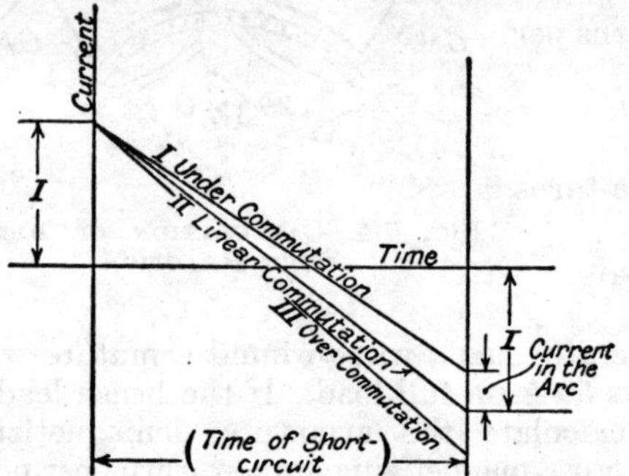

Fig. 7.6. Current Changes in a Short-circuited Element

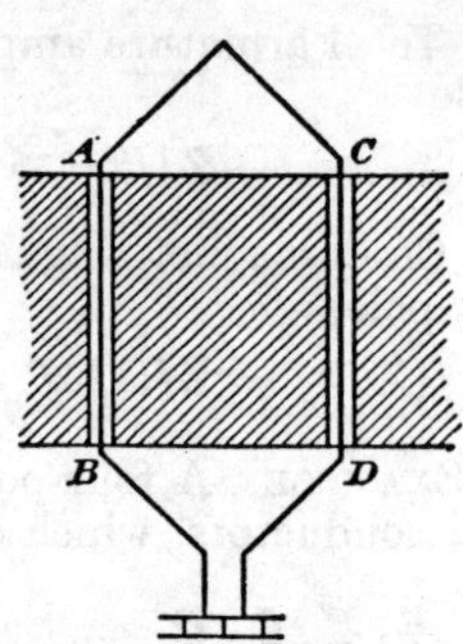

Fig. 7.7. To Illustrate Hobart's Rule

against time in such a case is shown in Fig. 7.6, curve I; what is required is a curve of current similar to curve II.

Reactance Voltage. The difficulty experienced by the current in attaining the full value in the reversed direction by the end of short-circuit is due to the fact that the rate of change of current is so great that the self-induction of the coil sets up a back e.m.f. which opposes the reversal. Since the current in the coil has to change from $+ I$ to $- I$, the total change is $2I$. If t is the time of short-circuit and L the self-induction of the coil, then the average value of the self-induced e.m.f. is

$$L(2I/t) = 2LI/t \text{ volts}$$

This is called the "reactance voltage." The self-induction, L, is here a composite quantity, being made up of the true self-induction of the short-circuited coil and the mutual inductions of its neighbouring coils. An approximate value for L can be determined by Hobart's Rule.* Fig. 7.7 shows an armature coil; the sides AB and CD are embedded in iron, whereas the triangular ends are free. For slots of normal proportions, Hobart has found by experiment that 1 A flowing through a coil of a single turn produces a flux of 4 lines/cm of embedded length,

* For more complete information the student should consult design texts, e.g. Clayton, A. E., *Performance and Design of D.C. Machines.* Pitman, London, 1938.

and 0·8 line/cm of free length. Hence, l_1 and l_2 are the total embedded and free lengths respectively, flux per ampere due to the whole coil

$$= 4l_1 + 0{\cdot}8l_2 \text{ for a single-turn coil}$$

and $$4nl_1 + 0{\cdot}8nl_2 \text{ for a coil of } n \text{ turns.}$$

With lengths expressed in metres and fluxes in webers these expressions become, respectively

$$(4l_1 + 0{\cdot}8l_2) \times 10^{-6} \text{ webers/ampere}$$
$$(4nl_1 + 0{\cdot}8nl_2) \times 10^{-6} \text{ webers/ampere}$$

The self induction of the coil is therefore

$$\text{Flux per ampere} \times \text{Number of turns}$$
$$= n^2(4l_1 + 0{\cdot}8l_2) \times 10^{-6} \text{ henries}$$

The effect of neighbouring coils can be taken into account by writing for the effective coefficient of self-induction of a short-circuited coil

$$L = \left(\begin{matrix}\text{total flux through coil per ampere,} \\ \text{due to self and neighbouring coils}\end{matrix}\right) \times \text{Number of turns}$$

Fig. 7.8 shows the disposition of the short-circuited coils in a lap winding. The flux per ampere through one coil due to the embedded length is twice that which would exist if the coil had no neighbours, but the flux due to the free length is not influenced by the neighbours.

∴ Total flux per ampere through a coil of n turns

$$= (8nl_1 + 0{\cdot}8nl_2) \times 10^{-6} \text{ webers}$$

and the effective coefficient of self-induction is

$$L = n^2(8l_1 + 0{\cdot}8l_2) \times 10^{-6} \text{ H}$$

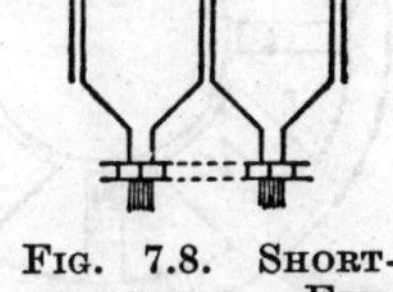

FIG. 7.8. SHORT-CIRCUITED ELEMENTS IN A LAP WINDING

This applies to the case of a brush short-circuiting only one coil at a time. If the brush width is such that it short-circuits m coils at a time, then, since these m coils will probably lie in the same slots, the flux through a coil will be increased m times. We then have

$$L = mn^2(8l_1 + 0{\cdot}8l_2) \times 10^{-6} \text{ henrys}$$

This expression shows that L is proportional to the square of the number of turns in an armature coil. For good commutation it is therefore advisable to use single-turn coils. This can be done in large machines where the large number of commutator bars required by such a winding does not make the bars too narrow, but it is obviously impossible with small machines.

It is obvious from Fig. 7.7 that the period of short-circuit is the time taken for a point on the commutator to move a distance equal to the brush width, less the thickness of the mica insulation between segments.

Calling these w_b and w_m respectively, and the peripheral velocity of the commutator v, we have

$$\text{Period of short-circuit } t = (w_b - w_m)/v \text{ second}$$

It is, of course, impossible to give maximum allowable values for the reactance voltage to cover all cases. In general, it can be taken that it should not exceed 0·6 or 0·7 V in non-interpole machines. For a large interpole machine it may be as high as 15 V.

EXAMPLE. Calculate the reactance voltage for a machine having the following particulars: number of commutator segments, 55; revolutions per second, 11·67; brush width in commutator segments, 1·74; coefficient of self-induction 0·000153 H; current per coil 27 A.

Time of commutation

$$= 1{\cdot}74/(55 \times 11{\cdot}67) = 2{\cdot}71 \times 10^{-3} \text{ sec}$$

$\therefore$ Average reactance voltage

$$= (2 \times 1{\cdot}53 \times 10^{-4} \times 27)/2{\cdot}7 \times 10^{-3} = 3{\cdot}06 \text{ V}$$

Interpoles. There are two methods of making the current in the short-circuited element attain its full value in the reversed direction by the end of short-circuit. They are known as e.m.f. and resistance commutation respectively. In e.m.f. commutation the short-circuited coil has a voltage induced in it which neutralizes the reactance voltage. This induced voltage must therefore be in the same direction as the final direction of the current. The magnetic field required to induce this e.m.f. is called the commutating field. One method is to shift the brushes so that they lie, not in the m.n.p., but in the fringe of the field produced by the next main pole farther ahead. This field will induce an e.m.f. in the required direction in the short-circuited coils, with the result that if the brushes are advanced sufficiently beyond the m.n.p. sparkless commutation will be obtained. This method is now obsolete because, since the effect of armature reaction is to weaken the field under the leading pole tips, and this weakening increases as the load increases, a very large brush shift is required when the load is heavy. Also the brush position has to be adjusted when the load varies, an obvious disadvantage. A better method of providing the commutating field is to make use of interpoles. These are small auxiliary poles placed in the geometric neutral planes, that is, midway between the main poles. Their polarity must, in the case of a generator, be that of the next main pole farther ahead, as shown in Fig. 7.9. Since the commutating field produced by them has to be proportional to the armature current, they are series excited (*see* Chapter 8). The neutralization of the

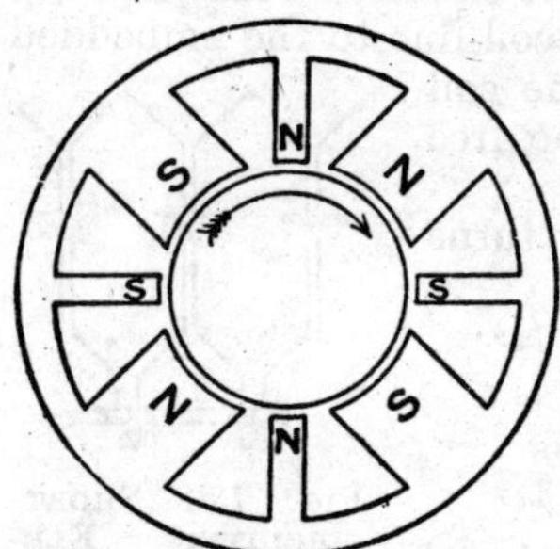

FIG. 7.9. FIELD FRAME WITH INTERPOLES

reactance voltage is thus rendered automatic when interpoles are used, and modern machines will operate between no load and 20 or 25 per cent overload with fixed brush position without appreciable sparking. The most important advantage of the use of interpoles is that they raise the sparking limit of a machine to about the same value as the heating limit, so that for a given output an interpole machine can be made smaller, and therefore cheaper, than a non-interpole machine.

It must be noticed that for a given armature current there is a proper value for the commutating field, and that it is possible for this field to be too strong. In such a case the reversed current in the short-circuited coil is forced to too high a value by the end of short-circuit, and sparking at the commutator takes place in the reversed direction. This is called over-commutation and is illustrated graphically by curve III, Fig. 7.6.

Let B_{ip} = flux density of the commutating field

l_{ip} = axial length of interpole

v_a = armature peripheral velocity

Then since each coil has two sides, the commutating e.m.f. is equal to $2nB_{ip}l_{ip}v_a$. If this is just equal to the reactance voltage E_r, we have

$$2nB_{ip}l_{ip}v_a = E_r$$
$$\therefore\ B_{ip} = E_r/(2nl_{ip}v_a)$$

EXAMPLE. A 6-pole machine has an armature diameter of 0·55 m. and an interpole axial length of 0·15 m: the speed is 450 r.p.m. The armature has single-turn coils and the reactance voltage is 1·86, Calculate the strength of the commutating field.

$$v_a = \pi D_a N/60 = \pi \times 0{\cdot}55 \times 450/60$$
$$= 12{\cdot}95 \text{ m/sec}$$
$$\therefore\ B_{ip} = 1{\cdot}86/(2 \times 1 \times 0{\cdot}15 \times 12{\cdot}95)$$
$$= 0{\cdot}478 \text{ webers/m}^2$$

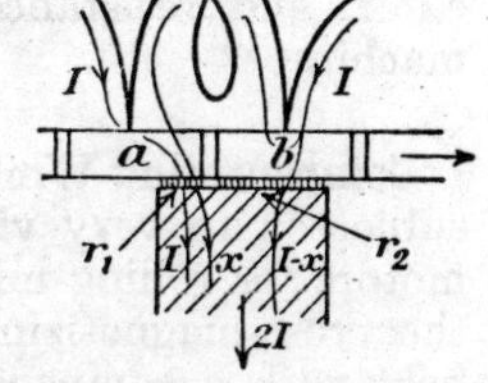

FIG. 7.10. EFFECT OF BRUSH CONTACT RESISTANCE

High-resistance Brushes. The second method of obtaining good commutation is to use high-resistance brushes. When the current flowing from the right (Fig. 7.10) reaches the commutator bar b, there are two parallel paths open to it. The first is straight across the bar b to the brush, the second, round the short-circuited element in a counter-clockwise direction and across the bar a. With brushes having a low contact resistance there is no inducement for the current to take the second path. With carbon brushes, which have a high contact resistance, more and more of the current flowing to the brush from the right hand will be shunted round the element as the bar b passes the brush, because the area of contact of bar b with the brush is diminishing, and its contact resistance, r_2, increasing, while the area of contact of the brush with bar a is increasing, and its resistance, r_1, therefore decreasing.

Carbon brushes have therefore almost entirely replaced copper brushes. The disadvantage of carbon brushes is that they can only be worked at a current density of about 40–50 A/in.², as compared with 150–200 for copper brushes. This necessitates a larger commutator. The properties of a few grades of carbon brush are shown below—*

	Max. Current Density, A/in.²	Max. Contact Resistance, Ω/in.²	Pressure on Commutator, lb/in.²
Copper . . .	200	0·003	1·5
Ordinary Carbon . .	40	0·04	2·0
Electrographitic . .	60	0·02	2·0

To illustrate the increase in contact resistance obtained by using carbon instead of copper brushes, take the case of the ordinary carbon brush. For the same area of brush,

$$\frac{\text{Contact resistance of carbon brush}}{\text{Contact resistance of copper brush}} = \frac{0{\cdot}04}{0{\cdot}003} = 13$$

But for the same current collected, the area of the carbon brush is 200/40 = 5 times the area of the copper brush. Hence, since the contact resistance is inversely proportional to the area, we have for the same current collected

$$\frac{\text{Contact resistance of carbon brush}}{\text{Contact resistance of copper brush}} = \frac{13}{5} = 2{\cdot}6$$

This is sufficient to give improved commutation. In practice, both e.m.f. and resistance commutation are used together in the same machine.

Compensating Winding. In the case of large direct-current machines subjected to very violent fluctuations in load, e.g. turbo-generators, motors for rolling mills, and colliery winders, it is usual to neutralize the cross-magnetizing effect of armature reaction by providing the field with a compensating winding. Whenever the load changes, the flux density in the air gap changes, as we have seen. A sudden change in load will produce a sudden change in flux threading the armature coils, and this in turn will set up a *statically*-induced e.m.f. in these coils. The magnitude of this induced e.m.f. will depend upon the rapidity of this change of load, and it will be very high if the change of load is very great and takes place almost instantaneously. This induced e.m.f. will appear as a voltage between consecutive commutator bars, and if it is great enough it will give rise to a flash over, that is, an arc completely encircling the commutator, and thereby short-circuiting the whole armature. Since the distortion of the field which

* P. Hunter-Brown. *Carbon Brushes for Electrical Machines.* Morgan Crucible Co., Ltd., London, 1923.

produces this phenomenon is caused by the armature conductors which lie under the pole faces, it can be eliminated by neutralizing the magnetic effect of those conductors. This is done by housing the compensating winding in the pole faces and connecting it in series with the armature in such a way that the currents in the armature and compensating windings are in opposite directions. The arrangement of these windings on a four-pole interpole generator is shown in Fig. 7.11.

Owing to the expense, and the room taken up, it will be clear that compensating windings can only be used in the case of machines of large output. They are required (*a*) if a machine is subjected to very violently fluctuating loads—the motor of a rolling mill, for example— or (*b*) where a generator has to deliver full current at a very low induced voltage, as for example, the generator of a Ward-Leonard set (*see* p. 242).

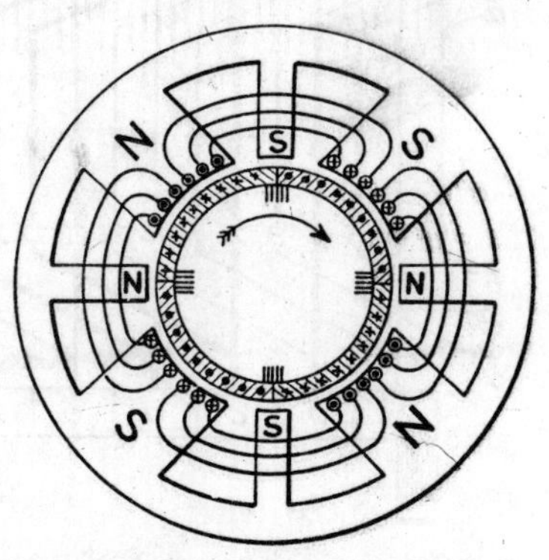

FIG. 7.11. MULTIPOLAR FIELD WITH COMPENSATING WINDINGS

Special Armature Windings. The methods discussed above for the prevention of sparking at the brushes are methods which are external to the armature itself. Obviously, if it is at all practicable, the armature itself should be wound in such a manner that its reactance voltage shall be as small as possible. This voltage is proportional to the self-induction of the coil undergoing commutation, and is therefore proportional to the square of the number of turns in the coil. Thus, the fewer the turns per coil the smaller will be the reactance voltage, for which reason it is preferable to use only single-turn coils whenever possible. Again, when a brush spans more than two segments, two or more neighbouring coils may be undergoing commutation at the same instant, in which case they will induce e.m.f.s in one another by the process of mutual induction. The total reactance voltage in any coil is thus the sum of its own self-induced e.m.f. and the mutually-induced e.m.f.s set up in it in virtue of the changes of current taking place in neighbouring coils.* It is, therefore, desirable that these mutually-induced e.m.f.s, as well as the self-induced e.m.f., shall be made as small as possible. One solution is to design the winding so that if adjacent conductors are undergoing commutation at the same instant, they do not lie side by side, but are separated by a small band of conductors carrying the full current in opposite directions. This can be done by using what is called a "short-chord" winding, that is, a winding in which the rear pitch, i.e. the coil width, is somewhat less than the pole pitch. Fig. 7.12 shows such a winding. In this winding

* See Clayton, *loc. cit.*, H. Cotton, *Design of Electrical Machinery*, or any standard text on the design of d.c. machines.

conductors 1 and 20 are undergoing commutation at the same time, but they are separated by the band of conductors numbered 24, 23, 22, and 21. Similarly with conductors 13 and 8.

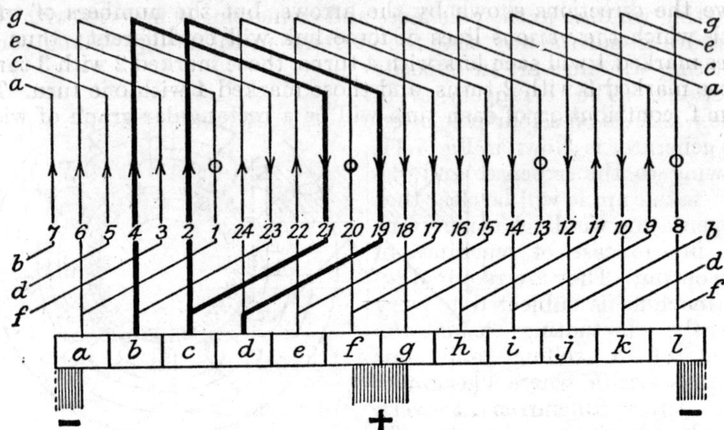

FIG. 7.12. EXAMPLE OF "SHORT-CHORD" WINDING

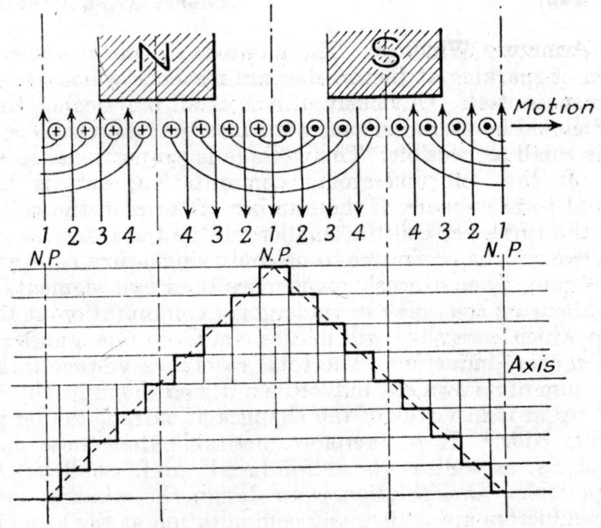

FIG. 7.13

Armature M.M.F. Diagram. Fig. 7.13 shows a portion of an armature winding with conductors represented in the conventional manner. For a normal full-pitch winding each conductor can equally well represent all the conductors in one slot, as the winding overhang does not make any contribution to the m.m.f. distribution within the core length. We can imagine that the conductors are joined by the series of end connections shown in the figure. In this way we can regard the armature winding as consisting of a series of distributed coils. The fluxes will have the directions shown by the arrows, but the numbers of turns with which the various lines of force link will be different: thus the lines marked 1 will each link with 4 turns, those marked 2 with 3 turns, those marked 3 with 2 turns, and those marked 4 with one turn. The m.m.f. contribution of each turn will be a rectangular graph of width

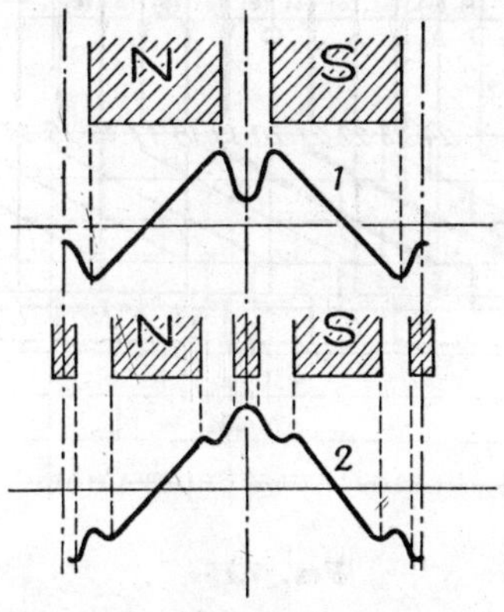

Fig. 7.14

equal to the turn width and of height proportional to the linkage. Inserting these rectangles we obtain the lower portion of Fig. 7.13, from which we see that the resultant m.m.f. distribution is the stepped triangular distribution shown.

With a large number of slots per pole the m.m.f. distribution would approximate to the dotted triangular distribution. With a constant air gap radial length, that is, with no interpolar gap, the flux distribution due to the armature currents above would tend even closer to this triangular shape, because the tendency to spread at places where there is a sudden change in m.m.f. would smooth out the steps. Owing to the presence of the interpolar gap the distribution would be like curve 1 of Fig. 7.14 in the case of a non-interpole machine, and like curve 2 for a machine with interpoles.

In the case of the non-interpole machine we see that the armature flux in the interpolar gaps becomes so small that compensation of the armature reaction can be effected by means of a winding embracing the pole arc only. This applies also to the interpole machine because, although the armature acting alone would produce a strong flux at the interpole axis this can be neutralized by an appropriate number of ampere-turns on the interpole itself. Thus a compensating winding spread over the pole arc, and not the whole pole pitch is sufficient in both cases. The m.m.f. diagram for a compensated machine is shown in Fig. 7.15. In an actual machine there would, of course, be a greater numbers of slots on both armature and pole face. By making the current loading of each pole-face slot equal to that of each armature slot it will be seen that there is complete compensation over the pole width, the cross-magnetizing effect thereby being neutralized. There is no

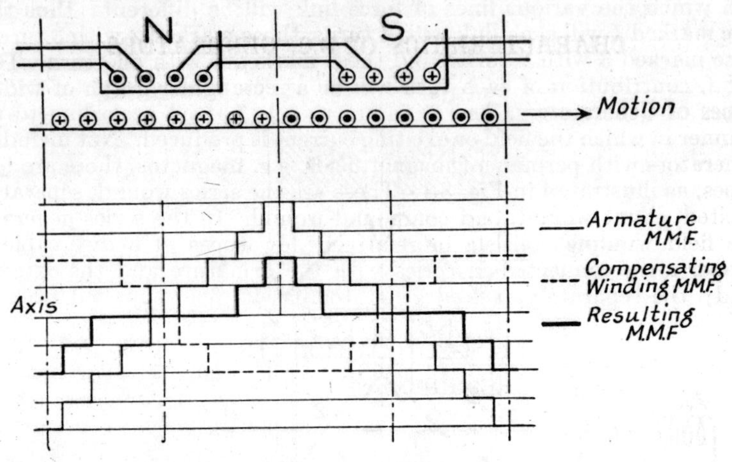

Fig. 7.15

compensation in the interpolar gap, but, as we have seen this is effected either by the gap itself, or by an appropriate number of interpole turns when interpoles are present.

Compensation of the armature m.m.f. is also important in the case of some a.c. commutator motors and this is considered in Chapters 25 and 26.

For further reading consult the books recommended at the end of Chapter 6.

CHAPTER 8

CHARACTERISTICS OF D.C. GENERATORS

Types of Generators. Generators are usually named according to the manner in which the field or exciting current is produced. Not including generators with permanent magnet fields, e.g. magnetos, there are four types, as illustrated in Fig. 8.1. These are the series wound, separately excited, shunt wound, and compound wound. In the series generator the field winding consists of relatively few turns of heavy cable or copper strip connected in series with the armature and the external load; the separately-excited generator has its field supplied from an

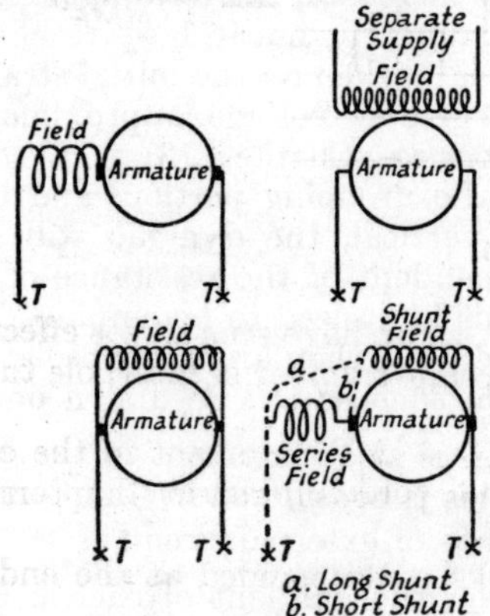

FIG. 8.1. METHODS OF EXCITATION

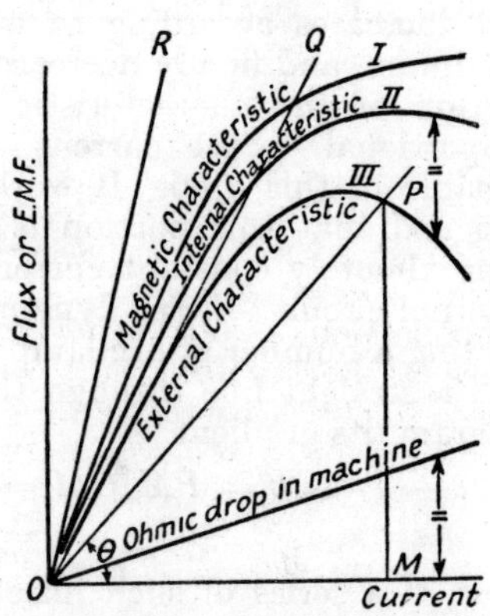

FIG. 8.2. CHARACTERISTICS OF SERIES GENERATOR

independent source; the shunt generator has its field connected across the armature terminals; and the compound generator has both series and shunt excitation.

Series-wound Generator. The current which flows through the armature also flows through the field winding, and this fact determines the nature of the characteristics. If the flux/pole is plotted against the current, the armature being driven at constant speed, the magnetic characteristic is obtained, its shape being given by curve I, Fig. 8.2. This curve starts a little way up the voltage axis due to the residual magnetism when the current is zero. If it were not for the demagnetizing effect of the armature reaction this curve would give the total flux when the machine is delivering current to the external load, but because of the armature reaction the actual curve lies somewhat below this. (Curve II.) This curve gives, to a different scale, the e.m.f.

generated in the armature, because this e.m.f. is proportional to the flux. It is called the internal or total characteristic. The terminal voltage of the generator is equal to the e.m.f. generated, E, less the ohmic drop of volts in the machine. If R_{se} and R_a are the resistances of the series field and armature respectively, then, since each carries the load current I, the ohmic drop is

$$(R_a + R_{se})I$$

and for the terminal voltage, V_t, we have

$$V_t = E - I(R_a + R_{se})$$

The ohmic drop is represented graphically by a straight line through the origin, and deducting ordinates of this lines from corresponding ordinates of curve II, we obtain the curve of terminal voltage, curve III. This is called the external characteristic. We see that as the current taken from a series generator is increased, the terminal voltage first increases according to an approximately linear law, reaches a maximum, and finally decreases. If it is worked on the initial straight portion of the characteristic it will give a voltage approximately proportional to the current. The booster (Chapter 13) is a series dynamo of this kind. If worked on the drooping portion, and if so designed that this portion is nearly vertical, the dynamo will give approximately constant current, independent of the resistance of the external circuit. Series dynamos of this kind used to be employed for lighting a number of arc lamps in series, but they are now obsolete.

If a straight line through the origin, such as OP, is drawn on the diagram, its gradient

$$\tan\theta = PM/OM = \text{(terminal voltage)/(current)}$$
$$= \text{resistance of external circuit}$$

Thus, if a series of such lines is drawn, the gradient of each corresponding to a definite resistance, the points of intersection with the external characteristic will give the terminal voltage and current which will be obtained when the external circuit has the corresponding resistance. Neglecting the initiate ordinate due to residual magnetism we can draw a straight line OQ which is tangential to the characteristic. The resistance represented by this line is called the *critical resistance*, since it is the greatest resistance with which the generator will be able to built up its magnetism. If the external resistance is greater than this, as represented by OR, then since OR has no point of intersection with the characteristic, the generator will not excite, and it will deliver no current.

Separately-excited Generator. The excitation is here independent of the load current, so that if there were no armature reaction the flux would be constant as indicated by curve I, Fig. 8.3. Because of armature reaction the curve of actual flux is slightly drooping (curve II). This second curve gives to a different scale the e.m.f. induced in the armature, and it is therefore the total characteristic. The terminal voltage on load is the total e.m.f. generated, less the ohmic drop in the armature (not in the field in this case); hence, if we draw the curve of armature drop, $R_a I$, curve III, and deduct its ordinates from those of curve II, we obtain the external characteristic, upper curve III. We see that in this case there is a slight decrease in terminal voltage as the load increases. This decrease can be easily neutralized by a slight increase in exciting current.

Shunt-wound Generator. When the external circuit is open the field winding can be regarded as being in series with the armature. The machine will therefore build up its own magnetism, and will give full

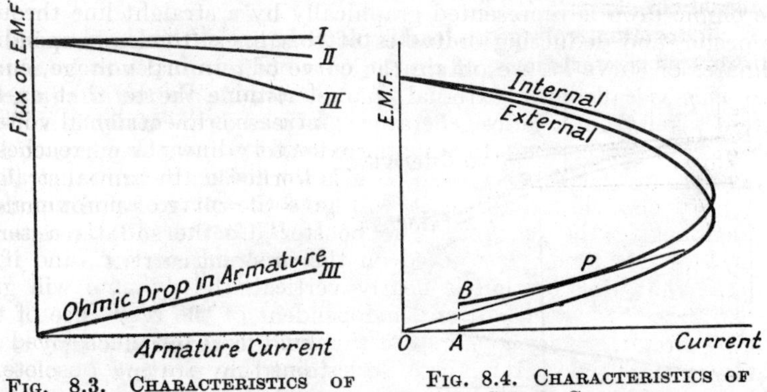

Fig. 8.3. Characteristics of Separately-excited Generator

Fig. 8.4. Characteristics of Shunt Generator

voltage on no load if the resistance of the field winding is less than the critical resistance. If the voltage across the shunt field did not fall as the external load increased, this generator would have a characteristic similar to that of a separately-excited generator, but since a fall in terminal voltage causes a decrease in exciting current, the total decrease in voltage is greater than if the machine were separately excited. The external characteristic therefore droops rather more than if the machine were separately excited. If the load on the machine is gradually increased by decreasing the resistance of the external circuit, a decrease in resistance when the current is small will cause an increase in current. At the same time the increased current will lower the terminal voltage, this, of course, tending to decrease the current. At first the effect of the decreased resistance predominates over the effect of decreased terminal voltage, but when the current has reached a certain value (much greater than the normal full load in modern machines) the load resistance shunts the field winding to such an extent that the terminal voltage decreases more rapidly than the load resistance. Then a further decrease in external resistance actually causes a decrease in current. The characteristic thus turns back, and when the armature is actually short-circuited it cuts the current axis at some point A (Fig. 8.4). The same thing happens to the total characteristic, except that in this case it stops at some point B, the ordinate AB

representing the voltage due to residual magnetism. Actually AB will, in such a case, be very small, because the magnetism will be almost completely neutralized by armature reaction: it may even be reversed. It is for this reason that shunt generators often fail to excite after they have been shut down through a severe short-circuit. If a tangent line OP to the total characteristic is drawn, the resistance represented by the gradient of this line gives the minimum external resistance for which the generator will excite if it is made to excite on load. If the external resistance is less than that represented by the line OP it will fail to excite, and therefore deliver no current. *There are thus two critical resistances for a shunt generator*, one for the field and the other for the external circuit.

If the terminal voltage on load is plotted against the load current the external characteristic is, of course, obtained. In order to be able to determine the total characteristic experimentally it is necessary to draw the curve of drop of volts in the armature, R_aI_a, and the curve of shunt current. If R_{sh} is the shunt resistance, then shunt current

$$I_{sh} = E_t/R_{sh}$$

and the required curve is obtained by plotting I_{sh} horizontally against E_t; it is, of course, a straight line through the origin, but because of the high resistance of the shunt field it has a very steep gradient, as shown in Fig. 8.5. Now the armature supplies both load and shunt currents, so that for the armature current we have

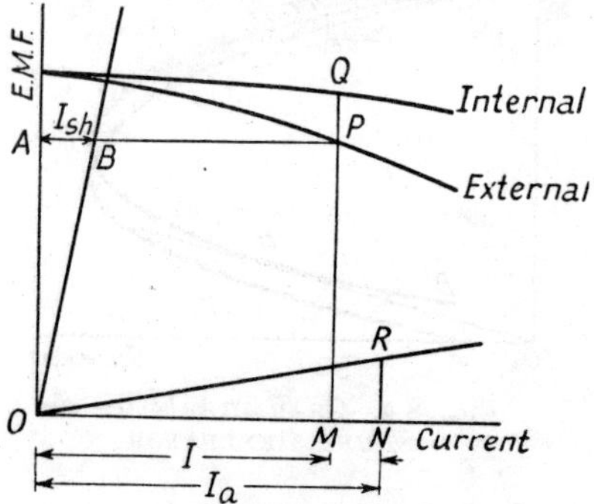

FIG. 8.5. CHARACTERISTICS OF SHUNT GENERATOR

$$I_a = (I + I_{sh})$$

where I is the load current.

If we take any point P on the external characteristic and draw the perpendicular PM, then for the given terminal voltage, $OM = I$. Draw PA horizontally, then $AB = I_{sh}$, and if we mark off $MN = AB$ then

$$ON = OM + MN = (I + I_{sh}) = I_a$$

Hence, the vertical RN is equal to the drop in the armature, and therefore if we produce MP to Q, making $PQ = RN$, the total length QM is the sum of the terminal voltage and total armature drop, that is the total e.m.f. generated. In this way a point Q on the total characteristic is obtained, and if other points are obtained in the same way, the total characteristic can be drawn.

The external characteristic can be pre-determined by the following graphical method. The open-circuit magnetization curve OM is drawn and the shunt resistance line OE_1, then drawn (Fig. 8.6). The no-load e.m.f. is E_1A_1 and the shunt ampere-turns per pole OA_1. For a given load current, the terminal PD is less than E_1A_1 because of (*a*) the armature volt drop and (*b*) the armature demagnetizing ampere-turns. Both of these are proportional to the armature current and can be calculated from the machine data. Set off the demagnetizing

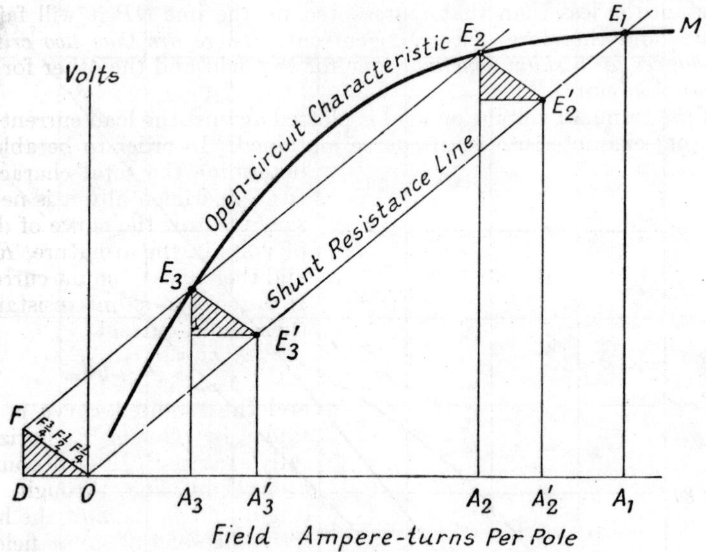

FIG. 8.6. CHARACTERISTICS OF SHUNT GENERATOR

ampere-turns for full-load current as OD (Fig. 8.6) and the armature volt drop as DF. Draw FE_2 parallel to OE_1. Then E_2A_2 and E_3A_3 are two possible values of the induced e.m.f. Parallels to FO from E_2 and E_3 locate the points E_2' and E_3'. Then, for the particular shunt circuit resistance which gives a no-load e.m.f. of E_1A_1, full-load current gives two possible induced e.m.f.s of E_2A_2 and E_3A_3 and two possible terminal p.d.s of $E_2'A_2'$ and $E_3'A_3'$. For any fraction of full load, say $\frac{1}{4}$, $\frac{1}{2}$, etc., the point F will have the positions indicated, and therefore by similar construction induced e.m.f.s and terminal p.d.s for corresponding loads can be determined. Curves similar to Fig. 8.4 can then be drawn.

EXAMPLE 1. A shunt generator has a total of 50 Ω resistance in the shunt circuit, this including the field winding as well as the regulator. Its terminal voltage is 25 V when run at 500 r.p.m., 113 V at 1000 r.p.m. and 200 V at 1500 r.p.m. Draw the magnetization characteristic for 1000 r.p.m. and hence determine the terminal voltage at 1000 r.p.m. if the resistance of the shunt circuit is reduced to 40 Ω.

The 50 Ω line is drawn first of all. With 200 V induced, the shunt current will be $200/50 = 4$ A, so that this line passes through the origin and the point (4, 200) as in Fig. 8.7. The three points of 25, 113, and 200 V lie on this line as indicated. Since the induced voltage on no

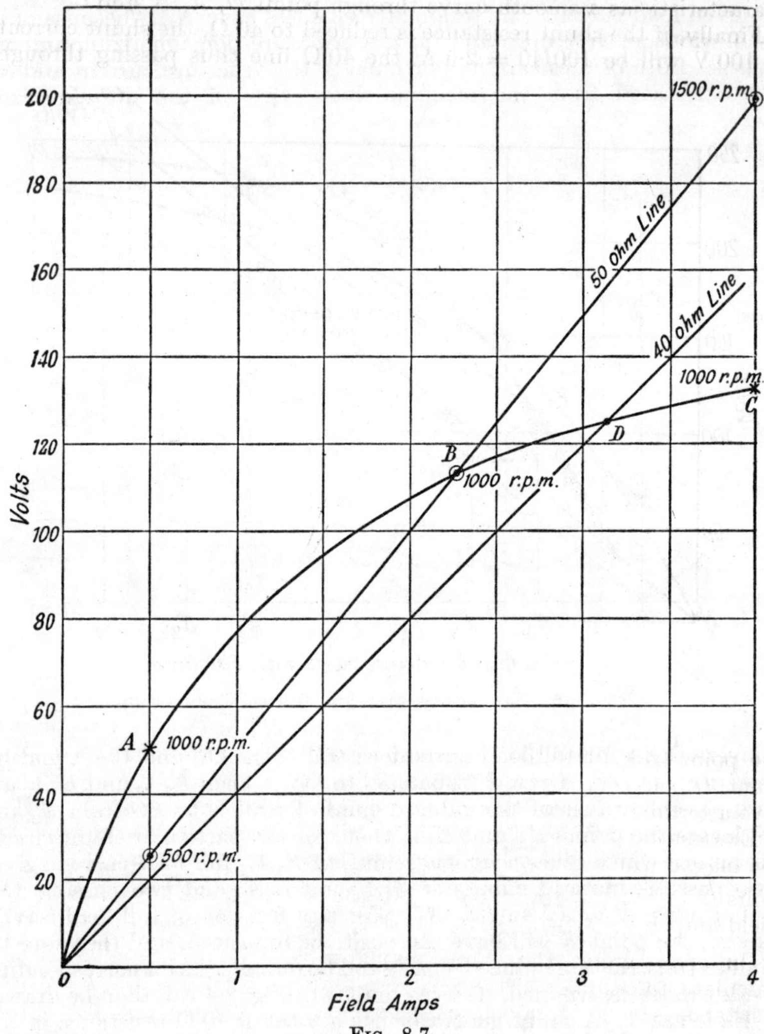

Fig. 8.7

load is proportional to the speed so long as the flux remains constant, the 25 V and 200 V points corrected to 1000 r.p.m. are

$$25 \times (1000/500) = 50 \text{ V}$$

and

$$200 \times (1000/1500) = 133{\cdot}3 \text{ V}$$

We thus have three points corresponding to 1000 r.p.m., viz. A, B, and C, and the origin makes a fourth point, so that we can draw the characteristic as a smooth curve through points O, A, B, and C.

Finally, if the shunt resistance is reduced to 40 Ω, the shunt current at 100 V will be 100/40 = 2·5 A, the 40 Ω line thus passing through

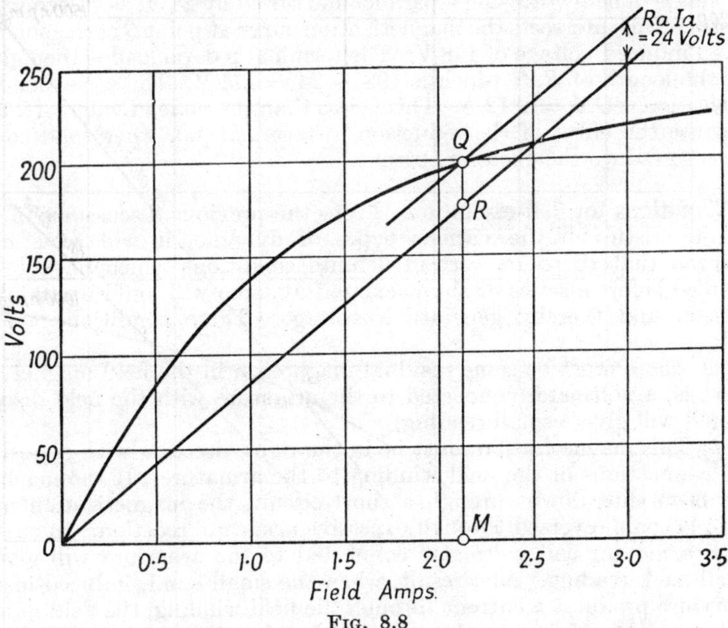

Fig. 8.8

the point (2·5, 100). Its intersection with the characteristic is at the point D, showing that at 1000 r.p.m. a shunt resistance of 40 Ω will give a terminal voltage of 126.

Example 2. When running at normal speed a shunt dynamo has the magnetization characteristic given by the following values—

Volts . . .	45	85	140	175	195	210	220	225
Field amperes	0·25	0·5	1·0	1·5	2·0	2·5	3·0	3·5

On open circuit the shunt regulator is adjusted to give 210 V. Calculate the total resistance in the shunt circuit. If the armature resistance is 0·095 Ω, calculate the terminal p.d. on load when delivering a current of 250 A.

The magnetization curve is given in Fig. 8.8. For a voltage is 210 V on no load the shunt current is 2·5 A, so the resistance of the shunt circuit is

$$210/2{\cdot}5 = 84\ \Omega$$

Since the load is in parallel with the shunt circuit, the resistance line is also a line of terminal voltage, since its ordinates give the p.d. applied to the shunt field. On load, to give the induced volts, this line has to be moved parallel to itself by an amount equal to the armature drop, namely

$$R_a(I + I_{sh}) = 0{\cdot}095(250 + 2{\cdot}5) = 24 \text{ V}$$

This is done by drawing a parallel line raised by 24 V, as shown in the figure. This intersects the magnetization curve at point Q corresponding to an induced voltage of 198 V. The terminal p.d. on load is then given by the length of RM, which is $198 - 24 = 174$ V. The corresponding excitation is $OM = 2{\cdot}12$ A. This is less than the no load value of 2·5 A, because the effect of the reduction in terminal p.d. when load comes on is to reduce the exciting current.

Conditions for Self-excitation. From the previous discussions of the characteristics of the various types of dynamo, it will have been realized that there are certain definite conditions which have to be fulfilled before a series or shunt-excited dynamo will build up its magnetism, and thereby generate a voltage. These conditions are as follows—

(1) There must be some residual magnetism in the field magnet. If there is, a voltmeter connected to the armature with the field disconnected will give a small reading.

(2) This magnetization must be in the right direction with regard to the connections of the field winding to the armature. If the machine has been shut down through a short-circuit, the magnetization may have become reversed through excessive armature reaction. In such a case, a moving-coil voltmeter connected to the armature will give a small back reading. As a result, when the small e.m.f. induced in the armature produces a current through the field winding, the field m.m.f. will strengthen the reversed flux and so build up the magnetization, and therefore the voltage, in the wrong direction. The remedy is to connect the field winding to a separate external source which will set up a new residual magnetization in the right direction.

(3) In the case of a series dynamo, the resistance of the circuit must be less than the critical resistance.

(4) In the case of the shunt dynamo, the field is *in series with the armature* with respect to the path taken by the exciting current, and consequently there is also a critical resistance for the field circuit of a shunt dynamo. There is almost invariably a shunt regulator in circuit, and failure to excite is often due to too much of this regulator being in circuit.

(5) In the case of the shunt dynamo there is also a lower limit for the resistance of the external load, below which the machine will fail to excite if started up with the load switched on to the armature.

The Compound-wound Generator. The shunt generator gives a terminal voltage which falls off somewhat with increase of load. It is usual to include an adjustable resistance called the shunt regulator in the field circuit, and to cut out some of this resistance when the load increases. This can be done by hand, and the terminal voltage kept constant if the fluctuations in voltage are fairly slow, as, for example, with a lighting load. When the fluctuations in load are very rapid, as with a traction load, then hand regulation is impossible and it is necessary to keep the voltage to the required value automatically. Again, it is usual to connect a generator to a pair of feeding points by a cable called a feeder. From these feeding points radiate other cables called distributors, across which the consumers are connected. The voltage across these feeding points is kept constant. Now if each conductor of a feeder has a resistance R, then, when delivering current I, there will be a drop in the feeder of $2IR$. Hence, if the constant voltage at the feeding points is E we have for the generator terminal voltage

$$V_t = E + 2IR$$

an equation which shows that the terminal voltage must *rise* with increase of load. This rise is obtained by providing a shunt generator with additional series excitation, the design being such that, over the working range, the series characteristic does not droop. The characteristics for the shunt and series turns separately are shown in Fig. 8.9, and that of the compound generator is obtained by adding ordinates of the two curves. The two component curves are somewhat concave downwards, and the resulting curve is therefore the same.

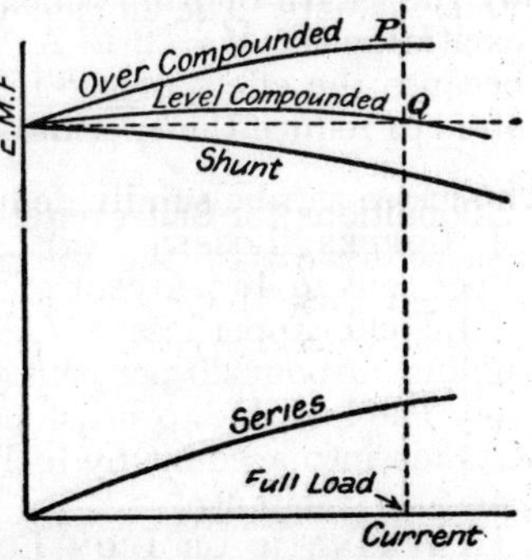

FIG. 8.9. CHARACTERISTICS OF COMPOUND GENERATOR

If the series excitation is such that the terminal voltage on full load is the same as on no load, the generator is "level" compounded. If the terminal voltage rises with load it is "over" compounded. The length PQ gives this rise, and if such a generator is to maintain constant voltage at a pair of feeding points we have, obviously,

$$PQ = 2IR$$

The length PQ expressed as a percentage of the voltage E at the feeding points is called the "percentage compounding." Example—

$$E = 500 \text{ V}, \; R = 0{\cdot}025 \; \Omega, \text{ full-load current}$$
$$I = 1000 \text{ A}$$

∴ Drop in feeder

$$2IR = 50 \text{ V}$$

∴ Full-load terminal voltage at the generator

$$V_t = 500 + 50 = 550 \text{ V}$$

∴ Percentage compounding

$$100(V_t - E)/E = 10 \text{ per cent}$$

There are two ways of connecting the shunt field in a compound generator, called short and long shunt. These are illustrated by the alternative connections shown dotted in Fig. 8.1. The short shunt is the more usual arrangement as it gives a somewhat higher voltage due to the fact that the shunt field has the full armature voltage across it. In the long shunt arrangement the voltage across the shunt is the armature voltage less the ohmic drop in the series field.

Efficiency of D.C. Machines. When calculating the efficiency of electrical machinery it is convenient to express power in electrical units, namely, watts. The efficiency of a generator is then

$$\eta = \frac{\text{output}}{\text{intake}} = \frac{\text{output}}{\text{output} + \text{losses}}$$

The losses can be subdivided as follows—

I. Copper Losses. (*a*) Armature copper loss $= I_a{}^2R_a$. About 30 per cent to 40 per cent of the total full-load losses.

(*b*) Field copper loss $= I_{sh}{}^2R_{sh}$ in a shunt winding; $I_{se}{}^2R_{se}$ in a series winding. About 20 per cent to 30 per cent of total losses at full load.

(*c*) The loss due to brush contact resistance. This can conveniently be taken into account by including the brush contact resistance in the armature resistance.

II. Magnetic or Iron Losses. (*a*) Hysteresis loss. Proportional to $\{(\text{Flux})^{1\cdot6} \times \text{Speed}\}$.

(*b*) Eddy-current loss. Proportional to $\{(\text{Flux})^2 \times \text{Speed}^2\}$.

These relationships apply only when the flux density varies sinusoidally with time and alternates on a single axis. These conditions do not apply in rotating electrical machines and, for flux densities below saturation the total loss, at constant frequency, is almost exactly proportional to B^2.* These two losses are therefore approximately constant in a machine whose flux is approximately constant, e.g. shunt or compound machines, but variable in the case of a series machine. The two together represent 20 per cent to 30 per cent of the total losses at full load.

* L. H. A. Carr. "The teaching of the principles of electrical machine design." *J. Inst. Elect. Engrs.* **94,** Pt. 2. 1947, 443.

The following is an extract from Carr's paper—

Analysis of tests of machines of different makes over the past thirty years leads to the following laws for total iron loss of machines built with ordinary dynamo sheet steel.

Iron loss at 50 c/s $= aB_c{}^2V_c + bB_t{}^2V_t$ watts, where B_c, B_t are the maximum values of apparent flux density in gauss $\times 10^{-4}$, in core and teeth respectively, and V_c, V_t are the corresponding values of actual iron in cubic centimetres. In the "tooth loss" is included the loss in the pole faces of salient-pole machines due to the passage of the teeth. The figures apply only if tooth-pulsation loss is not excessive.

Good average values for the constants are—

Induction motors	$a = 0{\cdot}036$	$b = 0{\cdot}050$
Synchronous machines (either synchronous induction or solient-pole)	$a = 0{\cdot}036$	$b = 0{\cdot}042$
Direct-current machines	$a = 0{\cdot}050$	$b = 0{\cdot}180$

III. Mechanical Losses. (*a*) Friction at bearings and commutator. (*b*) Windage of rotating armature.

Together 10 per cent to 20 per cent of total losses, independent of the load, and proportional to the speed for small variations in speed.

The magnetic and mechanical losses are often grouped together and called collectively the stray losses.

The efficiencies of generators depend, of course, upon the output, average values for the full-load efficiencies varying from 70 per cent for 1 kW machine to 95 per cent for a 1000 kW machine.

In the case of a shunt or compound machine the shunt copper loss and stray losses are constant. We can then write

$$\text{Total losses} = \text{armature copper loss} + \text{constant losses } (P_c)$$

$$= R_a(I + I_{sh})^2 + P_c$$

where I is the load current.

$$\text{Output} = VI$$

$$\therefore \eta = \frac{VI}{VI + R_a(I + I_{sh})^2 + P_c} = \frac{VI}{VI + R_a I^2 + P_c} \text{ approx.}$$

since I_{sh} is small compared with I.

$$\therefore \eta = 1/[1 + (R_a I/V + W_c/VI)]$$

Now the product of the terms in the brackets is a constant since V_t s approximately constant, I being the variable. Hence, their sum is a minimum, and the efficiency a maximum

when $R_a I/V = P_c/VI$

when $I = \sqrt{(P_c/R_a)}$

Hence, if the efficiency is plotted against the load current, a curve of the form shown in Fig. 8.10 will be obtained, the position of the maximum being defined by the above value of the load current.

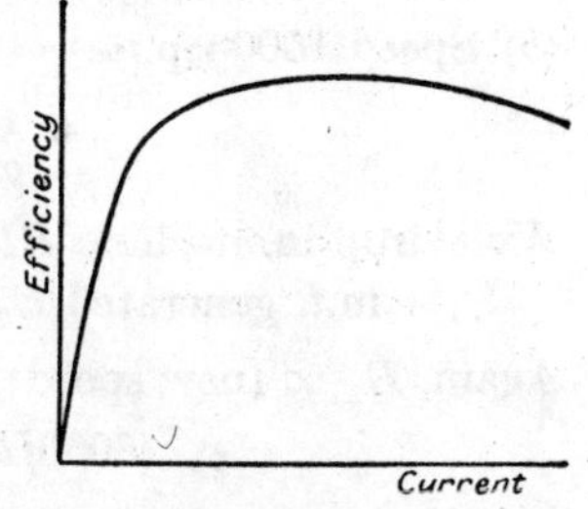

Fig. 8.10. Efficiency Characteristic

Example. A 400 V shunt generator has a full-load current of 200 A; its armature resistance is 0·06 Ω, and field resistance, 100 Ω; the stray losses are 2000 W. Find the h.p. of its prime mover when it is delivering full load, and find the load for which efficiency of the generator is a maximum.

$$I = 200 \text{ A on full load. } I_{sh} = 400/R_{sh} = 4 \text{ A}$$

$$\therefore \quad I_a = I + I_{sh} = 204 \text{ A}$$

$$\therefore \quad \text{Armature copper loss } I_a^2 R_a = 204^2 \times 0{\cdot}06 = 2500 \text{ W}$$

$$\text{Field copper loss } I_{sh}^2 R_{sh} \text{ or } I_{sh} V = 4 \times 400 = 1600 \text{ W}$$

$$\text{Constant losses } P_c = \text{field copper loss} + \text{stray losses} = 3600 \text{ W}$$

$\therefore$ Total full-load losses = 2500 + 3600 = 6100 W

Output VI = 400 × 200 = 80,000 W

Intake = 80,000 + 6100 = 86,100 W

$\therefore$ h.p. of engine = 86,100/746 = 115

η = 80,000/86,100 = 0·93, or 93 per cent.

The load at which η is a maximum is

$$I = \sqrt{(3600/0{\cdot}06)} = 245 \text{ A}$$

This is greater than the normal load because the armature resistance is on the low side for a 400 V generator of 80 kW output.

• EXAMPLE. A series generator of total resistance 0·5 Ω is running at 1000 r.p.m. and delivering 5 kW at a terminal p.d. of 100 V. If the speed is raised to 1500 r.p.m., and the load adjusted to 8 kW, find the new current and terminal p.d.

Assume that the machine is working on the straight portion of the characteristic; then the flux is proportional to the current, and the e.m.f. generated proportional to the product of flux and speed.

(*a*) Speed 1000 r.p.m.

Output 5 kW = 5000 W

V = 100 V $\therefore$ I = 5000/100 = 50 A

$\therefore$ Volt drop in machine IR = 50 × 0·5 = 25 V, and the total e.m.f. generated E_1 = 100 + 25 = 125 V

$\therefore$ 125 $\propto$ (speed × flux)

$\propto$ (speed × current), $\propto$ 1000 × 50 . . . (1)

(*b*) Speed 1500 r.p.m.

Let I = new current

$\therefore$ V = Output/I = 8000/I

Volt drop in machine RI = 0·5I

$\therefore$ e.m.f. generated E_2 = V + drop = 8000/I + 0·5I

Again, $E_2 \propto$ (new speed × new current)

$\therefore$ (8000/I) + 0·5I $\propto$ 1500 × I . . . (2)

Dividing equation (2) by (1),

$$\frac{(8000/I) + 0{\cdot}5I}{125} = \frac{1500I}{50{,}000}$$

$$I = 49{\cdot}6 \text{ A}$$

$$V = 8000/49{\cdot}6 = 161 \text{ V}$$

EXAMPLES ON CHAPTER 8

1. A shunt dynamo gave an external characteristic as follows—

Current	0	4	8	12	16	20
Terminal volts	89·5	88·5	87	84·4	82·2	79

Its armature and shunt resistances were 0·22 and 45 Ω respectively. Plot its total characteristic.

2. The above dynamo, when connected short-shunt compound, gave the following terminal voltages for the same load currents 89·5, 89·9, 90, 89·5, 88·2 V. The resistance of its series winding was 0·10 Ω. Deduce its total characteristic.

3. If a dynamo is level compounded at 1500 r.p.m., what will be the nature of its voltage characteristic at 1000 r.p.m., if the shunt regulator is not altered?

4. Given the magnetization curve of a self-excited, shunt-wound generator show how to determine the open-circuit voltage with a given total resistance in the shunt circuit. Show how to estimate the rise in voltage when the speed is increased by 10 per cent, and explain how the external characteristic of the machine may be estimated. (L.U.)

5. Find how many series turns per pole are needed on a 500 kW compounded dynamo required to give 500 V on no load and 550 V on full load, the requisite ampere-turns/pole being 7900 and 11,200 respectively. The shunt winding is designed to give 500 V at no load when its temperature is 20°C. The final temperature is 60°C. (C. and G.)

Ans. 4 series turns/pole.

6. The open-circuit characteristic of a d.c. generator run at 300 r.p.m. is as follows—

Field current (amperes)	0	2	3	4	5	6	7
Volts	7·5	93	135	165	186	202	215

Find the open-circuit characteristic for a speed of 375 r.p.m. and determine the voltage to which the machine will excite if the field circuit resistance is 40 Ω.

Ans. 264 V.

7. What additional resistance must be placed in series with the field circuit to reduce the open-circuit voltage of the machine in question to 230 V?

Ans. 7 Ω.

8. The open-circuit characteristic for a d.c. generator is as follows when the armature is driven at normal speed—

Field current (amperes)	0·5	1·0	1·5	2·0	2·5	3·0	3·5
Volts	60	120	138	145	149	151	152

The machine is connected as a shunt generator and driven at normal speed. The resistance of the shunt circuit being adjusted to 53 Ω, calculate (*a*) the open-circuit voltage, (*b*) the load current when the terminal voltage is 100, and (*c*) the terminal voltage when the load has a resistance of 0·4. Assume that the effect of armature reaction can be neglected.

Ans. 150·2 V, 450 A, 118 V.

9. A 6-pole generator armature has 1,000 conductors and is wave wound. If the flux per pole is 2×10^{-2} webers and the speed is 500 r.p.m., calculate the e.m.f. generated. If the above machine is self-excited and the armature and field resistances are 0·5 Ω and 250 Ω respectively, calculate the output current when the armature current is 40 A. (E.M.E.U.)

Ans. 500 V, 38·08 A.

10. The magnetization curve for a separately excited generator running at 800 r.p.m. and excited from 230 V mains is given by the following—

D.C. voltage	4	66	128	179	213	240	257	271	279	285	291
Field current (amperes)	0	0·2	0·4	0·6	0·8	1·0	1·2	1·4	1·6	1·8	2·0

Determine the field-circuit resistance required to give a no-load voltage of 275. What no-load voltage would be obtained if the same value of the field-circuit resistance were used when the generator was connected for shunt excitation? (E.M.E.U.)

Ans. 156 Ω, 286 V.

11. When running at its normal speed, the relationship between the voltage at the terminals of a generator on no-load, and the field current is as follows—

Field current (amperes)	1	2	3	4	5	6
Volts	188	336	427	484	516	530

The field resistance is 100 Ω. What series resistance must be used, and over what range must it be variable if the machine is to be self-excited as a shunt-machine to give a voltage of 400 to 500 V? (N.U.)

Ans. 10 to 50 Ω, 5 A rating.

12. A shunt generator has a full-load current of 100 A at 110 V. The stray losses are 500 W and the shunt resistance is 55 Ω. If the efficiency at full load is 88 per cent, find the armature resistance. Then find the efficiencies at one-half of full load and at 50 per cent overload, assuming that the voltage is kept constant at 110 V. At what load current will the efficiency be a maximum?

Ans. (*a*) 0·075 Ω; (*b*) 86 per cent; (*c*) 87 per cent; (*d*) 96 A.

13. A 300 kW shunt-excited generator has a full-load output current of 600 A at 500 V. The hot resistance of the armature winding is 0·0185 Ω, and of the interpole winding 4×10^{-3} Ω. The shunt current at no-load to maintain the rated voltage is 2·32 A, and at full-load 6·24 A. The iron loss is 1650 W at no-load and 2100 W at full load. Commutator friction loss 640 W and windage and friction loss 1500 W. The machine has 8 brush spindles, the brush contact drop per spindle being 1·1 V. Allowance for additional load loss can be made by deducting 1·25 per cent from the calculated efficiency at full-load. Assuming a linear law between fractional load and iron loss, shunt copper loss, and additional load loss respectively, calculate the efficiency at $\frac{1}{4}$, $\frac{1}{2}$, $\frac{3}{4}$, $\frac{4}{4}$ and $\frac{5}{4}$ of full load.

Ans. 92·3, 93·8, 94·0, 93·45, 92·7 per cent.

14. The relation between the excitation current and the e.m.f. generated by a d.c. shunt-wound generator running on open circuit at 850 r.p.m., is as follows—

Excitation (amperes)	2	3	4	5	6
E.M.F. (volts)	68	87	100·5	109	112·5

The shunt field resistance is 22·2 Ω. Find the voltage at the terminals of the machine when it runs at 850 r.p.m. self-excited. (L.U.)

Ans. 108·2 V.

CHAPTER 9

CROSS-FIELD DYNAMOS

The Rosenberg Constant-current Dynamo. This is a two-pole machine with two brush sets having their axes at right-angles, one along the field axis, the other in the normal directions at right-angles to it. The main brushes which supply the load are on the former axis, the auxiliarly brushes B_2B_2, which are short-circuited, along the latter. The poles are of abnormally small cross section, while the pole-shoes embrace an arc greater than is usual in normal machines, and are of considerable radial depth. This peculiarity is dictated by the necessity for a low reluctance path to the cross flux. These constructional

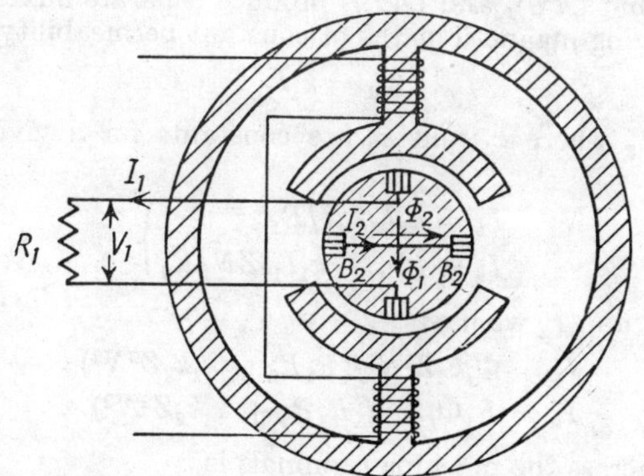

FIG. 9.1. GENERAL ARRANGEMENT OF CONSTANT-CURRENT DYNAMO

features are indicated in the diagram of Fig. 9.1, from which it will be seen that the excitation is by shunt-connected exciting coils.

Let $(AT)_f$ = ampere-turns due to the shunt field: this acts along the main brush axis B_1B_1

Let $(AT)_d$ = ampere-turns due to the load current I_1: owing to the orientation of the brush axes B_1B_1, this acts in direct opposition to $(AT)_f$.

Hence resultant $(AT)_r$ acting in the direction of the main field is

$$(AT)_r = (AT)_f - (AT)_d$$

Let Φ_1 = the flux produced by $(AT)_r$.

Let I_2 = the current flowing along the short-circuit B_2B_2

Let Φ_2 = the flux produced by the corresponding cross ampere-turns. This flux is required to be as large as possible, and this accounts for the shape of the pole shoes.

The short-circuit current I_2 is produced by the main flux Φ_1, while the main current I_1 is produced by the cross-flux Φ_2. As a result, the direction of the main current I_1 is independent of the direction of rotation. For suppose the direction to be reversed, then I_2 is reversed and this causes Φ_2 to reverse. Thus the e.m.f. producing the main current I_1 is induced by a reversed rotation in a reversed field, its direction therefore remaining unchanged.

Let R_1 = resistance of circuit to the main current I_1

and R_2 = resistance of the purely local circuit taken by the current I_2

Then, as the machine has two poles and therefore P/A = unity

$$I_1 = \Phi_2 ZN/R_1$$

and
$$I_2 = (\Phi_f - \Phi_d)ZN/R_2$$

Assuming that $(AT)_f$ and $(AT)_d$ produce separate fluxes of Φ_f and Φ_d, and assuming magnetic paths of constant permeability, we have

$$\Phi_d = k_1 I_1$$

and $\Phi_2 = k_2 I_2$ where k_1 and k_2 are constants for a given machine. Hence

$$\left.\begin{aligned} I_1 &= k_2 I_2 ZN/R_1 \\ I_2 &= (\Phi_f - k_1 I_1)ZN/R_2 \end{aligned}\right\} \quad . \quad . \quad . \quad (1)$$

Solving for I_1 and I_2 we have

$$I_1 = \Phi_f k_2 Z^2 N^2/(R_1 R_2 + k_1 k_2 Z^2 N^2) \quad . \quad . \quad . \quad (2)$$

$$I_2 = R_1 \Phi_f ZN/(R_1 R_2 + k_1 k_2 Z^2 N^2) \quad . \quad . \quad . \quad (3)$$

The voltage across the machine terminals is

$$\begin{aligned} V_1 &= R_1 I_1 \\ &= R_1(k_2 I_2 ZN/R_1) \\ &= k_2 ZN \times (\Phi_f - k_1 I_1)ZN/R_2 \\ &= (\Phi_f - k_1 I_1)k_2 Z^2 N^2/R_2 \quad . \quad . \quad . \quad (4) \end{aligned}$$

$$\therefore (\Phi_f - \Phi_d) = R_1 R_2 I_1/k_2 Z^2 N^2$$

Again
$$\Phi_f = [(R_1 R_2 + k_1 k_2 Z^2 N^2)/k_2 Z^2 N^2]I_1 \quad \text{from (2)}$$

$$\therefore (\Phi_f - \Phi_d)/\Phi_f = R_1 R_2/(R_1 R_2 + k_1 k_2 Z^2 N^2) \quad . \quad . \quad (5)$$

The following properties of the machine can be deduced from the above equations—

(1) If the quantity $k_1 k_2 Z^2 N^2$ is made large compared with $R_1 R_2$, then equation (2) reduces to

$$I_1 \simeq \Phi_f/k_1$$

the maximum value attained by I_1 being Φ_f/k_2. If the local resistance R_1 is kept constant then the voltage-speed characteristic is of the same shape as the current-speed characteristic, Fig. 9.2.

(2) If R_1 is varied over a moderate range we see from equation (2) that I_1 is not affected, provided N is kept constant. Thus under constant-speed, variable load-resistance conditions we have, within the limits of equation (2)

$$V_1 \propto I_1 R_1 \propto R_1$$

(3) From equation (3) we see that, for a given R_1, I_2 passes through a maximum value as N increases from zero, and then decreases with further increase in N, Fig. 9.3. Thus if there is any possibility of N

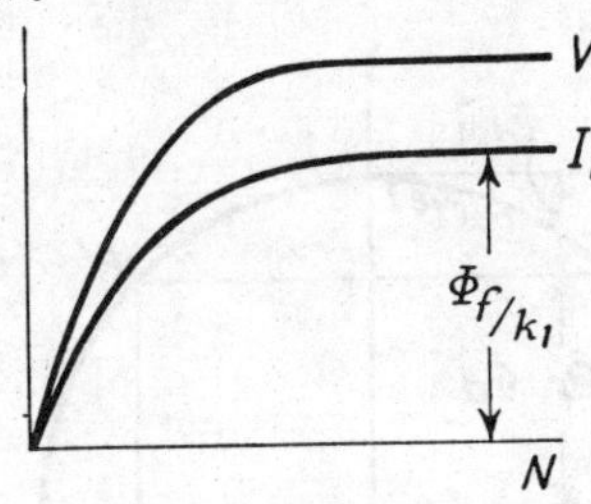

Fig. 9.2. Current-speed and Voltage-speed Characteristic

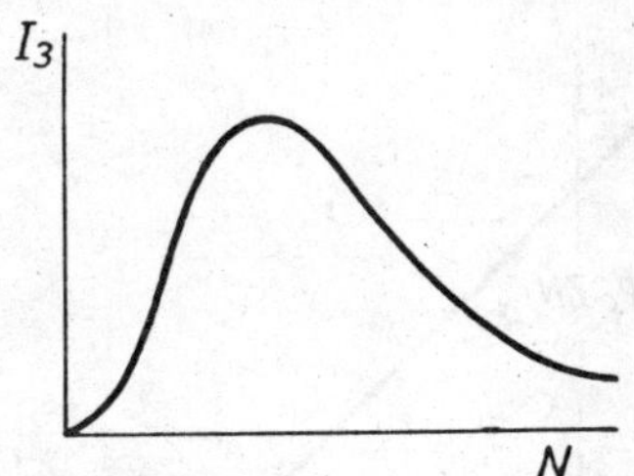

Fig. 9.3. Current in the Short Circuit

becoming too high, and I_2 therefore excessive, it is usual to insert a fuse in the short-circuit between B_2B_2.

(4) From the second of the equations (1) we see that the relationship between I_2 and I_1 is linear,

when $$I_1 = 0,\ I_2 = \Phi_f ZN/R_2$$

when $$I_2 = 0,\ I_1 = \Phi_f/k_1$$

We see from the characteristic Fig. 9.4, that I_2 becomes large if I_1 becomes small, e.g. if N becomes small.

(5) From equation (4) we see that if N is constant

$$V_1 \propto (\Phi_f - \Phi_d)$$

We can therefore determine the voltage-current characteristic, for a given value of N, as follows—

Draw the magnetization characteristic curve 1, Fig. 9.5 over the normal range of $(AT)_f$, and then set back $(AT)_d$ for any value of I_1, as shown, the difference thus being $(AT)_r$. Then

$$PN \propto V_1$$
$$O'N \propto (AT)_d \propto I_1$$

To obtain the required characteristic draw

$$O'N' = O'N, \text{ and } N'P' = NP$$

The point P' is the reflection of P in the line $O'Y$, showing that the voltage-current characteristic is the reflection of the magnetization characteristic. The final drooping portion beyond N' explains the constant-current property of the machine (provided the load resistance does not become too high).

(6) From equation (5) we see that since the numerator on the right-hand side must be small compared with the denominator

$$\Phi_f - \Phi_d \text{ must be small compared with } \Phi_f$$

and therefore Φ_d must be only slightly less than Φ_f Thus the machine has high demagnetizing ampere turns and also a high cross magnetizing

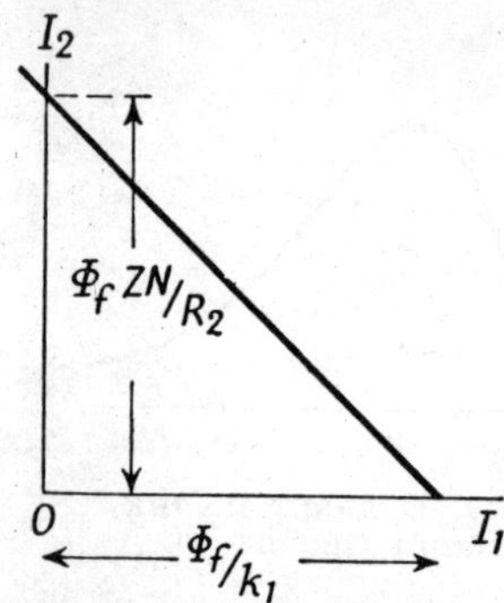

FIG. 9.4. LINEAR $I_2 - I_1$ CHARACTERISTIC

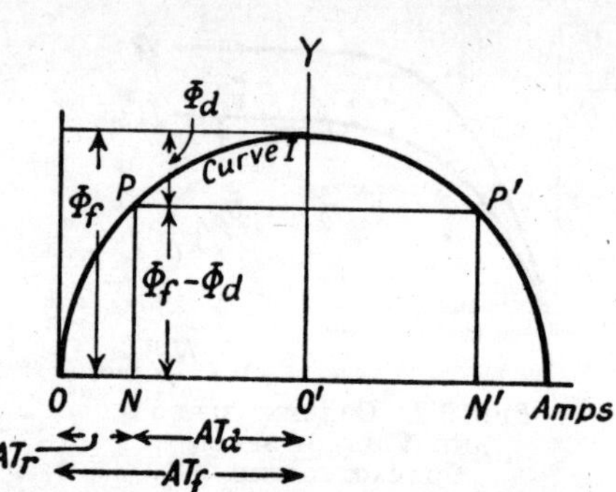

FIG. 9.5. VOLTAGE-CURRENT CHARACTERISTIC OF CONSTANT-SPEED DYNAMO

flux, so that it is essentially designed to make use of armature reaction, and not to suppress it, as in normal machines.

The Amplidyne. The amplidyne is a direct-current, high-gain, two-stage rotating amplifier with a quick speed of response; it is used for the control of large machinery and processes.

The stator of the amplidyne has four main-polar projections and two smaller interpoles, while the armature is of the normal two-pole construction. The control winding, which carries the input signal, is wound as shown in Fig. 9.6 to produce a flux vertically across the armature from the upper two N poles to the lower S poles. This flux induces an e.m.f. across the brushes A and B, which are shown, for clarity, bearing directly on the armature conductors; these brushes are short-circuited, causing a large current to flow in the armature, hence a large armature reaction m.m.f. will act horizontally across the machine. As the polar projections lie partly in the horizontal direction, the reluctance in this direction will be small and the armature m.m.f. will produce a large flux acting horizontally across the machine from the effective N poles on the right-hand side to the opposite S poles. An e.m.f. will therefore be produced across the brushes C and D, shown in Fig. 9.7, which are placed in the vertical axis. These brushes are connected to the load whose power input is being controlled.

With this simple arrangement, however, the armature reaction m.m.f. of the load current would annul the m.m.f. of the control field; it is therefore necessary to wind, in the pole faces, a compensating winding which carries the load current and overcomes the armature reaction of the load current. The arrangement of the polar projections

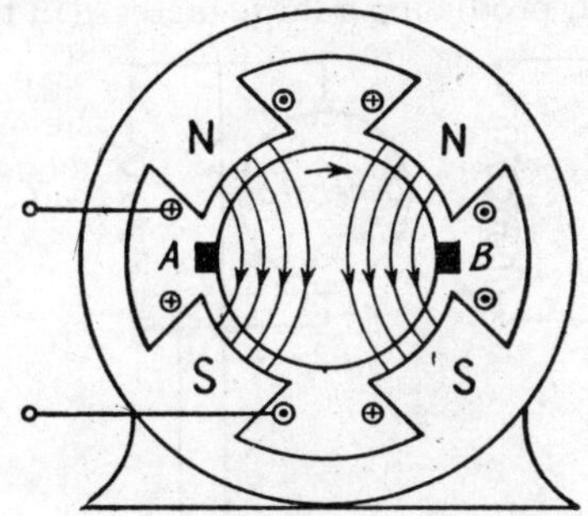

Fig. 9.6. Showing Amplidyne Control Winding and Flux

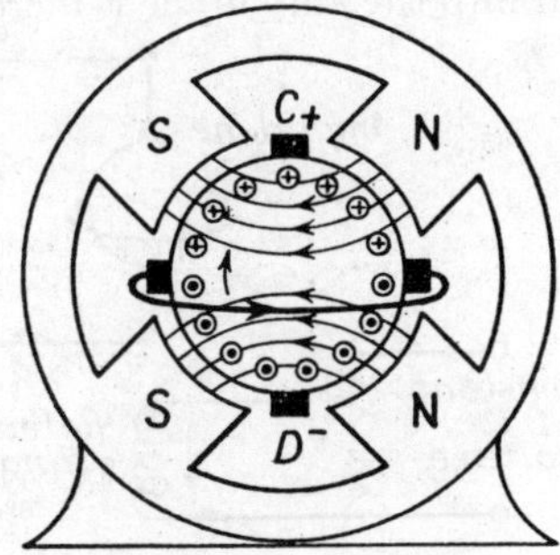

Fig. 9.7. Main Flux Distribution

enables commutation to take place in the interpolar spaces where flux is small; the commutation of the load current is also assisted by the interpoles, which are placed in the upper and lower spaces. The physical

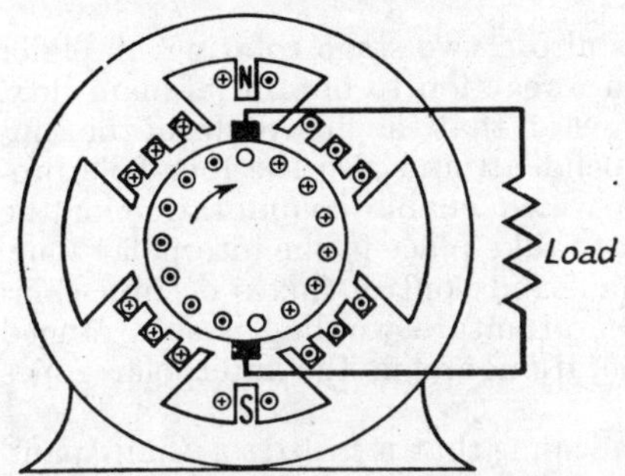

Fig. 9.8. Showing Compensating and Interpole Windings

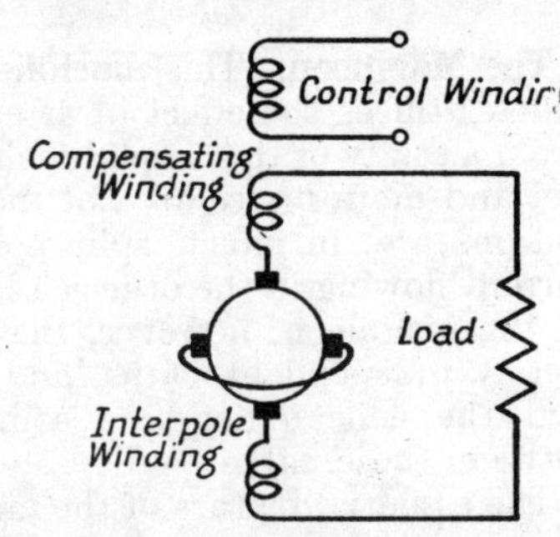

Fig. 9.9. Electrical Connections of Amplidyne

arrangement of these windings is shown in Fig. 9.8, the electrical connections being indicated in Fig. 9.9.

Application of the Amplidyne. The amplidyne is widely used where a form of "closed loop" control is required.* In this type of control system the controlled quantity is compared with a fixed reference quantity; the difference between the two quantities actuates the control system, causing an output that reduces the difference to a very small quantity.

Thus in the voltage control system shown in Fig. 9.10 the amplidyne

* *See* p. 215.

is wound with both a control and a reference field acting in opposition, the small difference between them being the required input signal. This input signal produces the large output required to supply the field of the generator; if the voltage of the generator tends to fall the input signal into the amplidyne is increased, producing a large increase in the

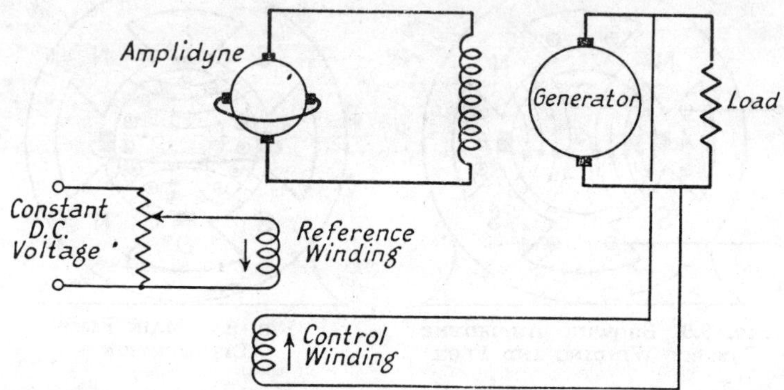

Fig. 9.10. Amplidyne Voltage-control System

generator field, which brings the voltage back to a value very close to its original setting.*

The Magnicon. This machine is also a two-stage rotating amplifier which utilizes the effect of armature reaction to obtain its main flux. From a study of the amplidyne it is seen that the flux paths of the control and main fluxes are not independent, and also the four-pole projections are, in effect, split poles which enable commutation of the current flowing in the other circuit to take place in the interpolar zone. In the magnicon, however, the main and control fluxes occupy completely independent paths, and the armature winding is so arranged that the coils undergoing commutation are in the interpolar zones between these paths.

The essential feature of the magnicon is that it utilizes a "half-span" armature winding, or that a four-pole lap-wound armature is used with a two-pole field system. This construction is seen from Fig. 9.11, where the control poles are shown; an e.m.f. is induced across the two brushes *A* and *B*, which are short-circuited, the armature current distribution then being as indicated. The armature conductors in the sectors *bc* and *da* carry currents in opposite directions and thus produce no resultant m.m.f.; in the sectors *ab* and *cd*, however, there will be a resulting m.m.f. which produces the main magnetic flux, in the low-reluctance path, horizontally across the armature, Fig. 9.12.

* The name "metadyne" is also given to the machine which is here called the "amplidyne," these being trade names adopted by different manufacturers. The development of the metadyne is discussed in a very illuminating manner in a paper by A. A. L. Bentall, "The metadyne generator in the service of industry." *Metro-Vic. Gaz.* (April–May, 1949). See also B. Adkins, "Amplidyne regulating systems." *J. Inst. Elect. Engrs.* (1947) 94, Pt. 2A.

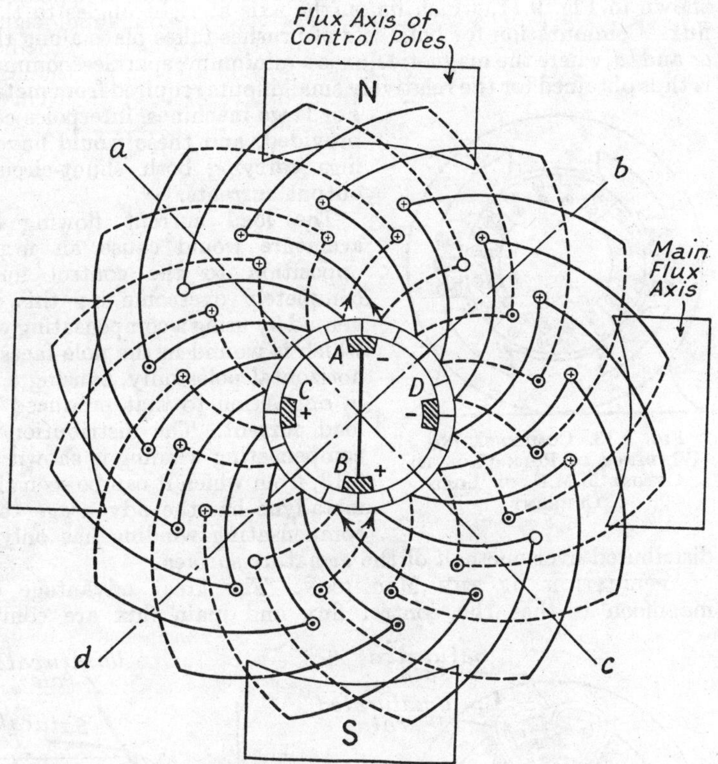

FIG. 9.11. THE MAGNICON

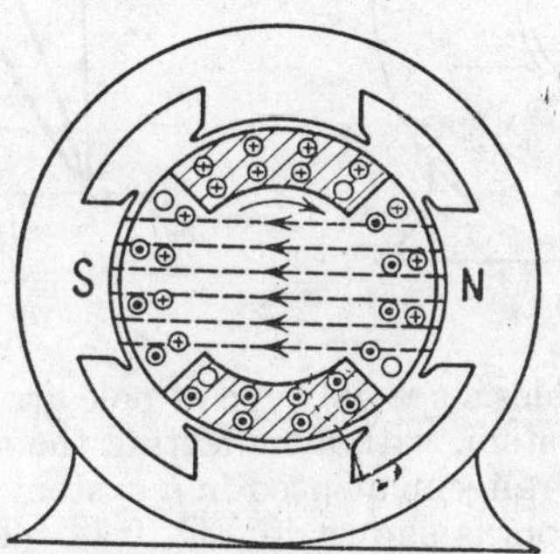

FIG. 9.12. MAIN MAGNETIC FLUX PRODUCED BY ARMATURE REACTION

This main flux induces the output e.m.f. across the brushes C and D, shown in Fig. 9.11, which lie on the axis at right angles to the main flux. Commutation for both sets of brushes takes place along the axes *ac* and *bd*, where the magnetic flux is a minimum; sparkless commutation is thus obtained for the relatively small inputs required from metadynes. For large machines, interpoles could be provided, and these would have windings carrying both shunt-circuit and output currents.

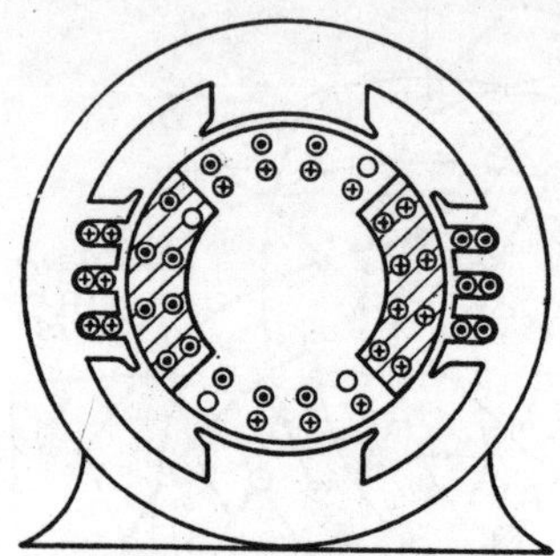

Fig. 9.13. Compensating Windings in Pole Face to Oppose M.M.F. of Load Current

The load current flowing in the armature would cause an m.m.f. in opposition to the control flux and completely overcome it; this is prevented by using a compensating winding which is wound in the pole faces of the horizontal poles only, causing a m.m.f. in opposition to that produced by the load current. The distribution of the compensating winding is shown in Fig. 9.13, from which it can be seen that the metadyne has the advantage that the compensating winding has only to be distributed over one-half of the armature surface.

Application of the Magnicon. The great advantage of the magnicon is that the control flux and main flux are completely

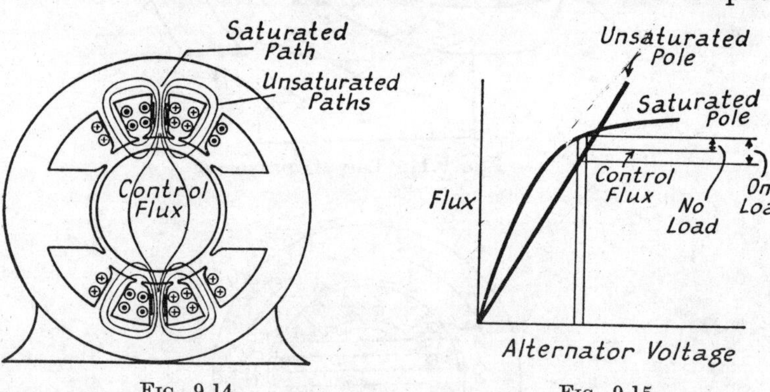

Fig. 9.14

Fig. 9.15

separate; this enables a special type of pole arrangement to be used for the control-field system, without affecting the main flux path.

One ingenious arrangement used in a system adapted for the voltage control of alternators is shown in Fig. 9.14. The control pole is split into three parts, the centre projection is highly saturated and has a winding carrying a current proportional to the alternator voltage. The outer projections also have windings carrying currents proportional to the alternator voltage; the iron in these projections is not saturated and the flux acts in opposition to the flux in the centre projection. The net control flux will be the difference of these two fluxes.

The operation of this control arrangement may be seen from Fig. 9.15. The alternator voltage is plotted as abscissae since the control current is proportional to it; the resultant control flux, which is the difference of the flux in the central, (or reference) projection and the outer projections, is therefore measured by the ordinate between the two curves. When the alternator is on no load the terminal voltage is V, and the control flux is sufficient to cause this voltage to be generated. If the alternator voltage tends to fall, the large difference in gradient between the two curves will cause the control flux to fall by an appreciable amount; this will result in the field excitation of the alternator increasing, thus bringing the terminal voltage of the alternator back nearly to its set value. The small drop in terminal volts v will cause sufficient increase in control flux to enable the magnicon to increase the alternator excitation to such an extent that the very large inherent regulation of the alternator will be compensated.

Basically, there is no difference in the fundamental principles of operation of the three special dynamos described above. In the amplidyne and metadyne, and in the magnicon, there is finite compensation, the amount of which can be varied to suit requirements. The performance is substantially affected by the amount. Thus with an amplidyne, 100 per cent compensation gives almost constant load voltage. Over compensation, by as little as 5 per cent, gives a steeply rising characteristic, and under compensation by the same amount a steeply falling one. With zero compensation—the case of the Rosenberg dynamo—the characteristic is almost a vertical straight line.

From the point of view of choice of type it can be said, in general, that amplidyne machines can be used for both current forcing and voltage forcing: up to the present the magnicon has been applied mainly to voltage control of both a.c. and d.c. machines.

Static amplifiers can do what rotary amplifiers do, but, except for small outputs, their size, when associated apparatus is taken into account, is prohibitive.

Further Reading

1. Say, M. G. (Editor). *Rotating Amplifiers*, Newnes (1954).
2. Tustin, A. *D.C. Machines for Control Systems*. E. and F. N. Spon (1952).
3. West, J. C. *Servo-Mechanisms*. English Universities Press, Ltd., London, 1953.

CHAPTER 10

THE D.C. MOTOR

ANY direct-current generator will run as a motor, that is, convert electrical power to mechanical power, if its field and armature are connected to a suitable electric supply. The essential construction of a motor is identical with that of a generator, and when there is any external difference in appearance this is due to the fact that whereas the frames of generators can as a rule be open, those of motors are either partly or totally enclosed because of the rougher (mechanical) usage to which the latter are subjected.

Consider first of all one conductor of a smooth-cored armature, i.e. an armature in which the conductors are not housed in slots, and in which they are, in consequence, acted on directly by the gap flux.

Fig. 10.1 (*A*) shows one conductor on the armature of a generator rotating clockwise under a N pole. Fleming's right-hand rule indicates that the induced e.m.f. is inwards, and therefore the current in that conductor is also inwards. This current sets up a magnetic field, the lines of force of which are concentric circles round the wire and in a clockwise direction (from the corkscrew rule). This field is seen to act in the same direction as the main field on the right-hand side, and in opposition on the left-hand side, the result being that the actual distribution of the lines of force in the resultant field is as indicated in Fig. 10.1 (*B*), some of the lines being bent round the conductor. Now magnetic lines of force are always in a state of tension, the bent lines of force setting up a mechanical force on the conductor much in the same way that the bent elastic of a catapult produces a mechanical force on the stone. In the case of the generator it will be noticed that this force is in opposition to the direction of motion. It is therefore called the magnetic drag, and it is this drag acting on all the conductors that the prime mover has to do work against. For a current of I amperes flowing in a conductor of l metres placed in a field of B webers/metre^2 the drag is given by

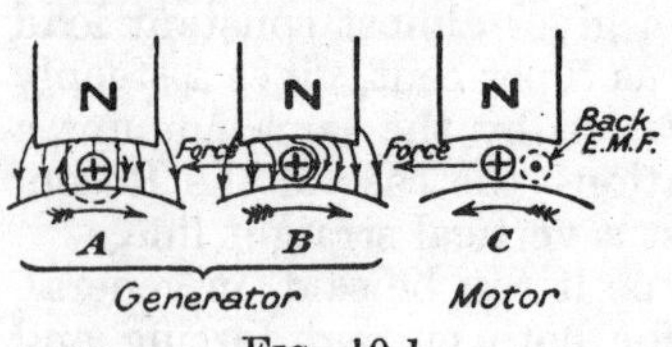

FIG. 10.1

$$f = BIl \text{ newtons}$$

The above explanation, although convenient for the calculation of the torque of the armature, is not a correct description of the mechanism of torque production. Hence let the conductors lie in slots as in Fig. 10.2. In figure *A* the main flux is shown concentrated into tufts which pass into the tops of the teeth, while the armature flux is shown by the dotted fluxes embracing the slots. The directions of these fluxes correspond with those of Fig. 10.1. We see that the effect of the armature flux is twofold. Firstly, it increases the flux in the left-hand halves of the teeth and reduces it in the right-hand halves, so that the distribution of flux density across the tooth section is no longer uniform. Secondly, it inclines the directions of the magnetic lines in the air gap so that the tufts of lines of force are not radial but are disposed somewhat after the manner of figure *B*. Since the lines of force are in a state of tension the tangential component of this tension will set up a torque on the armature, and from the figure we see that the direction

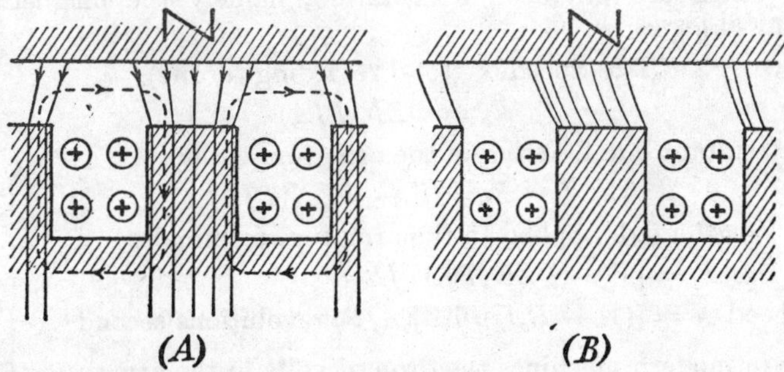

FIG. 10.2. MECHANISM OF TORQUE PRODUCTION IN A D.C. MACHINE

of this torque is counter-clockwise. Hence in an actual machine with slotted armature the torque is not due to mechanical forces on the conductors themselves, but to the tangential component of the magnetic pulls on the teeth.

Back E.M.F. Now suppose that the machine is uncoupled from the prime mover and that current is sent through the armature and field from an external source. If the excitation of the field and the directions of the armature currents are the same as before, the magnetic drag will be set up in the same direction as before, and under its influence the armature will rotate. The machine will now be running as a motor. We thus see that for the same direction of armature currents and the same excitation, the direction of rotation of a motor is opposed to that of a generator—Fig. 10.1 (*C*).

Now when the motor armature rotates, the armature conductors cut the lines of force of the field, and as a result, they have e.m.f.s induced in them. The direction of one such induced e.m.f. in an individual conductor is, of course, given by Fleming's right-hand rule, and applying this rule to the conductor in Fig. 10.1 (*C*) we see that the e.m.f. is outwards, that is, *opposed to the current*. This induced e.m.f. in the case of a motor is therefore called the "back e.m.f." The supply e.m.f. has to do work against this back e.m.f. in forcing the current through the armature. If the induced e.m.f. in an armature is opposite in direction to the current, then the machine is motoring; if it is in the same direction as the current, the machine is generating. The magnitude of the back e.m.f. is, of course, given by the expression for the generated e.m.f. Calling the back e.m.f. E_b, we have

$$E_b = \Phi ZN(P/A) \text{ volts}$$

Speed of a D.C. Motor. The speed of a motor automatically adjusts itself to the load so that the electrical power required to drive the current through the armature is equal to the mechanical power required to drive the load. The load here includes all those losses in the motor which are produced by rotation, namely the magnetic and mechanical losses.

Hence $E_b I \propto$ Speed $\times$ (total retarding torque).

Now $$E_b = \Phi ZN(P/A)$$

$\therefore$ Resultant e.m.f. acting in the armature circuit

$$V - E_b = V - \Phi ZN(P/A)$$

where V is the p.d. applied to the armature terminals.

$$\therefore\ I_a = [V - \Phi ZN(P/A)]/R_a$$

$\therefore$ Speed $N = [(V - R_a I_a)/\Phi Z](A/P)$ revolutions/second

Now in modern machines the drop of volts in the armature, $R_a I_a$, is small compared with V, so that we have, approximately,

$$N = (1/\Phi)(V/Z)(A/P)$$

$$\propto 1/\Phi \text{ so long as } V \text{ is constant}$$

Thus the speed of a d.c. motor is inversely proportional to the flux/pole.

Conditions for Maximum Power. We have

$$V = E_b + R_a I_a$$

$$\therefore\ VI_a = E_b I_a + R_a I_a^2$$

This is an equation of power, the various terms having the following meanings—

VI_a = total power supplied to the armature

$E_b I_a$ = power converted into mechanical power

$R_a I_a^2$ = power dissipated as heat due to the armature resistance

Hence, mechanical power of the motor

$$P_m = VI_a - R_a I_a^2$$

$$dP_m/dI_a = V - 2R_a I_a$$

$$= 0$$

and corresponds to a maximum when

$$R_a I_a = V/2 \text{ or } E_b = V/2$$

Thus a motor develops the maximum mechanical power when the armature current is such that the back e.m.f. is equal to one-half of the applied e.m.f. This is not attained in practice, since it necessitates a current well beyond the normal working range. Also an amount of power equal to the mechanical power developed would be wasted in heating the armature, so that, taking other losses into account, the efficiency would be well below 50 per cent.

Torque. If P = number of poles and A = number of parallel paths through the armature, the current in each individual conductor is

$$I = I_a/A$$

Consider that the flux per pole Φ is divided into separate fluxes Φ_1, Φ_2, Φ_3, etc., which embrace the individual conductors, as shown in Fig. 10.3.

Then force on conductor 1, using the first conception of torque production,

$$= BIl \text{ newtons}$$
$$= (\Phi_1/xl)(I_a/A)l$$
$$= \Phi_1 I_a/xA$$

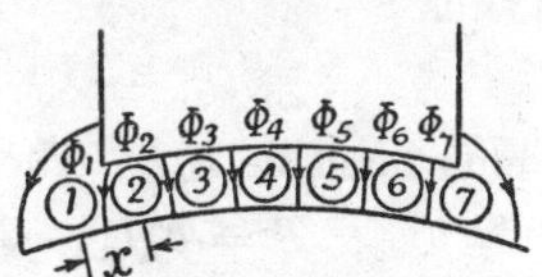

FIG. 10.3. CALCULATION OF TORQUE OF A MOTOR

force on conductor 2

$$= \Phi_2 I_a/xA$$

force on conductor 3

$$= \Phi_3 I_a/xA$$

$\therefore$ Total tangential force on all the conductors under one pole

$$= \Phi_1 I_a/xA + \Phi_2 I_a/xA + \Phi_3 I_a/xA + \ldots\ldots$$
$$= \Phi I_a/xA$$

$\therefore$ Torque due to this force

$$= \Phi I_a r/xA \text{ newton-metres, where } r = \text{radius}$$

Now the distance x

$$= 2\pi r/Z$$

$\therefore$ Torque due to all the conductors under one pole

$$= (\Phi I_a/A)(Z/2\pi r)r = (\Phi I_a Z/2\pi)(1/A)$$

Hence torque due to all the conductors under all the P poles

$$T = (\Phi I_a Z/2\pi)(P/A)$$
$$= 0{\cdot}159\Phi I_a Z(P/A) \text{ newton/metres}$$

EXAMPLE. A four-pole armature 50 cm in diameter has 1000 conductors, each of active length 25 cm. The pole span is two-thirds of the pole pitch, and the average flux density under the pole shoe is 6000 lines/cm². Calculate the torque in lb-in. when each conductor carries 10 A.

$$\Phi = \tfrac{2}{3}B_{av}Yl, \text{ where}$$
$$B_{av} = 0{\cdot}6 \text{ weber/m}^2$$
$$Y = \text{pole pitch} = \pi \times 0{\cdot}5/4 \text{ m}$$
$$l = \text{core length} = 0{\cdot}25 \text{ m}$$
$$\therefore \Phi = \tfrac{2}{3} \times 0{\cdot}6 \times 0{\cdot}5\pi/4 \times 0{\cdot}25 = 0{\cdot}0125\pi \text{ webers}$$

The current given is the current per path, so that the equation for torque becomes

$$T = (\Phi IZ/2\pi)P = (0{\cdot}0125 \times 10 \times 1000/2\pi) \times 4$$
$$= 250 \text{ newton-metres}$$

Now 1 newton-metre $= 0{\cdot}7373$ lb-ft $= 8{\cdot}848$ lb-in.

$$\therefore\ T = 250 \times 8{\cdot}848 = 2212 \text{ lb-in.}$$

The equation for the torque can also be derived directly from the back e.m.f. equation, as follows. The power required to overcome the back e.m.f. is that which provides the mechanical torque. If the torque is T, the work done per revolution is $T \times 2\pi$, since the work done by a torque is equal to the product of the torque and the angle turned through (in radians). The work done by the torque per second is thus

$$T \times 2\pi N \text{ watts}$$

which must be $E_b I_a$. Therefore

$$T = E_b I_a \div 2\pi N$$
$$= \Phi ZN \times (P/A) \times I_a \div 2\pi N$$
$$= (\Phi I_a Z/2\pi)(P/A) \text{ newton-metres, as before}$$

Armature Reaction and Commutation. Suppose that the brushes are placed first of all in the g.n.p. (Fig. 10.4), then if the armature is rotating clockwise the directions of the armature currents will be

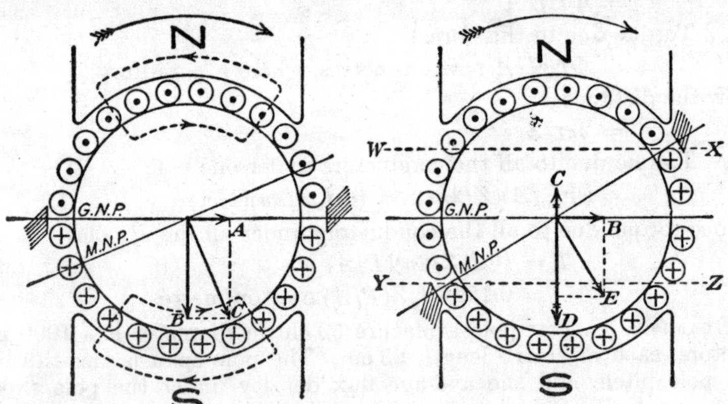

FIG. 10.4. ARMATURE REACTION OF A MOTOR

outwards under the N poles, and inwards under the S poles. The armature therefore acts like a solenoid whose axis coincides with the brush axis, as in the case of a generator, but in this case the armature m.m.f. OA is directed from left to right. OB is the main m.m.f., and OC the resultant m.m.f. We see that the magnetic neutral plane m.n.p., which is perpendicular to OC, is now shifted backwards, instead of forwards as in the case of the generator. Also on comparing the directions of the lines of force of the armature cross field with the main field, we see that the main field is strengthened under the leading pole tips and weakened under the trailing pole tips.

It is necessary to give the brushes a backward lead in order to bring the brush axis into the m.n.p. We can then divide the armature conductors into two groups, those lying above WX and below YZ, and those lying between these two planes. The first group produces a cross field OB whose m.m.f. acts from left to right, and the second group produces a m.m.f. OC which is in opposition to the main m.m.f. OD, and therefore sets up a demagnetizing effect. Thus when the brushes in a generator and a motor are set so as to be in the m.n.p., a demagnetizing action is set up by the armature in each case, but the cross-magnetizing fields are opposite in direction.

In some cases motors are required to run in both directions without sparking. In such a case it is obvious that the brushes must be placed in the g.n.p. In order to cut down the distortion of the field produced by the cross field to a minimum, the field magnets are designed to be nearly saturated. At the same time the air gap is made rather longer than in the case of a motor running in one direction only. Since each line of force of the cross flux has to cross the same air gap twice, this causes a considerable reduction in the total cross flux, thereby minimizing distortion. It is, of course, necessary to provide the excitation with additional ampere-turns to drive the flux across the longer air gaps, and such machines are said to have "stiff" magnetic fields, the field system being magnetically strong compared with the armature. Denoting the field ampere-turns per pole and the armature ampere-turns per pole by AT_f and AT_a respectively the ratio AT_f/AT_a varies from about 1·1 to 1·3 in the case of dynamos, and from 0·9 to 1·1 in the case of motors. The lower value for the motor is due to the fact that the terminal p.d. is maintained at the supply.

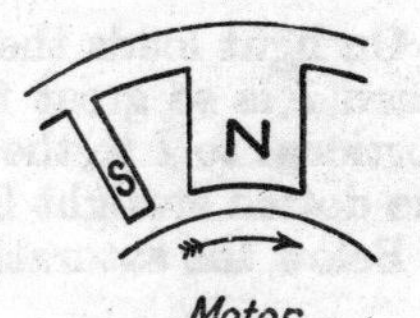

Fig. 10.5. Polarity of Interpoles

D.C. motors, like generators, are commonly fitted with interpoles in order to ensure good commutation. In either case their function is the same, namely, to force the current in a coil short-circuited by the brush to attain its full value in the reversed direction by the end of short-circuit. Since the currents in the armature conductors flow in the reverse direction to those of the generator when the directions of rotation are the same, it follows the polarity of the interpoles must be reversed. Their polarity with respect to the polarity of the main poles and to the direction of rotation is, for a motor, as indicated in Fig. 10.5.

Characteristics of D.C. Motors: The Series Motor. There are three ways of exciting the field of a d.c. motor; these are series, shunt and compound, and the characteristics of the motor are determined by the method of excitation. Consider first of all the series motor. The flux varies with the motor current, the relationship between the two being given by the magnetic characteristic (Fig. 10.6). If the flux due to armature reaction is deducted, the dotted curve in the figure will be obtained. Now we have seen that the speed is inversely proportional to the flux, and therefore, if we plot reciprocals of the flux against current, the speed-current curve will be obtained. From its shape we see that the series motor is essentially a variable speed motor, the speed being low on heavy load and dangerously high on light load. For this

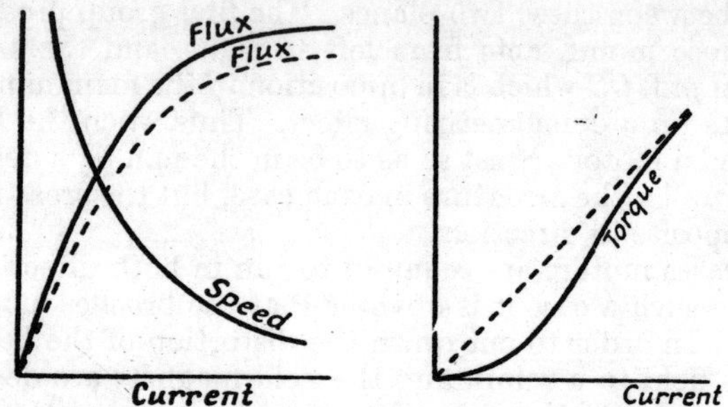

FIG. 10.6. CHARACTERISTICS OF SERIES MOTOR

reason the series motor is never run without some mechanical load on it, otherwise it might fly to pieces. If the saturation point is not reached, the flux Φ is approximately proportional to the current I.

Now torque $T \propto \Phi \times I$

But $\Phi \propto I$

$\therefore \quad T \propto I^2$

On light loads the torque-current curve is therefore a parabola, but when I is so great that Φ becomes nearly constant, T becomes proportional to I to the first power, and the curve becomes asymptotic to the dotted straight line through the origin, as shown in Fig. 10.6.

Below the saturation point we have Φ proportional to I

$$\therefore N \propto 1/\Phi$$
$$\propto 1/I$$

From the above we see that, so long as the iron is not saturated, the series motor exerts a torque proportional to the square of the current. Its starting torque is therefore very high, for which reason it is used in cases where heavy masses have to be accelerated quickly, e.g. for electric traction and for hoist work. The falling off in speed as the load increases is also a considerable advantage for such duties, since it automatically relieves the motor from having to carry excess loads. Thus, if a tram travelling up hill falls to half speed, the load on the motor is only half of what it would be if the tram raced up at full speed.

If the speed-current and torque-current curves are drawn on the same diagram, then by measuring ordinates on the two curves corresponding to a series of values of current, the speed-torque curve can be drawn. This curve is sometimes called the "mechanical characteristic," and for a series motor it is similar in shape to the speed-current curve.

EXAMPLE. A series motor takes 20 A at 400 V to drive a fan at 250 r.p.m. Its resistance is 1 Ω. If the torque required to drive the fan varies as the square of the speed, find the necessary applied voltage and current to drive the fan at 350 r.p.m.

Assuming a straight-line magnetic characteristic,

$$T \propto \Phi \times I \propto I^2. \text{ But } T \propto N^2$$

$\therefore I \propto N$. Hence, new current $= (350/250) \times 20 = 28$ A

Motor back e.m.f. at 250 r.p.m. $= 400 - (20 \times 1) = 380$ V

Motor back e.m.f. at 350 r.p.m. $= V - (28 \times 1)$

Now back e.m.f. $\propto \Phi \times N$, i.e. $\propto I \times N$

$$\therefore (V - 28)/380 = 28 \times 350/(20 \times 250)$$

$$\therefore V = 773 \text{ V}$$

The Shunt-wound Motor. For a constant applied p.d. the exciting current is constant. The flux will therefore have its maximum value at no load, and because of armature reaction will decrease slightly as the load increases. For most purposes this decrease can be neglected and the flux per pole, Φ, can be regarded as constant. We then have from the equation on p. 222, speed

$$N = (V - R_a I_a)/(\Phi Z)(A/P)$$
$$= K(V - R_a I_a) \text{ where } K \text{ is a constant}$$

Let N_o and I_o be the no load speed and armature current, and N and I, the speed and current at any other load. Then

$$N = K(E - R_a I)$$
$$N_o = K(E - R_a I_o)$$
$$\therefore N = N_o(E - R_a I)/(E - R_a I_o)$$
$$\therefore (N_o - N)/N_o = (R_a I - R_a I_o)/(E - R_a I_o)$$

This equation shows that since I_o is small the drop of speed of a shunt motor from no load to any other load is proportional to the drop of volts in the armature at that load. This, of course, assumes zero armature reaction; actually, since the armature exerts a demagnetizing action, Φ will decrease as I increases, and therefore the fall in speed will not be quite so great as indicated by the above expression. Since the full-load drop in the armature is small compared with the applied voltage, we see that the speed-current curve of a shunt motor is a slightly drooping curve, as shown in Fig. 10.7. It is instructive to deduce the drop of speed in another manner. Whatever the load on the motor may be, the back e.m.f. adjusts itself to such a value that sufficient armature current can pass to produce a torque equal to the total opposing torque.

Let the line AB represent the applied voltage and let the

A | ———————————————— $C_2\,C_1\,C$ | B

back e.m.f. be AC at a given load. Then BC is the resultant e.m.f. and represents to another scale the armature current. Since Φ is approximately constant, BC represents to still another scale the opposing torque. If the torque is doubled, C has a new position, C_1, where $BC_1 = 2BC$. The back e.m.f. is now AC_1, and, since the speed is proportional to the back e.m.f., the speed falls from AC to AC_1. If the torque is trebled, C moves to C_2 where $BC_2 = 3BC$, and the new current and speed are represented by BC_2 and AC_2 respectively. Thus, if the lengths of AC are plotted against the lengths BC, the speed-current characteristic is obtained.

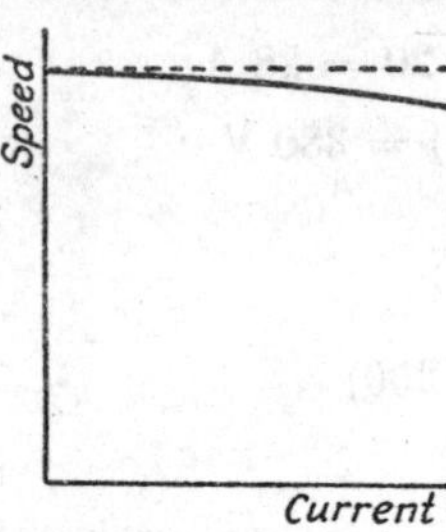

FIG. 10.7. CHARACTERISTICS OF SHUNT MOTOR

EXAMPLE. A six-pole lap wound shunt motor has poles 20 cm square and a flux density in the gap of 5000 gauss. The armature is wound with 500 conductors having a total length of wire of 24,000 cm and 0·07 cm² area. Find the speed of the motor with 100 V on the terminals and 120 A in the line. (London Univ.)

$$\Phi = 5000 \times 10^{-4} \times 20^2 \times 10^{-4} = 2 \times 10^{-2} \text{ webers}$$

Resistance of all the armature wires in series

$$= \rho l/a = (2/10^6) \times (24{,}000/0{\cdot}07)$$

$$= 0{\cdot}7\ \Omega$$

Hence resistance of armature with 6 parallel paths

$$R_a = 0{\cdot}7/6^2 = 0{\cdot}0194\ \Omega$$

Deducting, say, 5 A for the shunt current, the armature current will be 115 A, and the drop in the armature, $0{\cdot}0194 \times 115 = 2$ V, roughly.

$\therefore$ Back e.m.f.

$$E_b = 100 - 2 = 98 \text{ V}$$

But $\quad E_b = \Phi ZN(P/A)$

and $\quad P/A = 1$

$$\therefore N = 98/(2 \times 10^{-2} \times 500) = 9{\cdot}8 \text{ r.p.s.}$$

$$= 588 \text{ r.p.m.}$$

From the general equation for the torque of a d.c. motor, we see that, since Φ is approximately constant, the torque is proportional to the armature current. Hence, the torque-current characteristic is a straight line through the origin. The mechanical characteristic can be determined from the two electrical characteristics as before. It is a slightly drooping curve similar in shape to the speed-current curve.

Sometimes, when starting a shunt motor, it is found that it takes a very large current and runs at a high speed in the wrong direction. This phenomenon is due to a break in the shunt winding. The demagnetizing effect of the sudden rush of current at starting reverses the residual magnetism in the field, and the only flux present is that due to armature reaction. This flux is comparatively weak and in the reverse direction to the proper main field, thus causing the motor to run at high speed in the reversed direction.

Since the armature drop of volts on full load is proportional to the fall in speed from no load to full load, and since the full-load drop is only about 4 per cent of the applied voltage on the average, the shunt motor can be regarded as a constant-speed motor. If the drop in speed has to be made up, this can be effected by inserting resistance in the shunt field, thereby reducing the flux. Shunt motors are therefore suitable for driving light machine tools and for all purposes where an approximately constant speed is required.

Compound-wound Motor. The series field may be connected in the circuit so as to either help the shunt field or oppose it. In the first case the motor is "cumulatively" compounded, and in the second case, "differentially" compounded. If the series field of the differentially-compounded motor is so adjusted that the full-load decrease in flux produced by it is just sufficient to make the full-load speed equal to the speed on no load, then, for any other load within this range, its speed will be approximately constant. Its mechanical characteristic is shown in curve *A* (Fig. 10.8). This appears to be an advantage at first sight, but there are two disadvantages. In the first place, when the motor is being started, the shunt field will take some time to build up, and therefore the series field will be established first, owing to the initial rush of current through the armature and series field. The motor will therefore tend to start up the wrong way. When the shunt field is fully established the total field will be so small that there may not be sufficient torque to run the motor, and the armature will take an excessive current from the supply. This difficulty can be overcome by short-circuiting the series field during starting, and only putting it in circuit when the motor is under way. The second disadvantage is that if the motor becomes overloaded, the resulting decrease in flux will tend to force up the speed so that the motor will be very seriously overloaded. In fact, when the current attains a certain value, the motor becomes unstable, and it begins to race like a

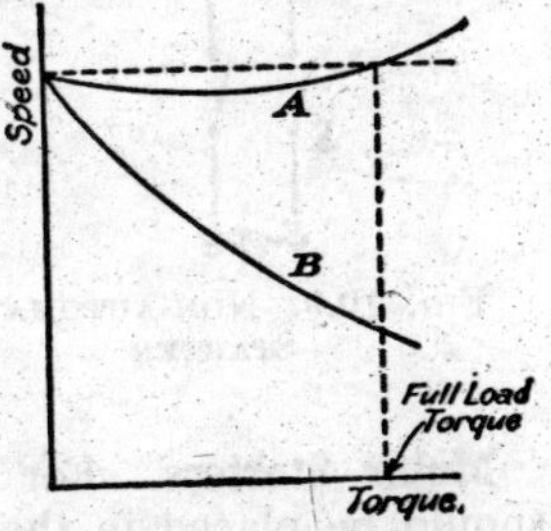

FIG. 10.8. CHARACTERISTICS OF COMPOUND MOTOR

series motor on no load. The motor is thus dangerous to use unless there is no possibility of the load exceeding the normal full load.

The cumulatively-compounded motor experiences an increase in flux as the load increases, with the result that the speed decreases, but not so rapidly as in the case of the plain series motor, because of the constant flux produced by the shunt field. The mechanical characteristic is therefore between those of the series and shunt motors, its nearness to one or the other depending upon the number of series turns (Curve *B*, Fig. 10.8). The falling off in speed affords relief to the motor if a sudden load comes on, and the motor is therefore used largely for driving heavy machine tools where sudden deep cuts may be taken. This motor is also suitable for driving continuous running rolling mills, for which duty it is coupled to a heavy flywheel. When a billet is inserted in the rolls the sudden increase in load causes the speed to decrease, with the result that the fly-wheel gives up some of its stored energy, thereby taking the strain of the sudden peak load off both motor and supply mains. When the load is light, the speed rises and the flywheel acquires the kinetic energy it gave up previously. In this way, provided that the successive peak loads do not come on before the speed has had time to reach the full value again, the fly-wheel, in conjunction with such a motor, acts as a load equalizer. Its function is to look after all the sudden fluctuations in the load and to leave the motor and supply to cope with the steady average load.

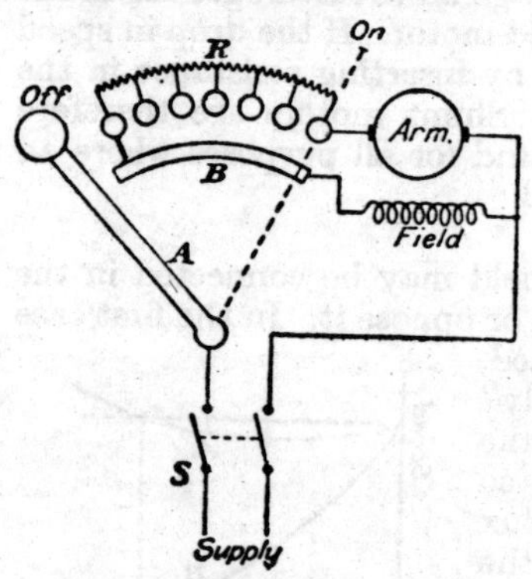

FIG. 10.9. NON-AUTOMATIC STARTER

Motor Starters. For all but fractional h.p. motors, a resistance should be placed in the armature circuit at the moment of starting, and then gradually cut out as the motor speeds up. The simplest arrangement of a starter for a shunt motor is shown in Fig. 10.9. The arm *A* makes contact with the studs connected to the starting resistance *R*, and also to a brass arc, *B*, by which the connection to the shunt field is made. This arrangement is perfectly satisfactory for starting, but it suffers from the disadvantage that when the main switch *S* is opened to shut down the motor, the arm *A* remains in the "on" position, and if the switch is again closed the armature is put right across the supply. It is thus obvious that when the switch is opened the arm should return to the "off" position. Also, for motors up to about 20 or 30 h.p. it is desirable to incorporate an "overload release" which will disconnect the motor from the supply in the event of an overload. For larger sizes it is better practice to leave this duty to a separate automatic circuit-breaker. A typical starter for a shunt or compound motor is shown in Fig. 10.10. The arm *A* is provided with a hinged armature, which is attracted by the coil of a "no-volt" release when in the "on" position. This release is connected in series with the shunt field. When the switch is opened the pull of this coil is removed and the arm is returned to the off position by a spring. The overload release consists of another coil, which carries the full line current. It attracts an armature which, when lifted, bridges two brass studs connected to the ends of the coil of the no-volt release, thereby short-circuiting this. The spring then pulls the arm back and stops the motor. The position of the armature of the overload release can be adjusted by a set screw, and the release thus set to operate at any desired overload.

Fig. 10.10. Starter with No-volt and Overload Releases

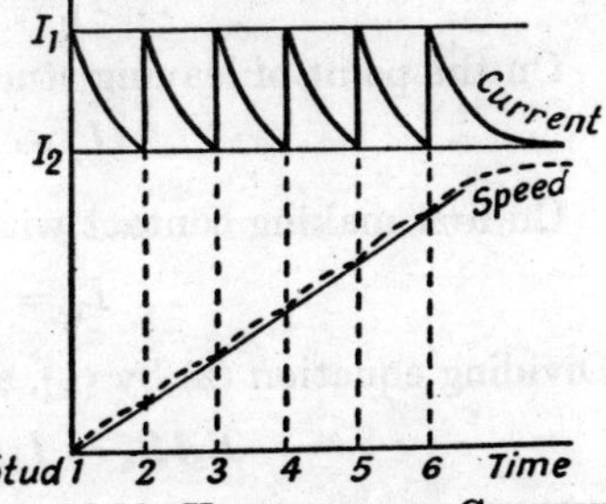

Fig. 10.11. Variations in Current and Speed During Starting

The grading of the steps of the starting resistance, and the time taken to move the arm across the row of contacts, should be such that, during the operation of starting, the motor current is kept within upper and lower limits. This is illustrated in Fig. 10.11. The stepped curve shows the variation of current. As soon as contact is made with any stud the diminution in resistances causes the current to increase initially; the motor then increases in speed with consequent increase in back e.m.f., so that the current decreases while the arm is held on that stud. When the current reaches the minimum value the arm is moved to the next stud, and the current again suddenly increases to the maximum value; and so on. The variations in speed are illustrated by the dotted curve. In order to calculate the various steps of the starting resistance, it is necessary to assume values for the upper and lower limits of the current; let these be I_1 and I_2 respectively. Let r_1, r_2, r_3, etc., be the various steps, and let R_1, R_2, R_3, etc. be the total resistances between the various studs and the far terminal of the armature, as illustrated in the scheme in Fig. 10.12. On making contact with any stud the

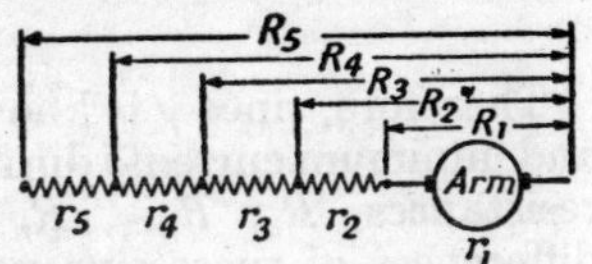

Fig. 10.12. Calculation of Shunt Motor Starter

current suddenly jumps to the value I_1, and when leaving any stud the current is I_2. Hence

when leaving stud 5, say, $I_2 =$ Resultant e.m.f.$/R_5$

$$= (V - E_b)/R_5$$

where E_b is the back e.m.f.

$$\therefore I_2 = (V - AN)/R_5 \quad . \quad . \quad . \quad . \quad (1)$$

where A is a constant.

On first making contact with stud 4 the current jumps to I_1, but the speed has not time to change from N,

$$\therefore I_1 = (V - AN)/R_4 \quad . \quad . \quad . \quad . \quad (2)$$

On the point of leaving stud 4 the speed has increased to say, N_1,

$$\therefore I_2 = (V - AN_1)/R_4 \quad . \quad . \quad . \quad . \quad (3)$$

On first making contact with stud 3,

$$I_1 = (V - AN_1)/R_3 \quad . \quad . \quad . \quad . \quad (4)$$

Dividing equation (2) by (1), and (4) by (3), we have

$$R_5/R_4 = I_1/I_2 \text{ and } R_4/R_3 = I_1/I_2$$

Hence, if we put $I_1/I_2 = \gamma$, we have, if there are n live studs, i.e. $(n - 1)$ sections in the resistance,

$$R_n/R_{n-1} = R_{n-1}/R_{n-2} = \ldots R_2/R_1 = \gamma$$

Multiplying all the terms together, we have

$$\gamma^{n-1} = R_n/R_1$$

$$\therefore \gamma = \sqrt[n-1]{(R_n/R_1)}$$

Now R_1 is the armature resistance, which is known. On making contact with the first live stud the speed is zero and the current is I_1. Hence

$$R_n = E/I_1$$

Therefore, since γ is known from the required ratio of the maximum and minimum currents during starting, n can be calculated. The various resistances, R_n, R_{n-1}, R_{n-2}, etc., can next be calculated, and the differences of successive pairs give the required resistances, r_1, r_2, r_3, etc.

Example. As an example of the calculation of the resistance steps of a shunt motor-starter, take the case of a 30 h.p., 220 V motor which takes 113 A when running normally on full load. The motor has to start up against full load and the starting current has not to exceed $1\frac{1}{2}$ times the normal full-load value. Resistance of armature circuit 0·02 Ω.

Since the motor has to start against full-load torque, the minimum current during starting will be the full-load current of 113 A. Hence $I_2 = 113$ A.

$$I_1 = 1\tfrac{1}{2} \times 113 = 169{\cdot}5 \text{ A}$$

and $$I_1/I_2 = \gamma = 1{\cdot}5$$

$$R_n = E/I_1 = 220/169{\cdot}5 = 1{\cdot}3\ \Omega$$

and $$R_1 = 0{\cdot}02, \text{ as stated.}$$

Now $$R_2/R_1 = \gamma = 1{\cdot}5$$

$$\begin{aligned}
\therefore \quad R_2 &= 1{\cdot}5 \times R_1 = 1{\cdot}5 \times 0{\cdot}02 = 0{\cdot}03\ \Omega\\
\text{Similarly,} \quad R_3 &= 1{\cdot}5 \times R_2 = 1{\cdot}5 \times 0{\cdot}03 = 0{\cdot}045\ \Omega\\
R_4 &= 1{\cdot}5 \times R_3 = 1{\cdot}5 \times 0{\cdot}045 = 0{\cdot}068\ \Omega\\
R_5 &= 1{\cdot}5 \times R_4 = 1{\cdot}5 \times 0{\cdot}068 = 0{\cdot}102\ \Omega\\
R_6 &= 1{\cdot}5 \times R_5 = 1{\cdot}5 \times 0{\cdot}102 = 0{\cdot}153\ \Omega\\
R_7 &= 1{\cdot}5 \times R_6 = 1{\cdot}5 \times 0{\cdot}153 = 0{\cdot}23\ \Omega\\
R_8 &= 1{\cdot}5 \times R_7 = 1{\cdot}5 \times 0{\cdot}23 = 0{\cdot}35\ \Omega\\
R_9 &= 1{\cdot}5 \times R_8 = 1{\cdot}5 \times 0{\cdot}35 = 0{\cdot}53\ \Omega\\
R_{10} &= 1{\cdot}5 \times R_9 = 1{\cdot}5 \times 0{\cdot}53 = 0{\cdot}8\ \Omega\\
R_{11} &= 1{\cdot}5 \times R_{10} = 1{\cdot}5 \times 0{\cdot}8 = 1{\cdot}2\ \Omega
\end{aligned}$$

The total was to be 1·3, so $R_{11} = 1{\cdot}2\ \Omega$ will be near enough for most purposes, but if it is desired to keep to the value 1·3, the adjustment can be made when the individual steps have been calculated. We have

$$\begin{aligned}
r_2 &= R_2 - R_1 = 0{\cdot}03 - 0{\cdot}02 = 0{\cdot}01\ \Omega\\
r_3 &= R_3 - R_2 = 0{\cdot}045 - 0{\cdot}03 = 0{\cdot}015\ \Omega\\
r_4 &= R_4 - R_3 = 0{\cdot}068 - 0{\cdot}045 = 0{\cdot}023\ \Omega\\
r_5 &= R_5 - R_4 = 0{\cdot}102 - 0{\cdot}068 = 0{\cdot}034\ \Omega\\
r_6 &= R_6 - R_5 = 0{\cdot}153 - 0{\cdot}102 = 0{\cdot}051\ \Omega\\
r_7 &= R_7 - R_6 = 0{\cdot}23 - 0{\cdot}153 = 0{\cdot}077\ \Omega\\
r_8 &= R_8 - R_7 = 0{\cdot}35 - 0{\cdot}23 = 0{\cdot}12\ \Omega\\
r_9 &= R_9 - R_8 = 0{\cdot}53 - 0{\cdot}35 = 0{\cdot}18\ \Omega\\
r_{10} &= R_{10} - R_9 = 0{\cdot}8 - 0{\cdot}53 = 0{\cdot}27\ \Omega\\
r_{11} &= R_1 - R_{10} = 1{\cdot}2 - 0{\cdot}8 = 0{\cdot}4\ \Omega
\end{aligned}$$

Thus there are 10 steps with 11 contacts, and, if desired, the deficit of 0·1 Ω can be divided equally among a few of the steps, say, r_7 to r_{11}, making these 0·097, 0·14, 0·20, 0·29, and 0·42 Ω respectively.

Finally, for a motor of this size it may be desirable to attain the first rush of current in two steps, the first step allowing 169·5/2, say, 85 A to pass. The total resistance in the starter must thus be 220/85 = 2·6 Ω, the extra step therefore being 2·6 − 1·3 = 1·3 Ω.

Starters for Series Motors. These are usually controllers, that is, they can be left in circuit for any length of time, since they also act as speed regulators. These controllers are in the form of a rotating cylinder carrying segments which made contact with fixed insulated "fingers." A blow-out coil is also provided, its function being to extinguish any arc formed when a segment breaks contact with a finger. This blow-out coil is simply an electro-magnet, the lines of force of which cross the space between the fingers and the segments, so that, when an arc is formed, it is acted on by a mechanical force perpendicular to the plane of the current and the flux, that is, in a vertical direction. The arc is thus blown out. The developed view of such a controller is shown in Fig. 10.13. It is arranged for both forward and reversed directions of rotation.*

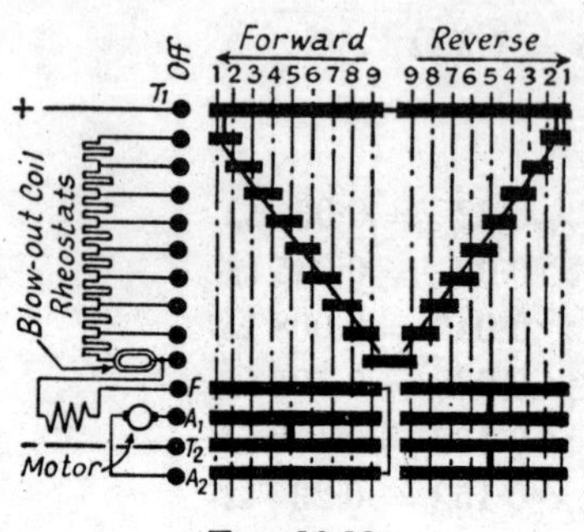

FIG. 10.13

Time of Acceleration. The time required for the motor to attain full speed can be calculated as follows. It will be seen by the speed curves of Fig. 10.11 that the speed, and therefore the angular velocity ω, is approximately proportional to the time if the starting handle is moved uniformly. Hence, the angular acceleration $d\omega/dt$ is constant. Let I be the mean value of the current *available for acceleration.* Then accelerating torque

$$T = (\Phi Z/2\pi)(P/A) \times I \text{ newton-metres}$$

If K is the moment of inertia of the moving masses in kilogramme-square metres, then

$$T = K(d\omega/dt) \text{ or } d\omega/dt = T/K = (\Phi Z/2\pi)(P/A)(I/K)$$

Hence, the time required to attain the full speed ω,

$$t = \omega/\text{angular acceleration}$$
$$= (2\pi/\Phi Z)(A/P)(K\omega/I)$$

EXAMPLE. A 50 h.p., 400 V, 500 r.p.m. shunt motor takes on full load 110 A. Its armature has a moment of inertia of 200×10^6 g-cm². Find the time taken to attain full speed if the maximum and minimum currents during starting are 150 and 120 A.

Full-speed angular velocity $\omega = 52{\cdot}5$ radians/sec
Torque on full load

$$= 33{,}000 \times \text{h.p.}/(2\pi \times N) = 525 \text{ lb-ft}$$
$$= 710 \text{ newton-metres}$$

* The calculation of a starter for a series motor is beyond the scope of this book. For a complete solution, *see* DOVER, A. T., *Electric Traction*, Pitman (London, 1954), SMITH, S. P., "Steps in the Starter of a Series Motor," *J. Inst. Elect. Engrs.*, **58**, p. 645, and DOVER, A. T., *Jnl. I.E E.*, **60**, (1922) 867.

The mean starting current is 135 A, and therefore, since 110 of these are required to produce the load torque, the other 25 A are avilable for acceleration. A current of 100 A produces a torque of 710 newton-metres, hence the current of 25 A produces an accelerating torque of

$$T = (25/110) \times 710 = 160 \text{ newton-metres}$$

$$\therefore d\omega/dt = T/K = 160/(200 \times 10^{-3} \times 10^{6} \times 10^{-4}) = 8$$

$$\therefore t = \omega/(d\omega/dt) = 52{\cdot}5/8 = 6{\cdot}6 \text{ sec}$$

Contactor Starters. When motors have to be started and stopped frequently, particularly if there is a definite duty cycle, and in the case of very large motors for which the starting gear is too heavy for manual operation, automatic starters are used. In these, the various

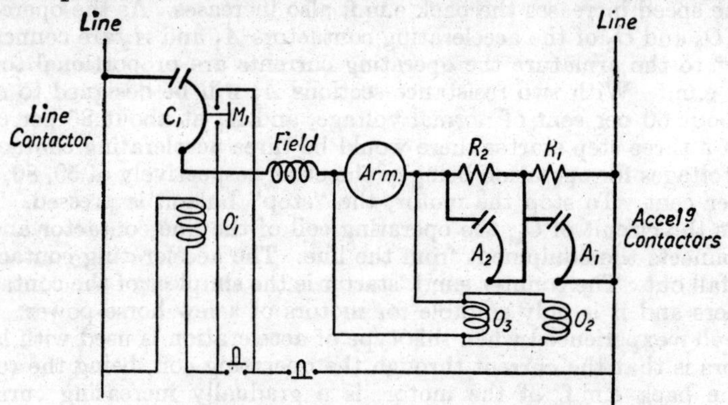

FIG. 10.14. CONNECTION SCHEME OF CONTACTOR STARTER

switching operations are performed by automatic switches called contactors. These switches can be operated (*a*) mechanically as, for example, by a cam, (*b*) pneumatically in localities where there is a supply of compressed air, and (*c*) electromagnetically, i.e. by means of a solenoid. This third method is almost invariably adopted for automatic starting gear.

When a motor is switched on to the line the following takes place—

(*a*) the speed increases and as a result, the back e.m.f., or counter e.m.f. increases;

(*b*) the current rises to a value decided by the circuit conditions at the moment, and then gradually decreases;

(*c*) the time progresses.

All three of these phenomena can be adopted as a means of controlling the acceleration of the motor, that is, by cutting out, or short-circuiting the various resistance steps.

(*a*) the counter e.m.f., or speed-limit method;

(*b*) the current limit method;

(*c*) the time limit method.

We will consider the operation of starters of type (*a*). Fig. 10.14

shows the connections scheme for a counter e.m.f. starter having two resistance sections R_1 and R_2. Three contactors are required, namely, a line contactor which connects the motor to the line with all the resistance in circuit, and two accelerating contactors whose function it is to short-circuit the resistance sections in sequence, first R_1 and then R_2. The control is effected by "start" and "stop" push-buttons which can be placed at any desired point, in fact there may be several of these so that the motor can be controlled from a number of points. When the "start" button is closed the operating coil O_1 of the line contactor C_1 is energized thereby closing this contactor and connecting the motor and resistances to the line. At the same time an auxiliary or maintaining contact M_1 is closed by the motion of the line contactor, thereby by-passing the "start" button. The motor now starts up and, as the speed increases the back e.m.f. also increases. As the operating coils O_2 and O_3 of the accelerating contactors A_1 and A_2 are connected direct to the armature the operating currents are proportional to the back e.m.f. With two resistance sections A_1 will be designed to close at about 60 per cent of normal voltage, and A_2 at about 85 per cent. With a three-step starter there would be three accelerating contactors the voltages for operation being of the order respectively of 50, 80, and 90 per cent. To stop the motor, the "stop" button is pressed. This opens the circuit of O_1, the operating coil of the line contactor and so disconnects the equipment from the line. The accelerating contactors also fall out. The counter e.m.f. starter is the simplest of the contactor starters and it is only suitable for motors of a few horse-power. The difficulty experienced when this type of acceleration is used with large motors is that the current through the operating coil, being the result of the back e.m.f. of the motor, is a gradually increasing current. Hence a heavy contactor operated in this manner closes only with light pressure instead of with a snap. The result is that such a contactor may chatter with consequent arcing, and the contacts may weld together.

The other types of contactor, such as the series and time-delay operated, are designed to overcome this difficulty, and are therefore more suited to large motors.*

Speed Control of D.C. Motors. I. RHEOSTATIC CONTROL. Let an adjustable resistance R be placed in series with the armature, making the total resistance in the armature circuit $(R + R_a)$, then the back e.m.f. for any armature current I_a is given by

$$E_b = V - (R + R_a)I_a$$

At no-load, and with no series resistance R in circuit the back e.m.f. is approximately equal to the applied p.d., E. Since, for constant excitation, the speed is proportional to the back e.m.f. we have, denoting the no-load speed by N_o—

* GARRAUD, C. C., *Electric Switch and Controlling Gear*. Benn (1927).
HARWOOD, P. B., *Control of Electric Motors*. McGraw-Hill (1936).
NORBURN, W. H. J., *Industrial Motor Control Gear*. Pitman (1933).

$$N/N_o = [V - (R + R_a)I_a]/V$$

$$\therefore\ N = N_o[1 - \{(R + R_a)/V\}I_a]$$

Now put $\quad R + R_a = R_t$

$$\therefore\ N = N_o[1 - (R_t/V)I_a]$$

For a given resistance R_t the speed is thus a linear function of the armature current I_a, the graph of N against I_a being a drooping straight line, as indicated in Fig. 10.15. The amount of the droop obviously depends upon the value of R_t, and therefore upon R, and obviously a whole family of speed-current curves can be drawn, each curve in the family corresponding to a definite value of R.

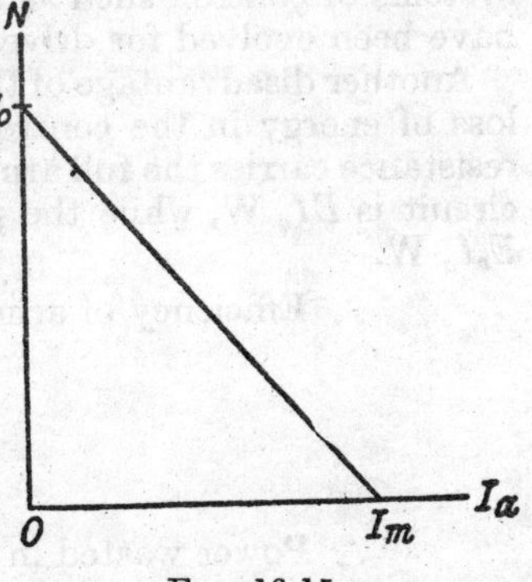

Fig. 10.15

We also see that for each value of R there will be a certain value, I_m, of the armature current which just stalls the motor. This value is given by

$$0 = N_o(1 - R_tI_m/V)$$

$$\therefore\ I_m = V/R_t$$

Example. A shunt-wound motor runs at 550 r.p.m. on a 200 V circuit. Its armature resistance is 0·5 Ω, and the current taken is 30 A, in addition to the field current. What resistance must be placed in series with the armature in order that the speed may be reduced to 300 r.p.m., the current in the armature remaining the same? If the load is changed so that with the inserted resistance the armature current is reduced to 15 A, what then will be the speed? (London Univ.)

From $\quad N = N_o[1 - (R_t/V)I_a]$

we have $\quad 300 = 500[1 - (R_t/200)30]$

$$\therefore\ R_t = 2{\cdot}667\ \Omega$$

$$\therefore\ R = R_t - R_a = 2{\cdot}667 - 0{\cdot}5 = 2{\cdot}167\ \Omega$$

If the current falls to 15 A, the resistance R_t remaining at 2·667 Ω we have

$$N = 500[1 - (2{\cdot}667/200)15]$$
$$= 400 \text{ r.p.m.}$$

We can also calculate the current which will just stall the motor. Thus with $R_t = 2{\cdot}667\ \Omega$, we have for this current

$$I_m = V/R_t = 200/2{\cdot}667 = 75 \text{ A}$$

The above example illustrates one of the disadvantages of this method of control, namely, that for a given value of the resistance the speed is not a constant, but is a function of the load current. Thus, with an external resistance of 2·167 Ω, the speed is 300 r.p.m. when the current is 30 A, but it rises to 400 r.p.m. when the current falls to 15 A. This means that if the speed is to be kept sensibly constant on a rapidly-changing load, the value of the controller resistance must be varied to suit those changes. For precise speed control it can be said that the speed must be a function of the controller resistance only, and not a function of the load current, and because of this, more elaborate systems of control, such as the Ward-Leonard control described later, have been evolved for drives where precision of control is a necessity.

Another disadvantage of the method is that there will be considerable loss of energy in the controller, particularly at low speeds, since the resistance carries the full armature current. The intake of the armature circuit is EI_a W, while the power converted into mechanical power is E_bI_a W.

$$\therefore \text{ Efficiency of armature circuit} = E_bI_a/VI_a$$
$$= (V - R_tI_a/V)$$
$$= [1 - (R_t/V)I_a]$$
$$= N/N_o$$

$$\therefore \text{ Power wasted in controller/Armature intake}$$
$$= (N_o - N)/N_o$$

In other words, the power wasted is proportional to the reduction of speed. Thus, if the speed is reduced 50 per cent, the power wasted will be 50 per cent, and the efficiency will be 50 per cent. If the speed is

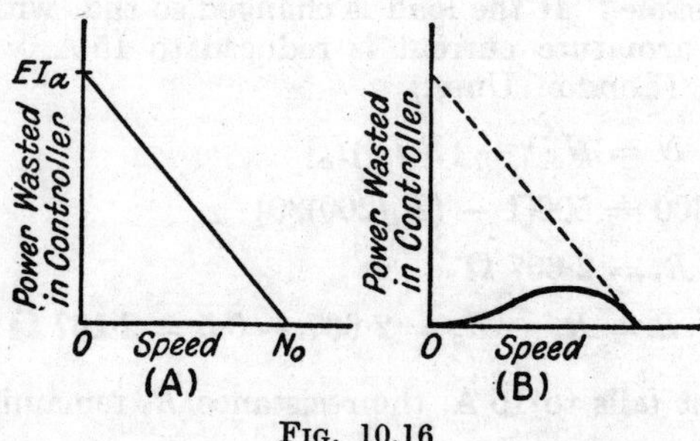

Fig. 10.16

brought down to zero, then the whole of the armature intake will be wasted in the controller. The graph of power wasted in the controller against speed is a straight line, as illustrated in Fig. 10.16 (*A*). The disadvantage of this inefficiency is that it not only entails a considerable loss of energy, but it also necessitates a large and expensive controller with elaborate arrangement for the dissipation of the heat produced in it.

In the above discussion we have assumed that the armature current has remained constant during the change in speed; in other words, that the load has been one of constant torque. In the case of appliances working on the centrifugal principle, such as fans, the torque is proportional to the square of the speed, and this makes a considerable difference to the efficiency of the rheostatic method.

Let $T = \text{torque}$

and $T_m = \text{maximum torque}$

At any speed $T \propto N^2$

But $I_a \propto T$

$$\therefore I_a \propto N^2 = AN^2, \text{ say,}$$

where A is a constant. At full speed N_o the torque is T_m, so that

$$I_m = AN_o^2$$

$$\therefore A = I_m/N_o^2$$

For the speed N at any torque T, we therefore have

$$N^2/N_o^2 = T/T_m = I_a/I_m$$

$$\therefore I_a = I_m(N/N_o)^2$$

Now the back e.m.f. is proportional to N, and we have

$$E_b = V(N/N_o)$$

Hence, drop of volts in the controller

$$= V - E_b = V - V(N/N_o)$$

$$= V(1 - N/N_o)$$

Hence, power wasted in the controller

$$P = (\text{drop along controller}) \times \text{current}$$

$$= V(1 - N/N_o) \times I_m(N/N_o)^2$$

$$= VI_m(1 - N/N_o)(N/N_o)^2$$

Now write $N/N_o = r$

$$\therefore P = VI_m(1 - r)r^2$$

$$= VI_m(r^2 - r^3)$$

when $N = N_o$, $r = 1$, and $P = 0$

when $N = 0$, $r = 0$, and $P = 0$

We thus see that in this case the power wasted in the controller is zero at both extreme limits of speed, and by giving r a series of values between 0 and 1·0 it is easy to see that the graph of power wasted is of the form shown in Fig. 10.16 (B).

Differentiating the expression for P with respect to r, we have

$$dP/dr = 2r - 3r^2$$

$$= 0 \text{ when } r = \tfrac{2}{3}$$

Hence, the maximum loss of power will take place when the speed is two-thirds of full speed. Again, EI_m is the full-load intake of the motor, P_m, say.

$$\therefore \text{ Maximum loss} = P_m[(2/3)^2 - (2/3)^3]$$
$$= 0{\cdot}148P_m$$

as against P_m, when the load is one of constant torque. This shows that when the load is one whose torque is a function of the speed, the rheostatic method is no longer excessively inefficient.

II. Field Control. We have seen that the speed of a d.c. motor is inversely proportional to the flux per pole, and hence, if this flux is varied, the speed will vary. This is accomplished by means of a shunt regulator in the case of a shunt motor, and a diverter in the case of a series motor, as shown in Fig. 10.17. This method is both convenient and economical, but obviously it will only give speeds greater than normal. By a combination of methods I and II, speeds below or above normal can be obtained. If a large range of speed by the field control is required, the motor must be fitted with interpoles, because, owing to the high speed, the reactance voltage will be high, and owing to the fact that the field is weakened, the commutating field, if it is the fringe of the main field, will also be weak. Because of the high reactance voltage at high speeds, the commutating field must be strong if the motor is to run sparklessly.

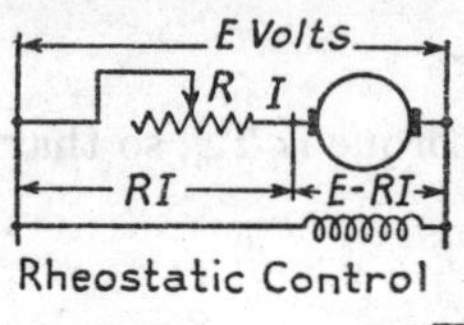

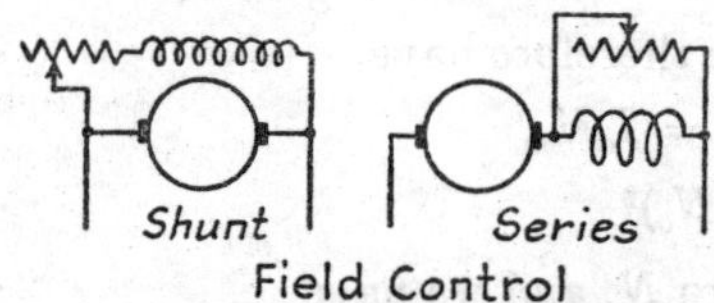

Fig. 10.17. Methods of Speed Control

The necessary variations in shunt resistance to obtain a given change in speed can only be determined when the magnetization characteristic of the machine is known. Since the back e.m.f. is proportional to the product of the flux and speed, we have

$$E_{b1} \propto \Phi_1 N_1$$
$$E_{b2} \propto \Phi_2 N_2$$
$$\therefore \Phi_2 = \Phi_1(E_{b2}/E_{b1})(N_1/N_2)$$

The conditions of loading of the motor will enable the back e.m.f.s to be calculated, and the new flux Φ_2 can then be determined from the above equation. A reference to the magnetization characteristic will then give the new excitation.

Example. A 230 V d.c. shunt motor whose magnetization curve is given by the figures below, runs at no-load at 1200 r.p.m. The resistance of the field magnet coils is 38·3 Ω. Find what resistance must be placed in series with the field magnet coils to increase the speed to 1400 r.p.m. at no-load. (London Univ.)

Amperes in magnet coils	1	2	3	4	5	6
Flux per pole in megalines	0·44	0·8	1·02	1·15	1·21	1·24

Since the armature resistance and current are not specified, we can take the back e.m.f. as being equal to the applied p.d., so that the expression simplifies to

$$\Phi_2 = \Phi_1(N_1/N_2)$$

The exciting current for full excitation is

$$I_{\cdot ex1} = 230/38{\cdot}3 = 6 \text{ A}$$

$$\therefore \Phi_1 = 0{\cdot}0124 \text{ webers}$$

$$\therefore \Phi_2 = 0{\cdot}0124(1200/1400) = 0{\cdot}01062 \text{ webers}$$

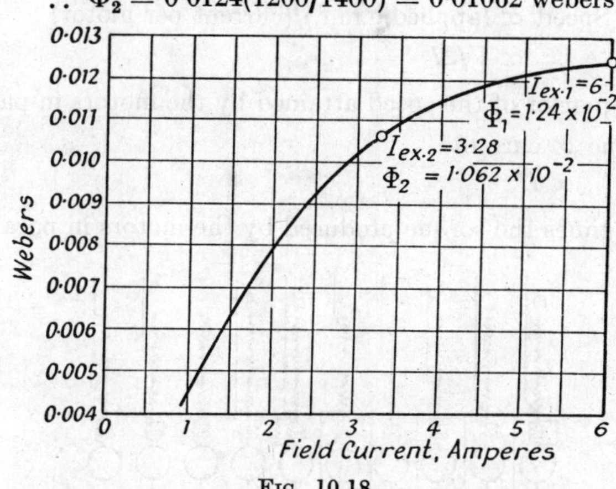

Fig. 10.18

The magnetization curve is drawn in Fig. 10.18, from which we see that the new value of the exciting current is

$$I_{ex \cdot 2} = 3{\cdot}28 \text{ A}$$

Hence, new total resistance in shunt circuit

$$= 230/3{\cdot}28 = 70{\cdot}2\ \Omega$$

$\therefore$ Resistance in shunt regulator to increase the speed from 1200 to 1400 r.p.m. $= 70{\cdot}2 - 38{\cdot}3 = 31{\cdot}9\ \Omega$.

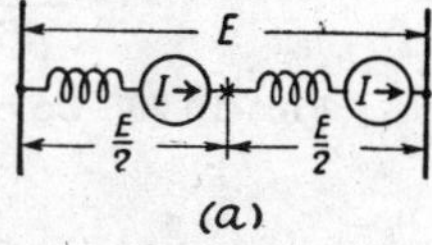

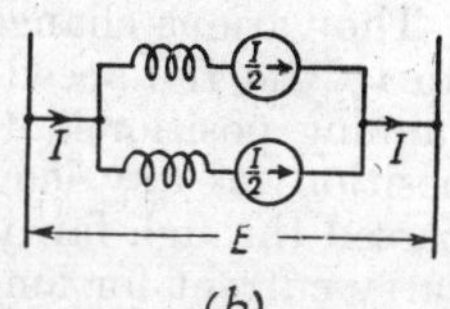

Fig. 10.19. Series-Parallel Control

III. Series Parallel Control. This method, in conjunction with auxiliary rheostatic control, is used for electric traction. It requires two motors mechanically coupled, this coupling, in the case of a tram or locomotive, being supplied by the adhesion of the wheels to the rails. Consider two series motors in parallel, as shown in Fig. 10.19 (*b*), and let them be taking a total current I.

Then, speed $\propto$ (back e.m.f.)/(flux)

$\propto$ (applied e.m.f.)/(current per motor)

approximately.

$$\propto V \div I/2 \text{ or } 2V/I$$

Again, torque $\propto$ (current $\times$ flux)

$$\propto (\text{current})^2$$

$$\propto I^2/4$$

Now consider the motors connected in series (Fig. 10.19 (*a*)), and let the pair take the same total current as before. Then the p.d. across each motor is $E/2$, so that

$$\text{Speed} \propto (\text{applied e.m.f.})/(\text{current per motor})$$

$$\propto V/2I$$

that is, one-quarter of the speed attained by the motors in parallel.

Also, torque $\propto$ current²

$$\propto I^2$$

that is, four times the torque produced by the motors in parallel.

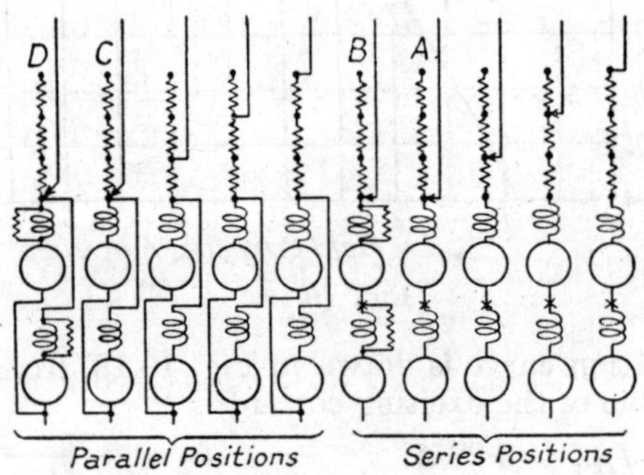

Fig. 10.20. Connections of Series-Parallel Controller for Different Speeds

The various changes in connection as the controller handle is moved up to the full-speed position are shown in Fig. 10.20. At the four running positions, *A*, *B*, *C* and *D*, the controller can be left in that position for any length of time, but the other positions have to be passed through fairly quickly, because the resistances are not rated to carry current for long periods.

IV. Ward-Leonard Control. This method is commonly used where a very delicate speed control over the whole range from zero to full speed is required, as, for example, with colliery winders. The method consists simply in working the motor with a constant excitation and applying to its armature sufficient voltage to give the speed required. A variable voltage supply is therefore required, and it is obtained from a motor-generator, or convertor, set. The scheme is shown in Fig. 10.21. *M* is the variable-speed motor. *A* is the motor, and *B* the generator of the convertor set. The variable voltage of the generator is obtained by varying its excitation by means of the shunt regulator R. Speeds in the reversed direction are obtained by reversing the generator excitation as indicated diagrammatically by the reversing switch (*R.S.*). The convertor set runs always in the same direction.

When applied to the control of very large motors, e.g. colliery

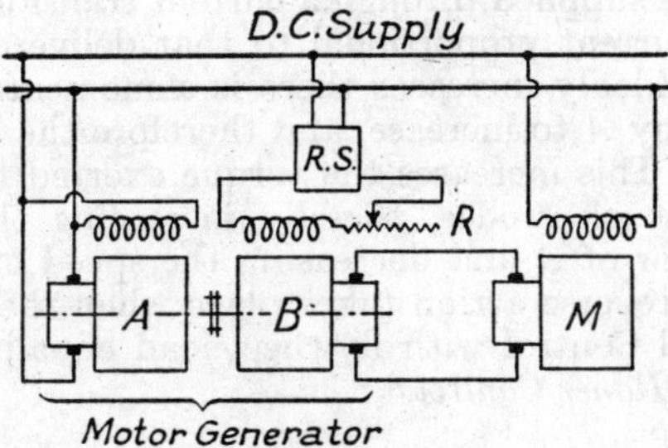

FIG. 10.21

winders or reversing rolling mills, the supply is nearly always an alternating one, and it is often desirable to incorporate a load equalizing arrangement in order to keep from the supply the violent peak loads which occur in such cases. The necessary modifications are shown in Fig. 10.22. The supply being alternating the motor A of the

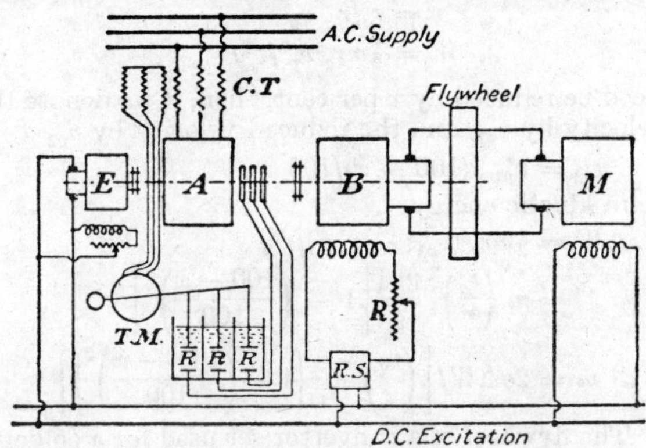

FIG. 10.22

convertor set is, of course, a three-phase motor, and since d.c. is required for the excitation of the variable speed motor M and variable voltage generator B, the convertor set has an exciter E direct coupled to it.* The flywheel for load equalizing is also coupled to the convertor set. Now if the flywheel is to keep the fluctuations in load from the mains, it must have its speed reduced when the peak load comes on, and have it raised again in times of light load. The necessary variations in speed

* It is not essential that the machine A shall be an electric motor, particularly as load equalization is not required. It can be a prime mover such as a steam turbine, a gas engine or an oil engine.

of the convertor set are obtained automatically by means of a "slip regulator," which puts resistance in the rotor circuit of the motor A when the load increases and cuts it out again when the load decreases. This resistance is generally a liquid resistance R, the electrodes of which balance the torque of a small "torque motor" ($T.M.$) at normal load. This motor is supplied through a current transformer ($C.T.$), and therefore carries current proportional to that delivered to the motor A. If the load suddenly increases there is a momentary tendency of the current taken by A to increase, and therefore the current in $T.M.$ tends to increase. This increases the torque exerted by $T.M.$, and it therefore raises the electrodes, thereby increasing the resistance in series with the rotor of A and decreasing the speed of the convertor-flywheel set. The reverse action takes place when the load decreases. The Ward-Leonard Control with flywheel load equalizer is known as the *Ward-Leonard-Ilgner Control.**

Calculation of Mass of Flywheel. A mass of m kilogrammes moving with a linear velocity of v metres/second has a kinetic energy of

$$W = \tfrac{1}{2}mv^2/g \text{ joules}$$

Let the mass be in the form of a rotating flywheel of radius r, peripheral velocity v_p, and radius of gyration r_g, then linear velocity at radius r_g and at normal speed

$$v = r_g/r \cdot v_p$$

$$\therefore \; W = \tfrac{1}{2}mr_g^2v_p^2/r^2g$$

Let the speed be reduced by x per cent, then, if we denote the initial peripheral velocity by v_{p1}, and the reduced velocity by v_{p2}

$$v_{p2} = v_{p1} \cdot (100 - x)/100$$

$\therefore$ Change in kinetic energy

$$\Delta W = \tfrac{1}{2}mr_g^2(v_{p1}^2 - v_{p2}^2)/r^2g$$

$$= m\left(\frac{r_g}{r}\right)^2 \frac{v^2_{p1}}{g}\left[1 - \left(\frac{100 - x}{100}\right)^2\right]$$

$$\therefore \quad m = 2g\Delta W\Big/\left[\left(\frac{r_g}{r}\right)^2 v^2_{p1}\left\{1 - \left(\frac{100 - x}{100}\right)^2\right\}\right]$$

EXAMPLE. The flywheel of a convertor set used for a colliery winder has to give up 3,000 h.p.-seconds when its speed is reduced by 15 per cent. Its peripheral velocity at full speed is 20,000 ft/min, and the shape is such that the radius of gyration is 0·78 of the outside radius. Calculate the mass of the flywheel.

$$\Delta W = 3{,}000 \text{ h.p.-sec}$$
$$= 3{,}000 \times 746 \text{ joules}$$
$$= 2{\cdot}238 \times 10^6 \text{ joules}$$

* For more complete description of Ward-Leonard equipments *see* Cotton, H. *Electricity Applied to Mining*, Chap. X (Pitman). Also *Electrical Equipment in Mines*, Chap IX, Ed. H. Cotton (Newnes).

$$g = 9{\cdot}81 \text{ m/sec}^2$$
$$r_g/r = 0{\cdot}78$$
$$v_{p1} = 20{,}000 \text{ ft/min}$$
$$= 20{,}000 \times 2{\cdot}54 \times 12/(100 \times 60)$$
$$= 101{\cdot}6 \text{ m/sec}$$

$$\therefore m$$
$$= 2 \times 9{\cdot}81 \times 2{\cdot}238 \times 10^6 /[(0{\cdot}78)^2 \times (101{\cdot}6)^2 \times (1 - 0{\cdot}85)^2] \text{ kg}$$
$$= 2{\cdot}51 \times 10^4 \text{ kg}$$
$$= 24{\cdot}4 \text{ tons}$$

Modified Ward-Leonard Control. The Ward-Leonard control can be modified in the following manner. Instead of obtaining the whole of its energy from the generator B, the motor M can have its armature

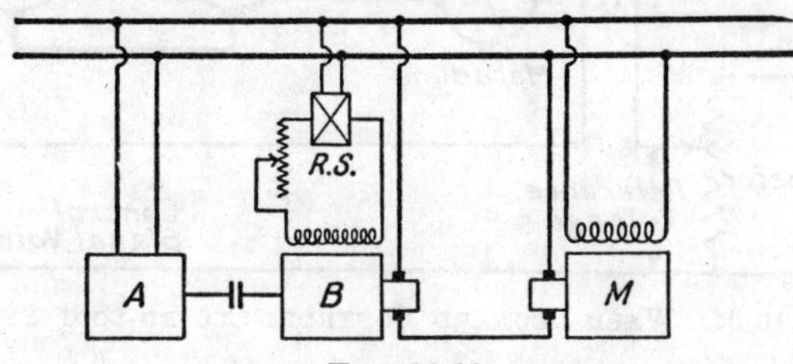

FIG. 10.23

connected in series with that of B, and with the supply, as shown in Fig. 10.23, with the result that the voltage applied to M is the resultant of the supply voltage and the voltage in B. Suppose that the excitation of B is adjusted so that its voltage is equal and opposite to that of the supply, then the resultant voltage will be zero and the motor will not rotate. As the excitation of B is reduced, the resultant voltage will increase, and the speed of M will therefore increase. With zero excitation of the generator the motor speed will be one-half of full speed. If, now, the exitation of B is reversed the generator will "boost" instead of "buck" the supply voltage, and the motor speed will increase still further. On full excitation when the generator voltage is equal to the supply voltage, the motor will be running at full speed.

This method differs in two important particulars from the Ward-Leonard control. In the first place, the two machines B and M in the Ward-Leonard control are wound for the same maximum voltage, and this voltage is independent of the supply voltage. In the second method the maximum voltage of B is equal to the supply voltage, while that of M is equal to twice the supply voltage. In the second place, the starting conditions for the Ward-Leonard control correspond to zero excitation of the generator B. Now the starting conditions are usually very heavy, which means that the generator has to be designed to fulfil the very difficult conditions of delivering a heavy current on a small excitation. This difficulty is overcome in the modified method, since the starting conditions correspond to full "buck" by B, and therefore to full excitation.

The Closed Loop System. There are many industrial drives in which very high precision of control is required over some particular aspect of the drive, e.g. of the speed or the position of some moving part. One way of achieving this is to take into account every factor which might have a bearing on the magnitude of the controlled quantity, and then to design with the utmost precision the electrical plant and associated equipment. For example, in the case of the driving motor of a colliery

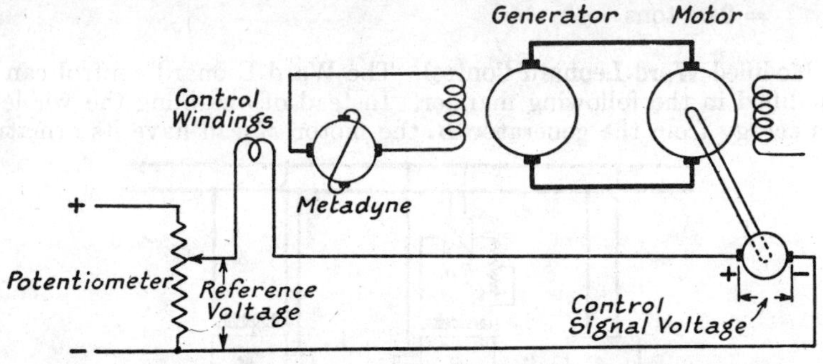

FIG. 10.24. WARD-LEONARD CONTROL: CLOSED-LOOP SYSTEM

winder, it is desirable that the speed of the motor shall depend on the position of the controller handle, and on nothing else. Suppose that the motor is controlled by the Ward-Leonard system then, even if the generator can be constructed to give an output voltage dependent on the field current only, and therefore on the controller position only, there is the volt-drop in the motor armature to allow for. Elaborate systems of compounding or of automatic voltage regulation can be designed to give the required characteristics under definitely specified conditions, but if the conditions during operation are not maintained exactly the same (temperature for example), a quite impossible requirement under normal industrial conditions, then the performance will deteriorate from the requirements. The above method belongs to the "open loop" systems of control. In the "closed loop" systems the quantity to be controlled, e.g. the speed, is compared with an appropriate analogue: thus the induced voltage of a tacho-generator may be set against an accurately maintained reference voltage. A difference between the two represents a departure from the desired operation, this difference therefore being called the error. This error is then amplified and applied to an appropriate controller which then receives just the right amount of energy to correct the error.

Reference has already been made on p. 218 to the control of the terminal voltage of an alternator by means of a rotary amplifier. Fig. 10.24 shows how a metadyne can be used as the amplifier in a closed-loop Ward-Leonard system, in which the controlled quantity is the speed. The reference voltage is obtained from a d.c. potentiometer, and the control voltage from a tacho-generator driven by the main motor. If there is any departure of the speed from the desired value there will be a difference voltage, the error voltage, which will produce a current in the control winding of the metadyne. The great advantage of the metadyne in this particular application is that, since a very small error voltage of even a fraction of a volt can produce the full output, the

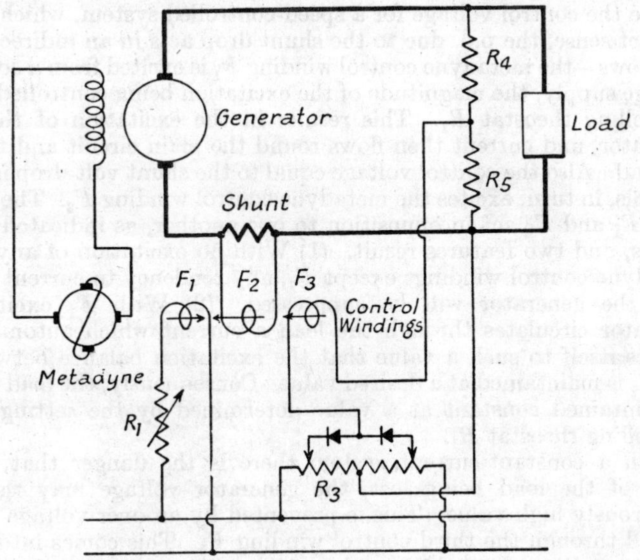

FIG. 10.25. METADYNE CONTROL OF CONSTANT-CURRENT SYSTEM

machine functions in the dual capacity of both regulating device and power amplifier.

The accuracy with which the controlled quantity is maintained depends on the accuracy of the regulating and amplifying systems. A possible weakness of the simple system given in Fig. 10.24 is the lack of constancy of the p.d. applied to the potentiometer from which the reference signal is derived, and, in some cases, it may be necessary to stabilize the supply to this potentiometer.

If rapid changes in load conditions give rise to rapid changes in the control signal then the control system must be quick-acting. To ensure accuracy under such conditions the tendency to hunt must be eliminated, and there must be sufficient damping to prevent overshooting of the mark.

Constant-current System.* When the control voltage is derived from a tacho-generator its value is proportional to the speed of the motor to

* *See* Bruce, G. A. "Diesel-electric Propulsion for Ships." *Met. Vic. Gaz.*, July, 1956.

be controlled, and consequently it is speed which is held at a predetermined value. Suppose that the control voltage is the volt-drop across a shunt in the armature circuit of the motor, then it is proportional to the motor current. As a result this current will be held at a constant value, the system thus becoming a constant-current system.

Fig. 10.25 shows the elements of the system. It will be seen that a metadyne is again used but that it now has three control windings. Unlike the control voltage for a speed-controlled system, which acts in a direct sense, the p.d. due to the shunt drop acts in an indirect sense, as follows—the metadyne control winding F_1 is excited from a constant-voltage supply, the magnitude of the excitation being controlled by the controlling rheostat R_1. This results in the excitation of the main generator, and current then flows round the main circuit and through the load. Also the control voltage equal to the shunt volt-drop is set up and this, in turn, excites the metadyne control winding F_2. The excitations F_1 and F_2 act in opposition to one another, as indicated by the arrows, and two features result. (1) With no excitation of any of the metadyne control windings except F_1, any tendency to current output from the generator will be suppressed. (2) With F_1 excited the generator circulates through the load a current which automatically adjusts itself to such a value that the excitation balance between F_1 and F_2 is maintained at a desired value. Consequently, the load current is maintained constant at a value determined by the setting of the controlling rheostat R_1.

With a constant-current system there is the danger that, in the event of the load being lost, the generator voltage may rise to a dangerously high value. This is prevented by an over-voltage control applied through the third control winding F_3. This comes into operation at a voltage dependent on the setting of the potentiometer taps on the centre tapped resistance R_3. By the combined use of this centre tapping and the double rectifier shown in the diagram, the voltage is limited in both positive and negative directions.

Examples on Chapter 10

1. Prove the expression for the force on a conductor carrying a steady current in a magnetic field. Hence, find the torque on an armature carrying a total of 50,000 ampere-conductors, the diameter of the core being 36 in., the length, 12 in. the pole arc being 70 per cent, and the flux density in the gap, 4500. (L.U.)

Ans. 2197 newton-metres.

2. The flux in each pole of a four-pole motor is 2×10^6 lines. What total number of ampere-conductors must be carried by the armature to produce a torque of 400 lb-ft? Prove any formulae you use for the calculation. (C. and G.)

Ans. 42,600.

3. Prove that when a current is taken through a magnetic field, work is done. Hence, find the pole flux of a 50 h.p. four-pole motor having 30,000 ampere-turns on the armature running at 800 r.p.m., the efficiency being 0·85. (L.U.)

Ans. $1{\cdot}28 \times 10^{-2}$ webers.
(assuming iron and friction losses to be one-third of total losses)

4. A shunt motor which is supplied with current at 440 V, runs when unloaded and at atmospheric temperature at a speed of 1000 r.p.m. After some hours its

temperature rises 30°C. Calculate the speed of the motor, assuming the voltage of the circuit falls to 435 V, that the armature current is 85 A, and that the armature resistance is 0·04 Ω. The temperature coefficient of the conductors may be taken as 0·44 per cent per °C. (L.U.)

***Ans.* 1130 r.p.m.**

5. A 220 V shunt motor running light takes 4 A. The field resistance is 220 Ω, and the resistance of the armature (hot) is 0·5 Ω. What will be the armature current when the motor is giving 6 h.p.? Explain what assumptions are made in calculating this result, and what errors these assumptions are likely to involve. (L.U.)

***Ans.* About 24·5 A.**

6. Show that a shunt motor when allowed to speed up with constant field without altering the resistance of the armature, will increase in speed according to an exponential law. (L.U.)

7. A 220 V shunt motor takes 3·4 A when running light. The field current is 1·0 A. When the armature is at rest it requires 6·2 V to pass 20 A through it. Find the approximate output and efficiency of the motor when the armature current is (*a*) 20 A, (*b*) 40 A. (L.U.)

***Ans.* (*a*) 5 h.p., 81 per cent; (*b*) 10·4 h.p., 86·2 per cent.**

8. State clearly what information you would need to enable you to calculate a metallic starter with a given number of steps for a series motor. Calculate the resistance steps for a starter of a 500-V shunt motor, given number of steps 12, maximum current during starting 20 A, resistance between armature terminals 1 Ω. (C. and G.)

***Ans.* 5·9, 4·5, 3·45, 2·6, 2·0, 1·54, 1·18, 0·9, 0·69, 0·52, 0·41, 0·31 Ω.**

9. Calculate the steps for the starter of a 10 h.p., 110 V, shunt motor, given that the current can rise to a maximum of 135 A. The motor is started against full load, the corresponding armature current being 90 A. The resistance of the armature circuit is 0·055 Ω.

***Ans.* 0·272, 0·186, 0·125, 0·086, 0·057, 0·041 and 0·010 Ω.**

10. A compound-wound continuous current generator is used as a motor without any change of connections. Discuss its speed characteristic on a moderate fluctuating load and its behaviour on very large loads. For what purpose would such a machine be suited? Would there be any advantage in fitting it with a flywheel? (L.U.)

11. Find an expression for the force on a conductor carrying a steady current in a magnetic field. Hence find the torque on an armature carrying a total of 25,000 ampere conductors, the diameter of the core being 36 in., the length 12 in., the polar arc 70 per cent, and the flux-density in the gap 6000. (L.U.)

***Ans.* 1460 newton-metres.**

12. What is meant by the "mechanical characteristic" of an electric motor? Sketch the form it takes in the case of a shunt-wound direct-current machine connected to mains which are kept at a constant voltage. A 20 h.p. direct-current shunt motor has a full load efficiency of 88 per cent when supplied with power at 200 V., its armature resistance being 0·06 Ω, and its shunt resistance 80 Ω. Find approximately, indicating the assumptions made, the percentage change in speed from no load to full load, the voltage of the supply mains being kept constant at 200 V. (L.U.)

***Ans.* 2·5 per cent.**

13. A four-pole series motor has a magnetization characteristic given by the following figures—

Amperes through field magnet—	10	20	30	40	50	60	70	80	90
Flux per pole (megalines)—	0·29	0·55	0·74	0·85	0·92	0·97	1·01	1·06	1·08

If the armature has 820 conductors and is wave wound, and the resistance of the armature and field magnet windings is 0·15 Ω, estimate the speed at which the motor will run when supplied from a constant p.d. of 230 V, when taking a current of (*a*) 45 A, (*b*) 85 A. Calculate also the torque in lb-ft which the motor will exert when carrying these currents. (L.U.)

***Ans.* (*a*) 920 r.p.m., 180 newton-metres; (*b*) 744 r.p.m., 237·5 newton-metres.**

14. A d.c. series motor has a magnetization characteristic similar to that in example 13. The armature resistance is 0·1 Ω and the field resistance 0·05 Ω. When connected to a 230 V supply and loaded to take 60 A, the speed of the motor is 750 r.p.m. What must be the resistance of a diverter connected in parallel with the field winding in order to increase the speed to 900 r.p.m., the total current remaining at 60 A?

***Ans.* 0·75 Ω.**

15. The armature of a 200 V, 4-pole series motor is lap-wound. There are 280 slots and each slot has 4 conductors. The current is 45 A and the flux per pole 1·8 megalines. The field resistance is 0·3 Ω, the armature resistance 0·5 Ω, and the iron and friction losses total 800 W. The pulley diameter is 16 in. Find the pull at the rim of the pulley. (L.U.)

***Ans.* 645 newtons.**

CHAPTER 11

OPERATION OF D.C. GENERATORS IN SERIES AND IN PARALLEL

It is usual to have several generators in a power station, so that the number of generators in operation at any time can be varied to suit the magnitude of the load on the station. There are two ways in which the generators can be connected to the load, namely, in parallel or in series with it. In the former, which is the method almost universally adopted, all the machines work at the same voltage, and the load on any individual machine is proportional to the current delivered by it. In the series arrangement, all the generators deliver the same current, and the load on any machine is proportional to its voltage.

Series Generators in Series. No difficulties are experienced in working any number of these machines in series (Fig. 11.1). The method is only rarely employed, the most important example being the constant current transmission system which supplies Lyons (France). In this system there are three generating stations in series, namely Bozel, 54,000 V; Moutiers, 30,000 V; and Fond de France, 36,000 V; the total voltage being therefore 120,000 V. The current is kept constant at 150 A, and if the load varies, the generated voltage is varied by rocking the brushes round the commutator. In times of light load the generators not required for service are taken out of action by short-circuiting them. The above constant-current series system is known as the Thury System. The supply of electricity by this system is now obsolete but the utilization by motors working at constant current, as described in the previous chapter has several advantages, particularly when the number of motors is greater than one. If any particular motor is not required it is unexcited and its armature short-circuited as in the case of generators in series.

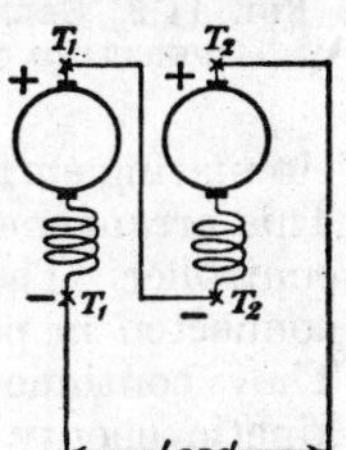

Fig. 11.1. Series Generators in Series

Series Generators in Parallel. Let E_1 and E_2 be the e.m.f.s induced in the two machines (Fig. 11.2), and let R be the resistance of either. Then they will share the load equally so long as E_1 and E_2 are equal. Suppose that E_1 becomes slightly greater than E_2, which, of course, is quite probable. Then a current given by

$$i = (E_1 - E_2)/2R$$

will circulate in a clockwise direction round the local path shown by the dotted line. The total current delivered by machine I will now be greater than that delivered by II, and therefore, because of the rising characteristics of series machines, the voltage E_1 will increase, while E_2 will decrease. The difference $(E_1 - E_2)$ will thus increase, and so also will the circulating current. The effect is thus cumulative, so that eventually, if there were no fuses or automatic switches in the circuit, the total current through II would be reversed. This would reverse E_2, and the resultant e.m.f. acting round the local path would be $(E_1 + E_2)$, and the circulating current $(E_1 + E_2)/2R$. The machines would then burn out. To prevent the possibility of reversal of either machine, it is obvious that the circulating current produced by inequalities in voltage must not be allowed to flow through the field windings; this is done by connecting a heavy copper bar of negligible resistance across the two machines, as shown in the second figure. The circulating current is now confined to the armatures and equalizing bar. This arrangement is used in the "braking" position of a locomotive controller. The two motors are disconnected from the line and are connected in parallel to the resistance. They continue to run because of the kinetic energy of the train, and they transform this kinetic energy into electrical energy which, in turn, is dissipated as heat in the resistances. The motors thus function as generators in the braking positions, and since they are series machines, it is necessary to interconnect them by an equalizing bar.

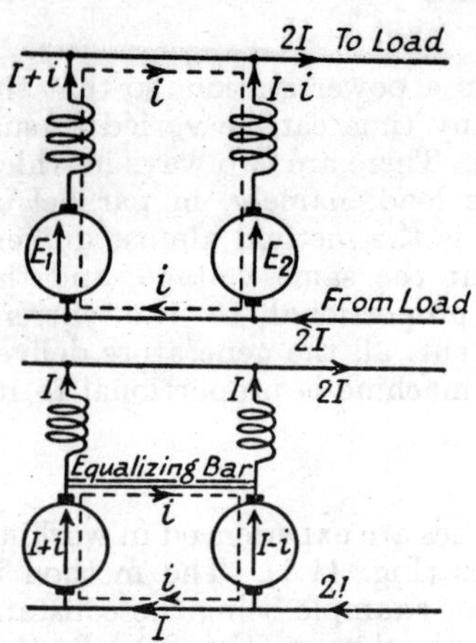

Fig. 11.2. Series Generators Operating in Parallel

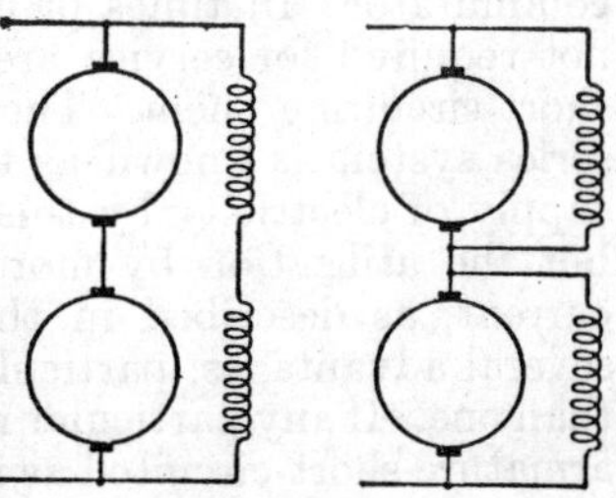

Fig. 11.3. Shunt Generators in Series

Shunt Generators in Series. When a high voltage is required it is fairly common practice to use two shunt generators, both mechanically coupled and electrically connected in series (Fig. 11.3). They then carry the same current, and they will share the load equally only so long as each gives the same voltage. If each shunt field is connected to its own armature, as shown in the first figure, any slight differences in voltage will cause differences in excitation, which, in turn, will increase the difference in voltage. There will thus be a tendency to throw more than a fair share of the load on to one machine. This is overcome very easily by connecting the two fields in series and applying the full voltage to them, as shown in the second figure. The excitations are now equal, and, since the speeds are equal, the machines being mechanically coupled, the voltages are equal and therefore the loads on the two machines are equal.

Shunt Generators in Parallel. When a number of generators work in parallel, all their + terminals and all their − terminals are connected to two heavy copper bars called bus-bars. In the case of d.c. plant these bars are placed behind the switch-board, and they may be regarded as the + and − terminals for the whole station. The generators are connected to these bars through main switches, S_1 and S_2, as shown in Fig. 11.4. Suppose that machine I is supplying the load and it is

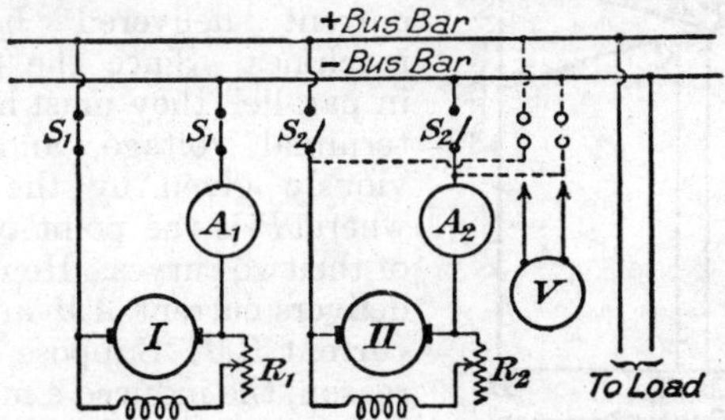

FIG. 11.4. PARALLEL OPERATION OF SHUNT GENERATORS

necessary to bring up II to share the load with it. The procedure is as follows. The incoming machine II is brought up to speed and its voltage then adjusted by means of its shunt regulator R_2 so as to be equal to, or 1 or 2 V greater than, the bus-bar voltage. The polarity must also be the same as the polarity of the bus-bars. This adjustment is made by means of a moving coil paralleling voltmeter V, which can be connected either to the incoming machine or to the bus-bars by means of the plug arrangement shown. When the adjustment is made, the switch S_2 is closed. The machine II will now either motor or generate, according as its induced e.m.f. is less than, or greater than the bus-bar voltage. Since the speed is kept constant by the engine governor, the induced e.m.f. depends on the excitation only. In order to make the machine generate, its excitation must therefore be such that the induced e.m.f. is greater than the bus-bar voltage. After closing S_2 the excitation of machine II is therefore gradually increased and during this operation the load will be gradually transferred from I to II, the ammeter reading A_1 decreasing while A_2 increases. Thus, by adjusting the excitations, either machine can be made to take any desired share of the total load. If it is desired to shut down one machine, say, No. I, it is not good practice to open its main switch, partly because of the violent sparking which will take place at the switch contacts, and also because the whole of the load will be suddenly thrown on to No. II. The excitation in No. I is gradually reduced until the whole of the current is taken by No. II; then the main switch, which is now carrying no current, is opened, and steam is finally shut off.* The machine then slows down gradually, and its exciting current gradually dies away. It is thus unnecessary to open the field circuit while it is carrying current. If this is done, excessively, high voltages will be set up because of the very high self-induction of the field winding of a large machine.

The slightly drooping character of the voltage characteristic of shunt generators renders the proper division of load between them when working in parallel, automatic. The voltage characteristics of two machines are shown in Fig. 11.5, the point A being the origin for curve I, and B the origin for curve II. The base AB represents the total current delivered by the two machines. Since the machines are in parallel, they must have the same terminal voltage, and this is obviously given by the length PM, where P is the point of intersection of the two curves. Hence, machine I delivers current AM and II delivers current BM. Suppose, that for some reason, the induced e.m.f. of machine I increases slightly, then its current will increase, say, to AM_1, and that taken by machine II will fall to BM_1. Now for this new condition to be stable, machine I should experience an increase in voltage, while II experiences a decrease; but just the reverse takes place, the voltage of I falling to QM_1 and that of II rising to RM_1. Such a state of affairs is thus impossible and the machines automatically return to the condition defined by their point of intersection P.

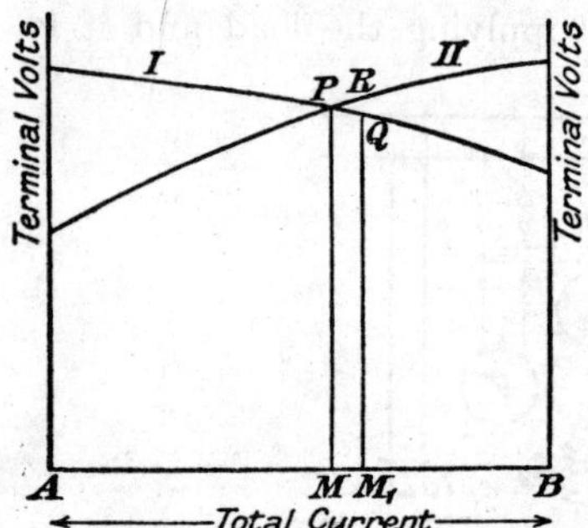

FIG. 11.5. PARALLEL OPERATION OF SHUNT GENERATORS

EXAMPLE. A shunt generator which gives a terminal voltage of 400 V at no load, and 360 V when delivering 100 A, is working in parallel with one which gives 400 V on no load and 350 V when delivering 100 A. If the voltage characteristics of each machine are approximately straight lines, find the common terminal voltage and the current in each when they are sharing a total load of 100 A.

If the two characteristics are drawn as in Fig. 11.5, and their equations are deduced, *both referred to A as origin,* we have

For machine I, $V = 400 - k_1 I$ where k_1 is a constant

$$\therefore\ 360 = 400 - k_1 \times 100;\ k_1 = 0{\cdot}4$$

For machine II, $V = 350 + k_2 I$

$$\therefore\ 400 = 350 + k_2 \times 100;\ k_2 = 0{\cdot}5$$

* In practice the excitations of both machines are adjusted in small steps; first a reduction in the excitation of No. I, then an increase in No. II. In this way large changes in bus-bar voltages are avoided.

Now let V be the common voltage and I the current in machine I. Then,

$$V = 400 - 0{\cdot}4I$$
$$V = 350 + 0{\cdot}5I$$

Subtracting $0 = 50 - 0{\cdot}9I \quad \therefore I = 55{\cdot}6\text{ A}$

$$\therefore V = 377{\cdot}8\text{ V}$$

and current in machine II

$$= 100 - 55{\cdot}6 = 44{\cdot}4\text{ A}$$

Compound Generators in Parallel. In all cases where generators having any series excitation have to be connected in parallel, an equalizing bar has to be used, since otherwise there is the danger that a circulating current through the bus-bars and machines may cause a reversal of one machine. Fig.

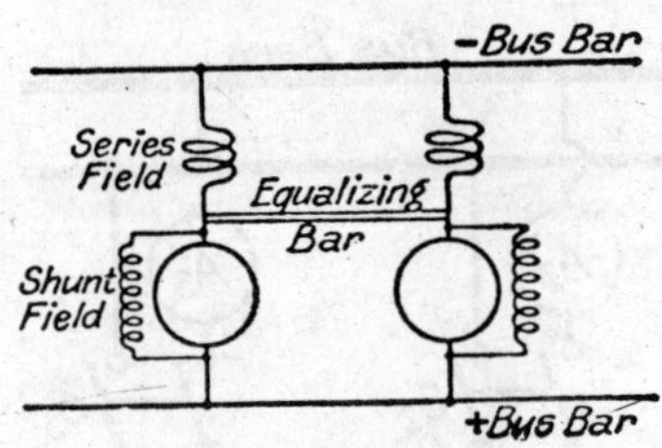

Fig. 11.6. Essential Connections for Compound Generators in Parallel

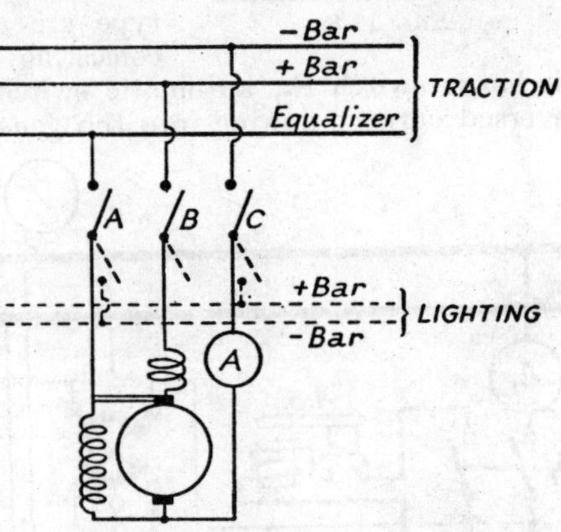

Fig. 11.7. Practical Circuit for Compound Generators in Parallel

11.6 is diagrammatic; it shows the essential connections without any switchgear or instruments, while in Fig. 11.7 the actual arrangement is shown. The main switch is a three-pole switch, and all three poles can be linked together so as to be closed or opened simultaneously, or the equalizer switch can be separate. In the latter case the equalizer switch is closed first and opened last. In stations supplying a combined traction and lighting load it is usual to have separate bus-bars for the two loads. Also, for the lighting load, the compounding is not required, the series fields therefore being left out of circuit. The bus-bars and connections for the lighting load are shown dotted, and it will be noticed that it is not the equalizer switch A, but the middle pole B, which is idle when the machine is thrown on to the lighting bars.

If the generators are provided with interpoles it is absolutely essential that the windings of these poles shall carry the same current as the armature. Hence, if compound interpole machines are worked in parallel, the equalizing bar is connected to the junction of the series and interpole fields, as in Fig. 11.8, so that, in the event of the armatures carrying any circulating current, this current will flow through the interpole windings and not be diverted from them by the equalizer bar.

GENERATOR AND BATTERY. When a battery is used to supply a load, either altogether, or as a standby, arrangements have to be made for charging, either when the voltage has fallen to 1·8 V per cell, or, more economically, while the load is actually on. Also, owing to the gradual fall in voltage during discharge, it is necessary to arrange that the number of cells in circuit can be gradually increased, so as to make up from time to time for this fall in voltage. The connections for an installation of this type are given in Fig. 11.9, the plant consisting of charging switch S_1, and discharge switch S_2, automatic switch AS for the prevention of a reversed current flowing into the generator, voltmeter switch S_4 for

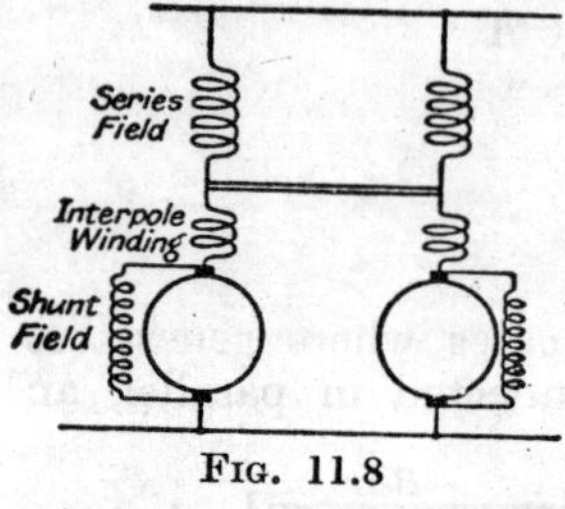

FIG. 11.8

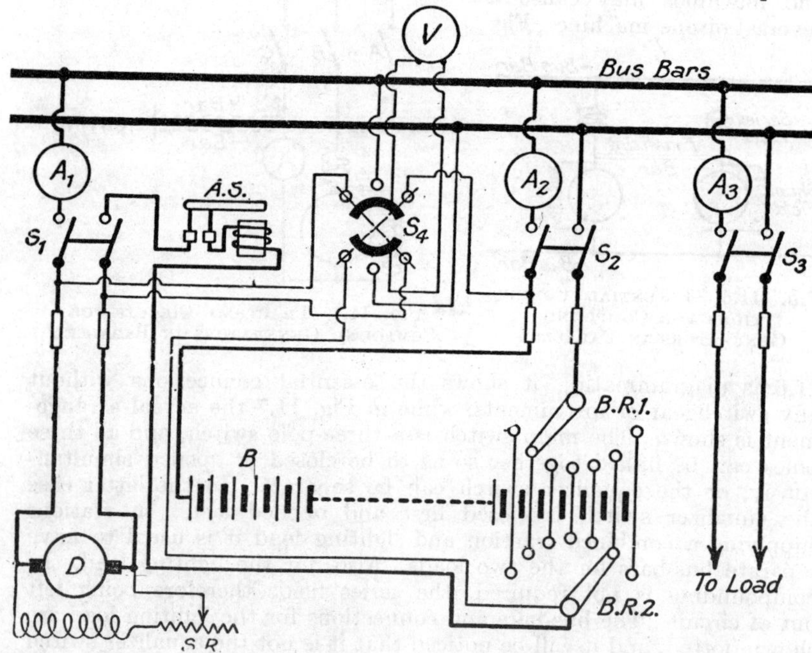

FIG. 11.9. SCHEME OF CONNECTIONS FOR BATTERY WITH HAND REGULATION DURING CHARGE AND DISCHARGE

paralleling the battery on to the generator, and the battery-regulating switch BR. This is a double selector switch, the bottom arm of which controls the numbers of cells in circuit during charge, and the top arm the number of cells during discharge.

The functions of the automatic switch are (1) to prevent the battery from discharging back into the generator, as explained above, and (2) to ensure that the battery and generator cannot be put in parallel until the generator voltage is in excess of the battery voltage by a small amount, usually 2 or 3 per cent. There are several switches of this type, and Fig. 11.10 shows the Neville type. The switch portion of this

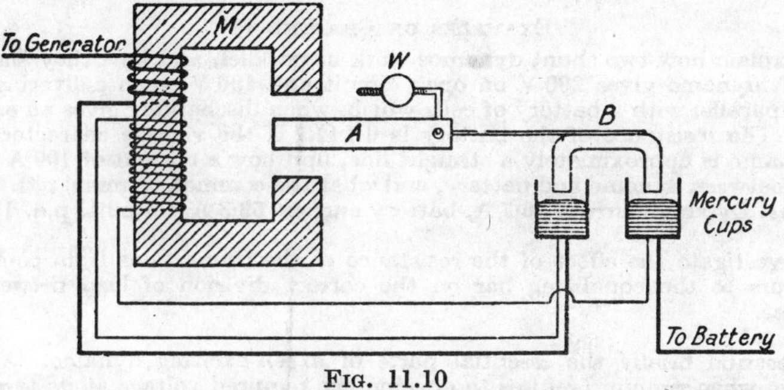

FIG. 11.10

consists of a copper fork which dips into two mercury cups. This forms one extremity of a pivoted beam, the other end of which is a magnetized armature, one pole of which can move in the air gap of a magnet wound with two coils. One coil is a series winding and the other a shunt winding, this latter being excited by the difference between the battery and generator voltages. The scheme is shown in Fig. 11.9.

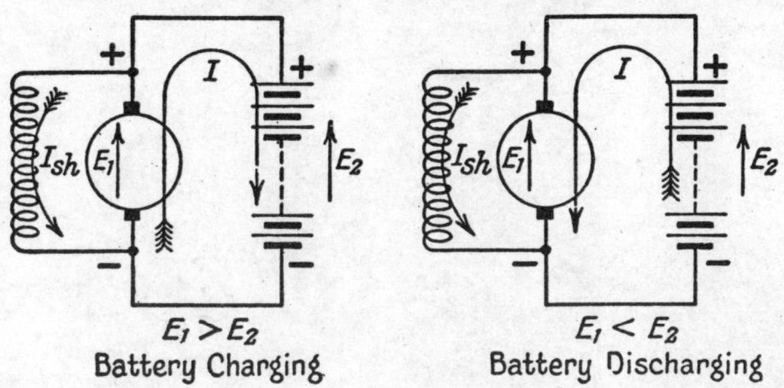

FIG. 11.11

If the battery voltage is in excess of the generator voltage, the current through the fine wire flows in such a direction as to hold the switch open. As soon as the generator voltage has exceeded the battery voltage by the required amount the reversed current in the shunt coil is of sufficient magnitude to close the switch, and the flow of current in the series coil then holds the switch closed. At the same time the shunt coil becomes short-circuited, thus preventing a wastage of energy in it. It will be seen that any failure of the current in the series winding releases the arm, and the counterweight lifts the copper fork out of the cups.

In connection with the charging of a battery of accumulators, it is necessary to note that the generator must be shunt wound, since this type only is immune from the risk of a reversal of polarity in the event of the direction of the current in the battery and armature circuit becoming reversed. This point is made clear by Fig. 11.11, which shows that no matter what the direction of the main current, the shunt current, and therefore the machine polarity, does not reverse.

Examples on Chapter 11

1. Explain how two shunt dynamos work in parallel, and how they share the load. A dynamo gives 200 V on open circuit and 190 V when delivering 50 A. It is in parallel with a battery of cells which, when discharging gives an e.m.f. of 196 V. The resistance of the battery is 0·1 Ω. If the voltage characteristic of the dynamo is approximately a straight line, find how a current of 100 A will be shared between dynamo and battery, and what the common terminal p.d. will be.

Ans. Dynamo current 46·7 A, battery current 53·3 A, terminal p.d. 190·7 V.

2. Investigate the effect of the resistance of the connections from compound generators to the equalizing bar on the correct division of load between the machines.

3. Describe briefly the essential parts of a self-exciting dynamo. A given machine, when running, refuses to develop the required voltage at its terminals. Enumerate possible causes of failure. (I.E.E.)

4. A shunt dynamo and a battery of accumulators are in parallel. The dynamo has an induced e.m.f. of 100 V and the battery has an open-circuit e.m.f. of 110 V. The dynamo resistance is 0·4 Ω and the battery resistance 0·5 Ω.

What is the common terminal p.d. and what is the current delivered by each if the output current is 40 A.

Ans. 95·6 V, dynamo 11·1 A, battery 28·9 A.

CHAPTER 12

TESTING OF D.C. MACHINES

Types of Tests. The methods of testing electrical machines can be divided into three classes, direct, indirect and regenerative.

In the direct method the motor or generator is put on full load and the whole of the power developed by it is wasted. For a generator the load usually consists of a water resistance, the output of the machine being used in heating the resistance. For a small motor a brake can be applied to a water-cooled pulley and the load adjusted until full load current is flowing. Then if W_1 and W_2 are the tensions in pounds on the tight and slack sides of the brake band, r the radius of the pulley in feet, and N the speed in r.p.m.,

$$\text{Motor output} = 2\pi N(W_1 - W_2)r/33{,}000 \text{ brake h.p.}$$
$$= 2\pi N(W_1 - W_2)r \times 746/33{,}000 \text{ W}$$

If I is the total current taken from the line, and E the voltage applied to the motor terminals,

$$\text{Intake} = VI \text{ W}$$

Hence,

$$\text{Efficiency, } \eta = \text{output/intake}$$
$$= 2\pi N(W_1 - W_2)r/VI \times 746/33{,}000$$
$$= 14{\cdot}2N(W_1 - W_2)r/VI \text{ per cent}$$

If a test of this kind is applied to a series motor the brake must be tight before the motor is switched on to the supply, otherwise the armature may fly to pieces.

A simple brake test like the above is suitable for small machines, but for a large motor it is preferable to direct couple it to a generator, and to load the generator by means of a resistance. The generator is carefully calibrated by determining the various losses in it beforehand, and its efficiency at different loads is thus known. Hence, if the generator output is measured by means of an ammeter and voltmeter, its intake can be calculated. This intake is obviously the motor output.

The testing of large machines by the direct method entails a considerable loss of power; in fact, in the case of a very large machine, there may be no facilities for a direct test. In some cases the generating machine is loaded back onto the mains by increasing its excitation to such an extent that its induced e.m.f. is greater than the mains voltage.

Indirect Methods. These consist in determining the losses and predetermining the efficiency from these. The amount of power required is that to supply the losses only, so that there is no difficulty in applying the method even to very large machines. The disadvantage is that the machine is running light during the test so that, although the efficiency can be calculated with fair accuracy from the results obtained, no indication is given as to the temperature rise on load, or to the commu tating qualities of the machine.

The simplest of the indirect tests is the Swinburne test, and it can only be applied to shunt or compound machines. The machine is run light as a motor, and the armature current I_o measured. The armature resistance and field resistance are also measured, and the temperature of these windings observed. Since the machine will be hot when running under normal conditions, the probable hot resistances are calculated, a temperature rise of 40°C being assumed. Then if R is the measured armature resistance, the resistance after a temperature rise of 40°C will be

$$R_a = R(1 + 40\alpha)$$

where the value of α, the coefficient of increase of resistance with temperature, will be fixed by the initial temperature.

Now let α_1 be the temperature coefficient corresponding to the initial temperature t_1. Then we have

$$\alpha_1 = \alpha_o/(1 + \alpha_o t_1) \text{ (\textit{see} p. 67)}$$

Thus, if we take the value $\alpha_o = 0{\cdot}00427$ for pure copper, we have for, say, $t_1 = 40$°C

$$\alpha_{40} = 0{\cdot}00427/(1 + 40 \times 0{\cdot}00427) = 0{\cdot}00364$$

and similarly for any other initial temperature.

If a copper coil of resistance 100 Ω at 40°C is heated to 100°C, its hot resistance will thus be

$$\begin{aligned} R_{100} &= R_{40}(1 + \alpha_{40} \times 60) \\ &= 100(1 + 0{\cdot}00364 \times 60) \\ &= 121{\cdot}8\ \Omega \end{aligned}$$

Similarly with the field winding. From the *cold* resistance of the armature the copper loss $I_o{}^2R$ is calculated, and from the cold resistance of the shunt field the field copper loss V^2/R_{sh} is calculated. The sum of these two losses deducted from the intake VI_o gives the friction and iron losses, that is, the stray losses. If now the calculated field copper loss V^2/R_{sh}, using the hot resistance, is added to the stray losses, the result will be the constant losses. Let these be P_c. Then, for any armature current I_a the armature copper loss will be $I_a{}^2R_a$, where R_a is now the hot resistance. Hence, total losses of all kinds

$$P_t = I_a{}^2R_a + P_c$$

The field current will be

$$I_{sh} = E/R_{sh}$$

and the line current will be

$$I_1 = I_a + I_{sh} \text{ if the machine is motoring}$$
$$I_2 = I_a - I_{sh} \text{ if the machine is generating}$$

Also, VI_1 is the intake when motoring, and VI_2 is the output when generating.

Hence, efficiency when acting as a motor

$$\eta = (\text{intake} - \text{losses})/(\text{intake}) = (VI_1 - P_t)/VI_1$$

and efficiency when acting as a generator

$$\eta = (\text{output})/(\text{output} + \text{losses}) = VI_2/(VI_2 + P_t)$$

Thus, if different armature currents I_a are assumed, the efficiency at different loads can be calculated.

If the machine under test is to function as a motor under normal conditions, its probable drop of speed on any load can be determined from the measured no-load speed N_o from the expression

$$N = N_o(V - R_aI_a)/(V - R_aI_o)$$

The drop in speed under normal working conditions will probably be somewhat less than this, because armature reaction on load will cause a decrease in total flux, and this will have a tendency to keep the speed up. The speed will also be altered if the brush position is different from that during the test.

EXAMPLE. A 500 V shunt motor takes a current of 5 A on no load. The resistance of the armature and field circuits are 0·22 Ω and 250 Ω respectively. Estimate the efficiency when the motor current is 100 A. What is the percentage change of speed between no load and full load? (C. and G., 1941.)

Accepting the data given and making no assumptions of any kind we have

$$I_{sh} = 500/250 = 2 \text{ A}$$

$$I_{ao} = 5 - 2 = 3 \text{ A, the no load armature current}$$

Field copper loss $= 500 \times 2 = 1000$ W

Back e.m.f. on no load

$$= 500 - 3 \times 0{\cdot}22 = 499{\cdot}3 \text{ V}$$

Stray losses $= 499{\cdot}3 \times 3 = 1498$ W

Armature current on load,

$$I_a = 100 - 2 = 98 \text{ A}$$

Armature copper loss as load

$$= (98)^2 \times 0{\cdot}22 = 2113 \text{ W}$$

Hence, total losses on load of 100 A intake

$$= 1000 + 1498 + 2113 = 4611 \text{ W}$$

Intake at this load $= 500 \times 100 = 50{,}000$ W

Output at this load $= 50{,}000 - 4611 = 45{,}389$ W

$$\eta = 100 \times 45{,}389/50{,}000 = 90{\cdot}78 \text{ per cent}$$

Back e.m.f. on load $= 500 - 98 \times 0{\cdot}22 = 478{\cdot}4$ V

Calling the no-load speed 100, we have

$$\text{Speed on load} = 100 \times 478{\cdot}4/499{\cdot}3 = 95{\cdot}8$$

Hence, percentage drop in speed

$$= 100 - 95{\cdot}8 = 4{\cdot}2 \text{ per cent}$$

We will now amplify the above example to make it conform more to actual practice. First allow 1 V drop at both positive and negative brushes, then

$$\text{Brush contact loss} = 2 \times 98 = 196 \text{ W}$$
$$\text{Modified total loss} = 4611 + 196 = 4807 \text{ W}$$
$$\text{Modified efficiency} = 100 \times 50{,}000 - 4807/50{,}000$$
$$= 90{\cdot}39 \text{ per cent}$$

Second, we deduct from this value an allowance which takes into account the additional load losses, that is, those losses which are not taken into account either by a no-load test or by resistance measurements: the most important in a small d.c. machine is the increase in iron loss due to the distortion of the armature flux distribution. For a machine of the type in the question we deduct 1·25 per cent* giving for the final value 89·14 per cent.

Separation of Losses. The above method does not give any indication as to how the stray losses are divided between friction, hysteresis, and eddy-current losses. These can be separated by the following test. The machine is run light as a motor, and a resistance R is placed in its armature circuit, so that any speed can be obtained. (*See* Fig. 12.1.) The field is separately excited to the normal value and different voltages applied to the armature, the speed N and the armature intake being observed for each value of the voltage. It is preferable, although not absolutely necessary, to measure the intake by means of a wattmeter, as shown. The armature current I_a is also observed, and the armature copper loss $I_a^2R_a$ deducted from the observed intake. This gives the total stray losses, say W. Now the various components of the stray losses are all functions of the speed, and can be written

Friction loss $= AN + BN^2$

Hysteresis loss $= CN$ so long as the excitation is constant

Eddy-current loss $= DN^2$ so long as the excitation is constant

A, B, C, and D are all constants. We therefore have

$$P = AN + BN^2 + CN + DN^2$$

or

$$P/N = (A + C) + (B + D)N$$

Hence, if P/N is plotted against the speed N, a straight line will be obtained (Fig. 12.2). By taking two points on this line, not necessarily observed results, and substituting the corresponding values of P/N

* British Standard Specification No. 269.

and N, simultaneous equations will be formed from which $(A + C)$ and $(B + D)$ can be calculated. The losses for any speed can then be calculated. So far the losses have been separated into those proportional to N, and those proportional to N^2, and such a separation is sufficient for most practical purposes. If it is required to determine each of the constants A, B, C, and D, then it is necessary to repeat the experiment with a reduced excitation. This will not affect the friction loss, but the constants C and D will be different, say, C' and D' respectively. Then the equation to the new curve of P/N against N is

$$P/N = (A + C') + (B + D')N$$

The terms $(A + C')$ and $(B + D')$ can then be determined as before, and by subtracting $(A + C')$ from $(A + C)$ and $(B + D')$ from $(B + D)$, the quantities $(C - C')$ and $(D - D')$ are found. Now C,

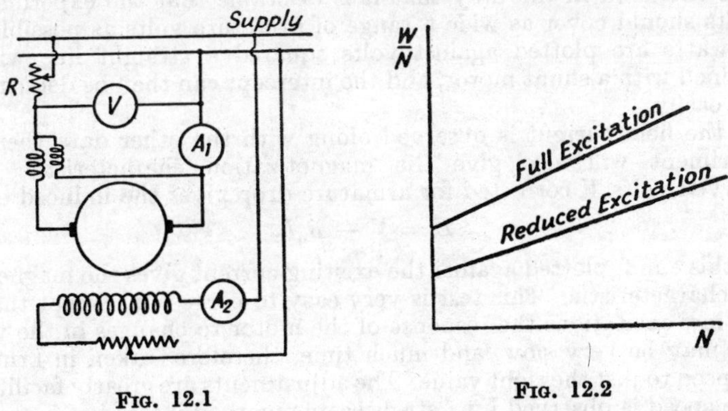

FIG. 12.1 FIG. 12.2

the coefficient for the hysteresis loss, is proportional to the flux Φ raised to the 1·6 power,* and D is proportional to Φ^2. Hence, calling Φ and Φ' the fluxes corresponding to the normal and reduced excitation, we have

$$C'/C = (\Phi'/\Phi)^{1 \cdot 6}; \quad D'/D = (\Phi'/\Phi)^2$$

Again, the ratio Φ'/Φ is given by the ratio of the back e.m.f.s in the two experiments *for any given speed*. Knowing the applied voltage and the armature current, the back e.m.f. can be calculated by deducting the armature drop from the applied voltage. The above ratios can thus be calculated and the constants A, B, C, and D then determined separately.

For works testing it is generally sufficient to separate the stray losses into iron (variable) and friction (constant) losses, without analysing the iron loss into its component hysteresis and eddy-current losses. In such a case the connection scheme of Fig. 12.1 is used, but the procedure is modified as follows: The motor is run at normal

* *See* footnote on p. 206.

speed with normal applied volts and excitation, and when the wattmeter reading has settled down to a steady value readings of armature watts, volts, amperes, and speed are taken. Next the excitation is reduced so that the speed increases, and the armature rheostat R is then adjusted to bring the speed back to the same value as before. Readings are again taken and the same procedure repeated over as wide a range of armature voltage as possible. The armature watts (corrected for armature copper loss) are then plotted against armature volts. Since the speed is kept constant the friction and windage loss will be constant, but since the excitation is gradually reduced, the flux per pole, and therefore the iron loss, will be gradually reduced. If the process could be continued until the armature volts were zero, then the flux would be zero and the iron loss zero. Hence if the graph is produced backwards to meet the watts axis the intercept on this axis will give the friction and windage loss. It is because the graph has to be extended in this way that it is desirable that the experimental results should cover as wide a range of armature volts as possible. If the watts are plotted against volts squared a straight line will be obtained with a shunt motor, and the intercept can then be determined very easily.

If the field current is observed along with the other data then the experiment will also give the magnetization characteristic. The observed volts V corrected for armature drop gives the induced e.m.f.

$$E = V - R_a I_a$$

and this e.m.f. plotted against the exciting current gives the magnetization characteristic. This test is very easy to carry out, except that at very low excitations the response of the motor to changes in the value of R may be very slow, and much time, therefore, taken in bringing the speed to just the right value. The adjustments are greatly facilitated if the speed is observed by a stroboscopic method instead of by means of an ordinary tachometer.

The "Retardation" or "Running Down" Test. This is another convenient method of separating the various losses. In this method the machine is run up to a little way beyond normal speed, and the supply is then switched off from the armature. As a result the armature slows down, its kinetic energy being drawn upon to supply the various losses produced by rotation. If I is the moment of inertia of the armature and ω its angular velocity at any instant, then losses due to rotation

$$\begin{aligned} P &= (d/dt)(\text{kinetic energy}) \\ &= (d/dt)(\tfrac{1}{2}I\omega^2) \\ &= I\omega(d\omega/dt) \end{aligned}$$

To calculate P it is therefore necessary to determine the curve of ω against t. There are several ways of doing this, the method involving the least amount of special equipment being to connect a very high resistance moving coil voltmeter across the armature terminals and to use it as a speed indicator. It is calibrated for speed as follows. When motoring at constant speed, the speed, the applied voltage, and armature current are observed. The back e.m.f. is calculated, and the ratio of observed speed to back e.m.f. gives the required constant of the voltmeter when used as a speed indicator. If the machine is a very large one, and takes a long time to slow down, the following procedure can be adopted. A circle of paper is pasted on the dial of a stop clock, and each time an observer watching the voltmeter gives a signal, a second observer marks the position of the seconds hand with a pencil. The first observer gives his signals each time the voltmeter needle passes one of the main divisions as the armature is slowing down. Such a method naturally necessitates a certain amount of skill, but when experience has been gained very consistent results can be obtained. In the case of small machines the fall in speed is too rapid to enable such a method to be adopted.

It is therefore preferable to observe, by means of a stop-watch, the time taken to slow down from a fixed vo tage somewhat higher than the normal voltage, to the various mainl divisions below this on the voltmeter, the motor being brought up to full speed for each observation. Thus, suppose that 200 V is taken as the reference mark on the voltmeter. The motor is brought up to such a speed that the voltmeter reads somewhat higher than this, and the armature supply is cut off. The voltmeter needle immediately begins to fall, and the time to fall from 200 V to 190 V is observed. The speed is then brought up to full value, supply cut off, and the time to fall from 200 V to 180 V is observed; and so on. The observed voltages are converted to speeds and the speed-time curve plotted as in Fig. 12.3. To find the gradient dN/dt at any point P, it is usual to draw a tangent to the curve and to measure the intercepts OM and ON. The change of curvature is so small that it is necessary to make several trials, the positions M and N being marked in pencil, but the tangent line not being drawn. If the means of the various lengths OM and ON are taken, we have

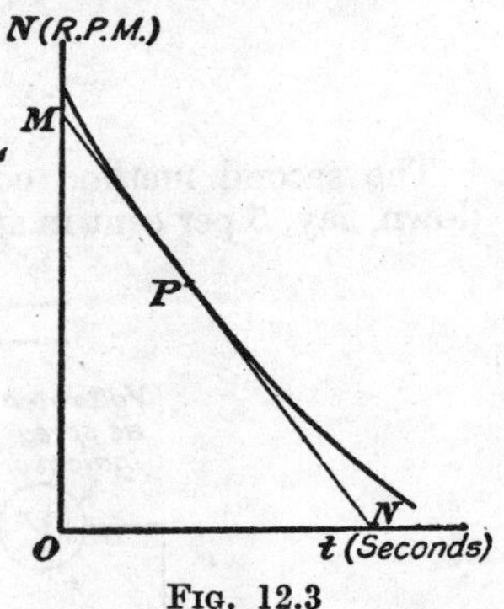

Fig. 12.3

$$dN/dt = \frac{OM \text{ (measured in units of } N)}{ON \text{ (measured in units of } t)}$$

Now, if the moment of inertia I is expressed in kilogramme-metres^2 and ω in radians/second, then

$$P = I\omega(d\omega/dt) \text{ watts}$$

Again, $\omega = 2\pi \times$ revolution/second

$$= (2\pi/60)N$$

and $d\omega/dt = (2\pi/60)(dN/dt)$

Hence, finally $$P = (2\pi/60)^2 IN(dN/dt)$$
$$= 0{\cdot}0109 IN(dN/dt)$$

so that, if I is known, the stray losses P can be calculated for any speed N.

There are several ways of finding I, the two most convenient being as follows. When the slowing-down curve has been obtained, a flywheel of known moment of inertia I' is keyed to the shaft and the curve obtained again. Naturally, the time of slowing down will be longer. For any given speed the rates of change (dN/dt) and $(dN/dt)'$ are determined as before. For the same speed the losses will be the same, since a smooth flywheel will not affect these appreciably, and we therefore have

$$0{\cdot}0109\, IN(dN/dt) = 0{\cdot}0109(I + I')N(dN/dt)'$$

$$\frac{I + I'}{I} = \frac{(dN/dt)}{(dN/dt)'}$$

$$\therefore\ I = I' \times \frac{(dN/dt)'}{(dN/dt) - (dN/dt)'}$$

The second method consists in observing the time taken to slow down, say, 5 per cent in speed, then applying a known retarding torque

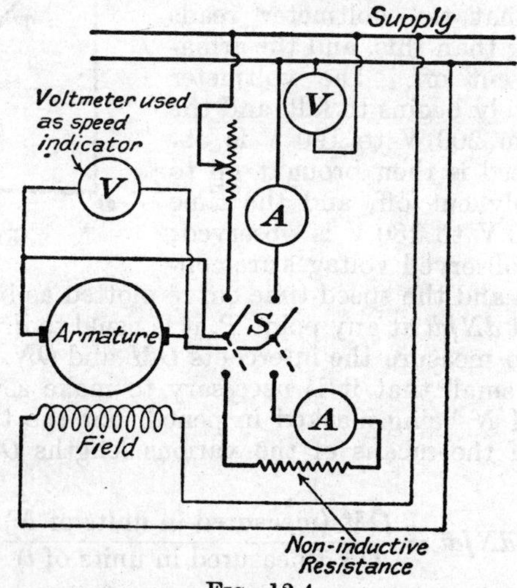

Fig. 12.4

and observing again the time taken to slow down the same amount. The retarding torque can be applied by means of a brake, but it is more accurate to apply it electrically, the method of doing so being indicated in Fig. 12.4. The armature is controlled by a double-throw switch S which is opened when the slowing down under normal conditions is to be observed. When applying an additional retarding torque, S is thrown right over so as to connect the armature to a non-inductive resistance as shown. The current I_a through this resistance produces an additional loss of $I_a^2(R + R_a)$. The current I_a will, of course, fall as the speed falls, but with a change in speed of not more than 5 per cent the change in I_a will be so small that it will be sufficiently accurate if a second observer takes its initial and final values, the mean value being used in the expression. Call this additional load P' watts and let the losses corresponding to the mean speed be P. Then, if δN is the change in speed and t and t' the times of slowing down without and with the extra load respectively,

$$P = 0{\cdot}0109IN(\delta N/t)$$

$$P + P' = 0{\cdot}0109IN(\delta N/t')$$

$$\therefore\ (P + P')/P = t/t'$$

$$P = P'[t'/(t - t')]$$

EXAMPLE. On breaking the armature circuit of a separately excited motor, the e.m.f. induced in the armature falls from 200 V to 190 V in 30 sec. If a current of 10 A is taken from the armature by connecting it to a resistance immediately after disconnecting it from the supply, the same fall in e.m.f. takes 20 sec. Find the stray losses.

We have $N = kE$, where k is a constant

$$\therefore\ P = 0{\cdot}0109IN(\delta N/t)$$

$$= 0{\cdot}0109k^2IE(\delta E/t)$$

The mean value of E is 195 V, and without the additional load $\delta E/t$ is $10/30 = \frac{1}{3}$.

$$\therefore\ P = 0{\cdot}0109k^2 \times \tfrac{1}{3} \times 195 \times I$$

The mean value of the additional load is $195 \times 10 = 1950$ W, and the corresponding rate of fall $\delta E/t$ is $\frac{1}{2}$.

$$\therefore\ P + 1950 = 0{\cdot}0109k^2 \times \tfrac{1}{2} \times 195 \times I$$

$\therefore$ Subtracting,

$$1950 = 0{\cdot}0109k^2 \times \tfrac{1}{6} \times 195 \times I$$

or $$k^2I = (6 \times 1950)/(0{\cdot}0109 \times 195)$$

$$\therefore\ P = 0{\cdot}0109 \times \tfrac{1}{3} \times 195 \times (6 \times 1950)/(0{\cdot}0109 \times 195)$$

$$= 3900 \text{ W}$$

Regenerative Methods. These methods require two machines, preferably identical. These are mechanically coupled, and are so adjusted electrically that one of them acts as a motor and the other as a generator. The motor supplies the mechanical power to drive the generator, while the electrical power developed in the generator is utilized in the motor. Thus, two machines of any size can be tested under full load conditions, and the power taken from the supply will be that required to overcome the losses only. The method is therefore invaluable where tests of long duration under full-load conditions have to be made on very large machines. Such tests are called "heat runs," because the object of the test is to determine the final temperature rise of the machine. Regenerative tests were first introduced by Hopkinson, for which reason they are often called *Hopkinson Tests*.

The method of procedure in the case of shunt machines is indicated in Fig. 12.5. Machine *I* is run up to full speed by means of the starting resistor, not shown, the main switch of machine *II* being open. This

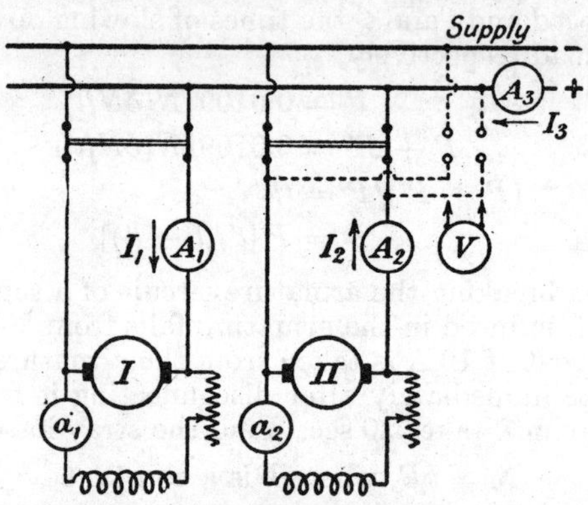

FIG. 12.5. HOPKINSON TEST

machine will therefore generate, and when its voltage has been adjusted equal to, and of the same polarity as, the bus-bar voltage, this adjustment being made by means of a paralleling voltmeter *V*, it can be switched on to the supply. Any required load can now be thrown on to the machines by adjusting the respective shunt regulators. The machine with the smaller excitation acts as a motor because, its back e.m.f. being less than that of the other, it admits a greater armature current, and therefore produces a greater torque. Assuming, first of all, that the shunt losses are the same in both machines, and that the armature copper losses and also the stray losses can be equally divided between them, we have

$$\text{Motor intake} = VI_1$$

$$\text{Generator output} = VI_2$$

Current from supply I_3, say, $= I_1 - I_2$

$$\therefore \text{ Motor intake} = V(I_2 + I_3)$$

$$\therefore \text{ Motor output} = \eta_1 V(I_2 + I_3) \text{ where } \eta_1 = \text{motor efficiency}$$

Again, generator output $= VI_2$

$$\therefore \text{ Generator intake} = VI_2/\eta_2 \text{ where } \eta_2 = \text{generator efficiency}$$

But the generator intake is equal to the motor output, so that

$$\eta_1\eta_2 = I_2/(I_2 + I_3)$$

By making the assumption of equal division of losses we also assume equal efficiencies. Hence, if we express either efficiency by η, we have

$$\eta = \sqrt{[I_2/(I_2 + I_3)]}$$

The error introduced by these assumptions will not be very important in the case of large machines, because the armature currents will not be very different, and the difference in the excitations required to give the necessary circulating current through the armatures will not greatly affect the iron losses. In the case of small machines the armature currents and the excitations are very different in the two machines, and the efficiency, as calculated above, is far from accurate. The greatest error is made in assuming that the armature copper losses are equal, because these losses are proportional to the square of the current. By calculating these losses separately, but assuming an equal division of stray losses, the error is reduced to a very small amount. The exciting currents are first adjusted till each of the ammeters A_1 and A_2 reads one-half of the ammeter A_3. The machines are now both motoring light, and if I_o is the current taken from the line, i.e. the reading of A_3, the total excitation and stray losses in the two machines are given by VI_o. We thus have for one machine

$$P_1 = VI_o/2$$

these losses being assumed constant. If the excitations are different the ammeters A_1 and A_2 will carry different currents. Calling these currents A_1 and A_2, and the shunt currents, a_1 and a_2, then

$$\text{Current in motor armature } I_M = (A_1 - a_1)$$

$$\therefore \text{ Motor armature copper loss } R_a I_M^2 = R_a(A_1 - a_1)^2$$

$$\therefore \text{ Total motor losses } P_M = P_1 + R_a(A_1 - a_1)^2$$

$$\text{Again, motor intake} = VA_1$$

$$\therefore \eta_M = (VA_1 - P_M)/VA_1$$

In the case of the generator the armature current is

$$I_G = (A_2 + a_2)$$

$\therefore$ Generator armature copper loss

$$R_a I_G^2 = R_a(A_2 + a_2)^2$$

$$\therefore \text{ Total generator losses } P_G = P_1 + R_a(A_2 + a_2)^2$$

$$\text{Again, generator output} = VA_2$$

$$\therefore \eta_G = VA_2/(VA_2 + P_G)$$

In testing compound machines, the series coils are very often left out of circuit; they can be kept in circuit if that on the motoring machine is reversed.

EXAMPLE. Two shunt motors loaded for the Hopkinson test take 15 A at 200 V from the supply. The motor current is 100 A and the shunt currents are 3 A and 2·5 A. If the resistance of each armature is 0·05 Ω, calculate the efficiency of each machine for its particular conditions of loading.

The motoring machine has the smaller shunt current, so that the distribution of current is as given in Fig. 12.6.

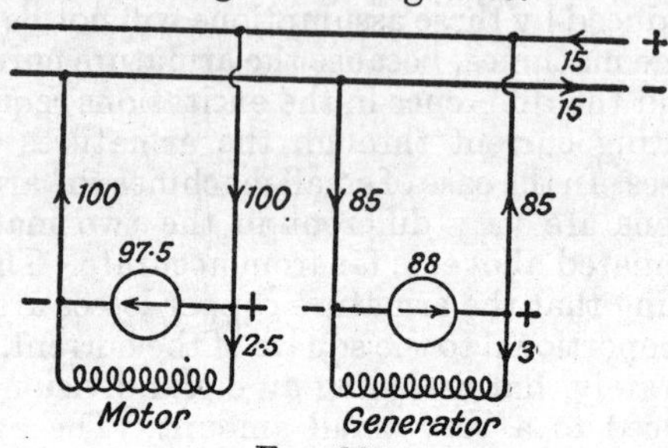

FIG. 12.6

Total intake of set	$= 15 \times 200$	$=$ 3000 W
Copper loss in motor armature	$= (97{\cdot}5)^2 \times 0{\cdot}05 =$	475 W
,, ,, generator armature	$= (88)^2 \times 0{\cdot}05$	$=$ 387 W
,, ,, motor field	$= 2{\cdot}5 \times 200$	$=$ 500 W
,, ,, generator field	$= 3 \times 200$	$=$ 600 W
	∴ Total copper losses	$=$ 1962 W

∴ Total stray losses $= 3000 - 1962$
$= 1040$, say

∴ Stray losses per machine $= 520$ W

Generator efficiency—

Output $= 200 \times 85 = 17{,}000$ W
Field copper loss $= 600$ W
Armature copper loss $= 387$ W
Stray losses $= 520$ W
∴ Total losses $= 1507$ W
∴ Intake $= 17{,}000 + 1507 = 18{,}507$ W

$$\therefore \eta_G = (17{,}000/18{,}507) \times 100 = 92 \text{ per cent}$$

Motor efficiency—

Intake $= 200 \times 100 = 20{,}000$ W
Field copper loss $= 500$ W
Armature copper loss $= 475$ W
Stray losses $= 520$ W
∴ Total losses $= 1495$ W
∴ Output $= 20{,}000 - 1495 = 18{,}505$ W

$$\therefore \eta_M = (18{,}505/20{,}000) \times 100 = 92{\cdot}55 \text{ per cent}$$

Series Motor Tests.* Because of the large variation in excitation of a series motor during normal working conditions, tests made with a fixed excitation are of little value. Again, no-load tests are impossible because of the dangerous speed attained. For these reasons the testing of series motors is much more difficult than that of shunt motors. If the motor is not too large, a brake test can be made on it, but in the case of large machines it is preferable to make a combined test with two machines. There is, as a rule, little difficulty in obtaining a pair of identical motors, since most series motors are used for traction work, and each tramcar has at least two. When a pair of machines is tested it is preferable to ensure that the iron losses are equal, if this is possible. This necessitates equal excitations. This condition is fulfilled in "Field's Test" by connecting the two fields in series. This arrangement

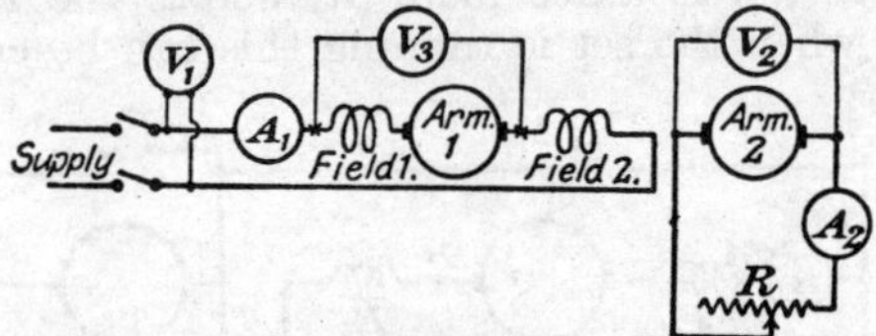

FIG. 12.7. CONNECTIONS FOR FIELD'S TEST

of the test is shown in Fig. 12.7. Machine I runs as a motor and drives machine II as a generator, the output of the latter being wasted in the adjustable load R. The connection of the generator and motor fields in series ensures that the stray losses in each machine will be the same. In making the test it is preferable to include no switch-gear in the connection between generator armature and load, since with this precaution there is no danger of the load being thrown off by mistake.

Let V_1 = supply voltage

I_1 = motor current

V_2 = voltage at terminals of load R

I_2 = load current

$\therefore$ Intake of whole set $= V_1 I_1$

Output $= V_2 I_2$

$\therefore$ Losses of all kinds $P_T = V_1 I_1 - V_2 I_2$

Total copper losses $P_C = \{(R_a + 2R_{se})I_1^2 + R_a I_2^2\}$

$\therefore$ Total stray losses $2P_S = P_T - P_C$

$\therefore$ Stray losses in one machine

$$P_S = (P_T - P_C)/2 \text{ watts}$$

The motor efficiency can now be determined as follows: the generator is working under such abnormal conditions that there is no object in calculating its efficiency.

* More complete information on the testing of series motors is given in *Electric Traction*, by A. T. Dover (Pitman) and *The Practical Testing of Dynamos and Motors*, by C. F. Smith (Scientific Publishing Co.).

Motor intake $= V_3I_1$ where V_3 is the reading of V_S

$$\text{Total motor losses} = (R_a + R_{se})I_1^2 + P_S$$
$$= P_M, \text{ say}$$
$$\therefore \eta_M = (V_3I_1 - P_M)/V_3I_1 \text{ per cent}$$

It is to be noted that the above test is not a regenerative test, since the output of the generator is wasted. A regenerative test on series motors is rather difficult to carry out, a good method being as follows. (*See* Fig. 12.8.) The generator during the test is loaded back on the supply, the amount of this loading being regulated by a load booster, this booster being a small low-voltage generator whose voltage can be varied between zero and a maximum. At the moment of starting it is necessary to have a resistance load, R, across the armature as in Field's test, but when the set is running this can be cut out and the

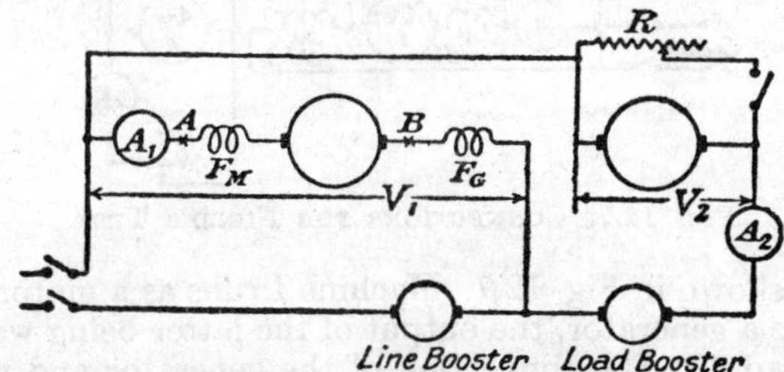

FIG. 12.8. REGENERATIVE TEST

load looked after by the booster. The generator and motor fields are in series, so that the stray losses in the two machines are the same. In order to keep the voltage across the motor terminals AB constant, an adjustable resistance can be connected in series with the motor, but if, as is usually the case, the supply is at normal motor voltage, it will be necessary to boost up the voltage in order to compensate for the drop in the generator field F_G. This is automatically carried out by a second booster, called the line booster, and this second booster is series excited.

Let V_1 = supply voltage and I_1 = motor current.

Let V_2 = generator voltage and I_2 = current loaded back on the supply.

$\therefore$ Total losses in the two machines $= V_1I_1 - V_2I_2$.

Total copper losses $= (R_a + 2R_{se})I_1^2 + R_aI_2^2$.

The calculation of the motor efficiency is thus the same as in Field's test and need not be proceeded with any further.

It will be realized that, by the use of boosters, the motoring machine in Field's test can be worked under normal conditions. In the Hopkinson test the motoring machine has its field weakened, and therefore neither machine is working under quite normal conditions. This can be avoided by keeping the normal voltage across the field of the motoring machine, and increasing the field of the generating machine by means of a field booster.

Where only a single series motor is available for test, the losses can be determined by methods similar to the Swinburne method as applied to a shunt motor. If there is a convenient source available, such as a battery or a low-voltage booster, the field can be separately excited and the losses then determined by running the armature on no load. In the case of the series motor it is necessary to make tests at a series of values of field current covering the range of the machine during normal operation, and the p.d. to be applied to the armature will, in each case, be the rated voltage less the drop of volts in the series winding.

By modifying the test as indicated in the connection diagram of Fig. 12.9 it is possible to dispense with the separate supply for the field winding. The motor is controlled by means of a rheostat R_2, and some of the intake current is diverted from the armature by means of the rheostat R_1. It is advisable to take a series of values for the field current and to calculate for each value the p.d. to be applied to the armature, this p.d. being given by the expression $(V - R_{se}I_{se})$. This p.d. is then plotted against the current. When starting up it is advisable to make R_1 small so that the whole of R_2 can be cut out without reaching a dangerous speed. The field current is now observed and R_2 is adjusted to give a reading of V corresponding to that on the curve. This may require one or two slight adjustments, after which the two currents, the voltage and the speed, are observed.

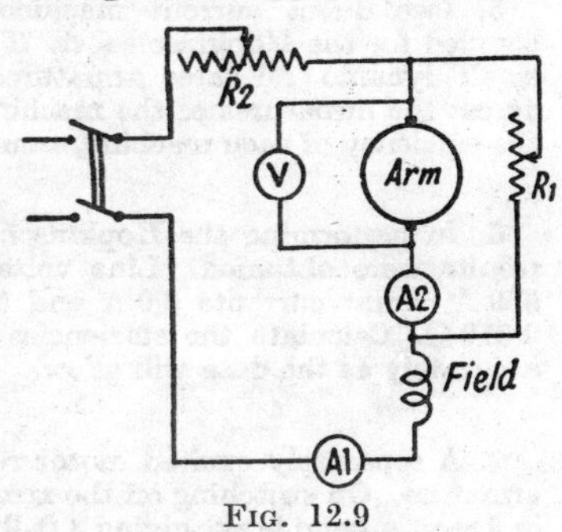

FIG. 12.9

Next, the resistance R_2 is increased, and the shunting resistance R_1 is adjusted until V again corresponds to the new value of the field current. Proceeding in this manner the whole range of operation of the motor is covered. The results are finally plotted so as to give curves of field current and armature watts, both against speed, as abscissae. Since the field resistance is known the field copper loss $I_{se}^2R_{se}$ can be determined, and also the armature copper loss $I_{se}^2R_a$, since during normal operation the series field and armature carry the same current. Thus the curve of total copper loss can be added, thereby providing all the data for the predetermination of the efficiency at any operating speed.

EXAMPLES ON CHAPTER 12

1. Describe in detail all the adjustments and tests you would make in determining the efficiency of a continuous-current shunt motor by the "loss" method at a given output. Deduce the formula for the current corresponding with a load

of *W* watts, supplied at *V* volts, assuming the loss tests have furnished the requisite data. (L.U.)

2. A shunt motor running light at 480 V, takes a current of 2·5 A. The resistance of its field winding is 800 Ω and of its armature 0·6 Ω. Determine the efficiency of the motor when loaded so that the current is 40 A, the terminal voltage being maintained at 480 V. (L.U.)

Ans. 88·9 per cent.

3. If you had to determine the efficiency of a very large motor, say, 200 h.p., and the only supply available was one from which about 20 h.p. could be taken, what procedure would you adopt?

4. You are required to determine the no-load magnetization curve of a shunt-excited 500 V, eight-pole, continuous-current motor intended to be run at 1000 r.p.m. The source of excitation available is only 250 V. The voltmeter available only reads to 250 V. Indicate how you would carry out the test, and give a diagram of connections. (L.U.)

5. Two direct current machines are separately excited and mechanically coupled for the Hopkinson test. The driving power is supplied electrically by a small dynamo, all three armatures being connected in series. If the voltages across the armatures of the machines under test are 200 V and 160 V, calculate the efficiency of each machine, assuming that they are the same.

Ans. 89·5 per cent.

6. In performing the Hopkinson test on two shunt machines, the following results were obtained. Line voltage 200 V, line current 50 A, motor current 500 A, shunt currents 6·0 A and 5·0 A. Armature resistance of each machine 0·012 Ω. Calculate the efficiencies of the motoring and generating machines as accurately as the data will allow.

***Ans.* Each 94·9 per cent approx.**

7. A separately excited motor ran at 1000 r.p.m. with 100 V applied to the armature. On switching off the armature supply the volts fell from 100 V to 90 V in 3 sec. A band brake giving 3 ft-lb was next applied and the same fall took place in 1·3 sec. Calculate the stray losses.

Ans. 325 W.

8. What are the difficulties in applying the regenerative test to two identical series motors? Describe the test in detail and state how these difficulties are overcome.

9. Two similar shunt machines are tested by the Hopkinson method, the intake being 40 A at 220 V. The motor armature current is 200 A, motor shunt current 6 A and dynamo shunt current 7 A. Each armature resistance is 0·05 Ω. Calculate the efficiency of each machine.

Ans. Motor 90 per cent, dynamo 89·4 per cent.

10. A test on two coupled similar tramway motors, with their fields connected in series, gave the following results where one machine acted as a motor and the other as a generator. Motor: armature current 56 A, voltage 550 V, volt drop across field winding 40 V. Generator: armature current 44 A, armature voltage 400 V, field volt drop 41 V. Resistance of each armature 0·3 Ω. Calculate the efficiency of the motor and gearing at this load and criticize the method of testing. (I.E.E.)

Ans. 74·3 per cent.

11. A 2000 kW turbo-alternator operates at 1200 r.p.m. on a 40 period, 5750 V line. The calculated kinetic energy of the complete machine is 7900 ft-tons, and when (*a*) unexcited, the speed falls 2 revs. per sec in 49 secs., and (*b*) when excited the speed falls 2 revs. per sec in 29 secs. Estimate the core loss. (I.E.E.)

Ans. 64·1 kW.

CHAPTER 13

DISTRIBUTION

A LOW-TENSION distribution consists of a network of cables which can be divided into three categories, viz. (*a*) the "feeders," which are the cables supplying power in bulk to a selected number of points called feeding points; (*b*) the "distributors," from which current is tapped for the various consumers, these cables generally having the main street for their route; (*c*) the "service mains," which are the small cables teed off from the distributors and taken into the premises of the various consumers. A very elementary scheme employing one feeder,

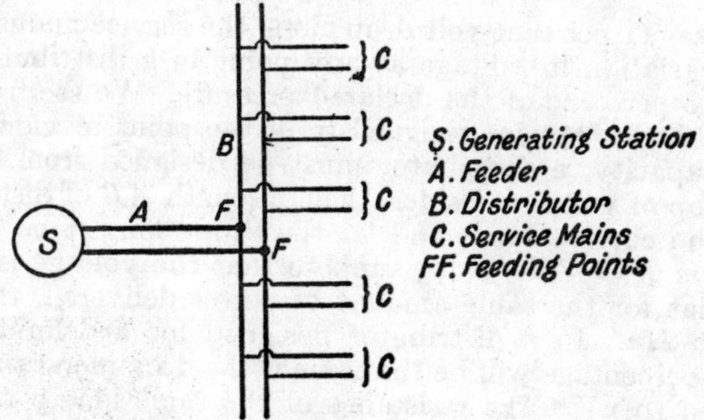

FIG. 13.1. ELEMENTARY DISTRIBUTION SCHEME

one distributor, and a number of service mains is given in Fig. 13.1. In this simple scheme the distributor is fed at one point only, but in actual schemes a long distributor may be fed at two points by means of two separate feeders, or alternatively the distributor may be arranged to form a closed circuit, and fed at any desired number of feeding points by means of a number of feeders radiating from the generating station. It will be shown on p. 290 that these arrangements give a much smaller drop of volts than a distributor fed at one end only.

Effect of Supply Voltage on the Size of Distributors and Feeders. By the size of a cable we always mean the cross section of the core carrying the current, and not the overall cross section. Consider first of all the case of a feeder. Since there are no consumers tapped off from a feeder, it is immaterial what the drop of volts along the feeder will be, provided that it is not outside the range of compounding of the generators in the power station. A feeder can thus be designed from the point of view of current-carrying capacity, with drop of volts as a secondary consideration. The allowable current density for a given type of cable laid in a given manner is not a constant, but decreases somewhat as the cable size increases; this is because the cooling facilities improve as the size is reduced. As a rough approximation we can, however, assume that the current density is a constant. Now, suppose, that the voltage of the system is increased n fold, then for a given power delivered the current is reduced to $1/n$th, and therefore the size of the cable is reduced to $1/n$th. There is thus a very appreciable saving in the price of the copper in the feeders when the supply voltage is increased.

Now take the case of a distributor. There is a Board of Trade regulation which states that the variation of voltage at a consumer's terminals must not exceed $\pm$ 4 per cent of the declared pressure.

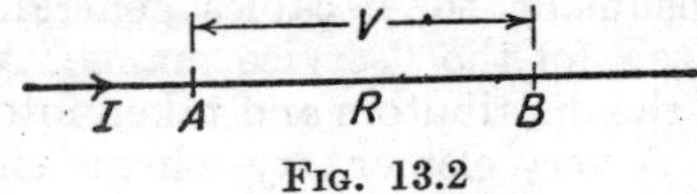

FIG. 13.2

Allowing, say, 1 per cent volt drop along the service mains, this means that the variation in voltage at any point in a distributor must not exceed $\pm$ 3 per cent of the declared pressure. We see from this that whereas a feeder can be designed from the point of view of current-carrying capacity, a distributor must be designed from the point of view of drop of volts. Consider, then, a length AB of distributor, Fig. 13.2, let the current be I, and let the drop along it be V. Then its resistance is $R = V/I$. Now, suppose that the voltage is increased n fold, so that for the same amount of power delivered, the current is reduced to I/n. In a distributor designed for a definite percentage drop, this percentage will be the same if the drop along the length AB is increased to nV. The resistance of the same length of distributor will now be

$$R^1 = nV/(I/n) = n^2R$$

Now, for a given length the cross section of a cable is inversely proportional to its resistance, showing that an increase in the working voltage of n times reduces the cross section, and therefore the weight of copper required to n^{-2}.

Effect of Pressure on the Efficiency of Transmission.

Let V_1 = voltage at station bus-bars

V_2 = voltage at feeding points

I = current in feeders

R = resistance of each side of a feeder

i.e. of both positive and negative sides.

Then, power put into feeder $= V_1I$

Power taken from feeder $= V_2I$

$\therefore$ Efficiency of transmission $= V_2I/V_1I = V_2/V_1$

Losses in feeder $= 2I^2R$

Suppose that 1000 kW have to be transmitted 1 mile, the feeder being worked at a current density of 1000 A/in.^{2} Suppose that the consumer's voltage V_2 is 100 V. Then,

$$I = \text{power delivered}/V_2 = 1{,}000{,}000/100 = 10{,}000 \text{ A}$$

Hence, cross section of copper in either positive or negative side of the cables

$$a = 10{,}000/1000 = 10 \text{ in.}^2$$

Distance l = 1 mile = 5280 × 12 in.

∴ Resistance of one side

$$R = \rho(l/a) = [2/(3 \times 10^6)](5280 \times 12/10)$$
$$= 0{\cdot}0042\ \Omega$$

∴ Drop in volts in feeders

$$= 2IR = 2 \times 0{\cdot}0042 \times 10{,}000 = 84 \text{ V}$$

∴ Station voltage

$$V_1 = 100 + 84 = 184 \text{ V}$$

∴ Efficiency of transmission

$$= 100/184 = 54{\cdot}4 \text{ per cent}$$

Now suppose that the voltage at the consumer's terminals is 1000 V instead of 100 V. Then $V_2 = 1000$ V and $I = 1000$ A.

∴ Cross section of copper in cables $a = 1$ in.^{2}

∴ $R = 0{\cdot}042$, and drop in feeders $2IR = 84$ V, as before.

∴ $V_1 = 1000 + 84 = 1084$ V and efficiency = 1000/1084 = 92 per cent.

There is thus a considerable increase in efficiency when the voltage is raised. Also, the cross section of copper in the cables is diminished in proportion to the increase in voltage, and this diminishes the cost of the copper in the cables in proportion. For these reasons it is desirable to transmit at a high voltage. Now in a d.c. system it is not possible to change the voltage between the feeders and distributors without using rotating machinery, and as this is undesirable, the feeders and distributors are worked at the same voltage. This means that the consumer's voltage is the same as the voltage at the feeding points. Suppose, for example, that the load consists of lamps. Then since 250 V is the highest suitable voltage for this purpose, the feeders must work at this voltage; 250 V is very low if a large amount of power has to be dealt with, especially if the feeders are long ones, and consequently with a simple two-wire system the efficiency of transmission will be low.

Three-wire System. The above difficulty is largely overcome by using the three-wire system, which consists of two "outers" and an earthed middle wire (Fig. 13.3). The consumer's apparatus or lamps are connected between one outer and the neutral, the result being that the transmitting voltage, i.e. the voltage between the outers, is twice that at the consumer's terminals. This results in a considerable increase in efficiency, and economy in copper. The middle wire or neutral has half the cross section of either outer. If the loads on the two sides are unequal, then the difference current, called the "out of balance" current, flows in the neutral. Thus, if the positive and negative outers carry I_1 and I_2 respectively, the neutral carries $(I_1 - I_2)$. If a three-wire and a two-wire feeder both transmit the same amount of power the same distance with the same voltage at the consumer's terminals, and with the same efficiency, then the three-wire feeder, if balanced, requires only five-sixteenths as much copper as the two-wire feeder. This is proved as follows.

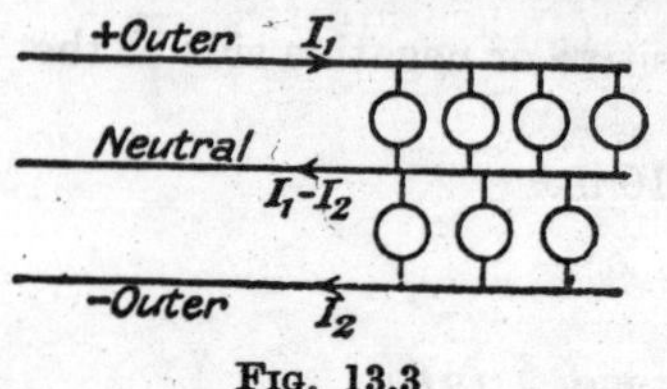

FIG. 13.3

Let P = power delivered to consumer in watts

V_1 = voltage at station

and V_2 = voltage at feeding points

Then, in the two-wire system, current $I = P/V_2$

∴ Drop in both positive and negative sides,

$$2IR_1 = 2PR_1/V_2$$

and
$$V_1 = V_2 + \text{drop} = V_2 + 2PR_1/V_2$$

where R_1 is the resistance of one conductor.

$$\therefore \text{ Efficiency} = V_2/(V_2 + 2PR_1/V_2)$$

In the three-wire system, the total voltage is doubled, and therefore, each outer will carry a current of $P/2V_2$ so long as the currents on the two sides are equal. Hence, if R_2 is the resistance of each outer,

$$\text{Efficiency} = 2V_2/(2V_2 + 2PR_2/2V_2)$$

Since the efficiencies are equal we have

$$\frac{V_2}{V_2 + 2PR_1/V_2} = \frac{2V_2}{2V_2 + PR_2/V_2} \text{ or } \frac{V_2}{V_2 + PR_2/2V_2}$$

Hence
$$R_2 = 4R_1$$

Now the cross section, and therefore volume, of a conductor of given length is inversely proportional to its resistance. If we represent the volume of the copper in the two-wire system by 100 we have 50 in each wire.

∴ Volume of each outer in the three-wire feeder = 50/4 = 12·5

and volume of middle wire = 12·5/2 = 6·25

$\therefore$ Total volume of copper in three-wire feed

$$= 12{\cdot}5 + 6{\cdot}25 + 12{\cdot}5 = 31{\cdot}25$$

$$\therefore \frac{\text{Copper in 3-wire feeder}}{\text{Copper in 2-wire feeder}} = \frac{31{\cdot}25}{100} = \frac{5}{16}$$

If the system is not balanced it is most convenient to compare the amounts of copper in the two cases by taking a numerical example.

EXAMPLE. A three-wire feeder whose outers have a cross section of 0·2 in.², is 1 mile long. It carries 220 A in the positive and 200 A in the negative outer, and the voltages across the two sides at the feeding points are 200 V. What size of two-wire feeder will transmit 420 A at 200 V with the same total loss? Compare the amounts of copper in the two feeders.

Taking the specific resistance of copper as $\frac{2}{3}$ microhm inch we have for the resistances of the cores of the three-wire feeder

Positive side, $R = \rho(l/a) = [2/(3 \times 10^6)](5280 \times 12/0{\cdot}2) = 0{\cdot}21$;

Negative side, $R = 0{\cdot}21$; neutral $R = 0{\cdot}42\ \Omega$

Hence, loss in feeder

$$= 220^2 \times 0{\cdot}21 + 200^2 \times 0{\cdot}21 + 20^2 \times 0{\cdot}42$$
$$= 18{,}770 \text{ W}$$

Let R_1 = resistance of each side of the two-wire feeder

$$\therefore 2 \times 420^2 \times R_1 = 18{,}770, \; R_1 = 0{\cdot}053\ \Omega$$

Now the amount of copper in a feeder of given length is inversely proportional to its resistance; hence, if we call the total amount of copper in the two-wire feeder 100, i.e. 50 per cent in each side, we have for the three-wire feeder

Copper in positive outer $= (0{\cdot}053/0{\cdot}21) \times 50 = 12{\cdot}6$ per cent

Copper in negative outer $=$ 12·6 per cent

Copper in neutral $= (0{\cdot}053/0{\cdot}42) \times 50 = 6{\cdot}3$ per cent

Total $= 31{\cdot}5$ per cent

Balancers. The generators supplying a three-wire feeder are all connected in parallel across the outers, and it is therefore necessary to fix the potential of the middle wire mid-way between that of the outers. If this is not done the voltages across the two sides will not be equal, unless the currents taken from the outers are equal.

The commonest form of balancer consists of two identical shunt machines mechanically coupled, and having their armatures and fields connected in series across the outers. The middle wire is connected to the junction of the armatures. Consider the arrangement shown in

Fig. 13.4. Let R_a be the resistance of each balancer armature. If the current I_1 is greater than I_2, then the positive side will be the more heavily loaded and the pressure on that side will tend to decrease. The armature of the balancer on the positive side will therefore act as a

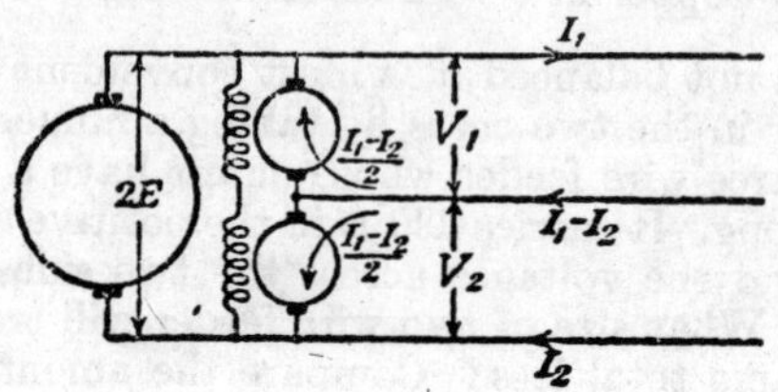

FIG. 13.4. BALANCE FOR THREE-WIRE SYSTEM

generator, while the armature on the less heavily loaded side, the negative, will motor. Thus the effect of the balancer is to convey power from the lightly loaded to the heavily loaded side.

Since the balancer armature on the positive side is generating, its terminal voltage is

$$V_1 = E - R_a(I_1 - I_2)/2$$

Since the balancer armature on the negative side is motoring, its terminal voltage is

$$V_2 = E + R_a(I_1 - I_2)/2$$

where $2E$ is the total voltage across the outers, and therefore twice the voltage *induced* in each armature.

∴ Difference of voltage between the two sides of the system,

$$V_2 - V_1 = R_a(I_1 - I_2)$$

Hence, this difference is proportional to: first, the "out of balance" current $(I_1 - I_2)$; second, the armature resistance of the balancer.

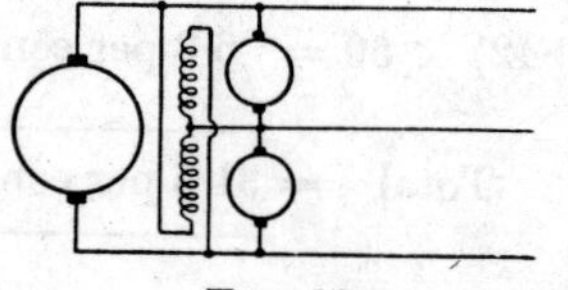

FIG. 13.5

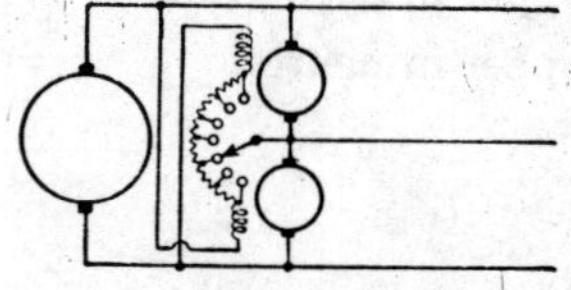

FIG. 13.6

For this reason balancers are designed with a very small armature resistance, and steps are always taken to arrange the loads on the two sides, so that the out of balance current will be as small as possible.

The difference $(E_2 - E_1)$ can be reduced by cross connecting the balancer fields, as shown in Fig. 13.5. This causes the generating machine to draw its excitation from the lightly loaded side, which is at the higher voltage, and the motoring machine to draw its excitation from the heavily loaded side.

Hence, the induced e.m.f. in the generating armature is greater than E, while that in the motoring armature is less than E. This increases V_1 and decreases V_2, the difference $(V_2 - V_1)$ being therefore decreased. Additional hand regulation of the voltage can be obtained by connecting an adjustable regulator in series with the balancer fields, as shown in Fig. 13.6.

Boosters. A booster is a generator whose voltage is added to, or injected into, a circuit, usually to compensate for a variation in voltage, as, for example, the drop of volts in a feeder due to its resistance. A booster which is used to compensate for the drop in a feeder is called a "feeder booster." Obviously it is a low voltage, heavy current machine. The booster in this case is a series generator connected in series with the feeder and driven at constant speed by a shunt motor. Since the drop in volts in a feeder is proportional to the current, the voltage injected by the booster must also be proportional to the current. Hence, the booster must work on the straight line portion of its voltage characteristic. The method of connecting the booster in the circuit is shown in Fig. 13.7. The advantage of using a booster instead of, or in addition to, over-compounding the generators is that each feeder can be regulated independently, a great advantage if the feeders are of different lengths. The arrangement of a three-wire station supplying one set of short

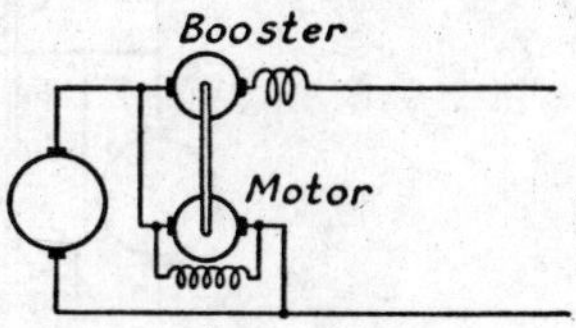

Fig. 13.7

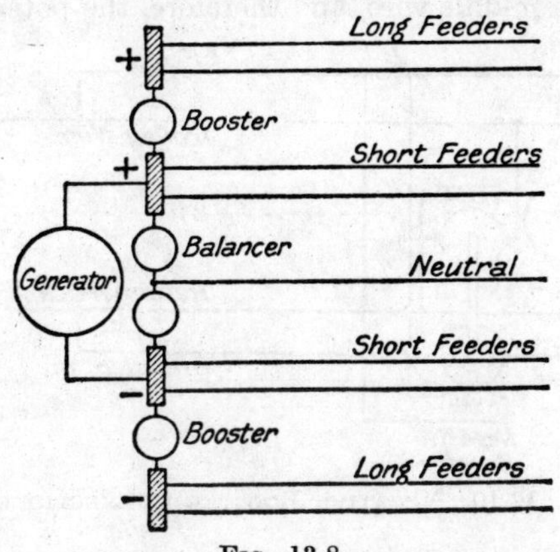

Fig. 13.8

and one set of long feeders is shown in Fig. 13.8. Notice that since there is a middle wire whose potential must be kept midway between the outers, it is necessary to have a booster on both the positive and negative sides for the regulation of the long feeders. The shaded rectangles in Fig. 13.8 represent the various bus-bars. For some cases, e.g. a tramway system, it is desirable to raise the voltage of the line at some distant point instead of merely at the generator end. This can be done by running a special feeder to the desired point and injecting a voltage into this feeder by means of a booster connected in series with it, as shown in Fig. 13.9.

Negative Booster. The negative booster subtracts instead of adds its voltage to the system. It is used in traction systems to prevent the potential of the return rail from becoming too high relative to the earth.

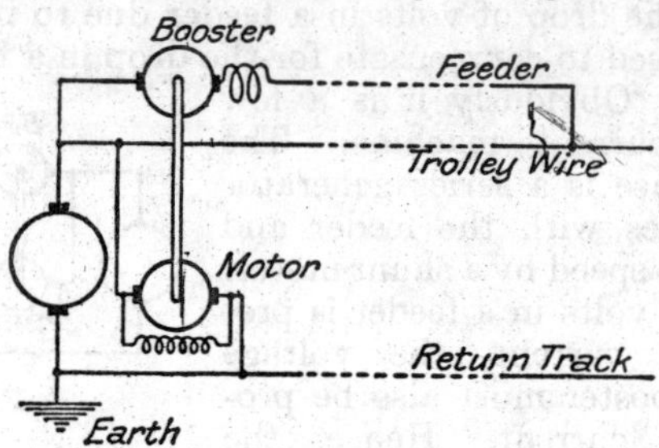

FIG. 13.9. BOOSTER CONNECTIONS

A Board of Trade regulation limits this potential to 4·2 V because, if it becomes much greater than this, the earth currents set up produce electrolysis which causes great damage to any iron pipes in the vicinity. The negative bus-bar at the station is always earthed, the traction system having no middle wire, and therefore, the potential of the rail

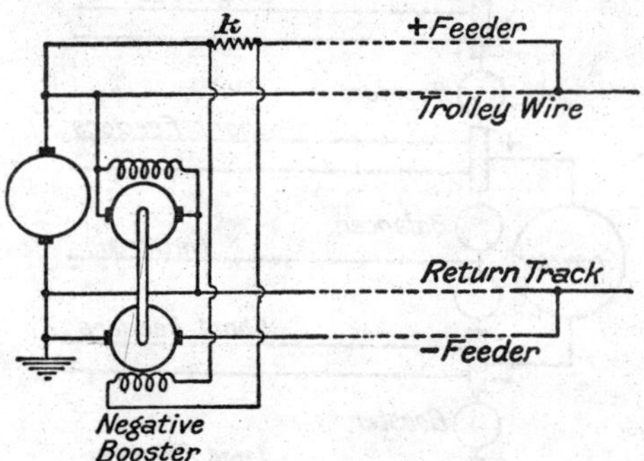

FIG. 13.10. NEGATIVE BOOSTER CONNECTIONS

will be positive relative to earth, and of increasing magnitude as the distance from the station is increased. It is kept within the prescribed limits by means of a booster connected to the rail, as shown in Fig. 13.10. A second Board of Trade regulation limits the potential difference between any pair of points on the return rail to 7 V, and to comply with regulation it is generally necessary to use several negative boosters. These negative boosters may be series excited, as in the case of feeder boosters, or they may be excited by having their fields connected across a resistance R in series with one of the feeders. The drop of volts in such a resistance will be proportional to the feeder current, and therefore, the excitation and the induced voltage of the booster will also be proportional to the feeder current. It is this second arrangement of exciting the booster field which is shown in Fig. 13.10.

Battery Boosters. In the case of d.c. systems subjected to violent variations in load, as, for example, a traction system, it is common practice to "float" a battery of accumulators across the bus-bars, as shown in Fig. 13.11. Let V_1 and V_2 be the generator and battery terminal voltages, and R the resistance of the circuit through generator and battery. Then generator current

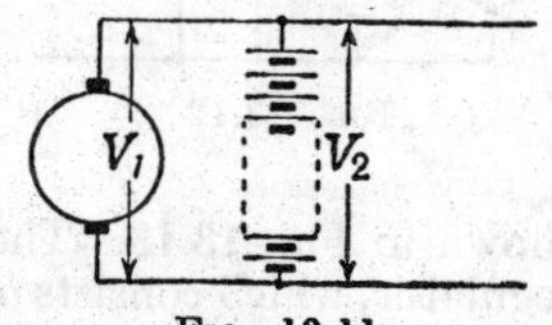

FIG. 13.11

$$I = (V_1 - V_2)/R$$

and if V_2 were constant, I would be constant, irrespective of any changes in the load. The battery would then constitute a perfect load equalizer, and the generators, working at constant load, would be working under ideal conditions. In times of heavy load the battery would discharge into the line by an amount equal to the difference between the line current and the current I, above. On light load the generator would charge the battery, the charging current being the difference between I and the line current. Unfortunately, the above simple arrangement is not practicable, because the battery terminal voltage V_2 is far from constant. Thus, when charging a battery it is necessary to apply about 2·5 V per cell, whereas when a battery is discharged it may not give more than 1·8 V per cell, so that there may be a 40 per cent variation in the terminal voltage of the battery. Suppose that the e.m.f. of a battery on no load is E; then when it is being charged, its e.m.f. will rise to, say, $E + e_1$. Hence, if I is the charging current, and R the battery resistance, the charging voltage to be applied to the battery will be

$$(V_2)_{\text{charge}} = E + e_1 + RI$$

On discharge the e.m.f. will fall to, say, $E - e_2$, and the terminal voltage for the same current I will be

$$(V_2)_{\text{discharge}} = E - e_2 - RI$$

Hence, $$(V_2)_{\text{charge}} - (V_2)_{\text{discharge}} = (e_1 + e_2) + 2RI$$

Now, assume that $e_1 = K_1I$ and $e_2 = K_2I$, then the difference between the terminal voltages on charge and discharge becomes

$$(K_1 + K_2 + 2R)I$$

that is, proportional to I. Hence, if a voltage is injected into the battery circuit and this voltage *reverses* with the battery current and is proportional to it, the terminal voltage will be kept constant. The booster which injects this voltage is called a "reversible battery booster." If the excess voltages e_1 and e_2 were rigorously proportional to I, and the characteristic of the booster were a perfect straight line through the origin, then the simple arrangement shown in Fig. 13.12 would give the necessary constant battery voltage E_2. Unfortunately, neither assumption is strictly true, and therefore the boosters used in practice have to be much more complicated.

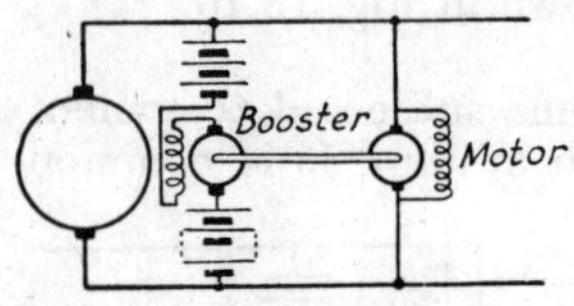

Fig. 13.12

There are several boosters which fulfil the necessary requirements, one form, the Entz booster, being taken as an example. The arrangement of the Entz booster is shown in Fig. 13.13. The most important part of this booster is the regulator, which consists of two piles of carbon plates pressed down by a lever L. This lever is pulled down on one side by a spring, W, and on the other side by a solenoid, S, in series with the line. The booster is separately excited, its field being connected direct to the exciter armature, while the exciter field is connected to the middle point of the regulator and the middle point of the battery. When the line current is equal to the normal full-load generator current, no help is needed from the battery, and the spring is adjusted so that, with this current, the resistances of the two carbon piles are equal, thus producing zero current in the exciter field and giving no voltage in the booster armature. If the line current increases, the solenoid is pulled down on the right-hand side, thus reducing the resistance of the right-hand pile. The exciter field now carries a current, and the e.m.f. induced in the booster armature enables the battery to discharge into the line. When the line current is less than normal, the pull on the solenoid relaxes and the left-hand pile has the smaller resistance. This reverses the direction of the current in the exciter field and so reverses the direction of the e.m.f. induced in the booster armature. The battery is then charged by the generator.

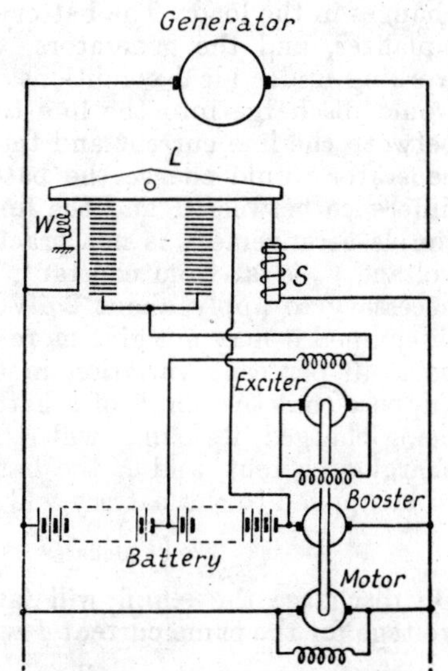

Fig. 13.13. Entz Booster

In the case of all boosters which have to carry a violently fluctuating current it is necessary that the voltage they produce shall follow, without any time lag, the variations in current. It is therefore necessary to laminate the whole of the field systems of such machines. If this were not done the rapid fluctuations in flux would induce eddy currents in the solid iron, which, apart from producing heat and loss of efficiency, would prevent the necessary rapid change in flux with current, and thus cause the voltage variations to lag behind those of the current.

If a booster is used merely to charge a battery it supplies current in one direction only, for which reason it is called an "Irreversible Battery Booster." The arrangement of such a booster is shown in Fig. 13.14. It will be seen that the booster adds its voltage to the bus-bar voltage, and its voltage range must therefore cover the variation of battery voltage between fully charged and discharged conditions. The booster in this case is shunt or separately excited, the necessary regulation being performed by hand. Sometimes a booster is used to charge individual cells if these have become discharged more than the other cells in the battery. Such a booster is called a "Milking Booster."

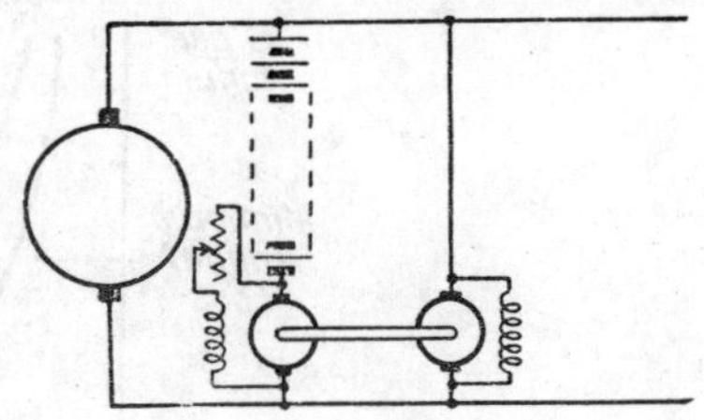

FIG. 13.14. IRREVERSIBLE BATTERY BOOSTER

Calculation of Feeders. KELVIN'S ECONOMY LAW. We have seen that the drop of volts is not a primary factor in the design of a feeder. A feeder can, therefore, be designed on the basis of current-carrying capacity and, where practicable, of minimum financial loss. The financial loss per annum occasioned by conductors of electricity is made up of (*a*) interest on the capital cost of the conductors, plus an allowance for depreciation, (*b*) the cost of the energy wasted in virtue of the ohmic resistance of the conductor. For a route of given length the weight, and therefore the cost, of the copper is proportional to the area of cross section. Hence, the annual value of the combined interest and depreciation is also proportional to the cross section, and can be written as £Pa,

where $P =$ a constant

$a =$ cross section of copper

The ohmic resistance is proportional to $1/a$, and therefore for a given curve of demand for a current throughout the year, the energy lost in the conductor will be proportional to the resistance, and therefore proportional to $1/a$. The annual value of this energy will, therefore, be proportional to $1/a$, and can be written £Q/a, where Q is another constant.

$$\therefore \begin{pmatrix}\text{Total annual}\\ \text{financial loss}\end{pmatrix} = £(Pa + Q/a)$$

If plotted against cross section, the graph of the loss Pa is a straight line, while that of Q/a is a rectangular hyperbola, as shown by curves 1 and 2 in Fig. 13.15. The graph of total annual loss is given by curve

3, and it exhibits a minimum at that value of a corresponding to the intersection of the two component curves. The above argument has not been concerned with any insulating covering in the case of cables, or of insulators in the case of bare conductors, from which we see that for bare conductors the most economical cross section is that which makes the annual value of the interest and depreciation on the conduc-

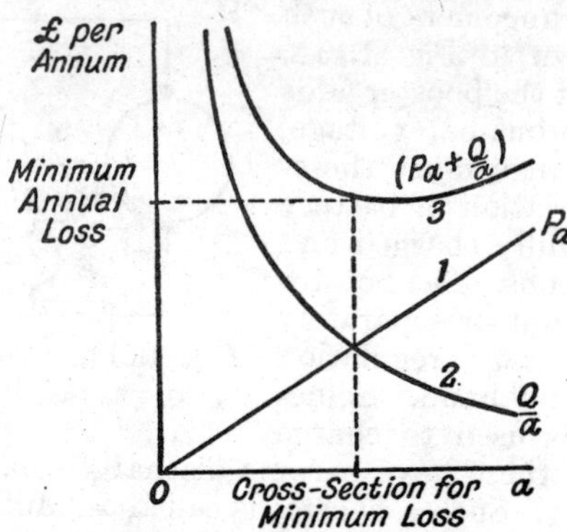

FIG. 13.15. ILLUSTRATING KELVIN'S LAW

tor equal to the annual value of the electrical energy loss due to the resistance of the conductor.

It is to be noticed that a feeder cross section, as calculated from the law, will not always be a practicable one, because it may be too small to carry the current.

EXAMPLE. If the cost of an overhead line is £600 × A (where A is the sectional area in square inches), and if the interest and depreciation charges on the line are 8 per cent, estimate the most economical current density to use for a transmission requiring full-load current for 60 per cent of the year. The cost of generating electrical energy is 0·5d./unit. The resistance of a conductor one mile long and 1 in.^{2} cross section is 0·43 Ω. (London Univ.)

The value for resistance should obviously be 0·043 Ω, and we will use this figure. Consider one mile, and let the full-load current be I, then in a two-conductor line in which each conductor is of resistance R, power lost at full current

$$= 2I^2R$$
$$= 2I^2 \times 0{\cdot}043 \times (1/A) \text{ watts}$$
$$= (0{\cdot}086I^2/A) \times 10^{-3} \text{ kilowatts}$$

The annual loss of energy is equal to that produced by the above loss taking place continually for 60 per cent of the year, i.e. for (0·6 × 365 × 24) hr. Hence, at £1/480 per unit, annual value of the energy lost

$$= £(0{\cdot}086I^2/A) \times 10^{-3} \times 0{\cdot}6 \times 365 \times 24 \times (1/480)$$
$$= £0{\cdot}00094(I^2/A)$$

The annual value of interest and depreciation

$$= 8 \text{ per cent of } £600A = £48A$$

Hence, for minimum annual loss

$$48A = 0{\cdot}00094(I^2/A)$$

$$\therefore \text{ Current density } I/A = \sqrt{(48/0{\cdot}00094)} = 226 \text{ A/in.}^2$$

We will now consider the influence of the insulation in the case of a covered cable. For a given type of cable, e.g. with vulcanized india-rubber, or impregnated paper insulation, for a given type of armouring, and for a given working voltage, the cost of the covering does not vary very much with the cross section of the cores, with the result that as a rough approximation we can regard the covering as adding a constant term to the cost of the cable. The annual value of the interest and depreciation, can therefore, be written $£(Pa + R)$, where R is another constant. The effect of this on the graph of interest and depreciation

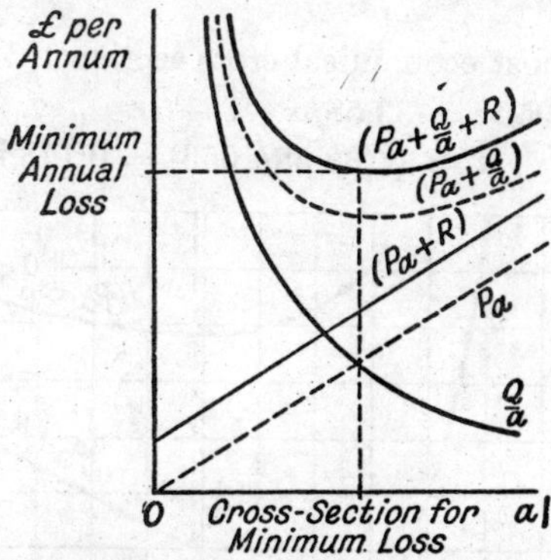

FIG. 13.16. KELVIN'S LAW APPLIED TO AN INSULATED CABLE

is to raise the graph of $£Pa$ vertically through a distance equal to $£R$. Similarly, the graph of total annual loss, namely, $£(Pa + Q/a + R)$ is merely the graph of $£(Pa + Q/a)$ raised vertically through a distance $£R$, without producing any horizontal displacement in the point of minimum cost. This is shown in Fig. 13.16. It follows from this that the insulation does not have any effect on the value of the cross section which gives the minimum annual loss, so that for an insulated cable the law can be stated as follows. The most economical cross section is that which makes the annual value of the interest and depreciation, due to the conductor in the cable, equal to the annual value of the energy lost.

EXAMPLE. A 500 V, 2-core feeder, two miles long supplies a maximum current of 200 A, and the demand is such that the copper loss per annum is such as would be produced by the full current flowing for six months. The resistance of a conductor one mile long of 1 in.² cross section is 0·046 Ω. The cost of the cable including installation is $£(6a + 1{\cdot}2)$/yd, and interest and depreciation charges 10 per cent. The cost of energy is 0·5d./unit. Find the most economical cross section, and plot curves of component and total annual losses.

For one mile, cost of conductors in cable

$$= £6a \times 1760$$
$$= £10{,}560a$$

∴ Interest and depreciation

$$= £1056a$$

Power lost per two cores, per mile, with maximum current flowing

$$= 2I^2R$$
$$= 2 \times (200)^2 \times 0{\cdot}046/a \text{ watts}$$

∴ Annual value of energy lost

$$= £\left(2 \times (200)^2 \times \frac{0{\cdot}046}{a} \times 10^{-3} \times \frac{365 \times 24}{2} \times \frac{1}{480}\right)$$
$$= £33{\cdot}58/a$$

Hence, for the most economical cross section

$$1056a = 33{\cdot}58/a$$
$$a = \sqrt{(33{\cdot}58/1056)} = 0{\cdot}178 \text{ in.}^2$$

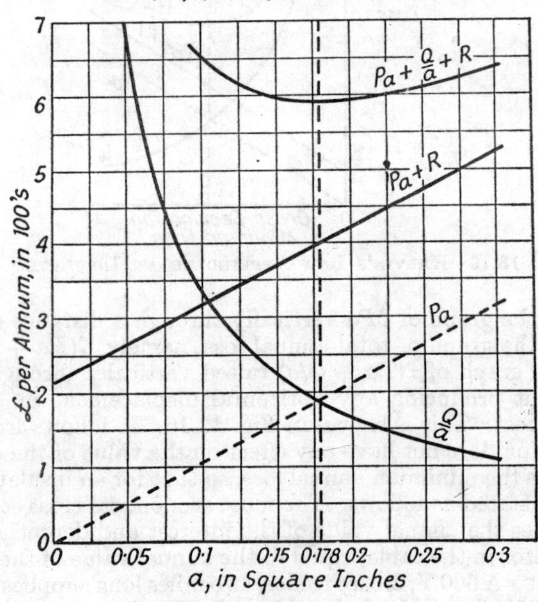

FIG. 13.17

From the above figures we have for the three constants $P = 1056$, $Q = 33{\cdot}58$, and $R = 211{\cdot}2$ for each mile of cable, so that the curves are as shown in Fig. 13.17.

Calculation of the Drop of Volts in Distributors. (*a*) Distributor Fed from One End. We have seen that the main consideration in the design of distributors is the drop of volts. This drop depends upon the nature of the current loading of the distributor, and also on whether it is fed from one end only or from both ends. Consider first of all a distributor with concentrated loads, and fed from one end. Let i_1, i_2, i_3, etc., be the currents tapped off, and I_1, I_2, I_3, etc., the currents in the distributor sections. Let r_1, r_2, r_3, etc. be the resistances of the distributor sections, and R_1, R_2, R_3, etc., the total resistances from the feeding point F to the successive points (Fig. 13.18).

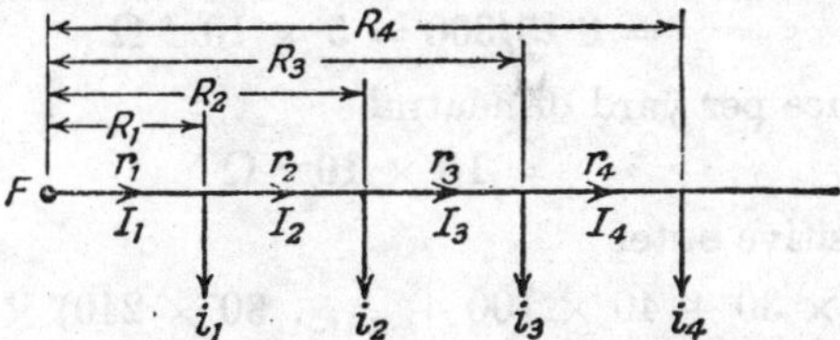

Fig. 13.18. Distributor Fed at One End

$$
\begin{aligned}
\text{Then total drop} &= r_1I_1 + r_2I_2 + r_3I_3 + \dots \\
&= r_1(i_1 + i_2 + i_3 + \dots) \\
&\quad + r_2(i_2 + i_3 + \dots) \\
&\quad + r_3(i_3 + i_4 + \dots) + \dots \\
&= i_1r_1 + i_2(r_1 + r_2) + i_3(r_1 + r_2 + r_3) + \dots \\
&= i_1R_1 + i_2R_2 + i_3R_3 + \dots \\
&= \Sigma(iR)
\end{aligned}
$$

Thus the drop *at the far end* of a distributor fed at one end is given by the sum of the moments of the various currents tapped off, about the feeding point. The drop at any intermediate point is equal to the sum of the moments of the currents up to that point, plus the moment of all the currents beyond that point assumed to be acting at that point. Thus the drop at the third tapping point will be

$$
\begin{aligned}
&i_1R_1 + i_1R_2 + i_3R_3 + (i_4 + i_5 + \dots)R_3 \\
&= i_1R_1 + i_2R_2 + I_3R_3
\end{aligned}
$$

and similarly for any other intermediate point. The total drop, reckoning both conductors, will, of course, be twice the above value.

(*b*) Three-wire Distributor Fed from One End. The drops in the two outers are calculated as above. In the case of the neutral, the currents flowing to it from the positive outer are reckoned positive, while the currents flowing from it to the negative outer are reckoned negative. If the total drop in the neutral is positive it is added to the positive drop and deducted from the negative drop; and vice versa.

Example. A three-wire distributor, 300 yd long, is fed at 230 V on each side, and is loaded as follows.

Yards from feeding point, + side	30	100	140	160	210	240
Load	20	40	50	10	25	30
Yards from feeding point, − side	60	80	120	180	260	
Load	30	25	35	60	25	

The outers each have a resistance of 0·15 Ω, and the neutral has a cross section of one-half each outer. Calculate the p.d. at the far end on each side.

Resistance per yard of outer

$$= 0{\cdot}15/300 = 5 \times 10^{-4}\ \Omega$$

∴ Resistance per yard of neutral

$$= 10 \times 10^{-4}\ \Omega$$

Drop in positive outer

$$= (20 \times 30 + 40 \times 100 + \ldots\ 30 \times 240) \times 5 \times 10^{-4}$$
$$= 25{,}650 \times 5 \times 10^{-4} = 12{\cdot}82\ \text{V}$$

Drop in negative outer

$$= (30 \times 60 + 25 \times 80 + \ldots\ 25 \times 260) \times 5 \times 10^{-4}$$
$$= 25{,}300 \times 5 \times 10^{-4} = 12{\cdot}65\ \text{V}$$

Drop along neutral

$$= \{(25{,}650) - (25{,}300)\} \times 10 \times 10^{-4}$$
$$= +\ 0{\cdot}35\ \text{V}$$

∴ Total drop on positive side

$$= 12{\cdot}82 + 0{\cdot}35$$
$$= 13{\cdot}2\ \text{V}$$

∴ p.d. on positive side at far end

$$= 230 - 13{\cdot}2 = 216{\cdot}8\ \text{V}$$

Total drop on negative side

$$= 12{\cdot}65 - 0{\cdot}35$$
$$= 12{\cdot}3\ \text{V}$$

∴ p.d. on negative side at far end

$$= 230 - 12{\cdot}3 = 217{\cdot}7\ \text{V}$$

(*c*) Distributor Fed at Both Ends. The potential of the conductor will gradually fall from one feeding point, say, F_1 (Fig. 13.19), reach a minimum at one of the tapping points, and will rise again as the other feeding point F_2 is approached. There is thus some tapping point M, at which the potential is a minimum. Between F_1 and M all the currents tapped off will be supplied from F_1, while between F_2 and M all the currents tapped off will be supplied from F_2. With regard to the point M itself, the current tapped off there will, in general, come partly from F_1 and partly from F_2, and we can denote these two parts by x and y respectively. If the distributor were cut in two at M, x amperes tapped off at the end served by F_1, and y amperes tapped off at the end served by F_2, then the potential distribution would be unchanged, showing that, from this point of view, we can regard the distributor as consisting of two separate distributors each fed from one

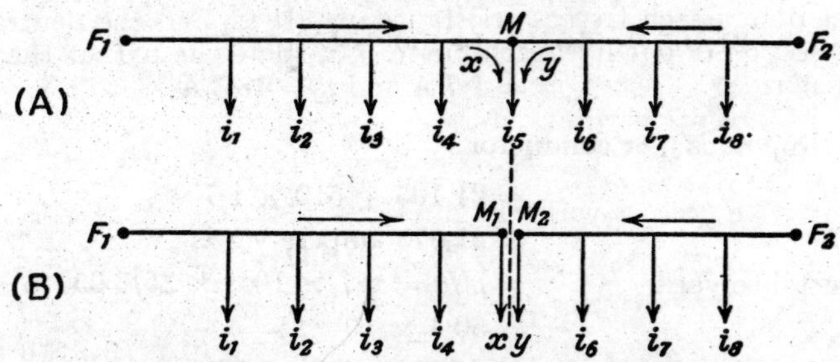

FIG. 13.19. DISTRIBUTOR FED AT BOTH ENDS

end only, as in Fig. 13.19 (*B*). In order to calculate the drop we thus require to locate M, and then calculate x and y. This is done by means of the pair of equations

$$\left.\begin{array}{c} x + y = i_5 \\ \text{Drop from } F_1 \text{ to } M_1 = \text{drop from } F_2 \text{ to } M_2 \end{array}\right\}$$

EXAMPLE. A distributor is fed at both ends. It is a two-core cable half a mile long and of cross section 0·05 in.² The loads tapped off are as shown in Fig. 13.20, the numbers along the distributor indicating

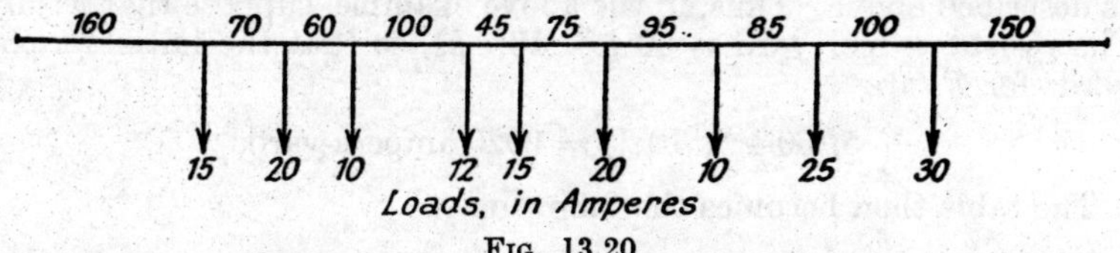

FIG. 13.20

yards. If the two ends F_1 and F_2 are at the same potential, which point will have the lowest potential, and what will be the drop at that point? Take the specific resistance of copper as 0·7 microhm inch.

The best method of locating the point of minimum potential is to take moments about the two ends, and by comparing the sum of these as the calculation proceeds, to make a guess at the point. This is done in the following table—

Moment about F_1 in Ampere-yards	Sums	Moment about F_2 in Ampere-yards	Sums
15 × 160 = 2400	2400	30 × 150 = 4500	4500
20 × 230 = 4600	7000	25 × 250 = 6250	10,750
10 × 290 = 2900	9900	10 × 335 = 3350	14,100
12 × 390 = 4680	14,580	20 × 430 = 8600	22,700
15 × 435 = 6525	21,105		

The table indicates that the sixth tapping point from F_1 is the required point. We then have for the two equations

$$x + y = 20$$

$$21{,}105 + 510x = 14{,}100 + 430y$$

$$\therefore\ x = 1{\cdot}7 \text{ A and } y = 18{\cdot}3 \text{ A}$$

$\therefore$ Drop at M per conductor

$$= 21{,}105 + 510 \times 1{\cdot}7$$

$$= 21{,}973 \text{ ampere-yards}$$

Resistance/yard $= \rho l/a = 0{\cdot}7 \times 10^{-6} \times 36/0{\cdot}05$

$$= 50{\cdot}4 \times 10^{-5}\ \Omega$$

$\therefore$ Drop per conductor $= 2{\cdot}1973 \times 10^4 \times 50{\cdot}4 \times 10^{-5}$

$$= 11{\cdot}1 \text{ V}$$

Hence, the total drop, reckoning both conductors, will be 22·2 V.

We have now to consider the effect of the feeding points F_1 and F_2 not being at the same potential, this condition being, of course, very probable in an actual case. Suppose that there is a difference of v volts between the points F_1 and F_2, F_1 being at the higher potential.

The quickest method is the following. We convert the v volts into ampere-yards; F_2 starts off with this drop below F_1, and this value appears in the column for F_2 as an initial drop. The procedure is then as described above. Thus, in the above example, suppose that v is 5 V. The resistance per yard is $50{\cdot}4 \times 10^{-5}\ \Omega$, so that the initial ampere-yards for F_2 are

$$5/(50{\cdot}4 \times 10^{-5}) = 9920 \text{ ampere-yards}$$

The table then becomes the following—

Moment about F_1	Sums	Moment about F_2	Sums
		Initial = 9920	9920
15 × 160 = 2400	2400	30 × 150 = 4500	14,420
20 × 230 = 4600	7000	25 × 250 = 6250	20,670
10 × 290 = 2900	9900	10 × 335 = 3350	24,020
12 × 390 = 4680	14,580	20 × 430 = 8600	32,620
15 × 435 = 6525	21,105		
20 × 510 = 10,200	31,305		

The dividing point is thus the same. We have

$$x + y = 20$$

and $$21{,}105 + 510x = 24{,}020 + 430y$$

giving $$x = 12{\cdot}2 \text{ and } y = 7{\cdot}8$$

The voltage drops can be calculated as before.

(*d*) Drop in a Uniformly-loaded Distributor

Let i = current tapped off per unit length
and r = resistance per unit length (Fig. 13.21)
∴ Drop at distance x from F

$$= \left(\begin{matrix}\text{Sum of moments}\\ \text{up to } x\end{matrix}\right) + \left(\begin{matrix}\text{Moment of the whole load beyond}\\ x\text{, assumed acting at } x\end{matrix}\right)$$

$$= \int_0^x idx \,.\, rx + i(l - x) \,.\, rx$$

$$= \tfrac{1}{2}irx^2 + irlx - irx^2$$

$$= irlx - \tfrac{1}{2}irx^2$$

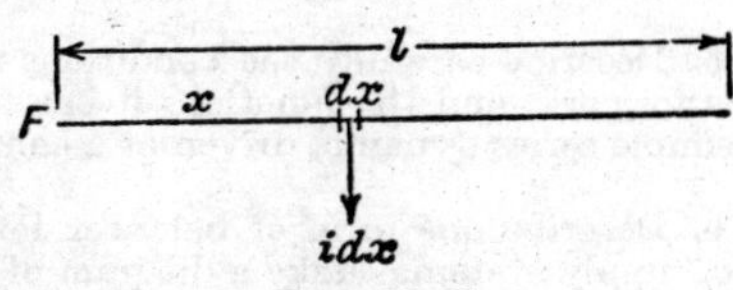

Fig. 13.21

At the far end $x = l$

$$\therefore \text{Drop} = \tfrac{1}{2}irl^2$$

$$= \tfrac{1}{2}(il) \times (rl)$$

$$= \tfrac{1}{2}IR$$

where I is the total current, and R the total resistance.

Thus a uniformly-loaded distributor fed at one end gives a total drop equal to that produced by the whole of the load assumed concentrated at the middle point.

Now let the distributor be fed at both ends. Then with both ends at the same potential the point of minimum potential is obviously the middle point. We can thus imagine the distributor cut into two at the middle point, thus giving two uniformly-loaded distributors fed at one end. The resistance of each is $R/2$, and the total load in each is $I/2$

$$\therefore \text{Drop at middle point} = \tfrac{1}{2}(I/2)(R/2)$$

$$= \tfrac{1}{8}IR$$

Further Reading

Russell, A. *Theory of Electric Cables and Networks.* Constable, London, 1935.

Cotton, H. *Transmission and Distribution of Electrical Energy.* English Universities Press, London, 1937.

Rapson, E. T. A. *Electrical Transmission and Distribution.* Oxford University Press, Oxford, 1933.

Sandford, F. *Electrical Distribution Fundamentals.* McGraw-Hill, New York, 1940.

Examples on Chapter 13

1. 600 h.p. at 600 V is to be delivered at a place 2 miles distant from an electrical generating station. Find the power and voltage of the generator required, if the loss in transmission is to be 14 per cent of the power generated. Also determine the cross section of the cable required. (C. and G.)

Ans. 520 kW at 698 V; 1·28 in.²

2. A three-wire feeder is half a mile long, the two outers having a cross section of 0·25 in.² and the neutral, 0·125 in.² If the voltage at the feeding points is to be 200 V on each side, and the loads on the positive and negative sides are 130 A and 100 A respectively, what voltages must be maintained on the two sides at the generating end?

Ans. 216·8 V and 203·6 V.

3. Calculate the power lost in the above feeder, and the efficiency of transmission. Find the resistance of a two-wire feeder which will transmit 230 A at 200 V and have the same loss as the three-wire feeder. Also compare the amounts of copper in the two- and three-wire feeders.

Ans. 2·55 kW; 94·6%; 0·024 Ω/conductor; 33·7 to 100.

4. What is the "time-constant" of an electric circuit? Explain its importance in direct current working, e.g. in reversible boosters, and in alternating current working. How does it vary with the saturation of the iron of the circuit? (L.U.)

5. Describe carefully the conditions under which a reversible battery booster has to work, and the functions it has to perform. Why is it not possible to use a simple series dynamo, driven at a constant speed, as such a booster?

6. Describe one form of balancer for balancing the voltages on a three-wire d.c. supply system. Make a diagram of connections, showing how it is connected to the mains. (L.U.)

7. Make a diagram of connections showing the necessary machines in the supply station of a three-wire continuous current system. What is the object of a balancer, and how does it act? If the supply voltage on each side of the three-wire system is kept at the same value at a feeding point, what is the effect on the voltage at terminals of lamps on each side, if the load on one side of the system is greater than on the other? (C. and G.)

8. Design a distribution cable, 500 yd long, to be fed with continuous current from both ends to supply the following loads at the stated distances from one end. In no case must the voltage drop exceed the feeder potential difference of 250 V. Specific resistance of copper 0·67 μΩ-in.

Consumer	Current	Yards	Consumer	Current	Yards
1	20	10	6	10	200
2	20	25	7	25	300
3	30	45	8	40	400
4	10	80	9	15	420
5	20	120	10	5	470

(C. and G.) *Ans.* 0·088 in.² cross-section.

9. A direct-current feeder consists of two copper conductors each 0·6 in.² in cross-section and 400 yd long. The average load on the feeder is 1·2 A/yd, and is uniformly distributed. The feeder is supplied with constant-pressure of 230 V at one end. Estimate the voltage between the conductors (*a*) at the centre of the feeder, (*b*) at the far end. (L.U.)

Ans. (*a*) 224 V, (*b*) 221·9 V.

10. A uniformly loaded distributor has a voltage between conductors of 215 V at one end and 210 V at the other: the conductor resistance is 0·08 Ω/yd. The loading is 0·25 A/yd. Find the position of the point at which the conductor current is zero.

Ans. 0·528 of the length reckoned from the 215 V end.

CHAPTER 14

ELECTROLYSIS—SECONDARY CELLS—PRIMARY CELLS

ELECTROLYSIS

EVERY appliance which carries an electric current must have at least one place where the current enters, and another where the current leaves. Between these points, and inside the appliance, the current takes a definite path, and by this statement we mean that electricity is transported along this path. In metals this transportation of electricity is due entirely to the valency electrons of the metal, and these electrons, breaking away from the atoms, as has been explained on p. 65, are able to move under the influence of the applied p.d. A current would also flow (*a*) if only positive electricity were transported along the current path, or (*b*) if both negative and positive electricity were transported at the same time, the movement of the negative electricity being against the potential gradient, and the movement of the positive electricity with the gradient.

We have now to consider how both kinds of electricity can take part in current flow at one and the same time. The negative flow is due to the movement of free electrons, but the positive flow is not due to the movement of free protons, because there are no such things. Consider again the case of a metal, and take a simple atom like that of sodium, which has for its electron array 2 electrons in the K shell, 8 in the L shell and only 1 in the M shell. It is *this lone electron* in the M shell which makes the sodium atom univalent. If it becomes detached from the atom then, under the influence of a potential gradient, it can move against that gradient and so make its contribution to current flow. By the removal of the valency electron the atom becomes an ion, having a positive charge numerically equal to that of one electron. Hence, if the sodium ions were free to move they would travel down the potential gradient and so make their contribution to current flow as well. However, in the metal in its normal state they are unable to do this because, at the most, they can oscillate about a mean position as the result of thermal agitation. Thus, if the sodium ion is to play a part in current flow, it cannot be an ion of metallic sodium in its normal state.

Now consider the important sulphuric acid molecule, H_2SO_4. In the absence of water the structure is

```
O     O—H
 \\  /
   S
 //  \
O     O—H
```

the molecule being completely covalent, that is, the valencies of all the constituent atoms are satisfied. If water is present, as in a water solution, then each molecule of acid reacts with two of water,

$$H_2SO_4 + 2H_2O = SO_4^{--} + H_3O^+ + H_3O^+$$

Thus the SO_4, sulphion group, is negatively charged with charge of magnitude $2e$, while each H_3O group is positively charged, of magnitude e, where e is the charge on the electron.

Now suppose that a solution of H_2SO_4 is placed in a container and conductors are immersed at either end of the container, the conductors being connected to a d.c. supply. The conductors are called electrodes, that connected to the positive side of the supply being the anode, and that connected to the negative side the cathode. The sulphion group SO_4^{--} is a negative ion and the "hydrated protons" H_3O^+ positive ions. There must clearly be an electrostatic attraction between positive and negative ions, but, as they are immersed in a medium (water) which has the extremely high permittivity of $\varepsilon = 81$, this force is extremely

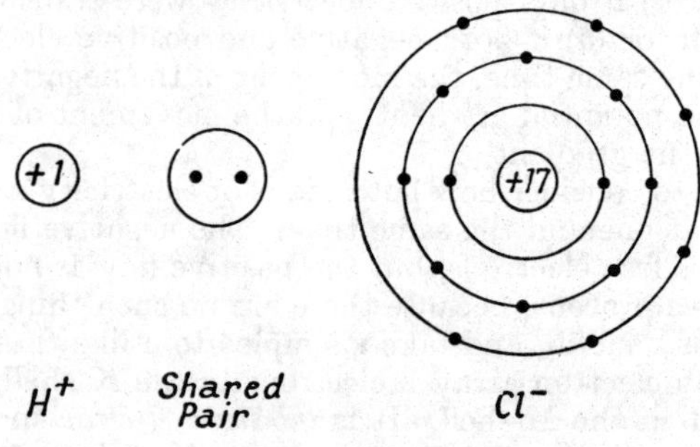

Fig. 14.1

small. Hence, under the influence of the potential gradient the positive ions move towards the cathode and the negative ions towards the anode. Thus there is current flow within the solution, or electrolyte, but it is now due to the movement of both negative and positive charges.

Before examining what takes place at the electrodes when the ions reach them consider a few more molecules. When two atoms combine to form a molecule a pair of electrons is shared by the participating atoms, one from one atom the other from the other atom. Thus, in the simple molecule of hydrochloric acid, HCl, the hydrogen atom has one electron only, the chlorine atom has the K and L shells completed, namely 2 and 8 electrons respectively, but it has 7 in the M shell, that is, one short of the full complement 8. In combining to form a molecule the two atoms share an electron somewhat as depicted in Fig. 14.1. In a solution in water the reaction is

$$HCl + H_2O = H_3O^+ + Cl^-$$

When ammonia, NH_3, is dissolved in water the reaction is

$$NH_3 + H_2O = NH_4^+ + OH^-$$

When copper sulphate is dissolved in water the reaction is

$$CuSO_4 = Cu^{++} + SO_4^{--}$$

Phenomena at the Electrodes. Since the flow of electricity is a continuous process with respect to the circuit as a whole it follows that if the motions of the positive and negative ions in the electrolyte constitute a current of I amperes then, since we are not considering any storage of electricity, this same current must be derived from the source, and must flow in the connecting wires. In these wires the current is due solely to the axial drift of electrons. Considering the case of a

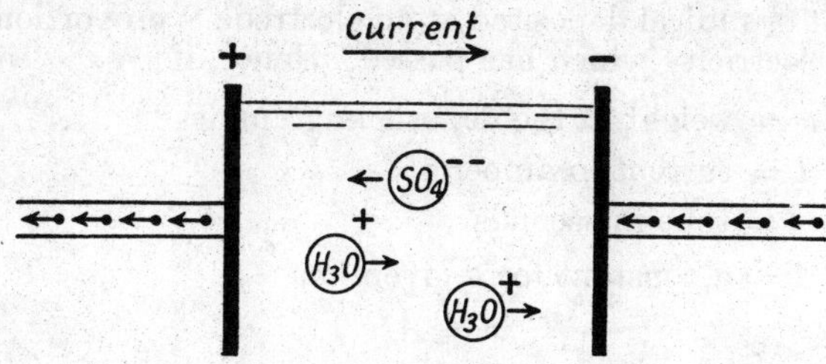

FIG. 14.2

solution of H_2SO_4, the picture is therefore that of Fig. 14.2. If the anode is inert, e.g. a carbon plate, then, when SO_4^{--} reaches it the reactions are as follows—

$$2SO_4^{--} + 2H_2O = 2H_2SO_4 + O_2 + 4e \quad . \quad . \quad (1)$$

$$2H_2SO_4 + 4H_2O = 4H_3O^+ + 2SO_4^{--} \quad . \quad . \quad (2)$$

The first equation shows that, for each SO_4^{--} ion reaching the anode, a molecule of oxygen is liberated, and also a charge of $2e$. This released charge maintains the potential of the anode, since it compensates for an equal charge on two free electrons which enter the connecting wire and so maintain the external current. Two sulphuric acid molecules so formed, together with additional water molecules, form additional SO_4^{--} and H_3O^+ ions. At the cathode, we have, remembering that there are two H_3O^+ ions to one SO_4^{--}, and writing the equation to correspond with equation (1) above

$$4H_3O^+ = 4H_2O + 2H_2 - 4e$$

the negative sign being used since the ionic charge is positive and therefore opposite to the electronic charge e of equation (1). Comparing this equation with equation (1) we see that there is a balance, and the electrolyte remains electrically neutral. Also this charge of $-4e$ compensates for the charge $4e$ on four electrons approaching the cathode from the outside circuit.

An important case is that of a solution of copper sulphate, $CuSO_4$, with copper electrodes. The ions are now Cu^{++} and SO_4^{--}. When the Cu^{++} ions reach the cathode there is a liberation of a charge $-2e$ and the ions become atoms of metallic copper which are deposited on the cathode. When the SO_4^{--} ions reach the copper anode the reaction is

$$SO_4^{--} + Cu = Cu^{++} + SO_4^{--} + 2e$$

the two electrode reactions thus being equal and opposite, as before. The Cu^{++}, being an ion, can enter the electrolyte, and thus there is as much copper removed from the anode as is deposited on the cathode.

Quantitative Laws of Electrolysis. The phenomena of electrolysis were thoroughly investigated by Faraday, who found that the weight of an element or radical deposited at an electrode is proportional to the quantity of electricity which has passed. Hence, if

w = weight of the deposit in grammes

I = current in amperes

t = time in seconds

A = a constant for a given ion

then
$$w = AIt$$

The constant A is composite, its value depending upon the atomic weight and the valency. It is obvious that for the electrolytic deposition of metals, or the liberation of gases, w will be proportional to the atomic weight, because each ion carries a definite quantity of electricity. Now the charge carried by an ion is proportional to its valency, for consider molecules of sulphuric acid and copper sulphate. The sulphion in the acid molecule is combined with two hydrogen atoms, whereas it is only combined with one copper atom in the copper sulphate molecule. From this it follows that the copper atom, which is divalent, has, when in the form of a copper ion, twice the charge of a hydrogen ion, which is univalent. In other words, in order to carry a given quantity of electricity, only one-half as many divalent copper ions as univalent hydrogen ions will be required. Hence, the greater the valency of the ion, the smaller the number of ions required to carry a given quantity of electricity, from which it follows that the weight deposited is inversely proportional to the valency. The equation can therefore be written

$$w = B(a/v)It$$

where a = atomic wt.; v = valency; B = another constant.

The constant B is equal to the number of grammes of hydrogen deposited by the passage of 1 coulomb (C) of electricity, namely, to 0·00001038 g/C. The quantity Ba, or $a \times 0{\cdot}00001038$, gives the number of grammes of a univalent substance of atomic weight a deposited by one coulomb. Thus, in the case of silver, the atomic weight is 107·6 and the weight/coulomb is $107{\cdot}6 \times 0{\cdot}00001038 = 0{\cdot}001118$. The quantity Ba/v gives the number of grammes of a substance of any valency v deposited by 1 C. Thus for divalent copper $a = 63{\cdot}2$, $v = 2$, and therefore, $Ba/v = 0{\cdot}000328$ g/C.

The quantity Ba/v is called the *electro-chemical equivalent*; hence, denoting it by z, we have

$$w = zIt$$

The quantity a/v is the *chemical equivalent*.

Polarization. The relationship between the current I through an electrolytic cell, and the applied p.d., V, is given by an equation of the form

$$V = \varepsilon + RI$$

where ε is the back e.m.f. or *polarization* e.m.f. of the cell, and RI is the ohmic drop due to its internal resistance. In the case of a cell in which the electrolyte is a salt of the metal of the electrodes, e.g. $CuSO_4$ with copper electrodes, the back e.m.f. is set up by concentrations of electrolyte round the electrodes. If the electrodes are not acted on by the electrolyte, e.g. platinum electrodes in a solution of H_2SO_4, it is due to deposits of gas. These concentrations of electrolyte, or deposits of gas, prevent the fresh arrivals of ions from reaching the electrodes, the result being that the ions are unable to give up their charges, and so set up a back e.m.f.

Multiplying both sides of the above equation by I, we have the energy equation

$$VI = \varepsilon I + RI^2$$

The term RI^2 represents the power used in producing heat, and the term εI represents the power required to effect the chemical changes taking place. The energy absorbed in effecting the chemical changes during a time t is, therefore, εIt, if I has remained constant. Otherwise, $\varepsilon\int Idt$. Consider the case of a dilute sulphuric acid electrolyte with platinum electrodes. Mixed oxygen and hydrogen will be liberated, and the potential energy of this mixture will of necessity be equal to the energy used in liberating it, e.g. to εIt. Now when one gramme of these mixed gases is exploded, 3780 g.cal of heat are evolved. The equivalent in watt-seconds or joules is $3780 \times 4{\cdot}2 = 15{,}870$. Again, the passage of 1 C of electricity liberates 0·00001038 g of hydrogen and $0{\cdot}00001038 \times 16/2$ g of oxygen; that is, $9{\cdot}35 \times 10^{-5}$ g of mixed gases. The quantity of electricity required to liberate 1 g of mixed gases is therefore $10^5/9{\cdot}35 = 1{\cdot}07 \times 10^4$ C. But we saw that the product εIt was equal to 15,870. Hence, for the back e.m.f. ε, in the case of such a cell, we have

$$\varepsilon = 15{,}870/(1{\cdot}07 \times 10^4) = 1{\cdot}47 \text{ V}$$

Hence, to electrolyze dilute sulphuric acid, or any electrolyte in which mixed oxygen and hydrogen are evolved, an applied p.d. greater than 1·47 V is necessary.

Secondary and Primary Cells

Accumulators. In the previous section we saw that an electrolytic cell could set up a back e.m.f. if there was a concentration of electrolyte round the electrodes, or if there was an accumulation of gas on the electrodes. This back e.m.f. can also exist if the electrodes themselves are changed chemically. The most striking example of this phenomenon occurs with lead electrodes in a dilute solution of sulphuric acid. The acid molecules split up into hydrogen and sulphion ions, associated with positive and negative charges respectively. On the application of a potential difference to the electrodes the hydrogen ions travel towards the cathode, where they give up their charges and are liberated as free hydrogen. The sulphion ions give up their charges at the anode, and they then have a choice of combining with either the lead of the anode, or the hydrogen of the water. Both reactions take place at the same time, but the most important is the combination with hydrogen of the water, so as to liberate oxygen. This oxygen at the moment of liberation is in the nascent, i.e. atomic form, and it therefore attacks the lead of the anode and forms in time a brown deposit of lead peroxide, PbO_2. On stopping the current and connecting a voltmeter to the electrodes it will now be found that the cell has an e.m.f. of 2 V. If the electrodes are joined by a wire the cell will produce a current which will flow through the electrolyte in the reverse direction to the previous current, and the cell will discharge. The following changes then take place. The hydrogen ions, carrying positive charges, now move towards the positive plate, which, as we have seen, is coated with PbO_2. After liberation of the charge, the hydrogen atoms combine with the oxygen of the PbO_2, while the acid attacks the lead, with the formation of $PbSO_4$.

$$PbO_2 + H_2 + H_2SO_4 = PbSO_4 + 2H_2O$$

At the negative plate, the sulphion ions, after losing their charge, combine with the lead to form $PbSO_4$.

$$Pb + SO_4 = PbSO_4$$

Thus if the cell is completely discharged, both plates will be coated with a whitish coating of lead sulphate. If the discharged cell is now charged again by passing current from an external source through it from positive to negative plate, the following changes will take place—

$$\text{At the negative plate } PbSO_4 + H_2 = Pb + H_2SO_4$$

$$\text{At the positive plate } PbSO_4 + SO_4 + 2H_2O = PbO_2 + 2H_2SO_4$$

From this it will be seen that on re-charging, the positive plate will re-acquire its brown coating of lead peroxide, while the negative plate will be reduced to grey metallic lead.

The capacity of such a cell, that is, the number of ampere-hours that can be obtained from it when discharging, will naturally be very small, because of the small surface exposed to the acid. With repeated charging and discharging, the lead plates will become "spongy," and will present a much greater effective surface to the acid, the result being that the capacity will be greatly increased. This process of producing accumulator plates is called "forming"; it is no longer used because of its high cost. A cheaper method is to use plates in the form of cast lead grids, the holes in which are filled with a paste of lead oxide. If red lead, Pb_3O_4, is used for the positive, and litharge, PbO, for the negative, both plates can be formed at the same time by immersing them in dilute sulphuric acid and passing current. The positive plate becomes further oxidized to PbO_2, while the litharge is reduced to metallic lead. Such plates are not mechanically strong, and the paste is liable to fall out if the cell is severely used. In more modern

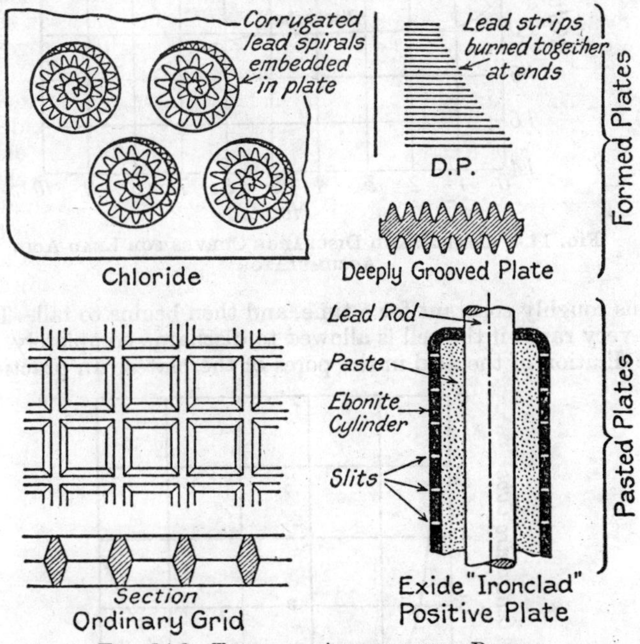

FIG. 14.3. TYPES OF ACCUMULATOR PLATES

processes the plates are formed by a very quick process. Thus, boiling in nitric acid to oxidize the lead, and then reducing the oxide electrically are possible. It is usual to perform these two operations together by adding "quick formation agents" to the electrolytic bath. These agents, of which nitrates, chlorates, and acetates are examples, form unstable compounds with the lead, these compounds being immediately formed into PbO_2 by the action of the current.

Various types of formed and pasted plates are illustrated in Fig. 14.3.

When a cell is charged the voltage gradually rises, remains in the neighbourhood of 2·3 V for some time, and then rises again, as shown by Fig. 14.4. At the same time a large quantity of gas is evolved. The final rise is due to a concentration of sulphuric acid in the pores of the positive plates. If the charged cell is now left standing this strong acid diffuses into the rest of the solution, and the voltage of the cell falls to about 2·2 V. If now the cell is discharged, the p.d. at the terminals

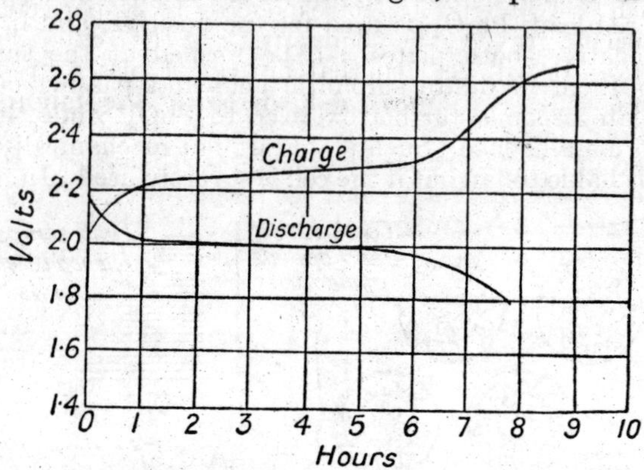

FIG. 14.4. CHARGE AND DISCHARGE CURVES FOR LEAD-ACID ACCUMULATOR

remains roughly constant for a time, and then begins to fall. The final fall is very rapid if the cell is allowed to discharge completely, because of the dilution of the acid in the pores of the plates. In practice a cell

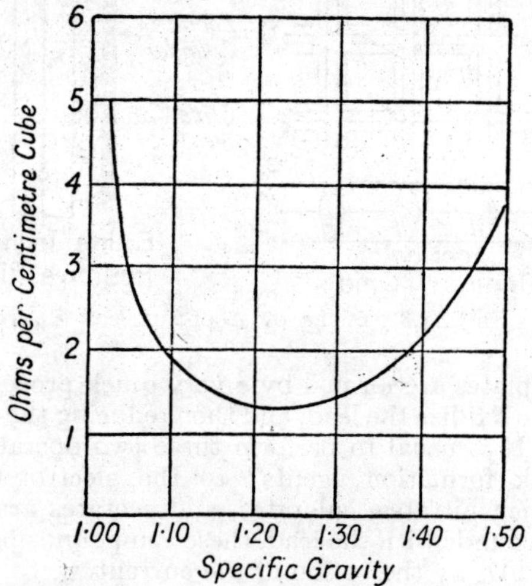

FIG. 14.5. VARIATION OF CONDUCTIVITY AND SPECIFIC GRAVITY

should not be allowed to discharge below 1·8 V, otherwise there will be an excessive formation of sulphate which will be difficult to remove. Also, the plates may tend to buckle and the paste become loosened. For this latter reason the charge and discharge currents should not be excessive, since too rapid chemical action is the main cause of buckling.

The specific gravity of the acid is 1·21 when fully charged, corresponding to 28 per cent acid by weight. When discharged to 1·8 V the s.g. is much less, about 1·18, corresponding to 25 per cent acid. With varying specific gravity, the conductivity of the acid varies, as shown in Fig. 14.5. It will be seen that an s.g. of just over 1·2 gives the minimum specific resistance, and therefore, the minimum internal resistance of the cell.

As stated before, the capacity of a cell is reckoned by the number of ampere-hours on discharge at the normal discharge rate. If the discharge rate is less than normal, the capacity will be greater, but with an excessive discharge rate the capacity will be very much reduced, as

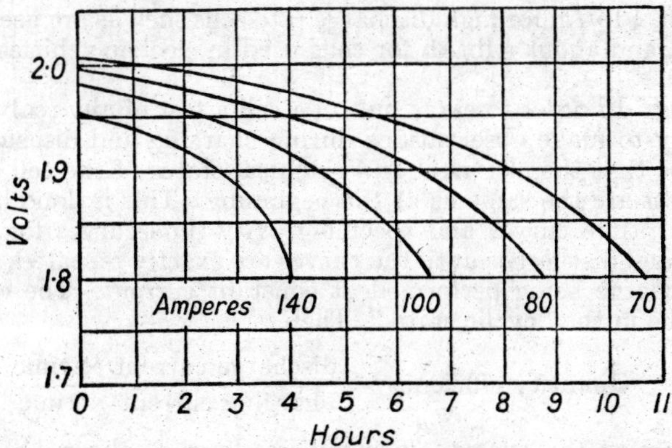

FIG. 14.6. VARIATION OF CAPACITY WITH DISCHARGE RATE

shown by Fig. 14.6. For a pasted positive plate the average capacity is about $\frac{1}{2}$ ampere-hour (Ah) for each square inch of positive surface (reckoning one side); the normal discharge rate, about $\frac{1}{6}$ A/in.^{2} of positive surface. Temperature has a great effect on the action of a cell, rise of temperature increasing the e.m.f. slightly, and the capacity, very considerably. Thus, a rise of temperature of 30°C will increase the capacity at normal discharge rate by 30 per cent.

The efficiency of a cell can be reckoned in two ways—

$$\text{Quantity efficiency} = \frac{\text{ampere-hours of discharge}}{\text{ampere-hours of charge}}$$

$$\text{Energy efficiency} = \frac{\text{watt-hours of discharge}}{\text{watt-hours of charge}}$$

If V is the terminal p.d. per cell, E_1 the e.m.f. per cell on charge, and E_2 the e.m.f. on discharge, then

$$V = E_1 + RI$$
$$V = E_2 - RI$$

The average terminal p.d. during charge is therefore greater than during discharge, the difference being twice the internal drop RI, plus the quantity $(E_1 - E_2)$. The e.m.f. during charging is greater than during discharging, because of the presence of the gases evolved. It therefore follows that the energy efficiency is less than the quantity efficiency. Increase in discharge rate, as we have seen, reduces the capacity and therefore the quantity efficiency to a certain extent, but it decreases the energy efficiency still further, since the greater the discharge rate the smaller the terminal p.d. of the cell. Average values are 90 per cent for the quantity efficiency and 75 per cent for the energy efficiency.

Cells of the lead-acid type weigh about $\frac{1}{2}$ lb/Ah for ordinary construction, $\frac{1}{3}$ lb/Ah for high discharge rate cells such as are used in power stations, and about $\frac{1}{5}$ lb/Ah for cells used in electric vehicles.

Testing. In order to carry out on a cell a test of any real value, it is necessary to make observations during charging and discharging, and to ensure that the electrical and chemical states of the cell at the end of the test are the same as at the beginning. This is done by plotting the e.m.f.-time curves and specific-gravity-time curves for successive charges and discharges until the curves are exactly repeated, the charge and discharge being performed at constant current. The cell is then said to be in the "cyclic state." Then

$$\text{Quantity efficiency} = \frac{\text{discharge current} \times \text{time}}{\text{charging current} \times \text{time}}$$

To determine the energy efficiency, readings of the p.d. have to be taken at frequent intervals and plotted against time. The area of the curve gives the volt-hours, which, when multiplied by the constant current of charge or discharge, gives the watt-hours.

$$\text{Energy efficiency} = \frac{\text{discharge current} \times \text{volt-hours of discharge}}{\text{charge current} \times \text{volt-hours of charge}}$$

If E_o is the open-circuit e.m.f. of a cell, and V_1, the terminal p.d. when delivering a current I, then

$$E_o - V_1 = \text{drop of volts in cell} = RI$$

$$\therefore \text{ Resistance of cell } R = (E_o - V_1)/I$$

In this way the resistance of a cell can be determined, but the method is not very accurate because $(E_o - V_1)$ is small. A better method is to balance the e.m.f. of the cell, C in Fig. 14.7, against the drop along a potentiometer AB, using a millivoltmeter (M.V.) in place of a galvanometer. The cell is then discharged at, say, I amperes through an adjustable resistance Rh, the millivoltmeter then giving directly the drop in the cell. Hence

$$R = (\text{reading of millivoltmeter})/(\text{current})$$

The Alkaline Accumulator. The active materials in this cell are nickel oxide on the positive, and iron oxide on the negative, plate. The electrolyte is a 21 per cent solution of caustic potash, KOH, with the addition of a small quantity of lithium hydrate. The action of the latter is not yet understood, but it very materially increases the capacity of the cell. The exact formula of the nickel oxide is not yet established, but the action of the cell can be followed by assuming the peroxide NiO_2, or its hydrated form $Ni(OH)_4$. On discharge the OH ions of the KOH travel to the negative, the iron therefore becoming oxidized. The K ions travel to the positive and reduce the $Ni(OH)_4$ to $Ni(OH)_2$. During charge the converse action takes place, so that the action can be represented by the reversible equation

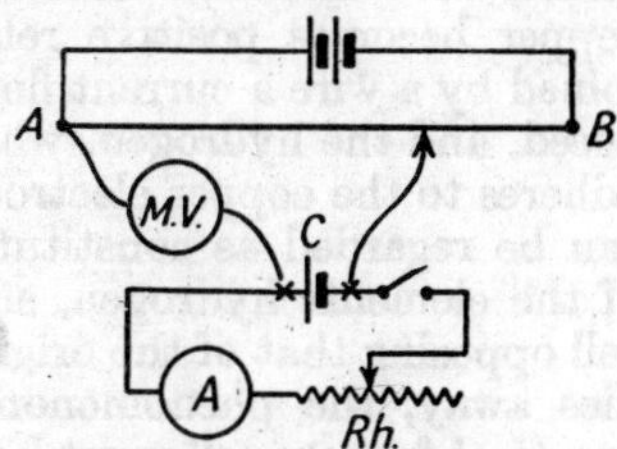

FIG. 14.7. DETERMINATION OF INTERNAL RESISTANCE

$$Ni(OH)_4 + KOH + Fe \rightleftharpoons Ni(OH)_2 + KOH + Fe(OH)_2$$

It will be noticed that the electrolyte acts merely as a vehicle for the transfer of OH from one plate to another, and it does not take part in any chemical change. As a result, the density does not change to the same extent as in the ordinary lead-acid cell.

The positive plate consists of a number of tubes made of perforated steel ribbon, wound spirally, and held together by steel rings. The tubes are very heavily nickel plated, packed with alternate layers of nickel hydroxide and flake nickel, and then clamped in a steel frame, which is also nickel plated. The flake nickel is added because the hydroxide is rather a poor conductor. The negative plate is made from finely perforated nickelled steel strip stamped into pockets, the pockets being filled with iron oxide. Here also the conductivity is not very good, and it is improved by adding a little mercury. In order to render the iron oxide susceptible to chemical changes in the cell, it is prepared by alternate reduction and oxidation of iron sulphide under KOH. The plates are separated from one another by hard rubber strips, and are held in a nickel-plated sheet steel container with welded seams.

The normal density of the electrolyte is 1·22, and this falls to about 1·16 during the first year's use. A fresh electrolyte of density 1·25 is then added, this higher density being to compensate for the old weak solution left in the plates.

The e.m.f. per cell when fully charged is 1·4 V, and it is usually allowed to fall to 1·1 V before recharging. Actually, the cell can be completely discharged, or even short-circuited, without damage, but it is not economical to allow the e.m.f. to fall below about 1 V. The efficiencies are lower than for lead-acid cells, average values being 80 per cent for the quantity efficiency and 60 per cent for the energy efficiency. The weight of the cell is about 1 lb for each 10 Ah.

The effect of an increase in temperature is to lower the e.m.f. slightly but to increase the capacity, and there is a critical temperature of 53°F below which the capacity falls off rapidly.

Primary Cells. The simplest cell consists of a vessel of dilute sulphuric acid in which electrodes of copper and zinc are placed. The copper becomes positive relatively to the zinc, and if the two are joined by a wire a current flows. The sulphuric acid is thereby decomposed, and the hydrogen, which travels in the direction of the current adheres to the copper electrode in the form of small bubbles. This gas can be regarded as constituting the positive of a new cell consisting of the elements hydrogen, sulphuric acid, and zinc, the e.m.f. of this cell opposing that of the original cell. The e.m.f. of the latter therefore dies away, the phenomenon being that of polarization. Since any practical form of cell must have a reasonably constant e.m.f., polarization must be prevented. The problem is thus the prevention of the formation of free hydrogen. There are two methods: a metal can be deposited on the cathode instead of hydrogen, or the hydrogen can be combined with some other substance the moment it is evolved.

The first method is used in the Daniell cell, in which two liquids, copper sulphate and dilute sulphuric acid, separated by a porous pot, are used. The positive plate, of copper, is immersed in the copper sulphate, and the negative plate, zinc, in the dilute acid. When current flows, the direction inside the cell is from zinc to copper, the hydrogen thus travelling to the copper sulphate solution. Here it deposits copper and forms sulphuric acid. The SO_4 ion attacks the zinc and forms zinc sulphate. This change is an oxidation and it supplies an amount of energy equal to the electrical energy supplied by the cell, plus the energy required to precipitate the copper. The electrodes are not changed, and therefore there is no polarization, the e.m.f. remaining constant at 1·07 V.

The second method is made use of in the Leclanché cell. The electrolyte is in this case a solution of ammonium chloride, NH_4Cl, in which a zinc rod (the negative electrode) is dipped. When the cell delivers current, zinc chloride, ZnCl, is formed, and the NH_4 group acts as the carrier of positive electricity. The positive electrode is a carbon plate placed in a porous pot and packed round with a mixture of manganese dioxide and carbon. This mixture becomes permeated with the electrolyte, and the NH_4 ions are able to travel to the carbon plate. There they are split up into ammonia and hydrogen gas, the latter being oxidized to water by the magnanese dioxide. The removal of hydrogen in this way is necessarily slow, and therefore the cell is only suitable for intermittent use. The energy in this cell is supplied by the formation of zinc chloride.

Further Reading

CRENNELL, J. T. and LEA, F. M. *Alkaline Accumulators.* Longmans, London, 1928.

CROSS, H. H. V. *The Battery Book.* Technical Press, London, 1937.

HUGHES, G. B. *Dry Batteries.* Hutchinson, London, 1940.

VINAL, G. W. *Storage Batteries.* Wiley, New York, 1940.

BROWN, H. G. *The Lead Storage Battery.* Locomotive Publ. Co., London, 1937.

FIELD, S. *Principles of Electrodeposition.* Pitman, London, 1949.

MANTELL, C. L. *Industrial Electrochemistry.* McGraw Hill, New York, 1940.

CHAPTER 15

A.C. CIRCUITS

AN alternating quantity is one which acts in alternate positive and negative directions, whose magnitude undergoes a definite series of numerical changes in definite intervals of time, and in which the sequence of changes while negative is identical with the sequence while

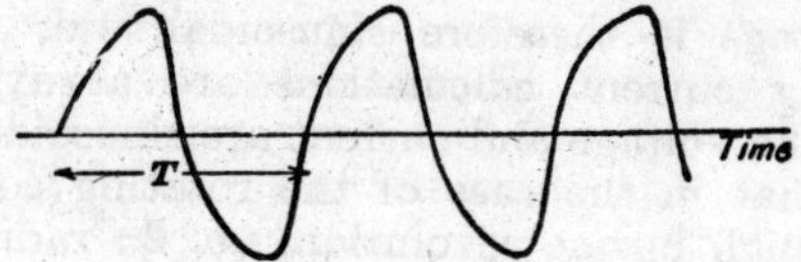

FIG. 15.1. GRAPH OF AN ALTERNATING QUANTITY

positive. The graph of such a quantity is shown in Fig. 15.1. It will be seen that the graph repeats after regular time intervals, and one repeat is called a complete "cycle." The time T of one cycle is called

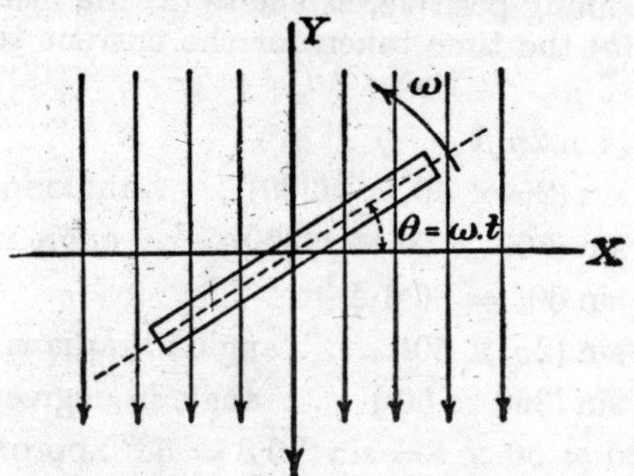

FIG. 15.2. A SIMPLE ALTERNATOR

the "periodic time," and the number of cycles per second, the "frequency," f. Hence,

$$T = 1/f \text{ and } f = 1/T$$

The Simple Alternator. Consider a coil rotating with angular velocity ω radians per second, as shown in Fig. 15.2, in a uniform magnetic field, and let time be reckoned from the instant the plane of the coil includes the OX axis. In this position the flux Φ linking with the coil has its maximum value Φ_{max}. Hence, if the linkage is expressed as a function of the time t,

$$\Phi N = \Phi_{max} N \cos \omega t,$$

N being the number of turns on the coil. The induced e.m.f. in the coil is equal to minus the rate of change of linkage. Denoting the instantaneous e.m.f. by e,

$$e = -(d/dt)(\Phi_{max} N \cos \omega t) \text{ volts}$$
$$= (\Phi_{max} N \omega) \sin \omega t$$

Obviously the quantity in the brackets is the maximum value of the e.m.f., E_{max}, and we can therefore write

$$e = E_{max} \sin \omega t$$

The induced voltage is therefore sinusoidal and, unless otherwise stated, alternating current calculations are always made on the assumption that the voltage and current are sinusoidal.

It is obvious that in the case of the rotating coil, one complete cycle is gone through in one revolution, i.e. 2π radians. Hence, the equation for the instantaneous voltage can also be written in the form

$$e = E_{max} \sin 2\pi ft$$

EXAMPLE. An alternating current of frequency 50 c/s has a maximum value of 100 A. Reckoning time from the instant the current is zero and is becoming positive, calculate (*a*) the instantaneous value after 1/300 sec; (*b*) the time taken for the current to reach 80 A for the first time.

(*a*) $i = I_{max} \sin 2\pi ft$

$= 100 \sin (2\pi \times 50 \times 1/300)$ . . . angle in radians

$= 100 \sin (360 \times 50 \times 1/300)$ . . . angle in degrees

$= 100 \sin 60° = 86{\cdot}6$ A

(*b*) $80 = 100 \sin (2\pi \times 50t)$ . . . angle in radians

$= 100 \sin (360 \times 50t)$ . . . angle in degrees

$\therefore 360 \times 50 \times t = \sin^{-1} 0{\cdot}8 = 53°$ approx.

$\therefore t = 53/(360 \times 50) = 0{\cdot}00295$ sec

Effective Value. The effective value of an alternating e.m.f. or current is given by that direct e.m.f. or current which, when applied to a given circuit for a given time, produces the same expenditure of energy as when the alternating e.m.f. or current is applied to the same circuit for the same time.

Consider an alternating current of any wave form (Fig. 15.3). Divide the base into a large number, n, equal intervals, each of (T/n) seconds, and let the mid-ordinates be i_1, i_2, i_3, etc. Let this current be flowing through a resistance of R ohms.

Then energy expended in 1st interval $= i_1^2 R \times (T/n)$ joules

" " 2nd " $= i_2^2 R \times (T/n)$ "

" " 3rd " $= i_3^2 R \times (T/n)$ "

etc.

Hence, total energy expended in time T

$$= T \times R[(i_1^2 + i_2^2 + i_3^2 + \ldots)/n]$$

Now let I be the effective current; then the energy expended in time

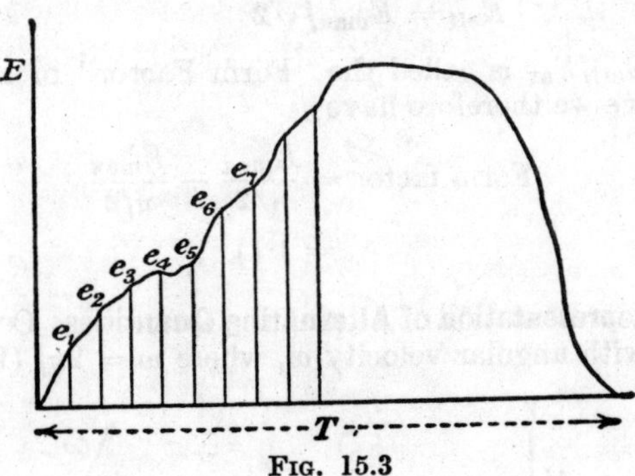

FIG. 15.3

T will be I^2RT joules. By definition these two expressions for the energy are equal

$$\therefore\ I^2RT = TR[(i_1^2 + i_2^2 + i_3^2 + \ldots)/n]$$

$$\therefore\ I = \sqrt{[(i_1^2 + i_2^2 + i_3^2 + \ldots)/n]}$$

Hence, the effective value is equal to the square root of the mean of the squares of successive ordinates. It is, therefore, sometimes called the "root mean square" (r.m.s.) value. It is also called the "virtual" value.

Average and Effective Values of a Sinusoidal Wave. We have

$$e = E_{\max} \sin \theta$$

where $\theta = \omega t$. A half-cycle is completed when θ varies from 0 to π radians.

Hence, average value

$$E_{av} = \frac{1}{\pi}\int_0^{\pi} E_{\max} \sin \theta d\theta$$

$$= -\frac{E_{\max}}{\pi}\Big[\cos\theta\Big]_0^{\pi}$$

$$= 2E_{\max}/\pi \text{ or } E_{\max}/(\pi/2)$$

Again,

$$E^2_{eff} = \frac{1}{\pi}\int_0^{\pi} E^2_{\max} \sin^2 \theta d\theta$$

$$= \frac{E^2_{\max}}{2\pi}\int_0^{\pi} (1 - \cos 2\theta)\, d\theta$$

$$= \frac{E^2_{max}}{2\pi}\left[\theta - \tfrac{1}{2}\sin 2\theta\right]_0^\pi$$

$$= \tfrac{1}{2} E^2_{max}$$

$$\therefore\ E_{eff} = E_{max}/\sqrt{2}$$

The ratio E_{eff}/E_{av} is called the "Form Factor" of a wave. For a sinusoidal wave we therefore have

$$\text{Form factor} = \frac{E_{max}}{\sqrt{2}} \div \frac{E_{max}}{\pi/2}$$

$$= 1{\cdot}11$$

Graphical Representation of Alternating Quantities. Consider a vector OP rotating with angular velocity ω, where $\omega = 2\pi f$ (Fig. 15.4). Let

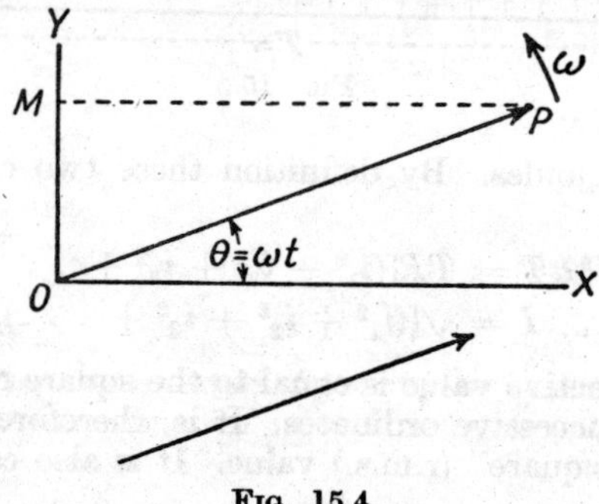

FIG. 15.4

the length of OP be E_{max}. Then at any instant the intercept on the Y axis will be

$$OM = OP \sin \omega t$$

$$= E_{max} \sin \omega t$$

Hence, $OM = e$, the instantaneous voltage.

Now the axes OX and OY are fixed in space, and it is not necessary to include them in the diagram: we are thus left simply with the vector OP. Finally, since the effective value E_{eff} bears a definite relationship to the maximum value, namely, $E_{max}/\sqrt{2}$, the length of the vector can be made equal to the effective value if desired.

We thus see that an alternating quantity which varies in magnitude sinusoidally with respect to time can be represented by a rotating vector of length equal to the maximum value. Its orientation in space at any instant is such that the length of its projection on to a fixed OY axis is equal to the instantaneous value at that particular instant. The r.m.s., or effective value can also be represented by a vector, but in this case, the vector has magnitude and direction, like a force vector for example, but it is not a rotating vector.

Phase Difference. Suppose two alternating quantities are represented at a given instant by two vectors, OP and OQ (Fig. 15.5). Then the angle ϕ between them is called the "phase" angle. With reference to the direction of rotation of the vectors, OP is in front of OQ, and OP is said to "lead" OQ, while OQ is said to "lag" behind OP. If the two quantities are also represented by wave diagrams, as shown in Fig. 15.10, the phase difference ϕ is given by the relative displacement of the curves along the angle, or time, axis. With reference to such a diagram the leading quantity is that which goes through its zero or crest value the first. The lag or lead obviously depends upon the assumed direction of rotation of the vectors, and in all the vector diagrams following the direction is counter-clockwise.

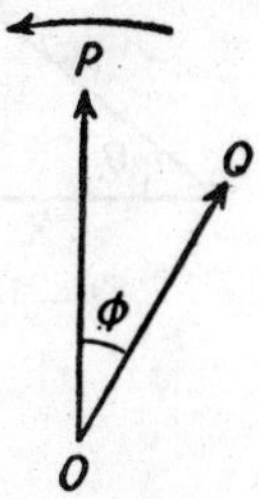

Fig. 15.5. Phase Difference

If the vectors are voltage vectors and the leading vector is represented by

$$e_1 = OP \sin \omega t$$

then the lagging vector will be represented by

$$e_2 = OQ \sin (\omega t - \phi)$$

If a number of alternating voltages act at the same time in a given circuit their resultant is given by their vector sum, just as the resultant of a number of forces acting at a point is given by the vector sum. The resultant voltage *at any instant* is given by the algebraic sum of the instantaneous values. This algebraic sum is obviously the same as the algebraic sum of the projections of all the vectors on the Y axis.

The Addition of Alternating E.M.F.s, or Alternating Currents. In the great majority of alternating current circuits it is necessary to consider the combined action of several e.m.f.s, all acting in the circuit at the same time, or in the case of a branched circuit, the action of several currents. One method of determining the resultant of a number of alternating e.m.f.s or currents is to regard their vectors as forces and to make the calculations exactly as in the case of a similar problem dealing with the effect of a number of forces acting at a point. For example, if there are two e.m.f.s acting at the same time, their equations being

$$e_1 = E_1 \sin \omega t$$

and

$$e_2 = E_2 \sin (\omega t + \phi)$$

the vectors are of lengths E_1 and E_2 (maximum values) and their phase displacement is ϕ, so that their resultant is given by

$$E^2_{\max} = E_1^2 + E_2^2 + 2E_1E_2 \cos \phi$$

$$\therefore E_{\text{eff}} = 0{\cdot}707 \times \sqrt{(E_1^2 + E_2^2 + 2E_1E_2 \cos \phi)}$$

Alternatively, we have, dealing in instantaneous values,

$$\begin{aligned} e = e_1 + e_2 &= E_1 \sin \omega t + E_2 \sin (\omega t + \phi) \\ &= E_1 \sin \omega t + E_2 (\sin \omega t \cos \phi + \cos \omega t \sin \phi) \\ &= (E_1 + E_2 \cos \phi) \sin \omega t + E_2 \sin \phi \, . \cos \omega t \end{aligned}$$

This shows how we can resolve the total e.m.f.s into two sinusoidal components in quadrature with one another, the magnitudes of these two components (maximum values) being $(E_1 + E_2 \cos \phi)$ and $E_2 \sin \phi$ respectively. Denoting these by X and Y respectively, we see that they will form the base and perpendicular of a right-angled triangle (Fig. 15.6). Let the base angle of this triangle be θ, then

FIG. 15.6

$$\sin \theta = Y/(X^2 + Y^2)^{\frac{1}{2}}$$

and $\cos \theta = X/(X^2 + Y^2)^{\frac{1}{2}}$

and $e = X \sin \omega t + Y \cos \omega t$

$$= (X^2 + Y^2)^{\frac{1}{2}}\left\{\frac{X}{(X^2 + Y^2)^{\frac{1}{2}}} . \sin \omega t + \frac{Y}{(X^2 + Y^2)^{\frac{1}{2}}} . \cos \omega t\right\}$$

$$= (X^2 + Y^2)^{\frac{1}{2}} (\sin \omega t . \cos \theta + \cos \omega t . \sin \theta)$$

$$= (X^2 + Y^2)^{\frac{1}{2}} . \sin (\omega t + \theta)$$

Thus the resultant e.m.f. is a sinusoidal e.m.f. of maximum value $(X^2 + Y^2)^{\frac{1}{2}}$, leading the e.m.f. e_1 by θ where $\theta = \tan^{-1} Y/X$.

For the effective value of the resultant, we have

$$E_{\text{eff}} = 0{\cdot}707(X^2 + Y^2)^{\frac{1}{2}}$$

$$= 0{\cdot}707[(E_1 + E_2 \cos \phi)^2 + E_2^2 \sin^2 \phi]^{\frac{1}{2}}$$

$$= 0{\cdot}707 \times \sqrt{(E_1^2 + E_2^2 + 2E_1E_2 \cos \phi)}, \text{ as before}$$

The special case in which E_1 and E_2 are of equal magnitude is of importance. Denoting each by E_1 we then have

$$e = e_1 + e_2 = E_1 \sin \omega t + E_1 \sin (\omega t + \phi)$$

$$= E_1 \{\sin \omega t + \sin (\omega t + \phi)\}$$

$$= 2E_1 \cos (\phi/2) \sin (\omega t + \phi/2)$$

showing that the resultant is of maximum value $2E_1 \cos \phi/2$, and therefore of effective $1{\cdot}414E_1 \cos \phi/2$, and that it leads the component $E_1 \sin \omega t$ by $\phi/2$.

Pure Resistance Circuit. A pure resistance circuit is one possessing neither inductance nor capacity. Hence, if a current passes through the circuit no back e.m.f. will be set up by any change in current. The applied voltage V has therefore to overcome the ohmic drop only as in a direct current circuit. We thus have, using effective values,

$$I = V/R \text{ in phase with } V \text{ (Fig. 15.7)}$$

The power, p, in the circuit at any instant is the product of the instantaneous voltage and instantaneous current. Hence, if the wave diagrams are drawn and the products of the ordinates at given instants plotted, the resulting curve will be the curve of power.

We have $$p = v \times i$$

Let $$v = V_{\max} \sin \omega t$$

Then $$i = I_{\max} \sin \omega t$$

and
$$p = V_{\max} I_{\max} \sin^2 \omega t$$
$$= \tfrac{1}{2} V_{\max} I_{\max}(1 - \cos 2\omega t)$$

since $$\sin^2 \omega t = \tfrac{1}{2}(1 - \cos 2\omega t)$$

It is clear that the power, p, has a constant part $\frac{1}{2} V_{\max} I_{\max}$ and a fluctuating part $-\frac{1}{2} V_{\max} I_{\max} \cos 2\omega t$, the latter averaging zero over a

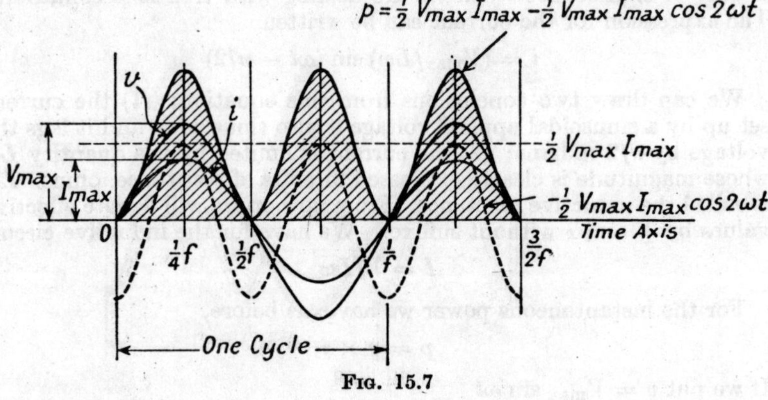

Fig. 15.7

complete cycle. Now the power in a circuit, as measured by a wattmeter for example, is the *average* of the instantaneous power, p, because power is a scalar and not a vector quantity. Hence average power is

$$P = \text{average of } p$$
$$= \text{steady part of } p$$
$$= \tfrac{1}{2} V_{\max} I_{\max}$$
$$= (V_{\max}/\sqrt{2})(I_{\max}/\sqrt{2}) = VI$$

Thus the power in a pure resistance circuit is given by the product of the effective voltage and current. The fluctuating power has twice the frequency of the voltage or current. This is illustrated by the curve of power on the wave diagram.

Purely Inductive Circuit. A pure inductive circuit is one which possesses inductance only, but no resistance or capacity. The nearest approach to such a circuit is obtained by winding a coil of heavy section copper wire on a laminated iron core. Such a coil is called a "choking coil." The magnetic field set up by the alternating current will be alternating; hence, its magnitude will be changing at every instant. Now a self-induced e.m.f. is set up whenever the magnetic flux, linking with a circuit, changes, and since there is no ohmic resistance, the applied voltage has to oppose the self-induced e.m.f. only. Hence, the applied voltage is equal and opposite to the self-induced e.m.f. at every instant.

$$\therefore\ v = -e' = +L(di/dt)$$

Put $$v = V_{max} \sin \omega t, \text{ as before}$$

Then $$di = (V_{max}/L) \sin \omega t \,.\, dt$$

$$\therefore\ i = -(V_{max}/L\omega) \cos \omega t$$

Neglecting the constant of integration, which, as is shown on p. 312, is only of significance when we are dealing with transient conditions. The expression for the current can be written

$$i = (V_{max}/L\omega) \sin (\omega t - \pi/2)$$

We can draw two conclusions from this equation: (1) the current set up by a sinusoidal applied voltage is also sinusoidal, and it lags the voltage by $\pi/2$ radians; (2) the current is limited by the quantity $L\omega$ whose magnitude is clearly expressed in units of resistance, ohms. $L\omega$ is called the inductive reactance. From now on we will denote effective values by V and I without suffixes. We have for the inductive circuit

$$I = V/L\omega$$

For the instantaneous power we have, as before,

$$p = v \times i$$

If we put $v = V_{max} \sin \omega t$

Then $$i = I_{max} \sin (\omega t - \pi/2) = -I_{max} \cos \omega t$$

$$\therefore\ p = -V_{max}I_{max} \sin \omega t \cos \omega t$$
$$= -\tfrac{1}{2} V_{max}I_{max} \sin 2\omega t$$

and average power

$$P = -\tfrac{1}{2} V_{max}I_{max} \times (\text{average of } \sin 2\omega t)$$
$$= 0$$

We therefore see that the total power is zero, a result which at first sight is surprising when it is realized that both V and I are finite. The power is, in fact, a pure sine wave of double frequency and maximum value $\tfrac{1}{2}E_{max}I_{max} = VI$.

The power is alternately positive and negative, the alternate lobes of the curve being of equal magnitude as shown in the wave diagram of Fig. 15.8. This is explained physically as follows. When the current is zero, there is zero magnetic field. As the current increases, the magnetic field increases, and since work has to be done to create a magnetic field, the circuit has to supply positive power. This goes on until the current is a maximum. The current, and therefore, the magnetic field, now begin to decrease. Since the field is decreasing, its potential energy also decreases; it is, in fact, returned to the circuit. This means that the coil is supplying power to the circuit, or in other words the power is negative. When the magnetic field has become zero again, the whole of its energy will have been returned to the circuit, and the total work done during the creation and destruction of the field will have been zero. Consequently, the *average* demand for power will have been zero.

An inspection of Fig. 15.8 shows that the maximum amount of energy which can be stored in the magnetic field is that given to the

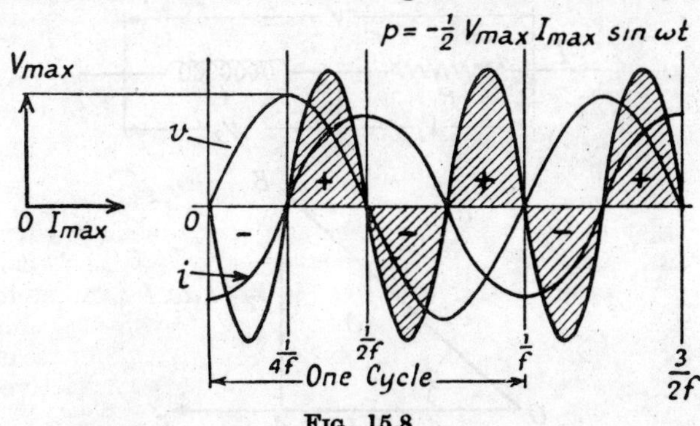

Fig. 15.8

field during a quarter-cycle of the applied voltage, which corresponds to a complete lobe of the power curve, for example, that corresponding to the time interval $\frac{1}{4}f$ to $\frac{1}{2}f$. The amount of this power is given by

$$P = \int_{1/4f}^{1/2f} -\tfrac{1}{2} V_{\max} I_{\max} \sin 2\omega t \,.\, dt$$

$$= \tfrac{1}{2} \cdot \frac{V_{\max} I_{\max}}{4\pi f} \Big[\cos 4\pi f t\Big]_{1/4f}^{1/2f}$$

$$= V_{\max} I_{\max}/2\omega$$

Again, $V_{\max} = L\omega I_{\max}$

$$\therefore\ P = \tfrac{1}{2} L I_{\max}{}^2 \text{ joules}$$

Resistance and Inductance in Series

Let V = total applied voltage (Fig. 15.9)

V_1 = voltage drop along R

and V_2 = voltage drop along L

Then $V_1 = IR$, in phase with I

and $V_2 = L\omega I$, leading I by $\pi/2$

Draw OI to represent the current I in phase and mark off $OA = V_1$. Then OA represents V_1 in magnitude and phase. Draw AB perpendicular to OI and equal to V_2. Then AB represents V_2 in magnitude and phase. Hence, OB, the vector sum of V_1 and V_2, is the applied voltage V. We therefore have

$$V = \sqrt{(OA^2 + AB^2)}$$
$$= I \times \sqrt{(R^2 + (L\omega)^2)}$$

or
$$I = V/\sqrt{(R^2 + (L\omega)^2)}$$

The quantity $\sqrt{(R^2 + (L\omega)^2)}$ is called the *impedance*, and it is measured in ohms. It is represented by the symbol Z.

Again, the vector OB represents the applied voltage in magnitude

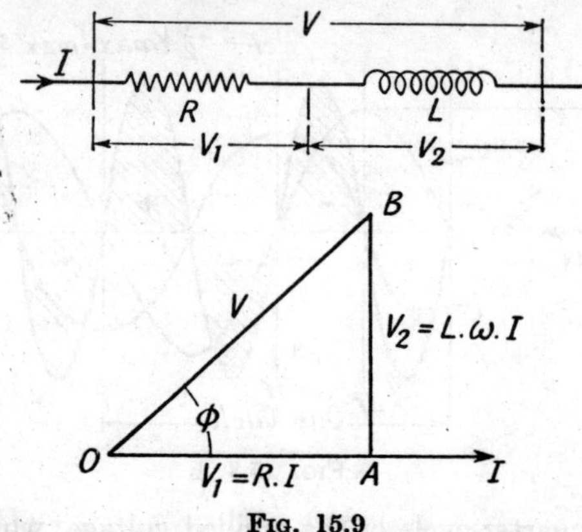

Fig. 15.9

and phase, whereas OI represents the current in phase. Hence, the current lags behind the voltage by an angle

$$\phi = \widehat{BOA} = \tan^{-1}(L\omega/R) \text{ or } \cos^{-1}(R/Z)$$

If each side of the triangle OAB is divided by I the new triangle so formed will have a base equal to R, perpendicular equal to $L\omega$, and hypotenuse equal to Z. We therefore have the relation

$$(\text{Impedance})^2 = (\text{resistance})^2 + (\text{reactance})^2$$

and this holds for any alternating-current circuit.

Now put $v = V_{max} \sin \omega t$

$$\therefore\ i = I_{max} \sin(\omega t - \phi)$$
$$\therefore\ p = v \times i = V_{max} I_{max} \sin \omega t \,.\, \sin(\omega t - \phi)$$
$$\therefore\ P = V_{max} I_{max} \times [\text{mean of } \sin \omega t \,.\, \sin(\omega t - \phi)]$$
$$= V_{max} I_{max} \times [\text{mean of } \tfrac{1}{2}\{\cos\phi - \cos(2\omega t - \phi)\}]$$
$$= (V_{max}/\sqrt{2})(I_{max}/\sqrt{2}) \cos\phi$$
$$= VI \cos\phi$$

The cosine of the angle of lag (or lead) of the current is called the "Power Factor," because it is the factor by which the apparent power

VI must be multiplied in order to obtain the true power. The presence of the term $\cos(2\omega t - \phi)$ indicates, as before, that the power has a periodic component of double frequency. In this case the positive lobes of the power curve (Fig. 15.10) are greater than the negative

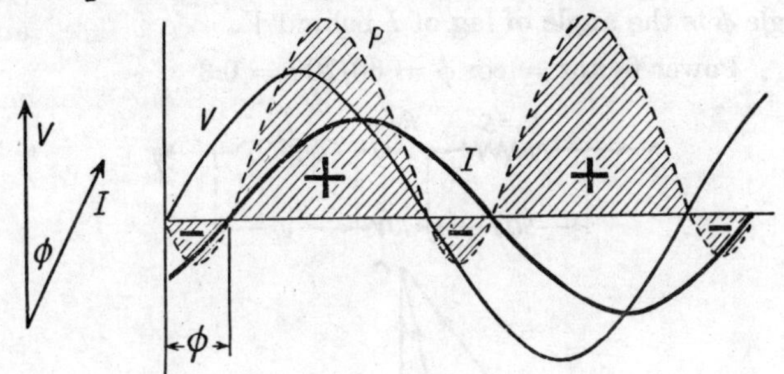

FIG. 15.10

lobes, because positive power has always to be drawn from the supply on account of the I^2R loss due to ohmic resistance.

It is useful to regard the above problem in another way. The current I lags behind V by the angle ϕ, as shown in Fig. 15.11. Resolve I into two components, one in phase with V and the other lagging $\pi/2$, or in "quadrature" with V. Then

Component in phase with V, OA $= I\cos\phi$

Power contributed by it $= V \times OA = VI\cos\phi$

Component in quadrature with V, $OB = I\sin\phi$

Power contributed by it $= 0$ (from p. 314)

$\therefore$ Total power $= VI\cos\phi$

The component OA is called the working, or "wattful," component; while OB is called the idle, or "wattless," component.

FIG. 15.11

EXAMPLE. An arc lamp (which may be regarded as being non-inductive) takes 10 A at 50 V. Calculate the impedance of a choke of 1 Ω resistance to be placed in series with it in order that it may be worked off a 200 V, 50 c/s supply. Find also the total power used and the power factor.

The circuit and vector diagram are given in Fig. 15.12. Resistance of arc $= 50/10 = 5\ \Omega$. Hence, total resistance $R = 6\ \Omega$. Total drop across this resistance $IR = 60$ V in phase with I. The total drop along the whole circuit is 200 V, and the drop, V_L, along the inductance L is in quadrature with I.

Hence, $\qquad V_L = \sqrt{[(200)^2 - (60)^2]} = 190{\cdot}8$ V

In the vector diagram, OA is the drop of 50 V across the arc, AB the drop of 10 V across the resistance in the choke, and BC the drop of

190·8 V across the inductance of the choke. The vector sum OC is the total applied voltage of 200 V, and AC is the total drop across the choke

$$= \sqrt{[(190{\cdot}8)^2 + 10^2]} = 191 \text{ V}$$

The angle ϕ is the angle of lag of I behind V.

$$\therefore \text{ Power factor} = \cos\phi = 60/200 = 0{\cdot}3$$

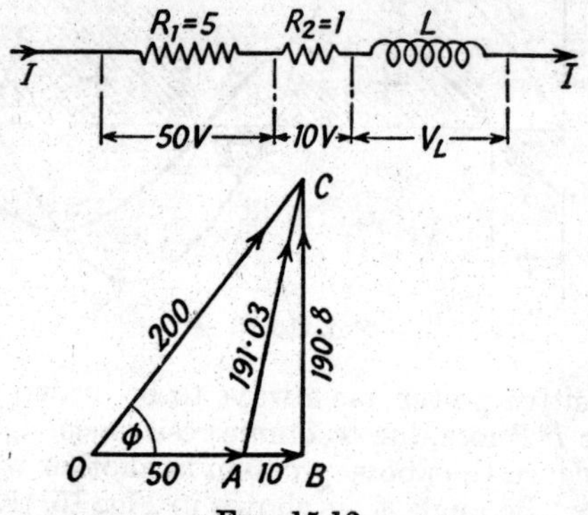

Fig. 15.12

$$\text{Power in circuit } VI\cos\phi = 200 \times 10 \times 0{\cdot}3 = 600 \text{ W}$$

Or, $= I^2R$, **where R is the total resistance**

$$= 10^2 \times 6 = 600 \text{ W}$$

Circuit Containing Capacitance Only

Let C = **capacitance in farads**

If the voltage at the terminals has at any moment a value v, then the quantity of electricity in the capacitor at that instant is

$$q = Cv \text{ coulombs}$$

Now quantity is the product of current and time, and if the current is varying, then

$$q = \int i\,.\,dt$$

or

$$dq = i\,.\,dt$$

$$\therefore\ i = dq/dt = C(dv/dt)$$

Put

$$v = V_{\text{max}} \sin \omega t$$

$$\therefore\ i = CV_{\text{max}}\,\omega\,.\cos\omega t$$

$$\therefore\ I_{\text{max}} = CV_{\text{max}}\,\omega$$

$$I = CV\omega \text{ or } \frac{V}{1/C\omega}$$

Hence, the reactance of a capacitor is given by $1/C\omega$.

Again, since the voltage is represented by $v = V_{\max} \sin \omega t$ and the current by $i = I_{\max} \sin (\omega t + \pi/2)$, we see that the current *leads* the applied voltage by $\pi/2$.

Again, since $$v = V_{\max} \sin \omega t$$

and $$i = I_{\max} \sin (\omega t + \pi/2) = I_{\max} \cos \omega t$$

$$\therefore\ p = V_{\max} I_{\max} \sin \omega t . \cos \omega t$$
$$= \tfrac{1}{2} V_{\max} I_{\max} \sin 2\omega t$$
$$\therefore\ P = \text{average of } p = 0$$

The total power in this case also is zero, and as before, it is periodic, with twice the supply frequency.

From the expressions for v and p we see that when v is zero and is on the point of becoming positive, so also is p, and consequently the maximum amount of energy will be supplied by the circuit to the capacitor during the quarter-period commencing at $t = 0$ and ending at $t = \frac{1}{4}f$.

$$\therefore\ P = \int_{o}^{1/4f} \tfrac{1}{2} V_{\max} I_{\max} \sin 2\omega t \,.\, dt$$
$$= \tfrac{1}{2} \cdot \frac{V_{\max} I_{\max}}{4\pi f} \Big[- \cos 4\pi f t \Big]_{o}^{1/4}$$
$$= V_{\max} I_{\max} / 4\pi f = V_{\max} I_{\max} / 2\omega$$

Now $$I_{\max} = V_{\max} C\omega$$
$$\therefore\ P = \tfrac{1}{2} V_{\max}^{2} C$$

Since a capacitor acts as an open circuit when connected in series with a direct-current circuit, there is often difficulty in realizing how it is that the capacitor can carry a current when placed in an alternating-current circuit. When a continuous p.d. is applied to the plates of a capacitor there is a momentary flow of electricity in the external circuit, which carries a sufficient quantity of electricity to the plates to make their p.d. equal to the applied p.d., after which all flow ceases. Similarly, if the plates of the charged capacitor are joined by a wire, the quantity of electricity originally given to the capacitor will flow in the opposite direction, thus producing a momentary current in the opposite direction, which ceases as soon as the plates are again at the same potential. This shows that by alternately charging and discharging a capacitor it is possible for an alternating current to flow in the capacitor circuit. But this current, regarded as a pure conduction current, is confined to the external circuit and does not flow through the dielectric. The mechanisms of the phenomena that take place at the plates and in the interior of the dielectric have been discussed on p. 297. With alternating p.d.s at the plates there is alternate storing up and releasing of energy by the dielectric.

The simple analogous hydraulic circuit of Fig. 15.13 will make this clear. The circuit consists of a tube in the form of a ring provided at one side with a piston, and at the opposite side with an expansion across which an elastic membrane is stretched. The whole tube is filled with water. If a steady pressure is applied to the piston in the direction indicated by the arrow there will be a displacement of water round the ring in a counter-clockwise direction, with consequent storing of water in the right-hand compartment and giving up of water by the left-hand compartment. Equilibrium will be established, and all flow of water then cease, when the reaction of the stressed membrane is equal to the applied pressure. The establishment of the stress in the membrane is thus accompanied by a momentary displacement of water, in exactly the same way that the establishment of an electrostatic stress in the dielectric of a capacitor is accompanied by a momentary flow, or displacement, of electricity. If the piston is released the membrane takes up its unstressed position, accompanied this time by a displacement of water in the opposite direction. Similarly when a capacitor is discharged there is a displacement of electricity in the reversed direction, this displacement ceasing when the p.d., and therefore the electrostatic stress, has been reduced to zero. This has been considered from a mathematical point of view on p. 37.

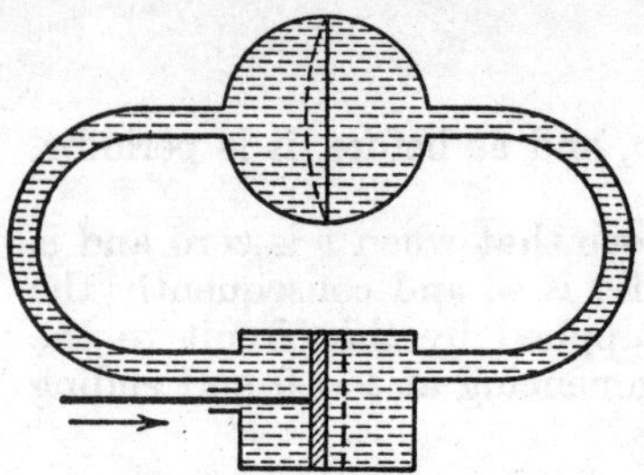

FIG. 15.13. HYDRAULIC ANALOGY

Now, suppose that the piston is given a simple harmonic displacement, then there will be an alternating flow of water in the circuit, but it is obvious that this flow will be bounded by the elastic membrane, the current in no sense flowing through the membrane. In spite of this the flow of water in the rest of the circuit is exactly the same as though the water did flow through the membrane, and similarly in the circuit of a capacitor to which an alternating p.d. is applied. From this point of view we thus see that it is justifiable to speak of the current as the capacitor current.

General Series Circuit. Fig. 15.14 represents a circuit with resistance, inductance, and capacitance all in series. We have

$V_R = RI$ in phase with I

$V_L = L\omega I$ in quadrature with I (leading)

$V_C = I/C\omega$ in quadrature with I (lagging)

Draw OI to represent the current in phase, and represent the above voltage drops by OA, OB, and OC respectively. The resultant of OB and OC is OD, where

$$OD = L\omega I - I/C\omega$$

Hence, for the total applied voltage we have

$$V = \sqrt{(OA^2 + OD^2)} = I\sqrt{[R^2 + (L\omega - 1/C\omega)^2]}$$

$$I = \frac{V}{\sqrt{[R^2 + (L\omega - 1/C\omega)^2]}}$$

The total impedance is now

$$Z = \sqrt{[R^2 + (L\omega - 1/C\omega)^2]}$$

and the reactance

$$X = (L\omega - 1/C\omega)$$

The capacitor reactance is reckoned negative, and therefore, if it is greater than the inductive reactance $L\omega$, the total reactance will be

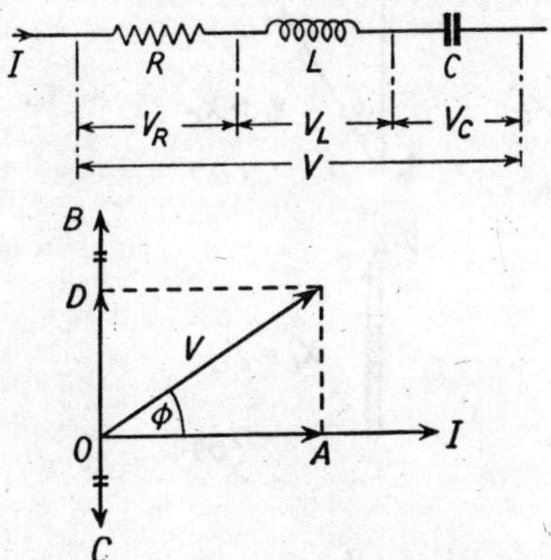

FIG. 15.14. GENERAL SERIES CIRCUIT

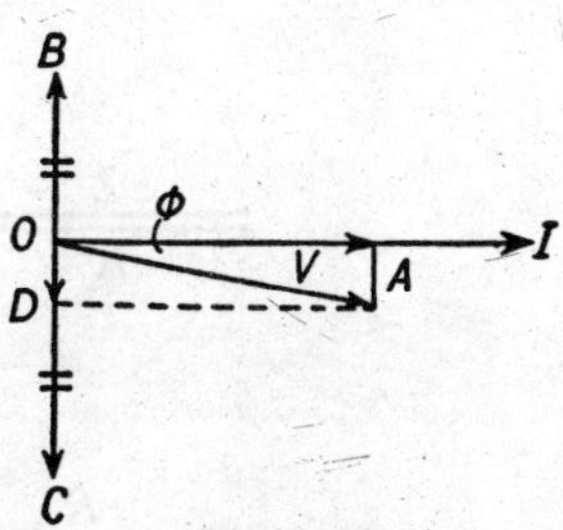

FIG. 15.15

negative, as in Fig. 15.15. If the total reactance is positive the current lags behind the applied voltage; if it is negative, the current leads.

For the angle of lag we have

$$\tan\phi = \frac{OD}{OA} = \frac{I(L\omega - 1/C\omega)}{RI}$$

$$\therefore \phi = \tan^{-1}\frac{L\omega - 1/C\omega}{R} = \tan^{-1}\frac{\text{reactance}}{\text{resistance}}$$

$$\text{Power factor} = \cos\phi = \frac{R}{\sqrt{[R^2 + (L\omega - 1/C\omega)^2]}} = \frac{\text{resistance}}{\text{impedance}}$$

If we denote the inductive reactance $L\omega$ by X_L and the capacitive reactance $1/C\omega$ by X_C we have

Total reactance $X = X_L - X_C$

Impedance $Z = \sqrt{(R^2 + X^2)} = \sqrt{[R^2 + (X_L - X_C)^2]}$

$$\cos\phi = R/Z$$

$$= R/[R^2 + (X_L - X_C)^2]^{\frac{1}{2}}$$

According to the values of X_L and X_C there are three cases—

(*a*) $X_L > X_C$. The inductive reactance predominates, and therefore if we replace X_L and X_C by a single equivalent reactance X, this will be inductive and the current will lag the applied voltage (Fig. 15.16 (*a*)).

(*b*) $X_L = X_C$. The total reactance is zero and therefore $Z = R$ numerically. Hence, $I = V/R$ and is in phase with V (Fig. 15.16 (*b*)).

(*c*) $X_L < X_C$. The capacitive reactance predominates, and therefore if we replace X_L and X_C by a simple equivalent reactance X, this will

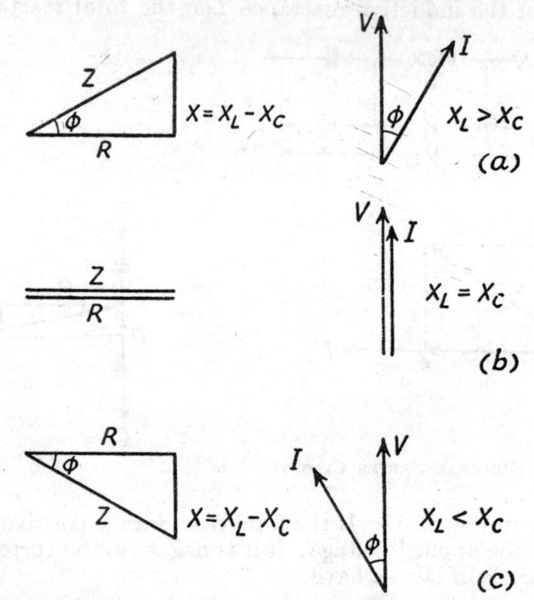

Fig. 15.16. Effect of Relative Values of X_L and X_C

be capacitive and the current will lead the applied voltage (Fig. 15.16 (*c*)).

Example. A R, L circuit has $R = 1\ \Omega$ and $L = 0{\cdot}00955$ H. Calculate the series capacitance which, converting the circuit to a R, L, C circuit will double the current. Hence calculate the inductance (or capacitance) of the equivalent simple R, L (or R, C) circuit. Supply frequency = 50 c/s.

$$X_L = L\omega = 0{\cdot}00955 \times 314 = 3\ \Omega$$

$$\therefore \text{original } Z = \sqrt{(1^2 + 3^2)} = 3{\cdot}1623\ \Omega$$

For the current to be doubled the new Z must be $1{\cdot}5812\ \Omega$

$$\therefore (1^2 + X^2) = 1{\cdot}5812^2 = 2{\cdot}5$$

$$\therefore X^2 = 1{\cdot}5$$

$$\therefore X = \pm 1{\cdot}2248$$

But

$$X = X_L - X_C$$

$$\therefore X_C = X_L - X = 3 \mp 1{\cdot}2248$$

$$= 1{\cdot}7752 \text{ or } 4{\cdot}2248$$

The necessary capacitance is thus

$$C = 1/X_C\omega = 1/(1{\cdot}7752 \times 314) = 1{\cdot}794 \times 10^{-3}\text{ F}$$

$$\text{or } = 1/(4{\cdot}2248 \times 314) = 0{\cdot}754 \times 10^{-3}\text{ F}$$

with $X_C = 1{\cdot}7752$, $X = 3 - 1{\cdot}7752 = +1{\cdot}2248\ \Omega$

The equivalent circuit is thus a R, L circuit with

$$L = 1{\cdot}2248/314 = 0{\cdot}0039\text{ H}$$

with $X_C = 4{\cdot}2248$, $X = 3 - 4{\cdot}2248 = -1{\cdot}2248\ \Omega$

The equivalent circuit is thus a R, C circuit with

$$C = 1/(1{\cdot}2248 \times 314) = 2{\cdot}60 \times 10^{-3}\text{ F}$$

Electrical Resonance. If the inductive and capacitive reactances, $L\omega$ and $1/C\omega$, are equal to one another in the general series circuit, then the total reactance is zero, and the current is given by the Ohm's Law value V/R, and is in phase with V. In order to understand what is taking place in the circuit when this condition is fulfilled, imagine that the resistance R is zero, that a constant voltage V is applied, and that the frequency is varied from zero to a very high value.

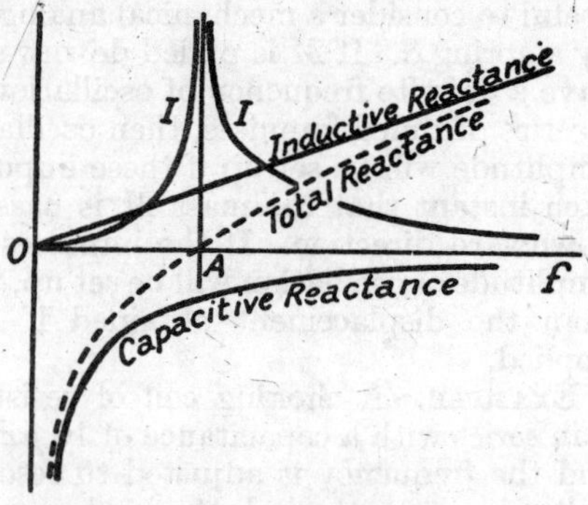

FIG. 15.17. SERIES RESONANCE WITH VARYING FREQUENCY, f

Then the inductive reactance $X_L = L\omega = 2\pi Lf$, and is represented graphically by a straight line through the origin (Fig. 15.17). The capacitive reactance

$$X_C = -1/C\omega = -1/2\pi Cf$$

It is therefore represented by a rectangular hyperbola in the fourth quadrant. The total reactance $X = X_L + X_C$, and the graph of X is also a hyperbola which crosses the frequency axis at some point A.

At this point the total reactance X is zero, and since the resistance is zero, the total impedance is zero. The current I is therefore theoretically infinite at this point, as shown. For frequencies greater than OA, the inductive reactance predominates and the current lags. For frequencies less than OA, the capacity reactance predominates, and therefore, the current leads.

The condition for zero reactance is

$$L\omega = 1/C\omega$$

$$\omega = \sqrt{(1/LC)} \text{ or } f = (1/2\pi)\sqrt{(1/LC)}$$

This condition is said to be the condition for *electrical resonance*, for the following reason. When a capacitor is discharged through an inductive external circuit of small resistance, the discharge is oscillatory, the frequency of the oscillations being

$$f = 1/[2\pi\sqrt{(LC)}]$$

When the capacitor is in an alternating-current circuit it is periodically charged and discharged, and if the frequency of the applied voltage is such that the total reactance is zero, then it coincides with the natural frequency of electrical oscillation of the circuit. In an actual circuit, the resistance is, of course, never zero, and there are also iron losses taking place in the inductor, and dielectric losses in the capacitor. The current therefore cannot become infinite, but when R is small it may reach a very high value. The voltages across the inductor and capacitor which then occur, namely,

$$V_2 = L\omega I, \text{ and } V_3 = I/C\omega$$

may be many times greater than the applied voltage. There may thus be the danger of break-down of the insulation.

It will be seen that resonance is the result of the coincidence of the applied frequency with the natural frequency of the circuit. It is useful to consider a mechanical analogy. Consider a mass M suspended by a spring S. If M is pulled downwards and released, the system will have a definite frequency of oscillation. If M is pushed downwards by a series of timed impulses, then oscillation will take place, but no large amplitude will be set up if these impulses are not so timed as to occur each instant that the mass M is passing its equilibrium position in a downward direction. If the impulses are so timed, then a very large amplitude of oscillation will be set up, the amplitude being much greater than the displacement obtained if a steady downward force were applied.

EXAMPLE. A choking coil of resistance 5 Ω and inductance 0·6 H is in series with a capacitance of 10 μF. If a voltage of 200 V is applied and the frequency is adjusted to resonance, find the current, and the voltages across the inductor and capacitor.

$$\begin{aligned}\text{Frequency for resonance } f &= (1/2\pi)\sqrt{(1/LC)}\\ &= (1/2\pi)\sqrt{[(10^6/(0{\cdot}6 \times 10)]}\\ &= 65 \text{ c/s}\end{aligned}$$

Current at resonance $I = V/R = 200/5 = 40$ A in phase with V

$$\begin{aligned}\text{Voltage across capacitor } I/C\omega &\\ &= 40 \times 10^6/(10 \times 2\pi \times 65)\\ &= 9800 \text{ V}\end{aligned}$$

$$\begin{aligned}\text{Voltage across inductor } &= I\sqrt{[R^2 + (L\omega)^2]}\\ &= 40 \times \sqrt{[25 + (0{\cdot}6 \times 2\pi \times 65)^2]}\\ &= 9800 \text{ V}\end{aligned}$$

Properties of a Resonant Circuit. For the general series circuit we have seen that the current is given by

$$I = \frac{V}{\sqrt{[R^2 + (L\omega - 1/C\omega)^2]}}$$

where $L\omega = X_L$ the inductive reactance

and $1/C\omega = X_C$ the capacitive reactance

$$\therefore\ I = V/[R^2 + (X_L - X_C)^2]^{\frac{1}{2}}$$

The p.d. across the inductive portion of the circuit is given by

$$V_L = L\omega I = X_L I$$

$$\therefore\ V_L = VX_L/\sqrt{[R^2 + (X_L - X_C)^2]}$$

Imagine that the inductive reactance X_L is varied over a wide range by varying L, the frequency f and applied voltage V meanwhile remaining constant. Differentiating with respect to X_L (since $X_L = \omega L$, and ω is now a constant)

$$dV_L/dX_L = \frac{[R^2+(X_L-X_C)^2]^{\frac{1}{2}}V - VX_L \times \frac{1}{2}[R^2+(X_L-X_C)^2]^{-\frac{1}{2}} \times 2(X_L - X_C)}{R^2 + (X_L - X_C)^2}$$

The maximum value of the voltage V_L is therefore given by the condition

$$\frac{[R^2+(X_L-X_C)^2]^{\frac{1}{2}}V - VX_L \,.\, \frac{1}{2}[R^2+(X_L-X_C)^2]^{-\frac{1}{2}} \,.\, 2(X_L - X_C)}{R^2 + (X_L - X_C)^2} = 0$$

$$\therefore\ [R^2 + (X_L - X_C)^2]^{\frac{1}{2}} V = VX_L[R^2 + (X_L - X_C)^2]^{-\frac{1}{2}} \,.\, (X_L - X_C)$$

$$R^2 + (X_L - X_C)^2 = X_L(X_L - X_C)$$

$$\therefore\ X_L = (R^2 + X_C^2)/X_C$$

It is only when $R = 0$ that this reduces to the resonance condition of $X_L = X_C$. Hence when the resistance is finite the condition for maximum voltage, V_L, is not the same as for resonance.

EXAMPLE. Consider a circuit in which $X_C = 200\ \Omega$. Let the supply voltage be 100. First of all give R a high value of $100\ \Omega$.

The condition for resonance is

$$X_L = X_C = 200, \text{ and } Z = R = 100$$

$$\therefore\ I = V/Z = V/R = 100/100 = 1\text{ A}$$

$$\therefore\ V_L = X_L I = 200 \times 1 = 200\text{ V}$$

The condition for maximum choke drop V_L is

$$X_L = (R^2 + X_C^2)/X_C$$
$$= (100^2 + 200^2)/200$$
$$= 250\ \Omega$$

$$\therefore I = V/[R^2 + (X_L - X_C)^2]^{\frac{1}{2}}$$
$$= 100/(100^2 + 50^2)^{\frac{1}{2}}$$
$$= 100/112 = 0{\cdot}893 \text{ A}$$
$$\therefore V_L = X_L I = 250 \times 0{\cdot}893$$
$$= 223{\cdot}3 \text{ V}$$

Now take the much lower resistance of $R = 10\ \Omega$. At resonance we still have the condition

$$Z = R \quad \therefore Z = 10$$
$$\therefore I = 100/10 = 10 \text{ A}$$
$$\therefore V_L = X_L I = 200 \times 10 = 2000 \text{ V}$$

The condition for maximum choke drop is now

$$X_L = (R^2 + X_C^2)/X_C$$
$$= (10^2 + 200^2)/200$$
$$= 200{\cdot}5\ \Omega$$
$$\therefore I = V/[R^2 + (X_L - X_C)^2]^{\frac{1}{2}}$$
$$= 100/[10^2 + (0{\cdot}5)^2]^{\frac{1}{2}}$$
$$= 9{\cdot}99 \text{ A}$$
$$\therefore V_L = X_L I$$
$$= 200{\cdot}5 \times 9{\cdot}99$$
$$= 1998 \text{ V}$$

This numerical example shows that for zero resistance or low values of resistance the condition for maximum choke drop is the same as the condition for resonance. For large resistances the two conditions are not the same, the divergence increasing with the magnitude of the resistance.

The Pass-band of a Resonant Circuit. The shape of the current-frequency curve for a general series circuit, often called the resonance curve, depends on the magnitude of R. If R is zero, then theoretically the current rises to infinity as shown by curve 1 of Fig. 15.18. For a low value of R the maximum current V/R is high, and the curve exhibits a very sharp peak as indicated by curve 2. As R increases the maximum current diminishes and the curve becomes flatter as shown by curves 3 and 4. When the peak is sharp it will be clear that the circuit has a selective property inasmuch as it prefers to pass a current of periodicity ω' (the resonant periodicity) than of any other periodicity, in fact with $R = 0$ the circuit can only pass current of periodicity ω', its impedance to all other frequencies being infinite. Suppose we draw a horizontal AB to cut any particular resonance curve, say No. 3, as shown, then the range of intersection will be from periodicity ω_1 to periodicity ω_2, the range $(\omega_1 - \omega_2)$ obviously depending on the height of AB. This range, the width of the shaded strip in the figure is called the *pass band*, and it is usual to define it in relation to a height of the horizontal of $V/(\sqrt{2}R)$ amperes.

At any pulsatance ω the current is

$$I = \frac{V}{\sqrt{[R^2 + (L\omega - 1/C\omega)^2]}}$$

while the pulsatance for resonance is

$$\omega' = 1/\sqrt{(LC)}$$

Since the pass band is defined in relationship to a current of $V/(\sqrt{2}R)$, we have for the corresponding frequency

$$\frac{V}{\sqrt{(2)}R} = \frac{V}{\sqrt{\{R^2 + (L\omega - 1/C\omega)^2\}}}$$

$$\therefore\ R^2 + (L\omega - 1/C\omega)^2 = 2R^2$$

$$L\omega - 1/C\omega \pm R = 0$$

$$\therefore\ \omega^2 \pm \omega R/L - 1/LC = 0$$

$$\therefore\ \omega = \frac{\pm R/L \pm \sqrt{(R^2/L^2 + 4/LC)}}{2}$$

$$\therefore\ \omega = \pm R/2L \pm \sqrt{(R^2/4L^2 + 1/LC)}$$

Now if a circuit is sharply selective, as for example in a tunedcircuit of a radio receiver, the resistance R has to be low and the term $R^2/4L$

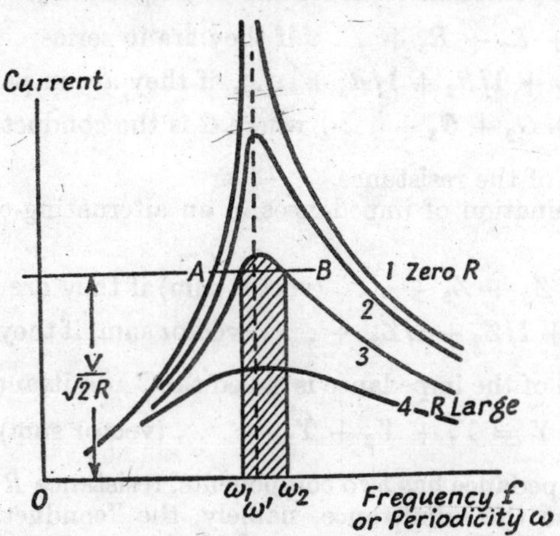

FIG. 15.18. THE *RLC* CIRCUIT AS A BAND SELECTOR

becomes negligible in comparison with the term $1/LC$. We then have

$$\omega = \pm R/2L \pm \sqrt{(1/LC)}$$

Again the term $1/\sqrt{(LC)}$ is the same as the resonant pulsatance ω' and therefore we have

$$\omega = \pm R/2L \pm \omega'$$

Since we are only concerned with one direction of rotation of the vector of applied voltage we have only the positive value of ω' to consider, this giving

$$\omega = \omega' \pm R/2L$$

Denoting the two values by ω_1 and ω_2 respectively, we have

$$\omega_1 = \omega' - R/2L$$

and
$$\omega_2 = \omega' + R/2L$$

and the bandwidth

$$\omega_2 - \omega_1 = 2R/2L = R/L \text{ radians/second}$$

If we want the bandwidth in terms of frequency we obviously have

$$f_2 - f_1 = (\omega_2 - \omega_1)/2\pi = R/2\pi L \text{ cycles/second}$$

The ability of a circuit to select any particular frequency is decided by the sharpness of the resonant circuit and therefore by the smallness of R. The selectivity is therefore inversely proportional to R, and it is arbitrarily defined as $\omega' L/R$. For any periodicity ω the term $\omega L/R$ is called the Q of the particular circuit.

Circuits in Parallel. In a direct-current circuit we have for the total resistance of a combination of resistances, R_1, R_2, R_3, etc.

$$R = R_1 + R_2 + R_3 + \ldots \text{ if they are in series}$$

$$1/R = 1/R_1 + 1/R_2 + 1/R_3 + \ldots \text{ if they are in parallel}$$

or $G = G_1 + G_2 + G_3 + \ldots$ where G is the conductance,

the reciprocal of the resistance.

For a combination of impedances in an alternating-current circuit, we have

$$Z = Z_1 + Z_2 + Z_3 + \ldots \text{ (vector sum) if they are in series}$$

$$1/Z = 1/Z_1 + 1/Z_2 + 1/Z_3 + \ldots \text{ (vector sum) if they are in parallel}$$

The reciprocal of the impedance is called the "Admittance," Y, so that

$$Y = Y_1 + Y_2 + Y_3 + \ldots \text{ (vector sum)}$$

Just as the impedance has two components, resistance R and reactance X, so also has the admittance, namely, the "conductance," G, and "susceptance," B. The impedance and admittance triangles (Fig. 15.19) are similar except that the admittance triangle is inverted, as shown. This is because B is negative for an inductive branch, and positive for a capacitive branch. The proof of this is given on p. 330.

$$\therefore\ B/G = \tan\phi = X/R$$

also
$$G^2 + B^2 = Y^2 = 1/Z^2$$
$$= 1/(X^2 + R^2)$$

The solution of these equations for G and B is

$$\left.\begin{aligned} G &= R/Z^2 \text{ mhos} \\ B &= X/Z^2 \text{ mhos} \end{aligned}\right\}$$

If the admittances of a number of circuits in parallel are represented on a vector diagram, as in Fig. 15.20, then

X component of total admittance

$$Y \cos \phi = Y_1 \cos \phi_1 + Y_2 \cos \phi_2 + Y_3 \cos \phi_3 + \ldots$$

$$\therefore\ G = G_1 + G_2 + G_3 + \ldots$$

Similarly, taking Y components,

$$B = B_1 + B_2 + B_3 + \ldots$$

Total admittance $\quad Y = \sqrt{(G^2 + B^2)}$

Total current $\quad I = V \times Y$

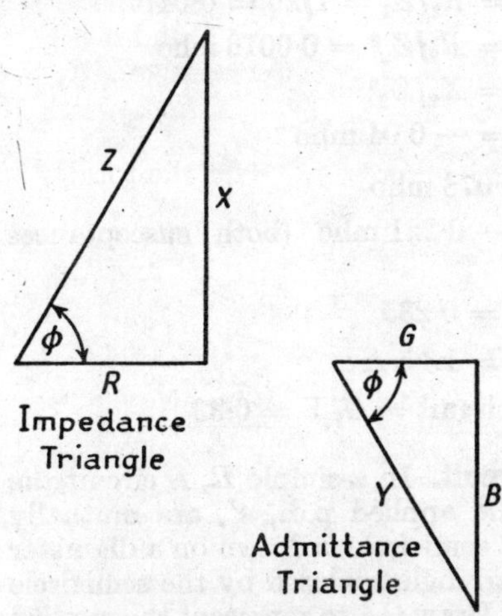

FIG. 15.19. IMPEDANCE AND CONDUCTANCE TRIANGLES FOR AN INDUCTIVE (R, L) BRANCH

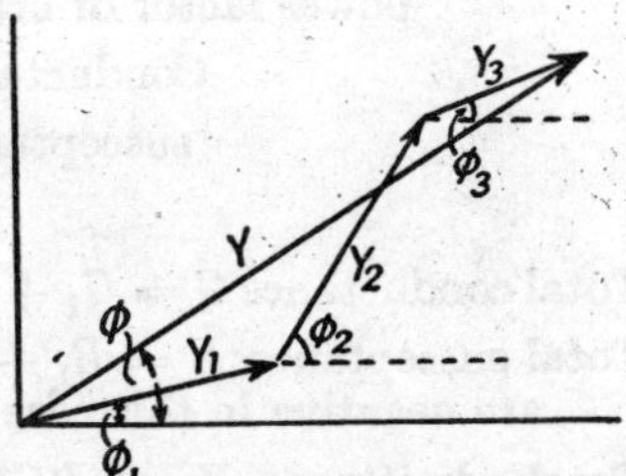

FIG. 15.20. VECTOR SUMMATION OF ADMITTANCES

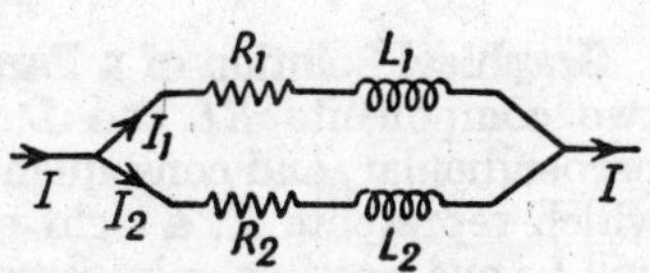

FIG. 15.21

where V is the voltage applied to the branched network, and ϕ the phase angle of the total current with respect to V

$$\phi = \tan^{-1}(B/G)$$

lagging if B is negative, and leading if G is positive.

EXAMPLE. Two coils, one of resistance 2 Ω and self-inductance 0·015 H, the other having a resistance of 1 Ω and a self-inductance of 0·08 H, are arranged in parallel on a 100 V, 50 c/s frequency circuit. Find the magnitudes and phases of the currents flowing in each circuit, and of the resultant current flowing through the whole system. (C. and G.)

This problem can be solved by considering each branch separately, calculating the current in each, and then determining their vector sum. In order to illustrate the method, the admittance method is used in the following calculation. Fig. 15.21 shows the circuit.

$\omega = 2\pi f$; $R_1 = 2$; $L_1 = 0{\cdot}015$; reactance $X_1 = L_1\omega = 4{\cdot}7\ \Omega$

Impedance $Z_1 = \sqrt{(2^2 + 4{\cdot}7^2)} = 5{\cdot}1$; current $I_1 = 100/5{\cdot}1 = 19{\cdot}6$ A

Power factor of branch 1 $= R_1/Z_1 = 2/5{\cdot}1 = 0{\cdot}39$.

$$\text{Conductance } G_1 = R_1/Z_1^2 = 0{\cdot}0765 \text{ mho}$$

$$\text{susceptance } B_1 = X_1/Z_1^2 = -\,0{\cdot}181 \text{ mho}$$

For branch 2,

$R_2 = 1$; $L_2 = 0{\cdot}08$; $X_2 = L_2\omega = 25\ \Omega$; $Z_2 = \sqrt{(25^2 + 1^2)} = 25$ very approximately.

$$\therefore\ I_2 = 100/25 = 4 \text{ A}$$

$$\text{power factor of branch 2} = R_2/Z_2 = 1/25 = 0{\cdot}04$$

$$\text{Conductance } G_2 = R_2/Z_2^2 = 0{\cdot}0016 \text{ mho}$$

$$\text{susceptance } B_2 = X_2/Z_2^2 = -\,0{\cdot}04 \text{ mho}$$

Total conductance $G = G_1 + G_2 = 0{\cdot}078$ mho

Total susceptance $B = B_1 + B_2 = -\,0{\cdot}221$ mho (both susceptances are negative in this case).

Total admittance $Y = \sqrt{(G^2 + B^2)} = 0{\cdot}235$

Total current $I = 100 \times 0{\cdot}235 = 23{\cdot}5$ A

$$\text{Power factor of whole circuit} = G/Y = 0{\cdot}33$$

Graphical Solution of a Parallel Circuit. In a simple R, L circuit the two components RI and $L\omega I$ of the applied p.d., V, are mutually perpendicular, and consequently, if a semicircle is drawn on a diameter which represents V, a right-angled triangle enclosed by the semicircle will be one possible voltage triangle. Draw OA to represent the applied p.d., V, and on it as diameter draw a semicircle as shown, Fig. 15.22. Draw a chord OB_1 such that

$$\tan \phi_1 = R_1/X_1, \text{ for branch No. 1}$$

Then, on joining AB_1 we have a voltage triangle OAB_1 in which OB_1 represents to scale the resistance drop R_1I_1, and B_1A represents the drop X_1I_1.

Draw in a similar manner the triangle OAB_2 such that

$$\tan \phi_2 = R_2/X_2, \text{ for branch No. 2}$$

Then OB_2 represents the drop R_2I_2 and B_2A the drop X_2I_2.

Now the current in any branch is equal to the resistance drop divided by the resistance, and it is also in phase with this drop. Hence, to obtain I_1, divide OB_1 (in volts) by R_1, and mark off OI_1 to represent I_1 thus found.

Similarly obtain OI_2, the current I_2 in the second branch, the triangle OB_2A being the voltage drop triangle for this branch. Then OI, the resultant of OI_1 and OI_2, gives the total current I. If there are more than two branches, then the individual branch currents are all found separately as above, the resultant of all of them being the total current.

Now triangle OBA is the drop of volts triangle for the whole branched circuit, so that the side OB in phase with I is the resistance drop, and

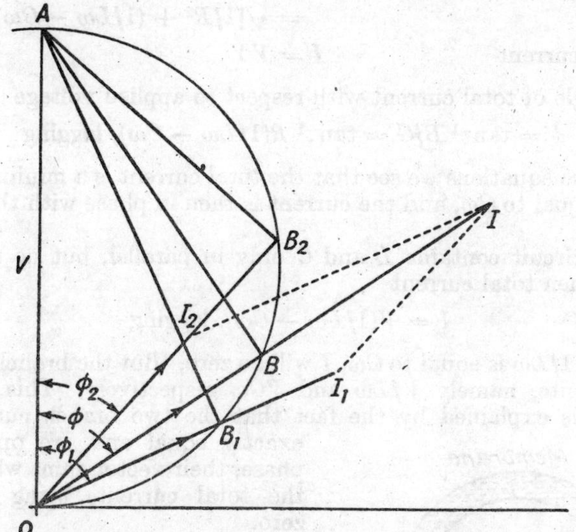

FIG. 15.22. GRAPHICAL SOLUTION OF BRANCHED CIRCUIT

BA, the reactance drop. Hence, if OB and BA are each measured on the voltage scale and divided by I, the equivalent single resistance R and reactance X are obtained.

Resonance in a Parallel Circuit. Consider a circuit consisting of R, L and C in parallel, as shown in Fig. 15.23.

Impedance of branch 1, $Z_1 = R$

Conductance of branch 1, $G_1 = R/Z_1^2 = 1/R$

Susceptance of branch 1, $B_1 = O$

Impedance of branch 2, $Z_2 = L\omega$

Conductance of branch 2, $G_2 = O$

Susceptance of branch 2, $B_2 = -L\omega/Z_2^2 = -1/L\omega$

Impedance of branch 3, $Z_3 = 1/C\alpha$

Conductance of branch 3, $G_3 = O$

R
L
C

FIG. 15.23

Susceptance of branch 3, $B_3 = +\dfrac{1/C\omega}{Z_3^2} = +C\omega$

$\therefore$ Total conductance $G = G_1 + G_2 + G_3$
$= 1/R$

Total susceptance $B = B_1 + B_2 + B_3$ (algebraic sum)
$= -1/L\omega + C\omega$

$\therefore$ Total admittance $Y = \sqrt{(G^2 + B^2)}$
$= \sqrt{[1/R^2 + (1/L\omega - C\omega)^2]}$ mhos

$\therefore$ Total current $I = VY$

Phase angle of total current with respect to applied voltage

$$\phi = \tan^{-1} B/G = \tan^{-1} R(1/L\omega - C\omega), \text{ lagging}$$

From these equations we see that the total current is a minimum when $1/L\omega$ is equal to $C\omega$, and the current is then in phase with the applied voltage.

If the circuit contains L and C only in parallel, but no resistance branch, then total current

$$I = V(1/L\omega - C\omega), \text{ lagging}$$

so that, if $1/L\omega$ is equal to $C\omega$, I will be zero. But the branch currents will be finite, namely $V/L\omega$ and $VC\omega$ respectively. This apparent anomaly is explained by the fact that the two branch currents are exactly equal and are opposite in phase, their vector sum, which gives the total current, being therefore zero.

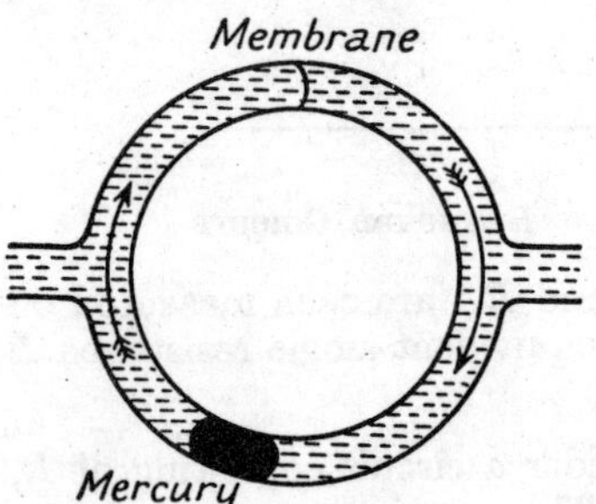

FIG. 15.24

This phenomenon is called "current resonance" or "parallel resonance" to distinguish it from the voltage resonance which takes place in series circuits. The condition for current resonance is the equality of inductive and capacity susceptances, namely,

$$1/L\omega = C\omega \text{ or } \omega = \sqrt{(1/LC)}$$

This is also the condition for voltage resonance in a series circuit. Current resonance does not produce a dangerous rise in pressure.

A useful hydraulic analogy of current resonance is shown in Fig. 15.24. A branched pipe has an elastic membrane across one branch and a column of mercury in the other. The elastic membrane possesses elasticity only, but no inertia, and it therefore corresponds to the capacitor branch. The mercury possesses inertia only and corresponds to the inductive branch. It is possible for an alternating current of water to exist in the circular tube as a whole since this closed system will have a natural period of oscillation of its own. This alternating current can exist when the water in the main pipe is stationary, a condition which corresponds with the branched circuit when current resonance takes place.

Now consider a two-branch circuit with branch 1 having R_1 in series with L, and branch 2 having R_2 in series with C. Referring back to series resonance in a R, L, C circuit, we found that the necessary condition for resonance was equality of the reactive voltages V_L and V_C. For many branched-circuit problems we can argue as for the series circuit but interchange current with voltage. In the case under

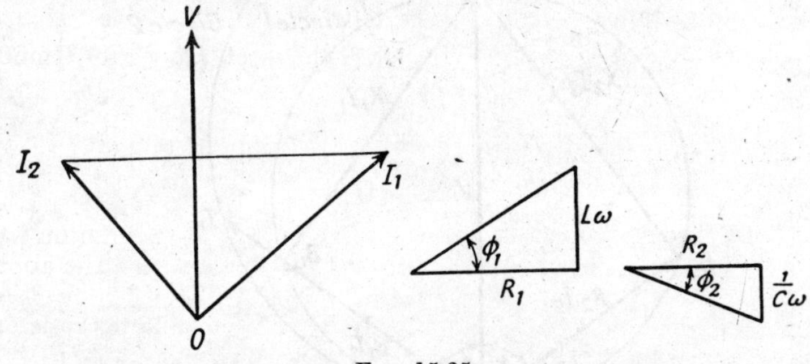

Fig. 15.25

consideration we can do the same and say that current resonance takes place when the reactive currents in the two branches are equal and opposite.

For branch 1,

$$I_1 = V/[R_1^2 + (L\omega)^2]^{\frac{1}{2}} \text{ (Fig. 15.25)}$$

Reactive component

$$\begin{aligned} I_{r1} &= I_1 \sin\phi_1 \\ &= V/[R_1^2 + (L\omega)^2]^{\frac{1}{2}} \times L\omega/[R_1^2 + (L\omega)^2]^{\frac{1}{2}} \end{aligned}$$

For branch 2,

$$I_2 = \frac{V}{[R_2^2 + (1/C\omega)^2]^{\frac{1}{2}}},$$

and reactive component

$$I_{r2} = I_2 \sin\phi_2 = \frac{V}{[R_2^2 + (1/C\omega)^2]^{\frac{1}{2}}} \times \frac{1}{C\omega[R_2^2 + (1/C\omega)^2]^{\frac{1}{2}}}$$

The condition for current resonance is thus

$$\frac{L\omega}{R_1^2 + (L\omega)^2} = \frac{1/C\omega}{R_2^2 + (1/C\omega)^2}$$

Thus the following quantities can be varied in order to bring about current resonance: C, L, R_1, R_2, and f $(= \omega/2\pi)$.

Consider the effect of varying L. Mark off OA to represent the applied p.d. V (Fig. 15.26), and on it as diameter draw circle 1. Since the constants of branch 2 are fixed we can draw the voltage triangle OB_2A for this branch, marking off the angle ϕ_2 to the left because the reactance

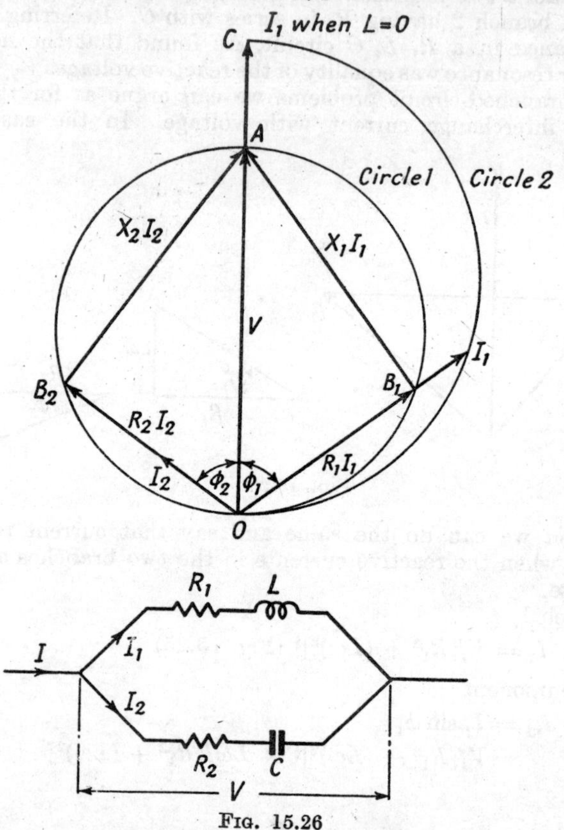

Fig. 15.26

of the branch is capacitive. Dividing OB_2 in volts by the known value of R_2, in ohms, we obtain the current I_2 which is marked off to an appropriate scale, as OI_2.

We have already seen that the triangle OB_1A is one possible voltage triangle for branch 1. If the branch current is OI_1 then there is a constant ratio between the lengths of OI_1 and OB_1 from which it follows that the locus of I_1 is another semicircle, whose diameter is OA produced to C, where point C corresponds to the condition L zero. Hence, on the current scale, $OC = V/R_1$.

Now I_2 is a fixed current, while I_1 is a variable current which is represented by circle 2 as locus. As the total current is the vector sum of I_1 and I_2, we move circle 2 bodily through a distance equal to, and parallel to, OI_2, the lower end of the diameter thus moving from O to point I_2. This gives circle 3 (Fig. 15.27) as the locus of the total current I. We can derive several conditions from the diagram—

(a) As the diagram is drawn there is one value of the total current, namely OP, which gives current resonance in the sense that the overall power factor is unity.

(b) If point Q were below point C, as at Q', there would be two

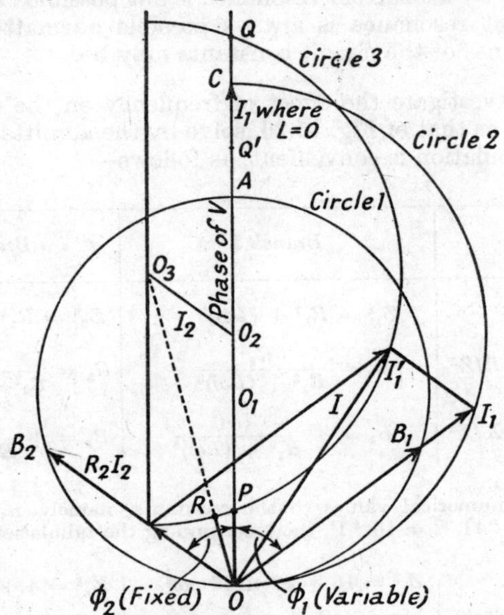

FIG. 15.27. LOCUS DIAGRAM FOR BRANCHED CIRCUIT WITH INDUCTANCE VARIED

different values of current for current resonance. The condition for two values in $V/2R_1 > I_2 \sin \phi_2$.

(c) If $V/2R_1 < I_2 \sin \phi_2$, circle 3 will not cut the perpendicular through O, and consequently there will be no resonance. This is the reverse of the series case in which the only condition for resonance is $L\omega = 1/C\omega$.

(d) With finite resistances the resonant condition is not the same as the condition of maximum current. This is shown in Fig. 15.27, in which the current of resonance is OP, but the maximum current is OR.

A similar locus diagram can be derived for the case in which C is varied. Now let R_1, R_2, L, and C be kept constant and let the frequency be varied. Solving the equation for current resonance in the branched circuit of Fig. 15.26 we have

$$\omega = [1/\sqrt{(LC)}]\,[(R_1{}^2C - L)/(R_2{}^2C - L)]^{\frac{1}{2}}$$

or

$$f = [1/2\pi\sqrt{(LC)}]\,[(R_1{}^2C - L)/(R_2{}^2C - L)]^{\frac{1}{2}}$$

(*a*) If $R_1 = 0$ and $R_2 = 0$, then the frequency for resonance is $1/2\pi\sqrt{(LC)}$, the same as for series resonance.

(*b*) If R_1 and R_2 are both very small then the frequency for resonance is very approximately the same as for case (*a*).

(*c*) If $R_1^2C < L$ while $R_2^2C > L$, or if $R_1^2C > L$ while $R_2^2C < L$, then the quantity in brackets is negative and its square root imaginary. In such a case current resonance is not possible. In the R, L, C series circuit resonance is always possible no matter what the individual values of the circuit constants may be.

In order to investigate the effect of frequency on the performance of such a circuit as that of Fig. 15.26, solve by the admittance method. Two column tabulation is convenient, as follows—

Quantity	*Branch* 1	*Branch* 2
(Impedance)², Z^2	$Z_1^2 = R_1^2 + (L\omega)^2$	$Z_2^2 = R_2^2 + (1/L\omega)^2$
Conductance, $G = R/Z^2$	$G_1 = \dfrac{R_1}{R_1^2 + (L\omega)^2}$	$G_2 = \dfrac{R_2}{R_2^2 + (1/L\omega)^2}$
Susceptance, $B = X/Z^2$	$B_1 = -\dfrac{L\omega}{R_1^2 + (L\omega)^2}$	$B_2 = +\dfrac{(1/C\omega)}{R_2^2 + (1/C\omega)^2}$

Suppose we give numerical values to the constants, namely $R_1 = 4\ \Omega$, $L = 7 \times 10^{-2}$ H, $R_2 = 2\ \Omega$, $C = 10^{-4}$ F, then, continuing the tabulation, we have

Z^2	$Z_1^2 = 16 + 49 \times 10^{-4}\omega^2$	$Z_2^2 = 4 + 10^8/\omega^2$
G	$G_1 = \dfrac{4}{16 + 49 \times 10^{-4}\omega^2}$	$G_2 = \dfrac{2}{4 + 10^8/\omega^2}$
B	$B_1 = -\dfrac{7 \times 10^{-2}\omega}{16 + 49 \times 10^{-4}\omega^2}$	$B_2 = +\dfrac{10^4/\omega}{4 + 10^8/\omega^2}$

By giving ω a series of values the characteristic curves similar to those of Fig. 15.16 for the series circuit can be drawn. First, determine the approximate value of f for resonance, on the assumption that R_1 and R_2 are small. The value is

$$f = 1/2\pi\sqrt{(LC)} = 1/2\pi\sqrt{(7 \times 10^{-2} \times 10^{-4})} = 60{\cdot}17 \text{ c/s}$$

The corresponding ω is

$$2\pi \times 60{\cdot}17 = 378{\cdot}1$$

It is therefore desirable to cluster the calculated points about this value of ω. The calculations are tabulated as shown on p. 337.

The graphs of admittance Y and overall power-factor are plotted in

ω	Z_1^2	$G_1 = R_1/Z_1^2$	$B_1 = -L\omega/Z_1^2$	Z_2^2	$G_2 = R_2/Z_2^2$	$B_2 = \frac{1/C\omega}{Z_2^2}$	$G = G_1 + G_2$	$B = B_1 + B_2$	$Y = (G^2 + B^2)^{\frac{1}{2}}$	$\cos\phi = G/Y$
0	16	0·25	∞	∞	0	0	0·25	$-\infty$	∞	1·0 lagging
100	65	0·0615	$1{\cdot}08 \times 10^{-1}$	10^4	2×10^{-4}	10^{-2}	6×10^{-2}	$-9{\cdot}8 \times 10^{-2}$	11×10^{-2}	0·545 ,,
200	212	$1{\cdot}89 \times 10^{-2}$	$6{\cdot}6 \times 10^{-2}$	2504	$7{\cdot}99 \times 10^{-4}$	2×10^{-2}	$1{\cdot}97 \times 10^{-2}$	$-4{\cdot}6 \times 10^{-2}$	5×10^{-2}	0·394 ,,
300	457	$8{\cdot}75 \times 10^{-3}$	$4{\cdot}59 \times 10^{-2}$	1115	$1{\cdot}79 \times 10^{-3}$	$2{\cdot}99 \times 10^{-2}$	$1{\cdot}05 \times 10^{-2}$	$-1{\cdot}6 \times 10^{-2}$	$1{\cdot}91 \times 10^{-2}$	0·550 ,,
350	619	$6{\cdot}46 \times 10^{-3}$	$3{\cdot}96 \times 10^{-2}$	820	$2{\cdot}44 \times 10^{-3}$	$3{\cdot}49 \times 10^{-2}$	$8{\cdot}9 \times 10^{-3}$	$-4{\cdot}7 \times 10^{-3}$	$10{\cdot}01 \times 10^{-3}$	0·890 ,,
375	705	$5{\cdot}67 \times 10^{-3}$	$3{\cdot}72 \times 10^{-2}$	715	$2{\cdot}80 \times 10^{-3}$	$3{\cdot}73 \times 10^{-2}$	$8{\cdot}47 \times 10^{-3}$	$+0{\cdot}1 \times 10^{-3}$	$8{\cdot}47 \times 10^{-3}$	1·0 leading
378·1	717	$5{\cdot}58 \times 10^{-3}$	$3{\cdot}69 \times 10^{-2}$	703	$2{\cdot}85 \times 10^{-3}$	$3{\cdot}76 \times 10^{-2}$	$8{\cdot}43 \times 10^{-3}$	$+0{\cdot}7 \times 10^{-3}$	$8{\cdot}43 \times 10^{-3}$	1·0 ,,
380	722	$5{\cdot}54 \times 10^{-3}$	$3{\cdot}68 \times 10^{-2}$	698	$2{\cdot}87 \times 10^{-3}$	$3{\cdot}77 \times 10^{-2}$	$8{\cdot}41 \times 10^{-3}$	$+0{\cdot}85 \times 10^{-3}$	$8{\cdot}46 \times 10^{-3}$	0·994 ,,
400	800	$5{\cdot}0 \times 10^{-3}$	$3{\cdot}6 \times 10^{-2}$	629	$3{\cdot}18 \times 10^{-3}$	$3{\cdot}97 \times 10^{-2}$	$8{\cdot}18 \times 10^{-3}$	$+3{\cdot}7 \times 10^{-3}$	$8{\cdot}98 \times 10^{-3}$	0·911 ,,
500	1241	$3{\cdot}22 \times 10^{-3}$	$2{\cdot}82 \times 10^{-2}$	404	$4{\cdot}95 \times 10^{-3}$	$4{\cdot}95 \times 10^{-2}$	$8{\cdot}17 \times 10^{-3}$	$+21{\cdot}3 \times 10^{-3}$	$22{\cdot}81 \times 10^{-3}$	0·358 ,,
700	2417	$1{\cdot}65 \times 10^{-3}$	$2{\cdot}03 \times 10^{-2}$	208	$9{\cdot}62 \times 10^{-3}$	$6{\cdot}88 \times 10^{-2}$	$11{\cdot}27 \times 10^{-3}$	$+48{\cdot}5 \times 10^{-3}$	$53{\cdot}38 \times 10^{-3}$	0·2111 ,,
1000	4900	$0{\cdot}82 \times 10^{-3}$	$1{\cdot}43 \times 10^{-2}$	104	$19{\cdot}2 \times 10^{-3}$	$9{\cdot}26 \times 10^{-2}$	$20{\cdot}2 \times 10^{-3}$	$+78{\cdot}3 \times 10^{-3}$	$80{\cdot}86 \times 10^{-3}$	0·250 ,,
∞	∞	0	0	4	2	∞	2	$+\infty$	∞	1·0 ,,
			all negative			all positive				

Fig. 15.28. If the applied p.d. V is kept constant then the line current given by $I = VY$, will be proportional to Y, with the result that the admittance curve is also, to a different scale, the curve of total current. We see that current resonance corresponds in this case to minimum total current, R_1 and R_2 being small. At the resonant frequency of

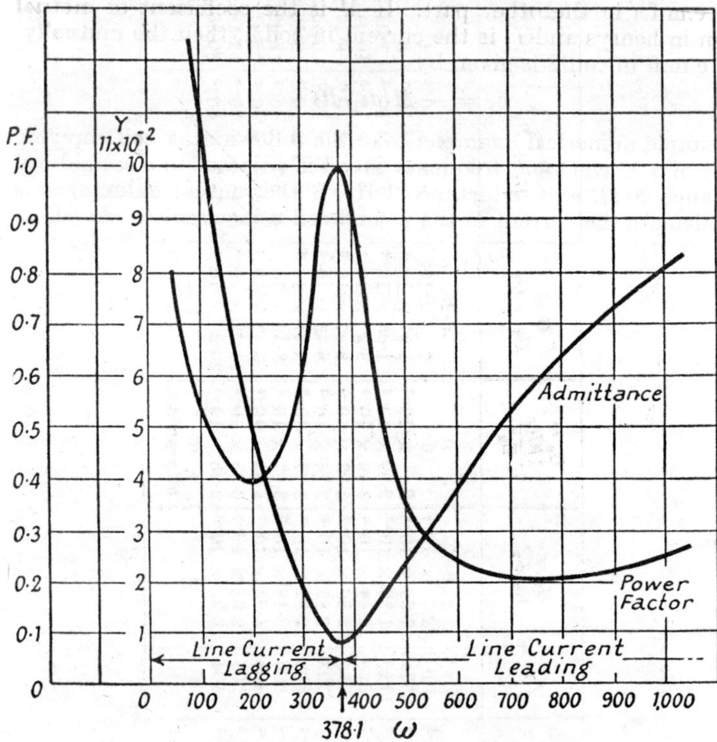

FIG. 15.28. EFFECT OF VARYING ω ON THE PERFORMANCE OF THE BRANCHED CIRCUIT OF FIG. 15.26

$378{\cdot}1/2\pi = 60{\cdot}18$ c/s, we see that the power-factor is unity, but that for values on either side the power factor falls away rapidly. The graphs then turn upwards, the power factor tending to unity both for $\omega = 0$ and for $\omega = \infty$. In the former case the branch current I_2 is obviously zero and the inductive reactance $L\omega$ is zero, the equivalent circuit at $\omega = 0$ thus being the single resistance R_1. When $\omega = \infty$ the branch current I_1 is obviously zero, and the capacitive reactance $-1/L\omega$ is zero, the equivalent circuit now being the single resistance R_2.*

* For a more complete treatment of the locus diagram *see* A. Hazeltine, "Current loci in the general linear a.c. network." *Elec. Eng.*, N.Y. **56** (1937) 325; A. T. Dover, *Theory and Practice of Alternating Currents*, Chap. VI. Pitman. London; M.I.T., *Electric Circuits*, Chap. IX. McGraw Hill, New York.

Circuit with Mutual Inductance. Since the magnetic field set up by any appliance may extend throughout a considerable space, it is inevitable that there must be many cases where the flux produced by one part of a circuit links with the windings in another part of the circuit. Thus changes in current in one part will be associated with induced e.m.f.s in the other part. If M is the coefficient of mutual induction in henrys and I_1 is the current in coil 1, then the mutually-induced e.m.f. in coil 2 is given by

$$e_2 = -M(di_1/dt)$$

As a simple numerical example, take the following: **a coil carrying a current of 5 A (r.m.s.) at frequency 50 c/s is adjacent to a second coil of resistance 20 Ω, and inductance 1 H. If the mutual inductance is 0·5 H, calculate the current in the second coil if its circuit is closed.**

$$i_1 = (I_1)_{\max} \sin \omega t$$

$$\therefore\ e_2 = -(d/dt)[(I_1)_{\max} \sin \omega t] \times M$$

$$= -(I_1)_{\max}\, \omega \cos \omega t \times M$$

$$\therefore\ (E_2)_{\max} = M\omega(I_1)_{\max}$$

$$\therefore\ E_2 = M\omega I_1$$

$$= 0{\cdot}5 \times 2\pi \times 50 \times 5$$

$$= 785 \text{ V}$$

$$\therefore\ I_2 = E_2/(R_2^2 + X_2^2)^{\frac{1}{2}}$$

$$= 785/(20^2 + (1 \times 314)^2)^{\frac{1}{2}}$$

$$= 2{\cdot}5 \text{ A}$$

EXAMPLE. Two coils, with terminals T_1T_2 and T_3T_4 respectively, are placed side by side. Measured separately, the inductance of the

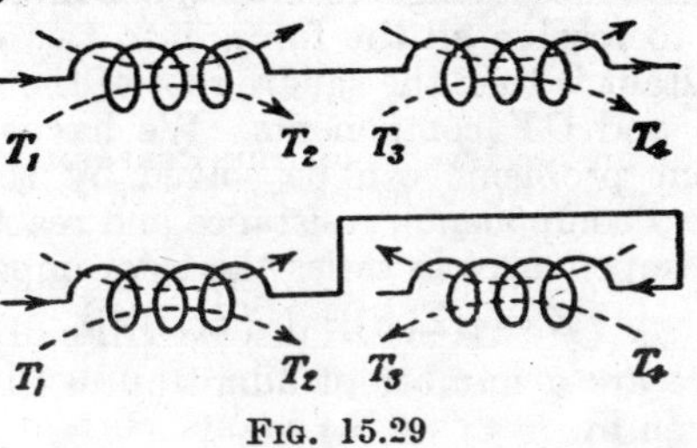

FIG. 15.29

first coil is 1200 μH, and that of the second coil is 800 μH. When T_2 is joined to T_3 the inductance between T_1 and T_4 is 2500 μH. What is the mutual inductance between the two coils, and what would be the inductance between T_1 and T_3 with T_2 joined to T_4? (*Lond. Univ.*)

Let L_1 and L_2 be the two self-inductances and M the mutual inductance, Fig. 15.29. Then, when the combination carries a current i, we have

Self-induced e.m.f. in coil 1 $= -L_1(di/dt)$

Mutually-induced e.m.f. in coil 1 due to change of current in coil 2 $= \mp M(di/dt)$

Self-induced e.m.f. in coil 2 $= -L_2(di/dt)$

Mutually-induced e.m.f. in coil 2 due to change of current in coil 1 $= \mp M(di/dt)$

The $-$ sign is given to $M(di/dt)$ when self and mutual fluxes are in the same direction, and the $+$ sign when they are in opposition.

$\therefore$ Total induced e.m.f. in the circuit

$$= -(L_1 \pm 2M + L_2)(di/dt)$$

$\therefore$ Equivalent self-induction of the whole circuit

$$= L_1 \pm 2M + L_2$$

the $+$ sign now being taken when self and mutual fluxes are in the same direction, and vice versa.

Since the total of $2500 > (1200 + 800)$ we have

$$1200 + 2M + 800 = 2500$$

$$\therefore M = 250\ \mu\text{H}$$

Hence, when the connections of coil 2 are reversed, total equivalent self-induction

$$= 1200 - (2 \times 250) + 800$$

$$= 1500\ \mu\text{H}$$

The Symbolic Method. When a number of forces acting at a point have to be combined into a single resultant, one method of determining this resultant is to resolve all the forces into two directions at right angles. The resultant is then the square root of the sum of the squares of the total OX and OY components. We have already seen that alternating-current problems can be solved by this manner. Thus impedance has two components, resistance and reactance, and if there are a number of impedances in series the total impedance is given by

$$Z = [\{\Sigma(R)\}^2 + \{\Sigma(X)\}^2]^{\frac{1}{2}}$$

Similarly, if there are a number of admittances in parallel the total admittance is given by

$$Y = [\{\Sigma(G)\}^2 + \{\Sigma(B)^2]\}^{\frac{1}{2}}$$

A more generally applicable method is to express a vector in terms of two components at right angles in the following manner. In Fig. 15.30, a vector of length z is resolved into two components x and y at right angles. In the symbolic notation the length of the vector is expressed in the form

$$Z = x + jy$$

where

$$j = \sqrt{(-1)}$$

The quantity j, although an imaginary quantity, is here used to indicate a real operation, namely, that of rotating a vector in a counter-clockwise direction through 90°, and the appearance of the operator j in front of the component y means that this component is leading the x component by 90°. The operator j^2 in front of a vector, e.g. j^2a, means that the a vector is 90° ahead of y, and therefore 180° ahead of x. Similarly if the operator is j^3, e.g. j^3b, the vector b will be 270° ahead of x. If the operator is j^4 the vector operated on will be in phase with x.

Hence we have, for any vector $\mathbf{Z}$,

$$j^2\mathbf{Z} = -\mathbf{Z}$$
$$j^3\mathbf{Z} = -j\mathbf{Z}$$
$$j^4\mathbf{Z} = \mathbf{Z}$$

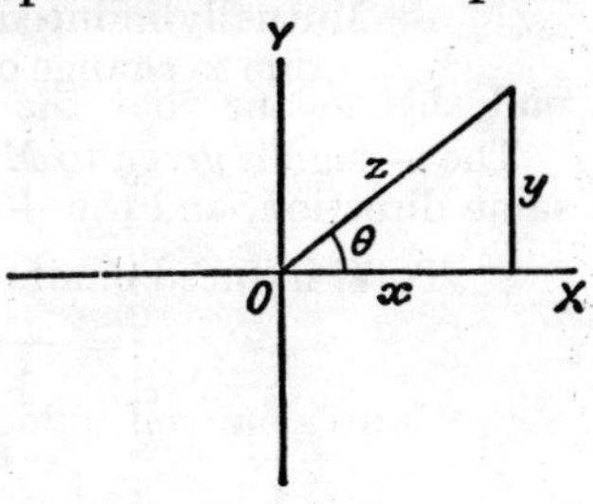

FIG. 15.30

Again, from the diagram

$$x = Z\cos\theta, \text{ and } y = Z\sin\theta$$

$$\therefore \mathbf{Z} = Z\cos\theta + jZ\sin\theta$$

This is the polar form and is commonly written in an abbreviated form

$$\mathbf{Z} = Z\underline{/\theta}$$

It can also be written in the exponential form

$$\mathbf{Z} = Ze^{j\theta}$$

the proof being as follows

$$e^{j\theta} = 1 + j\theta + \frac{(j\theta)^2}{1.2} + \frac{(j\theta)^3}{1.2.3} + \ldots \quad (1)$$

$$= 1 + j\theta - \frac{\theta^2}{1.2} - \frac{j\theta^3}{1.2.3} + \ldots$$

and
$$e^{-j\theta} = 1 - j\theta - \frac{\theta^2}{1.2} + \frac{j\theta^3}{1.2.3} + \ldots \quad (2)$$

Adding (1) and (2) and dividing by 2

$$\frac{e^{j\theta} + e^{-j\theta}}{2} = 1 - \frac{\theta^2}{1.2} + \frac{\theta^4}{1.2.3.4} - \ldots$$

$$= \cos\theta \quad (3)$$

Subtracting (1) and (2) and dividing by $2j$

$$\frac{e^{j\theta} - e^{-j\theta}}{2j} = \theta - \frac{\theta^3}{1.2.3} + \frac{\theta^5}{1.2.3.4.5} - \ldots$$

$$= \sin\theta \quad (4)$$

$$\therefore (e^{j\theta} - e^{-j\theta})/2 = j\sin\theta \quad (5)$$

Adding (3) and (5) we have

$$e^{j\theta} = \cos\theta + j\sin\theta$$

whence $$Z = Z\,\underline{/\theta} = Z\cos\theta + jZ\sin\theta = Z\,.\,e^{j\theta}$$

Suppose then that we have any vector quantity of magnitude P, then we can write this in the form

$$P = A\,(\cos\theta + j\sin\theta)$$

and this means that the operator $(\cos\theta + j\sin\theta)$ has rotated the

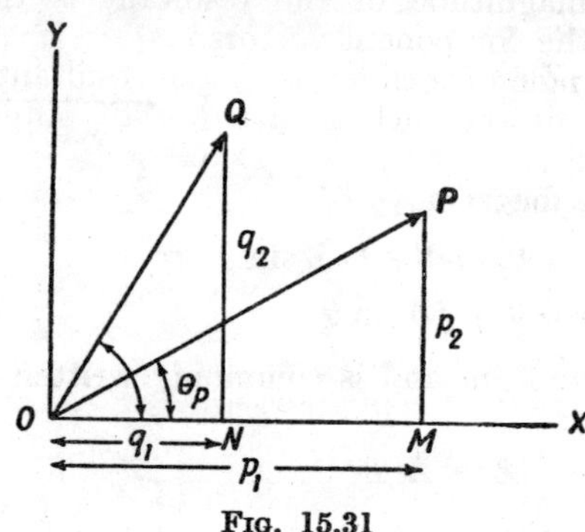

Fig. 15.31

vector through the angle θ in a *counter-clockwise* direction from the reference axis OX.

Multiplication of Vectors. Let the two vectors P and Q be written in the form

$$P = p_1 + jp_2, \text{ so that } P = \sqrt{(p_1^2 + p_2^2)}$$
$$Q = q_1 + jq_2, \text{ so that } Q = \sqrt{(q_1^2 + q_2^2)} \text{ (Fig. 15.31)}$$

then their product is

$$PQ = (p_1 + jp_2)(q_1 + jq_2)$$
$$= (p_1q_1 - p_2q_2) + j(p_1q_2 + p_2q_1)$$

This is a complex quantity also, and if we denote it by R we have

$$R = r_1 + jr_2$$

where $$r_1 = p_1q_1 - p_2q_2$$
$$r_2 = p_1q_2 + p_2q_1$$

The magnitude of R is therefore

$$R = \sqrt{(r_1^2 + r_2^2)}$$
$$= \sqrt{[(p_1q_1 - p_2q_2)^2 + (p_1q_2 + p_2q_1)^2]}$$
$$= \sqrt{[(p_1^2 + p_2^2)^2 \times (q_1^2 + q_2^2)^2]}$$
$$= PQ$$

Now express the vectors in the polar form

$$\boldsymbol{P} = P\,(\cos\theta_p + j\sin\theta_p)$$
$$\boldsymbol{Q} = Q\,(\cos\theta_q + j\sin\theta_q)$$
$$\therefore\ \boldsymbol{R} = \boldsymbol{PQ} = PQ[(\cos\theta_p\cos\theta_q - \sin\theta_p\sin\theta_q) + j(\sin\theta_p\cos\theta_q + \cos\theta_p\sin\theta_q)]$$
$$= PQ[\cos(\theta_p + \theta_q) + j\sin(\theta_p + \theta_q)]$$

or

$$\boldsymbol{R} = R\,[\cos(\theta_p + \theta_q) + j\sin(\theta_p + \theta_q)]$$

We thus see that when we multiply two vectors

(*a*) The magnitude of the resultant is the magnitude of the product of the component vectors.

(*b*) The angle of inclination of the resultant to the reference axis is the sum of the inclinations of the component vectors to the reference axis.

In the exponential form we have for the product

$$\boldsymbol{R} = \boldsymbol{PQ} = PQe^{j(\theta_p + \theta_q)}$$

Division of Vectors. The quotient is

$$\boldsymbol{S} = \frac{\boldsymbol{P}}{\boldsymbol{Q}} = \frac{P\,(\cos\theta_p + j\sin\theta_p)}{Q\,(\cos\theta_q + j\sin\theta_q)}$$
$$= \frac{P}{Q}\left(\frac{\cos\theta_p + j\sin\theta_p}{\cos\theta_q + j\sin\theta_q} \times \frac{\cos\theta_q - j\sin\theta_q}{\cos\theta_q - j\sin\theta_q}\right)$$
$$= \frac{P}{Q}\left[\frac{(\cos\theta_p\cos\theta_q + \sin\theta_p\sin\theta_q) + j\,(\sin\theta_p\cos\theta_q - \cos\theta_p\sin\theta_q)}{\cos^2\theta_q + \sin^2\theta_q}\right]$$
$$= \frac{P}{Q}[\cos(\theta_p - \theta_q) + j\sin(\theta_p - \theta_q)]$$

or $\boldsymbol{S} = S[\cos(\theta_p - \theta_q) + j\sin(\theta_p - \theta_q)]$

We thus see that when we divide two vectors

(*a*) The magnitude of the quotient is the same as the quotient of the two component vectors.

(*b*) The resultant is another vector quantity whose inclination to the reference axis is the *difference* of the inclination of the two components.

We thus have the alternative forms

$$\boldsymbol{S} = \frac{\boldsymbol{P}}{\boldsymbol{Q}} = \frac{P}{Q}\,\underline{/\theta_p - \theta_q}\,;\quad \boldsymbol{S} = \frac{P}{Q}e^{j(\theta_p - \theta_q)}$$

The Reciprocal of $(\cos\theta + j\sin\theta)$

$$\frac{1}{\cos\theta + j\sin\theta} = \frac{1}{\cos\theta + j\sin\theta} \times \frac{\cos\theta - j\sin\theta}{\cos\theta - j\sin\theta}$$
$$= \frac{\cos\theta - j\sin\theta}{\cos^2\theta + \sin^2\theta}$$
$$= \cos\theta - j\sin\theta$$
$$= \cos(-\theta) + j\sin(-\theta)$$

Hence the reciprocal is an operator which sets a vector back by the angle θ, or, in other words, produces an angular displacement θ in a negative sense.

LOGARITHM OF A COMPLEX NUMBER. Expressing the number in the exponential form we have

$$\boldsymbol{P} = P e^{j\theta}$$

$$\therefore \log_e \boldsymbol{P} = \log_e P + j\theta$$

Now $\quad P = \sqrt{(p_1{}^2 + p_2{}^2)}$ and $\theta = \tan^{-1}(p_2/p_1)$

$$\therefore \log_e \boldsymbol{P} = \log_e \sqrt{(p_1{}^2 + p_2{}^2)} + j\tan^{-1}(p_2/p_1)$$

showing that the logarithm is itself a complex number.

Application to Electric Circuit Calculations. In the case of an impedance Z made up of ohmic resistance R in series with reactance X, we have

$$\boldsymbol{Z} = R + jX = Ze^{j\theta}$$
$$= Z\underline{/\ \theta}$$

where $\quad \theta = \tan^{-1}(X/R)$

The admittance can be expressed symbolically in the following manner—

$$Y = 1/Z$$

$$\therefore \ Y = 1/Z = 1/(R + jX)$$
$$= (R - jX)/(R + jX)(R - jX)$$
$$= R/(R^2 + X^2) - jX/(R^2 + X^2)$$
$$= G - jb$$

$$\therefore \ Y = G - jb = Ye^{-j\theta} = Y\underline{/ -\theta}, \text{ or } Y\overline{\backslash\theta}$$

We will now consider several applications of the method to ordinary alternating-current calculations.

THE RESULTANT OF A NUMBER OF ALTERNATING E.M.F.S ALL ACTING AT THE SAME TIME IN A GIVEN CIRCUIT. Suppose there are two e.m.f.s represented by

$$e_1 = E_1 \sin \omega t, \text{ and } e_2 = E_2 \sin(\omega t + \phi)$$

Then taking $\boldsymbol{E}_1$ as the reference vector, we have

$$\boldsymbol{E}_1 = E_1 + j\,.\,0$$

and $\quad \boldsymbol{E}_2 = E_2 \cos\phi + j\,.\,E_2 \sin\phi$

$$\therefore \ \boldsymbol{E} = \boldsymbol{E}_1 + \boldsymbol{E}_2 = (E_1 + E_2\cos\phi) + j\,.\,E_2\sin\phi$$

Writing $\quad \boldsymbol{E} = X + j\,.\,Y$, we have

$$X = (E_1 + E_2\cos\phi)$$

and $\quad Y = E_2 \sin\phi$, thus giving

$$E = [(E_1 + E_2\cos\phi)^2 + E_2{}^2\sin^2\phi]^{\frac{1}{2}}$$

and $\quad \tan\theta = (Y/X) = E_2\sin\phi/(E_1 + E_2\cos\phi)$

Example (1) *Simple Series Circuit.* Fig. 15.32 shows a circuit consisting of two inductive resistances in series with one another, the resistances and reactances 5 and 6, 3 and 7 respectively. A sinusoidal p.d. of effective value 200 is applied to the combination. Calculate the current and power factor, and the voltage drops along the two coils.

The expression for impedance in symbolic notation is

$$Z = R + jX$$

$$\therefore\ Z = (5 + 3) + j(6 + 7) = 8 + j13$$

Put $V = 200 + j\,.\,0$

$$\therefore\ I = 200/(8 + j13)$$

$$= \frac{200(8 - j13)}{(8 + j13)(8 - j13)} = \frac{200(8 - j13)}{64 + 169} = 6{\cdot}88 - j11{\cdot}15$$

$$\therefore\ I = \sqrt{[(6{\cdot}88)^2 + (11{\cdot}15)^2]} = 13{\cdot}2\text{ A}$$

This is the effective value since we used the effective value of 200

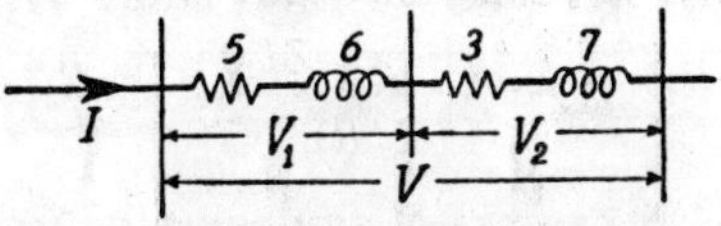

Fig. 15.32

for the applied p.d. Again, the minus sign in the expression for I shows that the current lags behind V, the angle, ϕ being

$$\phi = \tan^{-1}(11{\cdot}15/6{\cdot}88) = 58°\ 16'$$

and $$\text{p.f.} = \cos\phi = 0{\cdot}524$$

Again, for the voltage drops, we have

$$V_1 = IZ_1 = (6{\cdot}88 - j\,.\,11{\cdot}15)(5 + j\,.\,6)$$

$$= (101{\cdot}3 - j\,.\,14{\cdot}47)$$

$$\therefore\ V_1 = \sqrt{[(101{\cdot}3)^2 + (14{\cdot}47)^2]} = 102{\cdot}5\text{ V}$$

its phase angle with respect to V being

$$\phi_1 = \tan^{-1}(-\,14{\cdot}47/101{\cdot}3) = \tan^{-1}(-\,0{\cdot}1429) = -\,8°\ 13'$$

Similarly

$$V_2 = IZ_2 = (6{\cdot}88 - j\,.\,11{\cdot}15)(3 + j\,.\,7)$$

$$= 98{\cdot}7 + j\,.\,14{\cdot}5$$

$$\therefore\ V_2 = \sqrt{[(98{\cdot}7)^2 + (14{\cdot}5)^2]} = 99{\cdot}7\text{ V}$$

its phase angle with respect to V being

$$\phi_2 = \tan^{-1}(14{\cdot}5/98{\cdot}7) = \tan^{-1}0{\cdot}1471 = +\,8°\ 14'$$

We have also as a check

$$V = V_1 + V_2$$

$$= (101{\cdot}3 - j\,.\,14{\cdot}47) + (98{\cdot}7 + j\,.\,14{\cdot}5)$$

$$= 200 + j\,.\,0$$

EXAMPLE (2) *A Divided Circuit.* When making calculations on a divided circuit it is often convenient to reduce the circuit to the equivalent series circuit containing resistance and reactance. Thus let the symbolic impedances of the individual branches be $(R_1 + jX_1)$, $(R_2 + jX_2)$, $(R_3 + jX_3)$, etc., and the impedance of the whole branched circuit be $(R + jX)$. Then, since

$$I = I_1 + I_2 + I_3 + \ldots$$

$$\frac{V}{R+jX} = \frac{V}{R_1+jX_1} + \frac{V}{R_2+jX_2} + \frac{V}{R_3+jX_3} + \ldots$$

Dividing throughout by V and rationalizing, we have

$$\frac{R-jX}{R^2+X^2} = \frac{R_1-jX_1}{R_1^2+X_1^2} + \frac{R_2-jX_2}{R_2^2+X_2^2} + \frac{R_3-jX_3}{R_3^2+X_3^2} + \ldots$$

Equating real and imaginary parts, we have

$$\frac{R}{R^2+X^2} = \frac{R_1}{R_1^2+X_1^2} + \frac{R_2}{R_2^2+X_2^2} + \frac{R_3}{R_3^2+X_3^2} + \ldots$$

and

$$\frac{X}{R^2+X^2} = \frac{X_1}{R_1^2+X_1^2} + \frac{X_2}{R_2^2+X_2^2} + \frac{X_3}{R_3^2+X_3^2} + \ldots$$

In other words

$$G = G_1 + G_2 + G_3 + \ldots$$

and

$$B = B_1 + B_2 + B_3 + \ldots$$

the results thus being identical with those obtained by the admittance method.

As a numerical example consider the following circuit. A resistance of 10 Ω is in series with an inductive reactance of 6·28 Ω, and the combination is parallel with a circuit consisting of a resistance of 20 Ω in series with a capacitance of 100 μF. The above branched circuit is in series with a resistance of 15 Ω and an inductive reactance 15·7 Ω,

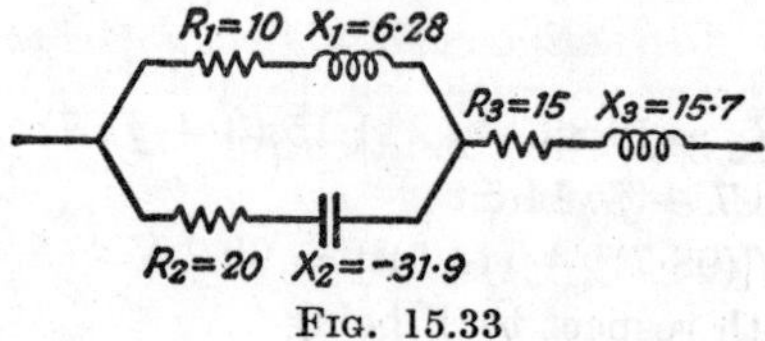

FIG. 15.33

Fig. 15.33. Calculate the equivalent resistance, reactance, and impedance of the whole circuit.

The reactance of the capacitor, assuming a 50 c/s supply, is

$$X_2 = -1/314C = -10^6/(314 \times 100) = -31{\cdot}9\ \Omega$$

The circuit is, therefore, as shown in Fig. 15.33.

For branch 1

$$Y_1 = 10/(10^2 + 6{\cdot}28^2) - j\,.\,6{\cdot}28/(10^2 + 6{\cdot}28^2)$$
$$= 0{\cdot}0718 - j \times 0{\cdot}045$$

For branch 2

$$Y_2 = 20/(20^2 + 31{\cdot}9^2) - j\,.\, - 31{\cdot}9/(20^2 + 31{\cdot}9^2)$$
$$= 0{\cdot}0141 + j \times 0{\cdot}0225$$

For the branched portion, we thus have

$$Y_{1\cdot2} = (0{\cdot}0718 + 0{\cdot}0141) - j(0{\cdot}045 - 0{\cdot}0225)$$
$$= 0{\cdot}0859 - j \times 0{\cdot}0225$$
$$\therefore\ R_{1\cdot2} = 0{\cdot}0859/(0{\cdot}0859^2 + 0{\cdot}0225^2) = 10{\cdot}87$$

and $X_{1\cdot2} = 0{\cdot}0225/(0{\cdot}0859^2 + 0{\cdot}0225^2) = 2{\cdot}86$ (and is positive)

Hence, for the whole circuit, we have

$$R = R_{1\cdot2} + R_3 = 10{\cdot}87 + 15 = 25{\cdot}87\ \Omega$$
$$X = X_{1\cdot2} + X_3 = 2{\cdot}86 + 15{\cdot}7 = 18{\cdot}56\ \Omega$$
$$Z = (25{\cdot}87^2 + 18{\cdot}56^2)^{\frac{1}{2}} = 31{\cdot}8\ \Omega$$

In order to determine the manner in which a current divides between two parallel branches, we have the two equations

$$\left.\begin{aligned} I &= I_1 + I_2 \\ I_1/I_2 &= Z_2/Z_1 \end{aligned}\right\}$$

and

These give

$$I_1/I = Z_2/(Z_1 + Z_2)$$
$$= (R_2 + jX_2)/[(R_1 + R_2) + j(X_1 + X_2)]$$
$$\therefore\ I_1/I = [(R_2{}^2 + X_2{}^2)/\{(R_1 + R_2)^2 + (X_1 + X_2)^2\}]^{\frac{1}{2}}$$

and similarly

$$I_2/I = [(R_1{}^2 + X_2{}^2)/\{(R_1 + R_2)^2 + (X_1 + X_2)^2\}]^{\frac{1}{2}}$$

EXAMPLE (3) *Alternating-current Bridge Networks.* Consider first of all a simple network such as that shown in Fig. 15.34. The condition for a balance with direct current is

$$R_1/R_2 = R_3/R_4$$

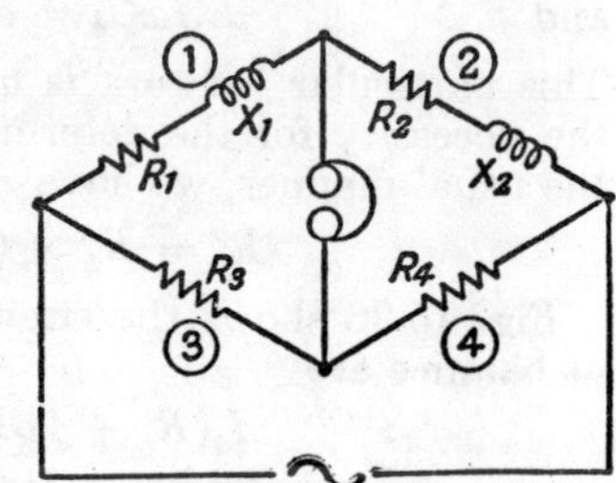

FIG. 15.34. SIMPLE A.C. BRIDGE NETWORK

The condition for a balance with alternating current is

$$Z_1/Z_2 = Z_3/Z_4$$
$$\therefore\ \frac{R_1 + jX_1}{R_2 + jX_2} = \frac{R_3 + j\,.\,0}{R_4 + j\,.\,0}$$
$$\therefore\ R_1R_4 + jX_1R_4 = R_2R_3 + jX_2R_3$$

Equating real and imaginary parts, we have

$$R_1/R_2 = R_3/R_4$$

as for direct current

and $$X_1/X_2 = R_3/R_4$$

Hence $$R_1/R_2 = R_3/R_4 = X_1/X_2$$

As a more complicated bridge network take the case of the Wien bridge shown in Fig. 15.35*. This consists of two non-inductive arms R_1 and R_3, a shunted capacitor for arm 2, and a capacitor in series,

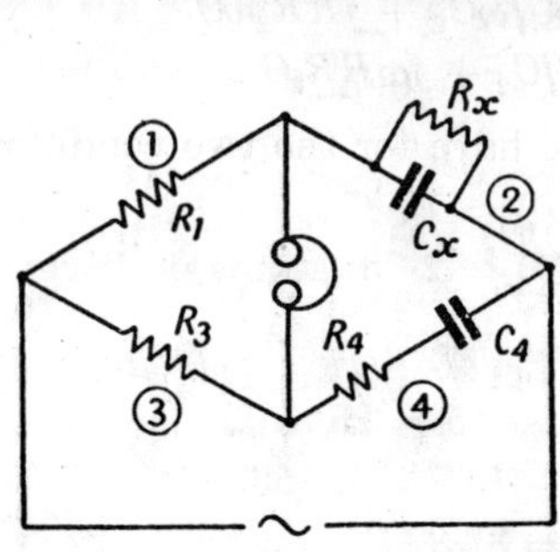

FIG. 15.35. WIEN A.C. BRIDGE

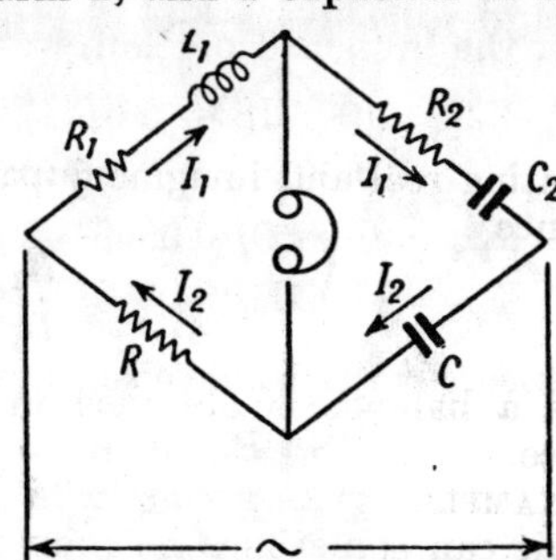

FIG. 15.36. OWEN'S BRIDGE

with a non-inductive resistance for arm 4. First of all reduce arm 2 to an equivalent series circuit of resistance R_2 and reactance x_2; we then have

$$R_2 = \frac{R_x}{1 + \omega^2 C_x^2 R_x^2} \text{ and } X_2 = -\frac{1}{\omega C_x(1 + 1/\omega^2 C_x^2 R_x^2)}$$

and for arm 4 $$X_4 = -1/\omega C_4$$

The condition for a balance is thus

$$\frac{R_1}{R_3} = \frac{R_x/(1 + \omega^2 C_x^2 R_x^2) - j \,.\, 1/\omega C_x(1 + 1/\omega^2 C_x^2 R_x^2)}{R_4 - j(1/\omega C_4)}$$

Equating real and imaginary parts, as before, this gives

$$\left.\begin{aligned} R_1 R_4 &= R_3 R_x/(1 + \omega^2 C_x^2 R_x^2) \\ \text{and} \quad R_1/\omega C_4 &= \omega C_x R_3 R_x^2/(1 + \omega^2 C_x^2 R_x^2) \end{aligned}\right\}$$

This particular network is used for the determination of R_x, without the necessity for the determination of C_x, hence if C_x is eliminated in the usual manner, we have

$$R_x = R_1 \times (1 + \omega^2 C_4^2 R_4^2)/\omega^2 R_3 R_4 C_4^2$$

Fig. 15.36 shows the circuit of the Owen's Bridge. The conditions at balance are

$$I_1(R_1 + j\omega L_1) = I_2 R$$
$$I_1(R_2 - j/\omega C_2) = I_2(-j/\omega C)$$

* *See* B. Hague, *A.C. Bridge Measurements*. Pitman, London.

$$\therefore \frac{R_1 + j\omega L_1}{R_2 - j/\omega C_2} = -\frac{R}{j/\omega C}$$

$$R_1 + j\omega L_1 = -\frac{R(R_2 - j/\omega C_2)}{j/\omega C}$$

$$= \left[\frac{R(R_2 - j/\omega C_2)}{j^2/\omega C}\right] \times -j$$

$$= R\omega C/\omega C_2 + jRR_2\omega C$$

$$= RC/C_2 + j\omega RR_2C$$

Equating real and imaginary parts we have for the two conditions of balance

$$R_1 = RC/C_2$$

$$L_1 = RR_2C$$

Thus a balance is obtained independently of the frequency of the source.

EXAMPLE (4) *Capacitor with Finite Dielectric Loss.* The loss in a circuit can sometimes be represented by the I^2R loss in a resistor of appropriate value and location. Thus, in the case of a capacitor with finite dielectric loss this loss can be replaced by that in a series resistor, or in a shunted resistor. Suppose that an actual capacitor is equivalent to (*a*) a capacitance C_1 in series with resistance R_1, or (*b*) a capacitance C_2 in parallel with resistance R_2. The two equivalents are shown in Fig. 15.37.

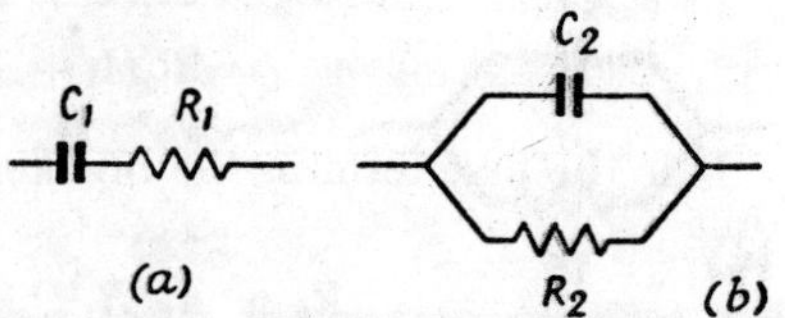

FIG. 15.37. EQUIVALENT CIRCUITS FOR CAPACITOR WITH FINITE DIELECTRIC LOSS

For the series equivalent we have

$$Z_1 = R_1 - j/C_1\omega$$

For the parallel equivalent we have

$$Z_2 = \frac{R_2 \times (-j/C_2\omega)}{R_2 + (-j/C_2\omega)} = \frac{R_2(1 - jR_2C_2\omega)}{1 + R_2^2C_2^2\omega^2}$$

If the two are electrically equivalent Z_1 and Z_2 must be equal

$$\therefore R_1 - j/C_1\omega = R_2/(1 + R_2^2C_2^2\omega^2) - jR_2^2C_2\omega/(1 + R_2^2C_2^2\omega^2)$$

Equating real and imaginary parts we have

$$R_1 = R_2/(1 + R_2^2C_2^2\omega^2) \quad . \quad . \quad . \quad (1)$$

and

$$1/C_1\omega = R_2^2C_2\omega/(1 + R_2^2C_2^2\omega^2) \quad . \quad . \quad (2)$$

$$\therefore R_1C_1\omega = 1/R_2C_2\omega \quad . \quad . \quad . \quad . \quad . \quad (3)$$

(by dividing (1) by (2))

Again, from (2)

$$C_1/C_2 = (1 + R_2^2C_2^2\omega^2)/R_2^2C_2^2\omega^2$$

$$\therefore\ (C_1 - C_2)/C_2 = 1/R_2^2C_2^2\omega^2$$

while from (3),

$$R_1^2C_1^2\omega^2 = 1/R_2^2C_2^2\omega^2 = (C_1 - C_2)/C_2$$

which gives $$C_2 = C_1/(1 + R_1^2C_1^2\omega^2)$$

This is as far as we can go unless we can make assumptions based on actual conditions. Fig. 15.38 shows the impedance triangle for the series arrangement, and, generally, the angle ϕ_1 is very nearly equal to 90°. The tangent

$$R_1 \div 1/C_1\omega = R_1C_1\omega$$

is therefore small, so that its square, $R_1^2C_1^2\omega^2$ is negligible in comparison with unity. The equation for C_1 thus reduces to $C_1 \simeq C_2$. From equation (1), we have

$$R_2 = R_1(1 + R_2^2C_2^2\omega^2)$$

$$= R_1(1 + 1/R_1^2C_1^2\omega^2) \quad \text{(from equation (3))}$$

Fig. 15.38

It is the first term in the brackets which is now negligibly small, so that

$$R_2 \simeq R_1/R_1^2C_1^2\omega^2 \simeq 1/R_1C_1^2\omega^2$$

Calculation of Power. We have

$$\boldsymbol{E} = E(\cos\alpha + j\sin\alpha)$$

and $$\boldsymbol{I} = I(\cos\beta + j\sin\beta)$$

$(\alpha - \beta)$ being the phase difference between E and I.

Multiplying these, we have

$$\boldsymbol{EI} = EI[(\cos\alpha\cos\beta - \sin\alpha\sin\beta) + j(\sin\alpha\cos\beta + \cos\alpha\sin\beta)]$$

$$= EI[\cos(\alpha + \beta) + j\sin(\alpha + \beta)]$$

The real part of this is $EI\cos(\alpha + \beta)$, and this obviously is not the power in the circuit, since this should be $EI\cos(\alpha - \beta)$. In addition, the imaginary part is $EI\sin(\alpha + \beta)$, and this also is not the correct expression for the reactive volt-amperes. Hence, when calculating power by the symbolic method, it is necessary to reverse the imaginary part of either $\boldsymbol{E}$ or $\boldsymbol{I}$. Thus, if we leave $\boldsymbol{E}$ as written above, but write

$$\boldsymbol{I} = I(\cos\beta - j\sin\beta)$$

then $$\boldsymbol{EI} = EI[\cos(\alpha - \beta) + j\sin(\alpha - \beta)]$$

The real term is the *mean* power, and the imaginary term gives the reactive volt-amperes.

Alternatively we can write the voltage and current in the form

$$\boldsymbol{E} = Ee^{j\alpha}; \; \boldsymbol{I} = Ie^{j\beta}$$

Reversing the sign of β, we have

$$\boldsymbol{I}_1 = I^{-}e^{j\beta}$$

$$\therefore \; \boldsymbol{EI}_1 = EIe^{j(\alpha-\beta)}$$

$$= EI\,[\cos(\alpha - \beta) + j\sin(\alpha - \beta)]$$

as before. The vector obtained by reversing the imaginary part of an original vector is sometimes referred to as the conjugate of the original vector.

As an example let

$$\boldsymbol{E} = 100 + j\,200$$

$$\boldsymbol{I} = 25 + j\,10$$

Then

$$E = \sqrt{(100^2 + 200^2)} = 223{\cdot}6$$

$$I = \sqrt{(25^2 + 10^2)} = 26{\cdot}93$$

$$\alpha = \tan^{-1}(200/100) = 63°\,26'$$

$$\beta = \tan^{-1}(10/25) = 21°\,48'$$

$$\therefore \; W = EI\cos(\alpha - \beta)$$

$$= 223{\cdot}6 \times 26{\cdot}93 \times 0{\cdot}7474$$

$$= 4500 \text{ W}$$

Alternatively, since the imaginary voltage component of 200 is in phase with the current component of 10, while the real voltage component of 100 is in phase with the current component of 25, we have (Fig. 15.39)

$$W = 200 \times 10 + 100 \times 25$$

$$= 4500 \text{ W}$$

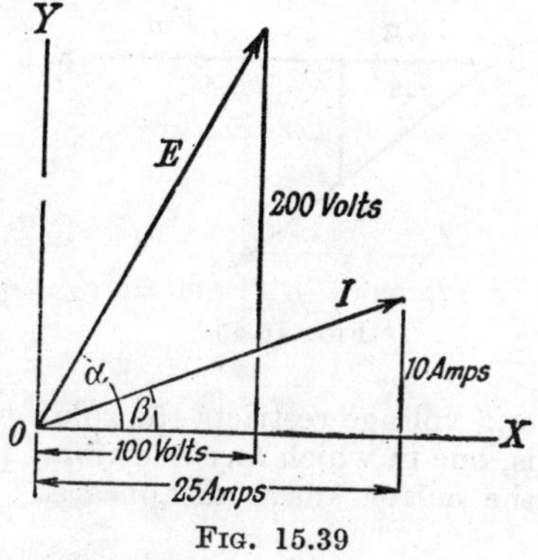

Fig. 15.39

Solution of the Differential Equation of the *R*, *L*, *C* Circuit. One method of solution affords an instructive application of the use of the operator j. In the R, L, C circuit the applied voltage v has three things to do, (1) supply the ohmic volt drop Ri, (2) overcome the self-induced e.m.f. $-L(di/dt)$, (3) supply the instantaneous p.d. at the capacitor plates, q/C. The component of v to overcome the second of these must obviously be $+L\,(di/dt)$, so that the voltage equation becomes

$$L(di/dt) + Ri + (q/C) = v$$

For a sinusoidal applied voltage we have

$$v = V_m e^{j0} = V_m e^{j\omega t}$$

time being reckoned from the instant the rotating vector crosses the OX-axis. The equation is now

$$L(di/dt) + Ri + (q/C) = V_m e^{j\omega t}$$

Differentiating with respect to t

$$L(d^2i/dt^2) + R(di/dt) + (i/C) = j\omega V_m e^{j\omega t}$$

For a sinusoidal current put $\qquad i = I_m e^{j\omega t}$

$$\therefore\ di/dt = j\omega I_m e^{j\omega t}$$

and $\qquad d^2i/dt^2 = -\omega^2 I_m e^{j\omega t}$

giving

$$-L\omega^2 I_m e^{j\omega t} + j\omega R I_m e^{j\omega t} + (I_m/C)e^{j\omega t} = j\omega V_m e^{j\omega t}$$

$$\therefore\ I_m = \frac{j\omega V_m}{j\omega R + (-L\omega^2 + 1/C)}$$

$$= \frac{V_m}{R + j(L\omega - 1/C\omega)}$$

We now proceed as before

$$I_m = \frac{V_m[R - j(L\omega - 1/C\omega)]}{R^2 + (L\omega - 1/C\omega)^2}$$

The denominator is the square of the impedance Z, while $(L\omega - 1/C\omega)$ is the total reactance X

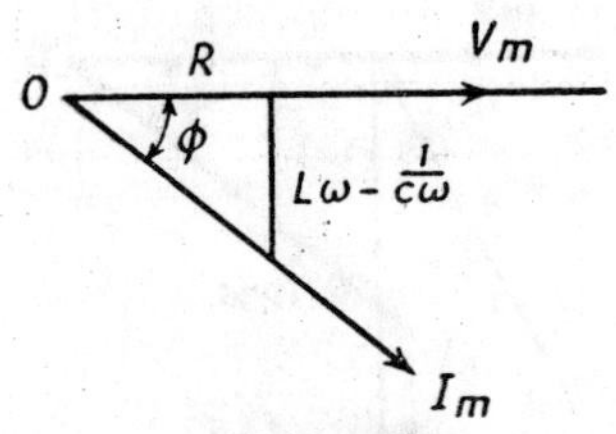

FIG. 15.40

$$\therefore\ I_m = V_m[R/Z^2 - j(X/Z^2)]$$

$$\therefore\ I_m^2 = V_m^2(R^2/Z^4 + X^2/Z^4)$$

$$\therefore\ I_m = V_m/Z$$

The relative phases of the vectors are given in Fig. 15.40, from which we see that I_m lags V_m by the angle $\phi = \tan^{-1}[(L\omega - 1/C\omega)/R]$.

The assumption of sinusoidal current and voltage restricts the solution to a steady-state condition only, that is, one in which any additional phenomena at the moment of starting the current have disappeared.

Starting Conditions. Consider a simple circuit having R and L in series, and let a sinusoidal p.d. of r.m.s. value V and frequency f be applied. Then, as we know, we have

$$I = V/[R^2 + (L\omega)^2]^{\frac{1}{2}} \text{ lagging } V \text{ by}$$

$$\phi = \cos^{-1}[R/\{R^2 + (L\omega)^2\}^{\frac{1}{2}}]$$

For the instantaneous values we have

$$v = V_m \sin \omega t$$

$$i = I_m \sin (\omega t - \phi)$$

The corresponding wave diagrams are given in Fig. 15.41. Suppose that the voltage is applied at the instant represented by point O, that is, when the voltage is zero, then we see that the current is finite and equal to OA. But the current must be zero at the instant of switching on, and furthermore it cannot attain any finite value suddenly because the circuit is inductive. Hence Fig. 15.41 does not represent initial conditions when the instant of switching on corresponds to the point

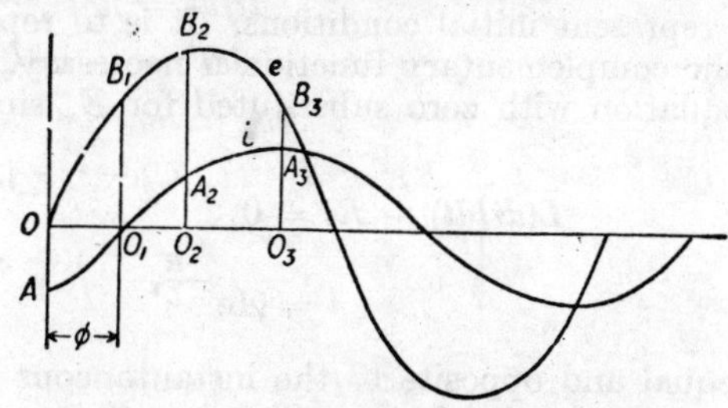

Fig. 15.41

O. Neither can it represent initial conditions when the instant of switching-on is, say O_2 or O_3, for these also correspond to finite current, namely O_2A_2 and O_3A_3 respectively. It can only represent initial conditions when the instant of switching on corresponds to point O_1, that is, to an instantaneous voltage of O_1B_1, for at that instant the current is zero. Hence, in general, the solutions we have used previously refer only to a circuit which has settled down to what are called "steady-state" conditions. At the instant of switching on, and for a short time afterwards, the circuit carries, in addition to the steady-state current, a transient current whose function it is to adjust the circuit conditions to the actual moment of starting. Thus, suppose that the voltage is switched on at the moment O in Fig. 15.41. The steady-state current is OA, which we will denote by $-I'$. Since the actual initial current is zero, the initial value of the transient current must be $+I'$. We can visualize this current rising instantaneously from zero to I' since initially there is no flux and therefore no induced e.m.f. This current only rises to the value I' because the flux associated with I' gives, with the equal and opposite flux associated with the steady state current of $-I'$, the circuit conditions appropriate to the moment of switching on. Having performed this function the transient current dies away, and as the circuit is inductive, it can only die away exponentially according to the law

$$i' = I' e^{-\frac{R}{L}t}$$

A more rigorous solution is as follows. The e.m.f. equation for the inductive circuit having R and L in series is

$$L(di/dt) + Ri = E_m \sin \omega t$$

Now the complete solution of this equation consists of two parts which are called the particular integral and the complementary function. The particular integral is the solution corresponding to the steady-state condition, namely

$$i = [E_m/\{R^2 + (L\omega)^2\}^{\frac{1}{2}}] \sin (\omega t - \phi)$$

This solution is obtained without any constants of integration, and cannot therefore represent initial conditions. It is to represent these conditions that the complementary function is necessary. This is the solution of the equation with zero substituted for $E_m \sin \omega t$, namely the solution of

$$L(di/dt) + Ri = 0$$

This is

$$i = Ae^{-\frac{R}{L}t}$$

where A is the equal and opposite to the instantaneous steady-state current at the moment of switching on. Since we wrote

$$e = E_m \sin \omega t$$

where e is the instantaneous e.m.f., we reckoned time from the instant $t = 0$. At any instant t' thereafter the instantaneous value of the steady-state current is

$$i' = I_m \sin (\omega t' - \phi)$$

and consequently the complementary function is given by

$$-I_m e^{-\frac{R}{L}(t-t')} \sin (\omega t' - \phi)$$

The complete solution is thus

$$i = I_m \sin (\omega t - \phi) - I_m e^{-\frac{R}{L}(t-t')} \sin (\omega t' - \phi)$$

The graph of current against time for an inductive circuit is shown in Fig. 15.42, from which it will be seen that the graph of total current gradually approaches that of the steady-state component, and ultimately merges with it.

An important case is that of a circuit in which R is so small that the steady-state current lags practically $\pi/2$ behind the voltage. In such a case L/R is so large that the transient component will die away very slowly. This case for switching at the instant $e = 0$ is shown in Fig. 15.43 and we see that the total current very nearly attains the value $2I_m$. This phenomenon is called current doubling.

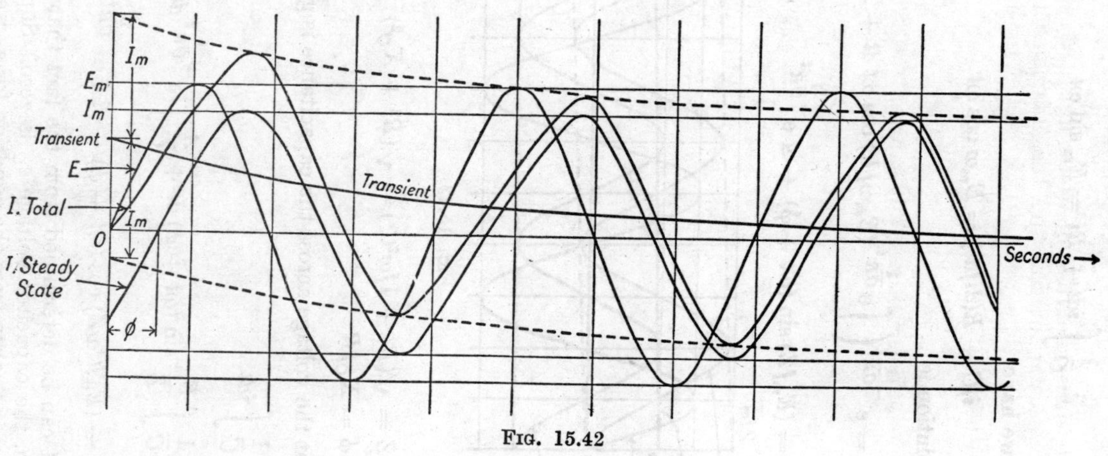

I_m
E_m
I_m
Transient
E
I. Total
I_m
O
I. Steady State
ϕ
Transient
Seconds →

Fig. 15.42

Now consider a circuit having resistance R in series with capacitance C. Then for a sinusoidally impressed e.m.f. the equation is

$$(q/C) + Ri = E_m \sin \omega t$$

$$\therefore \frac{1}{C}\int idt + Ri = E_m \sin \omega t$$

Differentiating, we have

$$i/C + R(di/dt) = E_m\omega \cos \omega t$$

The complete solution is

$$i = e^{-\frac{1}{CR}t}\left(\int e^{\frac{1}{CR}t}(E_m\omega/R)\cos \omega t\, dt + A\right)$$

$$i = (E_m/Z)\sin(\omega t + \phi) + A\, e^{-\frac{1}{CR}t}$$

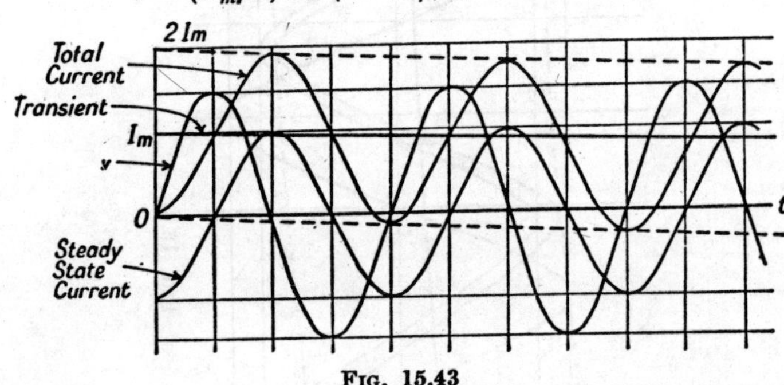

Fig. 15.43

where $\qquad Z = \sqrt{(R^2 + 1/\omega^2C^2)} = \sqrt{(R^2 + X_C{}^2)}$

and $\qquad \tan \phi = X_C/R$

At any instant the voltage across the capacitance is given by

$$e_C = \frac{q}{C} = \frac{1}{C}\int idt$$

$$= \frac{1}{C}\int \frac{E_m}{Z}\sin(\omega t + \phi)\, dt + \frac{A}{C}\int e^{-\frac{1}{CR}t}\, dt$$

$$= -(E_m/C\omega Z)\cos(\omega t + \phi) - RA\, e^{-\frac{1}{CR}t}$$

The constant A can be determined from the fact that at the instant of switching on, the capacitor voltage e_C is zero. Suppose then the switching instant is t', then from the equation for e_C we have

$$A = -(E_mX_C/RZ)e^{\frac{1}{CR}t'}\cos(\omega t' + \phi)$$

∴ For the complete solution we have

$$i = (E_m/Z)\sin(\omega t + \phi) - (E_m/Z)\,.\,(X_C/R)\,e^{-\frac{1}{CR}(t-t')}\cos(\omega t' + \phi)$$

$$i = I_m \sin(\omega t + \phi) - I_m X_C/R\,.\,e^{-\frac{1}{CR}(t-t')}\cos(\omega t' + \phi)$$

Thus the phenomenon is very similar to that for the inductive circuit, the total current being the resultant of a sinusoidal steady-state current and an exponentially decreasing direct-current transient.

The transients considered above are what are called "single energy" transients in that the energy associated with the particular transient is of one kind only. Thus in the inductive circuit it is magnetic energy seated in the magnetic field, whereas in the capacitive circuit it is electric energy stored in the dielectric. If a circuit has both inductance and capacitance the transient will be associated with energy of both kinds and will therefore be a "double-energy" transient.

Further Reading

Coulthard, W. B. *Transients in Electric Circuits.* Pitman, London, 1951.

Dover, A. T. *Theory and Practice of Alternating Currents.* Pitman, London, 1939.

Kutz, E. B. and Corcoran, G. F. *Introduction to Electric Transients.* Wiley, New York, 1935.

Russell, A. *Alternating Currents.* C.U.P., London, 1916.

M.I.T. *Electric Circuits.* Wiley, New York, 1940.

Tropper, H. *Electric Circuit Theory.* Longmans, London, 1949.

Carter, G. W. *Simple Calculations of Electric Transients.* C.U.P., London, 1944.

Ware, L. A. and Town, G. R. *Electrical Transients.* Macmillan, New York, 1954.

Examples on Chapter 15

1. An alternating current of frequency 50 c/s has a maximum value of 100 A. Calculate (*a*) its value 1/600 sec after the instant the current is zero; (*b*) in how many seconds after the zero value will the current attain the value of 86·6 A?

Ans. (*a*) 50 A; (*b*) $\frac{1}{300}$ sec.

2. Two currents represented by $i_1 = 50 \sin \omega t$ and $i_2 = 100 \sin (\omega t + \pi/4)$ are led into a common conductor. Find an expression for the total current in the form $i = I_{max} \sin (\omega t + \phi)$. If the conductor has a resistance of 10 Ω, calculate the number of calories produced in 20 min.

Ans. 28×10^6 g.cal.

3. Three voltages, represented by $e_1 = 20 \sin \omega t$, $e_2 = 30 \sin (\omega t - \pi/4)$, and $e_3 = 40 \cos (\omega t - \pi/6)$, act together in a circuit. Find an expression for the resulting voltage.

Ans. $E_{max} = 62{\cdot}6$ V.

4. A wooden ring of mean circumference 0·5 m and a circular section of 0·03 m diameter is uniformly wound with 400 turns of 1 mm copper wire. Calculate its resistance, inductance, and impedance at frequency 100 c/s. Take the specific resistance of copper to be $1{\cdot}7 \times 10^{-8}$ Ω-m.

Ans. (*a*) 0·82 Ω; (*b*) $2{\cdot}85 \times 10^{-4}$ H; (*c*) 0·83 Ω.

5. The air gap under the pole of a series motor is 0·06 in. in length, and the effective area is 80 in.² There are 5 turns of cable round the pole, having a resistance of 0·005 Ω. Assuming that the reluctance of the iron parts of the magnetic circuit is one-fifth of the reluctance of the air gap, calculate the voltage drop in a series coil when 150 A r.m.s. at 25 c/s are passing through it. (C. and G.)

Ans. 21 V.

6. A capacitor and an inductive conductor are placed in series across alternating current mains, whose voltage is kept constant while the frequency is varied. Show by means of a curve how the current varies with the frequency. What determines the maximum value reached by the current? If this maximum occurs at a frequency of 20 c/s, and if the capacitor has a capacitance of 50 μF, what is the value of the inductance? If the voltage of the mains contains a pronounced third harmonic, how will the curve connecting current with frequency be affected? (L.U.)

Ans. 1·27 H.

7. An alternating-current circuit includes two sections AB and BC in series. The section AB consists of two branches in parallel. The first of these is formed of a non-inductive resistance of 60 Ω in series with a capacitor of 50 μF, while the second consists of a resistance of 60 Ω having an inductance of 250 mH. The section BC consists of a resistance of 100 Ω having an inductance of 300 mH. The frequency of the current is 50 c/s. The voltage across the section AB is 500. What is the voltage across the section BC? (L.U.)

Ans. 960 V.

8. An air-cored choking coil is subjected to an alternating voltage of 100 V. The current taken is 0·1 A, and the power factor 0·2, when the frequency of the current is 50 c/s. Find the capacitance of a capacitor which, if placed in parallel with the coil, will cause the main current to be a minimum. What will be the impedance of this parallel combination (*a*) for currents of frequency 50 c/s, and (*b*) for currents of frequency 40 c/s? (L.U.)

Ans. 3·1 μF. (*a*) 5000 Ω; (*b*) 1940 Ω.

9. A choke coil of negligible resistance connected across a 500 V, 50 c/s circuit takes 1 A at 0·8 power factor. What capacitance must be placed in parallel with it in order to make the power factor of the combination equal to unity? (L.U.)

Ans. 3·82 μF.

10. A coil having a resistance of 5 Ω and an inductance of 0·02 H is arranged in parallel with another coil with a resistance of 1 Ω, and an inductance of 0·08 H. Calculate the current flowing through each coil when a pressure of 100 V at 50 c/s is applied to them. Find the total current passing, and estimate the resistance and inductance of a single coil which will take the same current at the same power factor. (L.U.)

Ans. 12·5, 3·97, and 15·9 A; 3·2 Ω and 0·0175 H.

11. A transmission line 40 miles long consists of two wires each 0·46 in. diameter, spaced 5 ft. If the conductors are short-circuited at the far end what voltage at frequency 25 c/s must be applied to produce a current of 200 A in the loop so formed?

Ans. 6140 V.

12. An alternating current passes through a non-inductive reactance R and an inductance L in series. Find the value of the non-inductive resistance which can be shunted across the inductance without altering the value of the main current. (L.U.)

Ans. $\omega^2 L^2/2R$.

CHAPTER 16

POLYPHASE CURRENTS

Up to the present, the alternating-current systems we have considered have possessed only one electrical circuit, as in the case of two-wire direct-current circuits. In alternating-current working it is possible to use two, three, or more individual circuits in the same apparatus or machine. The voltages and frequencies in the individual circuits are the same, but they have definite phase differences, the amounts of these differences depending on the number of circuits or "phases." The complication involved is only small, but the advantages of polyphase over single-phase working are enormous. They may be summed up as follows—

(*a*) For a given size of frame a polyphase generator or motor has a bigger output than a single phase.

(*b*) To transmit a given amount of power at a given voltage over a given distance, a polyphase transmission line requires less copper than a single-phase line.

(*c*) Polyphase motors have an absolutely uniform torque, whereas single-phase motors (except commutator motors) have a pulsating torque.

(*d*) Single-phase motors (except commutator motors) are not self-starting. Polyphase motors are self-starting.

(*e*) The pulsating nature of the armature reaction in single-phase alternators causes difficulty with parallel running unless the poles are fitted with exceedingly heavy dampers. Polyphase generators work in parallel without difficulty.

Relation of Output to Number of Phases. Imagine an armature with a uniformly distributed winding, and consider that portion of the winding lying in one pole pitch (Fig. 16.1). The points A and B are 180 electrical degrees apart, and therefore the phase difference in the e.m.f.s induced in the conductors at A and B is 180°. Also, the change of phase from A to B is uniform, and the vector polygon for a large number of conductors is a semicircle, as shown. For a given winding the maximum allowable current I is fixed. For simplicity we consider the case of unity power factor.

(*a*) Single Phase. All the winding is utilized and therefore the terminal voltage is the closing side of the whole polygon, that is, the diameter. Let this be E_1; then power generated $P_1 = E_1 I$.

(*b*) Two Phases. Each phase contains half the winding, and the closing sides for the two phases are the vectors AC and CB. Let these be E_2.

$$\therefore \text{ Total power generated, } P_2 = 2E_2 I = 2(E_1/\sqrt{2})I = \sqrt{2}P_1$$

Thus the power generated in the same frame when used two-phase is 41·4 per cent greater than when operated single-phase.

(c) THREE PHASE. The closing sides for the three phases are AD, DF, and FB. Calling these E_3, we have $E_3 = \frac{1}{2}E_1$. Hence total power

$$P_3 = 3E_3I = 1{\cdot}5E_1I = 1{\cdot}5P_1$$

The increase in power in this case is therefore 50 per cent.

(d) "m" PHASE, where m is a large number.

If the individual phase voltage is E_m, then

$$m \,.\, E_m = \text{circumference of semicircle} = (\pi/2)E$$

$$\therefore \text{ Power } P_m = mE_mI$$

$$= (\pi/2)E_1I = 1{\cdot}57P_1$$

The increase in power is therefore 57 per cent, and this is the maximum possible increase. Since it is only 7 per cent greater than the three-

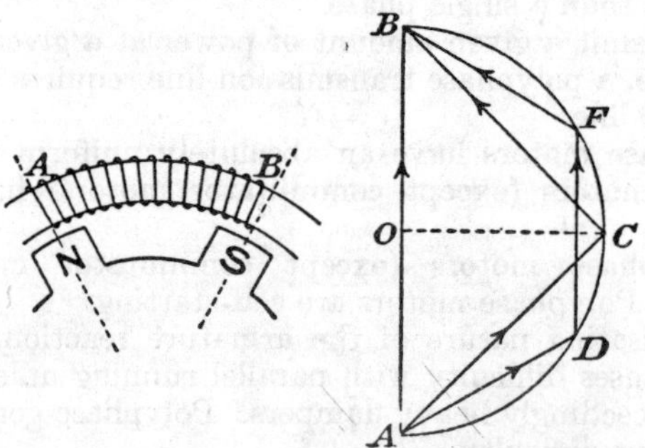

FIG. 16.1. RELATION OF OUTPUT TO NUMBER OF PHASES

phase power the increase does not justify the extra complication, and it is only in exceptional cases that more than three phases are used.

Two-phase Working. The armature has two distinct windings arranged so that there is a phase difference of $\pi/2$ in the e.m.f.s induced in them. The alternator can be connected to the receiving apparatus by either four or three conductors, as in Fig. 16.2, the circuit with four conductors being the same as two independent single-phase circuits.

Let V and I refer to line voltage and current, and V_p and I_p refer to phase voltage and current. Then obviously in the four-wire case,

$$V = V_p \text{ and } I = I_p$$

In the three-wire case the outers carry current $I, = I_p$, and the voltages between outers and middle wire are $V = V_p$. The middle wire carries the resultant of the currents I in the two outers, and since these are 90° apart, the current in the middle wire is $\sqrt{2}I$. Similarly the voltage between the two outers is $\sqrt{2}V$. In both four- and three-wire systems, the power

$$P = 2VI \cos \phi$$

where ϕ is the phase of I with respect to V.

Three-phase Working. The individual phases can be connected either in star (⅄), or in mesh or delta (Δ), as shown in Fig. 16.3. The

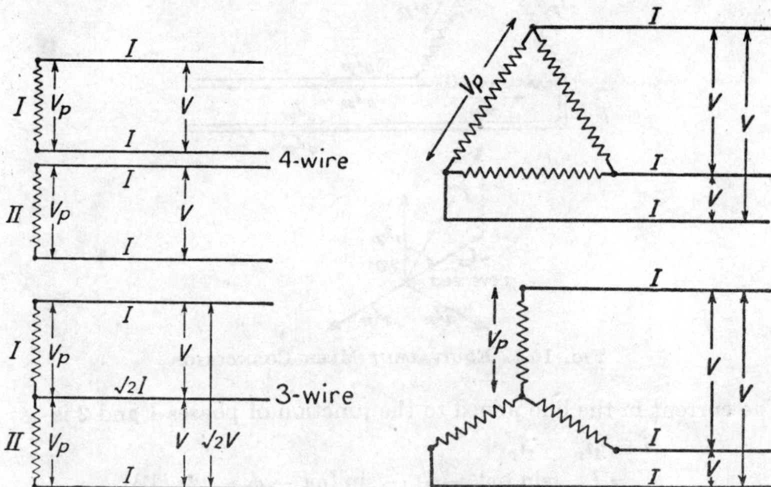

Fig. 16.2. Two-phase Working

Fig. 16.3. Three-phase Working

e.m.f.s in the individual phases have a mutual phase difference of 120°.

(*a*) Mesh Connection. The line voltages V are obviously equal to the phase voltages V_p. The line currents are the vector differences of the two phase-currents fed into a given line. This is easily understood by replacing the ordinary three-wire arrangement by a six-wire, the three phases being thus independent. This is shown in Fig. 16.4. Since each line wire in the actual circuit carries the currents flowing in a pair of wires in the six-wire arrangement, it follows that to obtain the line current the difference and not the sum of the phase currents must be taken. The phase difference of ${}_1I_p$ and ${}_2I_p$ reversed is 60°. Hence, in the case of a current balance, that is,

$${}_1I_p = {}_2I_p = {}_3I_p = I_p, \text{ say, we have}$$

$$I = \sqrt{3}\, I_p$$

Suppose the closed delta is an alternator armature winding delivering current to an external load. Let the three phases all operate at the same power factor of $\cos \phi$, which means that in any phase the current lags (or leads) the e.m.f. in the same phase by the angle ϕ. Thus we can write

$${}_1i_p = I_{mp} \sin (\omega t - \phi)$$

where I_{mp} is the maximum value of all the phase currents

$$_2i_p = I_{mp} \sin [\omega t - \phi - (2\pi/3)]$$
$$_3i_p = I_{mp} \sin [\omega t - \phi - (4\pi/3)]$$

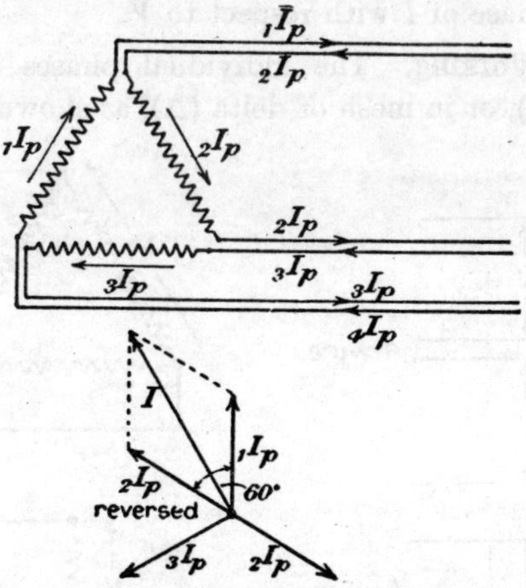

FIG. 16.4. EQUIVALENT MESH CONNECTION

The current in the line joined to the junction of phases 1 and 2 is

$$i = {}_1i_p - {}_2i_p$$
$$= I_{mp} [\sin (\omega t - \phi) - \sin \{\omega t - \phi - (2\pi/3)\}]$$
$$= \sqrt{3} I_{mp} \cos [\omega t - \phi - (\pi/3)]$$

Its effective value is

$$I = \sqrt{3}(I_{mp}/\sqrt{2}) = \sqrt{3} I_p, \text{ as above}$$

Similarly for the other phases taken in pairs.

$$\text{Power } P = 3 V_p I_p \cos \phi$$
$$= \sqrt{3}\, VI \cos \phi$$

We saw that the variations with respect to time of the power in a single-phase circuit follows a double frequency sine law, a double-frequency component being superposed on a steady value. The above treatment does not give any indication of the variation, if any, of three-phase power with respect to time. The total output power at any instant is

$$p = p_1 + p_2 + p_3$$

For a balanced circuit operating at a power factor $\cos \phi$

$$p_1 = {}_1V_p \,.\, {}_1I_p \sin \omega t \,.\, \sin (\omega t - \phi)$$
$$p_2 = {}_2V_p \,.\, {}_2I_p \sin [\omega t - (2\pi/3)] \sin [\omega t - (2\pi/3) - \phi]$$
$$p_3 = {}_3V_p \,.\, {}_3I_p \sin [\omega t - (4\pi/3)] \sin [\omega t - (4\pi/3) - \phi]$$

we have

$${}_1V_p = {}_2V_p = {}_3V_p = V_{mp} \text{ numerically}$$

and

$${}_1I_p = {}_2I_p = {}_3I_p = I_{mp} \text{ numerically}$$

$$\therefore\ p = 1{\cdot}5V_{mp}\,I_{mp}\cos\phi$$

which is a constant with respect to time. On the other hand, the various phases can be regarded as independent of one another as far as the production of power is concerned, and each phase shows the double-frequency component characteristic of single-phase working.

Although the delta is closed there will be no current flowing unless current is delivered to an external circuit, because of the balance of the induced e.m.f.s with respect to the closed circuit. Denote the phase e.m.f.s by e_1, e_2, and e_3 respectively. Then the total instantaneous e.m.f. acting round the circuit is

$$\begin{aligned}e_1 + e_2 + e_3 &= E_m \sin \omega t + E_m \sin[\omega t - (2\pi/3)] \\ &\quad + E_m \sin[\omega t - (4\pi/3)] \\ &= E_m[\sin\omega t + \sin\omega t \,.\, \cos(2\pi/3) - \cos\omega t \,.\, \sin(2\pi/3) \\ &\quad + \sin\omega t \,.\, \cos(4\pi/3) - \cos\omega t \,.\, \sin(4\pi/3)] \\ &= E_m(\sin\omega t - \tfrac{1}{2}\sin\omega t - \sqrt{3}/2 \,.\, \cos\omega t - \tfrac{1}{2}\sin\omega t \\ &\quad + \sqrt{3}/2 \cos\omega t) \\ &= 0\end{aligned}$$

(*b*) Star Connection. With this arrangement the line currents are obviously equal to the phase currents. The voltage V between any pair of lines is the vector difference of the voltages in the two phases supplying that pair. The phase difference between the voltage in one phase and the reversed voltage in the next phase is 60°, so that $V = \sqrt{3}\,V_p$.

Hence,

$$\begin{aligned}P &= 3V_pI_p\cos\phi \\ &= \sqrt{3}\,VI\cos\phi\end{aligned}$$

as in the case of the mesh connection.

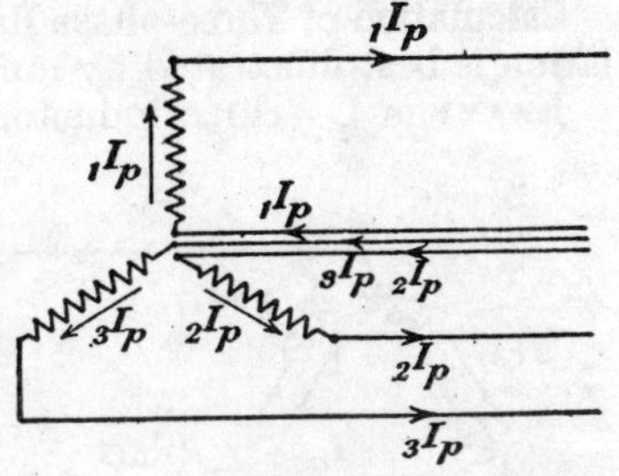

Fig. 16.5. Equivalent Star Connection

If a middle wire is connected to the junction of the three phases, called the "star" point, the current in this wire will be the vector *sum* of the three line currents, as can be seen by considering the three phases separated, as in Fig. 16.5. If the system is balanced, i.e. the three line currents equal and having the same phase angle with their respective phase voltages, then the line current vectors will be equal and will have a mutual phase difference of 120°. Hence, their vector sum is zero. The current in the neutral wire of a balanced star-connected system is therefore zero, for which reason the neutral wire is generally omitted.

Example. A 100 h.p., three-phase, star-connected motor works on a supply whose line voltage is 3000 V. The motor efficiency is 92 per cent, and power factor 90 per cent. Calculate the line current.

$$\text{Motor intake } P = (\text{output})/(\text{efficiency}) = 746 \times 100/0{\cdot}92$$
$$= 81{,}000 \text{ W}$$

$\therefore$ From $P = \sqrt{3}\, VI \cos\phi$

$$I = P/\sqrt{3}\, V \cos\phi = 81{,}000/(\sqrt{3} \times 3000 \times 0{\cdot}9)$$
$$= 17{\cdot}3 \text{ A}$$

Volt-amperes and Reactive Volt-amperes. For a balanced three-phase system the volt-amperes = three times the volt-amperes per phase.

$$\text{VA} = 3V_pI_p$$

For delta connection $V_p = V$ and $I_p = I/\sqrt{3}$

$$\therefore \text{VA} = \sqrt{3}\, VI$$

or

$$\text{kVA} = \sqrt{3}\, VI/1000$$

For a balanced three-phase system the reactive volt-amperes = three times the reactive volt-amperes per phase

$$\text{VAR} = 3V_pI_p \sin\phi$$
$$= \sqrt{3}\, VI \sin\phi$$
$$\therefore \text{kVAR} = \sqrt{3}\, VI \sin\phi/1000$$

Calculation of Three-phase Balanced Networks. The method of calculation is best illustrated by numerical examples.

EXAMPLE 1. Three inductors each of resistance 8 Ω and inductive

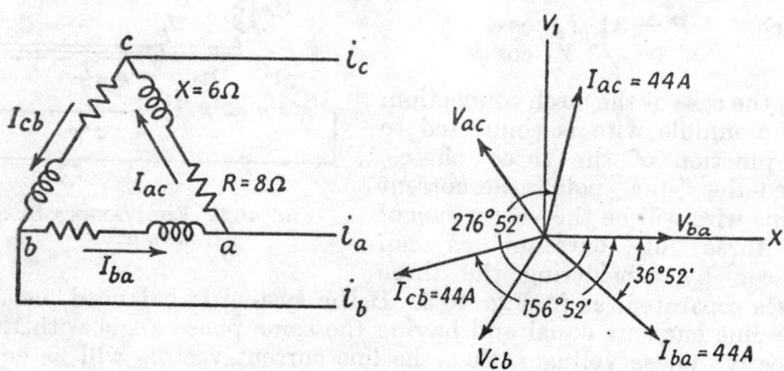

FIG. 16.6. BALANCED DELTA-CONNECTED SYSTEM AND VOLTAGE AND CURRENT VECTOR DIAGRAM

reactance 6 Ω are connected in closed delta, and supplied from a 440 V, 3-phase system. Calculate line and phase currents, the power-factor of the system and the intake in watts.

Using the notation of Fig. 16.6 we have

$$V_{ba} = 440 \underline{/0°} \text{ V}$$
$$V_{cb} = 440 \underline{/-120°} \text{ V}$$
$$V_{ac} = 440 \underline{/-240°} \text{ V}$$

For the impedance per phase

$$Z = 8 + j6$$
$$\therefore Z = \sqrt{(8^2 + 6^2)} = 10 \ \Omega$$

and the characteristic angle is

$$\theta = \tan^{-1} 6/8 = 36° \ 52'$$
$$\therefore I_{ba} = V_{ba}/Z_{ba} = 440/10 \underline{/36° \ 52'}$$

Using the rule for the division of vectors we have

$$I_{ba} = 44 \underline{/-36° \ 52'}, \text{ and } I_{ba} = 44 \text{ A}$$

Writing this in the form $x + jy$ we have

$$x = 44 \cos(-36° \ 52') = \quad 35{\cdot}2$$
$$y = 44 \sin(-36° \ 52') = -26{\cdot}4$$
$$\therefore I_{ba} = 35{\cdot}2 - j26{\cdot}4$$
$$I_{cb} = V_{cb}/Z_{cb} = 440 \underline{/-120°}/10 \underline{/36° \ 52'}$$
$$= 44 \underline{/-156° \ 52'}, \text{ and } I_{cb} = 44 \text{ A}$$
$$= -40{\cdot}47 - j17{\cdot}29$$
$$I_{ac} = V_{ac}/Z_{ac} = 440 \underline{/-240°}/10 \underline{/36° \ 52'}$$
$$= 44 \underline{/-276° \ 52'} \text{ and } I_{ac} = 44 \text{ A}$$
$$= 5{\cdot}26 + j43{\cdot}68$$

For the line currents we have

$$I_a = I_{ba} - I_{ac} = (35{\cdot}2 - j26{\cdot}4) - (5{\cdot}26 + j43{\cdot}68)$$
$$= 29{\cdot}94 - j70{\cdot}08$$
$$\therefore I_a = \sqrt{(29{\cdot}94^2 + 70{\cdot}08^2)} = 76{\cdot}21 \text{ A}$$
$$= \sqrt{3} \times 44 \text{ A}$$
$$I_b = I_{cb} - I_{ba} = (-40{\cdot}47 - j17{\cdot}29) - (35{\cdot}2 - j26{\cdot}4)$$
$$= -75{\cdot}67 + j9{\cdot}11$$
$$\therefore I_b = \sqrt{(75{\cdot}67^2 + 9{\cdot}11^2)} = 76{\cdot}21 \text{ A}$$
$$I_c = I_{ac} - I_{cb} = (5{\cdot}26 + j43{\cdot}68) - (-40{\cdot}47 - j17{\cdot}29)$$
$$= 45{\cdot}73 + j60{\cdot}97$$
$$\therefore I_c = \sqrt{(45{\cdot}73^2 + 60{\cdot}97^2)} = 76{\cdot}21 \text{ A}$$

EXAMPLE 2. The three inductors of example 1 are connected in tar to a 440 V, 3-phase system.

The magnitude of the phase voltage is $440/\sqrt{3} = 254$ V; using the notation of Fig. 16.7 we have

$$V_{na} = 254 \,\underline{/0°}\text{ V}$$

$$V_{nb} = 254 \,\underline{/-120°}\text{ V}$$

$$V_{nc} = 254 \,\underline{/-240°}\text{ V}$$

$$I_{na} = V_{na}/Z_{na} = 254 \,\underline{/0°}/10 \,\underline{/36° 52'}$$
$$= 25{\cdot}4 \,\underline{/-36° 52'}\text{ A}$$

$$I_{nb} = V_{nb}/Z_{nb} = 254 \,\underline{/-120°}/10 \,\underline{/36° 52'}$$
$$= 25{\cdot}4 \,\underline{/-156° 52'}\text{ A}$$

$$I_{nc} = V_{nc}/Z_{nc} = 254 \,\underline{/-240°}/10 \,\underline{/36° 52'}$$
$$= 25{\cdot}4 \,\underline{/276° 52'}\text{ A}$$

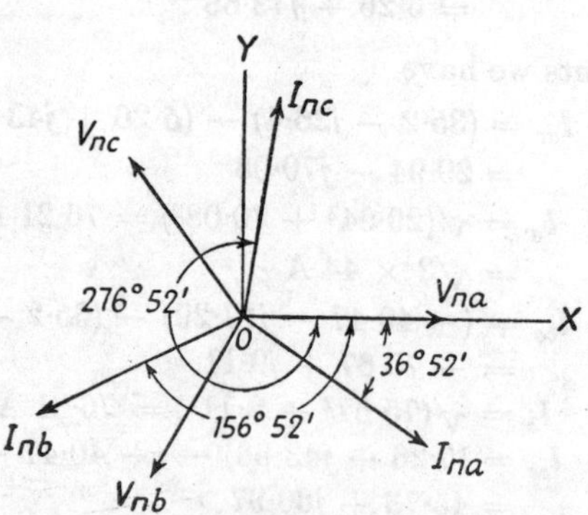

FIG. 16.7. BALANCED STAR-CONNECTED SYSTEM WITH VOLTAGE AND CURRENT VECTOR DIAGRAM

Measurement of Power in Three-phase Circuits. (*a*) THREE-WATTMETER METHOD. The wattmeters are arranged one in each phase as in

Fig. 16.8, the total power being the sum of the three readings. This method is only used in special cases. If the star point is not available when the alternator is star-connected, an "artificial star" can be made by connecting three high resistances in star to the three line conductors.*

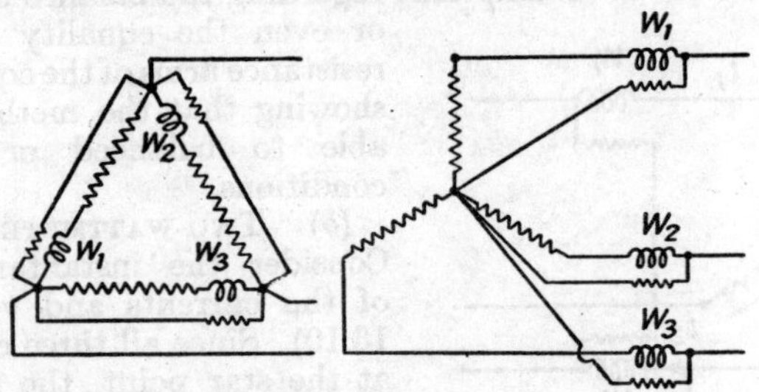

FIG. 16.8. MEASUREMENT OF POWER

Fig. 16.9 shows the circuit. Let e_1, e_2, and e_3 be the instantaneous voltage in the phases, and e_a, e_b, e_c the instantaneous voltages in the three arms of the artificial star resistance. Also let e_x be the instantaneous voltage difference between neutral points n and n'. Then the

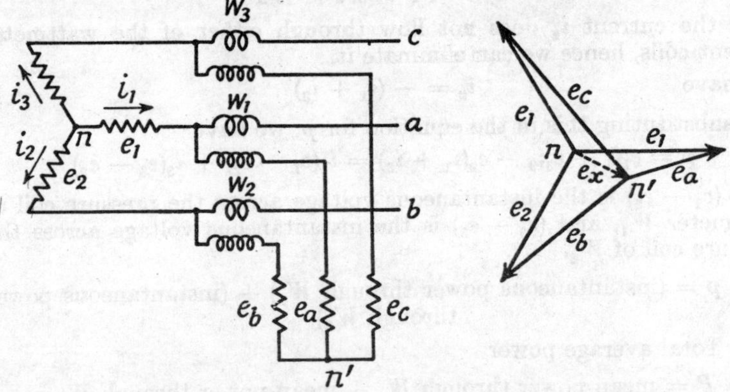

FIG. 16.9. THREE WATTMETER METHOD OF POWER MEASUREMENT WITH ARTIFICIAL NEUTRAL

instantaneous powers in the three wattmeters W_1, W_2, and W_3 are, respectively

$$p_1 = i_1e_a = i_1(e_1 + e_x)$$
$$= e_1i_1 + e_xi_1$$
$$p_2 = e_2i_2 + e_xi_2$$
$$p_3 = i_3e_3 + e_xi_3$$

* With low-voltage circuits it is sufficient to join the three pressure coils in star.

$$\therefore p_1 + p_2 + p_3 = (e_1 i_1 + e_2 i_2 + e_3 i_3) + e_x(i_1 + i_2 + i_3)$$

Now $p_1 + p_2 + p_3 = p$ the total instantaneous power

and $i_1 + i_2 + i_3 = 0$

We have made no assumptions regarding the balance of the phases, or even the equality of the three resistance arms of the equivalent star, showing that the method is applicable to balanced or unbalanced conditions.

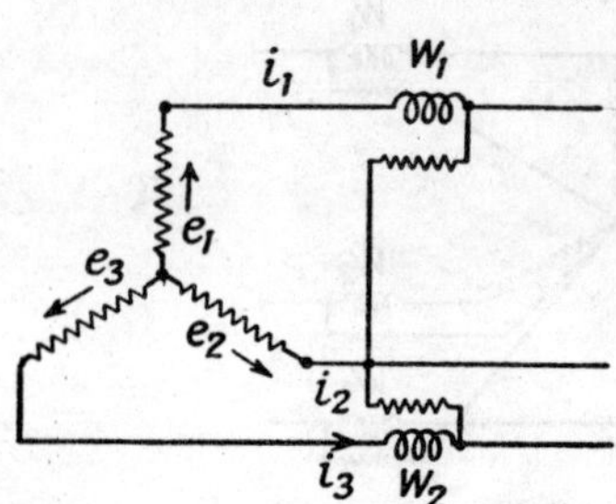

FIG. 16.10. TWO-WATTMETER METHOD OF MEASURING POWER

(*b*) TWO-WATTMETER METHOD. Consider the instantaneous values of the currents and voltages (Fig. 16.10). Since all three currents meet at the star point, the sum of their *instantaneous* values is zero, whether the system is balanced or not.

$$\therefore i_1 + i_2 + i_3 = 0$$

The instantaneous total power is the sum of the instantaneous powers in the three phases

$$p = e_1 i_1 + e_2 i_2 + e_3 i_3$$

Now the current i_2 does not flow through either of the wattmeter current coils, hence we can eliminate it.

We have $$i_2 = -(i_1 + i_3)$$

and substituting this in the equation for p, we have

$$p = e_1 i_1 + e_3 i_3 - e_2(i_1 + i_3) = i_1(e_1 - e_2) + i_3(e_3 - e_2)$$

Now $(e_1 - e_2)$ is the instantaneous voltage across the pressure coil of wattmeter W_1, and $(e_3 - e_2)$ is the instantaneous voltage across the pressure coil of W_2.

$\therefore p =$ (instantaneous power through W_1) + (instantaneous power through W_2)

$\therefore$ Total average power

$$P = \text{mean power through } W_1 + \text{mean power through } W_2$$
$$= \textit{algebraic} \text{ sum of the wattmeter readings}$$

This is the most important method of measuring power in a three-phase circuit, since only two wattmeters are required, and the method applies whether the system is balanced or not.

(*c*) ONE-WATTMETER METHOD. This method can only be used when the system is balanced. The current coil is connected in one line, as in Fig. 16.11, the pressure coil being connected alternately between this and the other two lines, and the readings taken. In each case the wattmeter is connected exactly like one of the wattmeters in the two wattmeter method, and therefore, the analysis following applies equally well to the two-wattmeter method when used on a balanced circuit. The current I through the wattmeter current coil is the vector difference of ${}_AI_B$ and ${}_CI_A$, i.e. the vector sum of ${}_AI_B$ and ${}_CI_A$ reversed, as shown in the vector diagram. In this diagram the phase currents

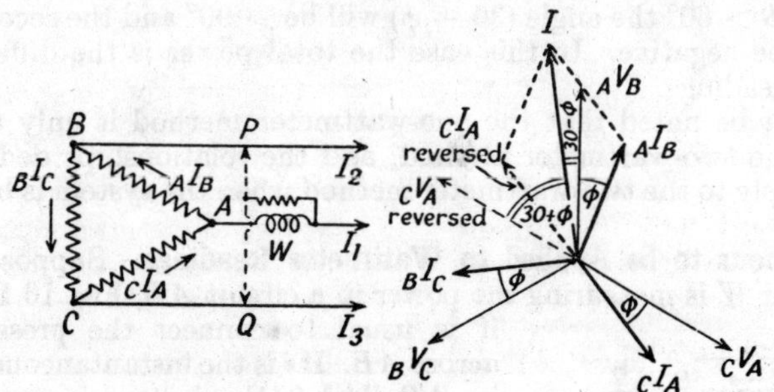

FIG. 16.11. MEASUREMENT OF POWER IN A BALANCED SYSTEM

are shown lagging behind the phase voltages at an angle ϕ, $\cos\phi$ therefore being the power factor of the load.

$$\therefore I = \sqrt{3}\, I_p$$

where I_p is the numerical value of each of the phase currents. When the wattmeter is connected to the point P, the pressure across it is ${}_AV_B$, the phase difference between this pressure and I being $(30° - \phi)$. Hence, if P_1 is the reading,

$$P_1 = {}_AV_BI\cos(30 - \phi)$$
$$= \sqrt{3}\, VI_p \cos(30 - \phi)$$

where V is the numerical value of each of the line or phase voltages.

When the pressure coil is disconnected from P and connected to Q the phase relation between the current and pressure in it is that between I and ${}_CV_A$ *reversed,* because to preserve a true cyclic rotation the pressure coil should have the end previously at P connected to A, and the end at A connected to Q, the reverse of what is actually done. The phase angle between I and ${}_CV_A$ reversed is $(30° + \phi)$. Hence, if P_2 is the new reading,

$$P_2 = {}_CV_AI\cos(30 + \phi) = \sqrt{3}\, VI_p\cos(30 + \phi)$$

Hence $\quad P_1 + P_2 = 3VI_p\cos\phi = \text{total power}$

Again, $\quad P_1 - P_2 = \sqrt{3}\, VI_p\sin\phi$

$$\therefore (P_1 - P_2)/(P_1 + P_2) = (1/\sqrt{3})\tan\phi$$

or $\quad \tan\phi = \sqrt{3} \times (P_1 - P_2)/(P_1 + P_2)$

In this way the phase angle of the load current, and therefore, the power factor, can be calculated.

For the second reading the phase angle in the wattmeter is $(30 + \phi)$; hence (*a*) If $\phi = 0$ both readings will be the same.

(*b*) If $\phi < 60°$ the second reading will be positive and the total power will be the sum of the readings.

(*c*) If $\phi = 60°$ the second reading will be zero.

(*d*) If $\phi > 60°$ the angle $(30 + \phi)$ will be $> 90°$ and the second reading will be negative. In this case the total power is the difference of the two readings.

It must be noted that the one-wattmeter method is only a special case of the two-wattmeter method, and the relationships deduced for it also apply to the two-wattmeter method when the system is balanced.

Corrections to be Applied to Wattmeter Readings. Suppose that a wattmeter W is measuring the power in a circuit AB, Fig. 16.12. Then it is usual to connect the pressure coil across AB. If i is the instantaneous current in AB, and i_1 the instantaneous current in the pressure coil $(i + i_1)$ is the instantaneous current in the current coil.

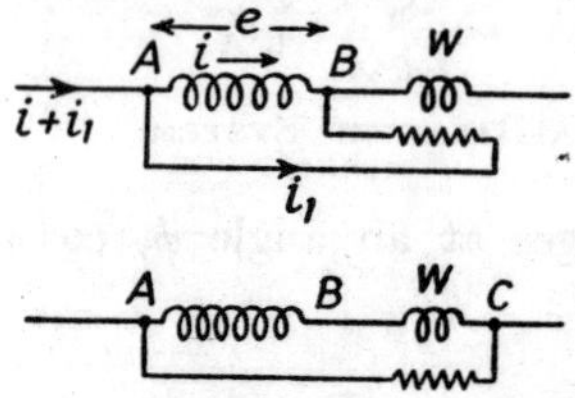

Fig. 16.12. Wattmeter Connections

$\therefore$ Torque at any instant, which is proportional to the product of the instantaneous currents in the pressure and current coils, is proportional to $i_1(i + i_1)$.

Now the current in the pressure coil is equal at any instant to e/R, since we can apply the ordinary Ohm's Law equation when dealing with instantaneous values.

$$\therefore \text{ Instantaneous torque } \propto (ei/R + ei_1/R) \propto (ei + ei_1)$$

$$\begin{aligned} \therefore \text{ Reading, which } &\propto \text{ average torque} \\ &\propto \text{ average of } (ei) + \text{ average of } (ei_1) \\ &\propto \text{ average power in } AB + \text{ average} \\ &\quad \text{power in pressure coil} \end{aligned}$$

A correction for the power used in the pressure coil has therefore to be applied. In standard wattmeters this correction is applied automatically by means of a compensating coil. This is a small coil placed with its axis along the axis of the current coil, and having the same number of turns as the current coil, but connected in series with the pressure coil. It is arranged so that its ampere-turns neutralize the extra ampere-turns in the current coil due to the current i_1.

When the current in the circuit AB is very small the pressure coil of the wattmeter is sometimes connected to the points A and C (Fig. 16.12). In this case the wattmeter reading

$$= \text{power in } AB + \text{power in current coil}$$

In addition to the above corrections, it is necessary to apply a second correction when the wattmeter is working on a circuit whose power factor is less than unity.

Let θ = angle of lag in the main circuit,

ϕ = angle of lag of current in the pressure coil behind the pressure across it,

i = instantaneous value of main current,

i_1 = instantaneous current in pressure coil.

Then from the phase relations shown in Fig. 16.13, in which the

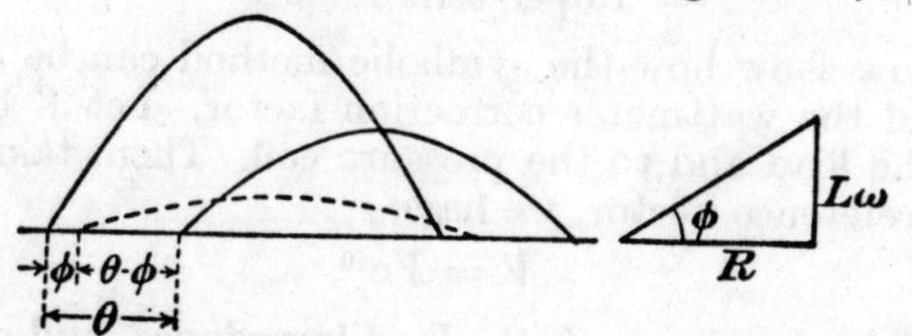

FIG. 16.13. PHASE RELATIONS IN WATTMETER CIRCUIT

current i_1 is represented by the dotted wave, we see that if

$$i = I_{\max} \sin \omega t; \text{ then } i_1 = I_{1\max} \sin (\omega t + (\theta - \phi))$$

$\therefore$ Wattmeter reading $\propto$ average torque

$\propto$ average of ii_1*

$\propto$ average of $I_{\max} I_{1\max} \times \frac{1}{2}[\cos (\theta - \phi) - \cos \{2\omega t + (\theta - \phi)\}]$

Now the average of $\cos \{2\omega t + (\theta - \phi)\}$ is zero, because it is a periodic quantity.

$$\therefore \text{ Reading} \propto (I_{\max}/\sqrt{2})(I_{1\max}/\sqrt{2}) \cos (\theta - \phi)$$
$$\propto I_{\text{eff}} \, . \, I_{1\text{eff}} \cos (\theta - \phi)$$
$$\propto I_{\text{eff}} \times [V_{\text{eff}}/\{R^2 + (L\omega)^2\}^{\frac{1}{2}}] \cos (\theta - \phi)$$

where R and L are the resistance and inductance of the pressure coil

$$\propto I_{\text{eff}} V_{\text{eff}} \times [R/\{R^2 + (L\omega)^2\}^{\frac{1}{2}}] \times (1/R) \cos (\theta - \phi)$$
$$\propto I_{\text{eff}} V_{\text{eff}} \cos \phi \cos (\theta - \phi)$$

omitting $1/R$ because it is constant.

But true reading should be proportional to $I_{\text{eff}} V_{\text{eff}} \cos \theta$.

$$\therefore \text{ Correcting factor} = (\text{true reading})/(\text{actual reading})$$
$$= \cos \theta / \cos \phi \cos (\theta - \phi)$$

or

$$= (1 + \tan^2 \phi)/(1 + \tan \theta \tan \phi)$$

In a standard instrument, ϕ should not be greater than 5′ or 10′. In a switchboard instrument, especially if used with instrument transformers, ϕ may be as high as 5° or 10°.

EXAMPLE. A wattmeter is measuring the power supplied to a circuit whose power factor is 0·7. The frequency is 50 c/s. The meter has a

* This applies to a compensated wattmeter in which an auxiliary coil, having the same number of turns as the current coil, is interwoven with the current coil but connected in series with the pressure coil. The m.m.f. due to its current i_1 neutralizes the m.m.f. due to the component in i_1 the current coil circuit.

shunt winding whose self induction is 0·4 H, and resistance 1000 Ω. Calculate its percentage error.

Power factor of load $\cos\theta = 0{\cdot}7 \quad \therefore\ \tan\theta = 1{\cdot}03$ approx.

$$\tan\phi = L\omega/R = 0{\cdot}4 \times 314/1000 = 0{\cdot}126$$

$$\therefore \text{Correcting factor} = (1 + (0{\cdot}126)^2)/(1 + 1{\cdot}03 \times 0{\cdot}126)$$
$$= 0{\cdot}9 \text{ approx.}$$
$$\therefore \text{Error} = 10 \text{ per cent}$$

We will now show how the symbolic method can be applied to the calculation of the wattmeter correction factor. Let V be the voltage applied to the load and to the pressure coil. Then, taking the vector of V as the reference vector, we have

$$\boldsymbol{V} = Ve^{j0}$$

Let I_1 be the load current, Z_1 the load impedance, and $\cos\theta$ its power factor, assumed lagging, then

$$\boldsymbol{I}_1 = I_1 e^{-j\theta}$$
$$= (V/Z_1)e^{-j\theta}$$

Similarly for the shunt circuit of the wattmeter we can write

$$\boldsymbol{I}_2 = (V/Z_2)e^{-j\phi}$$

Now the torque on the moving coil is proportional to the product of the currents I_1 and I_2, but when using the symbolic method we have to use the conjugate* of $\boldsymbol{I}_2$, namely, $\boldsymbol{I}_2'$, where

$$\boldsymbol{I}_2' = (V/Z_2)e^{+j\theta}$$

$$\therefore \text{Torque}, \propto \text{real part of } \boldsymbol{I}_1\boldsymbol{I}_2'$$
$$\propto \text{real part of } (V^2/Z_1Z_2)e^{j(\phi-\theta)}$$
$$\propto \text{real part of } (V^2/Z_1Z_2)\{\cos(\phi-\theta) + j\sin(\phi-\theta)\}$$
$$\propto (V^2/Z_1Z_2)\cos(\phi-\theta)$$

Now, if the shunt circuit acted like a pure resistance of value R_2, ϕ would be zero. The above expression would then reduce to

$$(V^2/Z_1R_2)\cos(-\theta) = (V^2/Z_1R_2)\cos\theta$$

Hence the correcting factor is

$$\{(V^2/Z_1R_2)\cos\theta\} \div \{(V^2/Z_1Z_2)\cos(\phi-\theta)\}$$
$$= Z_2\cos\theta/R_2\cos(\phi-\theta) = \cos\theta/\cos\phi\,(\cos\phi\cos\theta + \sin\phi\sin\theta)$$
$$= 1/\cos^2\phi\,(1 + \tan\phi\tan\theta) = \sec^2\phi/(1 + \tan\phi\tan\theta)$$
$$= (1 + \tan^2\phi)/(1 + \tan\phi\tan\theta)$$

which is the correction factor obtained previously.

Polyphase Vector Diagrams. Consider a star-connected alternator connected to a star-connected consuming device. Draw three vectors OA, OB, and OC to represent the phase voltages (Fig. 16.14). Then, on joining the ends, we obtain vectors AB, BC, and CA, which give the

* See p. 350.

line voltages at the alternator terminals. If the line and load are non-inductive, the line currents will be in phase with the alternator phase voltages, so that the directions of OA, OB, and OC will also give the directions of the line currents. The drop of volts in a non-inductive line will be in phase with the line currents, so that, if we deduct lengths AA_1, BB_1 and CC_1, where each length is equal to RI, R being the resis-

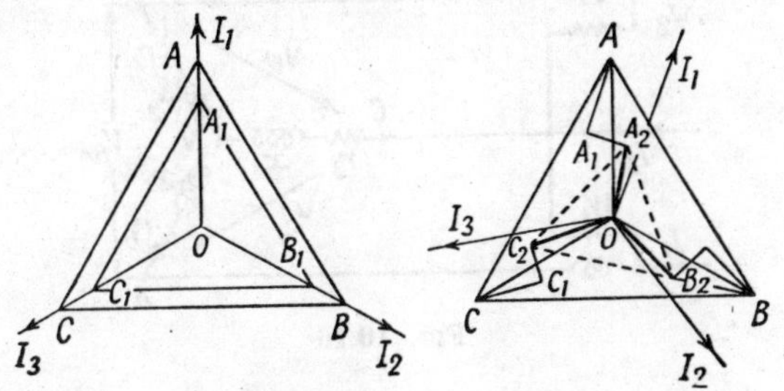

FIG. 16.14. THREE-PHASE VECTOR DIAGRAM

tance of each line, then OA_1, OB_1, and OC_1 are the phase voltages of the load, and A_1B_1, B_1C_1, C_1A_1, the line voltages at the load.

Now let the line possess resistance R and reactance X per phase, and let the load be inductive. The voltage triangle ABC for the alternator end is drawn as before, but the line current vectors OI_1, OI_2 and OI_3 will lag by equal amounts behind the phase voltages OA, OB and OC. Draw AA_1 parallel to OI_1 and equal to RI, where I is the line current. Then AA_1 is the resistance drop in line 1. Draw A_1A_2 perpendicular to OI_1 and equal to XI. Then A_1A_2 is the reactance drop in the line. The total line drop is therefore AA_2, and the phase voltage of the corresponding phase at the load end is OA_2. Similarly with the other phases. The line voltages at the load end are given by the sides A_2B_2, B_2C_2 and C_2A_2 of the triangle $A_2B_2C_2$.

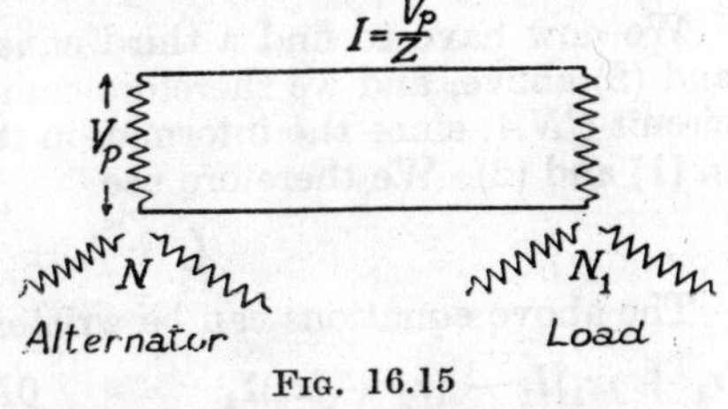

FIG. 16.15

In a symmetrical system, the alternator neutral N, and the load neutral N_1 (Fig. 16.15), will be at the same potential, so that they can be considered as joined by a cable of zero resistance and reactance. Hence, each phase can be considered separately. Let V_p be the phase voltage of the alternator, and Z the total impedance of one phase of the whole system, including alternator, line, and load. Then line current $I = V_p/Z$, and power factor $\cos \phi = R/Z$, R being the total resistance of one phase of the system.

Calculation of Unbalanced Three-phase Circuits. When making calculations on balanced three-phase circuits, as, for example, three-phase machines working under normal conditions, it is usually the best to work in terms of phase voltages and phase currents, and then perform the calculation exactly as for a single-phase circuit. With unbalanced circuits this is not possible, and the laws of networks have to be applied. Consider the star-connected network shown in Fig. 16.16.

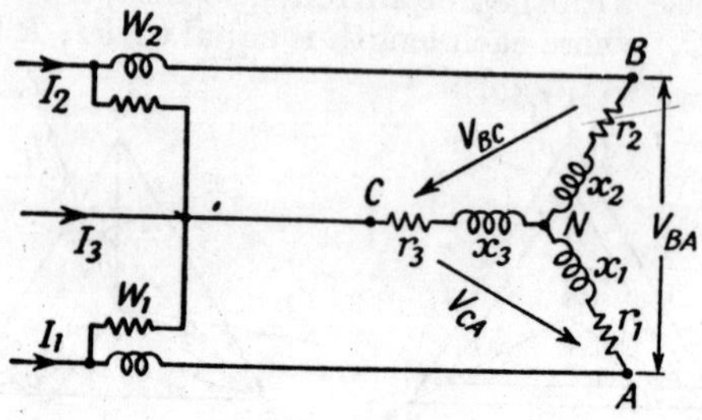

FIG. 16.16

Take the vector of V_{BA} as the reference vector, and let V be the numerical value of each line voltage, then we have

$$V_{AB} = V + j0$$
$$V_{BC} = -\,V/2 + j0{\cdot}866\,V$$
$$V_{CA} = -\,V/2 - j0{\cdot}866\,V$$

Hence, for the circuit ANB, we have

$$V + j0 = (r_1 + jx_1)I_1 - (r_2 + jx_2)I_2 \qquad . \quad . \quad (1)$$

For the circuit BNC we have

$$-\,V/2 + j0{\cdot}866\,V = (r_2 + jx_2)I_2 - (r_3 + jx_3)I_3 \qquad . \quad (2)$$

We now have to find a third equation which is independent of (1) and (2) above, and we therefore cannot use a similar equation for the circuit CNA, since the information it would give is already contained in (1) and (2). We therefore use

$$I_1 + I_2 + I_3 = 0 \qquad . \quad . \quad . \quad . \quad (3)$$

The above equations can be written in the form

$$\begin{array}{lllll} (r_1 + jx_1)I_1 - (r_2 + jx_2)I_2 & + & 0I_3 & - \;V & = 0 \\ \quad 0I_1 + (r_2 + jx_2)I_2 & - & (r_3 + jx_3)I_3 & + \;(V/2 - j0{\cdot}866\,V) & = 0 \\ \quad I_1 \qquad + \quad I_2 & + & I_3 & + \;0 & = 0 \end{array}$$

The solution can now be expressed in determinant form as follows

$$\frac{-\,I_1}{\begin{vmatrix} -(r_2 + jx_2) & 0 & -\,V \\ (r_2 + jx_2) & -(r_3 + jx_3) & \left(\dfrac{V}{2} - j0{\cdot}866\,V\right) \\ 1 & 1 & 0 \end{vmatrix}} = \frac{I_2}{\begin{vmatrix} 0 & -\,V & (r_1 + jx_1) \\ -(r_3 + jx_3) & \left(\dfrac{V}{2} - j0{\cdot}866\,V\right) & 0 \\ 1 & 0 & 1 \end{vmatrix}}$$

$$= \frac{-I_3}{\begin{vmatrix} -V & (r_1+jx_1)-(r_2+jx_2) & 0 \\ \left(\frac{V}{2}-j0{\cdot}866V\right) & 0 & (r_2+jx_2) \\ 0 & 1 & 1 \end{vmatrix}} = \frac{1}{\begin{vmatrix} (r_1+jx_1)-(r_2+jx_2) & 0 \\ 0 & (r_2+jx_2)-(r_3+jx_3) \\ 1 & 1 & 1 \end{vmatrix}}$$

Abbreviating this to the form

$$-I_1/D_1 = I_2/D_2 = -I_3/D_3 = 1/D_4$$

we will evaluate D_4, since this is necessary for the calculation of all three currents. On the other hand, it is preferable only to evaluate D_1, D_2 and D_3, when numerical values are given for the circuit constants. We then have

$$D_4 = (r_1 + jx_1)(r_2 + jx_2) + (r_2 + jx_2)(r_3 + jx_3) + (r_3 + jx_3)(r_1 + jx_1)$$

Suppose, for example, that

$$V = 100 \text{ V r.m.s.}$$
$$r_1 = 1,\ r_2 = 2,\ r_3 = 3$$
$$x_1 = 2,\ x_2 = 3,\ x_3 = 4$$

Then $D_1 = \begin{vmatrix} -(2+j3) & 0 & -100 \\ (2+j3) & -(3+j4) & (50-j86{\cdot}6) \\ 1 & 1 & 0 \end{vmatrix} = -(140{\cdot}2 + j723{\cdot}2)$

$$D_2 = \begin{vmatrix} 0 & -100 & 1+j2 \\ -(3+j4) & (50-j86{\cdot}6) & 0 \\ 1 & 0 & 1 \end{vmatrix} = -(523{\cdot}3 + j413{\cdot}4)$$

$$D_3 = \begin{vmatrix} -100 & 1+j2 & -(2+j3) \\ (50-j86{\cdot}6) & 0 & (2+j3) \\ 0 & 1 & 1 \end{vmatrix} = -(383 - j309{\cdot}8)$$

$$D_4 = (1+j2)(2+j3) + (2+j3)(3+j4) + (3+j4)(1+j2)$$
$$= (-15 + j34)$$
$$\therefore\ I_1 = (140{\cdot}2 + j723{\cdot}2)/(-15 + j34)$$
$$= 16{\cdot}4 - j11{\cdot}3$$
$$I_2 = -(523{\cdot}3 + j413{\cdot}4)/(-15 + j34)$$
$$= -4{\cdot}52 + j17{\cdot}4$$
$$I_3 = (383 - j309{\cdot}8)/(-15 + j34)$$
$$= -11{\cdot}85 - j6{\cdot}12$$
$$\therefore\ I_1 = 19{\cdot}8,\ I_2 = 18{\cdot}0,\ I_3 = 13{\cdot}3 \text{ A}$$

This gives for the power intake of the circuit

$$P = I_1^2 r_1 + I_2^2 r_2 + I_3^2 r_3$$
$$= 392 + 648 + 530 = 1570 \text{ W}$$

But we can also calculate this power symbolically; for example, suppose that the power is being measured by the two-wattmeter method, as indicated in the figure. Then, remembering that we have to use the conjugate of the current when calculating power, we have

$$P = P_1 + P_2$$

$$= \text{real part of } I_1 V_{AC} + \text{real part of } I_2 V_{BC}$$

with the necessary reversal of sign as explained previously

$$= \text{real part of } \{(50 + j86{\cdot}6)(16{\cdot}4 + j11{\cdot}3)$$
$$+ (-50 + j86{\cdot}6)(-4{\cdot}52 - j17{\cdot}4)\}$$
$$= 1570 \text{ W, as before}$$

Reduction of Mesh Connection to Equivalent Star Connection. In the case of a mesh-connected system the calculation is much simpler, as the voltage across each phase of the load and the phase currents can then be calculated independently of one another. For this reason it is often desirable to reduce an unbalanced star-connected system to the equivalent mesh-connected system, or alternatively, a mesh-connected system can be converted to the equivalent star. Let the systems indicated in Fig. 16.17 be electrically equivalent, then the measured impedances between the pairs of terminals must be the same for the two loads

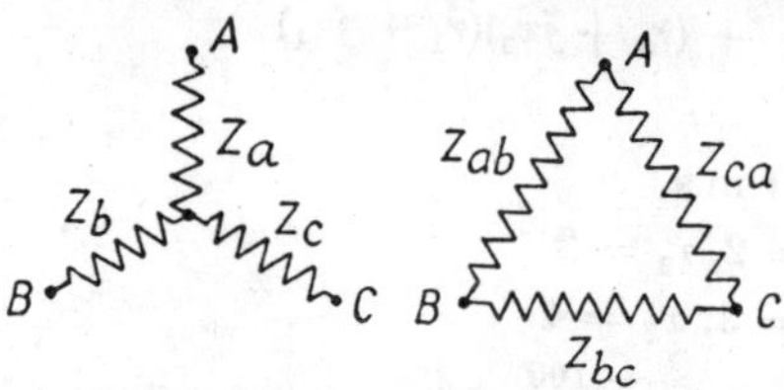

FIG. 16.17. EQUIVALENT STAR- AND MESH-CONNECTED SYSTEMS

$$\therefore \; Z_a + Z_b = Z_{ab}(Z_{bc} + Z_{ca})/(Z_{ab} + Z_{bc} + Z_{ca}) \quad . \quad (1)$$

$$Z_b + Z_c = Z_{bc}(Z_{ca} + Z_{ab})/(Z_{ab} + Z_{bc} + Z_{ca}) \quad . \quad (2)$$

$$Z_c + Z_a = Z_{ca}(Z_{ab} + Z_{bc})/(Z_{ab} + Z_{bc} + Z_{ca}) \quad . \quad (3)$$

These give

$$Z_a = Z_{ab} \, . \, Z_{ca}/(Z_{ab} + Z_{bc} + Z_{ca})$$
$$Z_b = Z_{bc} \, . \, Z_{ab}/(Z_{ab} + Z_{bc} + Z_{ca})$$
$$Z_c = Z_{ca} \, . \, Z_{bc}/(Z_{ab} + Z_{bc} + Z_{ca})$$

Thus, suppose we have a mesh-connected system as follows

$$Z_{ab} = 1 + j2$$
$$Z_{bc} = 2 + j3$$
$$Z_{ca} = 3 + j4$$

Then for the equivalent star-connected system we have

$$\therefore \; Z_{ab} + Z_{bc} + Z_{ca} = 6 + j9$$

$$\therefore \; Z_a = (1 + j2)(3 + j4)/(6 + j9) = 0{\cdot}513 + j0{\cdot}09$$
$$Z_b = (2 + j3)(1 + j2)/(6 + j9) = 0{\cdot}333 + j0{\cdot}666$$
$$Z_c = (3 + j4)(2 + j3)/(6 + j9) = 1 + j1{\cdot}33$$

The Drop of Volts in a Three-phase Line. We have seen* that the inductance of the loop formed by a pair of parallel conductors is

$$L = 14{\cdot}8 \times 10^{-4} \log_{10} (d/r) \text{ H/mile}$$

The inductance of a single conductor in the case of three conductors placed at the corners of an equilateral triangle of side d is $\frac{1}{2}L$. For consider the individual phase current to be flowing in three separate loops of a six-conductor line, as in Fig. 16.18. Let the currents in two of these loops be I_1 and I_2. Then the actual current, I_A, flowing in conductor A, is the vector sum of the currents I_1 and I_2. Again, the third loop is directly opposite to the conductor A, so that the flux produced by it will not link with A, and therefore will not induce a voltage in A. It is thus not necessary to consider the third loop when calculating the inductance of A, and the e.m.f. induced in A is due to the current I_1 in B and I_2 in C. The e.m.f. induced in both conductors of loop 1 is $L di_1/dt$. Hence, e.m.f. induced in one conductor only of loop 1, say conductor A, is $\frac{1}{2}L di_1/dt$. Similarly, the e.m.f. induced in conductor A by the current I_2 in loop 2 is $\frac{1}{2}L di_2/dt$.

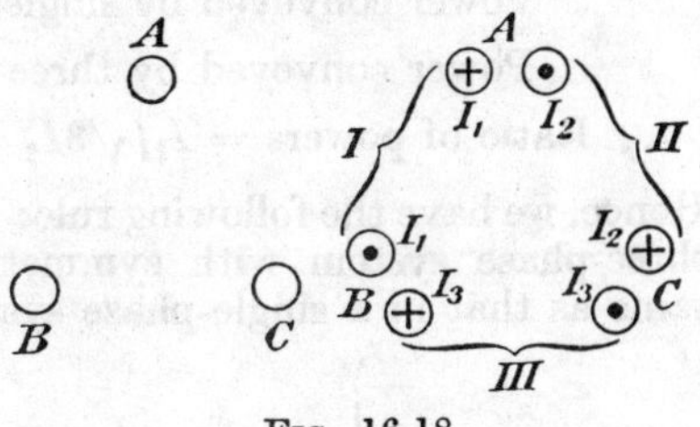

FIG. 16.18

$\therefore$ Total e.m.f. induced in conductor A

$$= \tfrac{1}{2}L(di_1/dt + di_2/dt)$$

Now $$i_A = i_1 + i_2$$

$$\therefore di_1/dt + di_2/dt = di_A/dt$$

$\therefore$ Total e.m.f. induced in conductor A

$$= \tfrac{1}{2}L di_A/dt$$

Hence, $\frac{1}{2}L$ is the equivalent self-induction of a single conductor.†

In a single-phase line formed by two of the conductors and carrying current I_1, the drop, reckoning both conductors, is

$$\{(2I_1R)^2 + (I_1L\omega)^2\}^{\frac{1}{2}} = I_1 \times \{4R^2 + (L\omega)^2\}^{\frac{1}{2}}$$

The drop per conductor in a three-phase line carrying a line current of I_2 is

$$\{(I_2R)^2 + (\tfrac{1}{2}I_2L\omega)^2\}^{\frac{1}{2}} = I_2 \times \{R^2 + \tfrac{1}{4}(L\omega)^2\}^{\frac{1}{2}}$$

$\therefore$ Drop in potential difference between two conductors of the three-phase line carrying current I_2, i.e. the drop in line voltage

$$= \sqrt{3} \times \text{drop per conductor}$$
$$= \sqrt{3}\, I_2 \times \{R^2 + \tfrac{1}{4}(L\omega)^2\}^{\frac{1}{2}}$$

* *See* p. 146.

† Compare this with the capacitance of a single conductor of a three-phase transmission line, p. 30.

If the drops in the single-phase and three-phase lines are equal, then

$$I_1 \times \{4R^2 + (L\omega)^2\}^{\frac{1}{2}} = \sqrt{3}\, I_2 \times \{R^2 + \tfrac{1}{4}(L\omega)^2\}^{\frac{1}{2}}$$

$$\therefore\ I_1/I_2 = \sqrt{3}/2$$

If the line pressures and the power factors are the same in the single- and three-phase lines, say, E and $\cos\phi$ respectively,

Power conveyed by single-phase line $= EI_1 \cos\phi$

Power conveyed by three-phase line $= \sqrt{3}\, EI_2 \cos\phi$

$\therefore$ Ratio of powers $= I_1/\sqrt{3}I_2 = \frac{1}{2}$

Hence, we have the following rule: The total drop of volts in a balanced three-phase system with symmetrically arranged conductors is the same as that in a single-phase system having conductors of the same diameter and spacing, operating at the same line voltage and the same power factor, and transmitting only half the amount of power transmitted by the three-phase system.

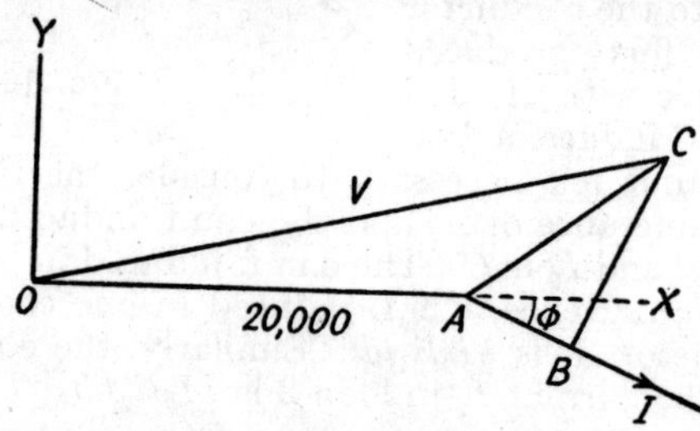

FIG. 16.19

EXAMPLE. A three-phase transmission line consists of $\frac{1}{2}$ in. conductors placed at the corners of an equilateral triangle of side 4 ft. The line is 20 miles long and supplies a load of 5000 kW at a voltage of 20,000 V and a power factor of 0·8. The frequency is 50 c/s. Find the drop of volts in the line.

Resistance of one conductor $R = 5\ \Omega$

$d/r = 48/\frac{1}{4} = 192:\ \log_{10}(d/r) = 2{\cdot}28$

$\therefore$ Inductance of a loop formed by two conductors

$$L = 14{\cdot}8 \times 10^{-4} \times 2{\cdot}28 \times 20 = 0{\cdot}068\ \text{H}$$

Reactance of loop $X = L\omega = 0{\cdot}068 \times 314 = 21{\cdot}3\ \Omega$

The current in a single-phase line using two of the conductors and transmitting 2500 kW would be

$$I = 2{,}500{,}000/(20{,}000 \times 0{\cdot}8) = 156\ \text{A}$$

$\therefore$ Resistance drop, AB in Fig. 16.19 $= 2 \times 5 \times 156 = 1560$ V

Reactance drop, $BC = 21{\cdot}3 \times 156 = 3330$ V

∴ Voltage at generating end V has for its X and Y components

$$(V)_X = 20{,}000 + 1560 \times 0{\cdot}8 + 3330 \times 0{\cdot}6$$
$$= 23{,}250$$
$$(V)_Y = 3330 \times 0{\cdot}8 - 1560 \times 0{\cdot}6$$
$$= 1730$$
$$\therefore\ V = \sqrt{\{(23{,}250)^2 + (1730)^2\}} = 23{,}300 \text{ V}$$

∴ Voltage drop = 3330 V. This is also the drop in the actual three-phase system.

Symmetrical Components of Unbalanced Voltages and Currents. In three-phase working the mutual phase difference between pairs of phase voltages or pairs of phase currents under balanced conditions

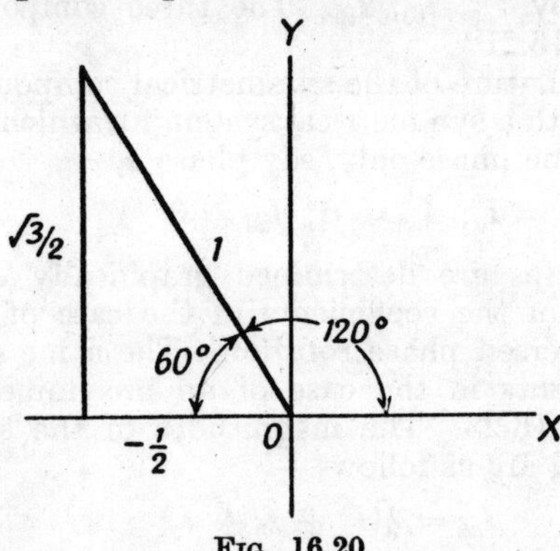

FIG. 16.20

is 120°. It is therefore possible to make three-phase calculations by means of an operator "a" which rotates a vector through 120° in a counter-clockwise direction, unlike the operator "j" which rotates a vector through 90°. Suppose a vector of length unity, originally oriented along the reference, OX, direction is operated on a succession of times by this operator a. Then, from Fig. 16.20, we have

$$a = e^{j2\pi/3} = -\tfrac{1}{2} + j\sqrt{3}/2$$
$$a^2 = e^{j4\pi/3} = -\tfrac{1}{2} - j\sqrt{3}/2$$
$$a^3 = e^{j2\pi} = 1$$
$$a^4 = e^{j8\pi/3} = a^3 \times a = a, \text{ and so on}$$

The following relationships hold

$$\left.\begin{aligned} 1/a &= a^2 \\ a + 1 &= -a^2 \\ a - 1 &= \sqrt{3}e^{j5\pi/6} = -j\sqrt{3}a^2 \\ a^2 - 1 &= j\sqrt{3}a \\ a^2 + a + 1 &= 0 \end{aligned}\right\}$$

The method of analysis is to reduce the unbalanced system into three separate symmetrical systems, as follows—

(1) A symmetrical rotating system of the same phase sequence as the original unsymmetrical system.

(2) A symmetrical rotating system of the inverse phase sequence.

(3) A non-rotating and, therefore, an alternating system in which the only symmetry is the equality of the three vectors.

The first of these systems is called the "positive phase sequence" system, the second the "negative phase sequence" system, and the third the "zero phase sequence" system. Consider a symmetrical three-phase current system with currents I_a, I_b, and I_c. Denote the vectors of the positive phase sequence system by I_{a1}, I_{b1}, I_{c1}; those of the negative phase sequence system by I_{a2}, I_{b2}, I_{c2}; those of the zero phase sequence system by I_{a0}, I_{b0}, I_{c0}. The three component systems are illustrated in Fig. 16.21.

To visualize the nature of the symmetrical components it is perhaps the best to start with a symmetrical system for which it will be obvious that, considering one phase only, say phase a,

$$I_{a1} = I_a; \; I_{a2} = 0; \; I_{a0} = 0$$

These relationships are determined graphically as shown in Fig. 16.22. The order of the coefficients in the case of $I_{a.2}$ are reversed because of the reversed phase rotation. The same graphical method gives the components in the case of an unsymmetrical system, as illustrated in Fig. 16.23. The magnitudes of the symmetrical components for phase a are as follows—

$$I_{a0} = \tfrac{1}{3}(I_a + I_b + I_c)$$
$$I_{a1} = \tfrac{1}{3}(I_a + aI_b + a^2I_c)$$
$$I_{a2} = \tfrac{1}{3}(I_a + a^2I_b + aI_c)$$

There are similar sets of equations for phases c and b, and if we combine them all we have

$$\begin{aligned}
I_a &= \tfrac{1}{3}(I_a + I_b + I_c) + \tfrac{1}{3}(I_a + aI_b + a^2I_c) + (\tfrac{1}{3}I_a + a^2I_b + aI_c) \\
&= I_{a0} + I_{a1} + I_{a2} \\
I_b &= \tfrac{1}{3}(I_a + I_b + I_c) + (a^2/3)(I_a + aI_b + a^2I_c) + (a/3)(I_a + a^2I_b + aI_c) \\
&= I_{a0} + a^2I_{a1} + aI_{a2} \\
&= I_{b0} + I_{b1} + I_{b2} \\
I_c &= \tfrac{1}{3}(I_a + I_b + I_c) + (a/3)(I_a + aI_b + a^2I_c) + (a^2/3)(I_a + a^2I_b + aI_c) \\
&= I_{a0} + aI_{a1} + a^2I_{a2} \\
&= I_{c0} + I_{c1} + I_{c2}
\end{aligned}$$

With two interconnected three-phase systems, say a star-connected alternator and a star-connected load, the only possibility of the flow of single-phase current is along a fourth wire joining the two neutrals. Thus with a three-wire system zero sequence components cannot exist.

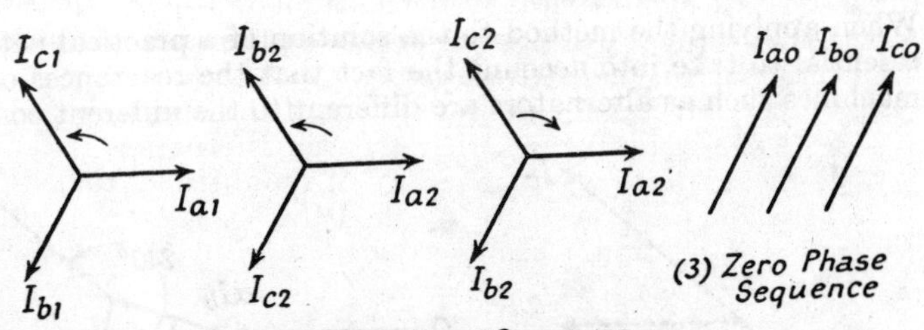

FIG. 16.21. SYMMETRICAL COMPONENTS OF A NON-SYMMETRICAL THREE-PHASE SYSTEM

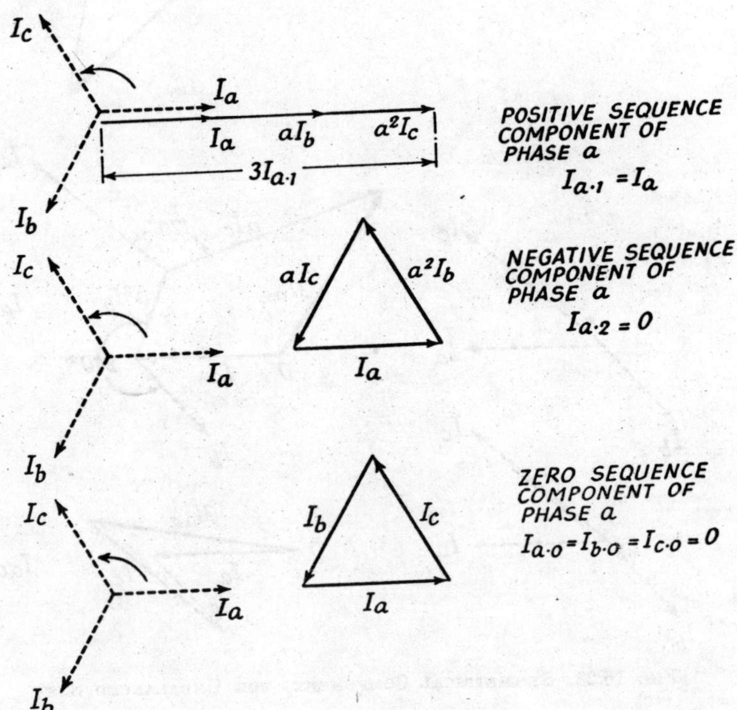

FIG. 16.22. SYMMETRICAL COMPONENTS FOR BALANCED SYSTEM

When applying the method to the solution of a practical network it is essential to take into account the fact that the reactances of rotating machines such as alternators are different to the different components.

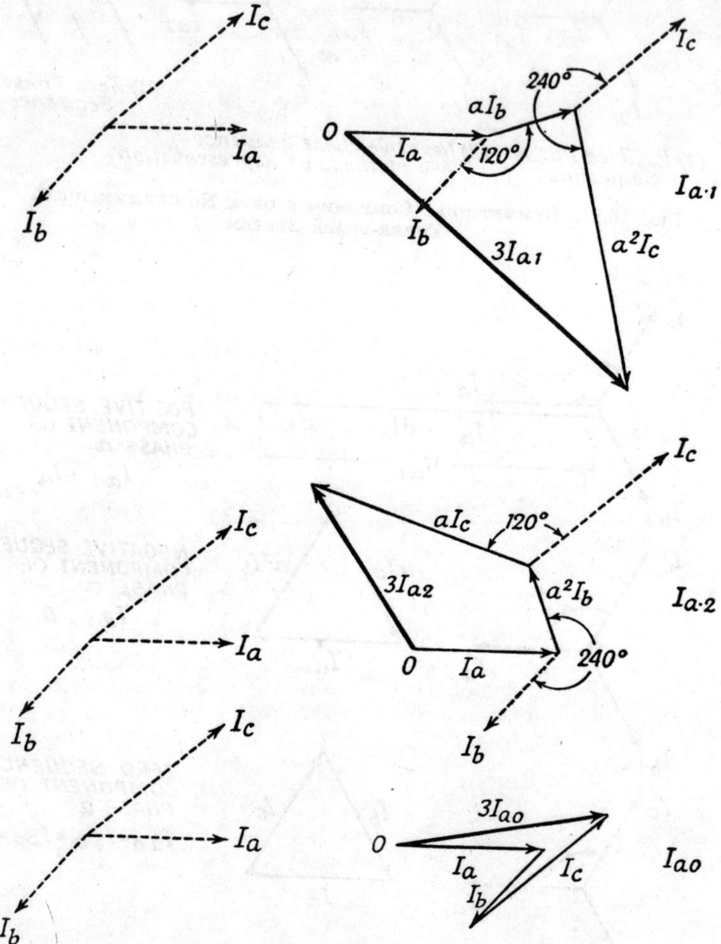

FIG. 16.23. SYMMETRICAL COMPONENTS FOR UNBALANCED SYST...

With transformers, transmission lines, and cables the reactances to the positive and negative sequence components are the same. In all cases the reactance to zero sequence components is different from the others.

EXAMPLE 1. Calculate the symmetrical components of a three-phase system in which $I_a = 10 \underline{/ 30°}$; $I_b = 20 \underline{/ -60°}$; $I_c = 15 \underline{/ -135°}$.

The vector diagram is shown in Fig. 16.24 and we have

$$I_{a0} = \tfrac{1}{3}(10\,\underline{/\,30°} + 20\,\underline{/-60°} + 15\,\underline{/-135°})$$
$$= \tfrac{1}{3}(8{\cdot}66 + j5{\cdot}0 + 10 - j17{\cdot}32 - 10{\cdot}67 - j10{\cdot}67)$$
$$= 0{\cdot}67 - j7{\cdot}67 = 7{\cdot}70\,\underline{/-85°}$$
$$\therefore\ I_{b0} = I_{c0} = 0{\cdot}67 - j7{\cdot}67 = 7{\cdot}70\,\underline{/-85°}$$

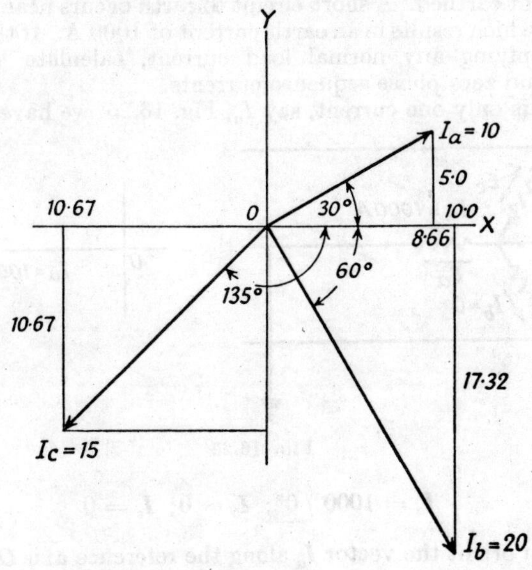

Fig. 16.24

For the positive sequence we have

$$I_{a1} = \tfrac{1}{3}(10\,\underline{/\,30°} + a20\,\underline{/-60°} + a^2 15\,\underline{/-135°})$$
$$= \tfrac{1}{3}(10\,\underline{/\,30°} + 20\,\underline{/\,60°} + 15\,\underline{/\,105°})$$
$$= \tfrac{1}{3}(8{\cdot}66 + j5{\cdot}0 + 10 + j17{\cdot}32 - 3{\cdot}88 + j14{\cdot}49)$$
$$= 4{\cdot}93 + j12{\cdot}27$$
$$= 13{\cdot}22\,\underline{/\,68°\,9'}$$
$$I_{b1} = I_{a1}\,\underline{/-120°} = 13{\cdot}22\,\underline{/-51°\,51'}$$
$$I_{c1} = I_{a1}\,\underline{/-240°} = I_{a1}\,\underline{/+120°} = 13{\cdot}22\,\underline{/\,188°\,9'}$$

For the negative sequence we have

$$I_{a2} = \tfrac{1}{3}(10\,\underline{/\,30°} + a^2 20\,\underline{/-60°} + a15\,\underline{/-135°})$$
$$= \tfrac{1}{3}(10\,\underline{/\,30°} + 20\,\underline{/\,180°} + 15\,\underline{/-15°}$$
$$= \tfrac{1}{3}(8{\cdot}66 + j5{\cdot}0 - 20 + j0 + 14{\cdot}49 - j3{\cdot}88$$

$$= 1{\cdot}05 + j0{\cdot}37$$
$$= 1{\cdot}11 \underline{/ 19^\circ\ 28'}$$
$$I_{b2} = I_{a2} \underline{/ 120^\circ} = 1{\cdot}11 \underline{/ 139^\circ\ 28'}$$
$$I_{c2} = I_{a2} \underline{/ -120^\circ} = 1{\cdot}11 \underline{/ -100^\circ\ 32'}$$

EXAMPLE 2. A three-phase alternator, star-connected with the neutral point earthed. A short-circuit to earth occurs near the terminal of phase a, which results in an earth current of 1000 A. If the alternator is not supplying any normal load current, calculate the positive, negative, and zero phase sequence currents.

As there is only one current, say I_a, Fig. 16.25, we have

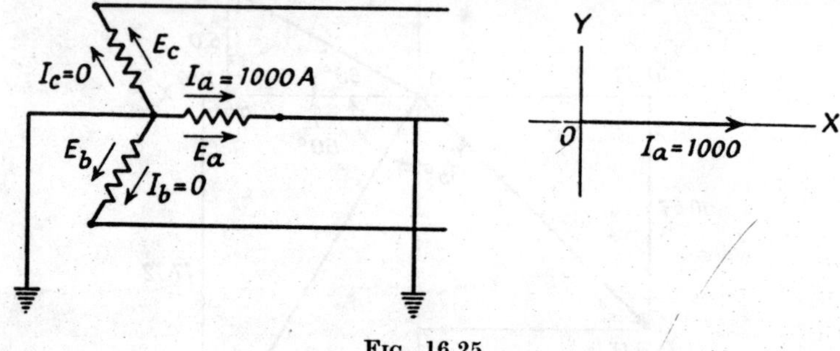

FIG. 16.25

$$I_a = 1000 \underline{/ 0^\circ};\ I_b = 0;\ I_c = 0$$

since we can orient the vector I_a along the reference axis OX.

$$\left.\begin{aligned} \therefore\ I_{a1} &= \tfrac{1}{3}(1000 \underline{/ 0^\circ} + 0 + 0) = 333{\cdot}3 \underline{/ 0^\circ}\text{ A} \\ I_{b1} &= 333{\cdot}3 \underline{/ -120^\circ}\text{ A} \\ I_{c1} &= 333{\cdot}3 \underline{/ -240^\circ}\text{ A} \end{aligned}\right\}\ \text{positive sequence}$$

$$\left.\begin{aligned} I_{a2} &= \tfrac{1}{3}(1000 \underline{/ 0^\circ} + 0 + 0) = 333{\cdot}3 \underline{/ 0^\circ}\text{ A} \\ I_{b2} &= 333{\cdot}3 \underline{/ -240^\circ}\text{ A} \\ I_{c2} &= 333{\cdot}3 \underline{/ -120^\circ}\text{ A} \end{aligned}\right\}\ \text{negative sequence}$$

$$I_{a0} = I_{b0} = I_{c0} = 333{\cdot}3 \underline{/ 0^\circ}\text{ A} \qquad \text{zero sequence}$$

EXAMPLE 3. An alternator supplying an unbalanced star-connected load: alternator and load star points both earthed.

The circuit is drawn in skeleton form in Fig. 16·26.

$$V_a = I_a Z_a = Z_a(I_{a1} + I_{a2} + I_{a0})$$
$$V_b = I_b Z_b = Z_b(I_{b1} + I_{b2} + I_{b0}) = Z_b(a^2 I_{a1} + a I_{a2} + I_{a0})$$
$$V_c = I_c Z_c = Z_c(I_{c1} + I_{c2} + I_{c0}) = Z_c(a I_{a1} + a^2 I_{a2} + I_{a0})$$

Hence

$$\begin{aligned} V_{a1} = \tfrac{1}{3}(V_a + aV_b + a^2V_c) &= (I_{a1}/3)(Z_a + Z_b + Z_c) \\ &\quad + (I_{a2}/3)(Z_a + a^2Z_b + aZ_c) \\ &\quad + (I_{a0}/3)(Z_a + aZ_b + a^2Z_c) \\ &= I_{a1}Z_0 + I_{a2}Z_2 + I_{a0}Z_1 \end{aligned}$$

where Z_1, Z_2 and Z_0 are positive, negative, and zero phase sequence components of the three impedances Z_a, Z_b and Z_c formed in a manner

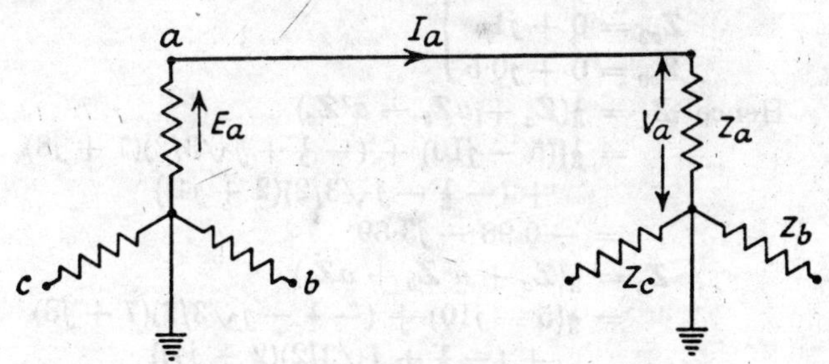

FIG. 16.26

analogous to that employed for currents and voltages. In this connection it is to be noted that these impedance components are not the impedances to positive, negative and zero phase sequence currents, for which, as pointed out previously, a machine such as an alternator, can have different values.

We obtain similar expressions for V_{a2} and V_{a0}, namely

$$V_{a2} = \tfrac{1}{3}(V_a + a^2V_b + aV_c) = I_{a1}Z_1 + I_{a2}Z_0 + I_{a0}Z_2$$
$$V_{a0} = \tfrac{1}{3}(V_a + V_b + V_c) = I_{a1}Z_2 + I_{a2}Z_1 + I_{a0}Z_0$$

Considering now the alternator, let the impedances of the windings to currents of positive, negative and zero sequence be Z_{g1}, Z_{g2}, and Z_{g0}. The positive sequence induced e.m.f. in phase a is E_a, and the induced e.m.f.s of the negative and zero sequence are each equal to zero. Hence

$$\begin{aligned} V_{a1} &= E_a - I_{a1}Z_{g1} \\ V_{a2} &= 0 \quad - I_{a2}Z_{g2} \\ V_{a0} &= 0 \quad - I_{a0}Z_{g0} \end{aligned}$$

On eliminating V_{a1}, V_{a2}, and V_{a0}, we obtain

$$\left.\begin{aligned} I_{a1}(Z_0 + Z_{g1}) + I_{a2}Z_2 + I_{a0}Z_1 &= E_a \\ I_{a1}Z_1 + I_{a2}(Z_0 + Z_{g2}) + I_{a0}Z_2 &= 0 \\ I_{a1}Z_2 + I_{a2}Z_1 + I_{a0}(Z_0 + Z_{g0}) &= 0 \end{aligned}\right\}$$

whence the values of the symmetrical components of the current in phase "a" may be calculated. The actual currents and line-to-neutral p.d.s may then be determined from the initial three equations of this section.

As a numerical exercise take the following values—

$$E_a = 230 + j0$$

$$\left.\begin{aligned} Z_a &= 5 - j10 \\ Z_b &= 7 + j8 \\ Z_c &= 2 + j4 \end{aligned}\right\}$$

$$\left.\begin{aligned} Z_{g1} &= 0 + j4 \\ Z_{g2} &= 0 + j1 \\ Z_{g0} &= 0 + j0{\cdot}5 \end{aligned}\right\}$$

Hence

$$\begin{aligned} Z_1 &= \tfrac{1}{3}(Z_a + aZ_b + a^2Z_c) \\ &= \tfrac{1}{3}\{(5 - j10) + (-\tfrac{1}{2} + j\sqrt{3}/2)(7 + j8) \\ &\qquad + (-\tfrac{1}{2} - j\sqrt{3}/2)(2 + j4)\} \\ &= -0{\cdot}98 - j3{\cdot}89 \\ Z_2 &= \tfrac{1}{3}(Z_a + a^2Z_b + aZ_c) \\ &= \tfrac{1}{3}\{5 - j10) + (-\tfrac{1}{2} - j\sqrt{3}/2)(7 + j8) \\ &\qquad + (-\tfrac{1}{2} + j\sqrt{3}/2)(2 + j4)\} \\ &= 1{\cdot}32 - j6{\cdot}78 \\ Z_0 + Z_{g1} &= 4{\cdot}67 + j4{\cdot}67 \\ Z_0 + Z_{g2} &= 4{\cdot}67 + j1{\cdot}67 \\ Z_0 + Z_{g0} &= 4{\cdot}67 + j1{\cdot}17 \end{aligned}$$

The symmetrical components of the current in phase "a" are now given by the equations

$$\left.\begin{aligned} (4{\cdot}67 + j4{\cdot}67)I_{a1} + (1{\cdot}32 - j6{\cdot}78)I_{a2} - (0{\cdot}98 + j3{\cdot}89)I_{a0} &= 230 \\ -(0{\cdot}98 + j3{\cdot}89)I_{a1} + (4{\cdot}67 + j1{\cdot}67)I_{a2} + (1{\cdot}32 - j6{\cdot}78)I_{a0} &= 0 \\ (1{\cdot}32 - j6{\cdot}78)I_{a1} - (0{\cdot}98 + j3{\cdot}89)I_{a2} + (4{\cdot}67 + j1{\cdot}17)I_{a0} &= 0 \end{aligned}\right\}$$

The solution can be written in determinant form* as follows—

$$I_{a1}$$

$$-\left|\begin{array}{ccc} (1{\cdot}32 - j6{\cdot}78) & -(0{\cdot}98 + j3{\cdot}89) & -230 \\ (4{\cdot}67 + j1{\cdot}67) & (1{\cdot}32 - j6{\cdot}78) & 0 \\ -(0{\cdot}98 + j3{\cdot}89) & (4{\cdot}67 + j1{\cdot}17) & 0 \end{array}\right.$$

* The expansion of a third order determinant can be performed very readily by means of a convenient trick usually termed the rule of Sarrus. This requires the repetition of the first two lines and cross-multiplication as indicated below—

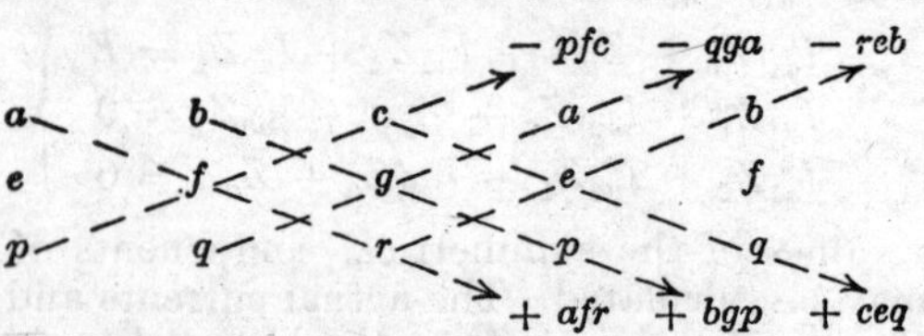

the products taken downwards being reckoned positive, and those taken upwards reckoned negative.

$$= \frac{I_{a2}}{+\begin{vmatrix} -(0{\cdot}98 + j3{\cdot}89) & -230 & (4{\cdot}67 + j4{\cdot}67) \\ (1{\cdot}32 - j6{\cdot}78) & 0 & -(0{\cdot}98 + j3{\cdot}89) \\ (4{\cdot}67 + j1{\cdot}17) & 0 & (1{\cdot}32 - j6{\cdot}78) \end{vmatrix}}$$

$$= \frac{I_{a0}}{-\begin{vmatrix} -230 & (4{\cdot}67 + j4{\cdot}67) & (1{\cdot}32 - j6{\cdot}78) \\ 0 & -(0{\cdot}98 + j3{\cdot}89) & (4{\cdot}67 + j1{\cdot}67) \\ 0 & (1{\cdot}32 - j6{\cdot}78) & -(0{\cdot}98 + j3{\cdot}89) \end{vmatrix}}$$

$$= \frac{1}{+\begin{vmatrix} (4{\cdot}67 + j4{\cdot}67) & (1{\cdot}32 - j6{\cdot}78) & -(0{\cdot}98 + j3{\cdot}89) \\ -(0{\cdot}98 + j3{\cdot}89) & (4{\cdot}67 + j1{\cdot}67) & (1{\cdot}32 - j6{\cdot}78) \\ (1{\cdot}32 - j6{\cdot}78) & -(0{\cdot}98 + j3{\cdot}89) & (4{\cdot}67 + j1{\cdot}17) \end{vmatrix}}$$

The expansion of the determinants in the numerical example gives

$$\left.\begin{aligned} I_{a1} &= 9{\cdot}37 - j12{\cdot}2 \\ I_{a2} &= -\,5{\cdot}26 + j13{\cdot}0 \\ I_{a0} &= 6{\cdot}68 + j13{\cdot}8 \end{aligned}\right\}$$

$$\therefore\ I_a = I_{a1} + I_{a2} + I_{a0} = 10{\cdot}79 + j14{\cdot}6$$

$$I_b = a^2 I_{a1} + a I_{a2} + I_{a0} = -\,17{\cdot}2 + j0{\cdot}72$$

$$I_c = a I_{a1} + a^2 I_{a2} + I_{a0} = 26{\cdot}5 + j26{\cdot}1$$

The alternator terminal p.d.s to neutral, that is, the p.d.s applied to each phase of the load are

$$V_a = I_a Z_a = (10{\cdot}79 + j14{\cdot}6)(5 - j10) = 200 - j35 = 203 \underline{/ -10^\circ}$$

$$V_b = I_b Z_b = (-17{\cdot}2 + j0{\cdot}72)(7 + j8) = -\,126 - j133 = 183 \underline{/ 226^\circ}$$

$$V_c = I_c Z_c = (26{\cdot}5 + j26{\cdot}1)(2 + j4) = -\,51 + j158 = 166 \underline{/ 108^\circ}$$

The vector diagram is given in Fig. 16·27, the angles indicated being measured off by means of a protractor. The calculated values are

phases a and c, $\phi = \tan^{-1}(10/5) = 63^\circ\ 26'$

phase b $\quad\phi = \tan^{-1}(8/7) = 48^\circ\ 45'$

The Varley Phase-sequence Indicator. The electrical phase sequence of a three-phase supply can be determined by the performance of an unbalanced network provided its behaviour is dependent on the phase

sequence. One arrangement commonly used is a star-connected device made up of two identical filament lamps and a choke, as shown in

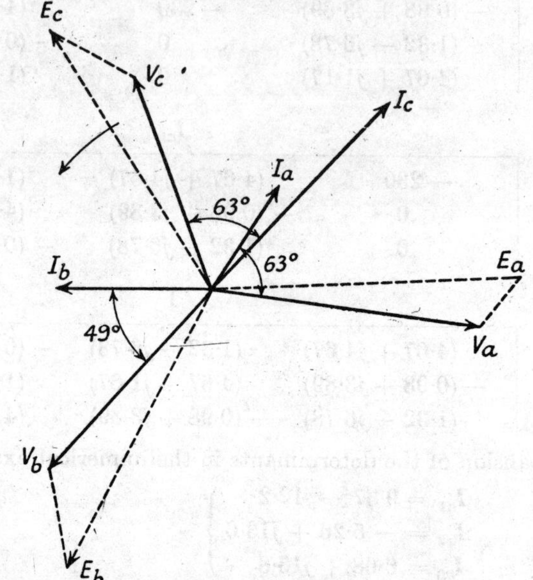

FIG. 16.27

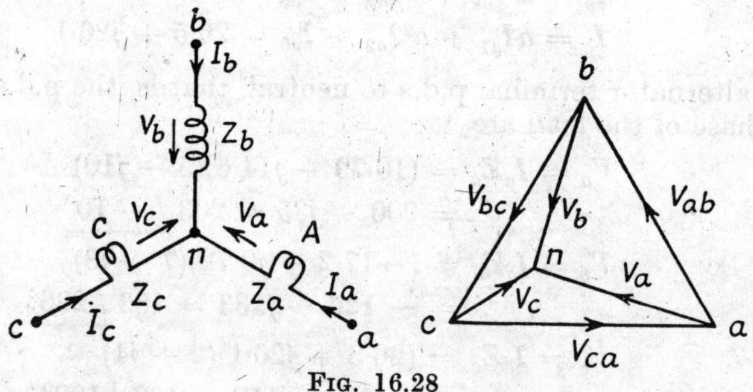

FIG. 16.28

Fig. 16.28.* Suppose that the electrical phase sequence is *ab*, *bc*, *ca*: Then

$$\boldsymbol{I_a + I_b + I_c = 0} \quad \ldots \quad (1)$$

$$\boldsymbol{V_{ab} = V_a - V_b = Z_aI_a - Z_bI_b} \quad . \quad (2)$$

$$\boldsymbol{V_{bc} = V_b - V_c = Z_bI_b - Z_cI_c} \quad . \quad (3)$$

From eqn. (1) $\boldsymbol{I_c = -(I_a + I_b)}$ . . . (4)

* *See* reference on p. 390.

$$\therefore \text{ From eqn. (3) } V_{bc} = Z_b I_b + Z_c(I_a + I_b)$$

$$\therefore V_{bc} = Z_c I_a + (Z_b + Z_c) I_b \quad . \quad . \quad (5)$$

$$\therefore I_b = \frac{V_{bc} - Z_c I_a}{Z_b + Z_c} \quad . \quad . \quad (6)$$

Substituting this value in eqn. (2), we have

$$Z_a I_a - \frac{Z_b(V_{bc} - Z_c I_a)}{Z_b + Z_c} = V_{ab}$$

$$\therefore I_a(Z_a Z_b + Z_a Z_c + Z_b Z_c) = (Z_b + Z_c)V_{ab} + Z_b V_{bc}$$

Multiplying both sides by Z_a and rearranging, we have

$$Z_a I_a = \frac{Z_a[(Z_b + Z_c)V_{ab} + Z_b V_{bc}]}{Z_a(Z_b + Z_c) + Z_b Z_c}$$

This gives the voltage across the lamp A: that across lamp C is given by

$$Z_c I_c = V_{ca} + Z_a I_a$$

These two voltages will be different, and which one is the greater will be determined by the phase sequence. If the sequence is *ab*, *bc*, *ca*, then lamp A will be the brighter, whereas if the sequence is *ab*, *ca*, *bc*, then lamp C will be the brighter.

Example. Let the line voltage be 200 V and let each of the three impedances be numerically equal. With a phase sequence of *ab*, *bc*, *ca*, we can write

$$V_{ab} = 200 \underline{/ 0°}, \quad V_{bc} = 200 \underline{/ -120°}, \quad V_{ca} = 200 \underline{/ -240°} \text{ volts}$$

also

$$Z_a = Z \underline{/ 0°}, \quad Z_b = Z \underline{/ 90°} \text{ (assuming a lossless choke)}$$

and

$$Z_c = Z \underline{/ 0°}$$

$$\therefore Z_a I_a$$

$$= \frac{Z \underline{/ 0°}[(Z \underline{/ 90°} + Z \underline{/ 0°})(200 \underline{/ 0°}) + (Z \underline{/ 90°})(200 \underline{/ -120°})]}{Z \underline{/ 0°}(Z \underline{/ 90°} + Z \underline{/ 0°}) + (Z \underline{/ 90°})(Z \underline{/ 0°})}$$

$$= \frac{(Z \underline{/ 90°} + Z \underline{/ 0°})(200 \underline{/ 0°}) + (Z \underline{/ 90°})(200 \underline{/ -120°})}{Z \underline{/ 90°} + Z \underline{/ 0°} + Z \underline{/ 90°}}$$

$$= \frac{386 \underline{/ 15{\cdot}0°}}{2{\cdot}24 \underline{/ 63{\cdot}45°}}$$

$$= 172{\cdot}6 \underline{/ -48° \ 27'}$$

This gives the p.d. across lamp A. For the p.d. across lamp C we have

$$Z_c I_c = V_{ca} + Z_a I_a$$

$$= 200 \underline{/ -240°} + 172{\cdot}6 \underline{/ -48° \ 27'}$$

$$= 46{\cdot}4 \underline{/ 71° \ 33'} \text{ V}$$

Hence lamp A is the brighter. By putting $V_{ab} = 200\underline{/0°}$, $V_{bc} = 200\underline{/-240°}$, $V_{ca} = 200\underline{/-120°}$, it can be shown in the same way that, for this particular phase sequence, lamp C is the brighter.

FURTHER READING

PAPERS

SWANN, S. A. "Varley test for phase sequence." *Electrician*, March 16, 1951.

PARTON, J. E. "A phase sequence indicator using back to back rectifiers." *J. Sci. Instrum.* **30** (1953) 375.

PARTON, J. E. "A sequence-component voltmeter." *Instrum. Pract.* **7** (1953) 888.

BOOKS

The works listed under Chapter 15, and the following.

KERCHNER, R. M. and CORCORAN, G. F. *Alternating Current Circuits.* Wiley, New York, 1951.

LYON, W. V. *Applications of the Method of Symmetrical Components.* McGraw Hill, New York, 1937.

WAGNER, C. F. and EVANS, R. D. *Symmetrical Components.* McGraw Hill, New York, 1933.

WARREN, A. G. *Mathematics Applied to Electrical Engineering.* Chapman & Hall, London, 1939.

EXAMPLES ON CHAPTER 16

1. A two-phase alternator having a voltage/phase of 1000 V and frequency 50 c/s, has an inductive resistance of 10 Ω and 0·1 H connected to phase I, and a capacitor of 50 μF capacitance connected to phase II. If the alternator is working on a three-wire circuit, find the current in the common return. Phase I leads phase II.

Ans. 38·2 A.

2. A single-phase and a two-phase three-wire system supply equal amounts of power to transmission lines of equal length. If the maximum voltage between any two conductors is the same in the two cases, and the conductors are all worked at the same current density, compare the amount of copper used in the two cases.

Ans. 1 to 1·21.

3. A 50 h.p., three-phase motor is supplied at a terminal voltage of 500 V. Its efficiency is 85 per cent, and power factor 0·8. Find (*a*) the line current; (*b*) the cost of running the motor at full load for 24 hr, the price of a kilowatt-hour being one penny.

Ans. 63·5 A, £4 8s. 0d.

4. Power in a balanced three-phase system is measured by the two-wattmeter method, and it is found that the ratio of the two readings is 2 to 1. What is the power factor of the system?

Ans. 0·866.

5. In testing the power supplied to a three-phase induction motor by two wattmeters connected across the line wires, one of the wattmeters reads backwards at light loads, though its readings are forwards at heavy loads. Examine the reason for this and find what is the true power reading. (L.U.)

6. A single wattmeter is used to ascertain the power supplied from three-phase mains. It is only possible to insert the current coil in one of the mains, but each of the other mains can in turn be connected with the pressure coil. Two wattmeter readings P_1 and P_2 are so obtained, the current coil being connected with one or other of the pressure coil terminals. Explain carefully why it is

necessary to alter the pressure coil connections for the second reading. Under what circumstances can the power and power factor be accurately obtained from the readings of P_1 and P_2? Prove in each case the formula to be used. (L.U.)

7. 500 kW at 11,000 V are received from a three-phase transmission line, each wire of which has a resistance of 1·2 Ω and a reactance of 1 Ω. Calculate the supply pressure when the power factor of the load is (*a*) unity; (*b*) 0·5 leading. (C. and G.)

Ans. (*a*) 11,055 V; (*b*) 10,985 V.

8. The intake of a balanced three-phase system is measured by means of two wattmeters. The lines are *A*, *B* and *C* and the current coils are connected in lines *A* and *C*. The two pressure coils each have one end connected to line *B*, but the other ends are connected: wattmeter *A* to line *C*, and wattmeter *C* to line *A*. Show that with this arrangement the sum of the readings gives the kVAR.

9. A three-phase system with unbalanced voltages supplies a three-phase load consisting of three identical reactors connected in star. If the three line voltages are represented by the sides of a triangle, show that the potential of the star point will be represented by the centre of gravity of the triangle.

10. A 3-phase, 440-V system with symmetrical voltages supplies a mesh-connected system in which the intakes of the three phases are: (1) 9·5 kVA at a p.f. of 0·5 lagging; (2) 15·0 kVA at a p.f. of 0·8 lagging; (3) 5·0 kVA at a p.f. of 0·9 leading. Calculate the three wire currents and their phases with respect to the p.d. across phase 1.

Ans. 44 A $\underline{/-20{\cdot}5^\circ}$; 57·3 A $\underline{/-156{\cdot}5^\circ}$; 39·9 A $\underline{/-172{\cdot}5^\circ}$.

11. The phase currents delivered by a three-phase system are $I_a = 22 + j0$; $I_b = -7 - j13{\cdot}3$; $I_c = -15 + j13{\cdot}3$. Calculate I_{a0}, I_{a1}, and I_{a2}.

Ans. $I_{a0} = 0$; $I_{a1} = 18{\cdot}66 + j2{\cdot}31$; $I_{a2} = 5{\cdot}33 - j2{\cdot}31$.

12. The phase currents delivered by a three-phase system are $I_a = 10 + j24$; $I_b = 14 - j2$; $I_c = -4 - j6$. Calculate the phase sequence components.

Ans.
$$\begin{aligned} I_{a0} &= I_{b0} = I_{c0} = 6{\cdot}67 + j5{\cdot}34 \\ I_{a1} &= 0{\cdot}52 + j14{\cdot}52 \\ I_{b1} &= 12{\cdot}32 - j7{\cdot}70 \\ I_{c1} &= -12{\cdot}82 - j6{\cdot}82 \\ I_{a2} &= 2{\cdot}82 + j4{\cdot}14 \\ I_{b2} &= -4{\cdot}98 + j0{\cdot}36 \\ I_{c2} &= 2{\cdot}16 - j4{\cdot}52 \end{aligned}$$

13. Three non-inductive resistors, each of 1000 Ω, are star-connected to a three-phase supply with 200 V between lines. What will be the reading on a voltmeter connected between one of the lines and the star point thus formed, if the voltmeter has a non-inductive resistance of 1000 Ω? (L.U.)

Ans. 86·6 V.

14. The impedances of the three phases of a star-connected load (no neutral wire) are $5 + j20$, $12 + j0$, and $1 - j10$ in order. The line voltage is 400, find the line currents. (L.U.)

Ans. $0{\cdot}5 - 29{\cdot}7j$, $16{\cdot}24 - 11{\cdot}5j$, $-16{\cdot}74 + 41{\cdot}2j$.

15. Use the method of symmetrical components to show that in the Varley phase-sequence test the voltage across the bright lamp is 3·73 times that across the dark lamp when the numerical impedances of the three legs of the circuit are equal.

16. In a test to determine the phase sequence of a 415-V, 50-c/s, 3-phase supply, a non-inductive resistor of 430 Ω was connected in series with a 5 μF

capacitor between two of the line terminals A and B. The voltage between the junction of the resistor and the capacitor and the third line terminal C was found to be 182 V. Draw the vector diagram and determine the phase sequence. If the phase sequence is reversed what will be the new reading on the voltmeter? (L.U.)

***Ans.* $A - C - B$, 557 V.**

17. The axes of three identical coils are in the same plane, are 120 degrees apart, and meet at a point. The inductance of each coil separately is 0·5 H, and its resistance is 200 Ω: the mutual inductance between each pair of coils is 0·2 H. What current will flow in each coil if the three are connected in star to a 3-phase, 50-c/s supply at 440 V? (L.U.)

***Ans.* 1·14 A.**

CHAPTER 17

THE TRANSFORMER

1. Construction. A transformer consists essentially of a magnetic circuit linking with two distinct windings, the primary and secondary. Fig. 17.1 shows the elementary scheme of a transformer. When the primary is connected to an a.c. supply an alternating flux will be set up in the core, and this flux linking with the secondary will induce an alternating e.m.f. in the secondary. The operation can thus be regarded as a case of mutual induction.

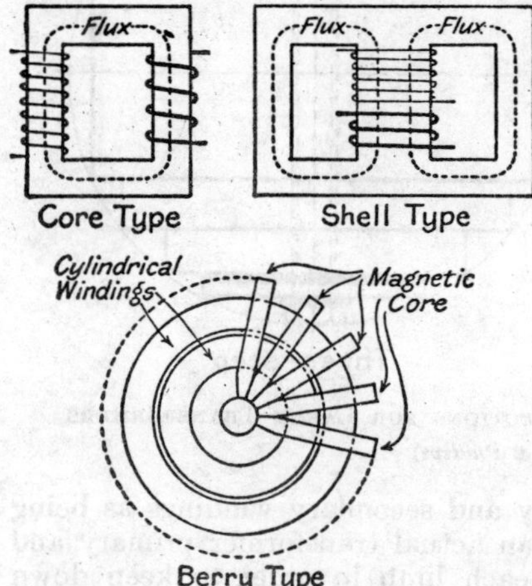

FIG. 17.1. TYPES OF TRANSFORMER

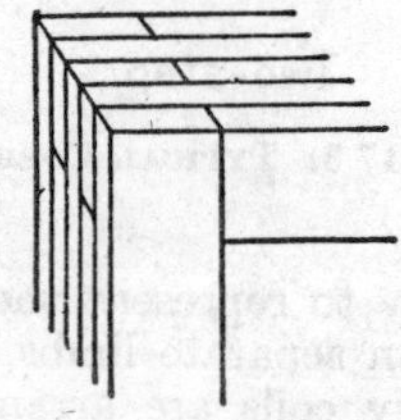

FIG. 17.2. CONSTRUCTION OF JOINTS

There are three main types of transformer, the type being decided by the disposition of the core. These are—

1. "Core Type" with single magnetic circuit.
2. "Shell Type" with double magnetic circuit.
3. "Berry Type" with distributed magnetic circuit.

These are shown in Fig. 17.1. The cores are built up of sheet iron, or alloyed steel, in order to keep down the eddy-current loss to a minimum. Average thicknesses are 0·5 mm for a frequency of 25 c/s and 0·35 mm for a frequency of 50 c/s. Fig. 17.2 shows the methods of arranging the core strips, and it will be seen that joints in alternate layers are staggered, so as to avoid the presence of narrow gaps right through the cross section of the core. Such joints are said to be "imbricated." The cross sections of small transformer cores are rectangular, but in large sizes it is common to adopt an approximately circular section since this section has the smallest perimeter for a given area, and therefore requires less copper than the rectangular section. In very large sizes the strips may be arranged in packets, as in armature cores, the ducts between them helping with the ventilation. Typical cross-sections are shown in Fig. 17.3.

The windings can be either of the "cylindrical" or "sandwich" type, as illustrated in Fig. 17.4, the former type being used mainly on core transformers and the latter on shell. In elementary diagrams it is

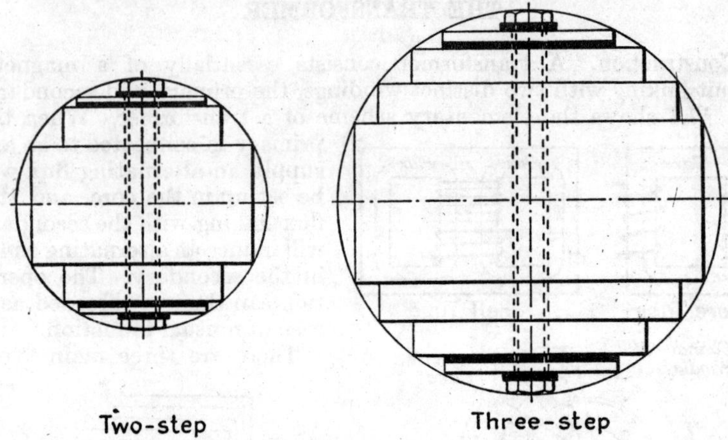

FIG. 17.3. TYPICAL CORE CROSS-SECTIONS FOR LARGE TRANSFORMERS
(*Johnson & Phillips*)

necessary to represent the primary and secondary windings as being placed on separate limbs, but in an actual transformer primary and secondary coils are arranged on each limb in order to keep down magnetic leakage.* This is explained more fully on p. 399.

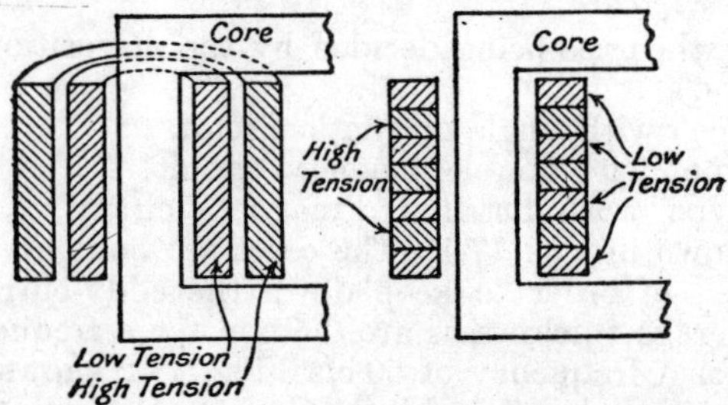

FIG. 17.4. TRANSFORMER WINDINGS

Transformation Ratio. The ratio

$$\frac{\text{Secondary induced e.m.f.}}{\text{Primary applied voltage}} = \frac{E_2}{V_1}$$

is called the *Transformation* ratio.

* A lengthy description of the methods of winding and insulating transformers is out of place here. For details *see* S. A. Stigant and H. M. Lacey, *The Johnson and Phillips Transformer Book*. Johnson and Phillips, London, and M. G. Say, *Performance and Design of A.C. Machines*, Ch. IV. Pitman, London.

Since the flux is alternating it can be written as

$$\Phi = \Phi_{max} \sin \omega t$$

$\therefore$ e.m.f. *induced* in the primary, instantaneous value,

$$e_1 = -(d\Phi/dt) \times N_1 \text{ volts}$$

where N_1 and N_2 are the numbers of turns on primary and secondary respectively.

$$\begin{aligned} \therefore\ e_1 &= -\Phi_{max}\,\omega\,.\cos \omega t\,.\,N_1 \\ &= -2\pi\Phi_{max}\,N_1 f\,.\cos \omega t \\ \therefore\ (E_1)_{max} &= -2\pi\Phi_{max}\,N_1 f \end{aligned}$$

and the effective value

$$\begin{aligned} E_1 &= -(2\pi/\sqrt{2})\Phi_{max}\,N_1 f \\ &= -4{\cdot}44\Phi_{max}\,N_1 f \text{ volts} \end{aligned}$$

Since the resistance of a transformer winding is very low, the ohmic drop is small and the applied primary voltage, V_1, has therefore only to oppose the induced primary e.m.f. E_1. Hence

$$V_1 = -E_1 = 4{\cdot}44\Phi_{max}\,N_1 f$$

The voltage induced in the secondary on no load is

$$E_{20} = 4{\cdot}44\Phi_{max}\,N_2 f \text{ volts}$$

by the same reasoning, and this is the voltage which appears at the secondary terminals on no load, i.e. on no load $E_{20} = V_2$.

Hence, no-load transformation ratio $E_{20}/V_1 = N_2/N_1$

Obviously, the induced voltages E_1 and E_{20} are in phase with one another, and they are therefore opposite in phase to the applied primary voltage V_1. Hence V_1 and E_{20} are opposite in phase.

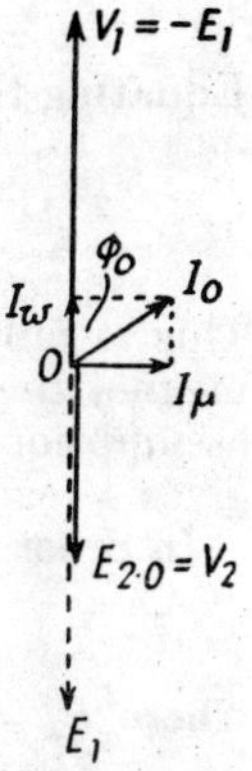

Fig. 17.5. No-load Vector Diagram

Primary No-load Current. On no load, i.e. with no current in the secondary winding, the primary carries a small current I_o, which has two components—

1. A magnetizing component I_μ which lags 90° behind V_1. This is an idle component and its function is to produce the magnetic flux.

2. A working component I_w which produces the necessary real power to supply the hysteresis and eddy-current losses in the iron. This component is in phase with V_1.

The no-load vector diagram is therefore as shown in Fig. 17.5. For high efficiency and good regulation I_μ and I_w must both be small. I_w is kept small by using low-loss iron and not working at too high a flux density. I_μ is kept small by having a closed magnetic circuit, i.e. one without an air gap. If an air gap were cut in the core, then the reluctance would be increased to such an extent that, in order to provide the necessary flux as given by the equation $V_1 = 4{\cdot}44\Phi_{max} N_1 f$, the magnetizing component I_μ would have to be abnormally large.

The separate components I_μ and I_w can be calculated as follows.

The magnetizing force produced by I_μ, is given by

$$H = NI_\mu/l$$

where l is the mean length in metres of the magnetic path. Hence flux density

$$B = \mu\mu_0 N_1 I_\mu/l$$

$$\therefore\ B_{max} = \mu\mu_0 N_1 I_{\mu max}/l$$

But

$$V_1 = \sqrt{2}\,.\,\pi\Phi_{max} N_1 f$$

$$= \sqrt{2}\,.\,\pi B_{max} A N_1 f$$

where A is the magnetic cross section of the core

$$\therefore\ (V_1)_{max} = 2\pi B_{max} A N_1 f$$

$$\therefore\ B_{max} = (V_1)_{max}/2\pi A N_1 f$$

Equating these two expressions for B_{max} we have

$$(I_\mu)_{max} = (V_1)_{max} \times l/2\pi\mu\mu_0 N_1^2 f A$$

$$\therefore\ I_\mu = V_1 l/2\pi\mu\mu_0 A N_1^2 f \text{ (effective values)}$$

This is only approximate, since it assumes a sinusoidal form of the magnetizing component and constant permeability of the iron. Neither assumption is correct, as will be shown in Chapter 28, but the equation gives a ready means of determining I_μ approximately.

In order to calculate the working component I_w, we have

$$V_1 I_w = P_h + P_e$$

where P_h = hysteresis loss and P_e = eddy-current loss.

$$\therefore\ I_w = (P_h + P_e)/V_1$$

For the hysteresis loss, we have

$$P_h = \eta B_{max}^{1\cdot6} f v \text{ watts}$$

where v is the volume of the iron in cubic metres. An average value for η is 500 for good quality transformer iron, and 191 for alloyed steel containing 5 per cent silicon.*

* *Note on Alternating Magnetization.* When the calculation of the losses in iron is based on the static hysteresis loop there is considerable discrepancy with respect to the measured losses. This is due to (*a*) the fact that the ratio B/H varies throughout the cycle, (*b*) the peak values of B and H at any finite depth below the surface are lower than the surface values and are also displaced in time in relation to them. This is due to the effect of eddy currents. It is therefore more logical to define the a.c. permeability as the ratio of δB to δH, where these two changes take place at the same instant of time. This ratio is called the *incremental permeability*. The hysteresis loops traversed by the various particles of iron are

In order to calculate the eddy-current loss imagine a portion of one core strip of thickness t metres, as shown in Fig. 17.6. Consider a rectangular prism of dimensions $t \times 1 \times 1$. The area of the front face is t square metres, and since the flux enters this face at right angles, the eddy currents will flow along paths parallel to the long sides. Consider two such paths each of width dx and distant x from the centre line.

Area of rectangle enclosed by paths
$= 2x$ metre^2

$\therefore$ Maximum flux entering the rectangle
$= 2B_{max}\, x$ lines

Regarding the rectangle as a single turn, the e.m.f. induced in it will be given by the transformer equation

$$E = 4{\cdot}44\Phi_{max}Nf$$
$$= 4{\cdot}44 \times (2B_{max}x) \times 1 \times f$$
$$= 8{\cdot}88B_{max}fx \text{ volts}$$

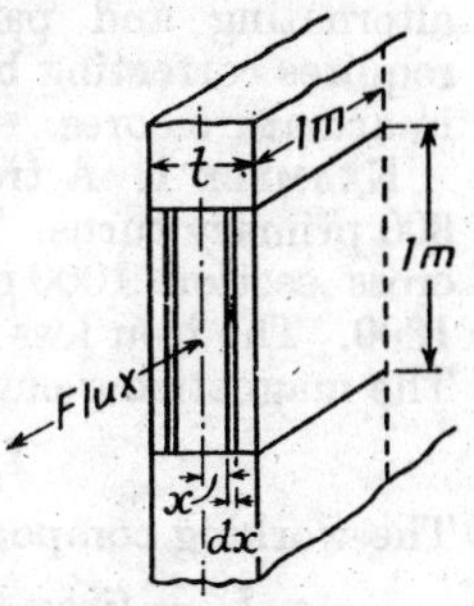

FIG. 17.6. CALCULATION OF EDDY-CURRENT LOSS

The current set up is confined to the two long sides, the total resistance of the path being therefore

$$\rho(l/a) = 2\rho/(dx \times 1) = 2\rho/dx$$

Hence, eddy-current loss in the two strips

$$E^2/R = [(8{\cdot}88)^2B^2_{max}f^2x^2/2\rho]\, dx$$

$\therefore$ Total eddy-current loss in the rectangular prism considered

$$= \int_o^{t/2} \frac{(8{\cdot}88)^2B^2_{max}f^2}{2\rho} \times x^2dx$$
$$= [(8{\cdot}88)^2B^2_{max}f^2/48\rho] \times t^3$$

Now the volume of the prism is $t \times 1 \times 1$ cubic metres. Hence, eddy-current loss per cubic metre

$$= [(8{\cdot}88)^2B^2_{max}f^2/48\rho] \times t^2$$

Taking ρ as 12×10^{-8} Ω-m at a temperature of 50°C the expression becomes

$$13{\cdot}6 \times 10^6B^2_{max}f^2t^2 \text{ watts/cubic metre}$$
$$\therefore\ P_e = 13{\cdot}6 \times 10^6B^2_{max}f^2t^2v \text{ watts}$$

---

not the same as the static loops having the same B_{max}, and, where eddy currents are considerable, the loops will be different for different positions in the specimen.

This problem is still under investigation, and from a great mass of literature the following, all of which have extensive bibliographies should be consulted—

GALL, D. and SIMS, L. G. A. *Jour. I.E.E.* **74** (1934) 453.
BUTLER, O. I. and MANG, C. V. *Jour. I.E.E.* **95** Pt. 2 (1948) 25; **95** Pt. 2 (1948) 410.
BRAILSFORD, F. *Jour. I.E.E.* **95** Pt. 2 (1948) 38.
BUTLER, O. I. and SARMA, M. R. *Jour. I.E.E.* **98** Pt. 2 (1951) 389.
BRAILSFORD, F. *Magnetic Materials.* Methuen.

Notice that the eddy-current loss is proportional to the square of the thickness of the core strips; hence, the necessity for using thin strips. It is to be noticed in passing that whereas the flux in a transformer core is a purely alternating one, the flux in armatures is partly alternating and partly rotating. The above expression therefore requires correcting before it can be used to calculate eddy-current loss in armature cores.

EXAMPLE 1. A transformer working on a 2000 V, 50 c/s circuit has 300 primary turns. The core has a mean magnetic path of 100 cm and cross section 1000 cm², the iron having a relative permeability of 1800. The iron loss is 400 W. Calculate the primary no-load current.

The magnetizing current

$$I_\mu = V_1 l / 2\pi\mu\mu_0 N_1^2 A f = 0{\cdot}31 \text{ A}$$

The working component

$$I_w = (\text{iron loss})/(\text{primary volts}) = 400/2000 = 0{\cdot}2 \text{ A}$$

Hence, no-load current

$$I_o = \sqrt{(0{\cdot}31^2 + 0{\cdot}2^2)} = 0{\cdot}37 \text{ A}$$

EXAMPLE 2. When a transformer is connected to a 1000 V, 50 c/s supply, the core loss is 1000 W, of which 700 are hysteresis and 300 are eddy-current loss. If the applied voltage is raised to 2000 V and the frequency to 100 c/s, find the new core loss.

The hysteresis loss P_h can be written in the form $PB_{\max}^{1{\cdot}6}f$, and the eddy-current loss P_e in the form $QB^2{}_{\max}f^2$: P and Q being constants. Now from the e.m.f. equation

$$V = 4{\cdot}44 B_{\max} A N_1 f$$

we see that $B_{\max} \propto V/f$

Hence, we can write for the hysteresis and eddy-current losses

$$P_h = PV^{1{\cdot}6}f^{-0{\cdot}6}; \quad P_e = QV^2$$

where the constants P and Q now have different values.

From the data given, we have

$$700 = P \times (1000)^{1{\cdot}6} \times (50)^{-0{\cdot}6}; \quad 300 = Q \times 1000^2$$

$$\therefore \; P = 700 \times 1000^{-1{\cdot}6} \times 50^{0{\cdot}6}; \quad Q = 300 \times 1000^{-2}$$

Hence, when the voltage is raised to 2000 V and the frequency to 100 c/s we have

$$P_h = (700 \times 1000^{-1{\cdot}6} \times 50^{0{\cdot}6}) \times 2000^{1{\cdot}6} \times 100^{-0{\cdot}6} = 1400 \text{ W}$$

$$P_e = 300 \times 1000^{-2} \times 2000^2 = 1200 \text{ W}$$

$\therefore$ Total core loss under the new conditions = 2600 W

Transformer on Load. Consider first of all a non-inductive load connected to the secondary terminals; then a secondary current I_2 will flow, and I_2 will be in phase with E_2. The secondary ampere-turns

N_2I_2 will tend to produce a secondary flux which, if allowed to exist, would disturb the flux conditions existing at no load. This would alter the primary induced e.m.f. E_1 and the balance between V_1 and E_1 would no longer exist. The presence of the secondary m.m.f. therefore necessitates the production of a primary m.m.f. equal in magnitude but opposite in direction. This is provided by a load current I_1' which flows from the supply through the primary. Since the m.m.f. produced by this is equal to the secondary m.m.f.,

$$1{\cdot}26N_1I_1' = 1{\cdot}26N_2I_2$$

$$\therefore\ I_1' = (N_2/N_1) \times I_2$$

The total primary current I_1 is therefore the vector sum of I_1' and I_o, and for all but light loads can be taken as equal to I_1'. We thus have

$$I_1 = (N_2/N_1) \times I_2 \text{ approx.}$$

or

$$I_2/I_1 = N_1/N_2$$

The current transformation ratio is thus the inverse of the voltage transformation ratio. The vector diagram for the transformer on non-inductive load is shown in Fig. 17.7 (*A*). A similar action takes place if the secondary load has any phase angle ϕ. The *induced* primary current I_1' is always opposite in phase to the secondary current I_2, and since I_o is small, the total primary current I_1 is almost exactly opposite in phase to I_2. (Fig. 17.7 (*B*).)

Since the induced primary and the secondary ampere-turns always neutralize one another, the flux on load is the same as the flux on no load. Hence, the iron losses are constant and are independent of the load.

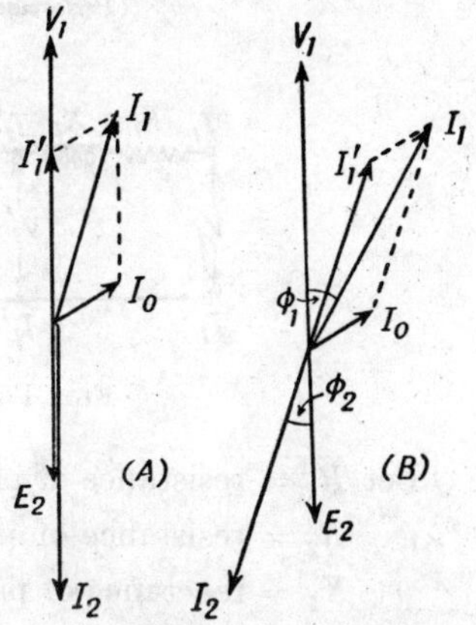

Fig. 17.7. Vector Diagram for Loaded Transformer

Effect of Resistance and Reactance in the Windings. So far we have assumed that the windings have no resistance and no reactance. In an actual transformer both are always present, the reactance being set up by the leakage fluxes. The paths of these fluxes for cylindrical and sandwich windings are as shown in Fig. 17.8. The primary leakage flux is defined as that flux which links with the primary, but not with the secondary. Similarly, the secondary leakage flux is the flux which links with the secondary but not the primary winding. Since a leakage flux links with only one winding, it produces a self-induced back e.m.f. in that winding, and it is therefore equivalent to a small choker in series with that winding.

The effects of resistance and reactance are best understood by considering the case of an ideal transformer having no resistance and reactance of its own, but having separate resistance and reactance coils connected in series with each winding, as shown in Fig. 17.9.

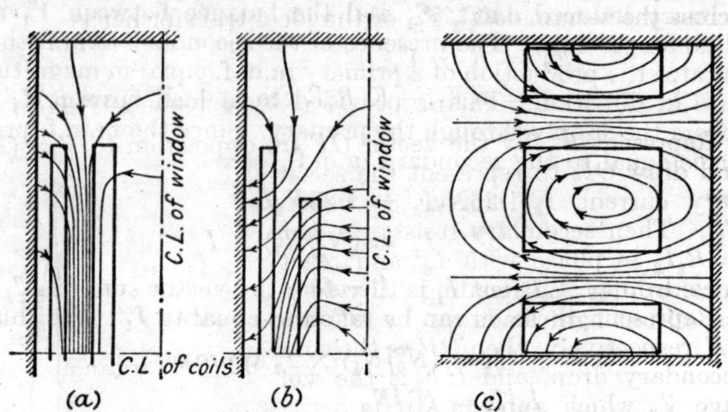

FIG. 17.8. MAGNETIC LEAKAGE IN TRANSFORMERS
(*a*) Cylindrical core-type coils of equal length. (*b*) Cylindrical core-type coils of unequal length. (*c*) Shell-type transformer with sandwich coils.
(*Performance and Design of A.C. Machines*)

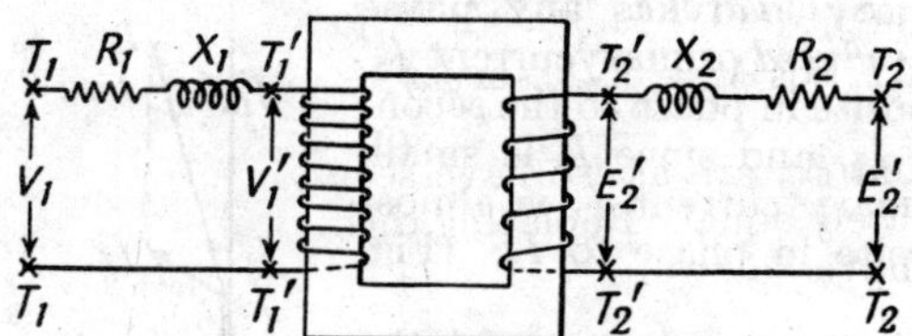

FIG. 17.9. EQUIVALENT CIRCUITS

Let R_1 = resistance of primary of actual transformer

R_2 = resistance of secondary of actual transformer

X_1 = reactance of primary of actual transformer

X_2 = reactance of secondary of actual transformer.

Represent the applied primary voltage by V_1 (Fig. 17.10). Let OI_1 represent the primary current I_1 in phase, lagging any angle ϕ_1 behind V_1. Then primary resistance drop = R_1I_1 in phase with I_1, and reactance drop = X_1I_1 leading I_1 by 90°. Represent these drops by OB and BC respectively. Then OC, the vector sum of OB and BC, is the total primary drop. Deducting OC from OA, we obtain $CA = V_1'$, the voltage across the primary terminals $T_1'T_1'$ of the ideal transformer.

Drop the perpendicular CD on to OA. Then since in an actual transformer OC is small compared with OA we have

$$V_1' = V_1 - OD, \text{ approx.}$$
$$= V_1 - (R_1I_1 \cos\phi_1 + X_1I_1 \sin\phi_1)$$

Let K = voltage transformation ratio.

Then the induced voltage in the secondary, namely, the voltage across the terminals $T_2'T_2'$ of the ideal transformer is

$$\begin{aligned} E_2' &= K \times V_1' \\ &= KV_1 - K(R_1 I_1 \cos\phi_1 + X_1 I_1 \sin\phi_1) \end{aligned}$$

Represent E_2' by the vector OF (in opposition to V_1') (Fig. 17.10), and draw OI_2 to represent the secondary current I_2, lagging ϕ_2 behind E_2'. Then secondary resistance drop $= R_2I_2$ in phase with I_2, and reactance drop $= X_2I_2$ leading I_2 by 90°.

Represent these drops by OG and GH respectively, then OH is the total secondary drop, and HF is the voltage V_2 which appears at the actual secondary terminals T_2T_2. Hence, as before,

$$\begin{aligned} V_2 &= E_2' - (R_2I_2 \cos\phi_2 + X_2I_2 \sin\phi_2) \\ &= KV_1 - K(R_1I_1 \cos\phi_1 + X_1I_1 \sin\phi_1) \\ &\quad - (R_2I_2 \cos\phi_2 + X_2I_2 \sin\phi_2) \end{aligned}$$

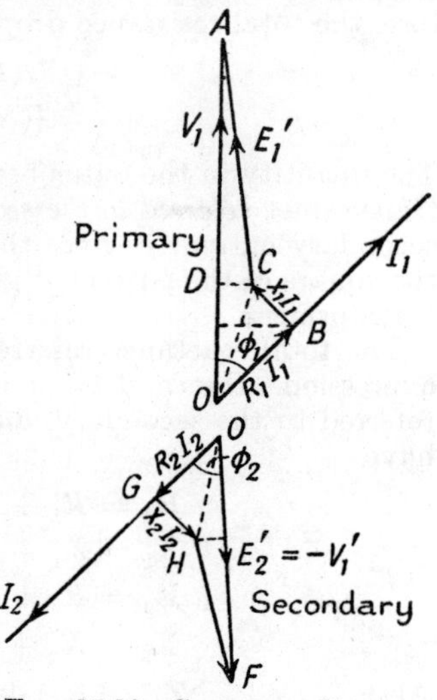

FIG. 17.10. COMPLETE VECTOR DIAGRAM

Now ϕ_1 and ϕ_2, are as we have seen, approximately equal. Replacing them by ϕ, we have

$$\begin{aligned} V_2 = KV_1 - \{&(KR_1I_1 + R_2I_2) \cos\phi \\ &+ (KX_1I_1 + X_2I_2) \sin\phi\} \end{aligned}$$

Again $\quad I_1 = KI_2$ approx.,

hence, the expression for the secondary terminal voltage, in terms of the applied primary voltage and the load current I_2, becomes

$$V_2 = KV_1 - \{(K^2R_1 + R_2) \cos\phi + (K^2X_1 + X_2) \sin\phi\}I_2$$

Now KV_1 is the secondary voltage on no load, E_{20} and therefore the expression

$$\{(K^2R_1 + R_2) \cos\phi + (K^2X_1 + X_2) \sin\phi\}I_2$$

is the drop of volts at the secondary terminals on load. Notice that the drop is proportional to the current I_2, and that it also depends upon the power factor, $\cos\phi$, of the load.

If the load is non-inductive, $\phi = 0$, $\cos\phi = 1$, $\sin\phi = 0$

$$\therefore \text{ Drop} = I_2(K^2R_1 + R_2)$$

It is therefore independent of the reactance.

If the load is purely inductive, $\phi = 90°$, $\cos\phi = 0$, $\sin\phi = 1$

$$\therefore \text{ Drop} = I_2(K^2X_1 + X_2)$$

In this case the drop is decided by the reactance and is independent of the resistance. The same applies when the load is a pure capacitive load, except that in this case the drop is negative, i.e. it is a rise of voltage.

Equivalent Resistance and Reactance. The primary resistance drop $= R_1I_1$, and this, when referred to the secondary, becomes KR_1I_1, or $K^2R_1I_2$. The secondary resistance drop is R_2I_2, and therefore, the total resistance drop in terms of I_2 is

$$(K^2R_1 + R_2)I_2$$
$$= \{(N_2/N_1)^2R_1 + R_2\}I_2$$

The quantity in the outer brackets is called the total resistance of the transformer *referred to the secondary*, because when multiplied by the secondary current it gives the total resistance drop. If multiplied by the square of the current it gives the total copper loss (I^2R loss) in the transformer.

The total reactance referred to the secondary is given by a similar expression. Hence, if $\mathbf{R}_2$ and $\mathbf{X}_2$ are the total resistance and reactance referred to the secondary, and $\mathbf{R}_1$ and $\mathbf{X}_1$ referred to the primary, we have

$$\mathbf{R}_1 = R_1 + (N_1/N_2)^2R_2$$
$$\mathbf{X}_1 = X_1 + (N_1/N_2)^2X_2$$
$$\mathbf{R}_2 = R_2 + (N_2/N_1)^2R_1$$
$$\mathbf{X}_2 = X_2 + (N_2/N_1)^2X_1$$

Also $$\mathbf{Z}_1^2 = \mathbf{R}_1^2 + \mathbf{X}_1^2 \text{ and } \mathbf{Z}_2^2 = \mathbf{R}_2^2 + \mathbf{X}_2^2$$

In the above discussion the primary and secondary leakage reactances X_1 and X_2 are treated as though they could exist independently of one another, in the sense that R_1 and R_2 are independent of one another. Actually this is not the case, because a leakage reactance can only exist physically when there is the appropriate amount of magnetic leakage. The primary and secondary magnetic leakages only take place when the primary is carrying the primary induced current I_1' and the secondary is carrying the corresponding output current I_2. Now the current I_2 cannot act alone, its condition of existence being that there shall be in the primary that induced current I_1' which restores the main magnetic circuit to the no-load condition. Hence it is not possible for the two leakage reactances to act independently of one another.

The total reactance can be measured by short-circuiting one winding of the transformer, and applying such a voltage to the other side (which for the purpose of the test can be regarded as the primary) that normal primary current I_1 flows. It then follows that the normal secondary current I_2 will also flow. Since magnetic leakage accompanies the load currents, and is not associated with the primary no-load current, the equivalent circuit for short-circuit conditions is that of Fig. 17.11.

Let V_s be the reduced primary voltage which causes normal current to circulate. In the figure the transformer is an ideal transformer which can be regarded as having no losses of any kind and therefore as having no internal volt drops. Starting from the secondary side, the terminal p.d. must be zero because of the short-circuit, and therefore, because of the condition of no volt drops, the secondary induced e.m.f. is also zero. Hence the primary inducing e.m.f. is also zero, and so is the primary terminal p.d. V_1', again because of the conditions of no internal volt drop. Consequently the whole of the primary applied voltage V_s is utilized in supplying the drop in the total impedance $\mathbf{Z}_1$ referred to the primary, which, in the equivalent circuit is represented as being external to the transformer.

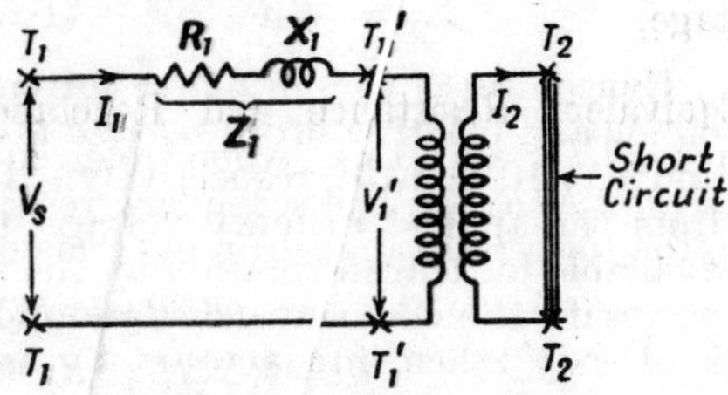

FIG. 17.11. EQUIVALENT CIRCUIT FOR THE SHORT-CIRCUIT TEST

$$\therefore \mathbf{Z}_1 = V_s/I_1$$
$$\therefore \mathbf{X}_1 = \sqrt{(\mathbf{Z}_1^2 - \mathbf{R}_1^2)}$$

If the power is measured during the test we have

$$P_s = \mathbf{R}_1 I_1^2 \quad \therefore \quad \mathbf{R}_1 = P_s/I_1^2$$

If the power factor during the test is $\cos\phi_s$ then we have, alternatively

$$\mathbf{R}_1 = \mathbf{Z}_1 \cos\phi_s$$

EXAMPLE. A transformer has a normal primary voltage of 1000 V and on open circuit it takes a current of 0·3 A at a power factor of 0·7. When the secondary is short-circuited and a reduced voltage of 50 V is applied to the primary, it takes 12 A at a power factor of 0·3. Calculate the efficiency and percentage drop of volts when delivering the full output of 10 kVA at a power factor of 0·8.

Iron losses = open circuit intake = $1000 \times 0{\cdot}3 \times 0{\cdot}7 = 210$ W

The *total* copper losses corresponding to a primary current of 12 A are equal to the intake of true power in the short-circuit test, that is, to $50 \times 12 \times 0{\cdot}3 = 180$ W. The full kVA capacity of the transformer is 10, and therefore, the primary current at this load (assuming 100 per cent efficiency) is $10 \times 1000/1000 = 10$ A. The total copper losses at this load are therefore $180 \times (10/12)^2 = 125$ W.

$$\text{Output in true power} = 10 \times 0{\cdot}8 = 8 \text{ kW} = 8000 \text{ W}$$

$$\therefore \text{Efficiency} = \frac{\text{output}}{\text{intake}} = \frac{8000}{8000 + 210 + 125} = 96 \text{ per cent}$$

Again, from the short-circuit test,

$$\mathbf{Z}_1 = 50/12 = 4{\cdot}16\ \Omega$$
$$\mathbf{R}_1 = \mathbf{Z}_1 \cos\phi_s$$

where $\cos \phi_s$ is power factor during the short-circuit test (Fig. 17.12)

$$= 4{\cdot}16 \times 0{\cdot}3 = 1{\cdot}25\ \Omega$$

$$\mathbf{X}_1 = \sqrt{\{(4{\cdot}16)^2 - (1{\cdot}25)^2\}} = 4{\cdot}0\ \Omega \text{ approx.}$$

Hence, since the primary current on ordinary load is 10 A, the ohmic and reactive drops *referred to the primary* are $1{\cdot}25 \times 10 = 12{\cdot}5$, and $4 \times 10 = 40$ V, respectively. If we deduct these drops vectorially from the applied primary voltage, the remainder will be the voltage available for transformation by the secondary winding. Also, it is not necessary to make any calculation for drop in the secondary since this is already taken into account by using the equivalent resistance and

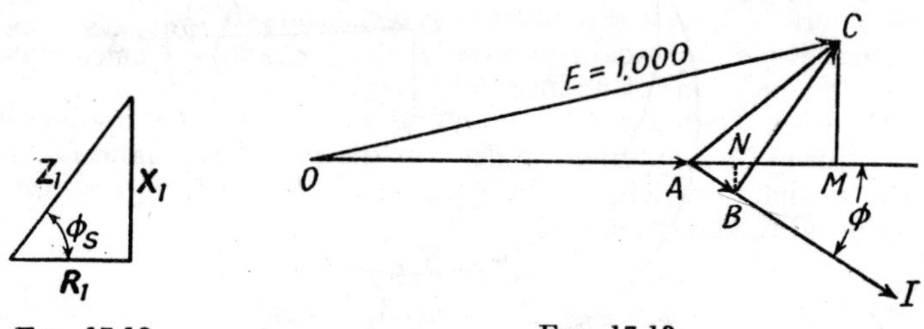

Fig. 17.12 Fig. 17.13

reactance referred to the primary. The vector diagram is given in Fig. 17.13. *OA* is the remainder of the voltage available for transformation; *AI* represents the phase of the current, this vector being therefore inclined at an angle of ϕ to *OC*, where $\cos \phi = 0{\cdot}8$. *AB* is the resistance drop of 12·5 V, in phase with *I*, and *BC*, the reactance drop of 40 V in quadrature with *I*. Hence, *OC* is the total applied voltage of 1000 V.

The voltage drop is $OC - OA$, which is approximately equal to *AM*. But

$$\begin{aligned} AM &= AN + NM = AB \cos \phi + BC \sin \phi \\ &= (12{\cdot}5 \times 0{\cdot}8) + (40 \times 0{\cdot}6) = 10 + 24 \\ &= 34 \text{ V} \end{aligned}$$

$$\therefore \text{ Percentage drop} = \mathbf{3{\cdot}4 \text{ per cent}}$$

Kapp Regulation Diagram. The *regulation* of a transformer, or an alternator, is the change of terminal voltage with load. Thus, if a transformer has a secondary induced e.m.f. of E_{20} on no load and a terminal voltage of V_2 on full load, the regulation is $(E_{20} - V_2)$. The percentage regulation is

$$(E_{20} - V_2)/E_{20} \times 100$$

This varies with the power factor of the load, as we have seen.

The graphical method of determining the voltage drop in a transformer is shown in Fig. 17.14. A vector *OA* is drawn to represent the secondary terminal voltage on load, and *OX* is drawn inclined at an angle ϕ to *OA*, where $\cos \phi$ is the power factor of the load. Then *OX* represents the phase of the secondary current. Draw AB parallel to OX and equal to the resistance drop referred to the secondary, namely, $\mathbf{R}_2 I_2$; and the perpendicular BC equal to the reactance drop referred to the secondary, namely, $\mathbf{X}_2 I_2$. Then the triangle CAB is the drop triangle referred to the secondary, and CA is the total voltage drop. Hence OC is the secondary no-load voltage E_{20}, and $(OC - OA)$,

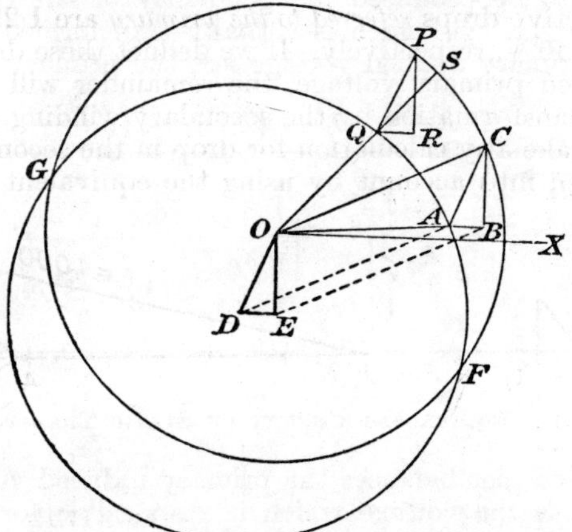

Fig. 17.14. Kapp Regulation Diagram

arithmetic difference, is the secondary drop from no load to full load. Now, if the triangle CAB is transferred to ODE, as shown, then $DA = OC = E_{20}$. Hence, for a given secondary current, the locus of C is a circle with centre O and radius E_{20}, while the locus of A is a circle with centre D and the same radius. In order to find the voltage drop on full load at any power factor $\cos \phi$, a radius OQS is drawn inclined at an angle ϕ to OX. If the impedance triangle is drawn in the position PQR, then $OP = OS$, is the no-load secondary voltage, and OQ the voltage on load. Hence, the length QS is the voltage drop. It is obvious that the triangle PQR need not be drawn, but simply the radius OQS. It will be seen that for all phase angles ϕ included between $\angle XOF$ (which is leading) and $\angle GOX$ (which is lagging), there is a drop of volts on load. For phase angles equal to either of these two limits the voltage on load is equal to that on no load. For phase angles between $\angle GOX$ and $\angle XOF$ (reckoning in a counter-clockwise direction) there is actually a rise of voltage on load.

The Kapp diagram affords a very convenient means of determining the variation of regulation with power factor, but it has the disadvantage that since the lengths of the sides of the impedance triangle are small compared with the radius of the circles, the diagram has to be drawn to a very large scale if accurate results are to be obtained.

Equivalent Circuit. On p. 399, the primary and secondary reactances were treated as though they had independent existences, like the primary and secondary resistances. Actually this is not the case, because, as we have seen, the leakage fluxes responsible for these reactances only exist when the windings carry load currents, and as it is impossible to have a load current in, say, the secondary without the corresponding induced current in the primary, it is equally impossible for one of the leakage fluxes to be produced independently of the other. On the other hand, the voltage which is actually transformed in the transformer is the primary applied voltage less the primary drop. This is

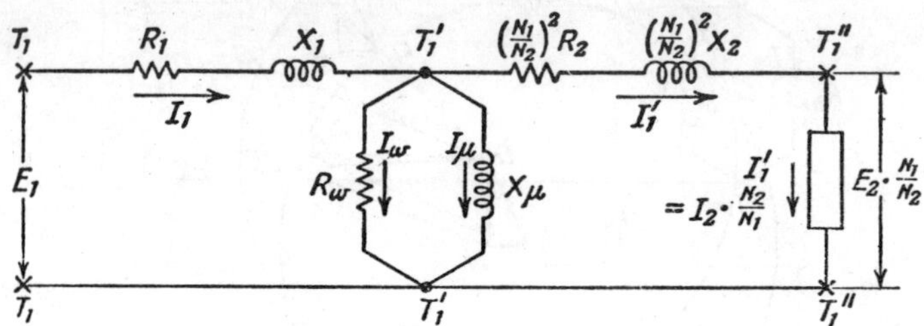

FIG. 17.15. EQUIVALENT CIRCUIT OF STATIC TRANSFORMER

the voltage which just balances the primary induced voltage V_1'. In other words, it is the voltage which is responsible for the iron loss current I_w, and the magnetizing current I_μ, and we can therefore represent these two components of the no-load current by the current taken by a non-inductive resistance R_w, and a pure reactance X_μ having the voltage V_1' applied to them, as indicated in Fig. 17.15. The rest of the equivalent circuit consists of the secondary resistance and reactance referred to the primary, and the primary induced load current I_1' is taken off at the terminals $T_1''T_1''$. Now the calculation of the regulation from this exact circuit is somewhat cumbersome, and a simplification can be made by transferring the resistance R_w and reactance X_μ to the terminals T_1T_1. The error introduced by doing this is, for a transformer of normal performance, quite negligible, because the currents I_w and I_μ are too small to affect the drop between T_1T_1 and $T_1'T_1'$. The simplified circuit is, therefore, as given in Fig. 17.16 (*A*). Now the resistance R_w and X_μ have no effect at all on the drop in the transformer when connected across T_1T_1, so that we can neglect them altogether, and we can make the further simplification of substituting the actual transformer current I_1 for the induced current I_1'. This is done in Fig. 17.16 (*B*). Alternatively, if the equivalent resistance and reactance are referred to the secondary side, we can use the equivalent circuit shown in Fig. 17.16 (*C*).

It is obvious that the circuits (*B*) and (*C*) lead to vector diagrams similar to that given in Fig. 17.13, but we will re-calculate the example on p. 403, using the symbolic method. The data are as follows—

$$V_1 = 1000 \text{ V}, \ I_1 = 10 \text{ A}, \ \mathbf{R}_1 = 1{\cdot}25, \ \mathbf{X}_1 = 4{\cdot}0$$

Denote the voltage V_1' (Fig. 17.16 (*B*)) by

$$x + j{\cdot}0$$

Since the current is at a power factor of 0·8 lagging, we have

$$\boldsymbol{I}_1 = 10 \times 0{\cdot}8 - j \times 10\sqrt{(1 - (0{\cdot}8)^2)}$$
$$= 8 - j6$$

∴ Impedance referred to primary side

$$\mathbf{Z}_1 = 1{\cdot}25 + j4$$

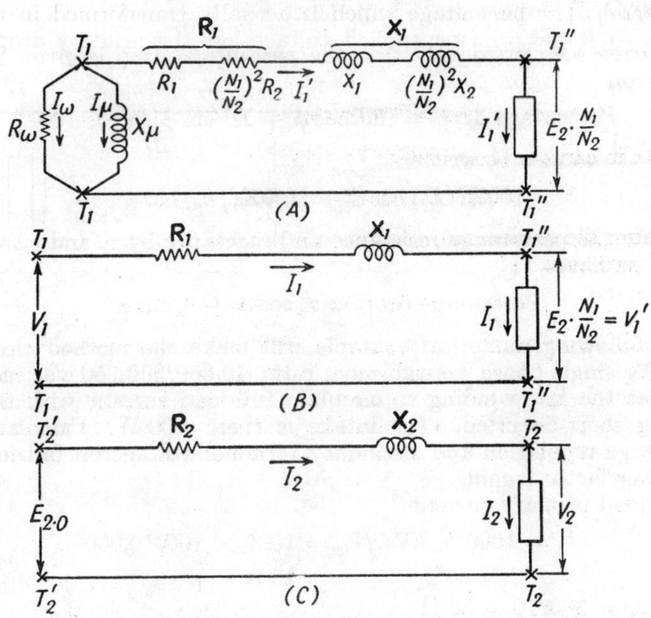

Fig. 17.16. Simplification of the Equivalent Circuit

∴ Drop of volts referred to primary side

$$\boldsymbol{I}_1\mathbf{Z}_1 = (8 - j6)(1{\cdot}25 + j4)$$
$$= 34 + j24{\cdot}5$$
$$\therefore\ V_1 = (x + j0) + (34 + j24{\cdot}5)$$
$$= (34 + x) + j24{\cdot}5$$

But $V_1 = 1000$

$$\therefore\ (34 + x)^2 + 24{\cdot}5^2 = 1000^2$$
$$\therefore\ x = 966 \text{ V, approx.}$$

∴ Drop of volts referred to primary = 1000 − 966 = 34

∴ Percentage drop = 3·4, as before

Percentage Resistance, Reactance, and Impedance. It is usual to express the resistance, reactance, and impedance of a transformer as percentages of the normal voltage. This is done in the following manner—

$100\mathbf{R}I/E_o$ = percentage resistance, where E_o is the normal voltage per phase of the winding on which the measurements or calculations are made

$100\mathbf{X}I/E_o$ = percentage reactance

and $100\mathbf{Z}I/E_o$ = percentage impedance

We have seen previously that the percentage drop is given by the expression

$$\text{Percentage drop} = (\mathbf{R}I\cos\phi + \mathbf{X}I\sin\phi)/E_o \times 100$$

so that it can also be written

$$(100\mathbf{R}I/E_o)\cos\phi + (100\mathbf{X}I/E_o)\sin\phi$$

Denoting the percentage resistance and reactance by v_r and v_x respectively, we have

$$\text{Percentage drop} = v_r\cos\phi + v_x\sin\phi$$

The following numerical example will make the method clear. A 100 kVA single-phase transformer, ratio 10,000/200, 50 c/s, requires 300 V at the h.t. winding to circulate full-load current with the l.t. winding short-circuited. The intake is then 1000 W. Calculate the percentage regulation and secondary terminal voltage on full load at 0·8 power factor lagging.

Full-load primary current

$$I_1 = 1000 \times \text{kVA}/V_1 = 1000 \times 100/10{,}000 = 10 \text{ A}$$

$$\therefore \mathbf{Z}_1 = 300/10 = 30\,\Omega$$

$$\mathbf{R}_1 = 1000/10^2 = 10\,\Omega$$

$$\mathbf{X}_1 = \sqrt{(30^2 - 10^2)} = 28{\cdot}3\,\Omega$$

$$\therefore v_r = (\mathbf{R}_1 I_1/V_1) \times 100 = 10 \times 10 \times 100/10{,}000 = 1$$

$$v_x = (\mathbf{X}_1 I_1/V_1) \times 100 = 28{\cdot}3 \times 10 \times 100/10{,}000 = 2{\cdot}83$$

$\therefore$ Percentage drop

$$= v_r\cos\phi + v_x\sin\phi$$

$$= 1 \times 0{\cdot}8 + 2{\cdot}83 \times 0{\cdot}6$$

$$= 2{\cdot}5$$

$$\therefore \text{Secondary drop} = 2{\cdot}5/100 \times 200 = 5 \text{ V}$$

$$\therefore V_2, \text{ on load} = 200 - 5 = 195 \text{ V}$$

"All-day" Efficiency. The ordinary or commercial efficiency is defined as the ratio

$$\frac{\text{Output in watts}}{\text{Intake in watts}}$$

The all-day efficiency is the ratio

$$\frac{\text{Output in kilowatt-hours}}{\text{Intake in kilowatt-hours}}$$

This second efficiency is thus measured on an energy basis. The reason for the introduction of this efficiency is that transformers used for distribution have their primary winding connected to the line for 24 hr/day. The core losses are thus going on for the whole 24 hr, whereas the copper losses go on only when the transformer is on load. Hence, if the load is not on the transformer for the whole of the time it is connected to the line, the all-day efficiency, for a given kVA loading, is less than the commercial efficiency.

EXAMPLE. A 20 kW lighting transformer of ordinary efficiency 95 per cent is on full load for 6 hr/day. Find the all-day efficiency if the full-load losses are equally divided between copper and iron.

Total losses on full load = 5 per cent of 20 kW = 1 kW

∴ Iron losses = 0·5 kW, and full-load copper losses = 0·5 kW

Output of 20 kW for 6 hr/day	=	120 kWh
Copper loss of 0·5 kW for 6 hr/day	=	3 kWh
Iron loss of 0·5 kW for 24 hr/day	=	12 kWh
Energy intake during 24 hr	=	135 kWh

∴ All-day efficiency = 120/135 = 89 per cent

Relation between Copper and Iron Loss.

$$\text{Output (watts)} = V_2 I_2 \cos\phi$$

$$\text{Primary copper loss} = R_1 I_1^2$$

$$\text{Secondary copper loss} = R_2 I_2^2$$

$$\therefore \text{Total copper loss} = (R_1 I_1^2 + R_2 I_2^2) = \mathbf{R}_2 I_2^2$$

$$\text{Iron loss} = P_i = \text{a constant}$$

$$\therefore \text{Efficiency} = \frac{\text{output}}{\text{intake}} = \frac{V_2 I_2 \cos\phi}{V_2 I_2 \cos\phi + \mathbf{R}_2 I_2^2 + P_i}$$

$$= \frac{V_2 \cos\phi}{V_2 \cos\phi + (\mathbf{R}_2 I_2 + P_i/I_2)}$$

For a given power factor the efficiency is a maximum when the expression in brackets is a minimum. Now the product of the two terms in the expression is a constant, and therefore, their sum is a minimum when they are equal. Hence, for maximum efficiency we have

$$\mathbf{R}_2 I_2 = P_i/I_2$$

$$\therefore \mathbf{R}_2 I_2^2 = P_i$$

The efficiency is thus a maximum at that load which makes the copper losses equal to the constant iron losses. If a transformer is intended to work most of the time on full load, it is designed to have the maximum efficiency at full load. If the load is variable it is usual to design it to have maximum efficiency at about three-quarters of full load.

EXAMPLE. A 100 kVA transformer has an iron loss of 1 kW and a copper loss on normal output current of 1·5 kW. Calculate the kVA loading at which the efficiency is a maximum, and the efficiency at this loading (*a*) at unity power factor, (*b*) at 0·8 power factor.

The current, and therefore the copper loss, is proportional to the square of the kVA loading. Hence, if the maximum efficiency occurs at "*m*" of full load, then

$$m^2 \times 1{\cdot}5 = 1 \text{ kW}$$

$$m = \sqrt{(1/1{\cdot}5)} = 0{\cdot}817$$

$\therefore$ kVA loading for maximum efficiency $= 0{\cdot}817 \times 100 = 81{\cdot}7$.

Total losses at this loading $= 2 \times$ iron loss $= 2$ kW

(*a*) unity power factor

$$\text{Output} = 81{\cdot}7 \text{ kW}$$

$$\text{Intake} = 83{\cdot}7 \text{ kW}$$

$$\eta = 100 \times (81{\cdot}7/83{\cdot}7) = 97{\cdot}61 \text{ per cent}$$

(*b*) 0·8 power factor

$$\text{Output} = 0{\cdot}8 \times 81{\cdot}7 = 65{\cdot}36 \text{ kW}$$

$$\text{Intake} = 67{\cdot}36 \text{ kW}$$

$$\eta = 100 \times (65{\cdot}36/67{\cdot}36) = 97{\cdot}03 \text{ per cent}$$

Division of Load between Two Transformers in Parallel. In Fig. 17.17 two transformers are shown connected in parallel on a pair of bus-bars, and their resistances and reactances, *referred to the secondary*, are, for the purposes of calculation, shown external to the secondary windings. If the transformation ratios are the same, the voltages across the secondary terminals of the two ideal transformers will be the same, and we can therefore imagine these connected to a second pair of bus-bars, as shown dotted. The voltage across these bars will not vary with load, but will remain constant. The equivalent simple circuit is therefore as shown in the second diagram, and the load current I will divide itself between the two transformers in the ratio of the branch currents I_1 and I_2 in the equivalent circuit.

Let $\mathbf{Z}_1 = \sqrt{(\mathbf{R}_1{}^2 + \mathbf{X}_1{}^2)}$, and $\mathbf{Z}_2 = \sqrt{(\mathbf{R}_2{}^2 + \mathbf{X}_2{}^2)}$

be the equivalent impedances of the two transformers referred to the secondaries. If the ratios $\mathbf{R}_1/\mathbf{X}_1$ and $\mathbf{R}_2/\mathbf{X}_2$ are equal, then the currents I_1 and I_2 will be in phase, and we can write

$$I_1/I_2 = \mathbf{Z}_2/\mathbf{Z}_1;\ I_1 + I_2 = I$$

$$\therefore\ I_1 = [\mathbf{Z}_2/(\mathbf{Z}_1 + \mathbf{Z}_2)] \times I, \text{ and } I_2 = [\mathbf{Z}_1/(\mathbf{Z}_1 + \mathbf{Z}_2)] \times I$$

If the ratios of resistance to reactance are not equal, then the graphical construction, as applied to branched circuits, can be used.

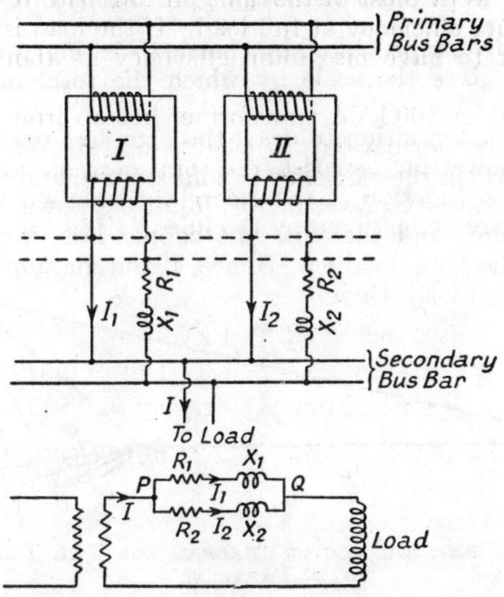

FIG. 17.17. TWO TRANSFORMERS IN PARALLEL

Assume any drop OA (Fig. 17.18) between the points P and Q in Fig. 17.17, and on OA draw a semicircle. Draw chords OB and OC inclined at ϕ_1 and ϕ_2, where $\phi_1 = \tan^{-1} \mathbf{X}_1/\mathbf{R}_1$, and $\phi_2 = \tan^{-1} \mathbf{X}_2/\mathbf{R}_2$. Then

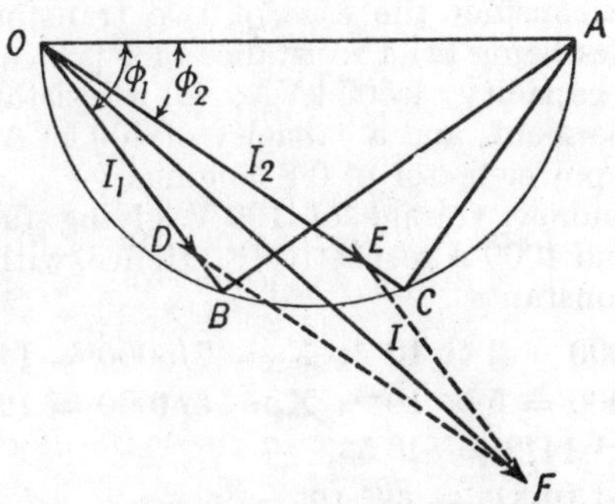

FIG. 17.18. GRAPHICAL CONSTRUCTION FOR DIVISION OF LOAD BETWEEN TWO TRANSFORMERS IN PARALLEL

these chords represent to scale the respective resistance drops, as before. Dividing them by the respective resistances $\mathbf{R}_1$ and $\mathbf{R}_2$, we obtain quotients proportional to the currents I_1 and I_2. Represent these quotients graphically by the lengths OD and OE, and find the resultant

OF. Then OF represents the total current delivered to the load. But this current is I, which is known. Hence we have

$$I_1 = (OD/OF) \times I \text{ and } I_2 = (OE/OF) \times I$$

The resistance drop represented by chord OB is then given numerically by $\mathbf{R}_1 I_1$. This gives the scale by which the total drop OA can be calculated.

We are now in a position to draw the complete vector diagram. If the load is of power factor $\cos\phi$, the total current vector OF will be inclined ϕ to the direction of the secondary terminal voltage V_2, this direction therefore being given by the line $O'V_2$ in Fig. 17.19. With A

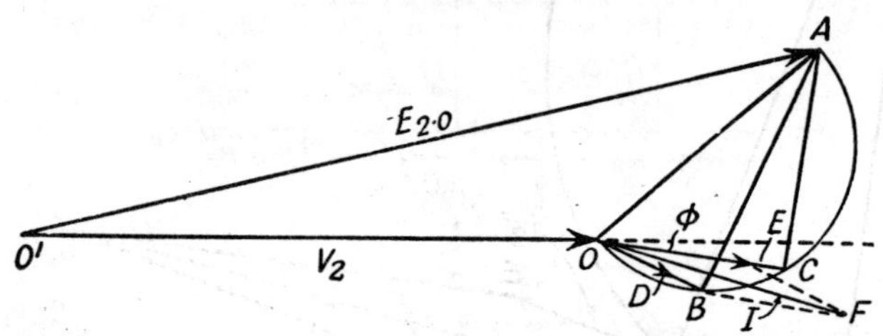

Fig. 17.19. Complete Vector Diagram for Two Transformers in Parallel

as centre and $E_{2.0}$, the secondary no-load voltage, as radius, mark off AO' equal to $E_{2.0}$; then the length $O'O$ gives the secondary voltage on load, and the voltage regulation of the "bank" of transformers is given by

$$[(O'A - O'O)/O'A] \times 100$$

As an example consider the case of two transformers in parallel, their characteristics being (a) a resistance of 1 per cent, a reactance of 7 per cent, and a capacity of 500 kVA; (b) a resistance of 2 per cent, a reactance of 5 per cent, and a capacity of 400 kVA. The load is one of 1000 kVA at a power factor of 0·8 lagging.

Assume a secondary voltage of 100 V giving full-load secondary currents of 5000 and 4000 A respectively. Hence, with this assumption, we have for the constants

$$\mathbf{R}_1 = 1/5000 = 2 \times 10^{-4}; \ \mathbf{X}_1 = 7/5000 = 14 \times 10^{-4}$$
$$\mathbf{R}_2 = 2/4000 = 5 \times 10^{-4}; \ \mathbf{X}_2 = 5/4000 = 12{\cdot}5 \times 10^{-4}$$
$$\phi_1 = \tan^{-1} 14/2 = 81^\circ\ 53'$$
$$\phi_2 = \tan^{-1} 12{\cdot}5/5 = 68^\circ\ 12'$$

The drop triangles are drawn to scale in Fig. 17.20, and working backwards as explained above, we see that the length OA in volts is equal to 6·68 V; with $E_{2.0}$ equal to 100, this gives 94·5 V for V_2, so that the percentage regulation of the bank is 5·5 per cent.

We also see that the currents I_1 and I_2 are 4940 and 5150 respectively, and the phases of these currents with respect to V_2 are $\alpha_1 = 45^\circ$

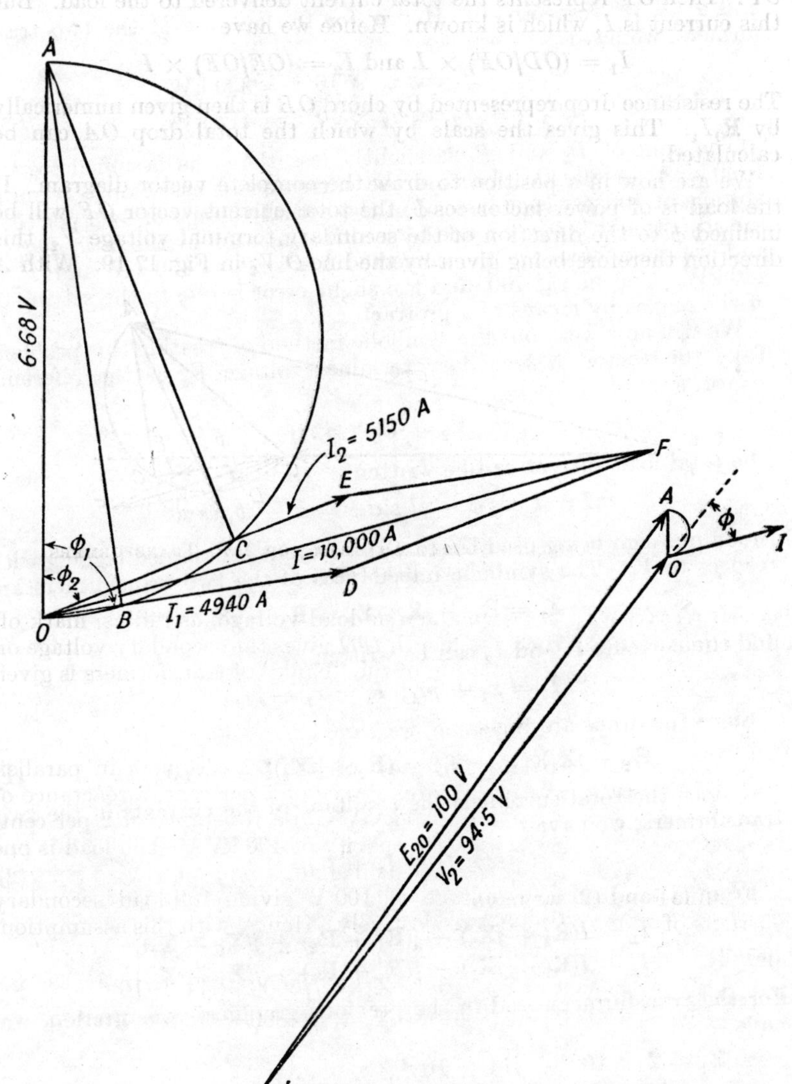
A
6·68 V
φ1
φ2
O
B
C
I2 = 5150 A
E
F
I = 10,000 A
D
I1 = 4940 A
A
O
φ
I
E20 = 100 V
V2 = 94·5 V
O′

FIG. 17.20

and $\alpha_2 = 32°$ respectively. Hence, for the outputs of the two transformers, we have

$$P_1 = V_2 I_1 \cos \alpha_1 \times 10^{-3} = 330 \text{ kW}$$
$$P_2 = V_2 I_2 \cos \alpha_2 \times 10^{-3} = 412 \text{ kW}$$

The sum of P_1 and P_2 is slightly less than the assumed load of 1000 kVA, or 800 kW, because the currents are based on a terminal voltage of 100 per cent instead of the actual voltage of 94·5 per cent; the adjustment is easily made by multiplying each by the ratio $E_{2\cdot 0}/V_2$, and we then have for the corrected loads $P_1 = 350$ kW and $P_2 = 440$ kW. The sum is 790, and the slight error is due to the setting off of the angles by means of a protractor.

We will now work out the symbolic method of treating this problem. Take the vector of secondary terminal voltage V_2 as the reference vector, so that

$$V_2 = V_2 + j \cdot 0$$

The total load current can be written

$$I = I \cos \phi - jI \sin \phi = a - jb, \text{ say,}$$

the minus sign being used because the current is assumed lagging with respect to V_2. The symbolic impedances of the two transformers are

$$Z_1 = \mathbf{R}_1 + j\mathbf{X}_1; \; Z_2 = \mathbf{R}_2 + j\mathbf{X}_2$$

and the currents I_1 and I_2 can be written

$$I_1 = x_1 + jy_1; \; I_2 = x_2 + jy_2$$

Since the drops are the same, we have

$$(\mathbf{R}_1 + j\mathbf{X}_1)(x_1 + jy_1) = (\mathbf{R}_2 + j\mathbf{X}_2)(x_2 + jy_2) \quad . \quad . \quad (1)$$

and since the total current is the resultant of the currents in the two transformers, we have

$$I_1 + I_2 = I \quad . \quad . \quad . \quad . \quad (2)$$

From (1) and (2) we have

$$I_1 = I(\mathbf{R}_2 + j\mathbf{X}_2) \div \{(\mathbf{R}_1 + \mathbf{R}_2) + j(\mathbf{X}_1 + \mathbf{X}_2)\}$$

and

$$I_2 = I(\mathbf{R}_1 + j\mathbf{X}_1) \div \{(\mathbf{R}_1 + \mathbf{R}_2) + j(\mathbf{X}_1 + \mathbf{X}_2)\}$$

For the transformers used in the previous graphical construction, we have

$$Z_1 = 2 \times 10^{-4} + j14 \times 10^{-4}$$

$$\therefore \; Y_1 = \frac{2 \times 10^{-4} - j14 \times 10^{-4}}{(2 \times 10^{-4})^2 + (14 \times 10^{-4})^2} = \frac{2 \times 10^{-4}}{200 \times 10^{-8}} - j\frac{14 \times 10^{-4}}{200 \times 10^{-8}}$$
$$= 10^2 - j7 \times 10^2$$

and
$$Y_2 = \frac{5 \times 10^{-4} - j12\cdot 5 \times 10^{-4}}{5^2 \times 10^{-8} + 12\cdot 5^2 \times 10^{-8}} = \frac{5 \times 10^{-4}}{180 \times 10^{-8}} - j\frac{12\cdot 5 \times 10^{-4}}{180 \times 10^{-8}}$$
$$= 2\cdot 76 \times 10^2 - j6\cdot 9 \times 10^2$$

Hence, for the whole of the branched circuit

$$Y = 10^2 - j7{\cdot}0 \times 10^2 + 2{\cdot}76 \times 10^2 - j6{\cdot}9 \times 10^2$$
$$= 3{\cdot}76 \times 10^2 - j13{\cdot}9 \times 10^2$$
$$\therefore Z = \frac{3{\cdot}76 \times 10^2 + j13{\cdot}9 \times 10^2}{3{\cdot}76^2 \times 10^4 + 13{\cdot}9^2 \times 10^4}$$
$$= 1{\cdot}81 \times 10^{-4} + j6{\cdot}69 \times 10^{-4}$$

Hence, voltage drop along the parallel path

$$ZI = (1{\cdot}81 \times 10^{-4} + j6{\cdot}69 \times 10^{-4})\,(10{,}000 \times 0{\cdot}8 - j10{,}000 \times 0{\cdot}6)$$
$$= 5{\cdot}47 + j4{\cdot}26$$

But $$V_2 = V_2 + j\,.\,0$$
$$\therefore E_{2\cdot 0} = V_2 + 5{\cdot}47 + j4{\cdot}26$$
$$\therefore E_{2\cdot 0} = \{(V_2 + 5{\cdot}47)^2 + 4{\cdot}26^2\}^{\frac{1}{2}}$$

But $$E_{2\cdot 0} = 100$$
$$\therefore V_2 = 94{\cdot}43 \text{ V}$$

For the calculation of the current carried by the two transformers, we have

$$I_1 = Y_1 \times \text{symbolic voltage drop}$$
$$= (10^2 - j7 \times 10^2)(5{\cdot}47 + j4{\cdot}26)$$
$$= 3{\cdot}53 \times 10^3 - j3{\cdot}40 \times 10^3$$
$$\therefore I_1 = 10^3(3{\cdot}53^2 + 3{\cdot}40^2)^{\frac{1}{2}}$$
$$= 4930 \text{ A}$$
$$I_2 = Y_2 \times \text{symbolic voltage drop}$$
$$= (2{\cdot}76 \times 10^2 - j6{\cdot}9 \times 10^2)(5{\cdot}47 + j4{\cdot}26)$$
$$= 4{\cdot}45 \times 10^3 - j2{\cdot}60 \times 10^3$$
$$\therefore I_2 = 10^3(4{\cdot}45^2 + 2{\cdot}60^2)^{\frac{1}{2}}$$
$$\therefore I_2 = 5150 \text{ A}$$

As a check $$I = 8 \times 10^3 - j6 \times 10^3$$
$$I_1 + I_2 = 3{\cdot}53 \times 10^3 - j3{\cdot}40 \times 10^3 + 4{\cdot}45 \times 10^3 - j2{\cdot}60 \times 10^3$$
$$= 7{\cdot}98 \times 10^3 - j6{\cdot}03 \times 10^3$$

Effect of an Inequality in the Turns Ratios. If the turns ratios are not exactly equal, then, with the primaries paralleled to the same supply, the secondary induced e.m.f.s will be unequal. This inequality will result in the production of a local, or circulating, current which will flow even inside no-load conditions. If, in addition, the transformers are sharing a load, then this circulating current will be superposed on the load currents. Consider the circuit of Fig. 17.21, and let the secondary induced e.m.f.s be E_1 and E_2, the suffixes referring to the transformer. Also let $E_1 > E_2$, then since E_1 and E_2 are in time phase, both being induced by the same applied primary voltage, the difference voltage acting round the local circuit dotted is

$$E_c = E_1 - E_2$$

$$\therefore\ I_c = (E_1 - E_2)/(Z_1 + Z_2)$$

FIG. 17.21. TRANSFORMERS WITH UNEQUAL TURNS RATIOS

lagging E_1 by the angle α where

$$\alpha = \tan^{-1}[(X_1 + X_2)/(R_1 + R_2)]$$

With normal ratios of reactance to resistance this will be a large angle of the order of 80°, showing that the circulating current is nearly a wattless current. Suppose that, apart from this circulating current,

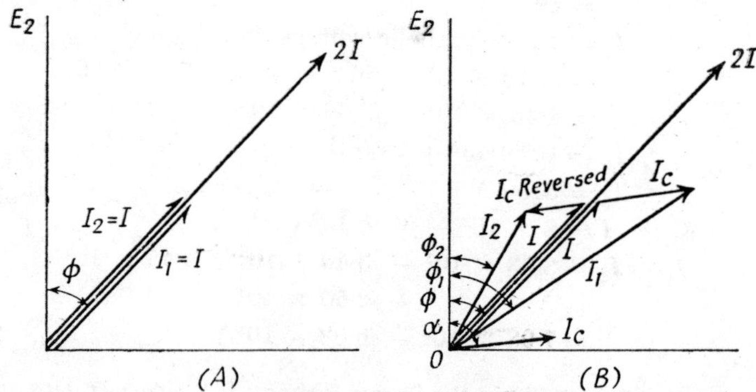

FIG. 17.22. EFFECT OF CIRCULATING CURRENT ON THE PERFORMANCE OF TRANSFORMERS IN PARALLEL

the two transformers are sharing a load current of $2I$ amperes equally, and suppose that their ratios of reactance to resistance are equal, then the contribution of each is I ampere at the same power factor as the load power factor, Fig. 17.22 (A). The circulating current I_c must be added vectorially to the load current of No. 1 and deducted vectorially from that of No. 2. This is done in Fig. 17.22 (*B*), and we see that the effect of the circulating current is to reduce the power factor of the transformer having the greater no-load induced e.m.f., and to increase the power factor of the other.*

EXAMPLE. A 100 kVA and a 50 kVA transformer are in parallel. The no-load secondary e.m.f.s are 1000 and 950 volts respectively; the resistances are 1·5 per cent and 2·0 per cent respectively, and the reactances 8 per cent and 6 per cent respectively. Calculate the no-load circulating current on the secondary side.

	No. 1. 100 kVA	No. 2. 50 kVA
Normal sec. current	$\frac{100{,}000}{1000} = 100$ A	$\frac{50{,}000}{950} = 52{\cdot}63$ A
$R_2 = \frac{E_{20}v_r}{100I_2}$	$\frac{1000 \times 1{\cdot}5}{100 \times 100} = 0{\cdot}15\ \Omega$	$\frac{950 \times 2}{100 \times 52{\cdot}63} = 0{\cdot}36\ \Omega$
$X_2 = \frac{E_{20}v_x}{100I_2}$	$\frac{1000 \times 8}{100 \times 100} = 0{\cdot}8\ \Omega$	$\frac{950 \times 6}{100 \times 52{\cdot}63} = 1{\cdot}08\ \Omega$

$$\therefore I_c = (1000 - 950)/\{(0{\cdot}15 + 0{\cdot}36)^2 + (0{\cdot}8 + 1{\cdot}08)^2\}^{\frac{1}{2}}$$
$$= 35{\cdot}7 \text{ A}$$
$$\alpha = \tan^{-1}(1{\cdot}88/0{\cdot}51) = 74° 50'$$

The following method of analysis applies whether the transformation ratios are equal or not. The circuit is given in Fig. 17.23, where Z_1 and

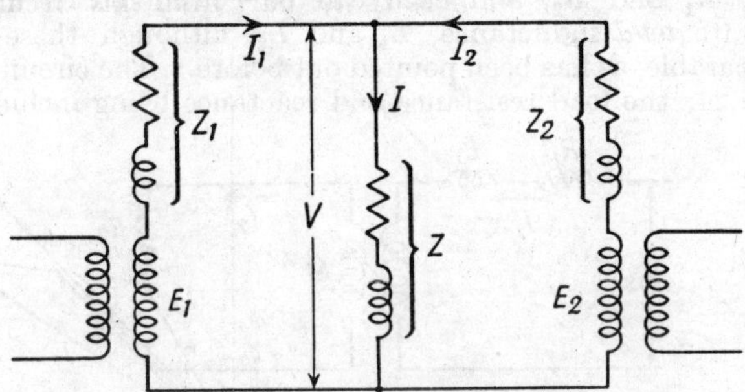

FIG. 17.23. TRANSFORMERS IN PARALLEL

Z_2 are the transformer impedances referred to the secondary sides, and Z the load impedance.

$$\left.\begin{aligned} E_1 &= I_1Z_1 + IZ = I_1Z_1 + (I_1 + I_2)Z \\ E_2 &= I_2Z_2 + IZ = I_2Z_2 + (I_1 + I_2)Z \end{aligned}\right\}$$

* A complete discussion of this problem is given by J. E. Parton, "Paralleled transformers with unequal ratios." *Elec. Journ.* (Sept. 3rd, 1954).

Solving these for I_1 and I_2 we have

$$I_1 = [E_1Z_2 + (E_1 - E_2)Z]/[Z_1Z_2 + (Z_1 + Z_2)Z]$$
$$I_2 = [E_2Z_1 - (E_1 - E_2)Z]/[Z_1Z_2 + (Z_1 + Z_2)Z]$$
$$\therefore\ I = [E_1Z_2 + E_2Z_1]/[Z_1Z_2 + (Z_1 + Z_2)Z]$$

Secondary, and load, terminal p.d.

$$V = IZ = [E_1Z_2 + E_2Z_1]/(Z_1Z_2/Z) + Z_1 + Z_2]$$

It is clear that if E_1 and E_2 are not equal, then the expressions for the currents must contain the circulating current. Comparing the expressions for I_1 and I_2 we see that the numerators have terms $(E_1 - E_2)Z$ and $-(E_1 - E_2)Z$ respectively, and the circulating current is given by

$$I_c = (E_1 - E_2)Z/[Z_1Z_2 + (Z_1 + Z_2)Z]$$
$$= (E_1 - E_2)/[Z_1 + Z_2 + (Z_1Z_2/Z)]$$

Under no-load conditions we must take the load impedance Z to be infinity, which then gives

$$I_c = (E_1 - E_2)/(Z_1 + Z_2)$$

the expression we obtained previously by a more elementary method. We see that the load itself has an effect on the magnitude of I_c because of the presence of the term Z_1Z_2/Z in the denominator.

The Transformer as a Mutually-inductive Circuit. Since practically the whole of the flux produced by a transformer links with both primary and secondary windings, the transformer is an appliance with very "closely coupled" circuits. Each winding has its own separate resistance, R_1 and R_2, and each can be considered as having its own separate *total* inductance, L_1 and L_2, although these are not strictly separable, as has been pointed out before.* The circuit is shown in Fig. 17.24, the load resistance and reactance being included in R_2

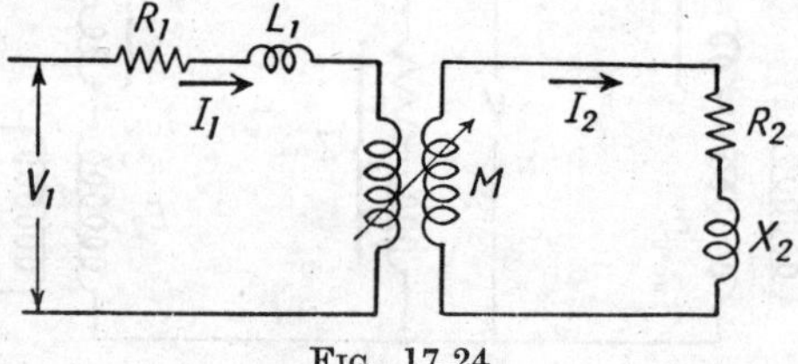

FIG. 17.24

and X_2 on the secondary side. The applied primary voltage has to do two things, viz. overcome the impedance drop in the primary and also overcome the mutually-induced primary e.m.f. due to the current variations in the secondary. Hence, using the symbolic notation,

$$V_1 = I_1(R_1 + jL_1\omega) + jM\omega I_2 \qquad . \quad . \quad . \quad (1)$$

* This case, however, is not identical with that discussed on p. 402, where the flux considered is the leakage flux, not the total flux.

The secondary induced e.m.f. is

$$\mathbf{E}_2 = -jM\omega\mathbf{I}_1$$

and this has only to supply the total impedance drop in the secondary circuit

$$\therefore \; -jM\omega\mathbf{I}_1 = \mathbf{I}_2(R_2 + jL_2\omega) \quad . \quad . \quad . \quad (2)$$

From equation (2) we have

$$\mathbf{I}_2/\mathbf{I}_1 = -jM\omega/(R_2 + jL_2\omega)$$
$$\therefore \; I_2/I_1 = M\omega/(R_2^2 + L_2^2\omega^2)^{\frac{1}{2}}$$

Substituting in equation (1) the value of $\mathbf{I}_2$ obtained from (2) we have

$$\begin{aligned}\mathbf{V}_1 &= \mathbf{I}_1(R_1 + jL_1\omega) + \mathbf{I}_1[M^2\omega^2/(R_2 + jL_2\omega)] \\ &= \mathbf{I}_1[(R_1 + jL_1\omega) + M^2\omega^2(R_2 - jL_2\omega)/(R_2^2 + L_2^2\omega^2)] \\ &= \mathbf{I}_1\left\{R_1 + R_2\frac{M^2\omega^2}{R_2^2 + L_2^2\omega^2} + j\omega\left(L_1 - L_2\frac{M^2\omega^2}{R_2^2 + L_2^2\omega^2}\right)\right\}\end{aligned}$$

Thus the quantity

$$\left(R_1 + R_2 \cdot \frac{M^2\omega^2}{R_2^2 + L_2^2\omega^2}\right) + j\omega\left(L_1 - L_2 \cdot \frac{M^2\omega^2}{R_2^2 + L_2^2\omega^2}\right)$$

is the equivalent impedance of the primary circuit.

$R_1 + R_2[M^2\omega^2/(R_2^2 + L_2^2\omega^2)]$ is the equivalent resistance,

and $L_1 - L_2[M^2\omega^2/(R_2^2 + L_2^2\omega^2)]$ is the equivalent inductance

Denote these by r_1 and l_1 respectively, then the equivalent primary reactance is

$$\therefore \; x_1 = l_1\omega$$
$$\therefore \; \mathbf{I}_1 = \mathbf{E}_1/(r_1 + jx_1)$$
$$\therefore \; I_1 = E_1/(r_1^2 + x_1^2)^{\frac{1}{2}}$$

But $\qquad I_2 = I_1 \times M\omega/(R_2^2 + L_2^2\omega^2)^{\frac{1}{2}}$

$$\therefore \; I_2 = [E_1/(r_1^2 + x_1^2)^{\frac{1}{2}}] \times [M\omega/(R_2^2 + L_2^2\omega^2)^{\frac{1}{2}}]$$

We see that $r_1 > R_1$ and $l_1 < L_1$, so that the effect of putting load on a transformer is to increase the apparent resistance and to decrease the apparent inductance.

In order to bring out the above property the problem was simplified by making the L's the total inductances, i.e. including both leakage flux and mutual flux. We will now give the more complete treatment in which the various fluxes are all regarded as quite separate. The solution applies to a transformer in which M is a constant, namely an air-core transformer. Since we have seen that the total flux of a constant-voltage, constant-frequency transformer is sensibly constant, the value of μ the permeability of the iron will be sensibly constant. Thus, the treatment applies very approximately to an iron-cored transformer. We will define the mutual inductance of the primary with respect to the secondary as $M_{1\cdot 2}$, and of the secondary with respect to the primary as $M_{2\cdot 1}$. Then if a current I_1 produces a flux through the secondary of Φ_1

$$M_{2\cdot 1} = N_2\Phi_1/I_1$$

Similarly

$$M_{1\cdot 2} = N_1\Phi_2/I_2$$

If we denote the reluctance of the common magnetic circuit by S, we have

$$\Phi_1/I_1 = N_1/S$$

and

$$\Phi_2/I_2 = N_2/S$$

Cross multiplying

$$N_2\Phi_1/I_1S = N_1\Phi_2/I_2S$$

or

$$N_2\Phi_1/I_1 = N_1\Phi_2/I_2$$

$$\therefore M_{2\cdot 1} = M_{1\cdot 2} = M \text{ say}$$

The various components of the primary applied voltage are therefore as follows—

(*a*) That to overcome the self-induced e.m.f. due to the true leakage reactance L_1, namely $j\omega L_1 \boldsymbol{I}_1$.

(*b*) That to overcome the induced e.m.f. due to the current change in the secondary, namely $+ j\omega M\boldsymbol{I}_2$.

(*c*) The self-induced e.m.f. due to the primary current I_1 in association with the main flux of the transformer. Since the induced e.m.f. due to changes in current I_2 is numerically equal to $j\omega M\boldsymbol{I}_2$, that due to changes in current I_1 is therefore $j\omega(M/\rho)\boldsymbol{I}_1$, where ρ is the turns ratio N_2/N_1. The sign of this must clearly be taken as positive since the applied voltage must have a component whose function it is to overcome it.

Hence for the e.m.f. equation of the primary we have

$$\boldsymbol{V}_1 = R_1\boldsymbol{I}_1 + j\omega L_1\boldsymbol{I}_1 + j\omega(M/\rho)\boldsymbol{I}_1 + j\omega M\boldsymbol{I}_2$$

The e.m.f. induced in the secondary is

$$\boldsymbol{E}_2 = - j\omega M\boldsymbol{I}_1$$

This has to overcome the self-inductance of the secondary $j\omega\rho M\boldsymbol{I}_2$ and the total impedance drop of both secondary and load. Denote the total resistance of secondary winding and load by R_2 and the sum of the secondary leakage reactance and load reactance by $L_2\omega$, then we have

$$\boldsymbol{E}_2 = - j\omega M\boldsymbol{I}_1 = R_2\boldsymbol{I}_2 + j\omega L_2\boldsymbol{I}_2 + j\omega\rho M\boldsymbol{I}_2$$

From the second equation we derive straight away the current relationship

$$\boldsymbol{I}_2 = -[j\omega M/(R_2 + j\omega L_2 + j\omega\rho M)]\boldsymbol{I}_1$$

or

$$- \boldsymbol{I}_1/\boldsymbol{I}_2 = \rho + (R_2 + j\omega L_2)/j\omega M$$

Substituting this in the equation for V_1 we have

$$V_1 = I_1\left[R_1 + j\omega L_1 + j\omega\frac{M}{\rho} + \frac{\omega^2M^2}{R_2 + j\omega(L_2 + \rho M)}\right]$$

$$= I_1\left[R_1 + j\omega\left(L_1 + \frac{M}{\rho}\right) + \frac{\omega^2M^2\{R_2 - j\omega(L_2 + \rho M)\}}{R_2{}^2 + \omega^2(L_2 + \rho M)^2}\right]$$

$$= I_1\left[\left(R_1 + \frac{\omega^2M^2R_2}{R_2{}^2 + \omega^2(L_2 + \rho M)^2}\right)\right.$$

$$\left. + j\omega\left(L_1 + \frac{M}{\rho} - \frac{\omega^2M^2(L_2 + \rho M)}{R_2{}^2 + \omega^2(L_2 + \rho M)^2}\right)\right]$$

The two terms in the "round" brackets represent the equivalent resistance of the whole circuit and the equivalent inductance of the whole circuit respectively.

EXAMPLE. An iron-cored transformer has

$R_1 = 1\ \Omega$; $L_1 = 0{\cdot}00637$ H

$R_2 = 50\ \Omega$; $L_2 = 0{\cdot}0795$ H (these include the load)

$M = 0{\cdot}5$ H

$\rho = 2$

$V_1 = 100$ V at $f = 50$ c/s

The no-load primary current will be given by putting $R_2 = \infty$. This gives

$$100 = I_{1\cdot0}[R_1 + 0 + j\omega\{L_1 + (M/\rho)\}]$$
$$= I_{1\cdot0}[1 + j314\{0{\cdot}00637 + (0{\cdot}5/2)\}]$$
$$= I_{1\cdot0}(1 + j80{\cdot}6)$$
$$\therefore\ I_{1\cdot0} = 100/(1 + j80{\cdot}6)$$
$$= 100(1 - j80{\cdot}6)/(1^2 + 80{\cdot}6^2)$$
$$= 0{\cdot}0154 - j1{\cdot}24$$
$$\therefore\ I_{1\cdot0} = 1{\cdot}24 \text{ A}$$
$$\phi_o = \cos^{-1}(0{\cdot}0154/1{\cdot}24)$$
$$= 89^\circ\ 17'$$

On load we put $R_2 = 50$. Working out the reactances separately we have

$$\omega L_1 = 314 \times 0{\cdot}00637 = 2\ \Omega$$
$$\omega L_2 = 314 \times 0{\cdot}0795 = 25\ \Omega$$
$$\omega M = 314 \times 0{\cdot}5 = 157\ \Omega$$

Substituting in the expression for I_1 we obtain

$$100 = I_1(11{\cdot}49 + j9{\cdot}5)$$
$$\therefore\ I_1 = 5{\cdot}2 - j4{\cdot}3$$

$$\therefore \; I_1 = 6{\cdot}75 \text{ A, and}$$
$$\phi_1 = 39° \, 30'$$
$$\therefore \; I_2 = [j\omega M/(R_2 + j\omega L_2 + j\omega pM)]I_1$$
$$= [j157/(50 + j25 + j2 \times 157)](5{\cdot}2 - j4{\cdot}3)$$
$$= [j157/(50 + j339)](5{\cdot}9 - j4{\cdot}3)$$
$$= j157(5{\cdot}9 - j4{\cdot}3)(50 - j339)/(50^2 + 339^2)$$
$$= 2{\cdot}65 - j1{\cdot}6$$

Of the total secondary impedance, suppose that of the load is $(45 + j20)$, then the secondary terminal voltage on load is

$$V_2 = (2{\cdot}65 - j1{\cdot}6)(45 + j20)$$
$$= 151 - j19$$
$$\therefore \; V_2 = 152\text{V}$$

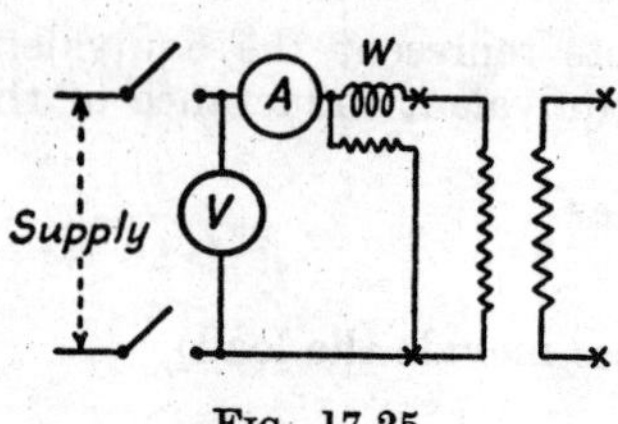

FIG. 17.25

Testing of Transformers. The various transformer tests are as follows—

(*a*) CORE LOSS TESTS. The primary is connected to a supply of normal frequency and voltage, through a wattmeter W, as shown in Fig. 17.25, and the secondary is left on open circuit. Since the core loss is independent of the load, and since the primary no-load current I_o is so small that the copper loss due to it can be neglected, we have

Core loss = wattmeter reading = P_i, say

Power factor $\cos \phi_o = P_i/VI_o$

Working component $I_w = I_o \cos \phi_o$

Magnetizing component $I_\mu = I_o \sin \phi_o$*

Since the power factor $\cos \phi_o$ will be low, it will be necessary to correct the wattmeter reading. (*See* p. 370.)

If the dimensions of the core and the number of primary turns are known, the hysteresis and eddy-current losses can be separated in the following way. We have, from the transformer e.m.f. equation,

$$B_{\max} = V/4{\cdot}44AN_1 f$$

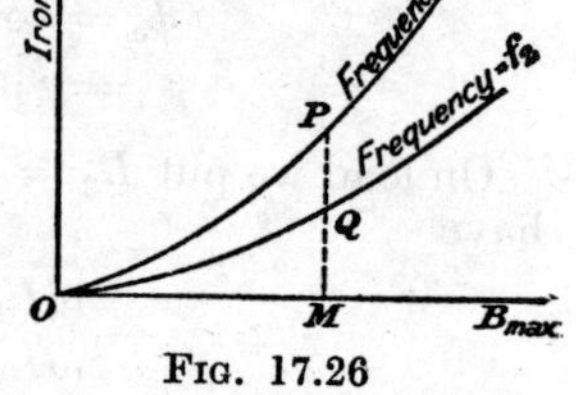

FIG. 17.26

The applied voltage V is varied over as wide a range as possible, the frequency being kept constant at, say, f_1. The iron loss, as measured by the wattmeter, is then plotted against the calculated values of $B_{\max}$ (Fig. 17.26). The test is then made at a second

* This elementary separation of the two components assumes that the current I_o is sinusoidal. In practice this is not the case, *see* Chap. 28, and the values of I_w and I_μ so obtained are only approximate.

frequency f_2, thus giving two curves. For a given value of B_{max} the hysteresis loss is proportional to the frequency, and the eddy-current loss to the square of the frequency. Hence, if we take any ordinate PQM, we have

Total loss at frequency f_1

$$P_1 = Af_1 + Bf_1^2$$

while total loss at frequency f_2

$$P_2 = Af_2 + Bf_2^2, \text{ } A \text{ and } B \text{ being constants}$$

From these equations A and B can be calculated. The separate losses at any frequency, *for the value* of B_{max} *equal to* OM, can then be calculated. By drawing a series of ordinates and making the above analysis for each, the losses at any value of B_{max}, as well as any value of the frequency, can be calculated.

It is usual to make the above test, not on the actual transformer itself, but on a test specimen built up of strips of transformer iron. In

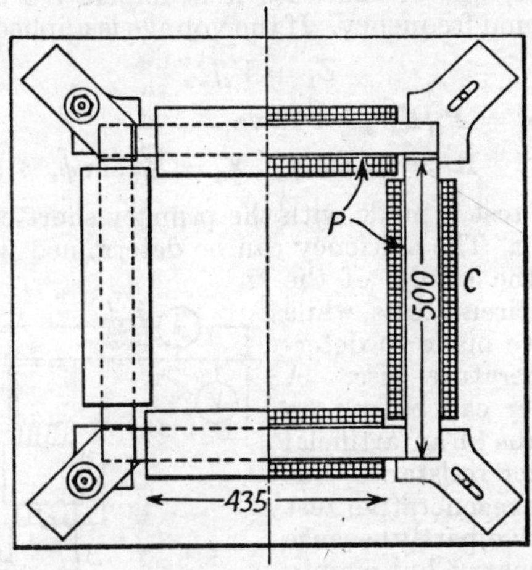

FIG. 17.27. EPSTEIN'S IRON-TESTING APPARATUS

Epstein's apparatus the specimen is arranged in a square, the standard dimensions of a strip being 50 cm × 3 cm. The total weight of strip used in building up the specimen is 10 kg. The arrangement of the specimen and the magnetizing coils is clearly shown in Fig. 17.27.

(*b*) IMPEDANCE AND COPPER LOSS TESTS. The secondary is short-circuited by a heavy conductor, as shown in Fig. 17.28, and a low voltage applied to the primary, the voltage being adjusted until the ammeter A indicates that the full-load current is flowing. The losses now taking place will be the full-load copper loss in both primary and secondary, and a small amount of iron loss. The voltage applied has to be so small that this iron loss is quite negligible. Hence, if P_s is the wattmeter reading, V_s the voltage applied, and I_s the current, then

$$\text{Full-load copper loss} = P_s$$

Remembering that the copper loss is proportional to the square of the current, the copper loss at any load can be calculated. If it is required to determine the copper loss only, it is not absolutely essential that the supply should be of normal frequency because the copper loss (I^2R) is independent of frequency. If the windings are of very heavy section it is preferable that normal frequency should be used, because eddy currents are set up in such windings, and the additional loss due to them is a part of the total copper loss.

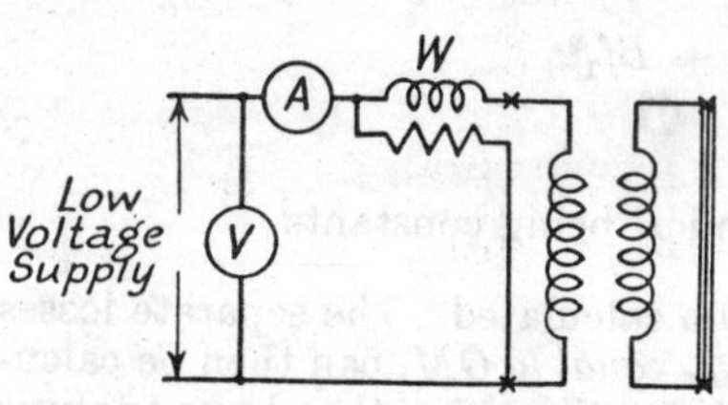

FIG. 17.28. CONNECTIONS FOR SHORT-CIRCUIT TEST

The impedance of the transformer can also be determined from the short-circuit test, but in this case it is imperative that the voltage should be of normal frequency. If the voltage is applied to the primary,

$$\mathbf{Z}_1 = V_s/I_s$$

Also, from $\cos \phi_s = P_s/E_sI_s$, we have

$$\mathbf{R}_1 = \mathbf{Z}_1 \cos \phi_s; \ \mathbf{X}_1 = \mathbf{Z}_1 \sin \phi_s$$

Similarly, if the test is made with the primary short-circuited.

(c) LOAD TEST. The efficiency can be determined with considerable accuracy from the results of the open- and short-circuit tests, while full-load tests are made to determine the temperature rise. A small transformer can be put on full load by means of an artificial load, say, a water resistance, but for large units a regenerative test is almost imperative, partly because of the saving in energy, but mainly because of the difficulty in arranging an artificial load to absorb very large amounts of power. The regenerative test on two transformers is called the Sumpner test, and sometimes the "back-to-back" test. Two similar transformers have their primaries connected to a supply of normal voltage and frequency as in Fig. 17.29. The wattmeter W_1 gives the total core losses. The two secondaries are connected together, care being taken to have their "polarities" in opposition. This is done as

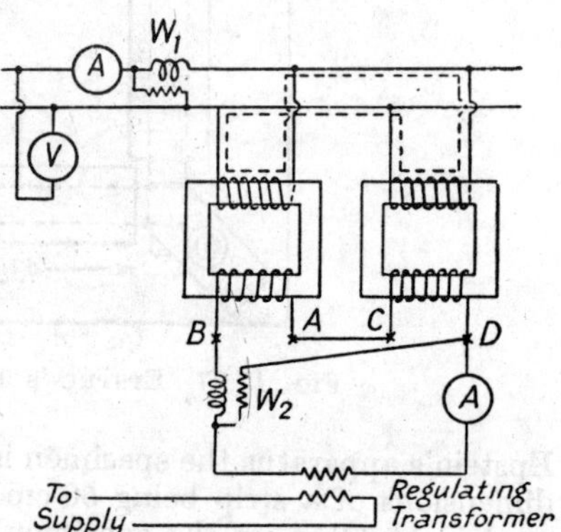

FIG. 17.29. CONNECTIONS FOR BACK-TO-BACK TEST

follows. Terminals *A* and *C* are connected together, and a voltmeter of range double the voltage of either transformer is connected across *BD*. If no reading is obtained, the voltages in the two secondaries are in opposition, and the terminals *BD* can be safely connected. If a reading is obtained, the voltages are acting in the same direction, so that it will be necessary to connect *A* to *D*. In the case of high-voltage transformers, it is common practice to use a bank of incandescent lamps in series instead of a voltmeter, care being taken, of course, to see that all the lamps are sound. If the two secondaries are connected in opposition no secondary current will flow, and to produce this current it is necessary to inject an e.m.f. into the secondary circuit by means of a regulating transformer. This injected voltage is adjusted until full-load secondary current is flowing. This will induce full-load currents in the primaries, but the primary currents will circulate through the bus-bars, as shown by the dotted path, and the reading of W_1 will be unaffected. The wattmeter W_2 will give the total copper losses, and thus, knowing both iron and copper losses, the efficiencies can be calculated. The advantage of the test is that two large transformers can be put under full-load conditions for several hours, so that the temperature rise can be measured, with an expenditure of energy equal to that required by the losses only.

The Sumpner test thus necessitates two transformers, and these are not always available. If a single transformer is to be tested, then provided another transformer of the same voltage and the same, or greater, capacity is available, the test can be carried out. Failing this, the single transformer can be tested as follows. One side is short-circuited and a voltage applied to the other side of such value that the transformer intake is equal to the sum of the normal copper and iron losses. It is obvious that with this method the losses are almost entirely copper losses, with the result that the temperature rise of the copper will be greater than normal, and that of the iron less than normal. On the other hand, the temperature rise of the oil in which the transformer is immersed will be of normal value, the test thus being sufficient for the checking of a specification giving a figure for the temperature rise of the oil, as is usual.

Other tests which can only be applied to three-phase transformers are described on p. 438.

Methods of Cooling Transformers. Since there are no rotating parts which induce a ventilating draught, transformers are more difficult to cool than rotating electrical machines. For small outputs, up to, say, 20 kW, the external surface is sufficient to enable the heat produced by the losses to be dissipated by radiation, but for larger sizes additional means of carrying away the heat have to be provided. The various methods are illustrated in Fig. 17.30. For outputs up to 500 kW it is sufficient to place the transformer in a tank of oil, and to provide as large a cooling surface as possible by (*a*) corrugating the sides of the tank, or (*b*) using a boiler-plate tank provided with external tubes. This second method is being used very extensively at the present time.

For still larger outputs the above method would necessitate a very large tank, and so it is common to use a small tank with a worm tube immersed in the oil, through which cooling water is passed. The disadvantage of this is that the water is under pressure, and if any flaws develop in the worm, water will find its way to the oil. It is therefore preferable to have the oil at higher pressure than the water, as in the

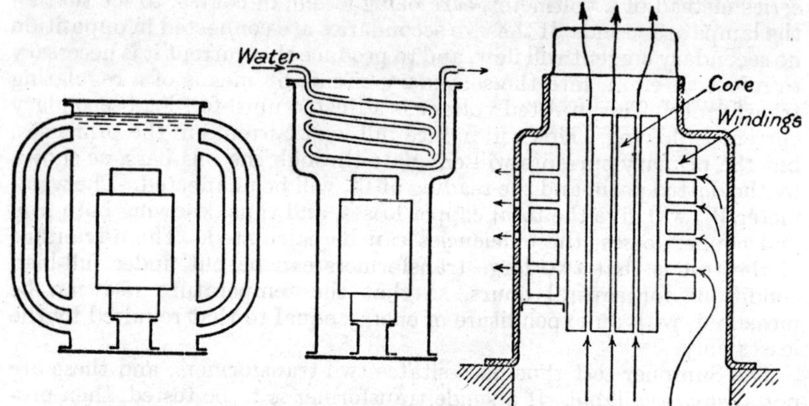

FIG. 17.30. METHODS OF COOLING TRANSFORMERS

(1) Tank with external tubes (natural cooling).
(2) Water cooling.
(3) Air-blast cooling.

case of the external oil cooler. The oil is pumped through a cooler very similar to a condenser, and is cooled by low-pressure water. Where a cheap water supply is not available, as is often the case in substations, the transformer can be cooled by an air blast (*c*). In such a case it is necessary to filter the air, or dust particles will in time clog up the ventilating ducts. The main disadvantage in air-blast cooling is that the very valuable increase in insulation strength due to immersion in oil is lost.*

Heating-time Curve. With very large transformers a very long time is required for the temperature to attain its final value, and it is not convenient to leave the transformer on test for the whole of this time. If the temperature is taken at regular intervals for a few hours, then the final temperature can be calculated from the initial rate of rise so determined.

Let P = full-load losses in copper and iron (watts)

H = amount of energy in joules required to raise the temperature of the whole transformer by 1°C

* For more complete information *see* R. O. Kapp, *Transformers*. Pitman, London; M. G. Say, *Performance and Design of A.C. Machines*. Pitman, London); S. A. Stigant and H. M. Lacey, *The J. and P. Transformer Book*. Johnson and Phillips, Ltd., London.

K = watts dissipated by the transformer and its cooling appliances per degree rise in temperature

θ_m = final temperature rise attained

Then $P = K\theta_m$ by Newton's law of cooling

Now at any temperature rise θ,

(Rate of production of heat) = (Rate of storage of heat) + (Rate of dissipation of heat)

$$\therefore \; P = H(d\theta/dt) + K\theta$$

the solution of which is

$$\theta = P/K + A\,.\,e^{-\frac{K}{H}}$$

where A is a constant and t is the time in seconds.

Now when

$$\theta = 0,\, t = 0, \quad \therefore \; A = -P/K$$

$$\therefore \; \theta = (P/K)(1 - e^{-\frac{t}{B}})$$

where $\qquad B = H/K$

B is called the "temperature-time" constant. Its components H and K can be calculated from the data of the transformer, and therefore, B is known. The equation for θ shows that the curve of θ against t is an

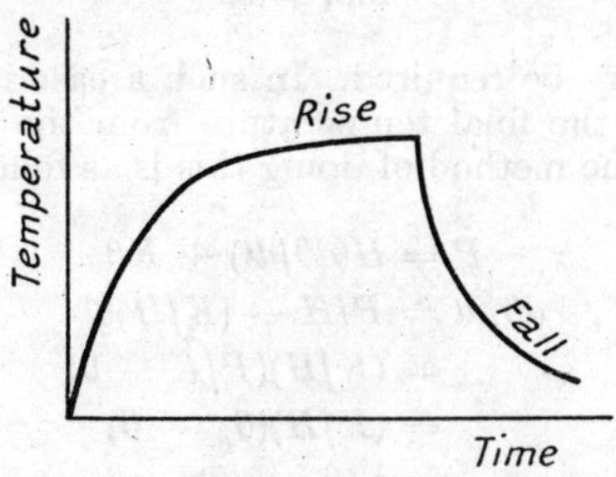

FIG. 17.31

exponential curve as shown in Fig. 17.31. The final temperature is theoretically attained when $t = \infty$. We therefore have

$$\theta_m = P/K$$

Differentiating the equation for θ with respect to t we have

$$d\theta/dt = (P/KB)e^{-\frac{t}{B}}$$

$\therefore$ Initial rate of rise of temperature

$$(d\theta/dt)_o = P/KB \text{ when } t = 0$$

$\therefore$ From $\theta_m = P/K$

We have $\theta_m = B \times (d\theta/dt)_o$

$= (\text{Temp.-time constant}) \times (\text{Initial rate of rise of temperature})$

If, after attaining its final temperature, the transformer is allowed to cool down, the curve of fall of temperature is complementary to the curve of rise of temperature. We then have

$$\theta = (P/K)e^{-\frac{t}{B}}$$

Now, when carrying out a heat run on a small machine, the final temperature will be attained in a comparatively short time, say five or six hours, but in the case of a very large machine twenty-four hours,

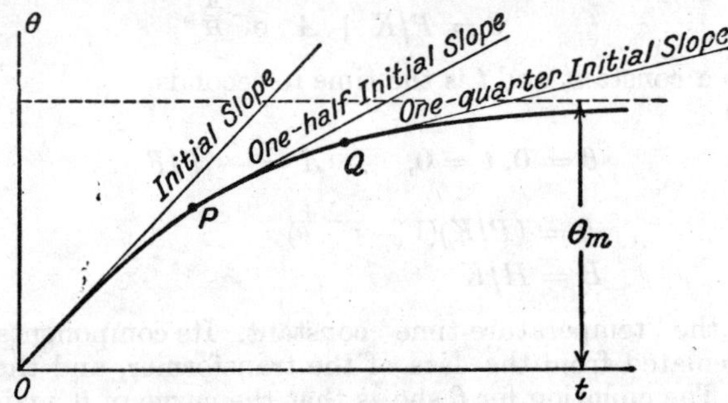

Fig. 17.32

or even longer, may be required. In such a case it is desirable to be able to determine the final temperature from the results of a run of short duration. One method of doing this is as follows. We have seen that

$$P = H(d\theta/dt) + K\theta$$

$$\therefore d\theta/dt = P/H - (K/H)\theta$$

$$= (K/H)(P/K - \theta)$$

$$= (K/H)(\theta_m - \theta)$$

$$\propto (\theta_m - \theta)$$

In other words, the slope of the curve at any instant is proportional to the difference between the actual temperature at that instant and the final temperature. Thus, in Fig. 17.32, the point P is taken at a temperature rise equal to half the final rise, and Q is taken at three-quarters of the final rise. The point O is at the commencement, for which there is no temperature rise. Hence, the slope of the tangent at O is proportional to the final rise θ_m, the slope of the tangent at P is proportional to $\frac{1}{2}\theta_m$, and that at Q is proportional to $\frac{1}{4}\theta_m$. The procedure is therefore as follows. Observe the initial rate of rise from the results of the test of short duration, and determine the slope at successive periods until a point is reached at which the slope is half the initial slope. Then the final rise of temperature is twice the rise of temperature reached at that point. As a check on the accuracy of the determination of the slopes at various instants, the curve of $(\theta - d\theta/dt)$ can be plotted against time, and this should be a straight line.

EXAMPLE. A 100 kVA transformer has a temperature-time constant of one hour. The ratio of iron loss to copper loss at full-load current is 1 to 2, and the temperature rise on normal full load is 50°C. What overload will produce a temperature rise of 60°C at the end of a run of 2 hr, starting from cold?

We have $$\theta = \theta_m(1 - e^{-\frac{t}{B}})$$

Hence under overload conditions

$$\begin{aligned} 60 &= \theta_m(1 - e^{-\frac{120}{60}}) \\ &= \theta_m(1 - e^{-2}) \\ &= 0{\cdot}865\theta_m \\ \therefore\ \theta_m &= 60/0{\cdot}865 = 69{\cdot}4°\text{C} \end{aligned}$$

But $$\theta_m = P/K$$

showing that θ_m is proportional to the total loss

$$\therefore \frac{\text{Total loss on overload}}{\text{Total loss on normal load}} = \frac{69{\cdot}4}{50} = 1{\cdot}39$$

Denote the full load by unity and the total load at overload by x, also let P_i = iron loss.

Then $$2P_i = \text{copper loss on normal load}$$

and $$2P_i \times x^2 = \text{copper loss on overload}$$

$$\therefore\ (P_i + 2P_i)/(P_i + 2P_i \times x^2) = 1/1{\cdot}39$$

$$3/(1 + 2x^2) = 1/1{\cdot}39$$

which gives $$x = 1{\cdot}26$$

Hence a 26 per cent overload will give a temperature rise of 60°C in 2 hr.

The Auto-Transformer. An auto-transformer has one winding only as shown in Fig. 17.33, the secondary voltage being obtained by tapping off at any convenient point. As in the ordinary transformer, the no-load transformation ratio is equal to the turns ratio

$$E_{2\cdot0}/V_1 = N_2/N_1$$

For the same capacity and voltage ratio the auto-transformer requires less copper than an ordinary transformer. The cross section of a conductor is proportional to the current carried, and the length of conductor in a winding is proportional to the number of turns. Hence, weight of copper in a winding is proportional to the product of the current and the number of turns.

(*a*) Ordinary Transformer

Weight of copper on primary $\propto N_1 I_1$

Weight of copper on secondary $\propto N_2 I_2$

Total weight of copper $\propto (N_1 I_1 + N_2 I_2)$

(*b*) Auto-Transformer

The bottom section of $(N_1 - N_2)$ turns carries current I_1

$\therefore$ Weight of copper in this section $\propto (N_1 - N_2) I_1$

The top section of N_2 turns carries a current of $(I_2 - I_1)$ because of the opposition of primary and secondary currents.

$\therefore$ Weight of copper in top section $\propto N_2(I_2 - I_1)$

$\therefore$ Total copper in auto-transformer $\propto$

$$(N_1 - N_2)I_1 + N_2(I_2 - I_1), \propto (N_1 - 2N_2)I_1 + N_2 I_2$$

$$\therefore \frac{\text{Weight of copper on auto-transformer}}{\text{Weight of copper on ordinary transformer}} = \frac{(N_1 - 2N_2)I_1 + N_2 I_2}{N_1 I_1 + N_2 I_2}$$

$$= \frac{N_1/N_2 - 2 + I_2/I_1}{N_1/N_2 + I_2/I_1}$$

Now $N_1/N_2 = I_2/I_1 = m$, say, where $m = 1/K$

$\therefore$ Ratio of weight of copper $= (2m - 2)/2m = (m - 1)/m$

$= 1 - K$

The current supplied to the load can be regarded as made up of two parts (*a*) that via the turns $(N_1 - N_2)$ without transformation, and (*b*) that due to transformation in the upper N_2 turns. The gain in the autotransformer follows from the fact that only a portion of the secondary output is the result of transformation.

Because of the saving of copper, auto-transformers are occasionally used where a transformation ratio not very different from unity is required; but where one side is at very high tension compared with the other, it is preferable to have the two sides electrically separate and to use ordinary transformers. Auto-transformers therefore have the biggest sphere of usefulness as regulating transformers. They are also sometimes used as boosters to raise the voltage in an alternating-current feeder.

The vector diagram of the auto-transformer is derived as follows—

Let the positive direction of the applied p.d. V_1 be upwards, that is from *a* to *c*: then the positive direction of I_1 round the primary circuit is counter-clockwise, and that of I_2 round the load circuit is also counter-clockwise. The positive directions of the induced voltage are downwards, that is from *c* to *b* and from *b* to *a*. The applied p.d. V_1 is balanced by the induced voltage in these two sections, plus the volt drops in them. The vector diagram is best started from the secondary side. First draw OA downwards, Fig. 17.33, to represent the secondary terminal p.d. on load, V_2. Then draw OB to represent the load current I_2, lagging V_2 by the load angle ϕ. The winding current in the section cb is given by

$$I_{cb} = [(m-1)/m]I_2 \text{ (namely } I_2 \dotplus I_1)$$

FIG. 17.33. AUTO-TRANSFORMER AND VECTOR DIAGRAM

and it is marked off as OC. The secondary volt drop triangle is ADE where

$$AD = R_2 I_{cb}; \quad DE = X_2 I_{cb}$$

The induced e.m.f. E_{cb} in the section cb is therefore given by the vector OE.

The induced e.m.f. in section ab is

$$E_{ba} = E_{cb}(m-1)$$

and therefore the component of the applied p.d., V_1 to balance this is

$$E_{ab} = -E_{ba} = -E_{cb}(m-1)$$

This is represented by OG, collinear with OE. The current I_1 is in phase opposition to I_2 and is equal to

$$I_1 = -I_2/m$$

It is represented by OF. The drop triangle for the section ab is therefore GHK where

$$GH = R_1I_1; \; HK = X_1I_1$$

The p.d. V_{ab}, across the section ab is therefore the vector sum of the e.m.f. OG, and the drop GK: it is therefore represented by OK. Finally, the applied p.d. V_1 is the vector sum of V_{ab} and V_{bc}, that is, V_{ab} plus V_{bc} where $V_{bc} = -V_{cb} = -V_2$. This is given by the vector OL.

Use of Transformers with Measuring Instruments. (*a*) When measuring large currents in a direct current circuit it is usual to pass the current through a shunt, and to measure the p.d. across the shunt by means of a millivoltmeter. This is not convenient with alternating current instruments, except hot-wire instruments, and it is therefore usual to pass the current to be measured through the primary of a "current" transformer, the secondary being connected to the ammeter (Fig. 17.34). Since ammeters are of very low resistance it will be seen

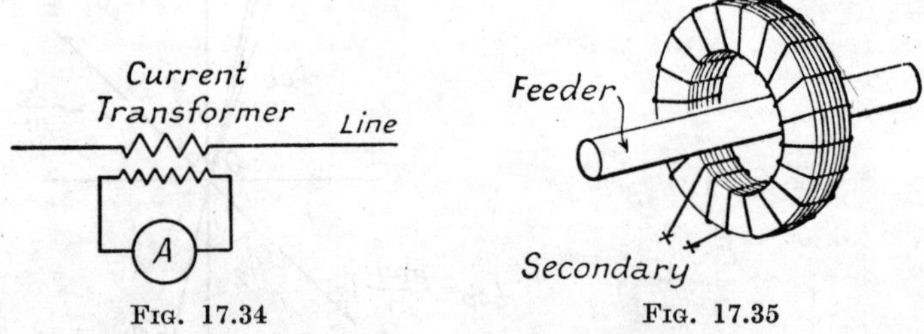

Fig. 17.34 Fig. 17.35

that a current transformer normally works short-circuited, and if for any reason the ammeter is taken from the circuit the secondary terminals must be short-circuited. If this is not done the primary winding will act like a choke connected in series with the line. The m.m.f. of the primary current will be unopposed and the core will carry an abnormally high flux, thus inducing high e.m.f.s in both primary and secondary windings. For the measurement of the very heavy currents carried by bus-bars or feeders the conductor itself constitutes the primary, as shown in Fig. 17.35.*

The ratio of a current transformer will not be exactly the inverse of the turns ratio, largely because of the effect of the magnetizing current. The primary current I_1 can be regarded as being made up of two components, the first of these being in exact phase opposition to the secondary current I_2, in order that the primary and secondary ampere turns may balance; the second being the magnetizing component I_0. The current vector diagram is therefore as shown in Fig. 17.36, in which a one-to-one ratio is assumed, since the balancing component of the primary current is $-I_2$. The component I_0, besides being responsible for a slight error in the current ratio, is also responsible for a phase

* The theory and use of instrument transformers are treated exhaustively in *Instrument Transformers*, by B. Hague. Pitman, London.

shift β, which is called the phase angle of the transformer. We have already seen that the phase angle between voltage and current in a wattmeter introduces errors in the reading. If the wattmeter is used in conjunction with a current transformer, as explained in section (*c*) then, in order to reduce such errors to a minimum the angle β must be

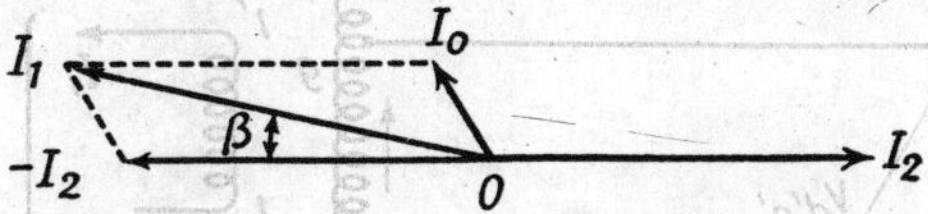

FIG. 17.36. CURRENT VECTOR DIAGRAM OF CURRENT TRANSFORMER

very small. Thus a current transformer which may be suitable for use with an ammeter may be quite unsuitable for use with a wattmeter, particularly at low power-factors.

(*b*) When measuring very high voltages, the voltage is stepped down by means of a "potential" transformer, the low-tension secondary of which is connected to the voltmeter. For switchboard use a voltmeter works practically all the time on one part of the scale, and if the transformer gives the correct ratio at this point, a variation of the ratio at other points is not of much account. On the other hand an ammeter has to be correct at all points on the scale, and therefore, the current ratio of a current transformer must be the same at all loads.

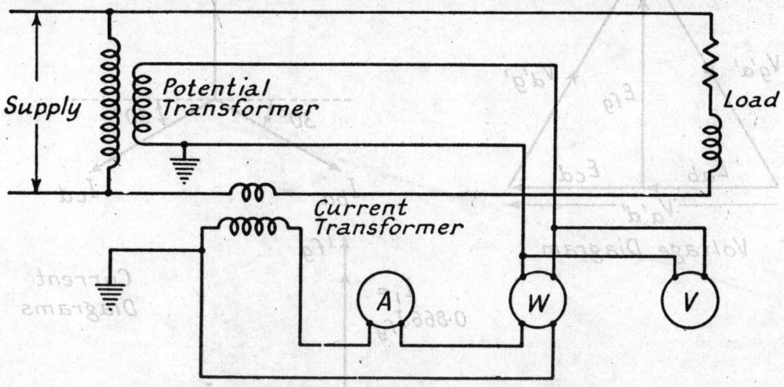

FIG. 17.37. TYPICAL CONNECTION SCHEME FOR INSTRUMENT TRANSFORMERS IN A SINGLE-PHASE CIRCUIT

(*c*) When measuring power on a high-voltage system both current and potential transformers are used, the way in which they are connected in the circuit being illustrated in Fig. 17.37. It is usual to ground (i.e. connect to earth) each transformer at one point.

Transformation from Three to Two Phase. SCOTT'S METHOD. In some cases it is desirable to work with two-phase current: examples are to supply electric furnaces, and the provision of low-voltage single-phase supplies in rural areas where the primary supply is derived from a three-phase medium voltage distributor. The method was first used for transformation from two-phase to three-phase at the first Niagara Falls power station, the generator then being two-phase. At

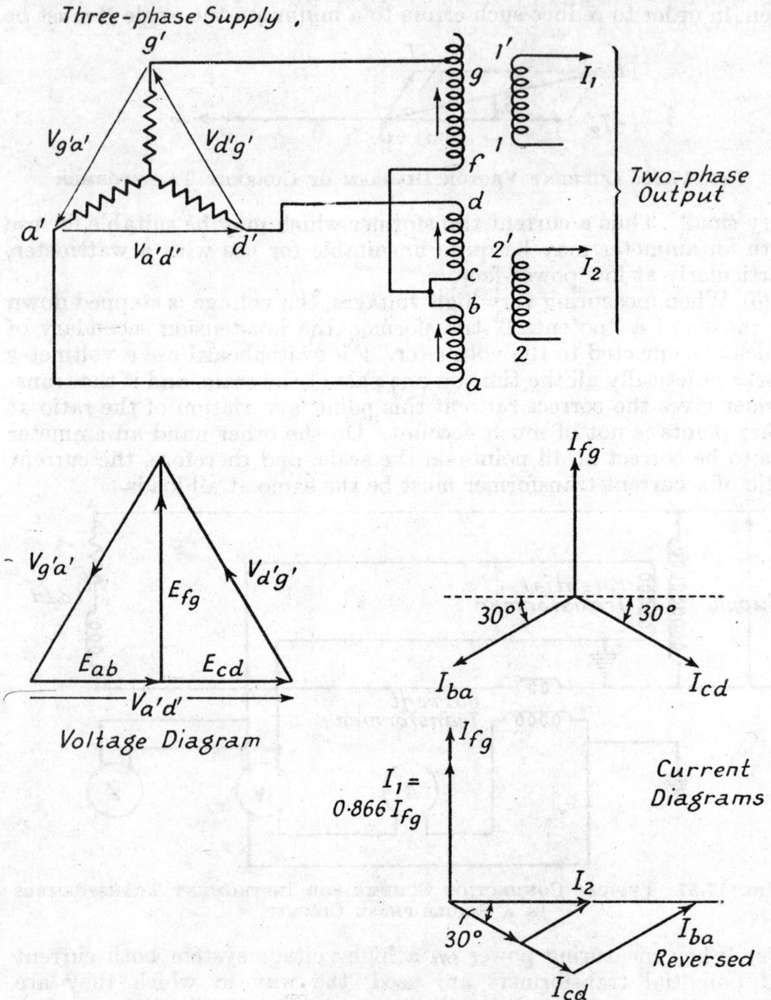

Fig. 17.38. Scheme of Scott Connections with Voltage and Current Diagrams

the present time bulk supplies are invariably three phase, and it is therefore necessary to convert from three to two phase in such cases. This can be done by means of two single-phase transformers connected in "T," as in Fig. 17.38.

Let the positive directions of the e.m.f.s E_{ab}, E_{cd}, and E_{fg} be as indicated. Then considering the line voltage $V_{d'g'}$ this has to be balanced by the voltages induced in *cd* and *fg* and therefore on traversing the circuit *d'dcfgg'* we see that the positive directions of E_{fg} and E_{cd} are in opposition so that we have

$$V_{d'g'} = E_{fg} - E_{cd}$$

Similarly
$$V_{g'a'} = -E_{fg} - E_{ab}$$

and
$$V_{a'd'} = E_{ab} + E_{cd}$$

Since *cb* is the mid point of the winding *ad* we have $E_{ab} = E_{cd}$, and therefore since the line voltage triangle is an equilateral triangle we see that the e.m.f. E_{fg} is in quadrature leading E_{ab} and E_{cd}. If the two secondary windings have equal numbers of turns, then in order that the induced e.m.f.s in them may be equal it is necessary that the number of turns in the winding *fg* shall be 0·866 of the turns in the complete winding *ad*. The transformer *ad* is usually called the main transformer and *fg* the *teaser* transformer.

In order to deduce the current relationship it is convenient to start with the primary side. With balanced conditions, and at unity power-factor the three line currents will be I_{ba}, I_{cd}, and I_{fg}, as shown. For simplicity, let each secondary phase have as many turns as the primary *ad* of the main transformer. Then for the balance of m.m.f.s in the teaser transformer we have

$$I_1 = 0{\cdot}866 I_{fg}$$

The currents in the two halves of the main transformer have a phase angle of 30°, as shown, and therefore since each half has half as many turns as the secondary, the m.m.f. balance is given by the relationship

$$\begin{aligned} I_2 &= 0{\cdot}5(I_{cd} + I_{ba} \text{ reversed}) \\ &= 0{\cdot}5 \times 2 \times 0{\cdot}866 I_{cd} \\ &= 0{\cdot}866 I_{cd} \\ &= I_1 \text{ numerically} \end{aligned}$$

and is in quadrature with I_1. Hence if the loadings on the two phases of the secondary are balanced, that on the three-phase side is balanced also.

If the two transformers are identical, then because of the 0·866 tapping on the teaser its output is only 0·866 of its rating. Because of the 30° phase angle, the output of the main transformer is also 0·866 of its rating so that, for the Scott-connected pair

$$\frac{\text{Output}}{\text{Rating}} = 0{\cdot}866$$

Obviously there is no point in providing the teaser transformer with turns which will not be used, and if it has 0·866 as many turns as the main transformer then its rating will be 0·866 that of the main transformer. We then have for the pair

$$\frac{\text{Output}}{\text{Rating}} = \frac{2 \times 0{\cdot}866}{1 + 0{\cdot}866} = 0{\cdot}926$$

Three-phase Transformation. The ordinary transformation of three-phase current at one voltage to three-phase current at another voltage can be effected either by means of a single three-phase transformer, or by three separate single-phase transformers, as shown in Fig. 17.39.

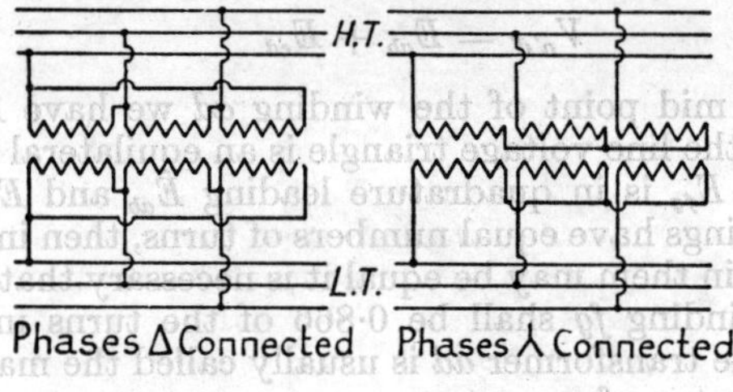

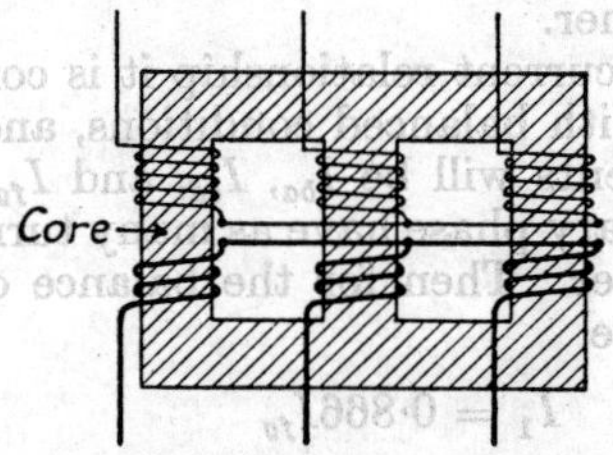

Fig. 17.39. Three-phase Transformer

The advantage of using three separate single-phase transformers is that only one-third of the total output need be kept as spare, an equal single-phase transformer being substituted for a faulty one, in the case of breakdown. With a three-phase transformer it is necessary to have another three-phase transformer of full output as a spare. If the three single-phase transformers are mesh connected, then, in the event of a fault developing in one of them, it can be removed from service, the two remaining transformers supplying the load in what is called "open delta" connection. This does not apply to the three single-phase transformers connected in star.

With one three-phase transformer, or a "bank" of three single-phase transformers, the primary may be star connected and the secondary mesh, or vice versa. If a number of three-phase transformers, or a number of banks of single-phase transformers, are worked in parallel, then there are certain methods of connection which cannot be employed. A few arrangements are tabulated below—

1st Transformer.	Primary Δ,	Secondary ⅄	Impossible
2nd Transformer.	Primary Δ,	Secondary Δ	

1st Transformer. Primary ⅄, Secondary ⅄ }
2nd Transformer. Primary Δ, Secondary Δ } Possible

1st Transformer. Primary ⅄, Secondary Δ }
2nd Transformer. Primary Δ, Secondary ⅄ } Possible

Consider for example a ⅄ — ⅄ transformer and a ⅄ — Δ transformer with their primary windings supplied from the same source. Let the

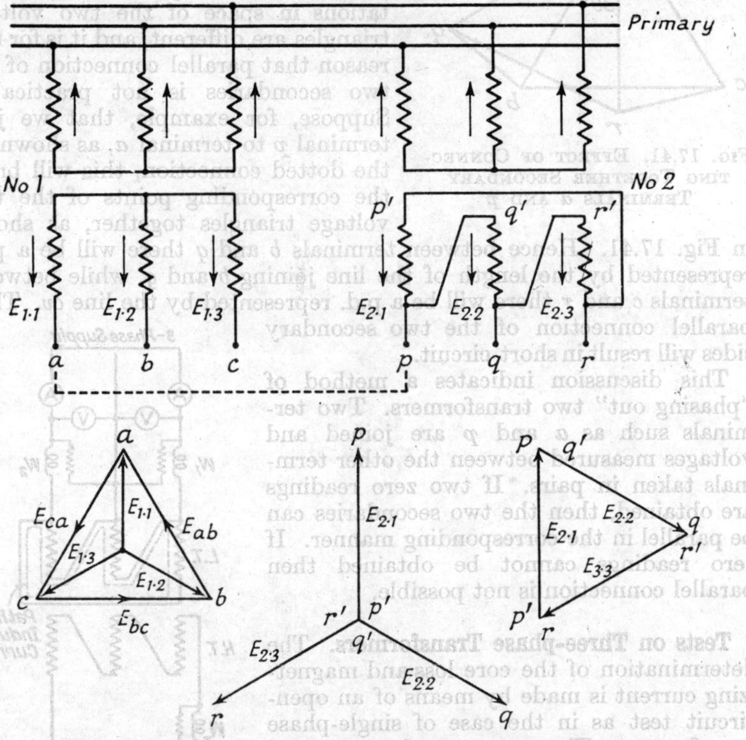

FIG. 17.40. VOLTAGE DIAGRAMS FOR ⅄ — ⅄ AND ⅄ — Δ TRANSFORMERS CONNECTED TO THE SAME SUPPLY

transformation ratios be equal, so that the voltage between any pair of secondary terminals of No. 1 is numerically equal to that between any pair of secondary terminals of No. 2. The connection scheme is indicated in Fig. 17.40. The corresponding phase voltages in the two primaries are identical, and therefore the corresponding phase voltage in the two secondaries are in phase with one another, thus $E_{1\cdot1}$ with $E_{2\cdot1}$, $E_{1\cdot2}$ with $E_{2\cdot2}$, and $E_{1\cdot3}$ with $E_{2\cdot3}$. The terminal voltages of transformer No. 1 are obtained by joining the ends of the three phase vectors as shown, the magnitude of the terminal voltage being, of course, $\sqrt{3}$

times the magnitude of the phase voltage. In transformer No. 2 the terminal voltages are numerically equal to the phase voltages because of the mesh connection, and therefore to obtain a terminal voltage diagram comparable with that of No. 1 we rearrange the three vectors in the form of a triangle. The line secondary voltages of No. 2 are $\sqrt{3}$ times the secondary phase voltage of No. 1, so that the two voltage triangles so formed have sides of equal length. We see from the figure that the orientations in space of the two voltage triangles are different, and it is for this reason that parallel connection of the two secondaries is not practicable. Suppose, for example, that we join terminal p to terminal a, as shown by the dotted connection, this will bring the corresponding points of the two voltage triangles together, as shown in Fig. 17.41. Hence between terminals b and q there will be a p.d. represented by the length of the line joining b and q, while between terminals c and r there will be a p.d. represented by the line cv. Thus parallel connection of the two secondary sides will result in short-circuit.

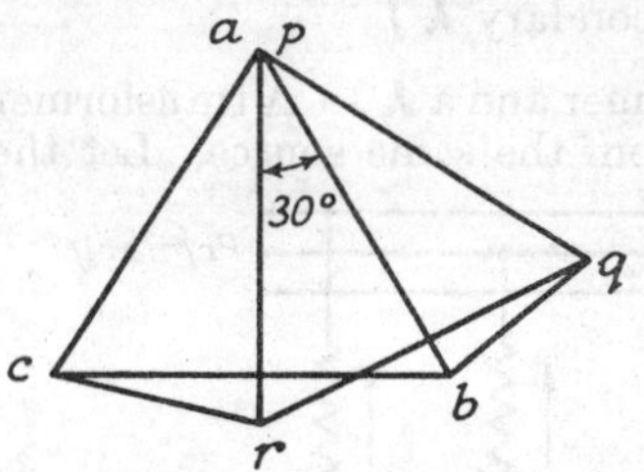

FIG. 17.41. EFFECT OF CONNECTING TOGETHER SECONDARY TERMINALS a AND p

This discussion indicates a method of "phasing out" two transformers. Two terminals such as a and p are joined and voltages measured between the other terminals taken in pairs. If two zero readings are obtained, then the two secondaries can be parallel in the corresponding manner. If zero readings cannot be obtained then parallel connection is not possible.

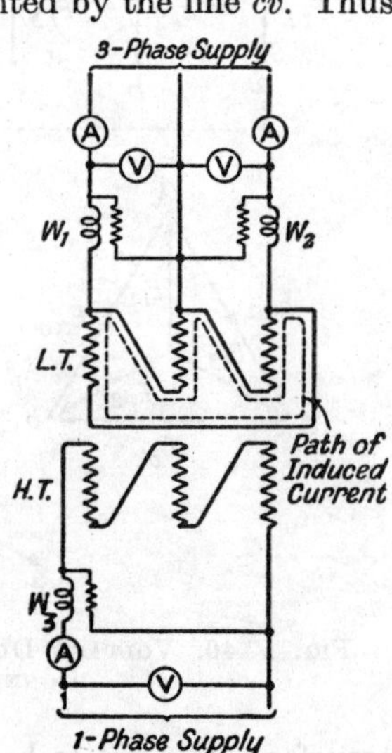

FIG. 17.42. CONNECTIONS FOR DELTA-DELTA RUN

Tests on Three-phase Transformers. The determination of the core loss and magnetizing current is made by means of an open-circuit test as in the case of single-phase transformers. The power can be measured by the two-wattmeter or the three-wattmeter method, or, alternatively, a single wattmeter used in conjunction with a transfer switch can be used. If the reader will work out the vector diagrams for the two-wattmeter and three-wattmeter methods, he will find that, owing to the low transformer power factor on open circuit, the *meter* power factor will be lower in the three-wattmeter than in the two-wattmeter method, and consequently the latter method is preferable.

The short-circuit test is also carried out by the two-wattmeter method, and the resistance, reactance, and impedance are worked out per phase.

For the purpose of a heat run the short-circuit test can also be used, the applied voltage being increased so that the transformer intake is equal to the sum of the normal iron and copper losses. As indicated in the section dealing with single-phase transformers, this method has the defect that the temperature rises of the iron and copper are not

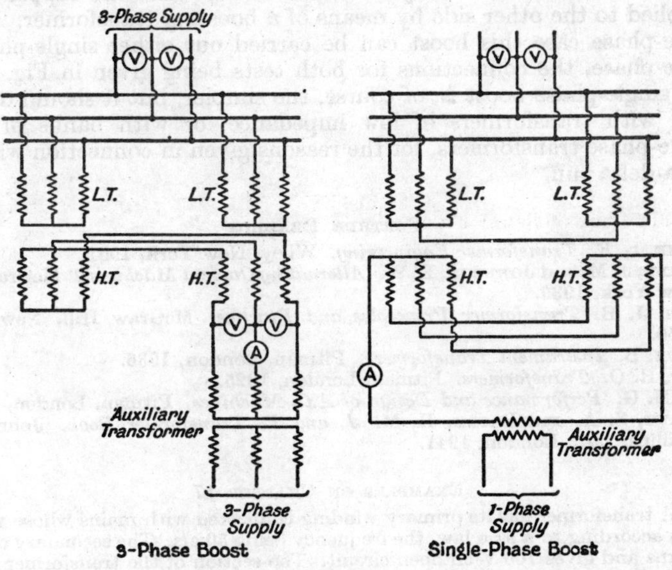

Fig. 17.43. Alternative Connections for Regenerative Test on Two Three-phase Transformers

normal. This difficulty can be overcome in the case of three-phase transformers by what is called a delta-delta run. Whatever the normal connections of the phases, both primary and secondary phases are connected in delta, and the iron losses are supplied from a three-phase supply of normal phase voltage and frequency, while the copper losses are derived from a single-phase supply. The connections for the test are shown in Fig. 17.42. Since the three-phase supply is at correct voltage and frequency, the sum of the wattmeter readings, W_1 and W_2, will be the iron loss. The high-tension winding is connected in open delta, and is connected to a single-phase supply, whose voltage, to produce normal full-load current, must be equal to three times the impedance voltage per phase referred to the high-tension side. The ammeter in this circuit will then indicate full-load current, and this current will induce full-load current in the low-tension winding, this induced current being a circulating current, as indicated by the dotted path, and therefore not affecting the readings of wattmeters W_1 and W_2. Hence, wattmeter W_3 will indicate the total copper loss. This test is suitable for low-reactance transformers, but not for high-reactance transformers, because the single-phase loading will set up abnormal leakage fluxes, which, in turn, will set up additional losses due to abnormal eddy currents in the windings, ironwork, and tank.

When two three-phase transformers are available, the Sumpner test can be carried out as with two single-phase transformers, the iron loss is supplied to one side, usually the low-tension, while the copper loss is supplied to the other side by means of a boosting transformer. In the three-phase case this boost can be carried out either single-phase or three-phase, the connections for both tests being given in Fig. 17.43. The single-phase boost is, of course, the simpler, but it should only be used with transformers of low impedance, or with banks of three single-phase transformers, for the reasons given in connection with the delta-delta run.

Further Reading

Blume, L. F. *Transformer Engineering.* Wiley, New York, 1951.

Bryant, J. M. and Johnson, E. W. *Alternating Current Machinery.* McGraw Hill, New York, 1939.

Gibbs, J. B. *Transformer Principles and Practice.* McGraw Hill, New York, 1950.

Hague, B. *Instrument Transformers.* Pitman, London, 1936.

Kapp, R. O. *Transformers.* Pitman, London, 1925.

Say, M. G. *Performance and Design of A.C. Machines.* Pitman, London, 1952.

Stigant, S. A. and Lacey, H. M. *J. and P. Transformer Book.* Johnson & Phillips, Ltd., London, 1941.

Examples on Chapter 17

1. A transformer has its primary winding connected with mains whose voltage varies according to a sine law, the frequency being 50 c/s. The secondary coil has 50 turns and gives 100 V on open circuit. The section of the transformer core is 20 in.^{2} Determine the maximum value of the induction density in the core. Prove the formula used. (L.U.)

Ans. 0·706 Wb/m.^{2}

2. A 100 kVA transformer with a secondary voltage of 400 V has a resistance referred to the secondary of 0·016 Ω, the iron loss of 1000 W. Plot its efficiency against secondary current for power factors of 1·0 and 0·8 lagging. Calculate the secondary current at which the efficiency is a maximum.

Ans. 250 A.

3. A 100 kVA transformer stepping down from 2000 to 400 V has a primary resistance of 0·17 Ω and a secondary resistance of 0·0068 Ω; the reactances are 0·25 Ω and 0·01 Ω respectively. Calculate the resistance, reactance, and impedance referred to the secondary, and hence find the per cent regulation on full secondary load of 250 A at a power factor of 0·8 lagging. Check the regulation by the Kapp diagram.

Ans. 0·0136 Ω, 0·02 Ω, 0·024 Ω, 1·43 per cent.

4. Give the theory of the open- and short-circuit method of testing a transformer, and show how from the measurement taken it is possible to calculate the efficiency and percentage drop of secondary voltage for a load of known magnitude and power factor.

A transformer to convert 10 kW from 2000 to 100 V when tested by the above method showed losses which for the open-circuit test at normal voltage were 200 W at power factor 0·7, and which, for the short-circuit test at full-load current, were 250 W at power factor 0·25. Calculate the efficiency, of the transformer for an output consisting of a full-load secondary current (lagging) at power factor 0·6. (L.U.)

Ans. 92·8 per cent.

5. Calculate the percentage regulation of the transformer in question 4 when working on the stated load.

Ans. 9·3 per cent.

6. Two 100 kW single-phase transformers are connected in parallel both on the primary and secondary. One transformer has an ohmic drop of $\frac{1}{2}$ per cent at full load and an inductive drop of 8 per cent at full-load current. The other has an ohmic drop of $\frac{3}{4}$ per cent and an inductive drop of 4 per cent. Show how they will share the following loads: (*a*) 180 kW at 0·9 power factor; (*b*) 120 kW at 0·6 power factor. (C. and G.)

Ans. 62·8 and 117·2 kW; (*b*) 44·8 and 75·2 kW.

7. A system supplies a load which varies in a period of 24 hr as follows: 6 a.m. to 12 noon, 1000 kW; 12 noon to 1 p.m., 100 kW; 1 p.m. to 6 p.m., 100 kW; 6 p.m. to 6 a.m., 300 kW. Energy is transmitted over a line in which the losses at full load are 5000 W, and is transformed by a transformer having a no-load loss of 6000 W, and a full-load copper loss of 8000 W. Calculate the total losses in transmission and conversion during the 24 hr. (C. and G.)

Ans. 301·2 kWh.

8. A 1000 kVA and a 500 kVA single-phase transformer are connected to the same bus-bars on the primary side. The secondary pressures at no load are 500 and 510 V respectively. The impedance voltage of the first transformer is 3·4 per cent and of the second 5 per cent. What cross current will pass between them when the secondaries are connected together in parallel? Assuming that the ratio of resistance to reactance is the same in each, what currents will flow in the secondary windings of the transformers when supplying a total load of 1200 kVA? (C. and G.)

Ans. 294, 1750, and 730 A (assuming a load of unity power factor).

9. Two electric furnaces are supplied with single-phase current at 80 V from a three-phase 11,000 V system by means of two single-phase Scott connected transformers, with similar secondary windings. When the load on one transformer is 500 kW and on the other, 800 kW, what current will flow in each of the three-phase lines (1) at unity power factor; (2) at 0·5 power factor? Neglect phase displacement in, and efficiency of, the transformers. (C. and G.)

Ans. (1) 78, 78 and 53 A; (2) 156, 156 and 105 A.

10. Two transformers *A* and *B* are joined in parallel to the same load. Determine the current delivered by each transformer, given: open-circuit e.m.f. 6,600 V for *A* and 6,400 V for *B*. Equivalent leakagei mpedance in terms of the secondary $= 0·3 + j3$ for *A*, and $0·2 + j1$ for *B*. The load impedance is $8 + j6$. (L.U.)

Ans. *A*, 195 A; *B*, 422 A.

11. Two 250 kVA transformers supplying a network are connected in parallel on both primary and secondary sides. Their voltage ratios are the same, the resistance drops are 1·5 and 0·9 per cent and their reactance drops 3·33 and 4 per cent respectively. Calculate the kVA loading of each transformer and its power-factor when the totall oad on the transformer is 500 kVA at 0·707 lagging power-factor. (C. and G.)

Ans. 263 kVA at 0·77 p.f.; 237 kVA at 0·63 p.f.

12. Two single-phase transformers rated at 500 kVA and 400 kVA respectively, are connected in parallel to supply a load of 1,000 kVA at 0·8 lagging power-factor. The resistance and reactance of the first transformer are 2·5 and

6 per cent respectively, and of the second transformer 1·6 and 7 per cent respectively. Calculate the kVA loading and the power-factor at which each transformer operates. (C. and G.)

Ans. 583 kVA at 0·84 p.f.; 417 kVA at 0·74 p.f.

13. Determine the kW output and the power-factor of the load on each of two transformers, *A* and *B*, operating in parallel on a load of 500 A at 0·8 power-factor, lagging.

No-load test, 11,000/3,000 V for each transformer
Short-circuit test with primary short-circuited
Transformer *A*, 150 V, 200 A, 10 kW
Transformer *B*, 150 V, 300 A, 12 kW (L.U.)

Ans. *A*, 498 kW at 0·83 p.f.; *B*, 702 kW at 0·78 p.f.

14. A three-phase transformer *A*, rated at 1,000 kVA, 33/11 kV, is connected in parallel with a similar transformer *B*, rated at 1,500 kVA, to supply a load of 2,000 kVA at 0·8 power-factor lagging at full-load the equivalent voltage drops per phase, referred to the secondary, due to resistance and leakage reactance are 2 and 5 per cent respectively for transformer *A*, and 1·7 and 6 per cent respectively for transformer *B*. Calculate the load on each transformer and the power-factor at which it operates. (I.E.E.)

Ans. *A*, 865 kVA at 0·83 p.f.; *B*, 1,135 kVA at 0·76 p.f.

CHAPTER 18

ROTATING MAGNETIC FIELDS

An alternating magnetic field is one which acts along a fixed axis, but whose direction of acting is alternatively positive and negative, and whose magnitude varies in a periodic manner. For example the field inside a solenoid excited by alternating current is an alternating field. A rotating magnetic field is one whose magnitude is constant but whose *axis of direction rotates in space.* A rotating field can be visualized very simply in terms of a field system with salient poles excited by direct current, for example, the field system of a d.c. machine. With a stationary field system the magnetic field is stationary in space, but if the field system can be rotated at a certain speed then its magnetic field will rotate with it at the same speed. In the alternator and synchronous motor we shall see that this is accomplished by making the field system the internal, rotating, member, the poles therefore projecting outwards instead of inwards. What corresponds to the armature, and carries the winding in which the e.m.f.s are induced, is the outer stationary member. This stationary member is called the stator, and the rotating member the rotor.

An important problem in the design of alternating-current machinery is the production of a rotating magnetic field by means of alternating currents flowing in the windings of the stator. We shall see that this necessitates polyphase currents.

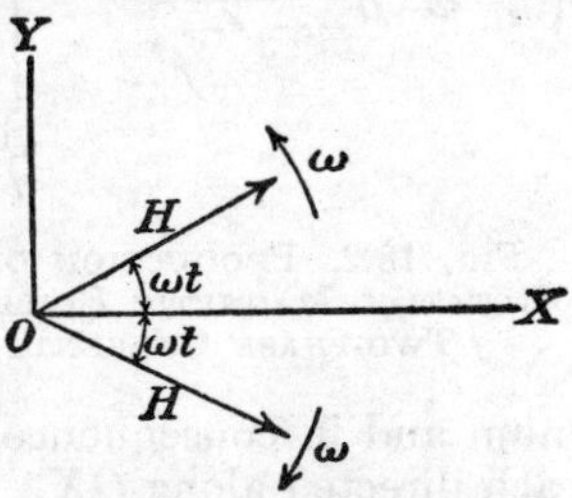

FIG. 18.1. COMPOSITION OF ROTATING FIELDS

Ferrari's Principle. If single-phase current is supplied to the stator it produces an alternating magnetic field which, by Ferrari's principle, can be split up into two rotating fields of half its amplitude, and travelling in opposite directions at synchronous speed. For, consider two such rotating fields both starting from the X axis at zero time, and rotating with angular velocity ω (Fig. 18.1). Then, after any time t

$$\begin{aligned}
&X \text{ component of No. 1} = H \cos \omega t \\
&X \text{ component of No. 2} = H \cos \omega t \\
\therefore\ &\text{Total } X \text{ component} = 2H \cos \omega t \\
&Y \text{ component of No. 1} = H \sin \omega t \\
&Y \text{ component of No. 2} = -H \sin \omega t \\
\therefore\ &\text{Total } Y \text{ component} = 0
\end{aligned}$$

Hence, the two rotating fields resolve into an alternating field,

$2H \cos \omega t$, of twice their amplitude. Conversely, an alternating field can be resolved into two rotating fields travelling in opposite directions at synchronous speed.

Two-phase Supply. Consider two coils placed 90° apart and connected, one to each phase of a two-phase supply, as in Fig. 18.2. Each coil sets up an alternating magnetic field along its own axis, and each of these fields can be split up into forward and backward travelling components, having an angular velocity of ω, where $\omega = 2\pi f$. Consider the instant the current in coil I is a maximum. Then its magnetic field is a maxi-

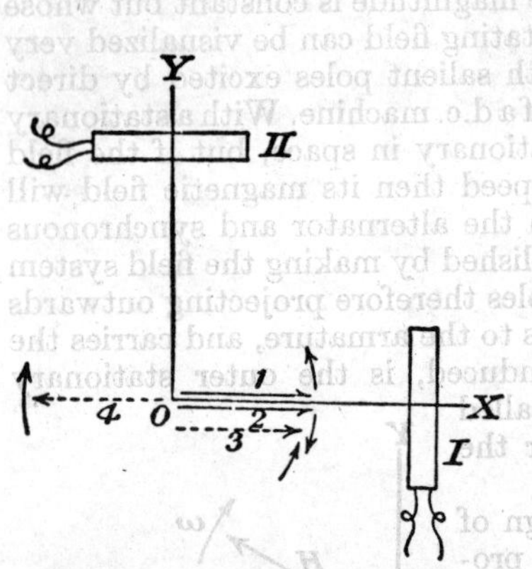

Fig. 18.2. Production of a Rotating Magnetic Field by Two-phase Currents

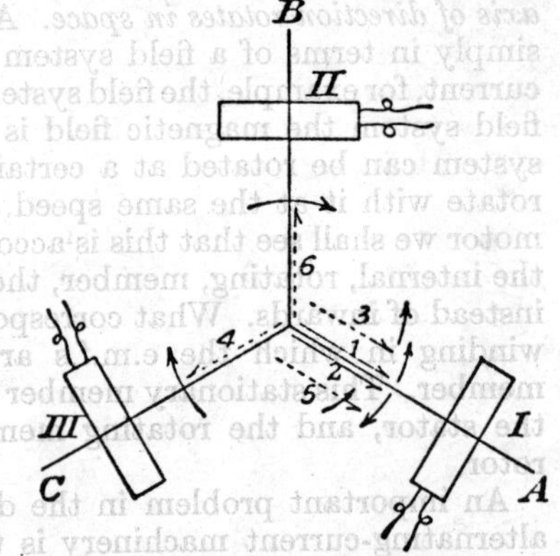

Fig. 18.3. Production of a Rotating Magnetic Field by Three-phase Currents

mum and in consequence the two rotating components, 1 and 2, are each directed along OX. At the same instant the current in coil II is 90° behind its maximum, so that its component fields, 3 and 4, have each to travel 90° before they point together along OY. Hence, they are along OX, but pointing in opposite directions. Now components 2 and 4 are both travelling in a clockwise direction with the same angular velocity. They are thus always opposed, and neutralize each other at every instant.

We are thus left with components 1 and 3, which combine together to produce a pure rotating field. It will thus be seen that the direction of rotation of the resulting field is the same as the direction of phase sequence in the two coils; in fact, the resulting field points along the axis of any coil at the instant the current in that coil is a maximum. Thus, to reverse the direction of the field, it is necessary to reverse the phase sequence, which can be done by reversing the connections to one of the coils.

Three-phase Supply. Now consider three coils arranged 120° apart, each supplied from one phase of a three-phase supply, as in Fig. 18.3. Consider the instant the current in coil I is a maximum. Then its two rotating fields, 1 and 2, will at that instant be directed along the axis OA. The current in coil II is 120° behind its maximum value, and therefore, each of its component rotating fields, 3 and 4, has to travel 120° before they point together along the axis OB. Hence, one of them will at that instant point along OA, and the other along OC, as shown. Similarly, the rotating fields 5 and 6, due to the current in coil III, have each to travel 240° before they point together along OC. Hence, field 5 is along OA, and 6 along OB.

Collecting the clockwise fields we see that these are 120° apart, and therefore neutralize one another at every instant. The counter-clockwise fields all point in the same direction and they therefore combine to produce a pure rotating field.

In this case also, the direction of rotation is the same as the direction of phase sequence, first coil I has its maximum current, then II, and then III. Thus, to reverse the direction of rotation of the field the phase sequence must be reversed. The three coils will be either star or mesh connected, and all that is necessary is to reverse two of the connections to the three terminals.

The above fields are bi-polar fields, that is, they have one N. and one S. pole. In a polyphase armature winding the various phases are separated by an angular distance in space equal to the electrical phase difference between the currents in them, so that such windings, when traversed by polyphase currents, produce a rotating field. In this case the rotating field has as many poles as the machines for which the winding is designed. It will thus be seen that the armature reaction in polyphase alternators consists of a rotating magnetic field which keeps pace with the poles of the rotor. The armature reaction is therefore uniform, instead of pulsating as in a single-phase alternator. Similarly, the torque produced by a polyphase synchronous motor is uniform, because the pure rotating field set up by the polyphase armature currents has no backward rotating component. This property of the motor enables it to be rendered self-starting by means of the dampers on the pole faces, the action of which in this connection is explained fully in Chapter 21.

Let each of the coils in Fig. 18.3 produce an alternating field of maximum strength H along its axes then, as with alternating current and voltage, we have the fundamental equation giving the instantaneous value with respect to time

$$h = H \sin \omega t$$

Take as zero time the instant that the rotating vector giving the field H_1 is directed along OA, then we have

$$\left.\begin{aligned} h_1 &= H_1 \sin \omega t \\ h_2 &= H_2 \sin (\omega t - 120°) \\ h_3 &= H_3 \sin (\omega t - 240°) \end{aligned}\right\}$$

or

$$\begin{aligned} h_1 &= H_1 \sin \omega t \\ h_2 &= -H_2[\tfrac{1}{2} \sin \omega t + (\sqrt{3}/2) \cos \omega t] \\ h_3 &= -H_3[\tfrac{1}{2} \sin \omega t - (\sqrt{3}/2) \cos \omega t] \end{aligned}$$

Now resolve these along OA and a perpendicular axis, and denote by $\bar{X}$ and $\bar{Y}$ the total components in these two directions. Then

$$\bar{X} = h_1 + h_2 \cos 120° + h_3 \cos 240°$$
$$\bar{Y} = 0 + h_2 \sin 120° + h_3 \sin 240°$$

Now put $H_1 = H_2 = H_3 = H$, and we have

$$\bar{X} = (3/2)H \sin \omega t$$
$$\bar{Y} = -(3/2)H \cos \omega t$$

Hence, for the resultant field, we have

$$H_r = (\bar{X}^2 + \bar{Y}^2)^{\frac{1}{2}}$$
$$= 3/2\, H$$

showing that the field is of constant strength, since its magnitude is independent of time.

For its direction θ with respect to the OA axis, we have

$$\tan \theta = \bar{Y}/\bar{X} = -\cos \omega t/\sin \omega t = \tan (\omega t - 90°)$$
$$\therefore\ \theta = \omega t - 90°$$

In other words, the angle θ is proportional to the time, and therefore the resultant field H_r must rotate in space with a uniform angular velocity of ω. In the example taken the direction of rotation of this field is counter-clockwise, because the sequence with which the component alternating fields go through their changes in magnitude is of order I, II, III. If it is required to reverse the direction of rotation in the field it is necessary to reverse this sequence, e.g. to change it to I, III, II. This is very easily accomplished by changing over two of the connections to the coils. Thus, if the coils are star connected, the leads to coils B and C can be changed over.

Rotating Field of a Three-phase Machine. So far we have dealt with the rotating fields produced by suitably disposed coils, arrangements which are sometimes employed in polyphase instruments. The disposition of the conductors in the stator winding of a polyphase machine is entirely different and therefore the field production by such a winding necessitates a separate description. Consider a simple two-pole three-phase machine, the stator winding of which consists of three coils disposed in space with the necessary angular spacing of 120°. A section of the machine is given in Fig. 18.4, the coils each having two conductors, or coil sides, AA' for phase I, BB' for phase II and CC' for phase III. Let these coils be supplied from a three-phase source, and consider first of all, the instant at which the current in phase I is zero. Then, with current vectors representing the maximum values of the phase currents, the diagram, at the instant considered will be oriented so that the vector I_1 points along OX. Projecting the other vectors on to OY we see that the instantaneous currents are

$$i_1 = 0;\ i_2 \text{ negative};\ i_3 \text{ positive};\ i_2 = i_3 \text{ numerically}$$

Adopt the convention that a current is positive when it flows inwards in conductors A, B, and C, and outwards in A', B', and C', and vice versa. Then at the instant considered the current i_2, being negative, flows inwards in conductor B' and outwards in conductor B. The current i_3, being positive, flows inwards in conductor C and outwards in C'. These directions are inserted in figure (*a*), and as far as the production of an m.m.f. is concerned we can imagine that the winding consists of a two-turn coil made up of turns $B'C'$ and BC, as shown by the dotted connectors. This coil produces an m.m.f. which acts vertically downwards, from A to A', and therefore the flux set up will be a two-pole flux with the north pole (where the lines of force leave the stator) centred at A, and the south pole (where the lines of force enter the stator) centred at A'.

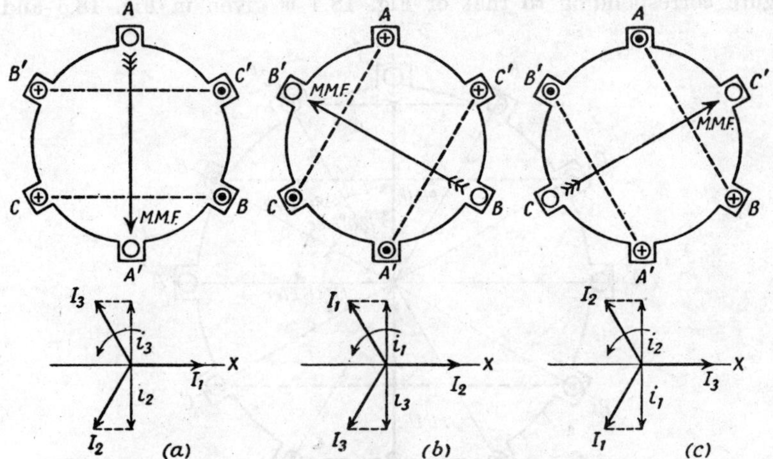

Fig. 18.4. Production of Rotating M.M.F. by Three-phase Stator Winding

One third of a cycle later the vector diagram will have rotated 120° in a counterclockwise direction so that for the instantaneous currents

$$i_2 = 0; \; i_1 \text{ positive}; \; i_3 \text{ negative}$$

Inserting the current directions as before and reducing the winding to the equivalent two-turn solenoid we obtain figure (*b*). The flux is again a two-pole flux but the m.m.f. now acts from B to B'.

After another one-third period we have

$$i_3 = 0; \; i_1 \text{ negative}; \; i_2 \text{ positive}$$

and the m.m.f. acts from C to C', as shown in figure (*c*).

After still another one-third period, that is, one period altogether from the commencement the conditions are again those of figure (*a*) showing that during one complete cycle the two-pole field has made one complete revolution.

$\therefore$ Speed of a two-pole field = one revolution in one periodic time
= one revolution in $1/f$ second
= f revolutions per second
= $60f$ revolutions per minute
= $120f/2$ revolutions per minute

Now consider a four-pole field. The simple armature will now have six coils, each coil will have an angular width of 90° instead of 180°, and opposite coils will be connected in pairs to form a phase. The figure corresponding to that of Fig. 18.4 is given in Fig. 18.5 and

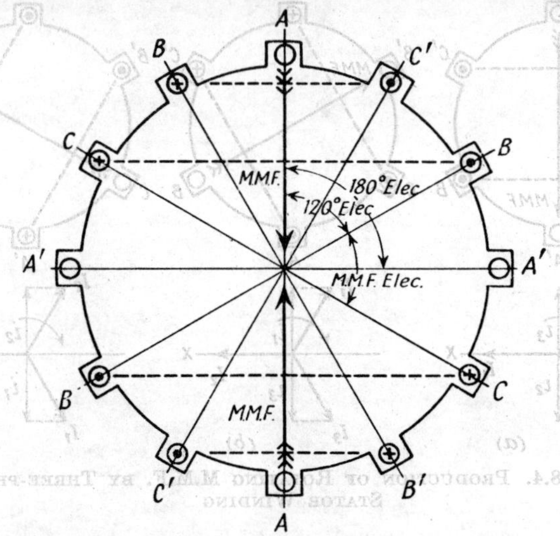

FIG. 18.5. PRODUCTION OF A 4-POLE M.M.F.

currents and m.m.f.s are drawn for instant 1 only, that is with $i_1 = 0$; i_2 negative; i_3 positive. There are, at this instant, eight conductors altogether which are carrying current and these can be regarded as making up two separate two-turn solenoids, one at the top, and one at the bottom, as indicated by the imaginary dotted connectors. We see from the current directions that the m.m.f. due to the top solenoid acts vertically downwards, while that due to the bottom solenoid acts vertically upwards. It follows that the fluxes set up by these m.m.f.s meet at the centre of the winding and are then diverted to either side, thus setting up a four-pole field.

Now after one complete cycle the distribution in space of both currents and fluxes must be the same as in Fig. 18.5, but this only necessitates that the flux shall rotate through one-half of a revolution. Obviously it cannot rotate through one whole revolution in this time otherwise there would be a repetition of the flux distribution after only one-half of a cycle, at which instant there could not be a repetition of the current distribution. Hence the speed of a 4-pole field is one-half of that of a 2-pole; the speed of a 6-pole field is one-third, and so on. Hence, in general the speed of a p-pole field is $2/p$ of that of a 2-pole field

$$\therefore \text{ Speed of } p\text{-pole field} = 2f/p \text{ revolutions per second}$$
$$= 120f/p \text{ revolutions per minute}$$

The direction of rotation corresponds to the electrical phase sequence. We considered an electrical phase sequence in the order 1—2—3, and on proceeding round the windings in this order we have to progress in a clockwise direction, the same as the direction of rotation of the field.

The Three-phase Power-factor Meter. This instrument is an application of the principle of the rotating field, one form of the instrument being shown in Fig. 18.6. Three fixed coils are arranged radially, each

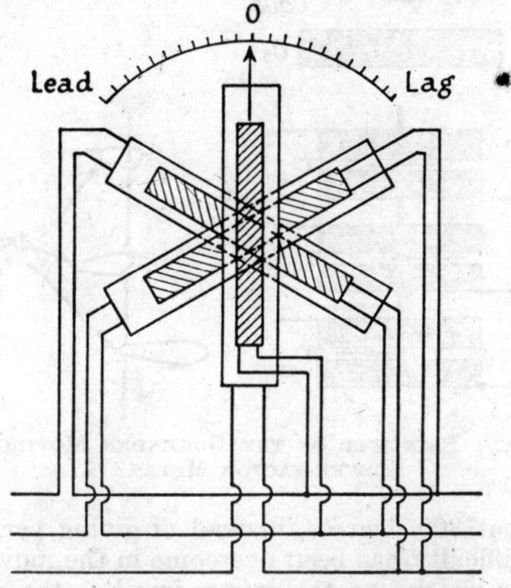

FIG. 18.6. POWER-FACTOR METER

coil being connected to one line conductor, if necessary, through current transformers. The currents flowing through these coils produce a rotating magnetic field, say, in a clockwise direction. Inside the fixed coils, and free to rotate, is a similar set of high-resistance pressure coils connected, each across one pair of lines, through a series resistance, or, if necessary, through a pressure transformer. The pressure coils also set up a rotating field which is in the same direction as the field produced by the fixed coils. Now if the pressure coils were clamped in the position in which they are drawn in the figure, there would be an angular distance between the two rotating fields equal to the angle of lag of the current. As the pressure coils are free to move, the two fields come into line, thus turning the pressure coils through an angle equal to the angle of lag of the current. If the current is leading, then the pressure coils will be turned in the opposite direction. If the current is in phase with the voltage, then, with the pressure coils in the position shown, the two rotating fields are already in phase, and there is no couple set up on the moving system, which remains in this position. Hence, the dial is graduated as shown.

A disadvantage of the instrument just described is that the ligament necessary for the leading of the current to the moving coils limits the

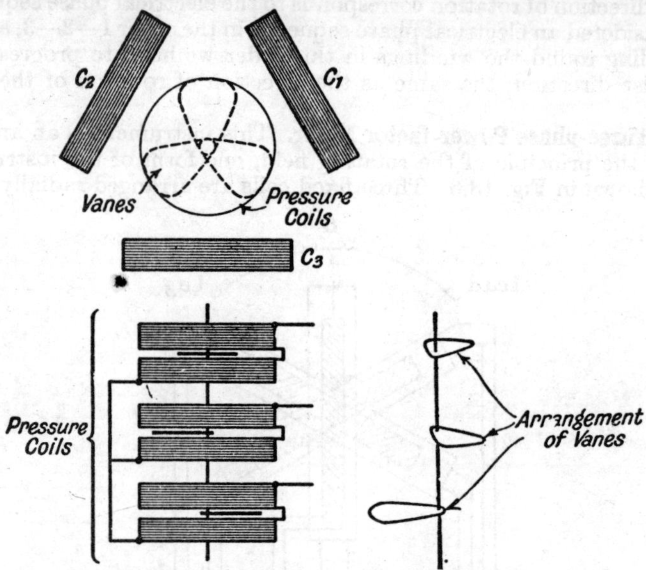

FIG. 18.7. PRINCIPLE OF THE OCKENDEN MOVING-IRON POWER-FACTOR METER

deflection to about 300 degrees, instead of giving perfect freedom of rotation. This difficulty has been overcome in the moving-iron power-factor meters in which, as the name implies, the moving system consists of iron vanes. In the Ockenden instrument, as manufactured by Messrs. Everett Edgcumbe & Co., Ltd., the rotating field principle is retained, the principle of the instrument being indicated by Fig. 18.7. C_1, C_2 and C_3 are the fixed current coils which, in virtue of their geometrical spacing, produce a rotating magnetic field as shown previously. The pressure coils are flat, fixed coils arranged in a stack with their axis perpendicular to the view of Fig. 18.7, so that in plan they appear as a circle, as shown. Each phase comprises two of these coils placed close together, and in the gap an iron vane can rotate. The pair of coils is wound so that like poles come together, the vane thus moving in an intense radial field. The result of this is that magnetic interference, both from neighbouring coils in the same instrument and also from external bodies, is eliminated.

CHAPTER 19

THE ALTERNATOR

ANY heteropolar rotating machine, that is, a machine with alternate north and south poles, is inherently an alternator, in the sense that the e.m.f. induced in each armature conductor is an alternating e.m.f. Thus the d.c. machine is an alternator in this sense, but since the field system is stationary in space, so also is the distribution of potential round the armature, with the result that fixed brushes which make contact with successive conductors in turn have between them a unidirectional p.d. corresponding to the angular positions of the brush axis with respect to the fixed potential distribution. If brushes are connected to points which are fixed with respect to the winding, and which therefore rotate with the winding, then, in a sense, the brush axis is rotating with respect to the fixed potential distribution, and the p.d. between the brushes is an alternating one. Thus two machines with identical field and armature systems can function, one as a d.c. machine if the armature winding is fitted with a commutator, the other as an alternator if tappings from fixed points on the armature winding are connected to slip-rings.

An alternator of this type is called a rotating-armature alternator: the construction is quite suitable for small outputs, but not for the very large outputs associated with alternators in modern power stations. With these the capacity of a single machine may be as high as 200,000 kVA, and obviously such an output could not be taken from rotating members such as slip-rings. Again, the terminal voltages of alternators range from 6600 to 33,000 volts, these values being quite unsuited to anything but terminals taken directly to the windings themselves, i.e. without the intermediary of rubbing contacts. Thus the only practicable construction for all but very small, low-voltage alternators, is the rotating-field, fixed-armature construction. The alternator of this type has a field system excited by direct current, and rotating inside a stator which carries the armature winding.

The Stator. The stator consists of (*a*) the iron core which carries the magnetic flux and (*b*) the stator winding which is housed in slots situated at the inner periphery of the stator. Since there is relative motion of flux to stator core, each particle of iron undergoes cycles of magnetization as in the case of the armature core of a d.c. machine, and consequently the core has to be laminated. For purposes of ventilation the core is arranged to have a series of radial ventilating ducts, say one duct 1·5 cm wide to each 7 cm of core length. This simple method of ventilation is suitable for slow and medium-speed machines, say up to 750 r.p.m., for which the ratio of core length to pole pitch is of the order of unity. The peripheral speeds are of the order 28 to 35 m/sec for slow speed, i.e. direct-coupled to reciprocating engines, 40 m/sec for medium speed, and as high as 150 m/sec for machines driven by turbines. The construction of a typical medium-speed machine is illustrated in Fig. 19.1.

The above simple ventilation scheme is not suitable for a turbo-alternator because of the great length of the core. Consider, for example a 50,000 kVA machine, speed 3000 r.p.m. Then for a rotor diameter of D metres, peripheral speed

$$V = \pi DN/60 \text{ metres/second}$$

Put $$V = 150 \text{ metres/second}$$

$$\therefore\ D \simeq 0{\cdot}95 \text{ metre, or 95 cms}$$

The output of an electrical machine can be expressed in the form

$$\text{kVA} = G \times D^2Ln \times 10^{-5}$$

where G is a constant for the particular type of machine under consideration and for the 50,000 kVA machine a suitable value is 24, when D and L are in centimetres, and n in r.p.s. The length of the core thus becomes

$$L = 50{,}000 \times 10^5 \times 60/(24 \times 95^2 \times 3000) = 460 \text{ cm}$$

With axial lengths of this, and greater, orders forced ventilation is essential. There are several alternative systems, and Fig. 19.2 shows the radial-axial system in which the radial ducts are intersected by axial ducts, and cooling air is forced into the machine from both ends, a system suitable for outputs up to 10,000 kVA at 3000 r.p.m.*

Frequency. Before considering the stator winding consider the frequency of the induced e.m.f. When a conductor is opposite the neutral plane, as at A, C and E (Fig. 19.3), its induced e.m.f. is zero. Opposite the middle of a pole, as at S, D, F, its induced e.m.f. is a maximum, its direction depending on the name of the pole influencing the conductor at any given instant. The e.m.f. induced in a conductor therefore goes through one complete cycle in an angular distance equal to twice the pole pitch.

$\therefore$ No. of cycles per revolution $= p/2$ where $p =$ No. of poles

Let $$N = \text{r.p.m.}$$

$$\therefore\ N/60 = \text{r.p.s.}$$

$\therefore$ Frequency in cycles per second

$$f = (p/2) \times (N/60) = Np/120$$

Converting this equation to the form

$$N = 120f/p \text{ r.p.m. or } 2f/p \text{ r.p.s.}$$

we see that the speed is the same as that of a rotating field produced in a p-pole winding by polyphase currents of frequency f. Hence, if the

* For very large turbo-alternators hydrogen is frequently used as the cooling medium (*see* p. 468).

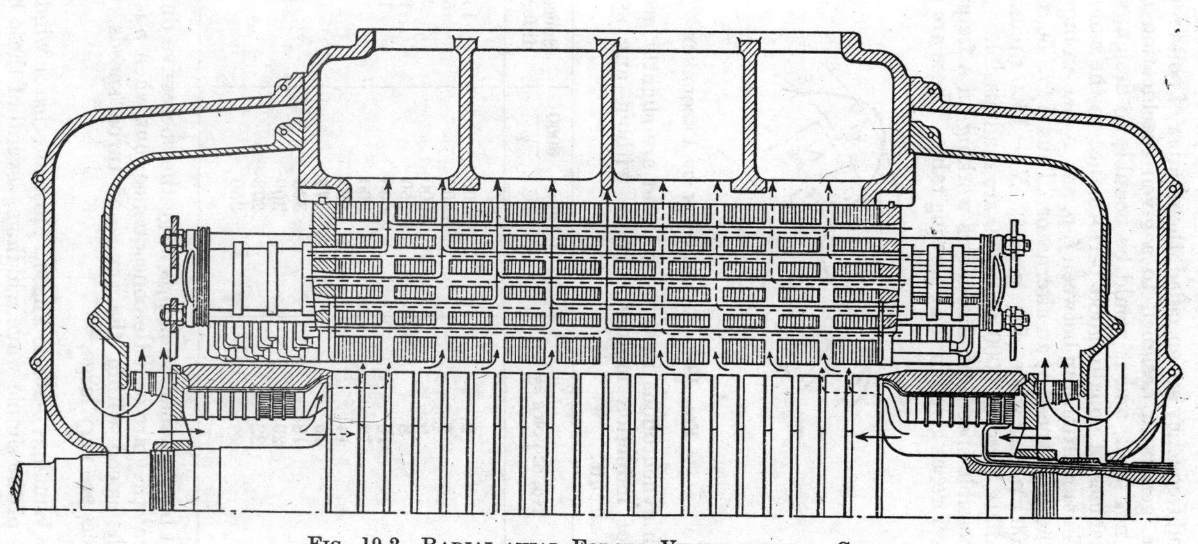

FIG. 19.2. RADIAL-AXIAL FORCED VENTILATION OF CORE
(*Courtesy Metropolitan Vickers Electrical Co., Ltd.*)

speed and frequency are specified, the number of poles required is fixed. In a d.c. generator designed to a given specification the number of poles is not fixed, and it would be possible to design, say, 10 and 12 pole machines of equally good performance to the same specification. This definite relationship between f, p, and N has an important bearing on the choice of speed of an alternator. Thus if $f = 50$, then for $p = 2$, $N = 3000$ r.p.m., while for $p = 4$, $N = 1500$ r.p.m., and speeds intermediate between 3000 and 1500 are inadmissible.

The possible speeds of alternators working on a frequency of 50 c/s are, therefore, as given in the following table. These are also the speeds

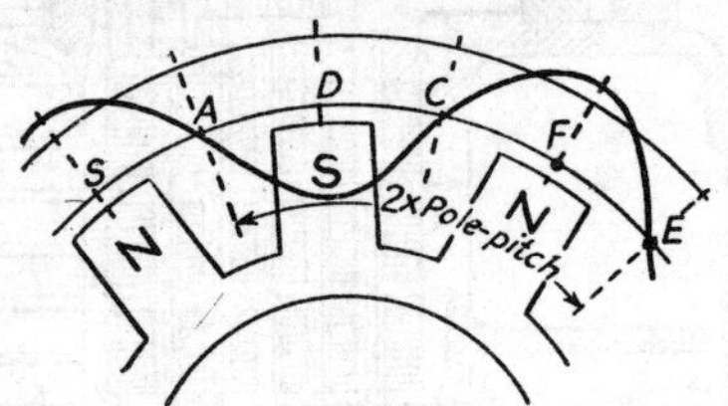

FIG. 19.3. CALCULATION OF FREQUENCY

at which synchronous motors run, and are slightly greater than the speeds of induction motors running without any speed-regulating plant in circuit.

No. of poles $= p$	$N = \dfrac{120f}{p} = \dfrac{6000}{p}$	Successive differences
2	3000	1500
4	1500	500
6	1000	250
8	750	150
10	600	100
12	500	72
14	428	53
16	375	42
18	333	33
20	300	27
22	273	23
24	250	

It will be seen that when p is large the successive differences are so small that they are of little consequence, but when p is so small as to bring the speed within the range of turbo speeds, the successive differences are very large.*

Stator Windings—Single Phase. Most alternator windings are modifications of the simple wave windings, several of these being shown in

* This is a disadvantage of a.c. working, since it is usual to adopt a high speed whenever possible. Thus a small 50-cycle alternator cannot run at a greater speed than 3000 r.p.m. and it may be necessary to interpose gearing so that the turbine driving it may run at a much higher speed.

Fig. 19.4. No. I is the simplest possible winding and can be referred to as a "skeleton" wave winding. In winding II, the single-turn coils of the skeleton winding are replaced by multi-turn coils, so that a higher

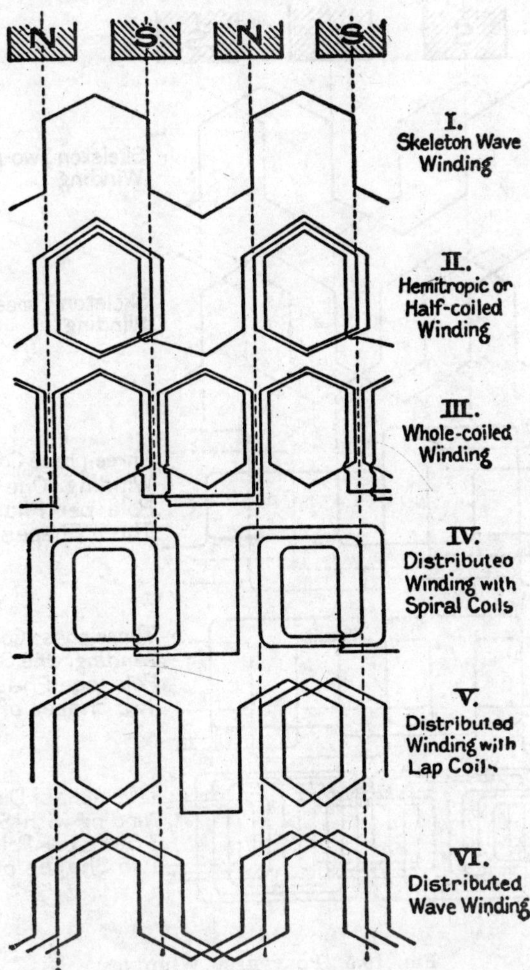

Fig. 19.4. Single-phase Stator Winding

e.m.f. can be obtained from the winding. This winding is called "hemitropic," or half-coiled, because the coils only cover one-half of the armature periphery. In winding III, the coils are distributed over the whole of the periphery, the winding therefore being called whole-coiled. All the above windings are "concentrated," because there is only one slot per pole. Concentrated windings give the maximum voltage for a given number of conductors, but the waveform of the voltage departs considerably from the desired sinusoidal form. Better waveform, at the sacrifice of a certain amount of output, is obtained

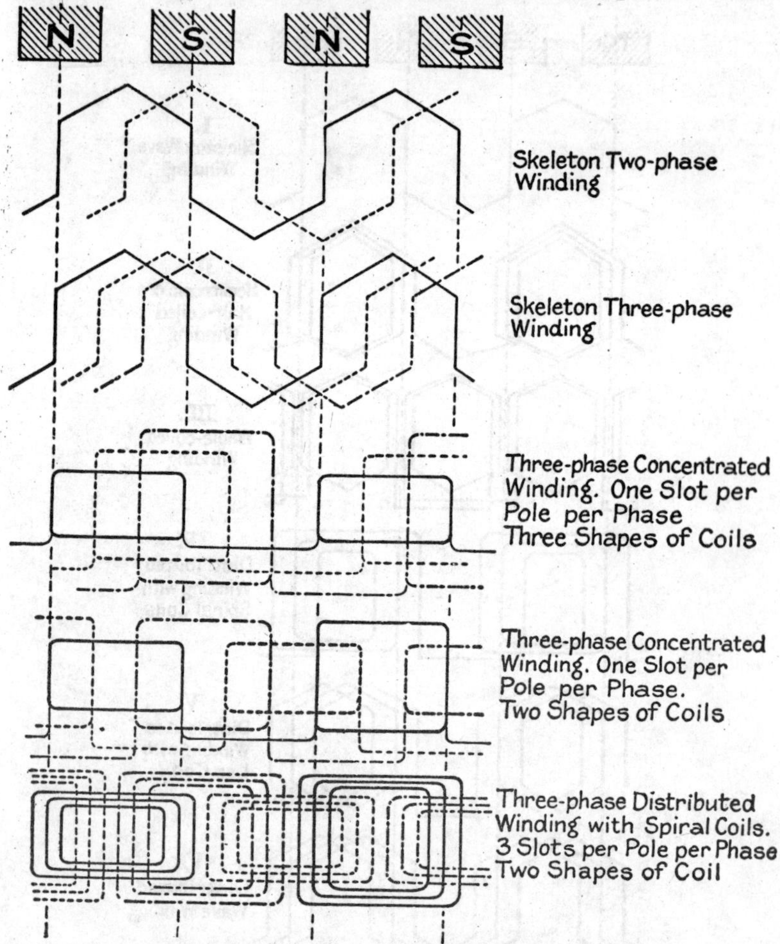

Fig. 19.5. Polyphase Windings

by distributing the winding in several slots per pole. The windings IV, V, and VI are distributed windings, IV and V having spiral and lap coils respectively: VI is carried out in wave fashion throughout

The above windings are "single-layer" windings, because each slot contains one coil side only. Two-layer windings arranged with two coil sides per slot, exactly like d.c. armature windings are frequently used in polyphase machines.

Polyphase Windings. These windings are arranged similarly to single-phase windings, the only difference being that in a two-phase alternator there are two separate single-phase windings, and in a three-phase alternator there are three separate windings. In a two-phase alternator the separate windings are 90 electrical degrees apart, thus giving the required phase difference. A skeleton two-phase winding is shown in Fig. 19.5. In a three-phase winding the individual windings are arranged 60° apart for convenience, so that the actual phase difference in the induced e.m.f.s are, between phases I and II, 60°; between II and III, 60°; between III and I, 240°. The second phase is therefore reversed when connected to the terminals, the correct phase relations in the external circuit being thus obtained. Several types of concentrated and distributed winding are illustrated in Fig. 19.5, their arrangements being the same as for the single-phase windings.

One of the most commonly used windings is the hemitropic winding

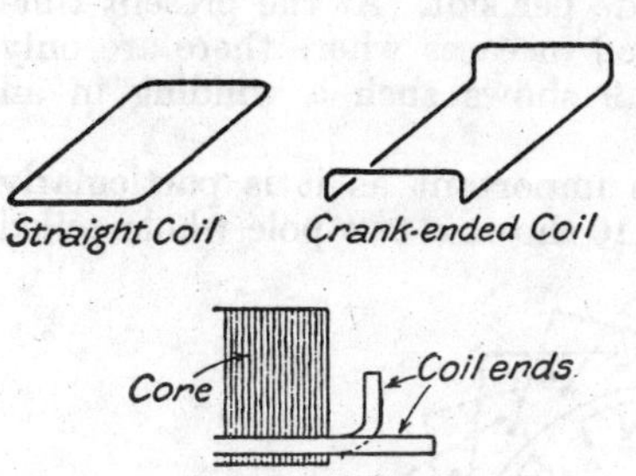

FIG. 19.6. ARRANGEMENT OF COILS

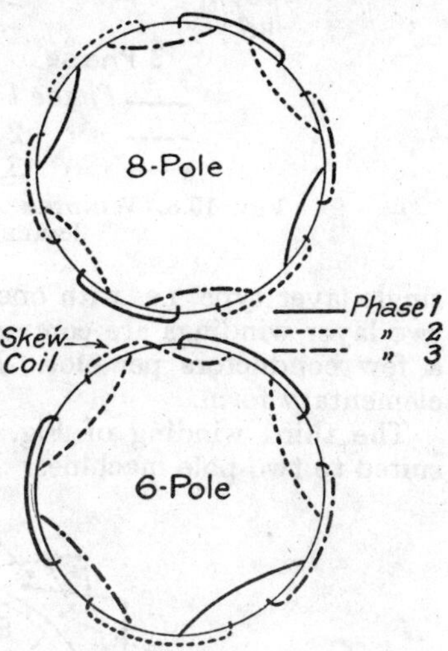

FIG. 19.7. WINDING WITH SKEW COIL

with spiral coils, as shown in the last example in Fig. 19.5. Two shapes of coil are required to carry out this winding, a plane coil and a crank-ended coil, as in Fig. 19.6. The coil ends for this winding are in two "ranges" to prevent fouling. If this winding is used with an alternator having an odd number of pairs of poles, then one of the coils has to be "skew" shaped, that is, one side straight and the other crank-ended. This is easily proved by tracing out the ends of the coils (Fig. 19.7).

A winding of this type can be used either as a three-phase or a two-phase winding. Consider the winding in Fig. 19.8. If the coil width is equal to the pole pitch, the winding is three-phase, each phase consisting of straight and crank-ended coils alternately. Suppose that the pole pitch is increased to $1\frac{1}{3}$ the coil width. Then all the straight coils can be joined together in one phase, and all the crank-ended coils, joined together in a second phase. If we consider any two consecutive coils in one phase, e.g. *A* and *B*, the e.m.f.s in them are not in phase, but since they are joined in series they combine to give an e.m.f. in phase with that which would be induced in a conductor placed in the position X. Similarly, the coils C and D in the second phase give together an e.m.f. in phase with that which would be induced in a conductor placed at Y. But Y is 90 electrical degrees from X, and therefore, the winding is a true two-phase winding.

At one time alternator windings were almost invariably of the

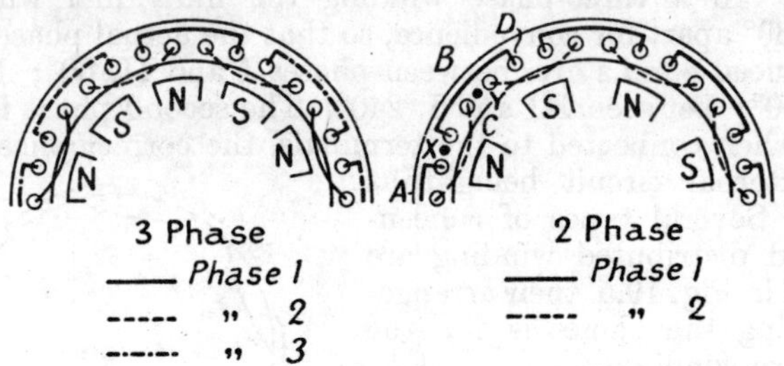

FIG. 19.8. WINDING ARRANGED FOR EITHER TWO- OR THREE-PHASE WORKING

single-layer type, i.e. with one coil side per slot. At the present time two-layer windings are commonly used in cases where there are only a few conductors per slot. Fig. 19.9 shows such a winding in an elementary form.

The third winding of Fig. 19.5 is important as it is particularly suited to two-pole machines. Fig. 19.10 shows a two-pole whole-coiled

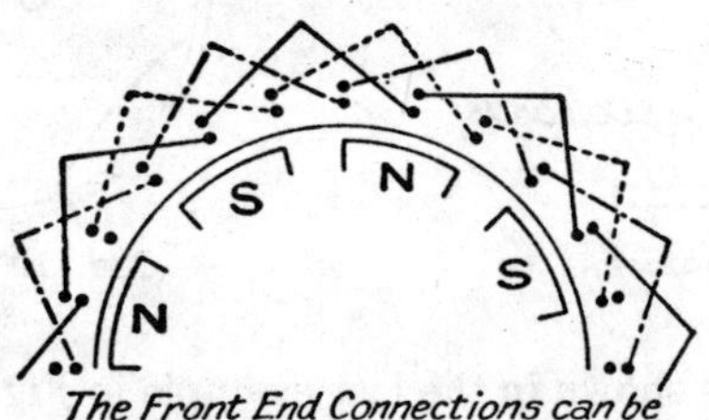

FIG. 19.9. TWO-LAYER WINDING

winding arranged in 24 slots, that is four slots per pole per phase. Actually the adjacent coil sides of a coil group occupy only two slots so that the winding is really a six-phase winding, the beginnings of the six phases being indicated by the terminals A_1, A_2, . . . A_6, and their "ends" by B_1, B_2, . . . B_6. In order to obtain a three-phase arrangement, opposite phases can be connected in either series or parallel, according to the voltage required, as indicated by the vector diagrams. In general it can be said that the single-layer windings are suited to high voltages because of the ease with which a large number of turns per coil can be secured. The winding of Fig. 19.9 is very suitable for medium-output turbo-alternators and high-speed, high-voltage induction motors.

Fractional Pitch Winding. All the windings considered so far are integral pitch windings, in that the number of slots per pole per phase is a whole number. With two-layer windings it is common to use a fractional number. We shall see that the use of a large number of slots per pole per phase tends towards a sinusoidal waveform of the terminal voltage—at a certain expense of output; a similar result can be obtained by the use of fractional pitches. With integral pitch windings the winding repeats itself after two poles, but with fractional-pitch

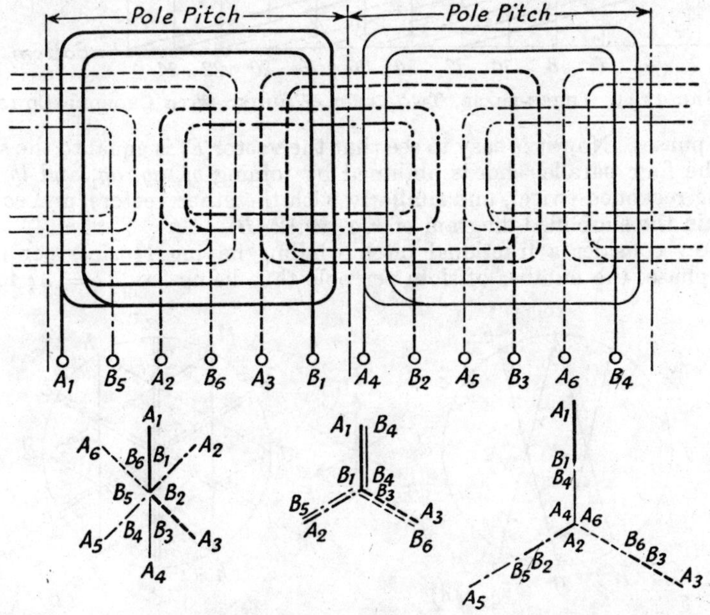

FIG. 19.10. TWO-POLE THREE-PHASE WINDING WITH THREE SHAPES OF COIL

arrangements several poles have to be passed over until the winding repeats. To understand this point it is a good plan to set out the coil ends. Consider first of all an integral-pitch winding with two slots per pole per phase, and let the coils be chorded by one slot, that is, the coil width less than the pole pitch by one slot pitch. Each coil spans five slots as shown in Fig. 19.11, and it will be seen that when two poles have been passed over, the winding repeats itself and has exactly the same dispositions with respect to the poles, so that we need consider two pole pitches only. Now if a winding could be distributed over an infinite number of slots the vector diagram for the conductors under two poles would be a circle. With distribution over a finite number of slots the phase and voltage of the e.m.f. induced in a coil is given by the length of the chord to the circle subtending at the centre an angle equal to the width of the coil in electrical degrees. Hence to obtain the vector diagram of the winding of Fig. 19.11, we draw a circle and mark off twelve equidistant points to represent the twelve slots per pole pair (Fig. 19.12). The four coil "throws" for phase I are represented by the chords joining *a* to *f*, *b* to *g*, *g* to *l*, and *h* to *a*. The resultant of these four e.m.f.s is E_1 acting in the direction *g* to *a*. Similarly we obtain vectors E_2 and E_3 for the e.m.f.s induced per pair of poles in the other

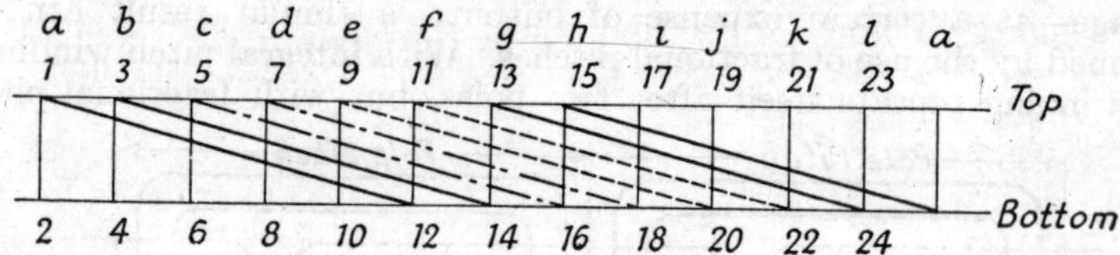

Fig. 19.11. Three-phase, Two-layer Winding with Chorded Coils

two phases. Now it is easy to see that the vector E_1 is equal to the sum of the four parallel chords obtained by joining *bf*, *ag*, *ag*, and *lh* (*ag* being reckoned twice), and similarly with the other vectors, and so we obtain the simplified diagram of Fig. 19.12 (*B*).

Now consider a fractional pitch winding having $1\frac{1}{4}$ slots per pole per phase, the number of slots per pole thus being $3 \times 1\frac{1}{4} = 3\frac{3}{4}$ for a

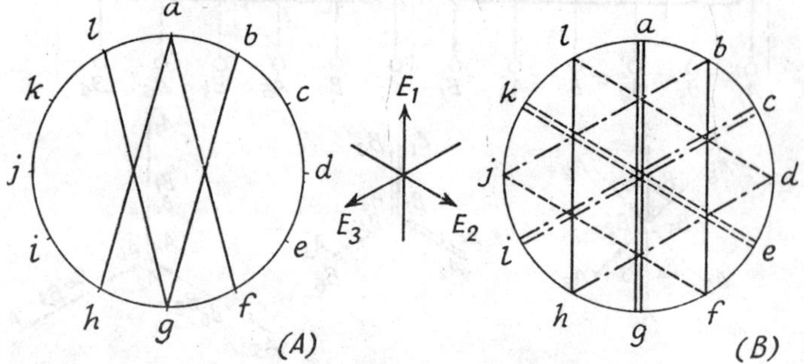

Fig. 19.12. Vector Diagram for Winding of Fig. 19.11

three-phase winding. The coil ends are drawn as before, Fig. 19.13, a coil throw of three slots being taken. We see now that we have to cover four pole-pitches before the winding repeats itself, showing straight away that such a winding can only be used for a pole number which is a multiple of four. The number of slots to four poles is 15 and so we divide the circle into 15 equal parts. The coil ends are drawn for a winding with each phase having two coils in a group, followed by three single coils, this being necessary to fill all the slots and retain three-phase symmetry. Each phase e.m.f. is given by the resultant of five parallel chords, one in each phase (*dh*, for phase one) being reckoned twice.

In some cases a winding does not repeat until many poles have been passed over. Consider, for example, $2\frac{2}{7}$ slots per pole per phase. The slots per pole for a three-phase machine is

$$2\tfrac{2}{7} \times 3 = 48/7$$

Hence, seven poles must be passed over to give a whole number of slots, but since a winding can only repeat with an even number of poles, the smallest number that can be used is 14.*

The e.m.f. Equation. Consider again Fig. 6.28, which can apply equally well to a rotating armature alternator as to a d.c. machine. Let there be Z' conductors per pole, as before, then there are $Z'/2$ turns

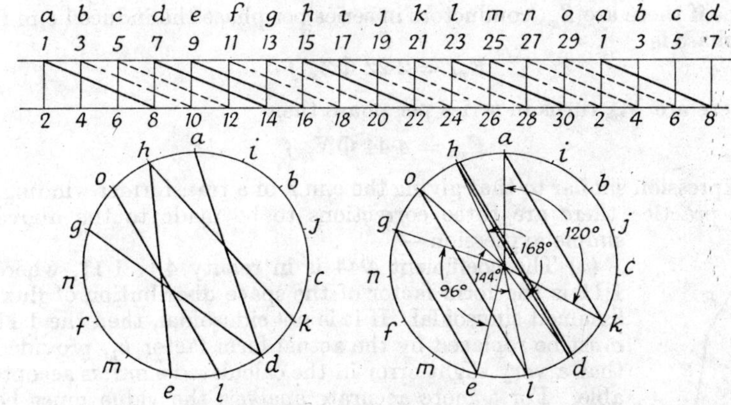

FIG. 19.13. FRACTIONAL-PITCH TWO-LAYER WINDING

per pole. Considering, first of all, a concentrated winding, the flux linking with a coil of angular width π electrical radians will be, at any angular position α, $2F(\alpha)$. Now let the flux distribution be sinusoidal in space, then

$$2F(\alpha) = 2 \times \Phi/2 \times \sin \alpha$$

$$= \Phi \sin \omega t$$

Again, for a concentrated winding, the e.m.f. in all the $Z'/2$ turns are in time phase, the total e.m.f. thus being

$$e = (Z'/2) \,.\, \Phi \,.\, d/dt\, (\sin \omega t)$$

neglecting the minus sign, since we are concerned with the magnitude only

$$\therefore \quad e = (Z'/2) \,.\, \Phi\omega \cos \omega t$$

$$= (Z'/2)\, \Phi\omega \cos \alpha$$

* An example of a fractional pitch winding is considered in detail on p. 465. For a very complete exposition *see* S. P. Smith, *The Theory of Armature Windings*. J.I.E.E. **59** (1916) 18. *See also* J. M. Bryant and E. W. Johnson *Alternating Current Machinery*, Chap. XIII, (McGraw Hill, New York, 1939), and C. C. Hawkins, S. P. Smith, and S. Neville, *Papers on the Design of Alternating Current Machinery*, Parts I and II, Pitman, 1919.

Hence for the r.m.s. value of this voltage

$$E' = (Z'/2)\ \Phi\omega \left[\frac{1}{\pi} \int_{-\pi/2}^{\pi/2} \cos^2 \alpha\, d\alpha \right]^{\frac{1}{2}}$$

also

$$\omega = 2\pi f$$

$$\therefore \quad E' = (\pi/\sqrt{2})\Phi Z'f$$

$$= 2{\cdot}22\ \Phi Z'f$$

Hence if there are Z_{ph} conductors in series per phase the induced e.m.f. per phase is

$$E_{ph} = 2{\cdot}22\ \Phi Z_{ph} f$$

If there are N_{ph} turns in series per phase then

$$E_{ph} = 4{\cdot}44\ \Phi N_{ph} f$$

an expression similar to that giving the e.m.f. of a transformer winding.

In practice there are three corrections to be made to the above simple expression—

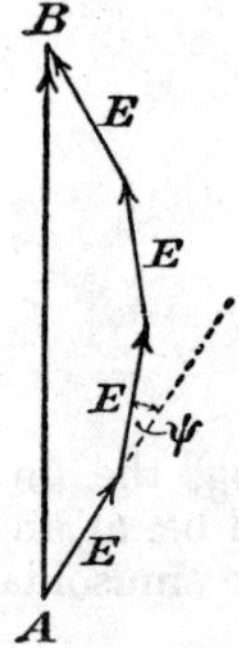

FIG. 19.14. CALCULATION OF BREADTH FACTOR

(*a*) The coefficient 4·44 is in reality 4 × 1·11, where 1·11 is the form factor of the space distribution of flux, assumed sinusoidal. If it is not sinusoidal, then the 1·11 must be replaced by the actual form factor k_1, provided that a very slight error in the calculated e.m.f. is acceptable. For a more accurate analysis the value must be kept at 1·11 and the fundamental and the various harmonics calculated separately.*

(*b*) Effect of distribution. With a distributed winding the e.m.f.s induced in the various coils constituting a group of coils are not in phase, so that the total e.m.f. is less than that due to a concentrated winding. It is therefore necessary to introduce a correcting factor k_2 which is less than unity. Fig. 19.14 gives the vector diagram for a coil group of four coils with a consecutive displacement of ψ electrical degrees. The coil group e.m.f. is the closing side AB of the polygon formed by the e.m.f.s of the individual coils. Let a group consist of m sections, and let E_c be the e.m.f. per coil, then the e.m.f. per group is

$$E_g = E_c \times \frac{\sin (m\psi/2)}{\sin (\psi/2)}$$

If the coil e.m.f.s had all been in phase the group e.m.f. would have been mE_c; hence by distributing the winding the e.m.f. is reduced in the ratio

$$k_2 = \frac{\sin (m\psi/2)}{m \sin (\psi/2)}$$

This ratio is called the "breadth factor" or "distribution factor."

* *See* Chap. 28.

PLATE II. ROTOR SLOT AND TEETH OF A TURBO-ALTERNATOR WITH HOLLOW CONDUCTORS
(*G.E.C.*)

PLATE II. ROTOR SLOT AND TEETH OF A TURBO-ALTERNATOR WITH HOLLOW CONDUCTORS
(*G.E.C.*)

(*c*) Effect of coil throw. We have seen that it is possible to construct a winding with a coil width less than the pole pitch. We shall see that this can have the effect of reducing harmonics in the voltage wave, and it also gives a saving in the amount of copper in the overhang. Such coils are said to be chorded, and they are commonly used, particularly with two-layer windings. In such a coil the e.m.f.s induced in the two

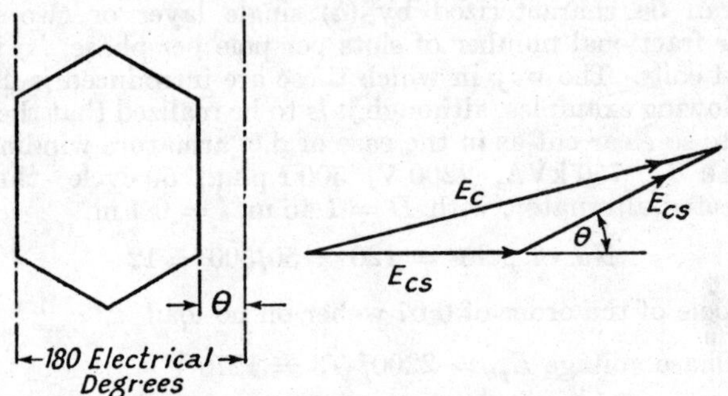

Fig. 19.15. Calculation of Chording Factor

coil sides are not in phase. Fig. 19.15 shows a coil chorded by θ degrees. With an e.m.f. of E_{cs} per coil side the coil e.m.f. is

$$E_c = 2E_{cs}\cos(\theta/2)$$

whereas it is $2E_{cs}$ for a full-pitch coil. Hence there is another correcting factor, the "chording factor" of magnitude

$$k_3 = \cos(\theta/2)$$

The final expression for the induced voltage per phase, that is the r.m.s. value of the fundamental and all the harmonics is, very nearly

$$E_{ph} = 2k_1k_2k_3\,\Phi Z_{ph}f$$

For extreme accuracy the sinusoidal components, i.e. the fundamental and the various harmonics, are calculated separately so that the value of Φ must be appropriate to the component being calculated, and so also must the values of k_2 and k_3; k_1 is then 1·11.

Example. A six-pole alternator rotating at 1000 r.p.m. has a single-phase winding housed in three slots per pole, the slots in groups of three being 20° apart. If each slot contains 10 conductors, and the flux per pole is 2×10^{-2} wb, calculate the e.m.f. generated, assuming a sinusoidal flux form.

$$\Phi = 2 \times 10^{-2} \quad Z = 18 \times 10 = 180 \text{ conductors}$$

$$f = Np/120 = 1000 \times 6/120 = 50$$

$$k_1 = 1{\cdot}11 \text{ for a sinusoidal field form}$$

$$k_2 = \frac{\sin(m\psi/2)}{m\sin(\psi/2)}, \text{ where } m = 3 \text{ and } \psi = 20$$

$\therefore k_2 = \sin 30/3 \sin 10 = 0{\cdot}96$

$\therefore E_o = 1{\cdot}11 \times 0{\cdot}96 \times 2 \times 2 \times 10^{-2} \times 180 \times 50$

$= 382$ V

Notes on the Choice of Stator Winding. We have seen that a stator winding can be characterized by (*a*) single layer or two layer (*b*) integral or fractional number of slots per pole per phase, (*c*) full-pitch or chorded coils. The way in which these are introduced is illustrated by the following examples, although it is to be realized that the decision is not quite so clear cut as in the case of d.c. armature windings.

EXAMPLE 1. 750 kVA, 2200 V, 500 r.p.m., 50 cycle, three-phase star-connected alternator, with $D = 1{\cdot}45$ m, $l = 0{\cdot}4$ m.

$$\text{No. of poles} = 120 \times 50/500 = 12$$

Flux per pole of the order of 0·07 weber on no load

$$\text{Phase voltage } E_{ph} = 2200/\sqrt{3} = 1270 \text{ V}$$

$$\text{Phase current } I_{ph} = 750{,}000/(\sqrt{3} \times 2200) = 197 \text{ A}$$

Giving the product $k_1k_2k_3$ the provisional value unity we have

$$Z_{ph} = E_{ph}/2\Phi f$$

$$= 1270/(2 \times 0{\cdot}07 \times 50) = 180, \text{ about}$$

$\therefore$ The total conductors $= 3 \times 180 = 540$, about

For a salient pole machine, if less than 1000 kVA, it is desirable that the ampere-conductors per slot shall be less than 1000.

$$\therefore \text{Slots} \not< 540 \times 197/1000 \not< 106$$

$\therefore$ Slots per pole per phase

$$\not< 106/(12 \times 3) \not< 2{\cdot}94$$

Thus 3 would be suitable, giving in all 108 slots and $540/108 = 5$ conductors per slot. Thus a single-layer winding like the bottom winding of Fig. 19.5 would be suitable: there would be no skew coil because the number of pairs of poles, six, is even.

If the number of conductors per slot had been an even number then a two-layer winding would have been possible. The voltage, 2200 V, is comparatively low, and the conductor current of about 200 A gives such a winding decided advantages from the constructional point of view. On the other hand, the two-layer winding is admirably suited to fractional pitches, and the following reasoning shows that in a 12-pole machine the main advantage of fractional pitches cannot be realized. For the suppression of harmonics it is desirable that coil sides which are consecutive in the scheme of the winding shall not occur in the same positions in the fields in which they lie, but that these positions should change progressively round the winding. Suppose that the same positions are not to recur in P' pole pairs, where $P' \not> P$, then, since

Plate III. English Electric 100 MW Turbo-alternator

The end covers are removed showing the passages for direct hydrogen cooling of the stator windings. This alternator is now installed in the Willington "A" Power Station of the Central Electricity Generating Board.

(*The English Electric Co. Ltd.*)

the angular displacements of the adjacent phases of a three-phase machine are $2\pi/3$ electrical radians, the condition to be fulfilled is

$$2\pi P'/3 = \pm 2\pi/3$$

This is possible when

$$P' = (3n \pm 1)$$

where n can have any but fractional values, including zero. Hence the following number of pole pairs can have a three-phase winding in which the positions of the coil sides in the field never recur*

$$P' = (3n \pm 1) = 1, 2, 4, 5, 7, 8, 10, 11, 13, 14, 16, \text{etc.}$$

In the present case $P = 6$, which is not included in the above series.

EXAMPLE 2. 2650 kVA, 3300 V, 187·5 r.p.m., 50 cycle, three-phase star-connected alternator, with $D = 2{\cdot}5$ m, $l = 0{\cdot}4$ m, and Φ of the order of 0·0845 weber per pole on no load.

$$p = 120 \times 50/187{\cdot}5 = 32 \text{ poles}$$

$$E_{ph} = 3300/\sqrt{3} = 1905 \text{ V}$$

$$I_{ph} = 2650 \times 1000/(\sqrt{3} \times 3300) = 463 \text{ A}$$

$$Z_{ph} \simeq E_{ph}/2\Phi f \simeq 1905/(2 \times 0{\cdot}0845 \times 50) \simeq 235$$

$\therefore$ Total conductors

$$\simeq 3 \times 235 \simeq 705$$

Suppose we try an integral number of slots per pole per phase, say 3, then

$$\text{Slots} = 3 \times 3 \times 32 = 288$$

$$\therefore \text{Conductors per slot} \simeq 705/288 \simeq 2{\cdot}45$$

This is just about midway between 2 and 3, the percentage change in both cases being so great that the flux per pole would be modified too much if either were used. This case is therefore one for a fractional pitch winding, and we can decrease the conductors per slot to 2, thus giving a two-layer winding.

$$\therefore \text{Slots } 705/2 \simeq 352$$

$$\therefore \text{Slots per pole per phase} \simeq 352/3 \times 32 \simeq 3{\cdot}66$$

A suitable number is therefore 3·5, so that there will be 336 slots altogether, each with 2 conductors per slot. The slot loading is thus $2 \times 463 = 926$ A which is suitable. The number of pole-pairs, 16, is included in the above series, showing the pole number to be suitable for a fractional pitch winding. If chorded coils are used then there is a further improvement in the magnitude of harmonic content and furthermore there will be a saving in the amount of copper in the overhang. Actually the current of 463 is rather high for a single-circuit winding, so that two circuits per phase, with 4 conductors per slot would be preferable. This, however, does not modify the scheme of the winding.

* *See* S. P. Smith, *loc. cit.*

With $10\frac{1}{2}$ slots per pole a full-pitch coil is impossible and a coil width of 9 slots, i.e. a throw of 1–10 will be suitable. The winding scheme and the vector diagram for one phase are given in Fig. 19.16A. Since there are 21 slots per pole pair the slot pitch is $360/21 = 17\frac{1}{7}$ electrical degrees. The second vector diagram, in which the vectors are replaced by parallel

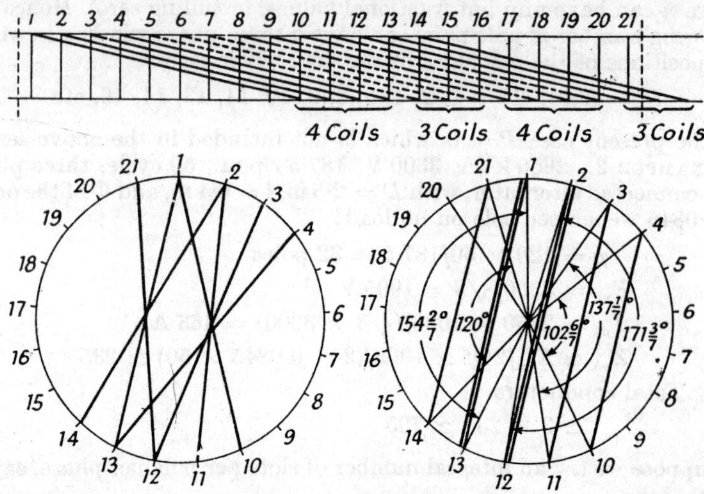

FIG. 19.16A. COIL ARRANGEMENT AND VECTOR DIAGRAMS FOR $3\frac{1}{2}$-SLOT PER-POLE-PER-PHASE THREE-PHASE WINDING

chords, shows how the product k_2k_3 can be determined in one operation. It is

$$k_2k_3 = (\sin 51\tfrac{3}{7} + \sin 68\tfrac{4}{7} + 2\sin 85\tfrac{5}{7} + 2\sin 77\tfrac{1}{7} + \sin 60)/7$$
$$= 0{\cdot}931$$

The m.m.f. analysis for this alternator is discussed on p. 481, and Fig. 19.16B shows how the ordinates for the stepped m.m.f. wave are obtained.

EXAMPLE 3. The turbo-alternator. A small machine will illustrate the problems encountered. 1250 kVA, 2750 V, 3000 r.p.m., 50 c/s, flux per pole of the order of 0·27 webers.

$$\text{No. of poles} = 120 \times 50/3000 = 2$$
$$E_{ph} = 2750/\sqrt{3} = 1590 \text{ V}$$
$$I_{ph} = 1250 \times 1000/(\sqrt{3} \times 2750) = 263 \text{ A}$$
$$Z_{ph} \simeq E_{ph}/2\Phi f \simeq 1590/(2 \times 0{\cdot}27 \times 50) \simeq 59$$

Total conductors

$$\simeq 3 \times 59 \simeq 177$$

With a turbo-alternator the slot loading can be greater than 1000 A, even in a small machine, because of the very efficient cooling, and this is desirable because of the saving in insulation resulting from the reduced slots. With 263 A per conductor, 4 conductors per slot will give 1052 A, and 6 conductors per slot 1598 A. Taking the latter figure the number of slots becomes

$$177/6, \text{ say } 30$$

which gives 5 slots per pole per phase. The single-layer winding of Fig. 19.10, with groups of five concentric coils per group, is suitable, or alternatively the winding can be two-layer, because of the even number of slots per pole. The second alternative is probably preferable (*a*) because of its mechanical simplicity, (*b*) because of the possibility of reducing copper in the overhang by chording.

An interesting problem in the turbo-alternator is the calculation of the ampere-turns per pole, since the rotor winding is distributed in slots. It is common with a two-layer stator winding to omit one conductor per phase, the number thus being 59 (not 60), the reason being that the starts are at one end of the machine and the finishes at the other, a device which simplifies the arrangement, and mechanical clamping, of the various connections. The distributed rotor winding gives $k_1 = 1{\cdot}11$, while for five slots per pole per phase $k_2 = 0{\cdot}914$. Hence, assuming full-pitch coils, with $k_3 = 1$, we have

$$\Phi = 1590/(2 \times 1{\cdot}11 \times 0{\cdot}914 \times 59 \times 50) = 0{\cdot}266 \text{ weber}$$

The core length is $l = 0{\cdot}76$ metre and mean gap diameter 0·508 metre, the pole pitch thus being

$$Y = \pi \times 0{\cdot}508/2 = 0{\cdot}798 \text{ m}$$

$$\therefore \; B_{\text{mean}} = 0{\cdot}266/(0{\cdot}76 \times 0{\cdot}798) = 0{\cdot}44 \text{ weber/m}^2$$

An approximate method of calculating the ampere-turns per pole is to assume a sinusoidal flux distribution of mean value equal to B_{mean}, that is, to 0·44 weber/m² in the present case.

$$\therefore \; B_{\text{max}} = (\pi/2) \times 0{\cdot}44 = 0{\cdot}69 \text{ weber/m}^2$$

The rotor has twenty slots altogether the angular distances being adjusted to give an m.m.f. as nearly sinusoidal as possible. These distances are given in Fig. 19.17. If the angular distance between the slot centres on either side of the wide central tooth is 2α, this tooth will carry the flux corresponding to the width 2α, and the average flux over this angle is

$$\frac{1}{\alpha}\int_o^\alpha B_{\text{max}} \cos\theta \, . \, d\theta = \frac{B_{\text{max}} \sin\alpha}{\alpha} = 0{\cdot}69 \times \frac{\sin 28^\circ}{28 \times (\pi/180)} = 0{\cdot}664$$

The angular width of this central tooth is 49·3°, so that the modified value of the average flux density is, approximately

$$[56 \times 2/(56 + 49{\cdot}3)] \times 0{\cdot}664 = 0{\cdot}705$$

This central tooth can now be regarded as a salient pole and contraction coefficients calculated as for a d.c. machine. These coefficients are

1·06 for stator teeth, 1·05 for stator ducts, and 1·02 for rotor ducts. Hence, using the method to extend the gap length, the actual length being $0{\cdot}89 \times 10^{-2}$ metre

$$l_g' = 0{\cdot}89 \times 10^{-2} \times 1{\cdot}06 \times 1{\cdot}05 \times 1{\cdot}02$$
$$= 1{\cdot}05 \times 10^{-2}\ \text{m}$$

Hence, ampere-turns per rotor pole for the gap

$$AT_g = H_g l_g'$$
$$= (B_g/\mu_0) l_g'$$
$$= [0{\cdot}705/(4\pi \times 10^{-7})] \times 1{\cdot}05 \times 10^{-2} = 5900$$

The Cooling of a Very Large Turbo-alternator. We have seen that for a given speed, the output of a rotating dynamo-electric machine is

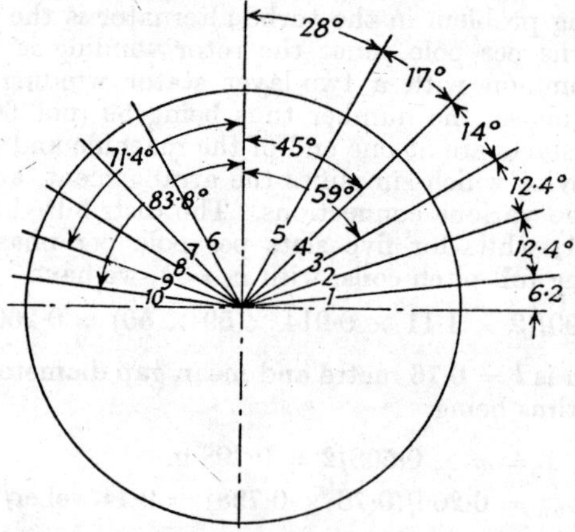

Fig. 19.17. Angular Position of Slots in 1250 kVA Two-pole Turbo-Alternator

proportional to D^2l, where D is the air-gap diameter and l the length. In the case of slow and medium-speed machines the division of D^2l into its components is decided by what experience has proved to be good proportions, as decided by manufacturing costs and desirable operating characteristics. In the case of the turbo-alternator, D is fixed by the allowable peripheral velocity, and with present-day materials this value is 150 m/sec. Hence the output is proportional to the length, and as output of single units increases, so the length increases. Thus the length of a 100,000 kVA (100 MVA) machine is about 5·5 m, and that of a 200 MVA machine about 7·5 m. With such great lengths the difficulty of removing the heat generated by the losses, particularly the slot losses, is so great that the methods of cooling adopted for machines of shorter length are quite inadequate. With air as the cooling medium, and with orthodox methods of construction of the windings, the upper limit of output for a single machine is 100 MVA. To increase this output two main modifications have been adopted (*a*) structural, (*b*) a change in cooling medium.

The structural change is the provision of a ventilating duct through the middle of each rotor and stator slot. The greatest difficulty is presented by the rotor slots, partly because of their great radial depth, and partly because there are so few, relative to the stator, that the ampere-conductors per slot and therefore the heat generated per slot are much greater than in the stator. If we assume for simplicity a

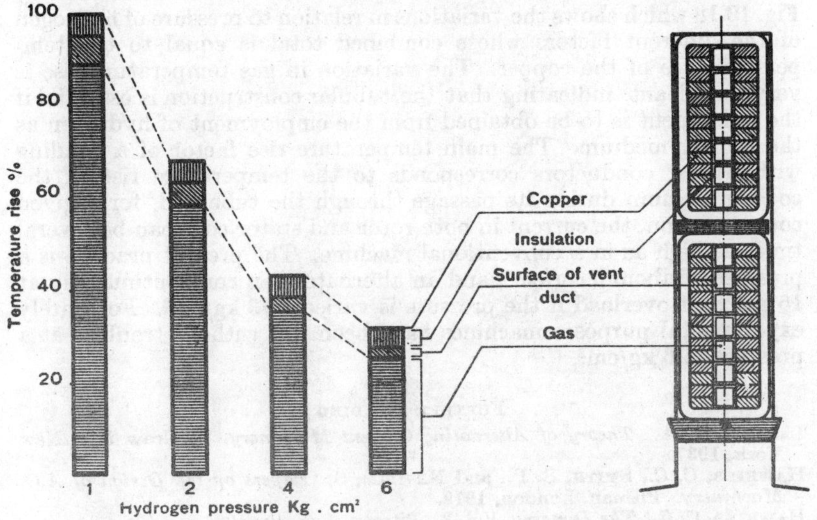

FIG. 19.18. VARIATIONS IN RELATION TO THE PRESSURE OF HYDROGEN OF THE VARIOUS FACTORS WHOSE COMBINED TOTAL IS EQUAL TO THE TEMPERATURE RISE OF THE COPPER OF THE WINDING
(*A.C.E.C.*)

radial heat flow, all the heat generated in the copper must first pass through the slot insulation, then through the surrounding metal, and finally to the heat dissipating surface where it has to traverse a gas film of poor conductivity before it is taken up by the main mass of the cooling medium. Thus the ideal solution is to evacuate the heat from the region where it is generated, usually inside the main conductor insulation, and this necessitates, either a single large vent duct serving the whole of the slot, or a separate duct to each conductor. Plate II (facing p. 464) shows the conductor arrangement in a rotor slot of this type.* The conductors are hard-drawn tubes of electrolytic copper alloyed with a small proportion of silver in order to improve the mechanical properties at elevated temperatures. For the main insulation, the traditional material, micanite, has been replace by a

* Easton, V., "The design of large turbo-alternators." *G.E.C. Journ.* (July, 1955).

synthetic resin bonded glass cloth laminate. It is able to withstand the electrical, mechanical and thermal stresses set up in operation, and also has a hard smooth surface able to withstand any frictional forces, as for example when fitting the conductor tightly into the slot.

The removal of heat from the stator slots is not so difficult as in the case of the rotor, and one large duct serves for the whole of the slot. A view of a stator winding having hollow conductors is given in Plate III (facing p. 465).

The cooling medium used with very large machines is hydrogen, its heat conductivity being 7 times that of air. The efficiency of the medium depends very largely on the pressure, as is well illustrated by Fig. 19.18 which shows the variations in relation to pressure of hydrogen of the different factors whose combined total is equal to the temperature rise of the copper. The variation in gas temperature rise is very significant, indicating that the tubular construction is essential if the full benefit is to be obtained from the employment of hydrogen as the cooling medium. The main temperature rise factor of a winding with hollow conductors corresponds to the temperature rise of the cooling medium during its passage through the tube and, for a given copper section, the current in both rotor and stator coils can be several times as high as in a conventional machine. The present practice is a pressure of about 2 kg/cm², and an alternator can run continuously at 15 per cent overload if the pressure is varied to 3 kg/cm². For purely experimental purposes machines have been run without troubles at a pressure of 6 kg/cm².

Further Reading

Langsdorf, A. *Theory of Alternating Current Machinery.* McGraw Hill, New York, 1937.

Hawkins, C. C., Smith, S. P., and Neville, S. *Papers on the Design of A.C. Machinery.* Pitman, London, 1919.

Hawkins, C. C. *The Dynamo*, Vol. 3. Pitman, London, 1925.

Kuhlmann, J. H. *Design of Electrical Apparatus.* Wiley, New York, 1950.

Say, M. G. *Performance and Design of A.C. Machines.* Pitman, London, 1952.

Examples on Chapter 19

1. If a single-phase alternator has 8 slots per pole uniformly spaced, but the winding is arranged with the middle two left empty, find the breadth coefficient.
Ans. 0·53.

2. A single-phase alternator of P poles has a large number S slots per pole, each containing one conductor. If the r.m.s. value of the e.m.f. induced in each conductor is 10 V, what is the total voltage induced?
Ans. $(20/\pi)\,SP$.

3. A single-phase alternator has an armature resistance of 0·1 Ω. When excited to give 50 V and then short-circuited, the short-circuit current is 200 A. To what induced voltage must the alternator be excited if it is to deliver 100 A at a power factor 0·8 lagging, with a terminal voltage of 200 V?
Ans. 222 V.

4. A single-phase alternator is excited to 1000 V at a frequency of 50 c/s. Its armature resistance is 0·8 Ω, and synchronous reactance, 3 Ω. Calculate what

length of unloaded cable, of capacitance $\frac{1}{3}$ μF/mile, must be connected to it terminals to produce resonance. Calculate the terminal voltage when this occurs If the flux per pole is 0·01 Wb, and the product of form factor and breadth factor 1·0, calculate the number of armature conductors.

Ans. 3200 miles; 3750 V; 1000 conductors.

5. The field form of an alternator measured from the neutral plane to the middle of a pole is as shown in the table. It is symmetrical about the centre.

Distance in m . .	0	0·01	0·02	0·03	0·04	0·05–0·1
Induction in gap, Wb/m²	0	0·04	0·21	0·45	0·66	0·80

The active length of conductor is 0·4m. All the 2000 conductors are housed in equally spaced slots 4 per pole, all the slots being filled. If the frequency is 50 c/s, find the induced e.m.f.

Ans. 8400 V.

6. The resultant air-gap flux of a 16-pole, 3-phase alternator is 6×10^{-2} webers per pole and is distributed sinusoidally over the pole. The stator has 2 slots per pole per phase, and 4 conductors per slot accommodated in two layers. The coil span is 150 electrical degrees. Calculate the e.m.f. generated in each phase when the machine runs at 375 r.p.m. (L.U.)

Ans. 795 V.

7. Derive an expression for the voltage induced in an alternator winding consisting of a number of similar full-pitch coils joined in series, the air-gap flux having sine distribution. Calculate the reduced voltage of a 6-pole, single-phase alternator running at 1200 r.p.m. The machine has 108 slots of which two-thirds are wound with full-pitch coils of 20 turns each; the field per pole is 5×10^{-2} wb. (C. & G.)

Ans. 7920 V.

8. The rotor of a 3-phase alternator has 16 poles and runs at 375 r.p.m. The flux distribution in the air-gap is symmetrical about the centre line of the pole, the values over half a pole pitch being as follows—

Angles (electrical degrees)	0	15	30	45	60	75	90
B wb/m²	0	0·225	0·585	0·790	0·880	0·9000	0·9000

The axial length of the stator core is 0·35 m, and the internal diameter 1·8 m. Determine the flux per pole. Plot to scale the e.m.f. waveform in a single-turn, full-pitch coil on the stator, and determine the frequency and r.m.s. value of this e.m.f. (L.U.)

Ans. 8×10^{-2} Wb, 18·5 V, 50 c/s.

CHAPTER 20

THE ALTERNATOR ON LOAD

THE d.c. machine on load experiences a change in flux owing to the effect of armature reaction, and an internal drop in volts due to the armature resistance. It is usual to keep the brush axis fixed and therefore the axis of the armature m.m.f., being coincident with the brush axis, is also fixed. If the brush axis were rotated with the machine under load, then the armature m.m.f. axis would conform to this. In the alternator with fixed armature winding the position in space of the armature reaction m.m.f. axis, at any particular instant in the voltage cycle is decided by the power factor. With an internal power factor of unity, that is, with the current in phase with the *induced* e.m.f., the space distribution of currents in the conductors is similar to that of e.m.f. induced in the conductors. If the power factor is altered, for example, if the current lags the induced voltage by γ electrical degrees then the armature will have to advance γ° with respect to the field before the m.m.f. axis is in the same position in space, as for internal unity power factor operation. With a rotating field machine this means that the field will have advanced γ° ahead of the position corresponding to unity power factor. Similarly, with a leading internal power factor of cos γ the field will be in retard of the position at unity power factor by the angle γ°, at the instant the armature m.m.f. axis is in the same direction as for unity power factor operation. Thus if we could imagine the power factor of an alternator to be varied progressively from unity to zero, this would be similar to a gradual shift of brush axis in a d.c. machine through 90 electrical degrees, the direction of shift being decided by the lag or lead of the current.

The internal volt drop in the alternator has two components, namely that due to the resistance of the winding and that due to the leakage reactance.

Rotating Field Produced by a Polyphase Winding. Since the total m.m.f. of a three-phase winding is made up of the separate effects of each of the phases it is necessary to consider, first of all, one phase by itself. Fig. 20.1 (a) shows three conductors of a single-phase concentrated winding having one slot per pole. Any point between conductors A and B is enclosed within a single turn and therefore the m.m.f. of the winding is constant from A to B. Similarly the m.m.f. is constant from B to C, of the same magnitude as before, but opposite in sign. Hence the m.m.f. of a single-phase concentrated winding has a rectangular distribution, as shown in the figure. It can thus be regarded as a rectangular wave, but it is a stationary wave because its position is fixed in space with respect to the winding itself. On the other hand, its amplitude varies sinusoidally with respect to time because the m.m.f.

at any instant is proportional to the magnitude of the current at that instant.

Fig. 20.1 (*b*) shows a single-phase winding distributed over two slots per pole. The winding consisting of conductors *A*, *B*, and *C* will produce the rectangular m.m.f. wave No. 1, while that consisting of conductors

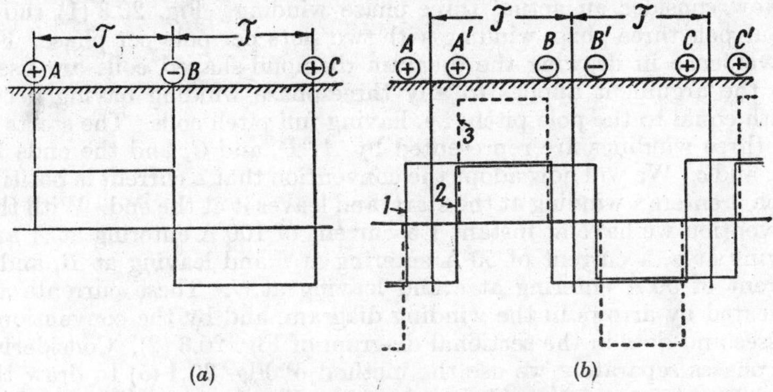

FIG. 20.1. M.M.F. OF A SINGLE-PHASE WINDING

A', *B'*, and *C'* will produce the m.m.f. wave No. 2. Adding the waves 1 and 2 together we obtain No. 3 for the total m.m.f. wave of the distributed winding. We see that the effect of distribution is to change the shape of the m.m.f. wave from rectangular to stepped.

In the case of a three-phase winding it is necessary to consider the state of affairs at several instants during one cycle, and we will now

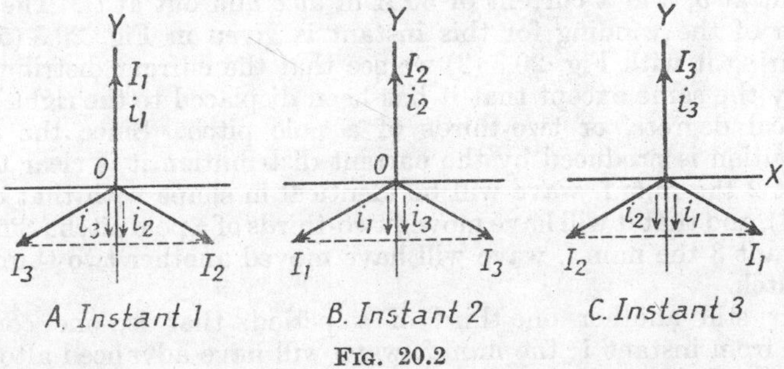

FIG. 20.2

choose as the first of these instants that at which the current in phase *I* is a maximum. Then, since we obtain the instantaneous value of a vector by projecting on to the *OY* axis, the vector of I_1 will be directed along *OY*, the diagram therefore being as in Fig. 20.2 (*A*). Suppose we call the maximum value of the current 100 per cent then, for the instantaneous current in the three phases we have

$$i_1 = +100, \; i_2 = -50, \; i_3 = -50$$

Fig. 20.2 (*B*) shows the state of affairs one-third of a period later, during which interval the vector diagram has rotated 120 degrees. At this instant we have

$$i_1 = -50, i_2 = +100, i_3 = -50$$

After another one-third period, we have, from Fig. 20.2 (*C*)

$$i_1 = -50, i_2 = -50, i_3 = +100$$

Now consider an actual three-phase winding. Fig. 20.3 (1) shows a four-pole three-phase winding with two slots per pole per phase. For convenience in drawing the diagram diamond-shaped coils are used, but the argument applies to any three-phase winding having a coil width equal to the pole pitch, i.e. having full pitch coils. The starts of the three windings are represented by *A*, *B*, and *C*, and the ends by *a*, *b*, and *c*. We will now adopt the convention that a current is positive when it enters a winding at the start, and leaves it at the end. With this convention we have at instant 1 a current of 100 A entering at *A* and leaving at *a*, a current of 50 A entering at *b* and leaving at *B*, and a current of 50 A entering at *c* and leaving at *C*. These currents are indicated by arrows in the winding diagram, and by the conventional crosses and dots in the sectional diagram of Fig. 20.3 (2). Considering the phases separately we use the method of Fig. 20.1 (*b*) to draw the three m.m.f. waves, these being shown in Fig. 20.3 (3). Adding these together we obtain for the total m.m.f. *at the instant considered* the wave shown in Fig. 20.3 (4). We see that this m.m.f. is not sinusoidal but that it is stepped, and a little reflection will show that the greater the number of slots per pole per phase, the less important will be the individual steps.

Now consider instant 2. At this instant a current of 50 A is flowing into phase *I* at *a* and out at *A*, a current of 100 A is flowing in at *B* and out at *b*, and a current of 50 A in at *c* and out at *C*. The cross-section of the winding for this instant is given in Fig. 20.3 (5), and comparing it with Fig. 20.3 (2) we see that the current distribution is exactly the same except that it has been displaced to the right by 120 electrical degrees, or two-thirds of a pole pitch. Since the m.m.f. distribution is produced by the current distribution it is clear that at instant 2 the m.m.f. wave will be identical in shape with that of Fig. 20.3 (4), and that it will have moved two-thirds of a pole pitch. Similarly at instant 3 the m.m.f. wave will have moved another two-thirds of a pole pitch.

After still another one-third of a period, that is, one complete period from instant 1, the m.m.f. wave will have advanced altogether six-thirds of a pole pitch, or two pole pitches. From p. 452 we have seen that this is the same as the speed of rotation of the rotor of the alternator, and this shows that the speed of the m.m.f. wave in space is the same as the speed of rotation of the rotor. Another way of expressing this fact is to say that the m.m.f. wave rotates at synchronous speed.

At the three intervals considered the current distributions in the conductors are identical and the m.m.f waves therefore identical in

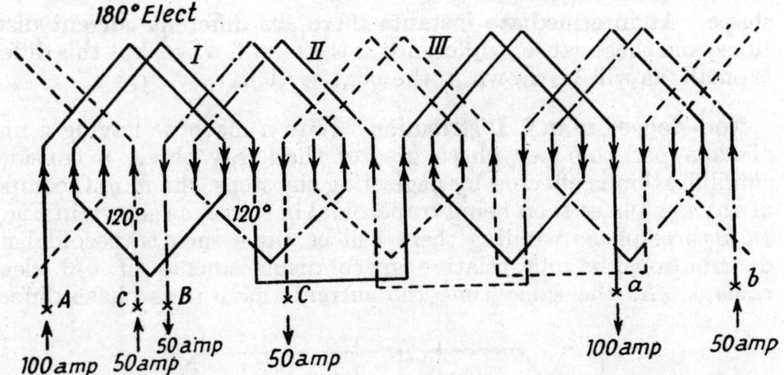

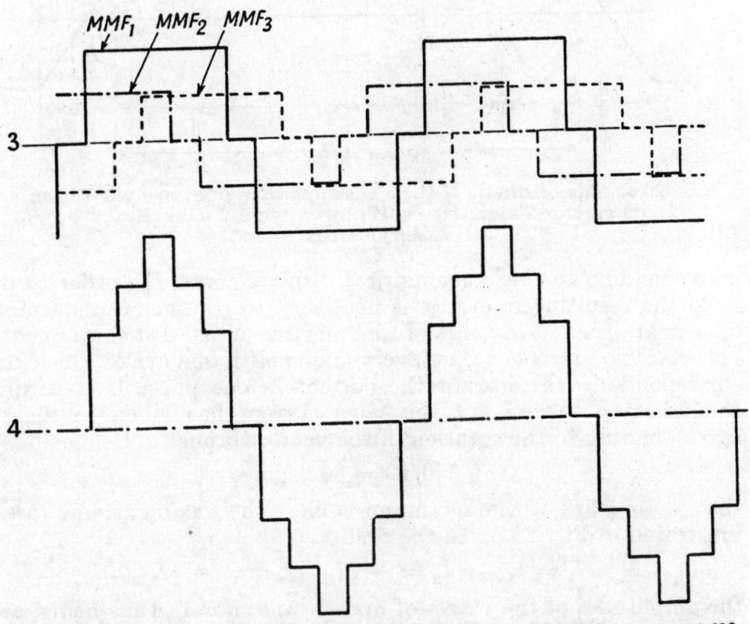

FIG. 20.3. M.M.F. WAVE OF A THREE-PHASE MACHINE

shape. At intermediate instants there are different current distributions, and these cause a difference in the m.m.f. wave, but this difference is small, as will be shown in the next section.

Non-stepped m.m.f. Distribution. With a machine having a number of slots per pole per phase greater than, say three, a considerable simplification is effected by neglecting the steps, the m.m.f. component of any one phase then being trapezoidal in shape, as shown in Fig. 20.4. In a three-phase winding there will be three such trapezoidal m.m.f. distributions having relative space displacements of $\pi/3$ electrical radians. At the same time the currents have time phase differences

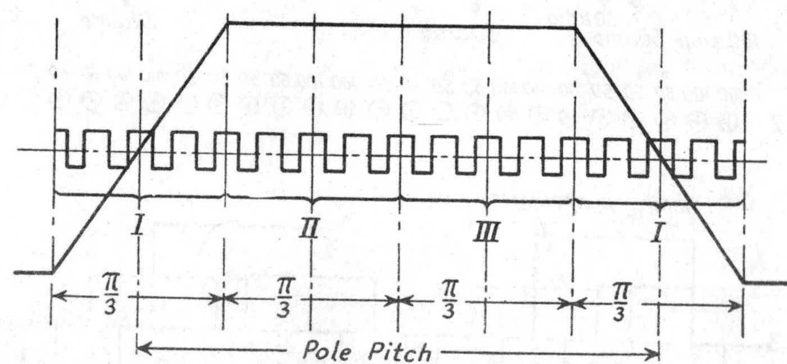

FIG. 20.4. TRAPEZOIDAL M.M.F. DISTRIBUTION DUE TO ONE PHASE OF A DISTRIBUTED THREE-PHASE WINDING WITH PHASE BELTS OF $\pi/3$ ELECTRICAL RADIANS

corresponding to these geometrical displacements. In order to determine the resulting m.m.f. it is necessary to consider the state of the system at different instants of time, for the actual distribution changes between two extreme values every one-twelfth of a cycle. One extreme corresponds to the instant the current in one phase is a maximum, the other to the instant the same current has fallen to the value corresponding to the rotation of the vector through

$$360/12 = 30 \text{ degrees}$$

that is, to $\sqrt{3}/2$ of the maximum value. These two extreme cases are illustrated in Fig. 20.5. In the first case

$$i_1 = I_{1\max},\ i_2 = -\tfrac{1}{2}I_{2\max},\ i_3 = -\tfrac{1}{2}I_{3\max}$$

the amplitudes of the waves of m.m.f.$_2$ and m.m.f.$_3$ thus being, at this instant, one-half of that of m.m.f.$_1$. In the second case we have

$$i_1 = 0{\cdot}866 I_{1\max},\ i_2 = 0,\ i_3 = -0{\cdot}866 I_{2\max}$$

In the first case the total m.m.f. distribution is a peaked wave whose centre coincides in space position with that of m.m.f.$_1$. In the second case the resultant is a flat-topped wave whose centre has advanced in space a distance of $Y/6$; in other words, the effect of the three component stationary waves is to produce a rotating m.m.f. wave which advances one-sixth of a pole-pitch in one-twelfth of a period, and therefore advances two pole-pitches in one period. This wave thus travels at synchronous speed, keeping pace with the rotor.

Let g denote the number of slots per pole per phase, and u the number of conductors (carrying full phase current) per slot. Then the height

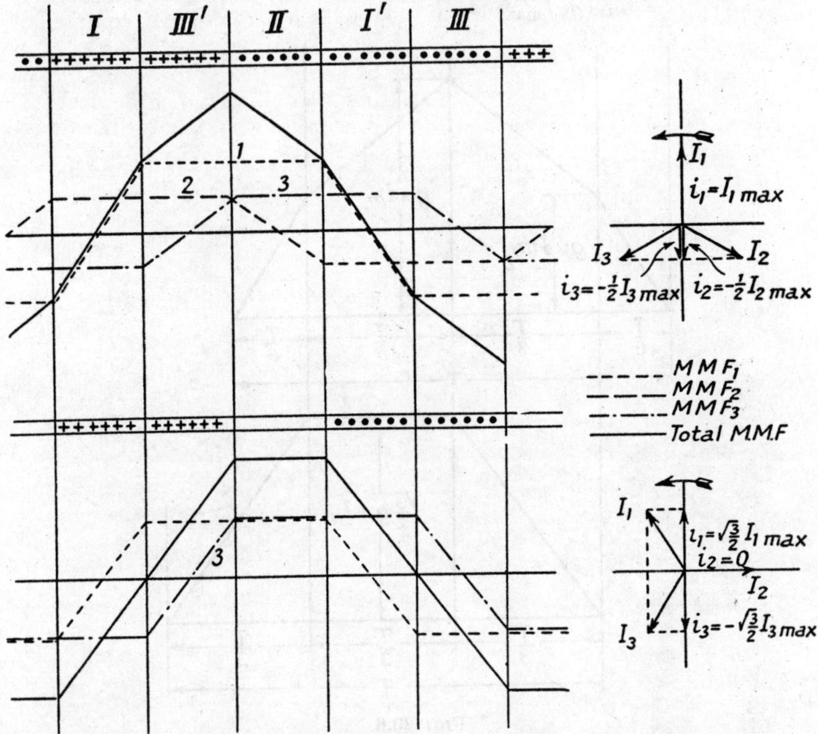

FIG. 20.5. EXTREME VALUES OF M.M.F.

of the m.m.f. wave due to one phase, at the instant the current in that phase is a maximum is $\frac{1}{2} \times 2guI_{\max}$. Let AT_a be the average value of the armature m.m.f. per pole at this instant, then from Fig. 20.6 (*top*), we have at this instant

$$AT_a \times Y = \tfrac{1}{2} \cdot \tfrac{1}{2}guI_{\max} \times 2 \times Y/6 + \tfrac{1}{2}(\tfrac{1}{2}guI_{\max} + guI_{\max}) \times 2 \times Y/3$$

$$= (7/12)guI_{\max}\, Y$$

$$\therefore\ AT_a = (7/12)guI_{\max}$$

In the other extreme case, Fig. 20.6 (*bottom*), we have

$$AT_a \times Y = \tfrac{1}{2}gu \times (\sqrt{3}/2)I_{\max} \times 2 \times Y/3 + (\sqrt{3}/2)guI_{\max} \times Y/3$$

$$= (1/\sqrt{3})guI_{\max}\, Y$$

$$\therefore\ AT_a = (1/\sqrt{3})guI_{\max}$$

The mean value for these two extremes is thus

$$AT_a = (\tfrac{1}{2})(\tfrac{7}{2} + 1/\sqrt{3})guI_{\max}$$
$$= 0{\cdot}58\, guI_{\max}$$

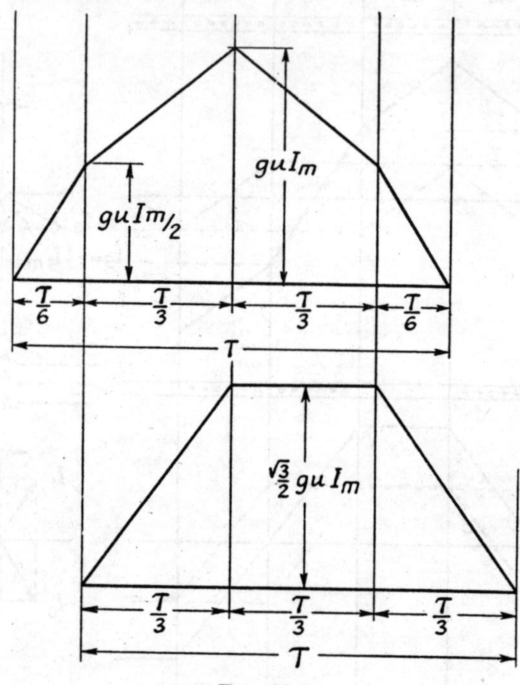

FIG. 20.6

In terms of the r.m.s. current per conductor this becomes

$$AT_a = 0{\cdot}82\, guI$$

If we replace the actual distribution by a sinusoidal distribution of the same average value we have

$$AT_{\max} = (\pi/2) \times 0{\cdot}82\, guI$$
$$= 1{\cdot}29\, guI$$

Fourier Analysis of the m.m.f. Wave. For the trapezoidal wave of Fig. 20.7, the analysis into sinusoidal components gives, for the m.m.f. at any position θ with respect to 0

$$\text{m.m.f.}_\theta = \frac{4}{\pi} \cdot \frac{\text{m.m.f.}_{\max}}{\alpha}$$
$$\left(\sin\alpha \sin\theta + \frac{1}{9}\sin 3\alpha \sin 3\theta + \ldots + \frac{1}{n^2}\sin n\alpha \sin n\theta\right)$$

For an armature winding with full-pitch coils, having a phase spread of $\pi/3$ electrical radians, the angle α for Fig. 20.6 (*bottom*) is $\pi/3$, so that

$$\text{m.m.f.}_\theta = \frac{4}{\pi} \cdot \frac{3\sqrt{3}\, guI_{\max}}{2\pi} \left(\frac{\sqrt{3}}{2} \sin\theta - \frac{\sqrt{3}}{2} \cdot \frac{1}{25} \sin 5\theta + \ldots + \frac{\sqrt{3}}{2} \cdot \frac{1}{n^2} \sin n\theta \right)$$

those terms in which n is a multiple of 3 disappearing. The above expression reduces to

$$\text{m.m.f.}_\theta = \frac{9}{\pi^2}\, guI_{\max} \left(\sin\theta - \frac{1}{25} \sin 5\theta + \ldots + \frac{1}{n^2} \sin n\theta \right)$$

FIG. 20.7

In terms of I(r.m.s.) this becomes

$$\text{m.m.f.}_\theta = \frac{9\sqrt{2}}{\pi^2}\, guI \left(\sin\theta - \frac{1}{25} \sin 5\theta + \ldots + \frac{1}{n^2} \sin n\theta \right)$$

$$= 1{\cdot}29\, guI \left(\sin\theta - \frac{1}{25} \sin 5\theta + \ldots + \frac{1}{n^2} \sin n\theta \right)$$

For the peaked m.m.f. waves of Fig. 20.6 (*a*) the Fourier analysis gives

$$\text{m.m.f.}_\theta = 1{\cdot}29\, guI \left(\sin\theta + \frac{1}{25} \sin 5\theta - \frac{1}{49} \sin 7\theta + \ldots \right)$$

The fundamental, namely $1{\cdot}29\, guI \sin\theta$, is the same in both cases and we see that it is, very approximately, the same as the sinusoidal distribution derived in a more elementary manner above. The harmonics decrease rapidly as n increases, and, in practice, it is usually sufficient to assume that the m.m.f. wave of a three-phase armature is of constant amplitude and rotates synchronously with the field system.

Now consider a single-phase winding. There is no gain in filling the whole of the periphery with conductors because of the very small increase in e.m.f. Suppose, for example, there is room in the pole-pitch for six equally spaced slots. The angular slot pitch is $\pi/6$ and therefore the vector polygon for a winding occupying all the slots is a six-sided polygon whose closing side is the diameter OA of the enclosing circle (Fig. 20.8). If one slot per pole is left empty the vector polygon for a coil group has five sides and the e.m.f. is OB. The ratio OB/OA is 0·866, showing that filling six slots instead of five increases the e.m.f. by 0·134 in 0·866, that is by 15·5 per cent, for a 20 per cent increase in copper. Thus there is little gain in making the group width more than $\frac{5}{6}$ of the pole pitch. The usual practice is to use a three-phase stator and house the winding in two-thirds of the slots.

Let the coil spread be α and let the symbol g now denote the number of slots within this range. Then when the current has its maximum

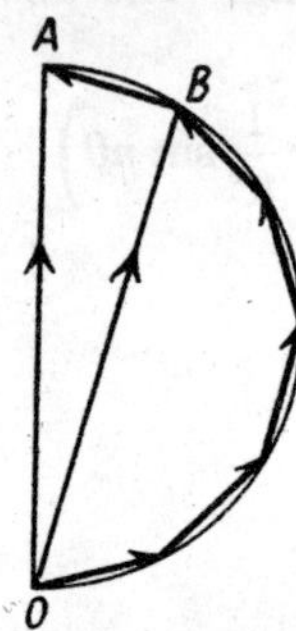

FIG. 20.8. E.M.F. OF A COIL GROUP OF A SINGLE-PHASE WINDING

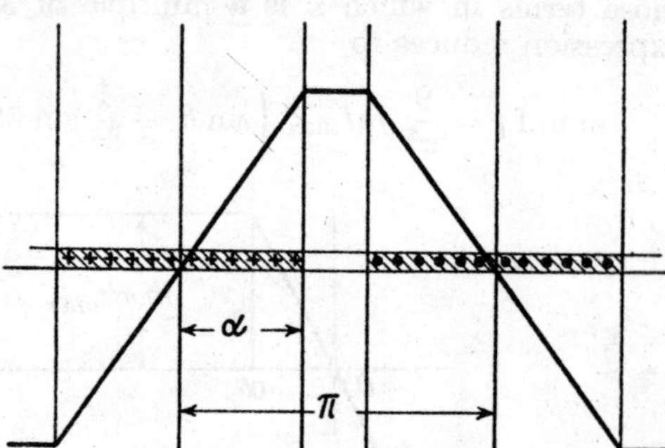

FIG. 20.9. M.M.F. OF SINGLE-PHASE WINDING

value I_{max} the m.m.f. distribution will be the trapezoid of Fig. 20.9, the amplitude being $\frac{1}{2}guI_{max}$, or 0·707 guI. The analysis of this gives

$$\text{m.m.f.}_\theta = \frac{4}{\pi} \cdot \frac{0{\cdot}707\, guI}{\alpha} \left(\sin\alpha \sin\theta + \frac{1}{9} \sin 3\alpha \sin 3\theta + \ldots \right)$$

The term in which n is a multiple of 3 will only disappear for $\alpha = \pi/3$ or $\alpha = 2\pi/3$, the former value being too small for a practicable winding, but the latter value possible. The amplitude of the fundamental is

$$(4/\pi)[(0{\cdot}707 \sin\alpha)/\alpha]\, guI$$

and this is now a stationary wave, as indeed are all the harmonics. The fundamental can be resolved, by Ferrari's principle, into forward and backward travelling components, each of amplitude

$$(2/\pi)[(0{\cdot}707 \sin\alpha)/\alpha]\, guI$$

One of these travels synchronously with the field system, and, in its effect on the performance under load of the machine, can be treated exactly like the m.m.f. wave of a three-phase machine. The backward component has a relative speed of twice synchronous speed with respect to the field system and its effect is to induce even harmonics in the e.m.f. In view of the amplitude of this backward-travelling component it is essential that it shall not penetrate the main field system, but shall be

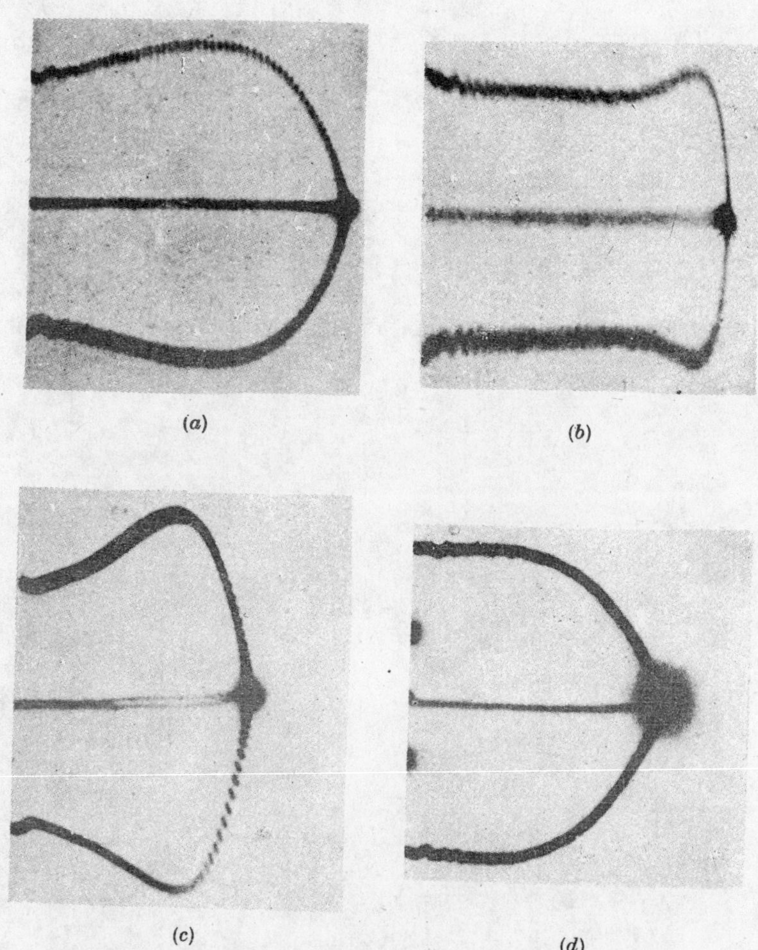

(a)

(b)

(c)

(d)

PLATE IV. TORQUE-SPEED CURVES FOR DIFFERENT INDUCTION MOTOR ROTOR CONSTRUCTIONS

(a) 4-pole single-cage rotor with relatively high starting torque.
(b) Double-cage (6-pole) rotor.
(c) 4-pole smooth-acceleration rotor.
(d) 6-pole high-slip rotor.

(*Metropolitan-Vickers*)

utilized in a manner which can do no harm. This is effected by providing the pole faces of single-phase alternators with a short-circuit "damping" winding. Under steady conditions the relative speed of this to the forward-travelling component is zero, so that no current is induced by this component. The backward component induces currents of twice normal frequency in the damping winding and the back m.m.f. set up by transformer action thus tends to neutralize the inducing m.m.f. The construction and disposition of the damping winding is illustrated in Fig. 20.9. This type of pole face winding is also supplied to three-phase alternators which are synchronized automatically, as well as to alternators driven by internal-combustion engines.

M.M.F. of a Fractional Pitch Winding. The exact analysis of the m.m.f. of such a winding is something in the nature of a mathematical recreation since it is doubtful if it could be used in actual alternator design. A very good method is to set out the coil ends, insert all the currents for some particular instant of time and then construct the stepped distribution. Consider, for example, the 2650 kVA alternator of p. 465. There are 32 poles, 336 slots each with two conductors carrying 463 A; slots per pole per phase $3\frac{1}{2}$. The coil ends have been set out in Fig. 19.16B. The maximum value of the full-load current is

$$I_{max} = \sqrt{2} \times 463 = 652$$

Taking the instant the current I_1 is a maximum and denoting this value by unity, the instantaneous phase currents are $i_1 = +1$, $i_2 = -\frac{1}{2}$, $i_3 = -\frac{1}{2}$. These are indicated in the end-connections in Fig. 19.16B (facing p. 466). Summing the vertical columns we obtain ordinates for the stepped m.m.f. wave, the maximum values of which are -6, or $6 \times 652 = 3912$, and $+6\frac{1}{2}$, or $6{\cdot}5 \times 652 = 4238$.

Let the value of β be 0·715.

Then slots per pole $= 0{\cdot}715 \times 10{\cdot}5 = 7{\cdot}5$

Area of positive half-wave under pole shoe

$$= (6\tfrac{1}{2} \times 2) + (5\tfrac{1}{2} \times 2) + (4\tfrac{1}{2} \times 2) + (3 \times 1{\cdot}5)$$
$$= 37{\cdot}5$$

Area of negative half-wave under pole shoe

$$= (6 \times 3) + (5 \times 2) + (4 \times 2) + (2\tfrac{1}{2} \times 0{\cdot}5)$$
$$= 37{\cdot}25$$

Mean $= 37{\cdot}375$

$\therefore AT_a = 37{\cdot}375 \times 652/7{\cdot}5 = 3250$

Effect of the Rotor Construction. (*a*) *Machine with cylindrical rotor, such as a turbo-alternator or synchronous-induction motor.* The whole of the armature ampere-turns per pole, namely $0{\cdot}82\,guI$, are available if the rotor is uniformly wound. With a turbo-alternator there is an unwound portion at the middle of each pole, so that the field m.m.f. distribution with d.c. excitations is trapezoidal, of constant amplitude. This is illustrated in Fig. 20.10, in which

$$\sigma = \text{(wound part of pole pitch)/(pole pitch)}$$

Let AT' be the number of ampere-turns required on each pole to neutralize the armature m.m.f. The average value of these will be given by

$$Y \times AT'_{av} = 2(\tfrac{1}{2} \times \tfrac{1}{2}\sigma Y \times AT') + Y(1 - \sigma)AT'$$
$$= (1 - \tfrac{1}{2}\sigma)AT'$$

Hence the additional ampere-turns per pole required to overcome the

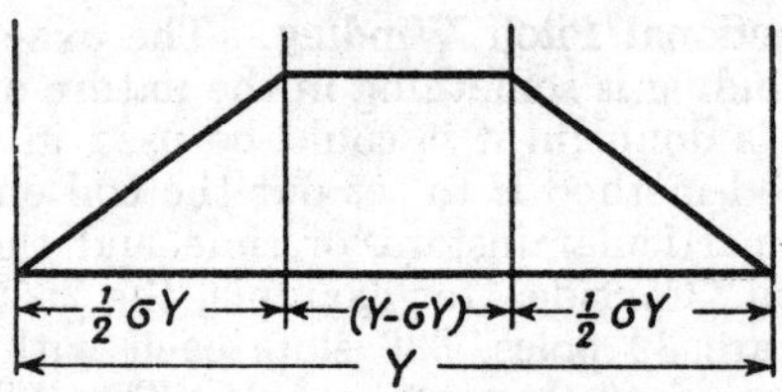

FIG. 20.10. ROTOR M.M.F. DISTRIBUTION FOR TURBO-ALTERNATOR

armature ampere-turns when these latter are directly demagnetizing, that is, at zero power-factor lagging,

$$AT' = [0{\cdot}82/(1 - \tfrac{1}{2}\sigma)]\, guI$$

(*b*) *Salient-pole machine.* The interpolar gaps interpose so much reluctance that, for all practical purposes, they can be regarded as cutting off those portions of the m.m.f. wave which project beyond the pole faces. It is now necessary to resolve the m.m.f. wave into direct and cross-magnetizing components, the simplest method being to work in terms of the fundamental sinusoidal component. We have seen that the average m.m.f. reckoned over the whole pole pitch is given by

$$AT_a = 0{\cdot}82\, guI$$

If we replace this by an equivalent sinusoidal distribution the maximum value will be

$$AT_{\max} = (\pi/2) \times 0{\cdot}82\, guI = 1{\cdot}29\, guI$$

The position of the crest of this travelling wave with respect to the polar axis is fixed by the internal phase angle γ. When γ is zero the crest is at the middle of the interpolar gap, and therefore for any finite angle γ the crest will be displaced γ electrical degrees from this position, as shown in Fig. 20.11. This sinusoidal distribution can be resolved into two sinusoidal components having their crest values $\pi/2$ electrical radians apart, so that if one of these has its crest at the polar axis the other will have its crest at the middle of the interpolar gap.

The former is the direct magnetizing component, the latter the cross-magnetizing component. The crest values are

Direct component $1{\cdot}29\ guI \sin \gamma$

Cross-magnetizing component $1{\cdot}29\ guI \cos \gamma$

We now take the pole span into account by calculating the area of each component curve which lies directly under the pole face. The two

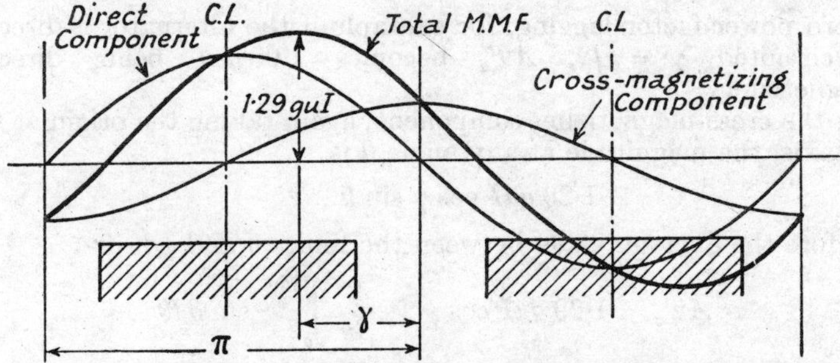

Fig. 20.11. The Two Components of the Armature M.M.F.

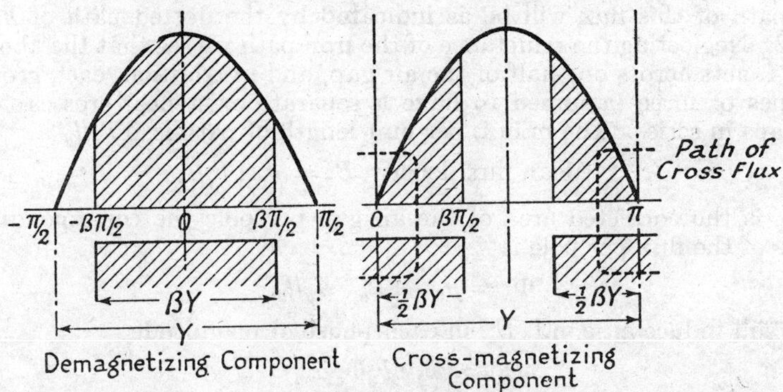

Fig. 20.12. Effective Positions of Armature M.M.F. Components in a Salient-pole Machine

cases are illustrated in Fig. 20.12. Taking as origin the centre of the pole face, the direct component at any angular position θ, is

$$1{\cdot}29\ guI \sin \gamma \cos \theta$$

Denoting the ratio pole span to pole-pitch by β, we average this expression between the limits $-\frac{1}{2}\beta\pi$ and $+\frac{1}{2}\beta\pi$, giving

$$AT_d = 1{\cdot}29\ guI \sin \gamma \times \frac{1}{\beta\pi}\int_{-\frac{1}{2}\beta\pi}^{+\frac{1}{2}\beta\pi} \cos \theta d\theta$$

$$= 1{\cdot}29\ guI \sin \gamma \times \frac{\sin (\beta\pi/2)}{\beta\pi/2}$$

Take as representative average value $\beta = \frac{2}{3}$, then

$$\frac{\sin(\beta\pi/2)}{\beta\pi/2} = \frac{\sin(\pi/3) \times 3}{\pi} = 0{\cdot}827$$

Hence direct component of the armature ampere-turns per pole

$$AT_a = 1{\cdot}29 \times 0{\cdot}827\, guI \sin\gamma$$
$$= 1{\cdot}06\, guI \sin\gamma$$

On zero power-factor lagging, for example if the alternator is directly short-circuited $\gamma = \pi/2$, AT_a becomes $1{\cdot}06\, guI$ being directly demagnetizing.

For the cross-magnetizing component, again taking the origin at the polar axis, the magnitude at any angle θ is

$$1{\cdot}29\, guI \cos\gamma \sin\theta$$

Therefore the average value between the limits 0 and $\frac{1}{2}\beta\pi$ is

$$AT_c = 1{\cdot}29\, guI \cos\gamma \times \frac{2}{\beta\pi}\int_0^{\frac{1}{2}\beta\pi} \sin\theta d\theta$$
$$= 1{\cdot}29\, guI \cos\gamma \times (2/\beta\pi)[1 - \cos(\beta\pi/2)]$$

If we imagine these ampere-turns as setting up a separate cross-flux, the path of this flux will be as indicated by the dotted path of Fig. 20.12. Neglecting the reluctance of the iron path we see that the above m.m.f. acts across one-half of the air gap, and in addition, each group of lines of force (assumed to have a separate existence) crosses two air-gaps in series. The m.m.f. per unit length of path is AT_c/l_g

$$\therefore \text{ Mean flux density } B_c = \mu_0 AT_c/l_g$$

If A_g is the corrected area of the air-gap per pole the corresponding value of the flux per pole is

$$\Phi_c = \mu_0 \,.\, AT_c \,.\, A_g/l_g$$

This will induce an e.m.f. E_c, in each phase of magnitude

$$E_c = 2k_1k_2k_3\Phi_c Z_{ph} f$$

Calculation of Leakage Reactance. In an alternating-current machine any flux set up by the load current which does not contribute to the useful flux of the machine is a leakage flux. We have already seen that such leakage flux exists in the static transformer, and that its effect is to set up a self-induced e.m.f. in the windings, thereby imparting leakage reactance to them. Similarly with an alternator. The leakage fluxes can be classified under four headings, the first three of which are indicated in Fig. 20.13.

(1) The slot leakage flux across the winding space: the lines of force, in general, link with part of the winding only.

(2) The slot leakage flux above the conductors: the whole of this flux links with the windings.

(3) The tooth head leakage.
(4) The coil-end, or overhang leakage.

The slot leakage fluxes and the overhang leakage flux take paths somewhat as indicated in Fig. 20.14, which shows a single coil of a

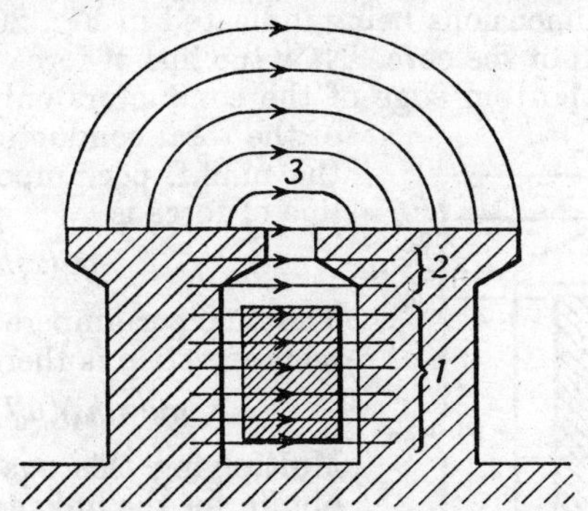

Fig. 20.13. Components of Armature Leakage Flux

single-layer winding. The determination of the overhang leakage flux presents an extremely complex problem and is beyond the scope of this book.* The calculation of the various components of the slot

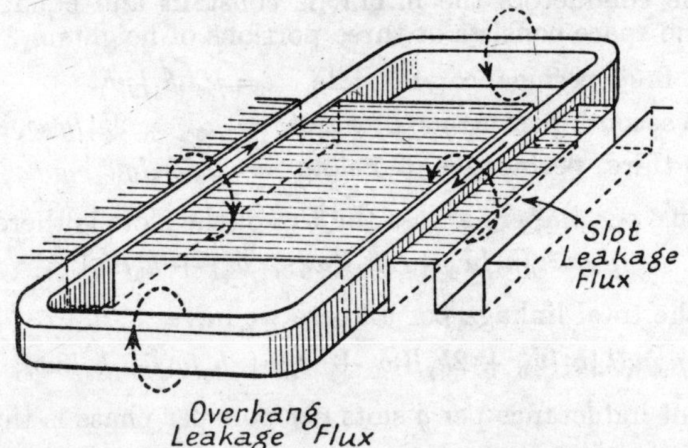

Fig. 20.14. Single Armature Coil Showing Paths of Leakage Flux

leakage flux affords a useful example in the application of first principles.

Consider first of all the slot leakage across the winding space: each line of force will cross g slots and will then pass through the iron below

* *See* A. Gray, *Electrical Machine Design*, p. 217. McGraw Hill, New York; H. Vickers, *The Induction Motor*. Pitman, London; P. L. Alger, "The calculation of the armature reactance of synchronous machines," *Trans. A.I.E.E.* **47**, No. 2 (1928) 493; L. A. Kilgore, "Calculation of synchronous machine constants," *Trans. A.I.E.E.* **50**, No. 4 (1931) 1201.

A useful empirical formula based on the results of tests is given by C. C. Hawkins, S. P. Smith and S. Neville, *Papers on the Design of A.C. Machinery*, p. 45. Pitman, London.

the slots, so that if we neglect the reluctance of all the iron paths, the permeance (inverse of reluctance) of an elementary path of depth dx will be

$$(\mu_0 \times \text{area})/\text{length} = \mu_0 l_i dx/gw_s$$

the various slot dimensions being indicated in Fig. 20.15, and l_i being the net iron length of the core. Now the line of force situated at a distance x from the bottom edge of the conductors only links with x/h_4 of the total conductor section, so that the m.m.f. per ampere acting on this line of force is

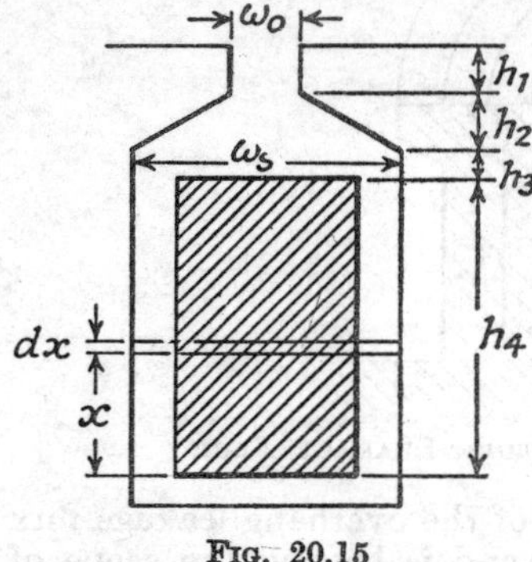

Fig. 20.15

$$gu(x/h_4)$$

The flux per ampere through the elementary strip is therefore

$$gu(x/h_4)(\mu_0 l_i dx/gw_s)$$

Multiplying this again by gux/h_4 we obtain for the linkages per ampere

$$\frac{\mu_0 gu^2 l_i}{w_s h_4^2}\int_0^{h_4} x^2 dx$$
$$= \mu_0 gu^2 l_i h_4/3w_s$$

Above the conductors the m.m.f. is constant and equal to gu per ampere. The space consists of three portions of heights h_1, h_2, and h_3.

For the first, permeance of g slots $= \mu_0 l_i h_1/gw_0$

For the second, permeance of g slots $= \mu_0 l_i \times 2h_2/g(w_s + w_0)$

For the third, permeance of g slots $= \mu_0 l_i h_3/gw_s$

The total flux per ampere across the top of the slots is therefore

$$\mu_0 gu^2 l_i[h_1/w_0 + 2h_2/(w_0 + w_s) + h_3/w_s]$$

Hence for the total linkage per ampere we have

$$\mu_0 gu^2 l_i[h_1/w_0 + 2h_2/(w_0 + w_s) + h_3/w_s + h_4/3w_s]$$

The total slot inductance per g slots per pole per phase is thus

$$L_s = 2\mu_0 gu^2 l_i[h_1/w_0 + 2h_2/(w_0 + w_s) + h_3/w_s + h_4/3w_s] \text{ H}$$

the coefficient 2 being introduced because each coil has two coil sides.

For the tooth-head leakage and overhang leakage we will assume recognized formulae. For the tooth-head leakage (Fig. 20.16)

$$L_{t.h} = \mu_0(2{\cdot}3/\pi)u^2 l_i[9 \log (Y/2y_s) + 3 \log \{1 + (\pi w_t/w_0)\} + 2{\cdot}1] \text{ H}$$

Finally, if we denote the inductance per phase due to the overhang leakage by L_0, we have for the reactance voltage per phase for a winding with n coils per phase

$$E_x = 2\pi fn(L_s + L_{t.h} + L_0)I \text{ volts}$$

EXAMPLE. A 750 kVA, 2200 V, 500 r.p.m., 50 cycle, star-connected alternator. $E_{ph} = 1270$ V, $I_{ph} = 197$ A.
Number of Poles: 12
Winding: single-layer with overhang in two ranges: 3 slots per pole per phase, 5 conductors per slot.
Dimensions:

Main Dimensions	Slots (*See* Fig. 20·17)	
$D = 1{\cdot}45$ m	$h_1 = 18{\cdot}8$ mm	$w_0 = w_s = 18$ mm
$l = 0{\cdot}4$ m	$h_2 = 0$	$w_t = 23{\cdot}7$ mm
$l_i = 0{\cdot}245$ m	$h_3 = 0$	$y_s = 41{\cdot}7$ mm
$Y = 0{\cdot}379$ m	$h_4 = 29{\cdot}9$ mm	

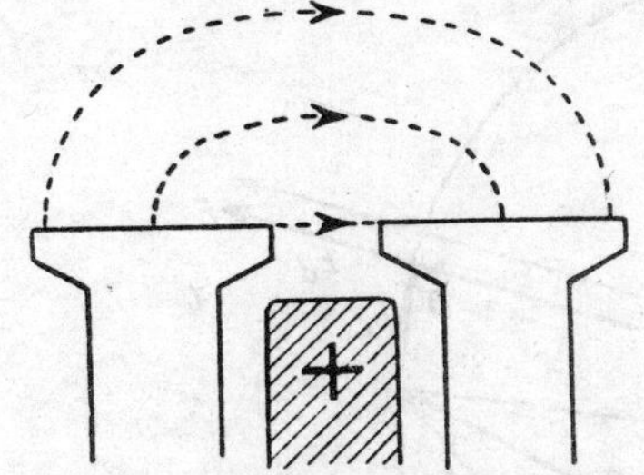

FIG. 20.16. TOOTH-HEAD LEAKAGE

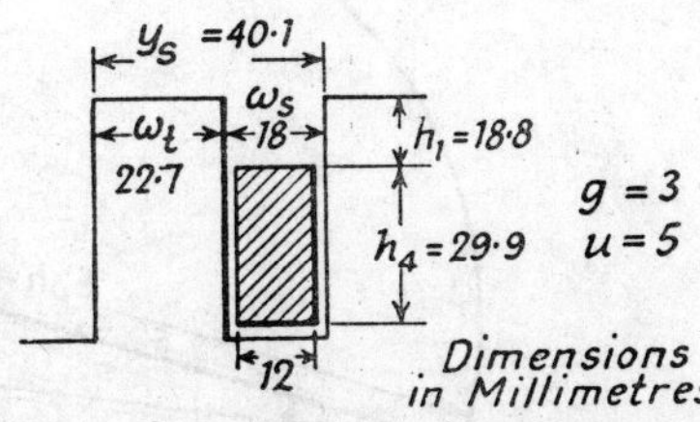

FIG. 20.17. SLOT PARTICULARS FOR ALTERNATOR DISCUSSED IN TEXT

Stator resistance per pole $5{\cdot}5 \times 10^{-2}\ \Omega$

Stator reactance per phase

(a) $L_s = 2\mu_0 g u^2 l_i (h_1/w_0 + 2h_2/(w_0 + w_s) + h_3/w_s + h_4/3w_s)$
$= 2 \times 4\pi \times 10^{-7} \times 3 \times 25 \times 0{\cdot}245(18{\cdot}8/18 + 29{\cdot}9/3 \times 18)$
$= 14{\cdot}8 \times 10^{-5}$

$$(b)\quad L_{t.h} = \mu_0 u^2 \times \left(\frac{2{\cdot}3}{\pi}\right) l_i \left[9 \log\left(\frac{Y}{2y_s}\right) + 3 \log\left\{1 + \left(\frac{\pi w_t}{w_0}\right)\right\} + 2{\cdot}1\right]$$

$$= 4\pi \times 10^{-7} \times \frac{2{\cdot}3}{\pi} \times 0{\cdot}245 \times 25 \left[9 \log\left(\frac{0{\cdot}379}{2 \times 41{\cdot}7 \times 10^{-3}}\right) + 3 \log\left\{1 + \left(\frac{\pi \times 23{\cdot}7}{18}\right)\right\} + 2{\cdot}1\right]$$

$$= 5{\cdot}7 \times 10^{-5}\ \text{H}$$

(c) The inductance due to overhang leakage is

$$L_0 = 7{\cdot}4 \times 10^{-5}\ \text{H}$$

(*d*) Reactance voltage. With a 12-pole machine there are $n = 6$ coils per phase

$$\therefore E_x = 2\pi \times 50 \times 6(14{\cdot}8 + 5{\cdot}7 + 7{\cdot}4) \times 10^{-5} \times 197$$
$$= 104 \text{ V}$$
$$= 8{\cdot}1 \text{ per cent of the phase voltage}$$

(*e*) Armature reaction m.m.f.

The direct component at zero power factor is

$$1{\cdot}06\, guI = 1{\cdot}06 \times 3 \times 5 \times 197$$
$$= 3140 \text{ AT/pole}$$

Predetermination of Voltage Regulation. (*a*) *The e.m.f. diagram.*

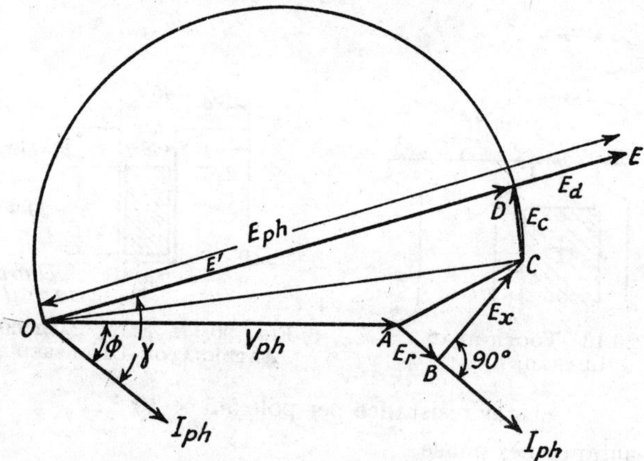

FIG. 20.18. VOLTAGE DIAGRAM FOR A THREE-PHASE ALTERNATOR ON LOAD

The complete voltage diagram for a three-phase alternator on load is shown in Fig. 20.18. In this diagram

$OA = V_{ph}$, the terminal p.d. per phase

$AB = E_r$, the armature resistance drop per phase

$BC = E_x$, the leakage reactance drop per phase

$CD = E_c$, the e.m.f. which would be induced by the cross flux per phase if this flux had a separate existence*

$DE = E_d$, the e.m.f. which would be induced by the direct component of the armature reaction per phase, if this could set up a separate flux

$OE = E_{ph}$, the total induced e.m.f. per phase

ϕ = external phase angle between I_{ph} and V_{ph}

γ = internal phase angle between I_{ph} and E_{ph}

* Note that in Figs. 20.18, 20.19, and 20.21 (*a*), the angle at *D* is in each case a right angle.

There is generally little difference between ϕ and γ, so that it can be assumed that V_{ph} is equal to its projection on OE. We then have, for an alternator which is excited to give E_{ph} volts per phase on no load

$$\text{Volt drop per phase} = E_{ph} - V_{ph} = E_d + E_r \cos\phi + E_x \sin\phi$$

$\therefore$ Percentage regulation up, that is, in terms of V_{ph}

$$100 \times [(E_{ph} - V_{ph})/V_{ph}] = 100 \times [(E_d + E_r \cos\phi + E_x \sin\phi)/V_{ph}]$$

To construct the diagram it is necessary to know E_c, the expression for

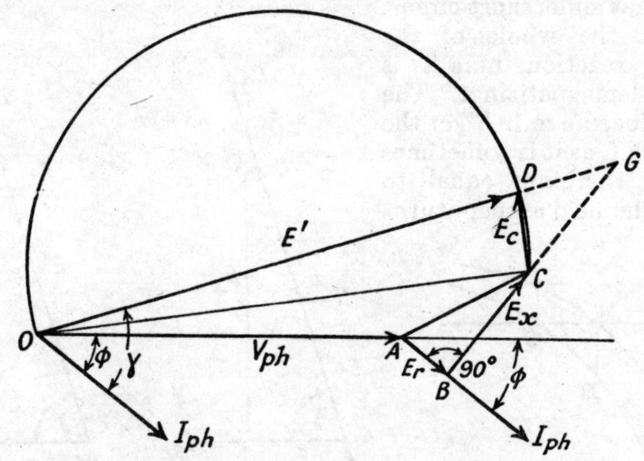

FIG. 20.19. CONSTRUCTION OF VOLTAGE DIAGRAM

which contains the angle γ, which is initially unknown. The difficulty is overcome as follows—

$$E_c = 2k_1k_2k_3Z_{ph}f \times (A_g/l_g) \times 1{\cdot}29\, guI \cos\gamma \times (2/\beta\pi)(1 - \cos\tfrac{1}{2}\beta\pi) \times \mu_0$$

$$\therefore\ E_c/\cos\gamma = 2k_1k_2k_3Z_{ph}f \times (A_g/l_g) \times 1{\cdot}29\, guI \times (2/\beta\pi)(1 - \cos\tfrac{1}{2}\beta\pi) \times \mu_0$$

The vector BC (Fig. 20.18) is extended to G, as in Fig. 20.19, the length CG being made equal to $E_c/\cos\gamma$. OG is joined and the intersection of OG with the semicircle on OC as diameter gives the point D. The length OD is the actual voltage E' which is induced per phase. The usual method of allowing for the direct component of the armature reaction is therefore to calculate the field ampere-turns necessary to induce the voltage E', and then add additional ampere-turns per pole equal to

$$AI_d = 1{\cdot}06\, guI \sin\gamma$$

It is to be noted that the regulation of an alternator can be expressed in two ways, namely, the percentage rise in voltage when full-load is thrown off, or the percentage drop in voltage when full-load is thrown on. The first is called "regulation up," and is more commonly used, the second is called "regulation down." Fig. 20.20 shows that the magnitudes of the two are different because, in each case, the initial voltage is the normal voltage.

(*b*) *Synchronous Impedance Method.* When an alternator is short-circuited the terminal voltage V_{ph} is zero and the vector diagram reduces to the form of Fig. 20.21 (*a*). Again, the resistance is so small in comparison with the leakage reactance that the angle γ is almost 90° and the cross-magnetizing component of the armature reaction negligibly small. This further simplifies the diagram to Fig. 20.21 (*b*), showing that under short-circuit conditions the whole of the armature reaction m.m.f. is directly demagnetizing. The actual induced e.m.f. E', or the air-gap e.m.f. as it is sometimes called, is therefore equal to AC, and the field ampere-turns

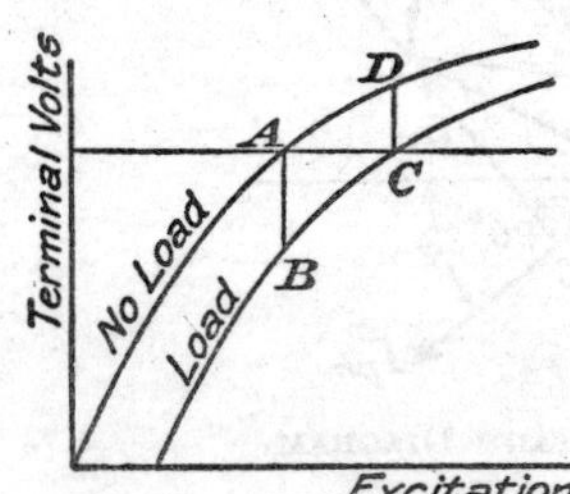

Fig. 20.20. Difference between "Regulation Up" and "Regulation Down"

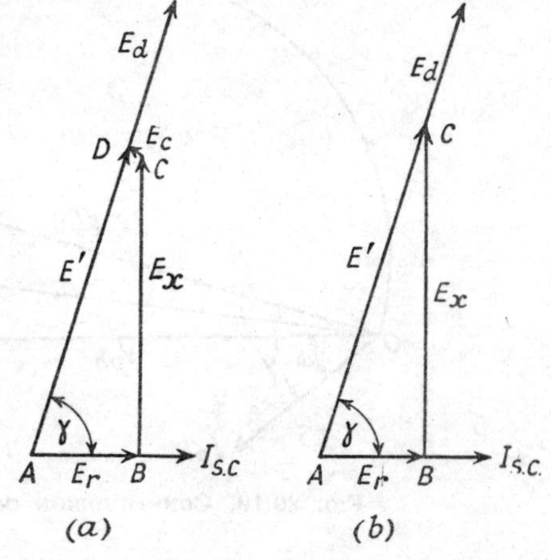

Fig. 20.21. Vector Diagrams for Short-circuit Conditions

have (*a*) to produce the flux which induces the e.m.f. E', and (*b*) to overcome the whole of the armature ampere-turns which, if they could act alone and produce a flux would set up an e.m.f. equal and opposite to E_d. Denoting the short-circuit current by I_{sc}, we have

$$E' = E_d + I_{sc}Z_a$$

where Z_a is the true impedance of the armature winding per phase.

If the excitation during short circuit is kept small enough to limit I_{sc} to the order of the normal current then the total m.m.f. will be very small and the iron paths unsaturated, so that we can write, in terms of the field ampere-turns AT_f

$$E' = k_1AT_f \text{ where } k_1 \text{ is a constant}$$

$$\therefore\ k_1AT_f = E_d + I_{sc}Z_a$$

But $\quad E_d = k_2I_{sc}$, where k_2 is another constant

$$\therefore\ I_{sc} = AT_f[k_1/(k_2 + Z_a)]$$

Thus the short-circuit current is, over the working range, proportional to the field ampere-turns, showing that the "short-circuit characteristic," the graph of I_{sc} against AT_f is a straight line through the origin. From the vector diagram of Fig. 20.21 we see that the voltage E_d and E' are in phase with one another and therefore, since γ is nearly a right angle E_x and E_d are very nearly in phase with one another. Hence if we imagine the true leakage reactance to be increased to such a value that the reactive drop is $(E_x + E_d)$, and if we also neglect the effect of armature reaction, then, under short-circuit conditions this fictitious machine will have the same performance as the actual machine. The reactance expressed by

$$X_s = (E_x + E_d)/I_{sc}$$

is called the synchronous reactance, and the impedance

$$Z_s = \sqrt{(R_{ph}^2 + X_s^2)}$$

the synchronous impedance.

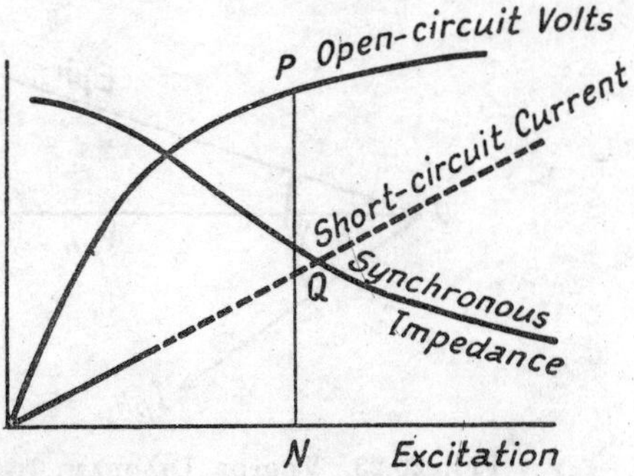

FIG. 20.22. ALTERNATOR CHARACTERISTICS

The voltage regulation can be predetermined experimentally, one method being to determine the open-circuit characteristic of induced volts against exciting current, or excitation ampere-turns, and the short-circuit characteristic of short-circuit current against exciting current. For this latter characteristic the terminals are short-circuited, one short circuit being through an ammeter, the machine run at normal speed and the excitation gradually increased until the stator current is, say 50 per cent above normal. Graphs similar to those of Fig. 20.22 are then obtained, the dotted line showing how the experimental characteristic has to be projected into the normal region. For any point P, we have

$$Z_s = PN/QN, \text{ where } PN \text{ is in volts and } QN \text{ in amperes}$$

where Z_s is now a combined effect called, at one time, the synchronous impedance.

This is the basis of the Behn-Eschenberg method of pre-determining the voltage regulation from the results of open-circuit and short-circuit tests. Draw a vector OA to represent the terminal p.d. on load (Fig. 20.23). Draw AI to represent the phase of the load current, I_{ph}, the load power factor being $\cos\phi$. Hence mark off $AB = R_a I_{ph}$. Erect $BC = X_s I_{ph}$ to represent the drop due to the synchronous reactance. Then OC represents the induced e.m.f. per phase, and

$$\text{Percentage regulation up} = 100 \times (OC - OA)/OA]$$

This simple method has several serious disadvantages—

(1) The armature reaction m.m.f. is only completely direct at zero power-factor. If the power-factor is gradually increased to unity the

direct component decreases gradually to zero while the cross-magnetizing component increases gradually from zero. Thus the magnitude of Z_s is only correct under the experimental condition of its determination.

(2) During the short-circuit test the iron paths are unsaturated, but under normal operating conditions much of the iron will be worked beyond the knee of the magnetization curve.*

Nevertheless, the conception of a constant impedance is valuable

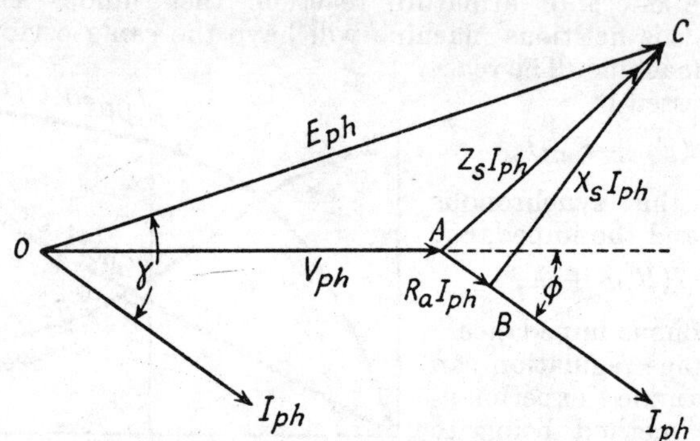

FIG. 20.23. VECTOR DIAGRAM, BEHN-ESCHENBERG METHOD

for the elucidation of certain operating characteristics. From Fig. 20.23 we have

$$E_{ph}^2 = V_{ph}^2 + (Z_s I_{ph})^2 - 2V_{ph} \times Z_s I_{ph} \cos [180 - (\psi - \phi)]$$

In general Z_s is so nearly equal to X_s that the resistance drop AB, can be neglected, the equation simplifying to

$$\begin{aligned} E_{ph}^2 &= V_{ph}^2 + (Z_s I_{ph})^2 - 2V_{ph} \times Z_s I_{ph} \cos (90 + \phi) \\ &= V_{ph}^2 + (Z_s I_{ph})^2 + 2V_{ph} Z_s I_{ph} \sin \phi \end{aligned}$$

Now $E_{ph} = Z_s I_{sc}$

where I_{sc} is the short-circuit current with normal excitation

$$\therefore (Z_s I_{sc})^2 = V_{ph}^2 + (Z_s I_{ph})^2 + 2V_{ph} Z_s I_{ph} \sin \phi$$

$$\therefore 1 = (V_{ph}/E_{ph})^2 + 2(V_{ph}/E_{ph})(I_{ph}/I_{sc}) \sin \phi + (I_{ph}/I_{sc})^2$$

The variables in this equation are V_{ph}, I_{ph}, and ϕ, and by taking different values for I_{ph} a family of V_{ph}/I_{ph} curves can be drawn, each curve corresponding to a different value of ϕ. It will be seen that whatever the value of ϕ all the curves will have the same intercept on the voltage axis, namely E_{ph}, and the same intercept on the current

* Harrison and Jones have shown that it is possible to run an alternator at low speeds under such conditions that the m.m.f. and flux distributions of a given full-load case are reproduced exactly. (D. Harrison and C. V. Jones. "A new method of predetermining the regulation of alternators at unity and lagging power factors." *J.I.E.E.* **95**, Pt. II, (1948) 374.)

axis, namely I_{sc}, the short-circuit current. A typical family of such voltage characteristics is given in Fig. 20.24, and we see that the terminal volt drop depends on ϕ as well as the magnitude of I_{ph}.

(c) *The m.m.f. Method.* Like the synchronous impedance method, this method necessitates the determination of the open-circuit and

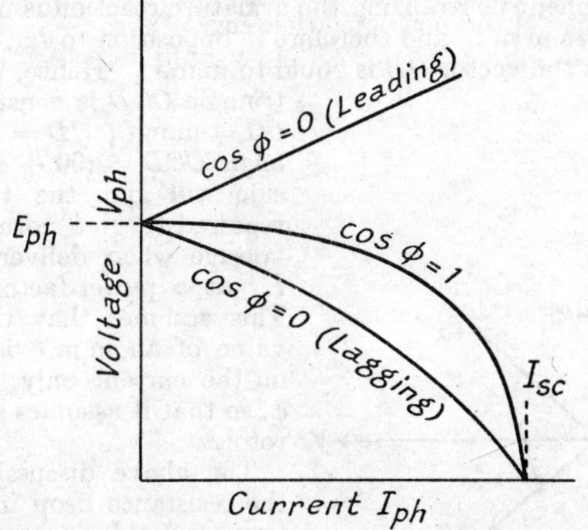

FIG. 20.24. FAMILY OF VOLTAGE CHARACTERISTICS

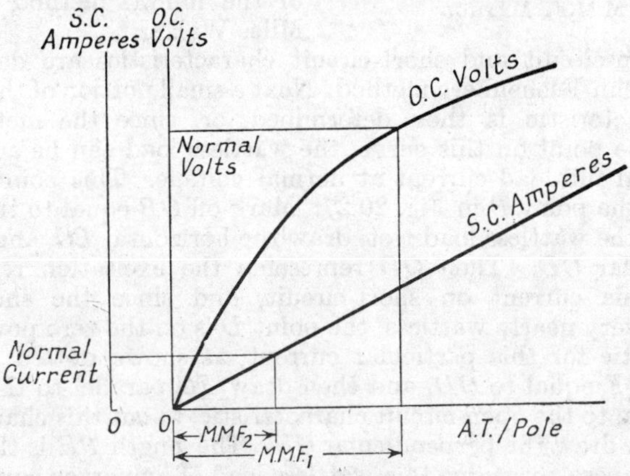

FIG. 20.25. M.M.F. METHOD

short-circuit characteristics. From Fig. 20.25 the m.m.f. per pole required to produce normal voltage on open circuit is m.m.f.$_1$, while that required to produce normal current on short-circuit is m.m.f.$_2$. This latter m.m.f. has to induce the impedance drop in the armature and also neutralize the armature reaction m.m.f. The m.m.f.$_2$ is in quadrature leading the induced voltage during the s.c. test. The two components of m.m.f.$_2$ are as follows. (i) The component which produces the flux necessary to induce the impedance drop E_x (neglecting the armature resistance drop) in quadrature, leading, with this drop. But E_x is also in quadrature leading with I_{ph}, so that this component of m.m.f.$_2$ is in phase opposition to I_{ph}, as indicated by OA in Fig. 20.26. (ii) The component neutralizing the armature reaction is in opposition to the armature m.m.f., and therefore in opposition to I_{ph}, as indicated by AB. Thus the vector OB is equal to m.m.f.$_2$. Hence, if the m.m.f. triangle OCD is constructed with OC = m.m.f.$_1$, CD = m.m.f.$_2$, and angle $OCD = (90 + \phi)$, the third side will give the total m.m.f. required to give normal terminal voltage when delivering current I_{ph} at a power-factor of cos ϕ. This assumes that the effective value of an m.m.f. is dependent on the current only, and not on ϕ, so that it assumes a cylindrical rotor.

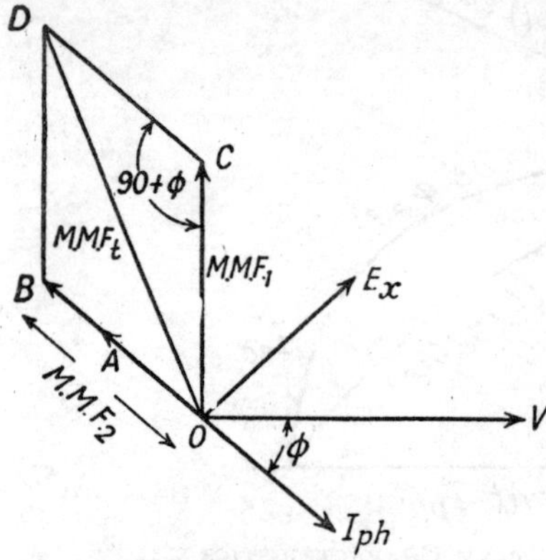

Fig. 20.26. Vector Diagram for M.M.F. Method

The above discussion neglects the resistance drop in the armature, and this is usually permissible. A convenient modification of the m.m.f. method is due to Miles Walker.*

The open-circuit and short-circuit characteristics are determined, as in the Behn-Eschenberg method. Next a small portion of the wattless load characteristic is then determined, or, since the method only requires one point on this curve, the wattless load can be adjusted to give normal full-load current at normal voltage. This point is represented by the point P in Fig. 20.27. Mark off OB equal to the current flowing in the wattless load test, draw the horizontal BC, and then the perpendicular CD. Then OD represents the excitation required to produce this current on short-circuit, and since the short-circuit current is very nearly wattless, the point D is on the zero power-factor characteristic for this particular current, as shown dotted. From P mark off PF equal to OD, and then draw FG parallel to the tangent at the origin to the open-circuit characteristic, to cut this characteristic at G. Then draw the perpendicular GH. The length PH is the demagnetizing ampere-turns due to a wattless load of amperage equal to that in the wattless current test, while GH is the reactance drop per phase in the armature. Hence, GH measured in volts, divided by the current, gives the reactance per phase of the machine. The excitation for the same current loading during short-circuit is OD, hence PH subtracted from OD (= FH) gives the excitation required to supply the reactance drop in the machine on short-circuit. The regulation is now determined

* *Specification and Design of Dynamo-electric Machinery*, p. 287, Longmans, London, 1917.

as follows. Draw OA, Fig. 20.28, to represent the normal voltage per phase, and set off AB equal to the reactance drop, and inclined ϕ to the perpendicular to OA; $\cos\phi$ is the power factor of the load at which

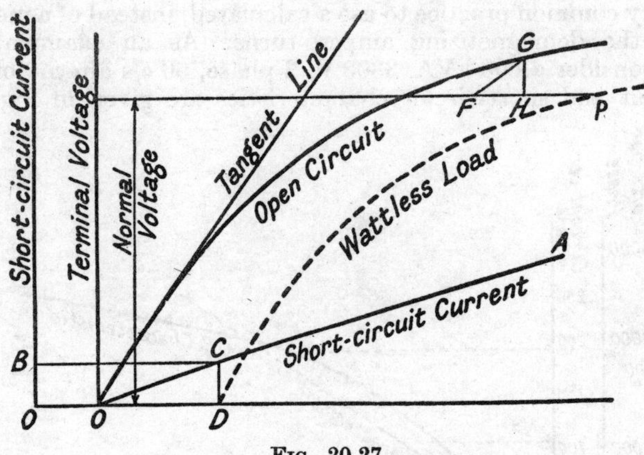

FIG. 20.27

the regulation is required. Next set off the perpendicular BC equal to the resistance drop per phase, Joining OC we obtain the e.m.f. generated per phase and, referring back to the open-circuit characteristic, we obtain the excitation required to generate this voltage. Mark this

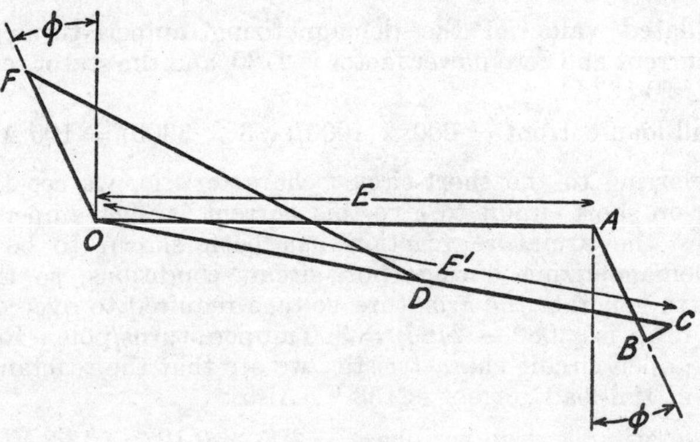

FIG. 20.28

excitation off along the direction of OC; let it be OD. Then mark off OF equal to the demagnetizing ampere-turns, or the equivalent exciting current (whichever is plotted in Fig. 20.27) as shown, and complete the triangle OFD. Then FD is the excitation required to maintain the normal terminal voltage at the load current used in the construction, and at a power factor equal to $\cos\phi$. Referring back to the open-circuit characteristic, the voltage corresponding to this excitation is read off. Calling this E_o, we have for the regulation up

$$\text{Percentage regulation up} = [(E_o - E)/E] \times 100$$

It is very common practice to use a calculated, instead of a measured, value of the demagnetizing ampere-turns. As an example of the method, consider a 600-kVA, 3300 V, 3-phase, 50 c/s alternator whose open-circuit and short-circuit characteristics are given in Fig. 20.29.

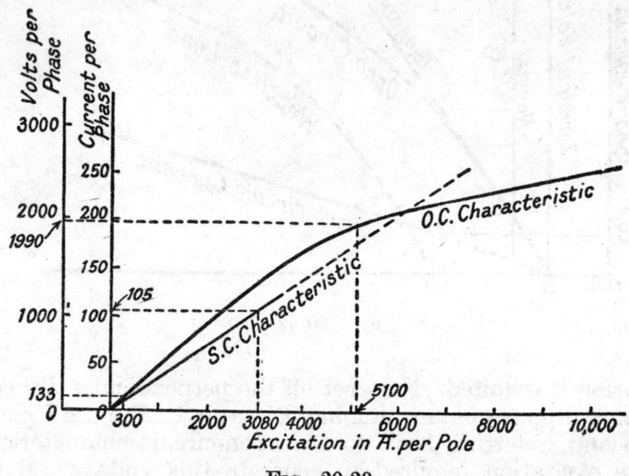

Fig. 20.29

The calculated value of the demagnetizing ampere-turns/pole at full-load current and zero power factor is 2780, and the stator resistance per phase is 0·127 Ω.

$$\text{Full-load current} = 600 \times 1000/(\sqrt{3} \times 3300) = 105 \text{ A}$$

and on referring to the short-circuit characteristic, we see that the excitation on short-circuit to give this current is 3080 ampere-turns/pole. Now the armature reaction has been shown to be almost entirely demagnetizing during short-circuit conditions, so that the excitation to generate the armature voltage required to overcome the reactance drop is (3080 − 2780) = 300 ampere-turns/pole. Referring this to the open-circuit characteristic, we see that the reactance drop per phase at full-load current is 133 V. Also

$$\text{Resistance drop per phase} = 105 \times 0{\cdot}127 = 13{\cdot}3 \text{ V}$$

and normal terminal volts per phase $= 3300/\sqrt{3} = 1910$ V

There is now sufficient data to make use of the construction of Fig. 20.28. The combined voltage and ampere-turn diagram is drawn to scale in Fig. 20.30 for a load current of the normal value of 105 A/phase, at a power factor of 0·8 lagging.

$$\therefore \phi = \cos^{-1} 0{\cdot}8 = 37° \text{ (very approx.)}$$

The diagram gives a generated voltage of 1990 V/phase, which by reference to the open-circuit characteristic, requires an excitation of 5100 ampere-turns. The demagnetizing ampere-turns of 2780 are set off along OF, as explained previously, the resultant excitation therefore being 7240 ampere-turns/pole. This corresponds to a voltage per phase of 2240 V when the load is thrown off, the regulation up, therefore, being

$$\frac{2240 - 1910}{1910} \times 100 = 17{\cdot}3 \text{ per cent}$$

Separation of Armature Reaction and Leakage Reactance Effects. This experimental method, due to Potier, necessitates the determination

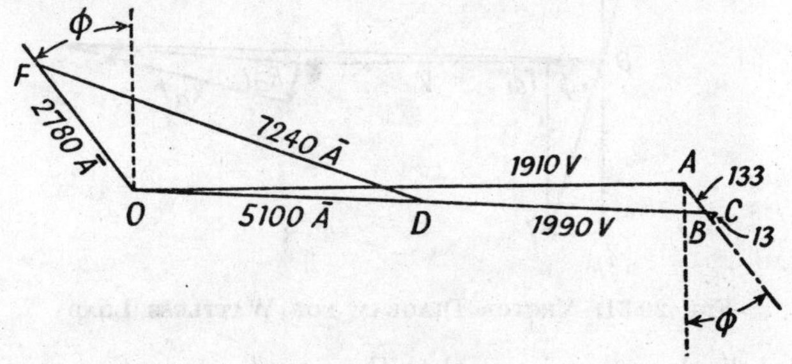

Fig. 20.30

of the open-circuit characteristic, and also the wattless load characteristic. This latter is the curve of terminal volts against excitation when the armature is delivering full-load current at zero power factor. Actually, this characteristic is very close to the characteristic for very low power factor, say not greater than 0·2, so that in practice the machine can be loaded by means of reactors with air gaps in the magnetic circuit so as to make the magnetizing current as large as possible. An alternative method is to supply an idle running synchronous motor whose excitation is reduced to as low a value as possible, without the machine actually falling out of step.

The vector diagram for wattless load conditions is shown in Fig. 20.31. The current I lags the terminal p.d. V by an angle ϕ nearly equal to 90°. Adding the resistance drop and reactance drop to this we obtain E the total e.m.f. per phase, and since ϕ is very nearly equal to 90° we have

$$E \simeq V + X_a I$$

From the m.m.f. triangle OCD, we have

$$\text{m.m.f.}_t \simeq \text{m.m.f.}_1 + \text{m.m.f.}_2$$

where m.m.f.$_1$ is that required to induce the terminal voltage V, and m.m.f.$_2$ that required to induce the drop $X_a I$ Now in both the o.c. and wattless-current characteristics volts are plotted as ordinates and excitation (m.m.f.) as abscissae. It therefore follows that two points P and Q on the o.c. and wattless load characteristics respectively for which the induced e.m.f. is the same will be so related that their

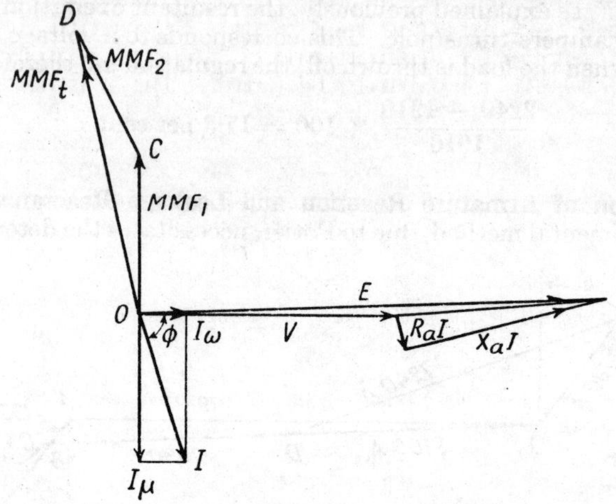

FIG. 20.31. VECTOR DIAGRAM FOR WATTLESS LOAD

horizontal displacement is equal to the armature reaction m.m.f.,$_1$ and their vertical displacement the leakage reactance drop. Thus, in Fig. 20.32, RQ is the armature reaction m.m.f. and PR the reactance drop.

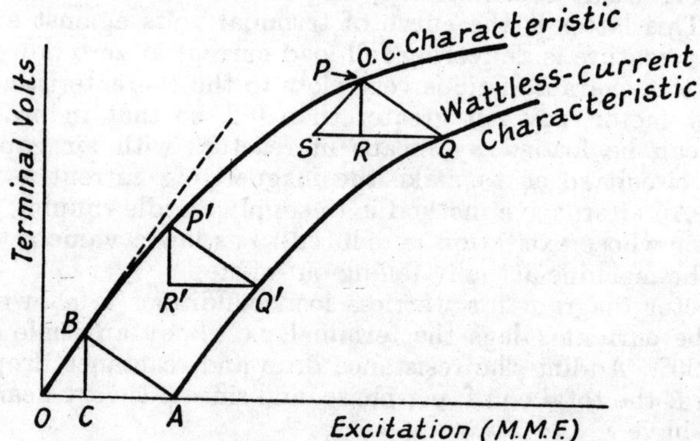

FIG. 20.32. ILLUSTRATING THE POTIER METHOD OF SEPARATING LEAKAGE REACTANCE AND ARMATURE REACTION EFFECTS

If a parallel $P'Q'$ is drawn in any position, the new triangle $P'R'Q'$ is equal to PRQ. In particular, if AB is drawn parallel to PQ and the perpendicular dropped from B on to the excitation axis, we have $CA = RQ$, and $BC = PR$, and therefore, since point A is given by the field m.m.f. required to circulate full-load current when the armature is short-circuited (since the terminal p.d. is zero), the triangle PRQ can be constructed from the determination of a single point Q on the wattless current curve. The procedure is as follows: From Q draw a horizontal $QS = AO$, and from S draw SP parallel to the initial portion OB of the o.c. characteristic. This gives P, from which the triangle PRQ can immediately be drawn. The reactance drop PR divided by the current flowing during the wattless-current test gives the leakage reactance per phase.

Transient Reactance of Synchronous Machines. We have seen that during normal steady-state operation there is an m.m.f. balance between the field m.m.f., demagnetizing component of the armature reaction m.m.f., and cross-magnetizing component. In addition there is an e.m.f. balance between the induced e.m.f., the terminal p.d. and the volt drops due to resistance and leakage reactance. If the impedance of the load is gradually reduced then there will be time for both of these balances to be maintained. If the machine is finally short-circuited the whole of the effective armature m.m.f. will be directly demagnetizing and the resultant m.m.f. will induce an e.m.f. equal to the internal voltage drop. Now suppose that a short-circuit is applied suddenly, then the armature m.m.f. will increase suddenly and, as the short-circuited armature is in the nature of a highly inductive circuit, this m.m.f. will be almost entirely demagnetizing. Owing to the very high inductance of the field winding, and its high ratio L/R, which in a large alternator may be of the order of 5 to 10, the armature m.m.f. cannot cause a sudden reduction in flux, but can only bring this about gradually. Hence, initially the short-circuit current is limited mainly by the leakage reactance. What the armature m.m.f. does do is to force more of the main flux to take field leakage paths (not to be confused with the paths of the armature leakage fluxes) so that there is a certain reduction in the induced e.m.f. with consequent reduction in the initial value of the short-circuit current. If the pole faces have damping windings, or if the poles are solid, the transformer action due to the rapid flux changes will tend to delay these changes and so to prolong the high initial value of the current.

Just as the effect of armature reaction and true leakage reactance can be combined to give the synchronous reactance, so also the effects of armature leakage flux and increased field leakage flux can be combined to give what is called the transient reactance.*

We have seen that the initial establishment of the current in an inductive circuit is associated with a rapidly diminishing d.c. component whose initial value is only zero if the circuit is closed at the instant the e.m.f. is a maximum. If a fully-excited alternator is short-circuited at the corresponding instant, then, owing to the above

* *See* R. H. Park and B. L. Robertson, "The reactances of synchronous machines," *Trans. A.I.E.E.*, **47**, No. 2 (1928) 514; B. Adkins, "Transient theory of synchronous generators connected to power systems," *J.I.E.E.*, **98**, Pt. II (1951) 510; R. E. Doherty and O. E. Shirley, "Reactance of synchronous machines and its applications," *Trans. A.I.E.E.*, **37**, Pt. 11 (1918) 1209; L. A. Kilgore, *loc. cit.*

phenomenon, the current rises suddenly to a high value and then dies down progressively to the value it would have if the machine were short-circuited gradually. The comparison of the two cases is illustrated in Fig. 20.33. If the envelopes of the curves, as determined by oscillographs, are drawn, it will be found that smooth curves will give intercepts OA, OA, but that the decrement of the first two, or perhaps

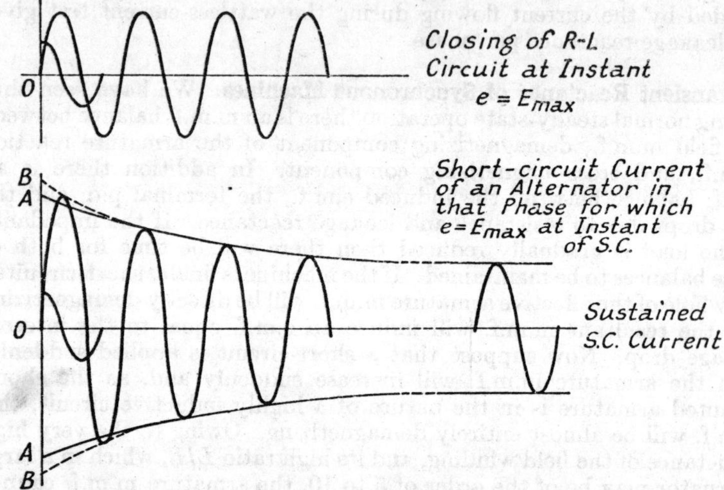

FIG. 20.33. SYMMETRICAL SHORT-CIRCUIT OF AN ALTERNATOR

three, cycles is so high that these require separate initial envelopes, as shown dotted, giving intercepts OB, OB. The ratio

$$\frac{\text{Max. value of excitation e.m.f.}}{OA \text{ (amperes)}} = \text{Transient Reactance}$$

and $$\frac{\text{Max. value of excitation e.m.f.}}{OB \text{ (amperes)}} = \text{Sub-transient Reactance}$$

The sub-transient reactance is due to closed field circuits such as damping windings and, in fact, any metal subjected to the effect of the field flux. This effect dies away rapidly, and the rest of the characteristic is due to the gradual reduction of main flux to that corresponding to the sustained steady conditions. It is for such reasons that the appropriate reactance has to be used in the calculation of the various symmetrical components of a short-circuit current.

FURTHER READING

The works recommended on p. 470.

CONCORDIA, C. *Synchronous Machines, Theory and Practice.* Wiley, 1951.

LAURENCE, R. R. and RICHARDS, H. E. *Principles of Alternating Current Machinery.* McGraw-Hill, 1953.

MCFARLAND, J. C. *Alternating Current Machines.* Van Nostrand, 1948.

SAH, PEN-JUNG. *Fundamentals of Alternating Current Machines.* McGraw-Hill, 1948.

EXAMPLES ON CHAPTER 20

1. A 3-phase synchronous machine, 2000 kVA, 11,000 V, has a two-layer winding housed in 6 slots per pole, 2 conductors per slot, with full pitch coils. The rotor is cylindrical. Draw the armature m.m.f. waves for the instants at which the crest value has its greatest and best values. Hence calculate the armature ampere-turns per pole at full load. *Ans.* 553 AT/pole.

2. A 750 kVA, 5000 V, 50 c/s, 3-phase, 250 r.p.m. alternator, has the following particulars: armature diameter, 230 cm; iron length of armature core, 55 cm; slot dimensions, $h_1 = 0$, $h_2 = 0{\cdot}4$ cm, $h_3 = 0{\cdot}3$ cm, $h_4 = 3{\cdot}8$ cm, $w_0 = w_s = 2{\cdot}0$ cm; slot pitch = 5·02 cm. Slots per pole per phase = 2; conductors per slot = 10 (single circuit winding). Calculate the inductance per phase due to (*a*) slot leakage, (*b*) tooth head leakage.

Ans. (*a*) $2{\cdot}72 \times 10^{-4}$ H; (*b*) $2{\cdot}5 \times 10^{-4}$ H.

3. Given that the inductance of the machine in question 2 due to overhang leakage is $1{\cdot}2 \times 10^{-4}$ H per phase, calculate the reactive drop per phase at full-load current. *Ans.* 165 V.

4. The open-circuit characteristic of a 3-phase, 50 c/s synchronous machine is given by—

Exciting current (amperes)	20	40	60	84	105	123
Terminal e.m.f. (volts)	850	1700	2460	3000	3300	3600

Determine the excitation necessary for full-load operation at 0·9 power factor, leading, on a 3000-V supply. When short-circuited and driven at normal speed an excitation of 50 amperes gives normal full-load stator current. The resistance drop is 2 per cent and the leakage reactance drop 15 per cent. (L.U.)

Ans. 76 A.

5. Deduce the exciting current for a 3-phase, 3300-V generator when supplying 1000 kW at 0·8 power factor, lagging, given the magnetization curve as open circuit—

Line voltage	3300	3600	3900
Exciting current (amperes)	80	96	118

There are 16 poles, 144 slots, 5 conductors per slot, single circuit full-pitched winding, star-connected. The stator winding has a resistance per phase of 0·15 ohm, and a leakage reactance of 1·2 Ω. The field coils have each 108 turns. (L.U.)

Ans. 124 A.

6. The following figures give the open-circuit and full-load zero power factor saturation curves for a 15,000 kVA, 11,000-V, 3-phase, 50 c/s, star connected turbo-alternator—

Field ampere-turns in 10^3	10	18	24	30	40	45	50
O.C. line kilovolts	4·9	8·4	10·1	11·5	12·8	13·3	13·65
Zero p.f. full-load Line kilovolts	—	0	—	—	—	10·2	—

Find the armature reaction, the armature reactance, and the synchronous reactance. Deduce the regulation for full-load at 0·8 power factor lagging. (*I.E.E.*)

Ans. 15,700 AT, 0·81 Ω, 6·15 Ω, 18·8 per cent.

CHAPTER 21

THE SYNCHRONOUS MOTOR

WE have seen that a d.c. machine can function as a generator or as a motor. Imagine such a machine connected to a supply of voltage V If the induced e.m.f. E, is greater than V, then the resultant e.m.f. in the armature circuit will be

$$E_r = E - V$$

and the armature current

$$I_a = (E - V)/R_a$$

its direction in the armature corresponding to the direction of E. If E is less than V, we have

$$E_r = E - V, \text{ and } I_a = (E - V)/R_a$$

as before, but the numerator will be negative, and the current therefore reversed, so that, in the armature circuit it is now in opposition to E.

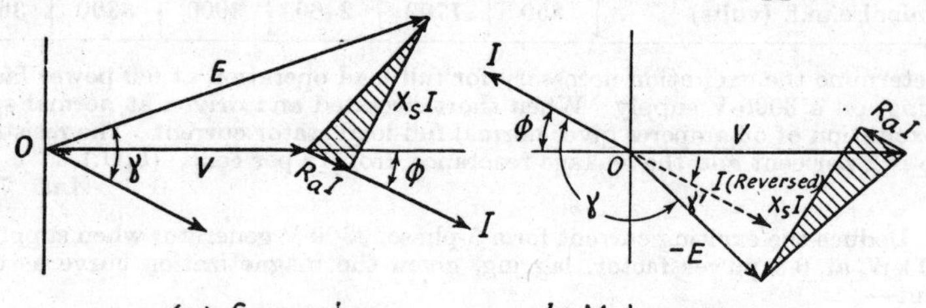

FIG. 21.1. VECTOR DIAGRAMS OF SYNCHRONOUS MACHINES
(a) Generating. (b) Motoring.

Thus, in general, a d.c. machine functions as a motor when the armature current flows in opposition to the induced e.m.f.

In the case of the a.c. synchronous machine, that is the machine with d.c. excitation of the rotor we have seen that the phase of the current can vary with respect to that of the induced e.m.f. If the current has a component in phase with the e.m.f. then the machine generates, but if it has a component in opposition, then the machine acts as a motor. Consider the vector diagrams of Fig. 22.1. Diagram (*a*) is the Behn-Eschenberg regulation diagram we have used before. For motor action, and for the same power factor the current vector is reversed, giving diagram (*b*). In the generator case the internal phase angle $\gamma < \pi/2$, in the motor case it is $> \pi/2$, and therefore if we reverse the current vector, the phase of this with respect to $-E$ is $\gamma' \ngtr \pi/2$. We can use this diagram to determine the nature of the space distribution of current with respect to that of air-gap flux in the two cases. Fig. 21.2 (*a*) shows a rotating armature machine: irrespective of whether the regime is generating or motoring, motion of the conductors to the right will lead to induced e.m.f.s acting inwards with conductors under a north pole. Hence in the belt of conductors AB the induced e.m.f.s, at the instant shown, will all be inwards. For generator action the current lags the induced e.m.f. by the angle γ and therefore this same belt of conductors must advance to the position $A'B'$, γ degrees in front of AB, before the current distribution in them corresponds to the e.m.f. distribution in Fig. 21.2 (*a*). Thus the current distribution *in space* is $\gamma°$ in advance of the e.m.f. distribution in space. Using the left-hand rule, as though the conductors were actually exposed to the main flux, as in a smooth-cored machine, we see that the tangential forces acting on the conductors are in opposition to motion in the case of those which carry inward current, and in the same direction as motion in the case of those which carry outward current. The former predominate and we have the expected result that the resultant torque acts in opposition to motion. To convert the conditions to those of a rotating field machine suppose an equal and opposite motion to the left, the field then moving to the left at synchronous speed but the armature being stationary. The relative motion is identical and therefore the field m.m.f. of a rotating field machine leads the armature m.m.f. by the angle $(\pi/2 - \gamma)$.

Now consider the motoring case: the reversed current leads the induced voltage by the angle γ' and therefore we reverse the armature current distribution and displace it γ' with respect to the induced e.m.f. distribution. This gives Fig. 21.2 (*c*) from which we see that the attraction between main field and armature now results in a torque in the same direction as motion, that is, a motoring torque. This is also shown by the contributions of the individual conductors, the greater of the two opposing torques being in the same direction as rotation. Using the same artifice to simulate the conditions of a rotating field machine we see that in the motor the field m.m.f. lags the armature m.m.f. with respect to motion by the angle $(\pi/2 - \gamma')$.

We can thus draw gap flux diagrams similar to those used for d.c. machines to demonstrate the differences between the two cases. This is done in Fig. 21.3.

Torque—Single-phase Machine. Consider first of all a single-phase machine, and let the field be rotating. Some of the rotor periphery may be left unfilled by conductors, so that the armature current distribution will be a rectangular block for each pole, the width of the block being fixed by the number of unfilled slots. This is illustrated in Fig. 21.4. As the field rotates this distribution will remain stationary in space, but the amplitude will vary sinusoidally with respect to time, the current distribution thus being in the nature of a stationary wave of rectangular shape. By Fourier analysis we can determine the sinusoidal components of this, and we will consider the fundamental only. This is a stationary wave of sinusoidal shape and amplitude

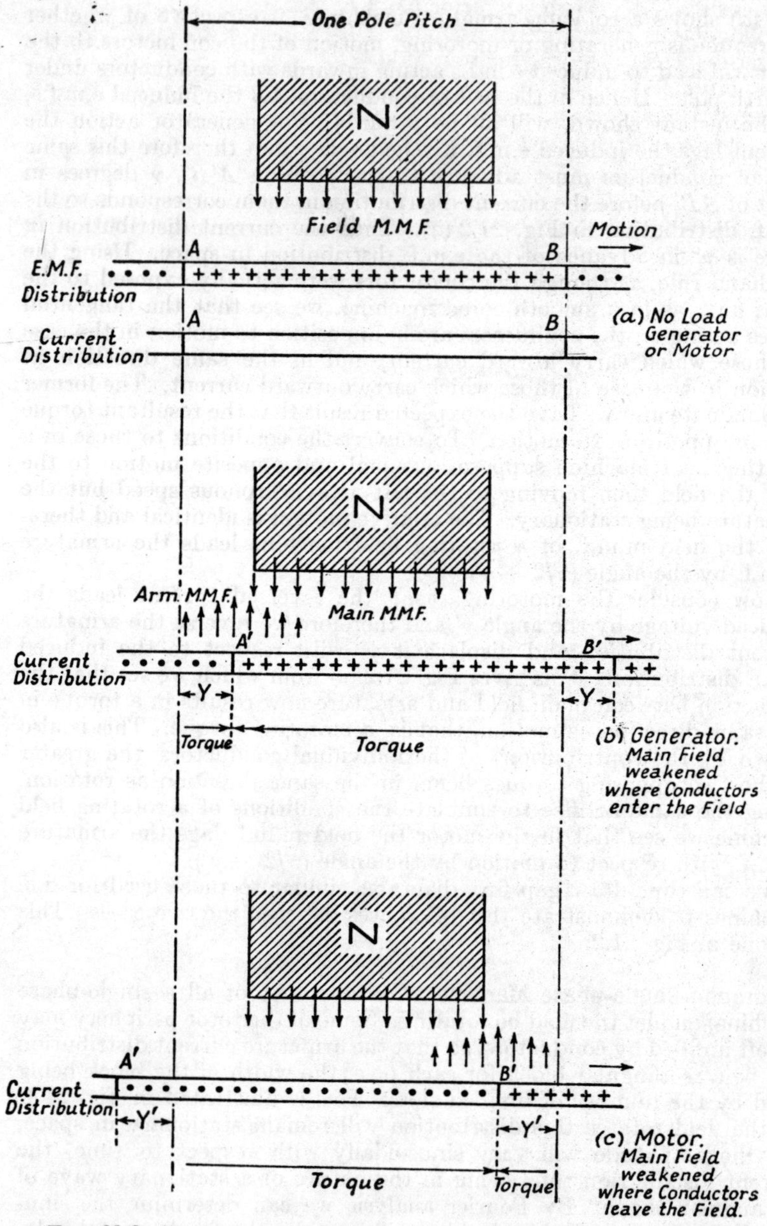

FIG. 21.2. MODES OF OPERATION OF THE SYNCHRONOUS MACHINE

$4I_{max} \cos \alpha/\pi$. By Ferrari's principle we can resolve this into two travelling waves, each of constant amplitude $2I_{max} \cos \alpha/\pi$, but rotating in opposite directions at synchronous speed.

The fundamental action of all dynamo-electric machines is the reaction of the magnetic flux set up by the field system on the current

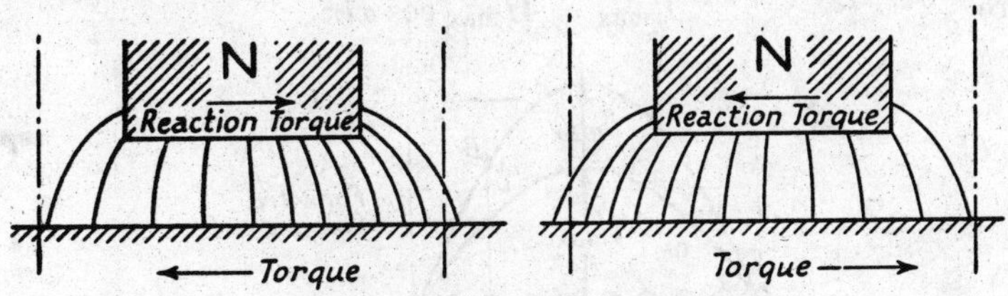

Fig. 21.3. Flux Distortion in a Synchronous Machine in Relation to its Mode of Operation

distribution in the armature winding. This reaction takes the form of a torque which must be unidirectional and this requires that the magnetic field and the armature current distribution must rotate in the same direction *and at exactly the same speed*, that is, they must keep

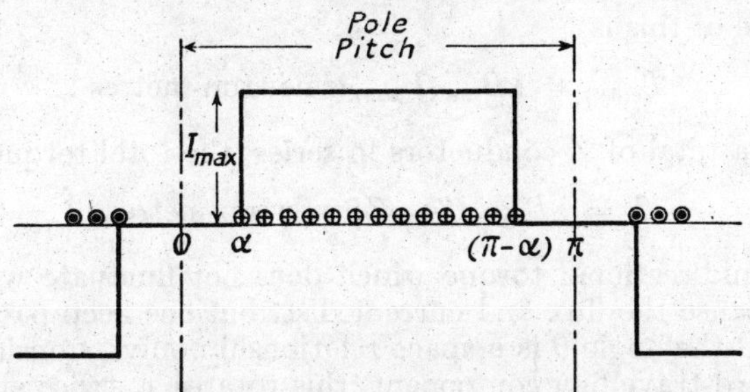

Fig. 21.4. Current Distribution of Single-phase Armature at Moment of Maximum Current

pace with one another. In the d.c. machine with its fixed field system it happens that speed of rotation of both magnetic field and armature current distribution is zero. This requirement indicates two things: (*a*) the backward rotating component of the current distribution does not set up the working torque, (*b*) the rotor must travel at exactly the same speed as the forward travelling component, that is, at synchronous speed. Consider then the forward-travelling component: the current distribution of this component (but not of the actual winding) is sinusoidal. Let the flux-density distribution be sinusoidal also, then, for unity power factor these two distributions will be in space phase, as shown in Fig. 21.5. Let angular distance be reckoned from the pole centre, then we can write

$$B = B_{max} \cos \theta$$

and $$i = I'_{max} \cos \theta$$

where $$I'_{max} = 4I_{max} \cos \alpha/\pi$$

FIG. 21.5. FLUX AND CURRENT RELATIONSHIP AT UNITY POWER-FACTOR

The torque due to a conductor at angular position θ, in a machine of conductor length l and radius r metres is

$$\begin{aligned} T_c &= Bilr \text{ newton-metres} \\ &= B_{max}I'_{max}lr \cos^2 \theta \\ &= \tfrac{1}{2}B_{max}I'_{max}lr(1 + \cos 2\theta) \end{aligned}$$

The average of this is

$$T_{c(av)} = \tfrac{1}{2}B_{max}I'_{max}lr \text{ newton-metres}$$

so that for a total of Z conductors in series, the total torque is

$$T = \tfrac{1}{2}B_{max}I'_{max}Zlr \text{ newton-metres}$$

This is a unidirectional torque which does not fluctuate with respect to time because the flux and current distributions keep pace with one another and the angle θ is a space relationship only. Considering now the backward-travelling component, this rotates at twice synchronous speed with respect to the flux and consequently it will set up an alternating torque of maximum value T and frequency $2f$. The total torque thus varies with time in accordance with the expression

$$T = \tfrac{1}{2}B_{max}I'_{max}Zlr(1 + \cos 2\omega t)$$

This shows that in the single-phase machine the torque fluctuates with a frequency of twice the supply frequency, and that periodically it falls to zero. In this respect the torque of a single-phase synchronous motor operating at unity power factor is analogous to the power in a single-phase circuit at unity power factor.

In the general case of an internal phase angle γ, lagging, the two components of the current distribution have a space phase of γ, with respect to the stationary wave, Fig. 21.6, so that, considering the forward component alone we have

$$T_c = B_{max}I'_{max}lr \cos\theta \,.\, \cos(\theta - \gamma)$$
$$= \tfrac{1}{2}B_{max}I'_{max}lr\,[\cos\gamma + \cos(2\theta - \gamma)]$$

The average of this over a pole pitch is

$$T_{c(av)} = \tfrac{1}{2}B_{max}I'_{max}lr \cos\gamma$$

Hence for Z conductors in series

$$T = \tfrac{1}{2}B_{max}I'_{max}\,Zlr \cos\gamma \text{ newton-metres}$$

Combining this with the torque due to the backward-travelling component we have for the total torque with respect to time

$$T = \tfrac{1}{2}B_{max}I'_{max}Zlr \cos\gamma\,[1 + \cos(2\omega t - \gamma)]$$

Thus in the general case the torque of a single-phase motor is analogous to the power in a single-phase circuit at a power factor of $\cos\gamma$.

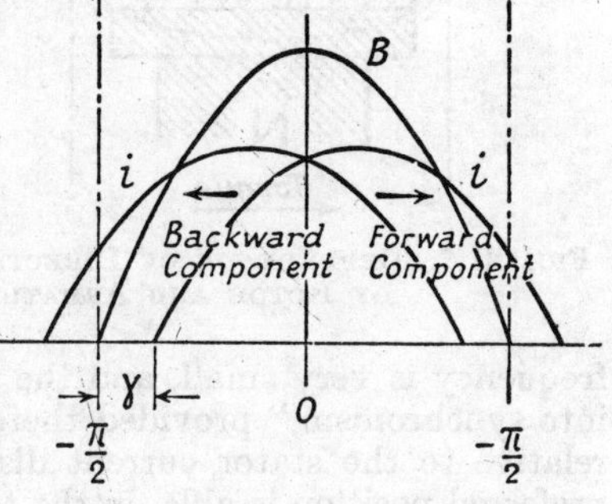

FIG. 21.6. FLUX AND CURRENT RELATIONSHIP

Three-phase Machine. For the three-phase machine we can consider the three current distributions separately. At any particular instant we see, from considerations similar to those of p. 362, that the three forward components are in space phase, whereas the backward components have a relative displacement of $2\pi/3$ electrical radians and therefore cancel out. The great advantage of the three-phase motor—or any motor having a number of phases greater than one, in fact—is therefore that the torque is uniform. For a phase spread of $2\pi/3$ radians the angle α is $\pi/6$, so that the amplitude of the fundamental of the stationary wave due to one phase alone is

$$4I_{max}\cos(\pi/6)/\pi = (2\sqrt{3}/\pi)I_{max}$$

Each travelling wave has an amplitude of one-half of this, and therefore the amplitude of the equivalent travelling wave, reckoning all three phases is $I'_{max} = 3\sqrt{3}I_{max}/\pi$.

General Properties of the Motor. Let the rotor be stationary, then, in the case of the single-phase machine it will be acted on by two equal and opposite travelling waves of current, the starting torque thus being zero. In the case of the three-phase machine there is only a forward-travelling wave of current. This wave will pass any given pole at synchronous speed, so that the direction of the torque set up will alternate according to the polarity of the wave. This is illustrated in Fig. 21.7, from which we see that the starting torque of the three-phase machine is an alternating torque, its average value thus being zero. If the rotor is unexcited then the flux will be that set up by the armature currents. Either a north or a south pole of this flux will attempt to grip a salient pole as it approaches it, but owing to the high relative speed of the armature flux with respect to the stationary rotor it is not possible to overcome the inertia of the latter and thus start the motor. Consequently, neither the single-phase nor the three-phase motor is capable of starting from rest by synchronous motor action, and so the motor must be run up to speed by some other means.

As the speed is increased from zero, the frequency of the torque decreases until, with a speed only a little less than synchronism the

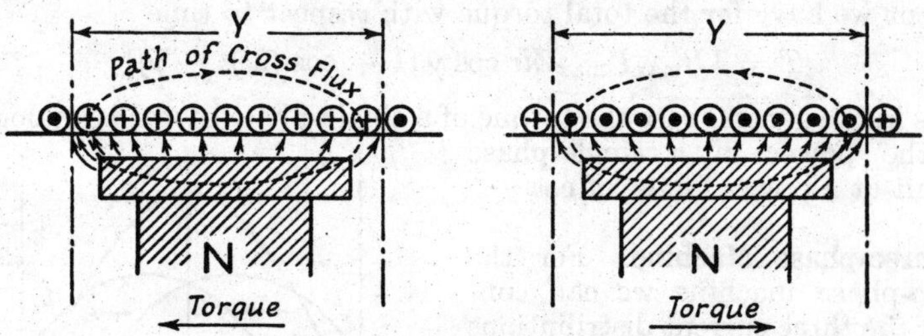

FIG. 21.7. DEPENDENCE OF DIRECTION OF TORQUE ON RELATIVE PORTIONS OF ROTOR AND ARMATURE CURRENT DISTRIBUTION

frequency is very small, and the rotor will attain full speed, or "pull into synchronism," provided there is a preferred position of the rotor relative to the stator current distribution, and if the torque at this preferred position is able, in the time available, to impart to the rotor the necessary kinetic energy. We will neglect the cylindrical rotor machine at the moment (*see* p. 540) and consider only the salient-pole machine with unexcited rotor. The most preferred position of a stator pole relative to the rotor is that of Fig. 21.8 (*A*). If the rotor speed is slightly less than synchronism then, after a very short interval of time the conditions will be those of Fig. 21.8 (*B*), in which the gap reluctance is considerably increased and the flux therefore decreased. It is clear that if the angular distance α exceeds a certain value the stator flux will leave the pole shown and will enter the next pole further ahead. Hence the condition for pulling into step, or into synchronism, is that the flux from any one stator pole shall be able to retain its hold on one rotor pole, and not slip from pole to pole.

Take the case of an 8-pole, 50 c/s motor, the synchronous speed thus being 750 r.p.m. Suppose that the starting device is able to bring the rotor up to 95 per cent of synchronism, the deficit of rotor speed below synchronism being

$$(5/100) \times 750 = 37{\cdot}5 \text{ r.p.m.}$$

The time taken for a stator pole to move forward one-half a pole pitch is 1/100 sec, and therefore the time taken for the stator pole to advance one-half a pole pitch relative to the rotor is

$$(100/5) \times (1/100) = 1/5 \text{ sec}$$

In this interval of time the rotor has to be accelerated from 712·5 r.p.m. to 750 r.p.m.

If the rotor could be brought up to 97·5 per cent of synchronism the

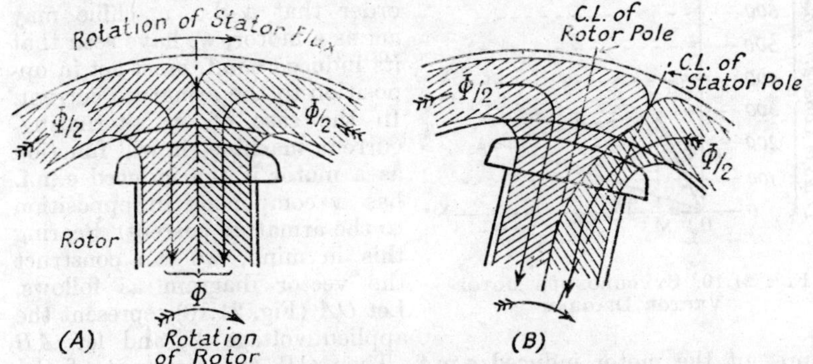

Fig. 21.8. Relative Positions of Rotor and Stator Poles

time available would be doubled and the speed would have to be increased by 18·75 r.p.m. instead of 37·5 r.p.m. This shows that the starting device should bring the rotor speed as near as possible to synchronism. When the motor has pulled into step the rotor can be excited, and the main flux then becomes the rotor flux. The armature m.m.f. then plays the normal role of armature reaction.

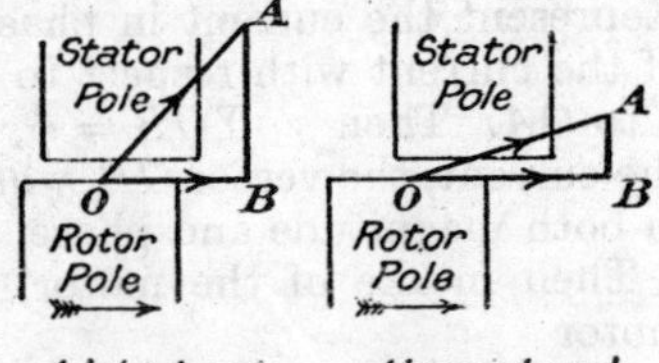

Fig. 21.9. Synchronous Motor under Varying Load

The d.c. motor and the induction motor adjust themselves to a change in load by an automatic change in speed. The synchronous motor *cannot* do this, and therefore its adjustment takes the form of a shift of the angular position of the rotor with respect to the position at which the flux enters the stator. A region where lines of force enter a mass of iron can be regarded as a south pole, and a region where they leave as a north pole, and therefore we can represent, in an elementary way, the adjustment to change in load by Fig. 21.9. The stator pole is supposed to be moving towards the right at synchronous speed. The torque is proportional to the tangential component of the total force of attraction, and therefore an increase in load must be accompanied by a falling back of the rotor pole with respect to the stator pole. It is obvious that as the load torque is gradually increased, a point will be reached at which the increasing distance between the rotor and stator poles causes the tangential component to decrease instead of increase. The motor then stops, since it can only run at synchronous speed or not at all. The torque at which this happens is called the "pull-out" torque.

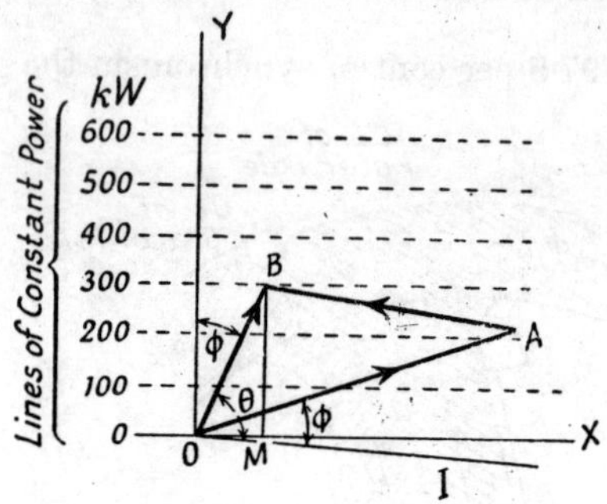

FIG. 21.10. SYNCHRONOUS MOTOR VECTOR DIAGRAM

Vector Diagram of the Motor. In order that a d.c. machine may act as a motor, we have seen that its induced e.m.f. must act in opposition to the armature current. In the case of an alternating-current machine, it will function as a motor if the induced e.m.f. has a component in opposition to the armature current. Bearing this in mind, we can construct the vector diagram as follows. Let OA (Fig. 21.10) represent the applied voltage V, and let AB represent the motor induced e.m.f. Then OB, the resultant of OA and AB, is the resultant e.m.f. acting in the armature circuit. Calling this resultant e.m.f. E_r, we have, current

$$I = E_r/Z, \text{ lagging an angle } \theta,$$

where $\tan\theta = X/R$, the various quantities referring to one phase.

Represent the current in phase by OI. Then $\angle AOI = \phi$, the phase of the current with respect to the applied voltage. Draw OY inclined θ to OA. Then $\angle YOB = \phi$, and since OB $(= E_r)$ is proportional to the current, the vector OB, when referred to OY, represents the current in both magnitude and phase.

Then intake of the motor, or intake per phase with a polyphase motor

$$P = VI\cos\phi$$
$$= V \times BM$$

where BM is the perpendicular dropped from B on to OX. But V is constant. Hence, intake is proportional to BM. Thus, if the motor is working with constant intake, the locus of the working point B is the straight line through B parallel to OX, since this line is at a constant distance from OX. Hence, we can draw a series of parallel lines on the diagram, each line corresponding to a definite value of the intake. These "power" lines are shown dotted in Fig. 21.10, and it is obvious that for equal increments of intake they will be equally spaced.

The length AB is the induced e.m.f., and since the speed is constant, this length also represents the flux per pole of the motor. We thus have

$$OA = \text{applied voltage}$$
$$AB = \text{flux per pole, a function of the excitation}$$

$$OB = \text{current}$$
$$\cos \angle YOB = \text{power factor}$$
$$\angle BAO = \text{angle in electrical degrees between the centre of the stator and rotor poles}$$

Perpendicular distance from working point B to the zero power line is proportional to the intake.

If the working point B is above the horizontal OA the machine will be working as a motor, but if it is below OA the machine will be generating. Since the angle θ is very nearly $\pi/2$ because R is so small, little error is made in taking θ equal to $\pi/2$. This is done in some subsequent diagrams.

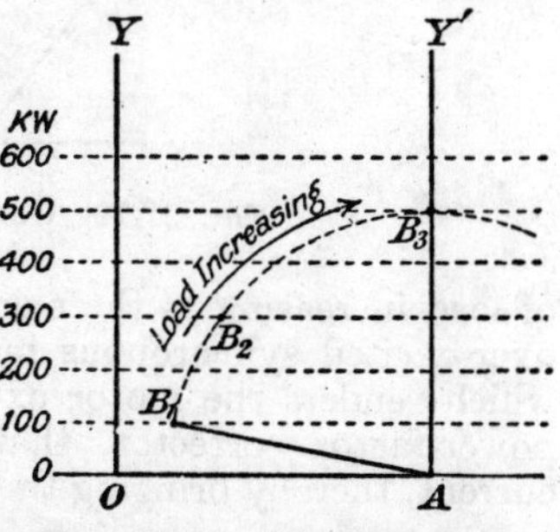

Fig. 21.11. Operation at Constant Excitation

Operation with Fixed Excitation. If the excitation is fixed, then AB is fixed, and the locus of B is a circle, shown dotted (Fig. 21.11), with A as centre and AB as radius. Suppose the load is such that the intake is 100 kW; then the working point will be at B_1. If the load is increased, the intake will have to increase, and therefore, the working point will have to reach a line of greater power. To do so it must travel along the circle in a clockwise direction. Thus it might travel to B_2, and it will remain there as long as the load remains at the new value. If the load is gradually increased, then the working point will travel round the circle until it reaches B_3, at which the intake is obviously a maximum. For a further increase in load, the point will travel on to a line of *smaller* power. Such a condition is impossible, since an increase in load naturally demands an increase in power, and the whole of the diagram to the right of AY' therefore represents unstable operation of the motor. The intake B_3A is thus the maximum intake for the given excitation, and if the motor losses are deducted from this, and the resulting output divided by the speed, the pull-out torque is obtained. We see that the pull-out torque depends upon the excitation, and the maximum possible value of this torque is fixed by that value of the exciting current which causes the maximum allowable temperature rise of the field windings.

Operation at Constant Power. The locus of the working point in this case is obviously one of the lines of constant power, say, 300 kW. The independent variable is the excitation. If the excitation is small, the working point will be at B_1, say (Fig. 21.12). The current OB_1 will be large and the angle of lag ϕ_1 will be large. The power factor will therefore be low. As the excitation is increased, the working point will move along the 300 kW line towards the left, and for one particular value of the excitation, will be at B_2 on OY. For the given power, the motor current is now a minimum, and the power factor is unity, the current being in phase with the applied voltage. A further increase in excitation will bring the working point to B_3, to the left of B_2. The current has now increased again, but it is leading instead of lagging, the angle

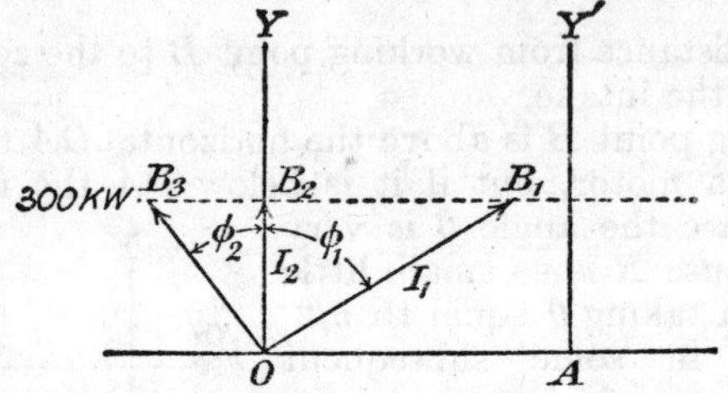

Fig. 21.12. Operation at Constant Power

of lead increasing as the excitation is increased. We thus see that an over-excited synchronous motor takes a leading current, a property which renders the motor extremely valuable as a phase advancer or power-factor corrector, that is, a machine which takes a leading current, thereby bringing the total current taken from the supply more

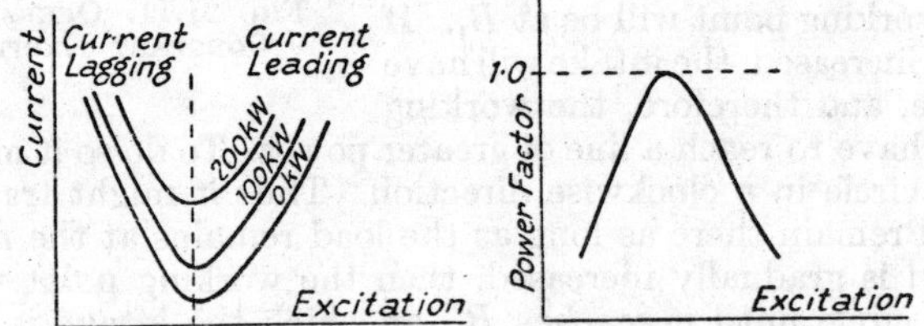

Fig. 21.13. Synchronous Motor Characteristics

nearly in phase with the voltage, in cases where the synchronous motor is working on the same supply as other plant which, by itself, takes current at a lagging power-factor.

If the motor current is plotted against excitation, the intake being constant, the characteristic obtained (Fig. 21.13) is called the "V" characteristic, from its shape. There is a family of such characteristics, each curve of the family corresponding to a definite intake. The curve of power factor against excitation is an inverted V, as shown.

Example. A single-phase alternator has an armature resistance of 0·5 Ω and synchronous reactance of 0·866 Ω. It is running as a synchronous motor on a 200 V supply, the mechanical load on the shaft, including iron and friction losses, being 6·5 kW. The current taken is 50 A. Find two possible phase angles of the current and two possible induced e.m.f.s.

Armature copper loss, $I^2R = 50 \times 50 \times 0{\cdot}5 = 1250$ W

$\therefore$ Intake $= 6500 + 1250 = 7750$ W

$$\therefore \text{ Power factor } \cos\phi = P/VI = 7750/(200 \times 50) = 0{\cdot}775$$
$$\therefore \phi = 39° \text{ lagging or leading}$$
$$\text{The angle } \theta \text{ (Fig. 21.10)} = \tan^{-1}(X/R) = \tan^{-1} 1{\cdot}732 = 60°$$
$$\therefore \text{ Phase angle, } \angle BOA = 60 - 39 = 21°; \text{ or } 60 + 39 = 99°$$

Also, resultant e.m.f., $OB = IZ = 50 \times \sqrt{(R^2 + X^2)} = 50$ V

$$\therefore AB^2 = OA^2 + OB^2 - 2\,OA \times OB\,.\cos \angle BOA$$
$$= 40{,}000 + 2500 - 20{,}000\,(\cos 21° \text{ or } \cos 99°)$$
$$\therefore AB = 154 \text{ V or } 214 \text{ V}$$

The method of drawing the "V" characteristic is as follows. The vector diagram with the lines of constant power is first drawn, and a series of concentric circles with centre at A then drawn to give a series of back e.m.f.s. The intersections of these circles with the lines of constant power will then locate the working points for specified powers and specified back e.m.f.s. The next step is to convert from back e.m.f.s to exciting current, and this necessitates a knowledge of the magnetization characteristic, a very convenient construction being to incorporate this characteristic in the diagram as shown in Fig. 21.14. Thus, for a power of 600 kW, the working points for the range of back e.m.f.s shown in the figure are B_1, B_2, B_3, . . . B_8, the current being OB_1, OB_2, OB_3, . . . OB_8; the back e.m.f. vectors are not drawn as there is no necessity to confuse the diagram by their presence. The various back e.m.f.s are projected on to the magnetization characteristic, giving the corresponding exciting currents. The above line currents are now plotted against these values of the exciting current, thereby giving the "V" characteristic for the particular power of 600 kW. Similarly, the characteristics for other powers can be drawn. If the angle of lag or lead of the various currents are measured the power factors can be calculated, and the curves of power factor against exciting current also drawn. By drawing a semicircle with O as centre and calling its radius 100, the intersection of the various current vectors with this semicircle can be projected on to OY, thereby giving the percentage power factors graphically, but this construction is not shown in Fig. 21.14.

The Blondel Diagram. This is an elaboration of the simple vector diagram of Fig. 21.10. Let V be the applied p.d. per phase and E the induced e.m.f. per phase. Let the power converted into mechanical power be P watts per phase, then

$$P = VI\cos\phi - RI^2$$
$$\therefore I^2 - (V/R)I\cos\phi + P/R = 0$$

This is the polar equation to a circle with its centre at some point O' on OY, Fig. 21.15, showing that if the excitation is varied while the power is kept constant, the working point B will move along this circle. Let the radius $O'B$ be r and the distance OO' be d, then

$$r^2 = I^2 - 2dI\cos\phi + d^2$$
$$\therefore I^2 - 2dI\cos\phi + (d^2 - r^2) = 0$$

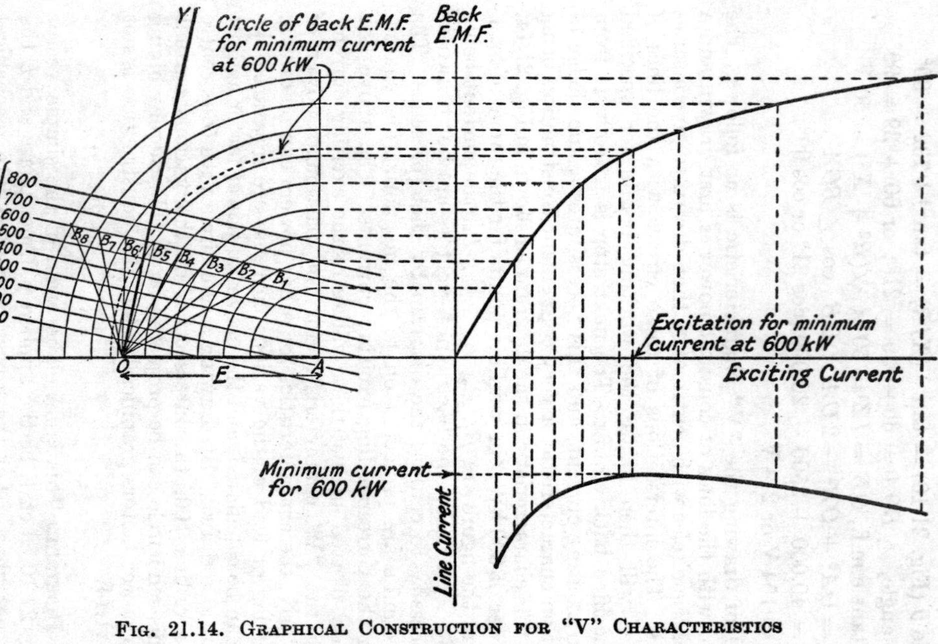

FIG. 21.14. GRAPHICAL CONSTRUCTION FOR "V" CHARACTERISTICS

Comparing this with the previous equation we have

$$OO' = d = V/2R$$

showing that the point O' is independent of P, and is a constant for a given motor operating at a fixed applied p.d. V.

Again
$$d^2 - r^2 = P/R$$
$$\therefore r = (d^2 - P/R)^{\frac{1}{2}}$$
$$= (V^2/4R^2 - P/R)^{\frac{1}{2}}$$
$$= (1/2R)(V^2 - 4PR)^{\frac{1}{2}}$$

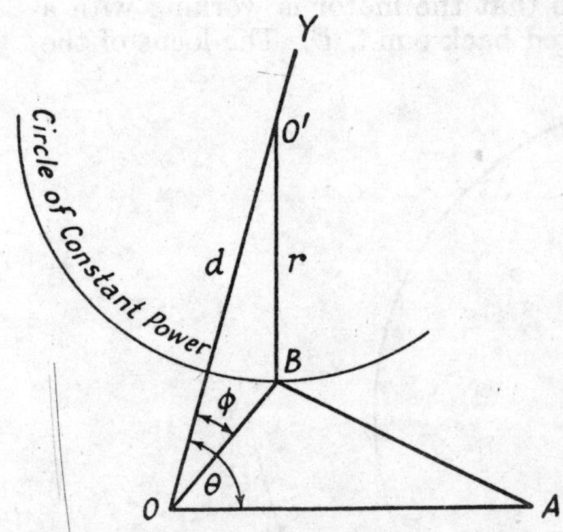

FIG. 21.15. DERIVATION OF THE BLONDEL DIAGRAM

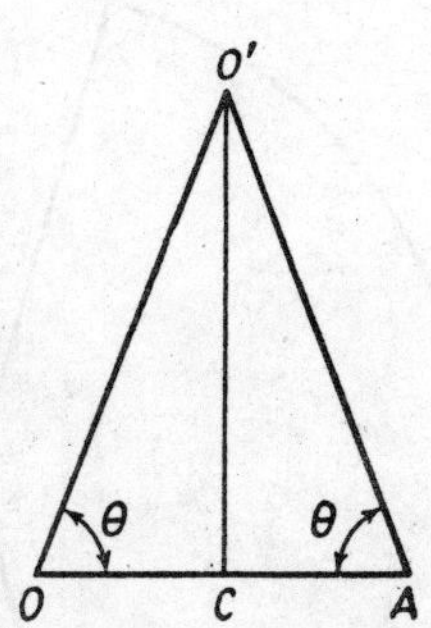

FIG. 21.16. SYMMETRY OF THE BLONDEL DIAGRAM

Drop the perpendicular $O'C$, Fig. 21.16, then

$$OC = OO' \cos \theta$$
$$= (V/2R) \times (R/Z)$$
$$= \tfrac{1}{2}(V/Z) = \tfrac{1}{2}OA$$

since the length OA must represent the current corresponding to V/Z when OB represents the current $I = E_r/Z$. The point C is thus the mid-point of OA, showing that the triangle $O'OA$ is isosceles, Fig. 21.16

Since the power P must be a real quantity we have the condition $4PR \ngtr V^2$, showing that the maximum possible power for phase is

$$P_{\max} = V^2/4R$$

and the radius of the circle of maximum power is zero: in other words this particular circle becomes the point O'. Again when $P = 0$ we have

$$r = V/2R = OO'$$

showing that the circle of zero power passes through the points O and A.

The radius for any power P whatsoever is

$$r = (V/2R)[1 - P/(V^2/4R)]^{\frac{1}{2}}$$
$$= (V/2R)(1 - P/P_{max})^{\frac{1}{2}}$$
$$= (V/2R)(1 - m)^{\frac{1}{2}}$$

where m is the power P expressed as a fraction of the maximum power P_{max}. The quantity $V/2R$ is the slant side OO' or d, so that

$$r = d(1 - m)^{\frac{1}{2}}$$

Conditions for Stability. Suppose that the motor is working with a fixed excitation and therefore a fixed back e.m.f. E. The locus of the

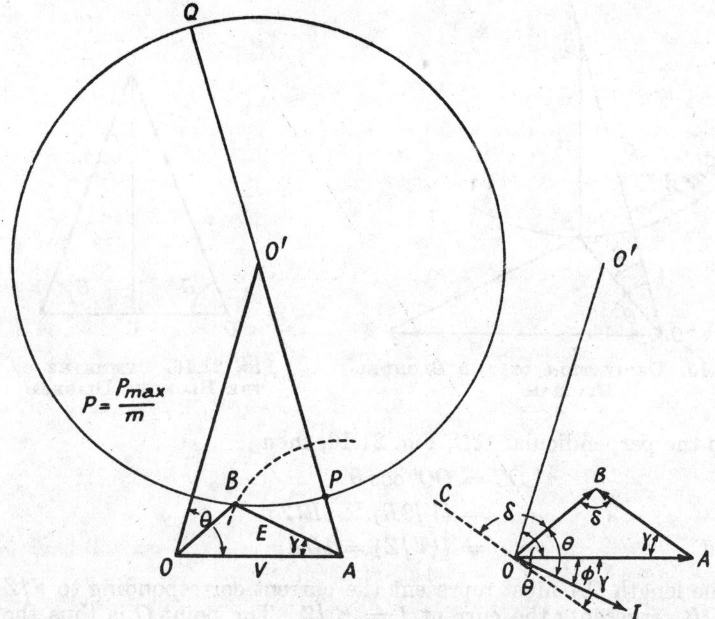

Fig. 21.17. Conditions for Stability

working point B will be the circle of radius E, the dotted circle of Fig. 21.17. If the load increases the point B must move on to a circle representing the new conditions, and since the radius of the power circle decreases with increase of power, this means that B must move clockwise for an increase of power, and vice versa. Thus an increase in power corresponds to an increase in γ, and the condition for stability is that $dP/d\gamma$ shall be positive. Draw OC parallel to AB, then

$$P = EI \cos(\pi - \angle COI) = -E(E_r/Z)\cos(\theta + \delta)$$
$$= -(EE_r/Z)(\cos\theta\cos\delta - \sin\theta\sin\delta)$$

We can eliminate E_r and functions of angles θ and δ as follows,

$$V/\sin\delta = E_r/\sin\gamma \quad \therefore \ \sin\delta = (V\sin\gamma)/E_r$$

$$E_r\cos\delta + V\cos\gamma = E \quad \therefore \ \cos\delta = (E - V\cos\gamma)/E_r$$

also

$$\cos\theta = R/Z \text{ and } \sin\theta = X/Z$$

Substituting these in the expression for P we have

$$P = -\frac{EE_r}{Z}\left(\frac{R}{Z}\cdot\frac{E - V\cos\gamma}{E_r} - \frac{X}{Z}\cdot\frac{V\sin\gamma}{E_r}\right)$$

$$= (VE/Z^2)(R\cos\gamma + X\sin\gamma) - E^2R/Z^2$$

Thus P is a function of two variables E and γ. For a fixed excitation E will be fixed, and so, differentiating with respect to γ, we have

$$dP/d\gamma = (VE/Z^2)(X\cos\gamma - R\sin\gamma)$$

$$= O \text{ when } X\cos\gamma = R\sin\gamma$$

i.e.

$$\text{when } \gamma = \tan^{-1}(X/R) = \theta$$

Thus for a given excitation the maximum power is obtained when $\gamma = \theta$.

Now differentiate with respect to E.

$$dP/dE = V\,.\,[(R\cos\gamma + X\sin\gamma)/Z^2] - 2R(E/Z^2)$$

$$= 0 \text{ when } E = V[(R\cos\gamma + X\sin\gamma)/2R]$$

If we now put $\gamma = \tan^{-1} X/R$, we have for the maximum possible power, by substituting the above values in the expression for P,

$$P_{\max} = V\,.\,(VZ/2R)(Z/Z^2) - (R/Z^2)V^2(Z^2/4R^2)$$

$$= V^2/4R, \text{ as obtained previously*}$$

EXAMPLE. A 1000 kVA, 3-phase, 2200-volt synchronous motor has the following data

$V = 2200/\sqrt{3} = 1270$ V/phase

$I = 1000 \times 1000/(\sqrt{3} \times 2200) = 263$ A/phase on full load

$R = 0{\cdot}2\ \Omega$/phase

$X = 3{\cdot}5\ \Omega$/phase

$Z \simeq 3{\cdot}5\ \Omega$/phase

The maximum power is obtained when

$$\gamma = \tan^{-1}(X/R) = \tan^{-1} 17{\cdot}5 = 86^\circ\ 44'$$

The maximum possible output per phase is

$$P_{\max} = V^2/4R = (1270)^2/(4 \times 0{\cdot}2) = 2{\cdot}02 \times 10^6 \text{ watts}$$

$$= 2020 \text{ kW}$$

* *See also* N. S. Thompson, "Maximum overload capacity of the synchronous motor," *Elec. Times*, **99** (1941) 343.

The maximum possible output for all three phases is thus 6060 kW, which is very much beyond the working range.

We have seen that

$$dP/d\gamma = (VE/Z^2)(X \cos \gamma - R \sin \gamma)$$

and that the condition for stability is that $dP/d\gamma$ shall be positive.

It is positive so long as

$$X \cos \gamma > R \sin \gamma$$

or

$$R \sin \gamma < X \cos \gamma$$

or

$$\tan \gamma < X/R < \tan \theta$$

Thus for the operation to be stable the point B must not move to the

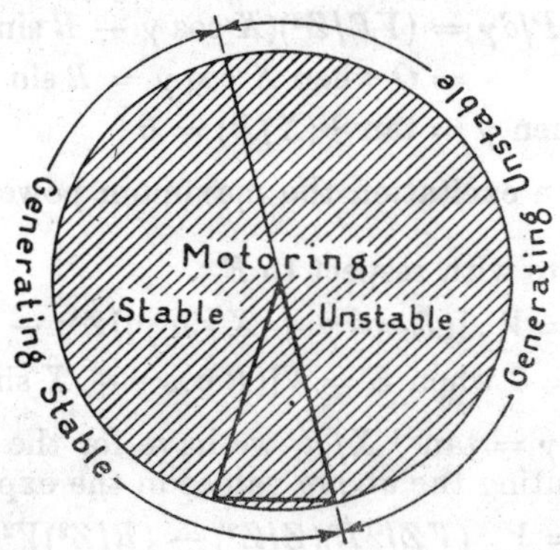

FIG. 21.18. MODES OF OPERATION OF THE SYNCHRONOUS MOTOR

right of the slant side $O'A$. Again, the circle of zero power is the circle passing through points O and A. For radii greater than this the power changes sign, which means that the regime is a generating one. The modes of operation can thus be represented by Fig. 21.18.

V Characteristics. Referring to Fig. 21.17 we have

$$OA^2 = AB^2 + OB^2 - 2AB \, . \, OB \cos \delta$$

$$V^2 = E^2 + I^2Z^2 - 2EIZ \cos [\angle (E, I) - \theta]$$

$$= E^2 + I^2Z^2 - 2EIZ [\cos \angle (E, I) \cos \theta + \sin \angle (E, I) \sin \theta]$$

But

$$EI \cos \angle (E, I) = - P$$

$$\therefore \cos \angle (E, I) = - P/EI$$

and

$$\sin \angle (E, I) = \pm (1 - P^2/E^2I^2)^{\frac{1}{2}}$$

again,

$$\cos \theta = R/Z, \text{ and } \sin \theta = X/Z$$

$$\therefore V^2 = E^2 + I^2Z^2 + 2EIZ[P(R/Z) \pm (E^2I^2 - P^2)^{\frac{1}{2}}(X/Z)] \times (1/EI)$$

$$= E^2 + I^2Z^2 + 2RP \pm 2X (E^2I^2 - P^2)^{\frac{1}{2}}$$

If we square and rearrange this equation, and also substitute $R^2 + X^2 = Z^2$, we have

$$Z^4I^4 - 2(Z^2V^2 - Z^2E^2 - 2RZ^2P + 2X^2E^2)I^2 + (V^2 - E^2 - 2RP)^2 + 4X^2P^2 = 0$$

This is a quadratic in I^2, the solution being

$$I^2 = (Z^2V^2 - Z^2E^2 - 2RZ^2P + 2X^2E^2)/Z^4 \pm (1/Z^4)\,[(Z^2V^2 - Z^2E^2 - 2RZ^2P + 2X^2E^2)^2 - Z^4\{(V^2 - E^2 - 2RP)^2 + 4X^2P^2\}]^{\frac{1}{2}}$$

The current must be real, and therefore I^2 must be both real and positive. The quantity under the root must never, therefore, become negative, so that

$$(Z^2V^2 - Z^2E^2 - 2RZ^2P + 2X^2E^2)^2 \not< Z^4[(V^2 - E^2 - 2RP)^2 + 4X^2P^2]$$

$$Z^2V^2E^2 \not< (R^2E^4 + 2Z^2RPE^2 + Z^4P^2)$$

or

$$ZVE \not< (RE^2 + Z^2P)$$

Hence the limiting values of the back e.m.f. E, for a given output P, are given by the equation

$$ZVE = RE^2 + Z^2P$$

or

$$E^2 - (VZ/R)E + Z^2P/R = 0$$

the solution of which is

$$E = \frac{\dfrac{VZ}{R} \pm \left(\dfrac{V^2Z^2}{R^2} - \dfrac{4Z^2P}{R}\right)^{\frac{1}{2}}}{2}$$

$$= \frac{VZ}{2R} \pm \frac{VZ}{2R}\sqrt{\left(1 - \frac{P}{V^2/4R}\right)}$$

$$= d[1 \pm (1 - m)^{\frac{1}{2}}]Z$$

Thus the limiting values of the back e.m.f. E, in order that the motor may develop an output equal to $1/m$ of the maximum possible output are, Fig. 21.19

$$AP = d[1 - (1 - m)^{\frac{1}{2}}]Z$$

$$AQ = d[1 + (1 - m)^{\frac{1}{2}}]Z$$

Consider again the above equations for I^2. If the motor is working on a constant output P, the only variables in the equation are E and I, of which E is the independent variable. Hence by giving a series of values to E, we can calculate the corresponding values of I and plot the curves of I against E for that particular value of P. We can repeat the process with different values of P thus obtaining a family of curves. We have also seen that for a given output P the limiting values of E are

$$E_{max} = (VZ/2R)[1 + (1 - m)^{\frac{1}{2}}]$$

and

$$E_{min} = (VZ/2R)[1 - (1 - m)^{\frac{1}{2}}]$$

Remembering these, no time will be wasted in trying to calculate points on curves which represent impossible conditions of operation.

For the 1000 kVA motor the equation for I^2 reduces to

$$I^2 = (120{,}500 + 0{\cdot}081E^2) \pm (1/150)(-815 \times 10^{12} + 882 \times 10^6 E^2)^{\frac{1}{2}}$$
$$= 120{,}500 + 0{\cdot}081E^2 \pm 190(1{\cdot}08E^2 - 10^6)^{\frac{1}{2}}$$

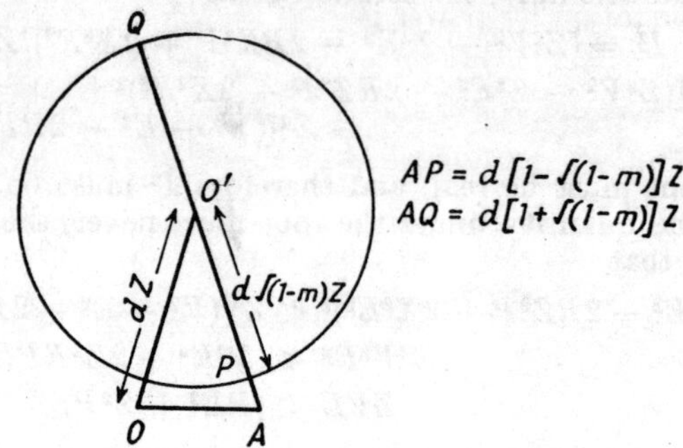

Fig. 21.19. Modes of Operation of the Synchronous Motor

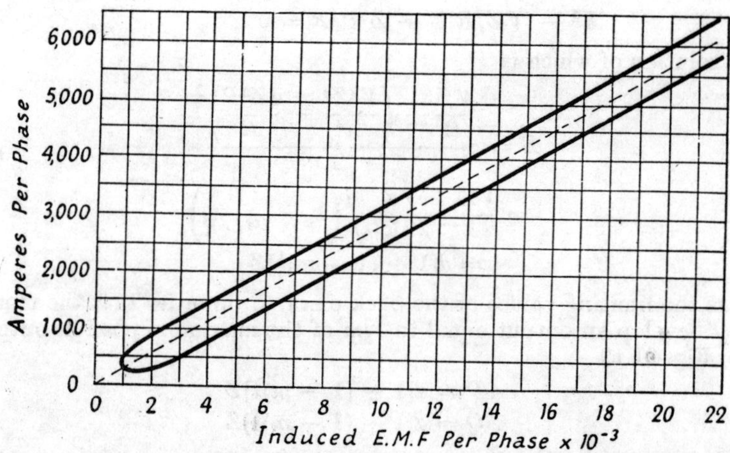

Fig. 21.20. V Characteristic for $m = 1$ for the 1000-kVA, 3-phase, 2200-volt Synchronous Motor of the Example

while the limiting values of E are $E_{max} = 21{,}200$ V and $E_{min} = 975$ V when working at the normal rated power of 333·3 kW per phase. The curve is drawn to scale in Fig. 21.20 for the case of the normal rated load, i.e. $m = 0{\cdot}165$. The fact that the rated terminal voltage per phase is 1270 V and that the induced e.m.f. per phase is unlikely to be greater than about 2,000 V shows that the V characteristic over the normal operating range is a very small portion of the total curve. Other V

curves for different values of power can be obtained by giving m a series of values and by thus obtaining new equations for I in terms of E.

Torque-angle Characteristic. The previous discussion of the Blondel diagram is based on the Behn-Eschenberg conception and is therefore more applicable to a cylindrical rotor machine than to the salient pole type. The equation for power is

$$P = (VE/Z^2)(R\cos\gamma + X\sin\gamma) - E^2R/Z^2$$

But $$R = Z\cos\theta, \text{ and } X = Z\sin\theta$$

$$\therefore\ P = (VE/Z)\cos(\theta - \gamma) - E^2R/Z^2$$

In an actual machine R is so small that the angle θ is only a few degrees short of $\pi/2$. We can then write

$$P = (VE/X)\sin\gamma$$

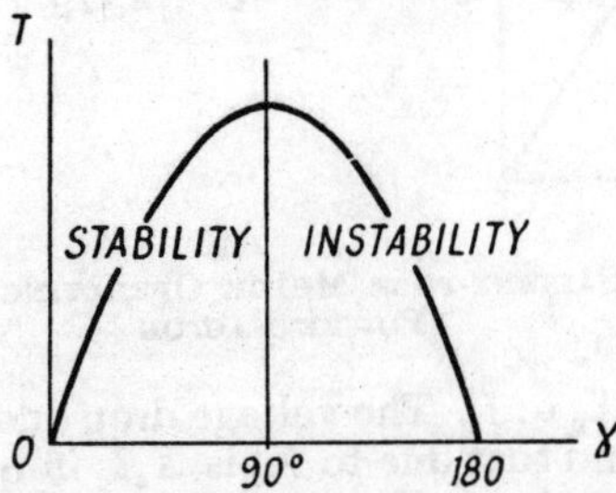

FIG. 21.21. TORQUE-ANGLE CHARACTERISTIC OF CYLINDRICAL ROTOR MACHINE

Since the speed is constant the torque is proportional to the power

$$\therefore\ T \propto (VE/X)\sin\gamma$$
$$\propto \sin\gamma \text{ when } E \text{ is constant}$$

The torque-angle characteristic of the cylindrical rotor machine is therefore the half sinusoid of Fig. 21.21.

We have seen that

$$dP/d\gamma = (VE/Z^2)(X\cos\gamma - R\sin\gamma)$$

With the above substitutions for R and X, this becomes

$$dP/d\gamma = (VE/Z)\sin(\theta - \gamma)$$
$$\simeq (VE/X)\cos\gamma \text{ when } R \text{ is small}$$

This is sometimes called the stiffness coefficient. It shows that for a given V and a given E (and therefore a given excitation) the coefficient is inversely proportional to the synchronous reactance X. Machines with long air-gaps have relatively small reactances, the coupling therefore being stiff. Such machines do not respond readily to rapid changes in shaft loading and under such conditions are therefore liable to high mechanical stresses. With a small air-gap and soft coupling the motor will respond to rapid load changes and the inertia of the rotating parts may result in values of γ such that there is a tendency to stray into the region of instability. A machine of this type also experiences large changes in the power factor as the load varies from zero-load to full-load.

Now consider the case of the salient-pole motor. We use the two-reaction theory and resolve the current into two components, the direct component I_d and the cross-magnetizing, or quadrature component I_c. The motor has two synchronous reactances which one can call X_d to

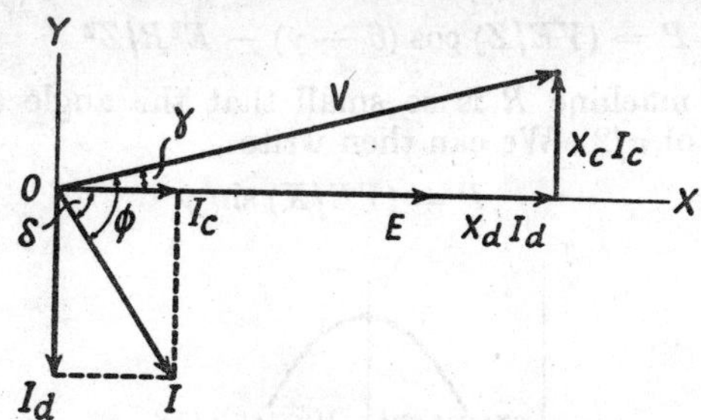

FIG. 21.22. SALIENT-POLE MOTOR OPERATING AT A LAGGING POWER FACTOR

component I_d, and X_c to I_c. The voltage drop due to X_d is X_dI_d in quadrature, leading I_d, and that due to X_c is X_cI_c in quadrature, leading I_c. The applied p.d. per phase, V, is the vector sum of the gap voltage E and these two drops, and this gives the vector diagram of Fig. 21.22, in which the small ohmic drops are neglected. Resolving along OX and OY we have, respectively

$$\begin{aligned} V\cos\gamma &= E + X_dI_d \\ &= E + X_dI\sin(\phi-\gamma) \\ V\sin\gamma &= X_cI_c \\ &= X_cI\cos(\phi-\gamma) \end{aligned}$$

$$\therefore\ V\cos\gamma - E = X_dI\sin\phi\cos\gamma - X_dI\cos\phi\sin\gamma$$

$$V\sin\gamma = X_cI\cos\phi\cos\gamma + X_cI\sin\phi\sin\gamma$$

We wish to eliminate terms in ϕ, hence multiply the first equation by $X_c\sin\gamma$ and the second by $X_d\cos\gamma$, then

$$X_cV\sin\gamma\cos\gamma - X_cE\sin\gamma = X_cX_dI\sin\phi\sin\gamma\cos\gamma - X_cX_dI\cos\phi\sin^2\gamma$$

$$X_dV\sin\gamma\cos\gamma = X_cX_dI\cos\phi\cos^2\gamma + X_cX_dI\sin\phi\sin\gamma\cos\gamma$$

Subtracting the first from the second

$$(X_d - X_c)V\sin\gamma\cos\gamma + X_cE\sin\gamma = X_cX_dI\cos\phi(\sin^2\gamma + \cos^2\gamma) = X_cX_dI\cos\phi$$

$$\tfrac{1}{2}(X_d - X_c)V\sin 2\gamma + X_cE\sin\gamma = X_cX_dI\cos\phi$$

Since we are neglecting resistance, power developed is equal to the intake

$$\therefore P = VI\cos\phi$$
$$= V[\{(X_d - X_c)V/2X_cX_d\}\sin 2\gamma + (E/X_d)\sin\gamma]$$
$$= (VE/X_d)\sin\gamma + \{V^2(X_d - X_c)/2X_cX_d\}\sin 2\gamma$$
$$\therefore T \propto [(VE/X_d)\sin\gamma + \{V^2(X_d - X_c)/2X_cX_d\}\sin 2\gamma]^*$$

This equation shows two very important differences between the salient-pole and nonsalient-pole machines. In the former there is a

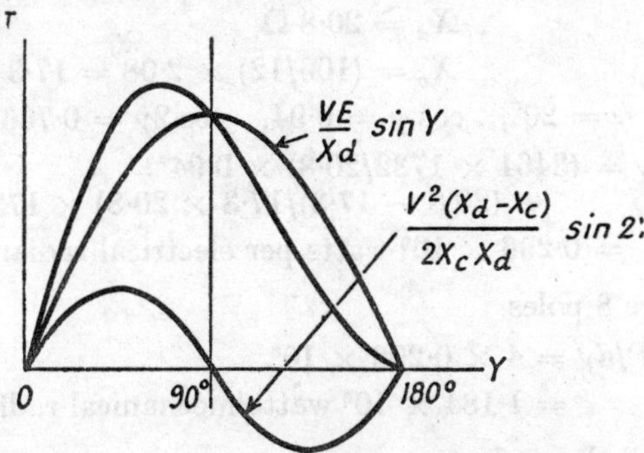

Fig. 21.23. Torque-angle Characteristic of Salient-pole Motor

second harmonic term in γ which results in a torque-angle characteristic like that of Fig. 21.23. The saliency results in an initial curve which rises more steeply than that of the nonsalient-pole machine and this assists the process of pulling into step. Secondly, even if E is zero, that is, the machine unexcited, there is still a finite torque, showing that it is possible, as has been discussed in a general manner on p. 508, for the nonsalient pole machine to pull into step even when unexcited.

The angle γ' at which the torque is a maximum can be obtained by differentiating the expression for P. We have

$$dP/d\gamma = (EV/X_a)\cos\gamma + [(X_d - X_c)/X_cX_d]V^2\cos 2\gamma$$

The equation for γ' is thus

$$2[(X_d - X_c)/X_cX_d]V^2\cos^2\gamma' + (EV/X_d)\cos\gamma' - [(X_d - X_c)/X_cX_d]V^2 = 0$$

Example. A 520 kVA, 3000 V, 50 c/s, 8-pole synchronous motor has a percentage direct-axis synchronous reactance of 120 and quadrature-axis reactance of 100. It is excited to an induced e.m.f.

* The development of this expression is dependent on unsaturated conditions. The correction for saturation is outside the scope of this text. *See* McFarland *Alternating Current Machines*, Van Nostrand (New York); also Robertson, Rogers and Dalziel, "The Saturated Synchronous Machine," *Trans.A.I.E.E.* **56**, 858.

of twice the rated voltage, and it is loaded to such an extent that the load angle is 20°. Calculate the restoring torque per mechanical radian displacement from the position of stability, the load angle at which the torque is a maximum, and the value of this maximum torque.

$$\text{Phase voltage } V = 3000/\sqrt{3} = 1732 \text{ V}$$

$$\therefore \text{ Induced phase e.m.f. } E = 3464 \text{ V}$$

$$\text{Normal full-load current} = 520 \times 1000/(\sqrt{3} \times 3000) = 100 \text{ A}$$

$$\therefore \; 100X_d = (120/100) \times 1732 = 2080$$

$$\therefore \; X_d = 20{\cdot}8 \; \Omega$$

and $$X_c = (100/12) \times 2{\cdot}08 = 17{\cdot}3 \; \Omega$$

with $$\gamma = 20°, \quad \cos\gamma = 0{\cdot}94, \quad \cos 2\gamma = 0{\cdot}766$$

$$\therefore \; dP/d\gamma = (3464 \times 1732/20{\cdot}8) \times 0{\cdot}94 + (20{\cdot}8 - 17{\cdot}3)/17{\cdot}3 \times 20{\cdot}8) \times 1732^2 \times 0{\cdot}766$$

$$= 0{\cdot}296 \times 10^6 \text{ watts per electrical radian}$$

Since there are 8 poles

$$dP/d\gamma = 4 \times 0{\cdot}296 \times 10^6$$

$$= 1{\cdot}184 \times 10^6 \text{ watts/mechanical radian}$$

Hence for all three phases

$$dP/d\gamma = 3{\cdot}552 \times 10^6 \text{ watts/mechanical radian}$$

Speed $$= 750 \text{ r.p.m.}$$

$$= 12{\cdot}5 \text{ r.p.s.}$$

$$\therefore \; \omega = 2\pi \times 12{\cdot}5 = 25\pi \text{ radian/sec}$$

$$\therefore \; T = 3{\cdot}552 \times 10^6/25\pi$$

$$= 0{\cdot}45 \times 10^5 \text{ newton-metres/radian}$$

To determine γ' put $\cos\gamma' = x$

$$\therefore \text{ Term in } x^2 = 2[(20{\cdot}8 - 17{\cdot}3)/20{\cdot}8 \times 17{\cdot}3] \, (1732)^2 x^2$$

$$= 0{\cdot}584 \times 10^6 x^2$$

$$\text{Term in } x = 3464 \times 1732/2{\cdot}08 \, x$$

$$= 2{\cdot}88 \times 10^6 \, x$$

Independent term

$$= [(2{\cdot}08 - 1{\cdot}73)/2{\cdot}08 \times 1{\cdot}73] \times 1732^2 = 0{\cdot}292 \times 10^6$$

$$\therefore \; 0{\cdot}584x^2 + 2{\cdot}88x - 0{\cdot}292 = 0$$

the solution of which is

$$\cos\gamma' = x = 0{\cdot}103$$

$$\therefore \; \gamma' = 84° \, 6'$$

and $$\sin\gamma' = 0{\cdot}9947 \text{ and } \sin 2\gamma' = 0{\cdot}2045$$

The power at this angle is

$$P_{\max} = \frac{1732 \times 3464}{2{\cdot}08} \times 0{\cdot}9947 + \frac{(1732)^2(20{\cdot}8 - 17{\cdot}3)}{2 \times 20{\cdot}8 \times 17{\cdot}3} \times 0{\cdot}2045$$

$$= 2{\cdot}87 \times 10^5 + 0{\cdot}03 \times 10^5$$

$$= 2{\cdot}90 \times 10^5 \text{ watts/phase} = 290 \text{ kW/phase}$$

Phase Swinging or Hunting. We have seen that when a synchronous motor is loaded, the rotor poles fall back a certain angle behind the poles of the forward rotating field, in order that sufficient tangential force may be set up to produce the necessary torque to cope with the load. If the load is suddenly thrown off, the rotor poles are pulled into almost exact opposition to the poles of the forward field, but because of the inertia of the rotor the rotor poles travel too far. They are then pulled back again, and so on, an oscillation thus being set up about the position of equilibrium corresponding to the new conditions of load. It will be seen that these oscillations will be set up whenever the load varies, and if the variations in load are periodic and synchronize with the natural period of oscillation of the rotor, mechanical resonance will be set up and the amplitude of the swing of the rotor poles relative to the poles of the rotating field will become so great that the motor will fall out of step. It is therefore evident that some damping couple must be introduced, so that, in the event of oscillations being set up, they will be immediately damped out and no large amplitude of swing obtained. These dampers take the form of heavy conducting grids housed in the pole faces. When the motion is uniform, there is no relative motion between the rotor and stator forward rotating poles, and no currents induced in the dampers. If hunting takes place, the relative velocity of the two sets of poles induces heavy currents in the dampers, and the kinetic energy of oscillation is thus damped down by being converted into heat energy in the dampers.

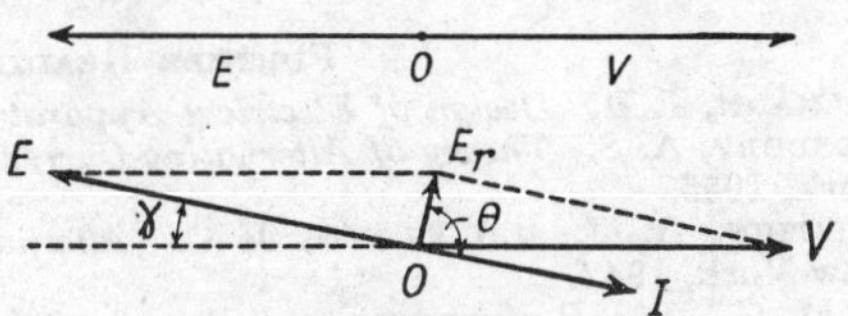

Fig. 21.24. Electrical Conditions During Phase Swinging

An expression for the period of swing of the rotor can be determined as follows. Suppose the motor is working on no load, and imagine for simplicity that there are no losses. Then the supply voltage V and the motor induced voltage E will be equal and opposite so long as the motion is steady. If for some reason the rotor is displaced relatively to the stator field by a small angle γ, then one vector will swing round through the angle γ, and a resultant voltage E_r will be set up (Fig. 21.24). For small values of γ, E_r is proportional to γ. We have, in fact, $E_r = V \times \gamma$.

Now E_r will set up a current through the armature, of magnitude $I = E_r/Z$, lagging very nearly $\pi/2$ behind E_r in the case of a modern machine with high reactance. Thus I will be nearly in phase with V, so that the power per phase conveyed by I will be

$$P = VI = V \times E_r/Z = V/Z \times V\gamma$$
$$= VI_s\gamma$$

where I_s is the short-circuit current corresponding to normal excitation. If the angular velocity of the rotor is ω radians per second, the torque set up by the current I will be

$$\text{Torque} = \frac{\text{power}}{\text{angular velocity}} = \frac{VI_s\gamma}{\omega} = A\,.\,\gamma, \text{ say}$$

If J is the moment of inertia of the rotating masses this torque will be utilized in accelerating the rotor so long as there is no damping.

$$\therefore\ J(d^2\gamma/dt^2) + A\gamma = 0$$

This is the differential equation of a simple harmonic oscillation, and comparing it with the equation

$$d^2x/dt^2 + n^2x = 0$$

we see that the frequency of the oscillations set up is

$$f = n/2\pi, \text{ or in the case of the motor}$$
$$f = (1/2\pi) \times (VI_s/J\omega)^{\frac{1}{2}}$$

This is the frequency of the oscillations when there is no damping.* If the pole faces are provided with damping windings, then there will be a retarding torque set up proportional to the angular velocity of phase swing. The effect of this torque is to damp out the oscillations and also to diminish their frequency.

These pole-face damping windings are also used in starting up the synchronous motor, but as their action is then that of an induction motor this method of starting will be dealt with in Chapter 23.

Further Reading

KUHLMANN, J. H. *Design of Electrical Apparatus.* Wiley, New York, 1950.

LANGSDORF, A. S. *Theory of Alternating Current Machinery.* McGraw Hill, New York, 1955.

PUCHSTEIN, A. F. and LLOYD, J. C. *Alternating Current Machines.* Wiley, New York, 1942.

SAY, M. G. *The Performance and Design of A.C. Machines.* Pitman, London, 1952.

Examples on Chapter 21

1. A single-phase synchronous motor rotates at 1000 r.p.m. on a 500 V supply. Its armature resistance is 0·1 Ω and reactance 0·5 Ω. If it can be excited up to an induced voltage of 2000 V, calculate its pull-out torque. Draw the vector diagram for the motor and from it deduce the "V" curves.

2. A single-phase synchronous motor for use on a 500 V circuit gives, when running as a generator, a short-circuit current of 150 A with no-load excitation applied. The armature resistance is 0·2 Ω. To what voltage must the motor be excited so as to develop 40 h.p. at unity power factor, the mechanical losses being 5 h.p.? What will be the armature current? *Ans.*—549 V, 69·0 A.

* This is a simplified treatment intended to show the factors on which the period of oscillations depends. For a more complete solution see pp. 543, 546.

3. A single-phase alternator, when driven at 1000 r.p.m., has a magnetization characteristic as follows—

Exciting current . .	30	60	90	120	150
Open-circuit volts . .	460	820	1040	1200	1300

A short-circuit current of 200 A is produced when the exciting current is 40 A. The armature resistance is 0·4 Ω. Plot the curve of current against exciting current with a constant intake of 100 kW at 1000 V when running as a synchronous motor at 1000 r.p.m.

4. The 3-phase, synchronous machine of question 4, p. 501, is run under the stated conditions as a synchronous motor. Calculate the required exciting current.

Ans. 121 A.

5. A 3-phase synchronous motor, when driven at normal speed as an alternator, has the following magnetization curve—

Exciting current . .	5	10	15	20	25
Terminal Voltage .	370	655	830	950	1050

An exciting current of 10 A produces a short-circuit current of 200 A. Calculate the power-factor of this machine when operating as a synchronous motor and taking a load of 75 kW at 800 V, the exciting current being 20 A. (C. & G.)

Ans. 0·67 p.f. leading.

6. Assuming constant synchronous impedance, show that the locus diagram of the current input to a synchronous motor is a circle if the excitation and supply voltage are constant. Draw the locus diagram and determine the current and power-factor for an input of 750 kW for a 3-phase, star-connected synchronous motor, the terminal voltage being 3,300, and the synchronous impedance 0·25 + $j2$ Ω per phase. The excitation corresponds to an induced e.m.f. of 3500 V. (L.U.)

Ans. 146 A, 0·9 p.f. leading.

CHAPTER 22

THE BEHAVIOUR OF ALTERNATORS DURING NORMAL OPERATING CONDITIONS

WE have seen that the terminal voltage V of an alternator on load is different from, and in general, less than, its no-load voltage. The difference is governed by the magnitude of the load current and also the power factor, and we have seen how it can be pre-determined. The normal method of operation in a central power station is to maintain the terminal voltage constant, and, under modern conditions, this necessitates that the regulation of the excitation, in order to maintain constant terminal voltage, shall be automatic: also the response must be rapid because of the violent fluctuations which are a common feature of an industrial load.

Automatic Voltage Regulators. Modern quick-acting voltage regulators are based on the "overshooting the mark" principle. When the load on the alternator increases and more excitation is required to keep the voltage constant, the regulator produces an increase in excitation more than is ultimately necessary. This is because the fluctuations in load may be very rapid, whereas the inductance of the alternator field will cause the flux to increase to the desired value very slowly, unless there is a large increase in exciting current. Before the voltage has time to increase to the value corresponding to the increased excitation, the regulator reduces the excitation again. There are two main types—

(*a*) The vibrating type, in which a fixed resistance is cut in and out of the exciter field circuit.

(*b*) The rotary type, in which a variable resistance is used.

The Tirrill Regulator. This is an example of the vibrating type. The essential parts are shown in Fig. 22.1. At the top there are two levers, that on the left operated by a solenoid energized in proportion to the exciter voltage, and that on the right by an a.c. magnet having both shunt and series excitations. This magnet is so adjusted that with normal load and voltage at the alternator, the pulls of the two coils are equal and opposite. In the event of an increase in load the series coil predominates, and so pulls down the a.c. magnet. This closes the contacts between the two levers and so de-energizes the relay horse-shoe magnet, which thereby releases its armature and short-circuits the rheostat in the exciter field. There is in consequence a sudden increase in excitation, which causes the alternator voltage to rise very rapidly. At the same time the excitation of the exciter control magnet is increased, thus pulling down the left-hand lever, energizing the relay, and re-inserting the field rheostat before the alternator voltage has had time to increase too far. The reverse action takes place when the load on the alternator decreases.

It will be seen that because of the overshooting the mark principle the terminal voltage does not remain absolutely steady, but oscillates

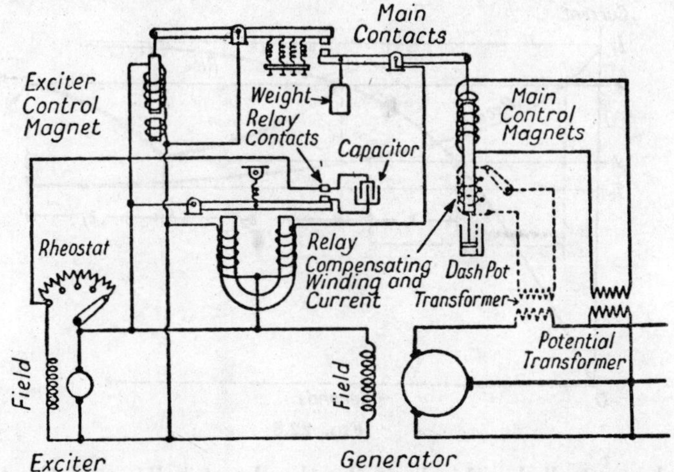

FIG. 22.1. CIRCUIT OF THE TIRRILL AUTOMATIC VOLTAGE REGULATOR

rapidly between maximum and minimum values. The regulator is so quick acting that the variation in voltage is less than 1 per cent.*

The exciter field circuit of an alternator controlled by a Tirrill regulator can be represented by the elementary diagram of Fig. 22.2.

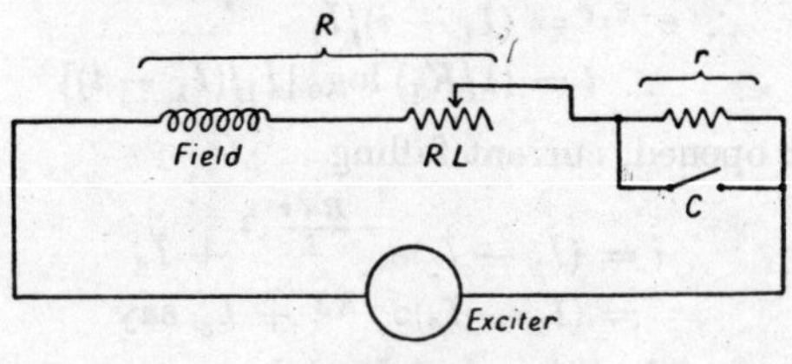

FIG. 22.2

When the contact C is open, the field circuit resistance is $R + r$; when C is closed the resistance is R. Thus the resistance oscillates between the extremes $(R + r)$ and R and the field current oscillates accordingly. If E is the voltage of the d.c. exciter, the field current will tend to a maximum of $I_1 = E/R$, and to a minimum of $I_2 = E/(R + r)$. Neither limit will be reached because of the speed of opening and closing of the contact C. The exciter field current graph will therefore consist of a succession of rise and decay curves for the particular

* For fuller information *see* C. C. Garrard, *Electric Switch and Control Gear*. Benn; and G. N. Patchett, *Automatic Voltage Regulator*. Pitman, London.

circuit conditions, and the maximum and minimum attained, x and y respectively, Fig. 22.3, will have to be such that the terminal voltage of the alternator is maintained at the correct value. There are, for a given value of r, two deciding factors, namely the value of R, which

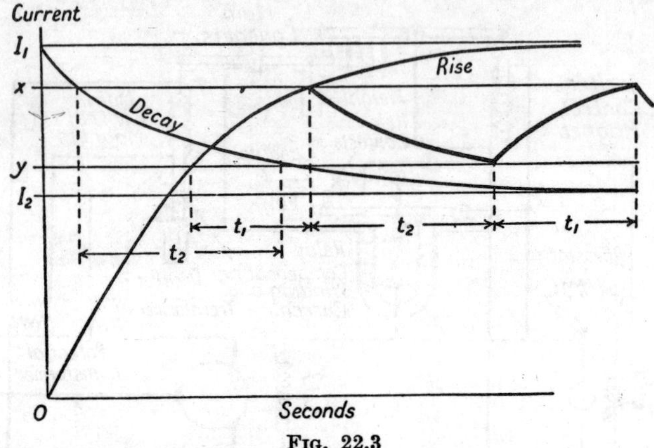

Fig. 22.3

can be controlled within limits by the rheostat RL, and the time of one cycle of operation of the contact C.

(*a*) Contact closed, current rising

$$i = I_1(1 - e^{-\frac{R}{L}\cdot t})$$
$$= I_1(1 - e^{-K_1t}) \text{ say}$$
$$\therefore\ e^{-K_1t} = (I_1 - i)/I_1$$
$$\therefore\ t = (1/K_1)\log_e [I_1/(I_1 - i)]$$

(*b*) Contact opened, current falling

$$i = (I_1 - I_2)e^{-\frac{R+r}{L}\cdot t} + I_2$$
$$= (I_1 - I_2)e^{-K_2t} + I_2, \text{ say}$$
$$\therefore\ e^{-K_2t} = (i - I_2)/(I_1 - I_2)$$
$$\therefore\ t = (1/K_2)\log_e [(I_1 - I_2)/(i - I_2)]$$

From the rising equation the time for the current to rise from 0 to x is therefore

$$t_x = (1/K_1)\log_e I_1/(I_1 - x)$$

while the time to rise from 0 to y is

$$t_y = (1/K_1)\log_e I_1/(I_1 - y)$$
$$\therefore\ t_1 = t_x - t_y$$
$$= (1/K_1)\log_e [(I_1 - y)/(I_1 - x)]$$

From the "falling" equation the time to fall from I_1 to x is

$$t_x' = (1/K_2)\log_e[(I_1 - I_2)/(x - I_2)]$$

while the time to fall from I_1 to y is

$$t_y' = (1/K_2)\log_e[(I_1 - I_2)/(y - I_2)]$$

$$\therefore t_2 = t_y' - t_x'$$

$$= (1/K_2)\log_e[(x - I_2)/(y - I_2)]$$

As an example let the exciter voltage be $E = 100$ V, and let $R = 50\ \Omega$ and $r = 25\ \Omega$. Then we have

$$I_1 = E/R = 100/50 = 2\text{ A}$$

$$I_2 = E/(R + r) = 100/75 = 1{\cdot}33\text{ A}$$

Suppose the desired current limits are $x = 1{\cdot}8$ A and $y = 1{\cdot}5$ A, and let the inductance of the exciter field winding be 8 H. While the current is rising we have

$$L/R = 8/50$$

while when the current is falling we have

$$L/(R + r) = 8/75$$

For the time t_1 we therefore have

$$t_1 = (8/50)\log_e[(2 - 1{\cdot}5)/(2 - 1{\cdot}8)]$$

$$= 0{\cdot}16\log_e 2{\cdot}5$$

$$= 0{\cdot}147\text{ sec}$$

For the time t_2 we have

$$t_2 = (8/75)\log_e[(1{\cdot}8 - 1{\cdot}33)/(1{\cdot}5 - 1{\cdot}33)]$$

$$= 0{\cdot}107\log_e 2{\cdot}76$$

$$= 0{\cdot}109\text{ sec}$$

Hence the time for one cycle of closing and opening is

$$0{\cdot}147 + 0{\cdot}109 = 0{\cdot}256\text{ sec}$$

Parallel Operation of Alternators. Before an incoming machine can be switched on to the bus-bars the following conditions have to be fulfilled—

(*a*) The voltage of the incoming machine must be the same as the bus-bar voltage.

(*b*) The phase of the machine voltage must be identical with the phase of the bus-bar voltage relative to the feeders, i.e. opposite in phase relative to the local circuit through the armatures and bus-bars. This circuit is shown dotted in Fig. 22.4.

(*c*) The frequency of the incoming machine must be the same as the bus-bar frequency.

Condition (*a*) is indicated by a voltmeter, and conditions (*b*) and (*c*) are both indicated by synchronizing gear.

Consider first of all the case of single-phase alternators. The simplest form of synchronizer consists of two lamps, *LL*, connected across the main switch as indicated in Fig. 22.4. If the frequencies of the alternators *A* and *B* are not equal, the phase angle between the voltages of *A* and *B* will be continually changing, and therefore, the current through the lamps and through the local circuit shown dotted will be changing. The resultant voltage will undergo changes similar in character to the beats produced when two sources of sound of slightly different frequencies are sounding together. This is indicated in Fig. 22.5. In consequence, the lamps will flicker, the alternations in brightness being rapid when there is a large difference in the frequencies, and slow when the frequencies are nearly equal. In the middle of a dark period the two voltages will be in opposition with respect to the local circuit. Hence, the speed of the incoming machine is adjusted until the lamps go in and out very slowly, the incoming voltage is adjusted equal to the bus-bar voltage, and then the switch is closed in the middle of a dark period. It is somewhat easier to judge the middle of the bright than the middle of the dark period, and some engineers prefer to synchronize "lamps bright." This necessitates the crossing over of the lamp connections, as in the case of machine *C*.*

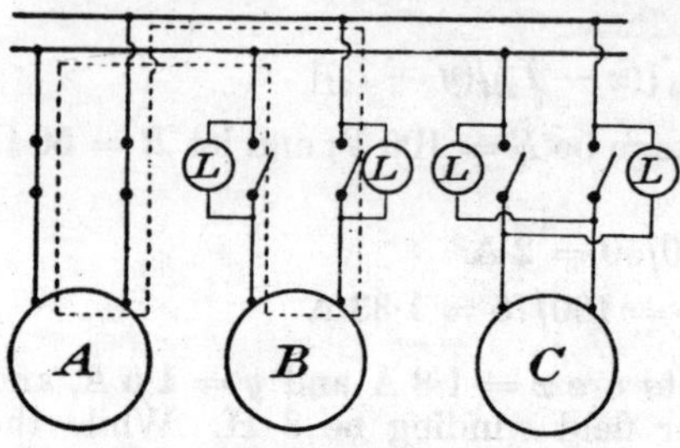

Fig. 22.4. Alternators in Parallel

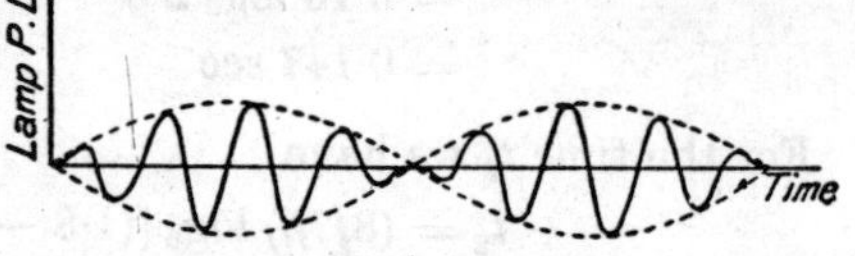

Fig. 22.5. Voltage Variations in Synchronizer

Conditions Necessary for Successful Parallel Operation. Since, with respect to the local circuit, the e.m.f. of an alternator is in phase opposition to the e.m.f. of another alternator with which it is working in parallel, the machines run as synchronous motors relative to one another. Hence, if one machine gets into difficulties, say, through a failure in steam supply, it must receive wattful motoring current from the other. Actually, with alternators driven by steam turbines or water turbines this is a very remote possibility. When the prime mover is a reciprocating engine, either steam or internal combustion, then the uneven torque results in cyclic variations in speed, with the result that the machine, if connected to a constant-frequency network, will

* For a discussion of the synchronizing equipment used in central power stations, see T. H. Carr, *Electric Power Stations*. Chapman & Hall, London.

alternately fall back from, and then run ahead of, its mean position.

(*a*) Consider two machines having resistance but no reactance. Their e.m.f.s E_1 and E_2 (Fig. 22.6) will be practically in phase opposition, so that their resultant E_r will be almost in quadrature with both E_1 and E_2. The synchronizing current I will be in phase with E_r and therefore, practically in quadrature with E_1 and E_2, so that it will be an idle current and will therefore convey no real power to the machine needing help.

(*b*) Suppose that the armatures have reactance only. Then the synchronizing current I will be in quadrature with E_r, and therefore, practically in phase with one of the machine voltages; E_2 in Fig. 22.6. Thus machine II will supply real power to machine I, so that the latter will keep running. This shows that for successful parallel operation, reactance in the armatures is absolutely necessary.

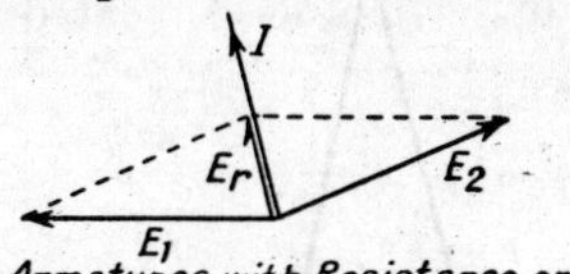

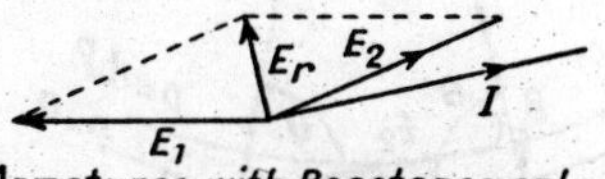

Fig. 22.6. Synchronizing Current

Now consider actual machines with both resistance and reactance. Let the angular phase difference of the two induced e.m.f.s be γ, and let the circulating current I_s lag an angle θ behind E_r, as shown in Fig. 22.7. Then so long as γ is small and the e.m.f.s are equal, we can write

$$E_1 = E_2 = E, \text{ and } E_r = E\gamma$$

$\therefore$ Circulating current

$$I_s = E_r/Z = E\gamma/Z$$

where Z is the combined impedance per phase of the two armatures.

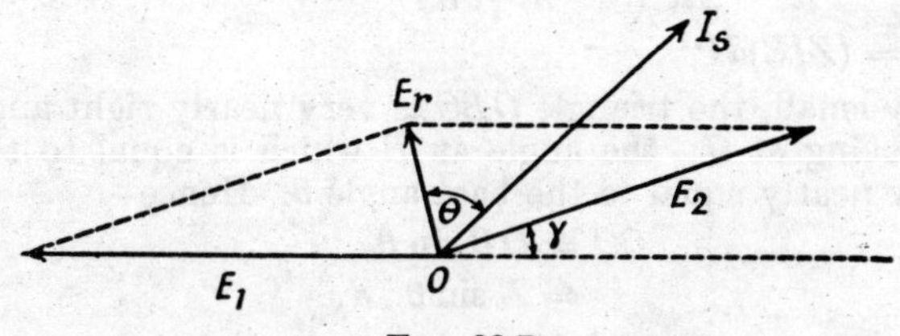

Fig. 22.7

Now so long as γ is small, the angle between OI_s and OE_2 is very nearly equal to $(90 - \theta)$, and therefore

$$\begin{aligned} \text{Synchronizing power } P_s &= E_2 I_s \cos(90 - \theta) \\ &= (E^2/Z)\,\gamma \sin\theta \\ \therefore dP_s/d\gamma &= (E^2/Z)\sin\theta \\ &= E^2/(R^2 + X^2)^{\frac{1}{2}} \times X/(R^2 + X^2)^{\frac{1}{2}} \\ &= E^2 \times X/(R^2 + X^2) \end{aligned}$$

and this is a maximum when $X = R$, showing that the maximum synchronizing power would be given with an armature reactance equal to the armature resistance. This condition is, of course, never fulfilled in actual practice, the resistance always being small in comparison with the reactance. Nor is it desirable that this condition should be fulfilled, since it would give rise to an excessively high restoring torque whenever a machine deviated from the steady-state angular position.

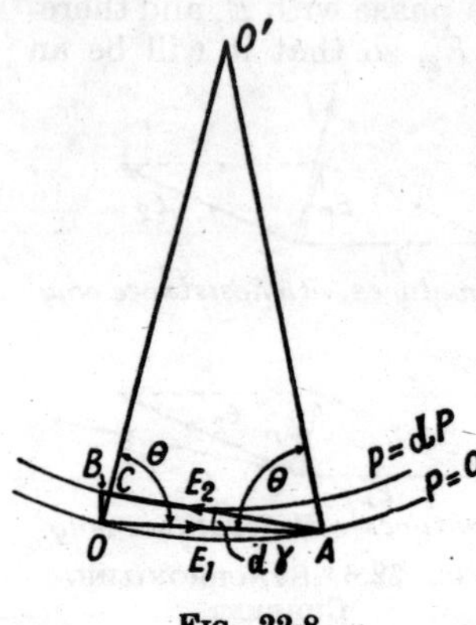

Fig. 22.8

The above relationship can be deduced in an instructive manner from the Blondel diagram. If γ is zero, then, with $E_1 = E_2$, the working point becomes the point O on the circle of $P = 0$, Fig. 22.8. A lag of E_2 with respect to E_1 by a small angle $d\gamma$ brings the working point to B on a circle of finite power dP. If C is the point of intersection of the circle of power dP with the slant side $O'O$, then

$$\begin{aligned} OC &= O'O - O'C \\ &= \left(\begin{matrix}\text{radius of the}\\ \text{circle of zero power}\end{matrix}\right) - \left(\begin{matrix}\text{radius of the}\\ \text{circle of power } dP\end{matrix}\right) \\ &= \frac{EZ}{2R} - \frac{EZ}{2R}\left(1 - \frac{dP}{E^2/4R}\right)^{\frac{1}{2}} \\ &= \frac{EZ}{2R} - \frac{EZ}{2R}\left(1 - \frac{dP}{E^2/2R}\right), \text{ approx.} \\ &= (Z/E)dP \end{aligned}$$

Now with $d\gamma$ small, the triangle OBC is very nearly right-angled, the right angle being at C; the angle at B which is equal to the angle $O'CA$ is very nearly equal to the base angle θ. Hence

$$\begin{aligned} OC &= OB \sin\theta \\ &= E \sin\theta \,.\, d\gamma \end{aligned}$$

Equating these two values of OC we have

$$(Z/E)dP = E \sin\theta \,.\, d\gamma$$

$\therefore\ dP/d\gamma = (E^2/Z)\sin\theta$, to which the value obtained previously on p. 521 reduces for the stated conditions.

Effect of a Change in Excitation. We have seen that when two d.c. shunt generators are operating in parallel, any desired division in load can be effected by making the necessary changes in the excitations, in fact, if load transfer by this method is carried out far enough the whole of the load can be thrown on to one machine. Now consider the case of two alternators in parallel, and let the operating conditions be initially identical, that is, each machine taking one-half of the kilowatt load and one-half of the kVAR load, the operating power factors thus being equal to the load power factor. This condition is illustrated in Fig. 22.9 (*a*), where the load current is indicated as $2I$. Now let the

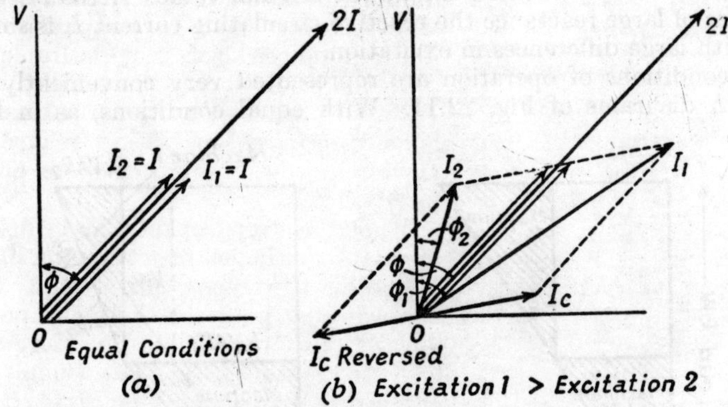

FIG. 22.9. EFFECT OF UNEQUAL EXCITATION ON ALTERNATORS IN PARALLEL

excitation of No. 1 be increased so that $E_1 > E_2$. The difference voltage $E_1 - E_2$ will set up a circulating current

$$I_c = (E_1 - E_2)/2Z$$

and this will lag E_1 by the angle θ. We can regard this current as being

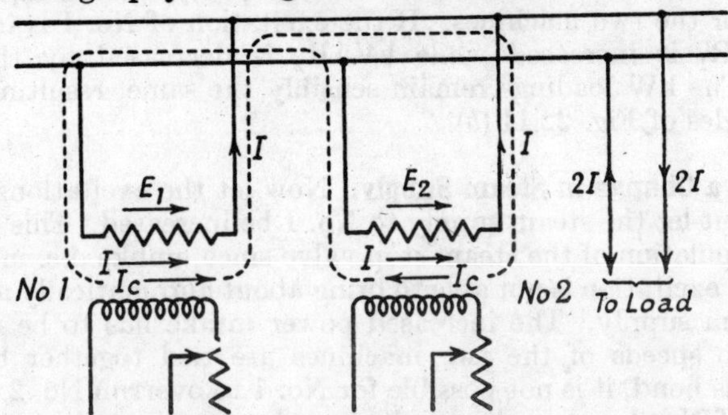

FIG. 22.10. PATH OF CIRCULATING CURRENT BETWEEN TWO ALTERNATORS IN PARALLEL

superposed on the original current distribution and since it is confined to the local path through the armature and round the bus-bars, as shown in Fig. 22.10, it follows that it must be added vectorially to the load current of No. 1, and subtracted from the load current of No. 2. The resulting currents I_1 and I_2 are now as in Fig. 22.9 (*b*) from which we see that the operating power factor, $\cos \phi_1$, of No. 1 is decreased, while that of No. 2, $\cos \phi_2$, is increased. Since the angle θ is very nearly

90° in actual machines it follows that the effect on the kW loading of the two machines is negligibly small, the main effect being to increase the kVAR output of No. 1 and to reduce that of No. 2. A secondary effect is to lower the bus-bar voltage V and slight excitation adjustments are necessary to bring this to the original value. Actually, with machines of large reactance the reactive circulating current I_c is small, even with large differences in excitation.

The conditions of operation are represented very conveniently by the kVA diagrams of Fig. 22.11. With equal conditions, as in Fig.

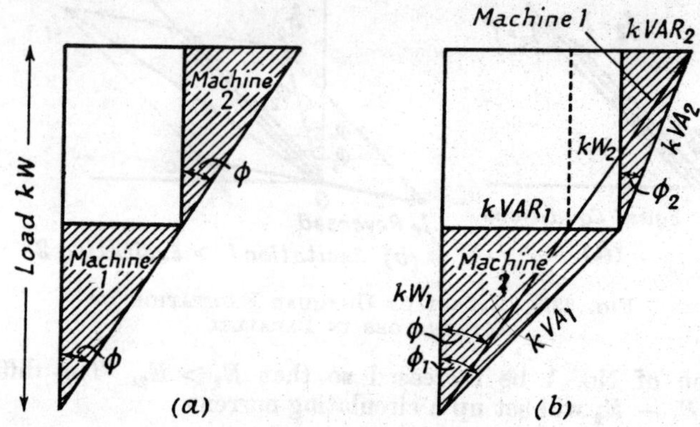

FIG. 22.11. KVA DIAGRAMS FOR ALTERNATORS IN PARALLEL

22.9 (*a*) the kW and kVAR are divided equally, giving equal kVA triangles for the two machines. If the excitation of No. 1 is increased then $kVAR_1$ is increased while $kVAR_2$ is decreased by the same amount. The kW loadings remain sensibly the same, resulting in the kVA triangles of Fig. 22.11 (*b*).

Effect of a Change in Steam Supply. Now let the excitations remain the same but let the steam supply to No. 1 be increased. This requires hand manipulation of the steam stop valve since, unlike d.c. machines, a change in excitation is not able to bring about automatically a change in the steam supply. The increased power intake has to be utilized, and, as the speeds of the two machines are tied together by their synchronous bond, it is not possible for No. 1 to overrun No. 2. Hence the rotor of No. 1 must advance its angular position with respect to No. 2, giving the conditions of Fig. 22.7, except that it is E_1 which is advanced by the angle γ. In this figure the vectors of E_1 and E_2 are drawn in opposition because the conditions are referred to the local circuit round the bus-bars. If they are to refer to the external circuit then E_2 must not be reversed with respect to E_1, so that the vector diagram, with E_1 in advance of E_2, becomes that of Fig. 22.12. Under the effect of the resultant e.m.f. E_r a current I_s flows, and it lags E_r by the characteristic angle θ which, being only slightly less than 90° brings this current into almost exact phase with E_1, so long as γ is small. The power per phase of No. 1 is thus increased by $E_1 I_s$ watts and, since the demand of the load is assumed unaltered, the power per phase of No. 2 is decreased by the same amount. The current I_s has

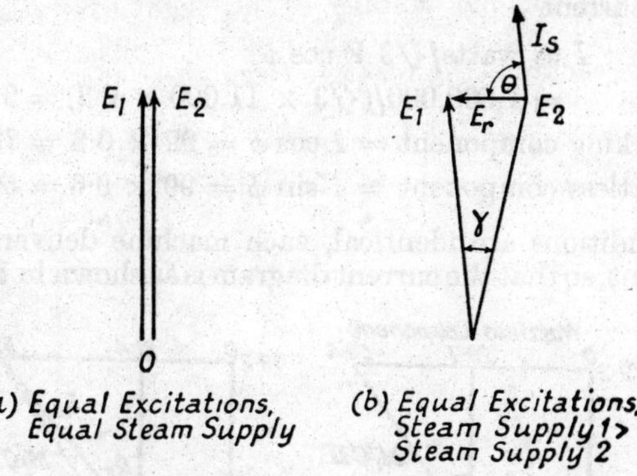

FIG. 22.12. EFFECT OF UNEQUAL STEAM SUPPLIES ON ALTERNATORS IN PARALLEL

no appreciable quadrature component, and therefore the increase in steam supply leaves the division of the kVAR almost the same as before, increases the kW output of No. 1 and reduces that of No. 2 by the same amount. This neglects the armature I^2R losses and any change in the iron losses, but these are relatively very small and do not materially affect the general result. If the two alternators are not connected to a constant voltage, constant-frequency, network there will have to be small adjustments of excitation, and an appropriate reduction in steam supply to No. 2 in order to keep the bus-bar voltage and the frequency the same as before.

The kVA diagram for equal excitations, but with a greater steam supply to No. 1 than to No. 2 is shown in Fig. 22.13.

FIG. 22.13. KVA DIAGRAM FOR TWO ALTERNATORS IN PARALLEL

Excitations are equal, but steam supply 1 is greater than steam supply 2.

EXAMPLE. Two identical three-phase alternators are coupled in parallel to a total load of 1500 kW at 11,000 V, power factor 0·8 lagging. The synchronous reactance of each machine is 60 Ω/phase, and resistance 2·8 Ω/phase. The power supplied by each machine being maintained the same, the excitation of the first alternator is adjusted so that its armature current is 45 A lagging. Calculate—

(*a*) the armature current of the second alternator,

(*b*) the power factor at which each alternator operates,
(*c*) the e.m.f. of the first alternator.

Total load current

$$I = \text{watts}/\sqrt{3}\,V\cos\phi$$
$$= 1{,}500{,}000/(\sqrt{3} \times 11{,}000 \times 0{\cdot}8) = 99\text{A}$$

Working component $= I\cos\phi = 99 \times 0{\cdot}8 = 79{\cdot}2$

Wattless component $= I\sin\phi = 99 \times 0{\cdot}6 = 59{\cdot}4$

When the conditions are identical, each machine delivers one-half of each component, so that the current diagram is as shown in Fig. 22.14 (*a*),

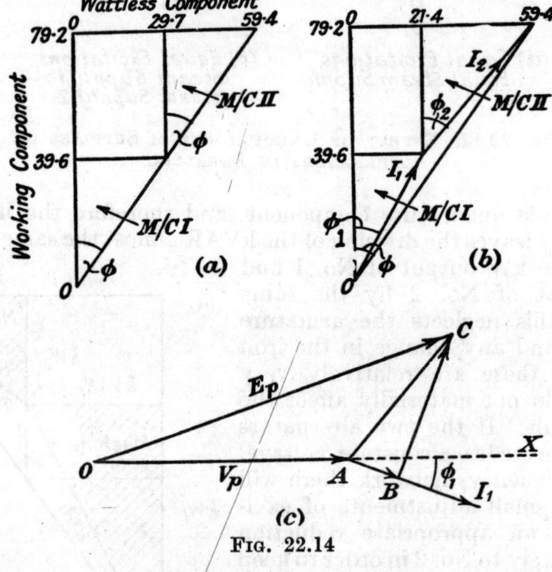

Fig. 22.14

and each machine has the same power factor as the load. Also each total machine current = 49·5 A.

Since the steam supplies are not altered, the working components will remain the same at 39·6 A/machine, but an adjustment of the excitations will alter the division of wattless current. The total current taken by machine I is reduced from 49·5 to 45 A and therefore its wattless current is reduced, the new value being

$$\sqrt{(45^2 - 39{\cdot}6^2)} = 21{\cdot}4 \text{ A}$$

Hence the wattless current delivered by machine II is

$$59{\cdot}4 - 21{\cdot}4 = 38 \text{ A}$$

The current diagram is now as in Fig. 22.14 (*b*), and we have

(*a*) $$I_2 = \sqrt{(39{\cdot}6^2 + 38^2)} = 55 \text{ A}$$

(*b*) $\cos\phi_1 = 39{\cdot}6/45 = 0{\cdot}88$

$\cos\phi_2 = 39{\cdot}6/55 = 0{\cdot}72$

(*c*) Terminal p.d. per phase

$$V_p = 11{,}000/\sqrt{3} = 6350\ \text{V}$$

For machine I

$$\text{Resistance drop per phase} = RI_1 = 2{\cdot}8 \times 45 = 126\ \text{V}$$
$$= AB,\ \text{Fig. 22.14 } (c)$$
$$\text{Reactive drop per phase} = XI_1 = 60 \times 45 = 2700\ \text{V}$$
$$= BC,\ \text{Fig. 22.14 } (c)$$

Denoting by X and Y the OX and OY components of the e.m.f. per phase, E_p, we have

$$X = V_p + RI_1\cos\phi_1 + XI_1\sin\phi_1$$
$$= 6350 + 126 \times 0{\cdot}88 + 2700 \times 0{\cdot}475$$
$$= 7741\ \text{V}$$
$$Y = XI_1\cos\phi_1 - RI_1\sin\phi_1$$
$$= 2700 \times 0{\cdot}88 - 126 \times 0{\cdot}475$$
$$= 2310\ \text{V}$$
$$\therefore\ E_p = \sqrt{(7741^2 + 2310^2)}$$
$$= 8060\ \text{V}$$
$$\therefore\ E = \sqrt{3} \times 8060$$
$$= 14{,}000\ \text{V}$$

Effect of Load on the Synchronizing Power. Since a change in the steam supply alters the phase relationship between the two e.m.f. vectors, we see that this is comparable in effect with a falling back of one machine with respect to the other. If this change is accidental, or is due to cyclic irregularity in the torque of the prime mover then the power which flows as a result of the change is called the synchronizing power. Now suppose that one machine is very large compared with the other, so large, in fact, that changes in the operating conditions of the smaller machine have no effect on its terminal voltage or frequency, then the busbar voltage V, and the frequency will remain constant. The smaller machine is then said to be operating on infinite bus-bars, and the impedance to any circulating current is its own impedance Z, and not $2Z$ as in the case of two equal machines not connected to any other system. For all practical purposes an alternator connected to an extensive supply system in which there are many alternators in parallel, or even power stations in parallel, can be regarded as connected to infinite bus-bars.

We have seen (p. 533) that the stiffness of coupling is given by

$$dP_s/d\gamma \simeq E^2[X/(R^2 + X^2)]\cos\gamma$$

and that $\quad P_s \simeq (E^2/Z)\sin\theta\sin\gamma$, neglecting terms in R.

These were developed on the assumption that $V_2 = E$, that is, they refer to no-load conditions. Now consider a cylindrical rotor machine, and use the single synchronous reactance. On load, the induced voltage will have to be increased by the internal volt drop ZI, according to the relationship

$$\boldsymbol{E} = \boldsymbol{V} + \boldsymbol{ZI}$$

where V is the bus-bar voltage. The vector of E is also displaced with respect to V by the angle COA, Fig. 22.15, which we have denoted by γ, so that the term $\sin\theta$ in the no-load case must be replaced by $\sin(\theta + \gamma)$ We have, under conditions of load

$$P_s = (EV/Z^2)\, X \sin\gamma \text{ per phase}$$

Again, since θ is very nearly 90° we can write

$$P_s \simeq (EV/Z) \sin\theta \sin\gamma^*$$

Owing to the very high reactances of modern alternators, the value of R has a quite negligible effect on the performance except, of course, on the efficiency. When an alternator is delivering no external load, then whatever output it produces is the same as the synchronizing power. From the expression for P_s we see that, over the operative range as an alternator, the synchronizing power is zero when γ is zero.

EXAMPLE. A 750 kVA, 11,000 V, 4-pole three-phase alternator has a resistance of 1·4 Ω/phase and a synchronous reactance of 30 Ω/phase. Calculate the synchronizing power per mechanical *degree* displacement (*a*) at no load, (*b*) on full load at power factor 0·8 lagging, the terminal p.d. being 11,000 V in both cases.

To illustrate the method we will use percentage values in this example·

Full load current per phase $I = 750{,}000/(\sqrt{3} \times 11{,}000) \simeq 40$ A

Phase voltage $E = 11{,}000/\sqrt{3} = 6350$ V

Resistance drop $RI = 1{\cdot}4 \times 40 = 56$ V

$\therefore$ Percentage resistance $= 56 \times 100/6350 \simeq 0{\cdot}9$

Synchronous reactance drop

$$XI = 30 \times 40 = 1200$$

$\therefore$ Percentage reactance $= 1200 \times 100/6350 = 18{\cdot}9$†

(*a*) No load

Percentage impedance $= (0{\cdot}9^2 + 18{\cdot}9^2)^{\frac{1}{2}} \simeq 18{\cdot}9$

$$\therefore \sin\theta = 1{\cdot}0 \text{ very nearly}$$

* For a more complete analysis *see* SAY, M. G., *Performance and Design of Alternating Current Machines*; Chap. XVII. Pitman, London.

† Note that the numerical values are given to illustrate the method of calculation only. In a modern machine the contribution of the leakage flux alone will give a reactance of the order given: the synchronous reactance will be much greater.

$$\gamma = 1° \text{ mechanical}$$
$$= 2° \text{ electrical for a 4-pole machine}$$
$$dP_s/d\gamma = (VE/Z) \sin\theta \simeq (E^2/X) \text{ W/mech. radian}$$
$$= (6350^2/30) \times (\pi/180) \times 2 \text{ W/mech. degree}$$
$$= 46{,}800 \text{ W/phase} = 46{\cdot}8 \text{ kW/phase.}$$

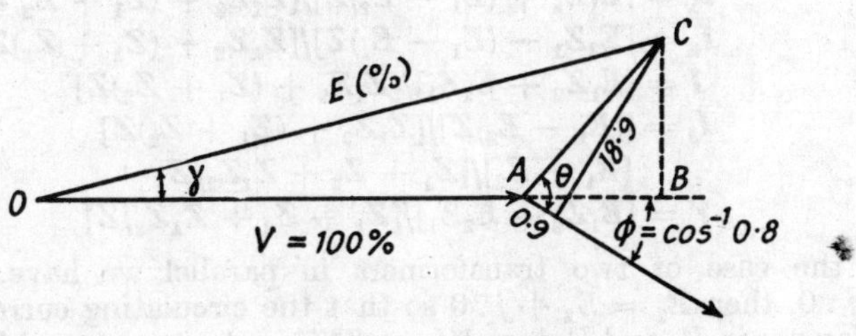

FIG. 22.15

(*b*) Full load, 0·8 power factor
Using percentage values, the regulation diagram is given in Fig. 22.15.

$$OB = 100 + 0{\cdot}9 \times 0{\cdot}8 + 18{\cdot}9 \times 0{\cdot}6$$
$$= 112{\cdot}1$$
$$BC = 18{\cdot}9 \times 0{\cdot}8 - 0{\cdot}9 \times 0{\cdot}6$$
$$= 14{\cdot}6$$
$$\therefore E(\%) = (112{\cdot}1^2 + 14{\cdot}6^2)^{\frac{1}{2}}$$
$$= 113$$
$$V = 6350 \text{ V/phase}$$
$$\therefore E = 1{\cdot}13 \times 6350 = 7176 \text{ V/phase}$$
$$\sin\gamma = 14{\cdot}6/113$$
$$\therefore \gamma = 7° 26'$$
$$dP_s/d\gamma = (VE/Z) \sin\theta \cos\gamma$$
$$= (6350 \times 7176/30) \sin 87° 17' \cos 7° 26'$$
$$= (6350 \times 7176/30) \times 0{\cdot}9989 \times 0{\cdot}9916$$
$$= 1{\cdot}5 \times 10^6 \text{ W/elect. radian}$$
$$= 1{\cdot}5 \times 10^3 \times (\pi/180) \times 2 \text{ kW/mech. degree}$$
$$= 52{\cdot}2 \text{ kW/mech. degree}$$
$$P_s = (750/3) \times 0{\cdot}8 \text{ kW/phase}$$
$$= 200 \text{ kW/phase.}$$

On full-load $P_s/\gamma \simeq dP_s/d\gamma$

$$= 200/7{\cdot}433 = 26{\cdot}8 \text{ kW/elect. degree}$$
$$= 53{\cdot}6 \text{ kW/mech. degree}$$

Solution by Symbolic Method. If E_1 and E_2 are the alternator induced e.m.f.s per phase, Z_1 and Z_2 the impedances per phase and Z the load impedance per phase we have, as with two transformers in parallel,

$$\left.\begin{aligned} E_1 &= I_1Z_1 + (I_1 + I_2)Z \\ E_2 &= I_2Z_2 + (I_1 + I_2)Z \end{aligned}\right\}$$

which lead to

$$I_1 = [E_1Z_2 + (E_1 - E_2)Z]/[Z_1Z_2 + (Z_1 + Z_2)Z]$$
$$I_2 = [E_2Z_1 - (E_1 - E_2)Z]/[Z_1Z_2 + (Z_1 + Z_2)Z]$$
$$I = [E_1Z_2 + E_2Z_1]/[Z_1Z_2 + (Z_1 + Z_2)Z]$$
$$I_s = [(E_1 - E_2)Z]/[Z_1Z_2 + (Z_1 + Z_2)Z]$$

or

$$[E_1 - E_2]/[Z_1 + Z_2 + Z_1Z_2/Z]$$
$$V = [E_1Z_2 + E_2Z_1]/[Z_1 + Z_2 + Z_1Z_2/Z]$$

In the case of two transformers in parallel we have, if $E_1 = E_1 + j\,.\,0$, then $E_2 = E_2 + j\,.\,0$ so that the circulating current due to a difference in E_1 and E_2 is wattless. This is also the case with alternators in which the angle γ is zero but the excitations are different, the circulating current then being the current previously denoted by I_c.

Now suppose that E_2 lags behind E_1 by the angle γ, then we can write

$$E_1 = E_1 + j\,.\,0;\; E_2 = E_2 \cos\gamma - jE_2 \sin\gamma$$

and the current I_s becomes the synchronizing current.

Division of Load between Two Machines. When a machine is switched on to the bus-bars its steam supply is small, so that it may be actually taking a small amount of motoring power from the bus-bars. At the most it will only supply a small amount of power to them. Its working point may therefore be in some position such as P_1 on Fig. 22.16, which, for this purpose can be the simplified Blondel diagram, in which the resistance is neglected and the output lines are therefore horizontal straight lines. To bring the working point into the generating half of the diagram, it is necessary to increase the steam supply. This causes the excitation vector BP_1 to swing round until the working point reaches, say, P_2. The machine is now generating, but its output may not be sufficient, and its power factor will be low because of the large angle of lag, $\angle P_2AZ$, of the current AP_2 behind the voltage. The next adjustment is therefore to increase the excitation until the power factor is in the neighbourhood

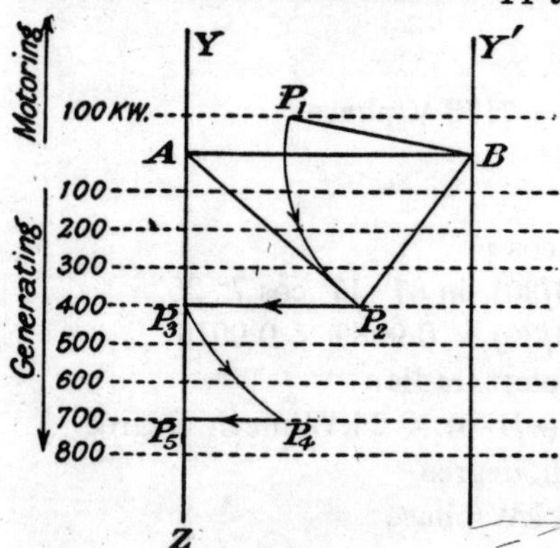

FIG. 22.16. To Illustrate Division of Load

of unity. During this adjustment the working point travels along a power line from P_2 to P_3. The steam supply is again increased, thus bringing the working point to P_4 on a line of greater output, and the excitation is then increased again. Thus the adjustments are made in steps, until the machine is delivering the required output. When adjusting the steam supply, the driver watches the wattmeter, whereas the adjustment of the excitation is made by means of the ammeter, or of a power-factor meter.

Compare this with the case of a d.c. shunt generator in which the load is adjusted by varying the excitation, the engine governor automatically admitting the required amount of steam.

Period of Free Oscillation of a Synchronous Machine.

Let $T' =$ torque/mechanical radian displacement

$mr^2 =$ inertia of the rotating masses

Then with torque expressed in newton-metres and inertia in kgm-metre² units the period of undamped free oscillation is given by

$$t = 2\pi\sqrt{(mr^2/T')} \text{ seconds}$$

Consider first of all a cylindrical rotor machine. The synchronizing power per phase has been shown to be

$$P_s = (E^2/Z)\sin\theta\sin\gamma \text{ for a machine on no load}$$

Since θ is very nearly 90° and γ must be small we have

$$P_s \simeq (E^2/Z)\,\gamma$$

where γ is in *electrical* radians.

$$\therefore\ P_s \text{ per electrical radian} \simeq E^2/Z$$

But one electrical radian is equal to $p/2$ mechanical radians, p being the number of poles, so that, approximately

P_s per mechanical radian

$$= (E^2/Z)(p/2)$$

$$T_s = \text{(watts)/(radians/second) newton-metres}$$

and $\omega = 2\pi N_s$ where $N_s =$ r.p.s.

$$\therefore\ T_s = pE^2/4\pi ZN_s \text{ newton-metres/mechanical radian}$$

For 3 phases

$$T_s = 3pE^2/4\pi ZN_s,\ E \text{ still being the e.m.f. per phase}$$

Now $E/Z = I_{sc}$ the short-circuit current

$$f = N_s p/2 \quad \therefore\ p/N_s = 2f/N_s^2$$

$$\therefore\ T_s = (3EI_{sc}/4\pi) \times (2f/N_s^2)$$

$$= 0{\cdot}478EI_{sc}f/N_s^2$$

$$\therefore\ t = 2\pi\sqrt{(mr^2/T_s)}$$

$$= 2\pi N_s\sqrt{(mr^2/0{\cdot}478EI_{sc}f)}$$

$$= 9{\cdot}11N_s\sqrt{(mr^2/EI_{sc}f)}$$

If the machine is on load, then

$$P_s/\gamma = (EV/Z) \sin\theta \sin\gamma/\gamma$$
$$\simeq (EV/Z) \sin\gamma/\gamma \quad . \quad . \quad . \quad . \quad (1)$$
$$dP_s/d\gamma = (EV/Z) \sin\theta \cos\gamma$$
$$\simeq (EV/Z) \cos\gamma \quad . \quad . \quad . \quad . \quad . \quad (2)$$

Now $(dP_s/d\gamma)_{av} = (EV/Z\gamma)\int_0^\gamma \cos\gamma\, d\gamma = (EV/Z) \sin\gamma/\gamma = P_s/\gamma$

Thus the expression (1) gives the average stiffness coefficient between no-load and the load in question. Expression (2) gives the stiffness coefficient at the load in question. As they are numerically much the same we have two alternative expressions for T_s, namely

$$T_s = (3pEV/4\pi ZN_s) \sin\gamma/\gamma$$

or
$$T_s = (3pEV/4\pi ZN_s) \cos\gamma$$

For the salient-pole machine we have

$$T_s = (3p/4\pi N_s\gamma)[(EV/X_d) \sin\gamma + \{(X_d - X_c)/2X_dX_c\}V^2 \sin 2\gamma]$$

or $T_s = (3p/4\pi N_s)[(EV/X_d) \cos\gamma + \{(X_d - X_c)/X_dX_c\}V^2 \cos 2\gamma]$

Example. A 2180 kVA, 6300 V, 50 c/s, 125 r.p.m. alternator is running in parallel with other alternators whose aggregate kVA is very large. Its transient short-circuit current is 5 times its normal current. It is driven by a gas-engine and has a direct-connected flywheel. The equivalent mr^2 of engine and alternator is 800 tons, and of the flywheel 1400 tons at one foot radius. Calculate the time of one free oscillation, on no load, given that X_c is 0·6 of X_d.

Normal current

$$I = 1000 \times 2180/\sqrt{3} \times 6300) = 200 \text{ A}$$
$$\therefore I_{sc} = 5 \times 200 = 1000 \text{ A}$$
$$E_{ph} = 6300/\sqrt{3} = 3637 \text{ V on no load}$$
$$\therefore X_d = 3637/1000 = 3{\cdot}64\ \Omega$$
$$\therefore X_c = 0{\cdot}6 \times 3{\cdot}64 = 2{\cdot}18\ \Omega$$

No. of poles $\quad p = 120f/N = 6000/125 = 48$

$$T_s = 3p/4\pi N_s[E^2/X_d + \{(X_d - X_c)/X_dX_c\}E^2]$$
$$= 3 \times 48/4\pi \times 60/125 \times 3637^2 \times (1/3{\cdot}64)[1 + (3{\cdot}64 - 2{\cdot}18)/2{\cdot}18]$$
$$= 3{\cdot}35 \times 10^7 \text{ newton-metres/mechanical radian}$$
$$mr^2 = 800 + 1400 = 2200 \text{ tons at one foot radius}$$
$$= 2200 \times 2240 = 4{\cdot}93 \times 10^6 \text{ lb-ft}^2 \text{ units}$$
$$= 4{\cdot}93 \times 10^6/[2{\cdot}2 \times (3{\cdot}28)^2] = 2{\cdot}08 \times 10^5 \text{ kg-m}^2 \text{ units}$$
$$\therefore t = 2\pi\sqrt{(2{\cdot}08 \times 10^5/3{\cdot}35 \times 10^7)}$$
$$= 0{\cdot}50 \text{ sec (undamped)}$$

Cyclic Irregularity. If an alternator is driven by an internal combustion engine, the cyclic variations in the torque of the engine, will cause cyclic variations in the angular velocity. Let ω_1 and ω_2 be the maximum and minimum values and ω_{av} the average, then

$$\text{Cyclic variation } \delta = (\omega_1 - \omega_2)/\omega_{av}$$

As a very rough value it can be said that this should not be greater than 1/200, but the permissible value depends on many circumstances, including the engine itself. If we define $1/\delta$ as the cyclic regularity, this

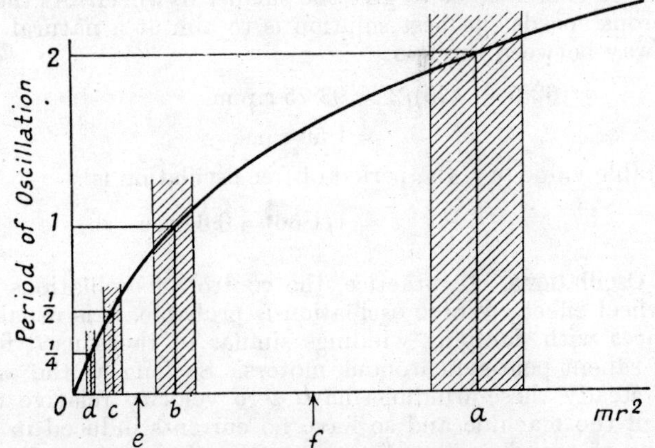

FIG. 22.17. FLYWHEEL EFFECT v. FREE OSCILLATION

is directly proportional to the mr^2, the so-called "flywheel effect" of the rotating masses. The period of free oscillation is, as we have seen, proportional to the square-root of mr^2, so that period plotted against mr^2 gives a parabolic curve, as shown in Fig. 22.17. Consider an alternator driven by a four-cycle engine, then there are oscillations corresponding to the main-shaft speed, which we will denote by unity and by the cam-shaft speed, of one-half: there are, in addition oscillations of one-quarter and of twice the shaft-speed. Thus there are four possibilities of resonance, and each spreads to either side to give danger zones, as shown by the shaded rectangles. If an alternator is working in parallel with others then the flywheel can be designed to give operation midway between two of the danger zones. From this point of view the chosen mr^2 could be either at e or at f, and, in practice, the former will be chosen because of the great mass of flywheel required for the other. It is to be noted that the above considerations do not apply to an alternator not in parallel with others, because such a machine does not constitute an oscillatory system, in that there is no restoring torque derived from the network to which it is connected, if there is a sudden change in the rotor displacement. In such a case there are no danger zones and the flywheel can be designed to give any desired value to the cyclic irregularity.*

* The mathematical discussion of this problem is beyond the scope of this book. *See* Miles WALKER, *Specification and Design of Dynamo Electric Machinery*, pp. 337–47. Longmans, London; R. Richter, *Elektrische Maschinen*, Vol. II, pp. 363–99. Springer, Berlin.

As an example consider the 2180 kVA alternator of p. 544. The main danger speeds are the cam-shaft speed of 62·5 r.p.m., and the main-shaft speed of 125 r.p.m. respectively. The danger zones can be regarded as extending to 20 per cent on either side of each. If there were only one dangerous speed, safety would be ensured by resonant speeds $\not>$ 80 per cent and $\not<$ 120 per cent of this, and the higher value would be chosen so as to give the smaller flywheel. As there are two dangerous speeds the best solution is to aim at a natural period about halfway between the two.

$$(62{\cdot}5 + 125)/2 = 93{\cdot}75 \text{ r.p.m.}$$

$$= 1{\cdot}56 \text{ r.p.s.}$$

Thus a suitable value of t, the period of free oscillation is

$$t = 1/1{\cdot}56 = 0{\cdot}64 \text{ sec}$$

Damped Oscillations. In practice, the control of oscillations is not left to flywheel effect. Where oscillation is probable, it is usual to fit the pole faces with damping windings similar to those used for the starting of salient-pole synchronous motors. So long as the angular velocity is steady these windings have zero velocity relative to the main field of the machine and so have no currents induced in them. In the event of any departure from exact synchronous speed, as for example when an oscillatory motion is superposed on the uniform rotation at synchronous speed, currents are induced and the kinetic energy of this oscillatory motion thereby converted into heat energy.

The damping winding functions in exactly the same manner as the squirrel-cage winding of an induction motor, and it is shown on page 559 that, for small relative angular velocities, the torque due to such a winding is directly proportional to the velocity. Provided there is no "forcing" of the oscillatory motion due to the application of a cyclic disturbance from an external source, the equation of motion is

$$J(d^2\gamma/dt^2) + T_d(d\gamma/dt) + T_s\gamma = 0$$

where T_d is the torque per unit angular velocity due to the damping winding, and T_s the synchronizing torque per unit displacement, the assumption being that γ is small enough for the conditions to be represented by the initial straight portion of the appropriate torque-angle characteristic. The above equation can be written

$$(d^2\gamma/dt^2) + (T_d/J)(d\gamma/dt) + (T_s/J)\gamma = 0$$

Put $\quad T_d/J = 2a \text{ and } T_s/J = b^2$

$$\therefore\ d^2\gamma/dt^2 + 2a(d\gamma/dt) + b^2\gamma = 0$$

Let $\gamma = A\,e^{\alpha t}$ be a solution, then

$$\alpha^2 + 2a\alpha + b^2 = 0$$

$$\alpha + a = \pm\sqrt{(a^2 - b^2)}$$

giving two values for α_1, namely

$$\alpha_1 = -a + \sqrt{(a^2 - b^2)}$$
$$\alpha_2 = -a - \sqrt{(a^2 - b^2)}$$

The complete solution is thus

$$\gamma = A\,e^{\alpha_1 t} + B\,e^{\alpha_2 t}$$

where $\alpha_1 = -T_d/2J + \sqrt{[(T_d/2J)^2 - T_s/J)]}$

and $\alpha_2 = -T_d/2J - \sqrt{[(T_d/2J)^2 - T_s/J]}$

The motion is aperiodic or is damped oscillatory according as $(T_d/2J)^2 >$ or $< T_s/J$. Normally the latter is the case and is therefore the only one which need be considered here. Now $a = T_d/2J$ and we can write

$$jb = j\sqrt{[T_s/J - (T_d/2J)^2]}$$

the symbol b now having a different meaning from before. The equation then becomes

$$\gamma = e^{-at}(A\,e^{jbt} + B\,e^{-jbt})$$

We must also add a constant whose value is decided by the initial conditions. Thus, suppose that a steady load torque T produces a steady displacement of γ_2, then we have

$$\gamma = e^{-at}(A\,e^{jbt} + B\,e^{-jbt}) + \gamma_2$$

where $\gamma_2 = T/T_s$

EXAMPLE. Consider again the example on p. 544.

$$T_s = 3{\cdot}35 \times 10^7 \text{ newton-metres/mechanical radian}$$
$$J = 2{\cdot}08 \times 10^5 \text{ kg-m}^2$$

Let $(T_d/2J)^2 = \frac{1}{2}(T_s/J)$, a possible value

$$\therefore\ T_d^2/(4 \times 2{\cdot}08^2 \times 10^{10}) = \tfrac{1}{2}[3{\cdot}35 \times 10^7/(2{\cdot}08 \times 10^5)]$$
$$T_d^2 = 13{\cdot}94 \times 10^{12}$$
$$T_d = 3{\cdot}73 \times 10^6$$

Let the initial conditions correspond to a momentary sudden overload applied to the machine when it is delivering full rated output.

$$T_{\text{full load}} = (\text{watts})/(\text{angular velocity}) = 2180 \times 1000/(125/60 \times 2\pi)$$
$$= 166{,}000 \text{ newton-metres}$$
$$\gamma_{\text{full load}} = T_{\text{full load}}/T_s \simeq 166{,}000/3{\cdot}35 \times 10^7$$
$$\simeq 0{\cdot}00495 \text{ mechanical radians}$$

If we use the symbol γ for the component of the displacement due to the suddenly applied overload, and γ_2 the displacement due to steady rated load then, at any instant, the total displacement is given by

$$\gamma' = \gamma + \gamma_2$$
$$= e^{-at}(A\,e^{jbt} + B\,e^{-jbt}) + \gamma_2$$
$$= e^{-at}(C\cos bt + D\sin bt) + \gamma_2$$

The initial condition is

$$\gamma = 0 \text{ when } t = 0 \quad \therefore C = 0$$

Let the magnitude of the suddenly applied overload be such that

$$\gamma_{max} = k\gamma_2$$

Hence, putting $\qquad d\gamma/dt = 0$

$$-a\,e^{-at}\,D\sin bt + b\,e^{-at}\,D\cos bt = 0$$

Hence $\quad \tan bt = b/a$ when $\gamma = \gamma_{max}$

$$\therefore \gamma_{max} = D\,e^{-at}\sin bt$$

$$D = k\gamma_2/e^{-at}\sin bt$$

$$a = T_d/2J = 9$$

$$b = \sqrt{[T_s/J - (T_d/2J)^2]} = \sqrt{(T_s/2J)} = \sqrt{80{\cdot}6} = 9$$

$$\therefore bt = \tan^{-1} 1 = \pi/4$$

$$\therefore \sin bt = 0{\cdot}707$$

$$e^{-at} = e^{-bt} = e^{-\pi/4} = 0{\cdot}456$$

We will now assume that $k = 0{\cdot}5$

$$\therefore D = 0{\cdot}5 \times 0{\cdot}00495/(0{\cdot}456 \times 0{\cdot}707) = 0{\cdot}00767$$

Hence $\qquad \gamma' = 0{\cdot}00767\,e^{-9t}\sin 9t + 0{\cdot}00495$

The solution of this is set out in tabular form below.

t	$9t$	$9t \times 180/\pi = \theta$	$\log \sin \theta$	$\log e^{-9t} = -3{\cdot}91t$	γ'
0·05	0·45	25° 48′	$\bar{1}$·6387	− 0·1954	+ 0·002128
0·10	0·90	51° 30′	$\bar{1}$·8935	− 0·3909	+ 0·002445
0·15	1·35	77° 18′	$\bar{1}$·9892	− 0·5865	+ 0·001938
0·20	1·80	103° 6′	$\bar{1}$·9885	− 0·7817	+ 0·001235
0·25	2·25	129° 0′	$\bar{1}$·8905	− 0·9775	+ 0·0006277
0·30	2·70	154° 42′	$\bar{1}$·6308	− 1·1726	+ 0·0002203
0·35	3·15	180° 30′	$\bar{3}$·9408	− 1·3700	− 0·0000028
0·40	3·60	206° 12′	$\bar{1}$·6449	− 1·5635	− 0·0000915
0·45	4·05	232° 0′	$\bar{1}$·8965	− 1·7600	− 0·0001051
0·50	4·50	258° 0′	$\bar{1}$·9904	− 1·954	− 0·0000834
0·55	4·95	284° 0′	$\bar{1}$·9869	− 2·150	− 0·0000527
0·60	5·40	309° 30′	$\bar{1}$·8874	− 2·345	− 0·0000267
0·65	5·85	335° 30′	$\bar{1}$·6177	− 2·543	− 0·0000091
0·70	6·30	361°	$\bar{2}$·2419	− 2·736	+ 0·0000002

From the above figures we see that the natural period of oscillation on full load with damping is very approximately 0·7 sec. This can be verified as follows—

$$\begin{pmatrix}\text{Natural period}\\ \text{with damping}\end{pmatrix} = \begin{pmatrix}\text{Undamped}\\ \text{period}\end{pmatrix} \times \sqrt{\left(\frac{T_s/J}{T_s/J - (T_d/2J)^2}\right)}$$

$$= \text{(Undamped period)} \times \sqrt{2}, \text{ in this case}$$
$$= 0{\cdot}50 \times \sqrt{2}$$
$$= 0{\cdot}705 \text{ sec}$$

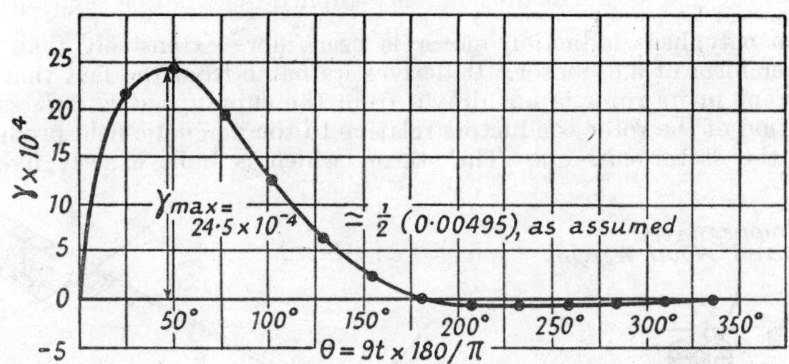

Fig. 22.18

agreeing with the above figure. The graph of angular deflection from the steady-state position, plotted against time is given in Fig. 22.18.

Further Reading

The works suggested on pp. 470 and 500. Also

Bewley, L. V. *Alternating Current Machinery.* Macmillan, 1949.

Crary, E. B. *Power System Stability* (2 vols.). Wiley, 1947.

Liwschitz-Garick, M. and Whipple, C. C. *Electric Machinery,* Vol. II. Van Nostrand, 1947.

CHAPTER 23

THE INDUCTION MOTOR

THE polyphase induction motor is used more extensively than any other form of a.c. motor. It derives its name from the fact that the current in its rotor is not drawn from the supply, but is *induced* by motion of the rotor conductors relative to the magnetic field produced by the stator currents. The stator, which is built exactly like an alternator stator, carries a polyphase winding similar to an alternator winding, and the conditions governing the choice of winding, as, for example between single-layer and two-layer are similar to those for alternators. Motors often work at the low-voltage of a.c. distribution systems and consequently large motors may require relatively few conductors and the winding will be a "bar" winding as illustrated in Fig. 23.1. For small motors operating at, say, 600 V a large number of small section conductors are necessary and a "mush" winding will be used. This, as shown in Fig. 23.2, can be regarded as consisting of hanks of insulated copper wire, the individual slot conductors being placed singly in the slots, the overhang then taped and shaped by hand.

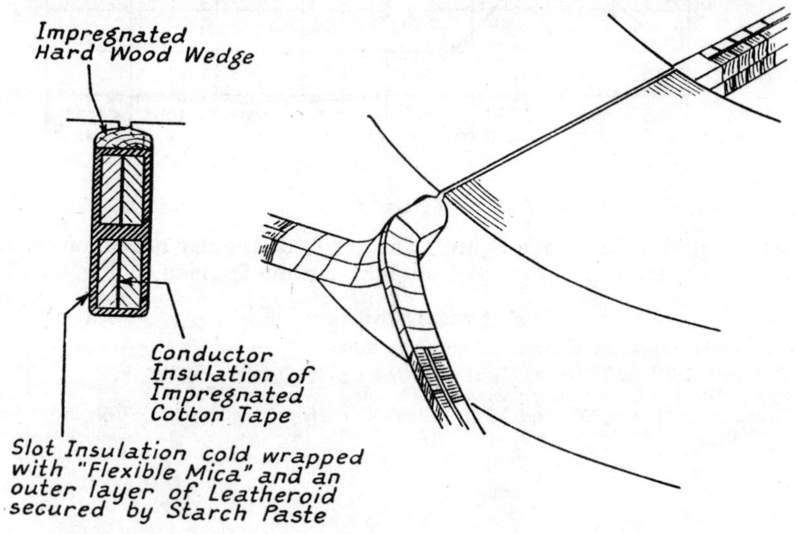

FIG. 23.1. BAR-WOUND STATOR OF INDUCTION MOTOR

The illustration shows a push-through type, two-layer winding, in semi-enclosed slots. The coil is inserted by pushing the bar through from one end of the slot, the straight end being afterwards bent to final shape.

(*Mather & Platt, Ltd.*)

The construction of a modern induction motor is shown in detail in Fig. 23.3. The stator winding is connected to the supply, and the polyphase currents circulating through it produce a magnetic field which rotates at synchronous speed. The lines of force of the stator field cut the rotor conductors and induce currents in them; hence, by Lenz's Law, the rotor follows after the stator field. Now it is on the relative velocity of the rotor and stator field that the magnitude of the induced rotor current, and therefore, of the torque, depends; hence, even on no

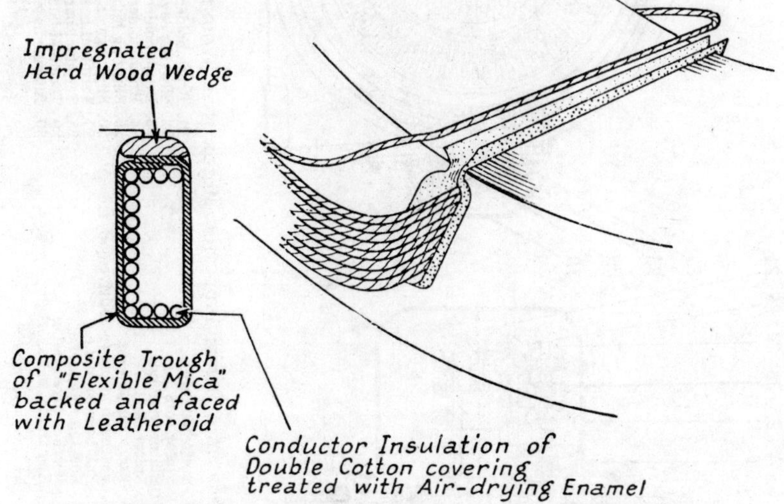

FIG. 23.2. WIREWOUND STATOR OF INDUCTION MOTOR

A mush-type winding being inserted into a semi-closed slot. The coil is inserted wire by wire, as shown.

load the rotor speed must be *less* than synchronous speed, because a torque must be set up to overcome the friction and iron loss torque. When extra load is put on the motor, the torque must increase in order to cope with it; hence, the induced rotor current must increase, and therefore, the relative velocity of rotor to stator field must increase. But the speed of the stator field is fixed by the frequency of the supply, and therefore, as the load on the motor increases, the actual speed of the rotor decreases. Its speed characteristic is therefore somewhat similar to that of a d.c. shunt motor.

Slip. For a unidirectional torque to be set up, the rotor field produced by the induced rotor currents must have the same angular velocity as the main stator field, and therefore, the lines of force "slip" past the rotor conductors.

Let ω_1 = angular velocity of stator field, and therefore of rotor field

ω_2 = actual angular velocity of rotor

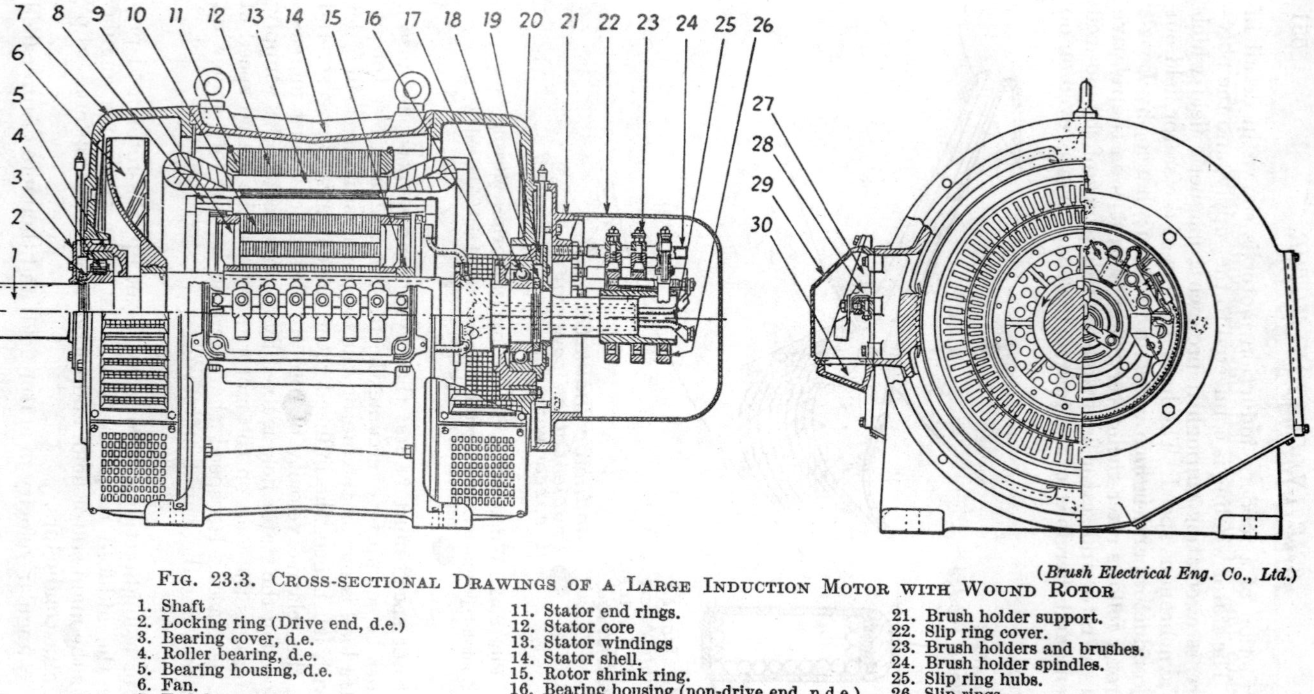

(*Brush Electrical Eng. Co., Ltd.*)

Fig. 23.3. Cross-sectional Drawings of a Large Induction Motor with Wound Rotor

1. Shaft
2. Locking ring (Drive end, d.e.)
3. Bearing cover, d.e.
4. Roller bearing, d.e.
5. Bearing housing, d.e.
6. Fan.
7. End frame, d.e.
8. Rotor windings.
9. Rotor end rings.
10. Rotor core.
11. Stator end rings.
12. Stator core
13. Stator windings
14. Stator shell.
15. Rotor shrink ring.
16. Bearing housing (non-drive end, n.d.e.)
17. Ball bearing, n.d.e.
18. Locking ring, n.d.e.
19. Bearing cover, n.d.e.
20. End frame, n.d.e.
21. Brush holder support.
22. Slip ring cover.
23. Brush holders and brushes.
24. Brush holder spindles.
25. Slip ring hubs.
26. Slip rings.
27. Terminal box.
28. Terminal bar and sockets.
29. Terminal box cover.
30. Intermediate chamber.

The difference $(\omega_1 - \omega_2)$ is called the "absolute" slip. The ratio $(\omega_1 - \omega_2)/\omega_1$ is called the "fractional" slip, and $100(\omega_1 - \omega_2)/\omega_1$ is the percentage slip.

The full-load slip of modern motors varies from about 4 per cent for small motors to 1·5 or 2 per cent for very large ones. The actual speed is thus very close to the synchronous speed under normal working conditions when the speed is unregulated.

When the rotor is stationary, the stator field sweeps past the rotor conductors with full synchronous speed, and the frequency of the rotor current is equal to the line frequency f. The motor then functions like a rotating-flux transformer, as distinct from an alternating-flux transformer, but with distributed windings and a narrow air gap in the magnetic circuit. When running at a speed less than synchronism the relative velocity of stator field to rotor is $\sigma\omega_1$, where σ is the fractional slip. But the frequency of the rotor currents is proportional to this relative velocity. Hence, the rotor currents have a frequency of σf.

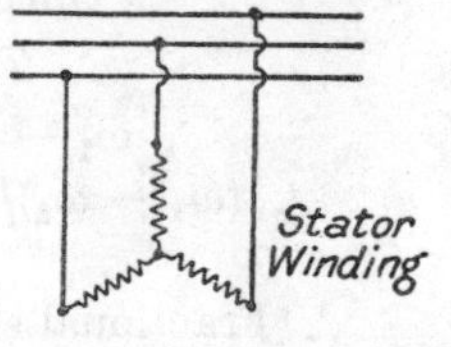

Phase-wound and Cage-type Rotors. There are two types of rotor winding, the "squirrel-cage" and the "phase-wound." The squirrel-cage rotor consists of a number of copper bars threaded through slots in a laminated rotor core, with their ends connected to stout copper or alloy end rings. These windings are thus permanently short-circuited on themselves, and external resistance cannot be connected in series for starting purposes.

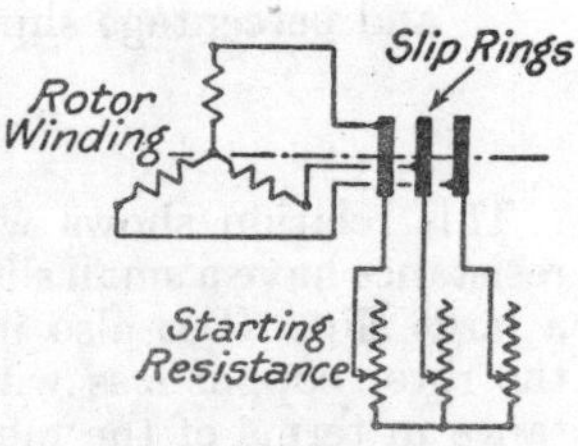

FIG. 23.4. INDUCTION MOTOR CONNECTIONS

The phase-wound rotor is provided with a distributed winding which is nearly always a two-layer wave winding in the case of large motors unless the voltage is so high that a single-layer "concentric" winding is preferable. The separate phases are generally connected in star, and the three free ends connected each to one of three insulated slip rings mounted on the motor shaft. When running normally these slip rings are short-circuited, but for starting purposes they are connected to a three-phase star-connected starting resistance, as in Fig. 23.4.

The squirrel-cage rotor is as a rule used only for comparatively small motors, because, as we shall see, its starting torque is very small and it takes a large current from the line at the moment of switching on. The rotor resistance is very low, and therefore, the motor is very efficient, and the simple construction of the rotor renders it mechanically robust. and comparatively cheap to manufacture. The small starting torque is the result of the low rotor resistance. When squirrel-cage motors are used for crane or hoist work, where a large starting torque is more important then high efficiency, because of the intermittent nature of the load, it is usual to increase the rotor resistance by making the end rings, and sometimes the bars as well, of high-resistance metal, e.g. bronze, and we shall see that, by this device, the starting torque is increased.

Relation between Slip and Rotor I^2R Loss. When an induction motor is operating normally, the slip is so small that the frequency of the magnetic reversals in the rotor core is only of the order of one or two per second, the rotor iron loss therefore being quite negligible. Hence

$$\text{(Power of rotating field)} = \text{Output} + \text{rotor copper loss}$$

If T is the motor torque in newton-metres

$$\begin{aligned}
\text{Power of rotating field} &= T\omega_1 \text{ watts} \\
\text{Output} &= T\omega_2 \text{ watts} \\
\therefore\ T\omega_1 &= T\omega_2 + \text{rotor copper loss} \\
\therefore\ \omega_1 - \omega_2 &= (\text{rotor copper loss})/T \\
\therefore\ (\omega_1 - \omega_2)/\omega_1 &= (\text{rotor copper loss})/T\omega_1 \\
&= (\text{rotor copper loss})/(\text{input power}) \\
\therefore\ \text{Fractional slip} &= (\text{rotor copper loss})/(\text{input power}) \\
\text{and percentage slip} &= \text{rotor copper loss/input power} \times 100 \\
&= \text{rotor copper loss expressed as a percentage} \\
&\qquad \text{of the input power of the rotor}
\end{aligned}$$

This relation shows why it is that motors with a very low rotor resistance have a small slip, while those with a high rotor resistance have a large slip. It is also important because it enables us to determine the rotor copper loss without measuring rotor current or rotor resistance in terms of the measured slip. Even with a slip-ring rotor this is an advantage and, in the case of a cage rotor it is the only possible method.

Torque. The mechanism of torque production in the induction motor is essentially the same as in the d.c. motor. Fig. 23.5 shows one rotor conductor which, for convenience, is shown midway between the stator and rotor surfaces. Assume the stator field is travelling in a clockwise direction, and therefore, since the rotor speed cannot be equal to the stator field speed, the rotor is travelling counter-clockwise *relative to the stator field.* Hence when applying the right-hand rule to the rotor conductor its motion must be considered as being from right to left. This shows that the e.m.f. in the conductor acts outwards, and therefore the current acts outwards. The directions of the lines of force in the two component fields are therefore as shown in the first figure, while the resultant field distribution is as given in the second figure. Thus lines of force are bent round the conductor and a torque set up in the same direction as the stator field. In this case also the forces which exist in an actual machine act on the tops of the rotor teeth and not on the conductors themselves, and the student is advised to draw the figure corresponding to Fig. 10.2.

In the d.c. motor the torque is proportional to the product of the flux per pole and the armature current, and similarly in the induction motor the torque is proportional to the flux per stator pole and the rotor current. There is, however, an additional factor to be taken into account, namely the rotor power factor. We have seen that the rotor frequency at any fractional slip σ is σf, and therefore if $X_{2\cdot 0}$ is the

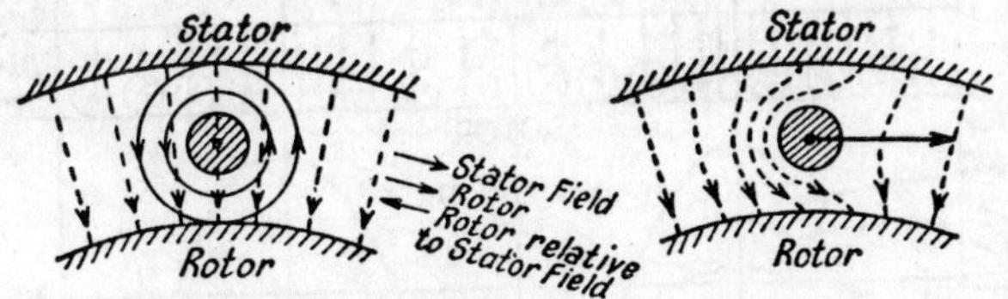

FIG. 23.5. TORQUE PRODUCTION IN AN INDUCTION MOTOR

rotor standstill reactance, i.e. its reactance to currents of line frequency, its reactance at any slip σ is $\sigma X_{2\cdot 0}$. First of all imagine the rotor at rest, then its reactance is $X_{2\cdot 0}$ and its impedance therefore

$$Z_2 = (R_2^2 + X_{2\cdot 0}^2)^{\frac{1}{2}}$$

while $$\cos\phi_2 = R_2/Z_2$$

Now in a normal motor $X_{2\cdot 0}^2$ is much larger than R_2^2 with the result that, at standstill

$$Z_2 \simeq X_{2\cdot 0}$$

and $\cos\phi_2$, is small.

Thus the current will lag the rotor induced voltage by an angle of the order of 90°.

Now imagine that the rotor could travel at exactly synchronous speed, then at this speed

$$\sigma = 0 \quad \therefore \ \sigma X_{2\cdot 0} = 0$$
$$\therefore \ Z_2 = R_2$$
$$\text{and } \cos\phi_2 = 1$$

Actually, an unregulated motor cannot run exactly at synchronism but the slip is so small that the condition $\cos\phi_2$ equal to unity is very nearly fulfilled. We will therefore consider the two extreme cases of $\cos\phi_2 \simeq$ unity and $\cos\phi_2 = 0$. Fig. 23.6 shows two pole pitches of an induction motor. We have already seen that the distribution in space of the m.m.f. of a three-phase winding is stepped and is not given by a smooth curve. But owing to the tendency of a magnetic flux to spread out laterally the flux density distribution produced by this m.m.f. will have the corners rounded off, so that when considering the general properties of the motor we can assume that the distribution of air-gap flux density in space is sinusoidal. This will result in a sinusoidal distribution of rotor induced e.m.f. in space, and therefore a sinusoidal distribution of rotor current in space in the case of a cage rotor, in which the various bars, not being connected in groups, can carry different currents. Hence for the case of unity power factor the gap density and rotor current distributions are in space phase. Now the force on a conductor is given by the product of its current and the gap density at its centre, and consequently by plotting the product of gap density and corresponding rotor current we obtain a curve giving the

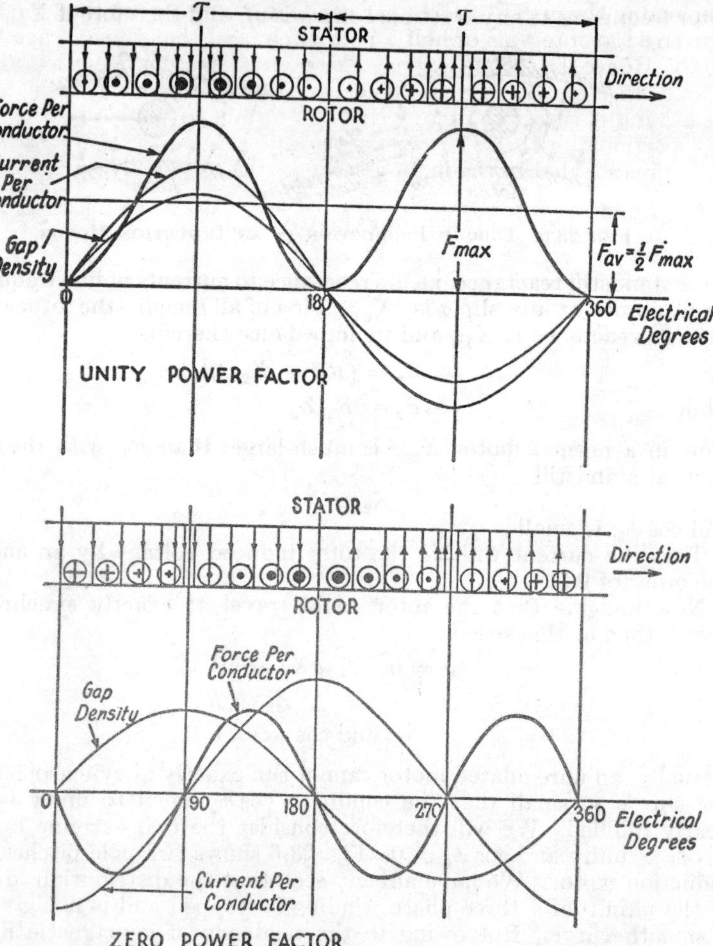

FIG. 23.6. EFFECT OF ROTOR POWER FACTOR ON THE TORQUE OF AN INDUCTION MOTOR

distribution of force round the rotor periphery. We see that for unity power factor this force is always positive but that at intervals of a pole pitch (Y) it falls to zero. The average force is one-half of the maximum force.

The second diagram of Fig. 23.6 shows the operating conditions for a rotor power factor of zero. The current lags 90 degrees behind the induced voltage, and therefore the rotor will have travelled 90 electrical degrees, or half a pole pitch, before the current in any particular bar will be the same as in the previous figure. In other words, the rotor current distribution will be shifted 90 electrical degrees to the right. On plotting the curve of force per conductor we now see that the force is reversed at intervals of half a pole pitch, and that the average force is zero. Hence the torque is zero. These curves are analogous with those of Figs. 15.7 and 15.8, giving the curves of power in a single-phase circuit, and it therefore follows that the torque developed by the rotor is proportional to the rotor power factor.

We thus see that in the induction motor the torque is proportional to

$$\begin{pmatrix}\text{Flux entering rotor}\\ \text{from stator}\end{pmatrix} \times (\text{rotor current}) \times \begin{pmatrix}\text{cos of angle between}\\ \text{flux and current}\end{pmatrix}$$

The flux is proportional to the stator applied voltage E_1, exactly as in a transformer.

$$\text{Rotor current per phase} = \frac{\text{rotor e.m.f.}}{\text{rotor impedance per phase}}$$

When the rotor is stationary, there will be a certain transformation ratio k between stator and rotor, and for the rotor e.m.f. we have

$$E_{p.2} = kE_{p.1}$$

When the motor is running, the relative velocity of rotor and stator field is proportional to the fractional slip σ, and since the fractional slip at standstill is unity, we have for the rotor e.m.f. when running,

$$E_{p.2} = \sigma k \,.\, E_{p.1}$$

Again, the frequency of the rotor current is σf; hence, if $X_{2\cdot 0}$ is the rotor reactance per phase to currents of line frequency, its reactance to currents of slip frequency will be $\sigma X_{2\cdot 0}$. Hence, the rotor current is given by

$$\sigma k E_{p.1}/(R_2^2 + \sigma^2 X_{2\cdot 0}^2)^{\frac{1}{2}}$$

The ratio k is equal to the turns ratio per phase N_2/N_1, if the stator and rotor have the same number of phases and the same values for the winding coefficients k_2 and k_3. The rotor can, however, have a different number of phases, since its performance depends on the rotating field of the machine and not on the manner in which this field is produced. Hence, in general, denoting the numbers of stator and rotor phases by m_1 and m_2 we have

$$k = m_2 N_2 k_{2\cdot 2} k_{2\cdot 3}/m_1 N_1 k_{1\cdot 2} k_{1\cdot 3}$$

Again. from the impedance triangle (Fig. 23.7), we have

$$\cos\phi = R_2/(R_2^2 + \sigma^2 X_{2\cdot 0}^2)^{\frac{1}{2}}$$

The expression for the torque, T, therefore becomes

$$\begin{aligned} T &\propto E_{p.1} \times \sigma k E_{p.1}/\sqrt{(R_2^2 + \sigma^2 X_{2\cdot 0}^2)} \times R_2/\sqrt{(R_2^2 + \sigma^2 X_{2\cdot 0}^2)} \\ &\propto \sigma k E_{p.1}^2 R_2/(R_2^2 + \sigma^2 X^2_{2\cdot 0}) \end{aligned}$$

If we start with the power developed, instead of the torque, then we can derive the same relationship and conveniently determine the factor which gives an actual torque equation. Let $E_{p.2}$ and $I_{p.2}$ be the rotor voltage and current per phase when on load, then the rotor power developed

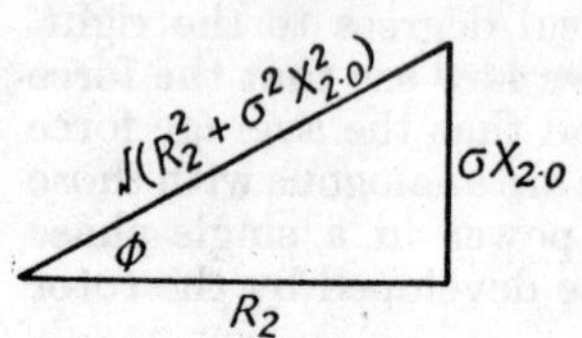

Fig. 23.7. Rotor Impedance Triangle

$$P_2' = m_2 E_{p.2} I_{p.2} \cos\phi_2/\sigma$$

Again $E_{p.2} = \sigma k E_{p.1}$

and $I_{p.2} = E_{p.2}/Z_{p.2}$

$$= k\sigma E_{p.1}/(R_2^2 + \sigma^2 X_{2\cdot0}^2)^{\frac{1}{2}}$$

$$\cos\phi_2 = R_2/(R_2^2 + \sigma^2 X_{2\cdot0}^2)^{\frac{1}{2}}$$

$$\therefore\ P_2' = m_2 k^2 E_{p.1}^2[\sigma^2 R_2/(R_2^2 + \sigma^2 X_{2\cdot0}^2)]$$

Rotor copper loss $= m_2 I_{p.2}^2 R_2$

$$= m_2 k^2 \sigma^2 E_{p.1}^2 R_2/(R_2^2 + \sigma^2 X_{2\cdot0}^2)\sigma P_2'$$

Hence power converted into mechanical power

$$P_2 = P_2' - \text{Rotor copper loss}$$

$$= m_2 k^2 E_{p.1}^2 \sigma R_2(1-\sigma)/(R_2^2 + \sigma^2 X_{2\cdot0}^2)$$

Now actual speed

$$N = N_s(1-\sigma) \text{ r.p.second}$$

$\therefore$ Angular velocity

$$\omega = 2\pi N_s(1-\sigma) \text{ radians/second}$$

But $P_2 = T\omega$

$$\therefore\ T = P_2/\omega$$

$$= m_2 k^2 E_{p.1}^2 \sigma R_2(1-\sigma)/[2\pi N_s(1-\sigma)(R_2^2 + \sigma^2 X_{2\cdot0}^2)]$$

$$= \frac{(m_2/2\pi)(1/N_s)k^2 E_{p.1}^2 \sigma R_2}{(R_2^2 + \sigma^2 X_{2\cdot0}^2)} \text{ newton-metres}$$

If N_s is expressed in r.p.m. and we require the torque in pound-feet units we have, since 1 newton-metre is equal to 0·7373 lb-ft unit

$$T = \frac{0{\cdot}7373(60m_2/2\pi)(1/N_s)k^2 E_{p.1}^2 \sigma R_2}{R_2^2 + \sigma^2 X_{2\cdot0}^2}$$

$$= \frac{7{\cdot}04(m_2/N_s)k^2 E_{p.1}^2 \sigma R_2}{R_2^2 + \sigma^2 X_{2\cdot0}^2} \text{ lb-ft units}$$

Several important properties of the motor can be deduced from the torque equation

(a) $T \propto (1/\sigma) \times \{\sigma k E_{p.1}/\sqrt{(R_2^2 + \sigma^2 X_{2\cdot0}^2)}\}^2 \times R_2$

The quantity in the brackets is the rotor current, and we therefore have

$$T \propto I_{p.2}^2 R_2/\sigma,$$

$$\propto \text{(rotor copper loss)/(slip)}$$

(h) When the motor is running under normal conditions σ is very small, and we can neglect $(\sigma X_{2\cdot0})^2$ in comparison with R_2^2. Hence, under these conditions,

$$T \propto \sigma E_{p.1}^2, \; k \text{ and } R_2 \text{ being constants}$$
$$\propto E_{p.1}^2(\omega_1 - \omega_2)$$

Now rotor intake $\propto T\omega_1$

$$\therefore \; (\omega_1 - \omega_2) \propto (\text{rotor intake})/E_{p.1}^2\omega_1$$
$$\propto \text{rotor intake}$$

so long as $E_{p.1}$ is constant. This shows that, under the assumed conditions, the *fall in speed* is proportional to the rotor intake, and therefore, very approximately to the output. We also see that the torque is proportional to the square of the applied voltage, a property of the motor which renders it very susceptible to changes in voltage. It is necessary that the voltage regulation of a line supplying induction motors should be very good.

Starting Torque. At the moment of starting, the rotor is stationary, and therefore, $\sigma = 1$. Hence

$$\text{starting torque} \propto R_2/(R_2^2 + X_{2\cdot0}^2)$$

This is a maximum when $R_2 = X_{2\cdot0}$, R_2 being regarded as the variable. Therefore, to obtain the maximum starting torque, sufficient resistance should be put in series with the rotor to make the total resistance per phase equal to the rotor reactance per phase, to currents of line frequency. The variation of starting torque with rotor resistance is shown graphically in Fig. 23.8. A squirrel-cage or other permanently short-circuited rotor winding has a small starting torque whenever the rotor resistance is very small, and it cannot be increased for starting purposes by the addition of an external resistance.

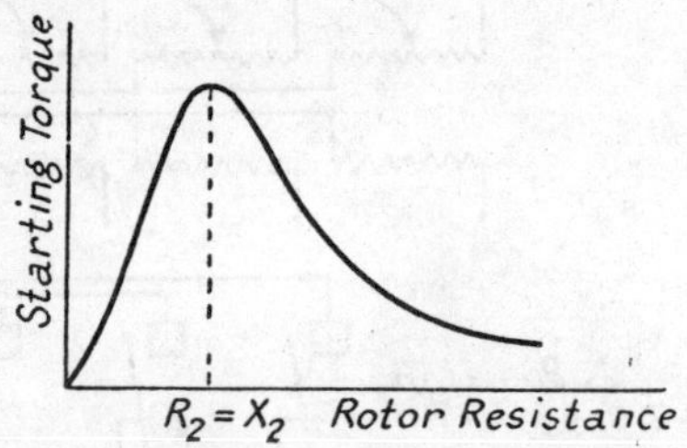

FIG. 23.8. EFFECT OF ROTOR RESISTANCE ON STARTING TORQUE

For this reason, cage motors whose duty is such that the starting performance is the most important characteristic, are generally provided with a rotor resistance greater than normal. This is accomplished by using for the end-rings a material of greater specific resistance than copper. This, of course, decreases the efficiency, increases the rotor temperature rise, and increases the slip, but, in cases where the motor is frequently started and stopped, these are out-weighed by the improved starting torque.

Methods of Starting Squirrel-cage Motors. An induction motor can be regarded as a transformer with a small air gap in the magnetic circuit. Hence, a squirrel-cage motor, when at rest, and therefore having no back e.m.f. induced in it, can be regarded as a transformer with a short-circuited secondary. Such a transformer would take a

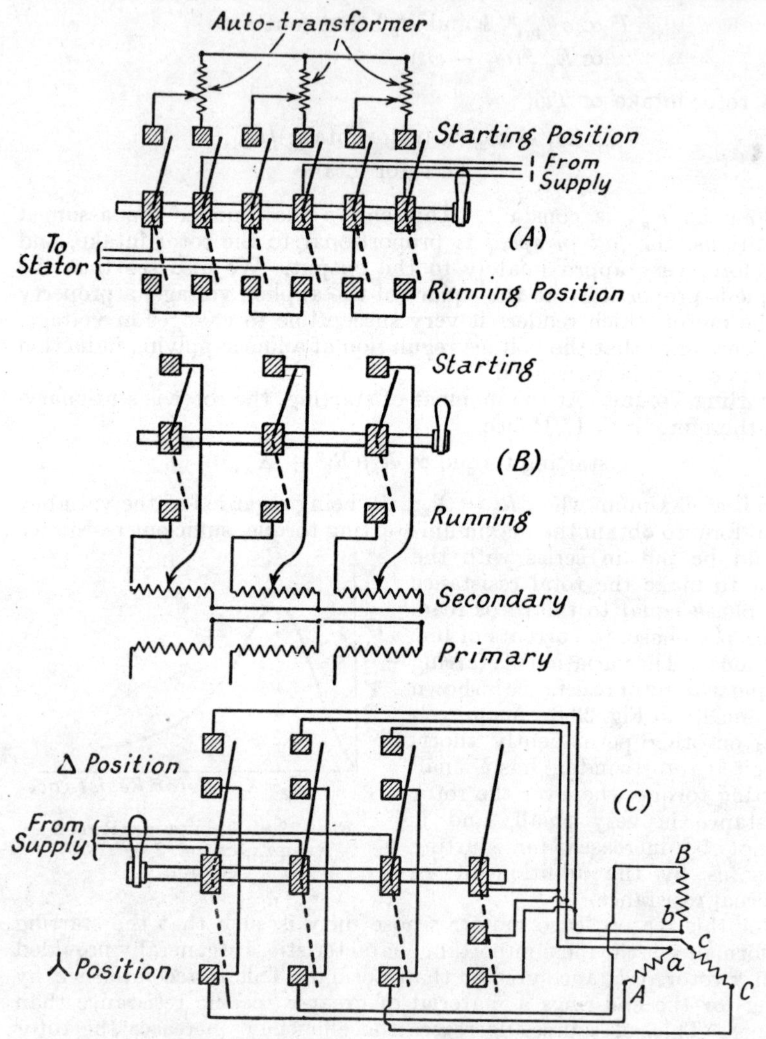

Fig. 23.9. Starters for Squirrel-cage Motors

very high current if the full voltage were applied to its primary winding, and therefore, for starting it is necessary to apply a reduced voltage. When the rotor is well under way, the full voltage can be applied. There are several methods of obtaining this reduced voltage.

In the auto-transformer starter, the reduced voltage is obtained by taking tappings from a three-phase auto-transformer, as shown in Fig. 23.9 (*A*). In the running position full line voltage is applied, and the auto-transformer left out of circuit.

If the motor is supplied through a step-down transformer, then the secondary of this transformer can be used as an auto-transformer as shown in Fig. 23.9 (*B*). This considerably simplifies the switchgear, but the extra connections from the transformer are a disadvantage.

In the "star-delta" method the motor is designed to operate normally with the stator phases mesh-connected, so that the normal phase pressure is equal to the line pressure. For starting, the phases are connected up in star, the pressure applied to each phase being thus $1/\sqrt{3}$ of the normal pressure. Fig. 23.9 (*C*) shows one arrangement of a star-delta starter and of the stator phases; the phases are each brought out to two terminals. A four-pole throw-over switch is used, the object of the fourth pole being to provide the star point when the phases are connected in star. It has no contacts on the delta side.

EXAMPLE. A squirrel-cage motor has a ratio of rotor standstill reactance to resistance per phase of 4 to 1, and the maximum torque is $2\frac{1}{4}$ times the normal full-load torque. Calculate (1) the full-load slip, (2) the ratio of starting torque to normal torque, (*a*) with direct-on starting, (*b*) with star-delta starting, (*c*) with an auto-transformer starter with 70 per cent tapping.

With normal voltage on the stator

$$T = k \times [\sigma R_2/(R_2^2 + \sigma^2 X_{2\cdot 0}^2)]$$

when $\quad T = T_{\max}, \sigma = R_2/X_{2\cdot 0}$

Let $\quad T_{\max} = 100$ per cent

$$\therefore\ 100 = \frac{kR_2^2/X_{2\cdot 0}}{R_2^2 + R_2^2 X_{2\cdot 0}^2/X_{2\cdot 0}^2}$$

$$\therefore\ k = 200X_{2\cdot 0}$$

(1) $\quad T_{\text{normal}} = T_{\max}/2\tfrac{1}{4} = 100/2\tfrac{1}{4} = 44{\cdot}4$

$$\therefore\ 44{\cdot}4 = 200X_{2\cdot 0}[\sigma R_2/(R_2^2 + \sigma^2 X_{2\cdot 0}^2)]$$

also $\quad X_{2\cdot 0} = 4R_2$

$$\therefore\ 44{\cdot}4 = 800R_2[\sigma R_2/(R_2^2 + 16\sigma^2 R_2^2)]$$

$$= 800\sigma/(1 + 16\sigma^2)$$

This gives $\sigma = 5{\cdot}7$ per cent or 107 per cent, showing that the motor will run normally with a full-load slip of 5·7 per cent. In the region of counter-current braking, the slip at normal full-load torque will be 107 per cent (Fig. 23.10).

(2) At starting $\quad \sigma = 1$

$$\therefore\ T_{\text{starting}} = kR_2/(R_2^2 + X_{2\cdot 0}^2)$$

Putting $\quad k = 200X_{2\cdot 0}$ and $X_{2\cdot 0} = 4R_2$

we have $\quad T_{\text{starting}} = 47$

(*a*) At full stator voltage, $T_{starting}$ is given by the above value of 47 per cent of the maximum possible torque

$$\therefore \; T_{starting} = 47/44{\cdot}4 = 1{\cdot}06T_{normal}$$

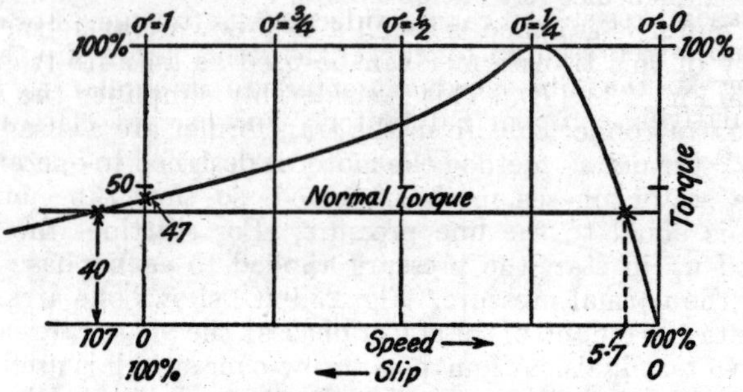

FIG. 23.10

(*b*) Star-delta starting gives $1/\sqrt{3}$ times full voltage, and therefore 1/3 times the torque at full voltage

$$\therefore \; T_{starting} = 1{\cdot}06/3 = 0{\cdot}35T_{normal}$$

(*c*)
$$T_{starting} = 1{\cdot}06 \times 0{\cdot}7^2 = 0{\cdot}52T_{normal}$$

Operation over the Normal Range. Under running conditions,

$$T \propto \sigma R_2/(R_2^2 + \sigma^2 X_{2\cdot0}^2)$$

This is a maximum when $R_2 = \sigma X_{2\cdot0}$, σ now being the variable. Substituting $R_2/X_{2\cdot0}$ for σ, we have for the maximum torque under running conditions

$$T_{max} \propto \frac{R_2^2/X_{2\cdot0}}{R_2^2 + R_2^2X_{2\cdot0}^2/X_{2\cdot0}^2} \propto \frac{1}{X_{2\cdot0}}$$

Since $X_{2\cdot0}$ is a constant for a given motor, we see that the maximum torque attainable is a constant and is independent of the rotor resistance. Hence, if we draw a family of torque v. slip or torque v. speed curves for a series of values of the total rotor resistance, they will all have the same maximum value. Such a family of curves is shown in Fig. 23.11. The perpendicular OY is drawn at unity slip or zero speed, and the intercepts on this axis give the starting torque of the motor with a rotor resistance equal to that corresponding to the particular curve. If R_2 is equal to $X_{2\cdot0}$ the maximum torque occurs at starting, as shown by curve 3, and for any other resistance the torque at starting is less than the maximum value. The shaded portion of the diagram represents the normal working range, and it will be seen that although a high resistance is an advantage for starting purposes, it is a disadvantage when running normally. Curve 1 is for a normal rotor of small resistance: this gives a greater torque than any of the others within the working range, and the maximum torque is not far removed from this range, a disadvantage in the case of a motor subjected to sudden violent overloads, even if of short duration. Curve 2 is for a high-resistance rotor, and we see that within the working range the increase in resistance causes a decrease in torque. For a given torque the motor represented by curve 2 suffers a larger slip than the normal motor, so that its speed characteristic is similar to that of a d.c. cumulatively-compounded motor. For certain classes of work this

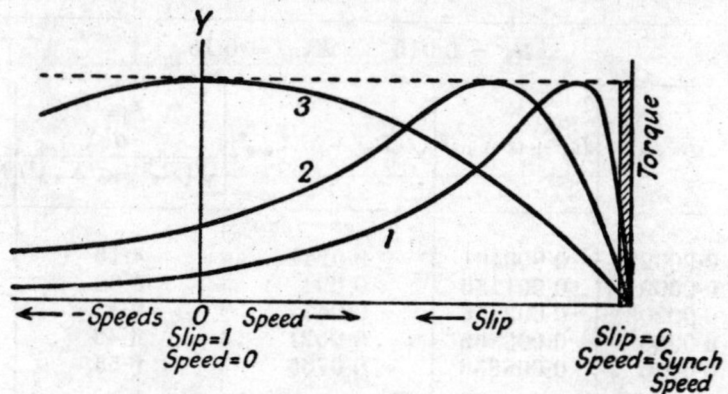

FIG. 23.11. TORQUE-SLIP CHARACTERISTICS

characteristic is an advantage, since it relieves the motor both electrically and mechanically when the opposing torque is heavy. Also, the position of the maximum torque is farther removed from the normal working range, so that the motor is not so liable to pull out in the event of a serious overload. For average motors the pull-out torque is about $2\frac{1}{2}$ times the normal full-load torque.

We have seen that T_{max} is dependent on (*a*) the square of the voltage $E_{2 \cdot 0}$, and therefore on the square of the supply voltage, (*b*) the frequency f, since $X_{2 \cdot 0} = 2\pi L_2 f$. This shows that a high maximum torque can be more easily attained in a low-frequency motor, (*c*) the rotor inductive reactance $X_{2 \cdot 0}$, showing that for a high torque the magnetic leakage must be made as small as possible. An important factor in deciding one of the components of the magnetic leakage, the zig-zag leakage (*see* p. 587) is the radial length of the air-gap, which should be made as small as possible.

As an actual example consider the case of a 250-b.h.p. induction motor having a rotor resistance per phase of 0·016 Ω and standstill reactance per phase of 0·15 Ω. We have seen that the torque and rotor current can be written

$$T = k_1 \frac{\sigma R_2}{[(R_2{}^2 + \sigma^2 X_{2 \cdot 0}{}^2)]}; \; I_2 = k_2 \frac{\sigma}{[(R_2{}^2 + \sigma^2 X_{2 \cdot 0}{}^2)^{\frac{1}{2}}]}$$

where k_1 and k_2 are constants which need not be determined. Substituting the numerical values for R_2 and X_2, we have

$$T = k_1[0{\cdot}016\sigma/\{(0{\cdot}016)^2 + (0{\cdot}15)^2\sigma^2\}]$$
$$I_2 = k_2[\sigma/\{(0{\cdot}016)^2 + (0{\cdot}15)^2\sigma^2\}^{\frac{1}{2}}]$$

We will take the values of k_1 and k_2 to be unity, the units in which T and I_2 are expressed then being arbitrary units. Taking a series of values of σ from zero up to 2·0 in steps of 0·1, we can tabulate as follows, this being more expeditious than making an individual calculation for each point on the curves.

It is not thought necessary to give the whole of the table, as the reader can make the necessary calculations for himself. The curves of

$R_2 = 0{\cdot}016 \qquad X_{2\cdot0} = 0{\cdot}15$

σ	σ^2	$\sigma X_{2\cdot0}^2$	$R_2^2+\sigma^2X_{2\cdot0}^2$	$\sqrt{(R_2^2+\sigma^2X_{2\cdot0}^2)}$	$I_2 = \dfrac{\sigma}{\sqrt{(R_2^2+\sigma^2X_{2\cdot0}^2)}}$	$T = \dfrac{\sigma}{R_2^2+\sigma^2X_{2\cdot0}^2}$
0·1	0·01	0·000225	0·000481	0·0244	4·16	208
0·2	0·04	0·0009	0·001156	0·034	5·88	180·4
0·3	0·09	0·00203	0·002256	0·0477	6·3	131
0·4	0·16	0·0036	0·003856	0·0621	6·42	101·6
0·5	0·25	0·0056	0·005856	0·0765	6·55	86

T and I_2 against σ are plotted in Fig. 23.12, and we see that the maximum torque occurs for a slip of 0·016/0·15 = 0·107 or 10·7 per

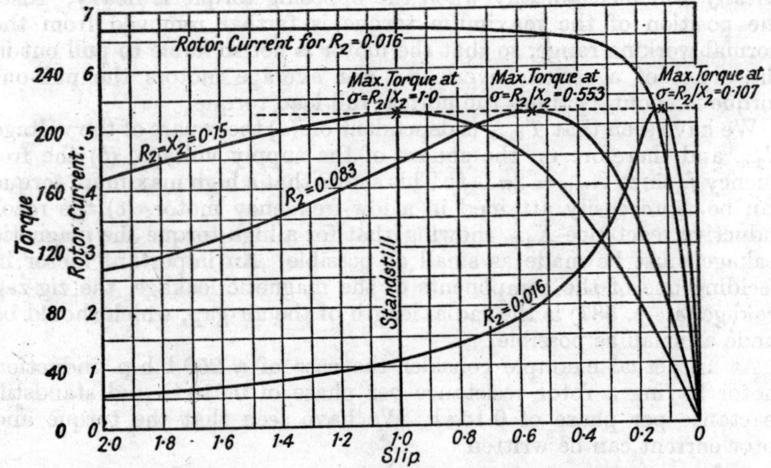

FIG. 23.12. INDUCTION MOTOR CHARACTERISTICS

cent. In the same figure the torque curves for a resistance R_2 of 0·083 and for a resistance of 0·15 Ω per phase are given, and it will be seen that these also exhibit a maximum at a slip of $R_2/X_{2\cdot0}$, namely, for $\sigma = 1$, when $R_2 = X_{2\cdot0}$, and for $\sigma = 0{\cdot}083/0{\cdot}15 = 0{\cdot}553$, when $R_2 = 0{\cdot}083$. For all three curves the magnitude of the maximum torque is the same.

Modes of Operation of the Motor. The normal operating range is within the limits of fractional slip $\sigma = 1$ (at starting) to σ very small, but still finite, during normal running. The particular torque-slip curve on which the motor is functioning depends only on the value of R_2, and in the case of a slip-ring motor it is possible to transfer to any desired curve by the switching into the rotor circuit of the appropriate amount of resistance.

Fig 23.11 shows the curves extended to the left of OY, the speed thus being negative, the slip greater than unity, and the torque still positive. This condition would apply if the rotor were reversed, the stator field travelling in the same direction as for the normal operating range. The conditions are identical if the stator field is suddenly reversed while the rotor is running in the forward direction. The torque will still be positive with respect to the new direction of the stator field, but it will be negative with respect to the old direction, and therefore it will be a braking torque which will decelerate the rotor. The sudden reversal of the stator field of a motor running with short-circuited rotor is called *plugging*, and is sometimes used as a method of control.* It is also occasionally used in connection with high-power hoists and colliery winders when these are lowering loads, but in such cases resistance is inserted since, as is shown by the curves, the amount of braking torque can be given any desired value by manipulation of the controller resistance.†

There is a third operating range to the right of the zero slip point, which means that the slip must be negative and therefore the rotor must over-run the stator field. This is merely an extension to the right-hand side of the curves for normal operation, so that the same torque equation applies but σ must now be given a negative sign. This shows that the torque is negative and therefore, as the direction of the stator field has not been reversed, the operation of the machine has changed from that of motor to that of alternator. The characteristic curves for this mode of operation are shown in Fig. 23.13, from which it will be seen that for any given torque, as indicated by the dotted torque line, there can be a series of resistance values with corresponding excess in speed above synchronism.

The "asynchronous" alternator, which is the name for the induction motor operated in this manner, is very different in its characteristics from the synchronous alternator studied in Chapters 19 and 20. The main differences, apart from the obvious structural differences are—

1. The synchronous alternator has d.c. magnetizing current supplied from an exciter and consequently it can function as an isolated machine. The asynchronous alternator draws its magnetizing current from the a.c. supply to which its stator is connected, and therefore it can only operate when connected to an existing a.c. supply.
2. In the synchronous alternator the rotor poles occupy fixed

* *See* H. Vickers, ***The Induction Motor***, Chap. V. Pitman, London.
† *See* H. Cotton, ***Electricity Applied to Mining***, Chap. X. Pitman, London.

positions with respect to the rotor structure because of the d.c. excitation. Hence, since in any dynamo-electric machine the rotor poles must travel in space at exactly the same speed as the stator poles, the machine can only operate at exact synchronism. The asynchronous alternator has a.c. excitation of the rotor and consequently the rotor field rotates in space at some velocity relative to the rotor itself which is determined by the negative slip at which

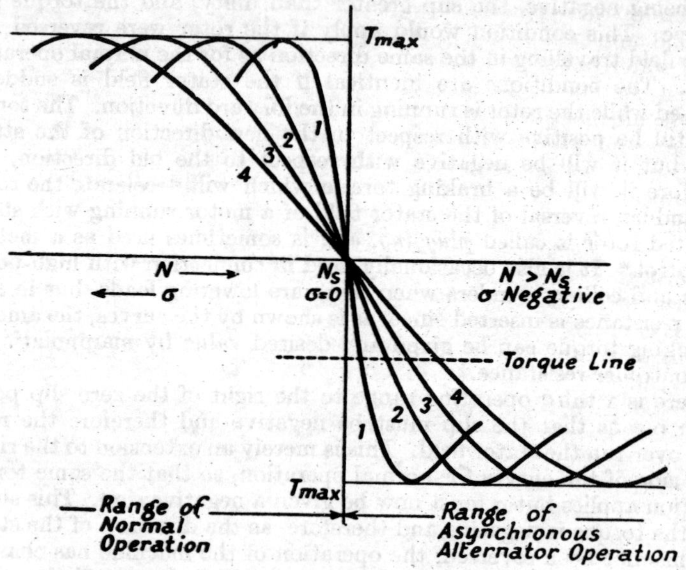

Fig. 23.13. Torque-slip Curves for Asynchronous Alternator Operation

the rotor is running. The regime is, in fact, merely an extension beyond zero slip of the normal motor operation. The field of the machine slips with respect to the rotor but it is now the rotor which has the greater velocity.

3. For a given rotor resistance the output increases with the excess of speed above synchronism, the relationship being almost linear. This output is fed back into the supply and thus there is no analogy with the case of the synchronous alternator since, even when working in parallel with others, an increase in steam supply does not result in an increase in speed.

4. Whatever the excess of rotor speed above synchronism the frequency of the output current remains the same as that of the supply voltage. The frequency of the rotor current, however, increases as the negative slip increases.

There are two important applications of this alternator. The first is when loads are being lowered by an a.c. driven hoist. Gravity can act as prime mover and a retarding torque will be set up if the speed is allowed to exceed synchronism. This kind of control requires that the period of time during this mode of operation shall be sufficiently long for the driver to maintain control.* Since the retarding torque ceases when there is no excess of speed above synchronism the braking from synchronous speed down to standstill must be accomplished by some other means. This is different from the counter-current method in which the braking effect exists right down to standstill.

The second application is for the utilization of small and variable amounts of water power which do not justify a normal power station. The simplest method is to arrange for a water wheel to drive an induction machine having a cage rotor. When there is sufficient water the machine is automatically switched on to an overhead line already energized from a normal station. If there is not sufficient water to drive the rotor above synchronism the equipment is automatically switched off.

More Accurate Analysis of the Torque-Slip Relationship. In the previous discussions it has been assumed that the stator voltage drops, R_1I_1 and X_1I_1 are negligibly small, and for large machines only, these assumptions may lead to little error. There will be appreciable error with small machines. The effect of the no-load current has also been neglected. This current is, in proportion, much larger than in a static transformer and its power factor is lower. The simplest way of taking it into account is to calculate first the wattless component of this, I_μ, and then the drop X_1I_μ due to it. This drop can be deducted directly from the supply voltage per phase, V_1, giving for the e.m.f. which is actually transformed in the machine

$$E_1' = V_1 - X_1I_\mu$$

We can now regard the secondary circuit as a distinct circuit fed at a voltage of

$$V_2' = kE_1' \text{ at standstill}$$

and

$$V_2' = k\sigma E_1' \text{ at any fractional slip } \sigma$$

The equivalent circuits for the rotor are, for the two cases, shown in Fig. 23.14. The changes in the coefficients of R_1, X_1 and X_2 from standstill to running conditions are due to the following—

R_2 is unchanged because it is an ohmic resistance in the secondary circuit.

k^2R_1 changes to σk^2R_1 because the voltage transformation ratio decreases in proportion to σ: similarly k^2X_1 changes to σk^2X_1.

The secondary leakage reactance $X_{2\cdot 0}$ changes to $\sigma X_{2\cdot 0}$ because of the decrease in frequency of the current in the secondary circuit.

Hence, although the standstill values k^2X_1 and $X_{2\cdot 0}$ are operated on equally by the factor σ, the physical phenomena associated with the

* *See* H. Cotton, *loc. cit.* p. 467.

operations are different, the first being the transformation of a line frequency voltage drop, the second a reduction in inductive reactance. From the equivalent circuit diagram for running conditions, we have

$$\sigma V_2' = [R_2 + \sigma k^2 R_1 + j(\sigma k^2 X_1 + \sigma X_2)]I_2$$

$$\therefore\ \sigma V_2' = I_2[(\sigma k^2 R_1 + R_2)^2 + \sigma^2(k^2 X_1 + X_2)^2]^{\frac{1}{2}}$$

$$I_2 = \sigma E_{2\cdot 0}/(R_2^{\,2} + \sigma^2 X_{2\cdot 0}^{\,2})^{\frac{1}{2}} \text{ as before}$$

$$\therefore\ E_{2\cdot 0} = (R_2^{\,2} + \sigma^2 X_{2\cdot 0}^{\,2})^{\frac{1}{2}} I_2/\sigma$$

$$= V_2'(R_2^{\,2} + \sigma^2 X_{2\cdot 0}^{\,2})^{\frac{1}{2}}/[(\sigma k^2 R_1 + R_2)^2 + \sigma^2(k^2 X_1 + X_{2\cdot 0})^2]^{\frac{1}{2}}$$

$$E_{2\cdot 0}^{\,2} = (V_2')^2[(R_2^{\,2} + \sigma^2 X_{2\cdot 0}^{\,2})/\{(\sigma k^2 R_1 + R_2)^2 + \sigma^2(k^2 X_1 + X_{2\cdot 0})^2\}]$$

Fig. 23.14. Equivalent Circuit of Rotor Winding

Substituting this more correct value in the previous expression for the torque, we have

$$T = \frac{(m_2/2\pi)(1/N_s)k^2(E_1'^2)\sigma R_2}{(\sigma k^2 R_1 + R_2)^2 + \sigma^2(k^2 X_1 + X_{2\cdot 0})^2}$$

Expanding the denominator we have the expression

$$\sigma^2 k^4 R_1^{\,2} + 2\sigma k^2 R_1 R_2 + R_2^{\,2} + \sigma^2(k^2 X_1 + X_{2\cdot 0})^2$$

In the numerator we have the variable σ; dividing the denominator by this we have

$$2k^2 R_1 R_2 + R_2^{\,2}/\sigma + \sigma[(k^2 X_1 + X_{2\cdot 0})^2 + k^4 R_1^{\,2}]$$

and this is a minimum, and therefore the torque a maximum when

$$\sigma[(k^2 X_1 + X_{2\cdot 0})^2 + k^4 R_1^{\,2}] = R_2^{\,2}/\sigma$$

when $$\sigma = R_2/((k^2 X_1 + X_{2\cdot 0})^2 + k^4 R_1^{\,2})^{\frac{1}{2}}$$

If we neglect R_1 and X_1 this reduces to the more approximate relationship $\sigma = R_2/X_{2\cdot 0}$ obtained previously.

Example. Compare the performances calculated by the approximate and accurate methods of the three-phase induction motor having the

following particulars: 380 V, star-connected, $R_1 = 0{\cdot}39\ \Omega$, $R_2 = 0{\cdot}151\ \Omega$, $X_1 = 1{\cdot}61\ \Omega$, $X_2 = 0{\cdot}416\ \Omega$, $k = 0{\cdot}57$, $I_\mu = 0{\cdot}6$ A, $\sigma = 5$ per cent at full load.

(*a*) *Approximate method*

$$T = A\,.\,[\sigma V_1^2 R_2/(R_2^2 + \sigma^2 X_{2\cdot0}^2)] \text{ where } A \text{ is a constant}$$

$$V_1 = 380/\sqrt{3} = 220 \text{ V/phase}$$

$$\therefore\ T_{\text{f.l.}} = A(1/20) \times 220^2 \times 0{\cdot}151/[0{\cdot}151^2 + (1/400) \times 0{\cdot}416^2]$$

$$= 15{,}700A$$

Fractional slip for maximum torque

$$\sigma' = R_2/X_{2\cdot0} = 0{\cdot}151/0{\cdot}416 = 0{\cdot}363$$

(*b*) *More accurate method*

$$E_1' = 220 - 1{\cdot}61 \times 0{\cdot}6 \simeq 220 \text{ V}$$

$$\therefore\ T = A\,(1/20) \times 220^2 \times 0{\cdot}151/[(1/20)\times 0{\cdot}57^2 \times 0{\cdot}39 + 0{\cdot}151]^2 + (1/400)(0{\cdot}57^2 \times 1{\cdot}61 + 0{\cdot}416)^2$$

$$= 13{,}600A$$

Fractional slip for maximum torque

$$\sigma' = 0{\cdot}151/[(0{\cdot}57^2 \times 1{\cdot}61 + 0{\cdot}416)^2 + 0{\cdot}57^4 \times 0{\cdot}39^2]^{\frac{1}{2}}$$

$$= 0{\cdot}16$$

Another Equivalent Circuit. Using the approximate equation for the rotor current we have

$$I_2 = \sigma E_{2\cdot0}/(R_2^2 + \sigma^2 X_{2\cdot0}^2)^{\frac{1}{2}} = E_{2\cdot0}/((R_2/\sigma))^2 + X_{2\cdot0}^2)^{\frac{1}{2}}$$

Thus we can imagine that the rotor resistance per phase is increased from R_2 to R_2/σ, that is, an increase of

$$R_2' = R_2/\sigma - R_2 = R_2(1 - \sigma)/\sigma \text{ per phase}$$

At the same time the rotor reactance retains the standstill value of $X_{2\cdot0}$. The power absorbed by this fictitious resistance R_2' is equal to the power developed by the rotor.

EXAMPLE. A three-phase motor has a rotor current per phase of 20 A when running with a slip of 4 per cent. Calculate the mechanical power developed given that the rotor resistance per phase is 0·06 Ω.

$$\sigma = 4/100 = 0{\cdot}04 \text{ or } 1/25$$

Resistance per phase representing the mechanical load

$$R_2' = R_2(1 - \sigma)/\sigma = 0{\cdot}06 \times 0{\cdot}96 \times 25$$

$$= 1{\cdot}44\ \Omega$$

Hence power developed in all three phases

$$P = 3 \times 20^2 \times 1{\cdot}44$$

$$= 1728 \text{ W}$$

An equivalent circuit diagram can be drawn in which this fictitious resistance R_2' takes the place of the load impedance in the circuit diagram of a transformer. Fig. 23.15 shows the diagram in which all quantities are referred to the primary (stator side), and the no-load current is represented by the current taken by the shunted admittance G_0, B_0. From the point of view of circuit relationships it is correct to place this admittance between the primary and secondary constants, but, for the purpose of analysis it is usual to place it across the supply

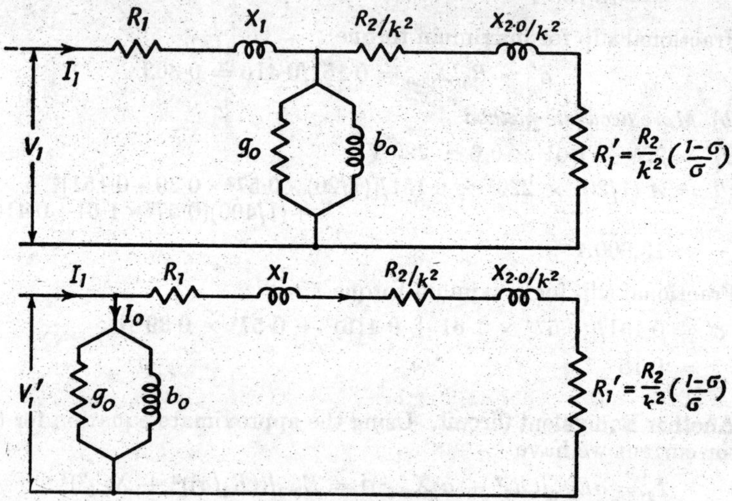

FIG. 23.15. EQUIVALENT CIRCUIT OF COMPLETE MOTOR

terminals so that it does not enter into expressions for the torque or power developed. This diagram can be used to deduce the circle diagram of the motor, as shown on p. 577.

The fictitious resistance R_2' can also be brought into the vector diagram of the motor. Fig. 23.16(a) shows the vector diagram of the motor at standstill with the rotor short-circuited. The primary impedance R_1, X_1, gives the hatched volt-drop triangle, the e.m.f. actually transformed in the motor being E_1. This induces the secondary e.m.f. E_2 in exact phase opposition to E_1. The secondary current I_2 lags E_2 by $\phi_{2\cdot 0}$ where

$$\tan \phi_{2\cdot 0} = X_{2\cdot 0}/R_2$$

and the component drops $R_2 I_2$ and $X_{2\cdot 0} I_2$ account for the whole of E_2, as shown. The current I_2 induces the primary current I_1' in exact phase opposition to it, and I_1' combined with the no-load current I_0 gives the total primary current I_1.

When running, Fig. 23.16 (b), the resultant rotor constants are R_2, $R_2(1 - \sigma)/\sigma$, and $X_{2\cdot 0}$ (not $\sigma X_{2\cdot 0}$), the power absorbed in the second of these representing the mechanical power developed. The two components of E_2 are now

$$I_2[R_2 + R_2(1 - \sigma)/\sigma] \text{ in phase with } I_2$$

and

$$X_{2\cdot0}I_2 \text{ in quadrature with } I_2$$

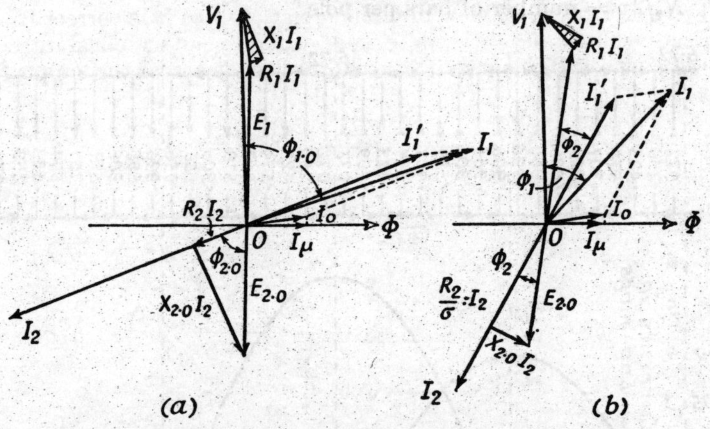

(a) (b)

FIG. 23.16. VECTOR DIAGRAMS OF INDUCTION MOTOR

and the first of these is equal to $(R_2/\sigma)I_2$. The angle of lag ϕ_2 is now given by

$$\tan \phi_2 = (\text{actual reactance})/(\text{actual resistance}) = \sigma X_{2\cdot0}/R_2$$

or

$$= \frac{\text{reactance in equivalent circuit}}{\text{resistance in equivalent circuit}}$$

$$= X_{2\cdot0}/(R_2/\sigma) = \sigma X_{2\cdot0}/R_2 \text{ as before}$$

Under running conditions $\phi_2 < \phi_{2\cdot0}$, becoming very small when σ is small.

The Resistance of the Cage Rotor. The winding structure of the cage rotor cannot be regarded as divided into a definite number of phases, as in the slip-ring motor; in fact, when the number of bars is prime to the number of poles, as is usually the case, each bar constitutes a separate phase. We will assume a sinusoidal flux density distribution round the air-gap, the current distribution in the rotor bars therefore being sinusoidal also, Fig. 23.17. This shows a rotor having 36 bars per pole pair, and, for simplicity, the maximum current per bar is taken to be unity. The currents in all the bars are indicated, and by regarding the end-rings as distributors and the bars as tappings from them, it follows that the variation of current round the rings is given by the algebraic sum of the currents tapped off as we proceed round the rings. In this way we see that the maximum current carried by the rings is 5·716 times the maximum current per bar, in the case of 36 bars per pole pair.

Let I_2 = r.m.s. value of the current in a bar

$\therefore \sqrt{2}I_2$ = maximum value

Let N_2 = number of rotor bars

and $2p$ = number of poles

$\therefore N_2/2p$ = number of bars per pole

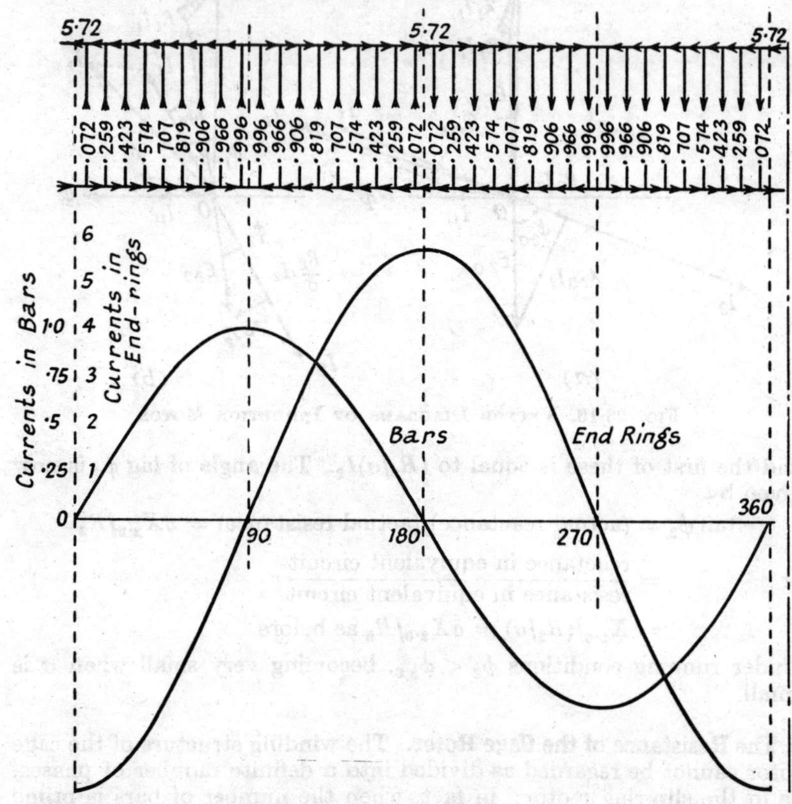

FIG. 23.17. CAGE ROTOR WITH 36 BARS PER POLE PAIR

Hence if all the bars carried the maximum current *at the same instant*, the maximum values of the current in an end ring would be

$$\tfrac{1}{2}(\sqrt{2}I_2N_2/2p)$$

But as the current distribution is sinusoidal, for which distribution the average value is $2/\pi$ of the maximum we see that the *maximum* current in an end ring is

$$(\tfrac{1}{2})(2/\pi)(\sqrt{2}I_2N_2/2p) = (1/\sqrt{2}\pi)(I_2N_2/p)$$

or

$$= 1/2\pi(\sqrt{2}I_2N_2/p)$$

Thus in the case of the motor with 36 bars per pole pair $N_2/p = 36$ and $\sqrt{2}I_2$ is the maximum bar current

$$\therefore \text{ maximum current per end ring} = (36/2\pi) \times (\text{max. current per bar})$$
$$= 5{\cdot}65 \times (\text{max. current per bar})$$

which agrees with the ratio 5·716 given above, when it is remembered that the individual currents in the diagram were only taken to two decimal places.

Now the r.m.s. value of a sinusoidal distribution is $1/\sqrt{2}$ of the maximum value

$$\therefore \text{ r.m.s. current in the ring} = (1/\sqrt{2})(1/2\pi)(\sqrt{2}I_2N_2/p)$$
$$= (1/2\pi)(I_2N_2/p)$$

Now let D_r = mean diameter of ring

a_r = cross-section of ring

σ_r = specific resistance of ring material

$\therefore$ Resistance of ring

$$R_r = \pi D_r\sigma_r/a_r$$

Reckoning both rings, the copper loss in these is given by

$$2(1/2\pi)^2(I_2N_2/p)^2 \times (\pi D_r\sigma_r/a_r) = I_2^2N_2^2 \times (1/2\pi) \times D_r\sigma_r/p^2a_r$$

Let l_b = length of bar

a_b = cross-section of bar

σ_b = specific-resistance of bar material

The total copper loss in N_2 bars is

$$N_2 \times (l_b\sigma_b/a_b) \times I_2^2 = (l_b\sigma_b/N_2a_b) \times I_2^2N_2^2$$

Hence total copper loss

$$= I_2^2N_2^2(l_b\sigma_b/N_2a_b + D_r\sigma_r/2\pi p^2a_r)$$

Hence, total rotor resistance, namely that resistance which, when carrying current I_2, would have the same copper loss as the whole rotor winding structure

$$R_2 = N_2^2(l_b\sigma_b/N_2a_b + D_r\sigma_r/2\pi p^2a_r)$$

This has now to be transferred to the stator, and as the rotor is in the nature of a perfectly distributed winding which cannot be split up into individual phases, we have to refer the resistance first of all to the whole of the stator.

(Total rotor resistance referred to the stator)

$$= N_2^2(l_b\sigma_b/N_2a_b + D_r\sigma_r/2\pi p^2a_r) \times [k_2k_2 \times (\text{stator conductors})/N_2]^2$$

Hence total rotor resistance referred to the stator per phase

$$= (N_2^2/m)(l_b\sigma_b/N_2a_b + D_r\sigma_r/2\pi p^2a_r) \times (k_2k_3 \times \text{stator conductors}/N_2)^2$$
$$= (1/m)(l_b\sigma_b/N_2a_b + D_r\sigma_r/2\pi p^2a_r) \times (k_2k_3 \times \text{stator conductors})^2$$

For the magnitude of the current in a rotor bar we make use of the equality of the rotor and the stator induced ampere-turns

$$N_2 I_2 = \text{Stator conductors} \times I_1'$$

$$I_2 = I_1' \times [(\text{stator conductors})/N_2]$$

EXAMPLE. A 6-pole, 3-phase induction motor has a stator winding housed in 3 slots/pole/phase, full-pitch coils. Conductors in series per phase 486, stator resistance per phase 1·23 Ω, full-load stator induced current per phase 11·2 A. Rotor with 57 round bars each $0{\cdot}95 \times 10^{-2}$ m in diameter and 0·16 m long. End rings 2×10^{-2} m, square section, mean diameter 0·23 m.

$$\sigma_b \text{ for copper bars, } 1{\cdot}7 \times 10^{-8} \text{ ohm-metre}$$

$$\sigma_r \text{ for brass rings, } 6{\cdot}8 \times 10^{-8} \text{ ohm-metre}$$

Current per bar at full-load

$$I_2 = 11{\cdot}2 \times (486 \times 3/57) = 292 \text{ A}$$

Current per ring at full-load

$$(1/2\pi) I_2 N_2/p = (1/2\pi) 292 \times 57/3 = 886 \text{ A}$$

$$a_b = 0{\cdot}7854 \times (0{\cdot}95 \times 10^{-2})^2 = 0{\cdot}71 \times 10^{-4} \text{ m}^2$$

$$a_r = 2 \times 10^{-2} \times 2 \times 10^{-2} = 4 \times 10^{-4} \text{ m}^2$$

Hence, total resistance referred to the stator per phase

$$= \frac{1}{3}\left(\frac{0{\cdot}16 \times 1{\cdot}7 \times 10^{-8}}{57 \times 0{\cdot}71 \times 10^{-4}} + \frac{0{\cdot}23 \times 6{\cdot}8 \times 10^{-8}}{2\pi \times 9 \times 4 \times 10^{-4}}\right) \times (0{\cdot}96 \times 1 \times 3 \times 486)^2$$

$$= 0{\cdot}9\ \Omega$$

Hence total resistance per phase referred to the stator

$$= 1{\cdot}23 + 0{\cdot}9 = 2{\cdot}13\ \Omega$$

$$\therefore \quad \text{Rotor } I^2R \text{ loss} = 3 \times 11{\cdot}2^2 \times 0{\cdot}9 = 340 \text{ W}$$

Crawling of Induction Motors. Squirrel-cage motors sometimes exhibit a tendency to run at a speed very much smaller than synchronous speed, usually one-seventh. This is due to the fact that since it is impossible to distribute the stator winding in more than 3 or 4 slots/pole/phase, even in large machines, the wave of stator flux, instead of being a pure sine wave, is stepped. This wave, when analysed, gives, in addition to the fundamental wave which rotates at synchronous speed, odd-numbered harmonics, third, fifth, seventh, and so on. The third harmonic waves produced by the three phases neutralize one another as will be seen later, and the most important of the remaining harmonics are the fifth and the seventh. It is shown below that the field set up by the fifth harmonic, considering this separately, rotates backwards, while that due to the seventh harmonic rotates forwards. Hence, the torque/slip characteristic will have three components, that due to the fundamental, that due to the fifth harmonic, and that due to the seventh. These separate characteristics are drawn in Fig. 23.18, the total torque being obtained by adding corresponding ordinates of the three separate curves.* The curve for the seventh harmonic crosses the slip axis at one-seventh fundamental speed. Similarly, the curve for the fifth harmonic crosses the slip axis at one-fifth fundamental synchronous speed, but in this case the point of intersection is to the left of the standstill axis. The curve of total torque is given by the heavy line, and from its shape we see that, as

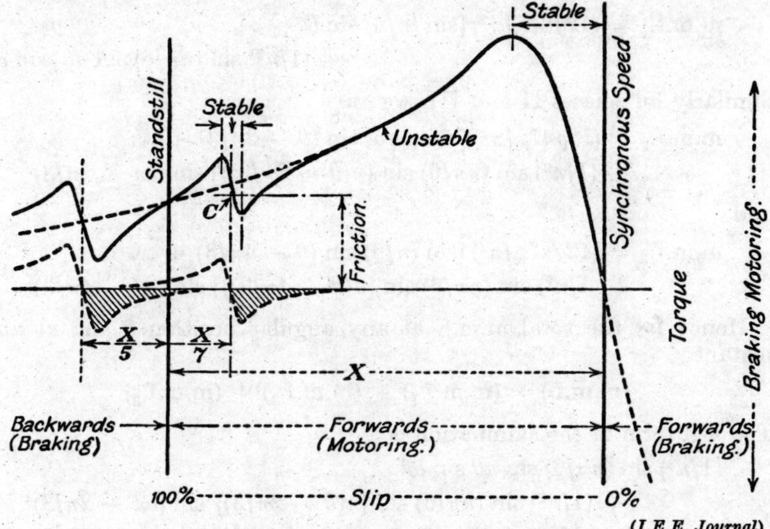

FIG. 23.18. EFFECT OF HARMONICS ON THE TORQUE-SLIP CURVE

the motor starts up from rest, the torque increases up to a maximum then decreases, and finally reaches another maximum. Also, the position of the first maximum is just below one-seventh synchronous speed. If the frictional torque of the motor is such that it requires the motor to develop a torque approaching that in the neighbourhood of one-seventh full speed, the motor will "crawl" at this speed and will be perfectly stable so long as the working point is in the neighbourhood of C. If the field form of the motor is such that a fairly large seventh harmonic is produced, then the only way to avoid this trouble is to make the rotor resistance sufficiently high to enable the motor to develop a torque at one-seventh speed greater than the frictional torque.

We have seen that the m.m.f. wave of one phase of a distributed winding can be regarded as of trapezoidal shape, and of maximum value $(1/2)guI_m$ where g = slots per pole per phase, and u = conductors

* *See* Catterson Smith, J. K., *J. Inst. Elect. Engrs.*, **49** (1912) 635.

per slot. Analysing this wave into a *height* Fourier series we have for the *height* at any angular position θ

$$\text{m.m.f.} = (2/\pi)(guI_m/(\pi/6))\,[\sin(\pi/6)\sin\theta + \ldots + (1/n^2)\sin(n\pi/6)\sin n\theta]$$

In the above series the even harmonics are absent because of the equality of the positive and negative half waves. Now the m.m.f. at any point in space is also varying sinusoidally, so that for the m.m.f. at any angular position θ, at any instant in time, we have

$$\text{m.m.f.}_1 = (12guI_m/\pi^2)\,[\sin(\pi/6)\sin\theta + \ldots + (1/n^2)\sin(n\pi/6)\sin n\theta]\sin\omega t$$

Similarly for phases II and III, we have

$$\text{m.m.f.}_2 = (12guI_m/\pi^2)\,[\sin(\pi/6)\sin(\theta - 2\pi/3) + \ldots + (1/n^2)\sin(n\pi/6)\sin\{n(\theta - 2\pi/3)\}]\sin(\omega t - 2\pi/3)$$

and

$$\text{m.m.f.}_3 = (12guI_m/\pi^2)\,[\sin(\pi/6)\sin(\theta - 4\pi/3) + \ldots + (1/n^2)\sin(n\pi/6)\sin\{n(\theta - 4\pi/3)\}]\sin(\omega t - 4\pi/3)$$

Hence for the total m.m.f. at any angular position θ and at any instant

$$(\text{m.m.f.}) = (\text{m.m.f.}_1) + (\text{m.m.f.}_2) + (\text{m.m.f.}_3)$$

The *n*th term of the summation is

$$(1/n^2)\sin(n\pi/6)\sin n\theta\sin\omega t + (1/n^2)\sin(n\pi/6)\sin\{n(\theta - 2\pi/3)\}\sin(\omega t - 2\pi/3) + (1/n^2)\sin(n\pi/6)\sin\{n(\theta - 4\pi/3)\}\sin(\omega t - 4\pi/3)$$

$$= (1/n^2)\sin(n\pi/6)\,[\sin n\theta\sin\omega t + \sin\{n(\theta - 2\pi/3)\}\sin(\omega t - 2\pi/3) + \sin\{n(\theta - 4\pi/3)\}\sin(\omega t - 4\pi/3)]$$

$$= (1/n^2)\sin(n\pi/6)\,(\tfrac{1}{2}[\cos(n\theta - \omega t) - \cos(n\theta + \omega t)] + \tfrac{1}{2}[\cos\{n\theta - \omega t - (n-1)2\pi/3\} - \cos\{n\theta + \omega t - (n+1)2\pi/3\}] + \tfrac{1}{2}[\cos\{n\theta - \omega t - (n-1)4\pi/3\} - \cos\{n\theta + \omega t - (n+1)4\pi/3\}])$$

We now put $n = 1$, $n = 3$, $n = 5$, etc.

For $n = 1$, we have

$$(1/1)\sin(\pi/6)[\tfrac{1}{2}\{\cos(\theta - \omega t) + \cos(\theta - \omega t) + \cos(\theta - \omega t)\} - \tfrac{1}{2}\{\cos(\theta + \omega t) + \cos(\theta + \omega t - 4\pi/3) + \cos(\theta + \omega t - 8\pi/3)\}]$$

$$= \tfrac{3}{4}\cos(\theta - \omega t)$$

For $n = 3$, the quantity inside the square brackets is zero, and similarly for $n = 9$, $n = 15$, etc., showing that harmonics whose numbers are multiples of 3 vanish.

For $n = 5$, we have

$$-(1/25)(\tfrac{3}{4})\cos(5\theta + \omega t)$$

and so on.

The height at any angular position θ, at any instant t, due to all three phases, is therefore given by

$$\text{m.m.f.} = 9guI_m/\pi^2\,[(1/1)\cos(\theta - \omega t) - (1/25)\cos(5\theta + \omega t) + (1/49)\cos(7\theta - \omega t) - \ldots]$$

The fact that the coefficient of ωt is unity in every term shows that the various harmonics rotate at a speed proportional to the reciprocal of their frequency number, because the general term $\cos(n\theta \pm \omega t)$ can be written in the form $\cos[n\{\theta \pm (\omega/n)t\}]$. The harmonics represented by the terms with the + sign, the general term for which is $\cos(n\theta - \omega t)$ include the fundamental, for which $n = 1$, the 7th, 13th, 19th, etc., and these all rotate in the same directions as the fundamental. The harmonics with the − sign, for which the general term is $\cos(n\theta + \omega t)$, namely the 5th, 11th, 17th, etc., rotate in the opposite direction to the fundamental, and this accounts for the displacement to the left of the standstill axis in Fig. 23.18, of the hump due to the 5th harmonic.*

The above phenomenon is not exhibited by motors with wound rotors, because the inserting of external resistance for starting purposes gives the motor a torque very much greater than the frictional torque. In fact, such motors will start against full-load torque without difficulty.

Magnetic Locking. The rotor of squirrel-cage motors sometimes exhibits a tendency to remain in one place, and not to rotate at all: this happens when the numbers of rotor and stator slots have a common factor. In such a case there will be certain rotor positions which make the air-gap reluctance a minimum, and it is in such positions that the rotor tends to remain. This is easily overcome by making the number of rotor slots prime to the number of stator slots.

The Circle Diagram of the Induction Motor. If an induction motor is run first of all at no load and then on load, the magnitude of which is gradually increased as far as possible, it will be found that the extremities of the line currents, plotted as vectors, will lie on a semicircle. This is the basis of the circle diagram which, in its completed form affords a valuable means of elucidating the properties of the motor.

Method 1. The rotor current is given by—

$$I_2 = \frac{k\sigma E_1'}{[(\sigma k^2 R_1 + R_2)^2 + \sigma^2(k^2 X_1 + X_{2\cdot 0})^2]^{\frac{1}{2}}}$$

$$\cos\phi_2 = \frac{\sigma k^2 R_1 + R_2}{[(\sigma k^2 R_1 + R_2)^2 + \sigma^2(k^2 X_1 + X_{2\cdot 0})^2]^{\frac{1}{2}}}$$

* H. M. Hobart, *Electric Motors*, Vol. II, Chap. XI. Pitman, London; Ottenstein, *E.T.Z.* (Nov. 13, 1919); T. F. Wall, "Crawling of 3-phase squirrel-cage induction motors." *World Power* (April, 1924).

$$\therefore \cos^2\phi_2 = \frac{\sigma^2k^4R_1^2 + 2\sigma k^2R_1R_2 + R_2^2}{\sigma^2k^4R_1^2 + 2\sigma k^2R_1R_2 + R_2^2 + \sigma^2(k^2X_1 + X_{2\cdot 0})^2}$$

$$\therefore \sin^2\phi_2 = 1 - \cos^2\phi_2$$

$$= \frac{\sigma^2(k^2X_1 + X_{2\cdot 0})^2}{\sigma^2k^4R_1^2 + 2\sigma k^2R_1R_2 + R_2^2 + \sigma^2(k^2X_1 + X_{2\cdot 0})^2}$$

Hence $$I_2 = [kE_1'/(k^2X_1 + X_{2\cdot 0})] \sin\phi_2$$

This is the polar equation to a circle passing through the origin, and of diameter $kE_1'/(k^2X_1 + X_{2\cdot 0})$. Now the induced stator current I_1' is in exact phase opposition to I_2, and is given by $I_1' = kI_2$. Hence the locus of the extremity of I_1' is also a circle, or, in practice, a semicircle, since the range of ϕ_2 is within 90°. Now the stator current I_1 is the vector sum of I_1' and I_0 and therefore we can displace the circle of I_1' bodily with respect to the origin, by the amount I_0.

EXAMPLE. 140 h.p., 3-phase, 3000 V, 50 c/s, 12-pole induction motor with star-connected stator and rotor windings.

$R_1 = 1{\cdot}41\ \Omega$, $R_2 = 0{\cdot}0204\ \Omega$

$X_1 = 5{\cdot}7\ \Omega$, $X_{2\cdot 0} = 0{\cdot}062\ \Omega$

$k = 1/9$, $I_0 = 12{\cdot}3$ A, $I_\mu \simeq 12{\cdot}3$ A, $I_w = 0{\cdot}85$ A, $\phi_0 = 86°$

$V_1 = 3000/\sqrt{3} = 1732$ V/phase

$I_\mu X_1 = 12{\cdot}3 \times 5{\cdot}7 = 70$ V/phase

$\therefore E_1' = 1662$ V/phase

$\therefore$ Diameter of circle of I_2

$= 1662/9(5{\cdot}7/81 + 0{\cdot}062) = 1400$ A

$\therefore$ Diameter of circle of I_1'

$= 1400/9 = 155{\cdot}6$ A

The maximum line current, i.e. the total current when the motor is stalled with full voltage applied is thus,

$$155{\cdot}6 + 12{\cdot}3 \simeq 168 \text{ A}$$

The circle diagram as derived in this manner is drawn in Fig. 23.19.

Method 2. In stator terms.

From the equivalent circuit of Fig. 23.15, we have for the load component of the stator current

$$I_1' = E_1'/[(R_1 + R_2/k^2 + R_1')^2 + (X_1 + X_{2\cdot 0}/k^2)^2]^{\frac{1}{2}}$$

lagging E_1', by ϕ_1', where

$$\sin\phi_1' = (X_1 + X_{2\cdot 0}/k^2)/[(R_1 + R_2/k^2 + R_1')^2 + (X_1 + X_{2\cdot 0}/k^2)^2]^{\frac{1}{2}}$$

Hence

$$I_1' = [E_1'/(X_1 + X_{2\cdot 0}/k^2)] \sin\phi_1'$$

This is the polar equation of a circle of diameter $E_1'/(X_1 + X_{2\cdot 0}/k^2)$, and in this way we can obtain the circle of the current I_1' directly. As the load is varied the extremity A of the I_1' vector travels round the semicircle, starting at O at no load and finishing at C, the standstill point, if the load is increased to such an extent that it stalls the motor

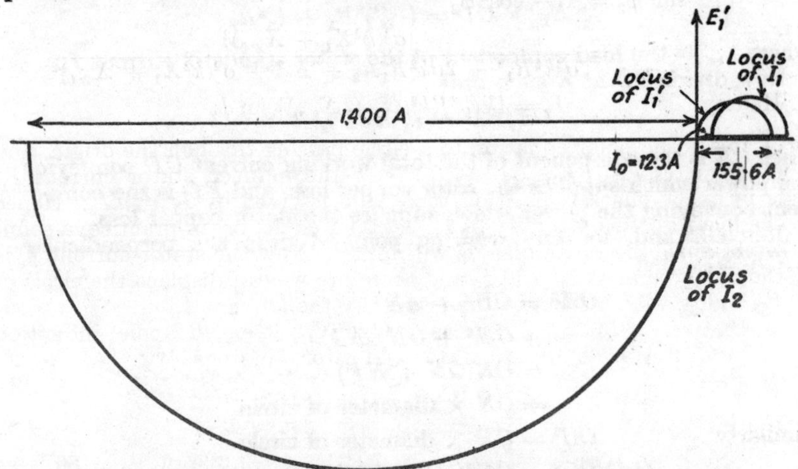

FIG. 23.19. LOCI OF ROTOR CURRENT I_2, STATOR INDUCED CURRENT I_1' AND TOTAL STATOR CURRENT I_1

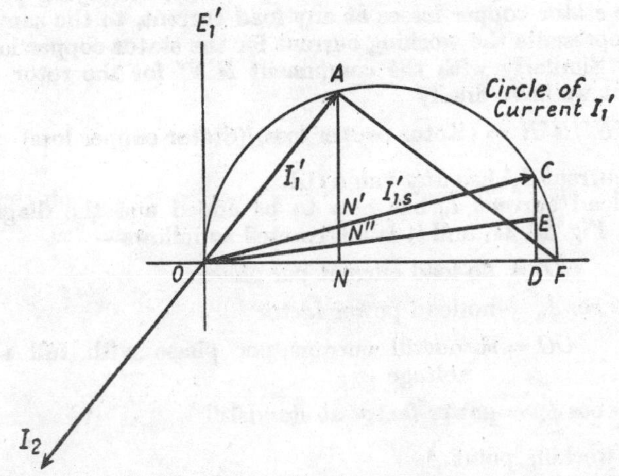

FIG. 23.20. CIRCLE DIAGRAM OF LOAD COMPONENT OF STATOR CURRENT

(*see* Fig. 23.20). The slip is then zero and so the equivalent load resistance R_1' is zero.

Drop the perpendicular CD, then CD is the working component of the standstill current. This current multiplied by the voltage E_1', therefore gives the sum of the stator and rotor copper losses. It does not include iron losses because the power for these is supplied by the working component, I_w, of the no-load current I_0 which, so far, has not been included in the diagram.

$$\therefore\ E_1' \times CD = (I_{1s}')^2 R_1 + I_2^2 R_2$$

where I_{1s}' is the load component of the stator standstill current I_s.

Now divide CD at E such that

$$CE/ED = I_2^2 R_2 / I'^2_{1s} R_1$$

Then CE is the component of the total working current CD conveying the power which supplies the rotor copper loss, and ED is the component conveying the power which supplies the stator copper loss.

Join OE and, for any working point A drop the perpendicular $AN'N''N$. Then

$$OA^2 = ON^2 + AN^2$$
$$= ON^2 + ON \,.\, NF$$
$$= ON(ON + NF)$$
$$= ON \times \text{diameter of circle}$$

Similarly $$OC^2 = OD \times \text{diameter of circle}$$

$$\therefore\ OA^2/OC^2 = ON/OD = N''N/ED$$

Hence as a copper loss is proportional to the square of the current it follows that $N''N$ represents the working current conveying power to supply the stator copper losses at any load current, to the same scale that ED represents the working current for the stator copper losses at standstill. Similarly with the component $N'N''$ for the rotor copper loss, so that we have finally

$$N'N''/N''N = (\text{Rotor copper loss})/(\text{Stator copper loss})$$

when the current I_1' has any value OA.

The no-load current I_0 has now to be added and the diagram is completed, Fig. 23.21, and it is interpreted as follows—

OO' = no-load current per phase

$\cos\phi_0$ = no-load power factor

OC = standstill current per phase with full applied voltage

$\cos\phi_s$ = power factor at standstill

For any working point A

OA = stator current per phase

$\cos\phi$ = power factor

AN = working component of stator current

$N'''N$ = component of AN required to supply the iron and friction losses

$N'N''$ = component of AN required to supply the rotor copper losses

$N''N'''$ = component of AN required to supply the stator copper losses

AN' = component of AN which conveys the output power

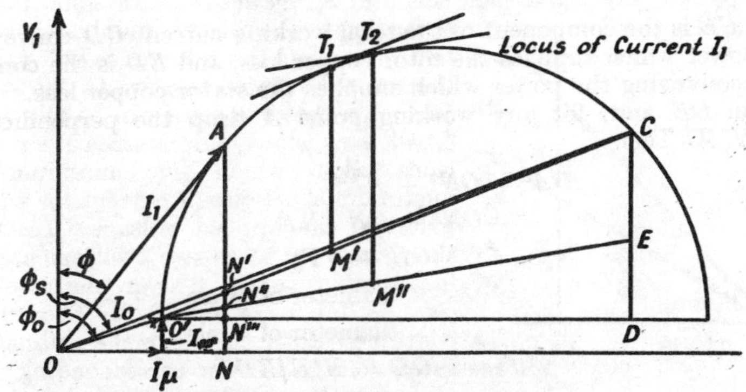

FIG. 23.21. COMPLETED CIRCLE DIAGRAM

$\therefore AN'/AN$ = efficiency

$N'N''/AN''$ = (rotor copper loss)/(rotor intake) = fractional slip

Intake = $\sqrt{3}\ V \times AN$ watts, where V = line voltage

Output = $\sqrt{3}\ V \times AN'$ watts

Torque = $(\sqrt{3}\ V/2\pi N_s) \times AN''$ newton-metres, N_s in r.p.s.

Starting torque = $(\sqrt{3}V/2\pi N_s) \times CE$, with short-circuited rotor

If tangent lines parallel to $O'C$ and $O'E$ are drawn to touch the circle at T_1 and T_2 respectively, then

$\sqrt{3}V \times T_1M'$ = maximum output in watts

$\sqrt{3}V \times T_2M''/2\pi N_s$ = maximum torque in newton-metres

Method 3. Construction from tests.

1. Run the motor on light load at the normal voltage and frequency, read the intake P_0, line volts, V_1 and the line current. If the stator is delta-connected, as is usual with cage motors, the phase current I_0 will be $1/\sqrt{3}$ of this. In what follows V and I refer to phase values, and the resulting circle diagram is for one phase of the motor. The power factor $\cos\phi_0$ is given by

$$\cos\phi_0 = P_0/3V_1I_0$$

2. Clamp the rotor and apply a reduced stator voltage, and then increase the voltage until a safe value of current is flowing. Owing to the thermal capacity of the motor, and the fact that the test is of very

short duration it is usually safe to go up to twice the rated current. The power factor is then

$$\cos \phi_s = P_s/3V_sI_s$$

It is a good plan to start with the maximum safe current and then gradually reduce V_s, taking a series of values of P_s, V_s and I_s. Then calculate $P_s/3V_s$, and plot this and I_s against V_s. In some motors, according to the relative numbers of stator and rotor slots the reluctance will vary as the rotor is rotated relative to the stator. In such a case the rotor can be allowed to crawl very slowly, or alternatively positions taken which give minimum and maximum short-circuit current for a given value of the applied voltage. Tests are then made for these two positions and the mean values taken. The graphs should be straight lines as in Fig. 23.22. They can be continued until they cut the ordinate of rated voltage V. The corresponding value of I_s can thus be determined, and that of $\cos \phi_s$ read off.

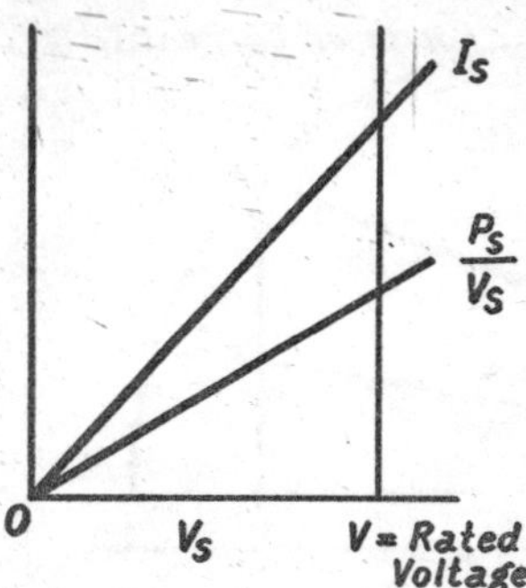

FIG. 23.22. SHORT-CIRCUIT CHARACTERISTICS

3. Determine the stator and rotor resistance per phase, although, actually, the diagram can be drawn by a measurement of R_1 only. At a short-circuit current I_s, equal to OC in the diagram

$$\text{Stator copper loss} = 3I_s^2R_1$$

$$\therefore \text{ Stator copper loss per phase} = I_s^2R_1$$

$$\therefore \text{ length } DE \text{ on the diagram} = I_s^2R_1/V_1$$

This automatically gives the length CE.

EXAMPLE. A 380-volt, 46 h.p., 50 c/s, 6-pole induction motor with cage rotor gave the following results on test. No-load current 16·0 A, no-load intake 1590 W, short-circuit current at rated voltage 280 A, intake 60,000 W. Stator star-connected with a resistance per phase of 0·10 Ω.

$$\cos \phi_0 = 1590/(\sqrt{3} \times 380 \times 16) = 0{\cdot}15$$

$$\therefore \phi_0 = 81° \, 23'$$

$$\cos \phi_s = 60{,}000/(\sqrt{3} \times 380 \times 280) = 0{\cdot}325$$

$$\therefore \phi_s = 71°$$

Rotor copper loss on short circuit

$$= P_s - 3R_1^2I_s^2$$

$$= 60{,}000 - 3 \times 0{\cdot}1 \times 280^2$$

$$= 36{,}500 \text{ W}$$

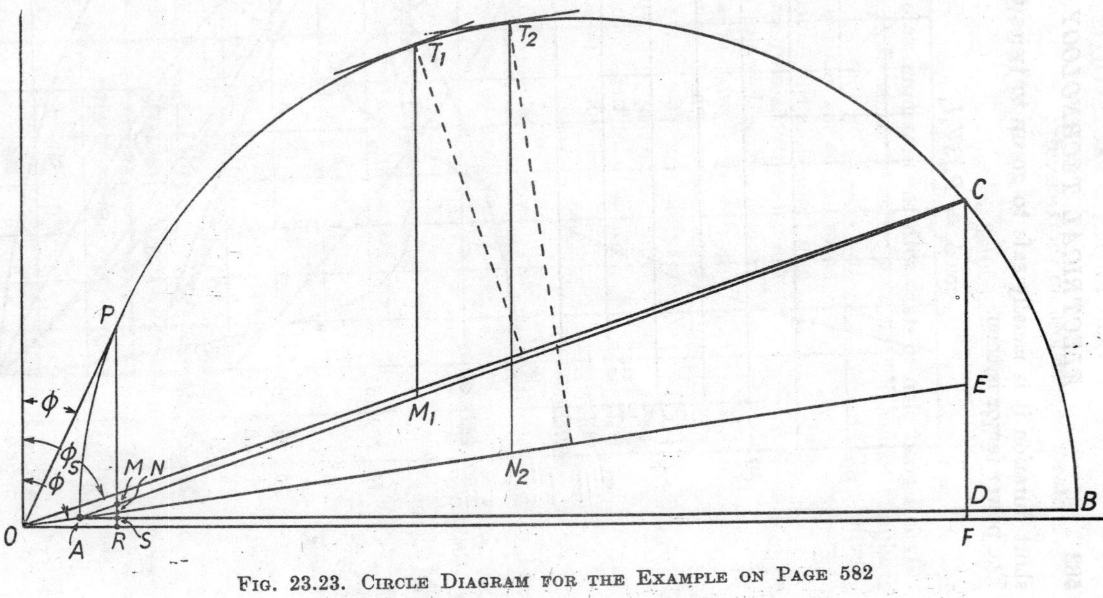

FIG. 23.23. CIRCLE DIAGRAM FOR THE EXAMPLE ON PAGE 582

$\therefore$ Stator copper loss on short-circuit

$$= 60,000 - 36,500 = 23,500 \text{ W}$$

$$\therefore \; CE/ED = 36,500/23,500$$

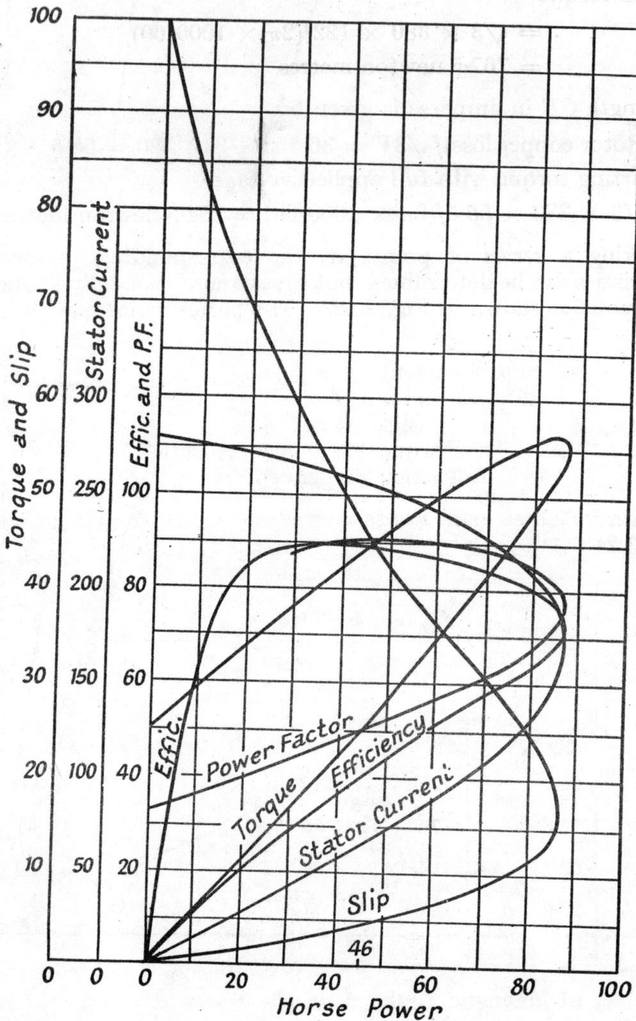

FIG. 23.24. PERFORMANCE OF INDUCTION MOTOR UNDER VARIOUS LOADS

The circle diagram is drawn to scale in Fig. 23.23 and, measured in amperes, we have

$$T_1M_1 = 99{\cdot}0 \text{ A}; \; T_2N_2 = 122 \text{ A}$$

$\therefore$ Maximum output in kilowatts

$$= \sqrt{3} \times 380 \times 99/1000 = 65{\cdot}2 \text{ kW}$$

Maximum torque

$$= \sqrt{3} \times 380 \times 122/(2\pi \times 1000/60)$$
$$= 767{\cdot}5 \text{ newton-metres}$$

The length CE in amperes is given by

$$(\text{Rotor copper loss})/\sqrt{3}V = 36{,}500/\sqrt{3} \times 380 = 55{\cdot}5 \text{ A}$$

$\therefore$ Starting torque with full applied voltage

$$= \sqrt{3} \times 380 \times 55{\cdot}5/(2\pi \times 1000/60) = 284{\cdot}3 \text{ newton-metres}$$

By taking a series of points A, the corresponding performance characteristics can be determined, but to save confusion only the normal full-load point is shown in Fig. 23.23. The performance on full-load is as follows—

Stator current 64 A
Power factor 0·905
Slip 4·4 per cent
Torque 301·0 newton-metres
Efficiency 89 per cent

The data for other loads are tabulated below and the results plotted in Fig. 23.24.*

Stator Current	Power Factor (%)	h.p. output	Slip (%)	Torque newton-metres	Efficiency (%)
16	15	0	0	0	0
50	89·6	35·3	3·55	262	89·7
75	90·7	52·6	5·61	399	87·8
100	89·0	66·7	7·95	532	85·3
125	86·0	77·6	10·5	620	82·0
150	82·4	84·8	13·45	700	78·0
163	77·8	87·2	16·87	751	73·7
196	72·3	84·0	21·4	763	67·6
225	62·7	71·9	29·7	731	58·0
250	52·2	50·6	42·5	631	44·1
275	37·9	15·8	74·0	435	17·2
280(s.c.)	32·5	0	100·0	340	0

The Effect of Magnetic Leakage on the Operation of the Induction Motor. The total stator flux produced by the stator current consists of two portions, namely—

Φ_1 = useful stator flux, i.e. the flux from the stator which enters the rotor

* *See* Carr, L. H. A. "The Circle Diagram of the Induction Motor," *Journ. I.E.E.* **66** (1928) 1174.

F_1 = leakage stator flux, i.e. the flux from the stator which does not enter the rotor.

Similarly, for the flux produced by the rotor currents, we have the useful rotor flux Φ_2, that rotor flux which enters the stator, and the leakage rotor flux F_2, that rotor flux which does not enter the stator. Then the total stator and rotor fluxes are $(\Phi_1 + F_1)$ and $(\Phi_2 + F_2)$ respectively, and we have

$$(\Phi_1 + F_1)/\Phi_1 = \lambda_1, \text{ the stator leakage factor}$$

$$(\Phi_2 + F_2)/\Phi_2 = \lambda_2, \text{ the rotor leakage factor}$$

Since the induction motor acts like a transformer, the total stator and rotor fluxes will be nearly, but not exactly, in phase opposition. Draw OA (Fig. 23.25) to represent the total stator flux, and mark off $OC = F_1$, $CA = \Phi_1$. Draw AB to represent the total rotor flux, making $AD = \Phi_2$ and $DB = F_2$. Then the resultant stator flux, which is equal to the vector sum of the total stator flux and the useful rotor flux, is equal to the vector sum of OA and AD, namely, OD.

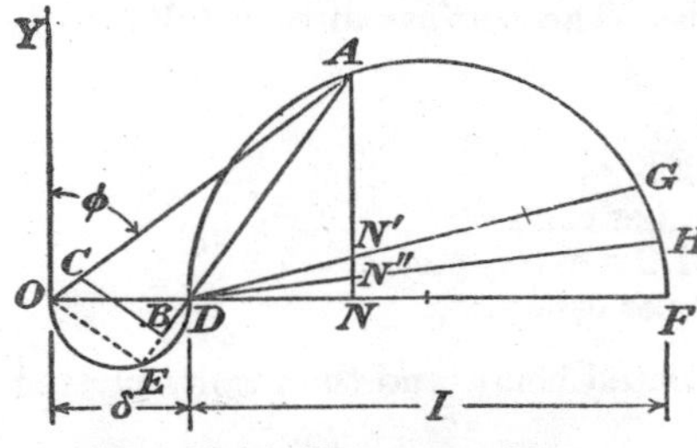

FIG. 23.25. INDUCTION MOTOR CIRCLE DIAGRAM

Now, as in a transformer, we are justified in assuming that the *resultant* flux remains constant. Also, since the wattless component of the no-load current is proportional to this flux, we see that OD represents, to another scale, the wattless or magnetizing component of the no-load stator current.

Again, the resultant rotor flux is the vector sum of the total flux produced by its own current and the flux entering the rotor from the stator. It is therefore the vector sum of AB and CA, that is, CB. Now the stator and rotor currents are in phase with the fluxes they produce, namely, OA and AB respectively. The rotor current is induced by the oscillations in the resultant rotor flux CB, and therefore, the rotor current is in quadrature with the flux CB. Hence, the angle ABC is a right angle. Producing AB to E, and drawing OE parallel to CB, we have, $\angle OED = 90°$, and therefore, the locus of E is a semicircle. Now

$$ED/DA = \lambda_1 . \lambda_2 - 1 = \text{a constant}$$

Hence, the locus of A is also a semicircle with its centre on OD produced. From this point the development of the diagram proceeds as before.

The quantity $(\lambda_1\lambda_2 - 1)$ is called the "dispersion coefficient," δ, say. Hence

$$OD/DF = ED/DA = \delta$$

Calling the diameter DF, unity, we therefore have $OD = \delta$. We have seen that for a given magnetizing current it is desirable that the diameter of the circle shall be as large as possible, and therefore, δ must be as small as possible. In order to keep δ small the magnetic leakage must be reduced to a minimum, and since the leakage depends largely on the radial length of the air gap, induction motors have an exceedingly narrow air gap.

The angle of lag is a minimum, and therefore, the power factor a maximum when the stator current vector is tangential to the circle, as at OT (Fig. 23.26).

$$\therefore \text{ Maximum power factor} = \cos\phi_{min}$$
$$= TM/MO = \tfrac{1}{2}/(\tfrac{1}{2} + \delta)$$
$$= 1/(1 + 2\delta)$$

Thus, to secure a good power factor it is necessary that δ should be small.

It will be seen that the characteristics of an induction motor depend very largely upon the value of the dispersion coefficient δ, and therefore, upon the stator and rotor magnetic leakage fluxes. Magnetic leakage in an induction motor can take place in several ways, as illustrated in Fig. 23.27.

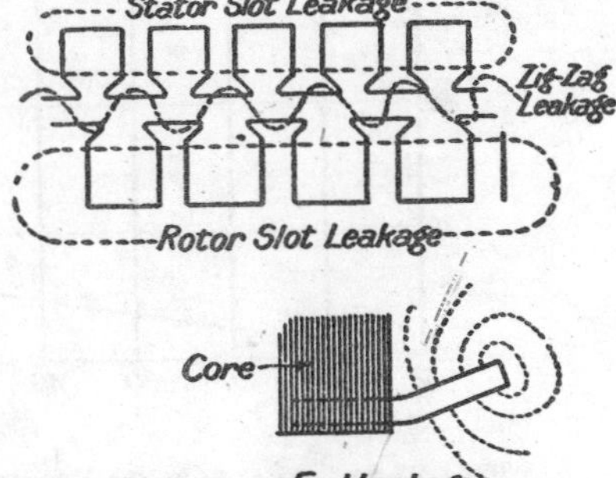

Fig. 23.26. Construction for Maximum Power Factor

Fig. 23.27. Leakage in Induction Motors

(a) *Slot leakage*. This flux passes across the slots and along the tops of the teeth.

(b) *Fluxes which zig-zag backwards and forwards between the tops of the stator and rotor teeth*. This leakage is called the "zig-zag" leakage.

(c) *End. leakage*. This takes place at the projecting ends of the coils.*

Testing of Induction Motors. The following tests are made on induction motors—

(a) *No-load Test at Normal Voltage and Frequency*. This gives the no-load current and power factor and also the sum of the core loss and friction loss.

* For a more complete discussion of induction-motor leakage, see any standard textbook on the design of electrical machinery.

If the no-load test is carried out at a series of applied voltages, after the manner of the no-load test on a d.c. shunt motor, it is possible to separate the iron and friction losses. Fig. 23.28 shows the results of a test made on a 5 h.p. 250 V squirrel-cage induction motor, the watts being plotted against $V^2/1000$, a very convenient unit for low-voltage motors. At low voltage the speed of an induction motor falls off considerably, with the result that the initial portion turns downwards. Also for voltages beyond normal, i.e. for $V^2/1000$ greater than 65 in

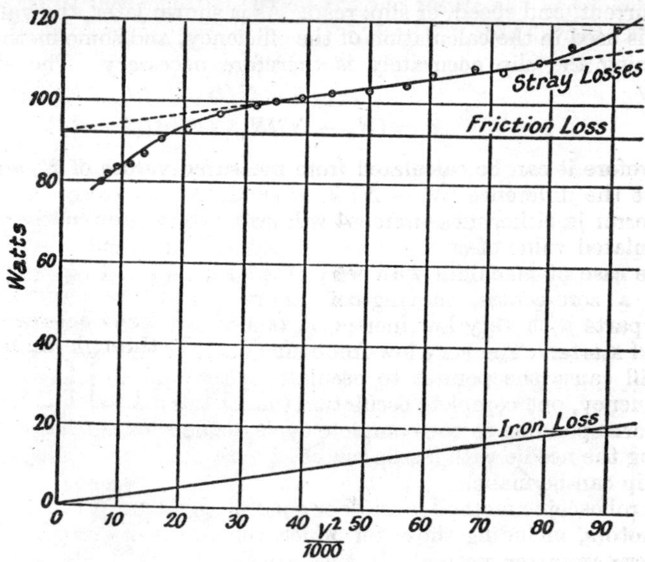

FIG. 23.28. ANALYSIS OF STRAY LOSSES IN INDUCTION MOTOR

the case of the 250 V motor, the curve turns upwards. There is, however, a sufficient range over which the graph is linear to enable a straight line to be drawn and the friction and windage loss to be thus determined with good accuracy.

(b) *Test at Reduced Voltage with Rotor Either Stalled or Allowed to Rotate Very Slowly.* This test gives the short-circuit current and power factor, so that the results of tests (a) and (b) along with resistance tests on the windings give sufficient data for the circle diagram to be drawn.

(c) *Heat Run.* It is possible to perform a regenerative test on two induction motors, but the test is difficult and is rarely carried out. The usual method is to couple the motor to a direct-current generator and to load up this machine by means of a resistance load, if of small size, or to adjust its excitation so that when paralleled on to a d.c. supply it returns energy to this supply. The excitation is, therefore, adjusted so that the induction motor takes its normal full-load current from the a.c. supply. This test gives the temperature rise of the different parts of the motor, usually by direct measurement with thermometer, and also information for the calculation of the efficiency. No attempt is made to calculate the efficiency directly from the output of the d.c. generator, since this method is liable to considerable errors. Instead, the various losses in the motor are determined separately, and the total losses then obtained by adding together the individual losses. This is called the segregation of losses method, and it is the best method for all classes of electrical machinery. The intake of the motor on load is determined by the two-wattmeter method, and the stator voltage, stator current, and speed, or slip, read. As is shown later, the value of the slip is used in the calculation of the efficiency, and some method of determining the slip accurately is therefore necessary. The slip is given by

$$\sigma = (N_s - N)/N_s$$

and therefore it can be calculated from measured values of N_s and N, but since the difference $(N_s - N)$ is so small in comparison with N_s, a small error in either measurement will lead to a considerable error in the calculated value of σ.

In the case of machines with very small slip, a good method is to connect a zero-centre, moving-coil ammeter, of the type having moving parts with very low inertia, in one of the leads between slip-rings and starter. The very low-frequency current through the instrument will cause the pointer to oscillate backwards and forwards at slip frequency, one complete oscillation (i.e. one "swing-swang") of the needle corresponding to one complete cycle of the rotor current. Thus, by timing the needle with a stop-watch, a very accurate determination of the slip can be made.

The stroboscopic method is another good method and it is applicable to all motors, including those for which the slip is too great for the centre-zero ammeter method. In this, a white disc with black sectors, the number of sectors being equal to the number of poles, is mounted on the motor shaft and is illuminated by light which is intermittently blotted out, as, for example, by being made to pass through a rotating disc with equidistant slits in the periphery. Suppose this disc is driven by a small synchronous motor, and that there are as many slits as the driving motor has poles, then the interval between the glimpses of the painted disc will be equal to the time taken for the synchronous motor to rotate an angular distance equal to the pole pitch, namely, half the periodic time of the supply. If the induction motor rotated at synchronous speed its disc would rotate through an angle equal to the spacing of the black sectors during this interval, so that at each glimpse the pattern would appear to be in the same position, i.e. stationary in space. But since the motor travels at a speed less than synchronism, the angular distance travelled by a sector during the period between two glimpses will be less than the sector pitch, and as a result the disc will have the appearance of rotating backwards slowly. The speed of the pattern is timed with a stop-watch, and this speed is the slip speed, the speed at which a synchronous motor would run if supplied at slip frequency.

$$\therefore \text{ Slip frequency} = \sigma f = np/120 \text{ c/s}$$

where n is the observed speed of the pattern in revolutions per minute.

If the disc can be screened from the light, the same phenomenon can be observed if the disc is illuminated by a number of neon lamps in parallel, worked from the stator supply. With either method observations are easy to make so long as the slip is small; but with a large slip the speed of rotation becomes so fast that, combined with the eyestrain caused by the intermittent illumination of the disc, the timing of the disc becomes difficult. This is overcome in the Drysdale slip

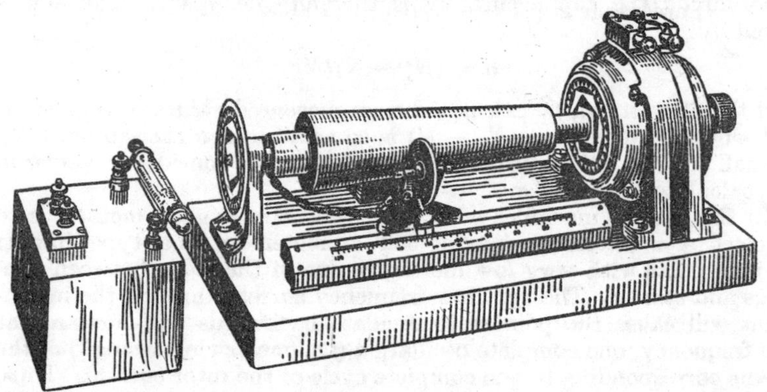

FIG. 23.29. DRYSDALE STROBOSCOPIC SLIP MOTOR
(*H. Tinsley & Co.*)

meter, in which the slotted disc, instead of being driven directly by the synchronous motor, is driven by a conical roller. When making an observation the speed of the disc is varied by sliding it along the roller until a position is reached at which the pattern appears stationary. The slip is then read directly from a scale on the instrument. This instrument is illustrated in Fig. 23.29.

The method of calculating the efficiency of the motor from the results of the test is best illustrated by a numerical example. It will be seen that the slip is required in order that the rotor I^2R loss may be separated from the total copper losses.

EXAMPLE. The results of no-load and full-load tests on a three-phase mesh-connected motor were as follows. The intake as measured by the two-wattmeter method gave 3000 and − 550 W on no-load, and 39,000 and 23,500 W on full load. The motor had four poles, the supply frequency was 50 c/s, full-load speed 1460 r.p.m., stator resistance per phase (hot) 0·13 Ω. No-load current 25 A, full-load current 92 A/line.

Stray losses + no-load stator I^2R loss

$$= 3000 - 550 = 2450 \text{ W}$$

$$\text{No-load stator } I^2R \text{ loss} = 3 \times (25/\sqrt{3})^2 \times 0{\cdot}13$$
$$= 82 \text{ W}$$
$$\therefore \text{ Stray losses} = 2450 - 82 = 2370 \text{ W (say)}$$
$$\text{Full-load intake} = 39{,}000 + 23{,}500 = 62{,}500 \text{ W}$$
$$\text{Stator } I^2R \text{ loss} = 3 \times (92/\sqrt{3})^2 \times 0{\cdot}13 = 1100 \text{ W}$$
$$\therefore \text{ Rotor intake} = 62{,}500 - (1100 + 2370)$$
$$= 59{,}030 \text{ W}$$
$$\text{Synchronous speed } N_s = 1500 \text{ r.p.m.}$$
$$\text{Actual speed } N = 1460 \text{ r.p.m.}$$
$$\therefore \text{ Percentage slip } (N_s - N)/N_s \times 100$$
$$= [(1500 - 1460)/1500] \times 100$$
$$= 2{\cdot}66$$
$$\therefore \text{ Rotor } I^2R \text{ loss} = 2{\cdot}66/100 \times 59{,}030$$
$$= 1570 \text{ W}$$
$$\therefore \text{ Output} = \text{Rotor intake} - \text{rotor } I^2R \text{ loss}$$
$$= 59{,}030 - 1570$$
$$= 57{,}460 \text{ W}$$
$$\text{Intake} = 62{,}500 \text{ W}$$
$$\therefore \text{ Percentage efficiency} = (57{,}460/62{,}500) \times 100$$
$$= 91{\cdot}9 \text{ per cent}$$

It will be seen that the above method necessitates that the motor shall be put on full load. If this is not possible, then it is necessary to draw the circle diagram from the results of no-load and locked tests, and to determine the efficiency from that.

Squirrel-cage Motors with High Starting Torque. The ordinary squirrel-cage motor suffers from the disadvantage that, owing to its very low rotor resistance, its starting torque is very low. We have seen that the starting torque can be improved by increasing the rotor resistance; a common method is to use end rings of alloy, another is to slot the end rings, thereby increasing the length of the mean path taken by the ring current. Many attempts have been made to produce a squirrel-cage motor with high starting torque. In the Boucherot motor the rotor has two separate squirrel-cages, one inside the other, as shown diagrammatically in Fig. 23.30. The outer one is of high-resistance metal, and the inner one of copper. The inner winding is much farther from the periphery of the core; its bars are surrounded by more iron, and therefore, its inductance is higher than that of the outer high-resistance winding. At the moment of starting, the rotor induced currents are at full line frequency, and therefore the reactance ($2\pi Lf$) of the inner cage will render its impedance so high that the rotor currents at starting will be confined to the outer cage, in spite of its greater resistance. The starting torque will therefore be high. As the motor speeds up, the frequency of the rotor current decreases, and the reactance of the windings becomes of less importance than the resistance. Hence, when running at full speed, when the rotor frequency is only of the order of one or two cycles per second, the impedance of the windings will depend only on their resistances. The rotor current at full speed is thus confined mainly to the inner low-resistance winding, and the efficiency of the motor is therefore high.*

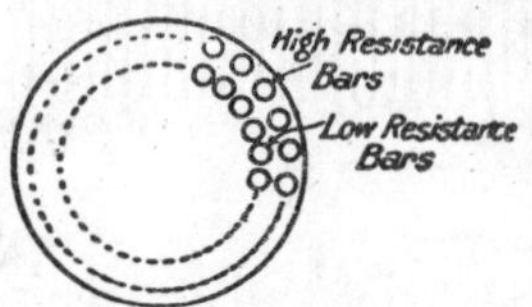

FIG. 23.30. BOUCHEROT ROTOR

If there were as many slots for the inner winding as for the outer the width of iron at the roots would be very small and it is common practice to have half as many slots for the inner winding. This leads to a rotor punching of the type of Fig. 23.31.

We can regard the two windings as acting independently of one another, each having its own torque-slip characteristic. The total torque-slip characteristic is obtained by adding corresponding ordinates of the two component curves, and considerable control over the resulting

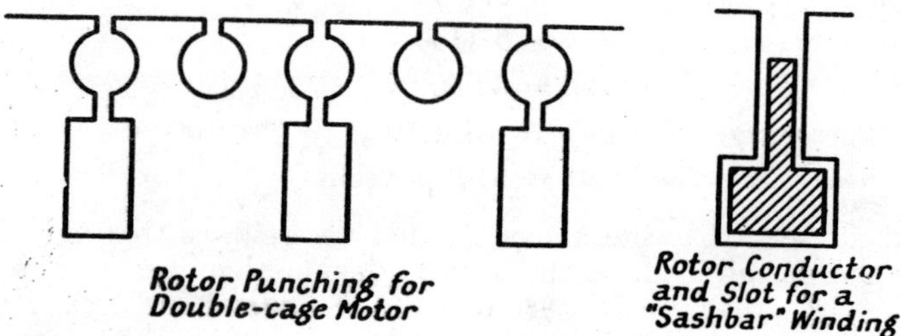

FIG. 23.31. MEANS OF OBTAINING A HIGH STARTING TORQUE

shape is possible by variations of the ratio $R/X_{2\cdot 0}$ for the two cages. Typical characteristics are shown in Plate IV. The characteristic of motor 3 is the smooth acceleration and absence of snatch due to torque variations, this being suitable for motors driving ring spinning frames in the textile industry. The great advantage of the double-cage motor is that "direct on" starting can be adapted because of the comparatively small starting current.

A similar effect is obtained by using a single-cage rotor with very deep, narrow, slots. If we regard the rectangular conductor as made up of a number of strips arranged on top of one another in the slot it is clear that there will be progressive increase in leakage inductance as we proceed from the mouth to the bottom of a slot. The effect is similar

* The theory of the double-cage motor is given in B. A. Behrend, *The Induction Motor*. (McGraw-Hill); F. Punga and O. Raydt, *Modern Polyphase Induction Motors*. Pitman, London; C. P. Steinmetz, *Theory and Calculation of Electrical Apparatus*. McGraw-Hill, New York. The method of determining the locus diagram for a double-cage motor is given in S. P. Smith and M. G. Say, *Electrical Engineering Design Class Manual* (Oxford).

in a solid conductor, the lower parts of which have a higher inductance than the upper. Hence, at starting, the current is high and the slot leakage therefore large, and the frequency is the line frequency, with the result that the lower parts of the bars carry little current. This has the effect of diminishing the effective cross-sections and so increasing the resistance. As the motor speeds up and the rotor frequency diminishes, the current distributes itself more equally over the cross-section until, at very small slip, the effective rotor resistance is practically the same as the ohmic value. To enhance this effect the conductors are sometimes specially shaped, as in the second figure of Fig. 23.31.

The manner in which the shape of the torque-speed characteristic can be controlled by the above methods is well illustrated by the four

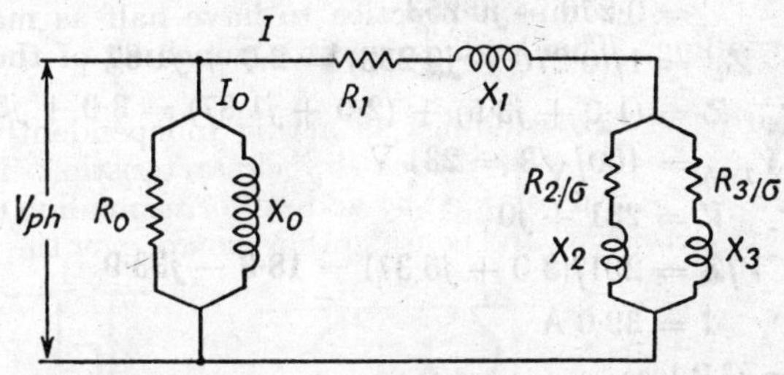

Fig. 23.32

curves of Plate IV (facing p. 481). That for the double-cage motor shows that an almost constant torque can be secured over a wide range of speeds: that for the high-slip motor shows the rapid falling off in torque chaıacteristic of a high rotor resistance. This characteristic is suited to the drive of presses, shearing machines, and similar duties where a drop in speed of 7 to 10 per cent at full load enables a direct connected flywheel to take up the load during peak conditions.

Determination of the Performance of the Double-cage Motor. The two cages are, in a sense, in parallel, so that the equivalent circuit is that of Fig. 23.32, in which all quantities are referred to the primary. The shunted impedance R_0, X_0, has no bearing on the torque and speed, and can thus be neglected when the calculation concerns these two characteristics only.

At standstill $\sigma = 1{\cdot}0$; the circuit is reduced to a single resistance and reactance in series and the current calculated. The rotor I^2R loss then gives the torque, the starting torque, in synchronous watts, that is, based on the synchronous speed N_s: thus the starting torque can be calculated.

At any speed, and therefore at any slip σ the circuit is again reduced to a single resistance and reactance in series, and the current again calculated. The rotor I^2R loss is based on the equivalent rotor resistance and on the synchronous speed N_s, so that the torque, at slip σ, can be calculated as before.

EXAMPLE. The standstill impedances per phase of a double-cage 400 V, 50 c/s, 4-pole, 3-phase motor are as follows, all values being referred to the stator—

$$\text{Stator, } \mathbf{Z}_1 = 1{\cdot}0 + j3{\cdot}5$$
$$\text{Cage 1, } \mathbf{Z}_2 = 3{\cdot}5 + j1{\cdot}2$$
$$\text{Cage 2, } \mathbf{Z}_3 = 0{\cdot}5 + j6{\cdot}0$$

(*a*) *Standstill.* Denoting the equivalent single impedances of the two cages by Z_2' we have

$$1/\mathbf{Z}_2' = 1/(3{\cdot}5 + j1{\cdot}2) + 1/(0{\cdot}5 + j6{\cdot}0)$$
$$= (0{\cdot}256 - j0{\cdot}088) + (0{\cdot}0138 - j0{\cdot}165)$$
$$= 0{\cdot}270 - j0{\cdot}253$$
$$\therefore \quad \mathbf{Z}_2' = 1/(0{\cdot}270 - j0{\cdot}253) = 2{\cdot}0 + j1{\cdot}87$$
$$\therefore \quad \mathbf{Z} = (1{\cdot}0 + j3{\cdot}5) + (2{\cdot}0 + j1{\cdot}87) = 3{\cdot}0 + j5{\cdot}37$$
$$V_{1\cdot ph} = 400/\sqrt{3} = 231 \text{ V}$$
$$\therefore \quad V = 231 + j0$$
$$\therefore \quad V/\mathbf{Z} = 231/(3{\cdot}0 + j5{\cdot}37) = 18{\cdot}3 - j33{\cdot}9$$
$$\therefore \quad I = 39{\cdot}0 \text{ A}$$

$\therefore$ Rotor I^2R loss

$$= 39^2 \times 2 = 3040 \text{ synchronous watts}$$
$$N_s = 25 \text{ r.p.s.} = 50\pi \text{ radians/sec}$$
$$\therefore \quad T \text{ starting} = 3040/50\pi = 19{\cdot}3 \text{ newton-metres.}$$

(*b*) *Normal full-load*, at which the slip is 3 per cent

$$\sigma = 0{\cdot}03$$
$$\therefore \quad \mathbf{Z}_2 = 3{\cdot}5/0{\cdot}03 + j1{\cdot}2 = 117 + j1{\cdot}2$$

and

$$\mathbf{Z}_3 = 0{\cdot}5/0{\cdot}03 + j6{\cdot}0 = 16{\cdot}7 + j6{\cdot}0$$
$$\therefore \quad 1/\mathbf{Z}_2' = 1/(117 + j1{\cdot}2) + 1/(16{\cdot}7 + j6{\cdot}0)$$
$$= (0{\cdot}0086 - j0{\cdot}0003) + (0{\cdot}0531 - j0{\cdot}0191)$$
$$= (0{\cdot}0617 - j0{\cdot}0194)$$
$$\therefore \quad \mathbf{Z}_2' = 1/(0{\cdot}0617 - j0{\cdot}0194) = 14{\cdot}85 + j4{\cdot}68$$
$$\therefore \quad \mathbf{Z} = (1{\cdot}0 + j3{\cdot}5) + (14{\cdot}85 + j4{\cdot}68)$$
$$= 15{\cdot}85 + j8{\cdot}18$$
$$\therefore \quad I = 231/(15{\cdot}85 + j8{\cdot}18) = 11{\cdot}6 - j5{\cdot}97$$
$$\therefore \quad I = 13{\cdot}05 \text{ A}$$
$$\therefore \quad \text{Rotor } I^2R \text{ loss} = 13{\cdot}05^2 \times 14{\cdot}85 = 2520 \text{ synchronous watts}$$
$$\therefore \quad T \text{ full-load} = 19{\cdot}3 \times 2520/3040$$
$$= 16{\cdot}1 \text{ newton-metres}$$

Control of Induction Motors. (*a*) STARTING. The slip-ring motor can be started by means of a resistance starter in series with the rotor, and in this way the torque, at the moment of starting, can be given any desired value up to the maximum the motor can develop. For small motors metallic resistances are generally used while for large motors the resistances can be a liquid (usually a water solution of washing soda), the ohmic value being varied by bringing a moving electrode progressively closer to a fixed electrode.

With the single-cage motor it is necessary, except with small sizes to apply a reduced voltage at starting since a sudden heavy rush of

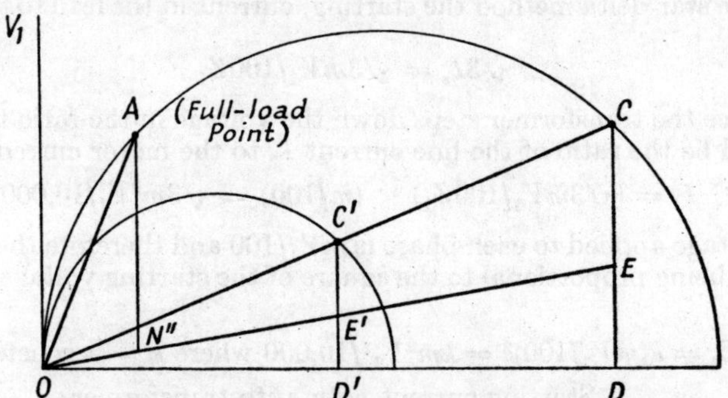

FIG. 23.33. STARTING PERFORMANCE AT REDUCED VOLTAGE

current at very low power-factor may disturb the voltage regulation of a low-capacity distribution system. We have seen that the diameter of the circular locus of primary current is equal to

$$E_1'/(X_1 + X_{2\cdot0}/k^2)$$

the denominator thus being a constant. Hence if we reduce the applied voltage to one-half, for example, we halve the diameter of the circle. Apart from that the main lines remain as before so that, neglecting the no-load current, the loci for full-voltage and half-voltage are as shown in Fig. 23.33. The standstill points are C and C' respectively, so that the starting currents are OC and OC' respectively, the power factors being the same.

The starting torque on full voltage is proportional to CE, but it is also proportional to the voltage, so that at full voltage

$$T_{\text{starting}} \propto V_1 \times CE$$

On one-half voltage we have

$$T_{\text{starting}} \propto V_1/2 \times C'E' = \tfrac{1}{4}V_1 \times CE$$

Thus the starting torque is reduced to one-quarter. Similarly for any other ratio of starting to normal voltage.

We have seen that the usual starters are the star-delta or the auto-transformer types. For very small motors a resistance starter in series with the stator windings is sometimes used. Since the starting torque is a function of the stator applied voltage it follows that an auto-transformer with a 57·7 per cent tapping will give the same torque as a star-delta starter, and the latter starter will therefore be used in general because it is the cheaper.

Let an auto-transformer have an "*m*" per cent tapping, then neglecting the motor magnetizing current and the transformer impedance, starting current per phase

$$I_s = mV_1/100Z_1$$

Assuming a delta-connected motor, in order to make a comparison with the star-delta method the starting, current in the lead to the motor is

$$\sqrt{3}I_s = \sqrt{3}mV_1/100Z_1$$

But since the transformer steps down the voltage in the ratio to $m/100$, this will be the ratio of the line current I_2 to the motor current

$$\therefore\ I_l = (\sqrt{3}mV_1/100Z_1) \times (m/100) = \sqrt{3}m^2V_1/10{,}000Z_1$$

The voltage applied to each phase is $mV_1/100$ and therefore the starting torque, being proportional to the square of the starting voltage, is given by

$$T_s = k(mV_1/100)^2 = km^2V_1^2/10{,}000 \text{ where } k = \text{a constant}$$

$$\therefore \frac{\text{Starting current with auto-transformer}}{\text{Starting current with star-delta}}$$

$$= \frac{\sqrt{3}m^2V_1/10{,}000Z_1}{V_1/\sqrt{3}Z_1}$$

$$= 3m^2/10{,}000$$

and
$$\frac{\text{Starting torque with auto-transformer}}{\text{Starting torque with star-delta}}$$

$$= \frac{km^2V_1^2/10{,}000}{kV_1^2/3}$$

$$= 3m^2/10{,}000$$

Since the ratio is the same in each case then the starting torque is proportional to the current in the line at the moment of starting, irrespective of the type of starter used. Referring to Fig. 23.33, the full-load point *A* is a fixed point whose position is decided by the rating of the machine, so that we have

$$AN'' \propto OA, \text{ i.e. Full-load torque} \propto \text{Full-load current}$$

We also have

$$C'E' \propto OC'$$

no matter what the reduction in diameter may be, since the inclinations of *OC* and *OE* are fixed by the parameters of the machine. Hence

$$\text{Starting torque} \propto \text{Starting current}$$

$$\therefore \frac{\text{Starting torque}}{\text{Full-load torque}} \propto \frac{\text{Starting current}}{\text{Full-load current}}$$

the relationship between these two ratios thus being linear.

EXAMPLE.

$I = 64$ A and $T = 310$ newton-metres at full load

$I = 280$ A and $T = 340$ newton-metres if full voltage is applied at the moment of starting

Hence torque ratio $= 340/310 \simeq 1{\cdot}1$

and current ratio $= 280/64 = 4{\cdot}38$

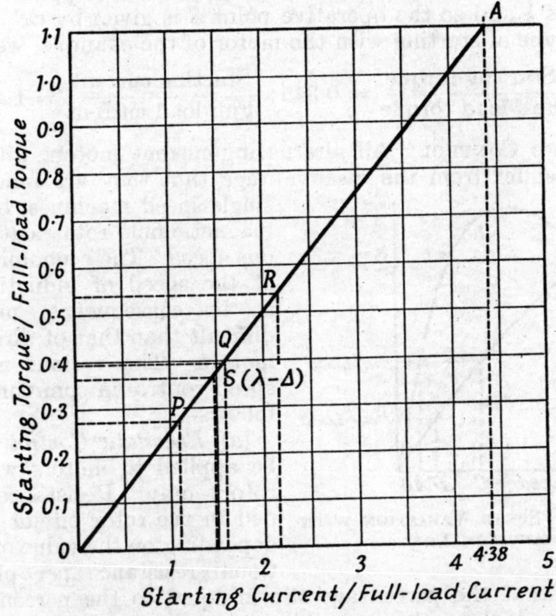

FIG. 23.34. STARTING CHARACTERISTICS WITH AUTO-TRANSFORMER STARTING

The characteristic is given in Fig. 23.34. We have seen that the torque is proportional to the square of the voltage, and thus we can mark off the points for the various percentage tappings along the graph OA. For example, suppose we decide on 50, 60, and 70 per cent tappings, then the ratios

$\frac{\text{Starting torque}}{\text{Full-load torque}}$ are 0·25, 0·36, and 0·49 times the value for this (ratio with full applied voltage), and so we mark off

$$OP = 0{\cdot}25 \text{ of } OA,\ OQ = 0{\cdot}36 \text{ of } OA,\ OR = 0{\cdot}49 \text{ of } OA$$

Reading off the corresponding ordinates and abscissae we have, for the motor of the example.

Percentage Tapping	$\dfrac{\text{Starting torque}}{\text{Full-load torque}}$	$\dfrac{\text{Starting current}}{\text{Full-load current}}$
50	0·27	1·1
60	0·39	1·6
70	0·54	2·18

For a star-delta starter we have a $100/\sqrt{3}$ or 57·7 tapping: the square of $1/\sqrt{3}$ is $\frac{1}{3}$ and so the operative point S is given by $OS = \frac{1}{3}$ of OA. For this type of starting with the motor of the example, we thus have

$$\frac{\text{Starting torque}}{\text{Full-load torque}} = 0{\cdot}345; \quad \frac{\text{Starting current}}{\text{Full-load current}} = 1{\cdot}49$$

(*b*) Speed Control. All alternating-current motors without commutators suffer from the disadvantage that they are fundamentally single-speed machines, because the magnetic field rotates at synchronous speed. The economical control of the speed of induction motors is, in consequence, much more difficult than that of direct-current motors. The various methods of speed control in common use are as follows—

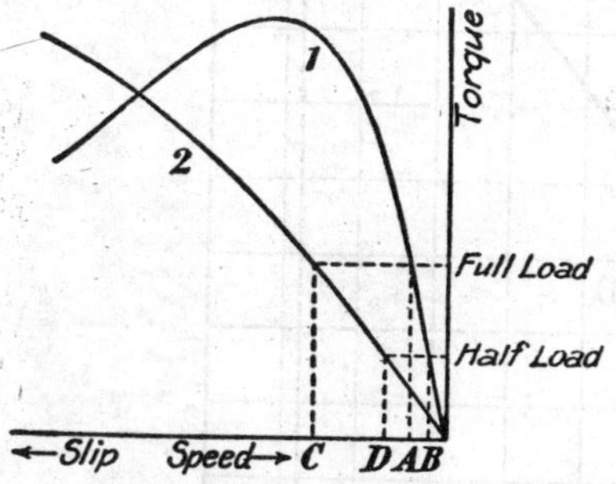

Fig. 23.35. Speed Variation with Change in Load

(*a*) *Rheostatic Control.* This can be applied to motors with wound rotors only. Resistance is included in the rotor circuit, the speed depending on the value of the additional resistance per phase. We have seen that the percentage slip is equal to the percentage rotor copper loss. If the motor is working on a constant torque, the current will be sensibly constant and the drop in speed below synchronism will be proportional to the extra resistance per phase. We thus see that the drop in speed is proportional to the power dissipated in the rheostat, the method therefore being very wasteful if speeds much below synchronism are required. Thus if the speed is half synchronous speed, half the power supplied to the motor will be wasted in the rheostat. A further disadvantage of the method is that the speed is only a function of the external resistance so long as the load torque remains constant. Consider the two torque v. speed characteristics in Fig. 23.35. Curve I corresponds to a motor running with the slip rings short-circuited. If the torque falls from, say, full-load, to half-load, then the change of speed will be AB. If so much resistance is added that the torque v. speed characteristic is represented by curve

II, then for the same change in torque the change in speed is CD, which is very much greater than AB.

For very large motors, resistance controllers are almost invariably of the liquid type, but for small and medium-sized machines metallic

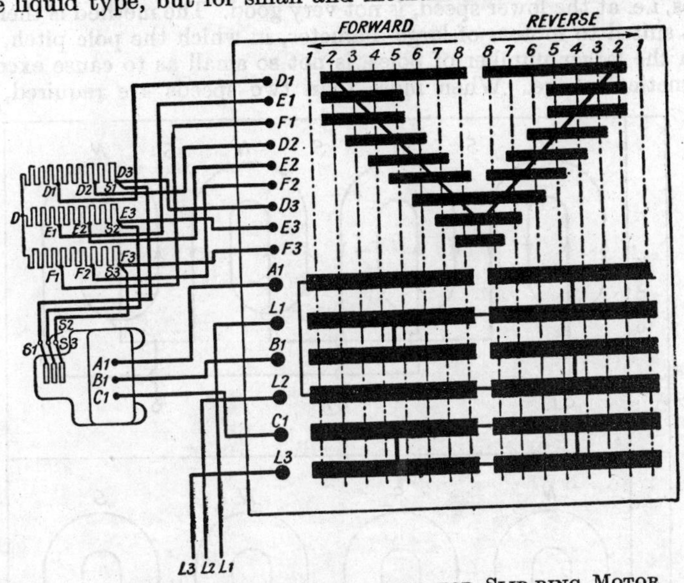

FIG. 23.36. CONTROLLER CONNECTIONS FOR SLIP-RING MOTOR

resistances with a controller of the barrel type are common. A typical connection scheme for such a controller is shown in Fig. 23.36.

It is essential that the student should appreciate the difference between a starter and a controller. A starter is in circuit for a short time, generally much less than one minute, and therefore its own thermal capacity will absorb much of the heat generated, a proportion only being radiated. A controller, however, may have some of its resistance in circuit for considerable periods, and so, when equilibrium is reached, all the heat generated has to be dissipated by radiation or some other means, such as an air-current in the case of metallic resistances, or water cooling of a liquid resistance.

(*b*) POLE CHANGING. The stator is wound so that the number of poles can be varied. If p_1 and p_2 are the numbers of poles, then there are two synchronous speeds, given by

$$n_1 = 120f/p_1, \text{ and } n_2 = 120f/p_2$$

It is easy to arrange for two numbers of poles in the ratio 1 : 2, Fig. 23.37 showing the method of changing from four to eight poles, a single-phase winding being shown for simplicity.* The larger number of

* For complete information on all methods of speed control of induction motors, *see* Miles Walker, *The Control of the Speed and Power Factor of Induction Motors*. Benn, London.

poles is obtained by making all the poles of one name, consequent poles. This introduces considerable magnetic leakage, the result being that the power factor, when running with the larger number of poles, i.e. at the lower speed, is not very good. The method is therefore best suited to motors of large diameter, in which the pole pitch, even with the larger number of poles, is not so small as to cause excessive magnetic leakage. When more than two speeds are required, it is

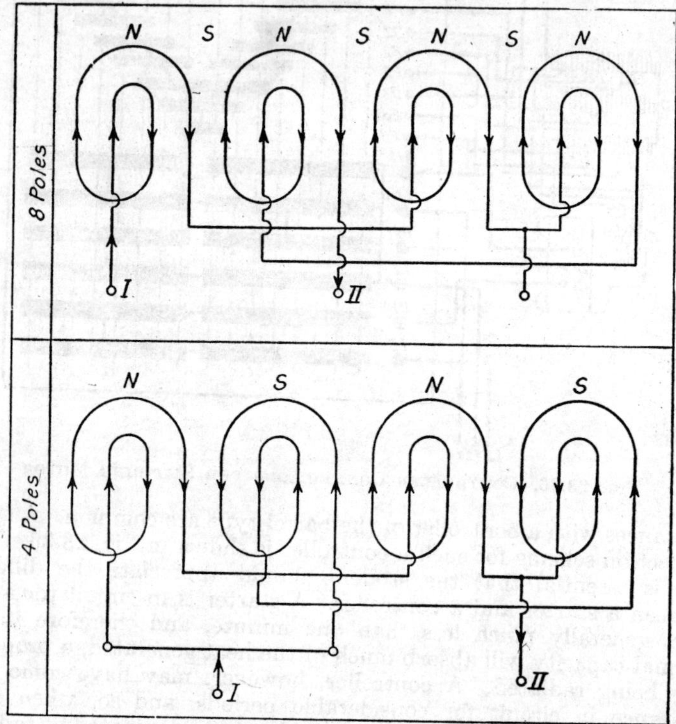

FIG. 23.37. PORTIONS OF SINGLE-PHASE WINDING FOR EIGHT AND FOUR POLES

possible to obtain these by re-arrangement of the same winding, but so many tappings are required that it is simpler to have two separate windings housed in the same slots. The rotor of these motors is usually of the squirrel-cage type, since such a rotor is suitable for any number of poles. If the rotor is phase-wound, then its number of poles must also be changed.

Additional speeds can be obtained by auxiliary rheostatic control, but this, of course, lowers the efficiency.

If Fig. 23.37 is examined it will be seen that, for 8 poles, all the coils carry current in the same sense, namely clockwise at the instant figured. For four poles, alternate coils carry currents in the opposite

direction, and consequently all that is required to halve the number of poles, and thus double the synchronous speed, is that the current in one-half of each phase shall be reversed. This is very easily accomplished

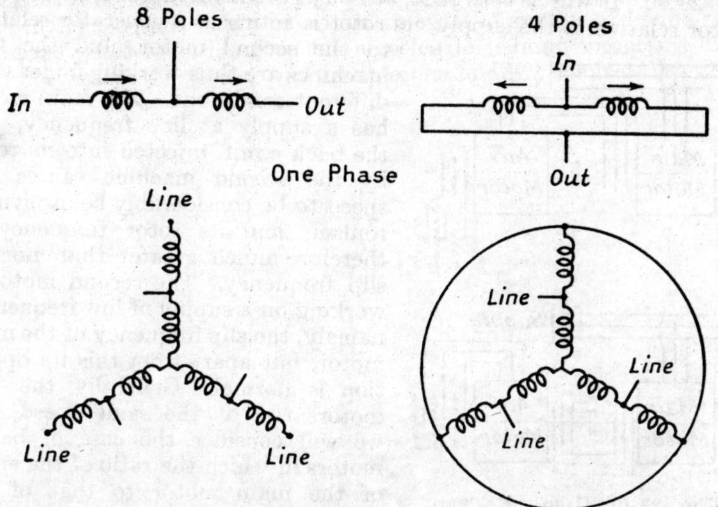

FIG. 23.38. CONNECTION SCHEME FOR TWO-SPEED POLE CHANGING WINDING

by bringing out a centre tapping from each phase. Fig. 23.38 shows the scheme both for one phase, and also for all three phases.

This simple arrangement suffers from the disadvantage that at the higher speed the two halves of each phase are in parallel, and therefore

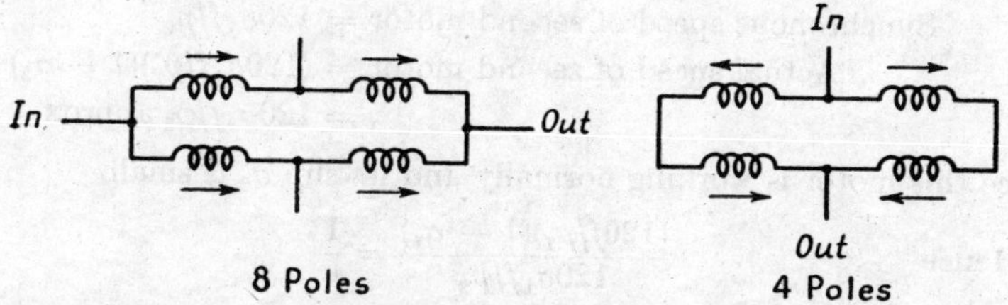

FIG. 23.39. METHOD OF RETAINING THE SAME NUMBER OF CONDUCTORS IN SERIES PER PHASE

the number of conductors in series per phase is halved. With very small motors this is not important, but with large motors it is necessary to keep the number unchanged. This is accomplished by arranging that for both numbers of poles each phase shall consist of two paths in parallel. This is illustrated in Fig. 23.39, which shows one phase only.

(*c*) CASCADE CONTROL. In this method two induction motors are mechanically coupled, and the rotor of the main motor is phase wound. The second rotor can be either squirrel-cage, or phase wound. The

stator of the first motor is connected to the supply, but the second motor receives its supply from the rotor of the first. Also the supply to this second motor can be taken either to its stator or rotor, as indicated in Fig. 23.40. It will be seen that, although the main motor is acting as a motor relative to the supply, its rotor is acting as a generator relative to the second motor, and the two machines are thus working under very different conditions. The main motor has a supply at line frequency, but the back e.m.f. injected into its rotor by the second machine causes the speed to be considerably below synchronism, and its rotor frequency is therefore much greater than normal slip frequency. The second motor is working on a supply of low frequency, namely, the slip frequency of the main motor, but apart from this its operation is normal. Generally the two motors run at the same speed, but we will consider the case of belted motors in which the ratio of the speed of the main motor to that of the second motor is $1/n$. Let the main motor have p_1, and the second motor p_2, poles. Let σ_1 and σ_2 be the respective fractional slips, and let the line frequency be f.

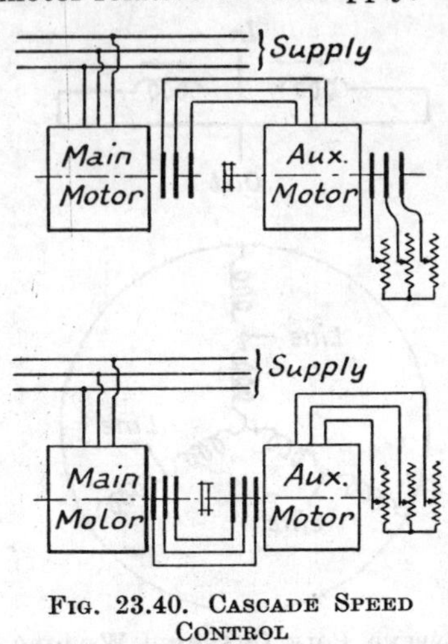

FIG. 23.40. CASCADE SPEED CONTROL

Then synchronous speed of main motor $= 120f/p_1$

Actual speed of main motor $= (120f/p_1)(1 - \sigma_1)$

Synchronous speed of second motor $= 120\sigma_1 f/p_2$

Actual speed of second motor $= (120\sigma_1 f/p_2)(1 - \sigma_2)$

$= 120\sigma_1 f/p_2$ approx.

since this motor is working normally and its slip σ_2 is small.

Hence
$$\frac{(120f/p_1)(1-\sigma_1)}{120\sigma_1 f/p_2} = \frac{1}{n}$$

$$\therefore \sigma_1 = \frac{1}{1 + 1/n \times p_1/p_2}$$

$$= np_2/(np_2 + p_1)$$

Thus, in the particular case in which the motors are solidly coupled, $n = 1$; also, if $p_1 = p_2$, then $\sigma_1 = \frac{1}{2}$, and the set runs at one-half synchronous speed. It is essential that the two motors shall be mechanically coupled, since if they are not, then, if mechanical load is put on any one motor it will stop and act like a transformer while the other motor will run light at full speed.

Substituting the above expression for the slip σ_1 in the expression for the speed of the main motor, we have

$$N_1 = (120f/p_1)(1 - \sigma_1)$$
$$= 120f \times [1/(p_1 + np_2)]$$

Hence, the main motor runs at the same speed as a single motor having $(p_1 + np_2)$ poles.

The usual method of applying the cascade method is to connect the two motors so that their torques act in the same direction, the speed then being obtained from the expressions developed above, and being less than the normal speed of the main motor. If a speed greater than normal speed is required, the torque of the second motor can be reversed by the simple expedient of changing over two of the leads to its stator winding. This is called differential cascading and the speed, for a direct-coupled set, is

$$N = 120f \times [1/(p_1 - p_2)]$$

This method of control is peculiar in that the torque developed by the machines varies very considerably with changes in speed. In particular, if the speed is the same as the synchronous speed of the main motor the torque will be zero, because of the fact that there will be no e.m.f. induced in the rotor of the main machine, and therefore no current in either stator or rotor winding of the rear machine. At speeds greater than the synchronous speed of the main machine, the slip of this machine is negative, and therefore the phase rotation of its rotor currents is reversed. This, combined with the fact that the connections to the stator of the second machine have been reversed with respect to normal cascading, explains why stable operation at speeds greater than synchronous speed is possible.

The main disadvantage of the cascade method, apart from the necessity of a second motor, is that the magnetizing current for both motors has to be drawn from the line, and since it is a wattless current it results in a poor power factor.

This problem was examined exhaustively by the author* who found that the performance in the case of two motors of the same size and same number of poles was characterized by—

1. A magnetizing current about twice that of a single motor.
2. A current locus of about one-half the diameter of that of a single motor.
3. A very rapid falling off in performance, e.g. power factor and torque for a given line current with increase in the dispersion coefficient $(\lambda_1\lambda_2 - 1)$.

The power factor curves are given in Fig. 23.41, and they illustrate very forcibly the necessity for very careful design of the magnetic circuits for this kind of control.

To minimize this lowering of the power factor, commercial cascade

* H. Cotton, "The operation of induction motors in cascade," *J. Inst. Elect. Engrs.*, **61** (1923) 284.

sets invariably have a small second motor with very few poles compared with the main motor, the magnetizing current required by the second motor being, therefore, small. The expression for the slip shows that in such a case the cascade speed is not very much below the normal synchronous speed of the main motor. If more than two speeds are

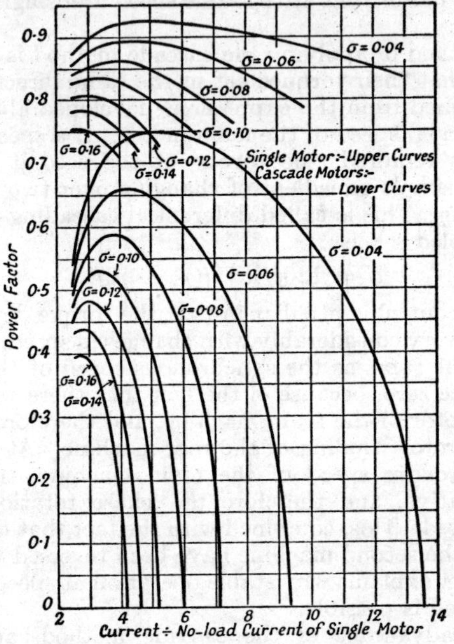

Fig. 23.41. Power Factor Curves

(In this figure, σ denotes the dispersion coefficient)

required, cascade control can be combined with pole changing; such sets were, at one time, in fairly extensive use for driving rolling mills and mine ventilation fans.

Because of the poor power factor and the limited number of economical speeds, it is usual, at the present time, to use some form of commutator motor as the second machine of a cascade-controlled set (*see* Chapter 26).

The Induction Motor run as a Synchronous Motor. At the present time it is common practice to use synchronous motors for drives where an absolutely fixed speed is not a disadvantage, e.g. for air compressors. The disadvantage of the ordinary synchronous motor is that it will not start under appreciable load. With an induction motor, starting against full-load torque is obtained, and when full speed is attained the motor can be converted to a synchronous motor by sending direct current through its rotor winding. This is done by switching over the rotor from the ordinary starting resistance to a d.c. exciter, usually direct coupled, as shown in Fig. 23.42. It will be seen that one rotor phase carries twice as much direct current as the other two, but although special windings have been evolved to avoid this, it is of little importance, because each rotor phase consists of conductors distributed over the whole periphery, the danger of unequal rotor heating therefore being avoided.

Now when the rotor carries only induced alternating currents, the rotor field slips past the rotor bars, and any motor speed is possible, according to the frequency of these currents. When the rotor carries direct current, the rotor poles are fixed relative to the rotor itself, and therefore, since the rotor field must travel at synchronous speed with the stator field, the rotor itself must also travel at synchronous speed. The induction motor action cannot bring the rotor quite up to synchronous speed, and therefore, when the rotor is switched over from the starting resistance to the exciter, the induction motor slip speed has to be made up. When running uniformly at synchronous speed the rotor will carry direct currents only, but if there is any departure from this speed the rotor will carry in addition induced alternating currents. The rotor being of low resistance, and the exciter also of low resistance, its winding acts as a damping winding, and it is not necessary to provide a separate damping winding, as in the salient-pole synchronous motor.

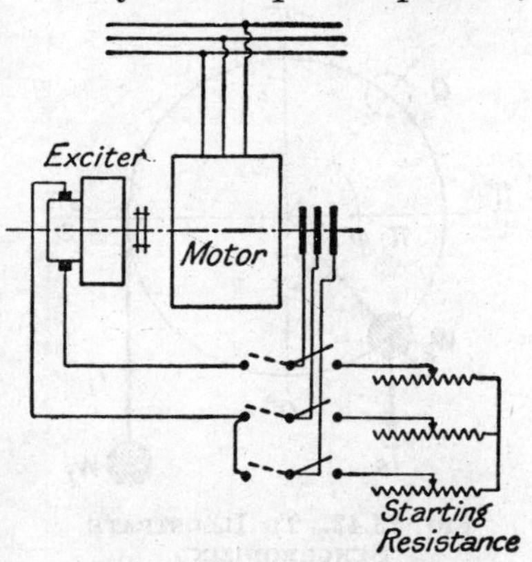

FIG. 23.42. SYNCHRONOUS INDUCTION MOTOR

Consider what happens when the direct-current excitation is switched on.* A synchronizing torque is suddenly set up, its magnitude being given by an expression $T_m \sin \theta$, where θ is the angle between the rotor and stator fields. Also, there is an induction-motor torque which, as we have seen, is proportional to the slip, $d\theta/dt$, as long as the slip is small. If the motor is started under load there is a constant load torque and finally there is the torque $Id^2\theta/dt^2$ required to accelerate the rotor. The variations in *slip* speed (not actual speed) are therefore represented by the variations in speed of the mechanical model represented in Fig. 23.43. A disc of moment of inertia I can rotate on a horizontal axis, and a weight W_1 is suspended by a string passed round it. This weight will set up a torque that will remain constant, no matter how fast the disc rotates, and this torque will represent the load torque T_l.

* For a complete account see H. Cotton, "The pulling into step of the synchronous induction motor," *Journ. I.E.E.*, **63** (1925) 211, and "The synchronous induction motor," *World Power*, **1** (New Series) (June, July, 1924). Also L. H. A. Carr, "The pulling into step of an induction-type synchronous motor," *Journ. I.E.E.*, **61** (1923) 692, and "Induction-type synchronous motors," *J. Inst. Elect. Engrs.*, **60** (1922) 165.

It will try to accelerate the disc, just as the load torque in the actual motor tries to accelerate the slip, and therefore, reduce the actual speed. If the disc is acted on by a frictional torque proportional to its speed, then this torque will represent the induction-motor torque in the actual motor. If these are the only external torques acting on the disc, then a uniform speed will be attained when they are equal and opposite. This uniform speed corresponds to the uniform slip speed of the induction motor before the direct-current excitation is switched on.

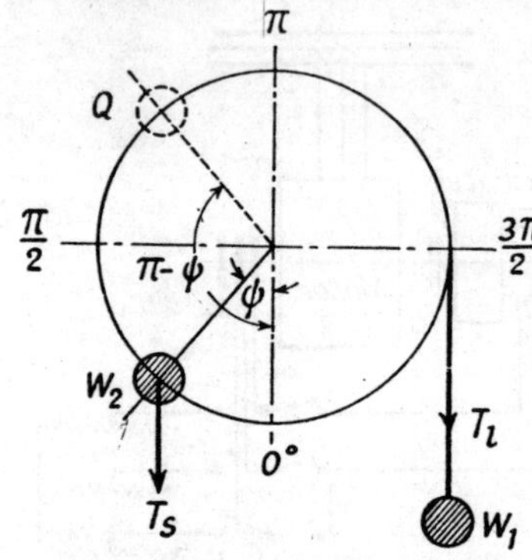

FIG. 23.43. TO ILLUSTRATE SYNCHRONIZING

Now suppose that another weight W_2 is suddenly clutched on to the periphery of the disc. This weight will set up another torque, and if we represent its angular position by the angle θ reckoned from the bottom position, it will have a maximum value of T_m when $\theta = \pi/2$. For any other angular position the torque can be written

$$T_s = T_m \sin \theta$$

and it represents the synchronizing torque in the actual machine due to the creation of the d.c. excitation. It will be seen from Fig. 23.43 that so long as $\theta < \pi$, and W_2 is therefore on the left-hand side of the centre the torque T_s acts in opposition to the torque T_l. It thus tends to reduce the angular velocity $d\theta/dt$ of the slip motion. If $\theta > \pi$ and $< 2\pi$, then T_s acts in conjunction with T_l and the slip velocity is increased.

For a certain angle $\psi < \pi/2$ the two torques balance, and we then have

$$T_s = T_m \sin \psi = T_l$$
$$\sin \psi = T_l/T_m$$

The point P at angular distance ψ is the point of stable equilibrium and if the conditions are such that the model comes to rest, and the motor succeeds in synchronizing, then P will be the position of rest of the weight W_2. There is also a point Q of unstable equilibrium its angular position being $(\pi - \psi)$. If after clutching on the weight W_2 the angular velocity becomes zero before the angular position $(\pi - \psi)$ is attained then the subsequent motion will be one of damped oscillations, as analyzed on p. 546, and the model will come to rest with W_2 at the position P. The angle ψ will thus represent the load angle of the machine.

If, after clutching on the weight W_2, the angular velocity does not become zero by the time the weight reaches the position Q of unstable equilibrium, then the weight will be carried over the position $\theta = \pi$ and down the right-hand side where, acting in conjunction with W_1, it will increase the slip velocity. By the time the position $\theta = 2\pi$ is reached the slip velocity will probably be so great that the model will continue to make whole revolutions, that is, the machine will fail to synchronize. Under these conditions, owing to the unbalance caused by the peripheral weight W_2 the angular velocity of the model will be far from uniform, since it will reach a minimum at the position $\theta = \psi$,

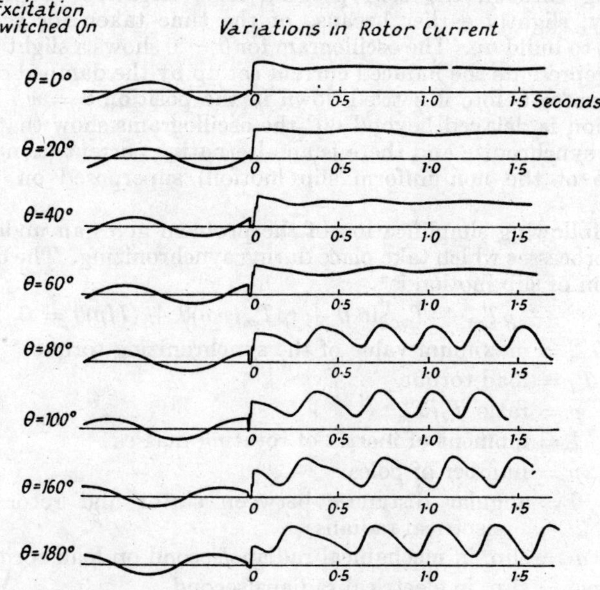

Fig. 23.44. Oscillograms of Rotor Current on Application of Excitation at Various Values of θ

$\theta = (2\pi + \psi)$ and so on, and reach a maximum at intermediate positions. In the motor the actual slip motion will correspond to this irregular motion of the model with the result that (*a*) the motor will be subjected to violent mechanical strains and (*b*) there will be violent oscillations of current and power factor in the line. It is thus imperative that the motor should synchronize as soon as possible after the switching on of the d.c. excitations. Since we must have the angle ψ within the allowable load angle of the machine we see that the first requirement is that T_m shall be sufficiently large with respect to T_l. A second requirement is that the synchronizing torque T_s shall, from the very moment of its application, be in opposition to T_l, and this requires that, in the model, the weight W_2 shall be clutched on at the bottom position, $\theta = 0$. In the motor it corresponds to the rotor poles set up by the d.c. excitation being directly opposite to stator poles of opposite polarity. Fig. 23.44 shows oscillograms of rotor current taken for various angular positions of application of the excitation. The initial portions correspond to the alternating slip-frequency current in the rotor and the final portions to the exciting current plus induced current, if any. It will be seen that at the position $\theta = 0$, the rotor induced current is zero and is on the point of changing over from negative to positive, where positive is taken as the direction of the d.c. excitation. Thus, if a moving-coil, zero-centre, ammeter is included in the rotor circuit, the moment of switching on the d.c. excitation will be when the needle is swinging through the zero position from negative to positive, or, actually, slightly earlier because of the time taken for the exciting current to build up. The oscillogram for $\theta = 0$ shows a slight oscillation which represents the induced current set up by the damped oscillations of the rotor before it settled down in the position $\theta = \psi$. When the excitation is delayed beyond 60° the oscillograms show that the rotor fails to synchronize and there is an alternating current (non-sinusoidal because of the non-uniform slip motion) superposed on the direct current.

The following simplification of the problem gives an understanding of the processes which take place during synchronizing. The differential equation of slip motion is*

$$-\rho T_m + T_m \sin\theta + (\rho T_m/p\omega)\dot{\theta} + (I/p)\ddot{\theta} = 0$$

where T_m = maximum value of the synchronizing torque
T_l = load torque
ρ = ratio T_l/T_m
I = moment of inertia of rotating masses
$2p$ = number of poles
θ = angular distances between stator and rotor fluxes in electrical radians
ω = slip, in mechanical radians/second on load torque T_l
$p\omega$ = slip, in electrical radians/second.

Assume that the d.c. excitation of the rotor is switched on at the instant $\theta = 0$, and that synchronism is reached by the time $\theta = \pi/2$, then the average value of the synchronizing torque is $(2/\pi)T_m$. The initial value of the asynchronous torque is, of course the load torque $T_l = \rho T_m$, and the final value when synchronism is achieved is zero, since $\dot{\theta}$ is then zero. Since the relationship between T_l and ρ is very nearly linear for the operative range of ω, the average value of the asynchronous torque is $\frac{1}{2}\rho T_m$. The load torque remains throughout at ρT_m, and thus for the torque available for acceleration, we have

$$T_{acc} = \frac{2}{\pi} T_m + \tfrac{1}{2}\rho T_m - \rho T_m$$

$$= \frac{2}{\pi} T_m - \tfrac{1}{2}\rho T_m$$

$$= \frac{2}{\pi} T_m - \tfrac{1}{2} T_l$$

* The mathematical solution of the equation of motion is discussed by L. H. A. Carr (*loc. cit.*) and H. Cotton (*loc. cit.*), also by Edgerton and Zak, "The pulling into step of a synchronous induction motor," *Journ. I.E.E.*, **68** (1930) 1205.

Let S_0 = fractional slip before synchronizing, so that $\frac{1}{2}S_0$ = average slip. A displacement of $\theta = \pi/2$ corresponds to one-quarter of a period of the induced rotor current, and since the average slip frequency is $\frac{1}{2}S_0 f$, the time during which the above accelerating torque acts is

$$t = 1/4(\tfrac{1}{2}S_0 f) = 1/(2S_0 f) \text{ second}$$

The increase in speed, with velocities in mechanical radians/sec, is

$$\delta\omega = \omega_s - \omega_s(1 - S_0) = S_0\omega_s$$

where ω_s = synchronous angular velocity.

$\therefore$ from $T_{acc} = I(\delta\omega/t)$

we have $\dfrac{2}{\pi} T_m - \frac{1}{2}T_l = IS_0\omega_s(2S_0 f) = 2I\omega_s f S_0^2$

$$\therefore \quad T_m = \pi I\omega_s f S_0^2 + 0{\cdot}78T_l$$

This equation indicates the conditions which must be fulfilled for easy synchronizing, namely—

1. The inertia of the rotating masses must be as low as possible: thus the high-speed motor of small diameter is indicated.
2. A high value of ω_s makes synchronizing difficult.
3. A high value of the supply frequency, f, makes synchronizing difficult.
4. A high value of the initial slip S_0 makes synchronizing very difficult, because T_m is proportional to S_0^2. Now S_0 is proportional to the rotor I^2R loss when running asynchronously, from which it follows that the rotor resistance must be made as low as possible, the amount of rotor copper therefore being greater than in a plain induction motor if the motor has to synchronize against a high load torque.
5. A high value of T_l makes synchronizing difficult, not only because of the term $0{\cdot}78T_l$ in the expression for T_m, but also because the value of T_l directly affects the value of S_0.

The induction motor run as a synchronous motor has the following advantages compared with the ordinary salient-pole synchronous motor—

(*a*) It will start and synchronize against greater than full-load torque, whereas the ordinary synchronous motor must be started under light-load conditions.

(*b*) Its air gap is not much greater than that of an ordinary induction motor, whereas the gap of a salient-pole motor is long. The exciter is not therefore of such large capacity as that for a salient-pole motor, but, on the other hand, it must be a low-voltage, heavy-current machine owing to the very low resistance of the rotor winding.

(*c*) The rotor winding combines the functions of exciting and damping winding: the salient-pole motor requires a separate damping winding.

(*d*) No special starting gear is required. The salient-pole motor must be started up, either by a separate starting motor or by applying a reduced voltage to the armature, either method involving complication and extra control gear.

The above comparison shows that if a mechanical load is to be driven, but at the same time the phase-advancing properties of the synchronous motor are to be utilized, then the synchronous induction motor is a good proposition. If the synchronous motor is to act as a phase advancer only, without doing any mechanical work, then the salient-pole type is the best, because owing to its longer air gap it is able to supply a greater amount of wattless leading kVA without becoming unstable.

In practice the choice between the salient-pole type and induction-motor type depends upon the conflicting requirements of high starting torque and high synchronizing torque, and stability during synchronous motor operation. The salient-pole motor cannot develop a starting torque equal to that of the induction-motor type unless elaborate and expensive modifications are made to the machine, and consequently where a high starting torque is necessary the induction-motor type is preferable. On the other hand, the salient-pole machine has a field system which is magnetically strong in comparison with the armature, because of the long air gap, with the result that it is very stable on violently fluctuating loads. The induction-motor type is at a disadvantage in this respect, as the very small air gap means a magnetically weak rotor, and in this respect it must be remembered that when a synchronous motor is operated at leading power factor the direct component of the armature reaction is demagnetizing. The best field of application of the synchronous induction motor is, therefore, for drives requiring a heavy starting torque, but in which the load torque is fairly steady.

There is another type of synchronous induction motor which is of salient-pole construction, but in which the simple damping winding of the synchronous motor is replaced by a complete three-phase winding connected to slip-rings. By this means the motor can be started up like an orthodox slip-ring motor, with ample starting torque, and after it is synchronized it runs with the stability of the ordinary salient-pole motor. Now in order that the induction-motor characteristics may be unimpaired it is necessary that the air gap shall be small, whereas, to retain the magnetic strength of the rotor with consequent stability on fluctuating load and on leading power factor, the air-gap must be large. For example, a medium-speed salient-pole machine of 2 m rotor diameter will have an air-gap of about 6 mm, whereas an induction motor of the same diameter will have an air-gap of only about 1·5 mm. In the salient-pole synchronous-induction motor this difficulty can be overcome by adopting the smaller air-gap and inserting non-magnetic liners between poles and hub, thereby increasing the reluctance of the main magnetic circuit by the required amount. This type of motor is valuable where the requirements of the drive are a high starting torque and also a high pull-out torque, as, for example, tube mills for making cement.

The comparison of the performances of the three types of synchronous motor is afforded by the following data. The figures are based on 800 h.p., 6600 V, 176·5 r.p.m., 50 c/s, unity power-factor machines, and they should be compared in the light of the previous discussion of the different types of synchronous motor.*

Type of Motor	Starting Torque (% Full load)	Starting Current (% Full load)	Efficiency (Full load)	Synchronous Pull-out Torque (% Full load)	Excitation	
					(A)	(kW)
Standard Salient-pole Motor	50	250	93·5	220	125	15
Synchronous Induction Motor	100	150	91·5	150	600	21
Salient-pole S.I. Motor	100	200	93·5	200	220	14·5

The values of 50 and 250 for the starting torque and starting current of the standard salient-pole motor are based on auto-transformer starting with a 60 per cent tapping.

The Induction Motor run as a Generator. If the rotor of an induction motor is driven by another motor, or by a prime mover, and at the same time the stator winding is connected to a three-phase supply, then so long as the rotor speed is less than synchronous speed, the rotor will develop a motoring torque which, for small values of the slip, is proportional to the slip. If it is driven at a gradually increasing speed, its torque will gradually decrease and will become zero at exactly synchronous speed. The rotor will then carry no induced current. If its speed is now increased above synchronism, its slip will be negative, the back e.m.f. induced in the stator by the pulsations in the useful rotor flux will be reversed, and will no longer have a component in direct opposition to the supply e.m.f. Hence, the induction motor will now be acting as a generator. The complete torque v. speed curve for speeds below and above synchronism is as shown in Fig. 23.45. When the induction motor functions as a generator, it is called an "asynchronous" generator. It differs from the ordinary synchronous generator in the following respects—

(*a*) It has no direct-current excitation.

(*b*) It will only generate when its stator is connected to a line of fixed frequency, its exciting current being the wattless magnetizing current drawn from the line. This current, as we have seen, produces a rotating field, and there is obviously no difference between a rotating magnetic field produced by polyphase alternating currents, and one produced by a direct-current excited system which is itself driven at synchronous speed.

* For a full discussion of the characteristics of the synchronous-induction motor *see* D. D. Stephens, "Salient-pole synchronous-induction motors." *B.T.H. Activities* (July-Aug., 1956).

(*c*) The frequency of the magnetizing current fixes the frequency of alternating current supplied by the induction motor. Thus the frequency is not affected by the speed at which the asynchronous generator is driven. There must be at least one synchronous alternator connected to the system in order to fix the frequency.

(*d*) No synchronizing is required since the machine cannot generate any e.m.f. until it is connected to the line.*

It will be seen that such generators have a somewhat limited application, and some uses have been examined on p. 566. They are also used in connection with the speed control of cage induction motors.

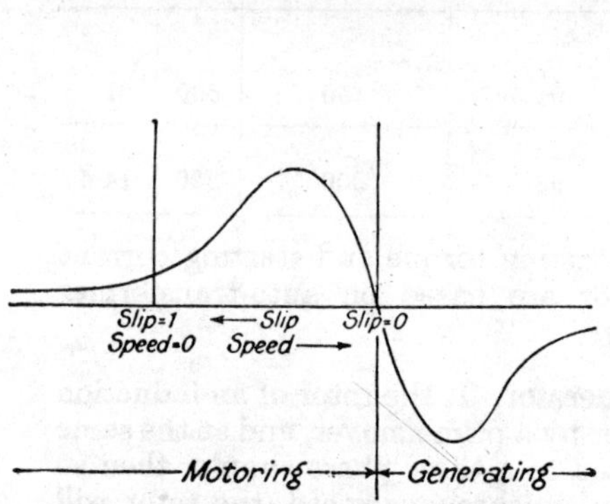

FIG. 23.45. TORQUE-SLIP CURVE

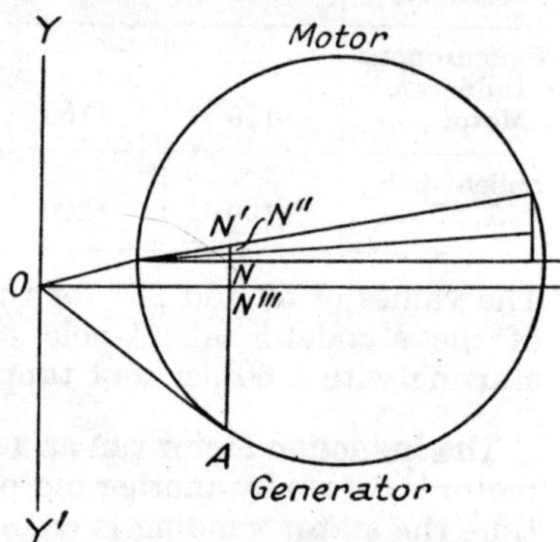

FIG. 23.46. CIRCLE DIAGRAM OF INDUCTION GENERATOR

Thus, if a commutator motor is worked in cascade with a main induction motor for speed-regulating purposes, it can be direct coupled to an asynchronous generator whose stator is connected to the supply, instead of being coupled to the main motor. In such a case the slip energy of the main rotor is returned (less losses) to the line, instead of being converted into mechanical energy.

Asynchronous generators, or induction generators as they are commonly called, are nearly always provided with squirrel-cage rotors.

The characteristics of the generator can be derived from the circle diagram, the working point being now below the horizontal axis. For any working point A (Fig. 23.46),

$$\text{Total power delivered to generator} = AN'$$
$$\text{Rotor copper loss} = N'N''$$
$$\text{Stator copper loss} = N''N$$
$$\text{Stray losses} = NN'''$$

* This property is extremely valuable in factories, where the same steam can be used both for the generation of electrical power and for process purposes. Where only part of the factory's power requirements are supplied in this way, the remainder being imported, the parallel operation of the two supplies offers no difficulty or complication. *See* F. W. Cox, "New practice in induction generation," *Metro-Vick. Gazette* (Nov./Dec., 1951).

$$\text{Output} = N'''A$$
$$\text{Slip} = N'N''/N''A$$
$$\text{Power factor} = \cos AOY'$$

We have seen that, for induction motor operation, the secondary current I_2, and power developed P_2, are given by

$$I_2 = k\sigma E_1/(R_2^2 + \sigma^2 X_{2\cdot 0}^2)^{\frac{1}{2}}$$
$$P_2 = m_2 k^2 E_1^2 \sigma R_2(1-\sigma)/(R_2^2 + \sigma^2 X_{2\cdot 0}^2)$$

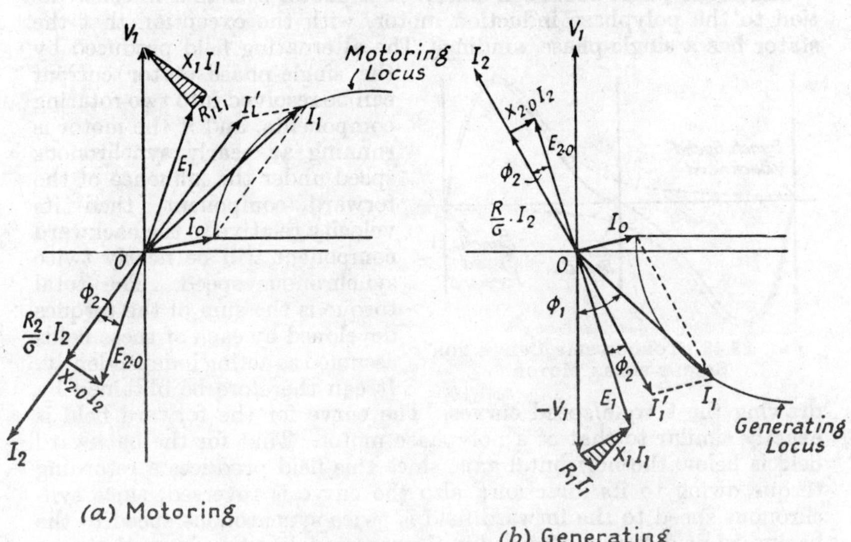

FIG. 23.47. VECTOR DIAGRAMS FOR MOTORING AND GENERATING REGIMES OF INDUCTION MOTOR

When operating as an asynchronous generator σ is negative, so that the corresponding expressions become

$$I_2 = -\,k\sigma E_1/(R_2^2 + \sigma^2 X_{2\cdot 0}^2)^{\frac{1}{2}}$$
$$P_2 = -\,m_2 k^2 E_1^2 \sigma R_2(1+\sigma)/(R_2^2 + \sigma^2 X_{2\cdot 0}^2)$$

The negative sign indicates, in each case, a reversal in direction so that, for a given numerical value of the slip, I_2 during generator operation is in opposition to the corresponding I_2 during motor operation. The negative sign for P_2 indicates that the rotor is now a receiver of power. The vector diagram for motor and generator operation are contrasted in Fig. 23.47. Note that in the motoring case V_1 is the vector sum of E_1, R_1I_1, and X_1I_1. In the generating case E_1 is the vector sum of V_1, R_1I_1, and X_1I_1.

During motoring action we have seen that it is possible to represent the power converted into mechanical power by the power absorbed by a fictitious resistance $R_2' = R_2[(1-\sigma)/\sigma]$. When generating this resistance becomes

$$R_2' = -R_2[(1+\sigma)/\sigma]$$

the negative sign indicating that R_2' now acts as a source of power. It is, in fact, the electrical equivalent of the power intake of the rotor via the shaft.

The Single-phase Induction Motor. This motor is similar in construction to the polyphase induction motor, with the exception that the stator has a single-phase winding. The alternating field produced by the single-phase stator current can be resolved into two rotating components, and if the motor is running at nearly synchronous speed under the influence of the forward component, then its velocity relative to the backward component will be nearly twice synchronous speed. The total torque is the sum of the torques developed by each of these fields assumed as acting independently. It can therefore be obtained by drawing the torque/speed curves. The curve for the forward field is exactly similar to that of a polyphase motor. That for the backward field is below the horizontal axis, since this field produces a retarding torque owing to its direction; also the curve is reversed, since synchronous speed to the forward field is twice synchronous speed to the backward field, while full speed in the reversed direction is synchronous speed to the backward field. The two component curves are shown dotted in Fig. 23.48, and the total torque is indicated by the full-line curve. It will be seen that this characteristic is similar in shape to that of a polyphase motor with low rotor resistance, with the exceptions that the torque is zero at a speed slightly less than synchronous speed, and it is zero again at zero speed. Consequently, the single-phase motor is not self-starting.

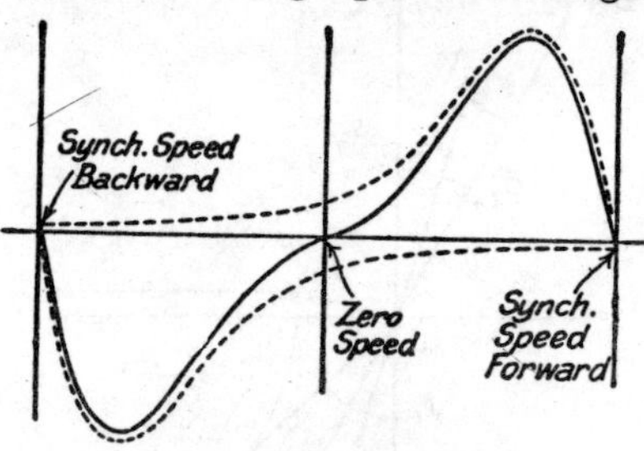

FIG. 23.48. TORQUE-SLIP CURVE FOR SINGLE-PHASE MOTOR

The mechanism of torque production is as follows: First of all let the motor be stationary. The single-phase stator winding produces a stator flux Φ_T which alternates with respect to time but always acts along the same axis: this flux is called the transformer flux. Generally, the rotor has a cage winding which is therefore short-circuited on itself, and we can therefore imagine pairs of conductors connected to form separate coils by means of cross-connectors, a few of which are shown dotted in Fig. 23.49. These short-circuited turns act like the secondary winding of a short-circuited transformer and therefore carry induced currents. Let the flux Φ_T be increasing then, by Lenz's law the induced current will set up an opposing flux, that is, one in an upward direction showing that, at the instant considered, the induced currents are outwards to the left of the axis and inwards to the right. Also, because of the differing flux linkage of the various coils

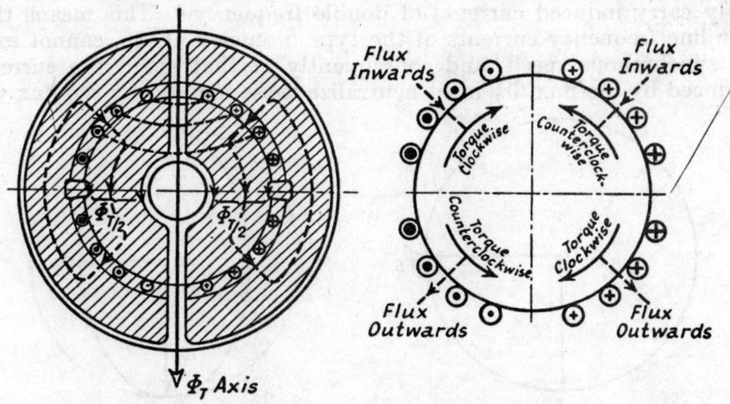

FIG. 23.49. ACTION OF THE SINGLE-PHASE MOTOR AT STANDSTILL

the magnitude of the currents will be graded, as indicated. For clarity these currents are transferred to the simple right-hand diagram, and, applying the left-hand rule we see that the four sectors of the rotor set up a series of torques which are self-cancelling, the total standstill torque then being zero.

Now let the rotor be travelling at synchronous speed, say in a counter-clockwise direction, as in Fig. 23.50. Then dynamically induced e.m.f.s will be set up in the conductors and, since the regions of maximum flux density are at the top and bottom, the space distribution of these e.m.f.s will be as indicated in Fig. 23.50. The currents due to these e.m.f.s will lag by 90° if we assume that the rotor resistance is negligibly small. Consider the instant at which Φ_T is a maximum and is then on the point of decreasing. After a rotation of 90° by the rotor the current distribution *in space* will correspond to the e.m.f. distribution of Fig. 23.50, but the flux Φ_T will have fallen to zero. Hence, no torque can be produced by the interaction of the currents with this particular flux Φ_T. The currents will set up a new flux, the speed flux Φ_S, along the quadrature axis, but we have already seen that the interaction of induced currents with the flux they produce does not

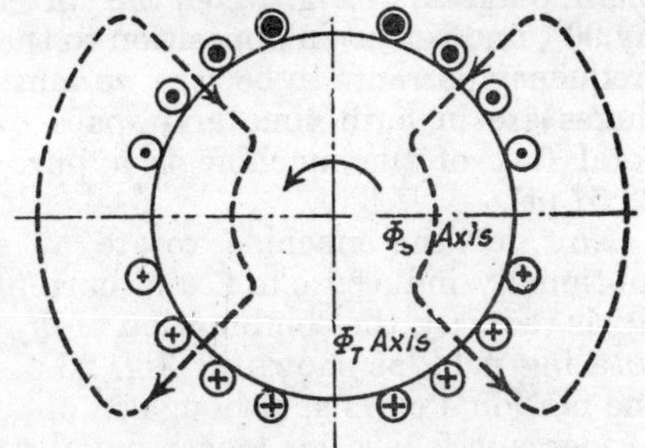

FIG. 23.50. ACTION OF THE SINGLE-PHASE MOTOR WHEN RUNNING AT SYNCHRONOUS SPEED

produce a torque either, and consequently at synchronous speed there is no torque due to either Φ_T or Φ_S. Now if a coil is rotated at synchronous speed in an alternating field, the field has reversed by the time the coil has made one-half of a revolution, and consequently it can only carry induced currents of double frequency.* This means that the line frequency currents of the type discussed above, cannot exist at synchronous speed, and consequently at that speed the currents induced by the flux Φ_T must neutralize those induced by the flux Φ_S.

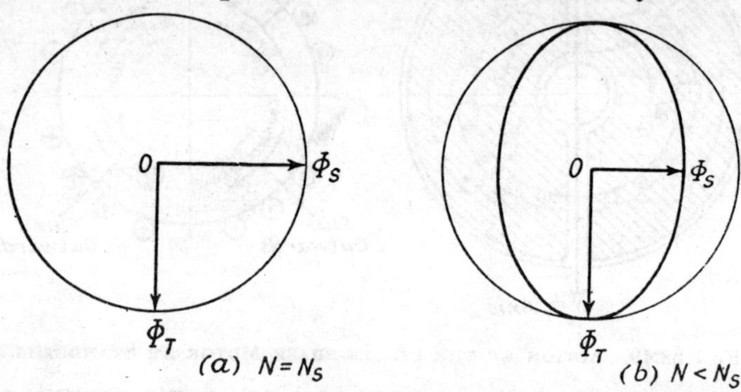

FIG. 23.51. RESULTANT MAGNETIC FIELD OF THE SINGLE-PHASE INDUCTION MOTOR

This is obvious from the fact that, at the instant represented by the e.m.f. diagram of Fig. 23.50, the current distribution in space is lagging by 90°, and is thus in opposition to that of Fig. 23.50. Thus for the line frequency currents to be zero we must have $\Phi_S = \Phi_T$. But these two fluxes are in both time and space quadrature and consequently the total flux of the machine is a pure, or circular rotating field, Fig. 23.51 (*a*).

Now let the machine rotate at some speed $N < N_S$, then the motionally induced e.m.f. and currents are reduced and consequently Φ_S is reduced, the combination of Φ_S with Φ_T now giving an elliptical rotating field, as shown in Fig. 23.51 (*b*). Again, the currents due to the motional e.m.f.s, although in opposition to those due to the transformer e.m.f., are no longer equal to them, so that there is a total current distribution similar to that of Fig. 23.49. This distribution reacting with the speed flux Φ_S produces a torque in the same direction as rotation, showing that motoring operation is possible so long as $N > 0$ and $< N_S$.

At standstill the only flux is Φ_T and the m.m.f. required to produce this draws a magnetizing current $I_{\mu \cdot 0}$ from the line. At synchronism there is an equal flux Φ_S and so the magnetizing current becomes $2I_{\mu \cdot 0}$. At any slip σ the flux Φ_S is reduced to $(1 - \sigma)\, \Phi_S$ and so the magnetizing current becomes

$$I_\mu = I_{\mu \cdot 0} + (1 - \sigma) I_{\mu \cdot 0} = (2 - \sigma) I_{\mu \cdot 0}$$

* *See* p. 619.

Now consider the alternating flux of the machine as equivalent to two rotating fluxes travelling in opposite directions, then denoting the corresponding induced rotor currents by $I_{2\cdot1}$ and $I_{2\cdot2}$ respectively we have, for the forward component

$$I_{2\cdot1} = \sigma E_{2\cdot0}/(R_2^2 + \sigma^2 X_{2\cdot0}^2)^{\frac{1}{2}}; \ \cos\phi_{2\cdot1} = R_2/(R_2^2 + \sigma^2 X_{2\cdot0}^2)^{\frac{1}{2}}$$

$$\begin{aligned}\therefore\ P_{2\cdot1} &= E_{2\cdot0} I_{2\cdot1} \cos\phi_{2\cdot1} \\ &= E_{2\cdot0} \times (\sigma E_{2\cdot0}/(R_2^2 + \sigma^2 X_{2\cdot0}^2)^{\frac{1}{2}}) \times (R_2/(R_2^2 + \sigma^2 X_{2\cdot0}^2)^{\frac{1}{2}}) \\ &= I_{2\cdot1}^2 (R_2/\sigma)\end{aligned}$$

Hence power converted into mechanical power

$$\begin{aligned}P_{2\cdot1}' &= P_{2\cdot1} - I_{2\cdot1}^2 R_2 \\ &= I_{2\cdot1}^2 R_2[(1-\sigma)/\sigma]\end{aligned}$$

The power due to the backward component is negative and can thus be written

$$P_{2\cdot2}' = -I_{2\cdot2}^2 R_2[(1-\sigma)/(2-\sigma)]$$

Again, if R_2 is very small, the currents $I_{2\cdot1}$ and $I_{2\cdot2}$ are both approximately equal to $E_{2\cdot0}/X_{2\cdot0}$ and are thus equal in magnitude. Denoting each by I_2 we have for the approximate power which is converted into mechanical power*

$$P = I_2^2 R_2[(1-\sigma)/\sigma - (1-\sigma)/(2-\sigma)]$$

The motor can be made self-starting by supplying the stator with an auxiliary winding for starting purposes, spacing this winding 90 electrical degrees from the main stator winding, and arranging that the current through it is as nearly in quadrature with the main current as possible. For starting purposes, the motor is thus converted temporarily into a two-phase motor. The most common method of obtaining the required phase difference is the split-phase method, and several alternatives are illustrated in Fig. 23.52.

In Fig. 23.52 (1) the starting winding S has a resistor R in phase with it, and this gives sufficient difference between the power factors of the main M, and starting winding S to enable the motor to start against no load. It is only suitable for small motors. When the motor has run up to speed a centrifugal switch, SW, cuts out the starting winding and the motor then continues to run single phase. The winding S is in circuit for such a short time that it need have very little thermal capacity and so can be made of a smaller gauge wire than the main winding.

In (2) the main winding has a choke in series with it and this is short-circuited and at the same time the starting winding open-circuited when sufficient speed has been attained. In (3) a capacitor is placed in series with the starting winding instead of the resistor of circuit (1). As capacitors can now be made very cheaply this is a very common method of starting.

* For a very complete analysis of the single-phase induction motor *see* E. Mallett, *Vectors for Electrical Engineers*. Chapman & Hall, London.

In all three cases above the split-phase device is used for starting only, the auxiliary circuits having a low thermal capacity, and are, therefore, what are called short-time rated. Method (3) can be modified by giving the auxiliary winding S the same rating as M, using a continuously rated capacitor C and omitting the centrifugal switch SW.

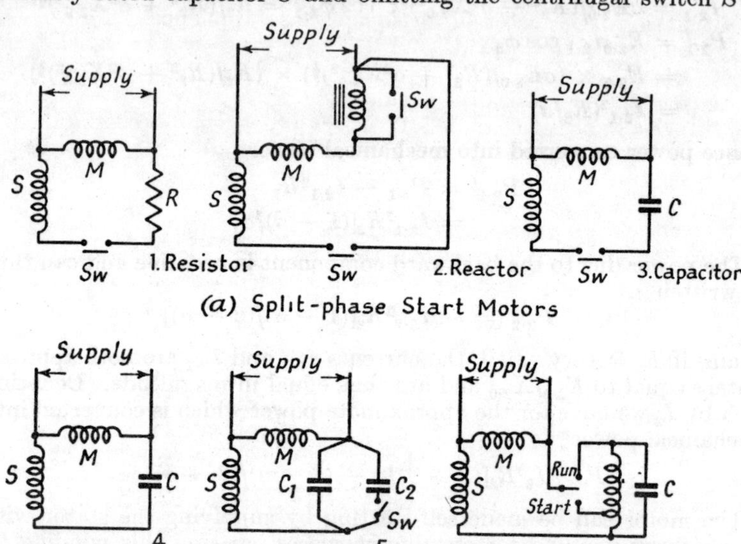

FIG. 23.52. ALTERNATIVE METHODS OF STARTING SINGLE-PHASE INDUCTION MOTORS

M—main winding. S—starting winding.

The motor then becomes a capacitor motor, as distinct from a capacitor-start motor, and it has the advantage of *running*, as well as starting, with a rotating field, although, owing to the impossibility of obtaining a phase difference of exactly 90° between the currents in M and S, the field is not truly circular. The simplest scheme is that of method (4). In order to obtain a better starting torque a short-time rated capacitor C_2 can be arranged in parallel with the main capacitor C_1, as in method (5), this additional capacitor being cut out when the speed is sufficiently high. A similar effect can be obtained by using an auto-transformer with the capacitor, as in method (6).

Capacitor-start motors are often of very small size, and the possible numbers of rotor and stator slots are therefore limited. Consequently, with slots parallel to the shaft there will be large variations in the gap reluctance as the rotor is turned with respect to the stator, with a consequent tendency to magnetic locking. This is overcome by skewing the rotor slots by one slot pitch.*

* *See* C. G. Veinott, *Fractional Horsepower Electric Motors*. McGraw-Hill, New York; E. K. Bottle, *F.H.P. Electric Motors*. Griffin, London.

The capacitor motor has appreciable advantages over the pure single-phase motor because the motor is virtually a two-phase motor. The advantages are mainly—

(a) MECHANICAL. The power intake of a purely single-phase motor fluctuates with respect to time at a frequency of twice line frequency, as does the power factor of all single-phase appliances. As a result, the torque is also fluctuating and this sometimes sets up considerable vibration.

(b) ELECTRICAL. The performance characteristics of the capacitor motor are better than those of the purely single-phase motor, the power factor, in particular, being considerably greater. For example, a 5 h.p., 4-pole single-phase motor will have a power factor and efficiency at normal full-load, both of about 76 per cent. The capacitor motor of the same rating and speed will have a power factor of 95 per cent and efficiency of 80 per cent.

Summing up, the presence of the backward-travelling field in the purely single-phase induction motor is a disadvantage in several respects.

(a) It reduces the total torque since it acts in opposition to the forward field.

(b) It renders the torque zero at starting so that an auxiliary winding has to be provided, thus increasing the size and cost of the motor.

(c) It requires a wattless magnetizing current just as the forward field does, the total magnetizing current being twice what it would be if the forward rotating field could be produced alone. This large magnetizing current gives the motor a very poor power factor.

(d) Owing to the greater magnetizing current, and the fact that the rotor current is not zero at synchronism, the copper losses in the single-phase motor are greater than in the polyphase motor. The rotor core losses are also not negligible.

Because of the increased losses and the space taken up by the starting winding it can be taken that, on the average, a frame which will give a certain horse-power at a given speed when used for a polyphase motor will only give about two-thirds of that horse-power when wound as a single-phase motor.

Rotating Coil in an Alternating Field. Let a single-turn coil of area A rotate with an angular velocity ω_a in an alternating field whose intensity varies sinusoidally with respect to time, so that

$$B = B_{\max} \sin \omega t \text{ where } \omega = 2\pi f$$

The angular position of the coil at any instant is

$$\theta = \omega_a t$$

Let $B = 0$ and $\theta = 0$ when $t = 0$, then at any instant the conditions are as shown in Fig. 23.53. The projected area of the coil is $A \cos \theta$, and therefore for the flux linkage we have

$$\Phi = B_{\max} \sin \omega t \,.\, A \cos \theta$$

Now the linkage varies both with respect to time and to angular position, and therefore we have, for the induced e.m.f.

$$e = -d\Phi/dt = -[(\partial\Phi/\partial\theta)(d\theta/dt) + \partial\Phi/\partial t]$$

$$\partial\Phi/\partial\theta = -B_{\max}A \sin \omega t \,.\, \sin\theta$$

$$\partial\Phi/\partial t = B_{\max}A\omega \cos \omega t \,.\, \cos\theta$$

$$d\theta/dt = \omega_a$$

$$\therefore\; e = B_{\max}A(-\omega \cos \omega t \,.\, \cos \omega_a t + \omega_a \sin \omega_a t \,.\, \sin \omega t)$$

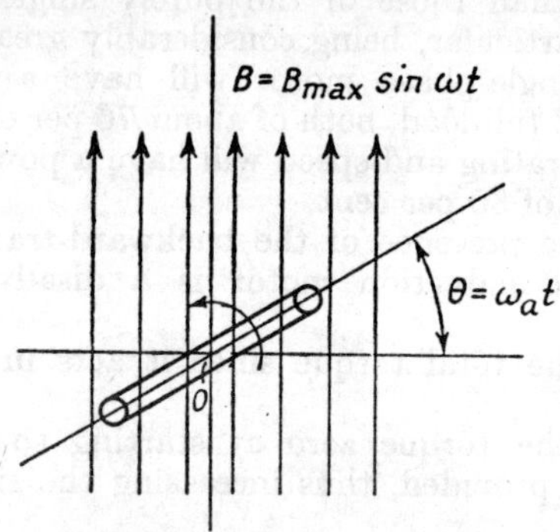

FIG. 23.53. ROTATING COIL IN ALTERNATING FIELD

The first component gives the transformer e.m.f., and the second the speed e.m.f. In the special case of $\omega_a = \omega$, that is the coil rotating at synchronous speed, then

$$e = B_{\max}A\omega\,(\sin^2 \omega t - \cos^2 \omega t)$$

$$= -B_{\max}A\omega \cos 2\omega t$$

showing that the e.m.f. is of twice line frequency. The two components are proportional to $\cos \omega t$ and $\sin \omega t$ respectively showing that they are in quadrature time phase. They are also proportional to $\cos \omega_a t$ and $\sin \omega_a t$ respectively, showing that they are also in quadrature space phase. Thus as the coil rotates one of the components waxes while the other wanes.

The actual mechanism is illustrated in Fig. 23.54. The flux density is a maximum and the direction of the flux is parallel to the plane of the coil. In this position the transformer e.m.f. is obviously zero because the rate of change of flux linkage is zero, but the speed e.m.f. is a maximum because the rate of cutting of the lines is a maximum. In Fig. 23.54 (*b*), which represents the conditions after one-quarter of a period, the flux is zero, the rate of cutting is zero and so the speed e.m.f. is zero. The *rate of change* of linkage is a maximum and the flux is about to reverse, with the result that the induced e.m.f. tends to circulate a current which will produce a flux in the original direction, as shown dotted. Thus the transformer e.m.f., with respect to the coil, is in

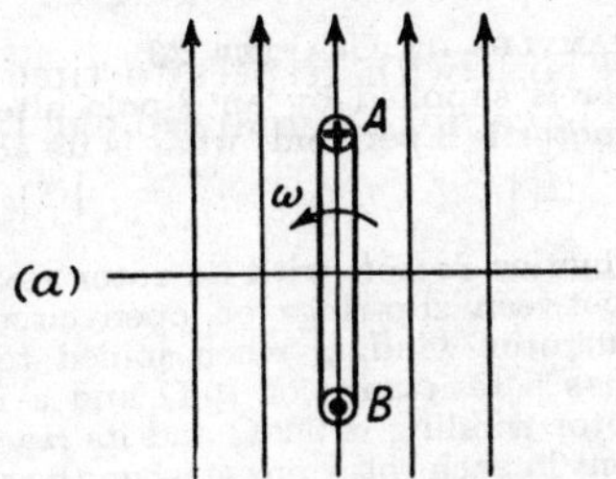

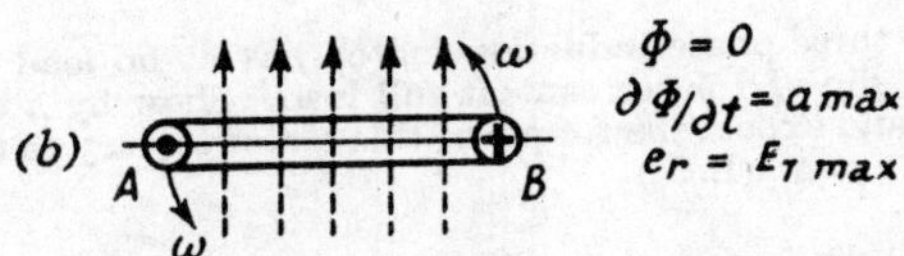

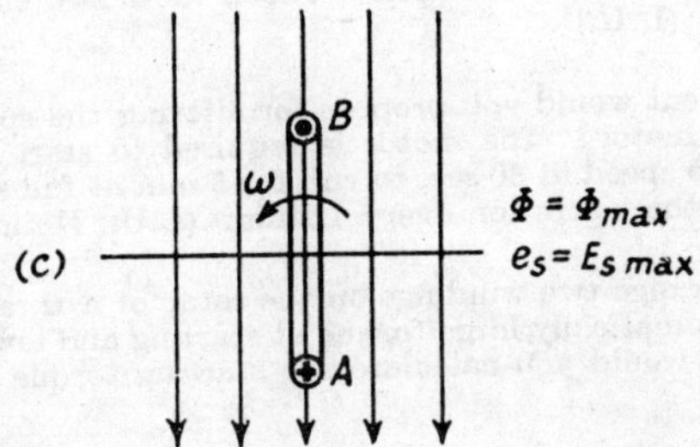

FIG. 23.54. ROTATING COIL IN AN ALTERNATING FIELD

the opposite direction. In (*c*) the flux has reversed and the coil has rotated through 180° so that the transformer e.m.f. is a maximum and in the same direction, with respect to the coil, as in case (*a*), and so on. We see that when the coil has made one complete revolution there have been two maxima and two reversed maxima of the induced e.m.f., which is therefore of *twice* the frequency $\omega/2\pi$.

FURTHER READING

ALGER, P. L. *The Nature of Polyphase Induction Machines.* Chapman & Hall, London, 1951.
ARCHIBALD, R. D. *Polyphase Induction Motor.* Chapman & Hall, London, 1926.
ASTON, K. *Design of Alternating Current Machines.* O.U.P., London, 1934.
CHAPMAN, F. T. *A Study of the Induction Motor.* Chapman & Hall, London, 1930.
GRIFFIN, J. *The Synchronous Induction Motor.* Macdonald, London, 1954.
JAMES, H. D. *Controllers for Electric Motors.* Wiley, New York, 1927.
JONES, H. H. *A.C. Motors of Fractional Horsepower.* Spon, London, 1938.
LAGRON, L. *Polyphase Induction Motors.* Blackie, London, 1931.
SAY, M. G. *Performance and Design of A.C. Machines.* Pitman, London, 1952.
SMITH, S. P. *Papers on the Design of A.C. Machinery.* Pitman, London, 1919.
VICKERS, H. *The Induction Motor.* Pitman, London, 1953.
WALKER, MILES. *Control of Speed and Power Factors of Induction Motors.* Benn, London, 1924.

Examples on Chapter 23

1. A 6-pole induction motor is supplied by an 8-pole alternator running at 750 r.p.m. If the slip of the motor is 3 per cent, what is its actual speed?

Ans. 970 r.p.m.

2. A three-phase, 50 c/s induction motor, with its rotor star connected, gives 100 V (r.m.s.) at standstill between slip-rings on open circuit. Calculate the current in each phase of the rotor winding when joined to a star-connected circuit, each limb of which has a resistance of 10 Ω and a reactance of 10 Ω. The resistance/phase of the rotor winding is 0·2 Ω and its reactance at standstill 10 Ω. Calculate also the current in each rotor phase when the slip-rings are short-circuited and the motor is running with a slip of 4 per cent. (L.U.)

Ans. 2·58 and 5·2 A.

3. A 200-h.p., three-phase induction motor has at no load a speed of about 500 r.p.m. and a slip of 1·5 per cent at full load. Show by a diagram drawn to scale how you would expect the torque of this motor to vary as the speed is varied from zero to 500 r.p.m. (L.U.)

4. Define the "slip" of a three-phase motor. Draw two curves connecting the torque of the motor with the slip for two different values of the resistance in the rotor circuits. State the connection between these two curves and give the theory of the curves. (L.U.)

5. What arrangement would you propose for altering the speed of a 600-h.p., three-phase haulage motor? The motor is required to start at twice full-load torque, to come up to speed in 30 sec, to run for 5 min at full speed, and then to stop. It is to repeat this operation every 12 min. (L.U., Mining.)

6. Show how to arrange two windings on the rotor of a three-phase induction motor to provide automatically large torque at starting and low rotor copper loss when running. How would you calculate the starting torque of such a motor? (C. and G.)

7. A 4-pole, 50 c/s, three-phase, 200 V induction motor is run on light load and takes 12·5 A/line at a power factor of 0·2. The same motor, when the rotor is allowed to rotate very slowly, takes 100 A when a pressure of 120 V between phases is applied, at a power factor of 0·25. Estimate the pull-out torque and maximum power factor at which the motor will work. (C. and G.) [Graphical.]

8. The speed of a 400-h.p. electrically driven fan is reduced by (*a*) a shunt motor with field control, (*b*) an induction motor with rotor resistance control. Compare the economy obtainable per hour in the two cases when the speed is reduced 15 per cent. Assume that the torque of the fan varies as the square of the speed and the cost of energy is 1d./kWh. Ignore losses in the motors. (C. and G.)

Ans. There is a saving of 2s. 8d./hr. by using a shunt motor.

9. Show how the losses in a three-phase slip-ring induction motor can be analysed from no-load and short-circuit tests taken when the power is supplied (*a*) to the stator and (*b*) to the rotor. (C. and G.)

10. An induction motor whose windings can be arranged for 12 or 24 poles can be run in cascade with a direct-coupled induction motor which can be arranged for 2 or 4 poles. With such an arrangement the cascade control is never used for the 24-pole speed. Why is this? Calculate the approximate speeds which can be obtained from the set, the supply frequency being 50 c/s.

Ans. 500, 250, 428, 375 r.p.m.

11. A 4-pole, 35-h.p., 50 c/s induction motor has a full-load slip of 2 per cent and a pull-out torque equal to $2\frac{1}{2}$ times the normal full-load torque. Draw to scale the torque/speed curve for all speeds between zero and full speed.

12. Draw the circle diagram for a 6-pole, star-connected, three-phase induction motor having the given data, and find the full-load power factor, stator current, and slip. Rated output 40 b.h.p., at 380 V, 50 c/s. No-load current 16 A. Power absorbed at no load 1550 W. Short-circuit (standstill) current 310 A at a power factor of 0·23. Resistance from terminal to terminal of stator winding (hot) 0·156 Ω.

Ans. 0·91, 60 A, 6 per cent.

13. From the circle diagram of the motor in question 12, plot the following characteristics against torque: speed, power factor, efficiency, output, and stator current.

14. Determine approximately the starting torque of an induction motor in terms of full-load torque when started by means of (*a*) a star-delta switch, (*b*) an autotransformer starter with 50 per cent tapping. Ignore the magnetizing current. The short-circuit current of the motor at normal voltage is 5 times the full-load current, and the full-load slip is 5 per cent. (I.E.E.)

Ans. (*a*) 0·14, (*b*) 0.31.

15. A 4-pole, 50 c/s, 3-phase induction motor has a slip-ring rotor with a resistance and standstill reactance of 0·04 Ω and 0·2 Ω respectively. Find the amount of resistance to be inserted in each rotor phase to obtain full-load torque at starting. What will be the approximate power-factor in the rotor at this instant? (L.U.)

Ans. 1·29 Ω.

16. Describe briefly how a high starting torque is obtained with a double-cage induction motor. The resistance and reactance (equivalent) values of such a motor for stator, outer and inner cages are 0·25, 1·0 and 0·15 Ω resistance and 3·5, zero, and 3·0 Ω reactance, respectively. Find the starting torque if the phase voltage is 250 V, and the synchronous speed is 1000 r.p.m. (I.E.E.)

Ans. 103 newton-metres.

17. The equivalent standstill impedances per phase of the first and second cages respectively of a double-cage induction motor are $5 + j1$ and $1 + j5$. Calculate the ratio of the torques of the two cages, (*a*) at standstill, (*b*) at a slip of 4 per cent. (*L.U.*)

Ans. (*a*) 5 to 1, (*b*) 1 to 4·8.

18. The short-circuit current of a 4 h.p. cage-type induction motor is 5 times its full-load current. The supply voltage is 400, and the full-load efficiency, power-factor, and slip, are respectively 0·84, 0·8, and 0·04. Ignoring magnetizing current, estimate the current taken from the mains at the instant of starting (*a*) when the motor is switched direct on to the mains, (*b*) when a star-delta starter is used. In each case calculate the starting torque in terms of full-load torque. (L.U.)

Ans. (*a*) 32 A, torque equal to full-load torque.
(*b*) 10·7 A, torque 0·11 of full-load torque.

CHAPTER 24

CONVERTING MACHINERY

Methods of Conversion. At the present time over 90 per cent of the total electrical energy generated is produced in a.c. stations. A large proportion of this is utilized as direct-current energy, and it is therefore necessary to convert from alternating to direct current. The most important methods of effecting this conversion are—

(*a*) Motor-generator sets. The motors are three-phase alternating current motors, either synchronous or asynchronous; and the generators are direct-current, shunt, or compound.

(*b*) Rotary convertors.

(*c*) Motor convertors. The motor convertor consists essentially of an induction motor and a d.c. generator mechanically coupled, and also electrically coupled by means of connections taken from the rotor winding of the motor to the armature winding of the generator. This machine is now quite obsolete, although there are a few still in commission. It will not be considered further.

(*d*) Mercury-arc rectifiers.

(*e*) Valve, and other rectifiers. Rectification by means of electronic devices is considered in Chapter 27.

The Rotary Convertor. This is essentially a direct-current machine having a direct-current armature of ordinary design, from which tappings are taken to slip-rings as well as the ordinary connections to the commutator. The field system is identical with that of an ordinary direct-current machine; it may be shunt or compound excited, and is usually fitted with interpoles. In all heteropolar machines, that is, machines with alternate N. and S. poles, whether for direct or alternating current, the e.m.f.s induced in individual armature conductors are alternating. The voltage across pairs of slip-rings will thus be alternating, while the voltage across the brushes at the commutator will be direct. The following are the rules for taking off the tappings to the slip-rings. If the armature is wave-wound, it has only two electrical circuits through it, whatever the number of poles. Hence, there will be only one tapping taken to each slip-ring. For single phase there will be two slip-rings, and there will be one-half the total number of armature conductors, in each path from one tapping point to another. For three phase, there will be three slip-rings and three tappings, the armature being divided into three equal parts. And so on.

If the armature is lap-wound, it has as many paths through it from brush to brush as there are poles. Hence, for any given potential relative to that of either of the brushes, there will be as many points in the winding having that potential, as there are pairs of poles. The number of tappings taken to each slip-ring is thus equal to the number of pairs of poles. The total number of tappings is equal to the product of the number of slip-rings and the number of pairs of poles. The

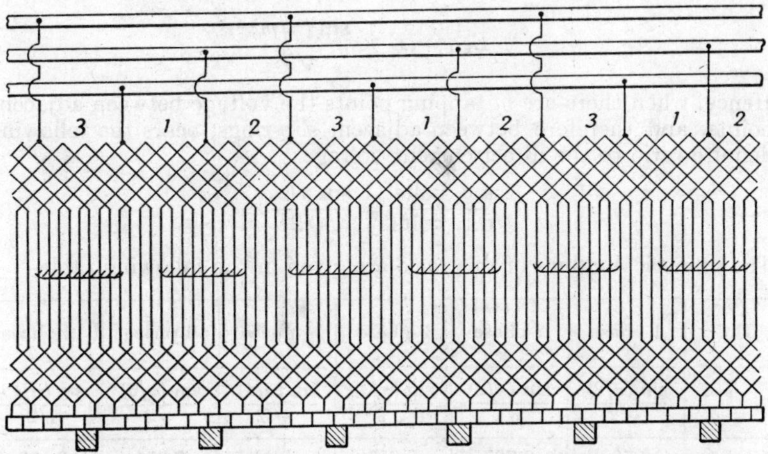

FIG. 24.1. ARRANGEMENT OF TAPPINGS ON ROTARY-CONVERTOR ARMATURE

arrangement of these tappings is shown in Fig. 24.1. The numbers of conductors in each path between two consecutive tappings must be equal. Hence, the winding must be such that it can be divided into the requisite number of identical paths. The existence of this condition shows that any direct-current winding is not necessarily suitable for use on a rotary convertor armature.

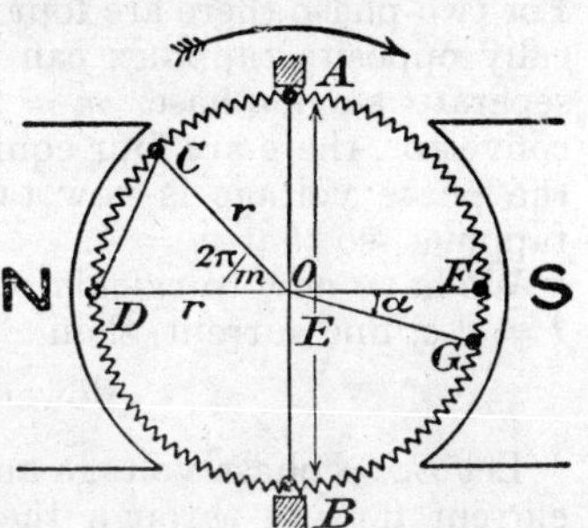

FIG. 24.2. CALCULATION OF VOLTAGE RATIOS

Voltage and Current Ratios. Fig. 24.2 represents an armature rotating in a two-pole field. The voltage at the direct-current side, namely, the voltage across the brushes AB, is obviously the same as the maximum value of the single-phase voltage, since in a single-phase rotary the tappings are diametrically opposite.

$$\therefore \text{ Max. a.c. volts} = \text{d.c. volts} = E, \text{ say}$$

$$\therefore \text{ Effective a.c. volts} = E/\sqrt{2}$$

Consider now an "m" phase convertor. Two consecutive tappings such as C and D will be an angular distance of $2\pi/m$ radians apart. If the direct-current voltage is represented by $2r$, the diameter AB of the circle, then the maximum value of the alternating-current voltage between the points C and D will be represented by the length of the chord CD.

$$\therefore E_{\max} = 2r \sin(\pi/m)$$
$$= E \sin(\pi/m)$$

$$\therefore E_{\text{eff}} = E \times \frac{\sin(\pi/m)}{\sqrt{2}}$$

Hence, when there are m tapping points the voltage between adjacent points, and therefore between adjacent slip-rings, bears the following definite ratio to the direct-current voltage.

$$\frac{\text{a.c. volts}}{\text{d.c. volts}} = \frac{\sin \pi/m}{\sqrt{2}}$$

The numerical values of this ratio are given in the following table—

D.C.	1-phase	2-phase	3-phase	4-phase	6-phase	12-phase
1	$\frac{\sin(\pi/2)}{\sqrt{2}}$	$\frac{\sin(\pi/2)}{\sqrt{2}}$	$\frac{\sin(\pi/3)}{\sqrt{2}}$	$\frac{\sin(\pi/4)}{\sqrt{2}}$	$\frac{\sin(\pi/6)}{\sqrt{2}}$	$\frac{\sin(\pi/12)}{\sqrt{2}}$
	0·707	0·707	0·612	0·5	0·354	0·183

The value of m taken in calculating the above ratios is only equal to the number of tappings when the voltage between adjacent tappings is required. Thus for single-phase there are two tappings and $m = 2$. For two-phase there are four tappings, but since each pair of diametrically opposite tappings can be regarded as being the terminals of a separate single-phase, $m = 2$ in this case also. In the four-phase convertor, there are four equidistant tappings as in the two-phase, but the phase voltage is now taken as the voltage between consecutive tappings, so that $m = 4$.

We have now to calculate the current ratios. If E = d.c. volts and I = d.c. line current, then

$$\text{Power on d.c. side} = EI$$

Let E_p = phase voltage and I_p = *phase* current, i.e. the alternating current flowing through the armature between two tapping points. Then if the power factor is $\cos\phi$,

$$\text{Power on a.c. side} = mE_pI_p \cos\phi$$

Now the a.c. line current I_l is the vector difference of the phase currents in two consecutive phases. Hence, from the vector diagram in Fig. 24.3, we have

$$I_l = 2I_p \cos(AOB/2)$$
$$= 2I_p \cos(\pi/2 - \pi/m)$$
$$= 2I_p \sin(\pi/m)$$

$$\therefore I_p = \frac{I_l}{2\sin(\pi/m)}$$

Also
$$E_p = E \times \frac{\sin(\pi/m)}{\sqrt{2}}$$

$$\therefore \text{ Power on a.c. side} = m \times \frac{E \sin(\pi/m)}{\sqrt{2}} \times \frac{I_l}{2 \sin(\pi/m)} \times \cos\phi$$
$$= (m/2\sqrt{2})EI_l \cos\phi$$

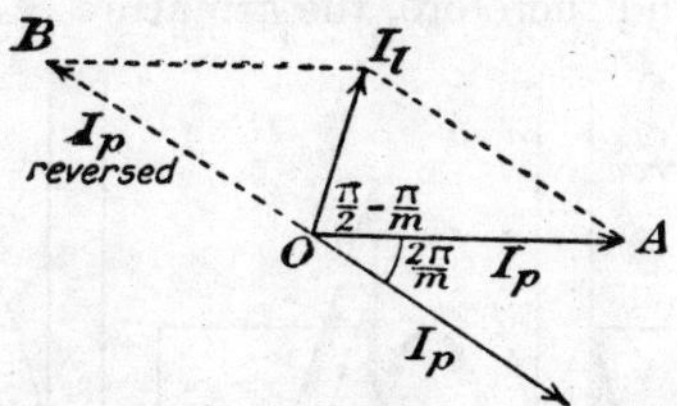

Fig. 24.3. Calculation of Current Ratios

Assuming 100 per cent efficiency and equating the expressions for the power on the d.c. and a.c. sides, we have

$$(m/2\sqrt{2})EI_l \cos\phi = EI$$
$$I_l/I = (2\sqrt{2}/m) \sec\phi$$

For unity power factor, $\sec\phi = 1$ and the current ratios are—

D.C.	1-phase	2-phase	3-phase	4-phase	6-phase	12-phase
1	1·414	0·707	0·943	0·707	0·472	0·236

Waveform of the Armature Current. Suppose that a rotary convertor is converting from alternating to direct current. Then relative to the alternating side it is motoring, while relative to the direct current side it is generating. The total current in the armature is thus the difference between the alternating and the direct-currents. Consider first of all a single-phase convertor; let A and B be the tapping-points (Fig. 24.2). For a point F midway between the tapping points, the alternating voltage has its crest value the instant F is opposite a pole centre; and if the current is in phase with the voltage, the current will go through its crest value at the same instant. The direct current is constant in magnitude and flows in one direction through the conductor at F while F is under the influence of the S. pole, and in the reverse direction while F is under the N. pole. It reverses each time the conductor passes under a brush. Hence, the direct current in one conductor can be regarded as an alternating current of rectangular waveform, which reverses each time the conductor passes under a brush. For the point F, mid-way between A and B, the alternating and rectangular waves are in phase, so long as the power factor is unity and the resulting waveform is as shown in Fig. 24.4 (A).

For a point G displaced any angle α from F, the alternating-current wave has the same form and is in phase with that through F, but G reaches a brush α degrees before F does, so that the rectangular wave for G reverses α degrees in front of the rectangular wave for F, thus modifying the resulting waveform as shown by Fig. 24.4 (B).

For a conductor at the tapping point the alternating and rectangular waves are 90° apart, the resulting waveform being as in Fig. 24.4 (C).

An examination of these curves shows that the effective value of the resultant current, and therefore, the armature I^2R loss, is the greatest

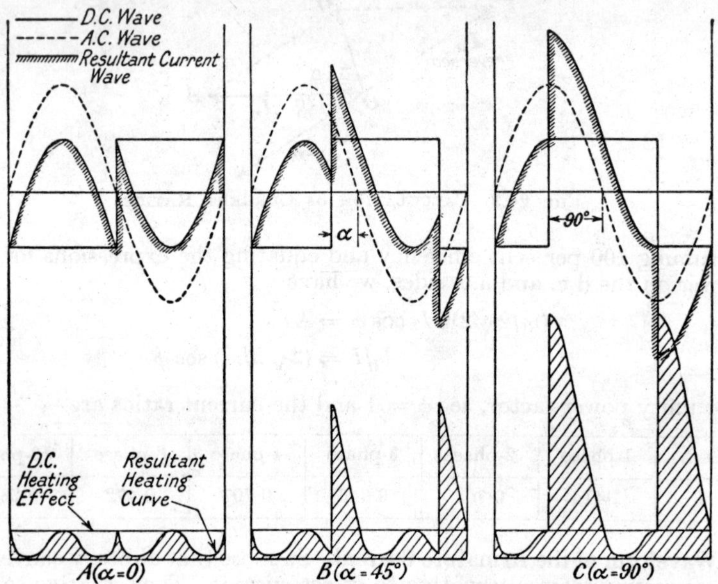

FIG. 24.4. CURRENT WAVEFORMS AND HEATING EFFECT IN A ROTARY CONVERTOR ARMATURE

at the tapping points and least at the mid-points. Since the effective value increases as the phase angle between the sinusoidal and rectangular waves increases, we see that the greater the number of phases, the smaller will be the heating effect, because, as the number of phases increases, the maximum possible phase difference between the two waves decreases.

The variation in the heating effect is best illustrated by curves of current squared, plotted against time for different values of the angle α. We will consider for simplicity a single-phase convertor, although the method can be extended to a machine with any number of phases. We require the relationship between I and I_p (not I_l) in order to draw the curves of resultant current. We have

$$mE_pI_p \cos\phi = EI$$

$$\therefore I_p = (1/m)(E/E_p)(I/\cos\phi)$$

For a single-phase machine $E/E_p = 1{\cdot}414$ and $m = 2$, hence, considering first of all a power factor of unity on the a.c. side, we have

$$I_p = \tfrac{1}{2} \times 1{\cdot}414 \times I$$
$$= I/\sqrt{2}$$

The direct current/conductor

$$I_a = I/2$$
$$\therefore\ I_p = \sqrt{2}I_a, \text{ and}$$
$$I_{p\cdot\max} = 2I_a$$

The current curves are drawn to scale in Fig. 24.4, the cases being A for a conductor half-way between tapping points, B for a conductor displaced 45 electrical degrees from A, and C for a conductor at a tapping point. In the lower half of the figure the curves of current squared are plotted to the same scale, and it is at once apparent how much greater is the heating at a tapping point than at the mid-point.

Now suppose that the power factor is less than unity, say 0·8, then for the same value of I_a, $I_{p\cdot\max}$ is increased in the ratio of sec ϕ to unity

$$\therefore\ I_{p\cdot\max} = 2I_a/0{\cdot}8 = 2{\cdot}5I_a$$

The angle ϕ is 37 degrees, very approximately, and as a result the maximum displacement between the direct- and alternating-current waves, which occurs at one of the tapping points can be $(90 + 37) = 127$ degrees. The corresponding curves for this case are drawn in Fig. 24.5, from which it is evident that the decrease in power factor produces a greater heating at one of the tapping points. It is true that the displacement at the other tapping point is now $(90 - 37) = 53$ degrees instead of 90 degrees, but this is offset by the fact that I_p is increased, so that on the whole the effect of a decrease in power factor is to produce a considerable increase in the heating.

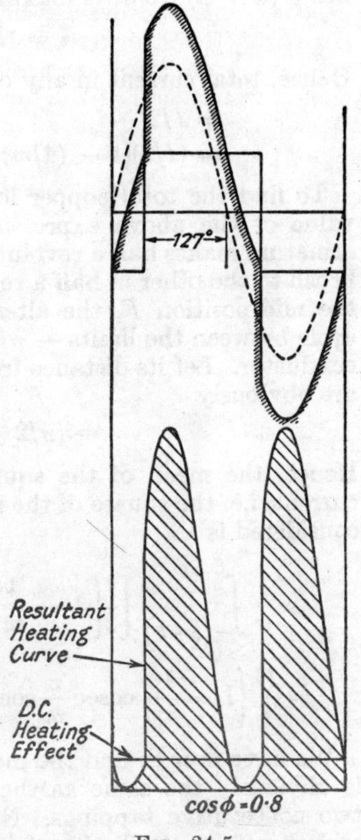

Fig. 24.5

A mathematical expression for the total I^2R loss in the armature is derived as follows—

We have seen that the alternating phase current

$$I_p = \frac{I_l}{2\sin(\pi/m)}, \text{ and } I_l = (2\sqrt{2}/m)\, I \sec\phi$$

$$\therefore\ I_p = (\sqrt{2}/m)\operatorname{cosec}(\pi/m)\sec\phi \times I$$

$$\therefore\ (I_p)_{\max} = (2/m)\operatorname{cosec}(\pi/m)\sec\phi \times I$$

Considering the two-pole armature of Fig. 24.2, the direct current flowing through any conductor G is $I/2$, and the alternating current, being $(\alpha - \phi)$ from its maximum value, is given by

$$i = (I_p)_{\max} \cos (\alpha - \phi)$$

Hence, total current in any conductor

$$= I/2 - i$$
$$= (I/2)[1 - (4/m) \operatorname{cosec} (\pi/m) \sec \phi \cos (\alpha - \phi)]$$

To find the total copper loss it is necessary to calculate the r.m.s. value of the above expression for one particular conductor as the armature makes half a revolution. The conductor at F passes from one brush to the other in half a revolution, and if angles are reckoned from the mid-position F, the alternating e.m.f. will go through one half-cycle between the limits $-\pi/2$ and $+\pi/2$. But we must consider any conductor. Let its distance from F be β. The limits for this conductor are obviously

$$-(\pi/2 + \beta) \text{ and } (\pi/2 - \beta)$$

Hence, the mean of the squares of the instantaneous values of the current, i.e. the square of the r.m.s. current in the particular conductor considered is

$$\frac{1}{\pi}\int_{-\left(\frac{\pi}{2}+\beta\right)}^{\frac{\pi}{2}-\beta} \left[\frac{I}{2}\left\{1 - \frac{4}{m}\operatorname{cosec}\frac{\pi}{m}\sec\phi\cos(\alpha-\phi)\right\}\right]^2 . d\alpha$$

$$= \frac{I^2}{4}\left\{1 - \frac{16}{\pi m}\operatorname{cosec}\frac{\pi}{m}\sec\phi\cos(\beta+\phi) + \frac{8}{m^2}\operatorname{cosec}^2\frac{\pi}{m}\sec^2\phi\right\}$$

We have now to find the mean of this for the whole armature. This is obviously the same as the mean for one armature path between two consecutive tappings. Now two such tappings are $2\pi/m$ apart, and therefore, the limits of integration are $-\pi/m$ and $+\pi/m$. The expression for the square of the r.m.s. value can be written

$$(I^2/4)\{1 - A\cos(\beta+\phi) + B\}$$

$$\therefore \text{ Mean value} = \frac{I^2}{4}(1+B) - \frac{I^2A}{4} \times \frac{m}{2\pi}\int_{-\pi/m}^{+\pi/m} \cos(\beta+\phi)\, d\beta$$

$$= \frac{I^2}{4}(1+B) - \frac{I^2A}{4} \times \frac{m}{\pi} . \sin\frac{\pi}{m}\cos\phi$$

Finally substituting the values of A and B, we have mean value of the square of the current for the whole armature

$$= (I^2/4)\{1 - 16/\pi^2 + (8/m^2)\operatorname{cosec}^2(\pi/m)\sec^2\phi\}$$

Hence, total I^2R loss in the armature

$$= 2 \times (I^2R/4)\{1 - 16/\pi^2 + (8/m^2)\operatorname{cosec}^2(\pi/m)\sec^2\phi\}$$

since we are considering a two-circuit winding, R being the resistance of each circuit. If the armature were carrying a direct current only, its external value being I, as before, the I^2R loss would be

$$2 \times (I/2)^2 R$$

Hence

$$\frac{I^2R \text{ loss when acting as a rotary}}{I^2R \text{ loss as a plain d.c. machine}} = 1 - \frac{16}{\pi^2} + \frac{8}{m^2} \operatorname{cosec}^2 \frac{\pi}{m} \sec^2 \phi$$

This shows that as the number of phases is increased, the I^2R loss decreases, and is always less than the loss in a plain d.c. machine, except in the case of a single-phase rotary. For a unity power-factor load, for which $\sec \phi = 1$, the square roots of the above ratio are

D.C.	$m = 2$	$m = 3$	$m = 4$	$m = 6$	$m = 12$
1	1·18	0·76	0·62	0·52	0·455

If the output of the rotary armature were limited only by the armature copper loss, then the ratios of the maximum outputs possible to the output when working as an ordinary d.c. machine, would be the reciprocals of the ratios in the above table, namely

D.C.	$m = 2$	$m = 3$	$m = 4$	$m = 6$	$m = 12$
1·0	0·85	1·33	1·63	1·92	2·20

This table shows how, in the case of all but a single-phase convertor, the output from a given armature increases as the number of phases increases. In practice it is usual to work rotary convertors six phase, since twelve-phase working involves too much complication.

The heating effects in the armature, at different positions with respect to the tapping points, and also in terms of the number of phases, are illustrated graphically in Fig. 24.6.

The above comparisons are based on the assumption that the power factor on the alternating-current side is unity. If it is not unity, then since the expression for the heating effect contains a term $\sec^2 \phi$, we see that as ϕ increases, the heating effect very considerably increases, the rotary convertor being no longer so very superior to the plain direct-current machine from the point of view of armature heating.

Armature Reaction in Polyphase Convertors. From the consideration of armature reaction in direct-current generators, we have seen that the m.m.f. due to the direct current is directed along the brush axis, and is therefore fixed in space. On the other hand, the armature is rotating at synchronous speed, so that we can also regard the armature m.m.f. due to the direct component of the current as rotating at synchronous speed relative to the armature itself. Its direction is at right angles to the main field. The alternating component of the armature current also sets up an armature reaction which also rotates at synchronous speed with respect to the armature, and in consequence these two components of the armature reaction have no relative velocity. Also, the m.m.f. due to the alternating component acts at right angles to the main field, but in a direction opposite to that due to the direct component, these two thus tending to cancel one another out. Actually they do not quite neutralize one another, the direct

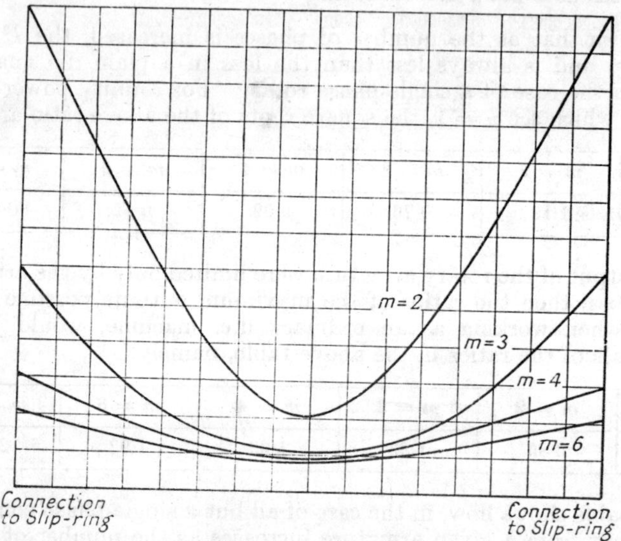

FIG. 24.6. RELATIVE HEATING BETWEEN TAPPING POINTS OF CONVERTOR ARMATURE HAVING m SLIP-RINGS

m.m.f. being slightly the greater, but the difference is so small that to all intents and purposes the rotary convertor can be regarded as having negligible armature reaction. As a result of this there is no distortion of the field form, and the machine can therefore withstand very heavy overloads without serious sparking at the commutator. In fact, when rotary convertors are required for traction purposes, one of the standard works tests is actually to short-circuit the machine momentarily.

Methods of Obtaining Six-phase Current. Since there is a fixed voltage ratio between the d.c. and a.c. sides, it will be seen that to give a definite voltage at the commutator, a definite voltage must be applied to the slip-rings, and this is almost sure to be different from the supply voltage. Transformers have therefore to be interposed between the alternating-current supply and the slip-rings, the necessary six-phase current being obtained by suitable arrangements of the secondary windings.

(*a*) DOUBLE-DELTA. Each phase of the secondary has two separate windings, which are arranged to form two independent mesh-connected secondaries, as shown in Fig. 24.7. By taking connections from points 1, 2, 3 on the first, and 1′, 2′, 3′ on the second, one of the two voltage triangles is reversed, relative to the other. Again, both windings are connected to a symmetrical armature, and this fixes the relative positions of the two voltage triangles, so that the lengths of the six

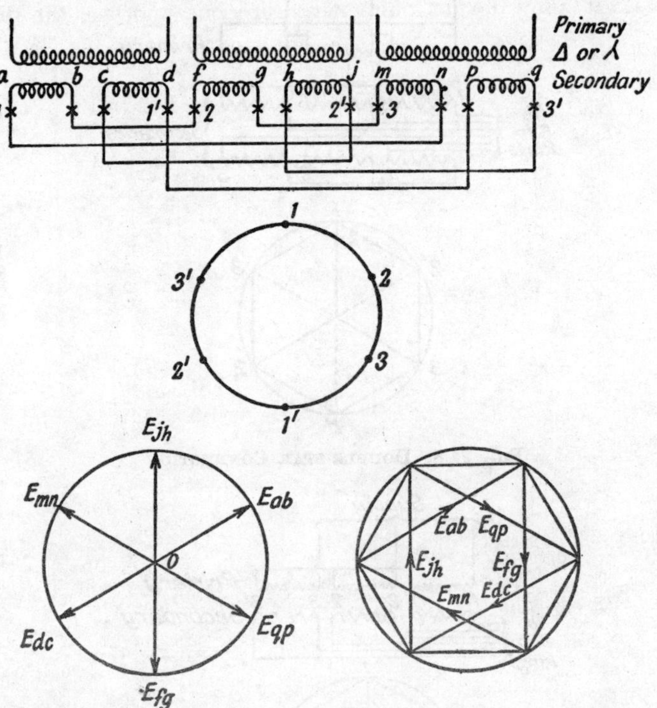

FIG. 24.7. DOUBLE-DELTA SYSTEM OF CONNECTIONS

vectors obtained by joining their corners are equal to one another. These vectors are the voltages applied to pairs of slip-rings, and since there is a phase difference of 60° between two consecutive vectors, we see that a true six-phase supply is obtained.

(*b*) DOUBLE-STAR. As before, there are two separate secondaries, but in this case they are star-connected (Fig. 24.8). The first secondary gives the ordinary three-phase voltage star 1, 2, 3, while the second gives the reversed star 1′, 2′, 3′, since the connections from it are at the opposite ends of the windings to the connections from the first. Also, the two star points must coincide because of the symmetry of the armature winding, and therefore, as before, the sides of the hexagon give the voltages of the six-phase supply.

(*c*) DIAMETRAL. In the double-star method, the two star points have the same potential and they could therefore be joined together if desired. If this were done, then each phase of the transformer secondary would have one winding only, the six ends being taken to the slip-rings, and the middle points, connected together. But the

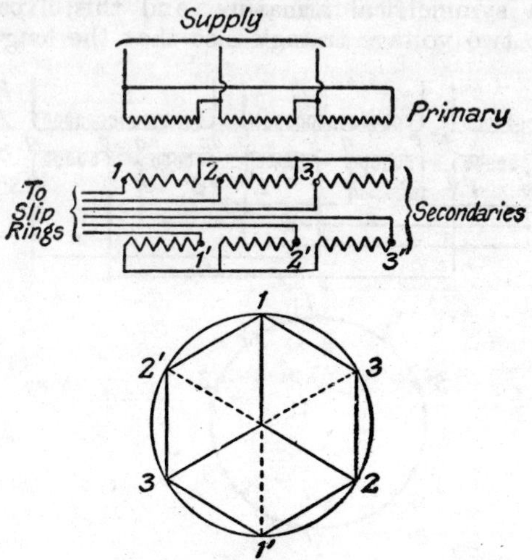

FIG. 24.8. DOUBLE-STAR CONNECTION

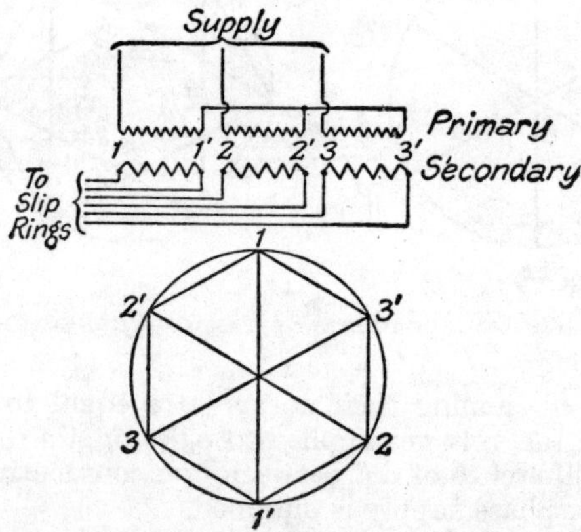

FIG. 24.9. DIAMETRAL CONNECTION

potential of the middle point is fixed by the armature of the convertor, and therefore, the star point need not be made, thus leaving the three separate phases of the secondary winding connected to the slip rings, as in Fig. 24.9. In this case the voltage induced in each phase of the secondary is given by the diameter of the circle, whereas in the double-star method, it is given by the radius. This method is the most commonly used, because it is the simplest. If the rotary convertor is supplying a three-wire system on the d.c. side, then the star point is formed, and the middle wire connected to it. This fixes its potential midway between the potentials of the positive and negative outers, and no extra balancing plant is required.

EXAMPLE. A star-connected three-phase alternator, giving 1500 V/phase, supplies six-phase current by means of three mesh-connected transformers, each provided with two equal secondaries. The current is passed into a rotary convertor supplying direct current at 550 V. Calculate the change ratio of the transformers and the current supplied by each of the secondaries, if 200 kW are taken from the convertor. Neglect magnetic leakage and loss in transformation. Show by diagrams how the six contact rings of the convertor are connected to the secondaries. (London Univ.)

The diagram of connections is given in Fig. 24.7. If E_a = r.m.s. voltage between adjacent slip-rings, and I_a = current supplied to each slip-ring, then

$$E_a = (E/\sqrt{2}) \sin (\pi/m) = (550/\sqrt{2}) \sin 30 = 194{\cdot}4 \text{ V}$$

$$I = 200{,}000/550 = 364 \text{ A}$$

$$\therefore I_a = (2\sqrt{2}/m) \times I = (2\sqrt{2}/6) \times 364 = 171{\cdot}4 \text{ A}$$

The voltage induced in each secondary winding is given by the length of the side of an equilateral triangle drawn in the six-phase voltage hexagon. Each side of the hexagon represents a voltage of 194·4 V. Hence, voltage in each secondary

$$= \sqrt{3} \times 194{\cdot}4 = 337 \text{ V}$$

$\therefore$ Transformation ratio of the transformer

$$K = 337/(\sqrt{3} \times 1500) = 0{\cdot}1295$$

$\therefore$ No. of turns on each secondary = 0·1295 × turns on each primary

Methods of Starting Rotary Convertors. In the great majority of cases, rotary convertors are supplied with alternating current, and deliver direct current. It is therefore unlikely that there will be any direct-current supply for starting, unless it is specially provided, so that the convertors in most cases are started up from the alternating-current side. If direct current is available, the procedure is as follows: The direct-current switch (Fig. 24.10) is closed, and the oil switch on the alternating-current side, left open. The rotary is then run up to speed like an ordinary shunt motor by means of the starting resistance. The shunt regulator is adjusted until synchronism is obtained, and the oil switch then closed. No voltage adjustment is necessary, as when synchronizing alternators, because the voltage ratio between the direct and alternating-current sides is fixed.

If the load on the direct-current side is rapidly fluctuating while synchronizing, it may be very difficult to decide the correct moment of synchronism, because of the variations in speed produced by the variations in voltage. In such a case it is better to run the rotary up to a speed well above synchronism, open the d.c. switch, and thus allow the machine to slow down. As soon as synchronous speed is reached, as indicated by the synchroscope, close the oil switch. Alternating current will then flow through the armature and keep it running at synchronous speed. There is no danger of a heavy rush of current on closing the oil switch, because no power is being drawn from the direct-current supply. At the worst, the armature will be violently

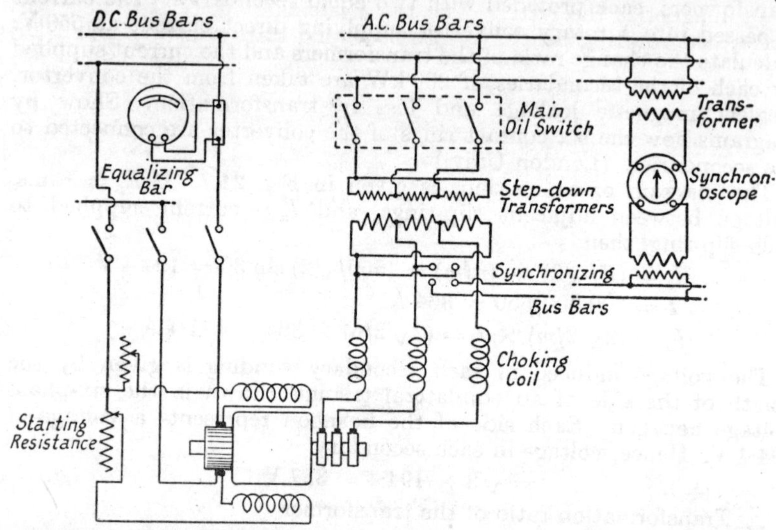

FIG. 24.10. CONNECTIONS FOR THREE-PHASE CONVERTOR TO BE STARTED FROM D.C. SIDE

pulled into step if the switch is not closed quite at the right moment. Finally the d.c. switch is closed.

There are three common methods of starting up from the alternating-current side—

(*a*) BY TAPS ON THE MAIN TRANSFORMER. This is a very commonly used method, the arrangement of the control gear being as indicated in Fig. 24.11. It will be noticed that no synchronizing gear is required, but a moving-coil zero-centre voltmeter V is connected across the brushes on the d.c. side, and the field winding is split up into a number of sections by means of a "field break-up switch." On the a.c. side, either auto-transformers are interposed between the slip-rings and the main transformer, or the secondaries of the main transformer are used as auto-transformers, this second arrangement being shown in the figure. When starting, the d.c. switch and the field break-up switch are opened, and the brushes lifted off the commutator. The a.c. switch is closed, the auto-transformer being arranged so that the smallest voltage is applied to the brushes. An alternating current flows through the armature and produces a rotating magnetic field, the lines of force of which sweep past the dampers on the pole faces, thereby inducing currents in them. Hence, the armature starts up like an inverted squirrel-cage induction motor. Now the rotating field produced by the armature rotates at synchronous speed *relative* to the armature, and therefore, as the speed of the armature increases, the actual speed, ir space, of the rotating field becomes smaller and smaller, until eventually its poles move so slowly past the salient poles of the field system, that they magnetize the latter. The two systems of poles then grip each other, the rotating field therefore becoming stationary in space. This

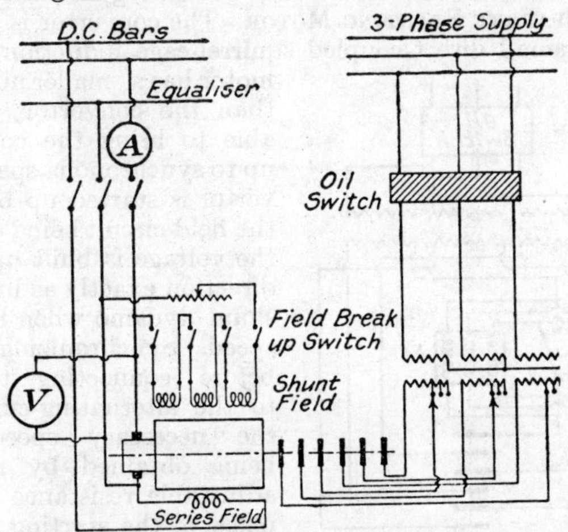

FIG. 24.11. STARTING BY TAPS ON MAIN TRANSFORMER

means that the armature must now be running at synchronous speed, the rotary having what is called "jumped into step." It is obvious that it is possible for the polarity of the d.c. side to come up in the wrong direction, because this polarity depends upon the ultimate polarity of the main poles, and therefore, on the relative positions of the main poles and the rotating field at the instant the convertor jumps into step. If the polarity on the d.c. side is wrong it will be indicated by the zero-centre voltmeter *V* being deflected in the wrong direction. It can be corrected by what is called "slipping a pole." The field switch is momentarily reversed, thus causing the armature field to be repelled by the main field, and as a result each pole of the armature field moves on towards the next main pole. This is indicated by the needle of the zero-centre voltmeter swinging over to the right side, the field break-up switch being then closed in the right position. Finally the convertor is connected to the d.c. system by closing the d.c. switch.

The field is split up into sections by the field break-up switch, because, at the moment of starting, the actual velocity in space of the armature field is the synchronous speed, and consequently, owing to the very large number of field turns, a very high voltage would be induced in the field winding. When the armature is running at synchronous speed, the armature field is stationary in space, and consequently no voltage is induced in the field winding, the sections of which can therefore be all connected in series. For the same reason, the brush gear on the d.c. side is lifted at starting, otherwise heavy currents would circulate across the brush faces and cause excessive sparking.

The full a.c. voltage is not applied to the slip-rings at starting, because of the very heavy rush of current which would be produced on the a.c. side. Also, this current would be very nearly wattless, so that it would probably interfere with the voltage regulation of the line.

(*b*) BY SEPARATE STARTING MOTOR. The convertor is started up by means of a small direct-coupled squirrel-cage induction motor. This motor has a smaller number of poles than the convertor, so that it is able to bring the convertor right up to synchronous speed. The convertor is started up by this motor, the field circuit being closed, so that the voltage is built up in the right direction exactly as in the case of a shunt dynamo when brought up to speed. Synchronizing is necessary before connecting the slip-rings to the alternating-current supply, the necessary speed regulation being obtained by means of an adjustable resistance in one stator phase of the starting motor. After synchronizing, the switch controlling the starting motor is opened and the convertor main switch on the d.c. side is closed.

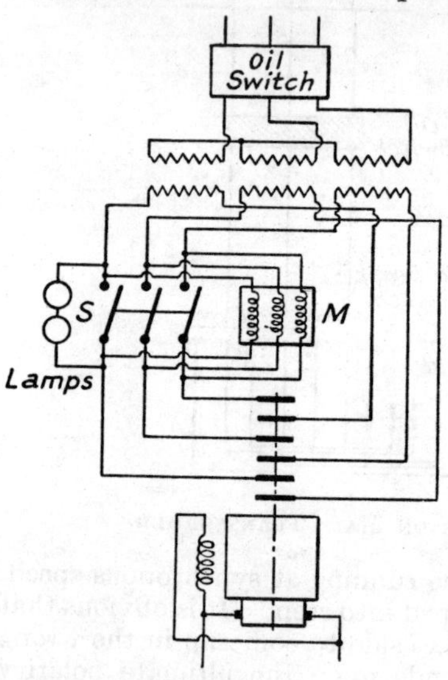

FIG. 24.12. SELF-SYNCHRONIZING METHOD

At the present time it is usual to build the starting motor with a rotor consisting simply of a solid iron cylinder, since there is no possibility of trouble with such a rotor as it is only in operation a very short time.

(*c*) SELF-SYNCHRONIZING METHOD. This method is due to Rosenberg. The convertor is started up by a small starting motor as in case (*b*), but the stator of this motor is in series with the convertor armature, as can be seen from Fig. 24.12. The main oil switch is first closed, the switch S across the starting motor being open. The shunt-field circuit is also closed, the regulator being set to the normal position. The starting motor M, having fewer poles than the convertor, brings the latter up to synchronous speed, and as soon as this speed is attained, the convertor jumps into step because its armature is also carrying the alternating current taken by the starting motor. There is very little possibility of reversed polarity, because the armature rotating field is produced by a very small current, and in consequence, it is not sufficiently strong to reverse the residual magnetism in the field. As soon as the convertor is in step the switch S is closed, thus short-circuiting the starting motor. Brush-lifting gear and field break-up switch are unnecessary, because the rotating field produced by the armature current during starting is not sufficiently strong to produce induced voltages of any importance.

Voltage Regulation of Rotary Convertors. There are four methods as follows—

(*a*) Hand Regulation. The main transformer is provided with tappings, so that the voltage applied to the slip-rings can be varied within limits. This is not a very satisfactory method, first, because it is not automatic; second, because the voltage can only be varied in jumps; third, because of excessive wear and tear on tappings from which very heavy currents are taken.

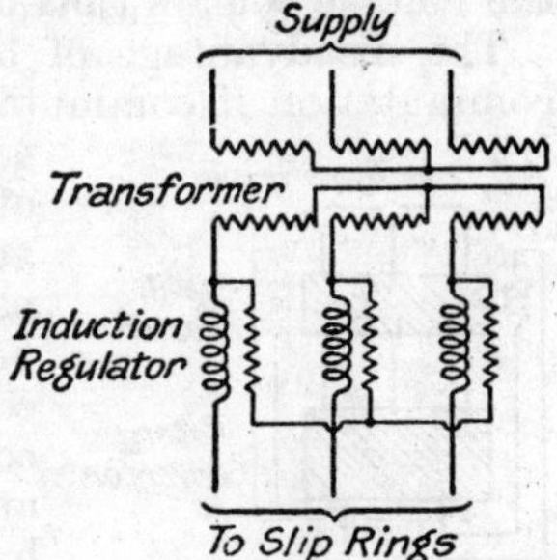

Fig. 24.13. Voltage Regulation by Induction Regulator

(*b*) By Induction Regulator. The method of connecting a three-phase induction regulator in the circuit is shown in Fig. 24.13. The regulator, which is very similar in construction to a three-phase induction motor, acts like a booster, and raises or lowers the total voltage applied to the slip-rings according to the position in which its rotor is placed relative to its stator. The regulator can be arranged for hand or automatic operation, and by means of it a variation of voltage on the d.c. side as high as 30 per cent or more can be obtained.

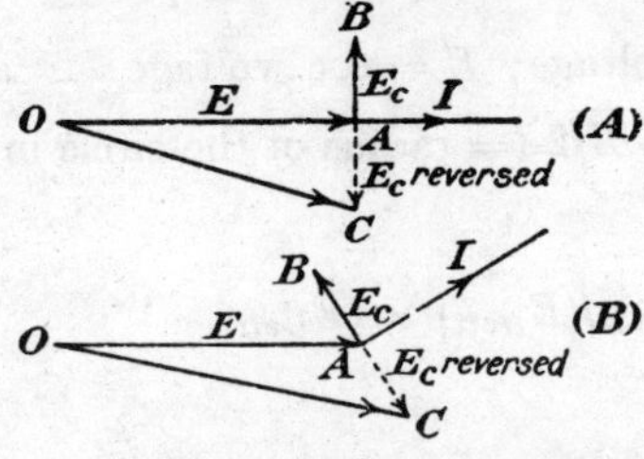

Fig. 24.14. Reactance Control

(*c*) By Choking Coils: Reactance Control. Choking coils are placed between the main transformer and the slip-rings, as in Fig. 24.10, and the convertor is compound excited, having both shunt and series excitation. The voltage at the slip-rings is obviously equal to the vector difference of the voltage E at the main transformer terminals, and the voltage E_c across the choker terminals, this latter voltage leading the current by 90°. On normal load the current I is in phase with the supply voltage OA (Fig. 24.14 (*A*)), the slip-ring voltage being therefore equal to OC. If the load increases, the convertor becomes over-excited due to the series excitation, and the current vector I and choke voltage vector OB are shifted round in a counter-clockwise direction, thus increasing the slip-ring voltage OC, as shown in Fig. 24.14 (*B*). This method is very good so long as a voltage regulation of not more than 15 per cent is required.

(*d*) Synchronous Booster Control. This method is often used where a wide range of voltage on the d.c. side is required, and where the fluctuations in load are very rapid as in traction work. The synchronous booster consists of an alternating-current generator, whose armature is mounted on the convertor shaft, and is connected electrically in series with the convertor armature, as shown diagrammatically in Fig. 24.15. The booster and convertor fields obviously have equal numbers of poles. If the booster field is supplied with series as well as separate excitation, the booster voltage will increase with the load, the voltage control thus being automatic.

The disadvantage of booster control is that it is liable to affect commutation if commutating poles are used. If the commutation is good when the voltage is about the middle of the range, then when the booster raises the voltage the commutating poles become magnetized by armature reaction, their effect then being too powerful. Conversely, when the convertor is giving the minimum voltage, the commutating poles are partially demagnetized and their effect is neutralized. This difficulty has been overcome by connecting diverters across the commutating poles when the convertor is delivering large d.c. voltages.

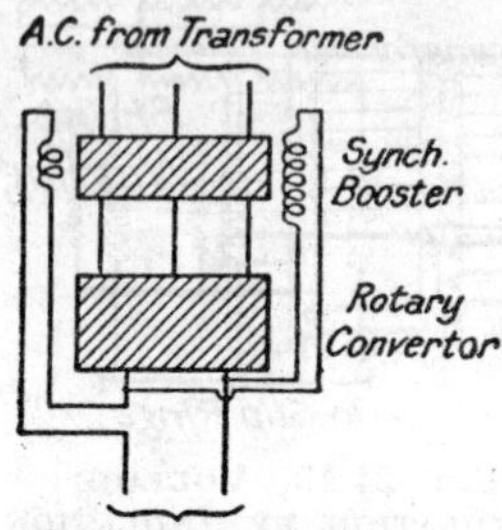

FIG. 24.15. VOLTAGE REGULATION BY SYNCHRONOUS BOOSTER

(*e*) SPLIT-POLE CONVERTORS. In the split-pole convertor the distribution of flux density under the poles is varied; this alters the waveform of the voltage across the slip-rings and so alters the transformation ratio.

Let E_l = line voltage; E_s = phase voltage; E = d.c. voltage

$\therefore$ for a six-phase machine $(E_s)_{max} = E/2$ (= radius of the circle in Fig. 24.2).

This can be put in the form

$$E = 2[(E_s)_{max}/(E_s)_{eff}] \times [(E_s)_{eff}/(E_l)_{eff}] \times (E_l)_{eff}$$
$$= 2K_1K_2 \times E_l$$

Thus $2K_1K_2$ is the transformation ratio.

Now for a sine wave $K_1 = \sqrt{2}$ and $K_2 = 1/\sqrt{3}$. As the waveform alters, K_2 remains practically constant, but K_1 varies very considerably, as shown below.

Sine Wave	Rectangular Wave	Triangular Wave
$K_1 = 1{\cdot}41$	$K_1 = 1$	$K_1 = 1{\cdot}74$
$K_2 = 1/\sqrt{3}$	$K_2 = 1{\cdot}04/\sqrt{3}$	$K_2 = 1{\cdot}006/\sqrt{3}$

Hence, we can say that the limits of K_1K_2 are 1 to 1·74, so that by varying the waveform, a very large change of d.c. voltage can be produced. This change is effected by altering the distribution of flux.

The pole consists of three limbs provided with main shunt, and auxiliary series windings as shown in Fig. 24.16. If the shunt winding produces, say, N. polarity, and the series windings tend to produce polarities of the order N., S., N., then the d.c. voltage will decrease. If the series windings tend to produce polarities of the order S., N., S., then the d.c. voltage will increase.

The objections to this method are: First, mecnanical difficulties of construction; second, difficulty with commutation, which limits its application to low frequencies. It is therefore only suitable for small capacity, 25 c/s machines. To simplify construction the split-pole convertor is often made with a two-part, instead of three-part, pole.

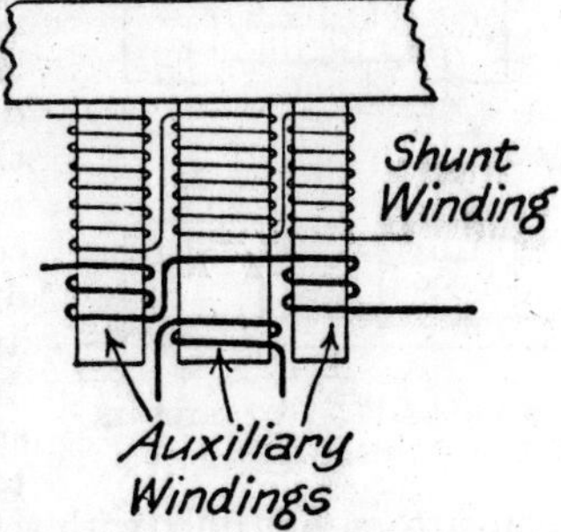

FIG. 24.16. SPLIT-POLE CONVERTOR

Inverted Rotary Convertors. Usually, rotary convertors change the supply from A.C. to D.C., and if they are worked in the opposite sense they are styled "inverted." With a convertor working normally, the speed is fixed at synchronism, because it is operating like a synchronous motor on the a.c. side; and this also applies to an inverted convertor if it shares the load on the a.c. side with a synchronous generator. If it does not share the load in this way, there is nothing to fix the speed, and the speed will vary as the flux per pole varies. Thus, if the load is highly inductive, the lagging alternating current will produce a powerful demagnetizing effect, thus causing a serious increase in speed. To prevent this, inverted convertors are always separately excited from a direct-coupled exciter. The increase in excitation due to a rise in speed will increase the flux/pole and so bring down the speed again.

Parallel Operation of Rotary Convertors. We have seen that two shunt generators work satisfactorily in parallel, because they have drooping voltage characteristics. Thus, if one machine takes too much load, its voltage drops, thereby automatically throwing the excess load on to the other machine. With convertors, the d.c. voltage is fixed because the a.c. voltage applied to the slip-rings is fixed. Therefore, if two convertors which are working from the same a.c. supply are also parallel on their d.c. sides, trouble due to unequal distribution of the load will result. This does not hold if they are connected to the a.c. supply via transformers, because the drop in the transformers gives the necessary drooping voltage characteristic. Often, to save the cost of transformers, two rotary convertors are supplied direct from a low-voltage alternator. In such a case, there are two methods of obtaining satisfactory parallel operation on the d.c. sides.

(*a*) The alternator can have two separate armature windings, one for each convertor.

(*b*) Balancing transformers can be used. The operation of these transformers will be understood by referring to Fig. 24.17, in which single-phase convertors are represented for simplicity. A single core has two windings, connected in series with each convertor on its a.c. side, as shown. They are arranged so that they produce fluxes through the core in opposition, so that when the load is balanced, the core will carry zero flux. If convertor *A* carries a greater load than *B*, its transformer coil will produce a greater flux than that of *B*, so that the transformer will now carry a finite resultant flux. This flux induces in the balancing coil of convertor *B* a current in the same direction as the ordinary load current, thus automatically equalizing the load.

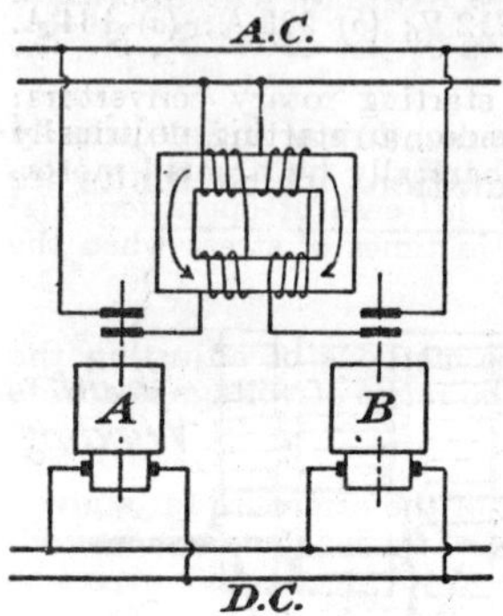

FIG. 24.17. ROTARIES IN PARALLEL

These balancing transformers are quite small, and therefore cheap, since they have to deal, not with the total intake of the convertors, but only with temporary differences in load.

FURTHER READING

HILL, E. P. *Rotary Convertors*. Chapman & Hall, London, 1927.

EXAMPLES ON CHAPTER 24

1. A six-phase rotary convertor is supplied from a transformer with the secondaries connected in double delta. If the convertor delivers 500 A to the d.c. load, what current will flow in the secondary winding of each transformer? *Ans.* 136 A.

2. A rotary convertor with six slip-rings is fed from the secondaries of a three-phase transformer with star-connected primaries. Each primary coil has ten times as many turns as the secondary. A load of 200 A at 500 V is taken from the continuous-current side. Draw carefully a diagram of the connections, and also a vector diagram showing the magnitude and phase relationship of the voltages of the line, of the transformer coils, and of the slip rings. Calculate the approximate voltages on the mains and the currents in the primary coils (assume the efficiency to be 100 per cent, and the power factor, unity). (L.U.)
Ans. Assuming diametrical secondary connections, line voltage 6100 V, current in primary coils 9·5 A, voltage at slip-rings 353 V.

3. A six-phase rotary convertor is supplying 100 A to the d.c. side. Draw graphs of the current in the armature (*a*) for a conductor mid-way between two tapping points, (*b*) for a conductor at a tapping point.

4. Explain the action of the rotary convertor, and explain why it is more economical to use a rotary convertor with a large number of phases. Show that in a six-phase rotary convertor working at unity power factor, the heating of a coil close to the tapping points is about twice as great as that of a coil mid-way between two tapping points. (L.U.)

5. A 6-ring, 12-pole, lap-wound synchronous convertor with diametral tappings runs at 600 r.p.m. and supplies 2000 A d.c. There are 1200 armature conductors and the flux per pole is 5×10^{-2} wb. The transformer is delta connected to a 6600 V, 3-phase supply. Assuming unity power factor and an armature efficiency

of 96 per cent calculate (*a*) the turn-ratio of the transformer and the voltage between adjacent rings, (*b*) the current per ring and (*c*) the current in the armature correction between a slip-ring and a tapping point. (C. and G.)

Ans. (*a*)15·55, 212 V; (*b*) 983 A; (*c*) 164 A.

6. Compare critically the following methods of starting rotary convertors: (*a*) starting electrically from the continuous-current side; (*b*) starting electrically from the alternating-current side; (*c*) starting mechanically by a small motor. Consider the matter under the following headings: (*d*) ease of operation; (*e*) time taken to come to synchronism; (*f*) reliability in times of stress when the generating voltage is unsteady. (C. and G.)

7. Examine critically the suitability of the various methods of adjusting the voltage on the d.c. side of a rotary convertor when the range of voltage is a wide one, say, from 440 to 550 V.

8. Explain why, in a rotary convertor, the heating of the armature for a given load is reduced as the number of phases on the a.c. side of the machine is increased. Show generally how the distribution of temperature in the armature changes as the excitation of the rotary convertor is changed. (C. and G.)

9. The speed of a rotary convertor, when used to convert direct-current power into alternating power, varies with the power factor, of the load on the alternating-current side of the machine. Explain this result, and describe an arrangement which may be used to reduce the speed variation. (L.U.)

10. Assuming an efficiency of 100 per cent, show that the ratio of the output of a synchronous convertor to that of the same machine run as a d.c. generator is 1·338 at unity power factor and 1·10 at 0·9 p.f. for a three-phase machine, and 1·937 at unity p.f. and 1·46 at 0·9 p.f. for a six-phase machine.

11. Prove that, in the case of a six-phase synchronous convertor, the ratio of the I^2R losses at the point of maximum and minimum temperature are 2·2 at unity p.f. and 4·9 at 0·9 p.f.

12. A six-pole machine has an armature with 162 slots and 4 conductors per slot. How would the winding (a lap winding) be tapped for operation as a three-phase convertor?

Ans. Pitch of tapping points in conductors, 72. Tappings to first ring at conductors 1, 217 and 433. Add 72 to each of these for tappings to second ring, and 144 to each for tappings to third ring.

CHAPTER 25

SINGLE-PHASE COMMUTATOR MOTORS

The e.m.f. of a Single-phase Commutator Motor. Let the armature rotate with an angular velocity of ω_a and let time be reckoned from the instant the flux is a maximum. We can consider two electrically opposite conductors as constituting a single coil, as on p. 179, but as this problem is more difficult to visualize than that of the d.c. machine,

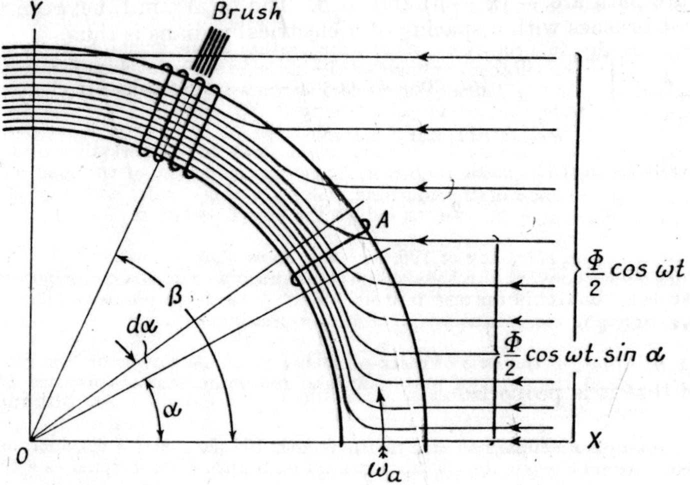

FIG. 25.1. E.M.F. OF A SINGLE-PHASE COMMUTATOR MOTOR

we will imagine that the machine has a ring winding, Fig. 25.1. The flux per half pole is, at any instant

$$(\Phi/2) \cos \omega t \text{ where } \Phi = \Phi_{max}, \text{ and } \omega = 2\pi f$$

If we assume a sinusoidal distribution of flux-density in the air-gap, a reasonable assumption with a distributed winding, the flux linkage with a single turn A at an angular position α with respect to OX, the main field axis, is

$$\phi = -(\Phi/2) \cos \omega t \sin \alpha^*$$
$$= -(\Phi/2) \cos \omega t \sin \omega_a t, \text{ since } \alpha = \omega_a t$$

where ω_a is the angular velocity of the armature. The negative sign is used because the flux linkage is decreasing with respect to time: this is because cos ωt decreases as t increases.

* Assuming, for simplicity, a two-pole field.

The e.m.f. induced in the single turn at A is therefore

$$e_A = -\,d\Phi/dt = (\Phi/2)(\omega_a \cos \omega t \,.\, \cos \alpha - \omega \sin \omega t \,.\, \sin \alpha)$$

With Z conductors altogether the number enclosed in an angular spread $d\alpha$ is

$$Z(d\alpha/2\pi)$$

the e.m.f. induced in this group of conductors thus being

$$de = (\Phi Z/4\pi)(\omega_a \cos \omega t \,.\, \cos \alpha - \omega \sin \omega t \,.\, \sin \alpha)d\alpha$$

Since we are reckoning angular distance in a counter-clockwise sense the limits of integration when calculating the total e.m.f. induced in one armature path are $-(\pi - \beta)$ and $+\beta$. The total e.m.f. between two adjacent brushes with a spacing of π electrical radians is thus

$$e = \int_{-(\pi-\beta)}^{+\beta} \frac{\Phi Z}{4\pi}(\omega_a \cos \omega t \,.\, \cos \alpha - \omega \sin \omega t \,.\, \sin \alpha)d\alpha$$
$$= (\Phi Z/2\pi)(\omega_a \sin \beta \cos \omega t + \omega \cos \beta \sin \omega t)$$

This shows that the induced e.m.f. has two components in quadrature,

$$e_S = (\Phi Z/2\pi)\omega_a \sin \beta \cos \omega t, \text{ the speed e.m.f.}$$
$$e_T = (\Phi Z/2\pi)\omega \cos \beta \sin \omega t, \text{ the transformer e.m.f.}$$

Since the first of these is a cosinusoidal function of ωt, the same as the flux, we see that this component is in time phase with the flux. Its effective value is

$$E_S = (\Phi Z/2\sqrt{2}\pi)\omega_a \sin \beta$$

We see that it is proportional to the speed of the armature but independent of the frequency of the supply. It is called the *speed* e.m.f. It is a maximum when $\beta = \pi/2$, that is, when the brush axis coincides with the magnetic neutral plane, as in a d.c. machine, and is zero when the brush axis coincides with the field axis.

If the armature rotated at the speed ω_a in a d.c. field system of flux Φ per pole, the induced e.m.f. would be

$$E_{dc} = \Phi Zn = \Phi Z\omega_a/2\pi^*$$
$$\therefore\ E_S/E_{dc} = \sin \beta/\sqrt{2} = 1/\sqrt{2} \text{ when } \sin \beta = 1$$

the relationship thus being the same as in a single-phase rotary convertor.

The second component E_T is an alternating e.m.f. of supply frequency f which lags E_S in time phase by $\pi/2$. Its magnitude is independent of the armature speed ω_a, showing that it is induced by variations in the flux linkage, and unassociated with the actual rotation. It is therefore called the transformer e.m.f. Its r.m.s. value is

$$E_T = (\Phi Z/2\sqrt{2}\pi)\omega \cos \beta$$
$$\therefore\ E_T/E_{dc} = (1/\sqrt{2})(\omega/\omega_a) \cos \beta$$
$$= (1/\sqrt{2}) \cos \beta, \text{ if } \omega_a = \omega$$

* This assumes a lap winding, the wave winding not being analogous to the elementary ring winding of Fig. 25.1.

The Single-phase Series Motor. If an alternating current is sent through an ordinary d.c. series motor, the field and armature currents reverse simultaneously, so that a unidirectional torque is produced and the armature rotates. The motor has to be modified very considerably for work with alternating currents. Thus, all the iron has to be thoroughly laminated both in armature and field.*

There are two e.m.f.s induced in an element of the armature winding. First, the dynamically-induced speed e.m.f. E_S, caused by the rotation of the armature conductors in the magnetic field. Second, the statically-induced transformer e.m.f., E_T, produced by the alternations

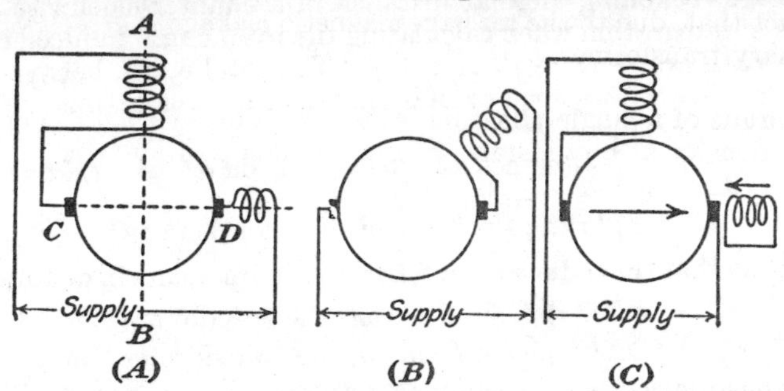

FIG. 25.2. COMPENSATED SERIES MOTOR

in the flux, independently of rotation. Whatever the brush position may be, there is always an e.m.f. induced in the short-circuited winding element. Thus, when the brushes are in the normal position, the short-circuited element has a transformer-induced e.m.f.; and when the brushes are perpendicular to the normal position, it has the dynamically-induced e.m.f. Hence, the motor shows a much greater tendency to sparking than a d.c. motor, and the elimination of sparking has been one of the most difficult problems in the evolution of the motor. Both excessive sparking and poor power factor are prevented by "compensating" the motor, the aim of which is to neutralize the effect of self-induction. This can only be done for the armature, since to neutralize the self-induction of the field means that the field flux must be neutralized, which is obviously impossible. The armature can be compensated by employing coils carrying the armature current, and placed in such a way as to neutralize the armature flux. Fig. 25.2 shows how this can be done. In Fig. 25.2 (A), the field flux is produced along the axis AB, and the armature produces its flux at right angles to this, namely, along the brush axis CD. Thus the compensating coil must produce its flux along CD and in a direction opposite to that of the armature flux at every instant. In the figure the field and compensating windings are shown as solenoids; in the actual motor they are distributed windings.

* For complete constructional details, *see* A. T. Dover, *Electric Traction.* Pitman, London. Also the works listed at the end of this chapter.

Since two fields at right angles produce a single resultant field, it is possible to combine the main field and compensating windings into a single winding producing its flux in the direction of this resultant, as shown in Fig. 25.2 (*B*).

The compensating coil, instead of being connected in series with the main field coil, can be short-circuited as in Fig. 25.2 (*C*). It then acts like the short-circuited secondary of a transformer, of which the primary is the armature winding. Since the fluxes produced by the primary and secondary of a transformer almost neutralize one another, we see that this method gives compensation which is good in practice, in spite of the fact that, due to the air gap, magnetic leakage is greater than in an ordinary transformer.

The Torque of a Single-phase Series Motor. Consider again Fig. 25.1. The flux density at any instant at any angular position α is

$$B = B_{\max} \cos \omega t \cos \alpha$$

The current in a conductor, when the operating power factor is $\cos \phi$ lagging, is

$$i_a = I_{c.\max} \cos (\omega t - \phi), \text{ where } I_c \text{ is the conductor current}$$

The tangential force on a conductor at angular position α is thus

$$f = BI_c l, \text{ where } l = \text{conductor length}$$
$$= B_{\max} I_{c.\max} l \cos \omega t \, . \cos (\omega t - \phi) \, . \cos \alpha$$

The maximum value of the flux per pole assuming, initially a two-pole machine

$$\Phi = B_{\max} \times 2lr, \text{ where } r = \text{radius}$$

since $2lr$ is the area occupied by the main flux when reckoned at right-angles to OX. It is thus necessary to use the area $2lr$ and not the angular pole pitch πlr, as in the case of a machine in which the radial air-gap density is uniform,

$$\therefore \; B_{\max} = \Phi/2lr$$

If we multiply the tangential force by r we obtain the torque contributed by one conductor

$$\therefore \; T_\alpha = (\Phi/2lr) \times lr \times I_{c.\max} \cos \omega t \, . \cos (\omega t - \phi) \cos \alpha$$

With Z conductors per pole pair, the number enclosed in an angular spread $d\alpha$ is $(Z/2\pi). \, d\alpha$, and their contribution to the total flux at instant t is

$$dT_\alpha = (1/4\pi)\Phi Z I_{c.\max} \cos \omega t \, . \cos (\omega t - \phi) \, . \cos \alpha \, d\alpha$$

This must be integrated between the limits defined by the angular brush positions, $-(\pi - \beta)$ and $+\beta$. It must also be integrated over a whole period of alternation of the current, the limits thus being 0 and

$2\pi/\omega$, where ω is again equal to $2\pi f$. As we require the average value of the torque we must also multiply by $\omega/2\pi$. This gives

$$T = \frac{1}{4\pi}\,\Phi Z I_{c.\max}\,\frac{\omega}{2\pi}\int_{-(\pi-\beta)}^{+\beta}\int_{0}^{2\pi/\omega}\cos\omega t\cos(\omega t-\phi)\cos\alpha\,.\,d\alpha\,dt$$

$$= (1/2\pi)\Phi Z I_{c.\max}\sin\beta\cos\phi$$

Now $I_{c.\max} = \frac{1}{2}\,I_{\max}$, where I = total brush current in a two-pole machine

$$= (\sqrt{2}/2)I \text{ (r.m.s.)}$$

$$\therefore\ T = (1/2\sqrt{2}\pi)\,.\,\Phi Z I\sin\beta\cos\phi$$

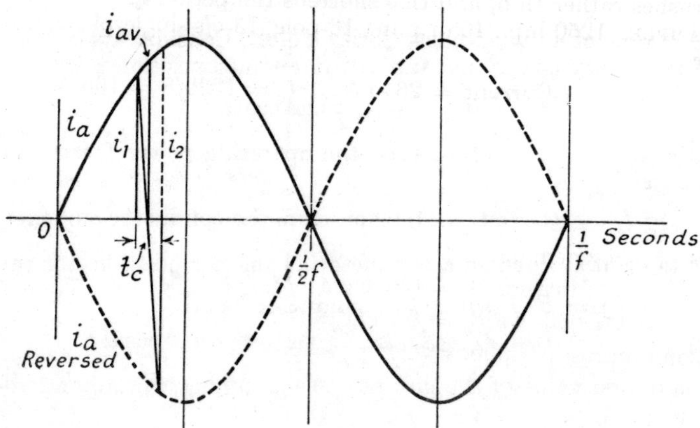

FIG. 25.3. COMMUTATION IN A SINGLE-PHASE COMMUTATOR MOTOR

As we have integrated over only one-half of the armature winding, the above expression must be doubled, giving

$$T = (1/\sqrt{2}\pi)\Phi Z I\sin\beta\cos\phi$$

We see that the torque is proportional to the power factor, $\cos\phi$, which we would expect, and it is also proportional to $\sin\beta$. Thus, other quantities being equal, it is a maximum when the brushes are set in the geometric neutral plane, and is zero when they are set in the field axis. This is another proof of the principle, explained in a different way on p. 614, that the interaction of a current with the flux produced by it does not necessarily result in torque production.

Commutation. In a d.c. machine the full conductor current I_c is reversed during the period of commutation t_c, so that the average rate of change of current in the short-circuited coil is $2I_c/t_c$ amperes/second. In the a.c. machine the instantaneous value of the current at the commencement of commutation can be anything from zero to $I_{c.\max}$. Suppose that the current is i_1 at the commencement and $-\,i_2$ at the end of commutation, Fig. 25.3. Let the commutation be linear, as represented by the diagonal joining the extremities of the ordinates of i_1 and $-i_2$. Then rate of change is

$$(i_1 + i_2)/t_c \simeq 2i_{av}/t_c$$

The maximum possible rate of change is thus only $\sqrt{2}$ times greater than in a d.c. machine whose I_a is the same as the r.m.s. value of I_a in the commutator motor, and for which t_c is the same. We shall see that it is necessary to keep the inductance of the short-circuited coil as low as possible and therefore single-turn coils are the rule, thereby requiring the maximum number of commutator segments for a given number of armature conductors. Thus the segments are narrow and the brushes rather thin, and this shortens the period t_c.

EXAMPLE. 1250-h.p., 465-r.p.m., 16-pole, 15-c/s single-phase traction motor

$$\text{Current} = 2640 \text{ A} \quad \therefore I_c = 1320/8 = 165 \text{ A}$$

$$I_{c.max} = 233 \text{ A}$$

$$Z = 1152 \text{ conductors}$$

$$\therefore N_2 = 576 \text{ comm. segments, single-turn coils}$$

$$D_2 = 0{\cdot}95 \text{ m, commutator diameter}$$

$$\therefore \text{Segment pitch} = \pi \times 0{\cdot}95/576 = 0{\cdot}517 \times 10^{-2} \text{ m}$$

$$\text{Mica width} = 0{\cdot}076 \text{ cm} = 0{\cdot}076 \times 10^{-2} \text{ m}$$

$$\text{Brush width} = 0{\cdot}9 \text{ cm} = 0{\cdot}9 \times 10^{-2} \text{ m}$$

$$\text{Commutator peripheral velocity} = \pi \times 0{\cdot}95 \times 465/60 = 23{\cdot}10 \text{ m/sec}$$

$\therefore$ Time during which a coil is short-circuited

$$t_c = (0{\cdot}9 - 0{\cdot}076) \times 10^{-2}/23{\cdot}10 = 3{\cdot}57 \times 10^{-4} \text{ sec}$$

With a frequency of 15 c/s the period of commutation thus corresponds to an angular interval of

$$15 \times 3{\cdot}57 \times 10^{-4} \times 360 = 1{\cdot}93 \text{ electrical degrees}$$

This very small value shows that there is complete justification for using the average current i_{av}.

The variations in the commuted current in the case of the 1250-h.p. motor are shown in Fig. 25.4. The width at the axis of each block of current is

$$1/(16 \times 15) = 1/240 \text{ sec; or } 360/16 = 22{\cdot}5 \text{ electrical degrees}$$

The maximum current is 233 A, and the r.m.s. current 165 A, so that the conditions in a comparable d.c. machine are as indicated by the dotted trapezoidal variation of current.

It will be quite clear that this difference between the a.c. and d.c. cases is not sufficient to make commutation much more difficult in the a.c. case. The real difficulty is the transformer e.m.f. induced in a winding section while it is undergoing commutation and therefore short-circuited by a brush. Fig. 25.5 shows that a short-circuited coil is linked by the whole of the main flux at this particular moment. This transformer e.m.f. is in time quadrature with the main flux, and therefore with the motor current, and it causes a current to circulate

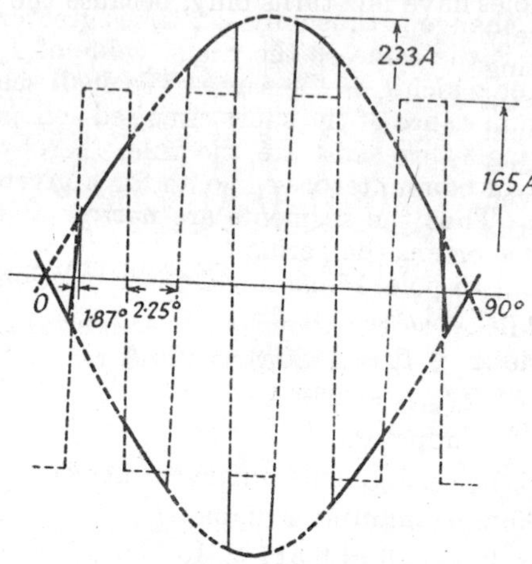

FIG. 25.4. COMMUTATION IN A 1250-H.P. SERIES COMMUTATOR MOTOR COMPARED WITH THAT IN COMPARABLE D.C. MACHINE

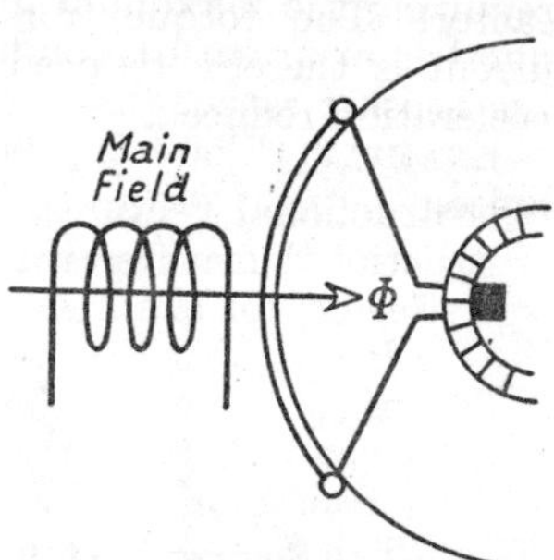

FIG. 25.5. PRODUCTION OF TRANSFORMER E.M.F. IN SHORT-CIRCUITED COIL

round the short-circuited coil and across the brush face. Correction cannot be effected by the use of a simple interpole as in the case of a d.c. machine for two reasons—

(*a*) at the moment of starting, the main current and therefore the main flux have their greatest value, so that the transformer e.m.f. is high. On the other hand the rotational e.m.f.s both with respect to the main flux and the interpole flux are zero,

(*b*) when running, the transformer e.m.f. and the rotational e.m.f. are in time quadrature, as explained on p. 645, and therefore with a simple series-connected interpole the effect is even worse than at starting, because these two e.m.f.s combine to give a resultant which is greater than either e.m.f. alone.*

The earliest method of suppressing the sparking resulting from inability to neutralize the transformer voltage was by the use of high resistance leads from the winding tapping points to the commutator segments, as shown in Fig. 25.6. It will be seen that such leads form high resistance paths for both main current and circulating current, and they cause inevitable I^2R loss. As the resistance is increased this loss due to the main current is increased, but that due to the circulating current is decreased, and it was the practice to choose a resistance such that the sum of these two losses was a minimum.

* F. Nouvion, "Electric locomotives on the Valenciennes-Thionville railway." *Proc. I.E.E.*, **101**, Pt. I (1954) 313.

This method is now rarely used by itself, a preferable method being to alter the phase of the interpole flux by shunting the interpole winding with a resistor.* By this expedient the rotational e.m.f. induced in the short-circuited coil has a component which is in phase opposition to the transformer e.m.f. The interpoles have few turns only, because the armature m.m.f. is almost completely neutralized by the compensating winding. In order to secure good commutation at starting it is usual to shunt the main field by a saturated reactor. The torque, for a given current is thereby reduced and the acceleration reduced.

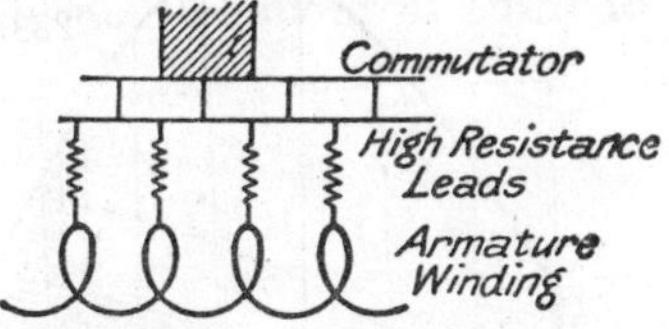

FIG. 25.6. RESISTANCE LEADS

Constructional Features. In addition to the above, there are certain constructional features characteristic of the machine, at any rate in large sizes as, for example, series motors for electric traction.

1. To keep down the transformer e.m.f. in the short-circuited coils the flux per pole must not be too great. Hence, for its output and speed the motor has a large number of poles. As a rough rule it can be taken that, for traction motors of this type there is a pair of poles for each 100 to 150 h.p.

2. The armature coils are single-turn coils and the brushes are narrow, rarely short-circuiting more than two coils at a time.

3. The compensating winding cannot be a full-pitch winding because of the peripheral space taken up by the interpoles. Now a full-pitch winding cannot be compensated by a chorded winding, and therefore it is usual to chord the armature winding. As explained on p. 193, this gives a zone of zero armature m.m.f. adjacent to the interpoles, therefore shortening the width of the m.m.f. distribution which has to be neutralized.

4. To secure a high operating power factor the inductance of the main field winding must be as low as possible: we have seen that the flux per pole is kept low by the use of a large number of poles, and the minimum possible turns per pole are also used. This reduction in flux per pole necessitates a corresponding increase in the armature conductors in order that the required torque may be developed, and armature e.m.f. set up. As a result the armature ampere-turns per pole are greater than the field ampere-turns per pole, a condition which could not be tolerated in a normal d.c. motor. It is made possible because of the compensation of the armature m.m.f.

5. The transformer e.m.f. is proportional to the frequency and therefore good commutation is more easily secured at low frequencies. Thus early motors were operated at 15 or $16\frac{2}{3}$ c/s, but they can now be operated satisfactorily at 25 or even 50 c/s.

6. The air-gap is very small, this being necessary because of the weak field. Thus the 1250-h.p. motor of the example has a rotor

* *See* F. Nouvion, "Electric locomotives on the Valenciennes-Thionville railway." *Proc. I.E.E.* **101**, Pt. I (1954) 313.

diameter of 150 cm, but an air-gap of only 0·3 cm. A d.c. machine of comparable diameter would have a gap length of about twice this.

7. Again, for the purpose of keeping down the inductance of all the windings of the machine, the operating voltage must be low, for a high voltage with proportionally low current would require a large number of turns to produce the same flux. Thus the 1250 h.p. motor of the

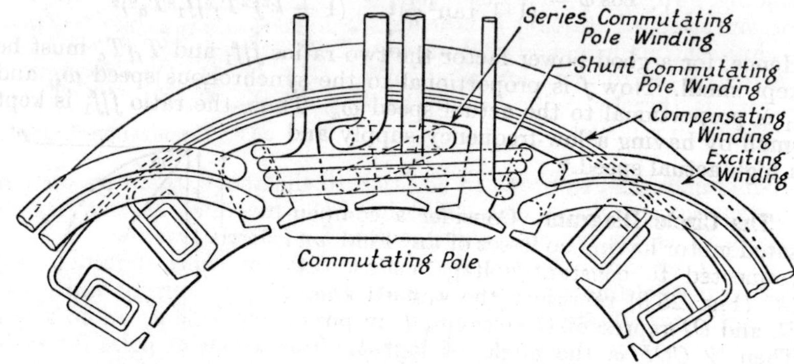

Fig. 25.7. Portion of a Railway Motor Stator

example is designed for a terminal voltage of 430, the armature induced e.m.f. when operating under the stated conditions being 398 V.

A portion of a railway motor stator is shown in Fig. 25.7. It will be seen that there is uniform slotting round the periphery, the interpole taking up three of the teeth. The reason for this is, that instead of shunting the interpole winding a similar effect is obtained by the use of an additional shunt interpole winding. The large interpole cross-section is necessary to prevent saturation since, if present, the relationship between current and flux would no longer be linear. With series excitation only, combined with the device of a non-inductive shunt, the interpole can consist of a single large tooth.

Conditions Necessary for a Good Power Factor. From the circle diagram, Fig. 25.8,

$$\tan\phi = \frac{CD}{OD} = \frac{\text{statically-induced e.m.f. in the field}}{\text{dynamically-induced e.m.f. in the armature}}$$

Let f = supply frequency

T_a = No. of turns on the armature

T_f = No. of turns on the field

Φ = flux (max. value)

Hence, statically-induced e.m.f. in the field

$$\propto \Phi T_f f$$

Now let the speed be such that the dynamically-induced e.m.f. in the armature is of frequency f_1.

$$\therefore \text{Dynamically-induced e.m.f.} \propto \Phi T_a f_1$$

$$\therefore \tan\phi \propto \Phi T_f f / \Phi T_a f_1$$

$$= k \times (f/f_1) \times (T_f/T_a), \text{ where } k \text{ is a constant}$$

$$\therefore \cos\phi = \frac{1}{(1+\tan^2\phi)^{\frac{1}{2}}} = \frac{1}{(1+k^2f^2T_f^2/f_1^2T_a^2)^{\frac{1}{2}}}$$

Hence, for a good power factor the two ratios f/f_1 and T_f/T_a must be kept small. Now f is proportional to the synchronous speed ω_s, and f_1 is proportional to the actual speed ω_a. Thus, the ratio f/f_1 is kept small by having a low-frequency supply and a high actual speed.*

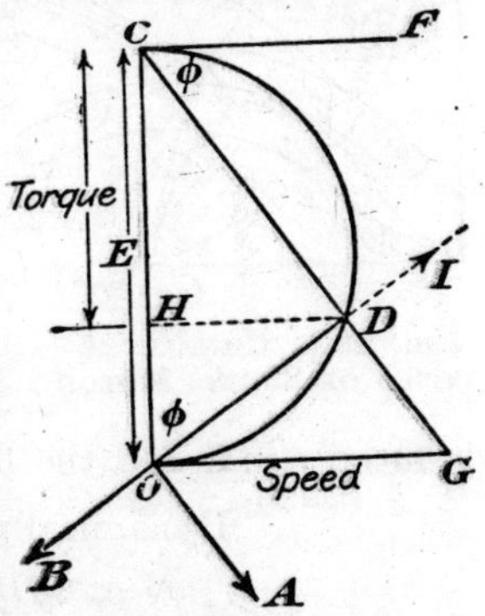

The Circle Diagram. Consider a compensated motor having no losses of any kind, and connected to constant voltage mains. Let OC (Fig. 25.8) represent the applied e.m.f., E, and OI represent the current, I, in phase. Then $\angle COI$ is the angle of lag, ϕ. The statically-induced e.m.f. due to self-induction of the field is equal to $L\omega I$, and lags 90° behind I. It is represented by vector OA on the diagram. The dynamically-induced e.m.f. in the armature is obviously the motor back e.m.f., and it is therefore represented by OB, in phase opposition to the current. Also, OB must have a component in phase opposition to E, since the machine is a motor. The applied voltage is equal and opposite to the resultant of OA and OB,

FIG. 25.8. CIRCLE DIAGRAM OF IDEAL MOTOR

$$\therefore E = OC = OB \text{ reversed} + OA \text{ reversed}$$

If we draw a semicircle on OC as diameter and join CD, we see that E is the resultant of OD and CD. Again, OD is in line with OB and DC is parallel to OA, the point D being the intersection of the current vector with the semicircle.

Hence, $OD = OB$, the dynamically-induced e.m.f.

and $CD = OA$, the statically-induced e.m.f., $L\omega I$

Now OC is constant, and therefore, the semicircle ODC is the locus of the intersection of these two vectors.

Again, $CD = L\omega I$, and since L and ω are constant, CD represents the current in magnitude. If now a perpendicular CF is drawn, then $\angle DCF = \angle COD = \phi$, and therefore, referred to CF, the vector CD

* Modern motors of this type can be operated satisfactorily at 50 c/s. *See* M. Garreau, "Electric traction using 50 c/s current." *Proc. I.E.E.* **101**, Pt. I (1954) 309.

represents the current both in magnitude and phase. If we assume that the iron is not saturated, but is worked on the straight part of the characteristic, a condition which approximately holds in order that the iron losses may not be too high, the flux is proportional to the current. Hence, CD also represents the flux.

The power factor $= \cos\phi = OD/OC$

is proportional to OD, since OC is constant.

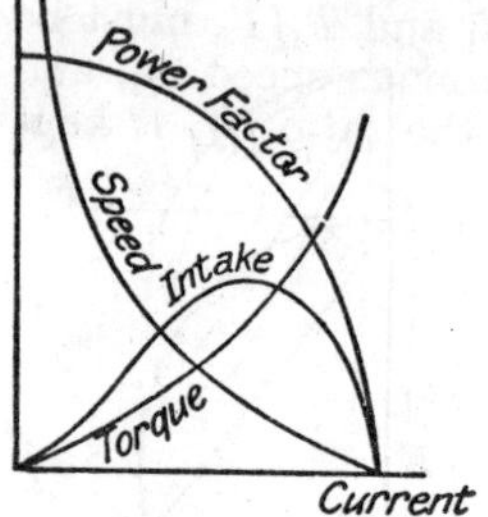

FIG. 25.9. CHARACTERISTICS OF SERIES MOTOR

Torque is proportional to the product of the components of current and flux in phase.

$$\therefore T \propto CD \times CD$$
$$\propto CH \times OC$$
$$\propto CH, \text{ since } OC \text{ is constant}$$
$$\text{Intake} = EI \cos\phi$$
$$\propto OC \times CD \times \cos\phi$$
$$\propto OC \times DH$$
$$\propto DH$$

Speed N is related to the dynamically-induced e.m.f. and the flux as follows—

$$\text{Dynamically-induced e.m.f.} \propto \text{flux} \times \text{speed},$$
$$\therefore N \propto OD/CD \propto \cot\phi \propto OG/OC \propto OG$$

If different positions of the point D are taken on the circle, and the above quantities measured off, the characteristics can be drawn. These are of the form shown in Fig. 25.9. Since the current is a maximum at standstill, the speed becomes zero when the current reaches its maximum value. The speed tends to become infinite when the current is very small, so that the mechanical characteristics of the motor are very similar to those of a d.c. series motor. The motor is thus very suitable for traction use, in which sphere it is very largely used. A motor designed for, say 25 c/s, will operate on 15 c/s, or on d.c., with improved performance. The supply frequency does not influence the speed, at any rate not directly, but with increased frequency there is a greater reactive voltage drop and therefore a slightly lower speed at a given current.

The Effect of Losses on the Diagram.

$$\text{Copper losses} = I^2R = I \times I \, . \, R$$

IR is the resistance drop, and is in phase with I. Hence, it can be represented by marking off a length DD' (Fig. 25.10) equal to IR.

$$\therefore \text{Copper losses} = CD \times DD'$$
$$= 2\,\Delta\, CDD'$$

If we neglect for the moment the iron and friction losses, we have

$$\begin{aligned} \text{Output} &= 2\,\Delta\,OCD - 2\,\Delta\,CDD' \\ &= 2\,\Delta\,OCD' \\ &\propto D'H' \end{aligned}$$

Again, the ratio $DD'/CD = IR/I = R$

$= \text{a constant},$

and therefore the angle DCD' is a constant, $= \alpha$, say; and the locus of D' is a circle whose centre is a little to the left of OC.

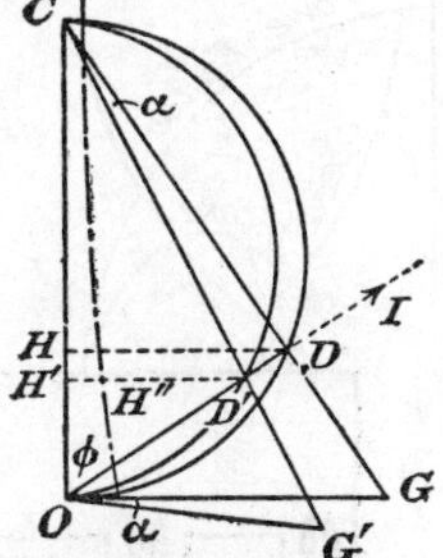

Fig. 25.10. Corrected Circle Diagram

$$\text{Speed} \propto \frac{\text{dynamically-induced e.m.f.}}{\text{flux}}$$

$$\propto (OD'/CD)\,(OD'/CD')$$

Draw OG' inclined at the fixed angle α to OG, meeting CD' produced in G'. Then

$$\text{Speed} \propto (OD'/CD')\,(OG'/OC)\,OG'$$

The iron and friction losses cannot be represented geometrically, because the speed and flux are both variables. If they are determined by a separate core loss test, or calculated from design data, a curve of these losses can be drawn as shown dotted. The output, taking losses of all kinds into account, is then $D'H''$,

$$\therefore \text{Efficiency} = D'H''/DH$$

Another Vector Diagram. Let suffixes 1, 2, 3, and 4, denote main field, armature, compensating winding and interpole winding respectively, Fig. 25.11. Then the applied voltage V consists of the following components—

1. Resistance drops R_1I, R_2I, R_3I and R_4I.
2. Inductive reactive drops X_1I, X_2I, X_3I and X_4I.
3. A component E' to overcome the rotational e.m.f. E of the armature. Now this e.m.f. is in time phase with the flux, and so is the current, with the result that the component E' is in phase with the current.

If we put $\quad R_t = R_1 + R_2 + R_3 + R_4$

and $\quad X_t = X_1 + X_2 + X_3 + X_4$

we have $\quad V = [(E' + R_tI)^2 + X_t^2I^2]^{\frac{1}{2}}$

and $\quad \cos\phi = (E' + R_tI)/[(E' + R_tI)^2 + X_t^2I^2]^{\frac{1}{2}}$

Now E' is a rotational e.m.f. and can therefore be written

$$E' \propto \Phi N$$

again, if we neglect saturation, we have

$$\Phi \propto I$$

$$\therefore\ E' = kIN \text{ where } k \text{ is a constant}$$

This gives

$$\cos\phi = (kN + R_t)/[(kN + R_t)^2 + X_t^2)^{\frac{1}{2}}$$

This shows that the power factor increases with the speed. Thus the 1250 h.p. motor has a power factor of 0·95 under the stated conditions

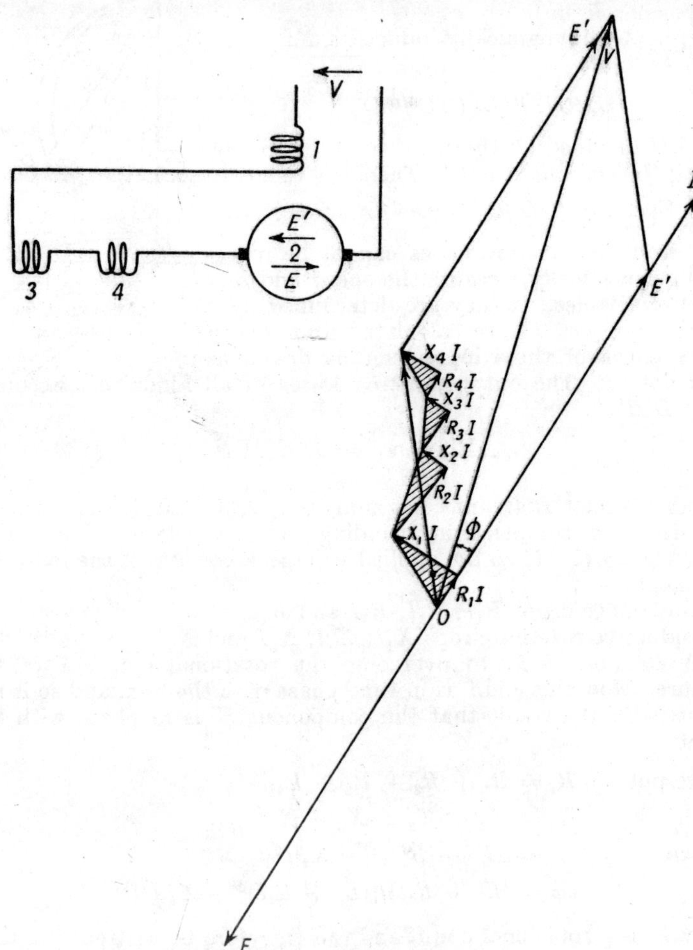

FIG. 25.11. COMPONENTS OF THE APPLIED P.D.

of load. At the moment of starting E' is zero and therefore the starting current and starting power factor with full applied voltage are given by

$$I_{\text{starting}} = V/(R_t^2 + X_t^2)^{\frac{1}{2}}$$
$$\cos\phi_{\text{starting}} = R_t/(R_t^2 + X_t^2)^{\frac{1}{2}}$$

The Repulsion Motor. This motor differs from the a.c. series motor in that no current is led into the armature, the brushes on the commutator being short-circuited, as shown in Fig. 25.12.*

Consider an ordinary d.c. armature with short-circuited brushes placed in a bi-polar field excited by alternating current.

(*a*) Let the brush axis be perpendicular to the main flux as in Fig. 25.12 (*A*). Then the e.m.f.s induced in the two halves of the winding shown will neutralize one another, and the armature will neither carry any induced current nor produce any torque.

(*b*) Let the brush axis be along the direction of the main flux, as in Fig. 25.12 (*B*). Then a current will be induced in the armature, the four quarters of which will produce four torques as indicated by the dotted arrows. These will neutralize one another, the total torque being again zero.

(*c*) Let the brush axis be inclined at any angle α to the direction of the main flux, as in Fig. 25.12 (*C*). Then calling the torques T_1 and T_2 as shown,

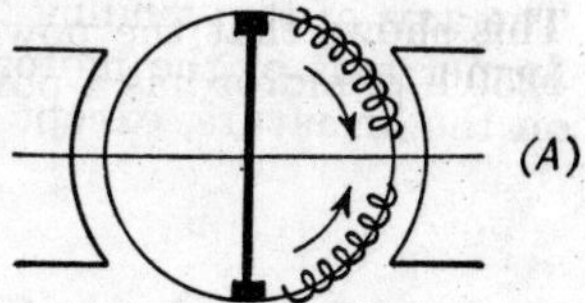

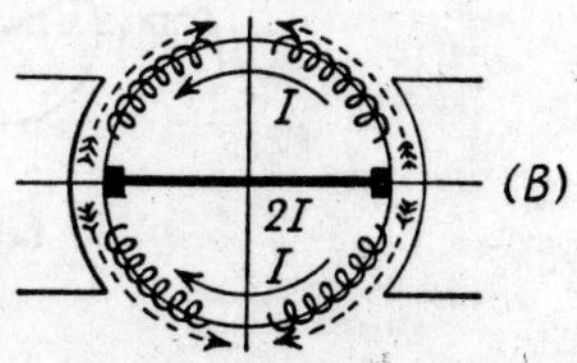

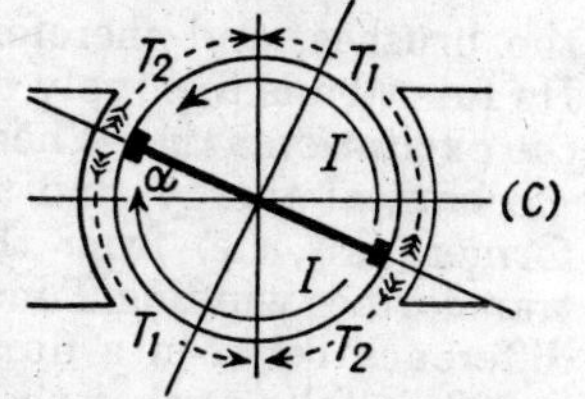

FIG. 25.12. ACTION OF REPULSION MOTORS WITH DIFFERENT BRUSH SETTINGS

$$\text{Resultant torque, } T = 2(T_1 - T_2)$$

The torque is also given by the expression

$$T \propto (\text{current between brushes}) \times (\text{component of main flux perpendicular to brush axis})$$

Now, current between brushes

$$\propto \text{component of main flux along brush axis}$$
$$\propto \Phi \cos\alpha$$
$$\therefore T \propto \Phi \sin\alpha \times \Phi \cos\alpha$$
$$\propto \sin\alpha \cos\alpha, \propto \sin 2\alpha$$

Hence, T is a maximum when $2\alpha = 90°$, i.e. when $\alpha = 45°$.

* For constructional information *see* A. T. Dover, *Electric Traction*, Pitman, London and *Electric Motors and Control Systems*, Pitman, London.

The actual arrangement of Fig. 25.13 (*A*) can be considered as replaced by that in Fig. 25.13 (*B*). If the total number of turns on the actual field is T_f, then

$$\text{Number of turns on winding } T = T_f \cos \alpha$$

$$\text{Number of turns on winding } E = T_f \sin \alpha$$

The axis of the winding *T*, namely, the brush axis, is called the transformer axis of the motor. The winding *E* has no transformer action on the armature, except on those coils which undergo short-circuit at

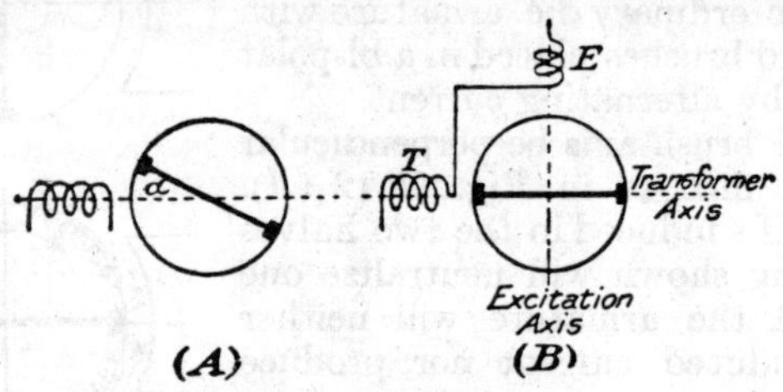

FIG. 25.13

the brushes, and therefore, it transmits no energy to the armature. Its function is to supply the excitation only, and the axis of *E* is called the excitation axis. When the motor is running, the armature develops mechanical energy, and this energy is imparted to it along the transformer axis, i.e. from the stator winding *T*. Also the flux in the transformer winding *T* induces a static e.m.f. in *T*; and since the phase difference between a flux and the statically-induced e.m.f. it sets up is 90°, it follows that the flux in winding *T* is in quadrature with the current, and therefore, with the flux along the excitation axis. But these two fluxes are 90° apart in space as well as in time, and therefore, the motor possesses

(*a*) A uniform rotating field, if the two fluxes are equal.
(*b*) An elliptical field, if they are different in magnitude.

In the armature there is a static e.m.f. along the energy axis set up by transformer action, and this is in quadrature with the flux producing it, i.e. with the flux through *T*. Again, the flux through *T* is in quadrature with the flux through *E*, and this latter flux produces the dynamically-induced e.m.f. in the armature, which is, of course, in phase with the flux through *E*. Hence, the armature dynamically-induced e.m.f. is in phase opposition, 180°, to the statically-induced e.m.f. The difference between these two armature e.m.f.s is equal to the armature drop in volts *RI*. If we assume an ideal motor with zero resistance, then the two are equal and opposite.

Let Φ_t = maximum flux produced by winding *T*

Φ_e = maximum flux produced by winding *E*

T_a = Number of armature turns

Then dynamic e.m.f. $E_a \propto \Phi_e T_a \times np$ . . . . (1)
and static e.m.f. $E_s \propto \Phi_t T_a f$ . . . . (2)

where n is the speed, and p the number of poles.

But E_d is equal to E_s

$$\therefore \Phi_e np = \Phi_t f$$

Again, if we call the synchronous speed r.p.m. n_s, then

$$f = n_s p/120, \propto n_s p$$
$$\therefore \Phi_e np \propto \Phi_t n_s p$$
$$\therefore \Phi_t/\Phi_e \propto n/n_s$$

Hence, the two fields are equal at synchronous speed only, and at this speed the total field is a pure rotating one.

At speeds below n_s $\Phi_t < \Phi_e$
At speeds above n_s $\Phi_t > \Phi_e$

Under these conditions the resultant field is elliptical, the major axis at speeds below n_s becoming the minor axis at speeds above n_s. At synchronous speed, the armature and the rotating field travel at the same rate, and therefore, there are no armature core losses. Also, there are no circulating currents in the coils undergoing commutation. At speeds above synchronism, the transformer flux increases rapidly with the speed, thereby resulting in large core losses and heavy circulating currents in the coils undergoing commutation. These conditions limit the operating speed of the motor to the neighbourhood of synchronous speed. If higher speeds are necessary, the transformer flux has to be weakened. On a low-frequency system the repulsion motor must therefore be a low-speed machine with relatively few poles, so that it will be considerably heavier than a series motor of the same power. This is because the fewer the poles, the greater must be the cross section of the magnetic paths for the same output. Thus, whereas series motors are suited to very low frequencies in the neighbourhood of 15 c/s repulsion motors are better suited for frequencies of about 25 c/s.

Mechanism of Torque Production. The mechanism of torque production in the repulsion motor is hardly apparent from the figures of Fig. 25.12, which are purely diagrammatic. It is similar to that for the single-phase induction motor, the main difference being that, whereas the induction motor rotor is circumferentially short-circuited, and is therefore unaffected by its angular position, the repulsion rotor is short-circuited along a definite axis, with consequent dependence for its performance on the positions of this axis. Fig. 25.14 shows a rotor with 24 conductors short-circuited on an axis at right-angles to the field axis, and the flux is increasing. The transformer e.m.f.s in the various conductors will have the directions indicated since any induced current will tend to set up a flux in opposition to the change in the main flux. Fig. 25.15 shows the rotor winding developed in the usual way. The brush short-circuits can be represented by the connection shown.

There are two directly short-circuited coils, namely 1–12 and 13–24. The rest of the armature winding automatically divides itself into two

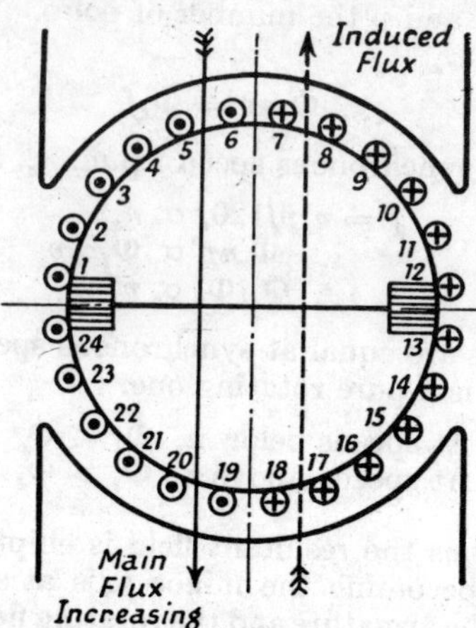

Fig. 25.14. Repulsion Motor with Brush Axis at Right Angles to Field Axis—E.M.F. Distribution

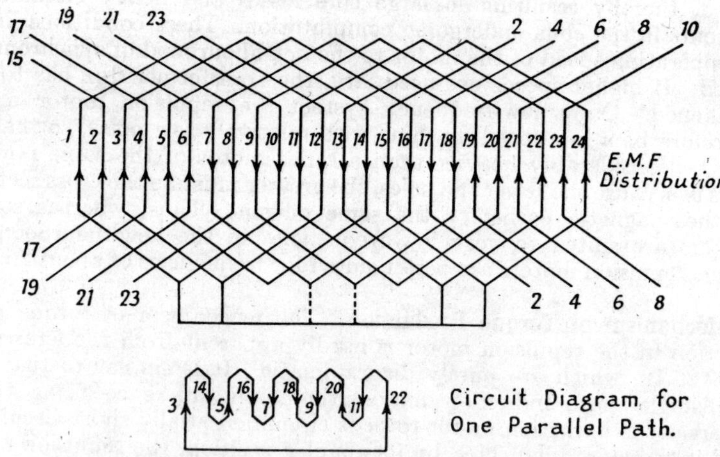

Fig. 25.15. E.M.F. Circulation in the Rotor Winding of Fig. 25.14

parallel paths, and one of these paths is shown separately in the form of an equivalent ring winding. It will be seen that there is a balance in the e.m.f. acting round the circuit, the resultant being zero and the current therefore zero. Consequently the torque is zero.

Now let the brush axis be shifted through 90°, this being equivalent to shifting the points of short-circuit of the winding to conductors 6 and 7 and conductors 18 and 19. The space distribution of induced e.m.f.s is the same as in Fig. 25.14, and we now see from Fig. 25.16 that the e.m.f.s in the two parallel paths act together thereby

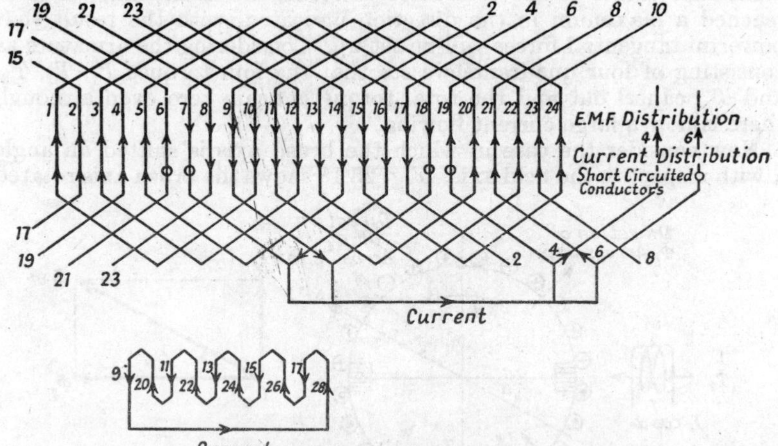

Fig. 25.16. E.M.F. and Current Circulation with Brush Axis In Line with Field Axis

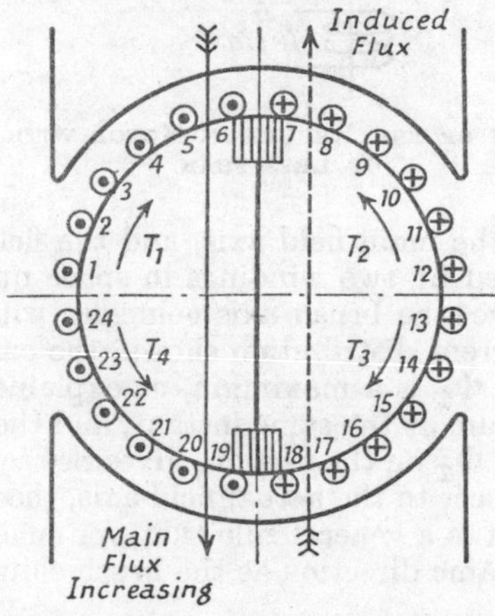

Fig. 25.17. Repulsion Motor with Brush Axis In Line with Field Axis

causing current to flow. The short-circuiting connection acts as a return path for both paths and therefore carries twice the conductor current. Now consider the time-phase relationships. Let the main flux be zero, its rate of change thus being a maximum, then the induced e.m.f.s will be a maximum. The short-circuit will be almost a purely inductive load and consequently the induced currents will lag behind the induced voltages by nearly a quarter of a period. In other words, by the time the space distribution of dots and crosses in Fig. 25.17 represents the maximum values of the conductor currents, the main flux will have reached a maximum in the direction indicated, with the result that powerful tangential forces will be set up. Considering the armature as consisting of four quadrants we see that the four torques T_1, T_2, T_3, and T_4, cancel out and the total torque is again zero even although there is now a large current flowing.

Now consider the case in which the brush axis is shifted an angle α with respect to the field axis. Fig. 25.18 shows the brush axis rotated

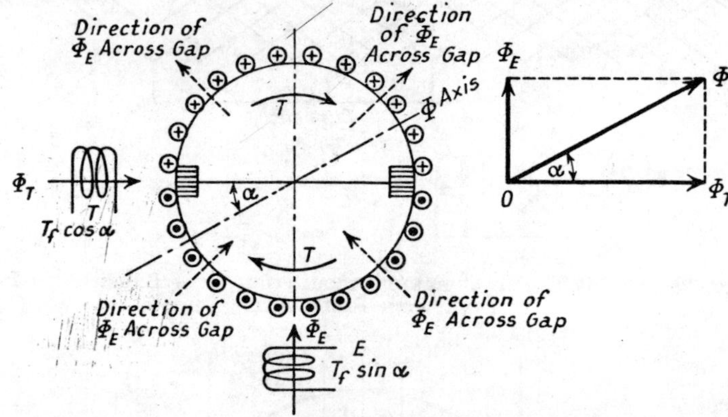

FIG. 25.18. ACTION OF THE REPULSION MOTOR WITH A BRUSH SHIFT α OF LESS THAN 90°

$\alpha°$ with respect to the main field axis, and the field, by the device of Fig. 25.13 is replaced by two windings in space quadrature T and E. Considering T above, the brush axis coincides with its field axis and gives rise to the current distribution shown, the current being a maximum when the flux Φ_T is a maximum, as explained above. The flux Φ_E is also a maximum at the same instant, and the interactions of the currents induced by Φ_T on the flux Φ_E gives rise to a clockwise torque. Note that, with respect to the actual field axis, the brushes are given a forward shift, and it is a general rule that, for small shifts, the torque always acts in the same direction as the brush shift reckoned from the actual field axis.

The Compensated Repulsion Motor. In the compensated motor, the field windings E, which set up the excitation flux Φ_e along the excitation axis are dispensed with, the excitation flux being set up by the armature itself. To do this, the armature has another set of brushes which have their axis along the excitation axis. (*See* Fig. 25.19.) Thus, the armature itself takes the place of the field E in the simple repulsion motor, two important advantages being derived from this arrangement.

(*a*) Since the main field is produced by the armature itself, the leakage which occurs between armature and field in all other motors is entirely eliminated.

(*b*) Practically complete compensation of the induced voltage in the excitation winding can be obtained at one particular speed, so that at this speed the motor operates at unity power factor. If there were no losses in the machine, this speed would be the synchronous speed, but because of losses it is slightly greater than this.

The Doubly Fed or Series Repulsion Motor. This is a modification of the repulsion motor to enable it to operate satisfactorily at speeds

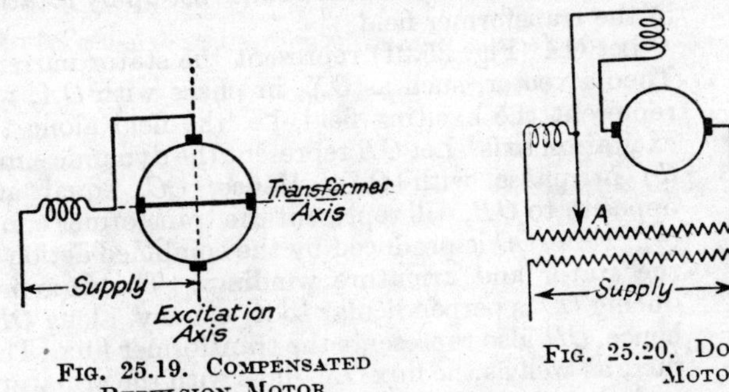

FIG. 25.19. COMPENSATED REPULSION MOTOR

FIG. 25.20. DOUBLE-FED MOTOR

greater than synchronous speed. One scheme of connections is shown in Fig. 25.20. We have seen that at high speeds the transformer flux increases rapidly with the speed, thereby setting up heavy circulating currents in the coils undergoing commutation, and also causing heavy core losses. In the doubly-fed motor this is overcome by weakening the transformer flux. The diagram indicates that the point A can be moved along the secondary of the main transformer, thus altering the voltage applied to the stator winding. At starting, the brushes are often short-circuited and the motor is thus brought up to speed as an ordinary repulsion motor. Then, when the speed is near to synchronism, the connections are automatically changed over to the doubly-fed arrangement. Thus, the doubly-fed motor combines some of the advantages of the series and the ordinary repulsion motor.

Vector Diagram of the Repulsion Motor. It is probable that the performance of the repulsion motor is more difficult to represent by means of a vector diagram than that of any other motor. There are many factors accounting for this, three of them being (*a*) the complex natures of the stator and rotor m.m.f. distributions, which depart considerably from the ideal sinusoidal distribution, (*b*) the difference between the shape of the two distributions and the fact that they can move relatively to one another, (*c*) magnetic saturation. The vector diagram is, of course, based on the assumed sinusoidal property of all the quantities represented, and thus it is possible for a diagram to give a useful qualitative picture of the performance, *but to be considerably in error if used quantitatively*.

Consider for simplicity an ideal motor having no losses. In the armature there are (*a*) the transformer e.m.f., (*b*) the dynamic e.m.f. We have seen that in an armature with zero resistance these are equal and opposite. Now in the compensated motor the armature produces its own exciting field, as previously explained, and therefore there are two more e.m.f.s to consider. These are (*c*) the statically-induced

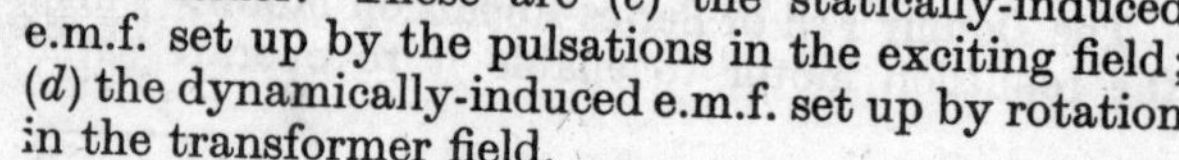

e.m.f. set up by the pulsations in the exciting field; (*d*) the dynamically-induced e.m.f. set up by rotation in the transformer field.

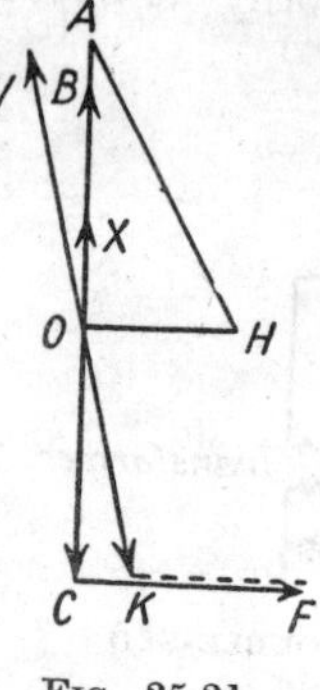

FIG. 25.21. VECTOR DIAGRAM OF COMPENSATED MOTOR

Let *OA* (Fig. 25.21) represent the stator current. Then a vector, such as *OX*, in phase with *OA*, will represent the exciting field, i.e. the field along the excitation axis. Let *OB* represent the dynamic e.m.f. (*b*) in phase with *OX*. Hence, *OC*, equal and opposite to *OB*, will represent the transformer e.m.f. (*a*). Now, *OC* is produced by the combined action of the stator and armature windings. The flux producing *OC* is perpendicular to it, namely, along *OH* hence, *OH* also represents the transformer flux. This flux, as well as the flux *OX*, links with the armature, and therefore, if *OH* is drawn proportional to the current producing this flux, the resultant of *OH* and *OA*, namely, *HA*, represents the total armature current.

The statically-induced e.m.f. (*c*) lags behind the main field, as shown by the vector *CF*, and therefore the dynamically-induced e.m.f. (*d*), which is in phase opposition to *CF*, will be represented by a vector such as *FK*. Now the supply e.m.f., *V*, is equal and opposite to the resultant of the e.m.f.s, *OC*, *CF*, and *FK*, namely, equal and opposite to *OK*. It is therefore represented by *OV* in the diagram.

FURTHER READING

ARNOLD, E. *Die Wechselstromtechnik*, Vol. 5 (2). Springer, Berlin, 1912.

GIBB, W. J. *Algebra of Electrical Machine Analysis*. B.T.H. Co., Technical Monograph.

HOBART, H. M. *Electric Motors*, Vol. 2. Pitman, London, 1923.

MAGGS, A. H. *Single-phase Commutator Motors*. B.T.H. Co. Technical Monograph.

MORRIS, D. "Characteristics of a single-phase shunt commutator motor with an improved field phase-correction circuit." *J. Inst. Elect. Engrs.*, **96**, Pt. II, 1949, p. 498

PRITCHARD, F. H. and KONN, F. "The modern single-phase motor for railway electrification." *Trans. A.I.E.E.*, **50**, 1931, p. 278.

PUCHSTEIN, A. F. and LLOYD, T. C. *Alternating Current Machines*. Wiley, New York, 1942.

CHAPTER 26

THE POLYPHASE COMMUTATOR MOTOR

Action of the Commutator. Imagine first of all an armature with a two-layer winding brought out to a commutator, a d.c. armature, in fact, and let this armature be rotated in a magnetic field. First of all let this field be stationary in space, then the e.m.f.s induced in individual armature conductors are alternating e.m.f.s, undergoing one complete cycle for an angular movement of two pole pitches. But the distribution of induced e.m.f.s in all the armature conductors is stationary in space, the e.m.f. at any particular point in space being proportional to the flux density at that particular point. Now, the action of the commutator is to connect the brushes which press on it to points in the winding which are fixed in space, and consequently if the magnetic field is stationary, the distribution of potential round the commutator is also stationary. Again, since each armature coil is connected to adjacent commutator segments (assuming a simple lap winding, as is usual), the potential distribution round the commutator is the coil e.m.f. integrated right round the armature. There will be a position of minimum potential, which for convenience can be called zero potential, and a pole pitch away there will be a point of maximum potential. Also there will be one point of zero potential and one of maximum potential for each pair of poles on the machine. Obviously, if the machine is to be used as a d.c. machine, these points will give the positions of the negative and positive brush positions respectively, so that the potential distributions and brush positions for two- and four-pole machines will be as shown in Fig. 26.1. The magnitude of the e.m.f. between brushes of opposite polarity will be

$$E = \Phi Z n, \text{ assuming a lap winding}$$

With a stationary magnetic field the external frequency is zero, i.e. the external circuit carries direct current, while the internal frequency is fixed by the speed n of the armature.

Now imagine that the magnetic field rotates in space, as would be the case in a machine consisting of a stator fed with polyphase currents after the manner of an induction motor, and having a rotor carrying a d.c. winding. Then, since the distribution of e.m.f. in the armature conductors is, at any instant, dependent on the position in space at that instant of the magnetic field, it follows that as the field rotates in space, the e.m.f. distribution round the winding will also rotate in space at the same speed and in the same direction. But obviously the potential distribution round the commutator will move whenever the e.m.f. distribution round the winding moves, so that we see, finally, that the potential distribution round the commutator rotates in space at the same speed, and in the same direction, as the magnetic field.

Thus it is obvious that the p.d. between two brushes pressing on the commutator will now be alternating, the frequency being determined by the speed of the magnetic field in space. But the internal frequency is determined by the relative speed of the armature to the stator rotating field, so that it follows that the commutator now acts as a frequency changer. Take, for example, the case of a 4-pole machine

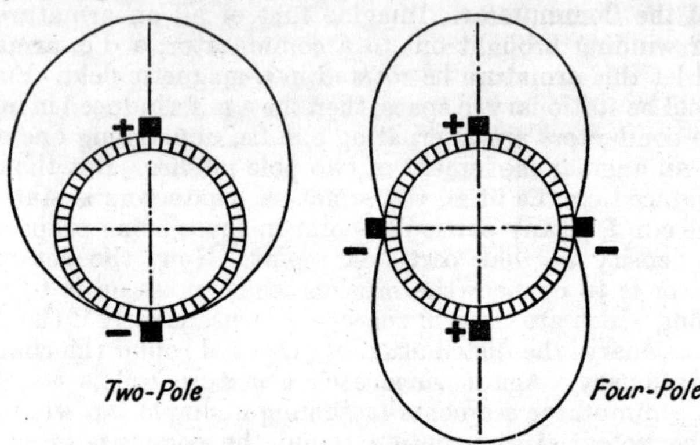

FIG. 26.1. POTENTIAL DISTRIBUTION ROUND COMMUTATOR

having its stator fed from a 50-c/s supply, and let the rotor travel at 1000 r.p.m. in the same direction as the stator field.

$$\text{Speed of stator field} = 120f/\text{poles} = 120 \times 50/4 = 1500 \text{ r.p.m.}$$

$$\therefore \text{Relative speed of stator field to rotor} = 1500 - 1000 = 500 \text{ r.p.m.}$$

External frequency, i.e. frequency of the alternating p.d. at the brushes

= frequency corresponding to field speed

= 50 c/s

Internal frequency, i.e. frequency of currents in the rotor conductors

= frequency corresponding to relative speed

$= 4 \times 500/120$

$= 16{\cdot}67$ c/s

It will be clear that (*a*) the p.d. between any two brushes will depend on their angular distance apart, and (*b*) that any number of phases of the external circuit can be arranged merely by providing the necessary number of brushes. Thus in the case of a 2-pole machine two diametrically opposite brushes will give single-phase, three equally spaced brushes will give three-phase, and so on. Thus the commutator can act, not only as a frequency changer, but as a phase convertor as well.

Suppose that the commutator is worked single-phase, then the brushes will be placed as in the case of a d.c. machine. The voltage given by the e.m.f. will be the maximum value, and therefore for the r.m.s. value (assuming a sinusoidal voltage) we have

$$E = \Phi Z n'/\sqrt{2}$$

where n' is the relative speed of stator field to rotor in r.p.m.

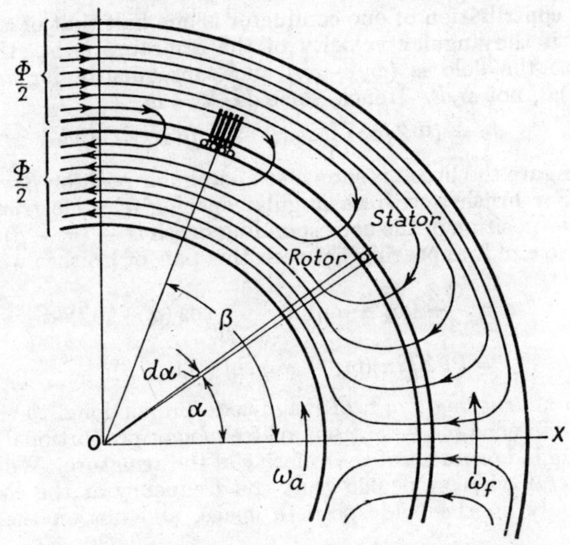

FIG. 26.2. ROTATING ARMATURE IN ROTATING FIELD

As in the case of the rotary convertor, we can regard the vector diagram per pair of poles as a complete circle and the voltage between two brushes of angular spacing α equal to the length of the chord which subtends an angle α at the centre. Thus, if the external circuit is three-phase, the angle α is 120° and we have to multiply the above value by

$$\sin(\pi/m) = \sin(\pi/3) = 0{\cdot}866$$

Hence for three-phase external supply the brush voltage is given by

$$E = (0{\cdot}866/\sqrt{2}) \times \Phi Z n' \text{ r.m.s. volts}$$

Induced e.m.f. Equation. In the position shown in Fig. 26.2 the flux of the pole shown is acting along OX. If Φ is the flux per pole the flux crossing the armature core at an angular distance α from OX is $(\Phi/2)\sin\alpha$. The linkage with a coil, one of whose coil sides is at A is therefore $\Phi \sin \alpha$. Hence if the field rotates with angular velocity ω_f and time is reckoned from the instant the field vector crosses the OX axis we have for the flux linking the coil

$$\phi = -\Phi \sin (\alpha - \omega_f t)$$

$$\therefore d\phi/dt = \omega_f \Phi \cos (\alpha - \omega_f t)$$

Let there be Z conductors altogether, then there are $(Z/2\pi)d\alpha$ conductors in the angle $d\alpha$, the e.m.f. induced in these being

$$de = +\tfrac{1}{2}(Z/2\pi)d\alpha(d\phi/dt) = (\Phi Z/4\pi)\omega_f \times \cos (\alpha - \omega_f t)d\alpha$$

since the contribution of one conductor is one-half that of a coil.

Again, if the angular velocity of the armature is ω_a, the velocity relative to the field is $(\omega_a - \omega_f)$ and consequently for $d\alpha$ we have $(\omega_a - \omega_f)dt$, not $\omega_a dt$. Hence, since $d\alpha/dt = \omega_a$

$$de = (\Phi Z/4\pi) [\cos (\alpha - \omega_f)](\omega_a - \omega_f)d\alpha$$

In the figure the brush is shown at an angular position β with respect to OX. For brushes with an angular spacing of π electrical radians, the angular position of the corresponding brush is $-(\pi - \beta)$ and therefore for the e.m.f. appearing between this pair of brushes we have

$$e = \frac{\Phi Z}{4\pi}(\omega_a - \omega_f)\int_{-(\pi-\beta)}^{\beta} \cos (\alpha - \omega_f t)d\alpha$$

$$= (\Phi Z/2\pi)(\omega_a - \omega_f) \sin (\omega_f t - \beta)$$

This is an alternating e.m.f. of magnitude proportional to the angular velocity difference $(\omega_a - \omega_f)$ but of frequency proportional to ω_f the latter being independent of the velocity of the armature. We found in a more general way on p. 665 that the frequency at the commutator depends only on the field speed in space, and not on the armature speed.

Now suppose that Φ were a unidirectional flux. In such a machine $\omega_f = -\omega_a$ as in a d.c. machine, so that to compare it with the a.c. machine we must imagine its armature running with an angular velocity of ω_f

Then $$E_{dc} = \Phi Z n$$

or, since $$n = \omega_f/2\pi$$

$$E_{dc} = \Phi Z \omega_f/2\pi$$

$$\therefore E_{ac}/E_{dc} = [(\omega_f - \omega_a/\omega_f)] \sin (\omega_f t - \beta)$$

When $\omega_f > \omega_a$ the e.m.f. is a maximum when

$$\omega_f t - \beta = \pi/2 \text{ (i.e. when } \omega_f t = \pi/2 + \beta)$$

that is, when the field vector is in quadrature with the brush axis.

The Three-phase Series Motor. As its name implies, the series motor has its rotor connected in series with the stator phases. The connection can be direct as in Fig. 26.3 (A) or via a current transformer as in

Fig. 26.3 (*B*). We shall see that the second method gives certain operating advantages but, fundamentally, the mode of operation is the same for both cases. The stator sets up a rotating field whose speed in space is the synchronous speed, and consequently it induces in the rotor an e.m.f. which, when transferred to the brushes is of line frequency. Thus series connection of stator and rotor is possible. The action of the motor is most easily followed by comparing the stator

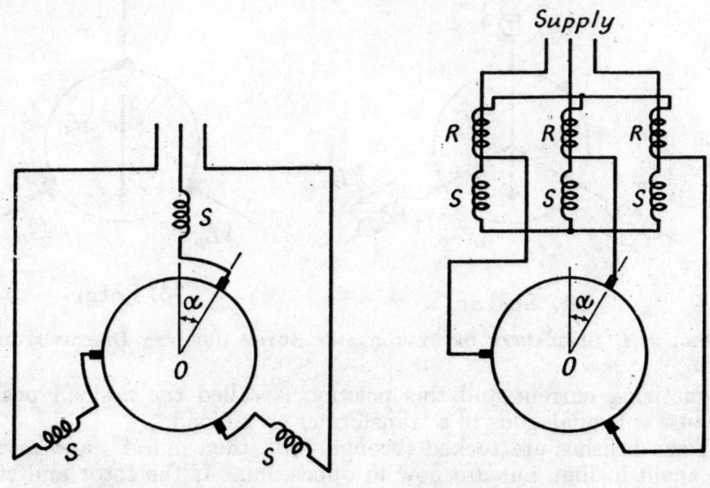

Fig. 26.3. Connection Schemes of Three-phase Commutator Motor

S—stator winding. *R*—secondaries of current transformers supplying the rotor.

and rotor m.m.f.s. Fig. 26.4 (*A*) shows the three stator phases diagrammatically and the instant represented is that for which the current in phase 1 is I_{max}. The instantaneous currents in phases 2 and 3 are therefore both $-I_{max}/2$, and we see that the stator m.m.f. therefore acts, at this instant, along the axis of phase 1. Fig. 26.4 (*B*) shows the rotor at the same instant. This is, of course, mesh-connected and therefore, at the instant considered, the winding currents are $\frac{1}{2}I_{max}$, $\frac{1}{2}I_{max}$, and zero. The rotor m.m.f. therefore coincides at this instant with the axis of brush B_1. The two m.m.f.s rotate at synchronous speed and therefore keep pace with one another. At any instant the position in space of m.m.f. 1 is fixed according to the position of the instant in the cycle, but that of m.m.f. 2 can be varied at will by shifting the brushes since, if the currents causing the distribution are fed in at the brushes, that distribution will be shifted, at any particular instant of time, by the same amount, and in the same direction as the brush shift.

When the angle α is zero we see from the figure that m.m.f.$_S$ and m.m.f.$_R$ are in line and act in the same sense. A magnetic field is set up but there is no torque because, as we have seen in connection with the operation of single-phase motors, the rotor m.m.f. must be in quadrature, or at any rate have a component in quadrature, with the stator m.m.f. for there to be a torque. The current is therefore a

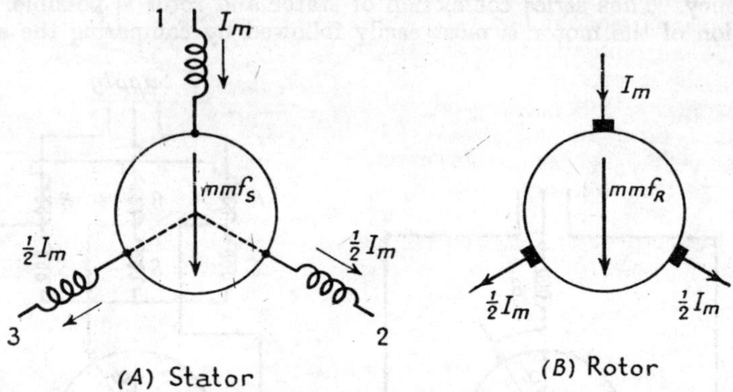

FIG. 26.4. SIMILARITY OF STATOR AND ROTOR CURRENT DISTRIBUTIONS

magnetizing current and this position is called the no-load position because it is analogous to a transformer on no-load.

If the brushes are rocked through 180°, then m.m.f.$_S$ and m.m.f.$_R$ are again in line, but are now in opposition. If the rotor and stator ampere-turns are equal, then when $\alpha = 180°$ the total m.m.f. is zero. This is called the short-circuit position, since it is analogous to a short-circuited transformer. The current is large but again there is no torque.

Now let the brushes be given any angular position, and put $\beta = (180° - \alpha)$, the angle β then being the displacement from the short-circuit position. Since we are assuming equal stator and rotor m.m.f.s we can replace the m.m.f.s by the stator and rotor currents I_1 and I_2, and the relationships are then shown in Fig. 26.5. The resultant of I_1 and I_2 is a magnetizing current I_μ which produces the motor flux Φ. Now Φ is in time phase with I_μ but it is in space quadrature with it, since, in space, the point of maximum flux density is displaced half a pole-pitch from the point of maximum m.m.f. Hence Fig. 26.5 is a space diagram only. Again, since torque is only produced by the mutual action of current (or m.m.f.) systems and flux systems which are in space phase, it follows that the energy component of the current is the median of the triangle, namely the component I_e.

Torque. Assume an unsaturated motor, then, if k_1, k_2, k_3 and k_4 are constants, we have

$$I_\mu = k_1\Phi$$

and

$$T = k_2 I_\mu I_e$$

Again, $I_\mu = 2I_e \tan(\beta/2)$

$$\therefore T = k_2 \times (k_1\Phi) \times (k_1\Phi/2 \tan(\beta/2))$$

or $$T = \tfrac{1}{2}k_1^2k_2\Phi^2 \cot(\beta/2)$$

Thus, as in other unsaturated series machines the torque is proportional to the square of the flux.

The current diagram shows that I_1 is the vector sum of I_e and $\frac{1}{2}I_\mu$, showing that only half the magnetizing current flows in the stator. If the machine had its brushes short-circuited it would become an induction motor and the whole of I_μ would then flow in the stator. Thus the power factor of the motor is better than that of the comparable induction motor.

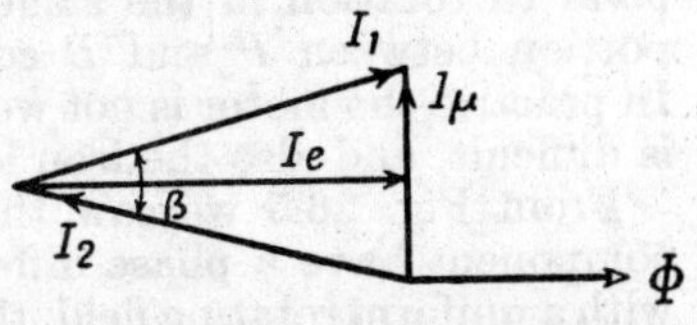

FIG. 26.5. SPACE RELATIONSHIPS OF CURRENTS AND FLUX

Let n_s be the synchronous speed and n the actual speed, then stator induced e.m.f.

$$E_1 = k_3\Phi n_s$$

While the rotor induced e.m.f.

$$E_2 = k_3\Phi(n_s - n)$$

$$\therefore E_2/E_1 = (n_s - n)/n_s = 1 - n/n_s$$

or $$n/n_s = (E_1 - E_2)/E_1$$

Substituting $\Phi = E_1/k_3n_s$ in the previous equation for the torque we have

$$T = \tfrac{1}{2}k_1^2k_2(E_1^2/k_3^2n_s^2)\cot(\beta/2)$$
$$= k_4E_1^2\cot(\beta/2)$$

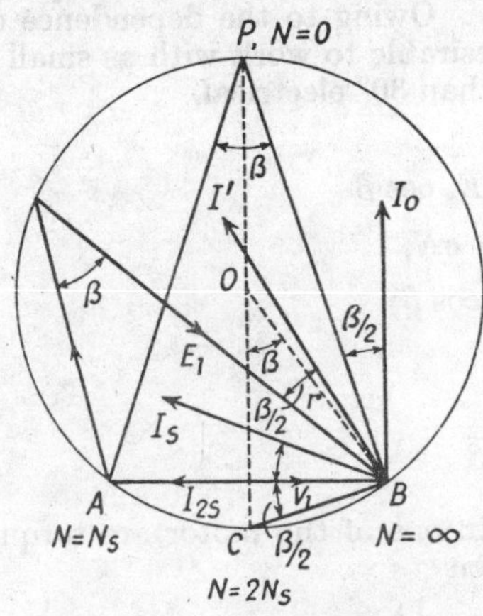

FIG. 26.6. VOLTAGE DIAGRAM FOR ONE PARTICULAR VALUE OF THE BRUSH ANGLE β

Speed. Let V_1 and β be maintained constant. When the machine is stationary the induced e.m.f.s E_1 and E_2 are numerically equal and the voltage diagram (Fig. 26.6) is similar to the current diagram of Fig. 26.5. The vector sum of E_1 and E_2 is the stator applied p.d. V_1. With β fixed, the angle between E_1 and E_2 is fixed and therefore the locus of the apex of the triangle is the circle circumscribing the equilateral triangle PAB. The radius of the circle is

$$r = V_1/2\sin\beta$$

To the left of point P the locus corresponds to speeds within the range standstill at P to N_s at A. This follows from the fact that

$$E_2/E_1 = 1 - n/n_s = 0 \text{ when } n = n_s$$

The rotor induced e.m.f. is then zero, and the stator induced e.m.f. is equal to V_1. For points between A and B along the arc ACB the speed is greater than n_s because the sense of E_2 is reversed. At C, E_1 and E_2 are numerically equal and the point thus corresponds to a speed of $2n_s$. At B, E_1 is zero and the speed becomes infinite. All the above correspond to rotation in the same direction as the field. The right-hand portion between P and B corresponds to rotation against the field. In practice the motor is not worked in this region because commutation is difficult, and also the iron losses are very high.

From Fig. 26.5 we saw that the total current I, and the energy component have a phase difference of $\beta/2$. In the case of a machine with a uniform rotating field, the space relationships also hold for current relationships. Now the energy component of the stator current is a current in phase opposition to E_1 and it therefore follows that the current vector is inclined $\beta/2$ to BP when inserted in Fig. 26.6. Hence if we denote the standstill current by I_0 this will be represented by the vector BI_0 lagging behind BP by $\beta/2$. The power factor is thus zero, which is to be expected as we are considering an ideal machine with no losses and no leakage reactance, so the current I_0 has only to produce a flux. As the motor starts up the working point P moves round the circle in a counter-clockwise direction and the current vector must rotate also in order to maintain the angle $\beta/2$ between stator induced e.m.f. and current. Thus for the point P' the current is OI' and the power factor $\cos \angle I'BA$. At synchronism E_1 becomes equal to V_1 and the current I_s lags V_1 by $\beta/2$ so that the power factor is then $\cos \beta/2$. At double synchronism, when the working point is at C the current I_{2s} is in phase with V_1 and the power factor is unity. Owing to the dependence of the power factor at the angle β it is desirable to work with as small a value of β as possible, say not greater than 30° electrical.

From Fig. 26.6 we also have

$$V_1^2 = E_1^2 + E_2^2 - 2E_1E_2 \cos \beta$$

Again,
$$E_2 = E_1[(n_s - n)/n_s] = \sigma E_1$$

$$\therefore\ V_1^2 = E_1^2(1 + \sigma^2 - 2\sigma \cos \beta)$$

$$\therefore\ \text{Torque } T = k_4 E_1^2 \cot (\beta/2)$$

$$= \frac{k_4 V_1^2 \cot (\beta/2)}{1 + \sigma^2 - 2\sigma \cos \beta}$$

Suppose that the normal rated conditions of the motor are torque T_n, stator e.m.f. E_n, brush angle β_n, then

$$T_n = k_4 E_n^2 \cot (\beta_n/2)$$

or
$$k_4 = \frac{T_n}{E_n^2 \cot (\beta_n/2)}$$

$$\therefore\ \frac{T}{T_n} = \frac{V_1^2 \cot (\beta/2)}{E_n^2 \cot (\beta_n/2)} \cdot \frac{1}{1 + \sigma^2 - 2\sigma \cos \beta}$$

By giving β a series of values the torque-slip characteristics can be drawn, and they take the form indicated in Fig. 26.7. It will be seen that for $\beta < 40°$ the curves are similar to those for a d.c. series motor, the operation of the motor being stable on any type of load. When $\beta > 40°$ the curves become progressively steeper and also bend back

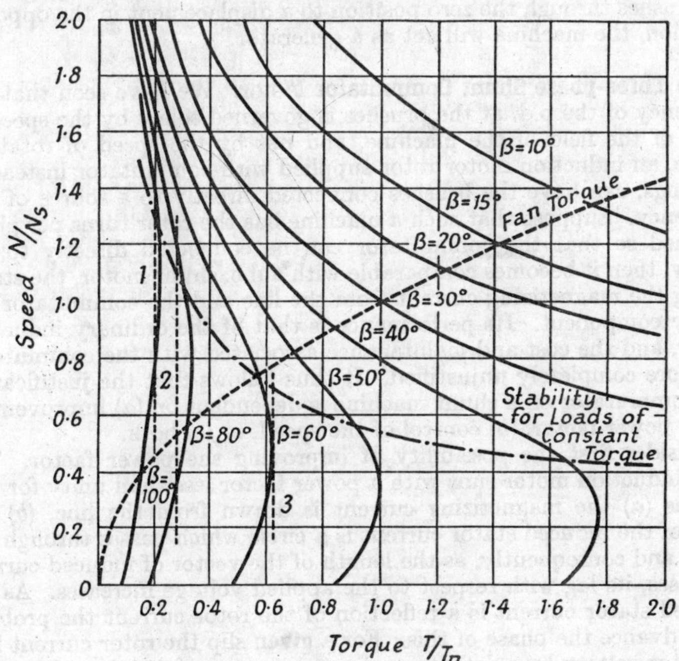

Fig. 26.7. Mechanical Characteristics of the Series Motor

on themselves. Consider the effect of this on stability. The dotted parabola gives the torque-speed curve of a fan whose output is chosen so that the curve passes through the point $T/T_n = 1$, $N/N_s = 1$. This cuts all the motor characteristics at a sufficiently large angle to ensure stability. Now let the load torque be independent of the speed. three such characteristics are indicated by the dotted vertical lines 1, 2 and 3. In case 1, the curve for $\beta = 100°$ is crossed twice and there is little difference between the motor and load characteristics showing that, under such conditions, the speed could wander over a very wide range. Conditions improve as the ratio T/T_n increases, but it is clear that for loads of constant torque, the area indicated is to be avoided.

In practice this instability can be largely overcome by supplying the rotor through a series transformer, as in Fig. 26.3 (*B*), the transformer being designed to take a relatively large magnetizing current. This gives the motor the characteristics of a heavily compounded d.c. machine and reduces the tendency to race at light load.

From Fig. 26.6 we see that when the motor is running at twice synchronous speed its power factor is unity. This means that it generates its own magnetizing current, that is, it is self-exciting. Hence, if the machine is braked, as for example, by connecting to resistances, or on the mains by reversing the flux sequence or by bringing back the brushes through the zero position to a displacement in the opposite direction, the machine will act as a generator.

The Three-phase Shunt Commutator Motor. We have seen that the frequency of the p.d. at the brushes is governed solely by the speed in space of the field of the machine, and not by the speed of rotation. Hence, an induction motor rotor supplied with commutator instead of slip-rings, can have the brushes connected directly to a source of line frequency. Suppose that such a machine has the rotor turns per phase designed so that the commutator can be connected directly to the supply, then it becomes comparable with a d.c. shunt motor, the stator taking the magnetizing current from the line and the commutator the energy component. Its performance is that of the ordinary induction motor, and the cost and maintenance associated with the commutator therefore completely unjustified. It thus follows that the justification of a commutator on a shunt machine is dependent on (*a*) improvement in the power factor, (*b*) control of the speed, or (*c*) both.

Consider first the possibility of improving the power factor. The plain induction motor runs with a power factor less than unity for two reasons (*a*) the magnetizing current is drawn from the line, (*b*) the locus of the induced stator current is a circle which passes through the origin and consequently, as the length of the vector of induced current increases, its lag with respect to the applied voltage increases. As the induced stator current is a reflection of the rotor current the problem is to advance the phase of this. For a given slip the rotor current lags the rotor voltage by a definite angle, and if an e.m.f. is injected into the rotor circuit so that the total rotor e.m.f. is advanced then, provided there is no change of slip, the rotor current will be advanced by the same amount. This is made clear by the vector diagrams of Fig. 26.8, in which (*a*) is for the uncompensated motor, and (*b*) for the compensated motor, the injected, or compensating e.m.f. being E_p.

It is clear that the frequency of E_p must always be the same as the slip frequency, and consequently it can only be supplied via a commutator.

Now consider speed control. In the rheostatic method some of the rotor induced e.m.f. is utilized externally in the form of a volt drop, and the speed correspondingly reduced. This is not only wasteful but it gives the motor a drooping speed characteristic which is often undesirable. There is a similar phenomenon with the d.c. motor and the difficulty can be overcome by using a series connected d.c. generator to inject an opposing voltage of the required amount. If this generator boosts, instead of "bucks," the armature voltage then the speed is increased. A similar effect is obtained if a voltage of the correct phase is injected via a commutator into the rotor circuit. The speed can be reduced or even raised above synchronism, and for all values of the boost or buck the motor will retain its shunt characteristic.

The Expedor. This is a device for phase compensation only. Consider an armature with d.c. winding rotating inside a stator without any windings, the stator being present merely to complete the magnetic

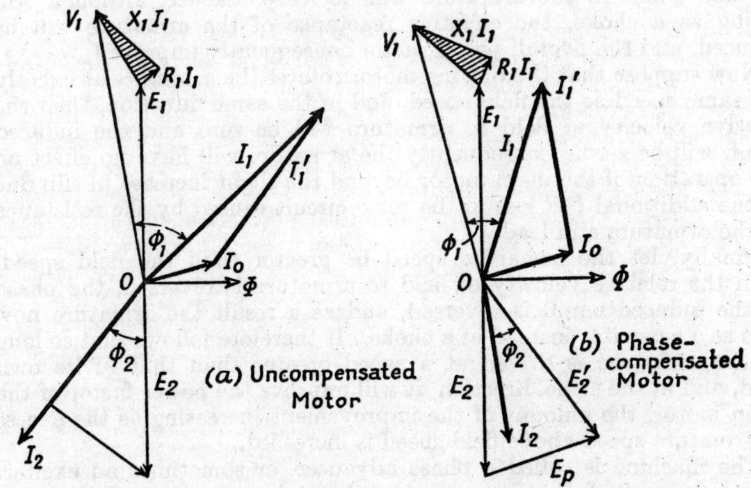

FIG. 26.8. PHASE COMPENSATION OF THE INDUCTION MOTOR

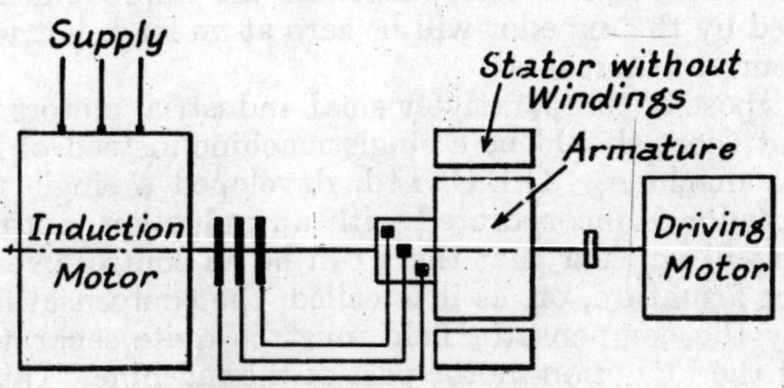

FIG. 26.9. THE EXPEDOR

path of the flux. Then, since the stator can produce no flux, this must obviously be produced by currents flowing in the armature winding. For example, if the armature is connected via the commutator to the slip-rings of an induction motor, as in Fig. 26.9, it will carry the current of the main motor and so set up a rotating field. The speed of this field in space will depend on the external frequency, as explained above, this external frequency being the rotor frequency of the main motor: the speed will be entirely independent of the speed of the armature.

First, suppose that the armature is stationary: then its conductors will be cut by the lines of force of its own rotating magnetic field, and in consequence it will act like a choke in series with the rotor of the main motor. As a result the power factor of the main motor will be reduced.

Now let the armature be rotated in the same direction as its own field in space; if its speed is less than the field speed, the relative velocity of field to armature will decrease, and consequently the induced e.m.f. in the armature will decrease. Hence, although still acting as a choke, the effective reactance of the armature will be reduced, and the overall power factor consequently improved.

Now suppose that the driving motor rotates the armature at exactly the same speed as the field speed, and in the same direction, then the relative velocity of field to armature will be zero and the induced e.m.f. will be zero. Consequently the armature will have no effect on the operation of the main motor beyond the slight increase in slip due to the additional I^2R loss in the rotor circuit caused by the resistance of the armature and leads.

Finally, let the armature speed be greater than the field speed, then the relative velocity of field to armature is reversed, the phase of the induced e.m.f. is reversed, and as a result the armature now acts as a capacitor instead of a choke. It therefore follows that so long as the armature is driven at a speed greater than that of its own field, and in the same direction, it will improve the power factor of the main motor, the amount of the improvement increasing as the excess of armature speed above field speed is increased.

The machine is called a phase advancer, or sometimes an exciter. As it carries the full rotor current of the main motor it is also a series exciter, and to this class of machine Miles Walker has given the name Expedor.* Because it is a series machine the improvement in power factor effected by the expedor will be zero at no load, but will increase as the load current increases.

For the purpose of comparatively small industrial motors it is clearly desirable that there should be a single machine instead of a combination of three machines. G.E.C. Ltd. developed a single machine in which an expedor is incorporated with an induction motor. In such a machine it will be clear that there can be no control over the speed of the exciter armature, or, as it is called, the compensating winding. Consequently the compensator field must be quite separate from the main field of the induction-motor part of the machine. This is accomplished by housing the compensating winding in entirely separate slots near the inner periphery of the rotor core. The core discs are therefore of the type shown in Fig. 26.10. The compensating winding is a lap winding brought out to a small commutator, and the electrical connections of main winding and compensating winding are the same as for Fig. 26.9.

* It is possible for the stator of the expedor to be provided with windings so that it can contribute to the total m.m.f., also interpole windings can be provided for the purpose of assisting commutation. *See* "Specification and Design of Dynamo-electric Machinery," Chap. 20.

The essential feature of this class of expedor is that the compensator field speed must be less than the actual speed of the compensator, and in the G.E.C. machine this is effected by arranging the compensator

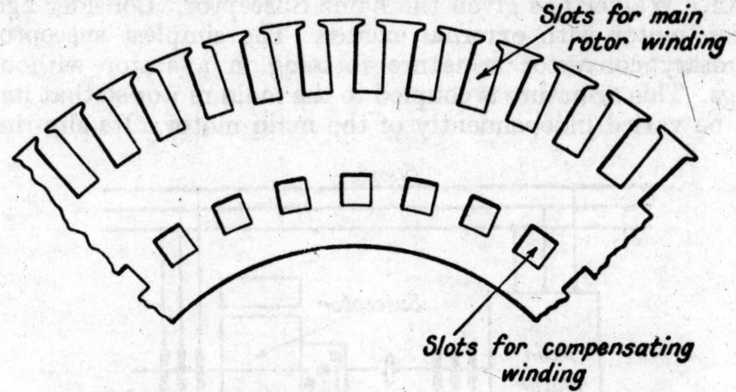

FIG. 26.10. CORE DISCS AND SLOTS IN THE EXPEDOR

to have a different number of poles from the main field. A numerical example will make this clear.

Let f = supply frequency

p_1 = number of main motor poles

p_2 = number of compensator poles

σ = fractional slip of motor

Then

$$\text{synchronous speed of motor} = 120f/p_1$$

and $$\text{actual speed of motor} = (120f/p_1)(1 - \sigma)$$

$$\text{slip frequency of motor} = \sigma f$$

$$\therefore \text{Speed of compensator field} = 120\sigma f/p_2$$

$$\therefore \frac{\text{motor speed}}{\text{compensator field speed}} = (120f(1 - \sigma)/p_1) \times (p_2/120\sigma f)$$

$$= p_2(1 - \sigma)/p_1\sigma$$

Thus if $p_1 = 6$ poles

$p_2 = 2$ poles

$\sigma = 5$ per cent, i.e. 1/20

then $$\frac{\text{motor speed}}{\text{compensator field speed}} = \frac{2(1 - 1/20)}{6/20}$$

$$= 6{\cdot}33$$

Hence under these conditions the compensating winding is rotated 6·33 times as fast as the field produced by it, thus fulfilling the necessary condition.

The Susceptor. In the expedor type of machine the compensating winding is in series with the main rotor. There is a class of machine in which the compensator is in the nature of a shunt exciter, and to this class Miles Walker has given the name Susceptor. Consider again an induction motor with external exciter. The simplest susceptor is a small rotary convertor armature rotating in a stator without any windings. This armature is coupled to the main motor so that its speed cannot be varied independently of the main motor. Its slip-rings are

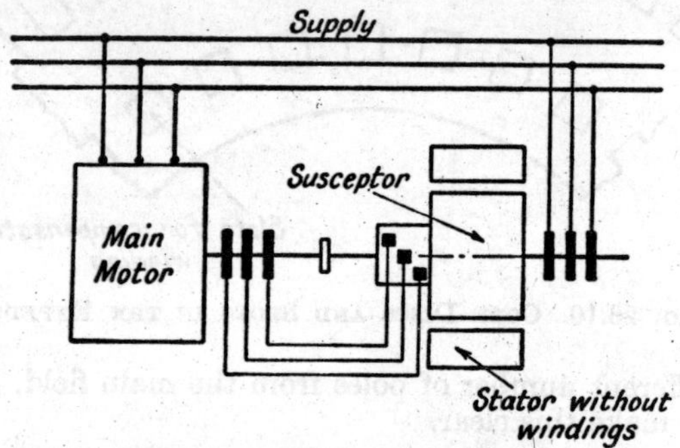

Fig. 26.11. The Susceptor

connected to the main supply, while its commutator is connected to the main rotor, the scheme being shown in Fig. 26.11.

The supply at line frequency to the susceptor will set up a rotating field at synchronous speed relative to the armature. Also since the slip-rings are connected to fixed points in the armature winding, this relative speed will be maintained whatever the actual armature speed may be. The connections are such that the rotations of susceptor and its field are in opposite directions, and therefore, since the actual speed is the same as the main motor speed, the speed of the field *in space* is the slip speed. We have seen that the external frequency on the commutator side corresponds to the actual field speed in space, and consequently this frequency is the same as the slip frequency of the main motor. It is for this reason that the commutator can be in direct electrical connection with the main slip-rings, as shown in Fig. 26.11.

Now, since the susceptor m.m.f. is produced by currents fed in at the slip-ring end, the position of the susceptor field in space at any instant will be entirely independent of the positions of the brushes on the commutator. But we saw on p. 665 that the potential distribution round the commutator rotates in space with the field, and always keeps pace with it. Hence the e.m.f. at the brushes at any particular instant depends on the relative position of the brushes and the commutator potential distribution. In other words, the phase of the e.m.f. at the commutator depends on the angular position of the brushes, and if the brush position is altered, the phase of this e.m.f. will be altered by a corresponding amount. It therefore follows that the phase of the e.m.f. injected into the main rotor circuit, and consequently the overall power factor of the equipment, can be controlled by rocking the brushes round the commutator.

Now, when considering the combination of induction motor and susceptor into a single machine, it will be realized at once that with an ordinary stator-fed induction motor this will be impossible, because

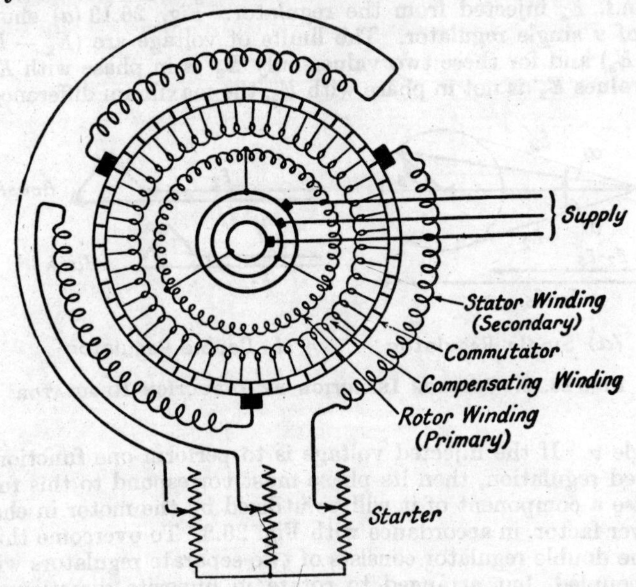

FIG. 26.12. THE COMPENSATED INDUCTION MOTOR—SUSCEPTOR-TYPE COMPENSATION

the motor field speed is the synchronous speed, N_s, whereas the susceptor field speed is σN_s. The difficulty is overcome by using a rotor-fed motor, i.e. a motor whose rotor is used as the primary and stator as the secondary. In such a motor the field speed is the slip speed σN_s, with the result that one and the same field can act as both main field and susceptor field. This further simplifies the construction, since both main rotor (primary) winding and compensating winding can be housed in the same rotor slots. The scheme of connections is therefore somewhat different from that of Fig. 26.11 and is given diagrammatically in Fig. 26.12.*

The Stator-fed Motor with Induction Regulator Control. This is an induction motor in which the variable voltage for speed control is supplied from an induction regulator. Imagine an induction motor with slip-ring rotor having the rotor fixed. An e.m.f. of supply frequency will be induced. If the rotor is given an angular shift then the

* This motor must not be confused with the Schrage motor described on p. 683.

e.m.f. will be constant in magnitude but its phase will change in accordance with the amount of the shift. This briefly, is the principle of the induction regulator. A means is provided for giving the rotor any angular position and therefore the e.m.f. can be given any phase with respect to the supply e.m.f. This e.m.f. can be injected into the rotor of the motor to be controlled via a commutator, and the resultant rotor voltage E_2' is then the vector sum of the induced voltage E_2 and the e.m.f. E_s injected from the regulator. Fig. 26.13 (*a*) shows the effect of a single regulator. The limits of voltage are $(E_2 - E_s)$ and $(E_2 + E_s)$ and for these two values only E_2' is in phase with E_2. For other values E_2' is not in phase with E_2, the maximum difference being

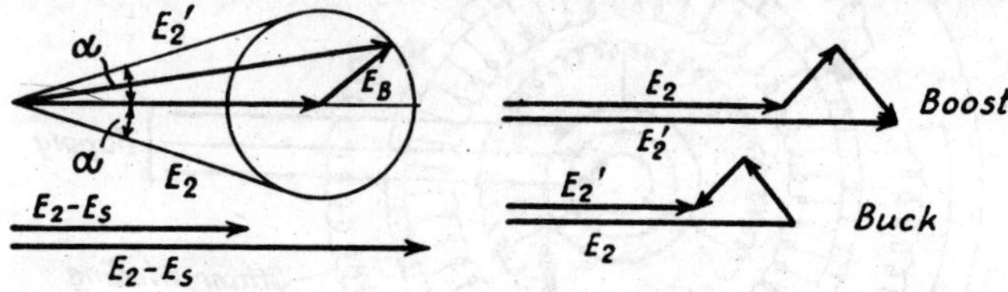

FIG. 26.13. VOLTAGE INJECTION BY INDUCTION REGULATOR

the angle α. If the injected voltage is to perform one function only, say speed regulation, then its phase must correspond to this function otherwise a component of it will be utilized by the motor in changing the power factor, in accordance with Fig. 26.8. To overcome this difficulty the double regulator consists of two separate regulators with the rotors coupled, but arranged to rotate in opposite directions. This gives the effect of Fig. 26.13 (*b*) and we see that E_2' is always in phase with E_2. Now it is desirable that there should be a certain amount of phase compensation from a device which is capable of providing this, and for this purpose it is merely necessary to ensure that the component of E_c in the direction of E_p of Fig. 26.8 is of the desired magnitude. The simplest way of doing this is to adjust the brush position at the commutator, this being possible because the potential distribution round the commutator is independent of the source from which E_c is derived. Fig. 26.14 (*a*) shows the effect of such a fixed-angle injection, the three cases of top speed, middle speed, and bottom speed being chosen. It will be seen that the phase compensating component E_p is variable in amount and that it reverses in phase when the motor is run through synchronism to hyper-synchronous speeds thus causing an undesirable lowering of power factor. This difficulty is overcome by using the voltage from the induction regulator solely as voltage E_s for the purpose of speed regulation and, at the same time injecting a constant phase compensating voltage E_p which is derived from a phase-compensating transformer. The vector diagrams are then as given in Fig. 26.14 (*b*) and Fig. 26.15 shows the circuit diagram.

Instead of the phase-compensating transformer indicated in Fig. 26.15 it is possible to use a boosting transformer, in which case its

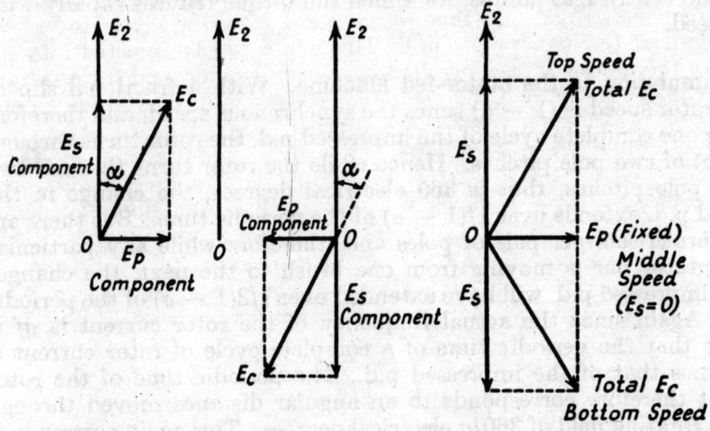

FIG. 26.14. SPEED AND POWER-FACTOR CONTROL OF STATOR-FED MOTOR

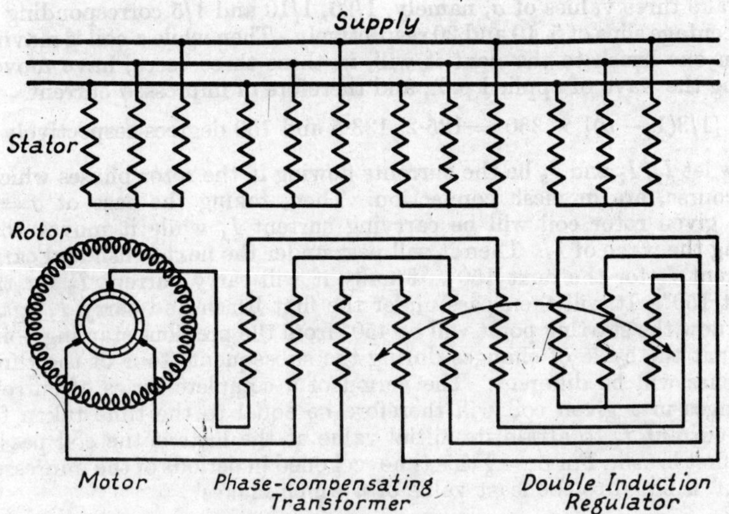

FIG. 26.15. CIRCUIT DIAGRAMS FOR STATOR-FED SHUNT COMMUTATOR MOTOR

primary windings are connected in series with the stator phases of the motor, and the secondary windings connected directly to the rotor brushes. With this arrangment it is possible to reduce the flux of the motor with reducing speed, by introducing the boosting voltage into the circuit in the necessary direction. Useful improvements in efficiency and power-factor are thereby obtained with such drives as fans and centrifugal pumps, for which the torque reduces rapidly with the speed.

Commutation in the Stator-fed Machine. With a fractional slip of σ the rotor speed is $(1 - \sigma)$ times the synchronous speed, and therefore during one complete cycle of the impressed p.d. the rotor turns through $(1 - \sigma)$ of two pole pitches. Hence while the rotor turns through two whole pole pitches, that is 360 electrical degrees, the change in the applied p.d. extends over $1/(1 - \sigma)$ of the periodic time. But there are three brush sets per pair of poles and, therefore while any particular commutator bar is moving from one brush to the next, the changes in the impressed p.d. will have extended over $1/3(1 - \sigma)$ of the periodic time. Again since the actual frequency of the rotor current is σf it follows that the periodic time of a complete cycle of rotor current is $1/\sigma$ times that of the impressed p.d. The periodic time of the rotor current therefore corresponds to an angular distance moved through by the rotating field of $360/\sigma$ electrical degrees. This again corresponds to an angular distance moved through by the rotor itself of

$$[(1 - \sigma)/\sigma] \times 360 \text{ electrical degrees}$$

Take three values of σ, namely, 1/20, 1/10 and 1/5 corresponding to percentage slips of 5, 10 and 20 respectively. Then while a coil is moving from one brush to the next it will, in these three cases, have moved along the wave of applied p.d., and therefore of impressed current

$$[1/3(1 - \sigma)] \times 360 = 125{\cdot}2,\ 133{\cdot}5 \text{ and } 150 \text{ degrees respectively}$$

Now let I_1, I_2 and I_3 be the currents flowing in the rotor phases which, of course, are in mesh connection. Then, taking the case of $\sigma = \frac{1}{5}$ any given rotor coil will be carrying current I_1 while it moves 150° along the wave of I_1. Then it will pass under the next brush and carry current I_2 for the next 150°. Finally it will carry current I_3 for the next 150°. It will then pass under the first brush and carry I_1 again but now the starting point will be 450° from the previous starting point so that the cycle of changes during the subsequent tour of the three brushes will be different. The period of a complete series of current changes in a given coil will therefore be equal to the time taken for the current I_1 to attain its initial value at the instant the coil passes the first brush. For $\sigma = \frac{1}{5}$ this time reckoned in periods of the impressed e.m.f. is equal to the least value of n which makes

$$n \times 360/450 \text{ a whole number}$$

This value is $n = 5$. To draw the curves of current changes in any coil for $\sigma = \frac{1}{5}$, it is therefore necessary to draw the three sinusoidal curves of I_1, I_2 and I_3, and to follow each for 150° and then change abruptly to the next. The curves for the above three values of σ are given in Fig. 26.16, only that for $\sigma = \frac{1}{5}$ being shown completely.

The change of current during the commutation period sets up a reactance voltage as in all machines with commutators: in addition there is the transformer voltage. This is proportional to the relative speed of rotor to rotating field and thus becomes zero at synchronism. If the speed is decreased the transformer voltage is set up, but the reactance voltage decreases so that good commutation is possible over a wide speed range below synchronism. For speeds above synchronism the transformer voltage is present and increases with excess of speed

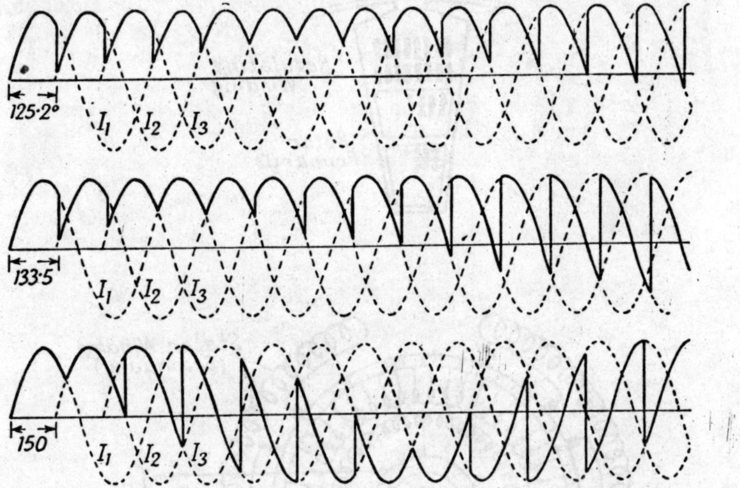

FIG. 26.16. COMMUTATION IN A THREE-PHASE COMMUTATOR MOTOR

above synchronism, and at the same time the reactance voltage increases, with the result that good commutation is not possible over such a wide range of hypersynchronous speeds.

The Schrage Motor. This is a motor in which power-factor correction and speed control are both available in one and the same machine. It has been explained that the speed of an induction motor is dependent on the total e.m.f. acting in the secondary circuit. If the secondary is short-circuited in the usual way, this e.m.f. is the induced e.m.f. and the motor runs at normal speed. If an e.m.f. of the correct frequency from some other source is injected into the secondary circuit, then if this injected e.m.f. has a component in direct opposition to the induced e.m.f., the speed will be below normal. On the other hand, if it has a component in phase with the induced e.m.f., the total e.m.f. acting in the secondary will be increased, and the speed will be above normal. In other words, if the injected e.m.f. bucks the induced rotor e.m.f., the motor will run at subsynchronous speeds, whereas if it boosts the induced e.m.f., the motor will run at hypersynchronous speeds. In addition, if this injected e.m.f. has a quadrature component which leads the induced rotor e.m.f., the power factor of the motor will be improved. The Schrage motor is a machine in which these adjustments are available. It is essentially an induction motor with integral compensator of the susceptor type, so that it is rotor-fed. It differs from the motor of Fig. 26.12 in that each brush set is duplicated, thus

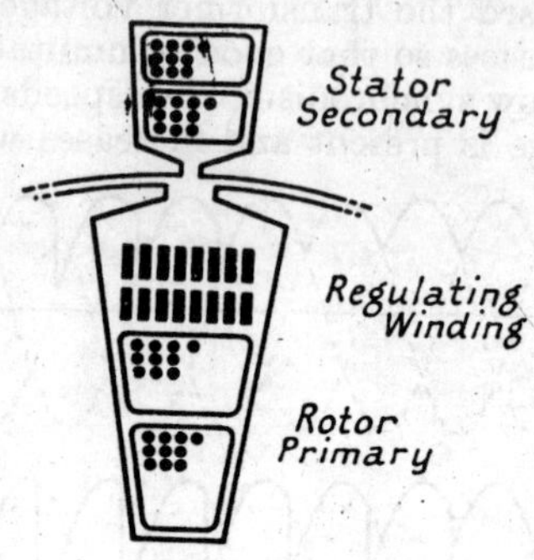

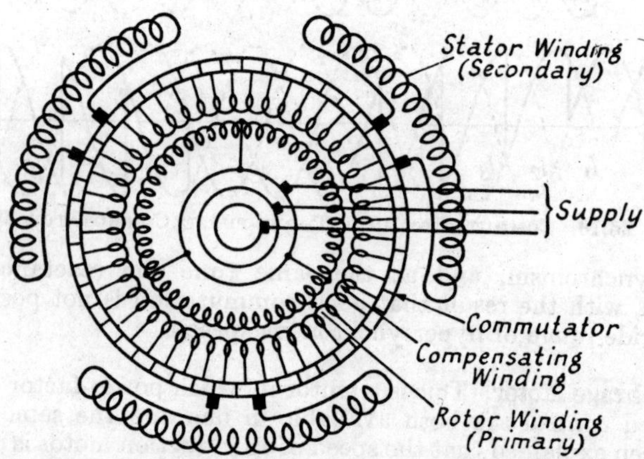

FIG. 26.17. CONNECTION SCHEME OF SCHRAGE MOTOR AND ARRANGEMENT OF STATOR AND ROTOR SLOTS

giving two separate brush sets, and these are arranged so that the brushes as a whole can be rocked round the commutator, or one set can be rocked relative to the other. Each phase of the stator (secondary) winding is connected to the brushes of one set according to the scheme of Fig. 26.17.

There are three possible modes of action illustrated in Figs. 26.18 (*a*), (*b*), and (*c*). In (*a*) the two brushes *a* and *b* of a set are in line with one another, so that at every instant they will be in direct electrical contact through a commutator bar. Hence each secondary phase is short-circuited and the motor will run like a plain induction motor.

In (*b*) the brush sets are opened out and the connections are such that the commutator p.d. between the points *a* and *b* will have a component in direct opposition to the secondary induced e.m.f., so that the motor will run at subsynchronous speed. In (*c*) the position of the brushes *a* and *b* is reversed, so that the e.m.f. picked off at the commutator is in phase opposition to that of (*b*). Hence this e.m.f. will have a component in phase with the rotor induced e.m.f., and the motor will run at hypersynchronous speeds. For any given setting of the brushes the motor will have a shunt characteristic.

Although the magnitude of the e.m.f. between the brushes *a* and *b* is dependent on their relative angular positions, the phase of the e.m.f.

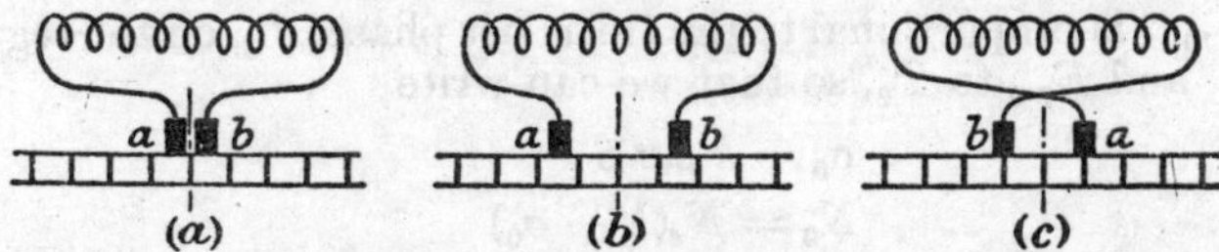

Fig. 26.18. Methods of Operation of the Schrage Motor

depends upon the position of the centre line with respect to the centre of the rotor winding. Hence if the brushes as a whole are rocked round the commutator, the phase of the e.m.f. injected into the secondary circuit will be varied relative to the rotor induced e.m.f., and therefore the power factor of the motor will be varied. Hence the speed depends on the angular distance between the individual brush sets *a* and *b*, while the power factor depends upon the angular position of the brushes as a whole.

In the case of Fig. 26.18 (*c*) there will be a certain angular distance between brushes *a* and *b* for which the motor runs at exactly synchronous speed. The field speed will then be zero, or, in other words, the field will be stationary in space as in a d.c. machine. The compensating winding will therefore act like the armature of a d.c. generator, and it will deliver a direct current to the secondary winding.

Theory of the Schrage Motor. Consider the machine at standstill. The rotating field will induce the following e.m.f.s—

(1) in the primary (rotor) an e.m.f. E_1 which in a machine without internal volt drops is equal and opposite to the applied p.d. V_1.

(2) A secondary e.m.f. $E_{2\cdot0}$.

(3) An e.m.f. E_k in the compensating, or regulating winding.

If the p.d. between brushes 180 electrical degrees apart is $E_{k.\max}$, then for any brush position β with respect to the neutral position, we have

$$E_k = E_{k.\max} \sin \beta$$

since, as with the rotary convertor, the p.d. between any pair of brushes is proportional to the length of the chord joining them.

Now let the rotor travel at any speed whatever. The relative speed of the motor field to primary and regulating winding remains equal to the synchronous speed and therefore E_1 and E_k are unchanged. The magnitude of E_2 changes, being given by

$$E_2 = \sigma E_{2\cdot 0} \text{ as in a plain induction motor}$$

When $\sigma = 0$, $E_2 = 0$, and when σ is negative, E_2 is reversed in phase.

Suppose that E_k is exactly equal and opposite to E_2, then no current flows in the secondary and this condition is therefore the light-load condition. Let σ_0 and N_0 be the light load slip and speed, then

$$\sigma_0 E_{2\cdot 0} = E_{k.\max} \sin \beta$$

$$\sigma_0 = (E_{k.\max}/E_{2\cdot 0}) \sin \beta$$

Now $E_{k.\max}$ is proportional to the turns per phase, T_3, on the regulating winding, and $E_{2\cdot 0}$ to T_2, so that we can write

$$\sigma_0 = k \sin \beta$$

and

$$N_0 = N_s(1 - \sigma_0)$$

$$= N_s(1 - k \sin \beta)$$

The lowest possible light-load speed is thus

$$N_{0.\min} = N_s(1 - k)$$

and is given by $\beta = 90°$. If the brushes are gradually brought together $\sin \beta$ decreases and so the speed increases until, when $\beta = 0$, and both brushes are at the neutral axis, $N_0 = N_s$. If the separation is now increased in the opposite sense, β becomes negative, and the speed becomes greater than synchronism. The maximum possible speed above synchronism is thus

$$N_{0.\max} = N_s(1 + k)$$

A very commonly designed range is from $N_{0.\min} = N_s/2$ to $N_{0.\max} = 1{\cdot}5\, N_s$, that is, a range of 1 : 3. We then have

$$(1 - h)/(1 + h) = \tfrac{1}{3} \quad \therefore\ k = \tfrac{1}{2}$$

If the motor is put on load a current I_2 flows in the secondary and regulating windings, and since these two are in series and form a closed circuit their m.m.f.s are in opposition. The combined ampere-turns are therefore

$$I_2(T_2 - T_3 \sin \beta)$$

The primary induced current I_1' is therefore given by

$$I_1'T_1 = I_2(T_2 - T_3 \sin \beta)$$

$$I_1' = I_2[(T_2 - T_3 \sin \beta)/T_1]$$

$$= I_2(T_2/T_1)[1 - (T_3/T_2) \sin \beta]$$

$$= I_2(T_2/T_1)(1 - k \sin \beta)$$

This current combined with the primary no-load current I_0, gives the total primary current I_1 as in a plain motor.

On no-load, since the resultant voltage in the secondary is zero, we have $E_2 = -E_k$. If σ_0 is the slip on no-load we also have $E_2 = \sigma_0 E_{2\cdot0}$. On load there is a resultant voltage given by

$$E_{\text{resultant}} = E_k + E_2 = -\sigma_0 E_{2\cdot0} + \sigma E_{2\cdot0}$$
$$\therefore \sigma - \sigma_0 = E_{\text{resultant}}/E_{2\cdot0}$$

Now the torque is proportional to I_2, and I_2 is proportional to $E_{\text{resultant}}$, also $E_{2\cdot0}$ is a constant. Hence the fall in speed is constant for a given torque and is not a function of σ_0. Thus if a motor is worked at constant torque its slip will be independent of the no-load speed to which it is adjusted.

In Fig. 26.19 the total angular separation, 2β, of the brushes is symmetrical with respect to the neutral axis with the result that the

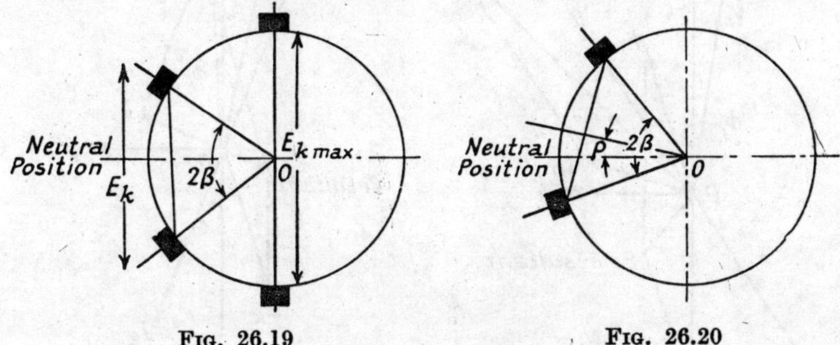

Fig. 26.19 Fig. 26.20

whole of the voltage $E_k = E_{k.\text{max}} \sin \beta$ is utilized for speed regulation If there is an angular shift ρ of the brush pair as a whole, as in Fig. 26.20 then the speed regulating component becomes

$$E_s = E_{k.\text{max}} \sin \beta \cos \rho$$

while the quadrature component

$$E_p = E_{k.\text{max}} \sin \beta \sin \rho$$

is a phase compensating component which acts exactly in accordance with the vector diagram of Fig. 26.8. Thus phase compensation is possible, but only at the expense of a reduction in the speed range.

The vector diagram of the motor for both sub- and hyper-synchronous speeds is given in Fig. 26.21. The difference between the two diagrams is due to the fact that as the speed passes through synchronism E_2 becomes zero and then reverses in phase, thus coming into phase with E_1 instead of in opposition to it as at subsynchronous speeds. Also, the frequency of the regulating voltage E_k changes, although the r.m.s. value remains constant. At synchronism the frequency is zero and the regulating winding acts like a d.c. generator and delivers direct current to the secondary. At hypersynchronous speed the frequency is, in a sense, negative, so that the phase sequence of the machine is reversed. To attain this condition the brushes are crossed over as shown in Fig. 26.18, and therefore the vector of E_r is also reversed. Owing to the negative slip the inductive reactance of the secondary becomes negative, and the current I_2 leads the resultant secondary voltage.

At exact synchronism the machine behaves, in a sense, like a synchronous motor, the regulating winding functioning as a d.c. exciter. There is, however, nothing to hold the machine in step because this mode of operation is only one of many, all of which can be passed through smoothly.* In the synchronous motor the exciter is external

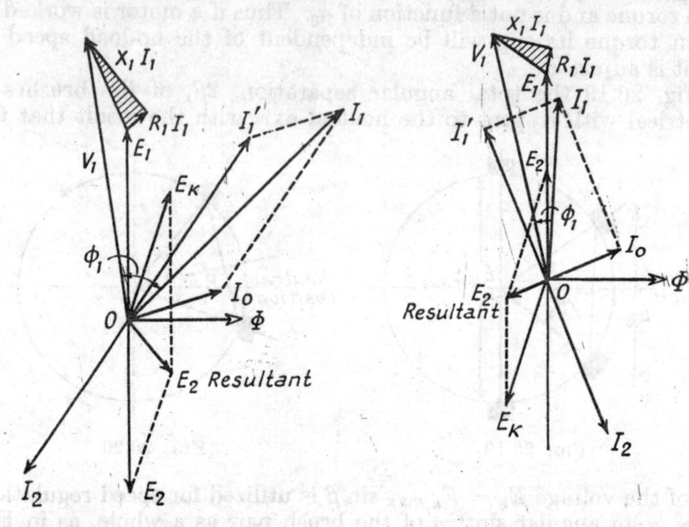

FIG. 26.21. VECTOR DIAGRAMS OF THE SCHRAGE MOTOR FOR SUBSYNCHRONOUS AND HYPERSYNCHRONOUS SPEEDS

and the poles are fixed with respect to the field system, even on a cylindrical rotor machine. In the Schrage motor, synchronous running is merely operation with $\sigma = 0$ and the slightest change in operating conditions will cause the field to slip in one direction or the other.

Cascade Control with Commutator Machine as the Rear Machine. (*a*) WITH ROTARY CONVERTOR. It will be seen that in the ordinary cascade control, speed variation is obtained by utilizing the slip energy of the main rotor in a second motor, instead of wasting it in a rheostat. In the Kramer method this slip energy is first converted into direct-current energy by means of a rotary convertor (Fig. 26.22), and it is

* This does not apply at synchronism at which the effects of space harmonics of flux and unbalance in the secondary circuits give rise to superimposed speed oscillations and primary-current hunting. *See* I. Thomas, "Second-order Torque Components in the Schrage Motor operating at Synchronous Speeds," *J. Inst. Elect. Engrs.*, Pt. C, **103,** 1956, p. 11.

then utilized in a direct-current motor which is direct coupled to the main motor. The speed is controlled merely by altering the excitation of the direct-current motor. This alters the back e.m.f., which, being approximately the same as the voltage at the commutator of the rotary alters the voltage at the slip-rings of the latter. This slip-ring voltage is injected into the rotor circuit of the main motor and causes the required reduction in speed. The advantage of this method is that any speed within the working range can be obtained, instead of only two or three speeds by the other methods. Also, if the rotary convertor is designed to operate with high excitation, it will take a leading current, and thus considerably improve the power factor of the set.

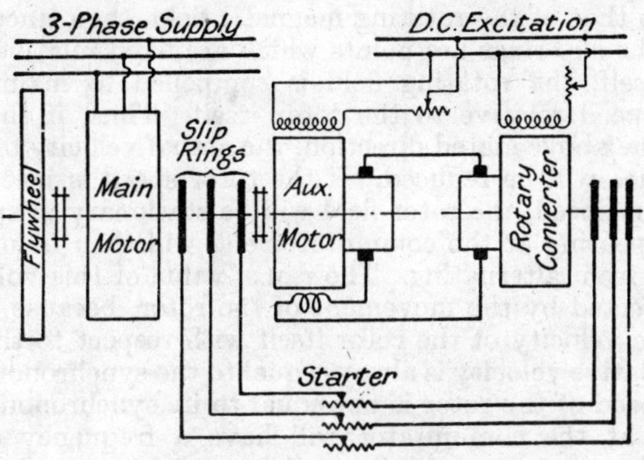

Fig. 26.22. Rotary Convertor in Cascade with Induction Motor

The Kramer system can be used with a flywheel load equalizer if means are provided to reduce the speed automatically when the load increases. This is done very simply by compounding the auxiliary d.c. motor. An increase in load causes an increase in rotor current in the induction motor, and this in turn passes through the rotary to the direct-current motor, whose back e.m.f., as a result of the increased series excitation, increases. This, as explained previously, brings down the speed, so enabling the flywheel to give up some of its stored energy. With a load equalizer of this kind the system is frequently used for driving continuously-running rolling mills. Without the load equalizer, and with a shunt-excited auxiliary motor, it is used for driving mine ventilating fans.

(*b*) With Auxiliary Polyphase Commutator Machine. Imagine an ordinary induction-motor rotor, without a stator, provided with the usual slip-rings and supplied with alternating current from an external source. If the rotor is stationary, a rotating magnetic field will be set up, this field travelling at synchronous speed with respect to the rotor, and therefore at synchronous speed in space. The field will induce in the rotor a back e.m.f. in a manner similar to the production of a back e.m.f. in the primary winding of a transformer on open circuit, and the current which flows will be practically wattless, and its magnitude will be limited almost entirely by the standstill reactance of the rotor windings. If the rotor is provided with a commutator as well as with slip-rings, the arrangement then being that of the rotor of the rotary convertor (*see* Chapter 24), there will be an e.m.f. of supply frequency between any brushes pressing on the commutator. This e.m.f. will not vary in magnitude if the brushes are rocked round the commutator and the angular spacing of the various brushes kept constant, but it will obviously vary in time-phase whenever such movement occurs. In fact, if the brushes are capable of movement throughout two pole pitches, phase displacement through 360 electrical degrees will be possible.

Now suppose that the rotor is made to rotate, say, in a direction opposite to that of the rotating magnetic field, then since the tapping points to the slip-rings are points which are fixed relative to the rotor winding itself, the rotating field is compelled to maintain its synchronous speed relative to the rotor itself. Thus, if the rotor itself moves in the above stated direction, the actual velocity of the rotating field in space will be reduced. If the rotor speed is just equal to the synchronous speed, the rotor field will be stationary in space, and the voltage appearing at the commutator end will then be unidirectional, as distinct from alternating. The r.m.s. value of this voltage is, however, unaffected by the movement of the rotor, because it is fixed by the relative velocity of the rotor itself with respect to the rotor field, and this relative velocity is always equal to the synchronous velocity.

If the speed of the rotor is not equal to its synchronous speed, then the e.m.f. at the commutator will have a frequency equal to the difference between the actual and the synchronous speeds, so that in this respect the appliance becomes a frequency convertor. In this form it can be connected in cascade with a large induction motor, and used for the purpose of speed regulation. There are three points to notice in connection with this frequency convertor—

(1) The frequency of the alternating voltage appearing at the commutator is not proportional to the speed of the convertor, but to the difference between its speed and the synchronous speed.

(2) For a given angular spacing of the brushes, e.g. three brushes spaced 120° apart for a three-phase two-pole machine, the e.m.f. is independent of the brush position.

(3) The time phase of this e.m.f. can be altered by rocking the brushes round the commutator.

Although we have only considered three-phase current fed to the slip-rings in the above discussion, the number of phases of converted-frequency current taken off at the commutator can be anything we like, being fixed simply by the number and spacing of the brushes.

Now consider a winding having a commutator and having the alternating-current supply brought to the commutator instead of to the slip-rings. It is now to be noted that, as long as the brushes are set in a definite position, the current is supplied to the rotor, not at points which are fixed relative to the rotor winding as when slip-rings are used, but at points which are fixed in space. This is so because, the brushes being fixed, the commutator puts them in electrical connection with the various rotor coils in succession. The result of this arrangement is that the rotor field always travels in space at synchronous speed, no matter what the actual speed of the rotor itself may be. It is therefore possible for a rotor fed in this manner to run at any speed inside a polyphase stator and still have a unidirectional torque acting on it, and this, in brief, is the mode of operation of the polyphase commutator motor. In fact, the rotor can run in either direction, because this direction does not influence the direction of the rotor field, which is decided only by the electrical phase sequence of the brushes pressing on the commutator.

The actual frequency of the currents flowing through the rotor windings is, of course, the slip frequency, this condition holding, no

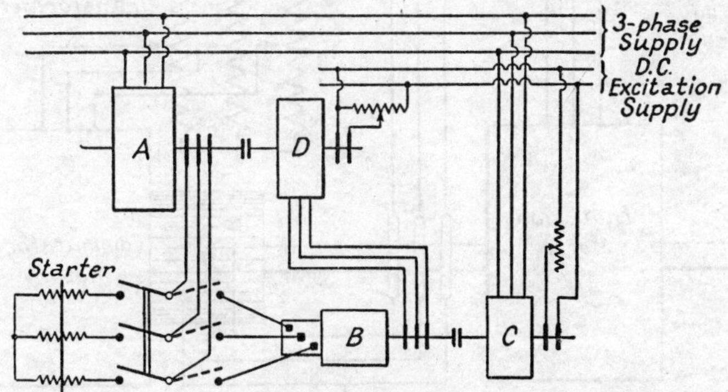

FIG. 26.23. SPEED CONTROL OF LARGE INDUCTION MOTOR BY FREQUENCY CONVERTOR

matter how the current is supplied to the rotor, i.e. whether induced as in the plain induction motor, or whether supplied to a commutator from an external source. Hence, with this method of supply also, the commutator acts as a frequency convertor.

It is also to be noted that in the case of the rotor of an induction motor, the commutator need not be part and parcel of the rotor construction, i.e. not integral with the machine as in the case of a direct-current armature and its commutator, since the action is exactly the same if the commutator and the winding connected to it constitute an entirely separate armature driven by a small motor. A separate frequency convertor of this type can be applied to the speed control of a large induction motor, one such scheme being given in Fig. 26.23. *A* is the main motor which, after being started up by an ordinary resistance starter, has its rotor winding connected to a frequency convertor *B*. This, as explained above, is a rotor provided with both commutator and slip-rings; a stator winding is not required, but the iron part of the stator is provided so as to give a low reluctance path for the magnetic flux. It will be seen that the frequency convertor is fed with alternating current from the rotor of the main motor at its commutator end. The current taken off at the slip-ring end of this convertor is used as supply for a synchronous motor *D*, which is direct-coupled to the main motor A. The frequency convertor is driven by another synchronous motor C. Suppose the main motor is running with a fractional slip of σ, then the frequency of the currents supplied to the commutator end of the frequency convertor is σf, with the result that the rotating field of this machine will rotate at a speed of σN_s. The actual rotor speed due to the synchronous motor drive is N_s, the relative speed of rotor to rotating field thus being $N_s(1 - \sigma)$. Hence the frequency of the current taken off at the slip-rings is $f(1 - \sigma)$

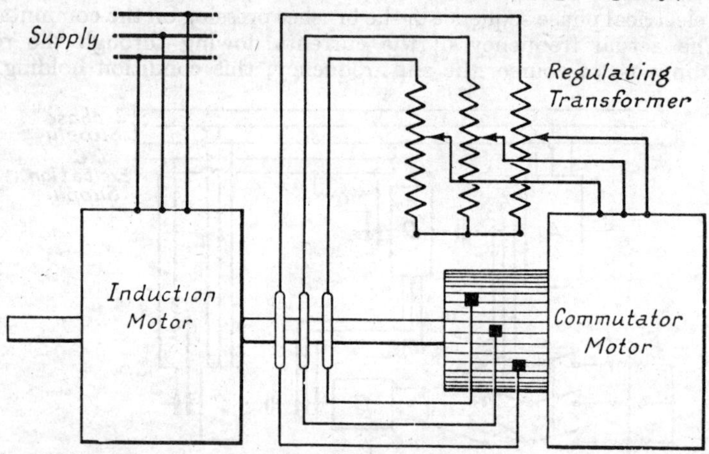

FIG. 26.24. INDUCTION MOTOR CASCADED WITH COMMUTATOR MACHINE (KRAMER SYSTEM)

and the speed of the synchronous motor D is $N_s(1 - \sigma)$. But this is also the speed of the main motor A to which it is coupled, showing that the frequency convertor used in the above manner supplies D with current of the correct frequency. The speed is obtained, as in the Kramer control, by varying the excitation of D, while power-factor control is obtained by varying the excitation of the synchronous motor C.

The stator-fed commutator motor, both with shunt-connected stator winding and series-connected winding, have already been discussed. Such motors can be cascaded with a main induction motor, and, for this application the stator can be shunt, series, or compound excited, as in a d.c. motor. The method of connecting a shunt commutator motor in cascade with an induction motor is shown in Fig. 26.24. It will be seen that the stator is supplied through a regulating transformer, and the speed of the set is adjusted by means of this transformer. The effect of an alteration of the number of turns on the secondary winding is to alter the back e.m.f. of the commutator motor, and thus to alter the speed of the set. The speed/torque characteristic of a set of this kind is fixed by the characteristics of the commutator motor, being similar to that of a shunt or compound motor, according as the commutator motor is shunt or compound wound.

(*c*) WITH SCHERBIUS MACHINE. The Scherbius machine is essentially a dynamo having an armature with commutator, and a stator carrying its own exciting winding. Irrespective of the speed of the armature the nature of the p.d. at the brushes is decided by the nature of the exciting current. If this is d.c. then the brush p.d. will be unidirectional, if it is a.c. then the brush p.d. will be alternating, of the same frequency. Hence if the machine is excited at the slip frequency of an induction motor the commutator can be connected to the slip-rings of the motor and the e.m.f. thereby injected into the rotor used for the purpose of speed control. The fundamental requirements for cascading are therefore—

(*a*) That the Scherbius machine shall be excited at the slip frequency of the main motor and that this condition shall prevail under all conditions of operation.

(*b*) That there shall be control over the magnitude of the injected e.m.f. in order that there may be control of the speed of the main motor. As this e.m.f. is a function of the exciting current it follows that this control is analogous to the control of the voltage of a d.c. generator by change in exciting current.

Compensation of the armature reaction m.m.f. is necessary for two reasons—

(*a*) This m.m.f., if operative, would induce a voltage in quadrature with the current, and this voltage, superposed on the output voltage would modify the operation partly because of its phase and partly because it would vary with the current.

(*b*) To limit the kVA supplied for excitation it is necessary to reduce the reactive volt drop in the exciting winding as far as possible. This drop depends on the total ampere-turns and therefore on the length of the air-gap, which must therefore be made as small as possible. This requires that ampere-turns required to neutralize the armature reaction must be provided by a compensating winding and not by the main exciting winding.

The brush positions of the Scherbius machine are not varied and therefore the reactance voltage can be neutralized by means of interpoles exactly as in the case of a d.c. machine. Because the brush positions round the commutator correspond to a spacing of 120 electrical degrees, the machine must have three interpoles per pole pair instead of two, as in a d.c. machine. This gives a pole pitch of 120 degrees instead of the usual 180 degrees. For the same reason the coil width of an armature coil is also 120 degrees. The arrangement of the machine is illustrated in Fig. 26.25.

As so far described the machine is used for speed control only when connected in cascade with a main induction motor, its output voltage therefore being the e.m.f. E_s. If phase compensation is also required the stator can be provided with still another exciting winding. This shifts the phase of the output voltage and therefore gives it a quadrature component corresponding to the e.m.f. E_p.

(*a*) *Single-range Regulation.* If speeds below synchronism only are required all that is necessary is that an e.m.f. E_s of slip frequency shall be injected into the rotor of the main motor in opposition to the induced e.m.f. E_2. This is provided by a Scherbius machine whose exciting winding is fed from the rotor of the controlled motor. The connection scheme is shown in Fig. 26.26 in which the Scherbius

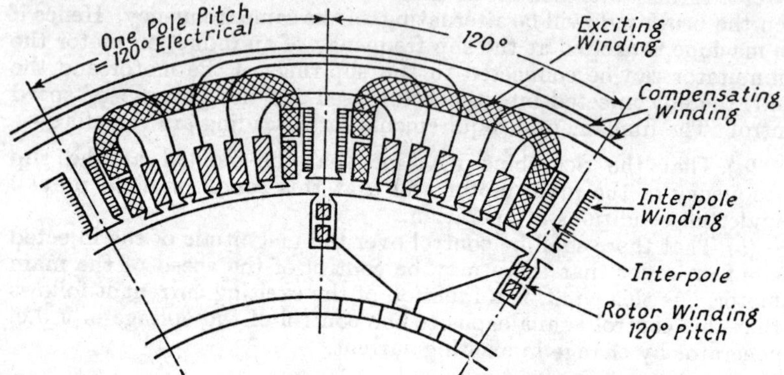

Fig. 26.25. Arrangement of Scherbius Machine

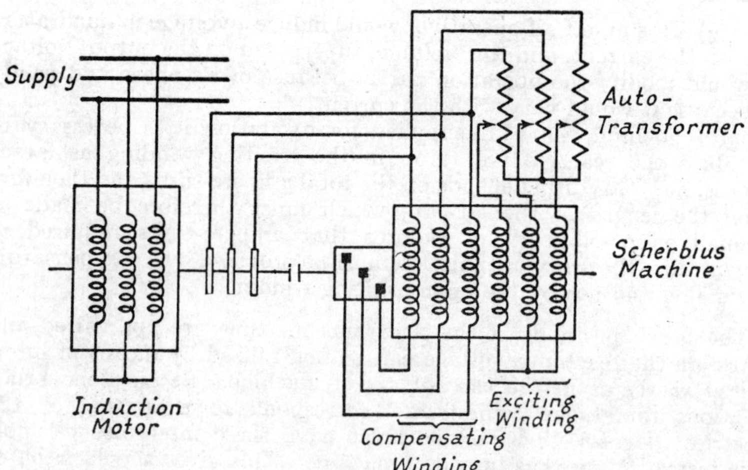

Fig. 26.26. Single-range Regulator

machine is shown direct-coupled to the main motor. An alternative arrangement is to separate the Scherbius machine and direct-couple it to a second induction motor whose stator is connected to the supply. The power received from the main induction motor rotor by the Scherbius machine causes it to drive the second induction machine above synchronism, the regulating power, less losses, thus being returned to the line. This is a common method since the Scherbius machine and the second induction machine can be high speed and therefore comparatively inexpensive machines. The speed is regulated by means of taps on the auto-transformer.

The terminal voltage of the Scherbius machine is clearly equal to the volt drop in the exciting winding, and owing to the preponderance of reactance over resistance in the exciting winding it follows that the armature induced voltage balances the reactive induced voltage in the exciting winding. This leads the current by 90° and therefore a phase displacement of the order of 90° is necessary. It is most easily accomplished by providing each pole with two exciting windings, as shown in Fig. 26.27 and connecting these so that the excitation for phase 1 is provided by phases 2 and 3, and so on. This is illustrated diagrammatically in Fig. 26.27, which shows the exciting windings only.

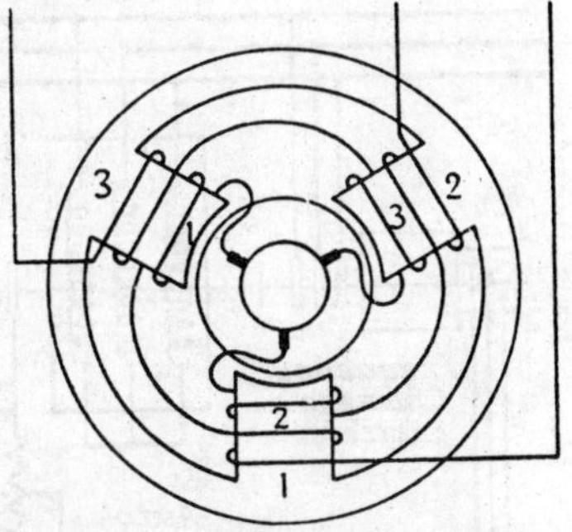

Fig. 26.27. Connection Scheme of Exciting Windings on Scherbius Machine

The exciting current is equal to the quotient of the main rotor induced e.m.f. and the exciting winding impedance. As this impedance is mainly reactive it follows that this quotient is sensibly constant except near to synchronism. Hence if the set is running at any particular speed, the induced e.m.f. of the Scherbius machine is constant for deviations on either side of this speed, while the main rotor e.m.f. is proportional to the slip. There is equilibrium at the slip for which both e.m.f.s are equal. Thus the Scherbius machine acts as a check on any tendency of the main motor to vary its speed with the result that speed variation with change in load is exceedingly small.

By providing series, as well as shunt, excitation the Scherbius machine can be compound wound. As this machine imposes its characteristics on the main motor it follows that, by compounding, a drooping speed characteristic of the main motor can be obtained and load equalization by flywheel adopted if desired.

(*b*) *Double-range Regulation.* Imagine a Scherbius regulated set working at constant power, the Scherbius machine being direct coupled to the induction motor as in Fig. 26.26. Suppose that the range of speed is from 250 to 125 r.p.m., then at the bottom speed each component provides one-half of the total power, the full-load power of the Scherbius machine thus being one-half the full-load power of the induction motor. Now suppose that the induction motor has a synchronous speed of 187·5 r.p.m., then the bottom speed of 125 r.p.m. can be attained by using a Scherbius machine of only half the power of the previous machine. If the induction motor can, at the same time, be run through synchronism and then brought up to 250 r.p.m. by this same Scherbius machine the same output and the same speed range have been obtained with a much smaller and cheaper

Scherbius machine. This is the justification for double-range regulation, which provides hyper-synchronous as well as sub-synchronous speeds.

In order that the motor may be run through synchronism it must first of all be brought up to synchronism, and the energy to do this must come from outside, that is, there must be something additional to the equipment of Fig. 26.26. An induction motor "slips" because of the rotor I^2R loss and therefore one function of the additional appliance is the supply of this loss: in this respect it acts as an ohmic-drop

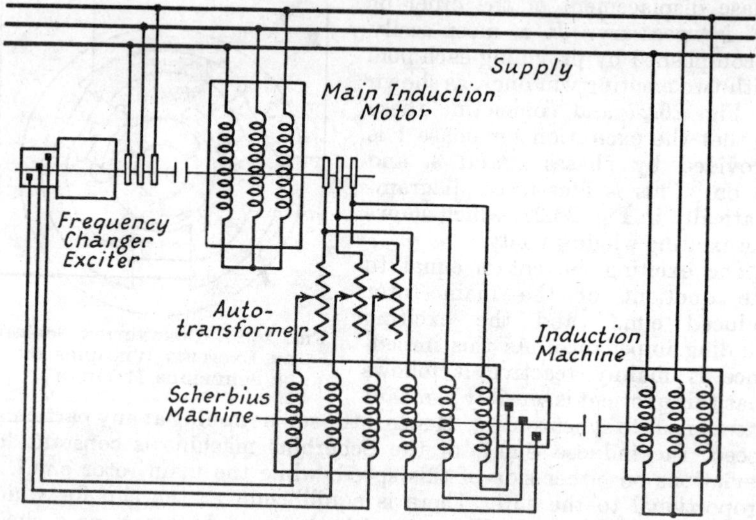

FIG. 26.28. CONNECTIONS FOR DOUBLE-RANGE REGULATION

exciter. A second function is the supply of exciting current to the Scherbius machine, since the rotor output of the induction motor is zero at synchronism. The appliance used is a frequency changer of the susceptor type driven from the main motor through gearing which steps up the speed in the ratio, poles on the frequency convertor to poles on the main motor. Its slip-rings are connected to the supply and its rotation is in the opposite direction to its own field, so that the field speed in space is the slip speed. Thus the p.d. at the commutator is of slip frequency. In order to inject the e.m.f. into the exciting winding the star-point of the latter is opened, as in Fig. 26.28, in which the Scherbius machine is shown driving an auxiliary induction machine. When running at sub-synchronous speeds the frequency convertor is cut out of circuit, the connections then being the same as for single-range operation. At hyper-synchronous speeds the Scherbius machine acts as a generator, supplying power to the main motor and consequently the auxiliary induction machine functions as a motor.

Control Equipment for Scherbius Machines. The circuit diagrams of Figs. 26.26 and 26.28 are reduced to their simplest terms and therefore give no indication of the type of equipment used for speed control by means of the Scherbius machine. Fig. 26.29* gives a schematic diagram of a single-range set in which the winding represented by the thick line is the main stator winding of the Scherbius machine, this winding being the distributed winding housed in the slotted pole faces. The winding

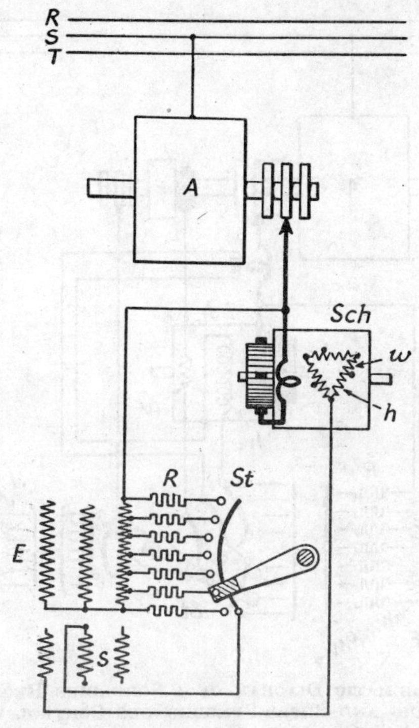

Brown Boveri)

FIG. 26.29. SCHEMATIC DIAGRAM OF A SCHERBIUS REGULATING SET FOR SUB-SYNCHRONOUS CONTROL, WITH SHUNT CHARACTERISTIC

The exciting transformer E is connected to the slip rings of the induction machine A, and feeds the exciting winding h of the Scherbius machine *Sch*. The speed is controlled simply by setting a suitable voltage on the exciting transformer, by means of the tapping switch *St*.

h is the exciting winding for the main poles, and w the exciting winding of the commutating poles. The slip-ring voltage of the main induction motor, which is equal to the terminal voltage of the Scherbius machine, is fed to the exciting transformer E, of the auto type, as explained previously. This has a large number of tappings, and the voltage tapped off is fed to the exciting windings. The p.d. applied to the exciting transformer is practically the same as the slip voltage,

* *Brown Boveri Rev.*, No. 5/6 (1951) 131; No. 7 (1952) 239; and No. 10/11 (1947) 191.

and since the magnitude of this voltage, and its frequency, are proportional to the slip, the ratio of voltage to frequency is a constant and so the flux of the transformer is practically constant.

In series with the secondary circuit of the exciting transformer is an additional secondary S. The windings connected in series are not

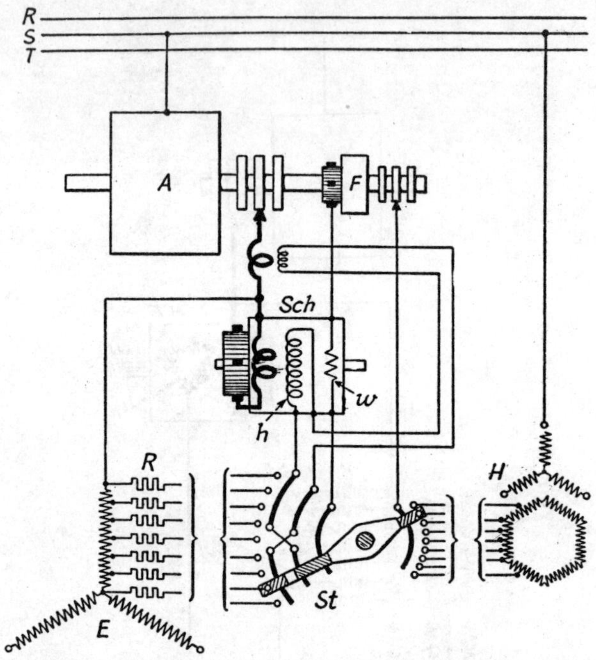

FIG. 26.30. SCHEMATIC DIAGRAM OF A SCHERBIUS REGULATING SET FOR SUB- AND SUPER-SYNCHRONOUS CONTROL WITH COMPOUND CHARACTERISTIC

The distinguishing feature here is the additional compound transformer T; this influences the excitation as a function of the rotor current.

on the same limb and the effect of S is therefore to cause phase displacement, its object being to impart the desired phase angle to the voltage fed to the Scherbius machine. A second method of producing a phase displacement is to insert a resistor R in the exciting circuit, the value of R being appropriate to the step to which it is connected. The losses in these resistors are very small. The above devices constitute a means of giving the main motor a high operating power-factor, and this is maintained over the whole of the range except near synchronism, where the slip e.m.f., and therefore the e.m.f. of the Scherbius machine, are so low that the effect on the main motor is very small.

Fig. 26.30 gives a schematic diagram of a Scherbius regulating set for sub- and super-synchronous speeds. The current transformer T, whose secondary is connected in series with the main exciting winding h, gives the main motor a compound characteristic; and the equipment is then suited to heavy-duty flywheel drives, e.g. continuous-running rolling mills. If this transformer is omitted then the main motor has a shunt characteristic. As in Fig. 26.29 the main additional equipment for double range operation is the frequency changer (ohmic drop exciter) F. The tapping switch St is a development of the simple switch of Fig. 26.29, since it has positions for sub- and super-synchronous speeds, the change-over from one to the other being through the reversing connections to the main exciting winding h. Again, the tapping switch has to select the appropriate voltage from the frequency-changer. This e.m.f. must correspond to the desired phase angle, and this is ensured by tappings on the polygon winding of an auxiliary transformer H which supplies voltage to the slip-rings of F. The resistors R in the tappings from the exciting transformer E are used for phase control and the additional secondary winding S of the single-range set is not required.

As explained previously, one advantage of the double-range over the single-range control is that a smaller Scherbius machine will suffice for the duty. There are two other alternative advantages: (*a*) the entire control range, which is fixed by the ratio of minimums to maximum speed is capable of enlargement, or (*b*) with a Scherbius machine of a certain capacity it is possible to control an induction machine of approximately double the output. There is an economic limit to the slip, and for a supply frequency of 50 c/s this is reached at a slip of about 30 to 40 per cent. Where the desired control range by single-range method reaches such a value the double-range method has to be employed.

Plate V (facing p. 481) shows the stator of a Scherbius machine, on which the various windings can be seen.

Design Limitations of 3-phase Variable-speed Commutator Motors. We have seen that commutating poles can be used in connection with certain types of a.c. commutator machines, both single-phase and polyphase, the reason being that the commutation zone is fixed in space. With the two general industrial types of variable-speed 3-phase motor, commutating poles are not possible: (*a*) in the stator-fed machine, this being controlled by a separate induction regulator, and (*b*) in the Schrage machine, because speed control involves the actual shifting of the brush axes.

Just as design limits were extended, and therefore increased outputs for a given weight of active materials were obtained for d.c. machines by the use of commutating poles, so also these advantages can be obtained for polyphase commutator machines if a means can be found of ameliorating the commutation conditions. In the case of d.c. machines the reactance voltage is neutralized by the rotationally induced e.m.f. due to the commutating field. This e.m.f. exists with commutator motors also, but, as we have seen, the current in a coil during the period of short-circuit by a brush, does not become reversed in value, but changed suddenly from one value to another, the magnitude of the change corresponding to the transfer of current from one phase to the next. In this respect an increase in the number of secondary phases is an advantage for two reasons; (*a*) because an increase in the number of phases reduces the magnitude of the secondary current for a given power, and (*b*) the sudden jump in instantaneous current is reduced. This second advantage is illustrated in Fig. 26.31 which shows

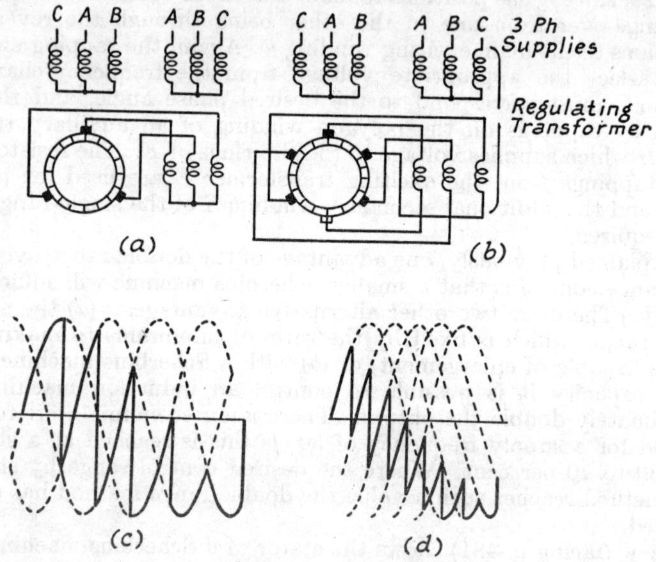

FIG. 26.31. IMPROVEMENT IN CURRENT COMMUTATION BY INCREASING THE SECONDARY PHASES

Stator-fed motor with its rotor connected: (*a*) 3-phase, and (*b*) 6-phase diametral. Instantaneous current in one conductor of the rotor when it is running at 1·33 × synchronous speed, (*c*) 3-phase, (*d*) 6-phase.
(J. E. Parton, *Trans. Inst. Engrs. & Shipbuilders in Scotland*, 1952.)

for simplicity a stator-fed machine (*a*) with three-phase secondary, (*b*) with six-phase, obtained by diametral connection.

There is a second difference (affecting commutation) between the d.c. machine and the three-phase commutator motor. In the d.c. machine the flux is stationary in space: in the commutator machine it rotates in space: (*a*) at synchronous speed in the case of the stator-fed machine, (*b*) at slip speed in the rotor-fed machine. Hence, since for any mode of operation, the brushes of either type are stationary, there exists in every armature coil, including those short-circuited by the brushes, a voltage induced by the main flux of the machine. In the short-circuited coil it is called the transformer voltage of commutation. The magnitude of this voltage is given by

$$E_t = \sqrt{2}\pi k_p \Phi f_a T \text{ volts, where,}$$

k_p = pitch-factor of the coil,

Φ = main flux in webers,

f_σ = frequency of main field relative to the commutator winding,

T = number of turns per segment.

Since, in the stator-fed machine the field speed in space is equal to the synchronous speed, the frequncy f_σ is proportional to the departure of the speed from synchronism. In the rotor-fed machine the value of E_t is practically constant since $f_\sigma = f$. In both machines it is practically independent of the load.

Assuming that the current changes uniformly from one value to the next during the period of short-circuit by the brush, the reactance voltage per segment is given by

$$E_c = L(\Delta I)$$

where L is the effective inductance of the short-circuited coil, and ΔI is the change in current, as in the case of a d.c. machine L is proportional to the square of the turns per coil, but the use of single-turn coils is more desirable in the case of the a.c. machine. The change of current, ΔI, is the brush current, and we see that E_c is proportional to the load current.

For 50 c/s machines the transformer voltage per turn in the case of a full-pitched winding varies from 1·11 V for a small machine of 1 h.p. per pole to 11·1 V for a large machine of 100 h.p. per pole*. It is this transformer voltage which causes the most difficulty, and the available means of combating it are: (*a*) by brush drops in the coil resistance, in the connections between coils and commutator bars (as explained in connection with single-phase motors) and, of course, in brush contact resistance; (*b*) by means of special auxiliary commutating windings in addition to the methods under (*a*). Resistance connections allow the use of a transformer voltage of about 2·6 per segment, but they make an important contribution to the armature heating and may therefore be dangerous if used with motors liable to heavy over-loads.†

The transformer voltage can be reduced by choosing a brush width so that not more than two armature coils are spanned. Hence the maximum possible number of segments is used, the segment pitch being of the order of 4 mm. The brush voltage drop per contact can be between 1 and 2 V.

The special commutating windings are of the nature of discharge or damping windings which consist of coils connected in parallel with the main coils, and so disposed that they are not themselves the seats of induced e.m.f.s. In other words they behave like the non-inductive discharge resistances connected in parallel with highly inductive circuits, such as d.c. field windings, when it is necessary to open the circuit while carrying current. The scheme is illustrated in Fig. 26.32.

* A. C. Lane, "Recent developments in armature winding of polyphase commutator motors," *Proc. B.T.H. Summer School*, (1948).

† *See also* p. 650.

Any energy associated with the inductance of the short-circuited coil, and not dissipated as brush-contact loss, is utilized in the production of I^2R loss in the discharge windings. These discharge windings are

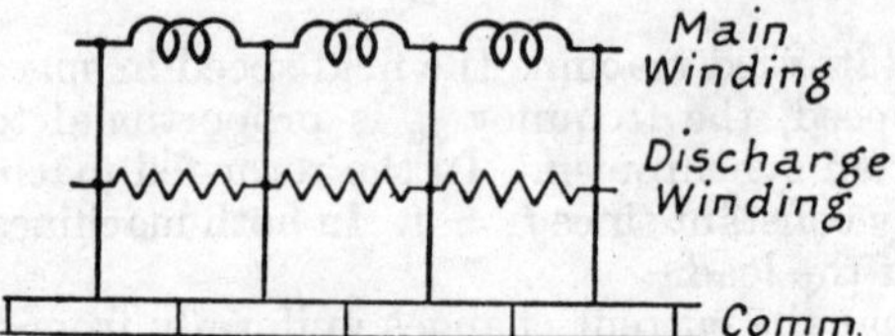

FIG. 26.32. DISCHARGE OR DAMPING WINDING

of three types, illustrated in Fig. 26.33 and described very completely in the paper by Lane.*

(1) *The Robinson discharge winding*, invented by P. W. Robinson. The main armature coils are short-pitched, and the discharge coils over-

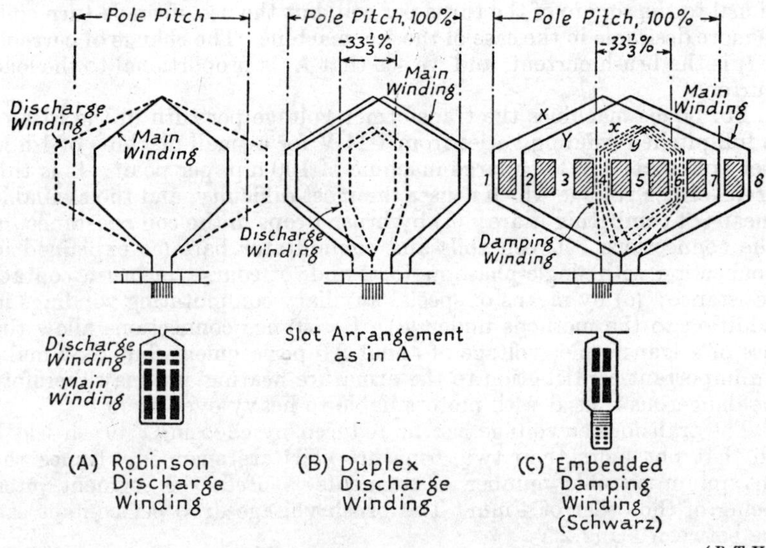

(*B.T.H.*)

FIG. 26.33. THREE TYPES OF DISCHARGE WINDING FOR POLYPHASE COMMUTATOR MOTORS

pitched by the same amount. Fig. 26.33 (*A*) shows the arrangement: it will be seen that the discharge winding is housed in the tops of the slots so that its inductance can be less than that of the main winding, whereas its resistance is made greater. Thus the complete winding consists of two ordinary two-layer windings, one on top of the other. It will be clear that the resultant motionally induced e.m.f., acting

* *See also* B. Adkins, "New auxiliary commutating windings in a.c. variable-speed commutator motors," *B.T.H. Activities*, (Apr., 1941).

round the local path made up of a main coil and discharge coil, is zero.

(2) *The duplex discharge winding*, Fig. 26.33 (*B*). The main winding is a two-circuit duplex winding consisting of full-pitch coils connected to alternate segments. This permits the use of a lage flux per pole without exceeding the permissible voltage between adjacent segments. The discharge winding has two-turn coils of only $33\frac{1}{3}$ per cent pitch, its location being such that the voltage induced by the main flux in one two-turn coil is equal to that induced in a single-turn coil of the main winding. The discharge winding is a simplex winding, so that it performs the additional function of equalizing the potential between the two circuits of the main winding.

(3) *Embedded damping winding*, Fig. 26.33 (*C*), invented by Dr. B. Schwarz.* The main winding is again a duplex winding with full-pitch coils, while the damping winding is simplex with $33\frac{1}{3}$ pitch coils. The damping winding is found at the bottoms of the slots and separated from the main winding by steel shims. Also the windings are so disposed that, whereas two consecutive main coils, like X and Y, lie in adjacent slots, the damping coils in parallel with them, x and y, lie in the same slots as one another. The effect of this arrangement is that every main coil is connected to another main coil in another pair of slots by a small transformer consisting of two auxiliary coils in the same slots. The second main coil thus acts as the discharge circuit for the first main coil, the connection being indirect, via the transformer. This has the advantage that the discharge coil, being a main coil, is the next coil to be commutated, the change of current in it prior to its commutation being such as to assist its commutation.

H.P. per Pole and Flux per Pole. In machines not subject to commutation difficulties, such as synchronous alternators and induction motors, there appears to be no limit, beyond a structural imit, to the h.p. per pole. Thus two-pole alternators of 100,000 kVA, and therefore of 66,700 h.p. per pole at unity power-factor, are in use, while 2-pole induction motors of 1000 h.p. and more are not uncommon. The presence of a commutator drastically reduces the possible h.p. per pole, particularly on a.c. machines. Consider for simplicity the speed regulation of a d.c. motor by means of a series-connected machine which can either buck or boost the supply voltage, V. Let this injected voltage be E_k, then, neglecting voltage drop in the armature circuit, the back e.m.f. is given by

$$E_b = V \pm E_k = V(1 \pm E_k V) = V(1 \pm k)$$

say, where k is the regulating fraction. The speed, being proportional to E_b in a constant flux machine is therefore given by

$$N = N_0(1 \pm k)$$

where N_0 = normal speed with $E_k = 0$.

For a given torque, and therefore for a given current, again in a constant flux machine, the power relationship is

$$P = P_0 \pm P_k = P_0\,(1 \pm k)$$

* *E.U.M.*, (Feb., 1934).

where P_0 is the power at normal speed N_0. Hence a regulating machine of power P_k can vary the motor power P and speed N between the limits $P_0(1 \pm k)$ and $N_0\ (1 \pm k)$, respectively, showing that the greater the speed range the greater the power of the regulating machine. We can write

$$P_{max} = P_0 + P_k$$

and

$$P_{min} = P_0 - P_k = P_{max} - 2P_k$$

$$\therefore \quad N_{min}/N_{max} = P_{min}/P_{max} = (P_{max} - 2P_k)/P_{max}$$

$$\therefore \quad P_k/P_{max} = \tfrac{1}{2}(1 - N_{min}/N_{max})^*$$

Giving N_{max}/N_{min} a series of values we have the following relationship

N_{max}/N_{min} .	1·2	1·5	2	3	5	10	20	50	∞
P_k/P_{max} .	0·083	0·167	0·25	0·333	0·40	0·45	0·475	0·49	0·50

Thus over a small speed range, say, up to 3 to 1, the value of P_k is approximately proportional to the speed range.

The h.p per pole is intimately connected with the allowable value of the transformer voltage E_t. The flux of a transformer is, for a given type, proportional to the kVA, and since an induction motor is fundamentally a transformer it is possible to employ an empirical expression based on this, namely

$$\text{flux per pole } \Phi = k_1\sqrt{\text{H.P.}/pf}$$

where H.P. = synchronous horse-power

p = number of poles

f = supply frequency in cycles per second

and

k_1 = a constant

This gives $\text{H.P.}/p = \Phi^2 f/k_1^2$

We can simplify the equation for E_t to

$$E_t = k_2\Phi f_\sigma$$

$$\therefore \quad E_t^2 = k_2^2\Phi^2 f_\sigma^2$$

$$\text{or} \quad \Phi^2 = E_t^2/(k_2^2 f_\sigma^2) = E_t^2/(k_2^2\sigma^2 f^2)$$

$$\therefore \quad \text{H.P.}/p = E_t^2/(k_2^2\sigma^2 f^2) \times (f/k_1^2)$$

$$= 1/(k_1k_2)^2 \times (E_t^2/\sigma^2 f)$$

The presence of the term f in the denominator shows that, as with single-phase motors, a low frequency is an advantage. The dominant

* J. E. Parton, "Polyphase commutator motors," *Trans. Inst. Eng. and Shipbuilders in Scotland*, Paper No. 1157 (1952).

factor is the term E_t^2, and the following table* shows how the values of H.P./p and of Φ are controlled by the allowable value of E_t. An examination of the figures shows that the value of H.P./p is approximately proportional to E_t^2

H.P. per pole	Φ (weber)	E_t per full-pitched turn
1	5×10^{-3}	1·11
2	7×10^{-3}	1·55
5	$1{\cdot}11 \times 10^{-2}$	2·46
10	$1{\cdot}57 \times 10^{-2}$	3·48
20	$2{\cdot}22 \times 10^{-2}$	4·92
50	$3{\cdot}5 \times 10^{-2}$	7·77
100	$5{\cdot}0 \times 10^{-2}$	11·1

Thielers† gives a somewhat different relationship, as follows—

$$\text{H.P.}_{\max}/p \propto (1 + 1/\sigma_{\max})m(I_2/2p)(E_t/f)$$

This can be deduced as follows—

$$\text{Transformer voltage } E_t \propto \Phi\sigma f$$

$$\text{H.P.} \propto D^2 lNB_{\text{mean}}(ac)$$

where D = rotor diameter

l = rotor axial length,

N = speed,

B_{mean} = mean gap density over pole pitch,

(ac) = ampere-conductors per unit length of periphery.

$$\therefore \quad \text{H.P.}_{\text{synch}} \propto D^2 lN_s B_{\text{mean}}(ac)$$

If $\sigma_{\max}$ is the maximum slip for speeds above synchronism

$$\text{H.P.}_{\max} \propto D^2 lN_s(1 + \sigma_{\max})B_{\text{mean}}(ac)$$

$$\propto (DlB_{\text{mean}})(D(ac))(1 + \sigma_{\max})(f/p)$$

$$DlB_{\text{mean}} = p\Phi$$

$$D(ac) \propto mZ_2 I_2$$

and, since for a given rotor voltage $Z_2 \propto 1/f$

$$D(ac) \propto mI_2/f$$

$$\therefore \quad \text{H.P.}_{\max}/p \propto \Phi(mI_2/f)(1 + \sigma_{\max})(f/p)$$

$$\text{again } \Phi \propto E_t/\sigma f$$

$$\therefore \quad \text{H.P.}_{\max}/p \propto (E_t/\sigma f)(mI_2/p)(1 + 1/\sigma_{\max})$$

$$\propto (1 + 1/\sigma_{\max})m(I_2/p)(E_t/f)$$

* A. C. Lane, *loc. cit.*

† G. Thielers, "The polyphase shunt commutator motor," *A.S.E.A. Journ.* 7 (1930) 98.

This expression also shows that most favourable conditions are given when f is low. The term $(1 + 1/\sigma_{max})$ shows that the output per pole is reduced as the range of speed is increased. The terms m and I must be taken together, since, as one increases the other decreases. The importance of an increase in m is that it decreases the current per brush arm, I/p, and also decreases the reactance voltage, as explained previously. The two expressions for H.P.$/p$, although dissimilar in form, are not contradictory, the first being based on a purely empirical expression in which the constant k_1 contains terms used in the development of the second.

By using special windings of the type described previously Lane states that a Schrage-type motor could be built up to 1000 h.p. with 12 poles, but that such a motor would not compete economically with other methods of obtaining the same speed range at the same output. He makes this comparison with the stator-fed machine—"In the stator-fed motor the transformer volts vary with the speed. Provided then that they do not exceed 2 volts in the working part of the speed range, it would be possible to use higher flux values than those indicated (in the above table). The shorter the range, i.e. the less the departure from synchronism, the higher the flux permissible. The limitation then becomes that set by the permissible transformer volts upon starting and running up to the working range. Due to the short duration of this operation, transformer volts up to 5 or 6 are able to be handled without damage. Therefore a stator-fed motor with a balanced speed range of 3 to 1 or less could use about twice as much flux per pole as a Schrage motor, and with corresponding ability to handle large powers at high speed. The larger the speed range the less is this permissible increase of flux per pole, until, with say 10 to 1 or greater range, there would be no permissible increase."

Further Reading

Adkins, B. and Gibbs, W. J. *Polyphase Commutator Motors.* C.U.P., London, 1953.

Arnold, E. *Die Wechselstromtechnik*, Vol. 5 (2). Springer, Berlin, 1912.

Charlton, O. B. *The Schrage Type A.C. Commutator Motor,* B.T.H. Co. Technical Monograph.

Franklin, P. W. "The Circle Diagram of the Polyphase Brush Shifting Commutator Motor (Schrage Type)." Trans. A.I.E.E., **66,** 1947.

Gibbs, W. J. "The Equations and Circle Diagrams of the Schrage Motor." *J. Inst. Elect. Engrs.,* **93,** Pt. II, 1946.

Liwschitz-Garik, H. and Whipple, C. C. *Electric Machinery,* Vol. II. Van Nostrand, N.Y., 1947.

Olliver, C. W. *A.C. Commutator Motors.* Chapman & Hall, London, 1927.

Richter, R. *Elektrische Machinen,* Vol. 5 (2 and 3). Springer, Berlin.

Teago, F. J. *The Commutator Motor.* Methuen, London, 1930.

CHAPTER 27

THE MERCURY-ARC RECTIFIER OR CURRENT CONVERTOR

The Mechanism of Conduction. Consider an enclosure provided with two electrodes as shown in Fig. 27.1. If a p.d. is applied to the electrodes then current will flow only if there are charged particles capable of

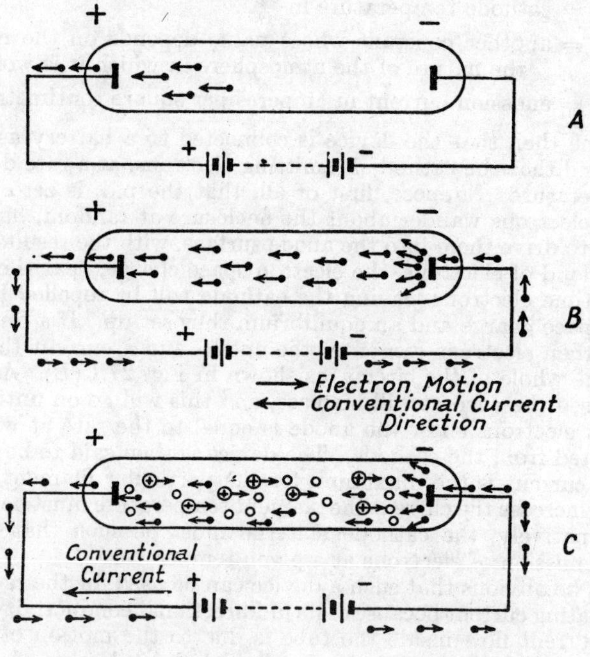

FIG. 27.1. MECHANISM OF CURRENT CONDUCTION IN A TWO-ELECTRODE ELECTRONIC DEVICE

producing current in virtue of their motion along the axis of the enclosure. If the enclosure is completely evacuated then there can be no current. Now let a number of electrons be introduced, as in Fig. 27.1 (*A*), then these electrons will move towards the anode, thereby setting úp a current whose conventional directions is opposite to the electron motion, that is, from anode to cathode inside the enclosure. When the electrons reach the anode they enter it and continue their motion as part of the electron gas of the electrical connections to the anode. Eventually all the electrons will have entered the anode and the tube current will then cease. In order that the tube current may persist it is therefore necessary that new electrons shall be introduced into the enclosure to take the place of those entering the anode, and clearly these must be emitted by the cathode. One method of fulfilling this requirement is to raise the cathode to a high temperature, the method of thermionic emission.

When the current is an "electron current" only, its magnitude is given by

$$I = AT^2 e^{-B/T}$$

where A = a constant of theoretical value 60·2 for most metals

T = cathode temperature in °K

B = another constant whose value depends on the metal and the nature of the atmosphere in which it is working

I = emission current in amperes per square centimetre

Imagine then that the device is connected to a battery as a source of p.d. and that the cathode is emitting electrons, at a rate decided by its temperature. Suppose, first of all, that the p.d. is zero, then the emitted electrons wander about the enclosure at random, but there is no force to drive them into the anode surface, with the result that they form a cloud of electrons, the electron space charge, near the cathode. After a time electrons leaving the cathode will be repelled back to it by the space charge and an equilibrium thus set up. If a finite p.d. is applied then electrons can enter the anode, and a current thus set up round the whole of the circuit, as shown in Fig. 27.1 (*B*). As the p.d. is increased the current will increase, and this will go on until the rate at which electrons enter the anode is equal to the rate at which they are emitted from the cathode. The device is then said to be saturated and the current is the maximum for the particular filament temperature. To increase the current the filament temperature must be increased or, alternatively, the cathode material must be such that it gives a greater emission of electrons at a given temperature.

It will be obvious that such a device can be used for the rectification of alternating current because of its unidirectional conductivity. So long as the current flow inside the tube is due to the motion of electrons only, then the current is very small and the tube impedance high. For the supply by rectification of direct current for traction purposes, or for industry, currents of the order of hundreds of amperes are required, and tube impedances must be reduced to the minimum possible value in order that high efficiency and good voltage regulation may be secured. For this purpose it is necessary to introduce a chemically inert gas or vapour, as, for example, mercury vapour. Now consider the application of a p.d. to a device containing mercury vapour and some free electrons. The mercury atoms in the normal state are electrically neutral, and consequently they have no tendency to move in any particular direction. Thus their motion is the random motion due to the thermal state of the enclosure and also to collisions with other particles. On the other hand the free electrons, acted on by the potential gradient, will acquire a progressively increasing velocity so long as their motion is not impeded. After travelling a certain distance, of average value decided largely by the vapour pressure, collision with a neutral atom will take place, and if the collision is of sufficient violence the valency electron of the atom will be removed, thereby converting the atom to a positive ion, and at the same time producing an additional electron.* There are then three charged particles, namely two electrons and one positive ion, in the place of the original electron, the conductivity of the path between the electrodes thus being greatly increased. The two electrons, acted on by the potential gradient, move towards the anode and when they have acquired sufficient velocity they will again take part in collisions with neutral atoms of sufficient violence to produce ionization. The ions will move towards the cathode at a much smaller velocity because of their mass but will nevertheless contribute to the total current. Thus, whereas the current in a metallic conductor is due to the axial drift of electrons only, the current in a gas or vapour filled electronic device is due to two components (*a*) the electron current and (*b*) the ion current. This is illustrated in an elementary way in Fig. 27.1 (*C*).

It will be obvious that, unless it is deliberately restricted, the process of ionization by collision is cumulative, the number of particles associated with either positive or negative charges increasing at a rapid rate. The phenomenon is called an "electron avalanche," a few of the stages being shown in Fig. 27.2. It is clear that the effect of the electron avalanche is to increase the current at a very rapid rate until conditions of near short-circuit are produced. If the electrodes are not able to deal with the corresponding emission, or acceptance, of electrons then the device will be destroyed, a well-known example being the behaviour of a small neon glow lamp when the limiting resistance is removed. The effect is to allow a rapid increase in conductivity which converts the discharge from a glow discharge to an arc, and so the lamp is destroyed. In the rectifier the maximum conductivity is encouraged. As explained previously, the electrons, when they reach the anode, enter it, and so become part of the external current. Where the enclosure contains electrons only a negative space charge is produced in the neighbourhood of the cathode and this has the effect of producing a saturation effect and so limiting the current. When a gas or vapour is present, some of the ions, when they encounter the electrons, re-combine with them to form neutral atoms again, and in this way tend to eliminate the effect of the negative space charge.†

Again, the ions, if they are in sufficient numbers can form a positive space charge in the immediate vicinity of the cathode, the effect being to pull the electrons from the surface layers of the cathode. The emission of electrons by this means is known as *field emission*, and it is

* Collision processes are discussed in greater detail on p. 823

† Recombination takes place mainly near the electrodes and the inner surface of the enclosure. In the other regions the effect of the negative space charge is partially offset by those electrons which pass sufficiently close to the positive ions.

so effective that the cathode can be operated at a considerably lower temperature than if the emission were due to cathode temperature only. Thus the so-called "cold-cathode" devices depend largely on field emission, but, where there is a gas or vapour present, the temperature may be raised sufficiently high to provide thermionic emission as well, even though the cathode is not deliberately heated by the passage

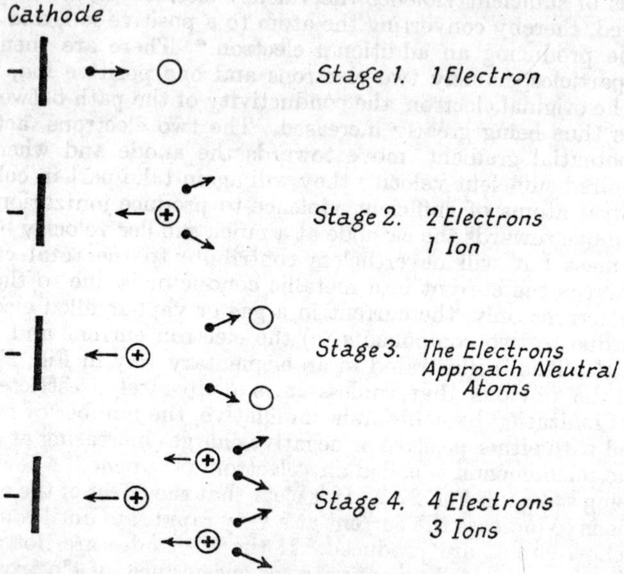

FIG. 27.2. STAGES IN THE PRODUCTION OF AN ELECTRON AVALANCHE

of a heating current. This is because those ions which do not recombine with electrons before they reach the cathode, fall into it at high velocity and thereby raise the temperature by impact. The ions then recombine with electrons at the cathode surface, they do not enter it in the way that electrons enter the anode.

From the above description it will be clear that, so long as the cathode can only emit electrons, but not accept them, and that the anode can only accept electrons but not emit them, the conductivity must be unidirectional and correspond to the conventional direction of current flow within the device from anode to cathode. It is this property which is utilized in the mercury-arc rectifier.

Influence of the Vapour Pressure. The vapour pressure decides the relative magnitudes of the positive ion concentration and electron concentration in the arc path. This relative concentration determines the anode voltage, and for a given anode current, the anode voltage is a minimum when the ion concentration and electron concentration are about equal, the reason being that this condition gives the optimum mutual space-charge neutralization. Fig. 27.3 shows in a general way the effect of vapour pressure on the performance of a gas-filled diode.* Curve 1 is for a vacuum, curve 2 for a very low vapour pressure, curve 3 for a higher vapour pressure. At a pressure giving the characteristic of curve 3, the anode voltage is almost constant and independent of the anode current. This is because of equal ion and electron concentrations, resulting in a small potential gradient in a large proportion of the arc path, extending from the anode almost up to the cathode. This region is called the *plasma* and it consists not only of ions and electrons in about equal concentrations, but also of neutral atoms, and of atoms which have accepted energy increments through collision, but not of sufficient value to produce ionization. These atoms are said to be in an excited state, and it is their return to the normal state which results in the emission of radiation from the plasma.†

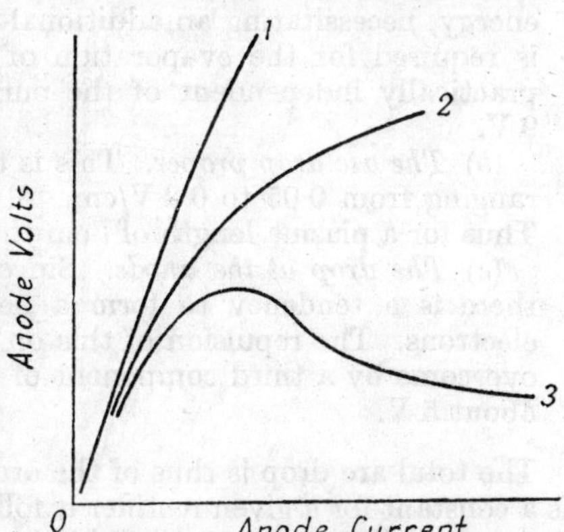

FIG. 27.3. ANODE-VOLTAGE/ANODE-CURRENT CHARACTERISTICS FOR A GAS-FILLED DIODE AT DIFFERENT GAS PRESSURES

The very small potential gradient along the plasma imparts to it a characteristic flexibility; in fact the plasma can be regarded as a flexible connection between anode and cathode, a property which is utilized in the multi-anode rectifiers discussed later in this chapter.

We have seen that ionization can only take place if the collisions of electrons with neutral atoms are sufficiently violent, i.e., if the electrons are accelerated to a sufficiently high velocity. This requires that the anode voltage shall be greater than the ionization voltage. This latter is a characteristic of the gas or vapour, a few values being: helium 24·56, neon 21·56, argon 15·75, krypton 14·00, xenon 12·13, mercury vapour 10·43. These values illustrate the great advantage of mercury vapour. Other advantages are (*a*) the great mass of the ion favours the emission of electrons by the process of cathode bombardment, (*b*) the cathode itself can be a mercury pool and thereby indestructible so long as the enclosure is not destroyed, (*c*) mercury evaporating from the cathode surface condenses on the walls of the enclosure and returns by gravity to the cathode pool. Thus, in contrast to "hot-cathode" devices, the electron emission does not diminish with time.

Arc Volt Drop. The total volt drop between anode and cathode is made up of three components, namely—

* *See* M. J. Druyvestign and F. M. Penning, "Electrical discharges in gases." *Rev. Mod. Phys.* **12** (1940) 87–174.

† *See* p. 822.

(*a*) *The drop at the cathode.* This is largely due to the high potential gradient in the immediate vicinity of the cathode due to the positive ionic space charge, and the product of this drop with the current gives a component of energy which has to be dissipated at the cathode. In addition, as the cathode is a mercury pool, additional energy, necessitating an additional component of the cathode drop, is required for the evaporation of mercury. The cathode drop is practically independent of the current, its value being from 7 to 9 V.

(*b*) *The arc drop proper.* This is the drop in the plasma, its value ranging from 0·05 to 0·2 V/cm. A good average value is 0·1 V/cm. Thus for a plasma length of l cm the plasma drop is 0·1 l volts.

(*c*) *The drop at the anode.* Since the electrons enter the anode there is a tendency to form a negative space charge of waiting electrons. The repulsion of this on approaching electrons has to be overcome by a third component of the total volt drop. Its value is about 5 V.

The total arc drop is thus of the order $(14 + 0{\cdot}1\,l)$ volts, and since l is a constant for a given rectifier it follows that the arc drop is sensibly constant in that it is practically independent of load current.

Construction. The enclosure can be either a glass bulb or a steel tank. The former type is used for relatively small outputs, a limit being set by the intricacy and cost of very large bulbs. The limiting output is therefore of the order of 500 A at 600 V. The cathode is a mercury pool and the anode head can be of fine-grain cast iron or of graphite. We have seen that there must be free electrons present before an arc can "strike" and consequently provision must be made for the necessary supply of these when the rectifier is started up. The process is called "ignition" and it consists merely in producing a small arc close to the cathode surface by means of a special ignition anode. This anode is brought into contact with the mercury pool thereby closing an auxiliary circuit and causing current to flow. This circuit is then broken at the mercury surface, the resulting arc "igniting" the rectifier. This can be performed in various ways—

(*a*) By tilting the rectifier so that the displaced mercury comes into contact with the ignition electrode, and then returning the rectifier to its normal position. This method is now practically obsolete.

(*b*) By lowering the ignition electrode to the mercury surface and then raising it. This is carried out by means of a solenoid.

(*c*) By the "pinch" effect, in which the electromagnetic forces produced by the current itself cause the required break in the circuit.

(*d*) By means of a small mercury jet thrown up from the mercury pool so as to make contact with a fixed ignition electrode.

In addition to ignition, provision has to be made for "excitation," that is the maintenance of an arc under conditions of no-load as far as the output circuit is concerned. This is provided by two auxiliary excitation anodes,* which enable the rectifier to supply a small single-phase load, thereby maintaining the conditions necessary for the creation of an arc to the main anodes when required. For ignition by the pinch effect the ignition electrode is situated close to the mercury pool cathode, as shown in Fig. 27.4, and the additional components of the ignition and excitation circuits are the exciter-ignition transformer, exciter choke and ignition relay. The exciter-ignition transformer has its primary connected to a tertiary winding on the main

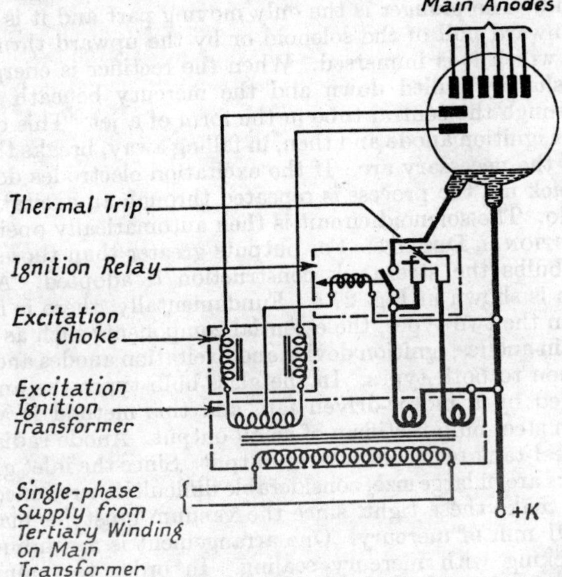

Fig. 27.4. Ignition and Excitation Circuits of Glass-bulb Rectifier

(*Nevelin Electric Supply Co., Ltd.*)

transformer and becomes energized immediately the set is switched on. Voltages are now induced in the two heavy-current secondary windings, one of which supplies the mercury pinch bridge between pool and ignition anode, via the ignition relay contact: the other feeds the thermal strip via the ignition relay and pinch bridge. These two windings cause a heavy current to flow in the pinch bridge in which the conductor, being mercury, and therefore in the fluid state, reduces its cross-section to such an extent that actual rupture takes place, resulting in a heavy spark which creates the necessary conditions for the striking of arcs to the excitation anodes. Before rupture, the mercury bridge can be regarded as made up of a large number of paths in parallel and all carrying current in the same direction at any one instant. Such conductors exert lateral attraction on one-another, and consequently if lateral motion is possible, as with a fluid conductor, the

* It will be seen from the connection scheme of Fig. 27.4 that these are connected as a full-wave rectifier.

cross-section is reduced until rupture takes place. The current now flows in the ignition coil, causing the contacts to open and disconnecting the pinch bridge from its source of supply. If, for any reason, the depth of mercury is too great for rupture to occur, the thermal strip becomes rapidly heated and, in a few seconds operates the trip mechanism and thereby opens the ignition relay contacts.

Mercury jet ignition is illustrated in Fig. 27.5 (*B*). It will be seen that, at the bottom of the mercury pool, immediately below the fixed ignition anode there is a tube with an annular iron plunger actuated by a solenoid. The plunger is the only moving part and it is acted on by the downward pull of the solenoid or by the upward thrust of the mercury in which it is immersed. When the rectifier is energized the plunger is sharply pulled down and the mercury beneath it forced upwards through the central tube in the form of a jet. This closes the circuit of the ignition anode and then, in falling away, breaks the circuit and sets up the necessary arc. If the excitation electrodes do not immediately pick up, the process is repeated through an excitation relay until they do. The solenoid circuit is then automatically opened.

CONSTRUCTIONAL DETAILS. For outputs greater than those possible with glass bulbs the steel-tank construction is adopted. A typical construction is shown in Fig. 27.5. Fundamentally, there is no difference between the two types, the essential components such as mercury cathode, main anodes, ignition device and excitation anodes and circuits being common to both types. In the glass-bulb types the anodes are usually cooled by a motor-driven fan, and this method is also used with modern steel-bulb rectifiers of small output. Anode radiators are used with steel-tank rectifiers of large output. Since the inlet glands for the electrodes are of large size, considerable difficulties were encountered in trying to make these tight, since the vacuum must be maintained at about 0·01 mm of mercury. One arrangement is a combination of asbestos packing with mercury sealing. In order to maintain the vacuum, the rectifier was, at one time, supplied with a direct-connected motor-driven air pump. It is now possible to build medium-sized steel tanks which will maintain their vacuum for several years without the necessity for frequent pumping. Then, where necessary, the vacuum can be restored by portable pumping units.* A second constructional difficulty was that of cooling the anodes. If the anodes reach too high a temperature due, say, to continuous overloading, they will send off electrons just as the cathode does. As a result, the current will flow in both directions, destroying the valve action and constituting a short-circuit. Thus, efficient cooling of the anodes is essential. The greatest difficulty has been the elimination of the possibility of short-circuits.

Internal short-circuiting can be caused by back-firing or cross-firing which is the result of a failure of the rectifying action. From what has been said, it will be clear that rectification can only take place if the arc continues to rotate round the ring of anodes, and to remain with each anode only for the correct length of time. If any of the anodes become

* Modern seals of vitreous enamel give vacuum-tight joints with adequate insulating properties.

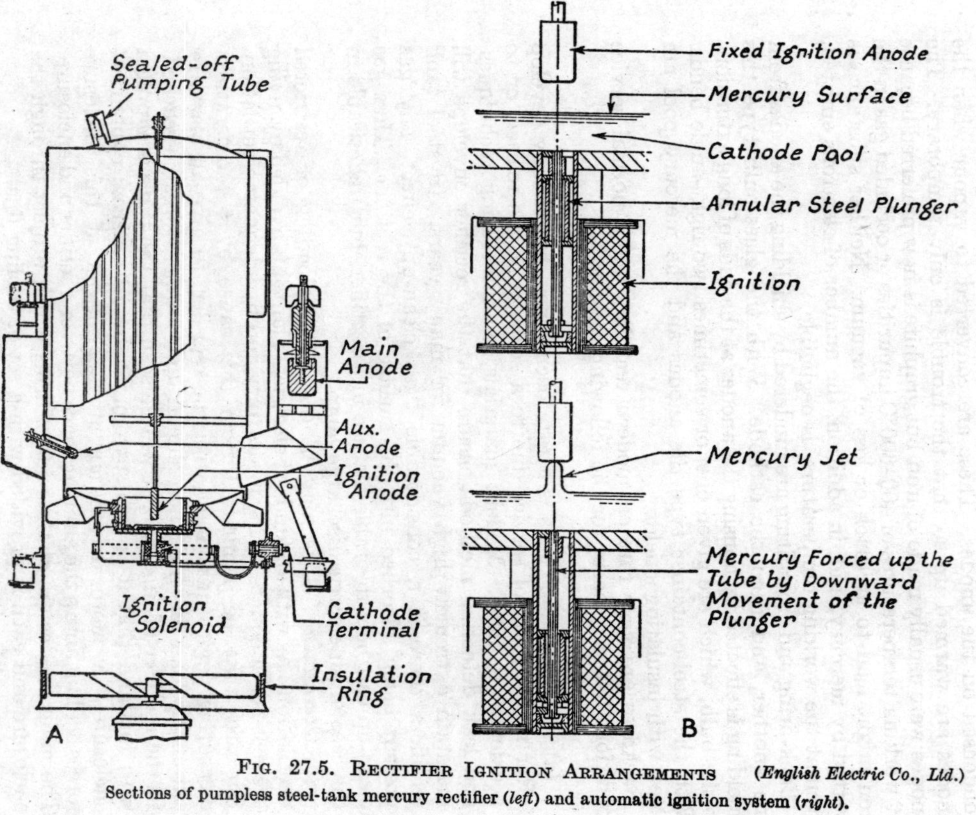

FIG. 27.5. RECTIFIER IGNITION ARRANGEMENTS (*English Electric Co., Ltd.*)
Sections of pumpless steel-tank mercury rectifier (*left*) and automatic ignition system (*right*).

overheated they may emit electrons themselves and thus try to act as cathodes. Irregular firing commonly takes place when starting up from cold, and this is due to small globules of mercury which have condensed on the anodes. These are converted to vapour when the anodes are warmed up so that the trouble is only temporary. The anodes were usually made of iron but graphite is now preferred because at working temperatures of 400–600°C impurities or occluded gas in the iron might lead to arc-back or loss of vacuum. Neither substance is wetted by mercury, and, in addition, the number of electrons emitted, even at the working temperature, is negligible.

Back-firing and cross-firing are hindered by shielding the anodes from one another, and from the cathode. With the glass-bulb type this shielding is effected by housing the anodes at the ends of long extensions to the bulb, which are given one, or sometimes two right-angle bends. With the steel-container type the anodes and the mercury pool are fitted with insulating shields.

Backfires are very rare on modern rectifiers under normal service conditions but may occur for the following reasons—

(*a*) *Poor vacuum.* This is a rare occurrence, but it may develop after many years of service due to a seepage or gas release, or to accidental damage. Modern materials, manufacturing techniques, and leak detecting methods ensure that the rectifier envelope will maintain extremely high vacuum for many years. In steel tank rectifiers there is a considerable "gas clean-up" effect, any gas present being adsorbed into the surface of the rectifier. Thus gas clean-up can be a desirable feature in a rectifier, but undesirable in the case of an electric discharge lamp.

(*b*) *Overheating.* This may be due to overloading or restricted ventilation. As with all kinds of electrical plant it is avoided by giving the rectifier adequate rating for its duty and by adequate ventilation. Backfires may be eliminated from this cause by a temperature-measuring device in contact with the rectifier, to give an alarm or automatic load disconnection should high temperatures be reached. In the rectifier of Fig. 27.5, it will be seen that the tank is supplied with fins and that it is also surrounded by an external cylindrical enclosure. A motor-driven fan at the bottom of this enclosure forces cooling air up the annular space between rectifier and enclosure. The main anodes are at the ends of arms which project through the enclosure and each is provided with a set of cooling fins.

(*c*) *Impurities in the graphite anodes.* This is most unlikely with the very high purity graphite now in use.

(*d*) *Mercury condensing in the anode areas.* This is avoided by designing for controlled temperature gradients and by the provision of drip shields for the anodes. In exceptional conditions, particularly for higher voltages, anode area heaters may be used. Every mercury-arc rectifier has an area segregated from the main arc path, in which the majority of the mercury evaporated at the cathode is condensed.

(*e*) Impact of jets of mercury vapour, drops of mercury, or radiation from the cathode. This is avoided by placing the anodes in separate bent arms, or, in some designs, in separate tubes projecting into the main cylinder with the inner openings screened.

The efficiency of the rectifier is high, the maximum value being at about three-quarters of full load, and only falling off appreciably below one-quarter. It is thus superior to any running machinery, even the rotary convertor.

If the current ceases for an interval of time as short as 1×10^{-5} sec, it is sufficient to cool the "cathode spot," which is the name given to the highly luminous cathode end of the arc, and thus stop the emission of electrons. Means must therefore be provided to prevent this occurring, the usual method being, as we have seen, to have a separate "exciting" anode worked from a separate external source. Cooling of the cathode spot can also be prevented by placing a choke in the d.c. circuit. This has the effect of prolonging the half-waves of current, with the result that the alternate half-waves overlap, and the resulting direct current never falls to zero. This is shown in Fig. 27.6 for a single-phase full-wave rectifier.

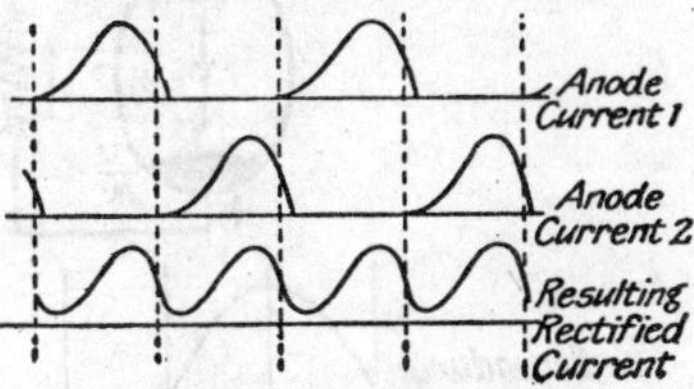

FIG. 27.6. WAVEFORMS OF RECTIFIED CURRENTS

Even without this choke the transfer of the electron stream from one anode to the next cannot be instantaneous because the current changes associated with the transfer, combined with the appreciable inductance of the circuits, result in an inductive drop which opposes the change. In addition, the ionized mercury vapour has a tendency to remain in the neighbourhood of an anode for an interval of time which, although very small, is comparable with the period of commutation. This effect sets an upper limit to the frequency at which this type of rectifier can be operated, this limit lying between 2000 and 5000 c/s.

Single-phase and Three-phase Rectification. The rectifier can be operated with any desired number of phases and Fig. 27.7 shows two methods of operation with single-phase supply. In the first figure there is only one anode, and, consequently, as current can only flow through the rectifier from anode to cathode, it follows that the external load will only carry current during those half-cycles when the anode is positive with respect to the cathode pool. In other words, only alternate half-waves are utilized, the rectifier therefore being called a "half-wave" rectifier.

In the second figure the load is taken to a centre tapping on the transformer secondary, with the result that each half-wave is utilized, the current stream being from anode A_1 to cathode K when A_1 is positive with respect to the cathode pool, and from A_2 to K when A_2 is positive with respect to it. In other words, the arc switches over from one anode to the other at the end of each half-cycle. Thus both half-waves are utilized and the rectifier becomes a "full-wave" rectifier.

As the voltage drop across the rectifier is practically constant, this sets a minimum to the magnitude of the a.c. input volts, and also means that, broadly speaking, the cathode is, at all instants during the con-

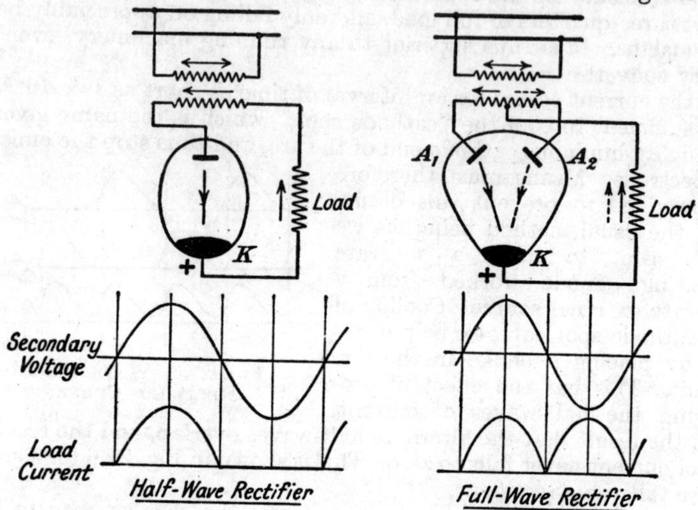

FIG. 27.7. HALF-WAVE AND FULL-WAVE SINGLE-PHASE RECTIFIERS

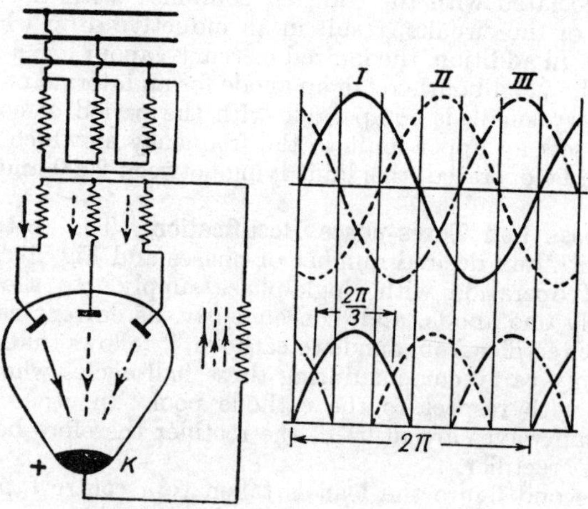

FIG. 27.8. THREE-PHASE RECTIFIER

duction period, at a potential of ($\sqrt{2}V$ − rectifier p.d.) above the neutral point of the transformer (where V is the r.m.s. input voltage). This potential therefore appears across the load "resistance" during conducting periods.

In the three-phase rectifier there are three anodes, Fig. 27.8, and since none of the phases has a centre tapping, the rectifier can be regarded as three half-wave rectifiers combined, the three components of the total load current having the usual three-phase mutual displacement of $2\pi/3$ radians. Now the arc will only pass between an anode and the cathode when that particular anode is at a higher potential than the others: as soon as the next anode in the sequence attains a potential higher than the first, the arc is immediately transferred to this second anode. Again, on examining the waves of secondary

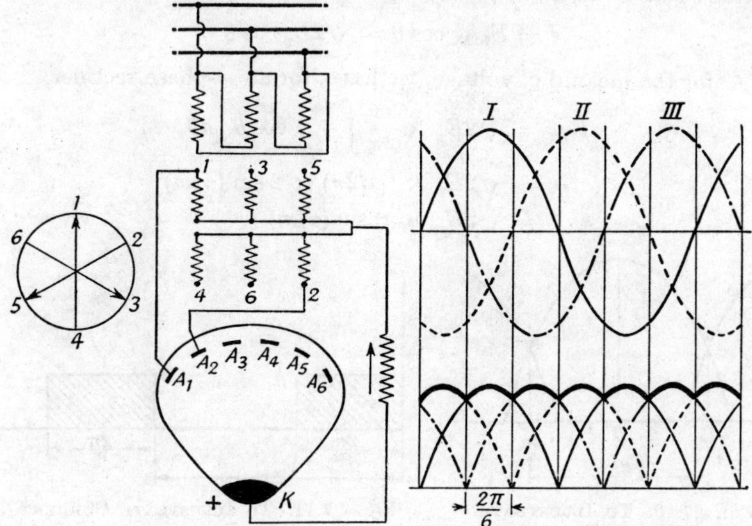

FIG. 27.9. THREE-PHASE FULL-WAVE RECTIFIER

voltage, we see that intersections occur at intervals of $2\pi/3$ radians, and this shows that each anode only carries current for $2\pi/3$, or one-third of a cycle, and not for one-half cycle as in the case of the single-phase rectifier.

If the anodes are arranged in a circle round the cathode, as is usual, then the rectifier can be regarded as a rotating switch which makes one complete revolution in each cycle and which maintains contact with each anode for one-third of a cycle per revolution. This is, in fact, the best way in which to regard the polyphase rectifier, and it is important to note that the rectifier is not in any way the seat of the energy supplied to the load. This energy comes from the transformer secondary, and therefore originally from the a.c. supply. On comparing the curve of load current with either of those obtained from a single-phase rectifier, we see that an increase in the number of phases very appreciably diminishes the percentage variation.

Still further improvement is effected by a three-phase full-wave rectifier, as shown in Fig. 27.9. The supply to this rectifier is shown as double star, which means that each phase is, in effect, centre-tapped, and, as a result, we can regard the rectifier, supplied in this manner, as a combination of three single-phase full-wave rectifiers. To obtain the curve of load current, we therefore draw the waves for the three phases each with the negative half reversed. This gives intersections $2\pi/6$ apart, showing that the period of commutation, i.e. the time during which the arc remains at any one anode, is now only one-sixth of the periodic time.

Voltage and Current Relationships. Let E_p = r.m.s. secondary voltage/phase. Then, with the origin at O, Fig. 27.10, we have

$$e = E_{\max}\cos\theta = \sqrt{2}E_p\cos\theta$$

$\therefore$ for the mean d.c. voltage, we have, for an m-phase rectifier,

$$E_{DC} = \sqrt{2}E_p \times \frac{m}{2\pi}\int_{-\pi/m}^{\pi/m}\cos\theta \, . \, d\theta$$

$$= \sqrt{2}E_p \times (m/2\pi) \times 2\sin(\pi/m)$$

$$= \sqrt{2}E_p(m/\pi)\sin(\pi/m)$$

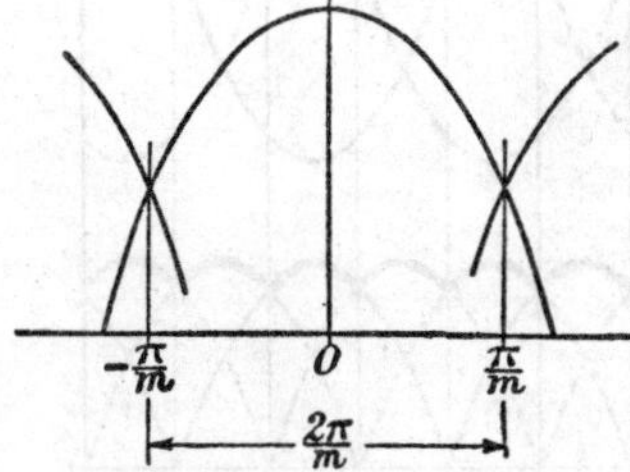

FIG. 27.10. TO ILLUSTRATE VOLTAGE RATIO

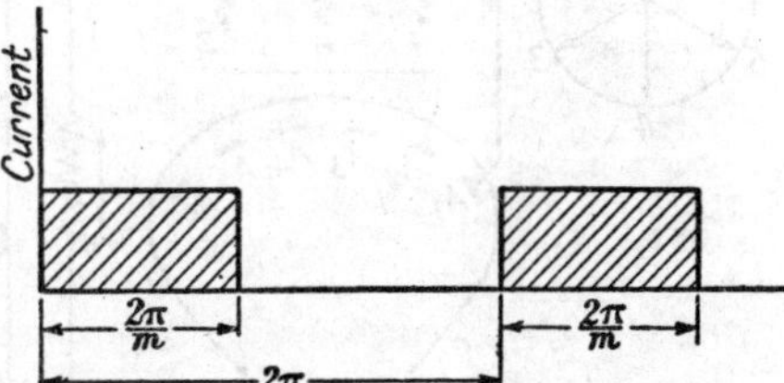

FIG. 27.11. TO ILLUSTRATE CURRENT RATIO

Actually the terminal voltage will be the above, less the sum of the drops in the rectifier and in the transformer.

EXAMPLE. The full-load terminal voltage on the output side of a six-phase rectifier is to be 1500 V, and the sum of all the drops is 25 V. Calculate the transformer secondary phase-voltage.

$$V_{DC} = \sqrt{2}E_p(m/\pi)\sin(\pi/m) - (\text{sum of drops})$$

$$\therefore 1500 = \sqrt{2}E_p(6/\pi)\sin 30 - 25$$

$$\sqrt{2}E_p(3/\pi) = 1525$$

$$E_p = 1525 \times \pi/3\sqrt{2} = 1130 \text{ V}$$

In calculating the current relationship, we have to note that the rectified current is supplied in turn by each secondary phase of the transformer, and that this current persists in each phase only for $2\pi/m$ of a cycle per cycle. The secondary current is thus in the nature of a series of intermittent blocks, as shown in Fig. 27.11. The mean current per phase is thus I_{DC}/m, while the r.m.s. current is $I_{DC}/\sqrt{m}$. Thus with a six-phase rectifier

$$I_{\text{mean}} = I_{DC}/6, \text{ while } I_{\text{RMS}} = I_{DC}/2{\cdot}45$$

Nature of the Anode Currents. In determining the expression for the current ratio we assumed that the anode current consisted of a series of rectangular blocks. The underlying assumption is that the output current is a perfectly steady current. Although this is not strictly correct, even with multi-anode rectifiers, it is a useful assumption to make when investigating the physical nature of the phenomena. In determining the nature of the currents in the transformer-rectifier combination we have to note that the rectified current is supplied by each secondary phase in turn. Consider first of all a 3-anode rectifier with delta-star transformer connections. The rectified current exists in each phase for only $2\pi/3$ of a cycle per cycle. This follows from Fig. 27.8. The anode currents, and therefore the secondary phase currents, are in the form of positive current blocks, as shown in Fig. 27.12. The directions of these are from anode to star point, as shown in the key diagram of Fig. 27.13, and each secondary current, $I_{2\cdot1}$, $I_{2\cdot2}$, and $I_{2\cdot3}$, while flowing, will be balanced by a corresponding primary current $I_{1\cdot1}$, $I_{1\cdot2}$, and $I_{1\cdot3}$. We will adopt the conventions that these currents are reckoned positive when they flow from the line to the transformer winding; and according to this, the primary phase currents also consist of successive blocks of positive direction current. This illustrates an objectionable feature of the method of transformer connection under consideration, namely, that it results in static magnetization of the core.

To determine the line currents, consider each of the three periods of $2\pi/3$ in turn. During the first period, Fig. 27.13 (2), anode 1 is conducting and the current $I_{1\cdot1}$ flows from the R line through phase 1 and directly to the B line. Hence during this period I_R is positive and I_B negative.

During the second period, in which anode 2 is conducting, Fig. 27.13 (3), we see that I_Y is positive and I_R negative. During the third period, Fig. 27.13 (4), I_B is positive and I_X negative. During the three periods we therefore have

Period 1. $I_R +$, $\quad I_Y = 0$, $\quad I_B -$

Period 2. $I_R -$, $\quad I_Y +$, $\quad I_B = 0$

Period 3. $I_R = 0$, $\quad I_Y -$, $\quad I_B +$

These give the diagram of primary line currents shown in the lower portion of Fig. 27.12.

Now consider the case of a 6-anode rectifier supplied by a delta-star transformer with diametrical connections, which gives full-wave rectification for each phase. The circuit diagram is given in Fig. 27.14, in which the voltage sequence is indicated by the clock diagram. During their conducting periods, anodes 1, 3 and 5 are at the potentials of the voltages V_1, V_2, and V_3, while anodes 4, 6, and 2 are at the potentials of V_1', V_2', and V_3', these being the reverse of the potentials V_1, V_2, and V_3 respectively. In the circuit diagram the terminals

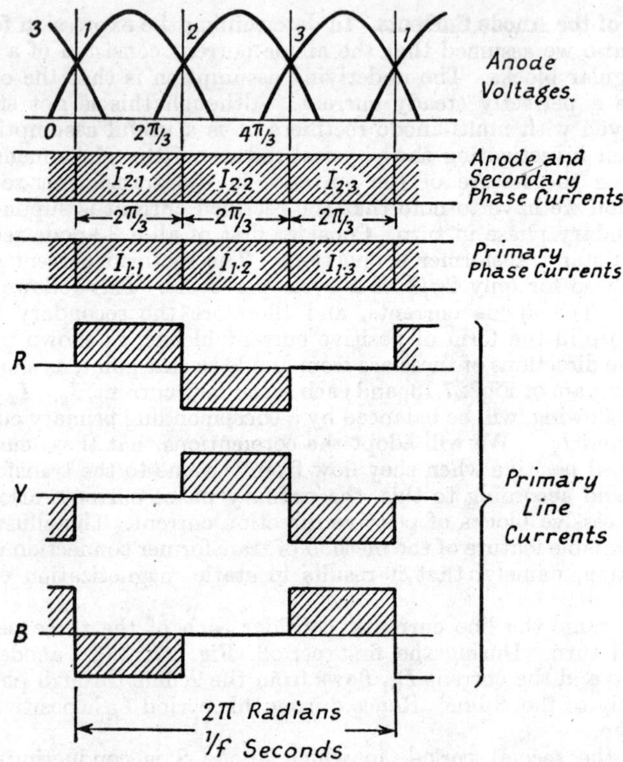

FIG. 27.12. CURRENT DIAGRAM FOR THE CIRCUIT OF A THREE-ANODE CONVERTOR WITH DELTA-STAR TRANSFORMER CONNECTIONS

1 1′, 2 2′, 3 3′ are the ends of the three secondary phases and therefore the anodes to which they are connected have to correspond with the numbering on the clock diagram: the anode numbers are enclosed in circles. The investigation of the current changes can now be made as before.

When anode (1) is conducting, the secondary current, $I_{2\cdot1}$, flows from N to terminal 1. Adopting the same convention as before the primary phase current $I_{1\cdot1}$ acts downwards and is therefore positive, thus making $I_R +$, and $I_B -$, while I_Y is zero.

When anode (2) is conducting, $I_{2\cdot3}'$ acts from N to the terminal 3′ with the result that the balancing current in the primary, $I_{1\cdot3}$ acts

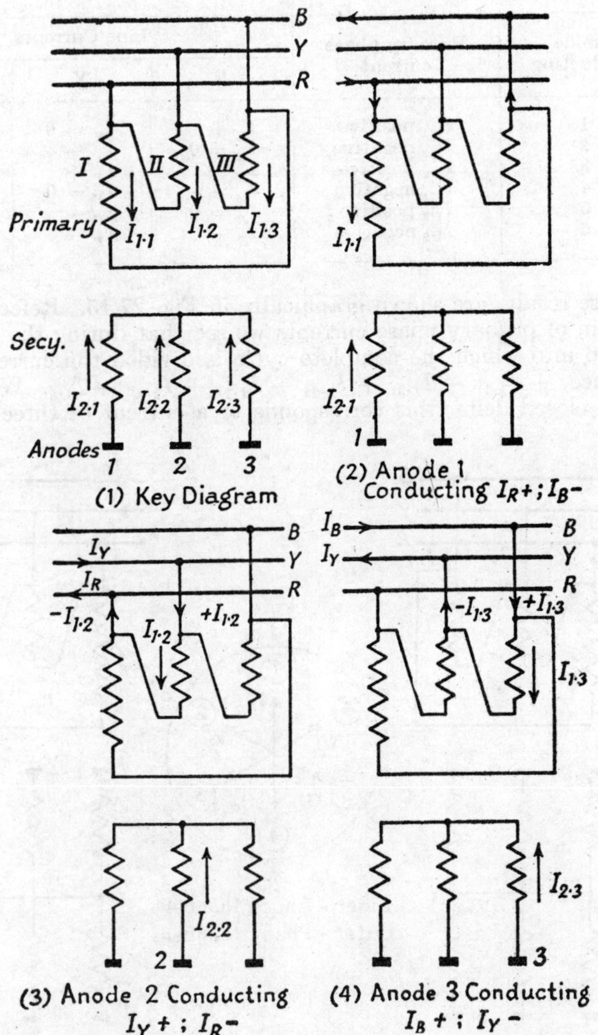

(1) Key Diagram

(2) Anode 1 Conducting I_R+; I_B-

(3) Anode 2 Conducting I_Y+; I_R-

(4) Anode 3 Conducting I_B+; I_Y-

FIG. 27.13. SEQUENCE OF CURRENT CHANGES DURING ONE CYCLE

upwards. By the same convention it is therefore reckoned negative, and we then have $I_R = 0$, $I_Y +$ and $I_B -$. Similarly when the other anodes are conducting. We can tabulate as follows

Anode conducting	Primary phase current	Line Currents		
		R	Y	B
1	$I_{1\cdot1}$ positive	$I_R +$	$I_Y = 0$	$I_B -$
2	$I_{1\cdot3}$ negative	$I_R = 0$	$I_Y +$	$I_B -$
3	$I_{1\cdot2}$ positive	$I_R -$	$I_Y +$	$I_B = 0$
4	$I_{1\cdot1}$ negative	$I_R -$	$I_Y = 0$	$I_B +$
5	$I_{1\cdot3}$ positive	$I_R = 0$	$I_Y -$	$I_B +$
6	$I_{1\cdot2}$ negative	$I_R +$	$I_Y -$	$I_B = 0$

These results are shown graphically in Fig. 27.15. Referring to the diagram of primary phase currents we see that during the six periods of $2\pi/6$ into which the complete cycle is divided the currents are, in sequence, $+I_{1\cdot1}$, $-I_{1\cdot3}$, $+I_{1\cdot2}$, $-I_{1\cdot1}$, $+I_{1\cdot3}$, $-I_{1\cdot2}$. With respect to the closed delta, this corresponds to a current of three times line

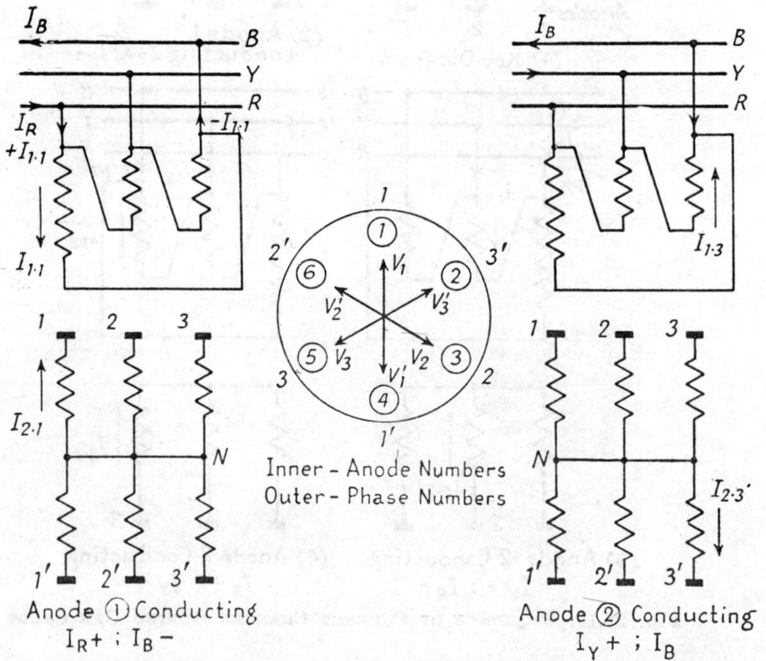

FIG. 27.14. CURRENT CHANGES IN A SIX-ANODE RECTIFIER WITH DIAMETRICAL CONNECTION, A FULL-WAVE STAR CONNECTION

The positive direction of the primary phase current is reckoned downwards, and the negative direction upwards.

frequency, showing that an alternating m.m.f. of triple frequency is superposed on the normal supply frequency m.m.f.

Effect of Transformer Reactance. In practice, the current blocks flowing in each anode circuit cannot have the rectangular shape of Fig. 27.11 because the leakage reactance of the transformer windings

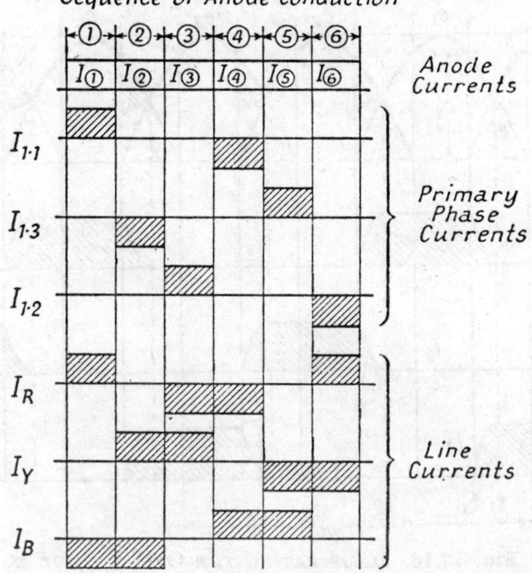

FIG. 27.15. CURRENT DIAGRAM FOR SIX-ANODE CONVERTOR WITH FULL-WAVE CONNECTION

(*a*) prevents an infinitely high rate of rise of current at the commencement and (*b*) prevents an infinitely high rate of fall of current at the end. Consider again a three-anode rectifier, for simplicity. The anode voltages overlap, as shown in Fig. 27.16. Now consider the instant t_1. This is the instant of overlap of anode voltages e_3 and e_1, so that without reactance, the current I_1 would be established suddenly, as shown by the dotted block of current I_1. Actually the current rises gradually. At the instant t_2, the voltage e_2 is on the point of becoming greater than e_1 so that, without reactance, i_1 should be extinguished immediately and i_2 established suddenly. In practice i_1 dies away gradually while i_2 is established gradually, with the result that there is a current overlap. This, of course, means that for the duration of this current overlap both anodes are carrying current, but that there is a gradual transfer from anode 1 to anode 2. Also during the overlap, the voltage at any instant will be the mean of the instantaneous values of e_1 and e_2. Thus at instant t_1 it will be e_1', while at instant t_3 it will be e_3', where e_3' is the mean of the instantaneous values of e_1 and e_3, for this particular intersection.

From the point of view of output voltage this means that the shaded triangular pieces are cut out of the voltage waves, the mean d.c. voltage thus being the mean of the heavy line curve. This has the effect of making the output voltage somewhat less than the value given by the e.m.f. equation.

Denote the angle of overlap by α electrical radians, then α is the angular extent of the shaded triangle in Figs. 27.16 and 27.17. During

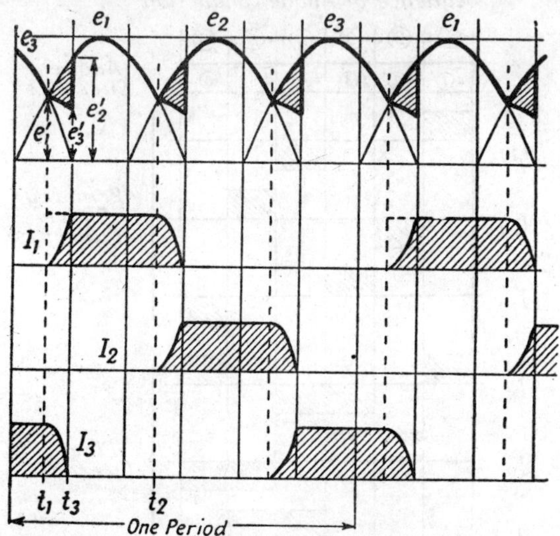

FIG. 27.16. ILLUSTRATING THE OVERLAP DUE TO TRANSFORMER REACTANCE

the period of overlap the mean rectified voltage is the mean of the voltages e_1 and e_2 in Fig. 27.17, namely $(e_1 + e_2)/2$. The voltages e_1 and e_2 now refer to the anode voltages and not to the particular instantaneous voltages of Fig. 27.16. During the remainder of the period $2\pi/m$, namely during $(2\pi/m - \alpha)$, the rectified voltage follows the anode voltage wave. Now the curve AB has the equation

$$e = \tfrac{1}{2}(e_1 + e_2) = \tfrac{1}{2}E_{\max}(\cos\beta + \cos\gamma)$$

the values of β and γ being decided by the point taken as the origin. We will take the point O, Fig. 27.17, corresponding to the point of intersection A of the waves of anode voltages e_1 and e_2. Then

$$e_1 = E_{\max}\cos(\theta + \pi/m)$$
$$e_2 = E_{\max}\cos(\theta - \pi/m)$$

This latter equation also refers to the period subsequent to overlap, namely from θ to $2\pi/m$. Hence the mean rectified voltage is given by

$$V_{dc} = \frac{m}{2\pi}E_{\max}\left[\int_0^{\alpha}\tfrac{1}{2}(e_1 + e_2)d\theta + \int_{\alpha}^{2\pi/m} e_2 d\theta\right]$$

$$= \frac{m}{2\pi} \cdot E_{\max} \left[\tfrac{1}{2} \int_0^{\alpha} \cos\left(\theta + \frac{\pi}{m}\right) d\theta + \tfrac{1}{2} \int_0^{\alpha} \cos\left(\theta - \frac{\pi}{m}\right) d\theta + \int_{\alpha}^{2\pi/m} \cos\left(\theta - \frac{\pi}{m}\right) d\theta \right]$$

$$= (m/2\pi)E_{\max}[\tfrac{1}{2}\,[\sin(\theta + \pi/m)]_0^{\alpha} + \tfrac{1}{2}\,[\sin(\theta - \pi/m)]_0^{\alpha} + [\sin(\theta - \pi/m)]_{\alpha}^{2\pi/m}$$

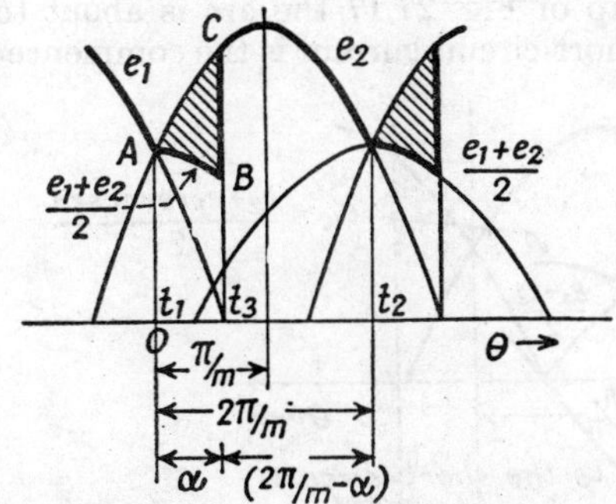

Fig. 27.17. Effect of Overlap

Fig. 27.18. Conditions during Short Circuit

This reduces to

$$\begin{aligned} &(m/2\pi)E_{\max} \cos\alpha \sin(\pi/m) + \sin(\pi/m) \\ &= 2(m/2\pi)E_{\max} \sin(\pi/m)(1 + \cos\alpha) \\ &= (m/\pi)E_{\max} \sin(\pi/m) \cos^2(\alpha/2) \\ &= (\sqrt{2}m/\pi)E_p \sin(\pi/m) \cos^2(\alpha/2) \end{aligned}$$

Summarizing, we see that the d.c. terminal voltage of a rectifier on load is less than the no-load value for the following reasons—

(1) The internal drops, made up of cathode drop, anode drop, and arc drop.

(2) The resistance of the transformer windings.

(3) The resistance of the smoothing chokes, not shown in the diagram of connections.

(4) The cutting out of the shaded portions of the voltage curve due to the effect of transformer reactance.

Calculation of the Overlap. It will be obvious that the angle α will be dependent on the transformer reactance: it can be determined as follows. The period of overlap is a period of short-circuit, since two anodes are conducting. Fig. 27.18 indicates that the voltage acting round the short-circuit path is $(e_1 - e_2)$. This voltage has to overcome the reactance voltage of two phases in series, showing that the reactance voltage per phase

$$e_x = \tfrac{1}{2}(e_1 - e_2)$$

$$= (E_{max}/2)\,[\cos(\theta - \pi/m) - \cos(\theta + \pi/m)]$$
$$= E_{max} \sin(\pi/m) \sin\theta$$

Hence short-circuit current

$$i_s = e_x/X$$
$$= (E_{max}/X) \sin(\pi/m) \sin(\theta - \pi/2)$$

its amplitude thus being $(E_{max}/X) \sin \pi/m$.

Note that for the shaded overlap of Fig. 27.17 the arc is about to move to anode 2 only, so that the short-circuit current is the commence-

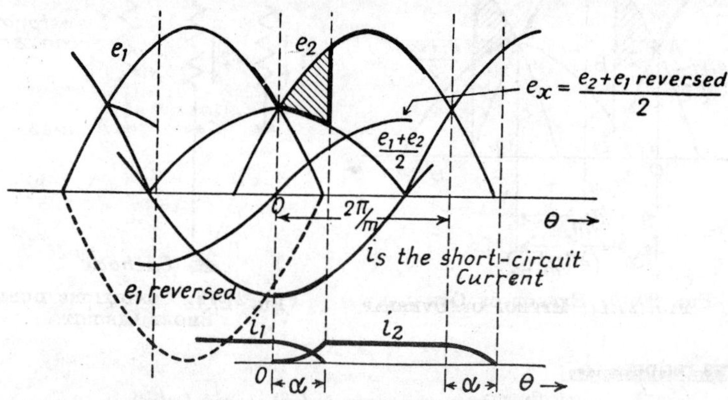

Fig. 27.19. Effect of Short-circuit Current on the Starting and Stopping of the Current at the Anode

ment of the block of current i_2. Hence we regard e_2 as positive and e_1 as negative; thus

$$e_x = \tfrac{1}{2}(e_2 + e_1 \text{ reversed})$$

The circuit of the short-circuit current is almost purely inductive, so that we can regard the current i_s as being in quadrature lagging with the voltage e_x. This gives the phase relationship shown in Fig. 27.19 and the portion of the wave of i_s within the angular width α is the initial portion of the current of anode 2. The overlap being the result, while the transformer reactance is the cause, we require an expression for α in terms of X_p the reactance per phase.

We can write

$$i_s = -(E_{max}/X) \sin(\pi/m) \cos\alpha^* = -I_{s.max} \cos\theta$$

During the periods of overlap the anode current can be regarded [illegible] being equal to the appropriate value of the short-circuit current [illegible]

* *See* H. Rissik, *Mercury Arc Current Convertors*, Chap. III. Pitman, Lo[illegible]ion.

a steady direct current superimposed. For the anode currents i_1 and i_2 we can thus write

$$i_1 = (A - i_s) \text{ and } i_2 = (B + i_s)$$

$$\therefore\ i_1 = A + i_{s.\max} \cos \theta$$

$$i_2 = B - i_{s.\max} \cos \theta$$

when $\theta = 0$ $\quad i_1 = I_{d.c}$ and $i_2 = 0$, showing that

$$A = I_{d.c} - i_{s.\max}$$

$$B = i_{s.\max}$$

$$\therefore\ i_1 = I_{d.c} - i_{s.\max} (1 - \cos \theta)$$

$$i_2 = i_{s.\max} (1 - \cos \theta)$$

Now when $\theta = \alpha$, $i_2 = I_{d.c}$

$$\therefore\ I_{dc} = i_{s.\max} - i_{s.\max} \cos \alpha$$

$$\cos \alpha = 1 - I_{dc}/i_{s.\max}$$

$$= 1 - I_{d.c}X/E_{\max} \sin (\pi/m)$$

Example. Calculate the drop in d.c. voltage in the case of a six-anode rectifier working with an overlap of 25°.

$$\cos \alpha = 1 - I_{dc}X/E_{\max} \sin (\pi/m)$$

$$\therefore\ I_{dc}X = E_{\max} (1 - \cos \alpha) \sin (\pi/m)$$

$$= \sqrt{2}E_p [\sin (\pi/6) - \sin (\pi/6) \cos \alpha]$$

$$= E_p \times \sqrt{2}(0{\cdot}5 - 0{\cdot}5 \times 0{\cdot}9061)$$

$$= 0{\cdot}066E_p$$

The average of this over the period of $2\pi/m$

$$= (0{\cdot}066 \times 6/2\pi)E_p = 0{\cdot}063E_p$$

Suppose for example that $E_p = 1130$ V, as in the previous example, and that there is an arc drop of 25 V, then sum of volt drops

$$= 25 + 0{\cdot}063 \times 1130$$

$$= 96{\cdot}2 \text{ V}$$

Regulation. Denote by E_{dc} the output voltage given by the equation on p. 720, that is, neglecting all the causes of volt drop. Then

$$E_{dc} = \sqrt{2}E_p(m/\pi) \sin (\pi/m)$$

Now, even on very light load there must be an arc with its associated drop, and if we denote by $V_{dc.0}$ the output voltage at the lowest current the rectifier can supply without ceasing to function

$$V_{dc.0} = E_{dc} - v_a$$

where $\quad v_a =$ the arc drop

There are two other volt drops, namely v_x, the drop due to transformer reactance, and v_r, that due to transformer resistance. Hence, on load

$$V_{dc} = (E_{dc} - v_a) - v_x - v_r$$
$$= V_{dc.0} - v_x - v_r$$

In the previous example

$$E_p = 1130 \text{ V}$$
$$\therefore\ E_{dc} = \sqrt{2} \times 1130 \times (6/\pi) \sin(\pi/6)$$
$$= 1525$$
$$\therefore\ V_{dc.0} = 1525 - 25 = 1500 \text{ V}$$

We have seen that $v_x = 71{\cdot}2$, and we will assume that $v_r = 5$. Then

$$V_{dc} = 1500 - 71{\cdot}2 \simeq 1429 \text{ V}$$

$\therefore$ Percentage regulation

$$= 100[(v_x + v_r)/(E_{dc} - v_a)]$$
$$= 100[71{\cdot}2/(1525 - 25)] = 4{\cdot}75 \text{ per cent}$$

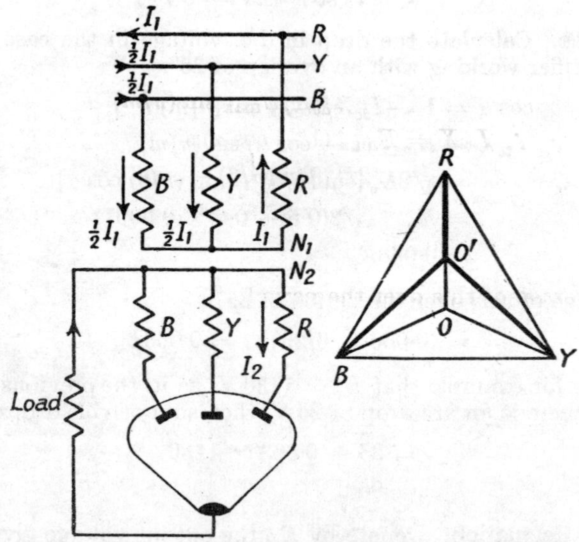

FIG. 27.20. TO ILLUSTRATE "SLIPPING OF THE NEUTRAL" WITH STAR-STAR CONNECTION

Star-connection of the Transformer Primary. So far we have dealt with simple star connection of the transformer secondary phases, but in each case with a delta-connected primary. Suppose that the primary is star connected. Consider the case of a three-anode rectifier, the circuit will then be as in Fig. 27.20. During the time that anode 1 is conducting, the primary phase R is loaded, while phases Y and B merely act as choking coils. Also the load current in phase R draws currents of one-half its value from the Y and B lines, since the resultant current at N_1 must be zero. These currents, flowing in the Y and B phases set up large potential drops in the phases because of the high impedances, the vector diagram of potential distribution therefore being modified by the displacement of the neutral point from O to O'. The voltage of the red phase is thereby reduced from the correct value OR to a much reduced value $O'R$, resulting in distortion of the primary current distribution and in poor voltage regulation.

This phenomenon is called "slipping of the neutral" and it is avoided by delta connection of the primary.

Phase Equalizing Circuits. The rectifier systems considered so far have all been of the "simplex" type, in which, in an m-anode rectifier, each anode conducts for $2\pi/m$ radians of the complete cycle of applied voltage with simplex working. The larger the number of anodes used the smaller the natural output voltage ripple, as is obvious from a comparison of Figs. 27.8 and 27.9, so that where smoothing is adopted, the problem is simplified by an increase in the number of anodes. However, as only one secondary phase is in use at any given instant the utilization of the transformer-rectifier system is poor.

The most satisfactory method of connection giving the greatest utilization of the plant and reduction in bulk of the transformer for a given overall rating, is that known as "multiplex." In such systems two, three, or four anodes operate simultaneously during current conducting periods which are greater in length than in the simplex case. To effect this economy, the m anodes of the rectifier are split into a number of star-connected groups of three, e.g. two groups for a six-anode and four groups for a twelve-anode. Each anode group has its own individual secondary winding, and the star points of the various groups are connected to the neutral point of the secondary system as a whole through an interphase transformer, or, to give it its functional name, a "phase equalizer." The common neutral point forms the negative pole of the system. The connection scheme for a six-anode rectifier is shown in Fig. 27.21. It is obvious that two anodes can only share a load successfully if they are maintained at the same potential relative to the cathode, and the function of the phase equalizer is the maintenance of this condition.

The two half-windings of the interphase transformer are identical, but are arranged to oppose each other on a common magnetic circuit. The difference in the potential between the d.c. output terminals and the instantaneous potential of that anode in a group which is conducting, appears across the corresponding half of the interphase transformer, and causes, by transformer action, a voltage of equal magnitude to be developed across the other half, but of opposite polarity. This is added algebraically to the instantaneous voltage across the conducting phase of the other group, thereby equalizing the potentials of the two groups. The load current is divided equally between the two groups (neglecting the effect of the small magnetizing current) and hence, since the two halves of the interphase transformer are wound in opposition, the nett magnetization of the core is reduced to a very small value. This results in an equalizing transformer of small physical dimensions.

The voltage variations across the individual phases during a complete cycle of the supply voltage are shown in Fig. 27.22. The mean

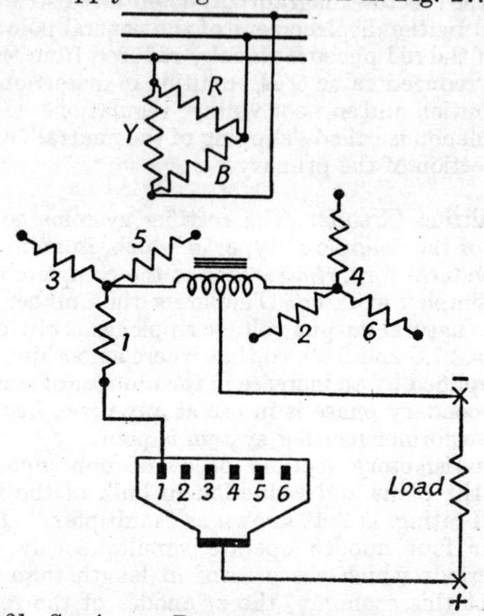

FIG. 27.21. SIX-ANODE RECTIFIER WITH PHASE EQUALIZERS
Only one anode is shown, for simplicity.

d.c. voltage is determined as follows. Let the origin correspond to the instant the voltage at anode 1 is a maximum, then

$$e = E_{max}\cos\theta$$
$$= \sqrt{2}E_p\cos\theta$$

Since anodes 1 and 2 are simultaneously conducting at this instant, the output voltage is the mean of the voltages of these two, namely

$$\tfrac{1}{2}[\sqrt{2}E_p\cos\theta + \sqrt{2}E_p\cos(\theta + \pi/3)]$$
$$= \sqrt{(3/2)}E_p\cos(\theta + \pi/6) \text{ averaged over the cycle}$$
$$\therefore E_{dc} = \frac{6}{\pi}\sqrt{\left(\frac{3}{2}\right)}\cdot E_p\int_0^{\pi/6}\cos\left(\theta + \frac{\pi}{6}\right)d\theta$$
$$= 1{\cdot}17E_p$$

which is identical with that obtained from a three-phase rectifier. The amplitude of the voltage ripple is clearly

$$\sqrt{2}E_p\cos(\pi/6) - (\sqrt{3/2})E_p\cos(\pi/6)$$
$$= E_p[\sqrt{2}\times(\sqrt{3}/2) - (\sqrt{3}/2)\times(\sqrt{3}/\sqrt{2})]$$
$$= 0{\cdot}162E_p$$

This is the same as for the simplex six-anode rectifier.

At the intersections of the anode voltage waves, the potentials of adjacent anodes are naturally equal, so that at the corresponding instants the phase equalizer has nothing to do and its windings have no e.m.f.s induced in them. Thus at six equally spaced instants during the cycle of applied voltage the equalizer e.m.f. is zero. Now consider the instant O. Anode voltage e_1 is equal to E_{max} while anode voltage

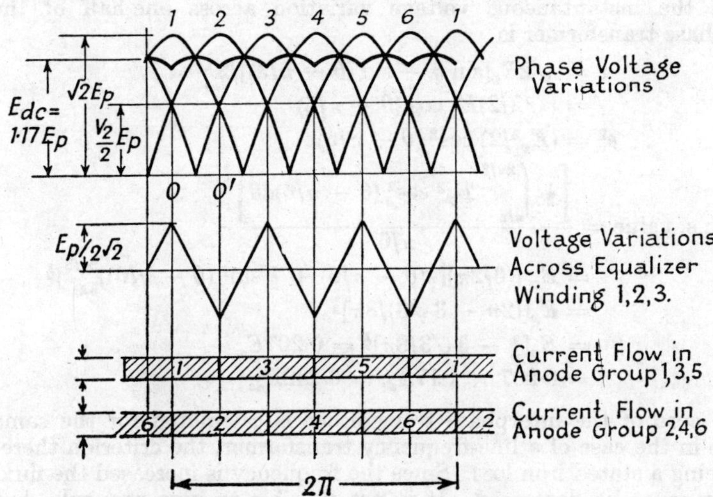

FIG. 27.22. VOLTAGE AND CURRENT VARIATIONS IN A SIX-ANODE RECTIFIER WITH PHASE EQUALIZING

e_2 is $\frac{1}{2}E_{max}$. For the purpose of load equalization the difference voltage $\frac{1}{2}E_{max}$ must exist in the phase equalizer windings, that is $\frac{1}{4}E_{max}$ in each half. This is accomplished by an increase in the current i_1 and a decrease in i_2 in order that the necessary unbalancing of the m.m.f. may exist, to set up the flux required for the inducing of the voltage $\frac{1}{4}E_{max}$ in each half winding. At instant O' a similar phenomenon takes place except that it is i_2 which must now increase, the e.m.f. of $\frac{1}{4}E_{max}$ in the corresponding half winding thus changing from $+\frac{1}{4}E_{max}$ to $-\frac{1}{4}E_{max}$ during the interval OO'. Again, the rate of change of voltage must be constant, and so we have across each half winding a triangular voltage of triple frequency, and of amplitude $\frac{1}{4}E_{max}$ or $E_p/2\sqrt{2}$. Thus the core of the interphase transformer is magnetized by a third-harmonic current producing a third-harmonic flux and inducing a third-harmonic e.m.f.

The above relationships enable the kVA rating of the interphase transformer to be calculated, as follows—

$$\text{Current loading} = \tfrac{1}{2}I_{dc}$$

The voltage per half winding, in terms of the mean rectified voltage E_{dc} is found as follows. The limits of the angle during which anode 1 and anode 2 are sharing the arc are, with reference to anode 1, $\pi/2$ and $2\pi/3$. The peak-to-peak voltage across the interphase transformer

$$2e = \begin{pmatrix}\text{Instantaneous volts}\\ \text{on anode 1}\end{pmatrix} - \begin{pmatrix}\text{Instantaneous volts}\\ \text{on anode 2}\end{pmatrix}$$
$$= \sqrt{2}E_p \sin\theta - \sqrt{2}E_p \sin(\theta - \pi/3)$$

Hence the instantaneous voltage variation across one-half of the interphase transformer is

$$e = \sqrt{2}E_p[\sin\theta - \sin(\theta - \pi/3)]/2$$
$$= (\sqrt{2}/2)E_p \cos(\theta - \pi/6)$$
$$\therefore\ e^2 = (E_p{}^2/2)\cos^2(\theta - \pi/6)$$
$$\therefore\ \text{r.m.s. value} = \frac{\left[\frac{1}{2}\int_{\pi/2}^{2\pi/3} E_p{}^2 \cos^2(\theta - \pi/6)d\theta\right]^{\frac{1}{2}}}{\pi/6}$$
$$= E_p\left[(6/2\pi)\{\tfrac{1}{2}(\theta - \pi/6) + \tfrac{1}{2}\sin^2(\theta - \pi/6)\}_{\pi/2}^{2\pi/3}\right]^{\frac{1}{2}}$$
$$= E_p[(2\pi - 3\sqrt{3})/8\pi]^{\frac{1}{2}}$$
$$= E_p[\tfrac{1}{4} - 3\sqrt{3}/8\pi]^{\frac{1}{2}} = 0{\cdot}207E_p$$
$$= 0{\cdot}207 \times 1{\cdot}17E_{dc} = 0{\cdot}242E_{dc}$$

The core of the interphase transformer is calculated for the same loss as in the case of a line-frequency transformer, the criterion therefore being a stated iron loss. Since the frequency is increased the flux-density must be decreased. It is not possible to give any rule, but one-half of the normal value for 50 c/s magnetization is appropriate. Now the volts per turn

$$V_t = B_{\max} Af$$
$$\propto B_{\max} f$$

Hence, since a 50 c/s transformer operates with about twice the flux density but only one-third the frequency of the interphase transformer, the volts per turn at supply frequency is only $\frac{2}{3}$ of the volts per turn at 150 c/s. The equivalent voltage rating is thus

$$(\tfrac{2}{3}) \times 0{\cdot}242E_{dc} = 0{\cdot}161E_{dc}$$
$$\therefore\ \text{kVA rating} = 0{\cdot}161E_{dc} \times 0{\cdot}5I_{dc}/1000$$
$$= 8{\cdot}05 \times 10^{-5}E_{dc}I_{dc}$$

Example. Output 1500 kW at 500 V.

$$E_{dc} = 500\ \text{V},\ I_{dc} = (1500/500) \times 10^3 = 3 \times 10^3\ \text{A}$$
$$\therefore\ \text{kVA rating} = 8{\cdot}05 \times 10^{-5} \times 5 \times 10^2 \times 3 \times 10^3$$
$$= 120{\cdot}8\ \text{kVA}$$

Current Changes During the Cycle. The current changes in the various circuits can be determined by the method previously described. Fig. 27.23 shows the conditions for two instants, one-sixth of a period apart. In the first figure anodes 1 and 2 are conducting together, with the result that I_{dc} is split into two halves, whose directions in the secondary windings are towards anode (1) and therefore towards terminal 1, and towards anode (2) and therefore towards terminal 3′, respectively. The balancing currents in the primary phases are $I_{1 \cdot 1}$

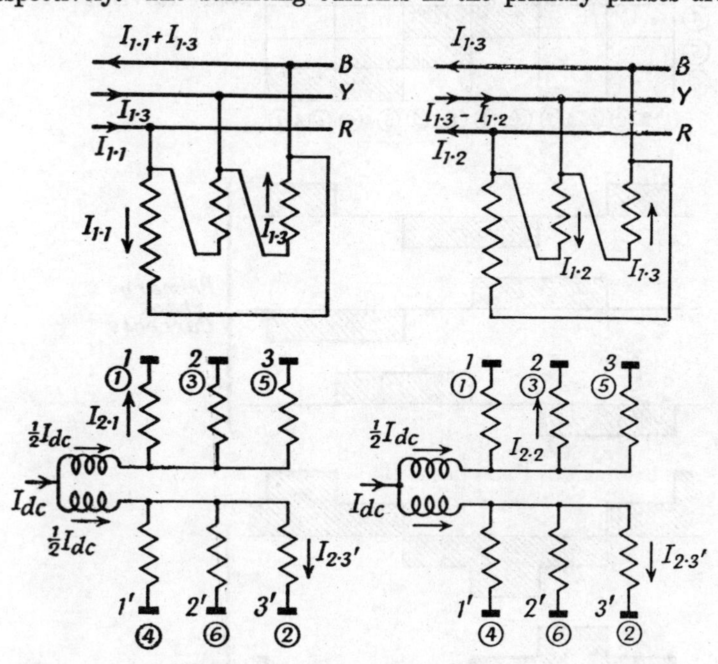

1. Anodes ① and ② Conducting   2. Anodes ② and ③ Conducting

FIG. 27.23. CURRENT CHANGES IN DOUBLE THREE-PHASE RECTIFICATION WITH INTERPHASE TRANSFORMER

and $I_{1 \cdot 3}$, the former being positive and the latter negative, according to the previously used convention. The line currents are thus—

$$I_R = I_{1 \cdot 1} \text{ and is positive}$$

$$I_Y = I_{1 \cdot 3} \text{ and is positive}$$

$$I_B = (I_{1 \cdot 1} + I_{1 \cdot 3}) \text{ and is negative}$$

In the second figure anodes (2) and (3) are conducting together, and by the same procedure we now have

$$I_R = I_{1 \cdot 2} \text{ and is negative}$$

$$I_Y = I_{1 \cdot 2} + I_{1 \cdot 3} \text{ and is positive}$$

$$I_B = I_{1 \cdot 3} \text{ and is negative}$$

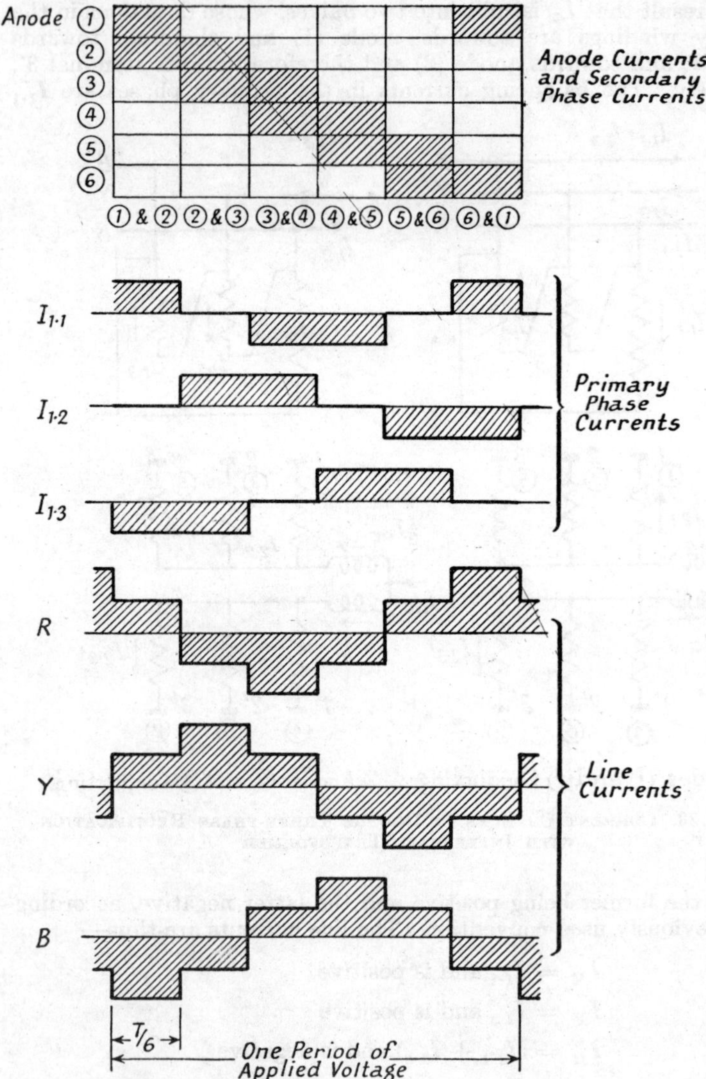

FIG. 27.24. CURRENT CHANGES DURING ONE COMPLETE CYCLE FOR DOUBLE THREE-PHASE RECTIFICATION WITH INTERPHASE TRANSFORMER

Similarly for the remaining instants in the cycle. All the currents are set out in graphical form in Fig. 27.24 and on comparing this figure with Figs. 27.12 and 27.15 we see that the transformer windings have a much better utilization, each winding carrying current for a greater fraction of the period.*

Grid Control of Mercury-arc Rectifiers. Imagine that a perforated grid is interposed between cathode and anode in such a way that the electrons can only reach the anode via the grid. Then it is quite clear that the motion of the electrons will now depend on the potential of the grid relative to that of the cathode. If the grid is positive, the

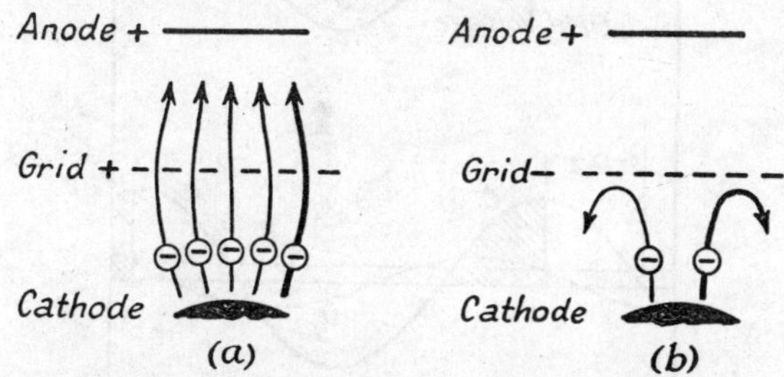

FIG. 27.25. TO ILLUSTRATE GRID CONTROL

electrons will be accelerated in their path to the anode, as in Fig. 27.25 (*a*), whereas if the grid is negative, the electrons will be decelerated. Provided this negative grid potential is large enough, the electrons will be turned back by the repulsion of the grid and will not be able to reach the anode at all, as in Fig. 27.25 (*b*). In such a case, the flow of current will cease. Rissik defines the function of grid control as follows:† "The function of any system of grid control is to permit the establishment of a current arc at the anodes of a current convertor at a predetermined instant in the anode voltage-cycle by the application of a positive potential to the individual control grids, and to prevent the re-establishing of the arc at some other instant by the subsequent application of a negative grid bias. The former potential is usually referred to as the *liberating* voltage, whilst the latter is generally known as the blocking potential." The liberating voltage varies between the limits 25 and 150, and the blocking potential between 100 and 300 V.

It is clear that the grid control voltage will have to be synchronous with the voltage supply to the anodes, and therefore the most obvious method of obtaining the control voltage is to use an alternating voltage of suitable value derived from the supply. The control is carried out by varying the phase of the grid voltage. Consider the first figure of Fig. 27.26. The grid voltage v_g, is in phase with the anode voltage.

* For the calculation of the supply, transformer, and rectifier loading with the various systems of connection, *see* H. Rissik, *loc. cit.*

† *Loc. cit.*, Chap. IX.

The horizontal ignition line of ordinate g represents the necessary voltage to be applied to the anode before ignition can take place, and therefore the point of intersection of this line with the grid voltage wave gives the moment at which the arc is struck. Now, once the arc is started, the grid has no further influence on the performance, the arc

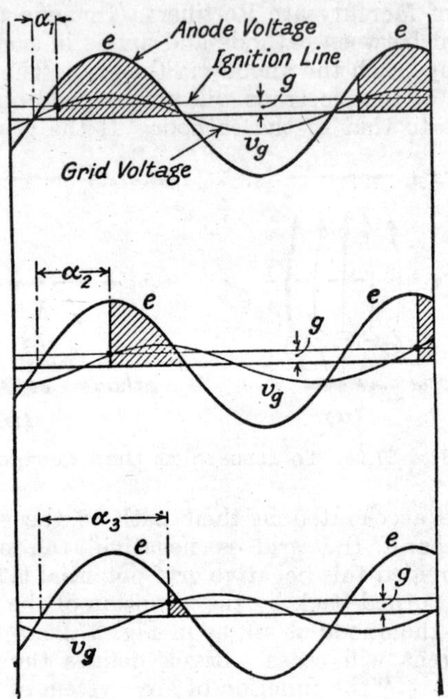

FIG. 27.26. PHASE-SHIFT CONTROL
(Mercury-arc Current Convertors, Pitman.)

continuing at the anode as in the case of a rectifier without grids. In the figure, this is until the anode voltage falls to zero, and the anode delivers current for practically half a cycle. The angle α, by which ignition is delayed from the zero of the anode voltage, is called the ignition angle. In the second figure, the ignition angle α_2 is considerably greater than α_1, with the result that the intersection of v_g with the ignition line occurs considerably later in the cycle, and the time of delivery of current by the anode is considerably reduced. In the third figure, the angle α_3 is still greater, and we see that when v_g is nearly in phase opposition to e the time of delivery of current by the anode is very small indeed. We have seen that the average direct voltage is the average of the anode voltage curve reckoned over the period of ignition, and we therefore see that grid control gives a means of varying the output voltage. Furthermore, since the grid voltage is inoperative once the arc has been started, this voltage need not be sinusoidal but can be in the form of almost instantaneous impulses, which can be obtained from a synchronously-driven alternator which gives a very peaky wave form.

The application of this method of voltage control in the case of a six-anode rectifier is illustrated by the figures in Fig. 27.27. In the

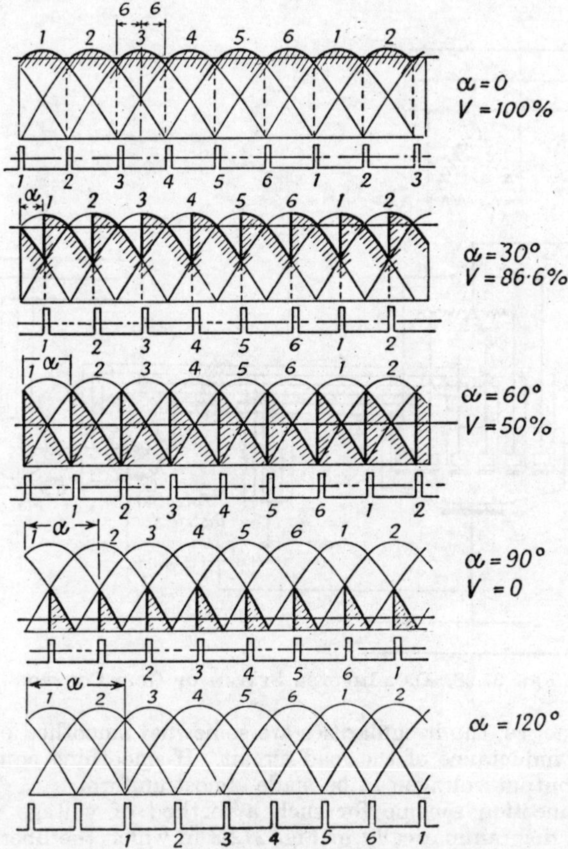

FIG. 27.27. VARIATION OF OUTPUT VOLTAGE WITH PHASE ANGLE OF ARC COMMUTATION

(English Electric Co., Ltd.)

first figure, the grid impulses synchronize with the intersections of the various anode voltage curves so that the grids try to start the arc at the moment it would have started without their aid. The grids thus have no effect and the output voltage is that of a six-phase rectifier, without grids: this voltage is taken to be 100 per cent. In the second figure, the instant of application of the grid impulses has been advanced an angle $\alpha = 30°$ with the result that the arc starts suddenly at an anode when the impulse is applied, and persists until the impulse has been applied to the next anode in the sequence. With $\alpha = 30°$ the average output voltage, as represented by the average under the shaded area, is reduced to 86·6 per cent. With $\alpha = 60°$ this average is reduced to 50 per cent, and so on, until when $\alpha = 90°$ the output voltage is reduced to zero.* The above curves of output voltage variation are theoretical curves corresponding to a pure resistance load,

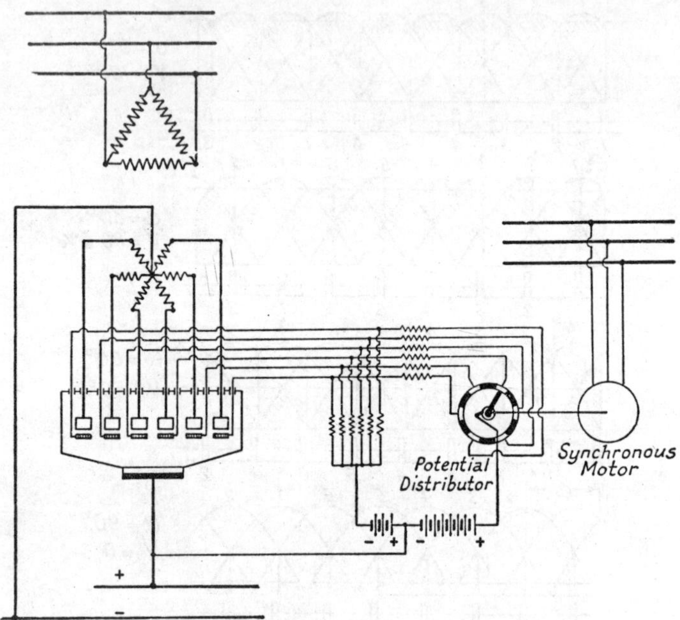

FIG. 27.28. D.C. IMPULSE SYSTEM OF GRID CONTROL

and, in practice, the irregularities are somewhat smoothed out by the inevitable inductance of the load circuit. If smoothing equipment is used, the output voltage can be made almost uniform.

The connection scheme for such a method of voltage control is illustrated diagrammatically in Fig. 27.28 in which the liberating and blocking voltages are shown as derived from batteries, and the grid impulses applied via a synchronously-driven distributor. In such a scheme the instant of ignition is controlled by varying the angular position of the distributor. There are other systems of voltage control, one being the saturated transformer, or "peaking" transformer, method. The circuit diagram for one system is given in Fig. 27.29.

The equipment comprises three peaking transformers, with series reactors for providing the necessary impulses for firing the rectifier

* For $\alpha = 90°$, $\cos \alpha = 0$. The shaded regions in the figure for $\alpha = 90°$ only give the positive halves. The sine wave of anode voltages when extended below the axis gives similar negative halves, the total being a saw-tooth wave of zero average value. $\alpha = 120°$ takes the device into the region of inverted working.

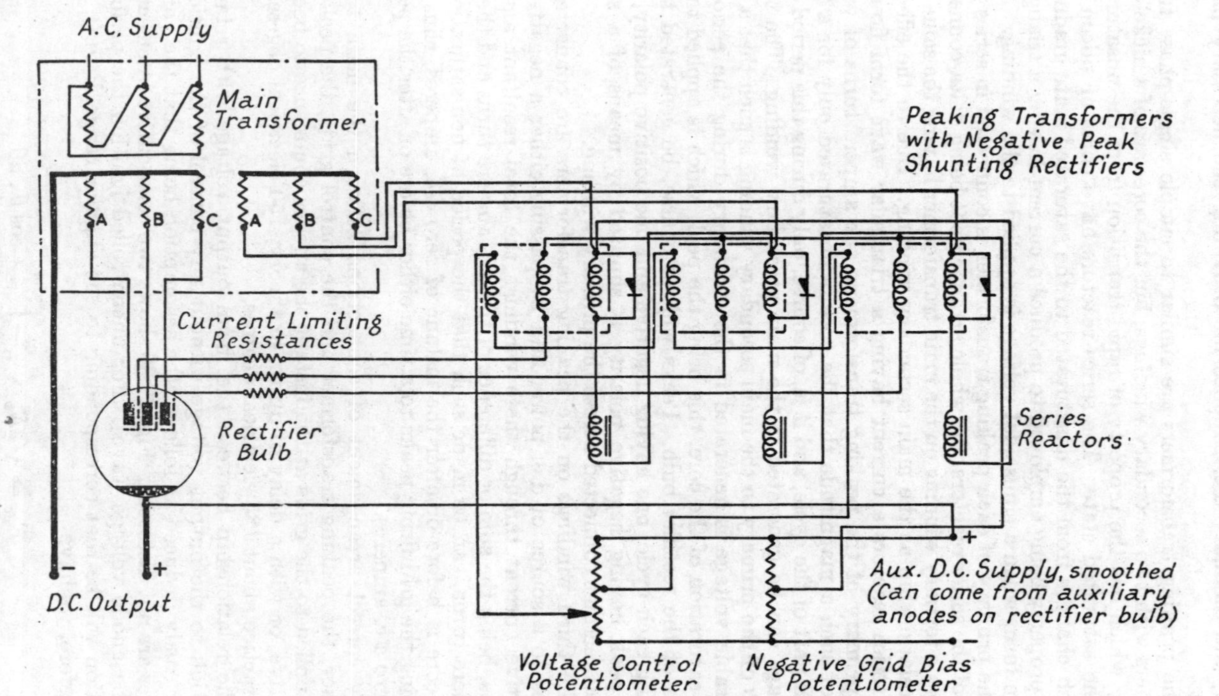

Fig. 27.29. Circuit Diagram for Impulse-type Grid Control

(*Nevelin Electric Supply Co., Ltd.*)

bulb, and auxiliary d.c. supplies for providing the necessary phase-shift.

The peaking transformers are similar to simple single-phase transformers with two secondary windings, but the cores are of a nickel-iron alloy which has the property of rapid transition from the unsaturated to the saturated state. The series reactors have cores of silicon iron which changes from the unsaturated to the saturated state gradually, this property being employed to produce a current having a triangular wave form when a sinusoidal voltage is applied to the winding.

The primary of each peaking transformer is connected in series with one of the series reactors, the whole arrangement being delta-connected across a tertiary winding on the main transformer having the same vector relationship as the main secondary winding. Due to the effect of the series reactors a current having a triangular wave form flows in the primary of the peaking transformers, the ampere turns of which are of such a magnitude that the core is unsaturated only for a very small part of the cycle, and it is, of course, only during this period that voltage can be generated in the main secondary winding. The turns ratio of the primary to the main secondary winding is such that a considerable voltage is generated in the secondary during the period of non-saturation of the core, this being the peak which is applied to the grids of the rectifier bulb. Peaks will, of course, be generated twice during each cycle, one having negative and one positive polarity, and the peak having negative polarity is shunted by means of a small metal rectifier connected across the primary winding.

The third windings on the peaking transformers are connected in series to a source of d.c. power and by passing either a negative or positive current through these windings the total resultant ampere turns, being the sum or difference of the d.c. ampere turns and the a.c. ampere turns, can be made such that the period of non-saturation of the core is before or after the instant of zero a.c. amperes, thus displacing the grid firing peaks to a time either before or after the period of zero a.c. amperes.

By correct arrangement of connections between the main transformer, the peaking transformer and the control grids, the peak can be applied to the grids at the instant the anodes commence to become positive, or with a delay of up to, in theory, 150 electrical degrees, or in practice about 135 electrical degrees.

The relationship between the mean output voltage with retarded ignition to the output voltage without any control of the ignition is very easily deduced as follows, the assumption being that the voltage waves are not distorted. Fig. 27.30 shows two consecutive intersections of the anode voltage waves. For uncontrolled ignition the period of ignition will be that corresponding to the angular range P to Q, and, as before, we have

$$E_{dc} = \sqrt{2}E_p \times \frac{m}{2\pi}\int_{-\pi/m}^{\pi/m} \cos\theta \, . \, d\theta$$

$$= \sqrt{2}E_p(m/\pi)\sin(\pi/m)$$

If the ignition is retarded an angle α by means of grid control, then the arc commences at the instant corresponding to the point R and changes over to the next anode in the sequence, at the instant corresponding to point S. The total period is the same as before, namely, $2\pi/m$, but the limits are now $-(\pi/m - \alpha)$ to $+(\pi/m + \alpha)$

$$\therefore E'_{dc} = \sqrt{2}E_p \times \frac{m}{2\pi}\int_{-\left(\frac{\pi}{m}-\alpha\right)}^{+\left(\frac{\pi}{m}+\alpha\right)} \cos\theta \, . \, d\theta$$

$$= \sqrt{2}E_p(m/\pi)\sin(\pi/m)\cos\alpha$$

$$\therefore E'_{dc}/E_{dc} = \cos\alpha$$

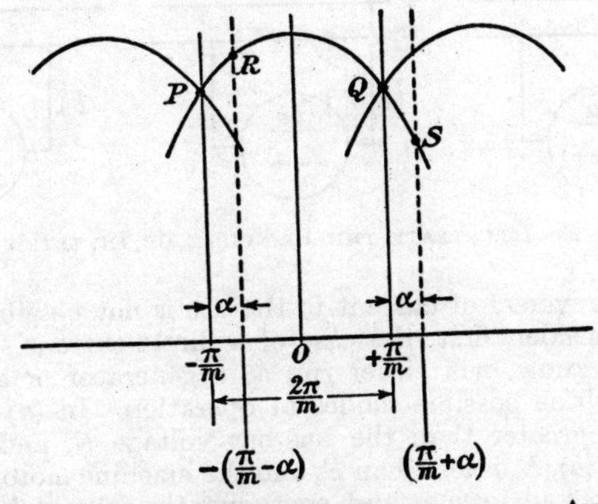

FIG. 27.30. TO ILLUSTRATE THE EFFECT OF THE ANGLE OF GRID EXCITATION

In other words, the ratio of the reduced output voltage, obtained by grid control, to the output voltage given with no such control is equal to the cosine of the angle of ignition retard.

The expression only holds for values of α up to $\pi/2 - \pi/m$. Thus in the case of a 6-anode rectifier the limiting value of α is

$$\pi/2 - \pi/6 = \pi/3 \text{ radians}$$
$$= 60°$$

EXAMPLE. Calculate the average value of the output voltage of a 6-anode rectifier with grids biassed to give a delay of 30°; the transformer has a 3-phase mesh-connected primary supplied at 3000 V, and a 6-phase star-connected secondary, the phase turns ratio being 150/11. Assume an arc drop of 25 V and neglect the effect of overlap. (I.E.E.)

$$E_p = 3000 \times 11/150 = 220 \text{ V}$$

Mean d.c. voltage without grid control

$$E_{dc} = \sqrt{2} \times 220 \times (6/\pi)\sin(\pi/6)$$
$$= 297{\cdot}0 \text{ V}$$

With grid control set to make $\alpha = 30°$

$$E'_{dc} = 297 \cos 30° = 257{\cdot}2 \text{ V}$$

Hence deducting the arc drop of 25 V

$$\text{Mean rectified output voltage} = 257{\cdot}2 - 25 = 232{\cdot}2 \text{ V}$$

Inverted Operation. The rotary convertor can operate either direct or inverted, but the rectifier, without grids, can only operate direct, because it is not possible to reverse the direction of current flow. With grid control, however, inverted operation is possible in spite of

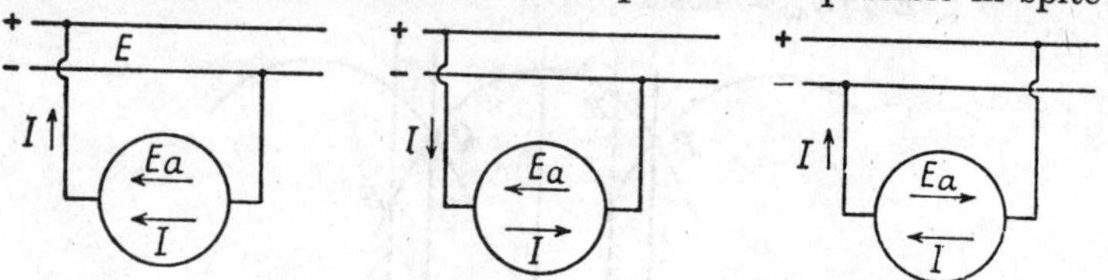

FIG. 27.31. TO ILLUSTRATE THE PRINCIPLE OF INVERTED OPERATION

the fact that reversal of current in the arc is not possible. To understand this, consider, first, the case of a direct-current shunt machine which, as we know, will either run as a generator or a motor. Fig. 27.31 shows three possible modes of operation. In (*a*) the armature voltage E_a is greater than the bus-bar voltage E, and the machine generates. In (*b*), E_a is less than E, and the machine motors, but neither voltage is reversed in sign and consequently current has to reverse. In (*c*), E_a is again less than E, but the connections to the bus-bars have been reversed, and E_a has been reversed in direction, with the result that motoring action has been secured without the reversal of current flow. The inverted operation of a mercury-arc rectifier is analogous to the above; the current flow cannot be reversed, and consequently the voltage must be reversed, and at the same time must be reduced. Now without grids only the positive portions of the anode voltage curves are utilized so that reversal is impossible, but with grid control the impulses can be timed so as to utilize negative portions of the voltage waves and also to give this reversed voltage the required reduced value. For inverted operation the leads to what is now the d.c. source must also be changed over as in the case of the direct-current machine of Fig. 27.31. Fig. 27.32 shows, in a diagrammatic way, the difference between normal and inverted running of the rectifier. The external load is shown as a direct-current machine which in the first case runs as a motor, and when the rectifier is functioning inverted, runs as a generator.

There are several practical applications of the mercury-arc invertor, one of these being Ward-Leonard control of a direct-current motor with two rectifiers taking the place of the motor-generator set, as shown diagrammatically in Fig. 27.33. The control of the speed is effected by grid control of the output voltage of a rectifier in place of excitation control of the generator voltage in the ordinary system. One of the rectifiers is used for forward running, while the other is used when the direction of rotation of the motor has to be reversed. For simplicity, a single-phase supply is shown in the diagram, the dotted

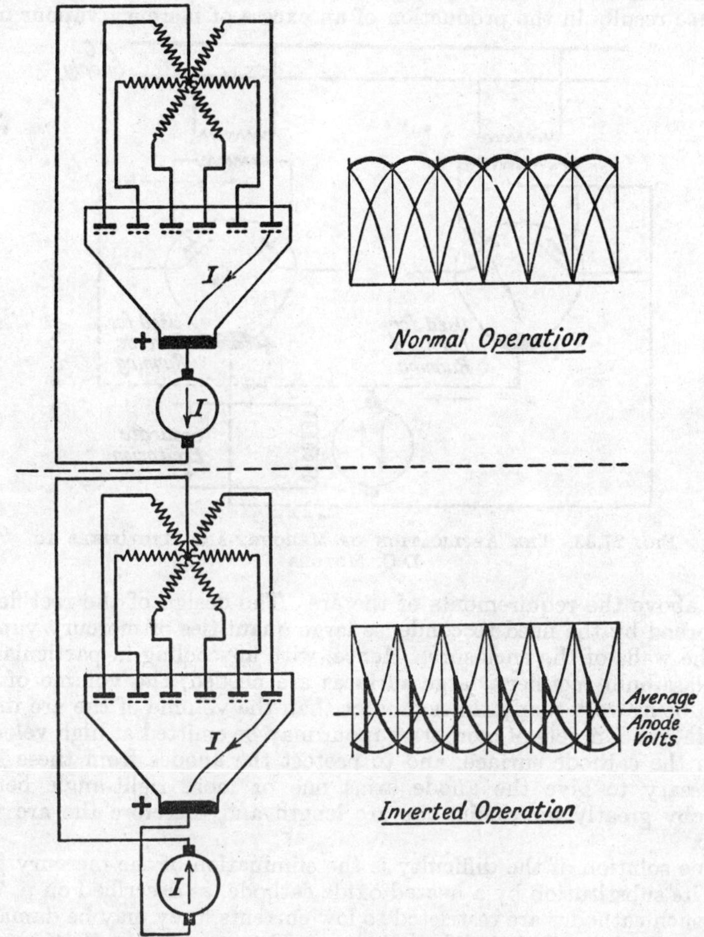

FIG. 27.32. INVERTED OPERATION OF MERCURY-ARC CONVERTOR

connections referring to the rectifier used for reverse running. So long as the driven machine is operating only as a motor, then neither rectifier will be used as an invertor, but if regenerative control is required, motor action will change over to generator action and one or other of the rectifiers will have to operate inverted, according to the direction of rotation.

The Single-anode Mercury-arc Rectifier. In the conventional type of rectifier described previously the arc terminates at a luminous cathode spot on the surface of the mercury pool, and because of uncontrolled thermal relationships this spot travels at high speed and in an erratic manner over the mercury surface. The necessary large cathode surface results in the production of an excess of mercury vapour over

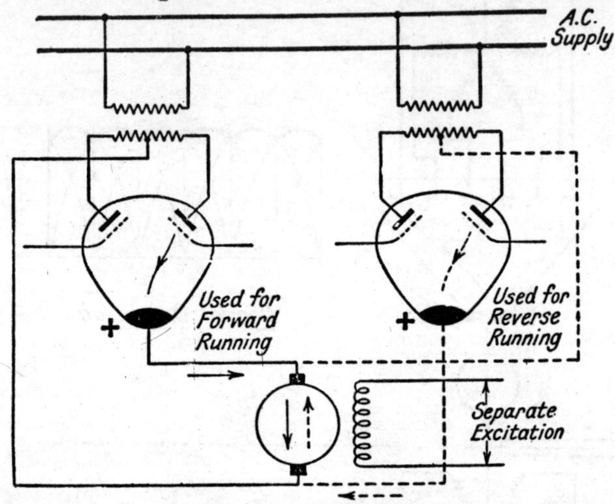

FIG. 27.33. THE APPLICATION OF MERCURY-ARC RECTIFIERS TO D.C. MOTORS

and above the requirements of the arc. The design of the rectifier is governed by the need to condense large quantities of mercury vapour at the walls of the enclosure. Hence, with air-cooling in particular as in glass-bulb rectifiers, large surfaces are needed, the volume of the bulb being very many times greater than the volume of the arc itself. Again, random jets of mercury vapour may be emitted at high velocity from the cathode surface, and to protect the anodes from these it is necessary to give the anode arms one or more right-angle bends, thereby greatly increasing the arc length and therefore the arc volt drop.

One solution of the difficulty is the elimination of the mercury pool and its substitution by a heated oxide cathode, as described on p. 749, but such cathodes are restricted to low currents, they may be damaged by overloads, and their life is limited. The current distribution over the cathode surface is also uneven, particularly in the larger sizes, and all these disadvantages impose an upper limit to the cathode current of about 20 A, r.m.s. Thus, in spite of the disadvantage of the conventional design, the mercury cathode is the best for power applications, and the ideals aimed for are (*a*) elimination of mercury evaporation, (*b*) an overload capacity comparable with that of the conventional type.*

* A more complete discussion is given by H. von Bertele, "Steady-vapour mercury-arc rectifiers." *Direct Current* (Sept., 1952) 43.

The former problem has been solved by the anchoring of the cathode spot. It has been known for some time that another metal protruding from the surface of the mercury pool tends to cause the spot to remain stationary at the junction between the mercury and that metal. The requirements are stated by Von Bertele as follows: ". . . certain rather complex relations were observed between the thermal field of the cathode and the general temperature level of the other parts of the rectifying vessel. It was noticed that with an anchoring system, for a given current, it was necessary for the thermal resistance of the anchoring device to be substantially lower than with equivalent free-moving spots; in the latter case unexpectedly low values of thermal resistance

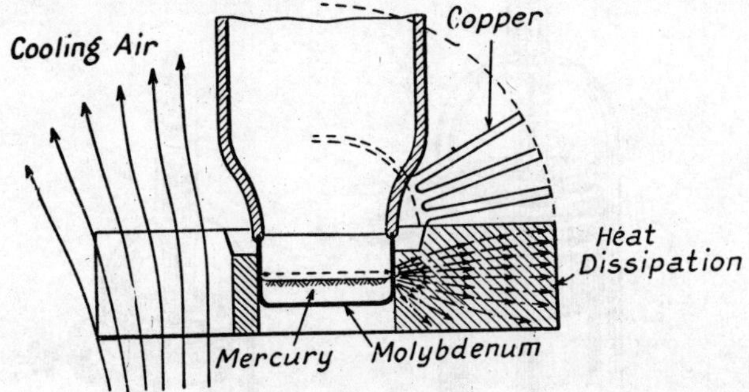

FIG. 27.34. SECTIONAL DIAGRAM OF THE CATHODE SYSTEM OF THE "NEVITRON" RECTIFIER
(Direct Current)

occur, and it was therefore difficult to devise anchoring systems of adequate current-carrying capacity, which possessed at the same time, not only the required low thermal resistance values, but also uniform heat conductivity along the full length of the anchor. This last point was important, because local differences in heat dissipation resulted in a tendency for the spot to concentrate at particular points on the anchor and the resultant heat developed at these points impaired the efficiency of the anchoring system." In the "Nevitron" rectifier the cathode spot is anchored at the periphery of the mercury pool, and is thereby in close proximity to external air-cooled fins. The mercury pool is contained in a molybdenum cup in intimate contact with an external copper ring of which the cooling fins are an integral part. The arrangement is shown in section in Fig. 27.34, which is taken from Von Bertele's paper. The general arrangement of the rectifier is illustrated in Fig. 27.35. The arrangements for ignition and excitation are similar to those used with other mercury-arc rectifiers, thus the ignition is by "dipper" electrode, the operating rod for which is located at the axis of the tube and passes through a cylindrical hole in the main electrode, as can be clearly seen in the views of Fig. 27.35.

Under light load conditions the cathode spot appears at the edge of the mercury pool and, as the load is increased, it extends to form ultimately a continuous luminous ring round the pool. As molybdenum is wetted by mercury a concave meniscus is formed, the mass of mercury at the actual cathode spot, or ring, thus being very small. As a result the vapour pressure within the tube does not vary widely with changes in load current, since it is not, as with the conventional rectifier, determined mainly by the condensation of large quantities of mercury.

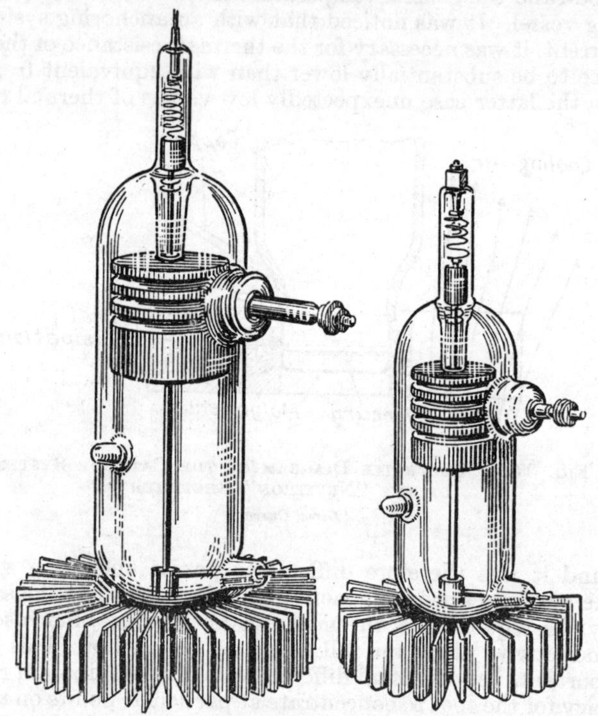

FIG. 27.35. SINGLE-ANODE RECTIFIERS
(*Nevelin Electric Supply Co., Ltd.*)

The absence of uncontrolled evaporation from the cathode makes it possible to locate the anode (and control grid when fitted) directly above the cathode without any intervening screen or baffle: the general layout of the valve is therefore comparable with that of a thyratron (p. 751).

This rectifier can be supplied with control grids, and for polyphase rectification the number required will correspond to the number of anodes on a single conventional type. A useful comparison with a six-anode grid-controlled 300 A rectifier is given below.*

The low arc-drop results in a higher overall efficiency, which is naturally most marked at low d.c. output voltages. Thus, for one

* H. von Bertele, *loc. cit.*

	Six Single-anode Valves with Anchored Cathode	Conventional Six-anode Glass Bulb
Specific valve volume (lit/A)	0·015	0·18
Enclosure volume (ft³)	3·0	42·0
Mercury content (kg)	1·2	11·0
Arc drop at normal load	12·5 V	21·5 V
Excitation power (VA)	600	360
Weight (kg)	7·5	30

particular capacity, the full-load efficiency of the anchored cathode rectifier is about 1·5 per cent higher than that of the conventional type when the output voltage is 500 V, but it is 5 per cent higher when the voltage is 100.

Hot-cathode Rectifiers. The mercury-arc rectifier previously described is a form of cold-cathode rectifier, in that the heating of the cathode necessary for electron emission is localized at the small cathode spot. In the hot-cathode rectifiers, the cathode is heated as a whole by means of a heating current from an external source. The enclosure is highly evacuated apart from a very small amount of mercury. This mercury does not form the cathode, its function being to supply mercury atoms to be converted into positively charged ions by collision, so that, for a given anode voltage, the current output is greater than that of the device used as a vacuum valve. The operation is fundamentally the same as that of the mercury-arc rectifier, the current passing through the tube as a discharge, but only with a possibility of one direction of conventional flow, namely from anode to cathode, because only the anode is capable of accepting electrons. With a single tube, half-wave rectification is obtained, while with two tubes and a centre-tapped transformer, the rectification is full wave. Comparison of the connection schemes of Fig. 27.36 with those of Fig. 27.7 will show that the hot-cathode rectifier and mercury-arc rectifier are essentially similar, the chief difference being that the cathode of the former is a filament which is heated by the passage of current derived from a small filament transformer.

This type of rectifier has several advantages as compared with the mercury-arc rectifier previously described, namely—

1. Smaller size for a given capacity.
2. Smaller arc length, leading to a lower arc drop and therefore a higher efficiency.
3. Lower drop at the cathode, this again increasing the efficiency.
4. Freedom from back-firing.

The cathode is usually on oxide-coated filament, since this type is very efficient from the point of view of electron emission and it is not necessary to raise its temperature to such a high value as a pure metal cathode.

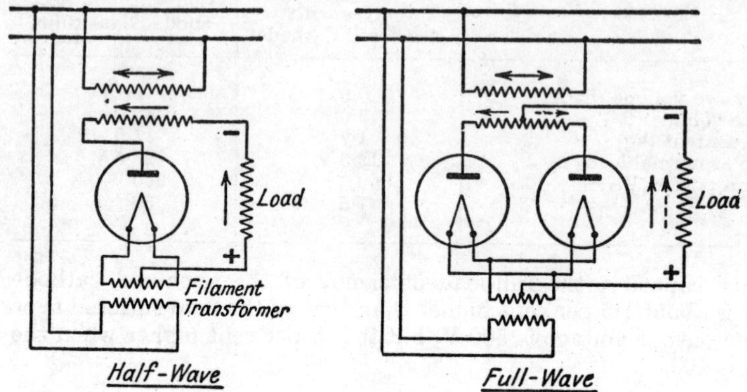

FIG. 27.36. HALF AND FULL-WAVE HOT-CATHODE RECTIFIERS

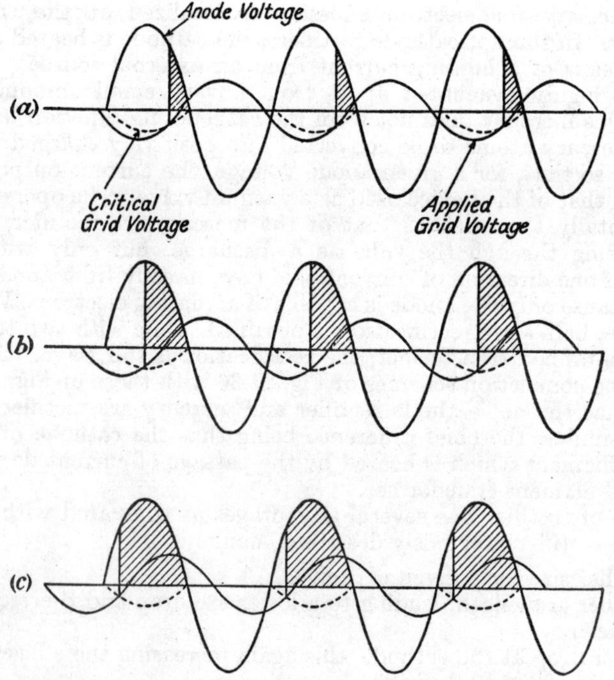

FIG. 27.37. PRINCIPLE OF THE THYRATRON

Unlike the mercury-arc rectifier, in which the vapour pressure is only a small fraction of a millimetre of mercury, the gas or vapour filled rectifiers operate at about 1 mm of mercury. A low anode potential of about 20 V will, under these conditions, accelerate the emitting electrons to the velocity required to produce ionization by collision.

The tungar rectifier is similar to the above, except that it has a tungsten filament and the tube contains an atmosphere of a rare gas, there being no mercury present. It is suited to low voltages and comparatively heavy currents, and is therefore commonly used for charging batteries and similar small power requirements. The kenotron is a highly-evacuated rectifier with no heavy gas molecules, such as

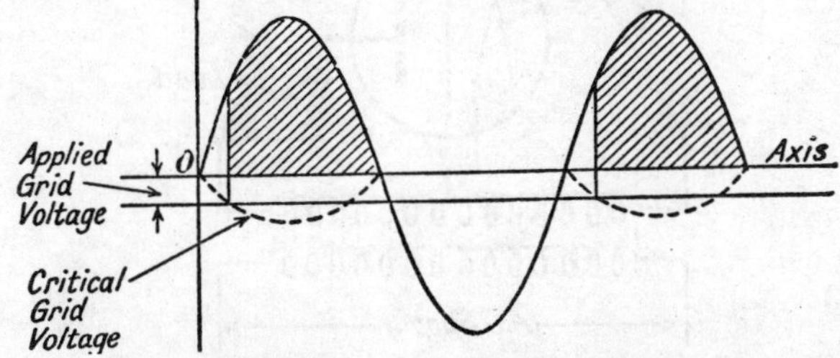

FIG. 27.38. PRINCIPLE OF THE THYRATRON

those of mercury vapour, present. It is therefore only suitable for very small currents, its application being to high-voltage rectification.

If the hot-cathode rectifier is provided with a control grid, then the rectifier becomes a gas-filled relay or thyratron. Just as the two-element hot-cathode rectifier is similar in operation to the mercury-arc rectifier, so is the thyratron similar in operation to the mercury-arc rectifier with grid control. Thus if an alternating potential of supply frequency is applied to the grid, the output will be dependent on the phase of this grid potential. The grid voltage necessary to prevent the anode current is called the *critical grid voltage*: it is not a constant, but varies with the anode voltage.* Hence, if the anode voltage is alternating, the critical grid voltage will also be cyclic (although not sinusoidal), as shown in Fig. 27.37. If the grid is given a steady bias, as represented by the horizontal line below the axis, Fig. 27.38, the thyratron will start to pass current at the intersection of this line with the critical grid voltage curve. By altering the grid bias, the grid line can be shifted either up or down, and can be made to coincide with the axis. There is no current flow during the negative half-cycle and the range of control by this method is from the utilization of the whole of the positive half-cycle to the utilization of half of the positive half-cycle. Thus the average current can only be varied from full to half value by this method.

By applying an alternating grid potential, and making use of phase control, then the average current can be varied from full value right down to zero, as illustrated by the curves of Fig. 27.37. The

* Note that the grid loses control on the start of conduction.

usual circuit for illustrating phase control of the grid voltage is shown in Fig. 27.39, in which it will be seen that the grid is connected to the junction of a capacitor C and a non-inductive resistor R in series. By varying C and R, the phase of the grid voltage can be varied, and therefore the output of the thyratron controlled.

There are many industrial appliances of thyratrons, one of these being illustrated in Fig. 27.40. The thyratron is here operating as a

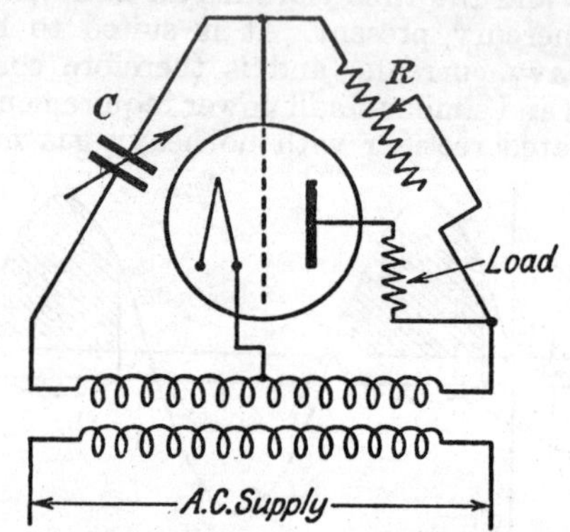

FIG. 27.39. THYRATRON CONTROL CIRCUIT

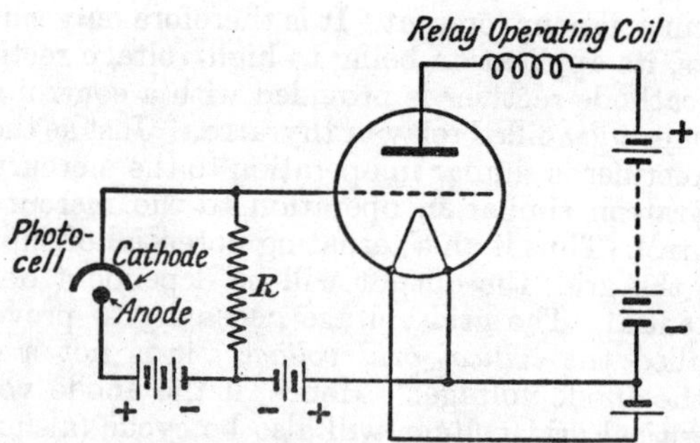

FIG. 27.40. AN APPLICATION OF THE THYRATRON

relay in conjunction with a photo-electric cell. When light is thrown on the cell, current flows through the cell and a potential difference is established across the ends of the high resistance R. This brings the grid of the thyratron within the ignition range and consequently an arc is struck and the thyratron anode current flows through the operating coil of the relay. In some cases the circuit is arranged so that the relay is operated when the light supply to the photo-cell fails. Typical examples of this application are the automatic lighting of street lamps with failing daylight, and the operation of burglar alarms. In the latter application the light beam consists of ultra-violet radiation only and is therefore invisible.

The anode characteristics of a vacuum triode valve, that is, the curves of anode current against anode voltage, are of the form shown in Fig. 27.41, there being a family of curves in which each corresponds to a definite grid voltage. Suppose that the grid is made very negative, corresponding to curve 1, then no anode current can flow until the anode voltage has the value V_{a1}, corresponding to point A. This point is called the cut-off point for the particular value of the grid-voltage, and the grid-voltage is called the cut-off bias. If the anode voltage is raised, say to V_{a2}, point B, then, for the same grid-voltage an anode

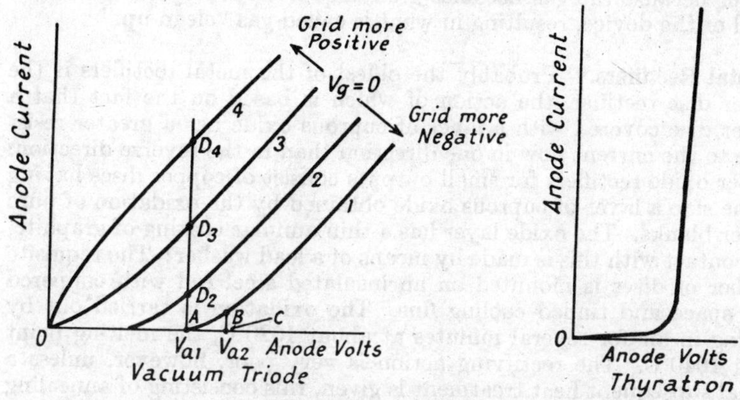

FIG. 27.41. CHARACTERISTICS OF VACUUM TRIODE AND THYRATRON CONTRASTED

current, given by the ordinate BC, will flow. Suppose that the anode voltage is kept at the valve V_{a1}, and that the grid is made progressively less negative then, for the particular grid voltages for which the curves are drawn the anode currents will be: 0, AD_2, AD_3, and AD_4 respectively. If, for the fixed anode voltage V_{a1} a curve is drawn of anode current against grid voltage this curve will be roughly linear, showing a progressively increasing anode current as the grid is made progressively more positive.

Now consider the thyratron: owing to the neutralization of the space charge, the anode-current/anode-voltage curve has the form shown, the whole characteristic, apart from the initial portion being almost a vertical line. Consider the device working with a fixed anode voltage and the grid made so negative that no current can flow. Now let the grid be made progressively less negative until the cut-off point is reached. The anode current then rises suddenly, owing to the shape of the characteristic, and its magnitude is limited only by the impedance of the external circuit. Any further increase in the "positiveness" of the grid has no effect whatever, for which reason the phenomenon is called the "firing" of the thyratron.

The grid potential necessary for ionization to take place, and consequently for the thyratron to fire, is very dependent on the gas or vapour pressure.* Thus as this pressure is increased the grid has to be made more negative. For this reason the thyratron with a mercury vapour filling is very susceptible to temperature, because there is sufficient metallic mercury present to ensure that the internal atmosphere is saturated. Where an inert gas is used instead of mercury this phenomenon is not present and the temperature effect consequently very small. Hence if a thyratron has to be operated over a wide temperature range and is required, at the same time, to have a consistent performance, then the inert gas is preferable to mercury vapour. The life is shorter because the gas becomes gradually adsorbed by the glass and metal of the device, resulting in what is called gas "clean up."

Metal Rectifiers.† Probably the oldest of the metal rectifiers is the copper disc rectifier, the action of which is based on the fact that a copper disc covered with a layer of cuprous oxide has a greater resistance to the current flow in one direction than in the reverse direction. Copper oxide rectifiers for small outputs consist of copper discs having on one side a layer of cuprous oxide obtained by the oxidation of pure copper blanks. The oxide layer has a thin annular coating of graphite, and contact with this is made by means of a lead washer: The requisite number of discs is mounted on an insulated steel rod with coppered steel space and tinned cooling fins. The oxidation is carried out by heating in air for several minutes at about 1020°C, the melting point being 1040°C. The rectifying action is very poor, however, unless a special subsequent heat treatment is given, this consisting of annealing at about 600°C, followed by rapid quenching. The surface layer of black cupric oxide, CuO, formed during cooling is removed by cyanide solution.

The forward average current rating for three-quarter inch diameter discs is about one-third of an ampere, and the maximum safe reverse voltage about 9 V. For large powers, of the order of several kilowatts, the large-plate type of element is used, consisting of a long oxidized copper strip about 2 in. wide. Series-parallel combinations involving standard elements can be arranged to meet any voltage and current requirements.

Another very important metal rectifier is the selenium type, first developed in Germany. With this rectifier, discs of steel are used with a specially prepared surface on which is deposited a thin layer of selenium, a hard grey-coloured element. Contact is made by a ring of special alloy apparently deposited by sputtering. Selenium rectifiers are very efficient, and for the same output are more compact than the copper oxide type.

* When the correct amount of gas or vapour is present in a thyratron the early current flow before firing is minute relative to the value after firing.

† It is not possible to describe the many forms of rectifier which have been devised for different purposes, e.g. the synchronously driven commutator, vibrating reed, electrolytic rectifier, and so on; very full particulars are given by L. B. W. Tolley, *Alternating Current Rectification and Allied Problems*. Chapman & Hall, London; H. Rissik, *The Rectification of Alternating Current*. E.U.P., London.

As with other types of rectifier, the metal rectifier can be used either for half-wave or for full-wave operation, the latter requiring a bridge arrangement of the elements, as shown in Fig. 27.42. Alternatively, full-wave rectification can be secured by means of two rectifiers and a centre-tapped transformer, exactly as with the other types of rectifier. Owing to its simplicity, and the fact that auxiliary batteries or filament supplies are not required, this type of rectifier is very widely used. In a variety of small sizes it is used in the mains units of much radio and electronic equipment (and in measuring instruments, while in larger sizes with large current output it is commonly used in the place of batteries if direct current is required where only an alternating-current supply is available.

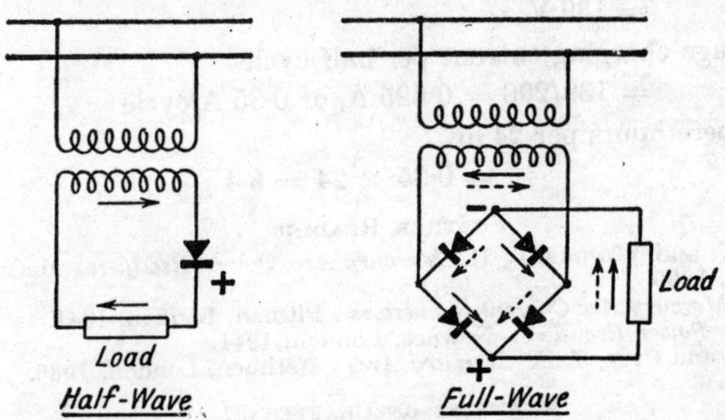

Fig. 27.42. Half- and Full-wave Rectifiers

Example. A half-wave rectifier is used to charge a 40 V battery from an alternating-current supply of r.m.s. voltage 200. The resistance of the rectifier to current in one direction is 200 Ω, and in the other direction very high. If the battery resistance is negligible, calculate the ampere-hours put into the battery in 24 hr.

The battery e.m.f. of 40 V is opposed to the supply voltage during the charging half-cycle, so that for the resultant voltage we have

$$e = E_{\max} \sin \theta - 40$$

Fig. 27.43

But charging cannot commence until the supply voltage has risen to 40 V, and charging ceases just before the end of the half-cycle at the instant the supply voltage has fallen to 40 V. The corresponding angles are

$$\sin^{-1}(40/E_{\max}) = \sin^{-1}(40/283) = 8° \, 7'$$

and

$$180 - 8° \, 7' = 171° \, 53' \text{ (Fig. 27.43)}$$

Taking the origin at the instant the applied voltage is a maximum, we have for the average charging e.m.f. over one-half period

$$\simeq \frac{1}{\pi}\int_{-81°53'}^{+81°53'} 283\cos\theta\, d\theta - 40$$

$$= (2/\pi) \times 283 \sin 81° 53' - 40$$

$$= (2/\pi) \times 283 \times 0{\cdot}99 - 40$$

$$= 179 - 40$$

$$= 139 \text{ V}$$

$\therefore$ Average charging current per half-cycle

$$= 139/200 = 0{\cdot}695 \text{ A, or } 0{\cdot}35 \text{ A/cycle}$$

Hence ampere-hours per 24 hr

$$= 0{\cdot}35 \times 24 = 8{\cdot}4$$

Further Reading

Marti, O. K. and Winograd, H. *Mercury Arc Power Rectifiers.* McGraw-Hill, New York, 1930.

Rissik, H. *Mercury Arc Current Convertors.* Pitman, London, 1941.

Rosslyn, J. *Power Rectifiers.* Newnes, London, 1941.

Teago, F. J. and Gill, J. F. *Mercury Arcs.* Methuen, London, 1936.

Examples on Chapter 27

1. A 3-anode mercury-arc rectifier has a direct full-load current of 100 A. Estimate the equivalent inductance per secondary transformer phase if the overlap angle on full load is to be limited to 45°. The load inductance may be considered large enough to maintain a sensibly constant output current. The transformer phase voltage is 400 V, and the frequency 50 cycles.

Ans. 4·6 mH.

2. A 3-anode mercury-arc rectifier has a full-load output of 20 A at 250 V. The arc-drop is 24 V, and the loss in connections, reactors, etc. is 50 W. A separate test on the transformer gave a core loss of 150 W on open circuit, and an I^2R loss of 40 W in each primary and each secondary phase when the (sinusoidal) output current was 11·55 A.

Calculate the full-load efficiency of the equipment for a highly-reactive load. The transformer has a mesh/star connection.

Ans. 85 per cent.

3. A six-anode, grid-controlled, mercury-arc rectifier equipment is supplied from a 440-V, 3-phase, 50-cycle system. The transformer has a 6-phase, diametrically-connected secondary and a 3-phase, mesh-connected primary. The phase-turn ratio is 110/53. Determine the mean output voltage when the grids are biassed to delay the ignition of the anodes by (*a*) 0, (*b*) $\frac{\pi}{4}$, (*c*) $\frac{\pi}{2}$; and (*d*) find the r.m.s. value of the output voltage and the fluctuation expressed as a percentage of the mean voltage in (*b*). Neglect arc-drop and leakage reactance.

Ans. (*a*) 288 V, (*b*) 203 V, (*c*) 0, (*d*) 212 V, 104 per cent.

4. A double-three-phase (or duplex-six-phase) mercury-arc rectifier gives a full-load d.c. output of 900 kW at 600 V.

Determine the required rating of the primary and secondary windings of the main transformer, and the interphase transformer, neglecting the losses in both the rectifier and transformers.

Assuming that the angle of overlap is 30 degrees, deduce the regulation drop at full load, and the effective leakage reactance of the transformer per phase.

State, briefly, the advantages which this form of circuit has over the Simplex 3- and 6-phase circuits.

Ans. 1440 kVA, 1010 kVA, 84·5 kVA, 41 V, 0·114 ohms.

CHAPTER 28

COMPLEX WAVEFORMS

UP to the present it has been assumed that alternating currents and voltages have been of simple sinusoidal form. Modern alternators are designed to give a terminal voltage which approaches very closely to a sine wave, but under certain conditions both current and voltage may be distorted very considerably. No matter what the waveform may be, the negative half-wave is an exact reproduction of the positive half, Fig. 28.1 (*A*) being a possible waveform, but not Fig. 28.1 (*B*). Since the waveform is repeated at regular intervals, it can be split up into fundamental and harmonic waves all of pure sinusoidal form. Also, because of the fact that the negative half is a reproduction of the positive half, there are no even harmonics. An alternating voltage of any waveform can therefore be represented by the general expression

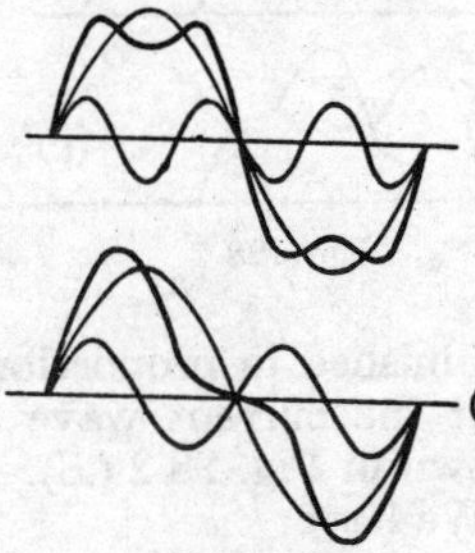

$$e = E_{1\max} \sin \omega t + E_{3\max} \sin (3\omega t + \phi_3) + E_{5\max} \sin (5\omega t + \phi_5) + \ldots$$

(A)

(B)

FIG. 28.1

where $E_{1\max}$, $E_{3\max}$, $E_{5\max}$, etc., are the amplitudes of the fundamental and harmonic waves, $\omega = 2\pi f$, f being the fundamental frequency, and the angles ϕ_3, ϕ_5, etc., the phases of the harmonics with respect to the fundamental.

Effective Value of a Complex Wave. Consider the voltage wave represented by

$$e = E_{1\max} \sin \omega t + E_{3\max} \sin (3\omega t + \phi_3) + E_{5\max} \sin (5\omega t + \phi_5) + \ldots$$

Then

$$E = \text{r.m.s. value} = (\text{average of } e^2)^{\frac{1}{2}}$$

Now

$$e^2 = E_{1\max}{}^2 \sin^2 \omega t + E_{3\max}{}^2 \sin^2 (3\omega t + \phi_3) + E_{5\max}{}^2 \sin^2 (5\omega t + \psi_5) + \ldots + 2E_{1\max}E_{3\max} \sin \omega t \,.\, \sin (3\omega t + \phi_3) + 2E_{1\max}E_{5\max} \sin \omega t \,.\, \sin (5\omega t + \psi_5) + \ldots$$

The average value of $\sin^2\alpha$, where α is any angle, is $\frac{1}{2}$. The average value of $\sin \alpha \sin \beta$, where α and β are angles which correspond to different frequencies (as, for example, ωt and $3\omega t + \phi_3$) is zero.

Hence

$$E^2 = \tfrac{1}{2}E_{1\max}{}^2 + \tfrac{1}{2}E_{3\max}{}^2 + \tfrac{1}{2}E_{5\max}{}^2 + \ldots = (E_{1\text{eff}})^2 + (E_{3\text{eff}})^2 + (E_{5\text{eff}})^2 + \ldots$$

$$\therefore\ E = \{(E_{1\text{eff}})^2 + (E_{3\text{eff}})^2 + (E_{5\text{eff}})^2\}^{\frac{1}{2}}$$

Hence, the effective value of a complex wave is equal to the square root of the sum of the squares of the effective values of the components. Or, in terms of the maximum values of the components, we have

$$E = 0{\cdot}707(E_1^2 + E_3^2 + E_5^2 + \ldots)^{\frac{1}{2}}$$

Influence of the Nature of the Circuit on the Shape of the Current Wave. Consider first a circuit possessing resistance only. Then we have

$$I_1 = E_1/R;\ I_3 = E_3/R;\ I_5 = E_5/R, \text{ etc.}$$

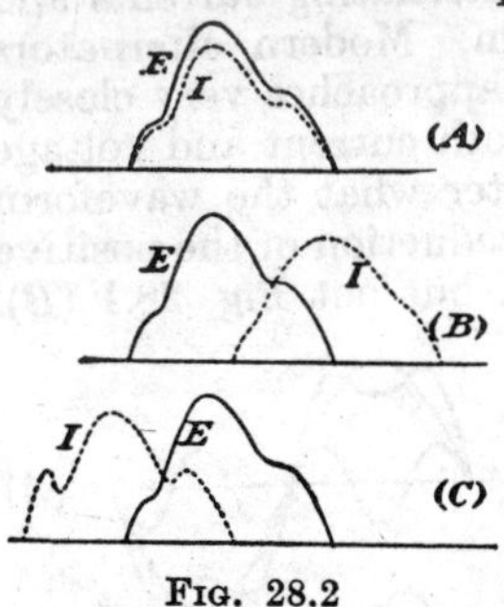

FIG. 28.2

Also, the various currents are in phase with the corresponding voltages. As a result the current wave is an exact reproduction of the voltage wave, and is in phase with it, as shown in Fig. 28.2 (*A*).

Consider now a purely inductive circuit of inductance *L*, then

$$I_1 = E_1/L\omega;\ I_3 = E_3/3L\omega;$$
$$I_5 = E_5/5L\omega, \text{ etc.}$$

Thus, the harmonics in the current wave are diminished in proportion to their frequency numbers, with the result that the current wave is much smoother than the voltage wave, as shown in Fig. 28.2 (*B*).

We have

$$E = 0{\cdot}707(E_{1max}^2 + E_{3max}^2 + E_{5max}^2 + \ldots)^{\frac{1}{2}}$$
$$I = 0{\cdot}707(I_{1max}^2 + I_{3max}^2 + I_{5max}^2 + \ldots)^{\frac{1}{2}}$$
$$= (0{\cdot}707/L\omega)(E_{1max}^2 + E_{3max}^2/9 + E_{5max}^2/25 + \ldots)^{\frac{1}{2}}$$

Hence the effective reactance

$$= E/I = L\omega \times \left(\frac{E_{1max}^2 + E_{3max}^2 + E_{5max}^2 + \ldots}{E_{1max}^2 + E_{3max}^2/9 + E_{5max}^2/25 + \ldots}\right)^{\frac{1}{2}}$$

Now consider a circuit containing capacitance only. Then $I_1 = E_{1max}C\omega$; $I_3 = 3E_{3max}C\omega$; $I_5 = 5E_{5max}C\omega$, etc., the effect of capacitance thus being to magnify the harmonics in the current wave. This is shown in Fig. 28.2 (*C*).

Again,

$$E = 0{\cdot}707(E_{1max}^2 + E_{3max}^2 + E_{5max}^2 + \ldots)^{\frac{1}{2}}$$
$$I = 0{\cdot}707(I_{1max}^2 + I_{3max}^2 + I_{5max}^2 + \ldots)^{\frac{1}{2}}$$
$$= 0{\cdot}707C\omega \times (E_{1max}^2 + 9E_{3max}^2 + 25E_{5max}^2 + \ldots)^{\frac{1}{2}}$$

Hence, effective reactance

$$\frac{E}{I} = \frac{1}{C\omega} \times \left(\frac{E_{1max}^2 + E_{3max}^2 + E_{5max}^2 + \ldots}{E_{1max}^2 + 9E_{3max}^2 + 25E_{5max}^2 + \ldots}\right)^{\frac{1}{2}}$$

Thus the effect of a complex waveform is to diminish the reactance of a capacitor and to increase the reactance of an inductive circuit.*

Selective Resonance. If the frequency of the supply is suitable, resonance may take place, not with the fundamental, but with one of the harmonics. This is called "selective resonance." For resonance with the fundamental, we have the condition $L\omega = 1/C\omega$. For resonance with the third harmonic, we have the condition $3L\omega = 1/3C\omega$; and so on. The effect of selective resonance is to produce enormous distortion of the current wave, in which the fundamental may be actually of smaller amplitude than the particular harmonic which causes resonance, although in the voltage wave the amplitude of the harmonic is much less than that of the fundamental.

EXAMPLE. A voltage wave having a fundamental of amplitude 500, and a third harmonic of amplitude 10, is applied to a circuit containing $R = 1\ \Omega$, $L = 1$ H, and $C = 1/50\ \mu$F in series. Find the frequency for resonance with the third harmonic, and draw the voltage and current waves for this frequency.

For resonance with the third harmonic,

$$\omega^2 = (1/9) \times 1/LC; \quad \therefore \omega = (1/3)1/(1/50 \times 10^{-6} \times 1)^{\frac{1}{2}} = 2360$$

$$\therefore f = 2360/2\pi = 375 \text{ c/s}$$

Impedance of the circuit to the fundamental,

$$\begin{aligned} Z_1 &= \{R^2 + (L\omega - 1/C\omega)^2\}^{\frac{1}{2}} \\ &= \{1 + (2360 - 50 \times 10^6/2360)^2\}^{\frac{1}{2}} \\ &= 19{,}000 \text{ approx.} \end{aligned}$$

$$\therefore I_1 = E_1/Z_1 = 500/19{,}000 = 0{\cdot}026 \text{ A}$$

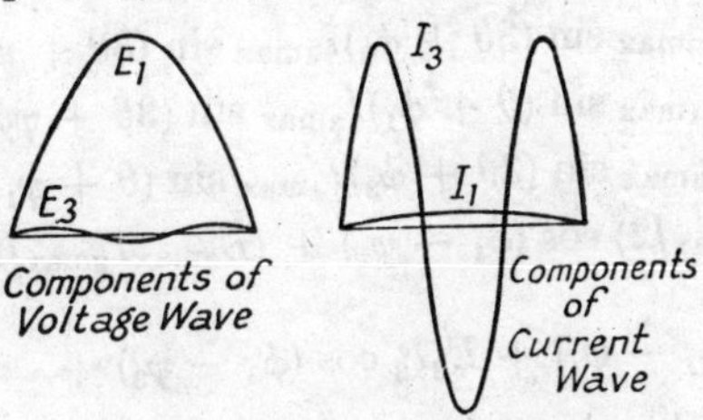

FIG. 28.3. TO ILLUSTRATE SELECTIVE RESONANCE

The impedance of the circuit to the third harmonic is the same as the resistance since there is resonance with this harmonic.

$$\therefore I_3 = E_3/R = 10 \text{ A}$$

The voltage and current waves are shown in Fig. 28.3; and it will be seen that the harmonic is so greatly magnified in the case considered, that to all intents and purposes the current is a triple-frequency current.

* *See also* E. W. Golding, *Electrical Measurements and Measuring Instruments*. Ch. XV, Pitman, London; P. Kemp, *A.C. Wave Forms*. Chapman & Hall, London.

Power Conveyed by Complex Waves. An alternating current can contribute power to a circuit only when it is flowing under the influence of a voltage of the same frequency. Hence if the voltage and current waves are assumed to be resolved into their components, the total power is made up of the following terms—

E_1 acting on I_1 produces $E_1 I_1 \cos \phi_1$ watts;

E_3 acting on I_3 produces $E_3 I_3 \cos \phi_3$ watts; and so on.

Hence, total power

$$P = E_1 I_1 \cos \phi_1 + E_3 I_3 \cos \phi_3 + E_5 I_5 \cos \phi_5 + \ldots$$

the various power factors being given by

$$\text{Cos } \phi_1 = R/Z_1; \ \cos \phi_3 = R/Z_3; \ \cos \phi_5 = R/Z_5; \text{ and so on}$$

A more general solution is as follows: Put $\theta = \omega t$, then

$$e = E_{1\max} \sin (\theta + \phi_1) + E_{3\max} \sin (3\theta + \phi_3) + \ldots$$

$$i = I_{1\max} \sin (\theta + \psi_1) + I_{3\max} \sin (3\theta + \psi_3) + \ldots$$

$$\therefore \ P = \frac{1}{2\pi} \int_0^{2\pi} ei d\theta$$

$$= \frac{1}{2\pi} \int_0^{2\pi} \{E_{1\max} \sin (\theta + \phi_1) + E_{3\max} \sin (3\theta + \phi_3) + \ldots\} \times \{I_{1\max} \sin (\theta + \psi_1) + I_{3\max} \sin (3\theta + \psi_3) + \ldots\}$$

$$= \frac{1}{2\pi} \int_0^{2\pi} \{E_{1\max} \sin (\theta + \phi_1) I_{1\max} \sin (\theta + \psi_1)\} + \{E_{3\max} \sin (3\theta + \phi_3) I_{3\max} \sin (3\theta + \psi_3)\} + \{E_{1\max} \sin (\theta + \phi_1) I_{3\max} \sin (3\theta + \psi_3)\} + \{E_{3\max} \sin (3\theta + \phi_3) I_{1\max} \sin (\theta + \psi_1)\} + \ldots\} d\theta$$

$$= (E_{1\max} I_{1\max}/2) \cos (\phi_1 - \psi_1) + (E_{3\max} I_{3\max}/2) \cos (\phi_3 - \psi_3) + \ldots$$

$$= E_1 I_1 \cos (\phi_1 - \psi_1) + E_3 I_3 \cos (\phi_3 - \psi_3) + \ldots$$

All the terms which are products of e.m.f.s and currents of different frequencies become zero when integrated between the limits 0 and 2π, the total power therefore being equal to the above summation.

With complex waves the power factor obviously cannot be denoted by the cosine of a phase angle, because the various components have, in general, different phase angles. It can be defined by the expression

$$\text{Power factor} = (\text{true power})/(\text{apparent power})$$

Now the true power in a circuit not containing motors, and in which there are no iron losses, is equal to the copper loss, i.e. to RI^2.

$$\therefore \ \text{Power factor} = RI^2/EI = RI/E$$

Again, $E = (E_1^2 + E_3^2 + E_5^2 + \ldots)^{\frac{1}{2}}$

$I = (I_1^2 + I_3^2 + I_5^2 + \ldots)^{\frac{1}{2}}$

But $I_1 = E_1/Z_1 = (E_1/R)(R/Z_1) = (E_1/R) \cos \phi_1$

Similarly, $I_3 = (E_3/R) \cos \phi_3$; $I_5 = (E_5/R) \cos \phi_5$; and so on

$$\therefore\ I = (1/R)(E_1^2 \cos^2 \phi_1 + E_3^2 \cos^2 \phi_3 + E_5^2 \cos^2 \phi_5 + \ldots)^{\frac{1}{2}}$$

Hence, power factor

$$= \left(\frac{E_1^2 \cos^2 \phi_1 + E_3^2 \cos^2 \phi_3 + E_5^2 \cos^2 \phi_5 + \ldots}{E_1^2 + E_3^2 + E_5^2 + \ldots}\right)^{\frac{1}{2}}$$

This expression shows that with a complex wave the power factor can never have the same value that it would have with sinusoidal waves of the same effective value, except in the case of a pure resistance circuit. Also, resonance cannot bring the power factor to unity, as it does with sinusoidal waves, because if one of the angles ϕ_1, ϕ_3, ϕ_5, etc. is zero, then all the others will be finite.

Effect of Iron on the Shape of the Current Wave. Consider an iron circuit, say, an anchor ring, of cross section A m². Let it be wound with a coil of N turns, and a sinusoidal voltage, applied. The apparatus is then similar to a transformer on no load and a sinusoidal flux given by

$$E = 4{\cdot}44\ B_{max}\ ANf$$

i.e.

$$B_{max} = E/4{\cdot}44\ ANf$$

will be produced. Also, this flux will lag 90° behind the applied voltage.

The magnetizing force H is proportional to the current, and therefore, if the B-H curve for the iron were a straight line through the origin, and there were no hysteresis, the current curve would be sinusoidal and in phase with the flux density curve. The current curve would thus be in quadrature (lagging) with the applied voltage, and the power would be zero. This is to be expected, since under these conditions the area of the B-H loop for a complete cycle would be zero, and we are not taking eddy-current loss into account. Actually, the B-H loop has a definite area, and therefore, the current lags less than 90° behind the voltage in order that there may be some real power to supply the hysteresis loss. Also, since the B-H curve is not a straight line, but is a curve of the well-known shape, the current is distorted from the pure sinusoidal form. In order to be able to draw the current curve it is necessary to calculate B_{max} and then draw the hysteresis loop corresponding to this value of B_{max}. (*See* Fig. 28.4.) Knowing B_{max}, and remembering that the flux density is sinusoidal, the curve of flux density can be very easily drawn as shown. Taking a series of points on this curve, projecting them on to the hysteresis loop, and reading off the corresponding value of the magnetizing current I, the current wave can be plotted. This wave contains a pronounced third harmonic whose amplitude, relative to the fundamental, increases as the saturation of the iron is increased. If ordinates are drawn, the mean of the products of corresponding ordinates on the current and voltage curves over a complete cycle will give the true power. Also, the effective (r.m.s.) value of the current can be calculated. The equivalent sinusoidal current wave is that current of the same

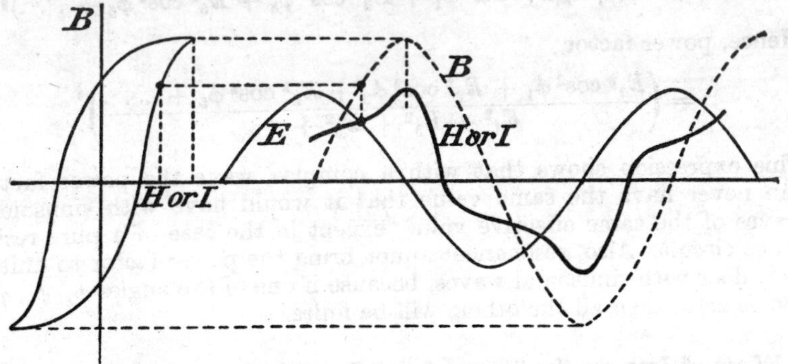

FIG. 28.4. WAVEFORM OF MAGNETIZING CURRENT

effective value, which, if acted on by the applied voltage, would produce the same power. If I is the effective value, as determined from the ordinates, and ϕ is the angle of lag of the equivalent sinusoidal wave, then

$$\text{Average power} = EI \cos \phi$$

$$\therefore \phi = \cos^{-1} (\text{average power}/EI)$$

The curve of the equivalent current wave thus leads the curve of flux density by an angle ψ

where $$\psi = 90 - \phi$$

The angle ψ is called the "Hysteric angle of lead."

It will be seen that the magnetizing current of a transformer will not be of pure sinusoidal form because of the hysteresis loss in the core; hence, the assumption of a sinusoidal form as on page 395 is only approximate.

The distortion produced by magnetic saturation is such that the current wave possesses a very pronounced third harmonic. This effect can be made use of to produce a triple-frequency current from a three-phase supply as follows. A transformer D (Fig. 28.5) has its primary winding in three separate parts, which are connected each in series with one of the chokers A, B, and C, to a three-phase supply. These chokers are designed to be very highly saturated, so that the primary currents each contain a large third harmonic. Now, while the mutual phase difference of the fundamentals is 120°, that of the harmonics is zero. As a result, there is no flux of supply frequency in the core of D, but there is a flux of three times normal frequency. This induces an e.m.f. of triple frequency in the secondary winding of D.

This method is sometimes used when a small amount of power for lighting is required from a three-phase low-frequency supply. A low frequency, although suitable for power work, is unsuitable for lighting, especially with metal filament lamps.

The above discussion of the effect of saturation of the iron applies to a circuit having a steady voltage of sinusoidal waveform applied to it. Now take the case of a coil with iron core, whose winding is connected in series with a circuit which is taking a sinusoidal current from

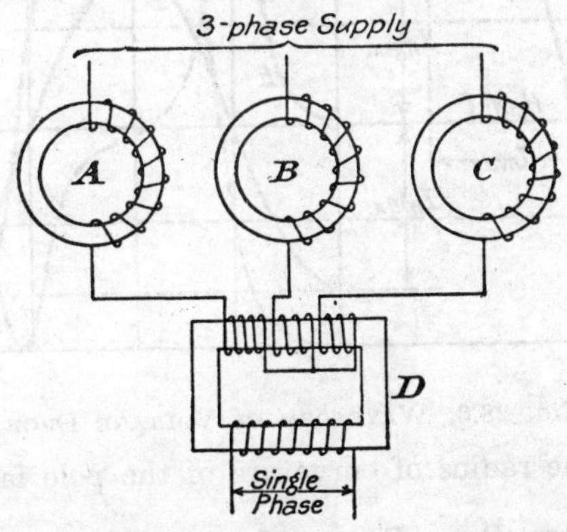

FIG. 28.5. TAYLOR'S METHOD OF FREQUENCY TREBLING*

the supply and whose impedance is so great that the introduction of the coil will not disturb this current. Then the curve of flux against time will no longer be sinusoidal, but will be flat-topped as is shown in Fig. 28.6. The flux wave is here obtained by projecting from the sinusoidal current wave on to the magnetization curve. Now the voltage induced in the coil, and therefore the voltage drop (neglecting the effect of resistance) is proportional to the rate of change of flux, the voltage drop curve thus being of the same form as the slope of the flux curve. This shows that the drop of volts is not sinusoidal, but is a peaked curve, the peaks being exceptionally pronounced when the iron is strongly saturated.

Waveform of an Alternator. The terminal voltage is the vector sum of the e.m.f.s in the individual conductors. Hence, to obtain a sinusoidal terminal voltage it is necessary that these individual e.m.f.s should be as nearly sinusoidal as possible. Each of these e.m.f.s has a waveform which is an exact reproduction of the distribution of flux round the air gap, so that the first essential for a good waveform is that this flux distribution should be as nearly sinusoidal as possible. There are three methods of obtaining this—

(*a*) ROUNDING OR CHAMFERING THE POLE FACE (Fig. 28.7). This

* The illustration does not show a fourth wire connected to the neutral of the three primary coils. This wire is essential, otherwise no third harmonic currents can flow as they are all co-phasal.

makes the air gap at the middle smaller than at the pole tips. Theoretically, the width of the air gap should be a sinusoidal function of the angular distance from the centre; but as this is not practicable, it is

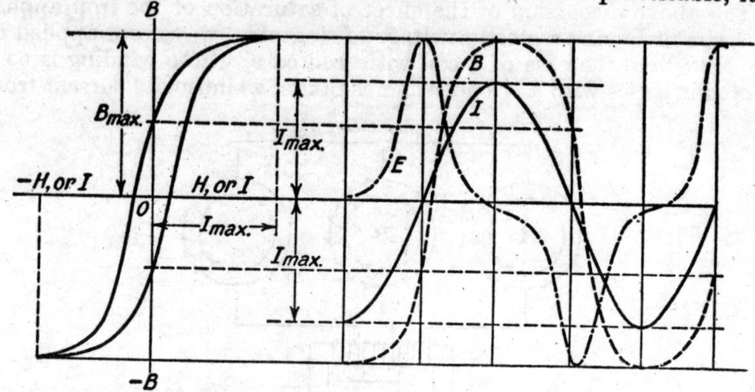

FIG. 28.6. WAVEFORM OF VOLTAGE DROP

usual to make the radius of curvature of the pole face about 0·7 of the air-gap radius.

(b) SKEWING THE POLE FACE. This causes the active length of conductor to vary from zero at the pole tip to a maximum at the centre. Theoretically, the axial length of the pole face should be a sinusoidal function of the angular distance from the pole centre, as in Fig. 28.8 (B). This is obviously impracticable, and therefore, the

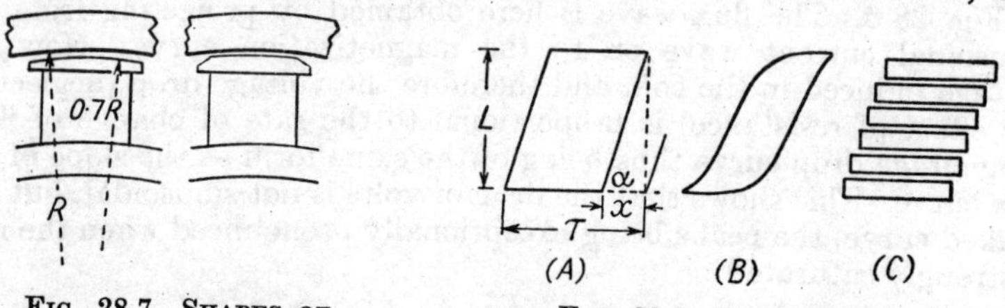

FIG. 28.7. SHAPES OF ALTERNATOR POLES

FIG. 28.8. POLE SHOES

rhomboidal shape of Fig. 28.8 (A) is adopted. Generally, the pole face is built up of laminations, and to enable stampings of the same shape to be used, the pole face is built up of separate packets of laminations, as in Fig. 28.8 (C).

The pole pitch $\tau = \pi D/p$; where D = diameter

Let β = ratio of pole span to pole pitch

$\therefore$ Distance $x = (\pi D/p)(1 - \beta)$

and the inclination of the pole edge is given by

$$\tan \alpha = L/x$$

An average value for the ratio β is 0·65 to 0·7.

(c) USE OF A CYLINDRICAL ROTOR WITH DISTRIBUTED FIELD WINDING. This method is now invariably adopted with turbo-alternators. With a salient-pole field system the flux distribution is approximately rectangular, as in Fig. 28.9 (*A*), and it therefore possesses a very pronounced third harmonic. With a distributed field winding and a non-salient-pole rotor, as in Fig. 28.9 (*B*), the m.m.f. distribution

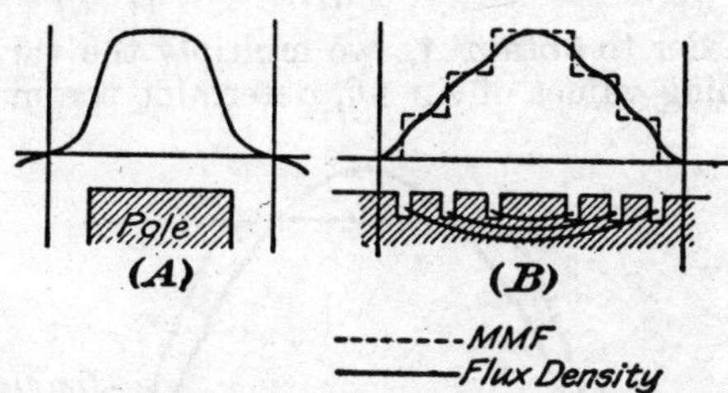

FIG. 28.9. COMPARISON OF SALIENT POLE AND CYLINDRICAL ROTORS

is stepped as shown dotted. Because of fringing, the flux distribution rounds off the corners, thus giving a rough sine wave for the flux density.

Determination of the Harmonics in the Field Form. We have already seen, p. 153, how the curve of gap density can be determined. This can be analysed into a Fourier series of general form

$$b = A_0 + A_1 \sin \omega t + A_2 \sin 2\omega t + A_3 \sin 3\omega t + \ldots$$
$$+ B_1 \cos \omega t + B_2 \cos 2\omega t + B_3 \cos 3\omega t + \ldots$$

or, in sine terms only

$$b = A_0 + C_1 \sin (\omega t + \alpha_1) + C_2 \sin (2\omega t + \alpha_2) + \ldots$$

where $C_1 = (A_1^2 + B_1^2)^{\frac{1}{2}}$; $C_2 = (A_2^2 + B_2^2)^{\frac{1}{2}}$; etc.

$$\alpha_1 = \cos^{-1} (A_1/C_1); \quad \alpha_2 = \cos^{-1} (A_2/C_2); \text{ etc.}$$

These angles being the angles between the various harmonics and the fundamental term of periodicity ω ($= 2\pi f$). In the case of the gap density distribution, the identical distribution is repeated under all the poles so that $A_0 = 0$, and all even harmonics are zero. Again, at no-load, the curve is symmetrical about the polar axis, so that the cosine terms in the original expression vanish, leaving

$$b = A_1 \sin \omega t + A_3 \sin 3\omega t + A_5 \sin 5\omega t + \ldots$$

or $$b = A_1 \sin \theta + A_3 \sin 3\theta + A_5 \sin 5\theta + \ldots$$

Suppose we require the amplitudes of the harmonics up to the eleventh, then the abscissae of flux distribution for one-half of the pole span is divided into 12 equal parts, namely, 7·5 electrical degrees, and the ordinates read at each division. We then have for the amplitude of the harmonic of the mth order

$$A_m = \frac{1}{\pi}\int_0^{2\pi} b \sin m\theta d\theta$$

showing that in order to obtain A_m we multiply the various ordinates by the corresponding values of $\sin m\theta$, determine the mean value and

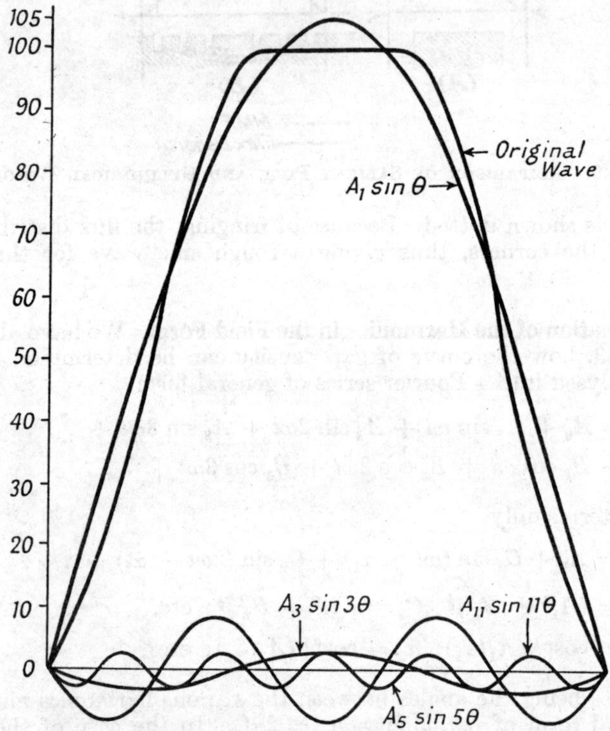

FIG. 28.10. ANALYSIS OF FIELD FORM

multiply by 2, the reason for this last step being that although the integration in the expression for A_m is taken over the range 0 to 2π, the integral is divided by π and not 2π. Again, as we are only taking ordinates up to 90° owing to symmetry, it is necessary to take only half the ordinate at this point. An actual flux distribution is shown in Fig. 28.10, the ordinates being

$\theta =$	0	7·5	15	22·5	30	37·5	45	52·5	60	67·5	75	82·5	90
$e =$	0	7·1	16·2	29	44	63	79	90·5	98	100	100	100	$\frac{100}{2} = 50$

$$\therefore\ A_1 = (2/12)(7{\cdot}1 \sin 7{\cdot}5 + 16{\cdot}2 \sin 15 + 29 \sin 22{\cdot}5 + \ldots + 50 \sin 90)$$

$$A_3 = (2/12)(7{\cdot}1 \sin 22{\cdot}5 + 16{\cdot}2 \sin 45 + \ldots + 50 \sin 270)$$

$$A_5 = (2/12)(7{\cdot}1 \sin 37{\cdot}5 + 16{\cdot}2 \sin 75 + \ldots + 50 \sin 450)$$

and so on.

These come to

$$A_1 = 104{\cdot}5,\ A_2 = -\,2{\cdot}95,\ A_5 = -\,8{\cdot}82$$

$$A_7 = -\,0{\cdot}65,\ A_9 = -\,0{\cdot}45,\ A_{11} = -\,2{\cdot}75$$

These are plotted in their correct relative amplitudes in Fig. 28.10.

Elimination of Harmonics by Special Arrangements of the Stator Windings. (*a*) USE OF SEVERAL SLOTS PER POLE PER PHASE. If a

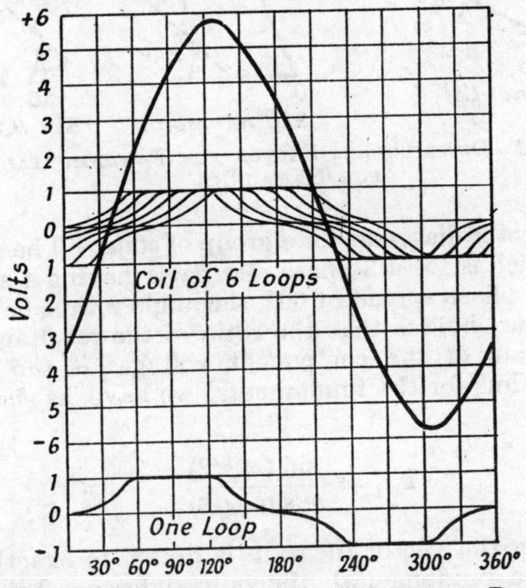

FIG. 28.11. TOTAL E.M.F. INDUCED IN A MULTI-TURN DISTRIBUTED COIL

number of non-sinusoidal e.m.f.s having a small mutual phase difference act in series, their resultant will, as shown by Fig. 28.11, be much more nearly sinusoidal than the components. Hence, if the winding is arranged so that the e.m.f.s in individual conductors in each phase have a slight phase difference, this will diminish the harmonics in the terminal voltage. This is done by adopting distributed instead of concentrated windings.

The e.m.f. induced in an individual armature conductor is of the same wave form as the air-gap flux density distribution and therefore contains the same harmonics in the same relative proportions. In a practical winding harmonics in the total e.m.f. can be reduced, and sometimes even eliminated.

The distribution of the winding over several slots per pole per phase displaces the various coils comprising a coil group in such a way that there is a progressive difference in phase between the various e.m.f.s induced in them. If the slot pitch is ψ electrical degrees, the phase difference of the components of fundamental frequency for two consecutive coils will be ψ, for the third harmonic it will be 3ψ, for the fifth harmonic 5ψ, and so on. Considering the various harmonics

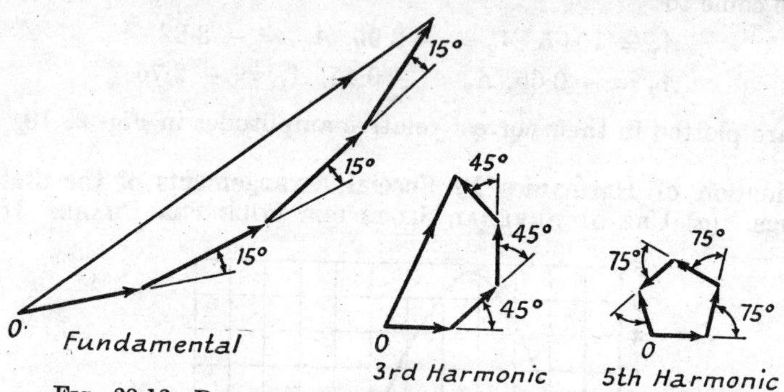

FIG. 28.12. DISTRIBUTION FACTOR FOR FUNDAMENTAL AND FOR HARMONICS

separately, the vector diagrams for a group of coils will be as illustrated in Fig. 28.12, which is for a 3-phase alternator having 4 slots per pole per phase, and a phase spread of 60°, the angle ψ thus being 15°. For any harmonic it is obvious that the ratio of the resultant voltage to the arithmetic sum of the component voltages is the distribution coefficient $k_{2.n}$. Thus for the fundamental we have, as shown on page 462,

$$k_{2\cdot1} = \frac{\sin(m\psi/2)}{m\sin(\psi/2)}$$

For a harmonic, the vector diagram is drawn in exactly the same manner as for the fundamental, the only difference being that the phase angle between the consecutive component voltage vectors is now $n\psi$ instead of ψ, with the result that the distribution factor for the nth harmonic now becomes

$$k_{2.n} = \frac{\sin(mn\psi/2)}{m\sin(n\psi/2)}$$

Consider the winding in which $m = 4$ and $\psi = 15°$

$$k_{2\cdot1} = \frac{\sin(4\times15/2)}{4\sin(15/2)} = 0\cdot96$$

$$k_{2\cdot3} = \frac{\sin(4\times3\times15/2)}{4\sin(3\times15/2)} = 0\cdot653$$

$$k_{2\cdot5} = \frac{\sin(4 \times 5 \times 15/2)}{4 \sin(5 \times 15/2)} = 0{\cdot}205$$

$$k_{2\cdot7} = \frac{\sin(4 \times 7 \times 15/2)}{4 \sin(7 \times 15/2)} = -\,0{\cdot}158$$

THE USE OF FRACTIONAL-PITCH WINDINGS. Generally, alternator windings are arranged with a whole number of slots per pole per phase. If the number is fractional, e.g. $3\frac{1}{2}$ slots per pole per phase, the winding is called a "fractional pitch" winding. With an ordinary distributed winding, the e.m.f.s in the conductors of one group of coils are given a small phase difference, but the successive groups occupy identical positions relative to the poles. With fractional-pitch windings, the various groups do not occupy identical positions with respect to the poles, with the result that phase differences are produced between groups of coils forming one phase, as well as between individual coil sides.

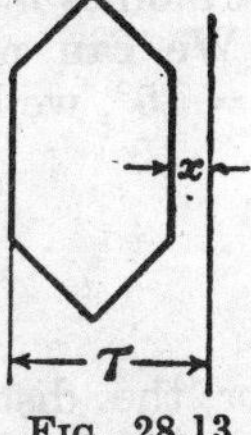

FIG. 28.13

(c) THE USE OF SHORT-CHORD WINDINGS. A "short-chord" winding is one in which the coil width is less than the pole pitch. There is thus a phase difference between the e.m.f.s in the two coil sides. By the use of such coils it is possible to eliminate one particular harmonic. Thus, if the distance x (Fig. 28.13) is $1/n$th of the pole pitch, then the nth harmonic will be eliminated, because the nth harmonic in the e.m.f.s in the two coil sides will be in phase opposition.

Now consider the effect of chording in a more general way. Let the deficiency of coil width against pole pitch be θ electrical degrees, giving rise to a chording factor of $k_{3\cdot1} = \cos \frac{1}{2}\theta$. With respect to any harmonic, say the nth, the phase difference will be $n\theta$, so that the chording factor for the nth harmonic becomes

$$k_{3.n} = \cos \tfrac{1}{2}n\theta$$

and therefore if the angle θ is such that $k_{3.n}$ is zero, that harmonic will vanish. Thus if $\theta = 60°$, the coil width then being $\frac{2}{3}$ pitch, the third, ninth, fifteenth, etc., harmonics will vanish, but at the expense of fundamental e.m.f., since a reduction of $\frac{1}{3}$ of the pole pitch is a large amount. We shall see that all harmonics which are multiples of 3 are eliminated from the terminal voltage wave, so that there is no point in eliminating these harmonics by chording. A $\frac{4}{5}$ pitch would eliminate the fifth harmonic and such a pitch would not be too great a reduction in coil width but it is only possible where the number of slots is suitable.

Since chording affects the amplitudes of all the harmonics it might be better to choose a value of θ which, instead of eliminating any one, considerably reduces more than one. Thus for a $\frac{5}{6}$ pitch, $\theta = 30°$, the chording factors are

$$k_{3\cdot1} = \cos(30/2) = 0{\cdot}966$$
$$k_{3\cdot3} = \cos(90/2) = 0{\cdot}707$$
$$k_{3\cdot5} = \cos(150/2) = 0{\cdot}259$$

$$k_{3\cdot 7} = \cos(210/2) = -0{\cdot}259$$
$$k_{3\cdot 9} = \cos(270/2) = -0{\cdot}707$$
$$k_{3\cdot 11} = \cos(330/2) = -0{\cdot}966$$

Such a pitch is very useful because harmonics which are multiples of 3 are not present in the terminal voltage wave in any case, the amplitude of the eleventh is bound to be small, and chording thus gives a very useful reduction in the fifth and seventh.

Example. The gap density distribution of a 3-phase, star-connected alternator can be expressed as

$$100 \sin\theta - 0{\cdot}5 \sin 3\theta - 9{\cdot}5 \sin 5\theta - 1{\cdot}8 \sin 7\theta$$

The winding is distributed over four slots per pole per phase and the coils are chorded by one slot. Calculate the relative amplitudes of the harmonics in the terminal voltage wave.

We can neglect harmonics which are multiples of 3, and since $\theta = 15°$, we have

$$k_{3\cdot 1} = \cos(15/2) = 0{\cdot}991$$
$$k_{3\cdot 5} = \cos(75/2) = 0{\cdot}793$$
$$k_{3\cdot 7} = \cos(105/2) = 0{\cdot}609$$

For the distribution factors $m = 4$ and $\psi = 15°$, giving the previously obtained values of $k_{2\cdot 1} = 0{\cdot}96$, $k_{2\cdot 5} = 0{\cdot}205$, $k_{2\cdot 7} = -0{\cdot}158$. The products of these factors and the amplitudes of the appropriate components of the gap density distribution are therefore

$$100 \times 0{\cdot}991 \times 0{\cdot}96 = 95{\cdot}1$$
$$-9{\cdot}5 \times 0{\cdot}793 \times 0{\cdot}205 = -1{\cdot}55$$
$$-1{\cdot}8 \times 0{\cdot}609 \times -0{\cdot}158 = 0{\cdot}17$$

The terminal voltage wave can thus be expressed as

$$e = 95{\cdot}1 \sin\theta - 1{\cdot}55 \sin 5\theta + 0{\cdot}17 \sin 7\theta$$

or, in terms of a 100 per cent fundamental

$$e = 100 \sin\theta - 1{\cdot}63 \sin 5\theta + 0{\cdot}18 \sin 7\theta$$

The fifth harmonic has been reduced to about one-sixth of its original value and the seventh harmonic to one-tenth.

Suppression of Tooth Ripples. As a pole moves past the stator teeth the configuration of the air gap is altered, its reluctance passing through a series of maximum and minimum values as shown in Fig. 28.14. These variations in reluctance set up a *stationary* wave of magnetism, one cycle of which corresponds to one slot pitch.

Let a = number of slots per pole

$\therefore\ 2a$ = number of slots per pair of poles

Hence, one pair of poles corresponds to $2a$ cycles of this stationary wave, whereas it only corresponds to one cycle of the fundamental e.m.f. The frequency of the stationary wave is therefore $2af$.

Now we have seen that a stationary wave can be split up into two travelling components which rotate in opposite directions. Hence, their angular velocities relative to the field are—

$$(2a + 1)\omega \text{ and } (2a - 1)\omega$$

ω being the synchronous angular velocity. As a result, they set up harmonics of frequencies $(2a + 1)f$ and $(2a - 1)f$.

EXAMPLE. If there are three slots per pole per phase $a = 9$, and the

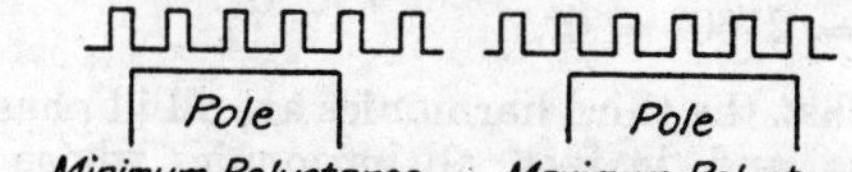

FIG. 28.14. VARIATION IN AIR-GAP RELUCTANCE

harmonics due to tooth ripples are the $(2 \times 9 + 1)$th and $(2 \times 9 - 1)$th, i.e. the 19th and 17th.

In order to suppress the tooth ripples it is necessary to reduce the variations of reluctance of the air gap to a minimum. This can be done—

(a) By making the pole span a whole number of slot pitches.

(b) By employing semi-enclosed or totally enclosed stator slots, as shown in Fig. 28.15. The placing of the windings in these slots is obviously more difficult than with open slots, and for this reason totally enclosed slots are rarely used except for low-voltage bar windings, in which each slot contains large section conductors which can be pushed into the slot from one end.

A B C.

A. Open
B. Semi-Enclosed
C. Totally Enclosed

FIG. 28.15. SHAPES OF SLOTS

Effect of the Interconnection of the Phases. Let the e.m.f. induced in phase I of a 3-phase alternator be given by

$$e_1 = E_1 \sin \theta + E_3 \sin 3\theta + E_5 \sin 5\theta + \ldots$$

Then for the e.m.f.s in phases II and III, we have

$$e_2 = E_1 \sin (\theta - 120°) + E_3 \sin 3(\theta - 120°) + E_5 \sin 5(\theta - 120°) + \ldots$$

and

$$e_3 = E_1 \sin (\theta - 240°) + E_3 \sin 3(\theta - 240°) + E_5 \sin 5(\theta - 240°) + \ldots$$

Hence, taking the direction of the e_1 vector as the reference axis at any instant, we have for the phases of the various components with respect to this axis

1. Fundamentals

   Phase I, 0°; Phase II, 120°; Phase III, 240°.

2. Third harmonics
   Phase I, 0°; Phase II, $3 \times 120° = 360° = 0°$; Phase III, $3 \times 240° = 720° = 0°$.
3. Fifth harmonics
   Phase I, 0°; Phase II, $5 \times 120° = 600° = 240°$; Phase III, $5 \times 240° = 1200° = 120°$.
4. Seventh harmonics
   Phase I, 0°; Phase II, $7 \times 120° = 840° = 120°$; Phase III, $7 \times 240° = 1680° = 240°$.
5. Ninth harmonics
   Phase I, 0°; Phase II, $9 \times 120° = 1080° = 0°$; Phase III, $9 \times 240° = 2160° = 0°$.

and so on. We see that the third harmonics are all in phase, as also are the ninth harmonics, and, in fact, all harmonics whose orders are a multiple of three. The fifth harmonics have the ordinary 3-phase displacements with the exception that the order is reversed from 0°, 120°, 240°, to 0°, 240°, 120°. The seventh harmonics are in the natural order of 0°, 120°, 240°.

Consider now a star-connected machine. The terminal voltage is the vector difference of the voltages of the two phases in series between the terminals, and consequently all the harmonics whose orders are a

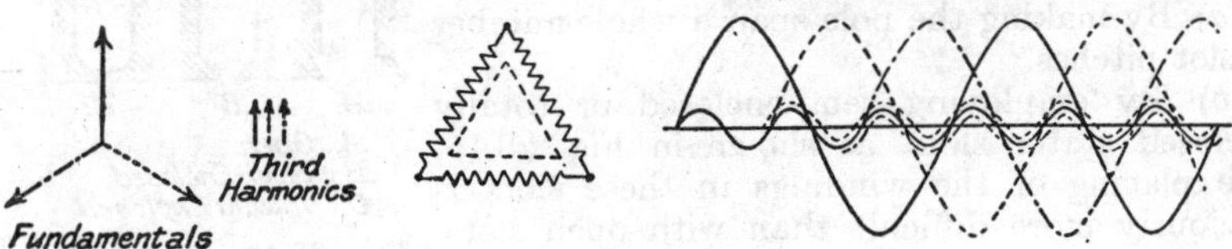

Fig. 28.16

multiple of three vanish from the terminal voltage. These harmonics are however, in phase with respect to the neutral, with the result that currents of corresponding frequencies will be set up if a number of alternators working in parallel have their neutrals connected together.

In the case of a delta-connected machine the voltage acting round the closed circuit is the vector sum of the voltages in the three phases, and we see that third harmonics and their multiples are in phase with respect to this circuit, Fig. 28.16. Consequently, local currents of corresponding frequencies are set up and these currents cause the appropriate voltage harmonics to be utilized internally in the form of volt drops. Thus the voltage harmonics whose orders are multiples of three are absent from the terminal voltage.

Hence, the terminal voltage of a given alternator will contain the same harmonics whether connected in star or in delta, but the wave forms will not be identical because of the phase displacement of the harmonics which the star connection brings about.

Effect of Waveform on Iron Losses. (*a*) Hysteresis Loss. Consider a transformer having a winding of N turns and a core of cross section A.

Then instantaneous applied voltage

$$e = N(d\Phi/dt)$$
$$= NA(dB/dt)$$
$$\therefore\ e\,.\,dt = NA\,.\,dB$$

Hence, the half-area of the voltage wave (Fig. 28.17)

$$S = \int_0^T e\,.\,dt$$
$$= NA\int_{-B_{\max}}^{B_{\max}} dB = 2NAB_{\max}$$

so that S is proportional to $B_{\max}$. But the hysteresis loss

$$W_h \propto B_{\max}^{1\cdot 6}$$
$$\therefore\ W_h \propto S^{1\cdot 6}$$

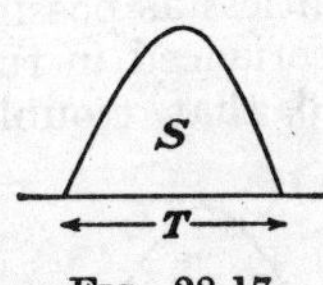

Fig. 28.17

Core Strips
e

Fig. 28.18

Now S is proportional to the average value of the applied voltage and

$$E_{\text{av}} = E/(\text{Form factor})$$
$$\therefore\ W_h \propto [E/(\text{Form factor})]^{1\cdot 6}$$
$$\propto [1/(\text{Form factor})]^{1\cdot 6}$$

if E is constant.

A peaked wave has a larger form factor than a sinusoidal wave, and consequently gives a smaller hysteresis loss for a given effective voltage. On the other hand, a flat-topped wave has a smaller form factor and therefore gives a greater hysteresis loss.

(*b*) Eddy-current Loss. Let R = resistance of a local eddy-current path (Fig. 28.18), and let e be the voltage acting round the path. Then power in the path = e^2/R. Now e is induced by the voltage E applied to the winding, and therefore the r.m.s. value of this voltage is proportional to E.

Hence, eddy-current loss is proportional to

$$E^2/R;\ \text{i.e. to } E^2$$

Hence, so long as E is constant, the eddy-current loss is constant, and is independent of the waveform.

Effect of Waveform on the Operation of a Synchronous Motor or on Two Alternators in Parallel. When two alternators are paralleled on the same bus-bars, their e.m.f.s are in opposition with respect to the local path round the busbars, and therefore, one machine acts as a synchronous motor relative to the other. Suppose the two e.m.f.s are sinusoidal, equal in magnitude and in exact phase opposition. Then there will be zero current circulating between the two machines. Suppose now that one waveform is sinusoidal, while the other possesses a harmonic, say, the third. Then although the fundamentals will be in opposition, the harmonic will have no e.m.f. of the same frequency to oppose it, so that it will produce an idle current round the local path.

If both waves possess a harmonic and the waveforms are identical, then, if the fundamentals are in phase opposition, so also will be the harmonics. Thus, no circulating current, either of fundamental or harmonic frequency, will flow.

If both waves possess, say, a third harmonic, but the waveforms are not identical, then if the fundamentals are in phase opposition, the harmonics will not be in opposition. They will therefore combine to produce an idle circulating current.

Hence, for satisfactory parallel operation the waveforms of the different alternators should be as nearly identical as possible. It is for this reason that difficulty is sometimes experienced in running alternators of different makes in parallel, and that trouble is almost invariably experienced when it is attempted to parallel old slow-speed sets with modern turbine-driven sets.

Components of Actual Wave — Components of Equivalent Sine Wave

Fig. 28.19

Effect of Waveform on the Drop in a Transformer. Let **R** and **X** be the effective resistance and reactance referred to the primary. Hence

$$\text{Drop} = I \times (\mathbf{R}^2 + \mathbf{X}^2)^{\frac{1}{2}}$$

Split up the actual current wave into its various components and also split up the equivalent sine wave as shown in Fig. 28.19. In the latter the components will be of the same amplitude as in the former, but their frequencies will be that of the fundamental. The drops produced by the various components can therefore be tabulated as follows—

Component	*Drops in Volts*	
	Actual Wave	Equivalent Sine Wave
Fundamental	$I_1 \times (\mathbf{R}^2 + \mathbf{X}^2)^{\frac{1}{2}}$	$(I_1^2 + I_3^2 + I_5^2 \ldots)^{\frac{1}{2}} (R^2 + X^2)^{\frac{1}{2}}$
3rd harmonic	$I_3 \times (\mathbf{R}^2 + 9\mathbf{X}^2)^{\frac{1}{2}}$	0
5th harmonic	$I_5 \times (\mathbf{R}^2 + 25\mathbf{X}^2)^{\frac{1}{2}}$	0
***n*th harmonic**	$I_n \times (\mathbf{R}^2 + n^2\mathbf{X}^2)^{\frac{1}{2}}$	0

Hence, for a sine wave and a complex wave of the same effective value, the complex wave will produce the greater total drop in volts.

Effect of Harmonics on the Form of the Rotating Field Produced by a Three-phase Winding. We have seen that the third harmonics neutralize one another, and we will therefore determine the effect of the fifth and seventh. The instantaneous m.m.f. of Phase I can be represented by the expression

$$m_1 = M_1 \sin \omega t + M_5 \sin 5(\omega t - \phi_5) + M_7 \sin 7(\omega t - \phi_7)$$

neglecting harmonics of orders higher than the seventh, and neglecting the third for the reason just stated. Hence, for the instantaneous m.m.f.s of Phases II and III, we have

$$m_2 = M_1 \sin (\omega t - 120°) \quad + M_5 \sin 5(\omega t - \phi_5 - 120°) + M_7 \sin 7(\omega t - \phi_7 - 120°)$$

$$m_3 = M_1 \sin (\omega t - 240°) \quad + M_5 \sin 5(\omega t - \phi_5 - 240°) + M_7 \sin 7(\omega t - \phi_7 - 240°)$$

Now $\sin 5(\omega t - \phi_5 - 120°) = \sin [5(\omega t - \phi_5) - 240°]$

$\sin 7(\omega t - \phi_7 - 120°) = \sin [7(\omega t - \phi_7) - 120°]$

$\sin 5(\omega t - \phi_5 - 240°) = \sin [5(\omega t - \phi_5) - 120°]$

$\sin 7(\omega t - \phi_7 - 240°) = \sin [7(\omega t - \phi_7) - 240°]$

$$\therefore m_1 = M_1 \sin \omega t + M_5 \sin 5(\omega t - \phi_5) + M_7 \sin 7(\omega t - \phi_7)$$

$$m_2 = M_1 \sin (\omega t - 120) \quad + M_5 \sin [5(\omega t - \phi_5) - 240°] + M_7 \sin [7(\omega t - \phi_7) - 120°]$$

$$m_3 = M_1 \sin (\omega t - 240) \quad + M_5 \sin [5(\omega t - \phi_5) - 120°] + M_7 \sin [7(\omega t - \phi_7) - 240°]$$

The sequence of the phase angles for the fundamental m.m.f.s in the three phases is 0°, 120°, and 240°. The sequence of the seventh harmonics is also 0°, 120°, and 240°, showing that the phase rotation of the seventh harmonics is the same as that of the fundamentals. On the other hand, the sequence of the phase angles for the fifth harmonic m.m.f.s in the three phases is 0°, 240°, and 120°. Hence, the phase rotation of this harmonic is opposite to that of the fundamental, and the seventh harmonic. The effect of this from the practical point of view has already been considered on page 574 in connection with the starting of squirrel-cage induction motors.

Experimental Determination of Waveform. It is possible to determine the waveform of an alternator by a "step-by-step" process, by means of a revolving contact connected to one point of the armature winding. This method is now obsolete except for instructional purposes, having been superseded by the oscillograph. The following description refers to the Duddell oscillograph. This instrument is essentially a permanent-magnet galvanometer in which the moving coil is in the form of a phosphor bronze loop, as shown in the scheme of Fig. 28.20. The loop carries a very light mirror M, and tension is applied as shown. This loop has an exceptionally small natural period of oscillation of its own (about $\frac{1}{2000}$ to $\frac{1}{10000}$ sec) and therefore, when an alternating current is passed through it, it can respond instantly to the changes in magnitude of the current. The deflection of the mirror at any instant is thus proportional to the instantaneous value of the current through the "vibrator" at that instant. The motion of the loop is damped and rendered aperiodic by immersing the loop and mirror in an oil bath. It is obvious that if a cinematograph film is moved at a uniform speed in a direction at right angles to the motion of the beam of light reflected from the mirror, this beam being at the same time focused on the film, a photographic record of the waveform will be obtained. Very often it is necessary to visualize the waveform on a screen. To do this it is necessary to impart to the beam of light reflected from the mirror a motion at right angles, this motion being such that the deflection is proportional to the time. To accomplish this the beam is reflected from M on to a long rocking mirror N (Fig. 28.21), and is then focused by a cylindrical lens L on to a screen. The mirror N is rocked by means of a face cam C, driven by means of a small synchronous motor. It is this motor which gives the deflection proportional to the time, since the motor speed is constant and is an exact multiple of the speed of the alternator whose waveform is required. At the end of one revolution of the cam, the mirror N has to be returned very quickly to the starting

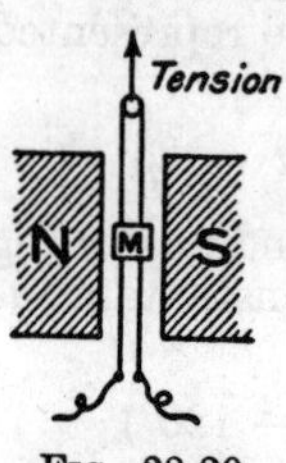

Fig. 28.20. Principle of Duddell Oscillograph

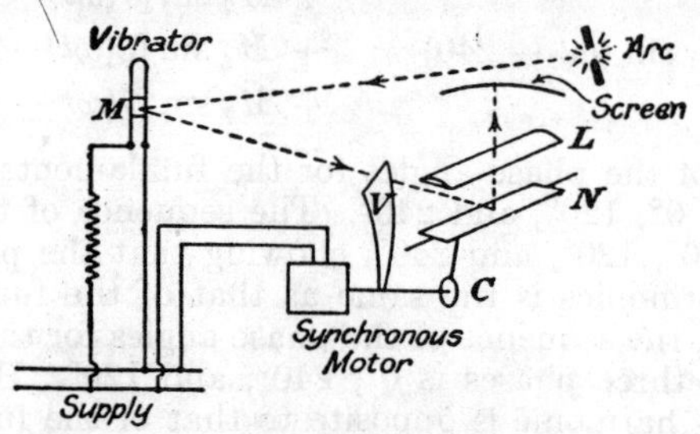

Fig. 28.21

point, and in order not to confuse the diagram, the light during this return motion is cut off by means of a vane V.

It is usual to provide the oscillograph with two vibrators side by side, so that both current and voltage waves can be determined at the same instant. The voltage vibrator is connected in series with a high non-inductive resistance, while the current vibrator, with a certain amount of series resistance, is connected across a non-inductive shunt. It is necessary to be careful in making the connections to the vibrators, otherwise the line potential may be thrown across the two vibrators, with the consequent danger of their sparking across to one another and so being destroyed. Fig. 28.22 shows the correct and incorrect methods of making the connections.

The Cathode-ray Oscillograph. When a high unidirectional p.d. is maintained between two electrodes in a highly evacuated tube, a stream of electrons passes from the cathode to the anode. As each of these has a negative charge, the stream of electrons is equivalent to an electric current, and consequently if a magnetic field is applied transversely to the stream, a deflection, whose direction is given by the left-hand rule, will take place. If the electron stream is brought to a focus at a fluorescent screen then the application of the magnetic field will produce a deflection of the light spot on the screen. Similarly, if the stream passes between a pair of plates between which a p.d. is maintained, a deflection will be obtained. In the first case the deflection is proportional to the strength of the magnetic field, while in the second case it is proportional to the strength of the electrostatic field. If the magnetic field is produced by a pair of current-carrying coils placed on either side of the tube, then the deflection will be proportional to the current.

E *I* *E Volts* *E Volts*

Correct *Incorrect*

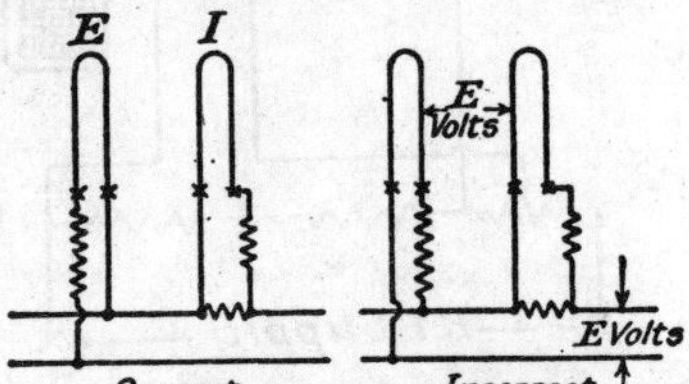

E. Voltage Vibrator
I. Current Vibrator

FIG. 28.22

The application of alternating currents or p.d.s will result in the spot of light being drawn out in a line, and, owing to the very small inertia of the electrons, the deflection at any instant will be proportional to the current, or voltage, at that instant. In other words the appliance can be used as an oscillograph. One type of tube is illustrated in Fig. 28.23, from which it will be seen that a pear shape is adopted, the wide end being coated on the inner surface with a fluorescent material.

The cathode C is indirectly heated by means of a heating element F which is supplied with current from a suitable mains transformer. Electrons ejected from C pass up the tube towards the screen, first through a cylindrical modulator M_1, and then through an axial hole in the anode A. The assembly comprising C, M_1 and A is called the electron gun. The anode is at a high positive potential with respect to the cathode, the p.d. between the two being called the accelerator voltage. Its magnitude varies from 200 or 300 for small tubes used for laboratory work, up to 15,000 or more for television tubes. The modulator, as will be seen from the method of connection to the h.t. supply is made negative with respect to the cathode: the reason for this is as follows: The tube shown in the figure is a gas-focused tube in which, after evacuation to a very high degree a rare gas, such as argon, is admitted at a low pressure of the order of 0·001 mm of mercury. High-speed electrons produce, by the collision process, described on p. 831, ions which, because of their relatively low mobility do not have the same tendency to wander from the tube axis; the attraction of the electrons by these ions thus produces a kind of focusing effect. This effect is enhanced by the negative potential of the modulator, which, by repelling electrons which may have a tendency to radial motion, causes a large proportion of the electrons to pass through the central hole in the anode. The modulator also intercepts

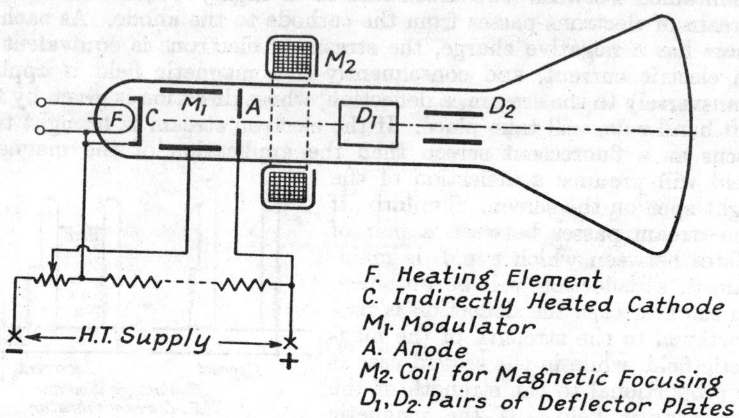

FIG. 28.23. MAGNETIC FOCUSING OF A CATHODE-RAY TUBE

the ions produced within it and so prevents ionic bombardment of the cathode.

The focusing arrangements described above are not effective at frequencies over, say 100 kc/s, and in addition there is loss of sensitivity at low deflecting voltages, resulting in what is called origin distortion. Focusing by magnetic or electrostatic means is usually employed.

Focusing can be effected magnetically by means of an axial magnetic field produced by an ironclad focusing coil M_2 (Fig. 28.23). This has the advantage that large angular deflection without loss of focus can be secured, enabling a shorter tube construction to be employed, which is of advantage in some applications.

Alternatively, focusing can be electrostatic, by means of what is called an electron lens.* Fig. 28.24 (*A*) shows a ray of light falling obliquely on to a curved air-glass surface, at which, after refraction, the angle $r°$ with the normal is less than the incident angle $i°$. The index of refraction for the particular combination of media is given by $n = \sin i°/\sin r°$. Fig. 28.24 (*B*) shows an electron path first through a region of low potential and second a region of high potential, the surface of separation being an equipotential surface. There is bending of the electron path at the surface and a law identical with the optical law holds, the index of refraction now being proportional to the electron velocity. In the optical case a converging lens is produced when the glass is enclosed between two spherical surfaces, giving a lens

* For a complete discussion *see* L. M. Myers, *Electron Optics*. Chapman & Hall, London.

which is thicker in the middle than at the edges. In the electrical case it is not possible to provide the abrupt change in medium characteristic of the air-glass surfaces of a lens, and a progressive increase of potential which is associated by equipotential surfaces of lens shape has to be employed. A common way of doing this is to arrange three anodes along the axis, the first of conventional plate form, the others of

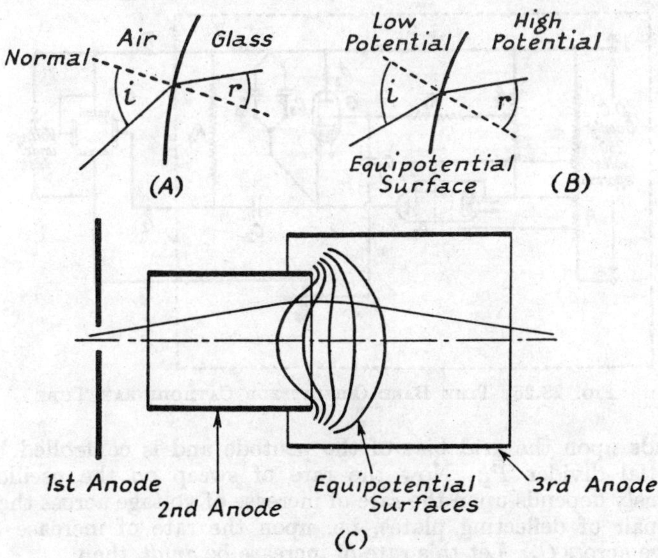

FIG. 28.24. ACTION OF THE ELECTRON LENS

cylindrical form. The second is made more positive than the first, and the third more than the second. The arrangement, and the resulting shapes of the equipotential surfaces, are shown in Fig. 28.24 (*C*).

In Fig. 28.23 two pairs of parallel plates, the deflecting plates, are shown. These are at right-angles to one another.

In order to obtain a wave trace, as distinct from a straight line on the viewing screen, it is necessary to give the electron stream a deflection at right angles, exactly as with the Duddell oscillograph, but in this case it must be given electrically and not optically. For this purpose an auxiliary appliance, called the linear time base or sweep circuit, is necessary. Fig. 28.25 shows such a circuit. P_e is a pentode and G a gas-filled triode (e.g. a thyratron). P_1 and P_2 are potential dividers, R_1 (about 50,000 Ω) is a resistance in the grid circuit of the triode, and R_2 (about 500 Ω) is a limiting resistance in its anode circuit. R_3 (about 2 MΩ) and C_2 (about 2 μF) are for the purpose of isolating the d.c. potentials so that these do not produce a steady deflection of the spot.

The action of the circuit is as follows: when the d.c. supply is first switched on, the capacitor C_4 (or C_3) acts momentarily as a short-circuit, and the full d.c. voltage is applied to the pentode. C_4 charges (at almost constant current, due to the shape of the anode current/anode volts characteristic of the pentode) and the voltage across it rises until it is sufficient for the triode G to discharge. The voltage required for discharge depends upon the grid bias of G, and is therefore controlled by the potential divider P_2. Again, the charging current of C_4

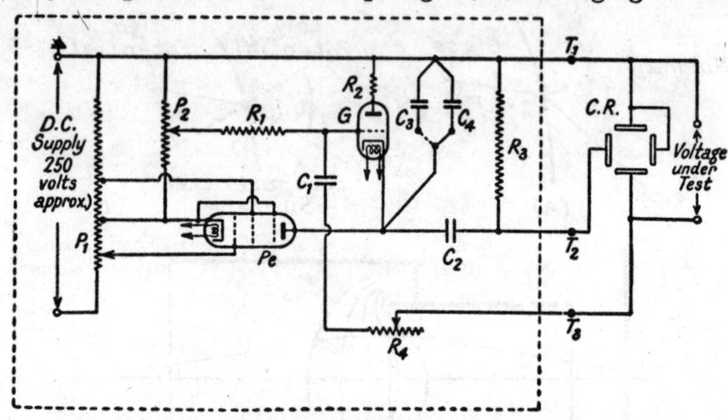

FIG. 28.25. TIME BASE CIRCUIT FOR CATHODE-RAY TUBE

depends upon the grid bias of the pentode and is controlled by the potential divider P_1. Now the rate of sweep on the oscillograph obviously depends upon the rate of increase of voltage across the time-base pair of deflecting plates, i.e. upon the rate of increase of the voltage across C_4. Let this rate of increase be dv/dt, then

$$dv/dt = (d/dt)(Q/C_4) = (1/C_4)(dQ/dt) = i/C_4$$

where Q is the quantity of electricity supplied to C_4 and i is the charging current. Since i is constant, the rate of sweep is constant and therefore the time base is linear.

The rate of sweep can be controlled by P_1 and the amplitude of the sweep by P_2. The capacitors C_3 and C_4 are of the order of 1 and 0·1 μF respectively.

In order to enable the sweep circuit to remain in synchronism with the voltage or current under examination, a synchronizing branch circuit consisting of C_1 (about 0·1 μF) and R_4 (about 2 MΩ) is added. This enables the sweep frequency to be locked to some sub-multiple of the test frequency, by applying to the grid circuit of the triode a small alternating voltage of this frequency.

Harmonic Analysis. This is the process of resolving a periodic, non-sinusoidal quantity into a series of sinusoidal components of ascending order of frequency. If $y = f(\theta)$ represents a periodic function of θ then

$$y = A_0 + A_1 \sin\theta + A_2 \sin 2\theta + \ldots + A_n \sin n\theta$$
$$+ B_1 \cos\theta + B_2 \cos 2\theta + \ldots + B_n \cos n\theta$$

Let the graph of y repeat when $\theta = 2\pi$, let the range $\theta = 0$ to $\theta = 2\pi$

be divided into m equal parts and ordinates erected at $\theta = 0, \theta = 2\pi/m$, $\theta = 4\pi/m \ldots \theta = 2\pi(m-1)/m$. Let the corresponding ordinates be $y_0, y_1, y_2, \ldots y_{m-1}$. The magnitudes of the coefficients of the various terms in the expanded form of $y = f(\theta)$ are

$$A_0 = 1/2\pi \int_0^{2\pi} y d\theta$$

$$A_n = 1/\pi \int_0^{2\pi} y \sin n\theta d\theta$$

$$B_n = 1/\pi \int_0^{2\pi} y \cos n\theta d\theta$$

If is clear that A_0 is the average value of y over the range $\theta = 0$ to $\theta = 2\pi$, and therefore, in terms of ordinates it is given by

$$A_0 = \frac{1}{m}(y_0 + y_1 + y_2 + \ldots y_{m-1})$$

that is, m ordinates in all.

The multiplying factors for A_n and B_n are $1/\pi$, but the limits of integration $\theta = 0$ to $\theta = 2\pi$. Hence A_n is twice the average ordinate of the graph of $y \sin n\theta$ against θ, while B_n is twice the average ordinate of the graph of $y \cos n\theta$ against θ. The equations for A_n and B_n are thus, in terms of ordinates

$$A_n = 2/m[y_0 \sin 0° + y_1 \sin 2n\pi/m + y_2 \sin 4n\pi/m + \ldots + y_{m-1} \sin 2n(m-1)\pi/m]$$

$$B_n = 2/m[y_0 \cos 0° + y_1 \cos 2n\pi/m + y_2 \cos 4n\pi/m + \ldots + y_{m-1} \cos 2n(m-1)\pi/m]$$

Now let the curve be drawn in polar form, instead of to rectangular coordinates, as shown in Fig. 28.26. We can draw a series of such diagrams, each corresponding to a definite value of n. Thus if $n = 1$ the angle between consecutive vectors is $2\pi/m$; if $n = 2$ it is $2 \times 2\pi/m$, and so on, the angle in the general case thus being $n \times 2\pi/m$.

Now $\quad y_0 \sin \theta° = 0 \quad$ and $\quad y_0 \cos \theta° = y_0$

$y_1 \sin 2n\pi/m$ is the vertical projection of y_1 when y_1 is inclined at an angle $2n\pi/m$ to OX, while $y_1 \cos 2n\pi/m$ is the horizontal projection of y_1, inclined at the same angle to OX. Similarly with the other ordinates. Hence, if polar diagrams corresponding to $n = 1$, $n = 2$, $n = 3$, etc., are drawn, the sums of the vertical projections, when multiplied by $2/m$, will give A_1, A_2, A_3, etc., while the sums of the horizontal projections, when multiplied by $2/m$, will give B_1, B_2, B_3, etc.

Again if we choose the value of m correctly the ordinates will arrange themselves into groups of 2, 3, 4, etc., according to the value of n, and thereby simplify the amount of computation to be done. For the waves common to electrical engineering problems $m = 12$ is a very useful value, giving $2\pi/m = \pi/6$ radian, or 30°. As we see from Fig. 28.26—

when $n = 1$, the radii are arranged in their numerical sequence.

when $n = 2$, they are arranged in groups of two per group.

when $n = 3$, they are arranged three per group, and so on.

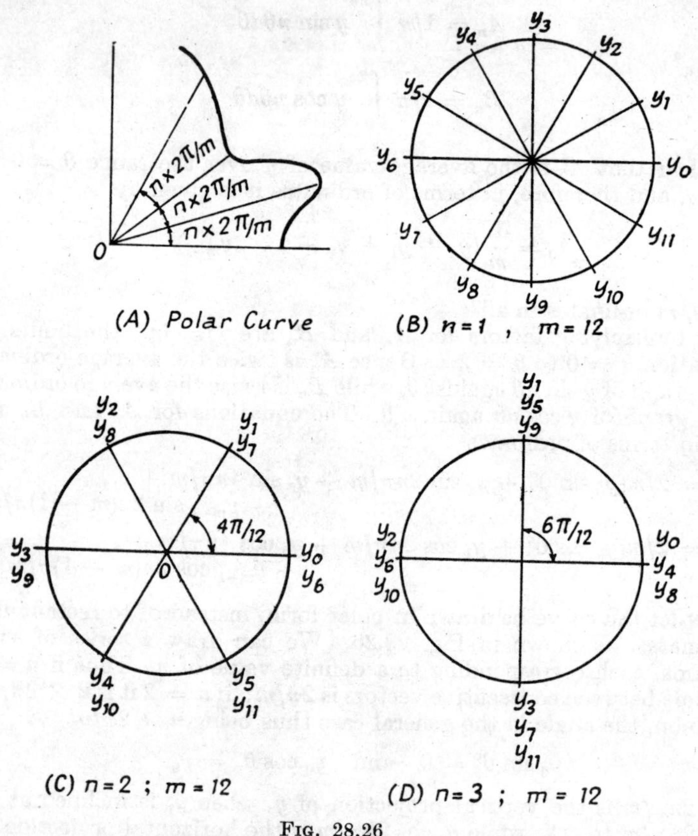

Fig. 28.26

In each diagram there are pairs of vectors, or pairs of vector groups which are in opposition; for example, with $m = 12$, y_1 is opposite to y_7 when $n = 1$

$(y_1 + y_7)$ is opposite to $(y_7 + y_{10})$ when $n = 2$.

$(y_1 + y_5 + y_9)$ is opposite to $(y_3 + y_7 + y_{11})$ when $n = 3$, and so on. This enables the number of groups to be reduced to one-half, as shown in Fig. 28.27.

Example. The periodic time for a certain periodic function is divided into 12 equal parts, the corresponding instantaneous values of the function being 3·370, 4·340, 5·315, 5·980, 5·315, 3·175, 1·187, 0·739, 1·260, 1·562, 1·710, 2·358.

Express the function as a Fourier series, then

(*a*) First evaluate A_0. Since $m = 12$, we have

$$A_0 = 1/12(3{\cdot}370 + 4{\cdot}340 + 5{\cdot}315 + \ldots + 2{\cdot}358) = 3{\cdot}025$$

(*b*) A_1 and B_1: Fig. 28.27 (*A*)

$$\begin{aligned} A_1 &= 1/6[2{\cdot}183 \sin 0^\circ + 3{\cdot}610 \sin 30^\circ + 4{\cdot}055 \sin 60^\circ \\ &\qquad + 4{\cdot}418 \sin 90^\circ + 3{\cdot}605 \sin 120^\circ + 0{\cdot}817 \sin 150^\circ] \\ &= 1/6[4{\cdot}427 \sin 30 + 7{\cdot}660 \sin 60 + 4{\cdot}418] \\ &= 2{\cdot}210. \end{aligned}$$

$$\begin{aligned} B_1 &= 1/6[2{\cdot}183 \cos 0^\circ + 3{\cdot}610 \cos 30^\circ + 4{\cdot}055 \cos 60^\circ \\ &\qquad + 4{\cdot}418 \cos 90^\circ + 3{\cdot}605 \cos 120^\circ + 0{\cdot}817 \cos 150^\circ] \\ &= 1/6[2{\cdot}183 + 2{\cdot}793 \cos 30^\circ + 0{\cdot}450 \cos 60^\circ] \\ &= 0{\cdot}804 \end{aligned}$$

(*c*) A_2 and B_2: Fig. 28.27 (*B*)

$$\begin{aligned} A_2 &= 1/6[-2{\cdot}985 \sin 0^\circ - 1{\cdot}946 \sin 60^\circ + 1{\cdot}042 \sin 120^\circ] \\ &= 1/6[0 - 0{\cdot}904 \sin 60^\circ] \\ &= -0{\cdot}132 \end{aligned}$$

$$\begin{aligned} B_2 &= 1/6[-2{\cdot}985 \cos 0^\circ - 1{\cdot}946 \cos 60^\circ + 1{\cdot}042 \cos 120^\circ] \\ &= 1/6[-2{\cdot}985 - 2{\cdot}988 \cos 60^\circ] \\ &= -0{\cdot}746. \end{aligned}$$

(*d*) A_3 and B_3: Fig. 28.27 (*C*)

$$\begin{aligned} A_3 &= 1/6[+1{\cdot}733 \sin 0^\circ + 0 \sin 90^\circ] \\ &= 0 \end{aligned}$$

$$\begin{aligned} B_3 &= 1/6[1{\cdot}733 \cos 0^\circ + 0 \cos 90^\circ] \\ &= 0{\cdot}289 \end{aligned}$$

And similarly with harmonics of higher orders. As far as the analysis has been made the expression for the function in the example is

$$\begin{aligned} y = 3{\cdot}025 + 2{\cdot}210 \sin \theta - 0{\cdot}132 \sin 2\theta + 0 \sin 3\theta \\ + 0{\cdot}804 \cos \theta - 0{\cdot}746 \cos 2\theta + 0{\cdot}289 \cos 3\theta \end{aligned}$$

Finally we can convert to an expression containing sine terms only, by the usual trigonomical methods. Thus

$$2{\cdot}210 \sin \theta + 0{\cdot}804 \cos \theta = (2{\cdot}210^2 + 0{\cdot}804^2)^{1/2} \sin (\theta + \alpha)$$

where $\alpha = \tan^{-1} (0{\cdot}804/2{\cdot}210)$

$$\therefore \quad 2{\cdot}210 \sin \theta + 0{\cdot}804 \cos \theta = 2{\cdot}291 \sin (\theta + 19^\circ\ 58')$$

and similarly with the higher orders.

With normal alternating voltage and current waveforms the ordinates repeat both in magnitude and sequence after each half

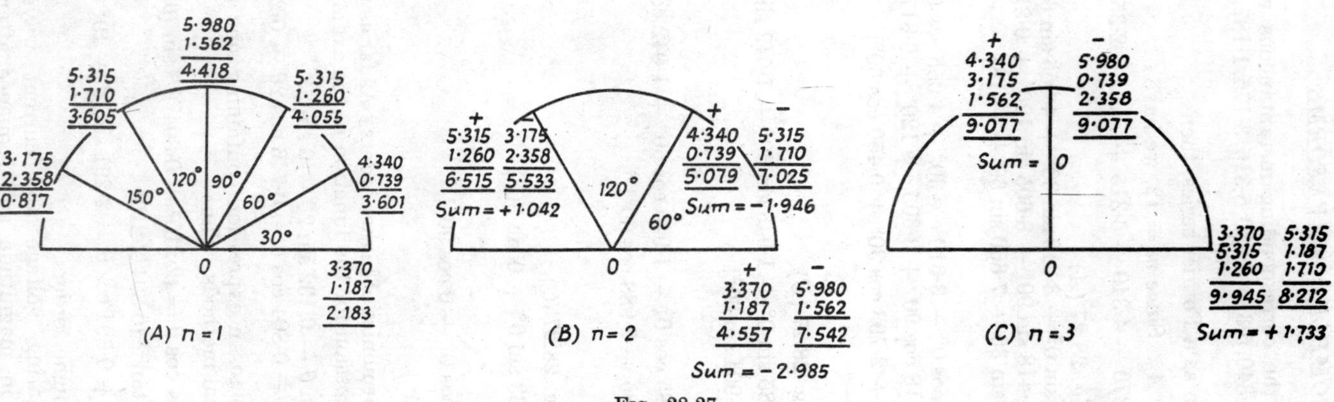

5·980
1·562
4·418
5·315
1·710
3·605
5·315
1·260
4·055
3·175
2·358
0·817
4·340
0·739
3·601
150°
120°
90°
60°
30°
0
3·370
1·187
2·183
(A) n = 1
+
5·315
1·260
6·515
3·175
2·358
5·533
Sum = + 1·042
120°
60°
+
4·340
0·739
5·079
5·315
1·710
7·025
Sum = − 1·946
0
+
3·370
1·187
4·557
5·980
1·562
7·542
Sum = − 2·985
(B) n = 2
+
4·340
3·175
1·562
9·077
5·980
0·739
2·358
9·077
Sum = 0
0
3·370
5·315
1·260
9·945
5·315
1·187
1·710
8·212
Sum = + 1·733
(C) n = 3

FIG. 28.27

period. Consequently (*a*) the constant term A_0 is zero, and (*b*) there are no harmonics of even orders. It is therefore sufficient to divide the half period into *m* equal parts, and if *m* is again equal to 12, the angular width of each part is now 15° apart from this the procedure is exactly as before.

Further Reading

Dover, A. T. *Theory and Practice of Alternating Currents.* Pitman, London, 1945.

Golding, E. W. *Electrical Measurements and Measuring Instruments.* Pitman, London, 1956.

Kemp, P. *Theory of A.C. Waveforms.* Chapman & Hall, London, 1952.

Maxfield, F. A. and Benedict, R. R. *Theory of Gaseous Conduction and Electronics.* McGraw-Hill, New York, 1941.

Reyner, J. H. *Cathode Ray Oscillographs.* Pitman, London, 1940.

Wilson, W. *The Cathode Ray Oscillograph in Industry.* Chapman & Hall London, 1943.

Examples on Chapter 28

1. An alternating voltage given by $100 \sin 314t + 10 \sin 1570t$ is applied to the terminals of a capacitor having a capacitance of 1 μF. Find the expression for the current flowing through the capacitor. (C. and G.)

Ans. $0{\cdot}0314 \cos 314t + 0{\cdot}0157 \cos 1570t$

2. Calculate the form factor of a triangular wave. If a sinusoidal voltage and then a triangular voltage wave are applied in turn to a transformer, the effective values and frequencies being the same, calculate the ratio of the hysteresis losses.

Ans. Form factor = 1·15. Hysteresis loss with triangular wave is 0·94 of loss with sinusoidal wave.

3. An electromotive force, $e = 2000 \sin \omega t + 400 \sin 3\omega t + 100 \sin 5\omega t$, is applied to a circuit consisting of a resistance of 10 Ω, a variable inductance, and a capacitance of 30 μF arranged in series with a hot-wire ammeter. Find the value of the inductance which will give resonance with the triple-frequency component of the pressure; and estimate the readings on the ammeter and on a hot-wire voltmeter connected across the supply when resonant conditions exist. $\omega = 300$. (L.U.)

Ans. 0·041 H; 1440 V; 31·6 A.

4. A current of 50 c/s, containing first, third, and fifth harmonics of crest values 100, 15, and 12 A respectively, is sent through an ammeter and an inductive coil of negligibly small losses. A voltmeter connected to the terminals shows 75 V. What will be the current indicated on the ammeter, and what is the exact value of the inductance of the coil expressed in henrys? (C. and G.)

Ans. 0·0027 H; 72 A.

5. A capacitor of 1·5 μF capacitance is supplied with a voltage having a waveform $e = 1000 \sin pt + 350 \sin 3pt + 270 \sin 5pt$, the frequency being 80 c/s. Calculate the current taken as measured on an ammeter. If energy is being taken from the supply circuit at the same time, how do the harmonics affect the power factor? (L.U.)

Ans. 1·06 A.

6. The waveform of an alternator is found to differ from that of a pure sine function. Assuming this to remain unchanged, show how to calculate the waveform of the current which the alternator would send (*a*) into a capacitor, (*b*) through a choking coil. (L.U.)

7. A voltage represented by $e = 500 \sin \omega t + 10 \sin 5\omega t$, is applied to a circuit containing 1 Ω, 1 H, and $\frac{1}{25}$ μF in series. Find the frequency for resonance with the 5th harmonic, and plot the voltage and current waves.

Ans. 159 c/s.

8. A potential difference represented by the formula

$$v = \sqrt{(2)} \,.\, 100 \,.\, \sin 2\pi \,.\, 50 \,.\, t + \sqrt{(2)} \,.\, 20 \,.\, \sin 2\pi \,.\, 150 \,.\, t$$

is applied to the terminals of a circuit made up of a resistance of 5 Ω, and inductance of 0·0318 H, and a capacitance of 12·5 μF, all in series. Calculate the effective current, and the power supplied to the circuit. (L.U.).

Ans. 0·547 A, 1·5 W.

9. The connection between the magnetizing current and the flux for a particular alternating-current electromagnet is shown by the following table, which gives values for one-half of the magnetization loop, the complete loop showing the usual symmetry with respect to the axes—

Flux (webers × 10−5) .	0	30	75	120	150	179	176	165	138	102	60	0
Magnetizing current (amperes) .	5·5	6·25	8	10·5	14·25	20·5	13·5	6·8	0	− 3	− 4·5	− 5·5

Plot the loop on squared paper, and by its aid deduce and plot the waveform of the magnetizing current when the flux follows a sine law, the amplitude of the flux being equal to the maximum value of the flux in the above magnetization loop. (L.U.)

10. Estimate the r.m.s. value of an alternating current of irregular wave shape in terms of the magnitudes of the fundamental component and the harmonics.

An alternating pressure is represented by $v = 1000 \sin \omega t + 250 \sin 3\omega t + 200 \sin 5\omega t$. Estimate the reading which will be given by an electrostatic voltmeter connected to the circuit. (L.U.)

Ans. 742 V.

11. The ordinates of a non-sinusoidal alternating voltage wave, measured in 15° intervals over the positive half wave are as follows—

0, 97, 152, 152, 134, 142, 176, 200, 189, 161, 133, 82

Determine the equation to the wave, up to the 5th harmonic.

Ans. $y = 200 \sin (\theta - 0° 50') + 40{\cdot}3 \sin (3\theta + 27° 42') + 20 \sin (5\theta - 48° 39')$.

12. A single-phase e.m.f. having a r.m.s. value of 400 V contains a third harmonic amounting to 20 per cent of the fundamental frequency component. This e.m.f. is applied to a circuit consisting of two parallel branches: (i) a resistance of 20 Ω in series with an inductance of 0·01 H; (ii) a resistance of 10 Ω in series with a capacitance of 50 μF.

Calculate the total current taken by the combination and the total power consumed, the fundamental frequency being 50 c/s. (E.M.E.U.).

Ans. 7·7 A, 2715 W.

13. A voltage whose instantaneous value is given by

$$340 \sin \omega t + 68\left(\sin 3\omega t + \frac{\pi}{4}\right) - 34 \cos 5\omega t$$

is applied to a circuit comprising a resistance of 60 Ω, an inductance of 0·015 H and a variable capacitance connected in series.

Calculate the value of the capacitance necessary to give resonance with the fifth harmonic and also the power and the power factor of the circuit in this condition. The fundamental frequency is 50 c/s. (N.U.)

Ans. 27 μF, 254 W, 0·50.

14. Denoting the ratio (effective value)/(maximum value) of an alternating waveform by $1/\sqrt{m}$, show that in the cases of rectangular, semi-circular, sinusoidal and triangular waves the values of m are 1, $1\frac{1}{2}$, 2, 3, respectively.

CHAPTER 29

PHASE ADVANCING

THE power factors of the great majority of commercial loads are considerably less than unity. This is mainly because the induction motor is the most widely used of the various alternating-current motors, and this motor works at less than unity power factor. This is partly because of the wattless magnetizing current it draws from the line. Similarly, all transformers take a wattless magnetizing current, but also because the combined effect of the circuit resistances and reactances is to impose on the extremity of the line current vector a circular locus of finite diameter. This results in a considerable falling off in power factor as the load is reduced. Electric lamps work practically at unity power factor, and therefore a system supplying a lighting load works at a much better power factor than one supplying power.

Rough average values for power factors under different conditions are as follows—

(*a*) System supplying lighting load only	0·9–0·95
(*b*) System supplying lighting and power, with the former predominating	0·8–0·85
(*c*) System supplying lighting and power with the latter predominating	0·75
(*d*) System supplying power only	0·65–0·7
(*e*) Single-phase system supplying power only	0·4–0·5

The Disadvantages of a Low Power Factor.

(*a*) The amount of true power is less than the kVA capacity of the central station, with the result that, although the generators may be fully loaded from the point of view of current output, and therefore of temperature rise, they will not be delivering their full load of true power. The same applies to the cables, to switchgear and transformers, in fact, to all apparatus and plant located between power station and consumer carrying the full current of the circuit in which it is connected. Therefore, if the power factor can be increased, the earning capacity of the system will be increased, without installing any new plant. Thus, if the power factor is raised from 0·6 to 0·9, the possible income will be increased 50 per cent.

(*b*) In addition to the earning capacity of the system being lowered by its presence, wattless kVA actually costs something to produce. From the examination of the data of running costs for a large number of Italian stations, Professor Arno has deduced the following expression for the running costs of a station—

$$\text{Cost} = \text{constant} \times \int_0^t \left(\frac{2}{3}\,\text{kW} + \frac{1}{3}\,\text{kVA}\right) dt$$

where t is the time. It will be seen that the cost has two components. One, the cost of the true energy (kilowatt-hours) delivered, and the other, the cost of the kilovolt-ampere-hours. The expression clearly refers to running costs only, since there are overhead charges, such as the annual cost of the central organization, capital charges associated with the land, and so on. The expression for running costs can also be modified to take into account the fact there are fixed charges associated with the size (kVA) of the plant irrespective of whether it is supplying energy or not. A more convenient expression is thus

$$\text{(Total annual cost, £)} = A + B \times \text{kVA} + C \times \text{kWh}$$

The constant C takes into account the cost of fuel, oil, wages and salaries, maintenance and everything which is automatically brought in when the plant delivers energy to the network.

It will be clear that, apart from the capital cost represented by $B \times$ kVA in the equation, the operational disadvantages of a low power factor increase the cost per kWh as the power factor is reduced. As the power factor of a station is a function of the characteristics of the load appliances, the supply authority is justified in framing a tariff in such a manner that the lower the power factor of a given installation the more the consumer pays for the energy (kWh). The possibility of reducing the annual account for electrical energy consumed is the justification for expenditure on the part of the consumer on apparatus which will improve the overall power factor of his plant.

(*c*) A poor power factor causes a large drop of volts in the alternators, partly because of the armature leakage reactance, and partly because of the demagnetizing effect of armature reaction. As a result the excitation on low power factors has to be much greater than on high power factors, in order that the terminal voltage may be maintained at the proper value. This necessitates larger exciters. In some cases when the power factor of the system has been lower than anticipated when ordering the alternators, the necessary overloading of the exciters and consequent overheating of the alternator field have caused serious trouble, and this in spite of the fact that the alternators were not delivering their maximum possible load of true power.

(*d*) The voltage drop of an overhead transmission line increases as the power factor decreases, with the result that it may be necessary to instal expensive appliances in order to keep up the voltage at the far end of the line. A good voltage regulation is specially important on circuits supplying induction motors, since the torque of these motors is proportional to the square of the applied voltage, and therefore, if there is a large sudden fall in voltage, the motors may stall.

Methods of Obtaining an Improved Power Factor. Imagine a system consisting of (*a*) power station, (*b*) long transmission line, feeding (*c*) distributors, supplying (*d*) individual consumers. There will, of course, be transformers at various places as, for example, for stepping up the voltage for transmission purposes, stepping it down for distribution, and possibly stepping it down still further for applications to the terminals of the consuming devices. The power factor of the outgoing supply at the power station is the resultant of the power factors of all the appliances connected to the distributors. It is possible to install phase-advancing apparatus at the power station and this, by neutralizing some of the lagging kVAR will reduce the total lagging kVAR and thereby reduce the kVA output for a given kW. Since the plant will have a definite kVA capacity it will then be able to increase the kW output while remaining within this definite kVA limit. This is illustrated in Fig. 29.1. The first figure shows the original load of

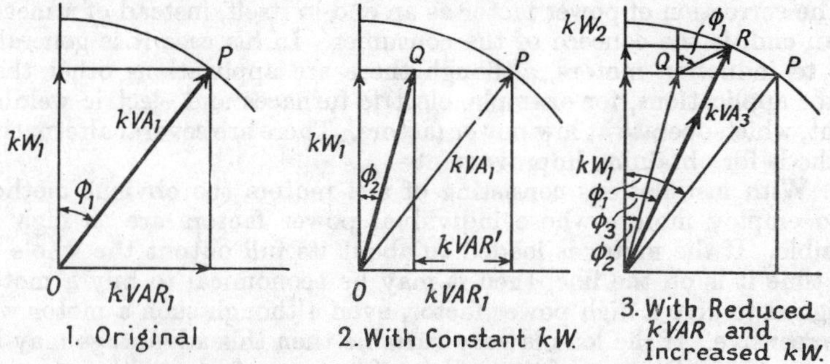

FIG. 29.1. POSSIBILITIES IN POWER FACTOR CORRECTION

kVA_1 at a lagging power factor of $\cos \phi_1$. The second figure shows the system supplied with kVAR′ of leading wattless kVA, which reduces the total kVA to kVA_2, and increases the power factor to $\cos \phi_1$. As kVA_2 is less than the output possible, as represented by the radius of the circle with centre at 0, it is now possible to put more load on the station. Assuming this additional load to be of the same power factor, $\cos \phi_1$, as before, it can be represented by QR in the third figure, the total kW output now being kW_2 and the overall power factor $\cos \phi_3$.

If this is done it will certainly improve conditions at the power station, and it will increase the earning capacity of the plant, but the cost will not be carried by the consumers whose apparatus is responsible for the trouble. It is therefore not practicable to install at the power station plant whose sole function is the improvement of the station power factor. We have seen that one of the operating disadvantages of low power factor is the high voltage drop in a long overhead transmission line, and it is not uncommon to install plant which, although power factor correcting in action, is regulation improving in function. The plant used for this purpose is a synchronous motor connected to the load end of the line. The motor runs idle and its excitation is controlled automatically in such a way that on full load in the transmission line the motor has its maximum excitation. The motor is designed to have the minimum losses, so that, having no mechanical output it operates at a power factor only slightly greater than zero. Thus a certain 20,000 kVA machine of this type has a total loss of only 1·4 per cent of 20,000 kW when fully excited and delivering full current, so that its leading power factor is then about 98·6 per cent. In its affect on the system it is to all intents and purposes, a capacitor, but a capacitor whose kVA can be varied at will and even reversed. It is therefore termed a synchronous capacitor. We see that, when the line is fully loaded the synchronous capacitor, having maximum excitation, delivers maximum leading kVAR, thereby increasing the line power factor and so reducing the volt drop. If the line is on no-load the distributed capacitance will actually cause the load end voltage to rise, and to combat this the synchronous capacitor is given a reduced excitation which causes it to take a small lagging current.

The correction of power factor as an end in itself, instead of a means to an end, is the concern of the consumer. In his case it is generally due to induction motors, although there are applications other than motor applications, for example, electric furnaces and electric welding plant, which operate at low power factors. There are several alternative methods for obtaining improvement—

1. With installations consisting of a.c. motors the obvious method is to employ motors whose individual power factors are as high as possible. If the motor is loaded to about its full output the whole of the time it is on the line, then it may be economical to buy a motor designed to give a high power factor, even although such a motor will be expensive. If the load is very variable then this advantage may be largely offset. Because of the falling off in power factor with reduction in load it is a mistake to instal an induction motor which is of too big a horse-power for the work it has to do.

With variable load the phase-compensated motor, or, in large sizes, the induction motor with separate phase advancer, should be considered.

2. If the characteristics are suitable then a synchronous motor can be used, and this is a particularly useful solution if there are one or two large motors along with a number of small motors, which will probably be plain induction motors. The synchronous motor will then have two functions, firstly the performance of work and secondly the production of leading wattless kVAR. The type of synchronous motor chosen will then be decided largely by (*a*) the desired leading power factor, (*b*) the starting and running conditions. We have seen that at a low leading power factor the armature reaction is largely demagnetizing, with the result that the salient-pole motor with its "stiff" magnetic field is desirable. For example in the extreme case of the synchronous capacitor a salient-pole machine is essential. If the load is violently fluctuating then, again, the stability of the salient-pole machine is an advantage. If starting conditions are severe and the proportion of leading kVAR to total motor kVA is small then the synchronous-induction motor is suitable.

3. The use of static capacitors. A capacitor is an appliance which takes leading kVAR from the line. The amount of this is fixed by the capacitance and consequently it cannot vary with varying load on the motor, or motors. It has no effect whatever on the motor operation. A most useful field of application is in a factory where capacitors can

be installed at various points to improve the power factors of groups of motors, or of single motors, if these are large enough. The capacitors can then be switched in or out according to the current demand for leading kVAR.

Consider the case of a single motor corrected by (*a*) a capacitor, (*b*) an expedor-type phase advancer, (*c*) a susceptor type. The capacitor takes a fixed amount of leading kVA and its effect is therefore a bodily

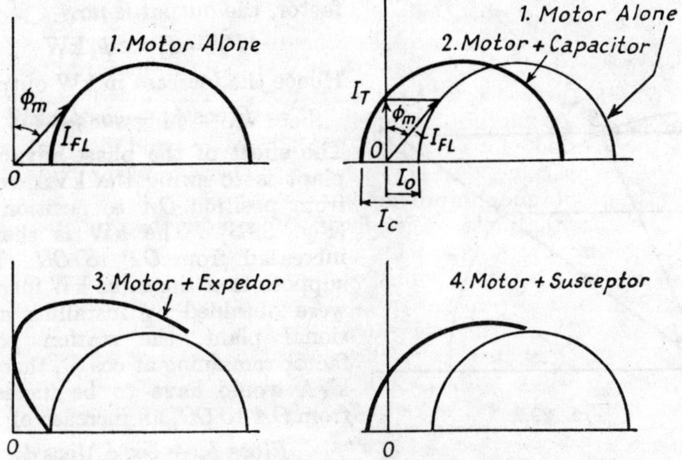

FIG. 29.2. COMPARISON OF DIFFERENT METHODS OF POWER-FACTOR COMPENSATION

shift to the left of the circle diagram of the motor. The motor functions exactly as normally but the locus of total line current is now the circle 2, Fig. 29.2, instead of circle 1. The capacitor current I_c per phase is, in the figure, of such magnitude that the overall power factor in full load is unity. At fractional loads the power factor is then less than unity and leading. The expedor type of phase advancer is a series machine whose flux increases with the rotor current of the motor, and therefore the injected e.m.f. increases as the load increases. The susceptor is a shunt machine and can therefore take leading kVAR when the main motor is running light. However, initially its effect is very similar to that of a capacitor.

Economical Limit of Power-factor Correction. The improvement of power factor can be justified for two entirely different reasons, although both are economic in the sense that the operation is carried out for the purpose of reducing expenditure. In the case of a generating plant the actual earning capacity is governed by the kW output, whereas the possible earning capacity is governed by the kVA output. Suppose, for example, that the kVA capacity of plant, switchgear and cables is equal to P, then if the power factor is $\cos\phi_1$, the output when the plant is fullv loaded is

$$P_1 = P\cos\phi_1 \text{ kW}$$

Suppose that the power factor at the station is improved to $\cos\phi_2$ by the use of phase-advancing apparatus which operates at zero power-factor leading, then, when the plant is fully loaded at this power factor, the output is now

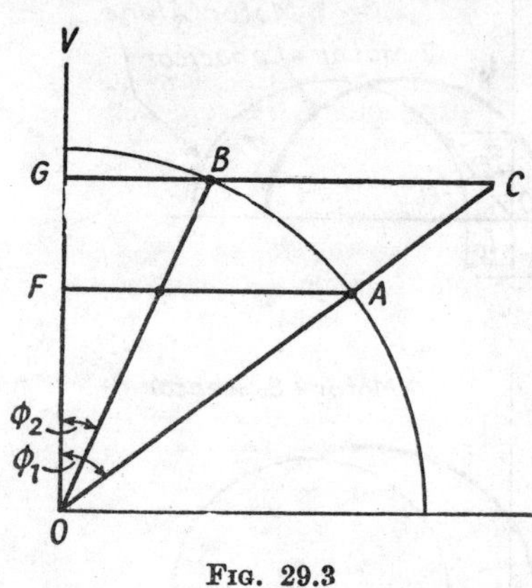

Fig. 29.3

$$P_2 = P\cos\phi_2 \text{ kW}$$

Hence the increase in kW output

$$= P(\cos\phi_2 - \cos\phi_1) \text{ kW}$$

The effect of the phase-advancing plant is to swing the kVA vector from position OA to position OB (Fig. 29.3). The kW is thereby increased from OF to OG. Now suppose that the same kW increase were obtained by installing additional plant, the station power factor remaining at $\cos\phi_1$, then the kVA would have to be increased from OA to OC, an increase of

$$P(\cos\phi_2 - \cos\phi_1)/\cos\phi_1$$

Hence, to effect this same increase but to keep the generating kVA down to the original value it is necessary that the phase-advancing apparatus shall have a kVAR of BC. This is equal to

$$OC\sin\phi_1 - OB\sin\phi_2$$
$$= P(\cos\phi_2/\cos\phi_1)\sin\phi_1 - P\sin\phi_2$$
$$= P(\tan\phi_1\cos\phi_2 - \sin\phi_2)$$

In the vector diagram (Fig. 29.3)

OA = original load kVA = generating plant kVA

OC = new load kVA

CB = kVAR of phase-advancing apparatus

OB = generating plant kVA under new conditions

= original value

Let the cost per kVA of generating plant, switchgear, and connections be £a, and that of the phase-advancing apparatus be £b per kVAR. If the problem is being assessed purely on a basis of capital cost then, as only one kind of apparatus is to be purchased, namely extra generating plant or phase-advancing apparatus, the two alternatives have to be compared one with the other. Phase-advancing apparatus will be cheaper, on this basis, so long as

$$£bP(\tan\phi_1\cos\phi_2 - \sin\phi_2) < £aP[(\cos\phi_2 - \cos\phi_1)/\cos\phi_1]$$

This reduces to

$$(a - b\sin\phi_2)/\cos\phi_2 < (a - b\sin\phi_1)/\cos\phi_1$$

and thus defines the minimum value to which the generating plant phase angle ϕ_2 can be taken by the use of phase-advancing plant before the cost of such plant becomes greater than that of additional generating plant, switchgear and connections all operating at the old phase angle of ϕ_1.

EXAMPLE 1. A system is working at its maximum kVA capacity with a lagging power factor of 0·707. An anticipated increase of load could be met (*a*) by raising the power factor of the system to 0·866 by the installation of phase advancers, (*b*) by installing extra generating plant, cables, etc. to meet the increased power demand. The total cost for the latter method is £10 per kVA. Estimate the limiting cost per kVA of phase-advancing plant which would justify its installation.

The limiting price is given by the value of *b* which satisfies the equation

$$(a - b\sin\phi_2)/\cos\phi_2 = (a - b\sin\phi_1)/\cos\phi_1$$

In the example

$$a = £10 \text{ per kVA}$$

$$\cos\phi_1 = 0{\cdot}707 \quad \therefore\ \sin\phi_1 = 0{\cdot}707$$

$$\cos\phi_2 = 0{\cdot}866 \quad \therefore\ \sin\phi_2 = 0{\cdot}5$$

$$\therefore\ (10 - 0{\cdot}5b)/0{\cdot}866 = (10 - 0{\cdot}707b)/0{\cdot}707$$

$$\therefore\ b = £6{\cdot}16 \text{ per kVA}$$

EXAMPLE 2. The load on the mains of a single-phase supply system is 100 kW at a power factor of 0·707, the current lagging behind the voltage. If phase-advancing apparatus is available for parallel connection, taking leading current at a power factor of 0·1, what must be its load in kVA if the power factor of the whole system is to be raised to (*a*) 0·8, (*b*) 0·9, (*c*) 0·95. (L.U.)

Since the method of calculation for all three parts is the same, consider only part (*a*). The vector diagram is shown in Fig. 29.4.

$$\cos\phi_1 = 0{\cdot}707 \quad \therefore\ \phi_1 = 45°$$

$$\cos\phi_2 = 0{\cdot}1 \quad \therefore\ \phi_2 = 84°\ 16'$$

$$\cos\phi_3 = 0{\cdot}8 \quad \therefore\ \phi_3 = 36°\ 52'$$

Hence in the triangle *OAC*, the angles at *O*, *A* and *C* are 8° 8′, 50° 44′, and 121° 8′ respectively.

$$\therefore\ \text{kVA}_2/\text{kVA}_1 = \sin 8°\ 8'/\sin 121°\ 8'$$

$$= 0{\cdot}1415/0{\cdot}8560 = 0{\cdot}165$$

$$\therefore\ \text{Phase-advancer kVA} = 0{\cdot}165 \times 100/0{\cdot}707 = 23{\cdot}3$$

The above discussion is of the case in which the earning power of an existing plant can be increased by an increase in the operating power factor of that plant, brought about by the use of phase-advancing apparatus. Thus the kW output of the plant has been increased although its kVA loading has remained the same. This case is the concern of the authority in charge of a generating plant in which such an improvement can be effected.

There is also the case of the consumer of electrical energy whose annual bill for electricity is based on a tariff which is framed to penalize the consumer whose plant operates at a low power factor. The tariff

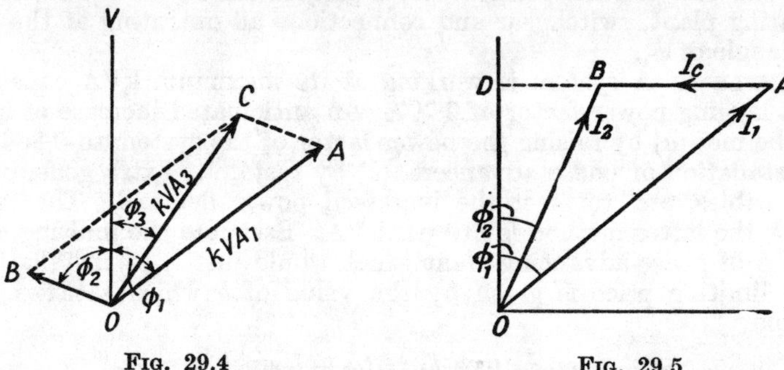

FIG. 29.4 FIG. 29.5

is almost invariably a two-part tariff made up of a fixed sum per year and a charge for the kWh consumed. The fixed sum is arranged so as to cover the cost incurred by the authority in making it possible for the consumer to use electricity at all, and consequently it is a justifiable charge whatever the actual kWh consumption per annum may be. The second, or running cost, covers the cost to the authority of supplying that particular amount of energy. The fixed charge is generally proportional to the consumers maximum demand, for example £5 per annum per kW, or kVA of maximum demand. It is by making this maximum demand charge on a basis of kVA, and not kW, that the consumer is penalized if the power factor of his consuming devices, motors for example, is low. By bringing his kVA demand more nearly equal to the kW demand he will thus reduce the annual bill. On the other hand there will be annual interest and also depreciation on the apparatus he purchases for the purpose of improving the power factor, and consequently his total annual outgoings are now the sum of (*a*) the bill for electrical energy, (*b*) the annual interest and depreciation on the phase-advancing apparatus. The criterion now is that this annual outgoing shall be a minimum.

Provided the phase-advancing apparatus has this one function only, static capacitors for example, then the supply is on the basis of constant kW. The vector diagram is shown in Fig. 29.5, in which the vectors are shown as currents.

OA = total load current I_1, = line current before correction

$DA =$ wattless component of I_1

$= I_1 \sin \phi_1$

$AB = I_c$ the current taken by the capacitor

$OB = I_2$ the line current after correction

The total wattless current after correction is

$$DB = I_2 \sin \phi_2$$

$$\therefore\ I_c = I_1 \sin \phi_1 - I_2 \sin \phi_2$$

Again,

$$I_2 \cos \phi_2 = I_1 \cos \phi_1$$

$$\therefore\ I_2 = I_1 (\cos \phi_1 / \cos \phi_2)$$

$$\therefore\ I_c = I_1(\sin \phi_1 - \cos \phi_1 \tan \phi_2)$$

If the plant kVA ($\sqrt{3} V I_1/1000$) is P then the necessary kVA of phase-advancing apparatus, operating at zero power factor, required to improve the overall power factor from $\cos \phi_1$ to $\cos \phi_2$ is

$$P(\sin \phi_1 - \cos \phi_1 \tan \phi_2)$$

In order to determine the economical limit to which this can be taken, start with the load kW, as this is a fixed value.

$$\text{Original kVA} = \text{kW}/\cos \phi_1$$

$$\text{Final kVA} = \text{kW}/\cos \phi_2$$

$\therefore$ Reduction in kVA

$$= \text{kW}(1/\cos \phi_1 - 1/\cos \phi_2)$$

The annual saving on a fixed charge of £a per kVA of maximum demand is thus

$$£\,\text{kW} \times a(1/\cos \phi_1 - 1/\cos \phi_2)$$

The necessary leading kVAR to effect this is

$$\text{kW}(\tan \phi_1 - \tan \phi_2)$$

The annual outgoings due to interest and depreciation amounting to £b per kVA of phase-advancing apparatus

$$= £\,\text{kW} \times b\,(\tan \phi_1 - \tan \phi_2)$$

The total annual saving is thus

$$£\,\text{kW} \times a(\sec \phi_1 - \sec \phi_2) - £\,\text{kW} \times b\,(\tan \phi_1 - \tan \phi_2)$$

$$= £\,\text{kW}\,[a \sec \phi_1 - a \sec \phi_2 - b \tan \phi_1 + b \tan \phi_2]$$

Differentiating this with respect to ϕ_2 and remembering that ϕ_1 is a fixed angle, we have

$$\text{kW}(-\,a \sin \phi_2 \sec^2\phi_2 + b \sec^2\phi_2)$$

This is equal to zero and corresponds to a maximum when

$$a \sin \phi_2 \sec^2\phi_2 = b \sec^2\phi_2$$

when $\sin \phi_2 = b/a$

This expression is independent of ϕ_1 showing that the most economical value for ϕ_2 depends only on the ratio of b to a.

EXAMPLE. A factory takes a steady load of 200 kVA at 0·707 power factor from 400 V, 50 c/s mains. The tariff is as follows: £5 per annum per kVA of maximum demand, plus 0·5d. per unit. It is required to reduce the annual outgoing for electricity to a minimum by installing static capacitors. If the installed cost of these is £4 10s. 0d. per kVA and interest and depreciation amount to 12 per cent per annum, calculate (*a*) the most economical power factor, (*b*) the capacitance of the capacitors.

$$a = \text{£5 per annum per kVA of maximum demand}$$
$$b = \text{12 per cent of £4 10s. 0d.} = \text{£0·54}$$
$$\therefore \sin\phi_2 = 0{\cdot}54/5 = 0{\cdot}108$$
$$\phi_2 = 6^\circ\, 12'$$
$$\cos\phi_2 = 0{\cdot}994$$

Referring to Fig. 29.5

$$I_1 = 200{,}000/(\sqrt{3} \times 400) = 288{\cdot}7 \text{ A}$$

In the triangle OAB

$$\angle BOA = \phi_1 - \phi_2 = 45^\circ - 6^\circ\, 12' = 38^\circ\, 48'$$
$$\angle OBA = (90 + \phi_2) = 96^\circ\, 12'$$
$$\therefore I_c = 288{\cdot}7 \times (\sin 38^\circ\, 48'/\sin 83^\circ\, 48')$$
$$= 288{\cdot}7 \times (0{\cdot}6266/0{\cdot}9942)$$
$$= 182 \text{ A}$$

Assuming delta-connected capacitors the capacitor current per phase

$$I_{c.ph} = 182/\sqrt{3} = 105{\cdot}1 \text{ A}$$
$$\therefore C_{ph} = I_{c.ph}/V_{ph}\omega = 105{\cdot}1/400 \times 314$$
$$= 8{\cdot}37 \times 10^{-4} \text{ F}$$

FURTHER READING

BOLTON, D. J. *Electrical Engineering Economics*. Chapman & Hall, London, 1950.

BURNHAM, T. H. *Engineering Economics*. Pitman, London, 1954.

CLAYTON, A. E. *Power Factor Correction*. Pitman, London, 1934.

ROGERS, P. L. *Power Factor Economics*. Wiley, New York, 1939.

STUBBINGS, G. W. *Power Factor Problems in Electricity Supply*. Spon, London, 1940.

TODMAN, J. C. *Power Economy in the Factory*. Pitman, London, 1932.

EXAMPLES ON CHAPTER 29

1. What are the disadvantages of a poor power factor in an alternating-current system? A single-phase motor takes 20 A at 200 V at a power factor of 0·8. Calculate the capacitance of the capacitor which, when shunted across the motor terminals, will bring the line current into phase with the voltage. Frequency of supply, 50 c/s.

Ans. 192 μF.

2. Explain clearly the action of the reactive component in a synchronous machine. A slow-speed alternator and a turbo-alternator working in parallel

supply 2500 kW at a power factor of 0·8. If the turbo-alternator gives 1000 kW at unity power factor, at what power factor must the slow-speed machine be working? (C. and G.)

Ans. 0·625.

3. Explain, by means of vectors, how an over-excited rotary convertor can compensate for a lagging power factor. A generating station supplies power for the following: Lighting 100 kW, induction motor of 400 h.p. having a power factor of 0·8 and efficiency 0·93, a rotary convertor giving 1000 A at 500 V at an efficiency of 0·94. What must be the power factor of the rotary convertor in order that the power factor at the supply station may be unity? (C. and G.)

Ans. 0·91 leading.

4. Define the term "power factor" as applied to alternating-current circuits. What power factor does one usually find for the following: (*a*) A water resistance, (*b*) the primary current of an alternating-current transformer when the secondary is unloaded, (*c*) a hand-feed alternating-current arc? Illustrate each case by means of a rectangular co-ordinate diagram of waveforms. Assume the wave of the alternator electromotive force to be sinusoidal.

5. A central station of 1000 kVA capacity supplies a load whose power factor is 0·8. How many wattless leading kVA must be supplied by phase-advancing apparatus in order to raise the power factor of the station to 0·95; (*a*) the true power supplied by the station remaining constant, (*b*) the kVA supplied by the station remaining constant?

Ans. (*a*) 336; (*b*) 288.

6. Describe a type of capacitor used to compensate for lagging currents in a power supply system. A 2000 V, 50 c/s motor installation has a maximum load of 300 kVA at a power factor of 0·6. Calculate the capacitance of a capacitor to raise the power factor to 0·95 at maximum load. What will the power factor be when the load is halved? (C. and G.)

Ans.—Assuming a three-phase system with the capacitors delta-connected, capacitance per phase = 48 μF. Power factor at half-load, 0·83 leading.

7. Explain by means of the vector diagram of the motor, why it is that an over-excited synchronous motor cannot supply so much wattless leading kVA when it is doing mechanical work as when it is running idle.

8. What methods would you recommend for the improvement of the power factor of (*a*) a central station and cable network, (*b*) a large induction motor, (*c*) a group of small motors? Give reasons for the various methods adopted.

9. A central station is working at a power factor of 0·7. Deduce an expression for the necessary reduction in wattless kVA to improve the power-factor to any given value. Hence, plot a curve with values of the reduction as ordinates against the new power factor as abscissae. Draw conclusions from the shape of the curve.

10. Show that the economical limit to which the power-factor of a lagging load can be raised is independent of the original value of the power-factor when the tariff consists of a fixed charge per kVA plus a flat rate per unit. A consumer is charged £3 per annum per kVA maximum demand, plus a flat rate per kVA. Phase advancing plant can be purchased for £3 per kVA, the cost for depreciation, housing, and interest on capital being 12 per cent. Find the most economical angle of lag to which the load can be improved under these conditions. (L.U.)

Ans. 7°.

11. What factors determine the economical limit of power-factor correction? A system is working at its maximum kVA capacity with a lagging power-factor of 0·707. An anticipated increase of load could be met by: (*a*) raising the power-factor of the system to 0·866 by the installation of phase advancers, (*b*) by in-

stalling extra generating plant, cables, etc., to meet the increased power demand. The total cost for the latter method is £8 per kVA. Estimate the limiting cost per kVA of phase advancing plant which would justify its installation. (I.E.E.)

***Ans.* £8 15s. 0d.**

12. A load has a maximum demand of 400 kVA at a power-factor of 0·8 lagging, and the tariff is £5 per kVA of maximum demand plus 0·5 pence per kWh. Calculate the saving in the maximum demand charge if the power-factor is corrected to 0·9 lagging. Calculate also the total saving if an annual charge of 10s. per kVA is allowed for the correcting apparatus.

***Ans.* £222; £180.**

13. A factory is charged for its electrical energy £4 per kW plus 0·4d. per kWh with an additional 5s. per kW of maximum demand for every 0·1 by which the power factor falls below 0·85. The maximum demand is 1200 kW at a power-factor of 0·6 lagging and the load-factor is 50 per cent. What is the maximum amount per kVA which could be economically spent on static condensers to improve the power-factor to 0·85? Take interest and depreciation as 10 per cent per annum.

***Ans.* £8 16s. per kVA.**

14. A large factory owns a generating station which has a maximum output of 10 MW. It is estimated that the maximum load will reach 18 MW in three years. In order to be able to supply the extra demand the management may either install two 5 MW turbo-alternator sets at a cost of £30 per kW, or obtain a bulk supply from the C.E.A. at £5 per annum per kW of maximum demand plus 0·3d. per kWh. Annual charges on new plant would be 12 per cent of capital cost, and the estimated generating costs would be £4 10s. per kW of the maximum demand plus 0·4d. per kWh. Assuming the estimated requirements in three years to be 80×10^6 kWh, and that for the bulk supply a load factor of 80 per cent would be maintained, which alternative would be the more economical?

***Ans.* Bulk supply £195,000**
Extension of sets £250,000.

CHAPTER 30

ILLUMINATION

Radiant Energy. Radiant energy is energy which can be transmitted through space without the necessity for some obvious transmitting agency, such as metallic conductors in the case of electrical energy, steam or hot-water pipes in the case of heat energy, and so on. It is quite clear that there must be some mechanism of transmission even in the case of radiant energy and it is convenient to assume, although by no means certain, that radiant energy is the energy content of a wave motion propagated through an all pervading medium, the ether. If some obstacle is interposed, for example the human body in front of a fire, or a thermojunction in the beam of light from a light source, then the energy incident on the obstacle will experience (*a*) reflection and generally scattering, the reflected energy again being in the form of radiant energy, and (*b*) absorption, in which case there will be a conversion of radiant to heat energy, with consequent rise in temperature of the obstacle.

If radiant energy is the energy content of a wave motion through a medium then it must be characterized by (*a*) a velocity of propagation, and (*b*) a wave length. Through free space, and very approximately through the atmosphere, the velocity of propagation of radiant energy is $2{\cdot}99776 \times 10^8$ m/sec, and obviously for the purpose of the engineer this can be simplified to 3×10^8 m/sec. Although the velocity of propagation is constant, the wavelength can be of any value, for, if we assume a structureless ether there is, apparently, no limit to the smallness, neither is there an upper limit. It is important to realize that radiations of differing wavelengths are of identical physical nature, for example, they all give rise to heat generation in an obstacle which has the property of absorbing incident radiation, a property which is possessed by all material bodies. On the other hand the wavelength of a radiation may have an important bearing on the suitability of that particular radiation for some particular purpose. Thus radio transmission can be carried out by radiations of wavelength 1 cm and upwards. The short wavelengths within the range $1{\cdot}5 \times 10^{-9}$ cm to 10^{-7} cm are the Röntgen or X-rays which are characterized by their ability to penetrate many materials normally regarded as opaque.

Owing to the physiological construction of the eye and its associated nerves, the physio-chemical processes which enable messages to be transmitted from the eye to the brain, and to the combined physiological and psychological processes which convert those messages to an impression, the range of wavelengths which are capable of producing the sensation of sight in an individual with normal vision is restricted to the very narrow band extending from about 4×10^{-5} to $7{\cdot}5 \times 10^{-5}$ cm. The word "about" is used deliberately here as the limiting wavelengths for vision vary with the individual, and also vary throughout life with the same individual. Thus a normal middle-aged person receives the sensation of light from radiations of wavelength 4×10^{-5} cm, but with young people the lower limit will be less than this, in some cases nearer 3×10^{-5} cm. There is not so much variation at the upper end, and $7{\cdot}5 \times 10^{-5}$ cm is a fairly good average.

As the centimetre is not a very convenient unit for the wavelength of "visible" radiation it is common to use a sub-multiple of the centimetre, namely 10^{-8} cm as the unit. This is called the *Angstrom unit* (Å)

$$1\ \text{Å} = 10^{-8}\ \text{cm}$$

and the average limiting wavelengths for vision are thus

$$4 \times 10^{-5}\ \text{cm} = 4 \times 10^{-5} \times 10^{8} = 4000\ \text{Å}$$

$$7{\cdot}5 \times 10^{-5}\ \text{cm} = 7{\cdot}5 \times 10^{-5} \times 10^{8} = 7500\ \text{Å}$$

Colour. Suppose that a beam of radiant energy is of one wavelength only, that is to say, it is a monochromatic radiation, then if that wavelength is within the above visible range it will give rise to the sensation of light: the colour of the sensation will be decided by the wavelength only. For example, if a monochromatic radiation is of wavelength 7000 Å it will give the sensation of red, if of 6000 Å the colour will be yellow, if 5500 Å green, if 4750 Å blue, and if 4000 Å violet. There are reds on either side of the 7000 Å wavelength, and similarly with the other colours, so that if there is a radiation including all the wavelengths within the visible range, then the spectrum of this radiation will include all the spectral colours, or hues, from extreme red to extreme violet. The sensation produced when radiation consisting of many wavelengths falls on the eye is the result of all the monochromatic components.

Relative Luminous Efficiency of Radiation. Imagine a beam of radiant energy consisting of all the wavelengths within the visible range, and let the relative values of the various components be such that for any narrow wavelength band $d\lambda$, the radiant energy within this bandwidth is a constant and independent of the wavelength λ. If the spectrum of this radiation is obtained by means of a dispersing system giving uniform spacing, then the radiant energy per constant bandwidth $d\lambda$ is also a constant. Hence if a slit is placed in front of the spectrum and traversed across it, then, no matter what its position the radiant energy coming from the slit is constant, and if this energy is determined by a thermojunction then the galvanometer deflection will be independent of the slit position. On the other hand if the slit is examined visually then, besides a change of colour as it traverses the spectrum, there will also be a change in the visual effect as indicated by the luminance of the slit. Starting at the extreme red the visual effect will be very small but as the wavelength is increased there will be first a slow and then a rapid increase until a maximum is reached at the green of 5540 Å. Further reduction in wavelength results in a diminution in visual effect, which becomes zero at about 4000 Å.

It is thus clear that the effect of visible radiation on the human eye is dependent on the wavelength, being a maximum (for a normal person) at 5540 Å and diminishing as the difference from 5540 Å increases, whether this difference corresponds to a greater or to a smaller wavelength. The graph of visual sensation against wavelength is shown in Fig. 30.1. It refers to normal intensities of illumination and

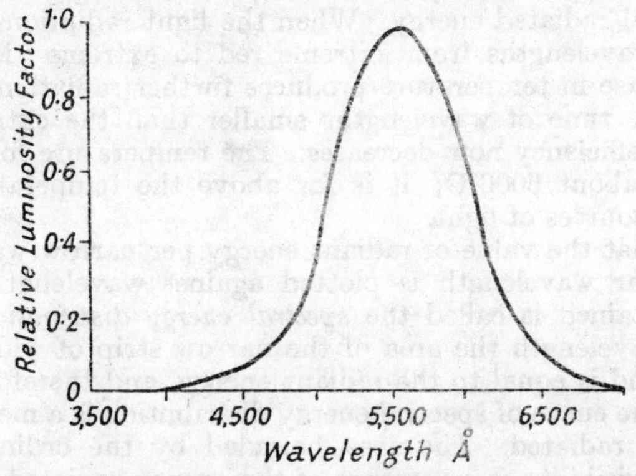

FIG. 30.1. RELATIVE LUMINOUS EFFICIENCY OF THE AVERAGE EYE

not to very low values: it is called the *relative luminous efficiency* of the average eye. Symbol V_λ.

Denoting the maximum value of this factor as unity, the curve can be plotted from the following data*—

λ	V_λ	λ	V_λ
3800	4×10^{-5}	5750	0·9154
4000	4×10^{-4}	6000	0·6310
4250	$7{\cdot}26 \times 10^{-3}$	6250	0·3210
4500	$3{\cdot}8 \times 10^{-2}$	6500	0·1070
4750	0·1226	6750	0·0232
5000	0·3230	7000	$4{\cdot}1 \times 10^{-3}$
5250	0·7932	7250	$7{\cdot}36 \times 10^{-4}$
5500	0·9950	7350	$3{\cdot}6 \times 10^{-4}$
5540	1·000		

Radiant Efficiency and Luminous Efficiency. When the temperature of a body is gradually increased, the body begins to radiate energy into space, the nature of the radiation depending on its temperature. Thus, when the temperature is low, only heat energy will be radiated, but when a certain temperature is reached, light will also be emitted and the body will then become luminous. At first the light radiated will consist of red waves only, but with increasing temperature more

* For complete data, *see* D. B. Judd; *Bureau Stds.* **6** (1931) 465.

and more of the shorter waves will be given off, until eventually the body is white hot. The heat waves still given off are identical in nature to the light waves, but as they produce no impression on the retina, it is obvious that from the point of view of light-giving, they represent so much wasted energy. The energy radiated as light, expressed as a percentage of the total energy radiated, is called the radiant efficiency of the body. As the temperature is increased above that at which light waves are first given off, the radiant efficiency will increase, because the energy of the light waves will increase in greater proportion than the total radiated energy. When the light radiations include all the visible wavelengths from extreme red to extreme violet, then a further increase in temperature produces further radiations which are invisible, this time of wavelengths smaller than the extreme violet. The radiant efficiency now decreases. The temperature for maximum efficiency is about 6000°C; it is far above the temperature usually attained by sources of light.

Suppose that the value of radiant energy per narrow wave band at any particular wavelength is plotted against wavelength, then the graph so obtained is called the *spectral energy distribution*. At any particular wavelength the area of the narrow strip of width equal to the wave band is equal to the radiant energy, and therefore the total area under the curve of spectral energy distribution is a measure of the total energy radiated. The area bounded by the ordinates at the extremes of visibility is a measure of the energy radiated in the form of visible light. But we have seen that the magnitude of the visual effect due to any monochromatic component is a function of the wavelength of that component, and consequently to obtain a measure of the light (as distinct from visible radiation) we have to multiply each ordinate of the spectral energy curve by the appropriate value of the relative luminosity factor. This is done in Fig. 30.2, and the area under the derived curve now gives the amount of light. It will be clear that the radiation in the form of light is considerably less than the total visible radiation.

Definitions.* *Radiant Flux.* Power emitted, transferred or received in the form of radiation. Symbol P or Φ_e.

Light. Radiant flux capable of stimulating the eye to produce visual sensation.

Luminous Flux. That quantity characteristic of radiant flux which expresses its capacity to produce visual sensation, evaluated according to the values of relative luminous efficiency for the light adapted eye adopted by the C.I.E.†. Symbol F or Φ.

Luminous Energy. The time integral of luminous flux. Symbol Q.

Illumination. At a point of a surface. The quotient of the luminous flux incident on an infinitesimal element of surface containing the point under consideration, by the area of that element. Symbol E.

* Extracted from B.S. 233 (1953) and reproduced by permission of the British Standards Institution.

† Commission Internationale de l'Eclairage.

Luminous Emittance from a point of a surface: the quotient of the luminous flux emitted from an infinitesimal element of surface containing the point under consideration, by the area of that element. Symbol H.

Luminous Intensity in a given direction: the quotient of the luminous flux emitted by a source, or by an element of a source, in an infinitesimal cone containing the given direction, by the solid angle of that cone. Symbol I.

Luminance at a point of a surface and in a given direction: the quotient of the luminous intensity in the given direction of an infinitesimal area of the surface containing the point under consideration,

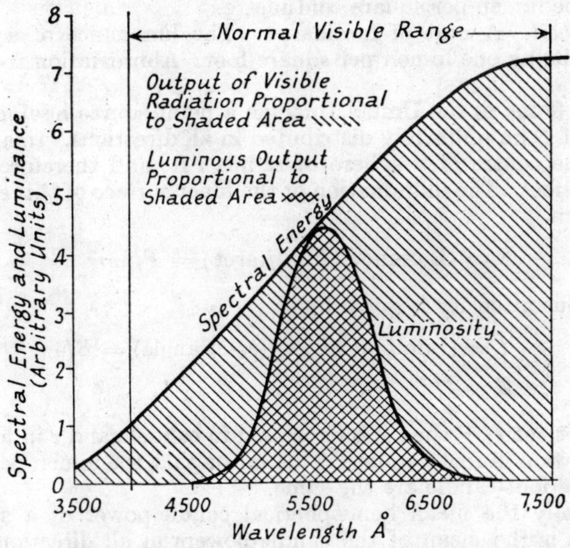

FIG. 30.2. ILLUSTRATING THE DIFFERENCE BETWEEN VISIBLE RADIATION AND LUMINOUS OUTPUT: FULL RADIATION AT 3400°K

by the orthogonally projected area of the element on a plane perpendicular to the given direction. Symbol L. The term luminance is now used in preference to brightness.

Luminous Efficiency of a radiation: the quotient of the luminous flux by the corresponding radiant flux. Symbol K; for monochromatic radiations of wavelength λ, K_λ; for monochromatic radiation for which the luminous efficiency is a maximum, K_m.

Lumen. The unit of luminous flux. The flux emitted in unit solid angle of one steradian by a point source having a uniform intensity of one candela (*see* below). Abbreviation lm.

Lumen per square foot. A unit of illumination. An illumination of one lumen per square foot. Abbreviation lm/ft² or lm/sq. ft. Note that this unit is the same as the *foot-candle*, a name which is *not* now recommended

Candela. The unit of luminous intensity. It is of magnitude such that the luminance of a full radiator (also known as a black-body radiator) at the temperature of solidifaction of platinum is 60 units of luminous intensity per square centimetre. Abbreviation cd.

Candle-power. The light-radiating capacity of a source in a given direction, in terms of the luminous intensity expressed in candelas.

Mean Spherical Candle-power. The average value of the candle-power of a luminous source in all directions.

Stilb. A unit of luminance. A luminance of one candela per square centimetre. Abbreviation sb.

Lambert. A unit of luminance. The luminance of a uniform diffuser emitting one lumen per square centimetre.

Foot-lambert. A unit of luminance. The luminance of a uniform diffuser emitting one lumen per square foot. Abbreviation ft-L.

Notes on Some of the Units. Imagine a point source having a luminous flux of F lm uniformly distributed in all directions. Imagine this to be at the centre of a sphere of radius r ft, and therefore of area $4\pi r^2$ square feet. The illumination of the inner surface of this sphere is, by definition

$$E = (\text{Luminous flux})/(\text{area}) = F/4\pi r^2$$

The luminous intensity is given by

$$I = (\text{Luminous flux})/(\text{solid angle}) = E/4\pi$$

$$\therefore\ E = I/r^2$$

but the quotient I/r^2 when I is expressed in candela and r in feet gives the illumination in foot-candles. This shows that the lumen per square foot and the foot-candle are the same.

Occasionally the mean hemispherical candle-power of a source is given: this is the mean of the candle-powers in all directions below the horizontal. The *reduction factor* is the ratio

$$f = (\text{m.s.c.p.})/(\text{m.h.c.p.})$$

The efficiency of a lamp is defined as the quotient of lumen output to the watts intake. The following relationships hold

$$\text{lumens per watt} = 4\pi/(\text{watts per m.s.c.p.})$$
$$= 4\pi f/(\text{watts per m.h.c.p.})$$
$$\text{watts per m.s.c.p.} = 4\pi/(\text{lumens per watt})$$
$$= (\text{watts per m.h.c.p.})/f$$
$$\text{watts per m.h.c.p.} = 4\pi f/(\text{lumens per watt})$$
$$= f \times (\text{watts per m.s.c.p.})$$

The following values are representative of the performance of modern gas-filled lamps.

Watts	Average Lumen Output Throughout Life			Average Efficiency Throughout Life lm/watt		
	Single-coil 110 V	Single-coil 230 V	Coiled-coil 230 V	Single-coil 110 V	Single-coil 230 V	Coiled-coil 230 V
25	133	113		5·32	4·52	
40	228	206		5·7	5·15	
60	449	330	389	7·48	5·5	6·48
75	759	584	665	10·12	7·79	8·87
100	1000	785	883	10	7·85	8·83
150	1400	1160	1270	9·33	7·73	8·47
200	2230	1970		11·15	9·85	
500	8950	7930		17·90	15·86	
1000	19,640	17,800		19·64	17·80	

It will have been realized that the lumen output of a source is dependent not only on the total visible radiation (radiant flux) emitted, but also on the wavelength content, in fact it is this which governs the form of the definition of luminous flux. Suppose that a source could emit monochromatic radiation of 5540 Å with no admixture of any other wavelength, visible or invisible, then this source would emit one lumen for each 0·0016 W of radiation, its radiant efficiency (not commercial efficiency as given in the above table) therefore being

$$1/0{\cdot}0016 = 625 \text{ lm/W}$$

If a source could emit monochromatic radiation only, but of some other wavelength λ, for which the relative luminous efficiency is V_λ, then the radiant efficiency would be 625 V_λ lm/W. For example the value of V_λ for the sodium D line (the double line whose components are of wavelengths 5890 and 5896 Å) is 0·75, showing that if the sodium lamp emitted the D line radiation only its luminous efficiency would be

$$0{\cdot}75 \times 625 = 468{\cdot}8 \text{ lm/W}$$

The commercial efficiencies of practical lamps are much below such values because (*a*) the lamps, particularly metal-filament lamps, emit large amounts of invisible radiation, particularly in the infra-red, and (*b*) because the visible radiation is usually spread over a wide band, thereby including much for which the value V_λ is very small. The high value of V_λ for the sodium D lines, combined with the fact that there is very little additional visible radiation, is one of the factors which gives the sodium vapour lamp the very high commercial efficiency, average throughout life, of 50–65 lm/W.

Laws of Illumination. The illumination of a surface receiving its flux from a source whose distance is sufficiently great to justify its being regarded as a point source, is inversely proportional to the square of the distance between the surface and the source.

The illumination of a surface at any point is proportional to the cosine of the angle between the normal at that point and the direction of the luminous flux. This is Lambert's Cosine Law. Fig. 30.3 represents a rectangular tube of flux F.

$$\text{Intensity of surface } ABCD = F/ABCD$$
$$\text{Intensity of surface } ABC'D' = F/ABC'D' = F\cos\theta/ABCD$$
$$\therefore (\text{Intensity of } ABCD)/(\text{Intensity of } ABC'D') = 1/\cos\theta$$

EXAMPLE. Two arc lamps of 1000 and 500 c.p. respectively (assumed the same in all directions) are suspended 40 ft above the ground and

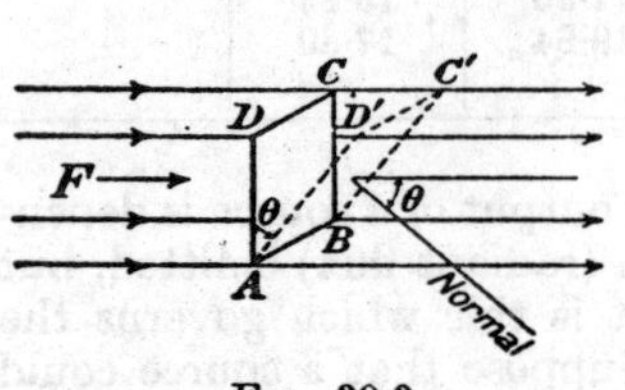

FIG. 30.3

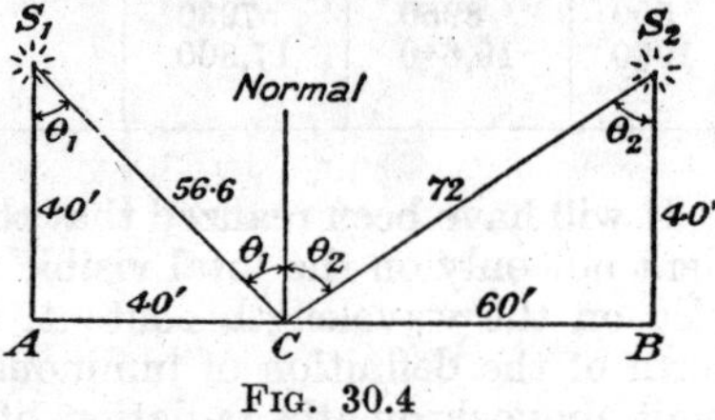

FIG. 30.4

are 100 ft apart. Find the intensity of illumination at a point on the ground in line with the two lamps and 40 ft from the base of the more powerful lamp.

From Fig. 30.4, we have

$$\text{Illumination at } C \text{ due to } S_1 = [1000/(S_1C)^2] \times \cos\theta_1 = [1000/(56{\cdot}6)^2] \times 0{\cdot}707 = 0{\cdot}22 \text{ lm/ft}^2$$
$$\text{Illumination at } C \text{ due to } S_2 = [500/(S_2C)^2] \times \cos\theta_2 = [500/(72)^2] \times 40/72 = 0{\cdot}05 \text{ lm/ft}^2$$
$$\therefore \text{Total illumination at } C = 0{\cdot}22 + 0{\cdot}05 = 0{\cdot}27 \text{ lm/ft}^2$$

Measurement of Luminous Intensity. In order to determine the luminous intensity of a source of light it is necessary to compare it with a source of known intensity. A standard lamp is therefore required. At one time the intensity of a candle of pure spermaceti wax, weighing $\frac{1}{6}$ lb and burning 120 gr/hr, was taken as the unit. It is clear that the candle, however closely it may conform to a definite specification, is a very inconvenient source to use as a primary standard. More modern standards are the Harcourt pentane lamp, which burns a mixture of pentane vapour and air, and has a candle power of about 10; the Heffner lamp, which burns amyl-acetate at a wick and has a candle power of 1; and the Carcel lamp, which burns colza oil and has a candle power of about 10. In practice, it is not convenient to use such standards, a more satisfactory way being to determine the candle power of an aged electric lamp by comparison with one of the above standards, and then to use this lamp as a secondary standard. It is usual to make periodic tests on a lamp which is to be used as such a standard, and when it is consistent, to remove the filament from the old bulb and place it in a new cylindrical bulb.

The most recent primary standard is known as a "black body" radiator,* and operates at the temperature of melted pure platinum.

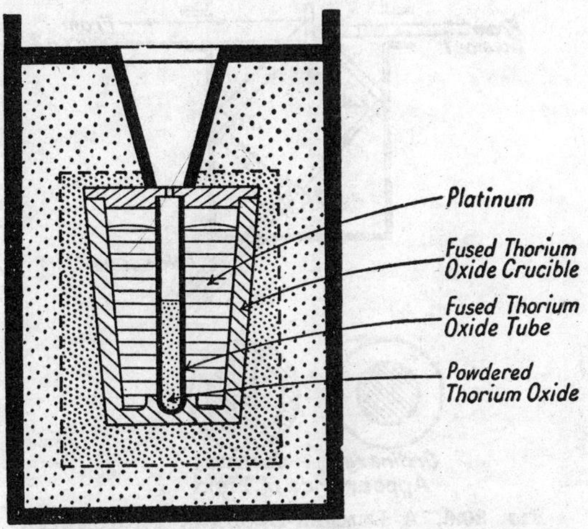

FIG. 30.5. THE PRIMARY STANDARD OF LIGHT

The only satisfactory refractory which will withstand this high temperature (1773°C) and at the same time remain chemically inert, thereby refraining from having any effect on the platinum, is fused thorium oxide. The black body consists essentially of a narrow tube of this material, filled about half-way with powdered thorium oxide. This tube is placed centrally in a larger crucible of the same material, the space between being nearly filled with platinum. The whole has a cover which locates the central tube, and there is a small central hole through which the radiation can pass. The crucible is embedded in thorium oxide, giving a construction somewhat as Fig. 30.5. The platinum is melted by means of a high-frequency induction furnace, and it is when the heating is removed and the platinum still molten that the black body is used as a standard. If temperature is plotted against time, there will be a fall in temperature until the platinum "freezing" temperature is reached, after which the temperature will remain stationary for a period until the platinum is solidified, after which the temperature falls again. During the period of steady temperature the brightness of the radiator is exactly 60 candles per cm^2.

The most convenient form of photometer for general purposes is the Lummer-Brodhun Photometer. This consists essentially of a plaster

* The modern preferred term is "full radiator" or "Planckian radiator."

of Paris screen S (Fig. 30.6), the two sides of which are illuminated by the standard lamp and the lamp under test, respectively. The light is scattered at the two surfaces, and the rays which leave at an angle of 45° reach the right-angled prisms A and B, where they are totally reflected to two more prisms, C and D. In one form of photometer a portion of the base of C is ground away, so that the contact area between C and D is a circle. Rays from prism A entering this circle pass through to the telescope, whereas those from B which enter the circle do not

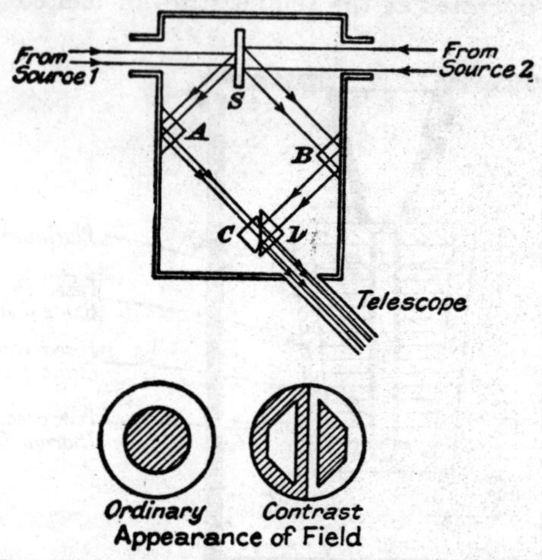

Fig. 30.6. A Lummer-Brodhun Photometer

reach the telescope. On the other hand, of the rays which reach the face of D outside this circle, only those from B reach the telescope. Hence, the field consists of an inner circle and an outer annular ring illuminated by the standard and the test lamp respectively. The position of the photometer is adjusted until the field is of uniform brightness and its distance from the two lamps read off. If CP_1 and CP_2 are the candle powers of the standard and test lamp, and d_1 and d_2 their respective distances from the photometer, then, from the inverse square law

$$CP_2/CP_1 = d_2^2/d_1^2 \text{ or } CP_2 = CP_1 \times (d_2/d_1)^2$$

In a modified form of Lummer-Brodhun Photometer, the area of contact between the two prisms C and D is so modified that the field of view consists of trapezoidal patches inside semicircles. With no further modification the instrument would still be an "equality of brightness" instrument with a different field pattern. To convert to "equality of contrast" thin strips of glass are placed in the positions indicated in Fig. 30.7, their thickness being such that they absorb about 8 per cent of the incident light. Hence the luminance of the trapezoid T_1 is lightly less than that of the background B_1, while that of trapezoid T_2 is slightly less than background B_2. Hence, when a balance is obtained, the contrast between trapezoid and background is the same in both halves of the field.

When the candle power of very powerful lamps, such as large discharge lamps, is being determined, the distance between the test and standard lamps must be very large, otherwise the photometer will have

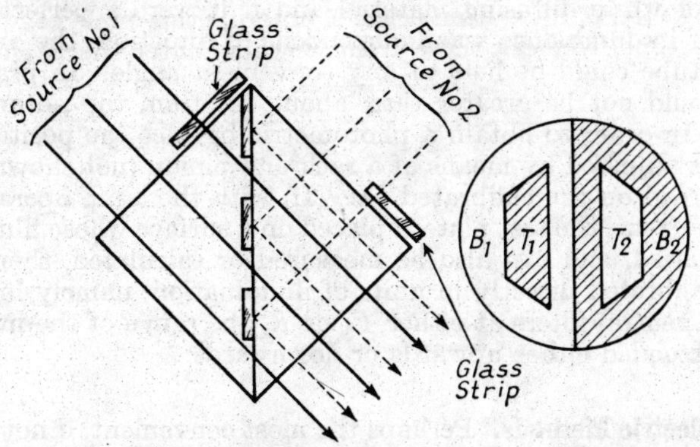

FIG. 30.7. PRISMS FOR THE EQUALITY OF CONTRAST PHOTOMETER HEAD

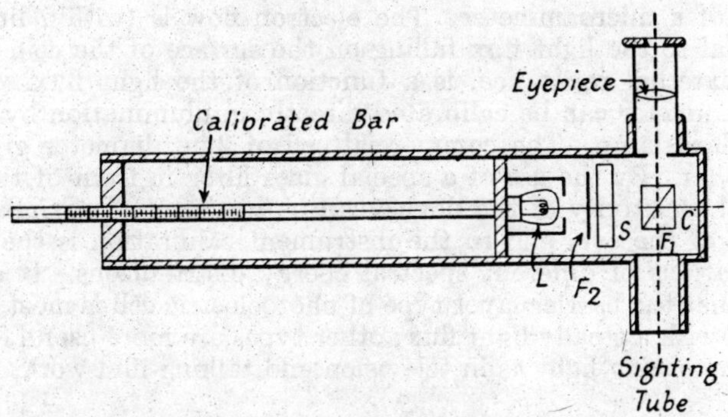

FIG. 30.8. PRINCIPLE OF MACBETH ILLUMINOMETER

to come so close to the standard that accurate observations are impossible. A photometer room is not large as a rule, consequently, in order to obtain the required distance, the test lamp can be slung from the ceiling and the light from it reflected by a mirror on to the photometer.

The Illumination Photometer. To enable the illumination of surfaces to be measured, special portable photometers have been evolved. One of these, the Macbeth Illuminometer, is illustrated in Fig. 30.8. The instrument is essentially a portable photometer having a Lummer-Brodhun cube C as the indicating device. One source is an opal glass screen S which is illuminated by a small lamp L and which, owing to its diffusing properties presents an area of uniform luminance to the cube *C*. The lamp is connected to a portable battery by leads not shown in the figure, and a rheostat and milliammeter enable the luminous intensity to be maintained at the correct value. The second source is a standard reflecting surface, called the test plate, which is placed on the surface for which the illumination is to be measured. This test plate is of white diffusing material and if it were a perfect diffuser, that is, if its luminance was independent of direction, the axis of the sighting tube could be held at any convenient angle. In practice the angle should not be greater than about 40° from the normal to the surface. In order to obtain a photometric balance the position of the lamp *L* is adjusted by means of a rack and pinion (not shown) and its position read on the calibrated bar. If, with the lamp operating at a definite current, the test plate is placed on a surface whose illumination can be varied, and can also be measured or calculated, then the bar can be calibrated directly in units of illumination, namely lm/ft^2. By inserting neutral filters at either F_1 or F_2 the range of the instrument can be extended either upwards or downwards.

Photoelectric Methods. Perhaps the most convenient, if not the most accurate, method of determining the illumination of a surface is to place a photovoltaic cell on the surface and to measure the cell current by means of a microammeter. The electron flow is (within limits) proportional to the light flux falling on the surface of the cell, and for a given external resistance, is a function of the light flux. Thus the micro-ammeter can be calibrated directly in illumination by means of a standard lamp. The commercial cell of 2 in. diameter gives about 450 μA/lm. By the use of a special glass filter in front of the cell the spectral sensitivity of the device can be made to approximate closely to that of the eye, and so the instrument calibration is the same for illuminations of different spectral energy distributions. It should be noted that the barrier-layer type of photoelectric cell is most suited for dealing with a steady light flux; other types are more useful for dealing with fluctuating light as in television and talking-film work.

Determination of M.H-S.C.P. and M.S.C.P. In order to determine the m.h.c.p., the candle power is measured in a horizontal direction, the lamp being rotated through, say, 10° about a vertical axis for each observation. The resulting candle powers plotted on a polar diagram give the horizontal curve of illumination. Such a curve is, roughly, circular for most lamps. The m.h-s.c.p. is the mean of the observed values. If these values differ considerably, so that the diagram departs from the circular form, it is preferable to plot candle power against angle with rectangular co-ordinate axes. The base is then divided into an even number of steps of equal width, and the ordinates measured. There will be an odd number. Applying Simpson's Rule, we have

$$\text{Area under curve} = \frac{h}{3}\left\{\begin{matrix} y_1 + y_n + 2(y_3 + y_5 + y_7 + \ldots) \\ + 4(y_2 + y_4 + y_6 + \ldots) \end{matrix}\right\}$$

where y_1, y_2, y_3, etc., are the various ordinates and h is the horizontal distance between consecutive ordinates. By dividing the area by the base the mean ordinate is found, and this is the required m.h.c.p. Usually, 10 steps, i.e. 11 ordinates, will be sufficient.

If the lamp is mounted so that it can be rotated in a vertical plane the resulting polar curve will exhibit two lobes, the cone of shadow about the position 0° being due to the holder. The m.s.c.p. can be determined from this polar diagram by Rousseau's construction. A circle is drawn on the diagram, as shown in Fig. 30.9, and an axis

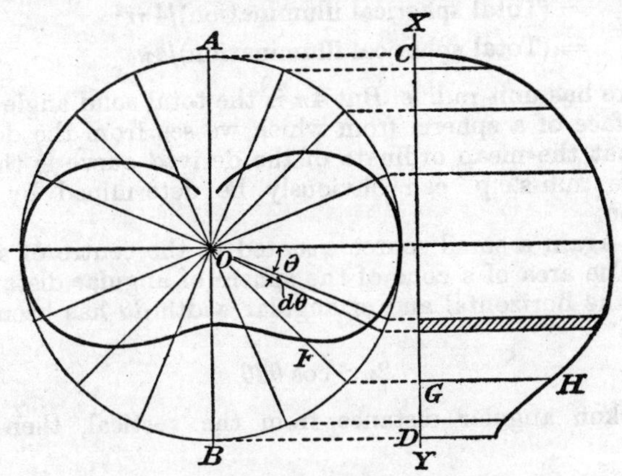

FIG. 30.9. THE ROUSSEAU DIAGRAM

XY drawn parallel to the vertical diameter AB. Points on the semicircle with uniform angular displacement are projected on to XY, and the lines produced beyond it by an amount equal to the candle power corresponding to the angular position. Thus, GH is made equal to OF, and so on. The mean ordinate of the curve so obtained (CD being regarded as its base) gives the m.s.c.p. For, suppose the intensity at any angular position θ is I, then, if an element of angular width $d\theta$ is taken and this element rotated about AB axis, it will describe a zone of a sphere of area

$$2\pi r \cos \theta \times r d\theta = 2\pi r^2 \cos \theta \, . \, d\theta$$

The illumination of the inner surface of this zone is therefore

$$2\pi I r^2 \cos \theta \, . \, d\theta$$

Hence, for the total spherical illumination, we have

$$2\pi \int_{-\pi/2}^{+\pi/2} I r^2 \cos \theta d\theta$$

I, of course, being a function of θ. Now, $r \cos \theta \, . \, d\theta$ is the width of the shaded strip in the figure, and therefore, $Ir \cos \theta \, . \, d\theta$ is the area of the shaded strip. The total area under the derived curve is therefore

$$\int_{-\pi/2}^{+\pi/2} Ir \cos \theta d\theta$$

$$= (\text{Total spherical illumination})/2\pi r$$

Now the base of this curve is of length $2r$. Hence, the length of the mean ordinate

$$= (\text{Total spherical illumination})/(2\pi r \times 2r)$$
$$= (\text{Total spherical illumination})/4\pi r^2$$
$$= (\text{Total spherical illumination})/4\pi$$

if the sphere has unit radius. But 4π is the total solid angle subtended by the surface of a sphere, from which we see from the definition of m.s.c.p., that the mean ordinate of the derived curve is the required value. The m.h-s.c.p. can obviously be determined by the same construction.

Consider again a small source situated at the centre of a sphere of radius r. The area of a zone of the sphere of angular distance θ with respect to the horizontal and of angular width $d\theta$ has been shown to be equal to

$$2\pi r^2 \cos \theta d\theta$$

Let us reckon angular distance from the vertical, then this area becomes

$$2\pi r^2 \sin \theta d\theta$$

Let the intensity in any direction θ be I_θ, then the luminous flux falling on the zone is

$$d\Phi = (I_\theta/r^2) \times 2\pi r^2 \sin \theta d\theta$$
$$d\Phi = 2\pi I_\theta \sin \theta d\theta$$

Hence total luminous flux from the source

$$\Phi = 2\pi \int_0^\pi I_\theta \sin \theta d\theta$$

Generally the integral cannot be evaluated, and therefore an obvious method is to plot $I_\theta \sin \theta$ against $d\theta$ and determine the enclosed area by means of a planimeter. Another method is the graphical method due to Rousseau described above.

A third method is the zonal method due to Russell. Let the sphere be divided into a series of zones of equal angular width α, so that the first zone extends from $\theta = 0$ to $\theta = \alpha$, the second from $\theta = \alpha$ to $\theta = 2\alpha$, etc., and let the angle α be chosen so that I_θ is sensibly constant over each zone. Then if we denote the mean intensities of the various zones by I_1, I_2, I_3, etc., we have for the fluxes in the various zones

$$\Phi_1 = 2\pi I_1 \int_0^\alpha \sin \theta d\theta = 2\pi I_1(1 - \cos \alpha)$$

$$\Phi_2 = 2\pi I_2 \int_{\alpha}^{2\alpha} \sin \theta d\theta = 2\pi I_2 (\cos \alpha - \cos 2\alpha)$$

$$\Phi_3 = 2\pi I_3 \int_{2\alpha}^{3\alpha} \sin \theta d\theta = 2\pi I_3 (\cos 2\alpha - \cos 3\alpha)$$

and so on. The flux in any zone, say the nth is thus

$$\Phi_n = 2\pi I_n [\cos (n - 1) \alpha - \cos n\alpha]$$

$$\therefore \Phi_n/I_n = 2\pi [\cos (n - 1) \alpha - \cos \alpha]$$

As an illustration of the method we will take a polar distribution of simple shape, justifying a comparatively small number of zones, say

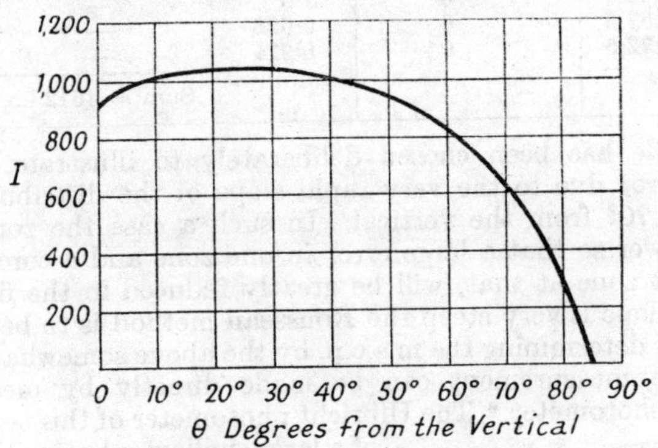

FIG. 30.10. INTENSITY DISTRIBUTION CURVE
(Rectangular co-ordinates)

12, so that $\alpha = 180/12 = 15°$. The values of I_θ are therefore read off at angles 7·5, 22·5, 37·5, etc., from the bottom. The intensity curve is given in Fig. 30.10 and the values of I_θ tabulated in column 1 of the table given below. For the ratio F_n/I_n we have

Zone 1, $n = 1$ $\therefore \Phi_n/I_n = 2\pi(1 - \cos 15°) = 0{\cdot}214$

Zone 2, $n = 2$ $\therefore \Phi_n/I_n = 2\pi (\cos 15 - \cos 30) = 0{\cdot}628$

Zone 3, $n = 3$ $\therefore \Phi_n/I_n = 2\pi (\cos 30 - \cos 45) = 1{\cdot}0$

and so on.

For the last zone,

Zone 12, $n = 12$ $\therefore \Phi_n/I_n = 2\pi (\cos 180 - \cos 165) = 0{\cdot}214$

Actually the values for the first six of the zones only need be calculated because Φ_n/I_n for zone 1 is the same as for zone 12, that for zone 2 is the same as for zone 11, and so on.

A typical intensity distribution for a filament lamp suspended inside a white diffusing bowl is illustrated in Fig. 30.10. The tabulation is given below—

θ	I_n (from curve)	Φ_n/I_n	$I_n \times \Phi_n/I_n$
7·5	1000	0·214	214
22·5	1115	0·628	700
37·5	1100	1·0	1100
52·5	920	1·3	1200
67·5	710	1·51	1072
82·5	200	1·63	326
97·5	0	1·63	—
112·5	0	1·51	—
127·5	0	1·3	—
142·5	0	1·0	—
157·5	0	0·628	—
172·5	0	0·214	—
			Sum = 4612 lm

This example has been chosen deliberately to illustrate a possible source of error due to the very rapid slope of the distribution curve after about 70° from the vertical. In such a case the zones can be made narrower so that a large error in one zone and a comparatively unimportant zone at that, will be greatly reduced in the final result. Where the slope is very steep the Rousseau method is to be preferred.

Instead of determining the m.s.c.p. by the above somewhat laborious method, the measurement can be made directly by means of an integrating photometer.* The Ulbricht photometer of this type consists of a large, hollow sphere with a smooth surface which is coated evenly with white paint. The diameter of the sphere should be large in comparison with the lamp to be tested, certainly not less than six times the diameter of the bulb of the lamp. When a lamp is placed inside the sphere the light is diffused in such a manner that the resulting illumination is uniform over the whole of the surface, the result of this being that, if a small window is arranged in the sphere, the illumination of this window is proportional to the m.s.c.p. of the lamp.

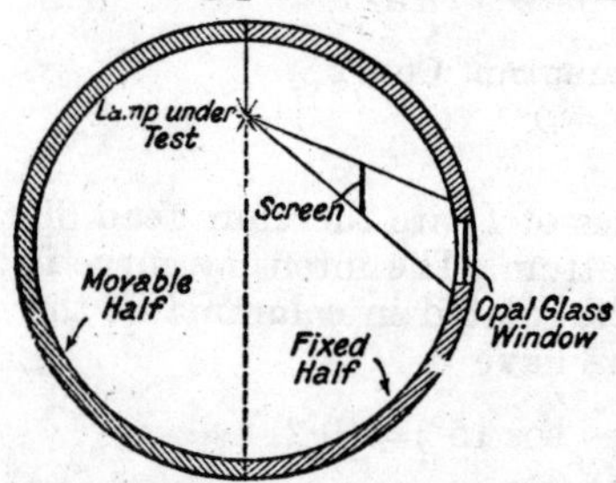

FIG. 30.11. PRINCIPLE OF SPHERICAL INTEGRATING PHOTOMETER

Consider the lamp placed at any position S, in the sphere (Fig. 30.11). The flux from S reaching any small area ΔS produces a certain illumination. Light is reflected and diffused at the surface ΔS, which in consequence becomes a secondary source of light. Let $\Delta\Phi$ be the flux reaching ΔS directly from the source, then the luminous intensity along the normal to ΔS is equal to $K.\Delta\Phi$, where K is a constant

* Integrating photometers of spherical and other shapes are discussed completely in J. W. T. Walsh, *Photometry*, Chap. VIII.

whose value is fixed by the nature of the interior surface of the sphere. Now consider any point P on the sphere. The illumination of a surface at P due to the reflected rays from ΔS is, for normal incidence,

$$(K \,.\, \Delta\Phi/x^2) \cos\theta$$

But as is shown in the figure, the normal to the surface at P makes an angle θ with the incident rays, so that for the actual illumination at P, due to the light reflected from ΔS, we have

$$\begin{aligned}&(K\,\Delta\Phi/x^2)\cos\theta \,.\, \cos\theta\\ &= (K\,.\,\Delta\Phi/x^2) \times (x/2r)^2\\ &= K\,.\,\Delta\Phi/4r^2\end{aligned}$$

This is independent of x, proving that all small areas such as ΔS which receive an equal flux from the source S produce identical illuminations at the point P. Hence, since P is any point, we see that the illumination at any point due to light reflected from the surface of the sphere is uniform. But by taking all the elementary surfaces ΔS we include the total flux of light from the source S, and therefore the illumination at any point of the sphere *due to reflected light* is proportional to the m.s.c.p.

The window where the observation is made must, therefore, be screened from direct rays, as indicated in Fig. 30.11. The sphere is made in two halves, one of which is fixed and the other movable on rails so that different lamps can be easily inserted and the surface inspected periodically.

The sphere photometer is naturally very expensive, especially when large, and therefore when very accurate results are not required a cubical box, as first suggested by Dr. Sumpner, can be used instead. The box has to be perfectly white on the inside and the surface uniform, as in the case of the sphere photometer, and it is used in exactly the same way. It is found that the cube gives very good results when the lamps under test have similar light distributions. If it is necessary to compare lamps with distributions which are not similar, it is necessary to obtain a correction factor for each distribution, and when this is done the measurements are of a high order of accuracy.

Incandescent Lamps. The incandescent lamp is a source from which the required radiation is obtained from a thin filament heated by the passage of an electric current. A source whose output is dependent on its temperature is called a *temperature radiator*.

There is a definite relationship between the diameter of a given filament and the current. Consider a filament working at a fixed temperature; then the power intake of the filament will be equal to the rate of dissipation of heat, chiefly by radiation.

$$\text{Power intake} = I^2R = I^2 \times (4\rho l/\pi d^2)$$

where l and d are the length and diameter respectively.

Now, the heat radiated per second

$$= \text{Surface area} \times K \times F(t)$$

where K is the emissivity of the filament. $F(t)$ is a constant only so long as the temperature t is constant.

Equating the intake to the rate of radiation of heat, we have

$$I^2 \times 4\rho l/\pi d^2 = \pi dl \times K \times F(t)$$

$$\therefore I^2 \times l/d^2 = A \times dl$$

where A is a constant.

$$\therefore d = B \times \sqrt[3]{I^2} \text{ or } I = C \times d^{1\cdot 5}$$

where B and C are constant. Note that similar expressions hold for the fusing current of a wire of given material under stated conditions.

Again, for filaments at the same temperature, the flux per unit area is the same. Hence

$$\text{lumen output} = D \times \text{surface area}$$

$$= D \times ld$$

where D is another constant.

In the evolutionary development of the filament lamp many materials have been used, carbon, the metals osmium and tantalum, and finally tungsten, and it seems unlikely that any material to replace tungsten is likely to be found in the forseeable future. The main reason for the superiority of tungsten in its ability to withstand a high operating temperature without undue evaporation of the filament. The necessity for a high temperature is illustrated in the curves of Fig. 30.12, which gives the spectral energy distributions of full radiators at five different temperatures, namely 2000, 2250, 2500, 3000 and 3400°K*. Ordinates are drawn at 4000 Å and 7500 Å to enclose the visible range. The total power of the radiation is the same in all cases, and so the curves are directly comparable. Two important facts are apparent.

1. The amount of visible radiation increases as the temperature is increased, with the result that the radiant efficiency also increases.

2. The relative values of the ordinates at various wavelengths, and therefore the colour of the radiation, is dependent on temperature.

Consider for example, the ordinates near the extreme red and near the extreme violet. At 2000°K there is no extreme violet and little blue or green as far as 5500 Å say, beyond this value the curve rises steeply indicating a predominance of yellow and red: as a result, and combined with the small amount of visible radiation (as indicated by the shaded area), the radiator has the visual appearance of dull red. At 3400°K the proportion of violet to red is very materially increased, the whole wavelength content in the visible range therefore being represented. The colour is therefore much closer to white (although not actually white) while the intensity, owing to the greater area under the curve, is very much greater.

To compare the relative intensities at the two temperatures each curve has to be operated on the relative luminous efficiency by multiplying the ordinate at any wavelength λ by the appropriate value V_λ.

* The symbol K denotes degrees on the absolute scale.

This has been done for the case of a temperature of 3400°K in Fig. 30.2, and a similar diagram for 2000°K is given in Fig. 30.13, with the luminous output curve at 3400°K for comparison. This shows the enormous increase in lumen output obtained by the increased temperature, and we have also seen that the colour is better (nearer to white).

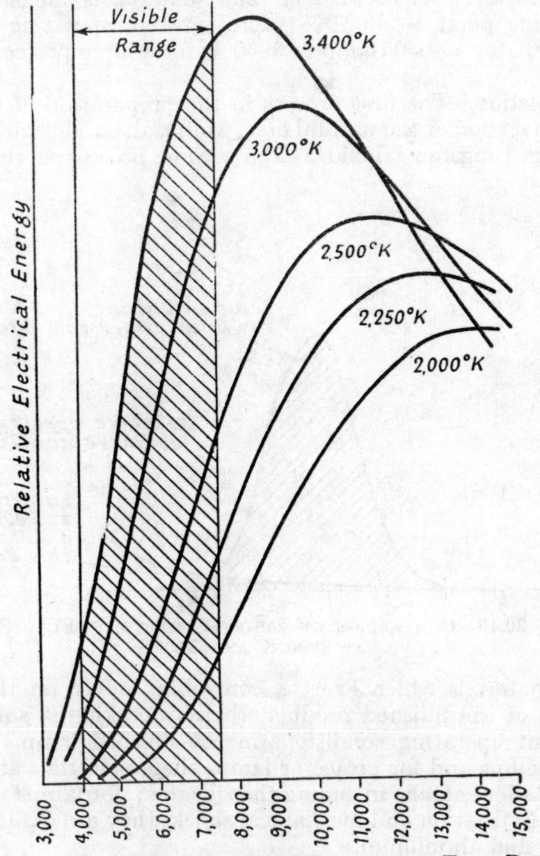

FIG. 30.12. SPECTRAL ENERGY DISTRIBUTION OF A FULL RADIATOR AT VARIOUS TEMPERATURES

As a matter of interest, a full radiator would require to be raised to a temperature of about 6000°K in order that the colour of its radiation should be of the quality accepted as white. The surface temperature of the sun has been computed to be of this order, that of the flame of a primitive oil lamp about 1870°K, and that of a batswing gas flame about 2160°K. It is therefore clear that the optimum condition of whiteness of the visible radiation cannot be obtained from any temperature radiator of a practicable kind. In this connection it is important to realize that over the visible range (but not beyond it, particularly in the U.V. region) the spectral energy *distribution* of a metal-filament lamp is almost identical with that of a full radiator.

For an electric lamp the best filament material is clearly that with the highest melting point, and the highest operating temperature assessed in terms of useful life. This material is tungsten, for which the melting point is 3660°K: representative operating temperatures are 2650°K for a 40-W lamp to 3000°K for a large projector lamp.

Construction. The first process in the preparation of tungsten wire is the conversion of the natural ores, wolframite and schielite, to chemically pure tungsten trioxide. The second process is the addition of

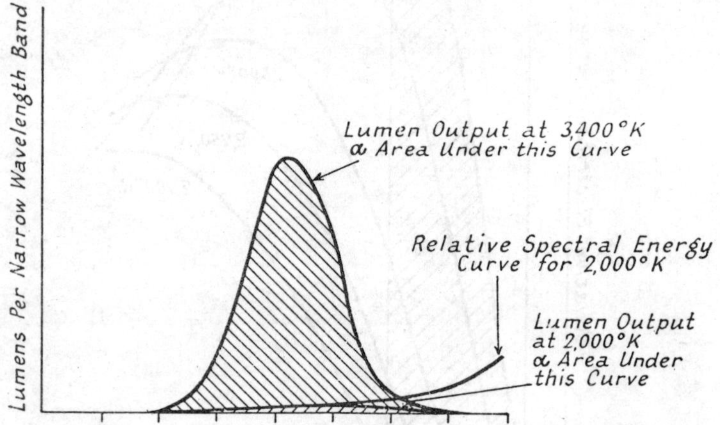

FIG. 30.13. COMPARISON OF LUMEN OUTPUTS OF FULL RADIATOR AT 2000°K AND 3400°K

certain materials which exert a controlling effect on the crystalline structure of the finished product, thereby making it suitable for the subsequent operating conditions in the finished lamp. For general-purpose lamps and for projector lamps these materials are compounds of the alkaline metals, including the silicates: for lamps which have to withstand vibration and mechanical shock, they are oxides of thorium, uranium, and aluminium.

The tungsten trioxide with its added compounds is heated in an atmosphere of hydrogen at a temperature of 1250°K, the final product from this process being the metal tungsten in the form of a grey powder. This is next pressed into bars of 40 cm length and 0·65 cm^2 cross section. These bars are "sintered" by the passage of a heavy current, current density 4000 A/in.^{2} The bar is sintered in an atmosphere of hydrogen to prevent oxidation. Before sintering the tungsten is in the form of a powder of density 11·23 g/cm^3, after sintering it is a hard, glass-brittle silver-bright metal of density 17·0.

In this form it is hammered into thin rods, then drawn on a chain-drawing bench to remove irregularities from the surface, and finally drawn through dies, firstly of tungsten carbide, and secondly of diamond, to finished wire of the required diameter.

The wire is coiled into spiral filaments which, in the case of the coiled-coil lamps are coiled a second time.

In the early lamps the filament was enclosed within an evacuated bulb in order to prevent the loss of heat from the filament to the gas filling that would otherwise have occurred. This led to rapid evaporation of the filament, with resulting blackening of the bulb and an early end to the life of the lamp. The preventatives were (*a*) the choice of filament material having a very slow evaporation, this leading to the

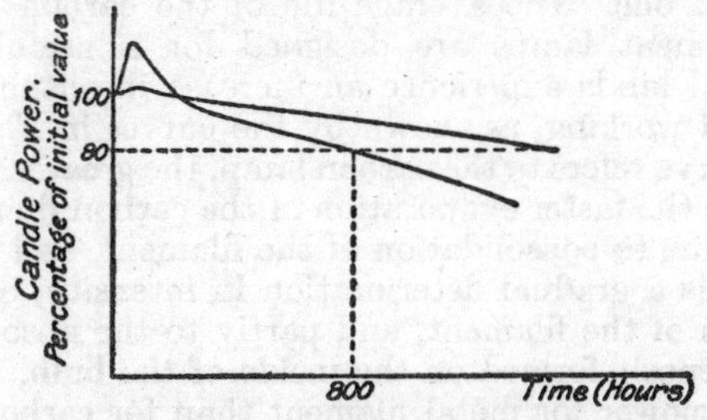

Fig. 30.14. Variation of Intensity with Time

abandonment of carbon for a metal of high melting point, and (*b*) a relatively low temperature, this, as we have seen, giving a low luminous efficiency and a considerable departure of the colour of the radiation from white.

Consequently, it became necessary to fill the bulbs with an inert gas. Nitrogen or argon, or a mixture of the two are used, they can be extracted from the atmosphere at reasonable cost, and their density is sufficient to prevent undue filament evaporation. Where the maximum efficiency is of more importance than cost, as for example the small miner's cap lamps and hand lamps, the temperature can be made higher than is usual in general-service lamps, evaporation being then prevented by the use of a gas of greater density, namely krypton.

The advantage of the coiled-coil filament is associated with the effect of a gas filling. It is found that there is a thin layer of gas in contact with the filament in which no appreciable gas movement takes place. Consequently the effect of this layer on a tightly-coiled spiral is to deny the free passage of gas through the individual turns and thus to give the effect of a short thick cylinder having a reduced effective cooling surface. The coiled-coil filament is an exploitation of this principle, the effective cooling surface being very much less than the surface of the filament itself. Thus an equal power intake of the filament (wattage) gives a higher temperature with this construction, or alternatively, a given temperature can be attained with a smaller wattage.

The diameter of the filament has an effect on the efficiency of gas filling. In the case of the very fine filaments used for low wattages and normal voltages (200–250 V), the surface area is so large in proportion to the filament volume that no advantage is derived. Hence, although the requirements of mass production impose coiled filaments in all sizes, it is only for 40 watts and upwards (in the 200–250 V range) that gas filling is adopted.

Life of Incandescent Lamps. The life of a carbon filament was ended when its luminous output fell to 80 per cent of the original value, because it was then cheaper to replace the lamp than continue working with the old one. The average life of the carbon lamp was 800 hr. Modern filament lamps are designed for a useful life of 1000 hr. Lamps of all kinds experience an increase in output during the first few hours of working, as shown by the curves in Fig. 30.14, in which the lower curve refers to the carbon lamp, the greater rate of diminution being due to the faster evaporation of the carbon filament. This initial increase is due to consolidation of the filament, and after it has taken place there is a gradual deterioration in intensity, due partly to slow vaporization of the filament, and partly to the absorption of light by the black deposit formed on the inside of the bulb. The falling off in intensity is slower for metal filament than for carbon filament lamps, and since the metal filament is more fragile, the life of these lamps is usually ended by breakage.

The term "smashing-point" is used to denote the percentage of initial intensity to which the lamp should be allowed to decline before it is replaced, the falling off in intensity beyond this point causing the efficiency to be so poor that it is cheaper to take the lamp from service and buy a new one. The smashing-point is generally taken as 80 per cent, although when tests have been made it has been found that lamps often fail before this point is reached.

Effect of Variations in Voltage on Operation. All of the operating characteristics of filament lamps are materially affected by a departure of the voltage from the rated value. After the initial heating up, which takes place very rapidly, owing to the low thermal capacity of the lamp, the power intake is steady, and, if the filament resistance were not dependent on temperature, the rate of generation of heat would be proportional to the square of the voltage. Because the filament has a positive temperature coefficient of resistance which is not small enough to be negligible, and because of the complex mechanism of heat transfer from filament to gas filling, the relationships between the various characteristics and the voltage must be empirical relationships based on experiment. For gas filled lamps they are as follows—

Let V_0 = normal volts; V = any voltage

lm_0 = normal lumens; lm = lumens at voltage V

P_0 = normal watts; P = watts at voltage V

I_0 = normal current; I = current at voltage V

Then
$$lm/lm_0 = (V/V_0)^{3.3}$$
$$P/P_0 = (V/V_0)^{1.43}$$
$$(lm/P)/(lm_0/P_0) = (V/V_0)^{2.0}$$
$$(\text{Life})/(\text{Life at normal voltage}) = (V/V_0)^{-13.3}$$

The characteristic curves are given in Fig. 30.15. The life characteristic is striking: small under voltages lengthens it appreciably, but an

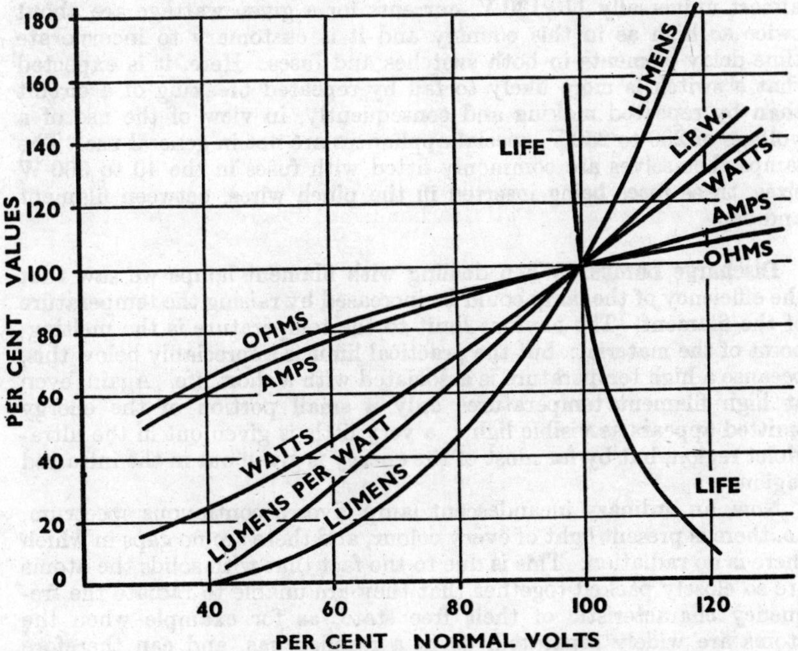

FIG. 30.15. CHARACTERISTIC CURVES FOR FILAMENT LAMP
("Electric Lamps," E.L.M.A. Lighting Service Bureau, E.D.A.)

overvoltage of only 5 per cent will shorten it to one-half of the normal value.

The luminance at normal voltage has the following values: for clear gas-filled general-purpose lamps 400 to 675 cd/cm² according to size: for pearl lamps, the area of maximum brightness, 6 for the 40 W lamp to 20 for the 150 W.

Starting Characteristics. The cold resistance of the tungsten filament is only one-fifteenth to one-seventeenth of the resistance at the working temperature and consequently there is a considerable rush of current at the moment of switching on. Actually the magnitude of this current rush is not given by the reciprocal of the above values because (*a*) the filament immediately heats up and thereby increases in resistance, (*b*) there is an initial voltage drop in the external circuit due to resistance (and reactance in the a.c. case) which is greater than normal, resulting in an initial applied voltage of less than normal. In practice the maximum current attained is of the order of rather more than one-half of the value decided by change in resistance above.

The maximum current is attained in about 1/300 sec, and normal current in about $\frac{1}{8}$ sec. In America, where the voltage for lighting is almost universally 110/130 V, currents for a given wattage are about twice as high as in this country and it is customary to incorporate time-delay elements in both switches and fuses. Here, it is expected that a switch is more likely to fail by repeated breaking of a circuit than by repeated making and consequently, in view of the use of a voltage of 200 to 250 V, special appliances are not in general use. The lamps themselves are commonly fitted with fuses in the 40 to 300 W sizes, these fuses being inserted in the pinch wires, between filament and cap.

Discharge Lamps. When dealing with filament lamps we saw that the efficiency of the lamp could be increased by raising the temperature of the filament. The obvious limit to the temperature is the melting-point of the material; but the practical limit is appreciably below this, because a high temperature is associated with a short life. Again, even at high filament temperatures only a small portion of the energy emitted appears as visible light: a very little is given out in the ultra-violet region, but by far most of the energy is given out in the infra-red region.

Now, an ordinary incandescent lamp gives a continuous spectrum, i.e. there is present light of every colour, and there are no gaps in which there is no radiation. This is due to the fact that with solids the atoms are so closely packed together that they are unable to radiate the frequency characteristic of their free state, as for example when the atoms are widely separated, as in a rarefied gas, and can therefore radiate without interference from one another. As a result of the continuous nature of the spectrum the ratio of the energy emitted as visible light to the total energy emitted is given by the ratio of the shaded portion to the total area under the curve. With a modern coiled-coil lamp this ratio is only about 10 per cent, showing that in order to obtain any appreciable gain in efficiency it is necessary to break away from the traditional heated-filament lamp and adopt some other principle.

This principle is available in the discharge lamp, in which the light is obtained from a luminous column of gas or vapour, sometimes rarefied, sometimes at a pressure of several atmospheres. Furthermore, this luminous column is not heated to a high temperature, but is excited electrically. To appreciate this, consider the nature of the electrical discharge through a rarefied gas or vapour. The phenomena are similar to those already studied in the case of the mercury-arc rectifier. Electrons travel from the cathode to the anode, and under the influence of the potential gradient they acquire a velocity, and therefore a kinetic energy, dependent upon the length of the mean free path. While in motion they will collide with gas or vapour atoms, and the result of the collision will depend on the kinetic energy of the electron just prior to the collision. One of three things can happen. Firstly, if the kinetic energy is small, the electron will bounce off the atom and there will be little result beyond a possible change of direction. Secondly, if the kinetic energy is sufficiently high, the impact will be sufficiently great to displace one of the electrons of the atom from its normal orbit to another orbit round the positive nucleus. The atom is then said to be in an excited state and radiation of a certain wavelength will be emitted on the return to the normal state, the wavelength depending on the change in the orbits of the electron round the nucleus. The lowest kinetic energy to produce excitation raises the atom to what is called the first excited state, and if the colliding electron has this amount of energy it will leave the atom with zero velocity. If the energy before collision is greater than this, then the electron will leave the atom with a velocity corresponding to the excess energy, or, if the excess energy is sufficiently great, the atom will be raised to what is called its second excited state and radiation of another wavelength will be emitted.

Hence the atom can only acquire definite increments of energy by such a collision process, and hence, the displaced electron can only be brought to one of a definite series of orbits. Regarding the possible orbits as circular paths round the nucleus, this means that there are definite radii to which the electron can be displaced, and an orbit of some radius in between two such consecutive radii is not possible. For a hydrogen atom the first six possible orbits are shown in Fig. 30.16. The radii of these orbits in Angstrom units (Å) are

$$r_1 = 0{\cdot}548 \text{ Å}$$
$$r_2 = 2{\cdot}112 \text{ Å}$$
$$r_3 = 4{\cdot}752 \text{ Å}$$
$$r_4 = 8{\cdot}448 \text{ Å}$$
$$r_5 = 13{\cdot}07 \text{ Å}$$
$$r_6 = 19{\cdot}008 \text{ Å}$$

Now if an electron is occupying any particular orbit say r_m, and it is displaced to another orbit r_n, of greater radius, then clearly work has to be done against gravitational attraction of the nucleus for the electron, and represented by the increased distance $(r_n - r_m)$. In the science of electronics it is usual to express this energy in "electron volts," the electron volt being the work done when an electron is moved in an electrostatic field from a point A to a point B such that the p.d. between A and B is one volt.

The Electron Volt. Let there be a p.d. of V volts between two points x m apart

Using m.k.s. units we have—

Potential gradient $X = v/x$ volts/metre, with x in metres.

$\therefore$ Force on an electron of charge e coulombs

$$\therefore \quad F = Xe$$
$$= Ve/x \text{ newtons}$$

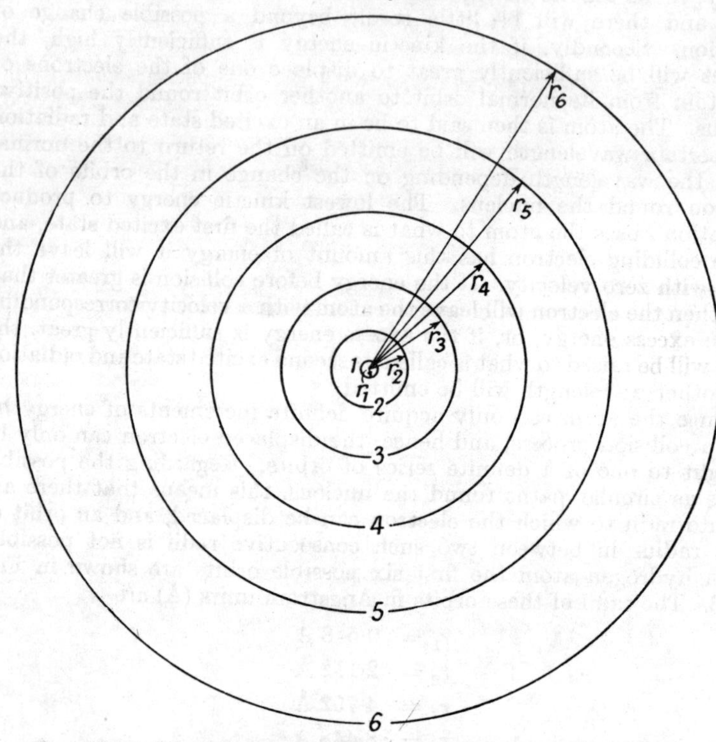

FIG. 30.16. POSSIBLE ORBITS OF THE VALENCY ELECTRONS OF A HYDROGEN ATOM

$\therefore$ Work done in moving the electron through the whole potential difference

$$\therefore \quad W = Fx$$
$$= Ve \text{ metre-newtons}$$
$$= e \text{ (numerically) when } V = 1 \text{ volt}$$

The charge on the electron is $1{\cdot}603 \times 10^{-19}$ coulombs

$$\therefore \quad W = 1{\cdot}603 \times 10^{-19} \text{ metre-newtons or joules}$$
$$\text{or} \quad = 1{\cdot}603 \times 10^{-12} \text{ ergs.}$$

Wavelength of Radiated Emission. Now suppose that an electron has been displaced from its normal orbit, or ground state, as it is called, the atom thereby being in an excited state. This state is a temporary state only, the duration being of the order of 10^{-8} sec. The electron then returns to an orbit of smaller radius, but not necessarily the ground state. It can be regarded as falling a definite distance under the gravitational attraction of the nucleus, and in consequence it gives up a definite amount of energy. This energy given up can appear only as radiation; there is no other form of energy to which it can be converted.

Let E_n = electron volt energy of the electron corresponding to the orbit of radius r_n

and E_m = electron volt energy of the electron corresponding to the new orbit to which it falls

Then $E_n - E_m$ = energy given up and converted into radiation.

It can be shown that this energy difference is equal to the product $h\nu$ where

h = a universal constant called Planck's constant

ν = frequency of the radiation emitted

In terms of electron volts, a difference of orbit corresponding to a difference of V volts will be equal to

$$1{\cdot}603 \times 10^{-19} \times V \text{ joules}$$

$$\therefore \quad h\nu = 1{\cdot}603 \times 10^{-19} \times V \text{ joules}$$

Now $$h = 6{\cdot}61 \times 10^{-34} \text{ joules}$$

$$\therefore \quad \nu = 1{\cdot}603 \times 10^{-19} \times V/(6{\cdot}61 \times 10^{-34})$$

$$= 2{\cdot}43 \times 10^{14} \times V$$

But frequency

$$\nu = c/\lambda$$

where c = velocity of radiation in free space (m/sec)

and λ = wavelength in m

$$\therefore \quad \lambda = c/\nu$$

Now $$c = 3 \times 10^{8} \text{ m/sec}$$

$$\therefore \quad \lambda = 3 \times 10^{8}/(2{\cdot}43 \times 10^{14} \times V)$$

$$= 1{\cdot}2336 \times 10^{-6}/V \text{ m}$$

$$= 1{\cdot}2336 \times 10^{-6} \times 10^{10}/V \text{ Å}$$

$$= 12{,}336/V \text{ Å}$$

This shows that the wavelength of the radiation emitted will bear an exact relationship to the energy given up. Hence, since only definite amounts of energy can be given up because of the definite series of possible orbits, only definite wavelengths can be emitted, and so the spectrum consists of a series of lines.

The Hydrogen Discharge. For a given atom, the attainment of any particular orbit of the displaced electron from the normal or ground state, will necessitate the acquisition of the definite amount of energy associated with that particular orbit. This energy is generally expressed in electron volts and is called the energy level of the orbit. By drawing a series of horizontal lines whose heights with respect to the ground state represent the energy levels of the various orbits we obtain the

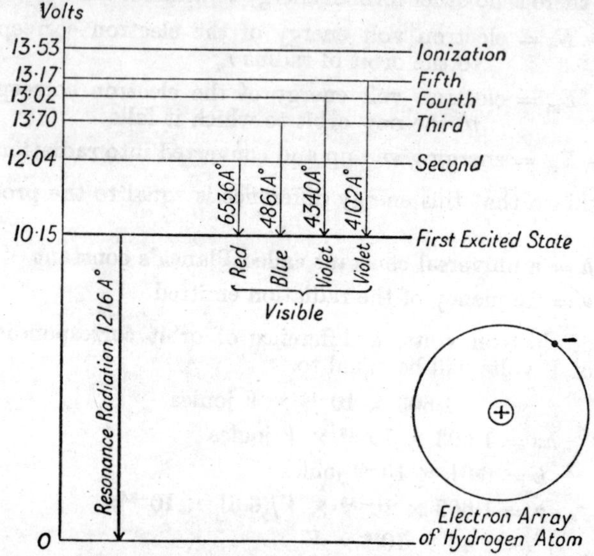

Fig. 30.17. Some of the Energy Levels of the Hydrogen Atom

energy-level diagram. That for the hydrogen atom is given in Fig. 30.17. The various levels are as follows—

1st	excited	state	10·15	electron volts
2nd	,,	,,	12·04	,,
3rd	,,	,,	12·70	,,
4th	,,	,,	13·02	,,
5th	,,	,,	13·17	,,

We see that the successive increments become progressively smaller, the energy levels thus becoming closer together as the order of excitation increases

There is a particular value of the energy, called the ionization level at which the valency electron of the atom is not merely displaced, but actually detached, the atom thereby losing an electron and becoming a positive ion. We will examine the effect of this later. For the hydrogen atom the ionization level is 13·53 V.

We can now investigate the nature of the radiation due to the passage of an electric current through hydrogen at very low pressure. It will be clear that with the vast number of collisions taking place simultaneously, these will be of all possible kinds, but the greatest number will be to the first excited state, that is, to the 10·15 level. The energy given up in the subsequent fall from this level to the ground state is thus

$$10{\cdot}15 - 0 = 10{\cdot}15 \text{ electron volts}$$

The wavelength of the radiation resulting from this fall is therefore

$$\lambda = 12{,}336/10{\cdot}15$$
$$= 1216 \text{ Å}$$

This is in the ultra-violet region and is therefore invisible, and so of no value as visible light. We see at once that an electric discharge lamp using hydrogen will be very inefficient as a light source because most of the radiant energy is the result of collisions which lead to the first excited state. In other words, a lamp whose light output is due solely to the changes in energy resulting from falls from one excited state to another, or from one excited state to ground state, will not be very efficient if the most frequent fall of all, that from the first excited state to ground state, gives rise to a radiation outside the visible spectrum. This particular radiation is called the resonance radiation.

In the hydrogen discharge the falls from all the possible levels to ground state result in ultra-violet radiation. Now consider the falls from the various excited states down to the first excited state. The energy changes are as follows—

2nd excited state to first	$12{\cdot}04 - 10{\cdot}15 = 1{\cdot}89$
3rd ,, ,, ,,	$12{\cdot}70 - 10{\cdot}15 = 2{\cdot}55$
4th ,, ,, ,,	$13{\cdot}02 - 10{\cdot}15 = 2{\cdot}87$
5th ,, ,, ,,	$13{\cdot}17 - 10{\cdot}15 = 3{\cdot}02$

The wavelengths of the radiations resulting from these are respectively

$$12{,}336/1{\cdot}89 = 6536 \text{ Å}$$
$$12{,}336/2{\cdot}55 = 4861 \text{ Å}$$
$$12{,}336/2{\cdot}87 = 4340 \text{ Å}$$
$$12{,}336/3{\cdot}02 = 4102 \text{ Å}$$

These are all in the visible region, the colours being respectively red, blue, violet, and extreme violet.*

When an electron has fallen from one state to the first excited state with resulting emission of radiation, it falls the rest of the way to ground state, with resulting emission of resonance radiation, which, for hydrogen, is in the ultra-violet. The light-giving cycles for hydrogen are therefore inefficient. Thus the collision resulting in the second excited state needs 12·04 electron volts of energy, but only 1·89 electron volts are required to give the 6536 Å radiation, and so the efficiency of the cycle for this particular radiation is

$$(1{\cdot}89/12{\cdot}04) \times 100 = 15{\cdot}7 \text{ per cent}$$

Similarly with the other cycles.

* The wavelengths given are not quite the same as the quotients, because the energy differences in the numerator are only given to three figures.

With the hydrogen discharge, falls from levels to the second excited state give rise to radiation of wavelengths 18,751, 12,818, and 10,938 Å, these being in the infra-red. Falls from higher levels to the third excited state also give rise to infra-red radiation of still greater wavelength. We thus see that the total radiation covers a very wide range of the spectrum, and the energy content of the few lines in the visible is only a small fraction of the total.

The Sodium Discharge. The sodium atom has eleven electrons arranged as follows: two in the inner or *K* shell, eight in the second

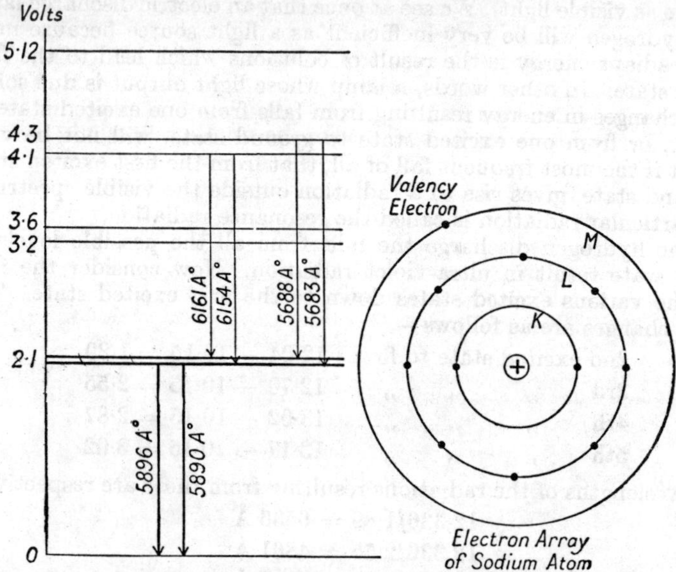

FIG. 30.18. ENERGY-LEVEL DIAGRAM FOR SODIUM

or *L* shell, and only one, the valency electron, in the third or *M* shell. It is to be noted that these shells are merely a convenient means of grouping the electrons, and are not to be confused with the orbits round which we assume the various electrons to rotate. The energy-level diagram is complicated but the relevant portion of it is shown in Fig. 30.18. The first excited state is at 2·1 V level, whereas the ionization state is at 5·12 V level. These values are very low in comparison with hydrogen. A second important point is that the various levels are actually in pairs having a very small difference, so small in fact that it cannot be shown on the diagram. We therefore draw two lines at 2·1 V level, to represent the two possible first excited states. A few of the other levels are 3·2, 3·6, 4·1 and 4·3 V, but as these are relatively unimportant we will regard them as single. Remembering the importance of the resonance radiation, that is, the radiation due to the fall from the first excited state to the ground state, we will calculate the wavelength. We have

$$\lambda = 12{,}336/2{\cdot}1$$

Remembering again the approximate value of the numerator, and the fact that there are actually two levels very close together we will state the actual values, which are

5890 and 5896 Å

These are in the visible range, the colour being yellow. Hence we see that the sodium discharge has the very desirable property of giving a visible resonance radiation, and that combined with this is the advantage of a low ionization level. There is another factor to take into account, namely the relative luminous efficiency V_λ. For the sodium yellow doublet this is 0·765. These factors lead to an extremely high efficiency and gives the sodium lamp a considerable industrial importance.

The spectrum of the sodium discharge shows a series of lines in pairs, or doublets, but the yellow colour is due almost entirely to the so-called D line, the 5890, 5896 doublet. Other doublets are the red, of wavelengths 6154, 6161 A, due to the falls from the 4·1 V level to the two first excitation, or resonance, levels, and the orange 5683, 5688 Å, due to the falls from the 4·3 to the resonance level. The falls from the 3·2 and 3·6 levels give rise to ultra-violet radiation.

It is of interest to note that sodium vapour is the only available vapour whose resonance radiation is in the visible range.

The Mercury Discharge. The mercury atom has a very complex electron array owing to the great number (80) of orbital electrons. The important point to note is that there are two valency electrons, and thus the various excited states are due to possible displacements of two electrons. This gives rise to a very complex energy-level diagram in which some of the levels are single and some in groups of three, the latter giving rise to triplet lines in the spectrum. A further complication arises from the fact that the component of a triplet line may have a considerable separation, unlike the sodium doublets, all of which have components very close together. From our point of view the important levels are

(*a*) The first excited state having three levels of 4·66, 4·86, and 5·43 electron volts respectively.

(*b*) The 7·7 V level.

(*c*) An 8·8 V level giving rise to single lines, or singlets.

(*d*) An 8·8 V level giving rise to triplet lines. These levels are shown in Fig. 30.19.

(*e*) The ionization level of 10·38 electron volts.

It would appear that with a triple first excitation level there should be a triplet resonance radiation, but, in fact certain lines are "forbidden," and there is only one resonance radiation, viz. that due to the fall from the 4·86 level.* The wavelength is

$$12{,}336 \div 4{\cdot}86 = 2537 \text{ Å}$$

This is in the ultra-violet. There is a less important ultra-violet line due to the fall from the 6·67 level, and of wavelength

$$12{,}336 \div 6{\cdot}67 = 1850 \text{ Å}$$

The visible lines are all due to falls from higher to lower excited states,

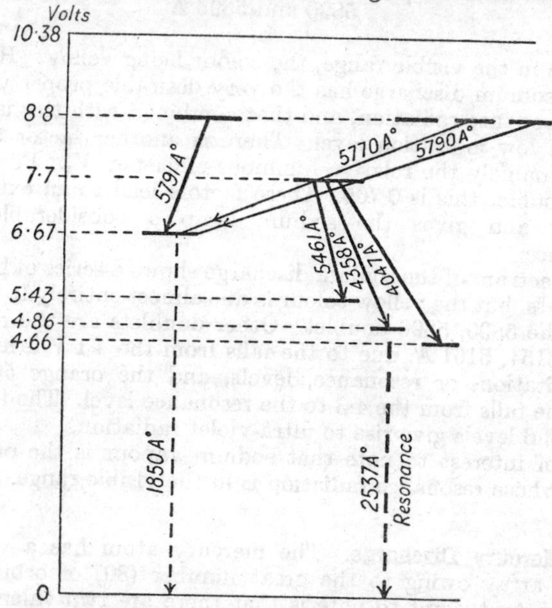

FIG. 30.19. ENERGY-LEVEL DIAGRAM FOR MERCURY

not to the ground state. These are of the following wavelengths, all of which are inserted in the diagram—

(*a*) 5770, 5790 and 5791 in the yellow. The two latter are so close together that the spectrum appears to have two yellow lines only,

(*b*) 5461 in the green. This is a very intense line, and, owing to the fact that the eye is most sensitive to green radiation, the line is of great importance,

(*c*). 4358 Å blue, and

(*d*) 4047 Å violet.

The value of V_λ for the mercury green line is the high value of 0·98, giving this line a very great importance.

* These phenomena, which are beyond the scope of this book, are discussed in the author's treatise *Electric Discharge Lamps*. Chapman & Hall, London.

In the mercury discharge, but not the sodium discharge there are certain excited states called *metastable states*. If there is a fall from a higher excited state to a metastable state, then, instead of this being followed by a subsequent fall to ground state, the atom waits, as it were, for another collision which will raise it again to a higher state. In the case of mercury the 4·66 and 5·43 levels of the first excited state are metastable states and it is for this reason that they do not give rise to resonance radiations, because there are no falls to ground state. In the visible range the falls from the 7·7 level to these two metastable levels are therefore 100 per cent efficient as far as energy exchange is concerned, and it is for this reason that the visible lines they produce, namely, 4047 A in the case of the 4·66 state, and 5461 A in the case of the 5·43 state, are important. The energy of the 4047 (violet) radiation is thus greater than that of the 5461 (green) radiation, but owing to its very high value of V_λ the contribution of the latter to the lumen output is the greater.

Ionization. We have seen that if a collision is sufficiently violent, then a valency electron will be detached from the atom, and the latter will become a positive ion. Now in order that the discharge path may be a conductor of electricity it is essential that there shall be both electrons and ions, but, since light emission is the function of a discharge lamp we only require sufficient conductivity to pass the current associated with the lumen output desired. Suppose that there were no check on the amount of ionization taking place then the conductivity would increase cumulatively and the tube would be destroyed. The mechanism, called by the physicist an "electron avalanche," would be somewhat as follows: An electron approaches a neutral atom and, the collision being sufficiently violent, detaches the valency atom, as shown in Fig. 30.20 (*a*). As a result of the collision there are three charged particles—two free electrons and one ion, instead of one charged particle and one neutral atom. Thus the conductivity of the path is very materially increased. Next, the ion moves towards the cathode and the two electrons towards the anode, and these encounter two more neutral atoms, as in Fig. 30.20 (*c*). Again ionization will take place if the collisions are sufficiently violent, and the result will be seven charged particles. Similarly with successive collisions. If the process is continued unchecked, the discharge path will become virtually a short-circuit and the apparatus destroyed because of the excessive current delivered by the electrodes. Now the possibility of ionization, as distinct from excitation or mere collision with no physical change, is decided by the velocity of the colliding electron, and this in turn is decided by the potential difference between anode and cathode. Hence if the rate of ionization becomes too great, i.e. if the current becomes too great, the p.d. between the electrodes must be reduced. This is effected by having a resistance or a choke in series with the lamp. If the current becomes too high the external volt drop increases and so the lamp voltage decreases. If the current falls, the external drop decreases and the lamp voltage increases. Thus there is an automatic control of the lamp current. Such an external resistance or choke is called a *ballast*. For d.c. working the ballast must obviously be a resistance, but for a.c. working it is usual to employ a choke because, for a given volt drop, the power loss in a choke is very small.

If the vapour pressure is very low, the running conditions are attained almost instantaneously, but for a pressure of the order of one atmosphere or more the starting characteristics of a discharge lamp are, in

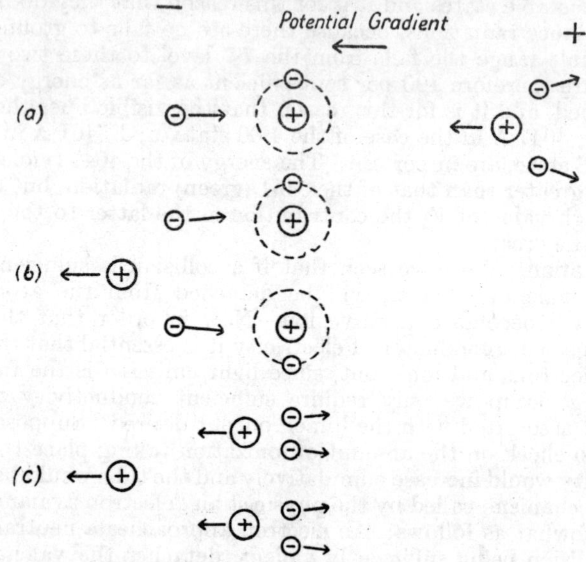

FIG. 30.20. TO ILLUSTRATE CUMULATIVE IONIZATION

some ways, analogous to those of a d.c. motor. At the moment of switching on, the back voltage of the arc is not established and the current is limited by the series choke. As ionization proceeds, the arc voltage rises and the current falls. Thus in the case of a 230 V, 400 W mercury-vapour lamp the starting current is 5·5 A and the arc voltage 15 V. After about seven minutes the lamp has settled down to steady state conditions, and the current has fallen to 3 A while the arc voltage has risen to 145 V.

As with all a.c. arcs an approximately sinusoidal current gives rise to a flat-topped arc voltage, but so far as the expression can be applied to two quantities of different waveforms, the arc voltage is in phase with the current. The drop across the choke is in quadrature with the current, and therefore the voltage triangle whose sides represent the applied voltage V, arc voltage V_a, and choke drop V_x, will be right-angled, having V as the hypotenuse. The vector diagram for starting and running conditions for the above 400 W lamp will therefore be as shown in Fig. 30.21.

The Sodium Vapour Lamp. The sodium lamp is essentially a low vapour-pressure lamp because an increase in pressure above a few mm of mercury does not increase the only visible radiant energy that matters, namely the yellow D line. This is because the D line radiation is the resonance radiation. This radiation, on colliding with an atom in the ground state can be completely absorbed and therefore, if the vapour pressure is too high, there will be little chance of any of the radiation reaching the outside. As it is, only that from the outer surface of the discharge manages to escape. Now the characteristic of all

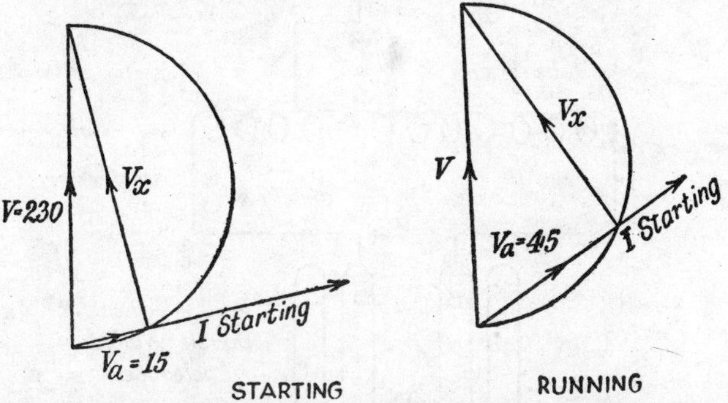

Fig. 30.21. Starting and Running Conditions for 400-w Medium-pressure Mercury-vapour Lamp

low-pressure discharges is the very low intensity, and consequently, to obtain a large lumen output, the lamp must be in the form of a long tube. With sodium there is a second factor, namely that the efficiency is very sensitive to changes in the tube temperature, and this is so critical that the lamp has to be enclosed in a double-walled jacket of the vacuum-flask type to protect it from the atmosphere. Clearly there is a limit to the practicable size of such a flask and therefore a limit to the length of lamp tube possible, even if the lamp is bent into a U so as to reduce the length of the enclosing jacket. The above characteristics are illustrated by the fact that the brightness is only 7·5 candles/cm^2.

A second factor is the fact that sodium is solid when cold. This means that the lamp cannot be started up as a sodium lamp because when cold there will be few sodium atoms in the discharge path. Hence there has to be a little neon gas in the tube and the lamp is, in fact, started up as a neon lamp giving the characteristic pink glow. The heat from the neon discharge gradually vaporizes the sodium, and after about 10 to 15 min the whole of the sodium is vaporized and the lamp fully run up.

The lamp is of the cold-cathode type, that is, the electrodes are not pre-heated, and therefore since the ionization voltage of neon is high, 21·5 V, a high voltage is necessary to start the discharge. Actually a very small amount of argon is introduced as well, its lower ionization voltage of 15·76 helping the discharge to start. The circuit diagram for the sodium lamp is therefore as shown in Fig. 30.22. It would appear at first sight that the lamp voltage will always be greater than the supply, and also that the circuit has no ballast. Actually the transformer is one in which leakage reactance is enormously increased by the use of magnetic shunts, so that the necessary characteristics can be obtained from a single transformer instead of using a transformer of the ordinary constant voltage type in conjunction with a separate choke. The characteristics of such a transformer are shown in Fig. 30.23. As an example

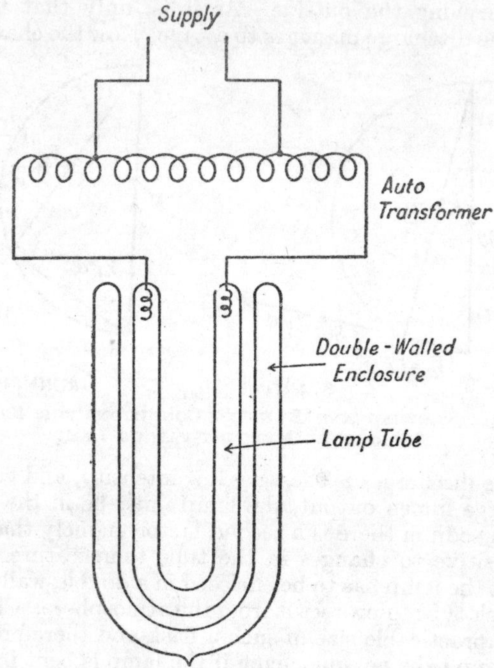

Fig. 30.22. Sodium-vapour Lamp and Auto-transformer

take the case of the 140 W sodium lamp. When fully run up the lamp voltage is 165 V and the lamp current 0·9 A. The starting voltage is 410 V. There is one other constructional feature of the sodium lamp which must be noted, and that is the glass used for the lamp tube. It is found that all ordinary glass, and even quartz, blacken rapidly under the effect of the sodium discharge, and therefore the sodium lamp only became a practical proposition when a glass which would withstand the sodium arc was discovered. There is now such a glass available, and it is applied as a thin coating to the inside of the lamp tube since, clearly, there is no necessity to make the whole of the tube of this very expensive material.

The Mercury-vapour Lamp. We have already seen that the low-pressure mercury discharge has a spectrum containing very few lines, the most important of which is a green of wavelength 5461 Å. As a result the colour is very objectionable. Furthermore, since, unlike the sodium lamp, none of the visible lines is a resonance radiation, the efficiency in terms of lumens per watt is very low. Again, in view of the low pressure, the lamp must be in the form of a long tube. The disadvantages due (*a*) to objectionable colour and (*b*) to low efficiency, are overcome in the case of the low-pressure lamp by coating the inside of the lamp tube with a fluorescent powder or *phosphor*, which has the property of absorbing radiation of one wavelength and re-emitting part of the energy so absorbed at a lower wavelength. Also

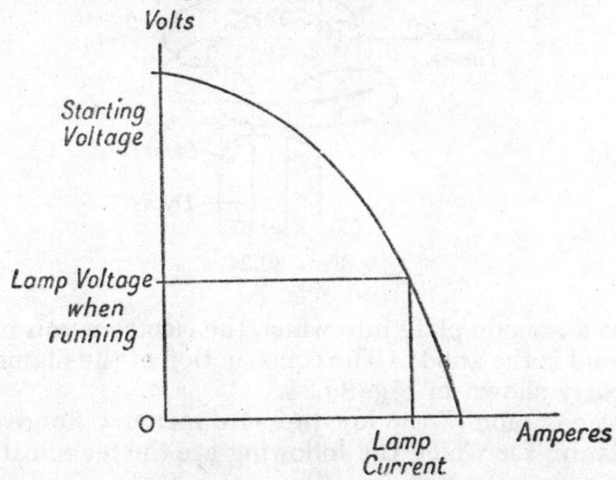

FIG. 30.23. CHARACTERISTICS OF LAMP TRANSFORMER

as is characteristic of all solid emitters of radiation, the energy re-emitted covers a band of wavelengths and is therefore no longer monochromatic. In the case of the mercury-vapour fluorescent lamp the phosphor absorbs the resonance radiation of 2537 Å, and re-emits a band characteristic of the particular material used. Thus the emitted ranges of a few commercial phosphors, excited by the mercury resonance line are as follows—*

Phosphor	*Colour*	*Emitted range* A
Calcium tungstate	Blue	3800–7000
Magnesium tungstate	Blue-white	3800–7200
Zinc silicate	Green	4500–6200
Zinc-beryllium silicate	Yellow-white	4500–7200
Cadmium silicate	Yellow-pink	4300–7200
Cadmium borate	Pink	4000–7000

By a suitable choice of phosphor, or mixture of phosphors it is thus possible to obtain a very wide range of colours, or what is most frequently required, a white.

It will be appreciated that most of the light output of a fluorescent lamp comes from the phosphor, and therefore the lamp can almost be regarded as a 2357 Å convertor.

* The commercial phosphors are crystalline materials containing a very small amount of a heavy impurity called the activator. The following have all been used as activators: silver, copper, bismuth, gold, and manganese.

In construction the lamp is merely a long tube coated on the inside with the phosphor. There is a spiral electrode at each end and this is coated with an electron emitting material very like the cathode of a thermionic valve. Now each end acts alternately as cathode and anode when a lamp is worked from an a.c. supply (as is usual) and a filament so coated does not readily act as an anode. Consequently at each end

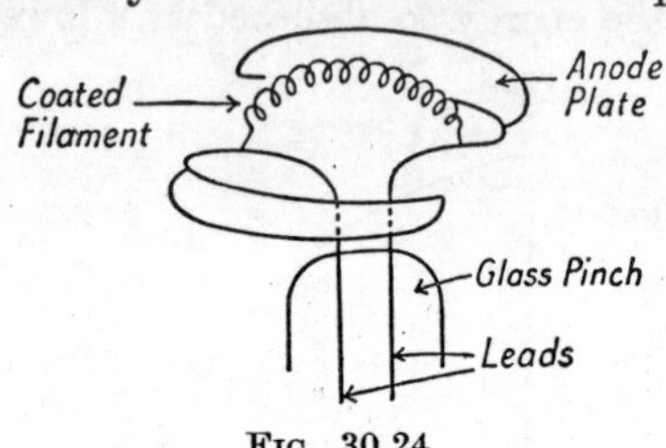

FIG. 30.24

there is also an anode plate into which the electrons can pass when that particular end is the anode. The construction of the filament and anode plate is clearly shown in Fig. 30.24.

The most common of the low-pressure mercury fluorescent lamps is the 80 W lamp, for which the following are the technical particulars—

Tube watts	80	Luminance	0·5 lm/cm²
Supply volts	200–250	Overall length	5 ft
Tube current	0·8 A	Tube diameter	38 mm
Initial efficiency	35 l/W	Arc length	1445 mm

The starting of this lamp is somewhat different from that of the sodium lamp. Instead of stepping up the voltage, a separate choke is connected in series with the lamp. This supplies the ballast when the lamp is running, but it also gives a voltage impulse for starting purposes. This is arranged as follows. The connections are made according to Fig. 30.25, the filaments being taken to a starter switch. This, in Fig. 30.25, is a small glow tube whose electrodes are mounted on bimetal strips. When cold, the electrodes are open and therefore when the main switch is closed the full supply voltage is applied to the starter. This sets up a glow discharge which warms the electrodes, thereby causing the bimetal strips to bend and the electrodes to touch. There is now a complete series circuit and a pre-heating current flows through the filaments thereby causing the emission of free electrons. At the same time the voltage at the starter falls to zero and the bimetal strips cool down. The electrodes of the starter switch then open again thus interrupting the current in the circuit, and, since the circuit is highly inductive owing to the presence of the choke, a voltage of the order of 1000 V is induced. As preparation for the discharge was previously made by the emission of electrons this voltage breaks down the long gap between the lamp electrodes and the lamp lights up immediately. Owing to the very small quantity of mercury used there is no appreciable running up period.

The overall intake of the lamp and choke is about 85 W. Thus, for a 220 V supply and tube current of 0·8 A, the power factor as given by the ratio of watts to volt-amperes is

$$85/(220 \times 0{\cdot}8) = 0{\cdot}48$$

For a very large installation this is too low, and it is usual to connect a capacitor for power-factor correction purposes across the supply, as shown at C in Fig. 30.25. According to the overall power factor required

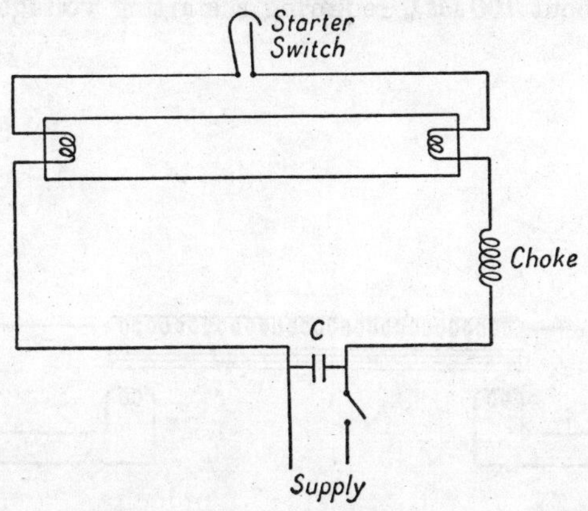

FIG. 30.25. CONNECTION SCHEME FOR TUBULAR FLUORESCENT LAMP

this varies from 4·5 μF for a power factor of 0·7 to 9·5 μF for a power factor of 0·95. The lamp current is not sinusoidal but can be regarded as a combination of sinusoidal and triangular components.* It is thus impossible to bring the power factor to unity.

The modern tendency is to do away with the starter switch in favour of what are called *quickstart* circuits. The starter switch depends for its operation on a high-voltage impulse after there has been a certain electron supply by the pre-heating of the electrodes. To do away with the necessity for the voltage impulse the electron supply must be greatly increased, a condition which is met by the continuous passage of heating current through the electrodes. To meet this condition the electrode construction of quickstart lamps is different from that of starter-start lamps. A lamp transformer is now necessary, two low voltage secondaries being provided for the heating of the two filaments. The transformer is usually of the low-reactance type, additional inductive ballast then being necessary as shown in Fig. 30.26.

The striking of the lamp without the assistance of a high-voltage impulse necessitates a more efficient electron supply, as explained above, but it is also assisted by two other factors.

1. An earthed strip in close proximity to the lamp, as illustrated in the figure. The action of this appears to be due to the presence of

* *See* Cotton *loc. cit.*

capacitance currents at the two ends of the lamp, which have the effect of extending the regions of ionizations from the immediate neighbourhood of the filaments along the whole length of the lamp.

2. The surface resistance of the lamp tube.

It is found that the starting voltage is greatly influenced by the resistance of the lamp surface, the worst conditions, corresponding to a resistance of about 100 MΩ, requiring a starting voltage of 2½ times

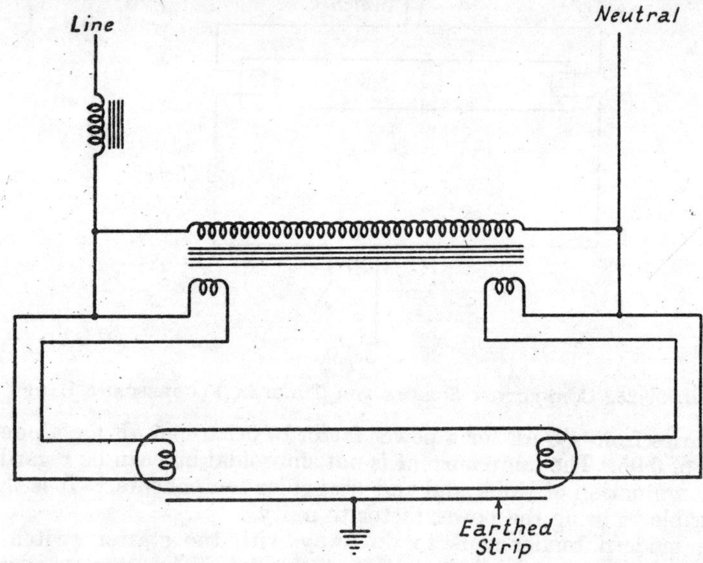

Fig. 30.26. Connections for Fluorescent Lamp with Quickstart Gear

the normal value. Thus there are two possible solutions; (*a*) a very low surface resistance, or (*b*) a very high surface resistance. The first can be secured by placing the earthed strip in direct contact with the glass, although there are some objections to this, (*b*) by treating the tube surface with a preparation which imparts an exceedingly high surface resistance.*

The High-pressure Mercury Lamp. If the vapour pressure of mercury is increased the energy per unit volume is increased and so the violence of the "average" collision is increased. This results in a greater proportion of excitations to the higher levels and so more and more spectral lines are brought in. However, the proportion of molecules to atoms increases and the phenomena are no longer purely atomic. The changes consist in the appearance of bands in certain places instead of the spectral lines characteristic of a purely atomic discharge, and at the same time the important lines becomes broadened. Thus as the

* J. W. Strange, "The dimming of fluorescent lamps." *Trans. Ill. Eng. Soc.* (*Lond.*) 15 (1950).

pressure is increased the gaps in the spectrum tend to become filled. Owing to the greatly increased energy per unit volume the luminance increases and at the same time, the supply voltage is not able to maintain a long arc, and so the arc length decreases. The following values illustrate this, all referring to a normal supply voltage of 200–250 V, a.c.

Type of Lamp	Average luminance lm/cm²	Efficiency (lumens per watt)	Arc length (mm)
80 W low-pressure lamp	0·5	35	1445
250 W lamp, about 2 atmospheres	140	36	120
250 W high-pressure air-cooled lamp, pressure about 20 atmospheres	8000	40 to 45	5
1000 W water-cooled lamp, pressure about 75 atmospheres	30,000	62·5	25

Another interesting phenomenon is the importance of the resonance radiation. At low pressure we have seen that this is of paramount

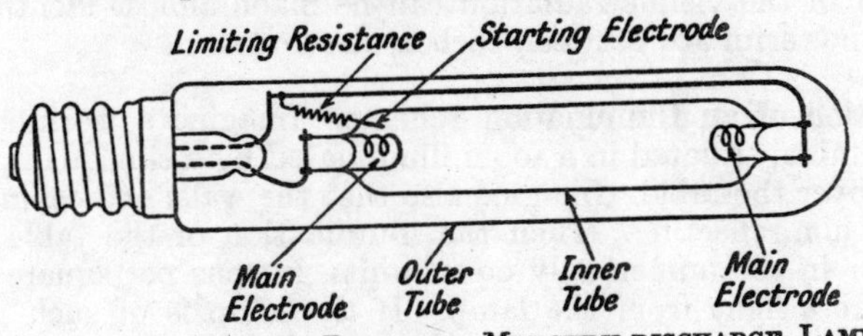

FIG. 30.27. MEDIUM PRESSURE MERCURY-DISCHARGE LAMP

importance. As the pressure is increased the proportion of radiant energy in the visible range rapidly increases. That of the resonance radiation decreases for two reasons. (*a*) The temperature becomes so high that a phosphor applied directly to the inside of the tube would be destroyed, and (*b*) the resonance radiation becomes absorbed in the arc itself, thereby becoming of relatively little importance at high pressures.

For industry the most commonly used of the high-pressure mercury lamps is one operating at about 2 or 3 atmospheres. It is a tubular lamp, and, because of the high arc temperature the tube has to be of very hard glass or quartz, the latter being essential for pressures above about five atmospheres. At this comparatively low pressure there is considerable ultra-violet radiation to which quartz is transparent, and therefore, as such radiation is extremely dangerous it is necessary to enclose the arc tube in a hard glass bulb (Fig. 30.27), which is opaque to ultra-violet. It will be seen that the lamp has a small auxiliary electrode called the starting electrode. When the lamp is switched on, the full voltage is applied between the starting and main electrodes, and as the gap is very small, an arc is struck. The current in this arc is limited to a very low value by the limiting resistance (of the order of 50,000 Ω) as the function of this arc is to set up ionization within the arc tube. The striking of this starting arc is facilitated by the introduction of a small amount of argon. The main arc between the main electrodes immediately follows but the initial light emission is very small, since most of the mercury in a cold lamp will be condensed. The heat of the discharge gradually vaporizes the mercury and the lamp is fully run up in about seven or eight minutes. The main electrodes are in the form of spirals of tungsten wire, each holding a pellet of an electron-emitting barium compound. There is no preheating as in the case of the low-pressure lamp.

Even at vapour pressures as high as 20 or 30 atmospheres the light from a mercury discharge is by no means similar to white light, since apart from a spectral energy distribution totally different from that of a full radiator, there is, even at this pressure, an appreciable deficiency in the red content. This can be overcome by amalgamating a small quantity of cadmium with the mercury, the light emission thus having cadmium as well as mercury lines. The most important of these is a red line of 6438 Å, and, by using the correct amount of cadmium the colour of the visible radiation can be made almost identical with that of a powerful arc between carbon electrodes.

Calculation of an Illumination Scheme. Imagine a working plane, such as a table, situated in a room illuminated by a single lamp placed centrally over the table. Imagine also that the walls and ceiling of the room are non-reflecting. Then the illumination of the table in foot candles, or in the numerically equal units, lumens per square foot, is due to direct light from the lamp. If the lamp is of such physical dimensions that it can be regarded as small in comparison with its distance from the table then the illumination at any point can be calculated from the two fundamental laws, the inverse square law and the cosine law.

Suppose that the average illumination of the table is E lumens per square foot and its area A square feet, then the table will receive EA lumens from the lamp. Even if the lamp is fitted with a reflector, the flux EA will be considerably less than the total luminous flux Φ emitted by the bare lamp. Thus if it is required that E shall have some specified value the necessary lumen output, Φ, of the lamp can only be calculated if we know the ratio, EA/Φ for the particular problem. In the case of a single lamp in a room with non-reflecting walls and ceiling this ratio can be easily calculated from first principles, but this particular case is very rarely met in practice.

Now let the walls and ceiling reflect some of the light they receive and, furthermore, let the source be provided with a reflector, or with a luminaire which gives a modified but non-mathematical polar distribution of intensity. The table will now receive light from all directions above the horizontal, but, except in very simple cases it will not be possible to calculate directly the illumination at the table top. The total lumens received, EA, will still be less than the total lumen output, Φ, of the bare lamp, and the ratio, EA/Φ will now depend on the dimensions of the room, the mounting height of the lamp above the table (or working plane as it is generally called) the reflecting powers of the walls and of the ceiling, and whether there is any specular reflection, and on the type of reflector or luminaire in use.

The ratio EA/Φ is called the coefficient of utilization: denoting it by C we have

$$\Phi = EA/C$$

This does not take into account any reduction in lamp output with time, as, for example, due to bulb blackening or accumulations of dust and dirt. A factor greater than unity, and called the depreciation factor, is therefore introduced. Denoting it by D, we have

$$\Phi = EAD/C$$

If A denotes the whole area of the working plane then Φ gives the total output of all the lamps, if there are more than one. If A is the area per lamp, then Φ is the lumen output per lamp.

It is preferable to have only one meaning to any symbol, so we will denote the whole area of the working plane by A. Then, if there are n lamps we have, finally

$$\Phi = EAD/nC \text{ lumens per lamp}$$

We will consider the various quantities in turn—

(*a*) The required illumination, E.

This naturally varies very considerably with the nature of the visual task. Very complete tables are available, but Fig. 30.28 will give some idea of the orders of magnitude involved. Taking column B as a mean we see that the necessary illumination varies from 300 lumens/ft² for very difficult visual tasks to as low as 3. A few recommended values for common situations are as follows—

Living rooms	7 lm/ft²
Work benches	15 ,,
General office work, typewriting, etc. . . .	20 ,,
Drawing office, on the boards	30 ,,
Reading tables	15 ,,

Tasks requiring illumination of the order of 100 are: assembling of very small parts, very fine bench and machine work, and jewellery and watch manufacturing.

(*b*) The area A is the total area of the working plane, and this is generally reckoned as the area of the floor plan. The height of the working plane is generally taken to be 2 ft 6 in.

(*c*) The depreciation factor, D.

Unless the situation is a very dirty one, it is usual to take D as equal to 1·4. In the case of installations using electric discharge lamps an allowance for depreciation is difficult to make, and it is usual for makers to give, not the initial lumens as in the case of filament lamps, but the average lumens throughout life When such an average value is used the value of D can, of course, be much lower, and a value of

1·25 is suitable. In extremely dirty situations, or where frequent cleaning is impossible, values of D higher than the above must be used,

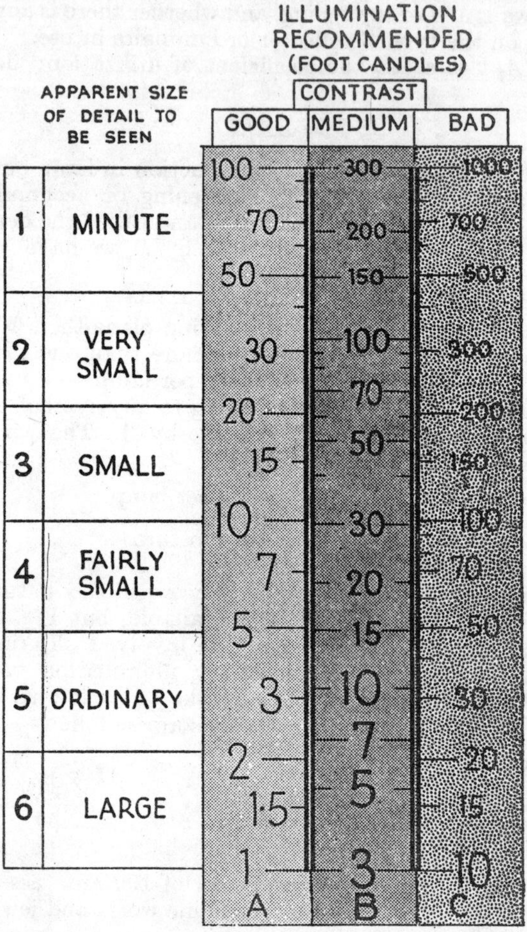

Fig. 30.28

and naturally these values can only be decided by experience with the particular location.

(d) The number of lamps, n.

The number of lamps and their spacing should be such as to give a reasonably uniform illumination over the working plane. With a 20 ft spacing, for a particular fitting and mounting height, the ratio

$$(\text{Maximum illumination})/(\text{Minimum illumination}) = 12{\cdot}5/3 = 4{\cdot}17$$

When the spacing is halved, the ratio becomes

$$14/12 = 1{\cdot}17$$

a value which, for all practical purposes, can be regarded as giving uniform illumination. Naturally this condition is only obtained by the use of a large number of units, and therefore, in practice, the economics of the problem may dictate a compromise between a high value like 4·17 and a desirable value like 1·17.*

The question of spacing is also linked with the geometrical arrangement of the units. Fig. 30.29 shows four commonly adopted arrangements. The dotted lines represent bays into which the plan view is automatically subdivided by such structural features as columns.

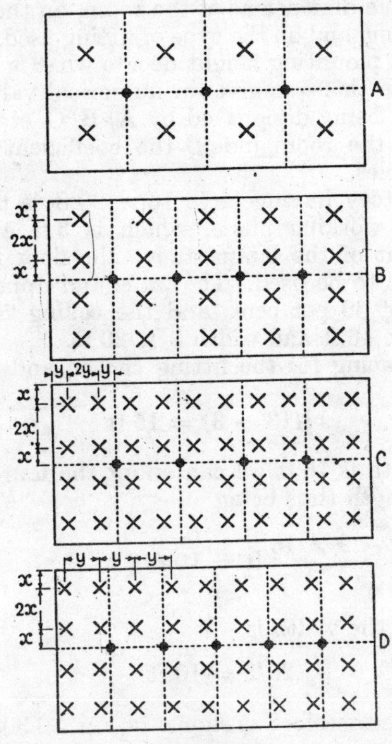

Fig. 30.29

Fig. *A* shows the simplest scheme with one unit per bay: obviously this is only possible when the bay size is less than the maximum permissible spacing, and in practice it is generally only suitable for high mounting heights. Fig. *B* shows two units per bay: this requires

* The recommended spacings for the more important types of fitting are as follows—

(1) Direct, semi-direct and general fittings, not more than 1½ times the mounting height above the working plane.

(2) Semi-indirect and indirect fittings, not more than 1½ times the ceiling height above the working plane.

(3) For direct fitting with globe enclosing the lighting unit, not more than 1¼ times the mounting height above the working plane.

rectangular bays with a width of less than two-thirds the length. Fig. *C* shows four units per bay: a very common arrangement with square bays of usual dimensions. Fig. *D* shows bay with alternate four and two units: this is suitable with bays which are not quite square. Obviously there are many other possible arrangements depending on the dimensions of the bays and the allowable spacing for the particular type of lighting unit chosen, and their mounting height.

(*e*) The coefficient of utilization, *C*.

This depends on the dimensions of the room, on the reflection factor of the walls and ceiling, and on the type of fitting used. The dimensions of the room and the mounting height decide what is called the "room index." At one time this was given a numerical value, but it is now only a classification, being designated by A, B, C, etc.

Having obtained the room index, the coefficient of utilization is determined from tables.

EXAMPLE. A factory having bays 16×20 ft is to be illuminated 7 lumens/ft^2 at the working plane, which is 3 ft above floor level. The mounting height of the lamps above the floor is 13 ft, and dispersive reflectors are to be used. The walls and upper structure have a reflection factor of 30 per cent, and the ceiling 70 per cent. The factory length is 6×16 ft and width 3×20 ft.

The maximum spacing for the fitting chosen and for its mounting height is

$$1\tfrac{1}{2}(13 - 3) = 15 \text{ ft}$$

Since the bay width is 16 ft we can adopt the four-two system, the spacing along the length thus being

$$\frac{2 \times 16}{3} \text{ ft} = 10 \text{ ft } 8 \text{ in.}$$

The spacing along the width is

$$20/2 = 10 \text{ ft}$$

The arrangement is therefore as shown in Fig. 30.30. So far we have $E = 7$; $A = 320$ ft^2 per bay, $D = 1{\cdot}4$, $n =$ average of 3 units per bay.

To determine the utilization factor *C*, extracts from tables of room index and utilization factor are given below.

ROOM INDEX

Room Width	Room Length	Room Index for Mounting Height 9 to $10\frac{1}{2}$ ft Above Working Plane
55–70	90–140	J

COEFFICIENT OF UTILIZATION

Lighting Equipment	Room Index	Ceiling	70%			50%			30%	
		Walls	50%	30%	10%	50%	30%	10%	30%	10%
Direct Standard Dispersive Reflectors	J		0·64	0·62	0·60	0·63	0·61	0·59	0·61	0·59

The value of C is thus 0·62

$$\therefore \ \Phi = EAD/nC = 7 \times (16 \times 20) \times 1{\cdot}4/(3 \times 0{\cdot}62)$$
$$= 1680 \text{ lumens per lamp}$$

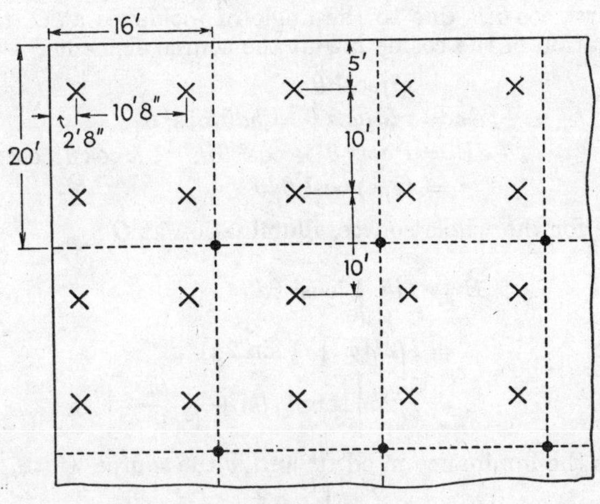

FIG. 30.30

Assuming a 230 V supply and filament lamps the average lumens throughout life for the larger sizes of lamp are—

Watts	*Single-coil*	*Coiled-coil*
75	785	883
100	1160	1270
150	1970	—
200	2725	—

Thus for the particular example the most suitable lamp would be the 150 W single-coil, the lumens "in hand" above the calculated value of 1680 being a valuable safeguard.

Illumination from Linear Sources. (*a*) Source Horizontal.

Let I = intensity per unit length normal to the surface illumination of a horizontal surface at point O (Fig. 30.31) directly underneath one end A, due to an element of length dx

$$dE = (I\,dx/r^2)\cos\theta \,.\, \cos\theta$$

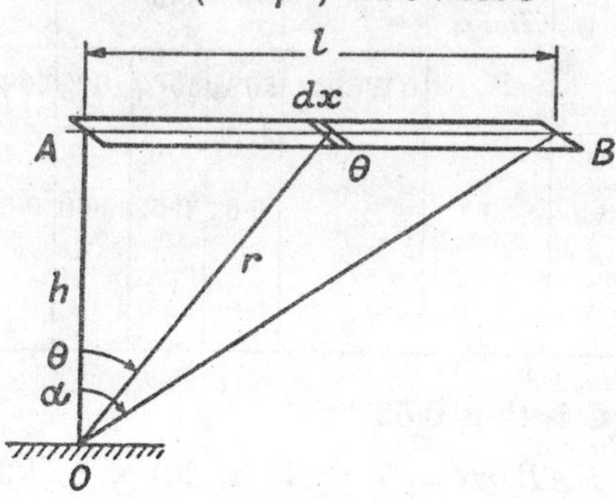

Fig. 30.31

The first $\cos\theta$ is due to the angle of incidence at O, the second to the operation of the cosine law at the source element

$$r^2 = h^2/\cos^2\theta$$

$$dx = r d\theta/\cos\theta = h d\theta/\cos^2\theta$$

$$\therefore \quad dE = I\cos^2\theta \times \cos^2\theta/h^2 \times h\cos\theta/\cos^2\theta$$

$$= (I/h)\cos^2\theta d\theta$$

Hence for the whole source, illumination at O

$$E = I/h \int_0^{\alpha} \cos^2\theta d\theta$$

$$= I/2h[\alpha + \tfrac{1}{2}\sin 2\alpha]$$

$$= I/2h\left[\tan^{-1} l/h + \frac{lh}{l^2 + h^2}\right]$$

If L is the luminance in cd/ft² and w the source width, $I = Lw$

$$\therefore \quad E = Lw/2h\left[\tan^{-1} l/h + \frac{lh}{l^2 + h^2}\right]$$

Example. A tubular source 10 ft long and 2 in. diameter is suspended 4 ft above a working plane. Calculate the illumination at a point directly underneath one end. The source has a uniform luminance of 500 cd/ft.²

Note that if the surface luminance is uniform, a tubular lamp can be regarded as a flat strip because of the operation of the cosine law.

Area of one foot of source = 1 × 2/12 = 1/6 ft².

$$\therefore \quad I = 500/6 = 83{\cdot}33 \text{ cd/ft.}^2$$

$$l = 10 \text{ ft};\ h = 4 \text{ ft}.$$

$$\therefore \quad \alpha = \tan^{-1} l/h = \tan^{-1} 10/4 = 68° \ 12'$$
$$= 68{\cdot}2° = 1{\cdot}190 \text{ radian}$$
$$\frac{lh}{l^2 + h^2} = \frac{10 \times 4}{116} = 0{\cdot}345$$
$$\therefore \quad E = \frac{83{\cdot}33}{2 \times 4}(1{\cdot}190 + 0{\cdot}345)$$
$$= 16{\cdot}09 \text{ lm/ft.}^2$$

If the point O is vertically underneath the source but located as in Fig. 30.32 (*a*) we can imagine that there are two sources of lengths l_1

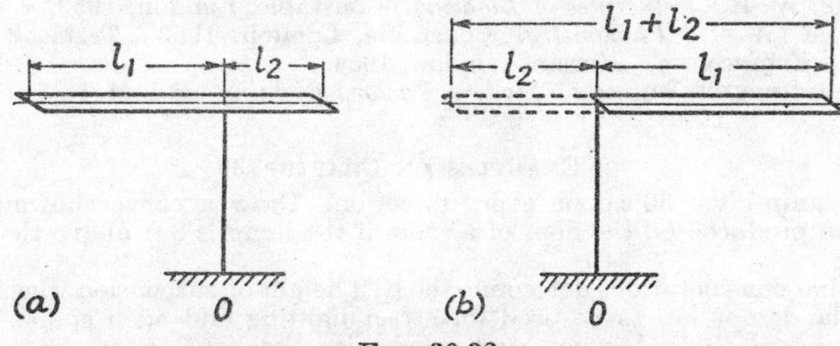

Fig. 30.32

and l_2, the total illumination thus being the sum of the two.

If the point O is outside the source as in (*b*) then we imagine that there is a source of length $(l_1 + l_2)$ and a *negative* source of length l_2, the total illumination then being the difference of the two.

(*b*) Source Vertical

If the perpendicular OA of Fig. 30.31 lies in the vertical plane (imagine the figure turned through 90°) then the angle of incidence of the ray from dx with this surface is now $(90 - \theta)$.

Hence $dE = (Idx/r^2) \times \cos(90 - \theta) \cos \theta$

$$\cos \theta = h/r; \ \sin \theta = x/r$$
$$\therefore \quad dE = (Idx/r^2) \sin \theta \cos \theta$$
$$\tan \theta = x/h$$
$$\therefore \quad x = h \tan \theta$$
$$\therefore \quad dx = h \sec^2 \theta d\theta$$
$$\text{also } r = h \sec \theta$$
$$\therefore \quad dE = Ih \sin \theta \cos \theta \sec^2 \theta d\theta / h^2 \sec^2 \theta$$
$$= (I/h) \sin \theta \cos \theta d\theta$$
$$\therefore \quad E = \int_0^\alpha (I/h) \sin \theta \cos \theta d\theta$$
$$= (I/2h) \sin^2 \alpha$$
$$= (I/2h) \cdot \frac{l^2}{l^2 + h^2} \quad \text{or} \quad (Lw/2h) \cdot \frac{l^2}{l^2 + h^2}$$

Further Reading

Light Sources

Atkinson, A. D. S. *Fluorescent Lighting.* Newnes, London, 1944.
Clarke, D. A. *Modern Electric Lamps.* Blackie, London, 1952.
Cotton, H. *Electric Discharge Lamps.* Chapman & Hall, London, 1946.
Lighting Service Bureau. *Electric Lamps.* E.L.M.A., London, 1956.

Principles of Illumination

Boast, W. B. *Illumination Engineering.* McGraw Hill, New York, 1953.
Jolley, L. B. W., *et alia. Theory and Design of Illuminating Engineering Equipment.* Chapman & Hall, London, 1930.
Moon, P. *The Scientific Basis of Illumination Engineering.* McGraw Hill, New York, 1936.
Stevens, W. R. *Principles of Lighting.* Constable, London, 1952.
Walsh, J. W. T. *Photometry.* Constable, London, 1953. *Textbook of Illuminating Engineering.* Pitman, London, 1954.
Lighting Service Bureau. *Modern Factory Lighting.* E.L.M.A., London, 1956.

Examples on Chapter 30

1. A lamp gives 30 c.d. in every direction. Draw a curve showing the illumination produced on the floor of a room if the lamp is 8 ft above the floor.

2. What considerations determine the best height of suspension when a number of similar lamps are to be used for street lighting and open spaces? A single arc lamp is such that, when suspended with the light source 22 ft above the ground, it produces an illumination I on the ground, which varies with the distance d, reckoned from the foot of the lamp as follows—

If d is 0, 10, 20, and 30 ft, the corresponding values of I are 1·8, 1·7, 1·5 and 1·2 f.c., while for values of d from 40 to 100, the values of I may be assumed given by the equation $I = 1{\cdot}4 - 0{\cdot}01d$.

Draw a curve for the above illumination, and another curve for it when the lamp is raised 8 ft. (L.U.)

3. Define (*a*) illumination, (*b*) luminous flux, and (*c*) brightness (luminance) of a luminous surface. State the units in terms of which these quantities are expressed. In what respect does the reflecting action of a diffusing surface differ from that of a mirror?

A large area is illuminated by a number of lamps each placed 12 ft above the ground on posts erected all over the area, so as to stand at the corners of squares of 50 ft side. The candle power of each lamp is 300 and may be assumed uniform. Calculate the illumination of the ground (*a*) at the base of each lamp, (*b*) at the centre of each square, and (*c*) as an average for the whole area. (L.U.)

Ans. (*a*) 2·26; (*b*) 0·35; (*c*) 0·75 f.c.

4. A metal-filament lamp has the following light distribution in a vertical plane, angles being reckoned from the vertical—

Angle	—	—	0	15	30	45	60	75	90	105	120	135	150	165
c.p.	—	—	8	10	26	35	44	52	49	46	40	30	14	0

Deduce its mean spherical candle power and mean hemispherical candle power for directions below the horizontal. If the candle power for all directions in the horizontal plane through the lamp is constant at 49, find also its reduction factor.

5. A light is placed 10 ft above the ground and its intensity is $100 \cos \phi$ in any downward direction, making an angle ϕ with the vertical. If P and Q are two points on the ground, P being vertically under the light and the distance PQ being 10 ft, calculate: (*a*) the illumination of the ground at P, and also at Q; (*b*) the total radiation sent down by the lamp. (I.E.E.)

Ans. (*a*) 1, 0·25 lumen/ft^2; (*b*) 100π lumens.

6. A hypothetical lamp has an intensity of 100 measured in any direction below the horizontal plane through the centre of the lamp, and zero above this plane. Draw the vertical polar curve. What is the output of the lamp in lumens? Calculate the intensity of illumination on a horizontal surface 10 ft below the lamp at a point 15 ft away from the vertical through the lamp. (I.E.E.)

Ans. 628 lumens; 0·171 lumen/ft^2.

7. If the filament of a 32 cd., 100 V lamp has a length l and diameter d, calculate the length and diameter of the filament of a 16 c.p., 200 V lamp, assuming that the two lamps are run at the same intrinsic brilliancy. (L.U.)

Ans. 1·25 l; 0·4 d.

8. Compare the effects of variable pressure of supply on glow lamps with carbon filaments, and on those with filaments of metal such as tungsten, as regards (*a*) fluctuations in light; (*b*) probable influence on the life; (*c*) fluctuations in the current flowing through. (L.U.)

9. Find the height at which a light having uniform spherical distribution should be placed over a floor in order that the intensity of horizontal illumination at a given distance from its vertical centre line may be the greatest. (L.U.)

Ans. $h = 0{\cdot}707\ l$, where l = distance from vertical centre line.

10. Describe in some detail the construction of either (*a*) a gas-filled incandescent-filament lamp, explaining what advantage is derived from coiling the filament, or (*b*) a gas-discharge lamp, mentioning any ancillary components which are used with this type.

The average illumination on the working plane of a shop, which is 150 ft long by 80 ft wide, is to be three foot-candles. Incandescent-filament lamps, each of 2500 lumens, in fittings having a utilization factor of 0·4 are to be used. Calculate a suitable number of lamps and suggest a spacing for them.

11. The floor plan of a workshop is 48 ft by 96 ft, and the roof is supported by a central row of columns 12 ft apart. The work is coil winding, requiring an illumination of 12 lumen/ft^2. The height from floor to roof trusses is 12 ft, the height of the benches 3 ft, and the "drop" of each lamp below the roof trusses is 1 ft. The walls can be assumed to have a reflectance of 10 per cent, and the ceiling 50 per cent. The fittings are direct, standard dispersive reflectors, the coefficiency of utilization under the stated conditions being 0·47. If the spacing is 12 ft in both directions, calculate the lumen output per lamp.

Ans. 5450 lumens.

(The standard output for filament lamps, assuming a 230 V supply are: 300 W, 4430; and 500 W, 7940. The 120 W high-pressure mercury lamps have an output of 5000 lumens. Thus mercury lamps could be used or, alternatively, the walls, ceiling, and structural work could be repainted. By so doing the coefficiency of utilization could be increased to 0·64 and the output per lamp reduced to 4000. The 300 W filament lamp would then do.)

12. A certain incandescent electric lamp has a filament 0·002 in. diameter and 40 in. long. It is required to construct another lamp of similar type to work at double the supply voltage and give half the candle power. Assuming that the new lamp operates at the same intrinsic brilliancy (luminance), determine suitable dimensions for the filament. (L.U.)

Ans. 50·4 in. long, 0·0008 in. dia.

13. A workroom 100 ft long by 40 ft wide is illuminated by means of 10 lamps suspended 10 ft above the working plane. Determine the mean spherical candle power of the lamps required to give an average illumination of 3 foot-candles at the work tables and the ratio, height of lamp/distance apart. Assume 25 per cent of the total light falls on the working plane. (I.E.E.)

Ans. 382 mscp, ratio 1/2.

14. Plot the polar curve of light distribution for a lamp and reflector from the following figures—

Downwards	0°	10°	20°	30°	40°	50°	60°	70° and over
Intensity .	800	850	610	530	460	200	75	0

The zone factors are—

Zone .	0–10°	10–20°	20–30°	30–40°	40–50°	50–60°	60–70°
Factor	0·095	0·283	0·463	0·628	0·774	0·897	0·993

Calculate the m.s.c.p. (C. & G.)

Ans. 101.

15. The relation between voltage and candle-power of a metal filament lamp is given by

$$\text{candle-power } I = \alpha V^{\beta}$$

where V is the lamp voltage. Explain how to determine the constants α and β, and calculate the percentage change in candle-power for a change in voltage of $\pm$ 6 per cent from the rated voltage if $\beta = 4{\cdot}5$. (I.E.E. Model Paper B.)

Ans. 28 per cent for an increase, 24 per cent for a decrease.

16. A show case is lighted by 10 feet of architectural tubular lamps arranged in a continuous line and placed along the top of the case. Determine the illumination produced on a horizontal surface 5 feet below the lamps in a position directly underneath the centre of the 10-ft. length of lamps, on the assumption that the tubular lamps emit 628 lumens per foot run. Neglect the effect of any reflectors which might be used. (C. & G.)

Ans. 16·4 lm/ft².

APPENDIX

TABLES SHOWING THE RELATIONS BETWEEN M.K.S., C.G.S. AND ENGLISH UNITS.

TABLE I

M.K.S. AND ENGLISH UNITS: NON-ELECTRICAL QUANTITIES

1 metre	= 3·281 ft	Length	1 ft	= 0·3048 metre
1 metre	= 39·37 in.	,,	1 in.	= 0·0254 metre
1 kilometre	= 0·6214 mile	,,	1 mile	= 1·609 kilometres
1 m^2	= 1,550 in.^{2}	Area	1 in.^{2}	= 6·452 × 10^{-4} m^2
1 m^2	= 1·973 × 10^9 circ. mils	,,	1 circ. mil	= 5·067 × 10^{-10} m^2
1 m^2	= 10·76 ft^2	,,	1 ft^2	= 0·0929 m^2
1 m^3	= 61,020 in.^{3}	Volume	1 in.^{3}	= 1·639 × 10^{-5} m^3
1 m^3	= 35·31 ft^3	,,	1 ft^3	= 0·02832 m^3
1 metre/sec	= 3·281 ft/sec	Velocity	1 ft/sec	= 0·3048 metre/sec
1 metre/sec	= 196·9 ft/min	,,	1 ft/min	= 5·08 × 10^{-3} metre/sec
1 metre/sec	= 2·237 m.p.h.	,,	1 m.p.h.	= 0·447 metre/sec
1 metre/sec^2	= 3·281 ft/sec^2	Acceleration	1 ft/sec^2	= 0·3048 metre/sec^2
1 metre/sec^2	= 2·237 m.p.h./sec	,,	1 m.p.h./sec	= 0·447 metre/sec^2
1 kg	= 0·0685 slug	Mass	1 slug	= 14·59 kg
1 kg	= 35·27 oz	,,	1 oz	= 0·02835 kg
1 kg	= 2·205 lb	,,	1 lb	= 0·4536 kg
1 kg	= 9·842 × 10^{-4} ton	,,	1 ton	= 1,016 kg
1 kg/m^3	= 0·06243 lb/ft^3	Density	1 lb/ft^3	= 16·02 kg/m^3
1 kg/cm^3	= 36·13 lb/in.^{3}	,,	1 lb/in.^{3}	= 0·02768 kg/cm^3
1 newton	= 7·233 poundals	Force	1 poundal	= 0·1383 newton
1 newton	= 0·2248 lb-wt		1 lb-wt	= 4·448 newtons
1 kNw	= 0·1004 ton-wt	,,	1 ton-wt	= 9·9635 kNw
1 Nw/m^2	= 1·450 × 10^{-4} lb/in.^{2}	Pressure	1 lb/in.^{2}	= 6,895 Nw/m^2
1 Nw/m^2	= 0·02089 lb/ft^2	,,	1 lb/ft^2	= 47·88 Nw/m^2
1 Nw/m^2	= 0·672 poundals/ft^2	,,	1 poundal/ft^2	= 1·488 Nw/m^2

SI/metric unit		Quantity	British unit	
1 kg/m²	= 0·2048 lb/ft²	Pressure	1 lb/ft²	= 4·882 kg/m²
1 kNw/m²	= 0·2953 in. Hg	,,	1 in. Hg	= 3·387 kNw/m²
1 kNw/m²	= 0·3344 ft water	,,	1 ft water	= 2·989 kNw/m²
1 mega-Nw/m²	= 0·06473 ton/in.²	,,	1 ton/in.²	= 15·45 mega-Nw/m²
1 joule	= 0·7376 ft-lb	Work or energy	1 ft-lb	= 1·3565 joules
1 joule	= 9·48 × 10^{-4} B.Th.U.	,,	1 B.Th.U.	= 1,055 joules
1 metre-kg	= 7·233 ft-lb	,,	1 ft-lb	= 0·1383 metre-kg
1 joule	= 8·851 in.-lb	,,	1 in.-lb	= 0·113 joule
1 watt	= 0·7376 ft-lb/sec	Power	1 ft-lb/sec	= 1·3565 watts
1 watt	= 44·26 ft-lb/min	,,	1 ft-lb/min	= 0·0226 watt
1 watt	= 0·05688 B.Th.U./min	,,	1 B.Th.U./min	= 17·58 watts
1 watt	= 1·341 × 10^{-3} h.p.	,,	1 h.p.	= 745·7 watts
1 kW	= 1·341 h.p.	,,	1 h.p.	= 0·7457 kW
1 kg-m/sec	= 7·233 lb-ft/sec	Momentum	1 lb-ft/sec	= 0·1383 kg-m/sec
1 kg-m/sec	= 0·2248 slug-ft/sec	,,	1 slug-ft/sec	= 4·448 kg-m/sec
1 newton-metre	= 0·7376 lb-ft	Torque	1 lb-ft	= 1·3565 newton-metres
1 kg-metre	= 7·233 lb-ft	,,	1 lb-ft	= 0·1383 kg-metre
1 newton-metre	= 8·851 lb-in.	,,	1 lb-in.	= 0·113 newton-metre
1 kg-m²	= 23·76 lb-ft²	Moment of inertia	1 lb-ft²	= 0·04208 kg-m²
1 kg-m²	= 0·7376 slug-ft²	,,	1 slug-ft²	= 1·3565 kg-m²
As "pressure" above		Young's modulus	As "pressure" above	
1 Nw/m	= 0·06852 lb/ft	Stiffness	1 lb/ft	= 14·59 Nw/m
1 Nw-sec/m²	= 0·02089 lb-sec/ft²	Viscosity	1 lb-sec/ft²	= 47·88 Nw-sec/m²
1 joule	= 9·48 × 10^{-4} B.Th.U.	Quantity of heat	1 B.Th.U.	= 1,055 joules
1 watt/m²/C°/m	= 6·937 B.Th.U./hr/ft²/F°/in.	Thermal conductivity	1 B.Th.U./hr/ft²/F°/in.	= 0·1442 watt/m²/C°/m

TABLE II

M.K.S. and C.G.S. Units: Non-electrical Quantities

M.K.S.	C.G.S. equivalent	Quantity	C.G.S.	M.K.S. equivalent
1 newton	= 10^5 dynes	Force	1 dyne	= 10^{-5} newton
1 newton	= 0·102 kg	,,	1 kg	= 9·807 newtons
1 newton/m²	= 10 dynes/cm²	Pressure	1 dyne/cm²	= 0·1 newton/m²
1 newton/m²	= 0·01 millibar	,,	1 millibar	= 100 newtons/m²
1 newton/m²	= 7·5 × 10^{-3} mm Hg	,,	1 mm Hg	= 133·3 newtons/m²
1 newton/m²	= 0·0102 g/cm²	,,	1 g/cm²	= 98·07 newtons/m²
1 joule	= 10^7 ergs	Work or energy	1 erg	= 10^{-7} joule
1 joule	= 0·2389 g-calorie	,,	1 g-calorie	= 4·186 joules
1 joule	= 0·102 metre-kg	,,	1 metre-kg	= 9·807 joules
1 joule	= 2·389 × 10^{-4} kilocalorie	,,	1 kilocalorie	= 4,186 joules
1 watt	= 10^7 ergs/sec	Power	1 erg/sec	= 10^{-7} watt
1 watt	= 10,200 cm-g/sec	,,	1 cm-g/sec	= 9·807 × 10^{-5} watt
1 watt	= 14·33 g-calories/min	,,	1 g-calorie/min	= 0·06977 watt
1 watt	= 0·2389 g-calories/sec	,,	1 g-calorie/sec	= 4·186 watts
1 watt	= 0·102 m-kg/sec	,,	1 m-kg/sec	= 9·807 watts
1 kW	= 1·36 metric h.p.	,,	1 metric h.p.	= 0·736 kW
1 kg-m/sec	= 10^5 g-cm/sec	Momentum	1 g-cm/sec	= 10^{-5} kg-m/sec
1 μNw-m	= 10 dyne-cm	Torque	1 dyne-cm	= 0·1 μNw-m
1 Nw-m	= 10^7 dyne-cm	,,	1 dyne-cm	= 10^{-7} Nw-m
1 Nw-m	= 10,200 g-cm	,,	1 g-cm	= 9·807 × 10^{-5} Nw-m
1 mNw-metre	= 10·2 g-cm	,,	1 g-cm	= 0·09807 mNw-m
1 kg-m²	= 10^7 g-cm²	Moment of inertia	1 g-cm²	= 10^{-7} kg-m²
1 Nw/m²	= 10 dynes/cm²	Young's modulus	1 dyne/cm²	= 0·1 Nw/m²
1 Nw/m	= 1,000 dynes/cm	Stiffness	1 dyne/cm	= 10^{-3} Nw/m
1 Nw-sec/m²	= 10 dyne-sec/cm²	Viscosity	1 dyne-sec/cm²	= 0·1 Nw-sec/m²
1 watt	= 10^7 ergs/sec	Sound energy flux	1 erg/sec	= 10^{-7} watt
1 joule/m³	= 10 ergs/cm³	Sound energy density	1 erg/cm³	= 0·1 joule/m³
1 watt/m²	= 1,000 ergs/sec/cm²	Sound intensity	1 erg/sec/cm²	= 0·001 watt/m²
1 M.K.S. Unit	= 0·001 C.G.S. Unit	Audibility	1 C.G.S. Unit	= 1,000 M.K.S. Units
1 M.K.S. acoustic ohm	= 10^{-5} C.G.S. acoustic ohm	Acoustic impedance	1 C.G.S. acoustic ohm	= 10^5 M.K.S. acoustic ohms
1 watt/m²/C°/m	= 2·389 × 10^{-3} g-cal/sec/cm²/C°/cm	Thermal conductivity	1 g-calorie/sec/cm²/C°/cm	= 418·6 watts/m²/C°/m

TABLE III

M.K.S. and C.G.S. Units: Electrical Quantities

Quantity	Symbol	M.K.S. Unit and Abbreviation		Number of C.G.S. Units in One M.K.S. Unit	
				Electromagnetic	Electrostatic
E.m.f. or p.d.	E or V	Volt	V	10^8	$(300)^{-1}$
Current	I	Ampere	A	10^{-1}	3×10^9
Resistance	R	Ohm	Ω	10^9	$(9 \times 10^{11})^{-1}$
Energy	W	Joule	J	10^7 (ergs)	10^7 (ergs)
Power	P	Watt	W	10^7	10^7
Resistivity	ρ	Ohm-metre	Ω-m	10^{11}	$(9 \times 10^9)^{-1}$
Current density	J	Ampere per square metre	A/m²	10^{-5}	3×10^5
Quantity of electricity	Q	Coulomb	C	10^{-1}	3×10^9
Conductance	G	Mho		10^{-9}	9×10^{11}
Conductivity	σ	Mho per metre		10^{-11}	9×10^9
Magnetic pole strength	m	Weber	Wb	$10^8 \div 4\pi$	$(1200\pi)^{-1}$
Magnetic space constant	μ_0	Henry per metre	H/m	$10^7 \div 4\pi$	$(36\pi \times 10^{13})^{-1}$
Magnetic moment	M	Weber-metre	Wb-m	$10^{10} \div 4\pi$	$(12\pi)^{-1}$
Magnetic flux	Φ	Weber	Wb	10^8 (maxwells)	$(300)^{-1}$
Flux density	B	Weber per square metre	Wb/m²	10^4 (gauss)	$(3 \times 10^6)^{-1}$
M.m.f.	F	Ampere-turn	AT	$4\pi \times 10^{-1}$ (gilberts)	$(12\pi \times 10^9)$
M.m.f. Gradient	H	Ampere-turn per metre	AT/m	$4\pi \times 10^{-3}$ (oersteds)	$12\pi \times 10^7$
Reluctance	S	Ampere-turn per weber	AT/Wb	$4\pi \times 10^{-9}$	$36\pi \times 10^{11}$
Permeance	Λ	Weber per ampere-turn	Wb/AT	$10^9 \div 4\pi$	$(36\pi \times 10^{11})^{-1}$
Intensity of magnetization	J	Weber per square metre	Wb/m²	$10^4 \div 4\pi$	$(12\pi \times 10^6)^{-1}$
Self inductance	L	Henry	H	10^9	$(9 \times 10^{11})^{-1}$
Mutual inductance	M	Henry	H	10^9	$(9 \times 10^{11})^{-1}$
Capacitance	C	Farad	F	10^{-9}	9×10^{11}
Electric space constant	ε_0	Farad per metre	F/m	$4\pi \times 10^{-11}$	$36\pi \times 10^9$
Surface density of charge	σ	Coulomb per square metre	C/m²	10^{-5}	3×10^5
Volume density of charge	ρ	Coulomb per cubic metre	C/m³	10^{-7}	3×10^3
Electric flux	Ψ	Coulomb	C	$4\pi \times 10^{-1}$	$12\pi \times 10^9$
Displacement	D	Coulomb per square metre	C/m²	$4\pi \times 10^{-5}$	$12\pi \times 10^5$
Electric intensity	$\mathscr{E}$	Volts per metre	V/m	10^6	$(3 \times 10^4)^{-1}$
Polarization	P	Coulomb per squ[illegible] metre	C/m²	10^{-5}	3×10^5

INDEX

ABSOLUTE systems, of units, 1
A.C.—
  bridge networks, 347
  commutator motors—
    polyphase, 665
    single-phase, 644
Acceleration, of d.c. motors, 234
Accumulators, 299
  alkaline, 304
  efficiency of, 303
  lead-acid, 300
  testing of, 304
Admittance, 329
Air, electric breakdown of, 61
Air-gap, calculation of, 153, 467, 481
All-day efficiency, 408
Alternating currents, 307
Alternating quantity, 307
  average value, 309
  effective value, 308
  graphical representation, 310
Alternating quantities, addition of, 311
Alternator, 451
  asynchronous, 565, 611
    circle diagram of, 612
    Vector diagram of, 613
  stator windings, 456, 459, 464
Alternators, parallel operation of, 531, 542
Ampere, 4, 85
Amplidyne, 214
Analysis, harmonic, 780
Ångstrom unit, 800
Annealing, magnetic, 113
Armature—
  ampere-turns, 186
  construction of, 160
  m.m.f. diagram, 194
  reaction, d.c., 185, 224
    in alternators, 472, 476, 481, 497
  windings, d.c., 163
Artificial neutral, 367
Atom, 6, 122
Atomic—
  number, 6
  weight, 6
Automatic voltage regulators, 528
Avalanche, electron, 709, 832
Average eye, 801

BACK e.m.f., 221
Ballistic method, 125
Battery booster, 283
Battery, charging of, 256
Balance, of e.m.f.s in armature windings, 175
Balancer, rotary, 279
Ballast, for discharge lamps, 832
Behn-Eschenberg, 491
Berry-type transformer, 393
Bitter patterns, 125
Blondel diagram, 513
Booster, 281
Boucherot motor, 592
Breadth factor, 462
Bridge, a.c., 347
Brush holders, 162
Brushes—
  high resistance, 191
  position of, 186

C.G.S. system, 1
Cage rotor—
  of induction motor, 553, 591
    resistance of, 571
Candela, 804
Candle power, 804
  mean spherical, 804
Capacitance, 18
  of coaxial cylinders, 21, 51
  of concentric spheres, 20
  of isolated sphere, 19
  of parallel conductors, 22
  of parallel plates, 20
  of three-phase line, 30
Capacitance, hydraulic analogy, 320
Capacitive reactance, 318
Capacitor—
  charge and discharge of, 37
  oscillatory charge and discharge of, 45
Capacitor motor, 618
Capacitor, with dielectric loss, 349
Capacitors, systems of, 24
Cascade control, 601, 688
Cathode-ray oscillograph, 777
  focusing of, 778
Characteristics, of d.c. generators, 197
Chording factor, 463
Circle diagram—
  of induction motor, 577, 586
  of single-phase series motor, 653
Circulating currents—
  in paralleled alternators, 535
  paralleled transformers, 415
Closed-loop system, 245
Coefficient of utilization, 841
Coefficients, potential, 27
Coercive force, 109, 112

Collision processes, 709
Colour, 800
Commutating poles, 190, 225, 652, 694
Commutation, 187, 224
Commutation—
  selective, 177
  in single-phase motor, 648
  in three-phase motor, 682, 700
Commutator—
  action of, 168, 665
  construction of, 162
Commutator, as frequency changer, 666
Commutator-motor—
  polyphase, 665
  single-phase, 644
Compensated a.c. series motor, 646
Compensated induction motor, 676, 679
Compensated repulsion motor, 662
Compensating winding, 192, 195
Complementary function, 370
Complex waveforms, 757
Compound-wound—
  generator, 204
  motor, 229
Conductance, 329
  specific, 66
Constant-current system, 247, 251
Contactor starters, 235
Contraction coefficient, 155
Control—
  of d.c. motors, 236
  of induction motors, 595
Convertor, rotary, 624
Cooling—
  of alternators, 468
  of transformers, 425
Copper-loss tests, of transformers, 423
Copper-oxide rectifier, 754
Core-type transformer, 393
Corona, 62
Coulomb's law, 7
Coulomb's theorem, 15
Crawling, of induction motors, 574
Cross-field dynamos, 211
Cross-magnetizing armature reaction, 187, 483
Cumulative compounding, 230
Curie temperature, 108
Current resonance, 332
Cycle, 317
Cyclic irregularity, 544
Cyclic state, of accumulator, 304

DAMPED oscillations, 46, 546
D.C.—
  machines, construction of, 151
  motor—
    control of, 220, 236
    speed of, 222
    torque of, 223
Decrement, logarithmic, 47
Delta connection, 361, 367, 376
Delta-star transformation, 73
Demagnetization curves, 115
Demagnetizing force, 118
Depreciation factor, 841
Design limitations, of commutator motors, 699, 703
Dielectrics, 52
  breakdown, 59
  flux density, 4
  phenomena, 55
  properties of, 53
Differential compounding, 229
Direct component, of armature reaction, 187, 483
Discharge lamps, 822
Distribution, 275
  factor, 462, 768
Distributor, volt drop in, 289
Division, of vectors, 343
Drop in volts, 289
  in three-phase line, 377
Domains, magnetic, 122, 124
Double-cage motor, 592
  performance of, 593
Double-range cascade control, 695
Doubly-fed motor, 663
Duddell oscillograph, 775

ECONOMICAL limits of p.f. correction, 791
Eddy-current loss, 206, 422
Effective value, 309, 757
Efficiency—
  of accumulator, 303
  of d.c. machine, 206
  of transformer, 408
  of transmission, 276
Electric current, 65
Electric field, 8
  of charged parallel plates, 13
  of charged sphere, 13
  of charged straight wire, 10
  of hollow spherical conductor, 14
  of uniformly charged ring, 11
Electric shielding, 29
Electric space constant, 1, 5
Electrical resistance, 65
Electrical resonance, 323
Electrification, 7
Electrodes, phenomena at, 297
Electrolysis, 294
  laws of, 298
Electron, 6, 65, 295
  avalanche, 709, 832
  current, 708
  lens, 779
  spin, 124
  volt, 823
Elliptical magnetic field, 616

E.m.f., dynamically induced, 135
equation—
of d.c. machine, 178
of an alternator, 461
of polyphase commutator motor, 667
of single-phase commutator motor, 644
of transformer, 405
fluctuations, 181
production of, 135
statically induced, 136
Emission—
field, 709
thermionic, 708
Energy, 2
division of stored, 35
in electric field, 31
in magnetic field, 103
loss of, 34
of charged system, 32
radiant, 799
Entz booster, 284
Epstein iron-testing apparatus, 423
Equalizer rings, 174
Equalizing bar, 252, 255
Equivalent circuit—
of induction motor, 568
of transformer, 405
Equivalent reactance—
of induction motor, 570
of transformer, 402, 419
Equivalent resistance—
of induction motor, 570, 572
of transformer, 402, 419
Excitation, of atoms, 823, 826, 828, 829
Expedor, 675

FEEDER, 275, 285
Ferrari's principle, 443, 505
Ferromagnetism, 104
Field control, of d.c. motor, 240
Field emission, 709
Field magnet, of d.c. machine, 151
Field's test, 272
Fleming's right-hand rule, 236
Fluctuations, in e.m.f. of d.c. machine, 181
Fluorescence, 835
Flux—
electric, 12
fringes, 153
magnetic, 4, 86
Flux density—
dielectric, 4
magnetic, 86
Flywheel, calculation of, 244, 545
Foot-lambert, 804
Force, 2
between parallel conductors, 85, 94
on a dielectric, 32
Force (*contd.*)—
magnetomotive, 94
Form factor, 310, 462
Fourier analysis, 478, 576
Fractional-pitch winding, 458, 465, 769
Frequency, 307, 452
convertor, 666, 691
of rotor oscillations, 526, 543, 546
tripling, 762
Fringing coefficient, 154
Frölich's equation, 105

GAS-FILLED lamps, 819
Gauss's theorem, 12
Generator—
and battery, 256
d.c., 197
in series and in parallel, 251, *et seq.*
Grid control, 737, 742, 751

HALF-WAVE rectifier, 717, 755
Harmonic analysis, 780
Harmonics, in alternator e.m.f. wave, 763
determination of, 765
elimination of, 767
Heating-time curve, 426
High-reactance transformer, 834
High-starting-torque motors, 591
Hobart's rule, 188
Hopkinson test, 268
Hot-cathode rectifier, 749
Hunting, 525, 543, 546
Hydrogen cooling of turbo-alternators, 468
Hydrogen discharge, 825
Hysteresis, 109
coefficient, 111
loop, 109
loss, 110, 206, 396, 422

ILLIOVICI permeameter, 128
Illumination, 799, 802
from linear source, 846
laws of, 805
photometer, 809
scheme, calculation of, 840
Impedance, 316, 321
synchronous, 490
tests on transformers, 423
Incandescent lamps, 815
construction of, 818
influence of voltage variations, 820
life of, 820
starting, 821
Inductance, mutual, 137, 145
of parallel conductors, 146
self, 137, 142

Induction motor, 550
 circle diagram of, 577, 586
 crawling, 574
 magnetic leakage in, 585
 magnetic locking, 577
 methods of starting, 559
 operating characteristics, 562, 565, 585
 rotors, 553, 591
 run as synchronous motor, 604
 single-phase, 614
 speed control of, 598
 starting characteristics, 595
 testing of, 587
 torque of, 554
 torque/slip characteristics, 558, 567, 592
 windings, 550
Induction regulator, 679
Inductive circuit, 138, 313
Inductive reactance, 314
Insulation resistance, 52
Integrating photometer, 814
Intensity—
 luminous, 803
 magnetic—
  due to circular current, 89
  due to short solenoid, 93
  due to short wire, 90
  due to solenoid, 91
  due to straight wire, 86
Interphase transformer, 731
Interpoles—
 on a.c. commutator motors, 652, 694
 on d.c. machines, 190, 225
Ion, 7
Ionization, 7, 831
Iron-testing apparatus, 423

KAPP regulation diagram, 404
Kelvin's double bridge, 80
Kelvin's economy law, 285
Kennelly's equation, 106
Kirchhoff's laws, 71
Kramer control, 688, 692

LAMBERT's cosine law, 804
Lamination—
 of armature core, 160
 of pole core, 151
 of transformer core, 393
Lamps, discharge, 822
 sodium vapour, 828, 832
 mercury vapour, 829, 834
Lamps, incandescent, 815
Lap winding, 164
Laplace's equation, 50, 69
Leakage—
 factor, 99
 magnetic, 98
Leakage (*contd.*)—
 reactance—
  of alternator, 484, 497
  of d.c. machine, 188
  of transformer, 399
Lifting power of electromagnet, 103
Light, 802
Linear source, 846
Load equalization, 243
Locking, magnetic, 577
Locus diagram, for parallel circuits, 333
Logarithmic decrement, 47
Losses—
 of power, 55, 110, 206
 separation of, 262
Lumen, 803
Luminance, 803
Luminous—
 efficiency, 800, 801, 803
 emittance, 803
 energy, 802
 flux, 802
 intensity, 803
Lummer-Brodhun photometer, 807

MAGNETIC—
 annealing, 113
 circuits, 97
  in parallel, 100
 field, rotating, 443
 flux, 4
 intensity—
  due to circular current, 89
  due to short solenoid, 93
  due to short wire, 90
  due to solenoid, 91
 leakage, 98
  in transformers, 400
  in induction motors, 585
 moment, 130
 pole, 95
 potentiometer, 128
 reluctance, 98
 reversals, 109, 113
 space constant, 1, 3, 5
Magnetization curves, 104
Magnetomotive force, 94
Magnicon, 216
Mean spherical candle power, 804, 810
Mechanical units, 1
Mercury—
 discharge, 829
 vapour lamp, 834
  high-pressure, 838
  quickstart, 837
  tubular fluorescent, 836
Mesh connection, 361, 367, 376
Metadyne, 216
M.K.S. system of units, 1

M.M.F.—
of a three-phase winding, 472
method, of alternator regulation determination, 493
wave—
Fourier analysis, 478, 576
of fractional pitch winding, 481
of single-phase winding, 479
Moment, magnetic, 130
Motor starters, 230, 559
Multiplication, of vectors, 342
Multi-turn windings, 169, 189
Mush winding, 550
Mutual induction, 137, 145, 339
Mutually induced e.m.f.s, 136

NEGATIVE booster, 282
Neutral—
artificial, 367
slipping of, 731
wire, 278, 363
Neutron, 6
Newton, 2
No-load current of transformer, 395
Non-linear resistor, 81

OCKENDEN p.f. meter, 450
Oersted, 4
One-wattmeter method, 368
Oscillations—
electric, 45
free and forced, 46, 525, 543, 546
of synchronous machines, 525, 543, 546
Oscillograph—
cathode-ray, 777
Duddell, 775
Overhang leakage, 485

PARALLEL—
circuit, 328
graphical solution of, 330
locus diagram, 333
resonance in, 331
conductors, inductance of, 146
connection—
of alternators, 531
of compound generators, 255
of series generators, 251
of shunt generators, 255
of transformers, 410
Particular integral, 354
Pass band, of resonant circuit, 326
Peaking transformer, 740
Percentage—
compounding, 206
impedance, 407
Periodic time, 307
Permanent magnets, 118
Permeability, of free space, 1, 3, 5
Permeameter, Illiovici, 128
Permittivity, 1, 8
Phase—
advancer, 675, 678
advancing, 787
difference, 311
equalizing circuits, 731
sequence indicator, 387
swinging, 525, 543, 546
Phosphor, 835
Photometer, 807, 809, 814
Pinch effect, 712
Pitch, winding, 163
Planck's constant, 825
Plasma, 711
Plugging, 565
Poisson's equation, 50
Polarization, 299, 306
Pole—
changing, 599
construction, of d.c. machines, 151
Polyphase currents, 359
Polyphase vector diagrams, 372
Potential, 15
coefficient, 27
due to charged sphere, 17
due to point charge, 16
gradient, 9, 16, 18, 25
Potentiometer, magnetic, 128
Potier method, 497
Power, 2, 68
Power in a.c. circuits, 313, 314, 316, 350
Power—
measurement, in polyphase circuits, 366
loss, 206
in controllers, 236
in transformers, 396
Power-factor, 316, 760
correction, economical limits of, 791
disadvantages of low, 787
meter, 449
methods of improving, 788
of single-phase series motor, 652
of three-phase series motor, 671
Practical units, 1
Primary cells, 306
Proton, 6
Pull-out torque, 510
Pulling into step, 508

QUICKSTART circuit, 837

RADIANT—
efficiency, 801
energy, 799
flux, 802
Rationalization of units, 4
Reactance voltage, 188
Reactance, transient, 499
Rectifiers, metal, 754

Rectifier, mercury arc, 707
 anode currents, 721, 734
 arc volt-drop, 711
 back firing, 716
 construction, 712, 714, 747
 current relationships, 720, 734
 effect of transformer reactance, 725
 effect of vapour pressure, 710
 excitation, 712
 full-wave, 717, 755
 grid control, 737, 742
 half-wave, 717, 754
 hot-cathode, 749
 ignition, 712
 interphase transformer, 731
 inverted operation, 744
 mercury-arc, 707
 overlap, 725, 727
 peaking transformer, 740
 phase equalizing circuits, 731
 short-circuit current, 727
 single anode, 746
 slipping of neutral, 730
 voltage regulation, 729
 voltage relationships, 720, 729
Reduction factor, 804
Regenerative tests, 267
Relative luminous efficiency, 800
Relative permittivity, 8
Reluctance, magnetic, 98
Reluctivity, 106
Repulsion motor, 657
 compensated, 662
 torque of, 659
Resistance—
 electrical, 65
 of cage rotor, 571
 leads for a.c. motors, 650, 701
 specific, 66
 variation with temperature, 67
Resistors, non linear, 81
Resonance—
 electrical, 323
 radiation, 827, 828, 830, 835, 838
 selective, 759
Resonant circuit, 325
 pass band, 326
Retardation tests, 264
Retentivity, 109, 112
Reversals, magnetic, 109, 113
Rheostatic control, 598
Right-hand rule, 136
RLC circuit, 48, 320
 differential equations of, 351
Room index, 844
Rosenberg dynamo, 211
Rotary convertor, 624
 armature—
  heating, 630
  reaction, 631
  windings, 625
Rotary convertor (*contd.*)—
 in cascade control, 688
 inverted operation, 641
 parallel operation, 641
 six-phase, 632
 split-pole, 640
 starting of, 635
 voltage and current ratios, 625
 voltage regulation, 639
 waveform of armature current, 627
Rotating coil—
 in alternating field, 619
 in uniform field, 307
Rotating magnetic field, 443
 effect of harmonics, 775
 of a three-phase machine, 446, 472
Rotor winding, 553, 591
Rotor-fed motor, 679, 683
Rousseau diagram, 811
Russell zones, 812

SAWTOOTH wave, 40
Scherbius machine, 693, 696
Schrage motor, 683
 theory of, 685
Scott method of transformation, 433
Selective commutation, 178
Selective resonance, 759
Selenium rectifier, 754
Self-induced e.m.f., 136
Self induction, 137, 142
Self-synchronizing, of rotary convertor, 638
Separately-excited generator, 198
Separation of losses, 262, 422
Series generators, 197
 in parallel, 251
 in series, 251
Series motor, d.c., 225
 polyphase, 668
 single-phase, 646
 tests, 271
  torque of, 647
Series-parallel—
 control, 241
 winding, 167
Shell—
 atomic, 122, 295
 type transformer, 393
Shielding, electric, 29
Short-chorded winding, 193, 459
Short-circuit current—
 of alternator, 490, 492, 499
 of induction motor, 582
 of transformer, 403
Shunt commutator motor, 674
Shunt motor, 227
Shunt-wound generator, 199
 self-excitation of, 204
Single-phase induction motor, 614
 magnetic field of, 612

Single-phase induction motor (*contd.*)—
  methods of starting, 617
Skew coils, 457
Slip, 551, 554
Slot leakage—
  in alternators, 484
  in d.c. machines, 188
  in induction motors, 587
Sodium discharge, 828
  vapour lamp, 832
Space charge, 708
Specific conductance, 66
Specific resistance, 66
Spectral energy distribution, 802, 817
Speed—
  of d.c. motor, 222, 226
  control—
    of d.c. motors, 236
    of induction motors, 598, 688
Spherical photometer, 814
Spin, of electron, 124
Split-phase motor, 617
Standard, of light, 807
Star connection, 363
Starters—
  for d.c. motors, 230
    contactor type, 235
    grading of resistances, 231
  for induction motors, 559
Starting—
  of d.c. motors, 230
  of induction motors, 553, 559
  of rotary convertors, 635
  of single-phase induction motors, 617
  of synchronous motors, 507
Stator windings—
  polyphase, 457, 550
  single-phase, 424
Steinmetz coefficient, 111
Stilb, 804
Stray losses, 207
Sumpner test, 424
Superposition theorem, 77
Susceptance, 329
Susceptor, 678
Symbolic method, 340, 344
Symmetrical components, 379
Synchronizing power, 539
Synchronous impedance, 490
Synchronous-induction motor, 604, 610
Synchronous motor, 502
  hunting, 525, 543, 546
  properties of, 507
  pull-out torque, 510
  stability, 516
  stiffness coefficient, 521
  torque, 503, 507
  torque-angle characteristic, 521
  V characteristics, 512, 518
  vector diagram, 510
System—
  of capacitors, 24
  of conductors, 70
Sweep circuit, 40, 779
Swinburne test, 260

TEETH, calculation of ampere-turns, 157
Temperature coefficient, 67
Testing—
  of accumulators, 304
  of d.c. machines, 259
  of induction motors, 587
  of transformers, 422, 438
Time constant, 140
Tirrill regulator, 528
Theory of magnetism, 122
Thermionic emission, 708
Thévenin's theorem, 78
Three-phase balanced networks, 364
Three-phase commutator motors—
  design limitations, 699
  series motor, 668
    characteristics, 673
    speed of, 671
    stability of, 673
  shunt commutator motor, 674
  special windings, 701
  unbalanced networks, 373
  working, 360
Three-wattmeter method, 366
Three-wire system, 277
Thyratron, 751
Tooth ripples, 770
Tooth-head leakage, 485
Toroid, 96
Torque—
  of d.c. motor, 223
  of repulsion motor, 659
  of single-phase series motor, 647
  of synchronous motor, 503, 507
Torque-angle characteristic, 521
Torque-slip characteristics of induction motor, 558, 562, 592
Transformation—
  three-phase, 436
  three- to two-phase, 433
Transformer, 393
  as mutually inductive circuit, 418
  auto, 429
  Berry-type, 393
  core-type, 393
  efficiency, 408
  e.m.f. of commutator motor, 645, 650
  equivalent circuit, 405
  heating-time curve, 426
  high reactance, 834
  interphase, 731
  magnetizing current, 395, 761
  methods of cooling, 425
  peaking, 740

Transformer (*contd.*)—
  percentage impedance, 407
  polyphase, 436
  regulation, 400, 404
  shell-type, 393
  short-circuit test, 403
  single-phase, 393
  testing, 422, 438
  transformation ratio, 394
  used with instruments, 432
  voltage regulation, 404
Transformers in parallel, 410, 415
Transients, 38, 46, 353
Transmission efficiency, 276
Tripling, of frequency, 762
Two-phase working, 360
Two-wattmeter method, 368

UNBALANCED polyphase systems, 373, 379
Units, 1
Utilization coefficient, 841

"V" CHARACTERISTICS of synchronous motor, 512, 518
Valency, 298
Varley phase-sequence indicator, 387
Vector diagram—
  compensated induction motor, 675
  polyphase, 372
    series motor, 671
  Schrage motor, 688
  single-phase series motor, 655
  synchronous alternator, 613
Vectors—
  addition of, 311
  division of, 343
  multiplication of, 342
  reciprocal of, 343
Ventilation, of alternators, 452
Voltage regulation—
  of alternator, 488

Voltage regulation (*contd.*)—
  of transformer, 400, 404
Voltage regulator, automatic, 528
Volt drop—
  in distributor, 289
  in three-phase line, 377

WARD-LEONARD control, 242, 245
Wattmeter corrections, 370
Wave winding, 165
Waveform—
  alternator, 763
  determination, of, 775
Wavelength, of radiation, 824
Waves, complex, 757
  effect on interconnection of phas 771
  effect on iron losses, 772
    operation of alternators in parall 773
    volt drop in transformer, 774
  effective value, 757
  influence of circuit on, 758, 761
  power conveyed by, 760
Weber, 4, 86
Wheatstone's network, 72
Winding—
  compensating, 192, 195
  fractional pitch, 459
  pitch, 162
  short-chorded, 193, 462
Windings—
  of alternators, 454, 767
  of d.c. machines, 162
  of induction motors, 550
  of variable-speed commutat motors, 701
  fractional pitch, 459, 769

YOKE, of magnet system, 151

ZONES, Russell, 812